Statistics in Practice

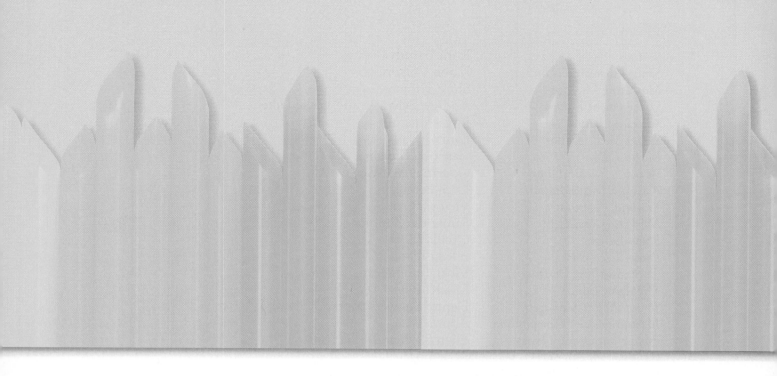

INSTRUCTOR'S EDITION

FIRST EDITION

Statistics in Practice

DAVID S. MOORE • **WILLIAM I. NOTZ** • **MICHAEL A. FLIGNER**

Purdue University   *The Ohio State University*   *University of California at Santa Cruz*

Instructor's material written by Jackie B. Miller, University of Michigan

W. H. Freeman and Company

A Macmillan Higher Education Company

Senior Publisher:	Ruth Baruth
Senior Acquisitions Editor:	Karen Carson
Marketing Manager:	Steve Thomas
Development Editors:	Andrew Sylvester and Leslie Lahr
Associate Editor:	Jorge Amaral
Senior Media Editor:	Laura Judge
Media Editor:	Catriona Kaplan
Associate Media Editor:	Courtney M. Way
Assistant Media Editor:	Liam Ferguson
Editorial Assistant:	Victoria Garvey
Marketing Assistant:	Samantha Zimbler
Photo Editor:	Cecilia Varas
Photo Researcher:	Eileen Liang
Cover and Text Designer:	Vicki Tomaselli
Managing Editor:	Lisa Kinne
Senior Project Manager:	Denise Showers, Aptara®, Inc.
Illustrations and Composition:	Aptara®, Inc.
Production Coordinators:	Ellen Cash and Julia DeRosa
Printing and Binding:	RR Donnelley

Library of Congress Control Number: 2014940824

Student Edition Hardcover (packaged with EESEE/CrunchIt! access card):
ISBN-13: 978-1-4641-5181-1
ISBN-10: 1-4641-5181-4

Student Edition Loose-leaf (packaged with EESEE/CrunchIt! access card):
ISBN-13: 978-1-4641-5238-2
ISBN-10: 1-4641-5238-1

Instructor Complimentary Copy:
ISBN-13: 978-1-4641-7451-3
ISBN-10: 1-4641-7451-2

Printed in the United States of America

First printing

W. H. Freeman and Company
41 Madison Avenue
New York, NY 10010
Houndmills, Basingstoke RG21 6XS, England
www.whfreeman.com

PREFACE TO THE INSTRUCTOR'S EDITION

As instructors of introductory statistics courses, we continually strive to improve our teaching. Instructors teaching an introductory statistics course using *Statistics in Practice* should find that the *Instructor's Edition* (IE) is a valuable resource with this mission. If you're an experienced instructor, you can use the *IE* as a source of additional, class-tested examples or simply to provide food for thought with respect to pedagogy. If you're a relatively new instructor, you can use the *IE* to help formulate lesson plans and enhance lecture notes. Regardless of your experience, I believe you will find examples, concepts, activities, pieces of advice, and discussions of pedagogy that can improve your class.

The *IE* introduces each chapter with the following resources:

- A chapter **Overview** highlights the most important chapter concepts and discusses some of the issues students often struggle with when confronted with those concepts. You will also find a discussion of course pedagogy in the Overview. In addition, there is discussion of how the material in the current chapter fits into the larger framework for the course. Where appropriate, scholarly articles have been referenced.

- The **Learning Objectives** section lists skills students should have after working through the chapter. These skills are then reinforced in the Part Review chapters of the text (Chapters 7, 10, 14, and 23).

- A **Teaching Suggestions and Additional Examples/Activities for the Classroom** section contains many examples, classroom activities, and exercises designed to introduce new concepts, as well as enhance classroom discussions. Included in this section are suggestions for real and compelling data sets that bring the material to life, yet make clear the concepts being taught. Here you will also find useful advice and additional discussion on pedagogy.

- The **Other Resources** section points to StatClips and Snapshots videos, case studies from the Electronic Encyclopedia of Statistical Examples and Exercises (EESEE), and applets, all of which are directly related to the chapter material and can be incorporated into class discussions and/or projects.

The activities and examples in the *IE* have already been used in class by many excellent teachers. These resources are often shared at conferences dedicated in large part to statistics pedagogy, and they are passed along from one colleague to another in discussions. Such conferences include the biennial U.S. Conference on Teaching Statistics (USCOTS) held in odd years, the quadrennial International Conference on Teaching Statistics (ICOTS), and the biennial Electronic Conference on Teaching Statistics (eCOTS), held in even years. Other activities can be found in the Resources section of the Consortium for the Advancement of Undergraduate Statistics Education (CAUSE, www.causeweb.org) or in the Mathematics and Statistics Section of the Multimedia Educational Resource for Learning and Online Teaching (MERLOT II, www.merlot.org).

It is in the spirit of sharing and our joint desire to make the learning experiences for our students the best they can be that I hope you will implement, improve on, or develop and share your own ideas for enhancing introductory statistics education. If there are any ideas, activities, data sets, or resources you would like to suggest for the instructor inserts for future editions, feel free to contact your local W.H. Freeman sales representative.

I would like to thank Bill Notz and Mike Fligner for entrusting this portion of the text to me. I am very happy to thank Patricia B. Humphrey for her helpful suggestions and dedication to helping me iterate what I wanted to communicate to instructors in the *IE*.

Jackie B. Miller
University of Michigan

BRIEF CONTENTS

---

*Starred material is not required for later parts of the text.

CONTENTS

---

*Starred material is not required for later parts of the text.

ix

Welcome to the first edition of *Statistics in Practice*. As the name suggests, this text provides an introduction to the practice of statistics that aims to equip students to carry out common statistical procedures and to follow statistical reasoning in their fields of study and in their future employment.

Statistics in Practice is designed to be accessible to college and university students with limited quantitative background—just "algebra" in the sense of being able to read and use simple equations. It is usable with almost any level of technology for calculating and graphing—from a $15 "two-variable statistics" calculator to a graphing calculator or spreadsheet program to full statistical software. Of course, graphs and calculations are less tedious with good technology, so we recommend making available to your students the most effective technology that circumstances permit.

Despite its rather low mathematical level, *Statistics in Practice* is designed to reflect the actual practice of statistics, where data analysis and design of data production join with probability-based inference to form a coherent science of data. There are good pedagogical reasons for beginning with data analysis (Chapters 1 to 7), then moving to data production (Chapters 8 to 10), and then to probability (Chapters 11 to 14) and inference (Chapters 15 to 30). In studying data analysis, students learn useful skills immediately and get over some of their fear of statistics. Data analysis is a necessary preliminary to inference in practice because inference requires clean data. Designed data production is the surest foundation for inference, and the deliberate use of chance in random sampling and randomized comparative experiments motivates the study of probability in a course that emphasizes data-oriented statistics. *Statistics in Practice* gives a full presentation of basic probability and inference (20 of the 30 chapters) but places it in the context of statistics as a whole.

Guiding Principles and the GAISE Guidelines

Statistics in Practice is based on three principles: balanced content, experience with data, and the importance of ideas. These principles are widely accepted by statisticians concerned about teaching and are directly connected to and reflected by the themes of the College Report of the Guidelines in Assessment and Instruction for Statistics Education (GAISE) Project.

The GAISE Guidelines include six recommendations for the introductory statistics course. The content, coverage, and features of *Statistics in Practice* are closely aligned to these recommendations:

1. Emphasize statistical literacy and develop statistical thinking. The intent of *Statistics in Practice* is to be modern and accessible. The exposition is straightforward and concentrates on major ideas and skills. One principle of writing for beginners is not to try to tell your students everything you know. Another principle is to offer frequent stopping points, marking off digestible bites of material. Statistical literacy is promoted throughout *Statistics in Practice* in the many examples and exercises drawn from the popular press and from many fields of study. Statistical thinking is promoted in examples and exercises that give enough background to allow students to consider the meaning of their calculations. Exercises often ask for conclusions that are more than a number (or "reject H_0"). Some exercises require judgment in addition to right-or-wrong calculations and conclusions. Statistics, more than mathematics, depends on judgment for effective use. *Statistics in Practice* begins to develop students' judgment about statistical studies.

2. Use real data. The study of statistics is supposed to help students work with data in their varied academic disciplines and in their unpredictable real-world employment. Students learn to work with data by working with data. *Statistics in Practice* is full of data from many fields of study and from everyday life. Data are more than mere numbers—they are numbers with a context that should play a role in making sense of the numbers and in stating conclusions. Examples and exercises in *Statistics in Practice*, though intended for beginners, use real data and give enough background to allow students to consider the meaning of their calculations.

3. Stress conceptual understanding rather than mere knowledge of procedures. A first course in statistics introduces many skills, from making a stemplot and calculating a correlation to choosing and carrying out a significance test. In practice (even if not always in the course), calculations and graphs are automated. Moreover, anyone who makes serious use of statistics will need some specific procedures not taught in a college statistics course. *Statistics in Practice* therefore tries to make clear the larger patterns and big ideas of statistics, not in the abstract, but in the context of learning specific skills and working with specific data. Many of the big ideas are summarized in graphical outlines. Formulas without guiding principles do students little good once the final exam is in the past, so it is worth the time to slow down a bit and explain the ideas.

4. Foster active learning in the classroom. Fostering active learning is the business of the teacher, though an emphasis on working with data helps. To this end, we have created interactive applets to our specifications that are available online. These are designed primarily to help in learning statistics rather than in doing statistics. Icons in the text call attention to comments and exercises based on the applets. We suggest using selected applets for classroom demonstrations, even if you do not ask students to work with them. The *Correlation and Regression*, *Confidence Interval*, and *P-value* applets, for example, convey core ideas more clearly than any amount of chalk and talk.

We also provide Web exercises at the end of each chapter. Our intent is to take advantage of the fact that most undergraduates are "Web savvy." These exercises require students to search the Web for either data or statistical examples and then evaluate what they find. Teachers can use these as classroom activities or assign them as homework projects.

5. Use technology to develop conceptual understanding and analyze data. Automating calculations increases students' ability to complete problems, reduces their frustration, and helps them concentrate on ideas and problem recognition rather than mechanics. At a minimum, students should have a "two-variable statistics" calculator with functions for correlation and the least-squares regression line as well as for the mean and standard deviation.

Many instructors will take advantage of more elaborate technology, as ASA/MAA and GAISE recommend. And many students who don't use technology in their college statistics course will find themselves using (for example) Excel on the job. *Statistics in Practice* does not assume or require use of software except in Parts V and VI, where the work is otherwise too tedious. It does accommodate software use and tries to convince students they are gaining knowledge that will enable them to read and use output from almost any source. Regular "Using Technology" sections appear throughout the text. Each of these presents and comments on output from the same three technologies, representing graphing calculators (the Texas Instruments TI-83 or TI-84), spreadsheets (Microsoft Excel), and statistical software (Minitab and CrunchIt!). The output always concerns one of the main teaching examples, so students can compare text and output.

6. Use assessments to improve and evaluate student learning. Within chapters, a few "Apply Your Knowledge" exercises follow each new idea or skill for a quick check of basic mastery—and also to mark off digestible bites of material. Each of the first four parts of the book ends with a review chapter that includes a point-by-point outline of skills learned, problems students can use to test themselves, and numerous supplementary exercises. (Instructors can choose to cover any or none of the chapters in Parts V and VI, so each of these chapters includes a skills outline.) The review chapters present supplemental exercises without the "I just studied that" context, thus asking for another level of learning. We think it is helpful to assign some supplemental exercises. Many instructors will find that the review chapters appear at the right points for pre-examination review. The "Test Yourself" questions provide multiple-choice, calculation, and short-answer questions to help students review, self-assess, and prepare for such an examination. In addition, assessment materials in the form of a test bank and quizzes are available online.

In this chapter we cover...

E ach chapter opener gives a brief look at where the chapter is heading, often with references to previous chapters, and includes a bulleted list of the major topics that will be covered.

CHAPTER 2

In this chapter we cover...

- Measuring center: the mean
- Measuring center: the median
- Comparing the mean and the median
- Measuring spread: the quartiles
- The five-number summary and boxplots
- Spotting suspected outliers*
- Measuring spread: the standard deviation
- Choosing measures of center and spread
- Using technology
- Organizing a statistical problem

EXAMPLE 2.9

GRADRATE

Comparing Graduation Rates

STATE: Law requires all states in the United States to use a common computation of on-time high school graduation rates for the 2010–11 school year. Previously, states chose one of several computation methods that gave answers that could differ by more than 10%. This common computation allows for meaningful comparison of graduation rates between the states.

We know from Table 1.1 (page 22), that the on-time high school graduation rates varied from 59% in the District of Columbia, to 88% in Vermont. The U.S. Census Bureau divides the 50 states and the District of Columbia, into four geographical regions, the Northeast (NE), Midwest (MW), South (S), and West (W). The region for each state is included in Table 1.1. Do these four regions of the country display distinct distributions of graduation rates? How do the mean rates compare?

PLAN: Use graphs and numerical descriptions to describe and compare the on-time high school graduation rates in the four regions of the United States.

SOLVE: We might use boxplots to compare the distributions, but stemplots preserve more detail and work well for data sets of these sizes. Figure 2.4 displays the stemplots with the stems lined up for easy comparison. The stems have been split to better display the distributions. The stemplots overlap, and some care is needed when comparing the four stemplots as the sample sizes differ, with some stemplots having more leaves than others. None of the plots show strong skewness, although the South has one low observation that stands apart from the others with this choice of stems. The Northeast and Midwest have distributions that are similar to each other as do the South and West. The graduation rates tend to be higher in the Northeast and Midwest and more spread out in the South and West. With little skewness and no serious outliers, we report $\bar{x}$ and s as our summary measures of center and spread/variability:

FIGURE 2.4
Stemplots comparing the distributions of graduation rates for the four census regions from Table 1.1.

```
Midwest        Northeast      South          West
8 | 66678     8 | 67        8 | 66        8 |
8 | 01334     8 | 33334     8 | 123       8 | 002
7 | 7         7 | 77        7 | 5688      7 | 6668
7 | 4         7 |           7 | 1124      7 | 4
6 |           6 |           6 | 7         6 | 88
6 |           6 |           6 |           6 | 23
5 |           5 |           5 | 9         5 |
```

Region	Mean Rate	Standard Deviation
Midwest	82.92	4.25
Northeast	82.56	3.47
South	75.93	7.36
West	73.58	6.73

CONCLUDE: The table of summary statistics confirms what we see in the stemplots. The Midwest and Northeast are quite similar to each other as are the South and West. The Midwest and Northeast have higher mean graduation rates as well as smaller standard deviations than the South and West. ■

4-Step Examples

In Chapter 2, students learn how to use the four-step process for working through statistical problems: State, Plan, Solve, Conclude. By observing this framework at work in selected examples throughout the text and practicing it in selected exercises, students develop the ability to solve and write reports on real statistical problems encountered outside the classroom.

Apply Your Knowledge

Major concepts are immediately reinforced with problems that are interspersed throughout the chapter (often following examples), allowing students to practice their skills as they work through the text.

Apply Your Knowledge

6.1 Video Gaming and Grades. The popularity of computer, video, online, and virtual reality games has raised concerns about their ability to negatively affect youth. The data in this exercise are based on a recent survey of 14- to 18-year-olds in Connecticut high schools. Here are the grade distributions of boys who have and have not played video games.[2] GAMING

	Grade Average		
	A's and B's	C's	D's and F's
Played games	736	450	193
Never played games	205	144	80

(a) How many people does this table describe? How many of these have played video games?

(b) Give the marginal distribution of the grades. What percent of the boys represented in the table received a grade of C or lower?

Using Technology

Located throughout the text, these special sections display and comment on the output from graphing calculators, spreadsheets, and statistical software in the context of examples from the text.

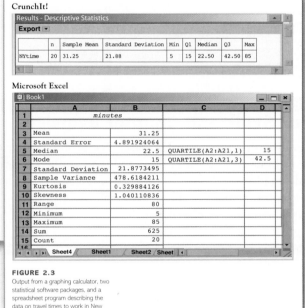

Using technology

Although a calculator with "two-variable statistics" functions will do the basic calculations we need, more elaborate tools are helpful. Graphing calculators and computer software will do calculations and make graphs as you command, freeing you to concentrate on choosing the right methods and interpreting your results. Figure 2.3 displays output describing the travel times to work of 20 people in New York State (Example 2.3, page 48). Can you find $\bar{x}$, s, and the five-number summary in each output? The big message of this section is: *Once you know what to look for, you can read output from any technological tool.*

The displays in Figure 2.3 come from a Texas Instruments graphing calculator, the Minitab and CrunchIt statistical programs, and the Microsoft Excel spreadsheet program. Minitab allows you to choose what descriptive measures you want, whereas the descriptive measures in the CrunchIt output are provided by default. Excel and the calculator give some things we don't need. Just ignore the extras. Excel's "Descriptive Statistics" menu item doesn't give the quartiles. We used the spreadsheet's separate quartile function to get Q_1 and Q_3.

Texas Instruments Graphing Calculator

```
1-Var Stats
x̄=31.25
Σx=625
Σx²=28625
Sx=21.8773495
σx=21.32340264
↓n=20
```

```
1-Var Stats
↑n=20
minX=5
Q₁=15
Med=22.5
Q₃=42.5
maxX=85
```

Minitab

Session

Descriptive Statistics: NYtime

Variable	N	N*	Mean	SE Mean	StDev	Minimum	Q1	Median	Q3	Maximum
NYtime	20	0	31.25	4.89	21.88	5.00	15.00	22.50	43.75	85.00

CrunchIt!

Results - Descriptive Statistics

Export ▾

	n	Sample Mean	Standard Deviation	Min	Q1	Median	Q3	Max
NYtime	20	31.25	21.88	5	15	22.50	42.50	85

Microsoft Excel

Book1

	A	B	C	D
1		*minutes*		
2				
3	Mean	31.25		
4	Standard Error	4.891924064		
5	Median	22.5	QUARTILE(A2:A21,1)	15
6	Mode	15	QUARTILE(A2:A21,3)	42.5
7	Standard Deviation	21.8773495		
8	Sample Variance	478.6184211		
9	Kurtosis	0.329884126		
10	Skewness	1.040110836		
11	Range	80		
12	Minimum	5		
13	Maximum	85		
14	Sum	625		
15	Count	20		

Sheet4 / Sheet1 / Sheet2 / Sheet

FIGURE 2.3

Output from a graphing calculator, two statistical software packages, and a spreadsheet program describing the data on travel times to work in New York State.

HE SAID, SHE SAID.

Height, weight, and body mass distributions in this book come from actual measurements by a government survey. Good thing that is. When *asked* their weight, almost all women say they weigh less than they really do. Heavier men also underreport their weight—but lighter men claim to weigh more than the scale shows. We leave you to ponder the psychology of the two sexes. Just remember that "say-so" is no substitute for measuring.

Statistics in Your World

These brief asides, found in each chapter, illustrate major concepts or present cautionary tales through entertaining and relevant stories, allowing students to take a break from the exposition while staying engaged.

CHAPTER 4 SUMMARY

Chapter Specifics

- To study relationships between variables, we must measure the variables on the same group of individuals.

- If we think that a variable x may explain or even cause changes in another variable y, we call x an **explanatory variable** and y a **response variable.**

- A **scatterplot** displays the relationship between two quantitative variables measured on the same individuals. Mark values of one variable on the horizontal axis (x axis) and values of the other variable on the vertical axis (y axis). Plot each individual's data as a point on the graph. Always plot the explanatory variable, if there is one, on the x axis of a scatterplot.

- Plot points with different colors or symbols to see the effect of a categorical variable in a scatterplot.

- In examining a scatterplot, look for an overall pattern showing the **direction, form,** and **strength** of the relationship and then for **outliers** or other deviations from this pattern.

- **Direction:** If the relationship has a clear direction, we speak of either **positive association** (high values of the two variables tend to occur together) or **negative association** (high values of one variable tend to occur with low values of the other variable).

- **Form: Linear relationships,** where the points show a straight-line pattern, are an important form of relationship between two variables. Curved relationships and **clusters** are other forms to watch for.

- **Strength:** The **strength** of a relationship is determined by how close the points in the scatterplot lie to a simple form such as a line.

- The **correlation** r measures the direction and strength of the linear association between two quantitative variables x and y. Although you can calculate a correlation for any scatterplot, r measures only straight-line relationships.

- Correlation indicates the direction of a linear relationship by its sign: $r > 0$ for a positive association and $r < 0$ for a negative association. Correlation always satisfies $-1 \leq r \leq 1$ and indicates the strength of a relationship by how close it is to -1 or 1. Perfect correlation, $r = \pm 1$, occurs only when the points on a scatterplot lie exactly on a straight line.

- Correlation ignores the distinction between explanatory and response variables. The value of r is not affected by changes in the unit of measurement of either variable. Correlation is not resistant, so outliers can greatly change the value of r.

Chapter Summary and Link It

Each chapter concludes with a summary of the chapter specifics, including major terms and processes, followed by a brief discussion of how the chapter links to material in both previous and upcoming chapters.

Link It

In Chapters 1 to 3, we focused on exploring features of a single variable. In this chapter we continued our study of exploratory data analysis but for the purpose of examining relationships *between* variables. A useful tool for exploring the relationship between two variables is the scatterplot. When the relationship is linear, correlation is a numerical measure of the strength of the linear relationship.

It is tempting to assume that the patterns we observe in our data hold for values of our variables that we have not observed—in other words, that additional data would continue to conform to these patterns. The process of identifying underlying patterns would seem to assume that this is the case. But is this assumption justified? Parts II to V of the book answer this question.

Check Your Skills and Chapter Exercises

Each chapter ends with a series of multiple-choice problems that test students' understanding of basic concepts and their ability to apply the concepts to real-world statistical situations. The multiple-choice problems are followed by a set of more in-depth exercises that allow students to make judgments and draw conclusions based on real data and real scenarios.

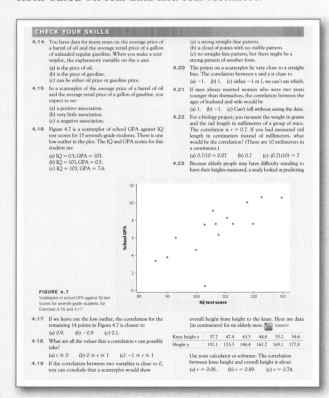

CHECK YOUR SKILLS

4.14 You have data for many years on the average price of a barrel of oil and the average retail price of a gallon of unleaded regular gasoline. When you make a scatterplot, the explanatory variable on the x axis

(a) is the price of oil.
(b) is the price of gasoline.
(c) can be either oil price or gasoline price.

4.15 In a scatterplot of the average price of a barrel of oil and the average retail price of a gallon of gasoline, you expect to see

(a) a positive association.
(b) very little association.
(c) a negative association.

4.16 Figure 4.7 is a scatterplot of school GPA against IQ test scores for 15 seventh-grade students. There is one low outlier in the plot. The IQ and GPA scores for this student are

(a) IQ = 0.5, GPA = 103.
(b) IQ = 103, GPA = 0.5.
(c) IQ = 103, GPA = 7.6.

(a) a strong straight-line pattern.
(b) a cloud of points with no visible pattern.
(c) no straight-line pattern, but there might be a strong pattern of another form.

4.20 The points on a scatterplot lie very close to a straight line. The correlation between x and y is close to

(a) -1.   (b) 1.   (c) either -1 or 1, we can't say which.

4.21 If men always married women who were two years younger than themselves, the correlation between the ages of husband and wife would be

(a) 1.   (b) -1.   (c) Can't tell without seeing the data.

4.22 For a biology project, you measure the weight in grams and the tail length in millimeters of a group of mice. The correlation is $r = 0.7$. If you had measured tail length in centimeters instead of millimeters, what would be the correlation? (There are 10 millimeters in a centimeter.)

(a) $0.7/10 = 0.07$ (b) 0.7   (c) $(0.7)(10) = 7$

4.23 Because elderly people may have difficulty standing to have their heights measured, a study looked at predicting

FIGURE 4.7
Scatterplot of school GPA against IQ test scores for seventh-grade students, for Exercises 4.16 and 4.17.

4.17 If we leave out the low outlier, the correlation for the remaining 14 points in Figure 4.7 is closest to

(a) 0.9.   (b) -0.9.   (c) 0.1.

4.18 What are all the values that a correlation r can possibly take?

(a) $r \geq 0$ (b) $0 \leq r \leq 1$ (c) $-1 \leq r \leq 1$

4.19 If the correlation between two variables is close to 0, you can conclude that a scatterplot would show

overall height from height to the knee. Here are data (in centimeters) for six elderly men: KNEEHT

Knee height x	57.7	47.4	43.5	44.8	55.2	54.6
Height y	192.1	153.3	146.4	162.7	169.1	177.8

Use your calculator or software: The correlation between knee height and overall height is about

(a) $r = 0.08$.   (b) $r = 0.89$.   (c) $r = 0.74$.

CHAPTER 4 EXERCISES

4.24 **Scores at the Masters.** The Masters is one of the four major golf tournaments. Figure 4.8 is a scatterplot of the scores for the first two rounds of the 2012 Masters for all the golfers entered. Only the 60 golfers with the lowest two-round total advance to the final two rounds (unless several people are tied for 60th place, in which case all those tied for 60th place advance). The plot has a grid pattern because golf scores must be whole numbers.   MASTR12

(a) Read the graph: What was the lowest score in the first round of play? How many golfers had this low score? What were their scores in the second round?
(b) Read the graph: Ben Crenshaw had the highest score in the second round. What was this score? What was Crenshaw's score in the first round?
(c) Is the correlation between first-round scores and second-round scores closest to $r = 0.3$, $r = 0.7$, or $r = 0.9$? Explain your choice. Does the graph suggest that knowing a professional golfer's score for one round is much help in predicting his score for another round on the same course?

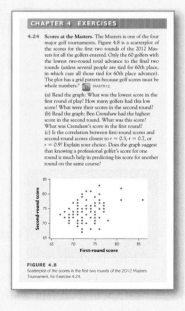

FIGURE 4.8
Scatterplot of the scores in the first two rounds of the 2012 Masters Tournament, for Exercise 4.24.

Web Exercises

A final set of exercises asks students to investigate data and statistical issues by researching topics online. These exercises tend to be more involved and provide an opportunity for students to dig deep into contemporary issues and special applications of statistics.

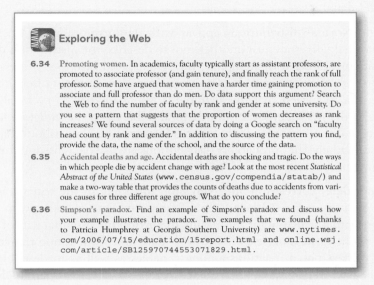

Exploring the Web

6.34 Promoting women. In academics, faculty typically start as assistant professors, are promoted to associate professor (and gain tenure), and finally reach the rank of full professor. Some have argued that women have a harder time gaining promotion to associate and full professor than do men. Do data support this argument? Search the Web to find the number of faculty by rank and gender at some university. Do you see a pattern that suggests that the proportion of women decreases as rank increases? We found several sources of data by doing a Google search on "faculty head count by rank and gender." In addition to discussing the pattern you find, provide the data, the name of the school, and the source of the data.

6.35 Accidental deaths and age. Accidental deaths are shocking and tragic. Do the ways in which people die by accident change with age? Look at the most recent *Statistical Abstract of the United States* (www.census.gov/compendia/statab/) and make a two-way table that provides the counts of deaths due to accidents from various causes for three different age groups. What do you conclude?

6.36 Simpson's paradox. Find an example of Simpson's paradox and discuss how your example illustrates the paradox. Two examples that we found (thanks to Patricia Humphrey at Georgia Southern University) are www.nytimes.com/2006/07/15/education/15report.html and online.wsj.com/article/SB125970744553071829.html.

Origins

Instructors who have seen our other texts will recognize that *Statistics in Practice* (*SIP*) has much in common with the 6th edition of the successful text *The Basic Practice of Statistics* (*BPS*). Although *Statistics in Practice* incorporates many new examples and exercises, the reorganization of several key topics distinguishes this text from *BPS*. There is no single best way to organize our presentation of statistics to beginners. That said, our choices reflect thinking about both content and pedagogy. Here are our reasons for several choices we have made about the order and selection of material in *Statistics in Practice*.

• **Why does the distinction between population and sample not appear in Part I?** There is more to statistics than inference. In fact, statistical inference is appropriate only in rather special circumstances. The chapters in Part I present tools and tactics for describing data—any data. These tools and tactics do not depend on the idea of inference from sample to population. Many data sets in these chapters (for example, the several sets of data about the 50 states) do not lend themselves to inference because they represent an entire population. John Tukey of Bell Labs and Princeton, the philosopher of modern data analysis, insisted that the population-sample distinction be avoided when it is not relevant. He used the word *batch* for data sets in general. We see no need for a special word, but we think Tukey was right.

• **Should we cover data production or data exploration first?** We prefer to begin with data exploration (Part I), as most students will use statistics mainly in settings other than planned research studies in their future employment. We place the design of data production (Part II) after data analysis to emphasize that data-analytic techniques apply to any data. However, it is equally reasonable to begin with data production—the natural flow of a planned study is from design to data analysis to inference. Because instructors have strong and differing opinions on this question, these two topics are now the first two parts of the book, with the text written so that it may be started with either Part I or Part II without loss of continuity.

• **How much probability should we cover?** The probability chapters are now contained in Part III. The instructor can choose to do the first chapter only, which gives a few basic definitions, or can cover one or both of the optional chapters, which go into greater detail.

• **Why do Normal distributions appear with data exploration?** Density curves such as Normal curves are just another tool to describe the distribution of a quantitative variable, along with stemplots, histograms, and boxplots. Professional statistical software offers to make density curves from data just as it offers histograms. We prefer not to suggest that this material is essentially tied to probability, as the traditional order does. And we find it helpful to break up the indigestible lump of probability that troubles students so much. Meeting Normal distributions early does this and strengthens the "probability distributions are like data distributions" way of approaching probability.

• **Why not delay correlation and regression until late in the course, as was traditional?** *Statistics in Practice* begins by offering experience working with data and gives a conceptual structure for this nonmathematical but essential part of statistics. Students profit from more experience with data and from seeing the conceptual structure worked out in relations among variables as well as in describing single-variable data. Correlation and least-squares regression are very important descriptive tools and are often used in settings where there is no population-sample distinction, such as studies of all a firm's employees. Perhaps most important, the *Statistics in Practice* approach asks students to think about what kind of relationship lies behind the data (confounding, lurking variables, association doesn't imply causation, and so on) without overwhelming them with demands of formal inference methods. Inference in the correlation and regression setting is a bit complex, demands software, and often comes at the end of the course. We find that delaying all mention of correlation and regression to that point means that students often don't master the basic uses and properties of these methods. We consider Chapters 4 and 5 (correlation and regression) essential and Chapter 24 (regression inference) optional.

• **Why is z for means with known variance not covered?** One of the justifications for covering the z procedures for means with known variance is that, under the additional assumption of Normality, it allows students to carry out explicit calculations of P-values, set up critical regions, compute confidence intervals, and calculate the power of the test using only the Normal table. There is no need to appeal to large-sample approximations and no need to estimate σ. Separating the initial reasoning of inference from messier practice enables students to focus on the basic ideas. But this takes a fair amount of time, and the ideas need to be rehashed with the introduction of t procedures. Many instructors face pressure from client departments to cover a large amount of material in a single semester. Eliminating coverage of the "unrealistic" z for means with known variance enables instructors to cover additional, more realistic applications of inference. To accommodate these instructors, we have chosen not to introduce inference using the z for means assuming Normality with known variance.

• **Why introduce inference with proportions rather than means?** Many instructors would agree that proportions are simpler and more familiar to students than means. Having made the decision not to use the unrealistic z procedures for means to introduce the basic ideas of inference, it makes the most sense to introduce the ideas of inference using the realistic setting of proportions. The use of a different standard error for the test and confidence interval does add an additional complication when dealing with proportions. Despite this, there are definite advantages to introducing inference with proportions, and using the familiar z distribution as an approximation outweighs the disadvantage of introducing inference using means, which involves introducing the t distribution at a very early stage.

• **Why divide sampling distributions into two chapters?** Sampling distributions represent one of the most difficult concepts in an introductory course, but without a good understanding of what they represent it is difficult for students to fully understand the central ideas of inference. Because we introduce inference in the context of proportions in Chapters 16 and 17, students only need the concept of sampling distributions and the specific results for the sampling distribution of $\hat{p}$ to understand these chapters. Once students have seen the application of sampling distributions to inference for a proportion, we return to the central limit theorem and the sampling distribution of $\bar{x}$ immediately before introducing inference for means. There is little overlap between the two sampling distribution chapters, and when covering the central limit theorem, the connection to proportions is made in an optional section. Reviewers have given positive feedback on the presentation of sampling distribution in two digestible chunks at appropriate locations in the text.

• **Why didn't you cover Topic X?** Introductory texts ought not to be encyclopedic. Including each reader's favorite special topic results in a text that is formidable in size and intimidating to students. We chose topics on two grounds: they are most commonly used in practice, and they are suitable vehicles for learning broader statistical ideas. Students who have completed the core of *Statistics in Practice*, Chapters 1–11 and 15–23, will have little difficulty moving on to more elaborate methods. There are, of course, seven chapters that discuss more advanced topics, three in this volume and four available online, to begin the next stages of learning.

ACKNOWLEDGMENTS

Once again, we have enjoyed the opportunity to rethink how to help beginning students achieve a practical grasp of basic statistics. What students actually learn is not identical to what we teachers think we have "covered," so the virtues of concentrating on the essentials are considerable. We hope that the first edition of *Statistics in Practice* offers a mix of concrete skills and clearly explained concepts that will help many teachers guide their students toward useful knowledge.

We are grateful to colleagues from two-year and four-year colleges and universities who commented on *Statistics in Practice*:

Dilshod Achilov, *East Tennessee State University*
Anna Bargagliotti, *Loyola Marymount University*
Jennifer Beineke, *Western New England University*
Pierre-Jérôme Bergeron, *University of Ottawa*
Monica Brown, *St. Catherine University*
Max Buot, *Xavier University*
Dennis L. Clason, *University of Cincinnati*
Amali Dassanayake, *Texas Tech University*
Christiana Drake, *University of California, Davis*
Robert Eby, *Blinn College, Bryan Campus*
Mark Gebert, *University of Kentucky*
Kim Gilbert, *University of Georgia*
Brenda Gunderson, *University of Michigan*
Gary Haefner, *University of Cincinnati*
Steve Howell, *Santa Fe College*
Pat Humphrey, *Georgia Southern University*
Kevin M. Iga, *Pepperdine University*
Pat A. Kan, *Cleveland State University*
Sheldon Lee, *Viterbo University*
Merrill Liechty, *Drexel University*
Antoinette M. Marquard, *Cleveland State University*
Catherine Matos, *Clayton State University*
Thomas McNamara, *Southwestern Oklahoma State University*
Jackie B. Miller, *University of Michigan*
Jack Morse, *University of Georgia*

Kathy Mowers, *Owensboro Community & Technical College*
Laura Patterson, *University of Colorado Boulder*
Maureen Petkewich, *University of South Carolina*
Nicole Radziwill, *James Madison University*
Deanna Robinson-Breidel, *McLennan Community College*
Hilary H. Seagle, *Southwestern Community College*
Shannon K. McClintock, *Emory University*
Shane Rollans, *Thompson Rivers University*
Robert Y. Shapiro, *Columbia University*
Manoj Sharma, *University of Cincinnati*
Kim Sheppard, *Cecil College*
Tom Short, *John Carroll University*
Murray Siegel, *Arizona State University*
Alla Sikorskii, *Michigan State University*
Mary Ann Teel, *University of North Texas*
Donna Tupper, *The Community College of Baltimore County–Essex*
Emiliano D. Vega, *Portland Community College*
Dottie Walton, *Cuyahoga Community College*
Carol Weideman, *St. Petersburg College–Gibbs Campus*
Lee Widmer, *College of Mount St. Joseph*
Jamie Wieland, *Illinois State University*
Jun Ye, *University of Akron*
Shiju Zhang, *St. Cloud State University*

We extend our appreciation to Ruth Baruth, Karen Carson, Andrew Sylvester, Leslie Lahr, Shona Burke, Lisa Kinne, Jorge Amaral, Liam Ferguson, Catriona Kaplan, Courtney Way, Laura Judge, Victoria Garvey, Cecilia Varas, Eileen Liang, Ellen Cash, Julia DeRosa, Denise Showers, and the other editors who have contributed to the development and cohesiveness of this book and its online resources.

Special thanks are due to Vicki Tomaselli and other design professionals whose talents were poured into the look and attractiveness of this book. We extend our appreciation to Denise Showers of Aptara®, Inc., who has tirelessly offered her knowledge, expertise, and patience throughout the production process.

We are indebted to our colleagues, Jackie B. Miller and Patricia B. Humphrey, for their many contributions, insights, and time. Their wisdom and experience in the classroom have added to a level of quality that students and instructors alike have come to expect.

Finally, we are indebted to the many statistics teachers with whom we have discussed the teaching of our subject over many years; to people from diverse fields with whom we have worked to understand data; and especially to students whose compliments and complaints have changed and improved our teaching. Working with teachers, colleagues in other disciplines, and students constantly reminds us of the importance of hands-on experience with data and of statistical thinking in an era when computer routines quickly handle statistical details.

David S. Moore, William I. Notz, and Michael A. Fligner

LaunchPad

W. H. Freeman's new online homework system, **LaunchPad,** offers our quality content curated and organized for easy assignability in a simple but powerful interface. We've taken what we've learned from thousands of instructors and hundreds of thousands of students to create a new generation of W. H. Freeman/Macmillan technology.

Curated Units. Combining a curated collection of videos, homework sets, tutorials, applets, and e-Book content, LaunchPad's interactive units give instructors a building block to use as is or as a starting point for their own learning units. Thousands of exercises from the text can be assigned as online homework, including many algorithmic exercises. An entire unit's worth of work can be assigned in seconds, drastically reducing the amount of time it takes to have a course up and running.

Easily customizable. Instructors can customize the LaunchPad units by adding quizzes and other activities from our vast wealth of resources. They can also add a discussion board, a dropbox, and RSS feed, with a few clicks. LaunchPad allows instructors to customize students' experience as much or as little desired.

Useful analytics. The gradebook quickly and easily allows instructors to look up performance metrics for classes, individual students, and individual assignments.

Intuitive interface and design. The student experience is simplified. Students' navigation options and expectations are clearly laid out at all times, ensuring they can never get lost in the system.

Assets integrated into LaunchPad include the following:

Interactive e-Book. Every LaunchPad e-Book comes with powerful study tools for students, video and multimedia content, and easy customization for instructors. Students can search, highlight, and bookmark, making it easier to study and access key content. And teachers can ensure that their classes get just the book they want to deliver: customize and rearrange chapters, add and share notes and discussions, and link to quizzes, activities, and other resources.

***LEARNING**Curve*   **LearningCurve** provides students and instructors with powerful adaptive quizzing, a game-like format, direct links to the e-Book, and instant feedback. The quizzing system features questions tailored specifically to the text and adapts to students' responses, providing material at different difficulty levels and topics based on student performance.

SolutionMaster   **SolutionMaster** offers an easy-to-use Web-based version of the instructor's solutions, allowing instructors to generate a solution file for any set of homework exercises.

New Stepped Tutorials are centered on algorithmically generated quizzing with step-by-step feedback to help students work their way toward the correct solution. These new exercise tutorials (two to three per chapter) are easily assignable and assessable.

Statistical Video Series consists of StatClips, StatClips Examples, and Statistically Speaking "Snapshots." View animated lecture videos, whiteboard lessons, and documentary-style footage that illustrate key statistical concepts and help students visualize statistics in real-world scenarios.

New Video Technology Manuals available for TI-83/84 calculators, Minitab, Excel, JMP, SPSS, R, Rcmdr, and CrunchIt!® provide brief instructions for using specific statistical software.

Updated StatTutor Tutorials offer multimedia tutorials that explore important concepts and procedures in a presentation that combines video, audio, and interactive features. The newly revised format includes built-in, assignable assessments and a bright new interface.

Updated Statistical Applets give students hands-on opportunities to familiarize themselves with important statistical concepts and procedures, in an interactive setting that allows them to manipulate variables and see the results graphically. Icons in the textbook indicate when an applet is available for the material being covered.

CrunchIt!® is a Web-based statistical program that allows users to perform all the statistical operations and graphing needed for an introductory statistics course and more. It saves users time by automatically loading data from *SIP,* and it provides the flexibility to edit and import additional data.

JMP **JMP Student Edition** (developed by SAS) is easy to learn and contains all the capabilities required for introductory statistics, including pre-loaded data sets from *SIP.* JMP is the leading commercial data analysis software of choice for scientists, engineers and analysts at companies throughout the globe (for Windows and Mac).

Stats@Work Simulations put students in the role of the statistical consultant, helping them better understand statistics interactively within the context of real-life scenarios.

EESEE Case Studies (*Electronic Encyclopedia of Statistical Examples and Exercises*), developed by The Ohio State University Statistics Department, teach students to apply their statistical skills by exploring actual case studies using real data.

Data files are available in ASCII, Excel, TI, Minitab, SPSS (an IBM Company),* and JMP formats.

Student Solutions Manual provides solutions to the odd-numbered exercises in the text. Available electronically within LaunchPad, as well as in print form.

Interactive Table Reader allows students to use statistical tables interactively to seek the information they need.

Instructor's Guide with Full Solutions includes teaching suggestions, chapter comments, and detailed solutions to all exercises. Available electronically within LaunchPad.

---

*SPSS was acquired by IBM in October 2009.

Test Bank offers hundreds of multiple-choice questions. Also available on CD-ROM (for Windows and Mac), where questions can be downloaded, edited, and resequenced to suit each instructor's needs.

Lecture PowerPoint Slides offer a detailed lecture presentation of statistical concepts covered in each chapter of *SIP*.

Additional Resources Available with *SIP*

Companion Web site www.whfreeman.com/sip This open-access Web site includes statistical applets, data files, supplementary exercises, and self-quizzes. The Web site also offers four optional companion chapters covering logistic regression, nonparametric tests, bootstrap methods and permutation tests, and statistics for quality control and capability. Instructor access to the Companion Web site requires user registration as an instructor and features all the open-access student Web materials, plus:

- Instructor version of **EESEE** with solutions to the exercises in the student version.

- **PowerPoint Slides** containing all textbook figures and tables.

- **Lecture PowerPoint Slides**

Special Software Packages Student versions of JMP and Minitab are available for packaging with the text. JMP is available at no additional cost within LaunchPad. Contact your W. H. Freeman representative for information or visit www.whfreeman.com.

Course Management Systems W. H. Freeman and Company provides courses for Blackboard, Angel, Desire2Learn, Canvas, Moodle, and Sakai course management systems. These are completely integrated solutions that you can easily customize and adapt to meet your teaching goals and course objectives. Visit macmillanhighered.com/Catalog/other/Coursepack for more information.

i-clicker i-clicker is a two-way radio-frequency classroom response solution developed by educators for educators. Each step of i-clicker's development has been informed by teaching and learning. To learn more about packaging i-clicker with this textbook, please contact your local sales rep or visit www1.iclicker.com.

ABOUT THE AUTHORS

David S. Moore is Shanti S. Gupta Distinguished Professor of Statistics, Emeritus, at Purdue University and was the 1998 president of the American Statistical Association. He received his A.B. from Princeton and his Ph.D. from Cornell, both in mathematics. He has written many research papers in statistical theory and served on the editorial boards of several major journals. Professor Moore is an elected fellow of the American Statistical Association and of the Institute of Mathematical Statistics and an elected member of the International Statistical Institute. He has served as program director for statistics and probability at the National Science Foundation. Professor Moore has made many contributions to the teaching of statistics. He was the content developer for the Annenberg/Corporation for Public Broadcasting college-level telecourse Against All Odds: Inside Statistics and for the series of video modules Statistics: Decisions through Data, intended to aid the teaching of statistics in schools. He is the author of influential articles on statistics education and of several leading texts. Professor Moore has served as president of the International Association for Statistical Education and has received the Mathematical Association of Americas national award for distinguished college or university teaching of mathematics.

William I. Notz is Professor of Statistics at The Ohio State University. He received his B.S. in physics from Johns Hopkins University and his Ph.D. in mathematics from Cornell University. His first academic job was as an assistant professor in the Department of Statistics at Purdue University. While there, he taught the introductory concepts course with Professor Moore, and as a result of this experience he developed an interest in statistical education. Professor Notz is a coauthor of EESEE (the *Electronic Encyclopedia of Statistical Examples and Exercises*) and coauthor of *Statistics: Concepts and Controversies*. Professor Notz's research interests have focused on experimental design and computer experiments. He is the author of several research papers and of a book on the design and analysis of computer experiments. He is an elected fellow of the American Statistical Association. He has served as the editor of the journal *Technometrics* and as editor of the *Journal of Statistics Education*. He has served as the Director of the Statistical Consulting Service, as acting chair of the Department of Statistics for a year, and as an Associate Dean in the College of Mathematical and Physical Sciences at The Ohio State University. He is a winner of The Ohio State University's Alumni Distinguished Teaching Award.

Michael A. Fligner is an Adjunct Professor at the University of California at Santa Cruz and a nonresident Professor Emeritus with The Ohio State University. He received his B.S. in mathematics from the State University of New York at Stony Brook and his Ph.D. from the University of Connecticut. He spent most of his professional career at The Ohio State University where he was Vice Chair of the Department for over 10 years and also served as Director of the Statistical Consulting Service. He has done consulting work with several large corporations in Central Ohio. Professor Fligner's research interests are in Nonparametric Statistical methods, and he received the Statistics in Chemistry award from the American Statistical Association for work on detecting biologically active compounds. He is coauthor of the book *Statistical Methods for Behavioral Ecology* and received a Fulbright scholarship under the American Republics Research program to work at the Charles Darwin Research Station in the Galapagos Islands. He has been an Associate Editor of the *Journal of Statistical Education*. Professor Fligner is currently associated with the Center for Statistical Analysis in the Social Sciences at the University of California at Santa Cruz.

Getting Started

What's hot in popular music this week? SoundScan knows. SoundScan collects data electronically from the cash registers in more than 14,000 retail outlets and also collects data on download sales from Web sites. When you buy a CD or download a digital track, the checkout scanner or Web site is probably telling SoundScan what you bought. SoundScan provides this information to *Billboard* magazine, MTV, and VH1, as well as to record companies and artists' agents.

Should women take hormones such as estrogen after menopause, when natural production of these hormones ends? In 1992, several major medical organizations said "Yes." In particular, women who took hormones seemed to reduce their risk of a heart attack by 35% to 50%. The risks of taking hormones appeared small compared with the benefits. But in 2002, the National Institutes of Health declared these findings wrong. Use of hormones after menopause immediately plummeted. Both recommendations were based on extensive studies. What happened?

Is the climate warming? Is it becoming more extreme? An overwhelming majority of scientists now agree that the earth is undergoing major changes in climate. Enormous quantities of data are continuously being collected from weather stations, satellites, and other sources to monitor factors such as the surface temperature on land and sea, precipitation, solar activity, and the chemical composition of air and water. Climate models incorporate this information to make projections of future climate change and can help us understand the effectiveness of proposed solutions.

SoundScan, medical studies, and climate research all produce data (numerical facts), and lots of them. Using data effectively is a large and growing part of most professions, and reacting to data is part of everyday life. In fact, we define statistics as **the science of learning from data.**

Although data are numbers, they are not "just numbers." *Data are numbers with a context.* The number 8.5, for example, carries no information by itself. But if we hear that a friend's new baby weighed 8.5 pounds at birth, we congratulate her on the healthy size of the child. The context engages our background knowledge and allows us to make judgments. We know that a baby weighing 8.5 pounds is a little above average, and that a human baby is unlikely to weigh 8.5 ounces or 8.5 kilograms (over 18 pounds). The context makes the number informative.

To gain insight from data, we make graphs and do calculations. But graphs and calculations are guided by ways of thinking that amount to educated common sense. Let's begin our study of statistics with an informal look at some aspects of statistical thinking.[1]

Where the data comes from matters

Although, data can be collected in a variety of ways, the type of conclusion that can be reached from the data depends on how the data were obtained. *Observational studies* and *experiments* are two common methods for collecting data. Let's take a closer look at the hormone replacement data to understand the differences.

EXAMPLE 0.1   Hormone Replacement Therapy

What's behind the flip-flop in the advice offered to women about hormone replacement? The evidence in favor of hormone replacement came from a number of observational studies that compared women who were taking hormones with others who were not. But women who choose to take hormones are very different from women who do not: they are richer and better educated and see doctors more often. These women do many things to maintain their health. It isn't surprising that they have fewer heart attacks.

Large and careful observational studies are expensive, but they are easier to arrange than careful experiments. Experiments don't let women decide what to do. They assign women to either hormone replacement or to dummy pills that look and taste the same as the hormone pills. The assignment is done by a coin toss, so that all kinds of women are equally likely to get either treatment. Part of the difficulty of a good experiment is persuading women to accept the result—invisible to them—of the coin toss. By 2002, several experiments agreed that hormone replacement does *not* reduce the risk of heart attacks, at least for older women. Faced with this better evidence, medical authorities changed their recommendations.[2] ■

Women who chose hormone replacement after menopause were on the average richer and better educated than those who didn't. No wonder they had fewer heart attacks. We can't conclude that hormone replacement reduces heart attacks just because we see this relationship in data. In this example, education and affluence are background factors that help explain the relationship between hormone replacement and good health.

Children who play soccer do better in school (on the average) than children who don't play soccer. Does this mean that playing soccer increases school grades?

Children who play soccer tend to have prosperous and well-educated parents. Once again, education and affluence are background factors that help explain the relationship between soccer and good grades.

Almost all relationships between two observed characteristics or "variables" are influenced by other variables lurking in the background. To understand the relationship between two variables, you must often look at other variables. Careful statistical studies try to think of and measure possible *lurking variables* to correct for their influence. As the hormone saga illustrates, this doesn't always work well. News reports often just ignore possible lurking variables that might ruin a good headline like "Playing soccer can improve your grades." The habit of asking, "What might lie behind this relationship?" is part of thinking statistically.

Of course, observational studies are still quite useful. We can learn from observational studies how chimpanzees behave in the wild or which popular songs sold best last week or what percent of workers were unemployed last month. SoundScan's data on popular music and the government's data on employment and unemployment come from *sample surveys,* an important kind of observational study that chooses a part (the sample) to represent a larger whole. Opinion polls interview perhaps 1000 of the 235 million adults in the United States to report the public's views on current issues. Can we trust the results? We'll see that this isn't a simple yes-or-no question. Let's just say that the government's unemployment rate is much more trustworthy than opinion poll results, and not just because the Bureau of Labor Statistics interviews 60,000 people rather than 1000. We can, however, say right away that some samples *can't* be trusted. Consider the following write-in poll.

EXAMPLE 0.2	Would You Have Children Again?

The advice columnist Ann Landers once asked her readers, "If you had it to do over again, would you have children?" A few weeks later, her column was headlined "70% OF PARENTS SAY KIDS NOT WORTH IT." Indeed, 70% of the nearly 10,000 parents who wrote in said they would not have children if they could make the choice again. Those 10,000 parents were upset enough with their children to write Ann Landers. Most parents are happy with their kids and don't bother to write. ■

Statistically designed samples, even opinion polls, don't let people choose themselves for the sample. They interview people selected by impersonal chance so that everyone has an equal opportunity to be in the sample. Such a poll showed that 91% of parents *would* have children again. *Where data come from matters a lot.* If you are careless about how you get your data, you may announce 70% "No" when the truth is close to 90% "Yes." Understanding the importance of where the data comes from and its relationship to the conclusions that can be reached is an important part of learning to think statistically.

Always look at the data

Yogi Berra, the Hall of Fame New York Yankee, said it: "You can observe a lot by just watching." That's a motto for learning from data. *A few carefully chosen graphs are often more instructive than great piles of numbers.* Consider the outcome of the 2000 presidential election in Florida.

EXAMPLE 0.3   ## Palm Beach County

Elections don't come much closer: after much recounting, state officials declared that George Bush had carried Florida by 537 votes out of almost 6 million votes cast. Florida's vote decided the 2000 presidential election and made George Bush, rather than Al Gore, president. Let's look at some data. Figure 0.1 displays a graph that plots votes for the third-party candidate Pat Buchanan against votes for the Democratic candidate Al Gore in Florida's 67 counties.

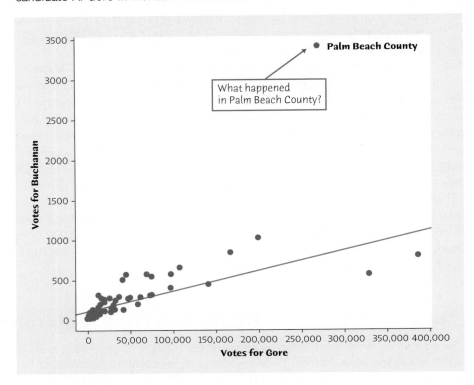

FIGURE 0.1

Votes in the 2000 presidential election for Al Gore and Patrick Buchanan in Florida's 67 counties. What happened in Palm Beach County?

What happened in Palm Beach County? The question leaps out from the graph. In this large and heavily Democratic county, a conservative third-party candidate did far better relative to the Democratic candidate than in any other county. The points for the other 66 counties show votes for both candidates increasing together in a roughly straight-line pattern. Both counts go up as county population goes up. Based on this pattern, we would expect Buchanan to receive around 800 votes in Palm Beach County. He actually received more than 3400 votes. That difference determined the election result in Florida and in the nation. ■

The graph demands an explanation. It turns out that Palm Beach County used a confusing "butterfly" ballot (see photo on page 5), in which candidate names on both left and right pages led to a voting column in the center. It would be easy for a voter who intended to vote for Gore to in fact cast a vote for Buchanan. The graph is convincing evidence that this in fact happened.

Most statistical software will draw a variety of graphs with a few simple commands. Examining your data with appropriate graphs and numerical summaries is the correct place to begin most data analyses. These can often reveal important patterns or trends that will help you understand what your data has to say.

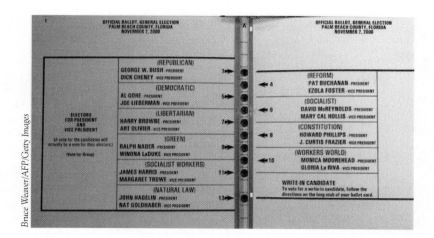

Bruce Weaver/AFP/Getty Images

Variation is everywhere

The company's sales reps file into their monthly meeting. The sales manager rises. "Congratulations! Our sales were up 2% last month, so we're all drinking champagne this morning. You remember that when sales were down 1% last month I fired half of our reps." This picture is only slightly exaggerated. Many managers overreact to small short-term variations in key figures. Here is Arthur Nielsen, former head of the country's largest market research firm, describing his experience:

> *Too many business people assign equal validity to all numbers printed on paper. They accept numbers as representing Truth and find it difficult to work with the concept of probability. They do not see a number as a kind of shorthand for a range that describes our actual knowledge of the underlying condition.*[3]

Business data such as sales and prices vary from month to month for reasons ranging from the weather to a customer's financial difficulties to the inevitable errors in gathering the data. The manager's challenge is to say when there is a real pattern behind the variation. We'll see that statistics provides tools for understanding variation and for seeking patterns behind the screen of variation. Let's look at some more data.

EXAMPLE 0.4   The Price of Gas

Figure 0.2 plots the average price of a gallon of regular unleaded gasoline each week from September 1990 to June 2013.[4] There certainly is variation! But a close look shows a yearly pattern: gas prices go up during the summer driving season, then down as demand drops in the fall. On top of this regular pattern, we see the effects of international events. For example, prices rose when the 1990 Gulf War threatened oil supplies and dropped when the world economy turned down after the September 11, 2001, terrorist attacks in the United States. The years 2007 and 2008 brought the perfect storm: the ability to produce oil and refine gasoline was overwhelmed by high demand from China and the United States and continued turmoil in the oil-producing areas of the Middle East and Nigeria. Add in a rapid fall in the value of the dollar, and prices at the pump skyrocketed to more than $4 per gallon. In 2010 the Gulf oil spill also affected supply and hence prices. The data carry an important message: because the United States imports much of its oil, we can't control the price we pay for gasoline. ■

Taylor Avelar/Bloomberg via Getty Images

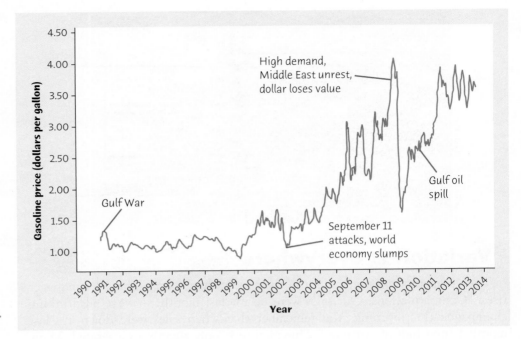

FIGURE 0.2

Variation is everywhere: the average retail price of regular unleaded gasoline, 1990 to mid 2013.

Variation is everywhere. Individuals vary; repeated measurements on the same individual vary; almost everything varies over time. One reason we need to know some statistics is that it helps us deal with variation and to describe the uncertainty in our conclusions. Let's look at another example to see how variation is incorporated into our conclusions.

EXAMPLE 0.5 **The HPV Vaccine**

Cervical cancer, once the leading cause of cancer deaths among women, is the easiest female cancer to prevent with regular screening tests and follow-up. Almost all cervical cancers are caused by human papillomavirus (HPV). The first vaccine to protect against the most common varieties of HPV became available in 2006. The Centers for Disease Control and Prevention recommend that all girls be vaccinated at age 11 or 12. In 2011, the CDC made the same recommendation for boys, to protect against anal and throat cancers caused by the HPV virus.

How well does the vaccine work? Doctors rely on experiments (called "clinical trials" in medicine) that give some women the new vaccine and others a dummy vaccine. (This is ethical when it is not yet known whether or not the vaccine is safe and effective.) The conclusion of the most important trial was that an estimated 98% of women up to age 26 who are vaccinated before they are infected with HPV will avoid cervical cancers over a 3-year period.

Women who get the vaccine are much less likely to get cervical cancer. But because variation is everywhere, the results are different for different women. Some vaccinated women will get cancer, and many who are not vaccinated will escape. Statistical conclusions are "on the average" statements only, and even these "on the average" statements have an element of uncertainty. Although we can't be 100% certain that the vaccine reduces risk on the average, statistics allows us to state how confident we are that this is the case. ◼

Because variation is everywhere, conclusions are uncertain. Statistics gives us a language for talking about uncertainty that is used and understood by statistically literate people everywhere. In the case of HPV vaccine, the medical journal used that language to tell us: "Vaccine efficiency . . . was 98% (95 percent confidence interval 86% to 100%)."[5] That "98% effective" is, in Arthur Nielsen's words, "shorthand for a range that describes our actual knowledge of the underlying condition." The range is 86% to 100%, and we are 95 percent confident that the truth lies in that range. We will soon learn to understand this language. We can't escape variation and uncertainty. Learning statistics enables us to live more comfortably with these realities.

What lies ahead in this book

The purpose of *Statistics in Practice* is to give you a working knowledge of the ideas and tools of practical statistics. We will divide practical statistics into three main areas.

- **Data analysis** concerns methods and strategies for looking at data; exploring, organizing, and describing data using graphs and numerical summaries. Your thoughtful exploration allows data to illuminate reality. Part I of this book (Chapters 1 to 6) discusses data analysis.

- **Data production** provides methods for producing data that can give clear answers to specific questions. Where data come from matters and is often the most important limitation on their usefulness. Basic concepts about how to select samples and design experiments are some of the most influential ideas in statistics. These concepts are the subject of Chapters 8 and 9.

- **Statistical inference** moves beyond the data in hand to draw conclusions about some wider universe. Statistical conclusions aren't Yes or No answers— they must take into account that variation is everywhere—variability among people, animals, or objects and uncertainty in data. To describe variation and uncertainty, inference uses the language of probability, introduced in Chapter 11. Because we are concerned with practice rather than theory, we need only a limited knowledge of probability. Chapters 12 and 13 offer more probability for those who want it. Chapters 15, 16, and 17 discuss the reasoning of statistical inference. These chapters are the key to the rest of the book. Chapters 18 to 22 continue our presentation of inference as used in practice to other common settings. Chapters 24 to 26 concern more advanced or specialized kinds of inference.

Because data are numbers with a context, doing statistics means more than manipulating numbers. You must **state** a problem in its real-world context, **plan** your specific statistical work in detail, **solve** the problem by making the necessary graphs and calculations, and **conclude** by explaining what your findings say about the real-world setting. We'll make regular use of this four-step process to encourage good habits that go beyond graphs and calculations to ask, "What do the data tell me?"

Statistics does involve lots of calculating and graphing. The text presents the techniques you need, but you should use technology to automate calculations and graphs as much as possible. Because the big ideas of statistics don't depend on any particular level of access to technology, *Statistics in Practice* does not require software or a graphing calculator until we reach the more advanced methods in Part IV of the text. Even if you make little use of technology, you should look at the "Using Technology" sections throughout the book. You will see at once that you can read and apply the output from almost any technology used for statistical calculations. The ideas really are more important than the details of how to do the calculations.

Unless you have access to software or a graphing calculator, *you will need a basic calculator with some built-in statistical functions.* Specifically, your calculator should find means and standard deviations and calculate correlations and regression lines. Look for a calculator that claims to do "two-variable statistics" or mentions "regression."

Although ability to carry out statistical procedures is very useful in academics and employment, the most important asset you can gain from the study of statistics is an understanding of the big ideas about working with data. *Statistics in Practice* tries to explain the most important ideas of statistics, not just teach methods. Some examples of big ideas that you will meet (one from each of the three areas of statistics) are "always plot your data," "randomized comparative experiments," and "statistical significance."

You learn statistics by doing statistical problems. As you read, you will see several levels of exercises, arranged to help you learn. Short "Apply Your Knowledge" problem sets appear after each major idea. These are straightforward exercises that help you solidify the main points as you read. Be sure you can do these exercises before going on. The end-of-chapter exercises begin with multiple-choice "Check Your Skills" exercises (with all answers in the back of the book). Use them to check your grasp of the basics. The regular "Chapter Exercises" help you combine all the ideas of a chapter. Finally, the four Part Review chapters (Chapters 7, 10, 14, and 23) look back over major blocks of learning, with many review exercises. At each step you are given less advance knowledge of exactly what statistical ideas and skills the problems will require, so each type of exercise requires more understanding.

The key to learning is persistence. The main ideas of statistics, like the main ideas of any important subject, took a long time to discover and take some time to master. The gain will be worth the pain.

CHAPTER 0 EXERCISES

0.1 **Observational studies and experiments.** Observational studies have suggested that vitamin E reduces the risk of heart disease. Careful experiments, however, showed that vitamin E has no effect. According to a commentary in the *Journal of the American Medical Association*:

Thus, vitamin E enters the category of therapies that were promising in epidemiologic and observational studies but failed to deliver in adequately powered randomized controlled trials. As in other studies, the "healthy user" bias must be considered; i.e., the healthy lifestyle behaviors that characterize individuals who care enough about their health to take various supplements are actually responsible for the better health, but this is minimized with the rigorous trial design.[6]

(a) Reread Example 0.1 (page 2) and the comments following it. Explain why observational studies suggest that vitamin E therapy reduces the risk of heart disease by describing some lurking variables.

(b) A randomized controlled trial is a type of experiment. How does "healthy user bias" explain how people who take vitamin E supplements have better health in observational studies but not in experiments?

0.2 **The price of gas.** In Example 0.4 (page 5) we examined the variation in the price of gasoline from 1990 to 2013. We saw both a regular pattern and the effects of international events. Figure 0.3 plots the average annual retail price of gasoline from 1929 to 1990.[7] Prices are adjusted for inflation. What overall patterns do you observe? What departures from the overall patterns do you observe? To what international events do these departures correspond?

0.3 **Online polls.** Ed Schultz is a liberal political commentator and host of *The Ed Show*. After an often lengthy and impassioned monologue from Mr. Schultz, viewers are asked to text their replies to a poll question. On June 23, 2013, after a monologue that included various issues that the Republicans could not be trusted on, Ed asked viewers to text in their responses to the question, "Do Republicans care about the personal struggles of undocumented immigrants?" Approximately 96.4% of those responding either by text or online said "No."

(a) This poll has some of the same problems as the Ann Landers' poll of Example 0.2 (page 3). Do you think that the proportion of Americans who feel this way is higher, lower, or close to 96.4%? Explain.

(b) For this poll, 868 people responded. Among those responding, 837 or 96.4% said "No." Do you think the results would have been more trustworthy if 2500 people had responded instead of 868? Explain.

0.4 **Traffic fatalities and 9/11.** Figure 0.4 provides information on the number of fatal crashes by month for the years 1996–2001.[8] The vertical blue line above each month gives the lowest to highest number of fatal crashes for that month for the years 1996–2000. For example, in January the number of fatal crashes for the 5 years from 1996 through 2000 was between about 2600 and 2900. The blue dots give the number of fatal crashes for each month in 2001. The numbers

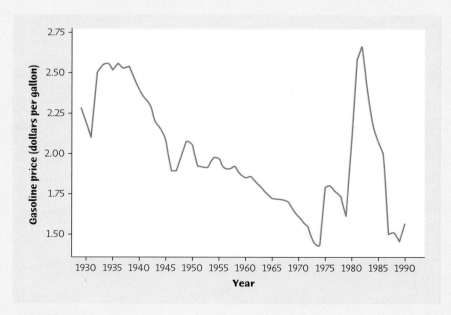

FIGURE 0.3

The average annual retail price of gasoline, 1929 to 1990. Prices are adjusted for inflation.

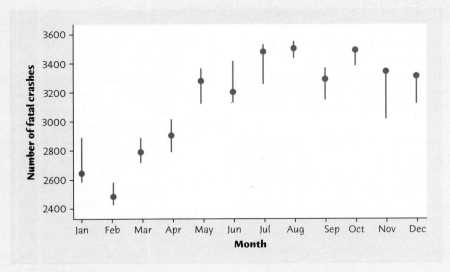

FIGURE 0.4

Number of fatal traffic accidents in the United States in 1996 through 2000 versus 2001. For each month, the blue lines represent the range of the number of fatal accidents from 1996 through 2000, and the blue dot gives the number of fatal accidents in 2001.

of fatal crashes from January through August of 2001 follow the general pattern for the 5 preceding years as we see the blue dots are well within the blue lines for each month.

(a) What happened in the last 3 months of 2001? The number of fatal crashes in October through December of 2001 are consistently at or above the values for the previous 5 years. How can you tell this from the graph?

(b) On September 11, 2001, terrorists hijacked four U.S. airplanes and used them to strike various targets on the East Coast. In part (a), we saw from the graph that fatal crashes seemed to be unusually high in the 3 months following the attacks. Did the terrorists cause fatal crashes to increase? Can you give a simple explanation for the apparent increase in fatal crashes during these months?

Exploring Data

"What do the data say?" is the first question we ask in any statistical study. *Data analysis* answers this question by open-ended exploration of the data. The tools of data analysis are graphs such as histograms and scatterplots and numerical measures such as means and correlations. At least as important as the tools are principles that organize our thinking as we examine data. The seven chapters in Part I present the principles and tools of statistical data analysis. They equip you with skills that are immediately useful whenever you deal with numbers.

These chapters reflect the strong emphasis on exploring data that characterizes modern statistics. Sometimes we hope to draw conclusions that apply to a setting that goes beyond the data in hand. This is *statistical inference*, the topic of much of the rest of the book. Data analysis is essential if we are to trust the results of inference, but data analysis isn't just preparation for inference. Roughly speaking, you can always do data analysis but inference requires rather special conditions.

One of the organizing principles of data analysis is to first look at one thing at a time and then at relationships. Our presentation follows this principle. In Chapters 1, 2, and 3 you will study *variables and their distributions*. Chapters 4, 5, and 6 concern *relationships among variables*. Chapter 7 reviews this part of the text.

Picturing Distributions with Graphs

Overview

The typical statistics application has three important components: collection of data, description of data, and inference. *Inferential statistics* concerns reaching conclusions (estimates, decisions, predictions, comparisons) about a population (or populations) based on information contained in a sample (or samples). *Descriptive statistics* are typically used to support this endeavor, though they are often useful outside any need for inference. The first chapters in this book discuss various descriptive statistical methods, and later chapters describe inferential methods. Probability (in Part III of the text) serves as a bridge between descriptive statistics and inferential statistics. Material on collecting data (experimental design, sampling methods, and related issues) in Part II of *Statistics in Practice* can be taught in its entirety prior to Part I if you prefer. The authors structured Part II of *SIP* so that you can make this change.

Chapter 1 concerns graphical displays of data for a single variable. Graphical displays can convey sophisticated or subtle messages that simple numbers cannot convey. Although there is no "correct" graph for a set of data, the best graph or plot can reveal surprising results that may inspire a different line of questioning, or it may reveal previously unknown truths; once data have been entered in a data analysis package, generating many graphs is quick and easy. Consider a famous early example of epidemiology: 19th-century physician John Snow plotted the addresses of cholera victims on a map of central London, revealing the location of a water pump that was the source of the epidemic. For the plot, and information about Snow and the epidemic, see http://en.wikipedia.org/wiki/John_Snow_(physician).

LEARNING OBJECTIVES**

- Identify the individuals and variables in a set of data.
- Identify each variable as categorical or quantitative. Identify the units in which each quantitative variable is measured.
- Recognize when a pie chart can and cannot be used, and when it is not desirable (for example, when there are too many categories).
- Make a bar graph to display the distribution of a categorical variable or to compare related distributions.
- Interpret pie charts and bar graphs.
- Make a histogram of the distribution of a quantitative variable.
- Make a stemplot of the distribution of a small set of quantitative observations.
- Look for the overall pattern and for major deviations from the pattern.
  - Assess from a histogram or stemplot whether the shape of a distribution is roughly symmetric, distinctly skewed, or neither.
  - Identify the center and variability/spread of the distribution in general terms (this will be covered more specifically in Chapter 2).
  - Recognize outliers and give plausible explanations for them.
- Recognize patterns such as trends and cycles in time plots.

**These learning outcomes appear later for the students in Chapter 7: Part I Review.

The type of graphical display we create depends on the type of variable one is graphing: we often use bar graphs or pie charts for categorical data and histograms or stemplots for quantitative data. Each of these types of graphical display is intended to help the investigator *explore* the data by revealing the *distribution* of the variable in question. A description of a quantitative variable's distribution involves three components: *shape, center,* and *variability/spread.* Graphical displays reveal all three of these features, as well as any potential *outliers,* and are almost always a first step in data analysis.

In Chapter 1, the authors refer to variability as "spread." Beginning with Chapter 2, the word "variability" is interspersed with "spread," so that instructors and students can remember that it is variability that is of import in statistics. This shift in wording comes from recent research in statistics education that has alerted statistics teachers to issues of lexical ambiguity in statistics. See "Lexical Ambiguity in Statistics: How Students Use and

Define the Words: Association, Average, Confidence, Random and Spread" (Kaplan, Fisher, and Rogness, *Journal of Statistics Education*, 18(2), 2010; http://www.amstat.org/publications/jse/v18n2/kaplan.pdf). If you have access to the journal *Teaching Statistics*, see also "Lexical Ambiguity: Making a Case against Spread" (Kaplan, Rogness, and Fisher, 34(2), pp. 56–60, Summer 2012; DOI: 10.1111/j.1467-9639.2011.00477.x).

Teaching Suggestions and Additional Examples/Activities for the Classroom

1. Provide a Context for Inference Right Away

At the beginning of a statistics course, many students will not grasp the relevance of the subject. On the very first day of class, bring the subject to life in the context of *making a decision in the face of uncertainty*. Some examples:

A. Ask students which of three medical treatments has the lowest average time to effect. Provide two cases of data obtained from some hypothetical experiments for them to compare (times given in minutes)

> **Case 1:** Treatment 1: 30 32 28 30 (mean = 30 minutes)
> Treatment 2: 36 37 35 36 (mean = 36 minutes)
> Treatment 3: 24 26 28 30 (mean = 27 minutes)
>
> **Case 2:** Treatment 1: 24 22 36 38 (mean = 30 minutes)
> Treatment 2: 30 34 38 46 (mean = 36 minutes)
> Treatment 3: 20 24 30 34 (mean = 27 minutes)

If you plot the data on number lines (e.g., a dotplot using colors or symbols to distinguish the three groups), it is clear that the evidence for Treatment 3's superiority is stronger in Case 1 than in Case 2, even though in both cases the treatment averages are the same. This example illustrates two important aspects of statistics problems: (1) We are not able to *know* (with absolute certainty) which option is best . . . but we *can* reach conclusions and make statements about the amount of faith we have in them (less faith in Case 2, more faith in Case 1) and (2) statistical decisions can only be made by considering variation in data . . . comparing means is not enough.

B. Tell your students that you are an 80% free throw shooter. Have them imagine conducting an experiment in which you attempt 100 free throws. For each of the following cases, ask whether they would believe your claim or conclude that you are not a 80% free throw shooter:

> **Case 1:** You make 82 of 100 shots. They should believe the claim.
>
> **Case 2:** You make 78 shots. They might believe the claim.
>
> **Case 3:** You make 75 shots. They might begin to doubt the claim.

Continue to decrease the number of shots made, and watch as students change their minds from "believe claim" to "disbelieve claim." Of course, they're using an uninformed guess for what values would be plausible if your claim was correct. Talk with your students about why they changed their minds and how much "evidence" was needed to change their minds. At this stage, you can show them a sampling distribution (don't call it that yet!) for the number of free throw shots made in 100 attempts by a real 80% free throw shooter. For this you can show your students the true binomial distribution with $n = 100$ trials and $p = 0.80$, which has mean 80 and standard deviation 4. Looking at the binomial distribution, it should be clear to students that virtually never would a true 80% shooter make fewer than about 68 shots. Another option is for you to use *The Reasoning of a Statistical Test* applet, as the applet uses the example of a free throw shooter. You can also simulate your own sampling distribution using R or Excel. Students should realize that the distribution provides a more objective framework for reaching an ultimately subjective decision. This example illustrates three important facts: (1) Variability in a distribution gives us a reference scale for what may or may not be unusual; (2) probability and chance play a strong role in decision making; and (3) all statistical decisions are subject to error and involve an element of subjectivity, even if they are rooted in objective benchmarks.

2. Emphasize Interpretation over Construction of Graphical Displays

Why would we want students to construct histograms by hand? Instead, students need to be able to look at a graph and extract the salient features (shape, center, variability/spread) and any important messages or lessons about the data represented. Rather than have students make graphs themselves, have them "predict" the shapes of various distributions: The heights of adult women (relatively symmetric), prices of houses in a large metropolitan area (right-skewed—there are many more lower-priced homes than higher-priced homes), exam scores for a very easy test (left-skewed), the lifetimes of a particular spider species (extremely right-skewed—most spiders don't live very long), etc.

3. Discuss Issues with Bad Graphical Displays

We want students to be savvy consumers of information they encounter in the media. We want them to ask questions about what they see. Exposing them to poorly designed graphical displays prepares them to view with skepticism graphs they encounter in the news. Besides, students enjoy seeing examples of very bad graphical displays. For some interesting examples, visit the Gallery of Data Visualization (http://www.datavis.ca/gallery/index. php). *USA Today* is usually a good source of these as well.

4. Illustrate Concepts with Simple but Good Graphical Displays

Use a simple data set for several different types of graphs. A good example is the set of launch-time temperatures (in degrees Fahrenheit) for the first 25 space shuttle missions (66, 70, 69, 80, 68, 67, 72, 70, 70, 57, 63, 70, 78, 67, 53, 67, 75, 70, 81, 76, 79, 75, 76, 58, 29, sorted chronologically). The 25th mission was the ill-fated *Challenger* disaster in January 1986. A histogram, stemplot, or dotplot will reveal just how large an outlier the 29 degree launch temperature was and will invite a "cause-and-effect" discussion. These data are observational, and no such cause-and-effect inference is possible . . . but at least the graph can point to a possible culprit for the accident. A stemplot can be created and can be shown again in Chapter 2, when the median, quartiles, and boxplots are discussed.

Other Resources (LaunchPad)

StatClips

> *Introduction to Statistics*
> *Types of Variables*
> *Summaries and Pictures for Categorical Data*
> *Exploratory Pictures for Quantitative Data*

Snapshots Videos

> *Introduction to Statistics*
> *Data and Distributions*
> *Visualizing and Summarizing Categorical Data*
> *Visualizing Quantitative Data.*

EESEE Case Studies

> *Historical Farm Data* (time plots)
> *How many poets wrote Cleanness?* (pie charts)
> *Nutrition and Breakfast Cereals* (histogram and shape)
> *Is Friday the 13th Unhealthy?* (graphing distributions of shoppers)

Applet

> *One-Variable Statistical Calculator*

Photograph by the U.S. Census Bureau, Public Information Office (PIO)

Picturing Distributions with Graphs

Statistics is the science of data. The volume of data available to us is overwhelming. For example, the U.S. Census Bureau's American Community Survey collects data from about 3,000,000 housing units each year. Astronomers work with data on tens of millions of galaxies. The checkout scanners at Walmart's 8000 stores in 15 countries record hundreds of millions of transactions every week, all saved to inform both Walmart and its suppliers. The first step in dealing with such a flood of data is to organize our thinking about data. Fortunately, we can do this without looking at millions of data points.

Individuals and variables

Any set of data contains information about some group of *individuals*. The information is organized in *variables*.

Individuals and Variables

Individuals are the objects described by a set of data. Individuals may be people, but they may also be animals or things.

A **variable** is any characteristic of an individual. A variable can take different values for different individuals.

A college's student database, for example, includes data about every currently enrolled student. The students are the individuals described by the data set. For each individual, the data contain the values of variables such as date of birth, choice of major, and grade point average (GPA). In practice, any set of data is accompanied by background information that helps us understand the data. When you plan a statistical study or explore data from someone else's work, ask yourself the following questions:

1. **Who?** What **individuals** do the data describe? **How many** individuals appear in the data?

2. **What?** How many **variables** do the data contain? What are the **exact definitions** of these variables? In what **unit of measurement** is each variable recorded? Weights, for example, might be recorded in pounds, in thousands of pounds, or in kilograms.

3. **Where?** Student GPAs and SAT scores (or lack of them) will vary from college to college depending on many variables, including admissions "selectivity" for the college.

4. **When?** Students change from year to year, as do prices, salaries, etc.

5. **Why?** What **purpose** do the data have? Do we hope to answer some specific questions? Do we want answers for just these individuals or for some larger group that these individuals are supposed to represent? Are the individuals and variables suitable for the intended purpose?

Some variables, like a person's sex or college major, simply place individuals into categories. Others, like height and GPA, take numerical values for which we can do arithmetic. It makes sense to give an average income for a company's employees, but it does not make sense to give an "average" sex. We can, however, count the numbers of female and male employees and do arithmetic with these counts.

statistics in Your World

WHAT'S THAT NUMBER?

You might think that numbers, unlike words, are universal. Think again. A "billion" in the United States means 1,000,000,000 (nine zeros). In Europe, a "billion" is 1,000,000,000,000 (twelve zeros). OK, those are words that describe numbers. But those commas in big numbers are periods in many other languages. This is so confusing that international standards call for spaces instead, so that an American billion is written 1 000 000 000. And the decimal point of the English-speaking world is the decimal comma in many other languages, so that 3.1416 in the United States becomes 3,1416 in Europe. So what is the number 10,642.389? Depends on where you are.

Categorical and Quantitative Variables

A **categorical variable** places an individual into one of several groups or categories.

A **quantitative variable** takes numerical values for which arithmetic operations such as adding and averaging make sense. The values of a quantitative variable are usually recorded in a **unit of measurement** such as seconds or kilograms.

EXAMPLE 1.1   The American Community Survey

At the U.S. Census Bureau Web site, you can view the detailed data collected by the American Community Survey, though of course the identities of people and housing units are protected. If you choose the file of data on people, the *individuals* are the people living in the housing units contacted by the survey. Over 100 variables are recorded for each individual. Figure 1.1 displays a very small part of the data.

⊠ eg01-01.csv							_ □ ✕
	A	B	C	D	E	F	G
1	SERIALNO	PWGTP	AGEP	JWMNP	SCHL	SEX	WAGP
2	283	187	66		6	1	24000
3	283	158	66		9	2	0
4	323	176	54	10	12	2	11900
5	346	339	37	10	11	1	6000
6	346	91	27	10	10	2	30000
7	370	234	53	10	13	1	83000
8	370	181	46	15	10	2	74000
9	370	155	18		9	2	0
10	487	233	26		14	2	800
11	487	146	23		12	2	8000
12	511	236	53		9	2	0
13	511	131	53		11	1	0
14	515	213	38		11	2	12500
15	515	194	40		9	1	800
16	515	221	18	20	9	1	2500
17	515	193	11		3	1	

eg01-01

FIGURE 1.1
A spreadsheet displaying data from the American Community Survey, for Example 1.1.

Each row in the spreadsheet contains data on one individual.

Each row records data on one individual. Each column contains the values of one *variable* for all the individuals. Translated from the U.S. Census Bureau's abbreviations, the variables are

SERIALNO   An identifying number for the household.
PWGTP   Weight in pounds.
AGEP   Age in years.
JWMNP   Travel time to work in minutes.
SCHL   Highest level of education. The numbers designate categories, *not* specific grades. For example, 9 = high school graduate, 10 = some college but no degree, and 13 = bachelor's degree.
SEX   Sex, designated by 1 = male and 2 = female.
WAGP   Wage and salary income last year, in dollars.

Look at the highlighted row in Figure 1.1. This individual is a 53-year-old man who weighs 234 pounds, travels 10 minutes to work, has a bachelor's degree, and earned $83,000 last year.

In addition to the household serial number, there are six variables. Education and sex are categorical variables. The values for education and sex are stored as numbers, but these numbers are just labels for the categories and have no units of measurement. The other four variables are quantitative. Their values do have units. These variables are weight in pounds, age in years, travel time in minutes, and income in dollars.

The *purpose* of the American Community Survey is to collect data that represent the entire nation to guide government policy and business decisions. To do this, the households contacted are chosen at random from all households in the country. We will see in Chapter 8 why choosing at random is a good idea. ■

Most data tables follow this format—each row is an individual, and each column is a variable. The data set in Figure 1.1 appears in a **spreadsheet** program that has rows and columns ready for your use. Spreadsheets are commonly used to enter and transmit data and to do simple calculations.

spreadsheet

Apply Your Knowledge

1.1 **Fuel Economy.** Here is a small part of a data set that describes the fuel economy (in miles per gallon) of model year 2013 motor vehicles:

Make and Model	Vehicle Type	Transmission Type	Number of Cylinders	City mpg	Highway mpg	Comb CO_2 (g/mile)
⋮						
Aston Martin Vantage	Two-seater	Manual	8	13	19	581
Honda Fit	Small station wagon	Automatic	4	28	35	286
Toyota Prius	Midsize	Automatic	4	51	48	179
Chevrolet Impala	Large	Automatic	6	18	29	413
⋮						

The Comb CO_2 measures the amount of CO_2 emitted by a vehicle's tailpipe in grams/mile for combined city and highway driving.

(a) What are the individuals in this data set?

(b) For each individual, what variables are given? Which of these variables are categorical, and which are quantitative? In what units are the quantitative variables measured?

1.2 **Students and Exercise.** You are preparing to study the exercise habits of college students. Describe two categorical variables and two quantitative variables that you might measure for each student. Give the units of measurement for the quantitative variables.

Categorical variables: pie charts and bar graphs

exploratory data analysis

Statistical tools and ideas help us examine data to describe their main features. This examination is called **exploratory data analysis.** Like an explorer crossing unknown lands, we want first to simply describe what we see. Here are two principles that help us organize our exploration of a set of data.

Exploring Data

1. Begin by examining each variable by itself. Then move on to study the relationships among the variables.

2. Begin with a graph or graphs. Then add numerical summaries of specific aspects of the data.

We will follow these principles in organizing our learning. Chapters 1 to 3 present methods for describing a single variable. We study relationships among several variables in Chapters 4 to 6. In each case, we begin with graphical displays, then add numerical summaries for more complete description.

The proper choice of graph depends on the nature of the variable. To examine a single variable, we usually want to display its *distribution.*

Distribution of a Variable

The **distribution** of a variable tells us what values it takes and how often it takes these values.

The values of a categorical variable are labels for the categories. The **distribution of a categorical variable** lists the categories and gives either the count or the percent of individuals who fall in each category.

EXAMPLE 1.2   Which Major?

Approximately 1.5 million full-time, first-year students enrolled in colleges and universities in 2011. What do they plan to study? Here are data on the percents of first-year students who plan to major in several discipline areas:[1]

MAJORS

Field of Study	Percent of Students
Arts and humanities	11.0
Biological sciences	10.9
Business	14.9
Education	5.9
Engineering	12.0
Physical sciences	3.7
Professional	14.9
Social sciences	12.1
Technical	1.0
Other majors and undeclared	13.5
Total	99.9

It's a good idea to check data for consistency. The percents should add to 100%. In fact, they add to 99.9%. What happened? Each percent is rounded to the nearest tenth. The exact percents would add to 100, but the rounded percents only come close. This is **roundoff error.** Roundoff errors don't point to mistakes in our work, just to the effect of rounding off results. ■

roundoff error

Columns of numbers take time to read. You can use a pie chart or a bar graph to display the distribution of a categorical variable more vividly. Figures 1.2 and 1.3 illustrate these displays for the distribution of intended college majors.

Pie charts show the distribution of a categorical variable as a "pie" whose slices are sized by the counts or percents for the categories. Pie charts are awkward to make by hand, but software will do the job for you. *A pie chart must include all the categories that make up a whole. Use a pie chart only when you want to emphasize each category's relation to the whole.* We need the "Other majors and undeclared" category in Example 1.2 to complete the whole (all intended majors) and allow us to make the pie chart in Figure 1.2.

pie chart

FIGURE 1.2

You can use a pie chart to display the distribution of a categorical variable. This pie chart shows the distribution of intended majors of students entering college.

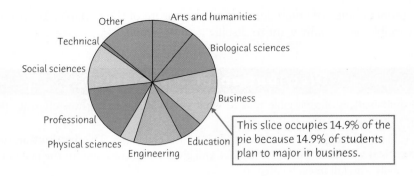

This slice occupies 14.9% of the pie because 14.9% of students plan to major in business.

bar graph   **Bar graphs** represent each category as a bar. The bar heights show the category counts or percents. Bar graphs are easier to make than pie charts and also easier to read. Figure 1.3 displays two bar graphs of the data on intended majors. The first orders the bars alphabetically by field of study (with "Other" at the end). It is often better to arrange the bars in order of height, as in Figure 1.3(b). This helps us immediately see which majors appear most often.

Bar graphs are more flexible than pie charts. Both graphs can display the distribution of a categorical variable, but a bar graph can also compare any set of quantities that are measured in the same units.

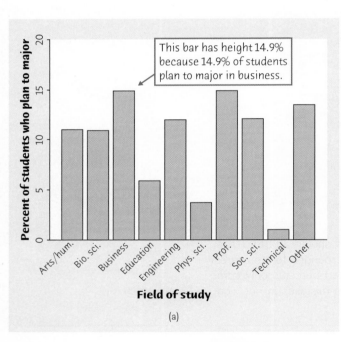

This bar has height 14.9% because 14.9% of students plan to major in business.

(a)

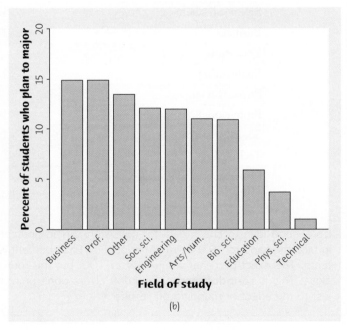

(b)

FIGURE 1.3

Bar graphs of the distribution of intended majors of students entering college. In (a), the bars follow the alphabetical order of fields of study. In (b), the same bars appear in order of height.

EXAMPLE 1.3   Smartphones Have Biggest Impact!

DEVICES

The rating service Arbitron asked Americans over 12 years old who used several digital platforms/devices to answer the question "How much of an impact on your life has (platform/device) had?" with 5 = Big impact and 1 = No impact at all. Here are the percents in decreasing order who said, "Big impact." Only those platforms/devices for which 15% or more said "Big impact" are reported.[2]

Platform/Device	Percent of Users Who Said "Big Impact"
Apple iPhone	53
Android Smartphone	50
Broadband Internet Access	43
BlackBerry	37
Cell Phone (not Smartphone)	35
Digital Video Recorder	27
Apple iPad	27
Apple iPod	23
Android Tablet	22
Windows Smartphone	19
eBook Reader	15

We can't make a pie chart to display these data. Each percent in the table refers to a different platform/device, not to parts of a single whole. Figure 1.4 is a bar graph comparing the 11 platforms/devices. We have again arranged the bars in order of height. ■

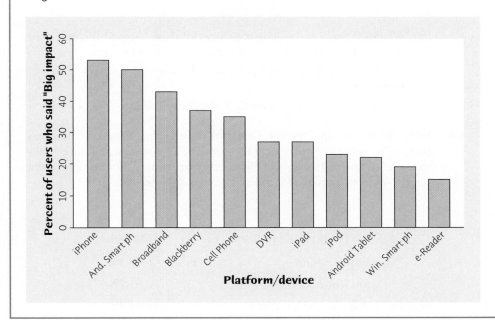

FIGURE 1.4
You can use a bar graph to compare quantities that are not part of a whole. This bar graph compares the percents of users who said, "Big Impact," when asked how much of an impact a certain platform/device had on their lives, for Example 1.3.

Bar graphs and pie charts are mainly tools for presenting data: they help your audience grasp data quickly. Because it is easy to understand data on a single categorical variable without a graph, bar graphs and pie charts are of limited use for data analysis. We will move on to quantitative variables, where graphs are essential tools.

Apply Your Knowledge

1.3 **Do You Listen to Country Radio?** The rating service Arbitron places U.S. radio stations into more than 50 categories that describe the kind of programs they broadcast. Which formats attract the largest audiences? Here are Arbitron's measurements of the share of the listening audience (aged 12 and over) at a given time for the most popular formats:[3] ▦ RADIO

Format	Audience Share
Country	14.1%
News/Talk/Information	12.1%
Adult Contemporary	8.8%
Pop Contemporary Hit	7.6%
Classic Hits	5.1%
Classic Rock	5.0%
Hot Adult Contemporary	4.4%
Urban Adult Contemporary	3.9%
All Sports	3.6%
Rhythmic Contemporary Hit	3.4%
Mexican Regional	3.0%

(a) What is the sum of the audience shares for these formats? What percent of the radio audience listens to stations with other formats?

(b) Make a bar graph to display these data. Be sure to include an "Other format" category.

(c) Would it be correct to display these data in a pie chart? Why or why not?

1.4   How Do Students Pay for College? The Higher Education Research Institute's Freshman Survey includes over 200,000 first-time, full-time freshmen who entered college in 2011.[4] The survey reports the following data on the sources students use to pay for college expenses. ▦ EXPENSE

Source for College Expenses	Students
Family resources	78.3%
Student resources	61.7%
Aid—not to be repaid	69.5%
Aid—to be repaid	52.5%

(a) Explain why it is *not* correct to use a pie chart to display these data.

(b) Make a bar graph of the data. Notice that because the data contrast groups such as family and student resources, it is better to keep these bars next to each other rather than to arrange the bars in order of height.

1.5   Never on Sunday? Births are not, as you might think, evenly distributed across the days of the week. Here are the average numbers of babies born on each day of the week in 2010:[5] ▦ BIRTHS

Day	Births
Sunday	7,110
Monday	11,662
Tuesday	12,821
Wednesday	12,629
Thursday	12,493
Friday	11,960
Saturday	8,007

Present these data in a well-labeled bar graph. Would it also be correct to make a pie chart? Suggest some possible reasons why there are fewer births on weekends.

Quantitative variables: histograms

Quantitative variables often take many values. The distribution tells us what values the variable takes and how often it takes these values. A graph of the distribution is clearer if nearby values are grouped together. The most common graph of the distribution of one quantitative variable is a **histogram.**

histogram

EXAMPLE 1.4   Making a Histogram

What percent of your home state's high school students graduate within four years? The No Child Left Behind Act of 2001 uses on-time high school graduation rates as one of its monitoring requirements. However, in 2001 most states were not collecting the necessary data to compute these rates accurately. The Freshman Graduation Rate (FGR) counts the number of high school graduates in a given year for a state and divides this by the number of ninth graders enrolled four years previously. Although the FGR can be computed from readily available data, it neglects high school students moving into and out of a state and may include students who have repeated a grade. Several alternative measures are available that partially correct for these deficiencies, but states have been free to choose their own measure, and the resulting rates can differ by more than 10%. Law now requires all states begin to use a common, more rigorous computation that tracks individual students. Table 1.1 presents the data for 2010–11, the first year in which states used a common formula and for which graduation rates could be compared between states.[6] Idaho, Kentucky, and Oklahoma received "timeline extensions" and were not required to file in 2010–11.

GRADRATE

The *individuals* in this data set are the states. The *variable* is the percent of a state's high school students who graduate within four years. The states vary quite a bit on this variable, from 59% in the District of Columbia to 88% in Iowa. It's much easier to see how your state compares with other states from a graph like a histogram than from the table. To make a histogram of the distribution of this variable, proceed as follows:

Step 1. Choose the classes. Divide the range of the data into classes of equal width. The data in Table 1.1 range from 59 to 88, so we decide to use these classes:

percent on-time graduates between 55 and 59

percent on-time graduates between 60 and 64

⋮

percent on-time graduates between 85 and 89

TABLE 1.1 PERCENT OF STATE HIGH SCHOOL STUDENTS GRADUATING ON TIME

STATE	PERCENT	REGION	STATE	PERCENT	REGION	STATE	PERCENT	REGION
Alabama	72	S	Louisiana	71	S	Ohio	80	MW
Alaska	68	W	Maine	84	NE	Oklahoma	—	S
Arizona	78	W	Maryland	83	S	Oregon	68	W
Arkansas	81	S	Massachusetts	83	NE	Pennsylvania	83	NE
California	76	W	Michigan	74	MW	Rhode Island	77	NE
Colorado	74	W	Minnesota	77	MW	South Carolina	74	S
Connecticut	83	NE	Mississippi	75	S	South Dakota	83	MW
Delaware	78	S	Missouri	81	MW	Tennessee	86	S
Florida	71	S	Montana	82	W	Texas	86	S
Georgia	67	S	Nebraska	86	MW	Utah	76	W
Hawaii	80	W	Nevada	62	W	Vermont	87	NE
Idaho	—	W	New Hampshire	86	NE	Virginia	82	S
Illinois	84	MW	New Jersey	83	NE	Washington	76	W
Indiana	86	MW	New Mexico	63	W	West Virginia	76	S
Iowa	88	MW	New York	77	NE	Wisconsin	87	MW
Kansas	83	MW	North Carolina	78	S	Wyoming	80	W
Kentucky	—	S	North Dakota	86	MW	Dist. of Columbia	59	S

It is equally correct to use classes 56 to 60, 61 to 65, and so on, or to use a different width for the classes. Just be sure to specify the classes precisely so that each individual falls into exactly one class. A state with an on-time graduation rate of 60% falls into the second class, whereas a state with an on-time graduation rate of 59% falls into the first.

Step 2. Count the individuals in each class. Here are the counts:

Class	Count	Class	Count
55 to 59	1	75 to 79	11
60 to 64	2	80 to 84	16
65 to 69	3	85 to 89	9
70 to 74	6		

Check that the counts add to 48, the number of individuals in the data set (the 47 states reporting and the District of Columbia).

Step 3. Draw the histogram. Mark the scale for the variable whose distribution you are displaying on the horizontal axis. That's the percent of a state's high school students who graduate within four years. The scale runs from 55 to 90 because that is the span of the classes we chose. The vertical axis contains the scale of counts. Each bar represents a class. The base of the bar covers the class, and the bar height is the class count. Draw the bars with no horizontal space between them unless a class is empty, so that its bar has height zero. Figure 1.5 is our histogram. An observation on the boundary of the bars, say 65, is counted in the bar to its right. So even though the last bar ends at 90, it only includes percents up to 89. ■

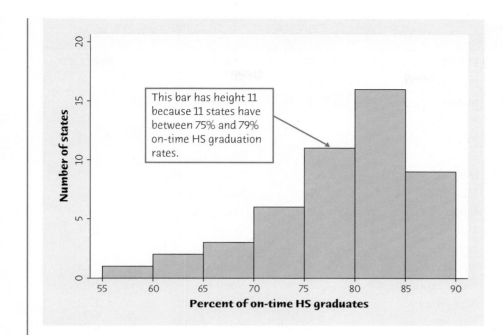

FIGURE 1.5
Histogram of the distribution of the percent of on-time high school graduates in 47 states and the District of Columbia, for Example 1.4.

This bar has height 11 because 11 states have between 75% and 79% on-time HS graduation rates.

Although histograms resemble bar graphs, their details and uses are different. A histogram displays the distribution of a quantitative variable. The horizontal axis of a histogram is marked in the units of measurement for the variable. A bar graph compares the sizes of different quantities. The horizontal axis of a bar graph simply identifies the quantities being compared and need not have any measurement scale. These quantities may be the values of a categorical variable, but they may also be unrelated, like the high-tech devices in Example 1.3 (page 18). Draw bar graphs with blank space between the bars to separate the quantities being compared. Draw histograms with no space, to indicate that all values of the variable are covered. A gap between bars in a histogram indicates that there are no values for that class.

Our eyes respond to the *area* of the bars in a histogram.[7] Because the classes are all the same width, area is determined by height, and all classes are fairly represented. There is no one right choice of the classes in a histogram. Too few classes will give a "skyscraper" graph, with all values in a few classes with tall bars. Too many will produce a "pancake" graph, with most classes having one or no observations. Neither choice will give a good picture of the shape of the distribution. You must use your judgment in choosing classes to display the shape. Statistics software will choose the classes for you. The software's choice is usually a good one, but you can change it if you want. The histogram function in the *One-Variable Statistical Calculator* applet on the text Web site allows you to change the number of classes by dragging with the mouse, so that it is easy to see how the choice of classes affects the histogram.

Apply Your Knowledge

1.6 Foreign Born. How are foreign-born residents distributed in the United States? The country as a whole has 12.5% foreign-born residents, but the states vary from 1.3% in West Virginia to 26.9% in California. Table 1.2 presents the data for all

TABLE 1.2 PERCENT OF STATE POPULATION BORN OUTSIDE THE UNITED STATES

STATE	PERCENT	STATE	PERCENT	STATE	PERCENT
Alabama	3.1	Louisiana	3.4	Ohio	3.8
Alaska	7.0	Maine	3.3	Oklahoma	5.1
Arizona	14.0	Maryland	12.8	Oregon	9.6
Arkansas	4.2	Massachusetts	14.3	Pennsylvania	5.5
California	26.9	Michigan	6.2	Rhode Island	12.7
Colorado	9.7	Minnesota	6.8	South Carolina	4.5
Connecticut	13.1	Mississippi	2.0	South Dakota	2.7
Delaware	8.4	Missouri	3.6	Tennessee	4.2
Florida	18.8	Montana	2.0	Texas	16.1
Georgia	9.4	Nebraska	5.9	Utah	7.8
Hawaii	17.3	Nevada	19.2	Vermont	3.3
Idaho	6.3	New Hampshire	5.2	Virginia	10.2
Illinois	13.5	New Jersey	20.2	Washington	12.2
Indiana	4.4	New Mexico	9.8	West Virginia	1.3
Iowa	3.9	New York	21.4	Wisconsin	4.5
Kansas	6.1	North Carolina	7.1	Wyoming	3.1
Kentucky	3.0	North Dakota	2.4	Dist. of Columbia	12.0

50 states and the District of Columbia.[8] Make a histogram of the percents using classes of width 5% starting at 0.0%. That is, the first bar covers 0.0% to 4.9%, the second covers 5.0% to 9.9%, and so on. (Make this histogram by hand, even if you have software, to be sure you understand the process. You may then want to compare your histogram with your software's choice.) FOREIGN

1.7 **Choosing Classes in a Histogram.** The data set menu that accompanies the *One-Variable Statistical Calculator* applet includes the data on foreign-born residents in the states from Table 1.2. Choose these data, then click on the "Histogram" tab to see a histogram.

(a) How many classes does the applet choose to use? (You can click on the graph outside the bars to get a count of classes.)

(b) Click on the graph and drag to the left. What is the smallest number of classes you can get? What are the lower and upper bounds of each class? (Click on the bar to find out.) Make a rough sketch of this histogram.

(c) Click and drag to the right. What is the greatest number of classes you can get? How many observations does the largest class have?

(d) You see that the choice of classes changes the appearance of a histogram. Drag back and forth until you get the histogram that you think best displays the distribution. How many classes did you use? Why do you think this is best?

Interpreting histograms

Making a statistical graph is not an end in itself. *The purpose of graphs is to help us understand the data.* After you make a graph, always ask, "What do I see?" Once you have displayed a distribution, you can see its important features as follows.

Examining a Histogram

In any graph of data, look for the **overall pattern** and for striking **deviations** from that pattern.

You can describe the overall pattern of a histogram by its **shape, center,** and **spread.**

An important kind of deviation is an **outlier,** an individual value that falls outside the overall pattern.

One way to describe the center of a distribution is by its *midpoint,* the value with roughly half the observations taking smaller values and half taking larger values. To find the midpoint, order the observations from smallest to largest, making sure to include repeated observations as many times as they appear in the data. First cross off the largest and smallest observations, then the largest and smallest of those remaining, and continue this process. If there were an odd number of observations initially, you will be left with a single observation, which is the midpoint. If there were an even number of observations initially, you will be left with two observations, and their average is the midpoint.

For now, we will describe the spread of a distribution by giving the *smallest and largest values.* We will learn better ways to describe center and spread in Chapter 2. The overall shape of a distribution can often be described in terms of symmetry or skewness, defined as follows.

Symmetric and Skewed Distributions

A distribution is **symmetric** if the right and left sides of the histogram are approximately mirror images of each other.

A distribution is **skewed to the right** if the right side of the histogram (containing the half of the observations with larger values) extends much farther out than the left side. It is **skewed to the left** if the left side of the histogram extends much farther out than the right side.

EXAMPLE 1.5   Describing a Distribution

GRADRATE

Look again at the histogram in Figure 1.5. To describe the distribution, we want to look at its overall pattern and any deviations. **Shape:** The distribution has a *single peak,* which represents states in which between 80% and 84% of students graduate high school on time. The distribution is *skewed to the left.* A majority of states have more than 75% of students graduate high school on time, but several states have much lower percents, so the graph extends quite far to the left of its peak. **Center:** Arranging the observations from Table 1.1 in order of size shows that 80% is the midpoint of the distribution. There are a total of 48 observations, and if we cross off the 23 highest graduation rates and the 23 lowest graduation rates, we are left with two graduation rates, both of which are 80%. The center is their average, which is 80%. **Spread:** The spread is from 59% to 88%, which shows considerable variability in graduation rates among the states.

Outliers: Figure 1.5 shows no observations outside the overall single-peaked, left-skewed pattern of the distribution. Figure 1.6 is another histogram of the same distribution, with classes of width 3% rather than 5%. Now there are three states that stand a bit apart to the left of the rest of the distribution, the District of Columbia at 58%, Nevada at 62% and New Mexico at 63%. Are these states outliers or just

FIGURE 1.6

Another histogram of the distribution of the percent of on-time high school graduates, with narrower class widths than in Figure 1.5. Histograms with more classes show more detail but may have a less clear pattern.

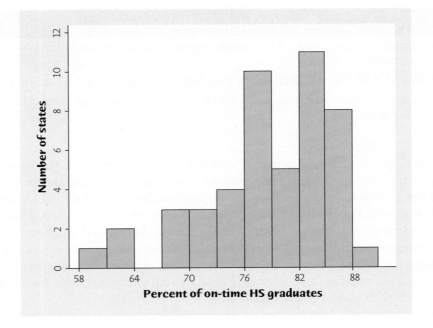

the smallest observations in a strongly skewed distribution? Unfortunately, there is no rule. Let's agree to call attention to only strong outliers that suggest something special about an observation—or an error such as typing 10.1 as 101. These states do not appear to be strong outliers. ■

Figures 1.5 and 1.6 remind us that interpreting graphs calls for judgment. We also see that *the choice of classes in a histogram can influence the appearance of a distribution.* Because of this, and to avoid worrying about minor details, concentrate on the main features of a distribution that persist with several choices of class intervals. Look for major peaks, not for minor ups and downs, in the bars of the histogram. For example, don't immediately conclude that Figure 1.6 shows two important peaks, one between 76% and 79% and the second between 82% and 85%. When you choose a larger number of class intervals, the histogram can become more jagged, leading to the appearance of multiple peaks that are close together. If Arizona, Delaware, and North Carolina had graduation rates of 79% instead of 78%, these states would have been in the class interval 79% to 81% rather than the class interval from 76% to 78%. This small change would have eliminated the second peak in between 76% and 79% in Figure 1.6, leaving only one peak as in Figure 1.5. Be sure to check for clear outliers, not just for the smallest and largest observations, and look for rough *symmetry* or clear *skewness.*

Here are more examples of describing the overall pattern of a histogram.

EXAMPLE 1.6   Iowa Test Scores

IOWATEST

Figure 1.7 displays the scores of all 947 seventh-grade students in the public schools of Gary, Indiana, on the vocabulary part of the Iowa Test of Basic Skills.[9] The distribution is *single-peaked* and *symmetric.* In mathematics, the two sides of symmetric patterns are exact mirror images. Real data are almost never exactly symmetric. We are content to describe Figure 1.7 as symmetric. The center (half above, half below) is close to 7. This is seventh-grade reading level. The scores range from 2.0 (second-grade level) to 12.1 (twelfth-grade level).

Notice that the vertical scale in Figure 1.7 is not the *count* of students but the *percent* of students in each histogram class. A histogram of percents rather than counts is convenient when we want to compare several distributions. To compare Gary with Los Angeles, a much bigger city, we would use percents so that both histograms have the same vertical scale. ■

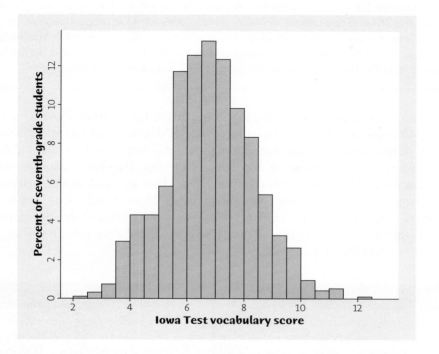

FIGURE 1.7
Histogram of the *Iowa Tests* vocabulary scores of all seventh-grade students in Gary, Indiana, for Example 1.6. This distribution is single peaked and symmetric.

EXAMPLE 1.7   Who Takes the SAT?

SATTAKER

Depending on where you went to high school, the answer to this question may be "almost everybody" or "almost nobody." Figure 1.8 is a histogram of the percent of high school graduates in each state who took the SAT Reasoning test.[10]

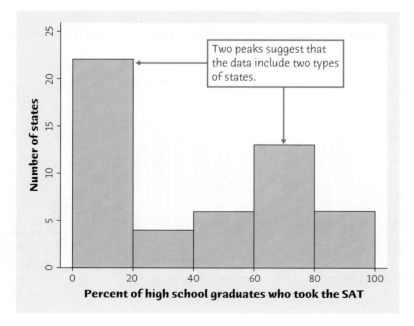

Two peaks suggest that the data include two types of states.

FIGURE 1.8
Histogram of the percent of high school graduates in each state who took the SAT Reasoning test, for Example 1.7. The graph shows two groups of states: ACT states (where few students take the SAT) at the left and SAT states at the right.

The histogram shows two peaks, a high peak at the left and a lower but broader peak centered in the 60% to 80% class. The presence of more than one peak suggests that a distribution mixes several kinds of individuals. That is the case here. There are two major tests of readiness for college, the ACT and the SAT. Most states have a strong preference for one or the other. In some states, many students take the ACT exam and few take the SAT—these states form the peak on the left. In other states, many students take the SAT and few choose the ACT—these states form the broader peak at the right.

Giving the center and spread of this distribution is not very useful. The midpoint falls in the 20% to 40% class, between the two peaks. The story told by the histogram is in the two peaks corresponding to ACT states and SAT states. ■

The overall shape of a distribution is important information about a variable. Some variables have distributions with predictable shapes. Many biological measurements on specimens from the same species and sex—lengths of bird bills, heights of young women—have symmetric distributions. On the other hand, data on people's incomes are usually strongly skewed to the right. There are many moderate incomes, some large incomes, and a few enormous incomes. Many distributions have irregular shapes that are neither symmetric nor skewed. Some data show other patterns, such as the two peaks in Figure 1.8. Use your eyes, describe the pattern you see, and then try to explain the pattern.

Apply Your Knowledge

1.8 Foreign Born. In Exercise 1.6 (page 23), you made a histogram of the percent of foreign-born residents in each of the 50 states and the District of Columbia, given in Table 1.2. Describe the shape of the distribution. Is it closer to symmetric or skewed? What is the center (midpoint) of the data? What is the spread in terms of the smallest and largest values? Are there any states with an unusually large or small percent of foreign-born residents? **FOREIGN**

1.9 Lyme Disease. Lyme disease is caused by a bacteria called *Borrelia burgdorferi* and is spread through the bite of an infected black legged tick, generally found in woods and grassy areas. There were 213,515 confirmed cases reported to the Centers for Disease Control (CDC) between 2001 and 2010, and these are broken down by age and sex in Figure 1.9.[11] Here is how Figure 1.9 relates

Kallista Images/SuperStock

FIGURE 1.9

Histogram of the ages of infected individuals with Lyme disease for cases reported between 2001 and 2010 in the United States, for males and females, for Exercise 1.9.

Confirmed Lyme disease cases by age and sex–United States, 2001-10

□ Male   ■ Female

(Cases vs Age(years) bar chart)

Reported cases of Lyme disease are most common among boys aged 5-9.

to what we have been studying. The individuals are the 213,515 people with confirmed cases, and two of the variables measured on each individual are sex and age. Considering males and females separately, we could draw a histogram of the variable age using class intervals 0–5 years old, 5–10 years old, and so forth. Look at the leftmost two bars in Figure 1.9. The light blue bar shows that approximately 9000 of the males were between 0 and 5 years old, and the dark blue bar shows slightly fewer females were in this age range. If we were to take all the light blue bars and put them side by side, we would have the histogram of age for males using the class intervals stated. Similarly the dark blue bars show females. Because we are trying to display both histograms in the same graph, the bars for males and females within each class interval have been placed alongside each other for ease of comparison, with the bars for different class intervals separated by small spaces.

(a) Describe the main features of the distribution of age for males. Why would giving a unique center and spread for this distribution be misleading?

(b) Suppose that different age groups of males spend differing amounts of time outdoors. How could this fact be used to explain the pattern that you found in part (a)? Remember to use your eyes to describe the pattern you see, and then try to explain the pattern.

(c) A 45-year-old male friend of yours looks at the histogram and tells you that he is planning on giving up hiking because this graph suggests he is in a high-risk group for getting Lyme disease. He will resume hiking when he is 65, as he will be less likely to get Lyme disease at that age. Is this a correct interpretation of the histogram?

(d) Comparing the histograms for males and females, how are they similar? What is the main difference, and why do you think it occurs?

Quantitative variables: stemplots

Histograms are not the only graphical display of distributions. For small data sets, a *stemplot* is quicker to make and presents more detailed information.

Stemplot

To make a **stemplot:**

1. Separate each observation into a **stem,** consisting of all but the final (rightmost) digit, and a **leaf,** the final digit. Stems may have as many digits as needed, but each leaf contains only a single digit.

2. Write the stems in a vertical column with the smallest at the top, and draw a vertical line at the right of this column. Be sure to include all the stems needed to span the data, even when some stems will have no leaves.

3. Write each leaf in the row to the right of its stem, in increasing order out from the stem.

EXAMPLE 1.8   Making a Stemplot

Table 1.2 (page 24) presents the percents of state residents who were born outside the United States. To make a stemplot of these data, take the whole-number part of the percent as the stem and the final digit (tenths) as the leaf. Write stems from 1 West Virginia to 26 for California. Now add leaves. Connecticut, 13.1%, has leaf 1 on

FOREIGN

the 13 stem. Illinois, at 13.5%, places leaf 5 on the same stem. These are the only observations on this stem. Arrange the leaves in order, so that 13|15 is one row in the stemplot. Figure 1.10 is the complete stemplot for the data in Table 1.2. ■

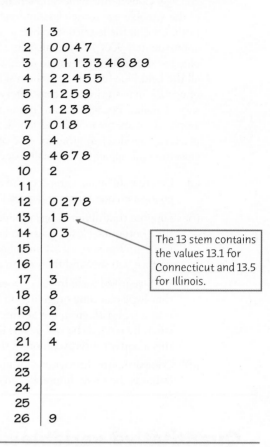

```
 1 | 3
 2 | 0 0 4 7
 3 | 0 1 1 3 3 4 6 8 9
 4 | 2 2 4 5 5
 5 | 1 2 5 9
 6 | 1 2 3 8
 7 | 0 1 8
 8 | 4
 9 | 4 6 7 8
10 | 2
11 |
12 | 0 2 7 8
13 | 1 5
14 | 0 3
15 |
16 | 1
17 | 3
18 | 8
19 | 2
20 | 2
21 | 4
22 |
23 |
24 |
25 |
26 | 9
```

The 13 stem contains the values 13.1 for Connecticut and 13.5 for Illinois.

FIGURE 1.10
Stemplot of the percents of foreign-born residents in the states, for Example 1.8. Each stem is a percent, and leaves are tenths of a percent.

THE VITAL FEW

Skewed distributions can show us where to concentrate our efforts. Ten percent of the cars on the road account for half of all carbon dioxide emissions. A histogram of CO_2 emissions would show many cars with small or moderate values and a few with very high values. Cleaning up or replacing these cars would reduce pollution at a cost much lower than that of programs aimed at all cars. Statisticians who work at improving quality in industry make a principle of this: Distinguish "the vital few" from "the trivial many."

A stemplot looks like a histogram turned on end, with the stems corresponding to the class intervals. The first stem in Figure 1.10 contains all states with percents between 1.0% and 1.9%. In this example, the stemplot is like a histogram with many classes. Compare the stemplot to Figure 1.11 (page 31), which is a histogram of the same data using class intervals 1.0% to 1.9%, 2.0% to 2.9%, and so forth. Although Figures 1.10 and 1.11 display exactly the same pattern, the stemplot, unlike the histogram, preserves the actual value of each observation.

 In a stemplot the classes (the stems) of a stemplot are given to you. Histograms are more flexible than stemplots because you can choose the classes more easily. In both Figures 1.10 and 1.11, California (26.9%) stands slightly apart from the long right tail of the skewed distribution. *Stemplots do not work well for large data sets, where each stem must hold a large number of leaves.* Don't try to make a stemplot of a large data set, such as the 947 Iowa Test scores in Figure 1.7.

When there are too many stems, there are often no leaves or just one or two leaves on many of the stems as in Figure 1.10. The number of stems can be reduced if we first **round** the data. In this example, we can round the data for each state to the nearest percent before drawing the stemplot. Here is the result:

rounding

```
0 | 1 2 2 2 3 3 3 3 3 3 4 4 4 4 4 5 5 5 5 6 6 6 6 7 7 7 8 8 9
1 | 0 0 0 0 2 2 3 3 3 4 4 4 6 7 9 9
2 | 0 1 7
```

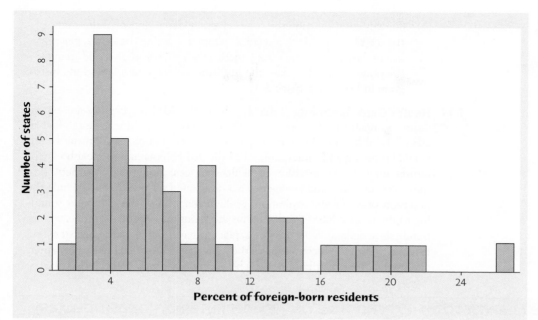

FIGURE 1.11
Histogram of the percents of foreign-born residents in the states, for Example 1.8. The class widths have been chosen to agree with the widths of the stems in the stemplot in Figure 1.10.

Now it seems that there are too few stems. You can also **split stems** in a stemplot to double the number of stems when all the leaves would otherwise fall on just a few stems, as occurred when we rounded to the nearest percent. Each stem then appears twice. Leaves 0 to 4 go on the upper stem, and leaves 5 to 9 go on the lower stem. If you split the stems with the data rounded to the nearest percent, the stemplot becomes

splitting stems

```
0 | 1 2 2 2 3 3 3 3 3 3 4 4 4 4 4
0 | 5 5 5 5 6 6 6 6 6 7 7 7 8 8 9
1 | 0 0 0 0 2 2 3 3 3 4 4 4
1 | 6 7 9 9
2 | 0 1
2 | 7
```

which makes the right skew pattern clearer. Rounding and splitting stems are matters for judgment, like choosing the classes in a histogram. Some data require rounding but don't require splitting stems, some require just splitting stems, and other data require both. The *One-Variable Statistical Calculator* applet on the text Web site allows you to decide whether to split stems so that it is easy to see the effect.

 Comparing Figures 1.11 (right-skewed) and 1.5 (left-skewed, page 23) reminds us that *the direction of skewness is the direction of the long tail, not the direction where most observations are clustered.*

Apply Your Knowledge

1.10 Graduation Rates. Table 1.1 (page 22) presents the data for the on-time high school graduation rates in each of the 50 states and the District of Columbia.

(a) Make a stemplot of the data as it appears in Table 1.1. Because the stemplot preserves the actual value of the observations, it is easy to find the midpoint and the spread. What are they? ▦ GRADRATE

(b) For this data, splitting the stems gives a clearer picture of the shape of the distribution. Make a second stemplot splitting the stems, placing leaves 0 to 4 on the first stem and leaves 5 to 9 on the second stem of the same value. How does the resulting stemplot compare to the histogram in Figure 1.5 (page 23)?

1.11 **Health Care Spending.** Table 1.3 shows the 2011 per capita total expenditure on health in 35 countries with the highest gross domestic product in 2011.[12] Health expenditure per capita is the sum of public and private health expenditures (in PPP, international $) divided by population. Health expenditures include the provision of health services, family-planning activities, nutrition activities, and emergency aid designated for health, but exclude the provision of water and sanitation. Make a stemplot of the data after rounding to the nearest $100 (so that stems are thousands of dollars and leaves are hundreds of dollars). Split the stems, placing leaves 0 to 4 on the first stem and leaves 5 to 9 on the second stem of the same value. Describe the shape, center, and spread of the distribution. Which country is the high outlier? [▮▮] **HEALTH**

TABLE 1.3 PER CAPITA TOTAL EXPENDITURE ON HEALTH (INTERNATIONAL DOLLARS)

COUNTRY	DOLLARS	COUNTRY	DOLLARS	COUNTRY	DOLLARS
Argentina	1434	India	141	Saudi Arabia	901
Australia	3692	Indonesia	127	South Africa	943
Austria	4482	Iran	929	Spain	3041
Belgium	4119	Italy	3130	Sweden	3870
Brazil	1043	Japan	3174	Switzerland	5564
Canada	4520	Korea, South	2181	Thailand	353
China	432	Malaysia	559	Turkey	1161
Colombia	618	Mexico	940	United Arab Emirates	1732
Denmark	4564	Netherlands	5123	United Kingdom	3322
France	4086	Norway	5674	United States	8608
Germany	4371	Poland	1423	Venezuela	659
Greece	2918	Russia	1316		

Time plots

Many variables are measured at intervals over time. We might, for example, measure the height of a growing child or the price of a stock at the end of each month. In these examples, our main interest is change over time. To display change over time, make a *time plot*.

Time Plot

A **time plot** of a variable plots each observation against the time at which it was measured. Always put time on the horizontal scale of your plot and the variable you are measuring on the vertical scale. Connecting the data points by lines helps emphasize any change over time.

EXAMPLE 1.9   Water Levels in the Everglades

Water levels in Everglades National Park are critical to the survival of this unique region. The photo shows a water-monitoring station in Shark River Slough, the main path for surface water moving through the "river of grass" that is the Everglades. Each day the mean gauge height, the height in feet of the water surface above the gauge datum, is measured at the Shark River Slough monitoring station. (The gauge datum is a vertical control measure established in 1929 and is used as a reference for establishing varying elevations. It establishes a zero point from which to measure the gauge height.) Figure 1.12 is a time plot of mean daily gauge height at this station from January 1, 2000, to December 31, 2011.[13] ■

WATERLEV

When you examine a time plot, look once again for an overall pattern and for strong deviations from the pattern. Figure 1.12 shows strong **cycles,** regular up-and-down movements in water level. The cycles show the effects of Florida's wet season (roughly June to November) and dry season (roughly December to May). Water levels are highest in late fall. If you look closely, you can see the year-to-year variation. The dry season in 2003 ended early, with the first-ever April tropical storm. In consequence, the dry-season water level in 2003 did not dip as low as in other years. The drought in the southeastern portion of the country in 2008 and 2009 shows up in the steep drop in the mean gauge height in 2009, whereas the lower peaks in 2006 and 2007 reflect lower water levels during the wet seasons in these years. Finally, in 2011, an extra-long dry season and a slow start to the 2011 rainy season compounded into the worst drought in the southwest Florida area in 80 years, which shows up as the steep drop in the mean gauge height in 2011.

Another common overall pattern in a time plot is a **trend,** a long-term upward or downward movement over time. Many economic variables show an upward trend. Incomes, house prices, and (alas) college tuitions tend to move generally upward over time.

Histograms and time plots give different kinds of information about a variable. The time plot in Figure 1.12 presents **time series data** that show the change in water level at one location over time. A histogram displays **cross-sectional data,** such as water levels at many locations in the Everglades at the same time.

cycles

Courtesy U.S. Geological Survey

trend

time series
cross-sectional

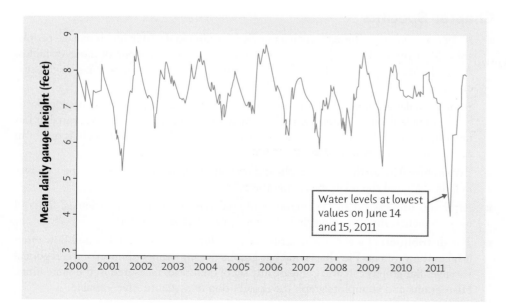

Water levels at lowest values on June 14 and 15, 2011

FIGURE 1.12
Time plot of average gauge height at a monitoring station in Everglades National Park over a twelve-year period, for Example 1.9. The yearly cycles reflect Florida's wet and dry seasons.

Apply Your Knowledge

1.12 The Cost of College. Here are data on the average tuition and fees charged to in-state students by public four-year colleges and universities for the 1980 to 2012 academic years. Because almost any variable measured in dollars increases over time due to inflation (the falling buying power of a dollar), the values are given in "constant dollars," adjusted to have the same buying power that a dollar had in 2012.[14] 📊 **COLLEGEX**

Year	Tuition	Year	Tuition	Year	Tuition	Year	Tuition
1980	$2227	1989	$3124	1998	$4559	2007	$6809
1981	$2273	1990	$3353	1999	$4621	2008	$6865
1982	$2423	1991	$3545	2000	$4652	2009	$7500
1983	$2633	1992	$3806	2001	$4862	2010	$8000
1984	$2703	1993	$4022	2002	$5213	2011	$8372
1985	$2801	1994	$4177	2003	$5787	2012	$8655
1986	$2959	1995	$4224	2004	$6201		
1987	$2990	1996	$4342	2005	$6439		
1988	$3051	1997	$4441	2006	$6534		

(a) Make a time plot of average tuition and fees.

(b) What overall pattern does your plot show?

(c) Some possible deviations from the overall pattern are outliers, periods when charges went down (in 2012 dollars) and periods of particularly rapid increase. Which are present in your plot, and during which years?

(d) In looking for patterns, do you think that it would be better to study a time series of the tuition for each year or the percent increase for each year? Why?

CHAPTER 1 SUMMARY

Chapter Specifics

■ A data set contains information on a number of **individuals.** Individuals may be people, animals, or things. For each individual, the data give values for one or more **variables.** A variable describes some characteristic of an individual, such as a person's height, sex, or salary.

■ Some variables are **categorical,** and others are **quantitative.** A categorical variable places each individual into a category, such as male or female. A quantitative variable has numerical values that measure some characteristic of each individual, such as height in centimeters or salary in dollars.

■ **Exploratory data analysis** uses graphs and numerical summaries to describe the variables in a data set and the relations among them.

■ After you understand the background of your data (individuals, variables, units of measurement), the first thing to do is almost always **plot your data.**

■ The **distribution** of a variable describes what values the variable takes and how often it takes these values. **Pie charts** and **bar graphs** display the distribution of a categorical variable. Bar graphs can also compare any set of quantities measured in the same units. **Histograms** and **stemplots** graph the distribution of a quantitative variable.

- When examining any graph, look for an **overall pattern** and for notable **deviations** from the pattern.

- **Shape, center, and spread** describe the overall pattern of the distribution of a quantitative variable. Some distributions have simple shapes, such as **symmetric** or **skewed.** Not all distributions have a simple overall shape, especially when there are few observations.

- **Outliers** are observations that lie outside the overall pattern of a distribution. Always look for outliers, and try to explain them.

- When observations on a variable are taken over time, make a **time plot** that graphs time horizontally and the values of the variable vertically. A time plot can reveal **trends, cycles,** or other changes over time.

Link It

Practical statistics uses data to draw conclusions about some broader universe. You should reread Example 1.1 (page 14), as it will help you understand this basic idea. For the American Community Survey described in the example, the data are the responses from those households responding to the survey, although the broader universe of interest is the entire nation.

In our study of practical statistics, we will divide the subject into three main areas. In exploratory data analysis, graphs and numerical summaries are used for exploring, organizing, and describing data so that the patterns become apparent. Data production concerns where the data come from and helps us to understand whether what we learn from our data can be generalized to a wider universe. And statistical inference provides tools for generalizing what we learn to a wider universe.

In this chapter we have begun to learn about data analysis. A data set can consist of hundreds of observations on many variables. Even if we consider only one variable at a time, it is difficult to see what the data have to say by scanning a list containing many data values. Graphs provide a visual tool for organizing and identifying patterns in data and are a good starting point in the exploration of the distribution of a variable.

Pie charts and bar graphs can summarize the information in a categorical variable by giving us the percent of the distribution in the various categories. Although a table containing the categories and percent gives the same information as a bar graph, a substantial advantage of the bar graph over a tabular presentation is that the bar graph allows us to visually compare percents among all categories simultaneously by means of the heights of the bars.

Histograms and stemplots are graphical tools for summarizing the information provided by a quantitative variable. The overall pattern in a histogram or stemplot illustrates some of the important features of the distribution of a variable that will be of interest as we continue our study of practical statistics. The center of the histogram tells us about the value of a "typical" observation on this variable, whereas the spread gives us a sense of how close most of the observations are to this value. Other interesting features are the presence of outliers and the general shape of the plot. For data collected over time, time plots can show patterns such as seasonal variation and trends in the variable. In the next chapter we will see how the information about the distribution of a variable can also be described using numerical summaries.

CHECK YOUR SKILLS

The multiple-choice exercises in Check Your Skills ask straightforward questions about basic facts from the chapter. Answers to all these exercises appear in the back of the book. You should expect almost all your answers to be correct.

1.13 Here are the first lines of a professor's data set at the end of a statistics course:

Name	Major	Total Points	Grade
ADVANI, SURA	COMM	397	B
BARTON, DAVID	HIST	323	C
BROWN, ANNETTE	BIOL	446	A
CHIU, SUN	PSYC	405	B
CORTEZ, MARIA	PSYC	461	A

The individuals in these data are

(a) the students.
(b) the total points.
(c) the course grades.

1.14 To display the distribution of grades (A, B, C, D, F) for all students in the course, it would be correct to use

(a) a pie chart but not a bar graph.
(b) a bar graph but not a pie chart.
(c) either a pie chart or a bar graph.

1.15 A description of different houses on the market includes the variables square footage of the house and the average monthly gas bill.

(a) Square footage and average monthly gas bill are both categorical variables.
(b) Square footage and average monthly gas bill are both quantitative variables.
(c) Square footage is a categorical variable, and average monthly gas bill is a quantitative variable.

1.16 A political party's data bank includes the zip codes of past donors, such as

47906   34236   53075   10010   90210   75204
30304   99709

Zip code is a

(a) quantitative variable.
(b) categorical variable.
(c) unit of measurement.

1.17 Figure 1.6 (page 26) is a histogram of the percent of on-time high school graduates in each state. The leftmost bar in the histogram covers percents of on-time high school graduates ranging from about

(a) 58% to 64%.   (b) 58% to 61%.   (c) 0% to 58%.

1.18 Here are the exam scores of 10 students in a statistics class:

50   35   41   97   76   69   94   91   23   65

To make a stemplot of these data, you would use stems

(a) 2, 3, 4, 5, 6, 7, 9.
(b) 2, 3, 4, 5, 6, 7, 8, 9.
(c) 20, 30, 40, 50, 60, 70, 80, 90.

1.19 How long must you travel each day to get to work? Here is a stemplot of the average travel times to work for workers in the 50 states and the District of Columbia who are at least 16 years of age and don't work at home.[15] The stems are whole minutes, and the leaves are tenths of a minute. TRAVEL

```
15 | 5 9
16 |
17 | 6 7 7 9
18 | 2 5
19 |
20 | 0 1 7 8 8 9
21 | 2 8
22 | 0 1 3 3 3 4 9 9
23 | 4 4 5 6 6 9
24 | 0 1 2 6 6
25 | 0 0 1 2 5 6 9
26 | 6 8 9
27 | 3 9
28 |
29 | 1 2
30 | 6 9
```

The state with the longest average travel time is New York. On average, how long does it take New Yorkers to travel to work each day?

(a) 30.69 minutes
(b) 309 minutes
(c) 30.9 minutes

1.20 The shape of the distribution in Exercise 1.19 is

(a) clearly skewed to the right.
(b) roughly symmetric.
(c) clearly skewed to the left.

1.21 The center of the distribution in Exercise 1.19 is close to

(a) 22 minutes.
(b) 23.4 minutes.
(c) 15.5 to 30.9 minutes.

1.22 You look at real estate ads for houses in Naples, Florida. There are many houses ranging from $200,000 to $500,000 in price. The few houses on the water, however, have prices up to $15 million. The distribution of house prices will be

(a) skewed to the left.
(b) roughly symmetric.
(c) skewed to the right.

CHAPTER 1 EXERCISES

1.23 Medical students. Students who have finished medical school are assigned to residencies in hospitals to receive further training in a medical specialty. Here is part of a hypothetical data base of students seeking residency positions. USMLE is the student's score on Step 1 of the national medical licensing examination.

NAME	MEDICAL SCHOOL	SEX	AGE	USMLE	SPECIALTY SOUGHT
Abrams, Laurie	Florida	F	28	238	Family medicine
Brown, Gordon	Meharry	M	25	205	Radiology
Cabrera, Maria	Tufts	F	26	191	Pediatrics
Ismael, Miranda	Indiana	F	32	245	Internal medicine

(a) What individuals does this data set describe?
(b) In addition to the student's name, how many variables does the data set contain? Which of these variables are categorical, and which are quantitative? If a variable is quantitative, what units is it measured in?

1.24 Buying a refrigerator. *Consumer Reports* is doing an article comparing refrigerators in their next issue. Some of the characteristics to be included in the report are the brand name and model; whether it has a top, bottom, or side-by-side freezer; the estimated energy consumption per year (kilowatts); whether or not it is Energy Star compliant; the width, depth, and height in inches; and both the freezer and refrigerator net capacity in cubic feet. Which of these variables are categorical, and which are quantitative? Give the units for the quantitative variables and the categories for the categorical variables. What are the individuals in the report?

1.25 What color is your car? The most popular colors for cars and light trucks vary with region and over time. In North America white remains the top color choice, with silver the top choice in South America and white the top choice worldwide for the second consecutive year. Here is the distribution of the top colors for vehicles sold in North America in 2012:[16] CARCOLOR

COLOR	POPULARITY
White	24%
Black	19%
Silver	16%
Gray	15%
Red	10%
Blue	7%
Beige, brown	5%
Other colors	

Fill in the percent of vehicles that are in other colors. Make a graph to display the distribution of color popularity.

1.26 Facebook, Twitter, and LinkedIn users. After years of explosive growth in users of social networking sites in all age ranges and demographics, it is hard to argue that social media haven't changed forever how we interact and connect online. Although Facebook is still the dominant player in social networking, both Twitter and LinkedIn have continued to increase their usage. Here is the age distribution of the users for the three sites at the end of 2010:[17] SOCIALNT

AGE GROUP	FACEBOOK USERS	TWITTER USERS	LINKEDIN USERS
18 to 22 years	16%	26%	6%
23 to 35 years	33%	34%	36%
36 to 49 years	25%	24%	32%
50 to 65 years	19%	13%	23%
Over 65 years	6%	4%	4%

(a) Draw a bar graph for the age distribution of Facebook users. The leftmost bar should correspond to "18 to 22," the next bar to "23 to 35," and so on. Do the same for Twitter and LinkedIn, using the same scale for the percent axis.
(b) Describe the most important difference in the age distribution of the audience for these three social networking sites. How does this difference show up in the bar graphs? Do you think it was important to order the bars by age to make the comparison easier? Why or why not?
(c) Explain why it *is* appropriate to use a pie chart to display any of these distributions. Draw a pie chart for each distribution. Do you think it is easier to compare the three distributions with bar graphs or pie charts? Explain your reasoning.

1.27 Deaths among young people. Among persons aged 15 to 24 years in the United States, the leading causes of death and number of deaths in 2011 were: accidents, 12,032; suicide, 4688; homicide, 4508; cancer, 1609; heart disease, 948; congenital defects, 429.[18]

(a) Make a bar graph to display these data.

(b) To make a pie chart, you need one additional piece of information. What is it?

1.28 Hispanic origins. According to the 2010 U.S. Census, 308.7 million people resided in the United States on April 1, 2010, of which 50.5 million (or 16%) were of Hispanic origin. What countries do they come from? Figure 1.13 is a pie chart to show the country of origin of Hispanics in the United States in 2010.[19] About what percent of Hispanics are Mexican? Puerto Rican? You see that it is hard to determine numbers from a pie chart. Bar graphs are much easier to use. (The U.S. Census Bureau includes the percents in many of its pie charts.)

Percent Distribution of Hispanics by Origin: 2010

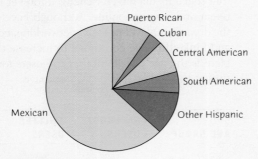

FIGURE 1.13

Pie chart of the national origins of Hispanic residents of the United States, for Exercise 1.28.

1.29 Canadian students rate their universities. The National Survey of Student Engagement asked students

UNIVERSITY	EXCELLENT RATING
Toronto	21%
York	18%
Alberta	23%
Ottawa	11%
Western Ontario	38%
British Columbia	18%
Calgary	14%
McGill	26%
Waterloo	36%
Concordia	21%

at many universities, "How would you evaluate your entire educational experience at this university?" Here are the percents of senior-year students at Canada's 10 largest primarily English-speaking universities who responded "Excellent":[20] CANADA

(a) The list is arranged in order of undergraduate enrollment. Make a bar graph with the bars in order of student rating.

(b) Explain carefully why it is not correct to make a pie chart of these data.

1.30 Do adolescent girls eat fruit? We all know that fruit is good for us. Many of us don't eat enough. Figure 1.14 is a histogram of the number of servings of fruit per day claimed by 74 seventeen-year-old girls in a study in Pennsylvania.[21] Describe the shape, center, and spread of this distribution. Are there any outliers? What percent of these girls ate six or more servings per day? How many of these girls ate fewer than two servings per day?

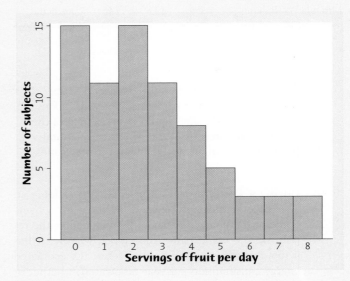

FIGURE 1.14

The distribution of fruit consumption in a sample of 74 seventeen-year-old girls, for Exercise 1.30.

1.31 IQ test scores. Figure 1.15 (page 39) is a stemplot of the IQ test scores of 78 seventh-grade students in a rural midwestern school.[22] IQ

(a) Four students had low scores that might be considered outliers. Ignoring these, describe the shape, center, and spread of the remainder of the distribution.

(b) We often read that IQ scores for large populations are centered at 100. What percent of these 78 students have scores above 100?

```
 7 | 2 4
 7 | 7 9
 8 |
 8 | 6 9
 9 | 0 1 3 3
 9 | 6 7 7 8
10 | 0 0 2 2 3 3 3 3 4 4
10 | 5 5 5 6 6 6 7 7 7 7 8 9
11 | 0 0 0 0 1 1 1 1 2 2 2 2 3 3 3 4 4 4 4
11 | 5 5 6 8 8 9 9 9
12 | 0 0 3 3 4 4
12 | 6 7 7 8 8 8
13 | 0 2
13 | 6
```

FIGURE 1.15

The distribution of IQ scores for 78 seventh-grade students, for Exercise 1.31.

1.32 **Returns on common stocks.** The return on a stock is the change in its market price plus any dividend payments made. Total return is usually expressed as a percent of the beginning price. Figure 1.16 is a histogram of the distribution of the monthly returns for all stocks listed on U.S. markets from January 1985 to December 2012 (336 months).[23] The extreme low outlier is the market crash of October 1987, when stocks lost 23% of their value in one month. The other two low outliers are 16% during August 1998, a month when the Dow Jones Industrial Average experienced its second-largest drop in history to that time, and the financial crisis in October 2008, when stocks lost 17% of their value. ▥ STOCKRET

(a) Ignoring the outliers, describe the overall shape of the distribution of monthly returns.

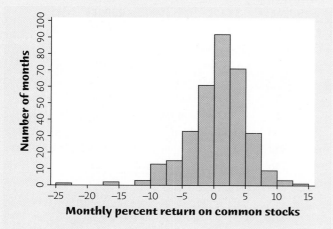

FIGURE 1.16

The distribution of monthly percent returns on U.S. common stocks from January 1985 to December 2012, for Exercise 1.32.

(b) What is the approximate center of this distribution? (For now, take the center to be the value with roughly half the months having lower returns and half having higher returns.)

(c) Approximately what were the smallest and largest monthly returns, leaving out the outliers? (This is one way to describe the spread of the distribution.)

(d) A return less than zero means that stocks lost value in that month. About what percent of all months had returns less than zero?

1.33 **Name that variable.** A survey of a large college class asked the following questions:

1. Are you female or male? (In the data, male = 0, female = 1.)
2. Are you right-handed or left-handed? (In the data, right = 0, left = 1.)
3. What is your height in inches?
4. How many minutes do you study on a typical weeknight?

Figure 1.17 shows histograms of the student responses, in scrambled order and without scale markings. Which histogram goes with each variable? Explain your reasoning.

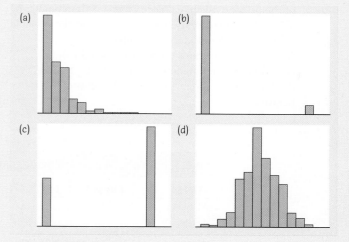

FIGURE 1.17

Histograms of four distributions, for Exercise 1.33.

1.34 **Food oils and health.** Fatty acids, despite their unpleasant name, are necessary for human health. Two types of essential fatty acids, called omega-3 and omega-6, are not produced by our bodies and so must be obtained from our food. Food oils, widely used in food processing and cooking, are major sources of these compounds. There is some evidence that a healthy diet should have more omega-3 than omega-6. Table 1.4 (page 40) gives the ratio of omega-3 to omega-6 in some common food oils.[24] Values greater than 1 show that an oil has more omega-3 than omega-6. ▥ FOODOILS

TABLE 1.4 OMEGA-3 FATTY ACIDS AS A FRACTION OF OMEGA-6 FATTY ACIDS IN FOOD OILS

OIL	RATIO	OIL	RATIO
Perilla	5.33	Flaxseed	3.56
Walnut	0.20	Canola	0.46
Wheat germ	0.13	Soybean	0.13
Mustard	0.38	Grape seed	0.00
Sardine	2.16	Menhaden	1.96
Salmon	2.50	Herring	2.67
Mayonnaise	0.06	Soybean, hydrogenated	0.07
Cod liver	2.00	Rice bran	0.05
Shortening (household)	0.11	Butter	0.64
Shortening (industrial)	0.06	Sunflower	0.03
Margarine	0.05	Corn	0.01
Olive	0.08	Sesame	0.01
Shea nut	0.06	Cottonseed	0.00
Sunflower (oleic)	0.05	Palm	0.02
Sunflower (linoleic)	0.00	Cocoa butter	0.04

(a) Make a histogram of these data, using classes bounded by the whole numbers from 0 to 6.
(b) What is the shape of the distribution? How many of the 30 food oils have more omega-3 than omega-6? What does this distribution suggest about the possible health effects of modern food oils?
(c) Table 1.4 contains entries for several fish oils (cod, herring, menhaden, salmon, sardine). How do these values support the idea that eating fish is healthy?

1.35 Where are the nurses? Table 1.5 gives the number of active nurses per 100,000 people in each state.[25]

(a) Why is the number of nurses per 100,000 people a better measure of the availability of nurses than a simple count of the number of nurses in a state?
(b) Make a histogram that displays the distribution of nurses per 100,000 people. Write a brief description of the distribution. Are there any outliers? If so, can you explain them? **NURSES**

TABLE 1.5 NURSES PER 100,000 PEOPLE, BY STATE

STATE	NURSES	STATE	NURSES	STATE	NURSES
Alabama	912	Louisiana	894	Ohio	1001
Alaska	756	Maine	1053	Oklahoma	712
Arizona	544	Maryland	869	Oregon	795
Arkansas	774	Massachusetts	1210	Pennsylvania	1017
California	641	Michigan	841	Rhode Island	1007
Colorado	761	Minnesota	1017	South Carolina	795
Connecticut	994	Mississippi	868	South Dakota	1215
Delaware	977	Missouri	958	Tennessee	894
Florida	814	Montana	748	Texas	662
Georgia	653	Nebraska	1010	Utah	625
Hawaii	753	Nevada	574	Vermont	912
Idaho	642	New Hampshire	970	Virginia	750
Illinois	812	New Jersey	907	Washington	774
Indiana	864	New Mexico	580	West Virginia	938
Iowa	990	New York	859	Wisconsin	905
Kansas	867	North Carolina	886	Wyoming	812
Kentucky	923	North Dakota	1097	Dist. of Columbia	1380

TABLE 1.6 ANNUAL CARBON DIOXIDE EMISSIONS IN 2009 (METRIC TONS PER PERSON)

COUNTRY	CO$_2$	COUNTRY	CO$_2$	COUNTRY	CO$_2$
Algeria	3.4740	Indonesia	1.9029	Poland	7.8147
Argentina	4.3611	Iran	8.2319	Russia	11.0048
Bangladesh	0.3471	Iraq	3.5489	South Africa	10.0302
Brazil	1.8999	Italy	6.6530	Spain	6.3156
Canada	15.2615	Japan	8.7011	Sudan	0.3375
China	5.7585	Kenya	0.3130	Tanzania	0.1464
Colombia	1.5602	Korea, South	10.6200	Thailand	3.9548
Congo	0.4932	Mexico	3.9831	Turkey	3.8672
Egypt	2.7113	Morocco	1.5431	Uganda	0.1075
Ethiopia	0.0972	Myanmar	0.2330	Ukraine	5.9538
France	5.8188	Nigeria	0.4546	United Kingdom	7.6977
Germany	8.9145	Pakistan	0.9456	United States	17.2239
India	1.6389	Philippines	0.7475	Vietnam	1.6370

1.36 Carbon dioxide emissions. Burning fuels in power plants and motor vehicles emits carbon dioxide (CO_2), which contributes to global warming. Table 1.6 displays the 2009 CO_2 emissions per person from countries with populations of at least 30 million in that year.[26] CO2EMISS

(a) Why do you think we choose to measure emissions per person rather than total CO_2 emissions for each country?

(b) Make a stemplot to display the data of Table 1.6. The data will first need to be rounded (see page 30). What units are you going to use for the stems? The leaves? You should round the data to the units you are planning to use for the leaves before drawing the stemplot. Describe the shape, center, and spread of the distribution. Which countries are outliers?

1.37 Fur seals on St. Paul Island. Every year hundreds of thousands of northern fur seals return to their haulouts in the Pribilof Islands in Alaska to breed, give birth, and teach their pups to swim, hunt, and survive in the Bering Sea. U.S. commercial fur sealing operations continued on St. Paul until 1984, but despite a reduction in harvest, the population of fur seals has

continued to decline. Possible reasons include climate shifts in the North Pacific, changes in the availability of prey, and new or increased interaction with commercial fisheries that increase mortality. Here are data on the estimated number of fur seal pups born on St. Paul Island (in thousands) from 1979 to 2010, where a dash indicates a year in which no data were collected:[27]

YEAR	PUPS BORN (THOUSANDS)	YEAR	PUPS BORN (THOUSANDS)
1979	245.93	1996	170.12
1980	203.82	1997	—
1981	179.44	1998	179.15
1982	203.58	1999	—
1983	165.94	2000	158.74
1984	173.27	2001	—
1985	182.26	2002	145.72
1986	167.66	2003	—
1987	171.61	2004	122.82
1988	202.23	2005	—
1989	171.53	2006	109.96
1990	201.30	2007	—
1991	—	2008	102.67
1992	182.44	2009	—
1993	—	2010	94.50
1994	192.10	2011	—
1995	—	2012	96.83

Make a stemplot to display the distribution of pups born per year. Describe the shape, center, and spread of the distribution. Are there any outliers? FURSEALS

©Arco Images GmbH/Alamy Images

1.38 Do women study more than men? We asked the students in a large first-year college class how many minutes they studied on a typical weeknight. Here are the responses of random samples of 30 women and 30 men from the class: STUDYTIM

WOMEN					MEN				
270	150	180	360	180	120	120	30	45	200
120	180	120	240	170	90	90	30	120	75
150	120	180	180	150	150	90	60	240	300
200	150	180	120	240	240	60	150	60	30
120	60	120	180	180	30	230	120	95	150
90	240	180	115	120	0	200	120	120	180

(a) Examine the data. Why are you not surprised that most responses are multiples of 10 minutes? What is the other common multiple found in the data? We eliminated one student who claimed to study 10,000 minutes per night. Are there any other responses you consider suspicious?

back-to-back stemplot (b) Make a **back-to-back stemplot** to compare the two samples. That is, use one set of stems with two sets of leaves, one to the right and one to the left of the stems. (Draw a line on either side of the stems to separate stems and leaves.) Order both sets of leaves from smallest at the stem to largest away from the stem. Report the approximate midpoints of both groups. Does it appear that women study more than men (or at least claim that they do)?

1.39 Fur seals on St. Paul Island. Make a time plot of the number of fur seals born per year from Exercise 1.37. What does the time plot show that your stemplot in Exercise 1.37 did not show? When you have data collected over time, a time plot is often needed to understand what is happening. FURSEALS

1.40 Marijuana and traffic accidents. Researchers in New Zealand interviewed 907 drivers at age 21. They had data on traffic accidents, and they asked the drivers about marijuana use. Here are data on the numbers of accidents caused by these drivers at age 19, broken down by marijuana use at the same age:[28]

	MARIJUANA USE PER YEAR			
	NEVER	1–10 TIMES	11–50 TIMES	51+ TIMES
Accidents caused	59	36	15	50
Drivers	452	229	70	156

(a) Explain carefully why a useful graph must compare *rates* (accidents per driver) rather than *counts* of accidents in the four marijuana use classes.

(b) Compute the accident rates in the four marijuana use classes. After you have done this, make a graph that displays the accident rate for each class. What do you conclude? (You can't conclude that marijuana use *causes* accidents, because risk takers are more likely both to drive aggressively and to use marijuana.)

1.41 Dates on coins. Sketch a histogram for a distribution that is skewed to the left. Suppose that you and your friends emptied your pockets of coins and recorded the year marked on each coin. The distribution of dates would be skewed to the left. Explain why.

1.42 El Niño and the monsoon. The earth is interconnected. For example, it appears that El Niño, the periodic warming of the Pacific Ocean west of South America, affects the monsoon rains that are essential for agriculture in India. Here are the monsoon rains (in millimeters) for the 23 strong El Niño years between 1871 and 2004:[29] ELNINO

628	669	740	651	710	736	717	698
653	604	781	784	790	811	830	858
858	896	806	790	792	957	872	

(a) To make a stemplot of these rainfall amounts, round the data to the nearest 10, so that stems are hundreds of millimeters and leaves are tens of millimeters. Make two stemplots, with and without splitting the stems. Which plot do you prefer?

(b) Describe the shape, center, and spread of the distribution. Are there any outliers?

(c) The average monsoon rainfall for all years from 1871 to 2004 is about 850 millimeters. What effect does El Niño appear to have on monsoon rains?

1.43 Watch those scales! Figures 1.18(a) and 1.18(b) page 43) both show time plots of tuition charged to in-state students from 1980 through 2012.[30]

(a) Which graph appears to show the biggest increase in tuition between 2000 and 2012?

(b) Read the graphs and compute the actual increase in tuition between 2000 and 2012 in each graph. Do you think these graphs are for the same or different data sets? Why?

The impression that a time plot gives depends on the scales you use on the two axes. Changing the scales can make tuition appear to increase very rapidly or to have only a gentle increase. The moral of this exercise is: Always pay close attention to the scales when you look at a time plot.

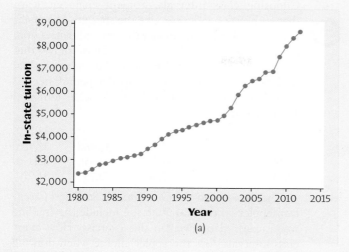

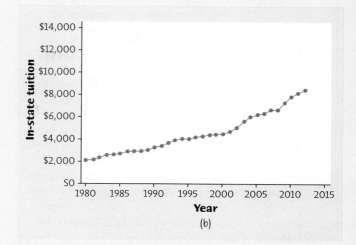

FIGURE 1.18

Time plots of in-state tuition between 1980 and 2012, for Exercise 1.43.

1.44 Housing starts. Figure 1.19 is a time plot of the number of single-family homes started by builders each month from January 1990 to April 2013.[31] The counts are in thousands of homes. 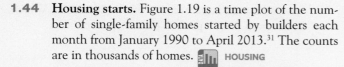 HOUSING

(a) The most notable pattern in this time plot is yearly up-and-down cycles. At what season of the year are housing starts highest? Lowest? The cycles are explained by the weather in the northern part of the country.

(b) Is there a longer-term trend visible in addition to the cycles? If so, describe it.

(c) The big economic news of 2007 was a severe downturn in housing that began in mid-2006. This was followed by the financial crisis in 2008. How are these economic events reflected in the time plot?

(d) How would you describe the behavior of the time plot since January 2011?

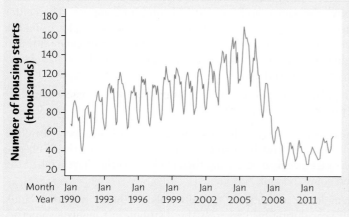

FIGURE 1.19

Time plot of the monthly count of new single-family houses started (in thousands) between January 1990 and April 2013, for Exercise 1.44.

1.45 Ozone hole. The ozone hole is a region in the stratosphere over the Antarctic with exceptionally depleted ozone. The size of the hole is not constant over the year but is largest at the beginning of the Southern Hemisphere spring (August–October). The increase in the size of the ozone hole led to the Montreal Protocol in 1987, an international treaty designed to protect the ozone layer by phasing out the production of substances, such as chlorofluorocarbons (CFCs), believed to be responsible for ozone depletion. The following table gives the average ozone hole size for the period September 7 to October 13 for each of the years from 1979 through 2012 (note that no data were acquired in 1995).[32]

YEAR	AREA (MILLIONS OF km²)	YEAR	AREA (MILLIONS OF km²)
1979	0.1	1996	22.7
1980	1.4	1997	22.1
1981	0.6	1998	25.9
1982	4.8	1999	23.2
1983	7.9	2000	24.8
1984	10.1	2001	25.0
1985	14.2	2002	12.0
1986	11.3	2003	25.8
1987	19.3	2004	19.5
1988	10.0	2005	24.4
1989	18.7	2006	26.6
1990	19.2	2007	22.0
1991	18.8	2008	25.2
1992	22.3	2009	22.0
1993	24.2	2010	19.4
1994	23.6	2011	24.7
1995	—	2012	17.9

To get a better feel for the magnitude of the numbers, the area of North America is approximately 24.5 million square kilometers (km^2). OZONE

The two parts of this exercise will have you draw two graphs of these data.

(a) First make a time plot of the data. The severity of the ozone hole will vary from year to year depending on the meteorology of the atmosphere above Antarctica. Does the time plot illustrate only year-to-year variation or are other patterns apparent? Specifically, is there a trend over any period of years? What about cyclical fluctuation? Explain in words the change in the average size of the ozone hole over this 34-year period.

(b) Now make a stemplot of the data. What is the midpoint of the distribution of ozone hole size? Do you think that the stemplot and the midpoint are a good description of this data set? Is there important information in the time plot that is not contained in the stemplot? When data are collected over time, you should always make a time plot.

1.46 Choosing class intervals. Student engineers learn that, although handbooks give the strength of a material as a single number, in fact the strength varies from piece to piece. A vital lesson in all fields of study is that "variation is everywhere." Here are data from a typical student laboratory exercise: the load in pounds needed to pull apart pieces of Douglas fir 4 inches long and 1.5 inches square.

33,190	31,860	32,590	26,520	33,280
32,320	33,020	32,030	30,460	32,700
23,040	30,930	32,720	33,650	32,340
24,050	30,170	31,300	28,730	31,920

The data sets in the *One-Variable Statistical Calculator* applet on the text Web site include the "pulling wood apart" data given in this exercise. How many class intervals does the applet choose when drawing the histogram? Use the applet to make several histograms with a larger number of class intervals. Are there any important features of the data that are revealed using a larger number of class intervals? Which histogram do you prefer? Explain your choice.

Exploring the Web

1.47 **Natural gas prices.** The Department of Energy Web site contains information about monthly wholesale and retail prices for natural gas in each state. Go to www.eia.doe.gov/naturalgas/data.cfm and then click on the link Monthly Wholesale and Retail Prices. Under Area, choose a state of interest to you, make sure the Period is monthly, and then under Residential Price click on View History. A window will open with a time plot covering approximately a 25-year period, along with a table of the monthly residential prices for each year.

(a) If you have access to statistical software, you should use the *Download Data (XLS file)* link to save the data as an Excel (.xls) file on your computer. Then enter the data into your software package, and reproduce the time series plot using the graphic capabilities of your software package. Be sure you use an appropriate title and axis labels. If you do not have access to appropriate software, provide a rough sketch of the time plot that is given on the Web site.

(b) Is there a regular pattern of seasonal variation that repeats each year? Describe it. Are the prices increasing over time?

1.48 **Hank Aaron's home run record.** The all-time home run leader prior to 2007 was Hank Aaron. You can find his career statistics by going to the Web site www.baseball-reference.com and then clicking on the Players tab at the top of the page and going to Hank Aaron.

(a) Make a stemplot or a histogram of the number of home runs that Hank Aaron hit in each year during his career. Is the distribution roughly symmetric, clearly skewed, or neither? About how many home runs did Aaron hit in a typical year? Are there any outliers?

(b) Would a time plot be appropriate for these data? If so, what information would be included in the time plot that is not in the stemplot?

Describing Distributions with Numbers

Overview

Graphical methods provide a visual impression of the data and often reveal features that summary numerical descriptive measures cannot. On the other hand, numerical summaries provide information that graphical displays cannot easily provide. Thus, we pair numerical descriptions with graphical displays to describe data more fully.

Numerical summaries are typically designed to provide insight about one of several different qualities of possible interest. There are measures of center, which may describe a data set's "typical" (or "middle") value or at its arithmetic average, and measures of "spread," which report the amount of variability in the data. Looking ahead to Chapter 3, we can also talk about measures of relative standing (e.g., z-scores), which measure how unusual a data value is. To describe the "center" of a data set, we typically use the mean or the median. The decision about which measure is "best" depends on properties of the distribution. If a data set is roughly symmetric, the mean and the median will be almost equal, so either will be fine. However, if a data set is skewed (or has outliers), the mean is influenced by the few extreme values in the longer tail of the distribution. In such cases, the median is the preferred measure of center. The median is resistant to the "pull" of a few extreme observations. Note that both of these measures require (at least) ordinal data for which addition (or sorting) makes sense. For purely categorical data, the only measure of center that makes sense is the mode (the most often occurring value).

In Chapter 2, we consider three measures of variability: the spread (typically called the range), the standard deviation, and the interquartile range (IQR). These mea-

sures of variability are sensitive to extreme values and skew in the same way that the mean and the median are. The range is most volatile, as it uses only the extreme values of the data set. The IQR is the measure of variability that is least impacted by extreme values, as it looks only at the variability of the middle 50% of the data. The standard deviation is an excellent measure of variability when the mean is the preferred measure of center. But, if a data set is strongly skewed or has outliers, variability should be measured with the IQR.

It is important for students to understand that we pair measures of center and variability: the mean and the standard deviation are reported together, and the median is paired with the IQR. The mean and standard deviation are preferable when the data are roughly symmetric; the median and IQR are preferred when the data are skewed (or have outliers). It is important to note, however, that with a bimodal distribution, no one measure of center (or of variability) can adequately describe the distribution.

When we examined graphical displays, we described the shape, center, and variability of a distribution. We also identified suspected outliers as values separated from the rest by gaps in a graph; in Chapter 2 we gain guidelines to determine whether such data values are truly unusual. The *boxplot*, based on the *five-number summary*, provides another graphical display of the data. Note too that the five-number summary is sometimes provided in lieu of the median and *IQR*. There are typically several reasonable choices for which numerical description we would use to answer a question of interest.

Teaching Suggestions and Additional Examples/Activities for the Classroom

1. Emphasize the Importance of Variability

Motivate the main ideas of this chapter with an example that needs a measure of variability in addition to a measure of center. Consider four factories, each trying to produce 20-ounce bottles of water. Based on the following samples, which factory is doing the *best* job?

Factory A: 18 19 21 22 ounces
Factory B: 19 20 20 21 ounces
Factory C: 17 18 18 19 ounces
Factory D: 18 20 22 24 ounces

Students in class should easily recognize that Factory B is best. Have them explain why they made this choice. Factories A and B both have "typical" values around 20 ounces, whereas Factory C is producing bottles with typically less than 20 ounces, and Factory D is producing bottles with typically more than 20 ounces. The data in Factory A is more variable than that of Factory B. With discussion, students will realize that measures of center and variability are equally important and measure different aspects. You can also emphasize the fact that variability does *not* depend on the center; Factories B and C have the same standard deviations but different means.

2. Work with Data You Have Used Before

You might revisit a data collection you used in Chapter 1 for graphical displays and use it to illustrate these ideas. For example, if you made a stemplot using the shuttle launch time temperatures mentioned in Chapter 1, you could use it to identify the five-number summary and construct a boxplot. Students see the elongated "whisker" of the boxplot, and that skew can be assessed. You can change the "29" to "50" or "−29" and discuss the impact on the mean and median.

3. Motivate Standard Deviation

Depending on the backgrounds of your students, you might consider having students "discover" the formula for standard deviation through an exploration. For example, you can present two simple samples: {2, 2, 8, 8} and {2, 4, 6, 8}. Ask students which data set is more variable. Some students will select the first set, if they think of variability in the sense of "larger differences." Others will select the second set, if they think of variability in the sense of "variety" (no repeated values). Still others may say there is no difference, as both sets range from 2 to 8. It is clear that we must first agree on which of these ways to think about variation is best for our purposes. (Refer to the instructor insert for Chapter 1 for resources on the current discussion of variability.) From here, you can graphically describe "deviations from the mean" and the idea of a "typical deviation from the mean." You will find (again, depending on the students you have) that students might suggest the "mean absolute deviation" to get around the fact that the sum of the deviations from the mean is always zero . . . and they will grasp that squaring deviations changes units (a motivation for the square root to come).

4. Talk about Why We Use $n - 1$ Instead of n

When students finally see the formula for standard deviation, they usually object to the use of $n - 1$ in the denominator. The easiest way to motivate the $n - 1$ is to remind your students that the sum of the $(x - \bar{x})$ terms is always 0, so this becomes a "restriction"—once you know $n - 1$ of the data values and the sample mean, you can calculate the value of the data point you do not know.

5. Address the Properties of Mean and Standard Deviation

Demonstrate to students that multiplying all numbers in a list by the same constant (rescaling the data) changes the mean, median, standard deviation, and IQR. However, when you add the same constant to all numbers in a list (shift the data), only measures of center change. Suppose we have 10 teachers with the following salaries:

$30,000	$30,750	$35,000	$35,000	$35,000
$40,000	$40,000	$45,000	$45,000	$64,250

The average teacher salary is $40,000, and the standard deviation is $9,992 (close to $10,000). The total salary for all 10 teachers is $400,000. Now suppose we have a 5% raise pool, which is $20,000. We can give each teacher a 5% raise or give each teacher a $2000 raise. Have your students calculate the new salaries under these two scenarios. Look at both sets of new salaries. Your students should be able to see that in addition to seeing increased variability in the salary distribution, the top paid teachers benefit more from the percentage raise than do the lower-paid teachers (and the top paid teachers get further from the rest).

Other Resources (LaunchPad)

StatClips

Exploratory Pictures for Quantitative Data
Summaries of Quantitative Data
Summaries of Quantitative Data: Example A (NC travel times)
Summaries of Quantitative Data: Example B (NY travel times)

Snapshots Videos

Summarizing Quantitative Data

EESEE Case Studies

Is Old Faithful Faithful?
Emissions from an Oil Refinery

Applet

Mean and Median

Logan Mock-Bunting/Getty Images

Describing Distributions with Numbers

We saw in Chapter 1 (page 15) that the American Community Survey asks, among much else, workers' travel times to work. Here are the travel times in minutes for 15 workers in North Carolina, chosen at random by the U.S. Census Bureau:[1]

30   20   10   40   25   20   10   60   15   40   5   30   12   10   10

We aren't surprised that most people estimate their travel time in multiples of 5 minutes. Here is a stemplot of these data:

```
0 | 5
1 | 0 0 0 0 2 5
2 | 0 0 5
3 | 0 0
4 | 0 0
5 |
6 | 0
```

The distribution is single peaked and right-skewed. The longest travel time (60 minutes) may be an outlier. Our goal in this chapter is to describe with numbers the center and spread of this and other distributions.

45

Measuring center: the mean

The most common measure of center is the ordinary arithmetic average, or *mean*.

The Mean $\bar{x}$

To find the **mean** of a set of observations, add their values and divide by the number of observations. If the n observations are $x_1, x_2, \ldots, x_n$, their mean is

$$\bar{x} = \frac{x_1 + x_2 + \cdots + x_n}{n}$$

or, in more compact notation,

$$\bar{x} = \frac{1}{n} \sum x_i$$

The Σ (capital Greek sigma) in the formula for the mean is short for "add them all up." The subscripts on the observations x_i are just a way of keeping the n observations distinct. They do not necessarily indicate order or any other special facts about the data. The bar over the x indicates the mean of all the x-values. Pronounce the mean $\bar{x}$ as "x-bar." This notation is very common. When writers who are discussing data use $\bar{x}$ or $\bar{y}$, they are talking about a mean.

EXAMPLE 2.1	**Travel Times to Work**

NCTRAVEL

The mean travel time of our 15 North Carolina workers is

$$\bar{x} = \frac{x_1 + x_2 + \cdots + x_n}{n}$$
$$= \frac{30 + 20 + \cdots + 10}{15}$$
$$= \frac{337}{15} = 22.5 \text{ minutes}$$

In practice, you can enter the data into your calculator and ask for the mean. You don't have to actually add and divide. But you should know that this is what the calculator is doing.

   Notice that only 6 of the 15 travel times are larger than the mean. If we leave out the longest single travel time, 60 minutes, the mean for the remaining 14 people is 19.8 minutes. That one observation raises the mean by 2.7 minutes. ■

Example 2.1 illustrates an important fact about the mean as a measure of center: it is sensitive to the influence of a few extreme observations. These may be outliers, but a skewed distribution that has no outliers will also pull the mean toward its long tail. Because the mean cannot resist the influence of extreme observations, we say

resistant measure   that it is not a **resistant measure** of center.

Apply Your Knowledge

2.1   *E. coli* **in Swimming Areas.** To investigate water quality, the *Columbus Dispatch* took water specimens at 16 Ohio State Park swimming areas in central Ohio. Those specimens were taken to laboratories and tested for *E. coli*, which are bacteria that can cause serious gastrointestinal problems. For reference, if a 100-milliliter specimen (about 3.3 ounces) of water contains more than 130 *E. coli* bacteria, it is considered unsafe. Here are the *E. coli* levels per 100 milliliter found by the laboratories:[2]

291.0	10.9	47.0	86.0	44.0	18.9	1.0	50.0
190.4	45.7	28.5	18.9	16.0	34.0	8.6	9.6

Find the mean *E. coli* level. How many of the lakes have *E. coli* levels greater than the mean? What feature of the data explains the fact that the mean is greater than most of the observations? 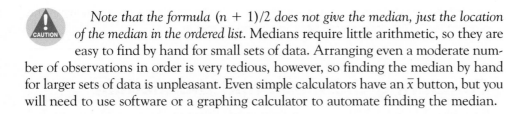 ECOLI

2.2 Health Care Spending. Table 1.3 (page 32) gives the 2011 health care expenditure per capita in 35 countries with the highest gross domestic product in 2011. The United States, at 8608 international dollars per person, is a high outlier. Find the mean health care spending in these nations with and without the United States included. How much does the one outlier increase the mean? HEALTH

Measuring center: the median

In Chapter 1, we used the midpoint of a distribution as an informal measure of center and gave a method for its computation. The *median* is the formal version of the midpoint, and we now provide a more detailed rule for its calculation.

The Median *M*

The **median *M*** is the midpoint of a distribution, the number such that half the observations are smaller and the other half are larger. To find the median of a distribution:

1. Arrange all observations in order of size, from smallest to largest.

2. If the number of observations *n* is odd, the median M is the center observation in the ordered list. If the number of observations *n* is even, the median M is midway between the two center observations in the ordered list.

3. You can always locate the median in the ordered list of observations by counting up (*n* + 1)/2 observations from the start of the list.

⚠ CAUTION
Note that the formula (n + 1)/2 does not give the median, just the location of the median in the ordered list. Medians require little arithmetic, so they are easy to find by hand for small sets of data. Arranging even a moderate number of observations in order is very tedious, however, so finding the median by hand for larger sets of data is unpleasant. Even simple calculators have an $\bar{x}$ button, but you will need to use software or a graphing calculator to automate finding the median.

EXAMPLE 2.2   Finding the Median: Odd *n*

What is the median travel time for our 15 North Carolina workers? Here are the data arranged in order:

5   10   10   10   10   12   15   **20**   20   25   30   30   40   40   60

The count of observations, *n* = 15, is odd. **The bold 20** is the center observation in the ordered list, with 7 observations to its left and 7 to its right. This is the median, *M* = 20 minutes.

Because $n = 15$, our rule for the location of the median gives

$$\text{location of } M = \frac{n+1}{2} = \frac{16}{2} = 8$$

That is, the median is the 8th observation in the ordered list. It is faster to use this rule than to locate the center by eye. ■

EXAMPLE 2.3   Finding the Median: Even *n*

Travel times to work in New York State are (on the average) longer than in North Carolina. Here are the travel times in minutes of 20 randomly chosen New York workers:

10 30 5 25 40 20 10 15 30 20 15 20 85 15 65 15 60 60 40 45

A stemplot not only displays the distribution but also makes finding the median easy because it arranges the observations in order:

```
0 | 5
1 | 0 0 5 5 5 5
2 | 0 0 0 5
3 | 0 0
4 | 0 0 5
5 |
6 | 0 0 5
7 |
8 | 5
```

The distribution is single peaked and right-skewed, with several travel times of an hour or more. There is no center observation, but there is a center pair. These are the bold **20** and **25** in the stemplot, which have 9 observations before them in the ordered list and 9 after them. The median is midway between these two observations:

$$M = \frac{20 + 25}{2} = 22.5 \text{ minutes}$$

With $n = 20$, the rule for locating the median in the list gives

$$\text{location of } M = \frac{n+1}{2} = \frac{21}{2} = 10.5$$

The location 10.5 means "halfway between the 10th and 11th observations in the ordered list." That agrees with what we found by eye. ■

Comparing the mean and the median

Examples 2.1 and 2.2 illustrate an important difference between the mean and the median. The median travel time (the midpoint of the distribution) is 20 minutes. The mean travel time is higher, 22.5 minutes. The mean is pulled toward the right tail of this right-skewed distribution. The median, unlike the mean, is *resistant*. If the longest travel time were 600 minutes rather than 60 minutes, the mean would increase to more than 58 minutes but the median would not change at all. The outlier just counts as one observation above the center, no matter how far above the center it lies. The mean uses the actual value of each observation and so will chase a single large observation upward. The *Mean and Median* applet is an excellent way to compare the resistance of M and $\bar{x}$.

Comparing the Mean and the Median

The mean and median of a roughly symmetric distribution are close together. If the distribution is exactly symmetric, the mean and median are exactly the same. In a skewed distribution, the mean is usually farther out in the long tail than the median.[3]

Many economic variables have distributions that are skewed to the right. For example, the median endowment of colleges and universities in the United States and Canada in 2012 was about $90 million—but the mean endowment was almost $491 million. Most institutions have modest endowments, but a few are very wealthy. Harvard's endowment was over $30 billion.[4] The few wealthy institutions pull the mean up but do not affect the median. Reports about incomes and other strongly skewed distributions usually give the median ("midpoint") rather than the mean ("arithmetic average"). However, a county that is about to impose a tax of 1% on the incomes of its residents cares about the mean income, not the median. The tax revenue will be 1% of total income, and the total is the mean times the number

 of residents. The mean and median measure center in different ways, and both are useful. *Don't confuse the "average" value of a variable (the mean) with its "typical" value, which we might describe by the median.*

Apply Your Knowledge

2.3 New York Travel Times. Find the mean of the travel times to work for the 20 New York workers in Example 2.3. Compare the mean and median for these data. What general fact does your comparison illustrate?

2.4 New House Prices. The mean and median sales prices of new homes sold in the United States in October 2011 were $212,300 and $242,300.[5] Which of these numbers is the mean and which is the median? Explain how you know.

2.5 Carbon Dioxide Emissions. Table 1.6 (page 41) gives the 2009 carbon dioxide (CO_2) emissions per person for countries with populations of at least 30 million. Find the mean and the median for these data. Make a histogram of the data. What features of the distribution explain why the mean is larger than the median? CO2EMISS

Getty Images/Flickr RF

Measuring spread: the quartiles

The mean and median provide two different measures of the center of a distribution. But a measure of center alone can be misleading. The U.S. Census Bureau reports that in 2011 the median income of American households was $50,054. Half of all households had incomes below $50,054, and half had higher incomes. The mean was much higher, $69,677, because the distribution of incomes is skewed to the right. But the median and mean don't tell the whole story. The bottom 20% of households had incomes less than $20,262, and households in the top 5% took in more than

 $186,000.[6] We are interested in the *spread* or *variability* of incomes as well as their center. *The simplest useful numerical description of a distribution requires both a measure of center and a measure of spread/variability.*

One way to measure spread is to give the smallest and largest observations. For example, the travel times of our 15 North Carolina workers range from 5 minutes to 60 minutes. These single observations show the full spread of the data, but they may be outliers. We can improve our description of spread by also looking at the spread of the middle half of the data. The *quartiles* mark out the middle half. Count up the ordered list of observations, starting from the smallest. The *first quartile* lies one-quarter of the way up the list. The *third quartile* lies three-quarters of the way up the list. In other words, the first quartile is larger than 25% of the observations, and the third quartile is larger than 75% of the observations. The second quartile is the median, which is larger than 50% of the observations. That is the idea of quartiles. We need a rule to make the idea exact. The rule for calculating the quartiles uses the rule for the median.

The Quartiles Q_1 and Q_3

To calculate the **quartiles**:

1. Arrange the observations in increasing order and locate the median, M, in the ordered list of observations.
2. The **first quartile, Q_1,** is the median of the observations whose position in the ordered list is to the left of the location of the overall median.
3. The **third quartile, Q_3,** is the median of the observations whose position in the ordered list is to the right of the location of the overall median.

Here are examples that show how the rules for the quartiles work for both odd and even numbers of observations.

EXAMPLE 2.4   **Finding the Quartiles: Odd *n***

Our North Carolina sample of 15 workers' travel times, arranged in increasing order, is

5   10   10   10   10   12   15   **20**   20   25   30   30   40   40   60

There is an odd number of observations, so the median is the middle one, the bold **20** in the list. The first quartile is the median of the 7 observations to the left of the median. This is the 4th of these 7 observations, so $Q_1 = 10$ minutes. If you want, you can use the rule for the location of the median with $n = 7$:

$$\text{location of } Q_1 = \frac{n+1}{2} = \frac{7+1}{2} = 4$$

The third quartile is the median of the 7 observations to the right of the median, $Q_3 = 30$ minutes. *When there is an odd number of observations, leave out the overall median when you locate the quartiles in the ordered list.*

The quartiles are *resistant* because they are not affected by a few extreme observations. For example, Q_3 would still be 30 if the outlier were 600 rather than 60. ■

EXAMPLE 2.5   **Finding the Quartiles: Even *n***

Here are the travel times to work of the 20 New Yorker workers from Example 2.3, arranged in increasing order:

5   10   10   15   15   15   15   20   20   20 | 25   30   30   40   40   45   60   60   65   85

There is an even number of observations, so the median lies midway between the middle pair, the 10th and 11th in the list. Its value is $M = 22.5$ minutes. We have marked the location of the median by |. The first quartile is the median of the first 10 observations because these are the observations to the left of the location of the median. Check that $Q_1 = 15$ minutes and $Q_3 = 42.5$ minutes. *When the number of observations is even, include all the observations when you locate the quartiles.* ■

Be careful when, as in these examples, several observations take the same numerical value. Write down all the observations, arrange them in order, and apply the rules just as if they all had distinct values.

The five-number summary and boxplots

The smallest and largest observations tell us little about the distribution as a whole, but they give information about the tails of the distribution that is missing if we know only the median and the quartiles. To get a quick summary of both center and spread, combine all five numbers.

The Five-Number Summary

The **five-number summary** of a distribution consists of the smallest observation, the first quartile, the median, the third quartile, and the largest observation, written in order from smallest to largest. In symbols, the five-number summary is

$$\text{Minimum} \quad Q_1 \quad M \quad Q_3 \quad \text{Maximum}$$

These five numbers offer a reasonably complete description of center and spread. The five-number summaries of travel times to work from Examples 2.4 and 2.5 are

North Carolina	5	10	20	30	60
New York	5	15	22.5	42.5	85

The five-number summary of a distribution leads to a new graph, the *boxplot*. Figure 2.1 shows boxplots comparing travel times to work in North Carolina and New York.

Boxplot

A **boxplot** is a graph of the five-number summary.
- ■ A central box spans the quartiles Q_1 and Q_3.
- ■ A line in the box marks the median M.
- ■ Lines extend from the box out to the smallest and largest observations.

Because boxplots show less detail than histograms or stemplots, they are best used for side-by-side comparison of more than one distribution, as in Figure 2.1. Be sure to include a numerical scale in the graph. When you look at a boxplot, first locate the median, which marks the center of the distribution. Then look at the spread. The span of the central box shows the spread of the middle half of the data, and the extremes (the smallest and largest observations) show the spread of the entire data set. We see from Figure 2.1 that travel times to work are in general a bit longer in New York than in North Carolina. The median, both quartiles, and the maximum are all larger in New York. New York travel times are also more variable, as shown by the span of the box and the spread between the extremes. Note that the boxes with arrows in Figure 2.1 that indicate the location of the five-number summary are *not* part of the boxplot, but are included purely for illustration.

FIGURE 2.1

Boxplots comparing the travel times to work of samples of workers in North Carolina and New York.

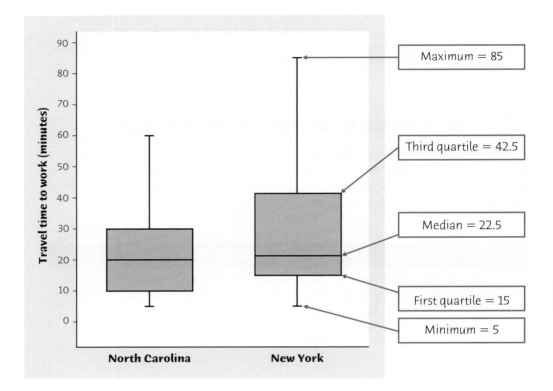

Finally, the New York data are more strongly right-skewed. In a symmetric distribution, the first and third quartiles are equally distant from the median. In most distributions that are skewed to the right, on the other hand, the third quartile will be farther above the median than the first quartile is below it. The extremes behave the same way, but remember that they are just single observations and may say little about the distribution as a whole.

Apply Your Knowledge

2.6 The Pittsburgh Steelers. The 2012–13 roster of the Pittsburgh Steelers professional football team included 8 defensive linemen and 9 offensive linemen. The weights in pounds of the defensive linemen were [STEELERS]

    280   324   307   285   300   288   325   348

and the weights of the offensive linemen were

    315   304   303   315   316   325   298   304   345

(a) Make a stemplot of the weights of the defensive linemen and find the five-number summary.

(b) Make a stemplot of the weights of the offensive linemen and find the five-number summary.

(c) Does either group contain one or more clear outliers? Which group of players tends to be heavier?

2.7 Fuel Economy for Midsize Cars. The Department of Energy provides fuel economy ratings for all cars and light trucks sold in the United States. Here are the estimated miles per gallon for city driving for the 173 cars classified as midsize in 2013, arranged in increasing order:[7] [MIDCARS]

11	11	11	11	12	14	14	14	14	15	15	15	15	15	16
16	16	16	16	16	16	16	16	16	16	16	17	17	17	17
17	17	17	17	17	17	17	18	18	18	18	18	18	18	18
18	18	18	18	18	18	18	18	18	18	18	19	19	19	19
19	19	19	19	19	19	19	19	19	19	19	19	19	19	20
20	20	20	20	20	20	20	20	20	21	21	21	21	21	21
21	21	21	21	21	21	21	22	22	22	22	22	22	22	22
22	22	22	22	22	22	23	23	23	24	24	24	24	24	24
24	24	25	25	25	25	25	25	25	25	25	25	25	26	26
26	26	26	26	26	27	27	27	27	27	27	27	27	27	27
27	27	28	28	28	28	28	28	28	28	29	29	30	30	30
31	40	40	40	43	45	47	51							

(a) Give the five-number summary of this distribution.

(b) Draw a boxplot of these data. What is the shape of the distribution shown by the boxplot? Which features of the boxplot led you to this conclusion? Are any observations unusually small or large?

Spotting suspected outliers*

Look again at the stemplot of travel times to work in New York in Example 2.3 (page 48). The five-number summary for this distribution is

$$5 \quad 15 \quad 22.5 \quad 42.5 \quad 85$$

How shall we describe the spread of this distribution? The smallest and largest observations are extremes that don't describe the spread of the majority of the data. The distance between the quartiles (the range of the center half of the data) is a more resistant measure of spread. This distance is called the *interquartile range*.

The Interquartile Range (*IQR*)

The **interquartile range** (*IQR*) is the distance between the first and third quartiles,

$$IQR = Q_3 - Q_1$$

For our data on New York travel times, $IQR = 42.5 - 15 = 27.5$ minutes. However, *no single numerical measure of spread, such as IQR, is very useful for describing skewed distributions*. The two sides of a skewed distribution have different spreads, so one number can't summarize them. That's why we give the full five-number summary. The interquartile range is mainly used as the basis for a rule of thumb for identifying suspected outliers. In some software, suspected outliers are identified in a boxplot with a special plotting symbol such as *.

The 1.5 × *IQR* Rule for Outliers

Call an observation a suspected outlier if it falls more than 1.5 × *IQR* above the third quartile or below the first quartile.

*This short section is optional.

EXAMPLE 2.6   Using the 1.5 × *IQR* Rule

For the New York travel time data, *IQR* = 27.5 and

$$1.5 \times IQR = 1.5 \times 27.5 = 41.25$$

Any values not falling between

$$Q_1 - (1.5 \times IQR) = 15.0 - 41.25 = -26.25 \quad \text{and}$$
$$Q_3 + (1.5 \times IQR) = 42.5 + 41.25 = 83.75$$

are flagged as suspected outliers. Look again at the stemplot in Example 2.3 (page 48): The only suspected outlier is the longest travel time, 85 minutes. The 1.5 × *IQR* rule suggests that the three next-longest travel times (60 and 65 minutes) are just part of the long right tail of this skewed distribution. ■

The 1.5 × *IQR* rule is not a replacement for looking at the data. It is most useful when large volumes of data are processed automatically.

Apply Your Knowledge

2.8 Travel Time to Work. In Example 2.1 (page 46), we noted the influence of one long travel time of 60 minutes in our sample of 15 North Carolina workers. Does the 1.5 × *IQR* rule identify this travel time as a suspected outlier?

2.9 Fuel Economy for Midsize Cars. Exercise 2.7 (page 52) gives the estimated miles per gallon (mpg) for city driving for the 173 cars classified as midsize in 2013. In that exercise we noted that several of the mpg values were unusually large. Which of these are suspected outliers by the 1.5 × *IQR* rule? Although outliers can be produced by errors or incorrectly recorded observations, they are often observations that differ from the others in some particular way. In this case, the cars producing the high outliers share a common feature. What do you think that is? MIDCARS

 HOW MUCH IS THAT HOUSE WORTH?

The town of Manhattan, Kansas, is sometimes called "the little Apple" to distinguish it from that other Manhattan, "the big Apple." A few years ago, a house there appeared in the county appraiser's records valued at $200,059,000. That would be quite a house even on Manhattan Island. As you might guess, the entry was wrong: the true value was $59,500. But before the error was discovered, the county, the city, and the school board had based their budgets on the total appraised value of real estate, which the one outlier jacked up by 6.5%. It can pay to spot outliers before you trust your data.

Measuring spread: the standard deviation

The five-number summary is not the most common numerical description of a distribution. That distinction belongs to the combination of the mean to measure center and the *standard deviation* to measure spread. The standard deviation and its close relative, the *variance*, measure spread/variability by looking at how far the observations are from their mean.

The Standard Deviation *s*

The **variance s^2** of a set of observations is an average of the squares of the deviations of the observations from their mean. In symbols, the variance of n observations $x_1, x_2, \ldots, x_n$ is

$$s^2 = \frac{(x_1 - \bar{x})^2 + (x_2 - \bar{x})^2 + \cdots + (x_n - \bar{x})^2}{n - 1}$$

or, more compactly,

$$s^2 = \frac{1}{n - 1} \sum (x_i - \bar{x})^2$$

The **standard deviation s** is the square root of the variance s^2:

$$s = \sqrt{\frac{1}{n - 1} \sum (x_i - \bar{x})^2}$$

In practice, use software or your calculator to obtain the standard deviation from keyed-in data. Doing an example step-by-step will help you understand how the variance and standard deviation work, however.

EXAMPLE 2.7   Calculating the Standard Deviation

Georgia Southern University had 2417 students with regular admission in its freshman class of 2010. For each student, data are available on their SAT and ACT scores (if taken), high school GPA, and the college within the university to which they were admitted.[8] In Exercise 3.49 (page 93), the full data set for the SAT Critical Reading scores will be examined. Here are the first five observations from that data set:

650   490   580   450   570

We will compute $\bar{x}$ and s for these students. First find the mean:

$$\bar{x} = \frac{650 + 490 + 580 + 450 + 570}{5}$$

$$= \frac{2740}{5} = 548$$

Figure 2.2 displays the data as points above the number line, with their mean marked by an asterisk (*). The arrows mark two of the deviations from the mean. The deviations show how spread out the data are about their mean. They are the starting point for calculating the variance and the standard deviation.

Observations x_i	Deviations $x_i - \bar{x}$	Squared Deviations $(x_i - \bar{x})^2$
650	$650 - 548 = 102$	$102^2 = 10{,}404$
490	$490 - 548 = -58$	$(-58)^2 = 3{,}364$
580	$580 - 548 = 32$	$32^2 = 1{,}024$
450	$450 - 548 = -98$	$(-98)^2 = 9{,}604$
570	$570 - 548 = 22$	$22^2 = 484$
	sum $= 0$	sum $= 24{,}880$

The variance is the sum of the squared deviations divided by one less than the number of observations:

$$s^2 = \frac{1}{n-1} \sum (x_i - \bar{x})^2 = \frac{24{,}880}{4} = 6220$$

The standard deviation is the square root of the variance:

$$s = \sqrt{6220} = 78.87 \quad ■$$

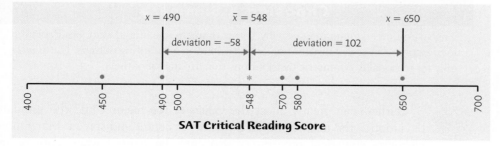

FIGURE 2.2
SAT Critical Reading scores for five students, with their mean (*) and the deviations of two observations from the mean shown.

Notice that the "average" in the variance s^2 divides the sum by one fewer than the number of observations, that is, $n - 1$ rather than n. The reason is that the deviations $x_i - \bar{x}$ always sum to exactly 0, so that knowing $n - 1$ of them determines the last one. Only $n - 1$ of the squared deviations can vary freely, and we **degrees of freedom** average by dividing the total by $n - 1$. The number $n - 1$ is called the **degrees of freedom** of the variance or standard deviation. Some calculators offer a choice between dividing by n and dividing by $n - 1$, so be sure to use $n - 1$.

More important than the details of hand calculation are the properties that determine the usefulness of the standard deviation:

- s measures *spread/variability about the mean* and should be used only when the mean is chosen as the measure of center.

- s is *always zero or greater than zero*. $s = 0$ only when there is no variability. This happens only when all observations have the same value. Otherwise, $s > 0$. As the observations become more spread out about their mean, s gets larger.

- s has the *same units of measurement as the original observations*. For example, if you measure weight in kilograms, both the mean $\bar{x}$ and the standard deviation s are also in kilograms. This is one reason to prefer s to the variance s^2, which would be in squared kilograms.

- Like the mean $\bar{x}$, s is *not resistant*. A few outliers can make s very large.

 The use of squared deviations renders s even more sensitive than $\bar{x}$ to a few extreme observations. For example, the standard deviation of the travel times for the 15 North Carolina workers in Example 2.1 (page 46) is 15.23 minutes. (Use your calculator or software to verify this.) If we omit the high outlier, the standard deviation drops to 11.56 minutes.

If you feel that the importance of the standard deviation is not yet clear, you are right. We will see in Chapter 3 that the standard deviation is the natural measure of spread for a very important class of symmetric distributions, the Normal distributions. The usefulness of many statistical procedures is tied to distributions of particular shapes. This is certainly true of the standard deviation.

Choosing measures of center and spread

We now have a choice between two descriptions of the center and spread/variability of a distribution: the five-number summary, or $\bar{x}$ and s. Because $\bar{x}$ and s are sensitive to extreme observations, they can be misleading when a distribution is strongly skewed or has outliers. In fact, because the two sides of a skewed distribution have different spreads, no single number describes the spread well. The five-number summary, with its two quartiles and two extremes, does a better job.

Choosing a Summary

The five-number summary is usually better than the mean and standard deviation for describing a skewed distribution or a distribution with strong outliers. Use $\bar{x}$ and s only for reasonably symmetric distributions that are free of outliers.

 Outliers can greatly affect the values of the mean $\bar{x}$ and the standard deviation s, the most common measures of center and spread. Many more elaborate statistical procedures also can't be trusted when outliers are present. *Whenever you find outliers in your data, try to find an explanation for them.* Sometimes the explanation is as simple as a typing error, such as typing 10.1 as 101. Sometimes a measuring device broke down or a subject gave a frivolous response, like the student in a class survey who claimed to study 30,000 minutes per night. (Yes, that really happened.) In all these cases, you can simply remove the outlier from your data. When outliers are "real data," like the long travel times of some New York workers, you should choose statistical methods that are not greatly disturbed by the outliers. For example, use the five-number summary rather than $\bar{x}$ and s to describe a distribution with extreme outliers. We will meet other examples later in the book.

!CAUTION   *Remember that a graph gives the best overall picture of a distribution.* If data have been entered into a calculator or statistical program, it is very simple and quick to create several graphs to see all the different features of a distribution. Numerical measures of center and spread/variability report specific facts about a distribution, but they do not describe its entire shape. Numerical summaries do not disclose the presence of multiple peaks or clusters, for example. Exercise 2.11 shows how misleading numerical summaries can be. **Always plot your data.**

Apply Your Knowledge

2.10 $\bar{x}$ and s by Hand. Radon is a naturally occurring gas and is the second leading cause of lung cancer in the United States.[9] It comes from the natural breakdown of uranium in the soil and enters buildings through cracks and other holes in the foundations. Found throughout the United States, levels vary considerably from state to state. Several methods can reduce the levels of radon in your home, and the Environmental Protection Agency recommends using one of these if the measured level in your home is above 4 picocuries per liter. Four readings from Franklin County, Ohio, where the county average is 9.32 picocuries per liter, were 5.2, 13.8, 8.6, and 16.8.

Getty Images/Photo Researchers

(a) Find the mean step-by-step. That is, find the sum of the 4 observations and divide by 4.

(b) Find the standard deviation step-by-step. That is, find the deviation of each observation from the mean, square the deviations, then obtain the variance and the standard deviation. Example 2.7 (page 55) shows the method.

(c) Now enter the data into your calculator and use the mean and standard deviation buttons to obtain $\bar{x}$ and s. Do the results agree with your hand calculations?

2.11 $\bar{x}$ and s Are Not Enough. The mean $\bar{x}$ and standard deviation s measure center and spread but are not a complete description of a distribution. Data sets with different shapes can have the same mean and standard deviation. To demonstrate this fact, use your calculator to find $\bar{x}$ and s for these two small data sets. Then make a stemplot of each and comment on the shape of each distribution. 📊 DATASET2

Data A:	9.14	8.14	8.74	8.77	9.26	8.10	6.13	3.10	9.13	7.26	4.74
Data B:	6.58	5.76	7.71	8.84	8.47	7.04	5.25	5.56	7.91	6.89	12.50

2.12 Choose a Summary. The shape of a distribution is a rough guide to whether the mean and standard deviation are a helpful summary of center and spread. For which of the following distributions would $\bar{x}$ and s be useful? In each case, give a reason for your decision.

(a) Percents of high school graduates in the states taking the SAT, Figure 1.8 (page 27)

(b) *Iowa Tests* scores, Figure 1.7 (page 27)

(c) New York travel times, Figure 2.1 (page 52)

Using technology

Although a calculator with "two-variable statistics" functions will do the basic calculations we need, more elaborate tools are helpful. Graphing calculators and computer software will do calculations and make graphs as you command, freeing you to concentrate on choosing the right methods and interpreting your results. Figure 2.3 displays output describing the travel times to work of 20 people in New York State (Example 2.3, page 48). Can you find $\bar{x}$, s, and the five-number summary in each output? The big message of this section is: *Once you know what to look for, you can read output from any technological tool.*

The displays in Figure 2.3 come from a Texas Instruments graphing calculator, the Minitab and CrunchIt statistical programs, and the Microsoft Excel spreadsheet program. Minitab allows you to choose what descriptive measures you want, whereas the descriptive measures in the CrunchIt output are provided by default. Excel and the calculator give some things we don't need. Just ignore the extras. Excel's "Descriptive Statistics" menu item doesn't give the quartiles. We used the spreadsheet's separate quartile function to get Q_1 and Q_3.

EXAMPLE 2.8   What Is the Third Quartile?

In Example 2.5, we saw that the quartiles of the New York travel times are $Q_1 = 15$ and $Q_3 = 42.5$. Look at the output displays in Figure 2.3. The Calculator, CrunchIt!, and Excel agree with our work. Minitab says that $Q_3 = 43.75$. What happened? *There are several rules for finding the quartiles. Some calculators and software use rules that give results different from ours for some sets of data.* This is true of Minitab and also Excel, though Excel agrees with our work in this example. Results from the various rules are always close to each other, so the differences are never important in practice. Our rule is the simplest for hand calculation. ■

Organizing a statistical problem

Most of our examples and exercises have aimed to help you learn basic tools (graphs and calculations) for describing and comparing distributions. You have also learned principles that guide use of these tools, such as "start with a graph" and "look for the overall pattern and striking deviations from the pattern." The data you work with are not just numbers—they describe specific settings such as water depth in the Everglades or travel time to work. Because data come from a specific setting, the final step in examining data is a *conclusion for that setting*. Water depth in the Everglades has a yearly cycle that reflects Florida's wet and dry seasons. Travel times to work are generally longer in New York than in North Carolina.

Texas Instruments Graphing Calculator

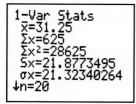

```
1-Var Stats
 x̄=31.25
 Σx=625
 Σx²=28625
 Sx=21.8773495
 σx=21.32340264
↓n=20
```

```
1-Var Stats
↑n=20
 minX=5
 Q₁=15
 Med=22.5
 Q₃=42.5
 maxX=85
```

Minitab

Session

Descriptive Statistics: NYtime

Variable	N	N*	Mean	SE Mean	StDev	Minimum	Q1	Median	Q3	Maximum
NYtime	20	0	31.25	4.89	21.88	5.00	15.00	22.50	43.75	85.00

CrunchIt!

Results - Descriptive Statistics

Export ▼

	n	Sample Mean	Standard Deviation	Min	Q1	Median	Q3	Max
NYtime	20	31.25	21.88	5	15	22.50	42.50	85

Microsoft Excel

Book1

	A	B	C	D
1		*minutes*		
2				
3	Mean	31.25		
4	Standard Error	4.891924064		
5	Median	22.5	QUARTILE(A2:A21,1)	15
6	Mode	15	QUARTILE(A2:A21,3)	42.5
7	Standard Deviation	21.8773495		
8	Sample Variance	478.6184211		
9	Kurtosis	0.329884126		
10	Skewness	1.040110836		
11	Range	80		
12	Minimum	5		
13	Maximum	85		
14	Sum	625		
15	Count	20		

Sheet4 / Sheet1 / Sheet2 / Sheet

FIGURE 2.3

Output from a graphing calculator, two statistical software packages, and a spreadsheet program describing the data on travel times to work in New York State.

Let's return to the on time high school graduation rates discussed in Example 1.4, page 21. We know from the example that the on time graduation rates vary from 59% in the District of Columbia to 88% in Iowa, with a median of 80%. State graduation rates are related to many factors, and in a statistical problem we often try to explain the differences or variation in a variable such as graduation rate by some of these factors. For example, do states with lower household incomes tend to have lower high school graduation rates? Or, do some regions of the country have lower high school graduation rates than others?

As you learn more statistical tools and principles, you will face more complex statistical problems. Although no framework accommodates all the varied issues that arise in applying statistics to real settings, we find the following four-step thought process gives useful guidance. In particular, the first and last steps emphasize that statistical problems are tied to specific real-world settings and therefore involve more than doing calculations and making graphs.

Organizing a Statistical Problem: A Four-Step Process

STATE: What is the practical question, in the context of the real-world setting?

PLAN: What specific statistical operations does this problem call for?

SOLVE: Make the graphs and carry out the calculations needed for this problem.

CONCLUDE: Give your practical conclusion in the setting of the real-world problem.

To help you master the basics, many exercises will continue to tell you what to do—make a histogram, find the five-number summary, and so on. Real statistical problems don't come with detailed instructions. From now on, especially in the later chapters of the book, you will meet some exercises that are more realistic. Use the four-step process as a guide to solving and reporting these problems. They are marked with the four-step icon, as the following example illustrates.

EXAMPLE 2.9   Comparing Graduation Rates

GRADRATE

STATE: Law requires all states in the United States to use a common computation of on-time high school graduation rates for the 2010–11 school year. Previously, states chose one of several computation methods that gave answers that could differ by more than 10%. This common computation allows for meaningful comparison of graduation rates between the states.

We know from Table 1.1 (page 22), that the on-time high school graduation rates varied from 59% in the District of Columbia, to 88% in Vermont. The U.S. Census Bureau divides the 50 states and the District of Columbia, into four geographical regions, the Northeast (NE), Midwest (MW), South (S), and West (W). The region for each state is included in Table 1.1. Do these four regions of the country display distinct distributions of graduation rates? How do the mean rates compare?

PLAN: Use graphs and numerical descriptions to describe and compare the on-time high school graduation rates in the four regions of the United States.

SOLVE: We might use boxplots to compare the distributions, but stemplots preserve more detail and work well for data sets of these sizes. Figure 2.4 displays the stemplots with the stems lined up for easy comparison. The stems have been split to better display the distributions. The stemplots overlap, and some care is needed when comparing the four stemplots as the sample sizes differ, with some stemplots having more leaves than others. None of the plots show strong skewness, although the South has one low observation that stands apart from the others with this choice of stems. The Northeast and Midwest have distributions that are similar to each other as do the South and West. The graduation rates tend to be higher in the Northeast and Midwest and more spread out in the South and West. With little skewness and no serious outliers, we report $\bar{x}$ and s as our summary measures of center and spread/variability:

Midwest	Northeast	South	West
8 | 6 6 6 7 8	8 | 6 7	8 | 6 6	8 |
8 | 0 1 3 3 4	8 | 3 3 3 3 4	8 | 1 2 3	8 | 0 0 2
7 | 7	7 | 7 7	7 | 5 6 8 8	7 | 6 6 6 8
7 | 4	7 |	7 | 1 1 2 4	7 | 4
6 |	6 |	6 | 7	6 | 8 8
6 |	6 |	6 |	6 | 2 3
5 |	5 |	5 | 9	5 |

Region	Mean Rate	Standard Deviation
Midwest	82.92	4.25
Northeast	82.56	3.47
South	75.93	7.36
West	73.58	6.73

FIGURE 2.4
Stemplots comparing the distributions of graduation rates for the four census regions from Table 1.1.

CONCLUDE: The table of summary statistics confirms what we see in the stemplots. The Midwest and Northeast are quite similar to each other as are the South and West. The Midwest and Northeast have higher mean graduation rates as well as smaller standard deviations than the South and West. ■

Apply Your Knowledge

2.13 Logging in the Rain Forest. "Conservationists have despaired over destruction of tropical rain forest by logging, clearing, and burning." These words begin a report on a statistical study of the effects of logging in Borneo.[10] Charles Cannon of Duke University and his coworkers compared forest plots that had never been logged (Group 1) with similar plots nearby that had been logged 1 year earlier (Group 2) and 8 years earlier (Group 3). All plots were 0.1 hectare in area. Here are the counts of trees for plots in each group: LOGGING

Group 1:	27	22	29	21	19	33	16	20	24	27	28	19
Group 2:	12	12	15	9	20	18	17	14	14	2	17	19
Group 3:	18	4	22	15	18	19	22	12	12			

To what extent has logging affected the count of trees? Follow the four-step process in reporting your work.

2.14 Diplomatic Scofflaws. Until Congress allowed some enforcement in 2002, the thousands of foreign diplomats in New York City could freely violate parking laws. Two economists looked at the number of unpaid parking tickets per diplomat over a five-year period ending when enforcement reduced the problem.[11] They concluded that large numbers of unpaid tickets indicated a "culture of corruption" in a country and lined up well with more elaborate measures of corruption. The data set for 145 countries is too large to print here, but look at the data file on the text Web site. The first 32 countries in the list (Australia to Trinidad and Tobago) are classified by the World Bank as "developed." The remaining countries (Albania to Zimbabwe) are "developing." The World Bank classification is based only on national income and does not take into account measures of social development. SCOFFLAW

Give a full description of the distribution of unpaid tickets for both groups of countries and identify any high outliers. Compare the two groups. Does national income alone do a good job of distinguishing countries whose diplomats do and do not obey parking laws?

© James Leynse/CORBIS

CHAPTER 2 SUMMARY

Chapter Specifics

■ A numerical summary of a distribution should report at least its **center** and its **spread** or **variability.**

■ The **mean** $\bar{x}$ and the **median** M describe the center of a distribution in different ways. The mean is the arithmetic average of the observations, and the median is the midpoint of the values.

■ When you use the median to indicate the center of the distribution, describe its spread by giving the **quartiles.** The **first quartile, Q_1,** has one-fourth of the observations below it, and the **third quartile, Q_3,** has three-fourths of the observations below it.

■ The **five-number summary** consisting of the median, the quartiles, and the smallest and largest individual observations provides a quick overall description of a distribution. The median describes the center, and the quartiles and extremes show the spread.

■ **Boxplots** based on the five-number summary are useful for comparing several distributions. The box spans the quartiles and shows the spread of the central half of the distribution. The median is marked within the box. Lines extend from the box to the extremes and show the full spread of the data.

■ The **variance s^2** and especially its square root, the **standard deviation s,** are common measures of spread about the mean as center. The standard deviation s is zero when there is no variability and gets larger as the spread increases.

■ A **resistant measure** of any aspect of a distribution is relatively unaffected by changes in the numerical value of a small proportion of the total number of observations, no matter how large these changes are. The median and quartiles are resistant, but the mean and the standard deviation are not.

■ The mean and standard deviation are good descriptions for symmetric distributions without outliers. They are most useful for the Normal distributions introduced in the next chapter. The five-number summary is a better description for skewed distributions.

■ Numerical summaries do not fully describe the shape of a distribution. Always plot your data.

■ A statistical problem has a real-world setting. You can organize many problems using the following four steps: **state, plan, solve,** and **conclude.**

Link It

In this chapter we have continued our study of exploratory data analysis. Graphs are an important visual tool for organizing and identifying patterns in data. They give a fairly complete description of a distribution, although for many problems the important information in your data can be described by a few numbers. These numerical summaries can be useful for describing a single distribution as well as for comparing the distributions from several groups of observations.

Two important features of a distribution are the center and the spread/variability. For distributions that are approximately symmetric without outliers, the mean and standard deviation are important numeric summaries for describing and comparing distributions. But if the distribution is not symmetric and/or has outliers, the five-number summary often provides a better description.

The boxplot gives a picture of the five-number summary that is useful for a simple comparison of several distributions. Remember that the boxplot is based only on the five-number summary and does not have any information beyond these five numbers. Certain features of a distribution that are revealed in histograms and stemplots will not be evident from a boxplot alone. These include gaps in the data and the presence of several peaks. You must be careful when reducing a distribution to a few numbers to make sure that important information has not been lost in the process.

CHECK YOUR SKILLS

2.15 The respiratory system can be a limiting factor in maximal exercise performance. Researchers from the United Kingdom studied the effect of two breathing frequencies on both performance times and several physiological parameters in swimming.[12] Subjects were 10 male collegiate swimmers. Here are their times in seconds to swim 200 meters at 90% of race pace when breathing every second stroke in front crawl swimming: **SWIMTIME**

151.6 165.1 159.2 163.5 174.8 173.2 177.6
174.3 164.1 171.4

The mean of these data is

(a) 165.10.   (b) 167.48.   (c) 168.25.

2.16 The median of the data in Exercise 2.15 is

(a) 167.48.   (b) 168.25.   (c) 174.00.

2.17 The five-number summary of the data in Exercise 2.15 is

(a) 151.6, 163.5, 167.48, 174.3, 177.6.
(b) 151.6, 163.5, 168.25, 174.3, 177.6.
(c) 151.6, 159.2, 168.25, 174.3, 177.6.

2.18 If a distribution is skewed to the right,

(a) the mean is less than the median.
(b) the mean and median are equal.
(c) the mean is greater than the median.

2.19 What percent of the observations in a distribution lie between the first quartile and the third quartile?

(a) 25%   (b) 50%   (c) 75%

2.20 To make a boxplot of a distribution, you must know

(a) all the individual observations.
(b) the mean and the standard deviation.
(c) the five-number summary.

2.21 The standard deviation of the 10 swim times in Exercise 2.15 (use your calculator) is about

(a) 7.4.   (b) 7.8.   (c) 8.2.

2.22 What are all the values that a standard deviation s can possibly take?

(a) $0 \leq s$ (b) $0 \leq s \leq 1$ (c) $-1 \leq s \leq 1$

2.23 The correct units for the standard deviation in Exercise 2.21 are

(a) no units—it's just a number.   (b) seconds.
(c) seconds squared.

2.24 Which of the following is least affected if an extreme high outlier is added to your data?

(a) The median   (b) The mean
(c) The standard deviation

CHAPTER 2 EXERCISES

2.25 **Incomes of college grads.** According to the U.S. Census Bureau's Current Population Survey, the mean and median 2011 income of people at least 25 years old who had a bachelor's degree but no higher degree were $49,648 and $61,691.[13] Which of these numbers is the mean and which is the median? Explain your reasoning.

2.26 **Saving for retirement.** Retirement seems a long way off, and we need money now, so saving for retirement is hard. Once every three years, the Board of Governors of the Federal Reserve System collects data on household assets and liabilities through the Survey of Consumer Finances (SCF). The most recent such survey was conducted in 2010, and the survey results were released to the public in April 2013. The survey presents data on retirement assets, which include defined contribution and Individual Retirement (IRA) balances. For married households the mean value per household is $123,968, but the median value is just $10,000. For single households, the mean is $33,585, and the median is $0.[14] What explains the differences between the two measures of center, both for married and single households? What does a median of $0 say about the percentage of single households with retirement assets?

2.27 **University endowments.** The National Association of College and University Business Officers collects data on college endowments. In 2012, 843 colleges and universities reported the value of their endowments. When the endowment values are arranged in order, what are the locations of the median and the quartiles in this ordered list?

2.28 **Pulling wood apart.** Exercise 1.46 (page 44) gives the breaking strengths of 20 pieces of Douglas fir. **WOOD**

(a) Give the five-number summary of the distribution of breaking strengths.
(b) Here is a stemplot of the data rounded to the nearest hundred pounds. The stems are thousands of pounds, and the leaves are hundreds of pounds.

```
23 | 0
24 | 1
25 |
26 | 5
27 |
28 | 7
29 |
30 | 259
31 | 399
32 | 033677
33 | 0237
```

The stemplot shows that the distribution is skewed to the left. Does the five-number summary show the skew? Remember that only a graph gives a clear picture of the shape of a distribution.

2.29 Comparing graduation rates. An alternative presentation to compare the graduation rates in Table 1.1 (page 22) by region of the country reports five-number summaries and uses boxplots to display the distributions. Do the boxplots fail to reveal any important information visible in the stemplots of Figure 2.4 (page 61)? Which plots make it simpler to compare the regions? Why? [ill] GRADRATE

2.30 How much fruit do adolescent girls eat? Figure 1.14 (page 38) is a histogram of the number of servings of fruit per day claimed by 74 seventeen-year-old girls.

(a) With a little care, you can find the median and the quartiles from the histogram. What are these numbers? How did you find them?

(b) You can also find the mean number of servings of fruit claimed per day from the histogram. First use the information in the histogram to compute the sum of the 74 observations, and then use this to compute the mean. What is the relationship between the mean and median? Is this what you expected?

(c) In general, you cannot find the exact values of the median, quartiles, or mean from the histogram. What is special about the histogram of the number of servings of fruit that allows you to do this?

2.31 Guinea pig survival times. Here are the survival times in days of 72 guinea pigs after they were injected with infectious bacteria in a medical experiment.[15] Survival times, whether of machines under stress or cancer patients after treatment, usually have distributions that are skewed to the right. [ill] GUINPIGS

43	45	53	56	56	57	58	66	67	73	74	79
80	80	81	81	81	82	83	83	84	88	89	91
91	92	92	97	99	99	100	100	101	102	102	102
103	104	107	108	109	113	114	118	121	123	126	128
137	138	139	144	145	147	156	162	174	178	179	184
191	198	211	214	243	249	329	380	403	511	522	598

(a) Graph the distribution and describe its main features. Does it show the expected right-skew?

(b) Which numerical summary would you choose for these data? Calculate your chosen summary. How does it reflect the skewness of the distribution?

2.32 Weight of newborns. Here is the distribution of the weight at birth for all babies born in the United States in 2010:[16]

WEIGHT (grams)	COUNT	WEIGHT (grams)	COUNT
Less than 500	5,980	3,000 to 3,499	1,566,755
500 to 999	22,015	3,500 to 3,999	1,055,004
1,000 to 1,499	29,846	4,000 to 4,499	262,997
1,500 to 1,999	63,427	4,500 to 4,999	36,706
2,000 to 2,499	204,295	5,000 to 5,499	3,964
2,500 to 2,999	744,181		

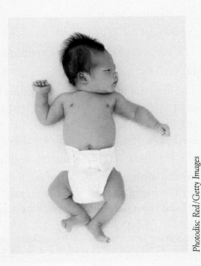

Photodisc Red/Getty Images

(a) For comparison with other years and with other countries, we prefer a histogram of the *percents* in each weight class rather than the counts. Explain why.

(b) How many babies were there?

(c) Make a histogram of the distribution, using percents on the vertical scale.

(d) What are the locations of the median and quartiles in the ordered list of all birth weights? In which weight classes do the median and quartiles fall?

2.33 More on study times. In Exercise 1.38 (page 42) you examined the nightly study time claimed by first-year college men and women. The most common methods for formal comparison of two groups use $\bar{x}$ and s to summarize the data. [ill] STUDYTIM

(a) What kinds of distributions are best summarized by $\bar{x}$ and s? Do you think these summary measures are appropriate in this case?

(b) One student in each group claimed to study at least 300 minutes (five hours) per night. How much does removing these observations change $\bar{x}$ and s for each group? You will need to compute $\bar{x}$ and s for each group, both with and without the high outlier.

2.34 **Making resistance visible.** In the *Mean and Median* applet, place three observations on the line by clicking below it: two close together near the center of the line and one somewhat to the right of these two.

(a) Pull the single rightmost observation out to the right. (Place the cursor on the point, hold down a mouse button, and drag the point.) How does the mean behave? How does the median behave? Explain briefly why each measure acts as it does.

(b) Now drag the single rightmost point to the left as far as you can. What happens to the mean? What happens to the median as you drag this point past the other two (watch carefully)?

2.35 **Behavior of the median.** Place five observations on the line in the *Mean and Median* applet by clicking below it.

(a) Add one additional observation *without changing the median*. Where is your new point?

(b) Use the applet to convince yourself that when you add yet another observation (there are now seven in all), the median does not change, no matter where you put the seventh point. Explain why this must be true.

2.36 **Never on Sunday: also in Canada?** Exercise 1.5 (page 20) gives the number of births in the United States on each day of the week during an entire year. The boxplots in Figure 2.5 are based on more detailed data from Toronto, Canada: the number of births on each of the 365 days in a year, grouped by day of the week.[17] Based on these plots, compare the day-of-the-week distributions using shape, center, and spread. Summarize your findings.

2.37 **Thinking about means.** Table 1.2 (page 24) gives the percent of foreign-born residents in each of the states. For the nation as a whole, 12.5% of residents are foreign-born. Find the mean of the 51 entries in Table 1.2. It is *not* 12.5%. Explain carefully why this happens. (*Hint:* The states with the largest populations are California, Texas, New York, and Florida. Look at their entries in Table 1.2.)

2.38 **Thinking about medians.** A report says that "the median credit card debt of American households is zero." We know that many households have large amounts of credit card debt. In fact, the mean household credit card debt is close to $8000. Explain how the median debt can nonetheless be zero.

2.39 **A standard deviation contest.** This is a standard deviation contest. You must choose four numbers from the whole numbers 0 to 10, with repeats allowed.

(a) Choose four numbers that have the smallest possible standard deviation.

(b) Choose four numbers that have the largest possible standard deviation.

(c) Is more than one choice possible in either (a) or (b)? Explain.

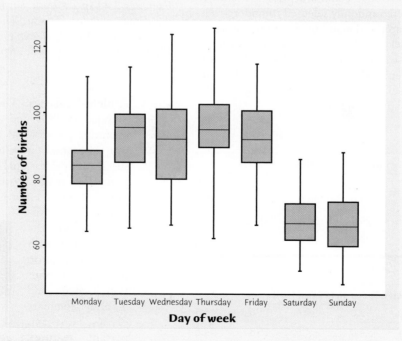

FIGURE 2.5
Boxplots of the distributions of numbers of births in Toronto, Canada, on each day of the week during a year, for Exercise 2.36.

2.40 **Test your technology.** This exercise requires a calculator with a standard deviation button or statistical software on a computer. The observations

$$10{,}001 \quad 10{,}002 \quad 10{,}003$$

have mean $\bar{x} = 10{,}002$ and standard deviation $s = 1$. Adding a 0 in the center of each number, the next set becomes

$$100{,}001 \quad 100{,}002 \quad 100{,}003$$

The standard deviation remains $s = 1$ as more 0s are added. Use your calculator or software to find the standard deviation of these numbers, adding extra 0s until you get an incorrect answer. How soon did you go wrong? This demonstrates that calculators and software cannot handle an arbitrary number of digits correctly.

2.41 **You create the data.** Create a set of 5 positive numbers (repeats allowed) that have median 7 and mean 10. What thought process did you use to create your numbers?

2.42 **You create the data.** Give an example of a small set of data for which the mean is smaller than the first quartile.

2.43 **Adolescent obesity.** Adolescent obesity is a serious health risk affecting more than 5 million young people in the United States alone. Laparoscopic adjustable gastric banding has the potential to provide a safe and effective treatment. Fifty adolescents between 14 and 18 years old with a body mass index (BMI) higher than 35 were recruited from the Melbourne, Australia, community for the study.[18] Twenty-five were randomly selected to undergo gastric banding, and the remaining twenty-five were assigned to a supervised lifestyle intervention program involving diet, exercise, and behavior modification. All subjects were followed for two years. Here are the weight losses in kilograms for the subjects who completed the study: GASTRIC

GASTRIC BANDING					
35.6	81.4	57.6	32.8	31.0	37.6
36.5	−5.4	27.9	49.0	64.8	39.0
43.0	33.9	29.7	20.2	15.2	41.7
53.4	13.4	24.8	19.4	32.3	22.0

LIFESTYLE INTERVENTION					
6.0	2.0	−3.0	20.6	11.6	15.5
−17.0	1.4	4.0	−4.6	15.8	34.6
6.0	−3.1	−4.3	−16.7	−1.8	−12.8

(a) In the context of this study, what do the negative values in the data set mean?

(b) Give a graphical comparison of the weight loss distribution for both groups using side-by-side boxplots. Provide appropriate numerical summaries for the two distributions and identify any high outliers in either group. What can you say about the effects of gastric banding versus lifestyle intervention on weight loss for the subjects in this study?

(c) The measured variable was weight loss in kilograms. Would two subjects with the same weight loss always have similar benefits from a weight-reduction program? Does it depend on their initial weights? Other variables considered in this study were the percent of excess weight lost and the reduction in BMI. Do you see any advantages to either of these variables when comparing weight loss for two groups?

(d) One subject from the gastric-banding group dropped out of the study, and seven subjects from the lifestyle group dropped out. Of the seven dropouts in the lifestyle group, six had gained weight at the time they dropped out. If all subjects had completed the study, how do you think it would have affected the comparison between the two groups?

*Exercises 2.44 to 2.49 ask you to analyze data without having the details outlined for you. The exercise statements give you the **State** step of the four-step process. In your work, follow the **Plan, Solve,** and **Conclude** steps as illustrated in Example 2.9 (page 60).*

2.44 **Athletes' salaries.** The Montreal Canadiens were founded in 1909 and are the longest continuously operating professional ice hockey team. They have won 24 Stanley Cups, making them one of the most successful professional sports teams of the traditional four major sports of Canada and the United States. Table 2.1 (page 67) gives the salaries of the 2012–2013 roster.[19] Provide the team owner with a full description of the distribution of salaries and a brief summary of its most important features.  HOCKEY

Oliver Samson Arcand/ NHLI via Getty Images

TABLE 2.1   SALARIES FOR THE 2012–2013 MONTREAL CANADIENS

PLAYER	SALARY	PLAYER	SALARY	PLAYER	SALARY
Andrei Markov	$5,750,000	Carey Price	$5,500,000	Brian Gionta	$5,000,000
Tomas Plekanec	$5,000,000	Tomas Kaberle	$4,250,000	Eric Cole	$4,000,000
Josh Gorges	$3,900,000	Rene Bourque	$3,000,000	Brandon Prust	$3,000,000
Alexei Emelin	$2,000,000	P. K. Suban	$2,000,000	Travis Moen	$1,850,000
Max Pacioretty	$1,750,000	Francis Bouillon	$1,500,000	Raphael Diaz	$1,200,000
Peter Budaj	$1,200,000	Lars Eller	$1,150,000	Colby Armstrong	$1,000,000
David Desharnais	$950,000	Alex Galchenyuk	$925,000	Yannick Weber	$900,000
Brendan Gallagher	$690,000	Ryan White	$687,500	Michael Blunden	$575,000
Petteri Nokelainen	$575,000				

2.45 **Returns on stocks.** How well have stocks done over the past generation? The Wilshire 5000 index describes the average performance of all U.S. stocks. The average is weighted by the total market value of each company's stock, so think of the index as measuring the performance of the average investor. Here are the percent returns on the Wilshire 5000 index for the years from 1971 to 2012: WILSHIRE

YEAR	RETURN	YEAR	RETURN	YEAR	RETURN
1971	16.19	1985	31.46	1999	24.23
1972	17.34	1986	15.61	2000	−10.89
1973	−18.78	1987	1.75	2001	−10.97
1974	−27.87	1988	17.59	2002	−20.86
1975	37.38	1989	28.53	2003	31.64
1976	26.77	1990	−6.03	2004	12.48
1977	−2.97	1991	33.58	2005	6.38
1978	8.54	1992	9.02	2006	15.77
1979	24.40	1993	10.67	2007	5.62
1980	33.21	1994	0.06	2008	−37.23
1981	−3.98	1995	36.41	2009	28.30
1982	20.43	1996	21.56	2010	17.16
1983	22.71	1997	31.48	2011	0.98
1984	3.27	1998	24.31	2012	16.06

What can you say about the distribution of yearly returns on stocks?

2.46 **Do good smells bring good business?** Businesses know that customers often respond to background music. Do they also respond to odors? Nicolas Guéguen and his colleagues studied this question in a small pizza restaurant in France on Saturday evenings in May. On one evening, a relaxing lavender odor was spread through the restaurant; on another evening, a stimulating lemon odor; a third evening served as a control, with no odor. Table 2.2 (page 68) shows the amounts (in euros) that customers spent on each of these evenings.[20] Compare the three distributions. Were both odors associated with increased customer spending? ODORS

2.47 **Daily activity and obesity.** People gain weight when they take in more energy from food than they expend. Table 2.3 (page 68) compares volunteer subjects who were lean with others who were mildly obese. None of the subjects followed an exercise program. The subjects wore sensors that recorded every move for 10 days. The table shows the average minutes per day spent in activity (standing and walking) and in lying down.[21] Compare the distributions of time spent actively for lean and obese subjects and also the distributions of time spent lying down. How does the behavior of lean and mildly obese people differ? OBESITY

2.48 **Good weather and tipping.** Favorable weather has been shown to be associated with increased tipping. Will just the belief that future weather will be favorable lead to higher tips? The researchers gave 60 index cards to a waitress at an Italian restaurant in New Jersey. Before delivering the bill to each customer, the waitress randomly selected a card and wrote on the bill the same message that was printed on the index card. Twenty of the cards had the message "The weather is supposed to be really good tomorrow. I hope you enjoy the day!" Another 20 cards contained the message "The weather is supposed to be not so good tomorrow. I hope you enjoy the day anyway!" The remaining 20 cards were blank, indicating that the waitress was not supposed to write any message. Choosing a card at random ensured that there was a random assignment of the customers to the three

TABLE 2.2 AMOUNT SPENT (EUROS) BY CUSTOMERS IN A RESTAURANT WHEN EXPOSED TO ODORS

NO ODOR

15.9	18.5	15.9	18.5	18.5	21.9	15.9	15.9	15.9	15.9
15.9	18.5	18.5	18.5	20.5	18.5	18.5	15.9	15.9	15.9
18.5	18.5	15.9	18.5	15.9	18.5	15.9	25.5	12.9	15.9

LEMON ODOR

18.5	15.9	18.5	18.5	18.5	15.9	18.5	15.9	18.5	18.5
15.9	18.5	21.5	15.9	21.9	15.9	18.5	18.5	18.5	18.5
25.9	15.9	15.9	15.9	18.5	18.5	18.5	18.5		

LAVENDER ODOR

21.9	18.5	22.3	21.9	18.5	24.9	18.5	22.5	21.5	21.9
21.5	18.5	25.5	18.5	18.5	21.9	18.5	18.5	24.9	21.9
25.9	21.9	18.5	18.5	22.8	18.5	21.9	20.7	21.9	22.5

TABLE 2.3 TIME (MINUTES PER DAY) ACTIVE AND LYING DOWN BY LEAN AND OBESE SUBJECTS

LEAN SUBJECTS			OBESE SUBJECTS		
SUBJECT	STAND/WALK	LIE	SUBJECT	STAND/WALK	LIE
1	511.100	555.500	11	260.244	521.044
2	607.925	450.650	12	464.756	514.931
3	319.212	537.362	13	367.138	563.300
4	584.644	489.269	14	413.667	532.208
5	578.869	514.081	15	347.375	504.931
6	543.388	506.500	16	416.531	448.856
7	677.188	467.700	17	358.650	460.550
8	555.656	567.006	18	267.344	509.981
9	374.831	531.431	19	410.631	448.706
10	504.700	396.962	20	426.356	412.919

experimental conditions. Here are the tip percents for the three messages:[22] **TIPPING**

Good weather report:	20.8	18.7	19.9	20.6	22.0	23.4	22.8	24.9	22.2	20.3
	24.9	22.3	27.0	20.4	22.2	24.0	21.2	22.1	22.0	22.7
Bad weather report:	18.0	19.0	19.2	18.8	18.4	19.0	18.5	16.1	16.8	14.0
	17.0	13.6	17.5	19.9	20.2	18.8	18.0	23.2	18.2	19.4
No weather report:	19.9	16.0	15.0	20.1	19.3	19.2	18.0	19.2	21.2	18.8
	18.5	19.3	19.3	19.4	10.8	19.1	19.7	19.8	21.3	20.6

Compare the three distributions. How did the tip percents vary with the weather report information?

2.49 Cholesterol levels and age. The National Health and Nutrition Examination Survey (NHANES) is a unique survey that combines interviews and physical examinations.[23] It includes basic demographic information,

questions about topics such as diet, physical activity, and prescription medications, as well as the results of a physical examination measuring a variety of variables, including blood pressure and cholesterol levels. The program began in the early 1960s, and the survey currently examines a nationally representative sample of about 5000 persons each year. You will work with the total cholesterol measurements (mg/dl) obtained from participants in the survey in 2009–2010. **CHOLEST**

To examine changes in cholesterol with age, we consider only the 3044 participants between 20 and 50 years of age and have classified them into the three age categories 20s, 30s, and 40s. The full data set is too large to print here, but here are the first ten individuals:

Age category:	30s	20s	20s	40s	30s	40s	20s	30s	30s	20s
Total cholesterol:	135	160	299	197	196	202	175	216	181	149

The first individual is in the 30s with a total cholesterol of 135, the second in the 20s with total cholesterol of 160, and so forth.

(a) Use graphical and numerical summaries to compare the three distributions. How does cholesterol change with age?

(b) The ideal range of total cholesterol is below 200 mg/dl. For individuals with elevated cholesterol levels, prescription drugs are often recommended to lower levels. Among the 3044 participants between 20 and 50 years of age, 4 individuals in their 20s, 24 individuals in their 30s, and 117 individuals in their 40s were taking prescription medications to reduce their cholesterol levels. How do you think your comparison of the distribution would be changed if none of the individuals were taking medication? Explain.

Exercises 2.50 to 2.53 make use of the optional material on the 1.5 × IQR rule for suspected outliers.

2.50 **Graduation rates.** In Exercise 1.10 (page 31) you were asked to use a stemplot to display the distribution of the percents of on-time high school graduates in the states. Stemplots help you find the five-number summary because they arrange the observations in increasing order. ⬛▦ GRADRATE

(a) Give the five-number summary of this distribution.
(b) Use the five-number summary to draw a boxplot of all the data. What is the shape of the distribution?

(c) Which observations does the 1.5 × IQR rule flag as suspected outliers? Is there a simple explanation for the outlier(s)?

2.51 **Carbon dioxide emissions.** Table 1.6 (page 41) gives the 2009 carbon dioxide (CO_2) emissions per person for countries with populations of at least 30 million in that year. A stemplot or histogram shows that the distribution is strongly skewed to the right. The United States and several other countries appear to be high outliers. ⬛▦ CO2EMISS

(a) Give the five-number summary. Explain why this summary suggests that the distribution is right-skewed.

(b) Which countries are outliers according to the 1.5 × IQR rule? Make a stemplot of the data or look at your stemplot from Exercise 1.36 (page 41). Do you agree with the rule's suggestions about which countries are and are not outliers?

2.52 **Foreign born.** Figure 1.10 (page 30) gives a stemplot of the percent of state residents who were born outside the United States for the 50 states and the District of Columbia. ⬛▦ FOREIGN

(a) Give the five-number summary of this distribution.
(b) Is California an outlier or just the largest observation in a strongly skewed distribution? What does the 1.5 × IQR rule say?

2.53 **Cholesterol for people in their 20s.** Exercise 2.49 gives the cholesterol levels of individuals in their 20s from the NHANES survey in 2009–10. The cholesterol levels are right-skewed, with a few large cholesterol levels. Which cholesterol levels are suspected outliers by the 1.5 × IQR rule? ⬛▦ CHOLES20

Exploring the Web

2.54 **Home run leaders.** The three top players on the career home run list are Barry Bonds, Hank Aaron, and Babe Ruth. You can find their home run statistics by going to the Web site www.baseball-reference.com and then clicking on the Players tab at the top of the page. Construct three side-by-side boxplots comparing the yearly home run production of Barry Bonds, Hank Aaron, and Babe Ruth. Describe any differences that you observe. It is worth noting that in his first four seasons, Babe Ruth was primarily a pitcher. If these four seasons are ignored, how does Babe Ruth compare with Barry Bonds and Hank Aaron?

2.55 **Crime rates and outliers.** The *Statistical Abstract of the United States* is a comprehensive summary of statistics on the social, political, and economic organization of the United States. It can be found at the Web site www.census.gov/compendia/statab/. Go to the section Law Enforcement, Courts and Prisons, and then to the subsection Crimes and Crime Rates. Several tables of data will be available.

(a) Open the Table on Crime Rates by State and Type for the latest year given. Why do you think they use rates per 100,000 population rather than the number of crimes committed? The District of Columbia is a high outlier in several crime categories.

(b) Open the Table on Crime Rates by Type for Selected Large Cities. This table includes the District of Columbia, which is listed as Washington, DC. Without doing any formal calculations, does the District of Columbia look like a high outlier in the table for large cities? Whether or not the District of Columbia is an outlier depends on more than its crime rate. It also depends on the other observations included in the data set. Which data set do you feel is more appropriate for the District of Columbia?

(c) Using the Table on Crime Rates by State and Type for the latest year given, choose a crime category, and give a full description of its distribution over the 50 states, omitting the District of Columbia. Your description of the distribution should include appropriate graphical and numerical summaries and a brief report describing the main features of the distribution. You can open the data as an Excel file and import it into your statistical software.

Overview

This chapter is primarily about using density curves to model the distribution of a random variable. In applications, researchers may use histograms to gain some insight into the shape of such a model and then use technology to fit a known distribution type to the histogram.

The Normal distribution is without question the most widely used distribution (and therefore most important) an introductory-level statistics student will encounter. It is true that some population distributions are Normal or can be approximated by the Normal distribution; however, the reason for the Normal distribution's importance usually has little or nothing to do with the populations from which we sample. Instead, Normal distributions attain their importance as large sample approximations to the sampling distributions of most of the important statistics we use for decisions, including sample means (and therefore proportions) and sums. Essentially, if the statistic you are using in an application is a sum or average, its sampling distribution is approximately Normal, provided the sample size is large enough. In this chapter, students develop the technical skills of obtaining percentiles or proportions under a Normal density curve. These skills will be critical later, when they study sampling distributions in Chapters 15 and 19.

At first, students will need to understand the relationship between a variable and a density curve. Students should make use of previous knowledge of graphical displays, namely histograms, to understand how a variable may be described using a density curve. We use this background knowledge to describe how the shape of a density function reveals relationships between the mean and median (e.g., a right-skewed density function describes a population for which the mean is larger than the median). Most students will prefer to think of density curves as "idealized" or "smoothed" histograms.

A common density curve that may be used to model many real-world variables is symmetric and bell-shaped (in fact, students are usually very familiar with the "bell curve"). This is the basic shape of a Normal random variable. A Normal distribution is fully described by its two parameters: μ and σ, the mean and standard deviation, respectively. To compute proportions or percentiles for an arbitrary Normal distribution, we use the important result that any Normal variable can be converted to a *standard* Normal variable, having mean 0 and standard deviation 1 through the equation $z = \dfrac{x - \mu}{\sigma}$. Students will encounter two varieties of applied problems: given values a and b find the proportion of the population

---

LEARNING OBJECTIVES**

■ Know that areas under a density curve represent a proportion of all observations and that the total area under a density curve is 1.

■ Approximately locate the median and the mean on a density curve.

■ Know that the mean and median both lie at the center of a symmetric density curve and that the mean moves farther toward the long tail of a skewed curve.

■ Recognize the shape of Normal curves and estimate by eye both the mean and standard deviation from such a curve.

■ Use the 68–95–99.7 rule and symmetry to state what percent of the observations from a Normal distribution fall between two points when both points lie at the mean or one, two, or three standard deviations on either side of the mean.

■ Find the standardized value (z-score) of an observation. Interpret z-scores and understand that any Normal distribution becomes standard Normal $N(0, 1)$ when standardized.

■ Given that a variable has a Normal distribution with a stated mean μ and standard deviation σ, calculate the proportion of values above a stated number, below a stated number, or between two stated numbers.

■ Given that a variable has a Normal distribution with a stated mean μ and standard deviation σ, calculate the point having a stated proportion of all values above or below it.

**These learning outcomes appear later for the students in Chapter 7: Part I Review.

between those values (note that a could be $-\infty$ and/or b could be ∞); and given a proportion, p, find the value a, such that p is the proportion of the population less (or greater) than a. Both problems will require access to a standard Normal distribution, through software, the *Normal Curve* applet, or a table.

Teaching Suggestions and Additional Examples/Activities for the Classroom

1. Introduce Density Functions Through a Uniform Distribution

When beginning to discuss density functions, work with one in class that is *not* Normal. Without evidence to the contrary, many students might conclude that everything has a Normal distribution. A Uniform distribution is a great starting example because students can easily compute areas under them; remind them that area = (length) $\times$ (width). Have your students try Exercise 3.2 on page 74 of the text.

2. Use a Normal Distribution the Students Can Relate To

After doing a non-Normal example, change gears and do an example for a mound-shaped histogram—ideally one that students can relate to. A good example is "batting averages" for all Major League Baseball players with at least (say) 150 at bats during the most recent season. This almost always generates a histogram very close to Normal. Use this example to illustrate that a smooth, bell-shaped density curve might fit well. In addition, point out the Normal distribution's key features: the "change of curvature" points (inflection points) lie one standard deviation above and below the mean, the mean and median are equal, the 68–95–99.7 rule fits the distribution well, and so on. You can view the collection of all batting averages at the CBS Sports Web site (http://www.cbssports.com/mlb/stats). Under "Sortable Stats," click on "Batting" under "Players." Once the players and their statistics come up, select "Regular Season." At the bottom of the page, you have the option to view "All" players. Click on "All." Remove the players with a handful of at bats by sorting by "AB" and copying only the batting averages of players with at least (say) 150 at bats. Paste these into the statistical software you are using, and create a histogram. Another example that college students are familiar with is SAT (or ACT) test scores; although the actual distribution varies from year to year, the tests are designed so the mean is approximately 500 and the standard deviation is approximately 100 (see Examples 3.5, 3.7, and 3.9, as well as Exercise 3.49).

3. Use the 68-95-99.7 Rule as a Guide

After introducing the Normal distribution, ask your students for various proportions directly tied to the 68–95–99.7 rule. For example, if batting averages have mean .270 and standard deviation .030, what proportion of players have an average of .300 or higher? Students should be able to answer questions like this without the formulaic computation of a z-score. The key is that they naturally relate values to the 68–95–99.7 rule. This will make using z-scores later intuitively appealing by having students discover this link. Using the 68–95–99.7 rule is also a good way for students to check the reasonability of their answers later. Only after students feel comfortable answering questions for values 1, 2, or 3 standard deviations away from the mean, ask them about values that are not so conveniently placed. This motivates use of software, the *Normal Curve* applet, or a table. Making pictures of the distribution and area of interest is highly recommended in terms of deciding if an answer is reasonable.

4. Any Value Can Be Standardized, but Not All Distributions are Normal

It is important to note to your students that although a value from any distribution can be standardized using the equation $z = \dfrac{x - \mu}{\sigma}$, the resulting distribution is only standard

Normal if the initial distribution is Normal. At the same time, it might help students to know that they can standardize values from different distributions to examine relative standing. For example, suppose a student earns 83 on an exam in history and 67 on an exam in chemistry. Although it appears at first that the student did better on the history exam, we don't know that unless we know the relative standing of the student on both exams. If the history exam had a mean of 80 and standard deviation 5, and the chemistry exam had a mean of 63 and standard deviation 2, the student did relatively better on the chemistry exam ($z = 2.0$ for chemistry; $z = 0.6$ for history).

5. Begin to Make Connections

Think ahead to help your students make the connections required later in the course. For example, emphasize that Z (as a random variable) is a "reserved letter" in the statistical world; it indicates the standard Normal distribution—a Normal distribution with mean 0 and standard deviation 1.

Other Resources (LaunchPad)

StatClips
The Normal distribution (and examples a–c)

Snapshots Videos
Normal Distributions

Applet
Normal Curve

ImageSource/Photolibrary

The Normal Distributions

We now have a tool box of graphical and numerical methods for describing distributions. What is more, we have a clear strategy for exploring data on a single quantitative variable.

Exploring a Distribution

1. Always plot your data: make a graph, usually a histogram or a stemplot.
2. Look for the overall pattern (shape, center, spread) and for striking deviations such as outliers.
3. Calculate a numerical summary to briefly describe center and spread/variability.

In this chapter, we add one more step to this strategy:

4. Sometimes the overall pattern of a large number of observations is so regular that we can describe it by a smooth curve.

Density curves

Figure 3.1 is a histogram of the scores of all 947 seventh-grade students in Gary, Indiana, on the vocabulary part of the *Iowa Tests of Basic Skills*.[1] Scores of many students on this national test have a quite regular distribution. The histogram is

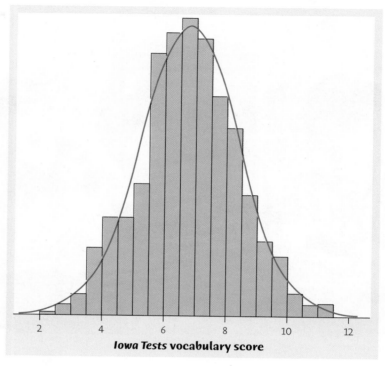

FIGURE 3.1

Histogram of the *Iowa Tests* vocabulary scores of all seventh-grade students in Gary, Indiana. The smooth curve shows the overall shape of the distribution.

symmetric, and both tails fall off smoothly from a single center peak. There are no large gaps or obvious outliers. The smooth curve drawn through the tops of the histogram bars in Figure 3.1 is a good description of the overall pattern of the data.

EXAMPLE 3.1   From Histogram to Density Curve

IOWATEST

Our eyes respond to the *areas* of the bars in a histogram. The bar areas represent proportions of the observations. Figure 3.2(a) is a copy of Figure 3.1 with the left-most bars shaded. The area of the shaded bars in Figure 3.2(a) represents the students with vocabulary scores of 6.0 or lower. There are 287 such students, who make up the proportion 287/947 = 0.303 of all Gary seventh-graders.

Now look at the curve drawn through the bars. In Figure 3.2(b), the area under the curve to the left of 6.0 is shaded. We can draw histogram bars taller or shorter by adjusting the vertical scale. In moving from histogram bars to a smooth curve, we make a specific choice: adjust the scale of the graph so that *the total area under the curve is exactly 1.* The total area represents the proportion 1, that is, all the observations. We can then interpret areas under the curve as proportions of the observations. The curve is now a *density curve*. The shaded area under the density curve in Figure 3.2(b) represents the proportion of students with score 6.0 or lower. This area is 0.293, only 0.010 away from the actual proportion 0.303. The method for finding this area will be presented shortly. For now, note that the areas under the density curve give quite good approximations to the actual distribution of the 947 test scores. ■

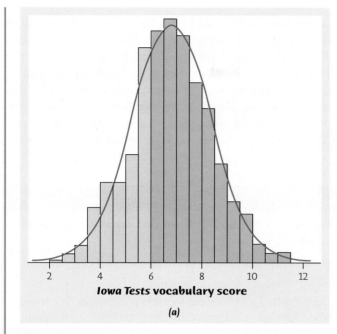

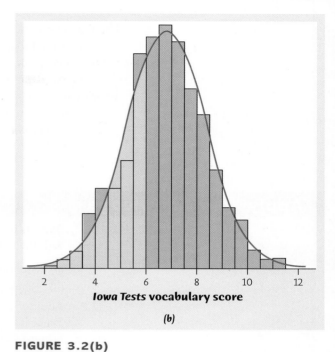

FIGURE 3.2(a)
The proportion of scores less than or equal to 6.0 in the actual data is 0.303.

FIGURE 3.2(b)
The proportion of scores less than or equal to 6.0 from the density curve is 0.293. The density curve is a good approximation to the distribution of the data.

Density Curve

A **density curve** is a curve that

■ Is always on or above the horizontal axis.

■ Has area exactly 1 underneath it.

A density curve describes the overall pattern of a distribution. The area under the curve and above any range of values is the proportion of all observations that fall in that range.

Density curves, like distributions, come in many shapes. Figure 3.3 shows a strongly skewed distribution, the survival times of guinea pigs from Exercise 2.31 (page 64).

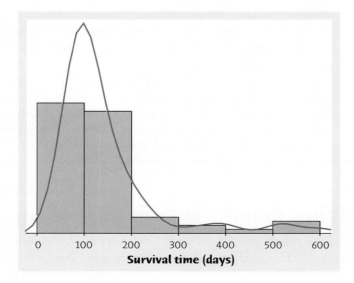

FIGURE 3.3
A right-skewed distribution pictured by both a histogram and a density curve.

The histogram and density curve were both created from the data by software. Both show the overall shape and the "bumps" in the long right tail. The density curve shows a single high peak as a main feature of the distribution. The histogram divides the observations near the peak between two bars, thus reducing the height of the peak. A density curve is often a good description of the overall pattern of a distribution. Outliers, which are deviations from the overall pattern, are not described by the curve. *Of course, no set of real data is exactly described by a density curve. The curve is an idealized description that is easy to use and accurate enough for practical use.*

Apply Your Knowledge

3.1   Sketch Density Curves. Sketch density curves that describe distributions with the following shapes:

(a)   Symmetric, but with two peaks (that is, two strong clusters of observations)

(b)   Single peak and skewed to the left

3.2   Accidents on a Bike Path. Examining the location of accidents on a level, 5-mile bike path shows that they occur uniformly along the length of the path. Figure 3.4 displays the density curve that describes the distribution of accidents.

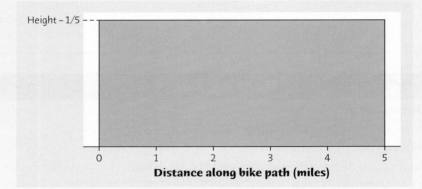

FIGURE 3.4

The density curve for the location of accidents along a 5-mile bike path, for Exercise 3.2.

(a)   Explain why this curve satisfies the two requirements for a density curve.

(b)   The proportion of accidents that occur in the first mile of the path is the area under the density curve between 0 miles and 1 mile. What is this area?

(c)   There is a stream alongside the bike path between the 0.8 mile mark and the 1.3 mile mark. What proportion of accidents happen on the bike path alongside the stream?

(d)   There are roads at the two ends of the bike path, but the remainder of the bike path follows a paved path through the woods. What proportion of accidents happen more than 1 mile from either road? (Hint: First determine where on the bike path the accident needs to occur to be more than 1 mile from either road, and then find the area.)

Describing density curves

Our measures of center and spread apply to density curves as well as to actual sets of observations. The median and quartiles are easy. Areas under a density curve represent proportions of the total number of observations. The median is the point with half the observations on either side. So *the median of a density curve is the equal-areas point,* the point with half the area under the curve to its left and the remaining half of the area to its right. The quartiles divide the area under the curve into quarters. One-fourth of the area under the curve is to the left of the first quartile, and three-fourths of the area is to the left of the third quartile. You can roughly locate the median and quartiles of any density curve by eye by dividing the area under the curve into four equal parts.

Because density curves are idealized patterns, a symmetric density curve is exactly symmetric. The median of a symmetric density curve is therefore at its center. Figure 3.5(a) shows a symmetric density curve with the median marked. It isn't so easy to spot the equal-areas point on a skewed curve. There are mathematical ways of finding the median for any density curve. That's how we marked the median on the skewed curve in Figure 3.5(b).

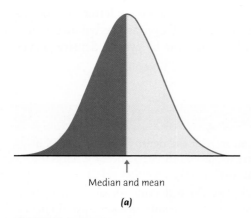

↑
Median and mean

(a)

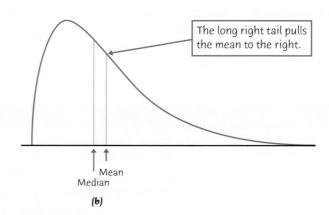

The long right tail pulls the mean to the right.

↑ ↑
     Mean
Median

(b)

FIGURE 3.5(a)
The median and mean of a symmetric density curve both lie at the center of symmetry.

FIGURE 3.5(b)
The median and mean of a right-skewed density curve. The mean is pulled away from the median toward the long tail.

What about the mean? The mean of a set of observations is their arithmetic average. If we think of the observations as weights strung out along a thin rod, the mean is the point at which the rod would balance. This fact is also true of density curves. *The mean is the point at which the curve would balance if made of solid material.* Figure 3.6 illustrates this fact about the mean. A symmetric curve balances at its center because the two sides are identical. *The mean and median of a symmetric density curve are equal,* as in Figure 3.5(a). We know that the mean of a skewed distribution is pulled toward the long tail. Figure 3.5(b) shows how the mean of a skewed density curve is pulled toward the long tail more than is the median. It's hard to locate the balance point by eye on a skewed curve. There are mathematical ways of calculating the mean for any density curve, so we are able to mark the mean as well as the median in Figure 3.5(b).

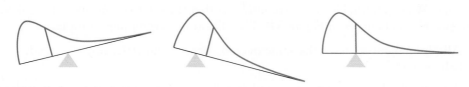

FIGURE 3.6
The mean is the balance point of a density curve.

Median and Mean of a Density Curve

The **median** of a density curve is the equal-areas point, the point that divides the area under the curve in half.

The **mean** of a density curve is the balance point at which the curve would balance if made of solid material.

The median and mean are the same for a symmetric density curve. They both lie at the center of the curve. The mean of a skewed curve is pulled away from the median in the direction of the long tail.

Because a density curve is an idealized description of a distribution of data, we need to distinguish between the mean and standard deviation of the density curve and the mean $\bar{x}$ and standard deviation s computed from the actual observations. The usual notation for the **mean of a density curve** is μ (the Greek letter mu). We write the **standard deviation of a density curve** as σ (the Greek letter sigma). We can roughly locate the mean μ of any density curve by eye, as the balance point. There is no easy way to locate the standard deviation σ by eye for density curves in general.

mean μ
standard deviation σ

Apply Your Knowledge

3.3   **Mean and Median.** What is the mean μ of the density curve pictured in Figure 3.4 on page 74? (That is, where would the curve balance?) What is the median? (That is, where is the point with area 0.5 on either side?)

3.4   **Mean and Median.** Figure 3.7 displays three density curves, each with three points marked on them. At which of these points on each curve do the mean and the median fall?

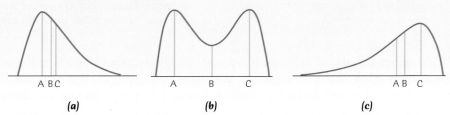

FIGURE 3.7
Three density curves.

Normal distributions

Normal curve
Normal distribution

One particularly important class of density curves has already appeared in Figures 3.1 and 3.2. They are called **Normal curves.** The distributions they describe are called **Normal distributions.** Normal distributions play a large role in statistics, but they are rather special and not at all "normal" in the sense of being usual or average. We capitalize Normal to remind you that these curves are special. Look at the two Normal curves in Figure 3.8. They illustrate several important facts:

■ All Normal curves have the same overall shape: symmetric, single-peaked, bell-shaped.

■ Any specific Normal curve is completely described by giving its mean μ and its standard deviation σ.

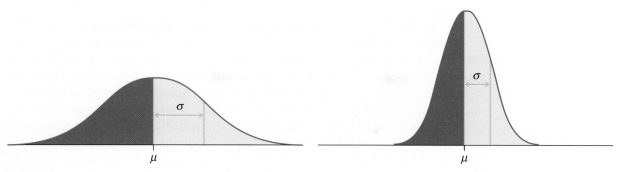

FIGURE 3.8
Two Normal curves, showing the mean μ and standard deviation σ.

- The mean is located at the center of the symmetric curve and is the same as the median. Changing μ without changing σ moves the Normal curve along the horizontal axis without changing its spread/variability.
- The standard deviation σ controls the spread of a Normal curve. Curves with larger standard deviation are more spread out.

The standard deviation σ is the natural measure of spread/variability for Normal distributions. Not only do μ and σ completely determine the shape of a Normal curve, but we can also locate σ by eye on a Normal curve. Here's how. Imagine that you are skiing down a mountain that has the shape of a Normal curve. At first, you descend at an ever-steeper angle as you go out from the peak:

Fortunately, before you find yourself going straight down, the slope begins to grow flatter rather than steeper as you go out and down:

The points at which this change of curvature takes place are located at distance σ on either side of the mean μ. You can feel the change as you run a pencil along a Normal curve, and so find the standard deviation. *Remember that μ and σ alone do not specify the shape of most distributions,* and that the shape of density curves in general does not reveal σ. These are special properties of Normal distributions.

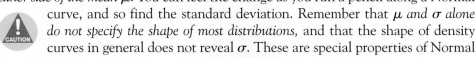

> ### Normal Distributions
>
> A **Normal distribution** is described by a Normal density curve. Any particular Normal distribution is completely specified by two numbers, its mean μ and standard deviation σ.
>
> The mean of a Normal distribution is at the center of the symmetric Normal curve. The standard deviation is the distance from the center to the change-of-curvature points on either side.

Why are the Normal distributions important in statistics? Here are three reasons. First, Normal distributions are good descriptions for some distributions of *real data.* Distributions that are often close to Normal include scores on tests taken by many people (such as *Iowa Tests* and SAT exams), repeated careful measurements of the same quantity, and characteristics of biological populations (such as lengths of crickets and yields of corn). Second, Normal distributions are good approximations to the results of many kinds of *chance outcomes,* such as the proportion of heads in many tosses of a coin. Third, we will see that many *statistical inference* procedures based on Normal distributions work well for other roughly symmetric distributions. However, many sets of data do not follow a Normal distribution. Most income distributions, for example, are skewed to the right and so are not Normal. Non-Normal data, like non-normal people, not only are common but also are sometimes more interesting than their Normal counterparts.

The 68–95–99.7 rule

Although there are many Normal curves, they all have common properties. In particular, all Normal distributions obey the following rule.

The 68–95–99.7 Rule

In the Normal distribution with mean μ and standard deviation σ:

■ Approximately **68%** of the observations fall within σ of the mean μ.
■ Approximately **95%** of the observations fall within 2σ of μ.
■ Approximately **99.7%** of the observations fall within 3σ of μ.

Figure 3.9 illustrates the 68–95–99.7 rule. By remembering these three numbers, you can think about Normal distributions without constantly making detailed calculations.

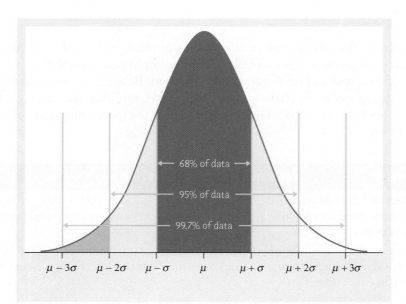

FIGURE 3.9
The 68–95–99.7 rule for Normal distributions.

EXAMPLE 3.2   *Iowa Tests* **Scores**

IOWATEST

Figures 3.1 and 3.2 (see pages 72 and 73) show that the distribution of *Iowa Tests* vocabulary scores for seventh-grade students in Gary, Indiana, is close to Normal. Suppose that the distribution is exactly Normal with mean $\mu = 6.84$ and standard deviation $\sigma = 1.55$. (These are the mean and standard deviation of the 947 actual scores.)

Figure 3.10 applies the 68–95–99.7 rule to the *Iowa Tests* scores. The 95 part of the rule says that approximately 95% of all scores are between

$$\mu - 2\sigma = 6.84 - (2)(1.55) = 6.84 - 3.10 = 3.74$$

and

$$\mu + 2\sigma = 6.84 + (2)(1.55) = 6.84 + 3.10 = 9.94$$

The other 5% of scores are outside this range. Because Normal distributions are symmetric, half of these scores are lower than 3.74 and half are higher than 9.94. That is, about 2.5% of the scores are below 3.74 and 2.5% are above 9.94. ■

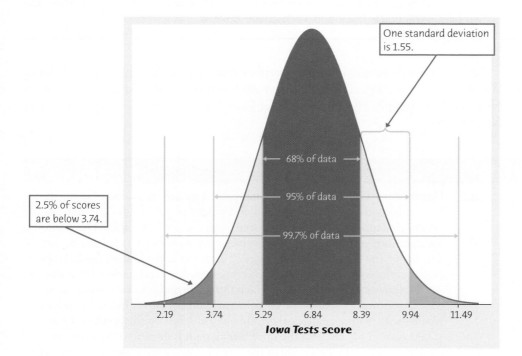

FIGURE 3.10

The 68–95–99.7 rule applied to the distribution of *Iowa Tests* scores for seventh-grade students in Gary, Indiana, for Example 3.2. The mean and standard deviation are $\mu = 6.84$ and $\sigma = 1.55$.

The 68–95–99.7 rule describes distributions that are exactly Normal. Real data such as the actual Gary scores are never exactly Normal. For one thing, *Iowa Tests* scores are reported only to the nearest tenth. A score can be 9.9 or 10.0, but not 9.94. We use a Normal distribution because it's a good approximation and because we think the knowledge that the test measures is continuous rather than stopping at tenths.

How well does our work in Example 3.2 describe the actual *Iowa Tests* scores? Well, 900 of the 947 scores are between 3.74 and 9.94. That's 95.04%, very accurate indeed. Of the remaining 47 scores, 20 are below 3.74 and 27 are above 9.94. The tails of the actual data are not quite equal, as they would be in an exactly Normal distribution. Normal distributions often describe real data better in the center of the distribution than in the extreme high and low tails.

EXAMPLE 3.3   *Iowa Tests* **Scores**

Look again at Figure 3.10. A score of 5.29 is one standard deviation below the mean. What percent of scores are higher than 5.29? Find the answer by adding areas in the figure. Here is the calculation in pictures:

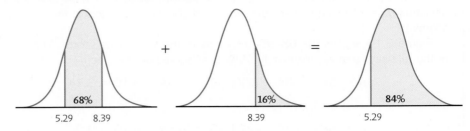

percent between 5.29 and 8.39 + percent above 8.39 = percent above 5.29

         68%   +   16%   =   84%

So approximately 84% of scores are higher than 5.29. Be sure you understand where the 16% came from. We know that 68% of scores are between 5.29 and 8.39, so 32% of scores are outside that range. These are equally split between the two tails, 16% below 5.29 and 16% above 8.39. ■

Because we will mention Normal distributions often, a short notation is helpful. We abbreviate the Normal distribution with mean μ and standard deviation σ as $N(\mu, \sigma)$. For example, the distribution of Gary *Iowa Tests* scores is approximately $N(6.84, 1.55)$.

Apply Your Knowledge

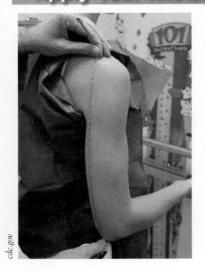

FIGURE 3.11

Correct tape placement when measuring upper arm length, for Exercise 3.5.

cdc.gov

3.5   **Upper Arm Lengths.** Anthropomorphic data are measurements on the human body that can track growth and weight of infants and children and evaluate changes in the body that occur over the adult life span. The resulting data can be used in areas as diverse as ergonomics and clothing design. The upper arm length of females over 20 years old in the United States is approximately Normal with mean 35.8 centimeters (cm) and standard deviation 2.1 cm. Draw a Normal curve on which this mean and standard deviation are correctly located. (*Hint:* Draw an unlabeled Normal curve, locate the points where the curvature changes, then add number labels on the horizontal axis.)

As seen in Figure 3.11, the upper arm length is measured from the acromion process, the highest point of the shoulder, down the posterior surface of the arm to the tip of the olecranon process, the bony part of the mid elbow.[2]

3.6   **Upper Arm Lengths.** The upper arm length of females over 20 years old in the United States is approximately Normal with mean 35.8 centimeters (cm) and standard deviation 2.1 cm. Use the 68–95–99.7 rule to answer the following questions. (Start by making a sketch like Figure 3.10.)

(a)   What range of lengths covers almost all (99.7%) of this distribution?

(b)   What percent of women over 20 have upper arm lengths below 33.7 cm?

3.7   **Monsoon Rains.** The summer monsoon rains bring 80% of India's rainfall and are essential for the country's agriculture. Records going back more than a century show that the amount of monsoon rainfall varies from year to year according to a distribution that is approximately Normal with mean 852 millimeters (mm) and standard deviation 82 mm.[3] Use the 68–95–99.7 rule to answer the following questions.

(a) Between what values do the monsoon rains fall in 95% of all years?

(b) How small are the monsoon rains in the driest 2.5% of all years?

The standard Normal distribution

As the 68–95–99.7 rule suggests, all Normal distributions share many properties. In fact, all Normal distributions are the same if we measure in units of size σ about the mean μ as center. Changing to these units is called *standardizing*. To standardize a value, subtract the mean of the distribution and then divide by the standard deviation.

Standardizing and *z*-Scores

If x is an observation from a distribution that has mean μ and standard deviation σ, the **standardized value** of x is

$$z = \frac{x - \mu}{\sigma}$$

A standardized value is often called a **z-score**.

A z-score tells us how many standard deviations the original observation falls away from the mean, and in which direction. Observations larger than the mean are positive when standardized, and observations smaller than the mean are negative.

 HE SAID, SHE SAID.

Height, weight, and body mass distributions in this book come from actual measurements by a government survey. Good thing that is. When *asked* their weight, almost all women say they weigh less than they really do. Heavier men also underreport their weight—but lighter men claim to weigh more than the scale shows. We leave you to ponder the psychology of the two sexes. Just remember that "say-so" is no substitute for measuring.

EXAMPLE 3.4   Standardizing Women's Heights

The heights of women aged 20 to 29 in the United States are approximately Normal with $\mu = 64.2$ inches and $\sigma = 2.8$ inches.[4] The standardized height is

$$z = \frac{\text{height} - 64.2}{2.8}$$

A woman's standardized height is the number of standard deviations by which her height differs from the mean height of all women aged 20 to 29. A woman 70 inches tall, for example, has standardized height

$$z = \frac{70 - 64.2}{2.8} = 2.07$$

or 2.07 standard deviations above the mean. Similarly, a woman 5 feet (60 inches) tall has standardized height

$$z = \frac{60 - 64.2}{2.8} = -1.50$$

or 1.50 standard deviations less than the mean height. ■

We often standardize observations from symmetric distributions to express them in a common scale. We might, for example, compare the heights of two children of different ages by calculating their z-scores. The standardized heights tell us where each child stands in the distribution for his or her age group.

If the variable we standardize has a Normal distribution, standardizing does more than give a common scale. It makes all Normal distributions into a single distribution, and this distribution is still Normal. Standardizing a variable that has any Normal distribution produces a new variable that has the *standard Normal distribution*.

Standard Normal Distribution

The **standard Normal distribution** is the Normal distribution $N(0, 1)$ with mean 0 and standard deviation 1.

If a variable x has any Normal distribution $N(\mu, \sigma)$ with mean μ and standard deviation σ, then the standardized variable

$$z = \frac{x - \mu}{\sigma}$$

has the standard Normal distribution.

Apply Your Knowledge

Spencer Grant/PhotoEdit

3.8 SAT versus ACT. In 2012, when she was a high school senior, Idonna scored 670 on the mathematics part of the SAT.[5] The distribution of SAT math scores in 2012 was Normal with mean 514 and standard deviation 117. Jonathan took the ACT and scored 26 on the mathematics portion. ACT math scores for 2012 were Normally distributed with mean 21.1 and standard deviation 5.3. Find the standardized scores for both students. Assuming that both tests measure the same kind of ability, who had the higher score?

3.9 Men's and Women's Heights. The heights of women aged 20 to 29 in the United States are approximately Normal with mean 64.2 inches and standard deviation 2.8 inches. Men the same age have mean height 69.4 inches with standard deviation 3.0 inches.[6] What are the z-scores for a woman 6 feet tall and a man 6 feet tall? Say in simple language what information the z-scores give that the original nonstandardized heights do not.

Finding Normal proportions

Areas under a Normal curve represent proportions of observations from that Normal distribution. There is no formula for areas under a Normal curve. Calculations use either software that calculates areas or a table of areas. Most tables and software calculate one kind of area, *cumulative proportions*. The idea of "cumulative" is "everything that came before." Here is the exact statement.

Cumulative Proportions

The **cumulative proportion** for a value x in a distribution is the proportion of observations in the distribution that are less than or equal to x.

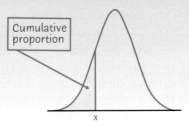

Cumulative proportion

x

The key to calculating Normal proportions is to match the area you want with areas that represent cumulative proportions. If you make a sketch of the area you want, you will almost never go wrong. Find areas for cumulative proportions either from software or (with an extra step) from a table. The following example shows the method in a picture.

EXAMPLE 3.5   Who Qualifies for College Sports?

The National Collegiate Athletic Association (NCAA) uses a sliding scale for eligibility for Division I athletes.[7] Those students with a 2.5 high school GPA must score at least 820 on the combined mathematics and critical reading parts of the SAT to compete in their first college year. The combined scores of the almost 1.7 million high school seniors taking the SAT in 2012 were approximately Normal with mean 1010 and standard deviation 214. What percent of high school seniors meet this SAT requirement of a combined score of 820 or better?

Here is the calculation in a picture: the proportion of scores above 820 is the area under the curve to the right of 820. That's the total area under the curve (which is always 1) minus the cumulative proportion up to 820.

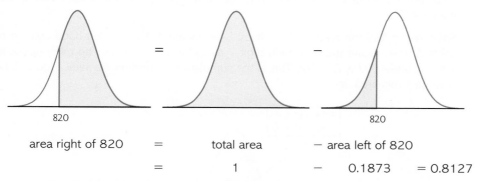

area right of 820	=	total area	− area left of 820
	=	1	− 0.1873 = 0.8127

About 81% of all high school seniors meet this SAT requirement of a combined math and reading score of 820 or higher. ■

There is *no* area under a smooth curve and exactly over the point 820. Consequently, the area to the right of 820 (the proportion of scores > 820) is the same as the area at or to the right of this point (the proportion of scores ≥ 820). The actual data may contain a student who scored exactly 820 on the SAT. That the proportion of scores exactly equal to 820 is 0 for a Normal distribution is a consequence of the idealized smoothing of Normal distributions for data.

To find the numerical value 0.1873 of the cumulative proportion in Example 3.5 using software, plug in mean 1010 and standard deviation 214 and ask for the cumulative proportion for 820. Software often uses terms such as "cumulative distribution" or "cumulative probability." We will learn in Chapter 10 why the language of probability fits. Here, for example, is Minitab's output:

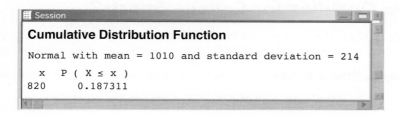

The *P* in the output stands for "probability," but we can read it as "proportion of the observations." The *Normal Density Curve* applet is even handier because it draws pictures as well as finding areas. If you are not using software, you can find cumulative proportions for Normal curves from a table. This requires an extra step.

Using the standard Normal table

The extra step in finding cumulative proportions from a table is that we must first standardize to express the problem in the standard scale of z-scores. This allows us to get by with just one table, a table of *standard Normal cumulative proportions*. Table A in the back of the book gives cumulative proportions for the standard Normal distribution. The pictures at the top of the table remind us that the entries are cumulative proportions, areas under the curve to the left of a value z.

| EXAMPLE 3.6 | The Standard Normal Table |

What proportion of observations on a standard Normal variable z take values less than 1.47?

Solution: To find the area to the left of 1.47, locate 1.4 in the left-hand column of Table A, then locate the remaining digit 7 as .07 in the top row. The entry opposite 1.4 and under .07 is 0.9292. This is the cumulative proportion we seek. Figure 3.12 illustrates this area. ■

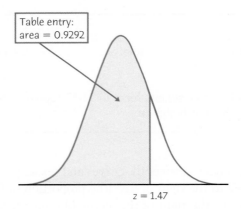

FIGURE 3.12
The area under a standard Normal curve to the left of the point $z = 1.47$ is 0.9292. Table A gives areas under the standard Normal curve.

Now that you see how Table A works, let's redo Example 3.5 using the table. We can break Normal calculations using the table into three steps.

| EXAMPLE 3.7 | Who Qualifies for College Sports? |

Scores of high school seniors on the SAT follow the Normal distribution with mean $\mu = 1010$ and standard deviation $\sigma = 214$. What percent of seniors score at least 820?

Step 1. Draw a picture. The picture is exactly as in Example 3.5. It shows that

area to the right of 820 = 1 − area to the left of 820

Step 2. Standardize. Call the SAT score x. Subtract the mean, then divide by the standard deviation, to transform the problem about x into a problem about a standard Normal z:

$$x \geq 820$$

$$\frac{x - 1010}{214} \geq \frac{820 - 1010}{214}$$

$$z \geq -0.89$$

Step 3. Use the table. The picture shows that we need the cumulative proportion for $x = 820$. Step 2 says this is the same as the cumulative proportion for $z = -0.89$. The Table A entry for $z = -0.89$ says that this cumulative proportion is 0.1867. The area to the right of -0.89 is therefore $1 - 0.1867 = 0.8133$. ■

The area from the table in Example 3.7 (0.8133) is slightly less accurate than the area from software in Example 3.5 (0.8127) because we must round z to two decimal places when we use Table A. The difference is rarely important in practice. Here's the method in outline form.

Using Table A to Find Normal Proportions

Step 1. State the problem in terms of the observed variable x. **Draw a picture** that shows the proportion you want in terms of cumulative proportions.

Step 2. Standardize x to restate the problem in terms of a standard Normal variable z.

Step 3. Use Table A and the fact that the total area under the curve is 1 to find the required area under the standard Normal curve.

EXAMPLE 3.8   Who Qualifies for College Sports?

Recall that the NCAA uses a sliding scale for eligibility for Division I athletics. Students with a 2.5 GPA must have a combined SAT score of 820 or higher to be eligible. Students with lower GPAs will require higher SAT scores for eligibility, whereas students with higher GPAs can have a lower SAT score and still be eligible. For example, students with a 2.75 GPA are only required to have a combined SAT score that is at least 720. What proportion of all students who take the SAT would meet an SAT requirement of at least 720, but not 820?

Step 1. State the problem and draw a picture. Call the SAT score x. The variable x has the $N(1010, 214)$ distribution. What proportion of SAT scores fall between 720 and 820? Here is the picture:

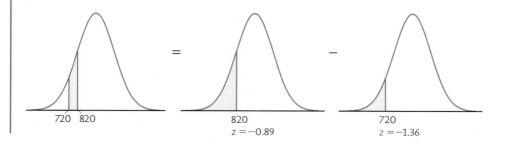

Step 2. Standardize. Subtract the mean, then divide by the standard deviation, to turn x into a standard Normal z:

$$720 \leq x < 820$$
$$\frac{720 - 1010}{214} \leq \frac{x - 1010}{214} < \frac{820 - 1010}{214}$$
$$-1.36 \leq z < -0.89$$

Step 3. Use the table. Follow the picture (we added the z-scores to the picture label to help you):

area between -1.36 and -0.89 = (area left of -0.89) − (area left of -1.36)
= 0.1867 − 0.0869 = 0.0998

About 10% of high school seniors have SAT scores between 720 and 820. ■

Sometimes we encounter a value of z more extreme than those appearing in Table A. For example, the area to the left of $z = -4$ is not given directly in the table. The z-values in Table A leave only area 0.0002 in each tail unaccounted for. For practical purposes, we can act as if there is zero area outside the range of Table A. Specifically, we act as if there is zero area below $z = -3.5$ and zero area above $z = 3.5$.

Apply Your Knowledge

3.10 Use the Normal Table. Use Table A to find the proportion of observations from a standard Normal distribution that satisfies each of the following statements. In each case, sketch a standard Normal curve and shade the area under the curve that is the answer to the question.

(a) $z < -0.76$ (b) $z > -0.76$
(c) $z < 1.45$ (d) $-0.76 < z < 1.45$

© John Henry Claude Wilson/Robert Harding Picture Library Ltd/Alamy

3.11 Monsoon Rains. The summer monsoon rains in India follow approximately a Normal distribution with mean 852 millimeters (mm) of rainfall and standard deviation 82 mm.

(a) In the drought year 1987, 697 mm of rain fell. In what percent of all years will India have 697 mm or less of monsoon rain?

(b) "Normal rainfall" means within 20% of the long-term average, or between 682 mm and 1022 mm. In what percent of all years is the rainfall normal?

3.12 The Medical College Admissions Test. Almost all medical schools in the United States require students to take the Medical College Admission Test (MCAT).[8] The exam is composed of three multiple-choice sections (Physical Sciences, Verbal Reasoning, and Biological Sciences). The score on each section is converted to a 15-point scale so that your total score has a maximum value of 45. The total scores follow a Normal distribution, and in 2012 the mean was 25.2 with a standard deviation of 6.4. There is little change in the distribution of scores from year to year.

(a) What proportion of students taking the MCAT had a score over 30?

(b) What proportion had scores between 20 and 25?

Finding a value given a proportion

Examples 3.5 to 3.8 illustrated the use of software or Table A to find what proportion of the observations satisfies some condition, such as "SAT score above 820." We may instead want to find the observed value with a given proportion of the observations above or below it. Statistical software will do this directly.

EXAMPLE 3.9   Find the Top 10% Using Software

Scores on the SAT reading test in recent years follow approximately the $N(504, 111)$ distribution. How high must a student score to place in the top 10% of all students taking the SAT?

We want to find the SAT score x with area 0.1 to its *right* under the Normal curve with mean $\mu = 504$ and standard deviation $\sigma = 111$. That's the same as finding the SAT score x with area 0.9 to its *left*. Figure 3.13 poses the question in graphical form. Most software will tell you x when you plug in mean 504, standard deviation 111, and cumulative proportion 0.9. Here is Minitab's output:

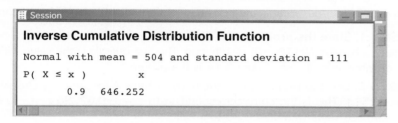

Minitab gives $x = 646.252$. So scores above 647 are in the top 10%. (Round up because SAT scores can only be whole numbers.) ■

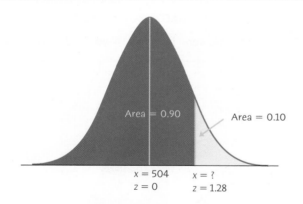

FIGURE 3.13
Locating the point on a Normal curve with area 0.10 to its right, for Examples 3.9 and 3.10.

Without software, use Table A backward. Find the given proportion in the body of the table and then read the corresponding z from the left column and top row. There are again three steps.

EXAMPLE 3.10   Find the Top 10% Using Table A

Scores on the SAT verbal test in recent years follow approximately the $N(504, 111)$ distribution. How high must a student score to place in the top 10% of all students taking the SAT?

Step 1. State the problem and draw a picture. This step is exactly as in Example 3.9. The picture is Figure 3.13. The x-value that puts a student in the top 10% is the same as the x-value for which 90% of the area is to the left of x.

Step 2. Use the table. Look in the body of Table A for the entry closest to 0.9. It is 0.8997. This is the entry corresponding to $z = 1.28$. So $z = 1.28$ is the standardized value with area 0.9 to its left.

Step 3. Unstandardize to transform z back to the original x scale. We know that the standardized value of the unknown x is $z = 1.28$. This means that x itself lies 1.28 standard deviations above the mean on this particular Normal curve. That is,

$$x = \text{mean} + (1.28)(\text{standard deviation})$$
$$= 504 + (1.28)(111) = 646.08$$

A student must score at least 647 to place in the highest 10%. ■

EXAMPLE 3.11   **Find the First Quartile**

High levels of cholesterol in the blood increase the risk of heart disease. For 14-year-old boys, the distribution of blood cholesterol is approximately Normal with mean $\mu = 170$ milligrams of cholesterol per deciliter of blood (mg/dl) and standard deviation $\sigma = 30$ mg/dl.[9] What is the first quartile of the distribution of blood cholesterol?

Step 1. State the problem and draw a picture. Call the cholesterol level x. The variable x has the $N(170, 30)$ distribution. The first quartile is the value with 25% of the distribution to its left. Figure 3.14 is the picture.

Step 2. Use the table. Look in the body of Table A for the entry closest to 0.25. It is 0.2514. This is the entry corresponding to $z = -0.67$. So $z = -0.67$ is the standardized value with area 0.25 to its left.

Step 3. Unstandardize. The cholesterol level corresponding to $z = -0.67$ lies 0.67 standard deviations below the mean, so

$$x = \text{mean} - (0.67)(\text{standard deviation})$$
$$= 170 - (0.67)(30) = 149.9$$

The first quartile of blood cholesterol levels in 14-year-old boys is about 150 mg/dl. ■

FIGURE 3.14

Locating the first quartile of a Normal curve, for Example 3.11.

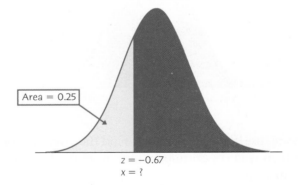

Area = 0.25

$z = -0.67$
$x = ?$

Apply Your Knowledge

3.13 Table A. Use Table A to find the value z of a standard Normal variable that satisfies each of the following conditions. (Use the value of z from Table A that comes closest to satisfying the condition.) In each case, sketch a standard Normal curve with your value of z marked on the axis.

(a) The point z with 65% of the observations falling below it

(b) The point z with 20% of the observations falling above it

3.14 The Medical College Admissions Test. The total scores on the Medical College Admission Test (MCAT) follow a Normal distribution with mean 25.2 and standard deviation 6.4. What are the median and the first and third quartiles of the MCAT scores? What is the interquartile range?

CHAPTER 3 SUMMARY

Chapter Specifics

- We can sometimes describe the overall pattern of a distribution by a **density curve.** A density curve has total area 1 underneath it. An area under a density curve gives the proportion of observations that fall in a range of values.

- A density curve is an idealized description of the overall pattern of a distribution that smooths out the irregularities in the actual data. We write the **mean of a density curve** as μ and the **standard deviation of a density curve** as σ to distinguish them from the mean $\bar{x}$ and standard deviation s of the actual data.

- The mean, the median, and the quartiles of a density curve can be located by eye. The **mean μ** is the balance point of the curve. The **median** divides the area under the curve in half. The **quartiles** and the median divide the area under the curve into quarters. The **standard deviation σ** cannot be located by eye on most density curves.

- The mean and median are equal for symmetric density curves. The mean of a skewed curve is located farther toward the long tail than is the median.

- The **Normal distributions** are described by a special family of bell-shaped, symmetric density curves, called **Normal curves.** The mean μ and standard deviation σ completely specify a Normal distribution $N(\mu, \sigma)$. The mean is the center of the curve, and σ is the distance from μ to the change-of-curvature points on either side.

- To **standardize** any observation x, subtract the mean of the distribution and then divide by the standard deviation. The resulting **z-score**

$$z = \frac{x - \mu}{\sigma}$$

says how many standard deviations x lies from the distribution mean.

- All Normal distributions are the same when measurements are transformed to the standardized scale. In particular, all Normal distributions satisfy the **68–95–99.7 rule,** which describes what percent of observations lie within one, two, and three standard deviations of the mean, respectively.

- If x has the $N(\mu, \sigma)$ distribution, then the **standardized variable** $z = (x - \mu)/\sigma$ has the **standard Normal distribution** $N(0, 1)$ with mean 0 and standard deviation 1. Table A gives the **cumulative proportions** of standard Normal observations that are less than z for many values of z. By standardizing, we can use Table A for any Normal distribution.

Link It

When exploring data, some data sets can be shown to closely follow the Normal distribution. When this is true, the description of the data can be greatly simplified without much loss of information. We can calculate the percentage of the distribution in an interval for *any* Normal distribution if we know its mean and standard deviation. This also shows why the mean and standard deviation can be important numerical summaries. For distributions that are approximately Normal, these two numerical summaries give a complete description of the distribution of our data. It is important to remember that not all distributions can be well approximated by a Normal curve. In these cases, calculations based on the Normal distribution can be misleading.

Normal distributions are also good approximations to many kinds of chance outcomes such as the proportion of heads in many tosses of a coin—this setting will be described in more detail in Chapter 15. And when we discuss statistical inference in Part IV of the text, we will find that many procedures based on Normal distributions work well for other roughly symmetric distributions.

 THE BELL CURVE?
Does the distribution of human intelligence follow the "bell curve" of a Normal distribution? Scores on IQ tests do roughly follow a Normal distribution. That is because a test score is calculated from a person's answers in a way that is designed to produce a Normal distribution. To conclude that intelligence follows a bell curve, we must agree that the test scores directly measure intelligence. Many psychologists don't think there is one human characteristic that we can call "intelligence" and can measure by a single test score.

CHECK YOUR SKILLS

3.15 Which of these variables is most likely to have a Normal distribution?

   (a) Income per person for 150 different countries
   (b) Sale prices of 200 homes in Santa Barbara, CA
   (c) Lengths of 100 newborns in Connecticut

3.16 To completely specify the shape of a Normal distribution, you must give

   (a) the mean and the standard deviation.
   (b) the five-number summary.
   (c) the median and the quartiles.

3.17 Figure 3.15 shows a Normal curve. The mean of this distribution is

   (a) 0.   (b) 2.   (c) 3.

3.18 The standard deviation of the Normal distribution in Figure 3.15 is

   (a) 2.   (b) 3.   (c) 5.

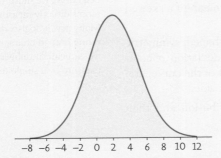

-8 -6 -4 -2 0 2 4 6 8 10 12

FIGURE 3.15
A Normal curve, for Exercises 3.17 and 3.18.

3.19 The length of human pregnancies from conception to birth varies according to a distribution that is approximately Normal with mean 266 days and standard deviation 16 days. About 95% of all pregnancies last between

   (a) 250 and 282 days.   (b) 234 and 298 days.
   (c) 218 and 314 days.

3.20 The scores of adults on an IQ test are approximately Normal with mean 100 and standard deviation 15. The organization MENSA, which calls itself "the high IQ society," requires an IQ score of 130 or higher for membership. What percent of adults would qualify for membership?

   (a) 95%   (b) 5%   (c) 2.5%

3.21 The scores of adults on an IQ test are approximately Normal with mean 100 and standard deviation 15. Clara scores 132 on such a test. Her z-score is about

   (a) 2.13.   (b) 2.80.   (c) 8.47.

3.22 The proportion of observations from a standard Normal distribution that take values greater than 1.45 is about

   (a) 0.9265.   (b) 0.0735.   (c) 0.0808.

3.23 The proportion of observations from a standard Normal distribution that take values less than -1.25 is about

   (a) 0.1151.   (b) 0.1056.   (c) 0.8849.

3.24 The scores of adults on an IQ test are approximately Normal with mean 100 and standard deviation 15. Clara scores 132 on such a test. She scores higher than what percent of all adults?

   (a) About 10%   (b) About 90%
   (c) About 98%

CHAPTER 3 EXERCISES

3.25 **Understanding density curves.** Remember that it is areas under a density curve, not the height of the curve, that give proportions in a distribution. To illustrate this, sketch a density curve that has a tall, thin peak at 0 on the horizontal axis but has most of its area close to 1 on the horizontal axis without a high peak at 1.

3.26 **Daily activity.** It appears that people who are mildly obese are less active than leaner people. One study looked at the average number of minutes per day that people spend standing or walking.[10] Among mildly obese people, minutes of activity varied according to the $N(373, 67)$ distribution. Minutes of activity for lean people had the $N(526, 107)$ distribution. Within what limits do the active minutes for about 95% of the people in each group fall? Use the 68–95–99.7 rule.

3.27 **Low IQ test scores.** Scores on the Wechsler Adult Intelligence Scale (WAIS) are approximately Normal with mean 100 and standard deviation 15. People with WAIS scores below 70 are considered intellectually disabled when, for example, applying for Social Security disability benefits. According to the 68–95–99.7 rule, about what percent of adults are intellectually disabled by this criterion.

3.28 **Standard Normal drill.** Use Table A to find the proportion of observations from a standard Normal distribution that falls in each of the following regions. In each case, sketch a standard Normal curve and shade the area representing the region.

   (a) $z \leq -2.15$ (b) $z \geq -2.15$ (c) $z > 1.57$
   (d) $-2.15 < z < 1.57$

3.29 **Standard Normal drill.**

(a) Find the number z such that the proportion of observations that are less than z in a standard Normal distribution is 0.3.

(b) Find the number z such that 35% of all observations from a standard Normal distribution are greater than z.

3.30 **Fruit flies.** The common fruit fly *Drosophila melanogaster* is the most studied organism in genetic research because it is small, is easy to grow, and reproduces rapidly. The length of the thorax (where the wings and legs attach) in a population of male fruit flies is approximately Normal with mean 0.800 millimeters (mm) and standard deviation 0.078 mm.

Plastique1/Dreamstime.com

(a) What proportion of flies have thorax length less than 0.6 mm?

(b) What proportion have thorax length greater than 0.9 mm?

(c) What proportion have thorax length between 0.6 mm and 0.9 mm?

3.31 **Acid rain?** Emissions of sulfur dioxide by industry set off chemical changes in the atmosphere that result in "acid rain." The acidity of liquids is measured by pH on a scale of 0 to 14. Distilled water has pH 7.0, and lower pH values indicate acidity. Normal rain is somewhat acidic, so acid rain is sometimes defined as rainfall with a pH below 5.0. The pH of rain at one location varies among rainy days according to a Normal distribution with mean 5.43 and standard deviation 0.54. What proportion of rainy days have rainfall with pH below 5.0?

3.32 **Runners.** In a study of exercise, a large group of male runners walk on a treadmill for 6 minutes. Their heart rates in beats per minute at the end vary from runner to runner according to the $N(104, 12.5)$ distribution. The heart rates for male nonrunners after the same exercise have the $N(130, 17)$ distribution.

(a) What percent of the runners have heart rates above 135?

(b) What percent of the nonrunners have heart rates above 135?

3.33 **A milling machine.** Automated manufacturing operations are quite precise but still vary, often with distributions that are close to Normal. The width in inches of slots cut by a milling machine follows approximately the $N(0.8750, 0.0012)$ distribution. The specifications allow slot widths between 0.8725 and 0.8775 inch. What proportion of slots meet these specifications?

3.34 **Body mass index.** Your body mass index (BMI) is your weight in kilograms divided by the square of your height in meters. Many online BMI calculators allow you to enter weight in pounds and height in inches. High BMI is a common but controversial indicator of overweight or obesity. A study by the National Center for Health Statistics found that the BMI of American young men (ages 20 to 29) is approximately Normal with mean 26.8 and standard deviation 5.2.[11]

(a) People with BMI less than 18.5 are often classified as "underweight." What percent of men aged 20 to 29 are underweight by this criterion?

(b) People with BMI over 30 are often classified as "obese." What percent of men aged 20 to 29 are obese by this criterion?

Miles per gallon. *In its* Fuel Economy Guide *for 2013 model vehicles, the Environmental Protection Agency gives data on 1108 vehicles. There are a number of high outliers, mainly hybrid gas-electric vehicles. If we ignore the vehicles identified as outliers, however, the combined city and highway gas mileage of the other 1093 vehicles is approximately Normal with mean 21.9 miles per gallon (mpg) and standard deviation 5.1 mpg. Exercises 3.35 to 3.38 concern this distribution.*

3.35 **I love my bug!** The 2013 Volkswagen Beetle with a five-cylinder 2.5L engine and automatic transmission has combined gas mileage 25 mpg. What percent of all vehicles have better gas mileage than the Beetle?

3.36 **The top 10%.** How high must a 2013 vehicle's gas mileage be to fall in the top 10% of all vehicles?

3.37 **The middle half.** The quartiles of any distribution are the values with cumulative proportions 0.25 and 0.75. They span the middle half of the distribution. What are the quartiles of the distribution of gas mileage?

3.38 **Quintiles.** The quintiles of any distribution are the values with cumulative proportions 0.20, 0.40, 0.60, and 0.80. What are the quintiles of the distribution of gas mileage?

3.39 **What's your percentile?** Reports on a student's test score such as the SAT or a child's height or weight usually give the percentile as well as the actual value of the variable. The percentile is just the cumulative proportion stated as a percent: the percent of all values of the variable that were lower than this one. The upper arm lengths of females in the United States are approximately Normal with mean 35.8 cm and standard deviation 2.1 cm, and those for males are approximately Normal with mean 39.1 cm and standard deviation 2.3 cm.

(a) Larry, a 60-year-old male in the United States, has an upper arm length of 37.2 cm. What is his percentile?

(b) Measure your upper arm length to the nearest tenth of a centimeter, referring to Exercise 3.5 (page 80) for the measurement instructions. What is your arm length in centimeters? What is your percentile?

3.40 **Perfect SAT scores.** It is possible to score higher than 1600 on the combined mathematics and reading portions of the SAT, but scores 1600 and above are reported as 1600. The distribution of SAT scores (combining mathematics and reading) in 2012 was close to Normal with mean 1010 and standard deviation 214. What proportion of SAT scores for these two parts were reported as 1600 (that is, what proportion of SAT scores were actually higher than 1600?)

3.41 **Heights of women.** The heights of women aged 20 to 29 follow approximately the $N(64.2, 2.8)$ distribution. Men the same age have heights distributed as $N(69.4, 3.0)$. What percent of women aged 20 to 29 are taller than the mean height of men aged 20 to 29?

3.42 **Weights aren't Normal.** The heights of people of the same sex and similar ages follow a Normal distribution reasonably closely. Weights, on the other hand, are not Normally distributed. The weights of women aged 20 to 29 in the United States have mean 161.9 pounds and median 149.4 pounds. The first and third quartiles are 126.3 pounds and 181.2 pounds. What can you say about the shape of the weight distribution? Why?

3.43 **A surprising calculation.** Changing the mean and standard deviation of a Normal distribution by a moderate amount can greatly change the percent of observations in the tails. Suppose a college is looking for applicants with SAT math scores 750 and above.

(a) In 2012, the scores of men on the math SAT followed the $N(532, 119)$ distribution. What percent of men scored 750 or better?

(b) Women's SAT math scores that year had the $N(499, 113)$ distribution. What percent of women scored 750 or better? You see that the percent of men above 750 is more than twice the percent of women with such high scores. Why this is true is controversial. (On the other hand, women score higher than men on the new SAT writing test, though by a smaller amount.)

3.44 **Grading managers.** Some companies "grade on a bell curve" to compare the performance of their managers and professional workers. This forces the use of some low performance ratings so that not all workers are listed as "above average." Ford Motor Company's "performance management process" for this year assigned 10% A grades, 80% B grades, and 10% C grades to the company's managers. Suppose Ford's performance scores really are Normally distributed. This year, managers with scores less than 25 received C's and those with scores above 475 received A's. What are the mean and standard deviation of the scores?

3.45 **Osteoporosis.** Osteoporosis is a condition in which the bones become brittle due to loss of minerals. To diagnose osteoporosis, an elaborate apparatus measures bone mineral density (BMD). BMD is usually reported in standardized form. The standardization is based on a population of healthy young adults. The World Health Organization (WHO) criterion for osteoporosis is a BMD 2.5 standard deviations below the mean for healthy young adults. BMD measurements in a population of people similar in age and sex roughly follow a Normal distribution.

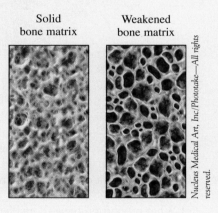

Solid bone matrix   Weakened bone matrix

(a) What percent of healthy young adults have osteoporosis by the WHO criterion?

(b) Women aged 70 to 79 are of course not young adults. The mean BMD in this age is about −2 on the standard scale for young adults. Suppose the standard deviation is the same as for young adults. What percent of this older population have osteoporosis?

In later chapters we will meet many statistical procedures that work well when the data are "close enough to Normal." Exercises 3.46 to 3.50 concern data that are mostly close enough to Normal for statistical work, whereas Exercise 3.51 concerns data for which the data are not close to

Normal. These exercises ask you to do data analysis and Normal calculations to investigate how close to Normal real data are.

3.46 **Normal is only approximate: IQ test scores.** Here are the IQ test scores of 31 seventh-grade girls in a Midwest school district:[12] **MIDWSTIQ**

114	100	104	89	102	91	114	114
103	105	108	130	120	132	111	128
118	119	86	72	111	103	74	112
107	103	98	96	112	112	93	

(a) We expect IQ scores to be approximately Normal. Make a stemplot to check that there are no major departures from Normality.

(b) Nonetheless, proportions calculated from a Normal distribution are not always very accurate for small numbers of observations. Find the mean $\bar{x}$ and standard deviation s for these IQ scores. What proportions of the scores are within one standard deviation and within two standard deviations of the mean? What would these proportions be in an exactly Normal distribution?

3.47 **Normal is only approximate: ACT scores.** Composite scores on the ACT test for the 2012 high school graduating class had mean 21.1 and standard deviation 5.3. In all, 1,666,017 students in this class took the test. Of these, 159,259 had scores higher than 28 and another 54,167 had scores exactly 28. ACT scores are always whole numbers. The exactly Normal $N(21.1, 5.3)$ distribution can include any value, not just whole numbers. What is more, there is *no* area exactly above 28 under the smooth Normal curve. So ACT scores can be only approximately Normal. To illustrate this fact, find

(a) the percent of 2012 ACT scores greater than 28 using the actual counts reported.

(b) the percent of 2012 ACT scores greater than or equal to 28, using the actual counts reported.

(c) the percent of observations that are greater than 28 using the $N(21.1, 5.3)$ distribution. (The percent greater than or equal to 28 is the same, because there is no area exactly over 28.)

3.48 **Are the data Normal? Acidity of rainfall.** Exercise 3.31 (page 91) concerns the acidity (measured by pH) of rainfall. A sample of 105 rainwater specimens had mean pH 5.43, standard deviation 0.54, and five-number summary 4.33, 5.05, 5.44, 5.79, 6.81.[13]

(a) Compare the mean and median and also the distances of the two quartiles from the median. Does it appear that the distribution is quite symmetric? Why?

(b) If the distribution is really $N(5.43, 0.54)$, what proportion of observations would be less than 5.05? Less than 5.79? Do these proportions

suggest that the distribution is close to Normal? Why?

3.49 **Are the data Normal? SAT critical reading scores.** Georgia Southern University (GSU) had 2417 students with regular admission in their Freshman class of 2010. For each student, data is available on their SAT and ACT scores, if taken, high school GPA and the college within the university to which they were admitted.[14] Here are the first 20 SAT critical reading scores from that data set: **SATCR**

650	490	580	450	570	540	510	530
510	560	560	590	470	690	530	570
460	590	530	490				

The complete data is in the file SATCR, which contains both the original scores and the ordered scores.

(a) Make a histogram of the distribution (if your software allows it, superimpose a normal curve over the histogram as in Figure 3.1). Although the resulting histogram depends a bit on your choice of classes, the distribution appears roughly symmetric with no outliers.

(b) Find the mean, median, standard deviation, and quartiles for these data. Comparing the mean and the median and comparing the distances of the two quartiles from the median suggest that the distribution is quite symmetric. Why?

(c) In 2010, the mean score on the critical reading portion of the SAT for all college-bound seniors was 501. If the distribution were exactly Normal with the mean and standard deviation you found in (b), what proportion of regularly admitted GSU freshman scored above the mean for all college-bound seniors?

(d) Compute the exact proportion of regularly admitted GSU freshman that scored above the mean for all college-bound seniors. It will be simplest to use the ordered scores in the SATCR file to calculate this. How does this percentage compare with the percentage calculated in part (c)? Despite the discrepancy, this distribution is "close enough to Normal" for statistical work in later chapters.

3.50 **Are the data Normal? Monsoon rains.** Here are the amounts of summer monsoon rainfall (millimeters) for India in the 100 years 1901 to 2000:[15] **MONSOON**

722.4	792.2	861.3	750.6	716.8	885.5	777.9	897.5	889.6	935.4
736.8	806.4	784.8	898.5	781.0	951.1	1004.7	651.2	885.0	719.4
866.2	869.4	823.5	863.0	804.0	903.1	853.5	768.2	821.5	804.9
877.6	803.8	976.2	913.8	843.9	908.7	842.4	908.6	789.9	853.6
728.7	958.1	868.6	920.8	911.3	904.0	945.9	874.3	904.2	877.3
739.2	793.3	923.4	885.8	930.5	983.6	789.0	889.6	944.3	839.9
1020.5	810.0	858.1	922.8	709.6	740.2	860.3	754.8	831.3	940.0
887.0	653.1	913.6	748.3	963.0	857.0	883.4	909.5	708.0	882.9
852.4	735.6	955.9	836.9	760.0	743.2	697.4	961.7	866.9	908.8
784.7	785.0	896.6	938.4	826.4	857.3	870.5	873.8	827.0	770.2

(a) Make a histogram of these rainfall amounts. Find the mean and the median.

(b) Although the distribution is reasonably Normal, your work shows some departure from Normality. In what way are the data not Normal?

3.51   Are the data Normal? Weight of females in their 20s. Many body measurements of people of the same sex and similar ages such as height and upper arm length follow a Normal distribution reasonably closely. Weights, on the other hand, are not Normally distributed. The NHANES survey of 2009–10[16] includes the weights of a representative sample of 548 females in the United States aged 20 to 29. The mean of the weights was 161.58 pounds, and the standard deviation was 48.96 pounds. Figure 3.16 gives a histogram of the data along with a smooth curve representing an $N(161.58, 48.96)$ distribution. From the figure, the Normal curve does not appear to follow the pattern in the histogram that closely. Because of this, the use of areas under the Normal curve may not provide a good approximation to incomes in various intervals. FEMWEIGH

(a) Using the data file on the text Web site, what proportion of females aged 20 to 29 weighed under 100 pounds? What percent of the $N(161.58, 48.96)$ distribution is below 100?

(b) What proportion of females aged 20 to 29 weighed over 250 pounds? What percent of the $N(161.58, 48.96)$ distribution is above 250?

(c) Based on your answers in (a) and (b), do you think it is a good idea to summarize the distribution of weights by an $N(161.58, 48.96)$ distribution?

The Normal Curve applet allows you to do Normal calculations quickly. It is somewhat limited by the number of pixels available for use, so that it can't hit every value exactly. In the following exercises, use the closest available values. In each case, make a sketch of the curve from the applet marked with the values you used to answer the questions.

3.52   How accurate is 68–95–99.7? The 68–95–99.7 rule for Normal distributions is a useful approximation. To see how accurate the rule is, drag one flag across the other so that the applet shows the area under the curve between the two flags.

(a) Place the flags one standard deviation on either side of the mean. What is the area between these two values? What does the 68–95–99.7 rule say this area is?

(b) Repeat for locations two and three standard deviations on either side of the mean. Again compare the 68–95–99.7 rule with the area given by the applet.

3.53   Where are the quartiles? How many standard deviations above and below the mean do the quartiles of any Normal distribution lie? (Use the standard Normal distribution to answer this question.)

3.54   Grading managers. In Exercise 3.44, we saw that Ford Motor Company once graded its managers in such a way that the top 10% received an A grade, the bottom 10% a C, and the middle 80% a B. Let's suppose that performance scores follow a Normal distribution. How many standard deviations above and below the mean do the A/B and B/C cutoffs lie? (Use the standard Normal distribution to answer this question.)

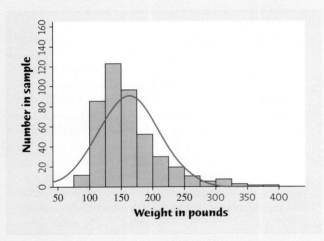

FIGURE 3.16

Histogram of the weights of 548 females aged 20 to 29 in the 2009–10 NHANES survey, with a Normal curve superimposed, for Exercise 3.51.

 Exploring the Web

3.55   Are the data Normal? Comparing quartiles. The Web site http://professionals.collegeboard.com/data-reports-research/sat presents data for high school seniors who participated in the SAT Program from both the current year, as well as previous years. Click on the link for *College Bound*

Seniors for the most recent year given. In the window that opens, click on the link for *Total Group Report* for this year. The Total Group Profile will open and contains several tables, each giving different summary information. Go to the table for Overall Mean Scores. How many students took the Critical Reading portion of the SAT? What were the mean and standard deviation of the scores? Assuming the distribution of scores is Normal with the mean and standard deviation given in the Overall Mean Scores table, what would be the first and third quartiles of the distribution. Now go to the table for Percentiles for the Total Group, and compare the actual first and third quartiles from the data to the values obtained from the Normal curve. Does this give any evidence that the distribution of Critical Reading scores is not Normal?

3.56 **Are the data Normal? Comparing proportions.** The Web site `http:// professionals.collegeboard.com/data-reports-research/sat` presents data for high school seniors who participated in the SAT Program from both the current year, as well as previous years. Click on the link for *College Bound Seniors* for the most recent year given. In the window that opens, click on the link for *Total Group Report* for this year. The Total Group Profile will open and contains several tables, each giving different summary information. Go to the Table for overall mean scores. How many students took the Critical Reading portion of the SAT? What were the mean and standard deviation of the scores? Now go to the table on the Score Distribution. The scores are broken into intervals 200–290, 300–390, and so on. Using the Total column, find the actual percentage of students who scored in each of the six intervals reported for the Critical Reading portion of the SAT. Now, assuming the distribution of scores is Normal with the mean and standard deviation given in the Overall Mean Scores table, find the area under the Normal curve for each interval. Because of the discreteness of the actual scores, take the interval 200–290 as the interval 200–300, the interval 300–390 as the interval 300–400, and so on, when finding the areas under the Normal curve. How do the actual percentages compare to the areas under the Normal curve? Does this give any evidence that the distribution of Critical Reading scores is not Normal?

Scatterplots and Correlation

Overview

Statistical model building is certainly one of the primary research tools of the social sciences, life sciences, physical sciences, and engineering. This chapter sets a foundation for the study of linear regression, one of the most basic and widely used statistical models. Before studying regression, students must first understand the concept of how and to what degree two quantitative variables may be related to one another, and we focus on linear relationships here. The most important concepts in this chapter will not only be used in subsequent chapters of this book, but will also play a key role in any future study of statistical model building.

An important first step in assessing the relationship between two quantitative variables (for example, "education level" and "income") is to determine which is the explanatory variable and which is the response variable. In most applications, this is clear from the context of the problem. For example, if the researcher is interested in predicting income from knowing how many years of schooling a person has, then "income" is the response variable, and "education level" is the explanatory variable. Even though scatterplots can be generated and correlation can be calculated regardless of which variable is explanatory and which is the response, it is important to have students think ahead to linear regression, where the roles are of utmost importance.

> ### LEARNING OBJECTIVES**
> ■ Identify the explanatory and response variables in situations where one variable explains or influences another.
>
> ■ Make a scatterplot to display the relationship between two quantitative variables measured on the same subjects. Place the explanatory variable (if any) on the horizontal scale of the plot.
>
> ■ Add a categorical variable to a scatterplot by using a different plotting symbol or color.
>
> ■ Describe the direction, form, and strength of the overall pattern of a scatterplot. In particular, recognize positive or negative associations and linear (straight-line) patterns. Recognize outliers in a scatterplot.
>
> ■ Judge whether it is appropriate to use correlation to describe the relationship between two quantitative variables. Find the correlation r.
>
> ■ Know the basic properties of correlation: r measures the direction and strength of only straight-line relationships; r is always a number between -1 and 1; $r = \pm 1$ only for perfect straight-line relationships; r moves away from 0 toward ± 1 as the straight-line relationship gets stronger.
>
> **These learning outcomes appear later for the students in Chapter 7: Part I Review.

When students begin to examine the relationship between two quantitative variables, they should always begin by plotting the data in a *scatterplot*. Scatterplots are intended to provide a simple visualization of the relationship (or lack thereof) between the two variables. The scatterplot may exhibit a linear relationship or some other kind of relationship. If there is a strong linear relationship—the kind of relationship required for simple linear regression studied in the next chapter—then it could indicate a positive association, where high (or low) values of the explanatory variable tend to be associated with high (or low) values of the response variable; or a negative association, where high (or low) values of the explanatory variable tend to be associated with low (or high) values of the response.

An important numerical value that summarizes both the direction and strength of the linear relationship between two quantitative variables is *correlation*, denoted by r. If there is a linear relationship between the variables, then the sign of the correlation (positive or negative) will agree with the direction of the association (positive or negative) as depicted in the scatterplot. The magnitude of correlation indicates the strength of the linear relationship between two quantitative variables (strength increases as the correlation moves away from 0 and toward $+1$ or -1). That is, a correlation of $+1$ or -1 indicates that the two variables have a perfect linear relationship—the points in the scatterplot perfectly fit a line with positive or negative slope, respectively. A correlation near 0 does not indicate that there is *no* relationship between the two variables; rather, it indicates that there is no *linear* relationship (see Exercise 4.8, for example).

Teaching Suggestions and Additional Examples/Activities for the Classroom

Note: If you prefer to wait to teach scatterplots, correlation, and linear regression until later in the course, you can move Chapters 4 and 5 to where they best fit your curriculum.

1. Use Data Collected in Class

Consider using data from your class to demonstrate construction of a scatterplot, preferably using software. A student survey given at the beginning of the course may provide some examples you can prepare ahead of time. Some possible examples include "age of student's car" (use "year of manufacture" for a negative association) versus "number of miles on odometer," or "number of hours studied in an average week" versus "number of credit hours taken."

2. Relate Correlation to Scatterplots

Show students scatterplots, along with associated values of correlation. Demonstrate that outliers can have a big impact on correlation. Demonstrate that curved relationships are not described by correlation (plot the cosine function, using several points, and show that correlation is very small, even though there is a very strong relationship between X and Y). Show the students several scatterplots for which there is increasing strength in the linear relationship between the two variables, and demonstrate that as the data become more linear, correlation moves closer to $+1$ or -1. All of these ideas can be demonstrated in class using the *Correlation and Regression* applet that accompanies the textbook. You may choose to work on Exercises 4.40 and/or 4.41 with your students in class or to come up with your own scenarios for the students to try with the applet.

3. Motivate Correlation with a Geometric Interpretation

The formula for correlation given in its definition text box in this chapter has an intuitively appealing geometric interpretation, which makes it clear that correlation measures only linear association between two quantitative variables. Make sure that your students are able to grasp this interpretation before doing the following. Start with a scatterplot in which there is an obvious positive or negative linear association. Ask students to plot the point $(\bar{x}, \bar{y})$—the point of average x, and average y. Sketch a new system of axes with new origin at the point $(\bar{x}, \bar{y})$. Consider the four quadrants defined by the new system of axes. Points in the first quadrant (the "upper-right" quadrant) are above average in x and above average in y and so have positive z-scores in both variables. Go through all four quadrants and ask students whether the products $\left(\dfrac{x - \bar{x}}{s_x}\right)\left(\dfrac{y - \bar{y}}{s_y}\right)$ for points in the quadrants are positive or negative. Show them that the formula for correlation essentially averages these products. If the first and third quadrants dominate the points, then correlation will be positive. If the second and fourth quadrants dominate, then correlation will be negative. If there is no linear association, then correlation will be close to zero because these products cancel in the sum. This presentation can also be deferred until later in the course, when regression is revisited in the context of inference. And, remind your students that we will use technology to calculate correlation for us.

4. Discuss the Impact of Averages on Correlation

Scatterplots and correlation are widely misunderstood and misused—intentionally and unintentionally—by practitioners. Demonstrate to students that a plot of "average age" versus "average height" in children is highly linear, but a plot of "age" versus "height" for individual children is far less linear. That is, correlations computed on averages instead of on individuals are highly misleading. These misleading correlations are sometimes referred to as "ecological" correlations.

Other Resources (LaunchPad)

Snapshots Videos

Correlation and Causation

EESEE Case Studies

Blood Alcohol Content

State of the SAT (scatterplots, correlation, and ecological correlation)

Fears in Children (scatterplots, correlation, adding categorical variables)

Applets

Two-Variable Statistical Calculator

Correlation and Regression

© Reinhard Dirscherl/age fotostock

Scatterplots and Correlation

A medical study finds that short women are more likely to have heart attacks than women of average height, whereas tall women have the fewest heart attacks. An insurance group reports that heavier cars have fewer deaths per 10,000 vehicles registered than do lighter cars. These and many other statistical studies look at the *relationship between two variables.* Statistical relationships are overall tendencies, not ironclad rules. They allow individual exceptions. Although smokers on the average die younger than nonsmokers, some people live to 90 while smoking three packs a day.

To understand a statistical relationship between two variables, we measure both variables on the same individuals. Often, we must examine other variables as well. To conclude that shorter women have higher risk from heart attacks, for example, the researchers had to eliminate the effect of other variables such as weight and exercise habits. In this and the following chapter, we study relationships between variables. One of our main themes is that the relationship between two variables can be strongly influenced by other variables that are lurking in the background.

Explanatory and response variables

We think that car weight helps explain accident deaths and that smoking influences life expectancy. In each of these relationships, the two variables play different roles: one explains or influences the other.

Response Variable, Explanatory Variable

A **response variable** measures an outcome of a study. An **explanatory variable** may explain or influence changes in a response variable.

You will often find explanatory variables called *independent variables* or *predictor variables* and response variables called *dependent variables*. The idea behind this language is that the response variable depends on the explanatory variable. Because "independent" and "dependent" have other meanings in statistics that are unrelated to the explanatory–response distinction, we prefer to avoid those words.

It is easiest to identify explanatory and response variables when we actually set values of one variable to see how it affects another variable.

EXAMPLE 4.1	Beer and Blood Alcohol

How does drinking beer affect the level of alcohol in our blood? The legal limit for driving in all states is 0.08%. Student volunteers at The Ohio State University drank different numbers of cans of beer. Thirty minutes later, a police officer measured their blood alcohol content. Number of beers consumed is the explanatory variable, and percent of alcohol in the blood is the response variable. ■

When we don't set the values of either variable but just observe both variables, there may or may not be explanatory and response variables. Whether there are depends on how we plan to use the data.

EXAMPLE 4.2	College Debts

A college student aid officer looks at the findings of the National Student Loan Survey. She notes data on the amount of debt of recent graduates, their current income, and how stressed they feel about college debt. She isn't interested in predictions but is simply trying to understand the situation of recent college graduates. The distinction between explanatory and response variables does not apply.

A sociologist looks at the same data with an eye to using amount of debt and income, along with other variables, to explain the stress caused by college debt. Now amount of debt and income are explanatory variables, and stress level is the response variable. ■

AFTER YOU PLOT YOUR DATA, THINK!

The statistician Abraham Wald (1902–1950) worked on war problems during World War II. Wald invented some statistical methods that were military secrets until the war ended. Here is one of his simpler ideas. Asked where extra armor should be added to airplanes, Wald studied the location of enemy bullet holes in planes returning from combat. He plotted the locations on an outline of the plane. As data accumulated, most of the outline filled up. Put the armor in the few spots with no bullet holes, said Wald. That's where bullets hit the planes that didn't make it back.

In many studies, the goal is to show that changes in one or more explanatory variables actually *cause* changes in a response variable. Other explanatory–response relationships do not involve direct causation. Nations with more television sets per person have greater life expectancies, but shipping many television sets to Botswana won't *cause* life expectancy to increase.

Most statistical studies examine data on more than one variable. Fortunately, statistical analysis of several-variable data builds on the tools we used to examine individual variables. The principles that guide our work also remain the same:

■ Plot your data. Look for overall patterns and deviations from those patterns.

■ Based on what your plot shows, choose numerical summaries for some aspects of the data.

Apply Your Knowledge

4.1 Explanatory and Response Variables? You have data on a large group of college students. Here are four pairs of variables measured on these students. For each pair, is it more reasonable to simply explore the relationship between the two variables or to view one of the variables as an explanatory variable and the other as a response variable? In the latter case, which is the explanatory variable, and which is the response variable?

(a) Amount of time spent studying for a statistics exam and grade on the exam

(b) Weight in kilograms and height in centimeters

(c) Hours per week spent online using Facebook and grade point average (GPA)

(d) Score on the SAT Writing exam and score on the SAT Critical Reading exam

4.2 Coral Reefs. How sensitive to changes in water temperature are coral reefs? To find out, scientists examined data on sea surface temperatures and coral growth per year at locations in the Red Sea.[1] What are the explanatory and response variables? Are they categorical or quantitative?

© George Holland/age fotostock

4.3 Beer and Blood Alcohol. Example 4.1 describes a study in which college students drank different amounts of beer. The response variable was their blood alcohol content (BAC). BAC for the same amount of beer might depend on other facts about the students. Name two other variables that could influence BAC.

Displaying relationships: scatterplots

The most useful graph for displaying the relationship between two quantitative variables is a *scatterplot*.

EXAMPLE 4.3   State SAT Mathematics Scores

Figure 1.8 (page 27) reminded us that in some states most high school graduates take the SAT test of readiness for college, and in other states most take the ACT. Who takes a test may influence the average score. Let's follow our four-step process (page 60) to examine this influence.[2]

MATHSAT

STATE: The percent of high school students who take the SAT varies from state to state. Does this fact help explain differences among the states in average SAT Mathematics score?

PLAN: Examine the relationship between percent taking the SAT and state mean score on the Mathematics part of the SAT. Choose the explanatory and response variables. Make a *scatterplot* to display the relationship between the variables. Interpret the plot to understand the relationship.

SOLVE (make the plot): We suspect that "percent taking" will help explain "mean score." So "percent taking" is the explanatory variable, and "mean score" is the response variable. We want to see how mean score changes when percent taking changes, so we put percent taking (the explanatory variable) on the horizontal axis.

Figure 4.1 is the scatterplot. Each point represents a single state. In Colorado, for example, 17% took the SAT, and their mean SAT Math score was 581. Find 17 on the *x* (horizontal) axis and 581 on the *y* (vertical) axis. Colorado appears as the point (17, 581) above 17 and to the right of 581.

CONCLUDE: We will explore conclusions in Example 4.4. ■

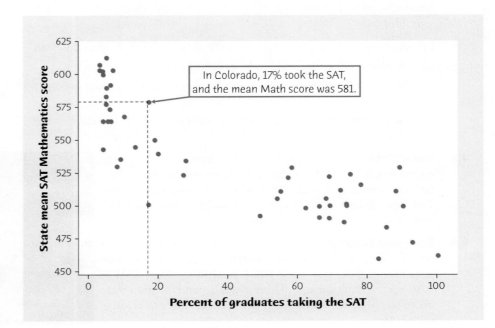

FIGURE 4.1
Scatterplot of the mean SAT Mathematics score in each state against the percent of that state's high school graduates who take the SAT, for Example 4.3. The dotted lines intersect at the point (17, 581), the data for Colorado.

Scatterplot

A **scatterplot** shows the relationship between two quantitative variables measured on the same individuals. The values of one variable appear on the horizontal axis, and the values of the other variable appear on the vertical axis. Each individual in the data appears as the point in the plot fixed by the values of both variables for that individual.

Always plot the explanatory variable, if there is one, on the horizontal axis (the *x* axis) of a scatterplot. As a reminder, we usually call the explanatory variable *x* and the response variable *y*. If there is no explanatory-response distinction, either variable can go on the horizontal axis.

Apply Your Knowledge

4.4   Do Heavier People Burn More Energy? Metabolic rate, the rate at which the body consumes energy, is important in studies of weight gain, dieting, and exercise. We have data on the lean body mass and resting metabolic rate for 12 women who are subjects in a study of dieting. Lean body mass, given in kilograms, is a person's weight leaving out all fat. Metabolic rate is measured in calories burned per 24 hours, the same calories used to describe the energy content of foods. **METAB**

Mass	36.1	54.6	48.5	42.0	50.6	42.0	40.3	33.1	42.4	34.5	51.1	41.2
Rate	995	1425	1396	1418	1502	1256	1189	913	1124	1052	1347	1204

The researchers believe that lean body mass is an important influence on metabolic rate. Make a scatterplot to examine this belief. (The *Two-Variable*

Statistical Calculator Applet provides an easy way to make scatterplots. Click "Data" to enter your data, then "Scatterplot" to see the plot.)

4.5   Outsourcing by Airlines. Airlines have increasingly outsourced the maintenance of their planes to other companies. A concern voiced by critics is that the maintenance may be less carefully done, so that outsourcing creates a safety hazard. In addition, flight delays are often due to maintenance problems, so one might look at government data on percent of major maintenance outsourced and percent of flight delays blamed on the airline to determine if these concerns are justified. This was done, and data from 2005 and 2006 appeared to justify the concerns of the critics. Do more recent data still support the concerns of the critics? Here are data from 2011:[3] AIRLINE

Airline	Outsource Percent	Delay Percent	Airline	Outsource Percent	Delay Percent
AirTran	70.7	12.75	Hawaiian	78.6	5.43
Alaska	53.3	9.61	JetBlue	59.7	23.87
American	24.4	10.73	Southwest	59.0	22.20
Continental	46.6	19.14	United	51.5	17.95
Delta	38.0	15.96	US Airways	57.8	14.92
Frontier	0.30	19.10			

Make a scatterplot that shows the relation between delays and outsourcing.

Interpreting scatterplots

To interpret a scatterplot, adapt the strategies of data analysis learned in Chapters 1 and 2 to the new two-variable setting.

Examining a Scatterplot

In any graph of data, look for the **overall pattern** and for striking **deviations** from that pattern.

You can describe the overall pattern of a scatterplot by the **direction, form,** and **strength** of the relationship.

An important kind of deviation is an **outlier,** an individual value that falls outside the overall pattern of the relationship.

EXAMPLE 4.4   Understanding State SAT Scores

We continue to explore the state SAT Mathematics scores by interpreting what the scatterplot tells us about the variation in scores from state to state.

SOLVE (interpret the plot): Figure 4.1 shows a clear *direction:* the overall pattern moves from upper left to lower right. That is, states in which higher percents of high school graduates take the SAT tend to have lower mean SAT Mathematics scores. We call this a *negative association* between the two variables.

The *form* of the relationship is roughly a straight line with a slight curve to the right as it moves down. What is more, most states fall into two distinct **clusters.** As in the histogram in Figure 1.8 (page 27), the ACT states cluster at the left and the SAT states at the right. In 23 states, fewer than 20% of seniors took the SAT; in another 24 states, more than 50% took the SAT.

The *strength* of a relationship in a scatterplot is determined by how closely the points follow a clear form. The overall relationship in Figure 4.1 is moderately strong:

clusters

states with similar percents taking the SAT tend to have roughly similar mean SAT Math scores.

CONCLUDE: Percent taking explains much of the variation among states in average SAT Mathematics score. States in which a higher percent of students take the SAT tend to have lower mean scores because the mean includes a broader group of students. SAT states as a group have lower mean SAT scores than ACT states. So average SAT score says almost nothing about the quality of education in a state. It is foolish to "rank" states by their average SAT scores. ■

Douglas Faulkner/Science Source

Positive Association, Negative Association

Two variables are **positively associated** when above-average values of one tend to accompany above-average values of the other, and below-average values also tend to occur together.

Two variables are **negatively associated** when above-average values of one tend to accompany below-average values of the other, and vice versa.

Of course, not all relationships have a clear direction that we can describe as positive association or negative association. Exercise 4.8 gives an example that does not have a single direction. Here is an example of a strong positive association with a simple and important form.

EXAMPLE 4.5   The Endangered Manatee

MANATEE

STATE: Manatees are large, gentle, slow-moving creatures found along the coast of Florida. Many manatees are injured or killed by boats. Table 4.1 contains data on the number of boats registered in Florida (in thousands) and the number of manatees killed by boats for the years between 1977 and 2012.[4] Examine the relationship. Is it plausible that restricting the number of boats would help protect manatees?

PLAN: Make a scatterplot with "boats registered" as the explanatory variable and "manatees killed" as the response variable. Describe the form, direction, and strength of the relationship.

TABLE 4.1 FLORIDA BOAT REGISTRATIONS (THOUSANDS) AND MANATEES KILLED BY BOATS

YEAR	BOATS	MANATEES	YEAR	BOATS	MANATEES	YEAR	BOATS	MANATEES
1977	447	13	1989	711	50	2001	944	81
1978	460	21	1990	719	47	2002	962	95
1979	481	24	1991	681	53	2003	978	73
1980	498	16	1992	679	38	2004	983	69
1981	513	24	1993	678	35	2005	1010	79
1982	512	20	1994	696	49	2006	1024	92
1983	526	15	1995	713	42	2007	1027	73
1984	559	34	1996	732	60	2008	1010	90
1985	585	33	1997	755	54	2009	982	97
1986	614	33	1998	809	66	2010	942	83
1987	645	39	1999	830	82	2011	922	88
1988	675	43	2000	880	78	2012	902	81

SOLVE: Figure 4.2 is the scatterplot. There is a positive association—more boats goes with more manatees killed. The form of the relationship is **linear.** That is, the overall pattern follows a straight line from lower left to upper right. The relationship is strong because the points don't deviate greatly from a line.

linear relationship

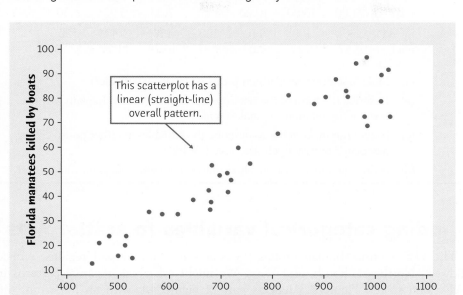

FIGURE 4.2
Scatterplot of the number of Florida manatees killed by boats in the years 1977 to 2012 against the number of boats registered in Florida that year, for Example 4.5. There is a strong linear (straight-line) pattern.

CONCLUDE: As more boats are registered, the number of manatees killed by boats goes up linearly. Data from the Florida Wildlife Commission indicate that in 2012 boats accounted for 21% of manatee deaths and 30% of deaths whose causes could be determined. Although many manatees die from other causes, it appears that fewer boats would mean fewer manatee deaths. ■

⚠ **CAUTION**

As the following chapter will emphasize, *it is wise to always ask what other variables lurking in the background might contribute to the relationship displayed in a scatterplot.* Because both boats registered and manatees killed are recorded year by year, any change in conditions over time might affect the relationship. For example, if boats in Florida have tended to go faster over the years, that might result in more manatees killed by the same number of boats.

Apply Your Knowledge

4.6   Do Heavier People Burn More Energy? Describe the direction, form, and strength of the relationship between lean body mass and metabolic rate, as displayed in your plot for Exercise 4.4. 📊 METAB

4.7   Outsourcing by Airlines. Does your plot for Exercise 4.5 show a positive, negative, or no association between maintenance outsourcing and delays caused by the airline? One airline is a low outlier in delay percent. Which airline is this? Aside from the outlier, does the plot show a roughly linear form? If it does, is the relationship very strong? 📊 AIRLINE

4.8   Does Fast Driving Waste Fuel? How does the fuel consumption of a car change as its speed increases? Here are data for a British Ford Escort. Speed is measured in kilometers per hour, and fuel consumption is measured in liters of gasoline used per 100 kilometers traveled.[5] 📊 **FASTDR**

Speed	10	20	30	40	50	60	70	80
Fuel	21.00	13.00	10.00	8.00	7.00	5.90	6.30	6.95

Speed	90	100	110	120	130	140	150
Fuel	7.57	8.27	9.03	9.87	10.79	11.77	12.83

(a) Make a scatterplot. (Which is the explanatory variable?)

(b) Describe the form of the relationship. It is not linear. Explain why the form of the relationship makes sense.

(c) It does not make sense to describe the variables as either positively associated or negatively associated. Why?

(d) Is the relationship reasonably strong or quite weak? Explain your answer.

Adding categorical variables to scatterplots

The U.S. Census Bureau groups the states into four broad regions, named Midwest, Northeast, South, and West. We might ask about regional patterns in SAT exam scores. Figure 4.3 repeats part of Figure 4.1, with an important difference. We have plotted only the Midwest and Northeast groups of states, using circles for the Midwest states and squares for the Northeast states.

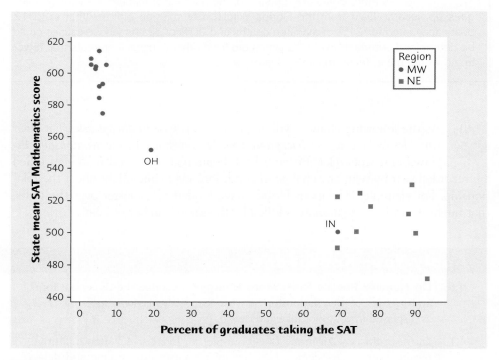

FIGURE 4.3
Mean SAT Mathematics score and percent of high school graduates who take the test for only the Midwest (blue) and Northeast (red) states.

The regional comparison is striking. The 9 Northeast states are all SAT states—at least 69% of high school graduates in each of these states take the

SAT. The 12 Midwest states are mostly ACT states. In 11 of these states, fewer than 8% of high school graduates take the SAT. One Midwest state is clearly an outlier within the region: Indiana is an SAT state (69% take the SAT) that falls close to the Northeast cluster. Ohio, where 19% take the SAT, also lies outside the Midwest cluster.

Dividing the states into regions introduces a third variable into the scatterplot. "Region" is a categorical variable that has four values, although we plotted data from only two of the four regions. The two regions are identified by the two different plotting symbols.

Categorical Variables in Scatterplots

To add a categorical variable to a scatterplot, use a different plot color or symbol for each category.

Apply Your Knowledge

4.9 **Do Heavier People Burn More Energy?** The study of dieting described in Exercise 4.4 collected data on the lean body mass (in kilograms) and metabolic rate (in calories) for both female and male subjects: METAB2

Sex	F	F	F	F	F	F	F	F	F	F
Mass	36.1	54.6	48.5	42.0	50.6	42.0	40.3	33.1	42.4	34.5
Rate	995	1425	1396	1418	1502	1256	1189	913	1124	1052

Sex	F	F	M	M	M	M	M	M	M
Mass	51.1	41.2	51.9	46.9	62.0	62.9	47.4	48.7	51.9
Rate	1347	1204	1867	1439	1792	1666	1322	1614	1460

(a) Make a scatterplot of metabolic rate versus lean body mass for all 19 subjects. Use separate symbols to distinguish women and men.

(b) Does the same overall pattern hold for both women and men? What is the most important difference between women and men?

Measuring linear association: correlation

A scatterplot displays the direction, form, and strength of the relationship between two quantitative variables. Linear (straight-line) relations are particularly important because a straight line is a simple pattern that is quite common. A linear relation is strong if the points lie close to a straight line and weak if they are widely scattered about a line. Our eyes are not good judges of how strong a linear relationship is. The two scatterplots in Figure 4.4 depict exactly the same data, but the lower plot is drawn smaller in a large field. The lower plot seems to show a stronger linear relationship. Our eyes can be fooled by changing the plotting scales or the amount of space around the cloud of points in a scatterplot.[6] We need to follow our strategy for data analysis by using a numerical measure to supplement the graph. *Correlation* is the measure we use.

FIGURE 4.4
Two scatterplots of the same data. The straight-line pattern in the lower plot appears stronger because of the surrounding space.

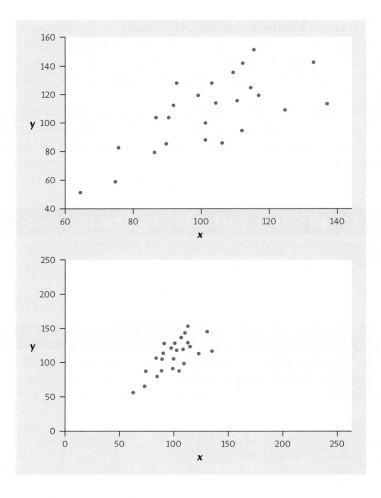

Correlation

The **correlation** measures the direction and strength of the linear relationship between two quantitative variables. Correlation is usually written as r.

Suppose that we have data on variables x and y for n individuals. The values for the first individual are x_1 and y_1, the values for the second individual are x_2 and y_2, and so on. The means and standard deviations of the two variables are $\bar{x}$ and s_x for the x-values, and $\bar{y}$ and s_y for the y-values. The correlation r between x and y is

$$r = \frac{1}{n-1}\left[\left(\frac{x_1 - \bar{x}}{s_x}\right)\left(\frac{y_1 - \bar{y}}{s_y}\right) + \left(\frac{x_2 - \bar{x}}{s_x}\right)\left(\frac{y_2 - \bar{y}}{s_y}\right) + \cdots + \left(\frac{x_n - \bar{x}}{s_x}\right)\left(\frac{y_n - \bar{y}}{s_y}\right)\right]$$

or, more compactly,

$$r = \frac{1}{n-1}\sum\left(\frac{x_i - \bar{x}}{s_x}\right)\left(\frac{y_i - \bar{y}}{s_y}\right)$$

DEATH FROM SUPERSTITION?

Is there a relationship between superstitious beliefs and bad things happening? Apparently there is. Chinese and Japanese people think that the number 4 is unlucky because when pronounced it sounds like the word for "death." Sociologists looked at 15 years' worth of death certificates for Chinese and Japanese Americans and for white Americans. Deaths from heart disease were notably higher on the fourth day of the month among Chinese and Japanese but not among whites. The sociologists think the explanation is increased stress on "unlucky days."

The formula for the correlation r is a bit complex. It helps us see what correlation is, but in practice you should use software or a calculator that finds r from keyed-in values of two variables, x and y. Exercise 4.10 asks you to calculate a correlation step-by-step from the definition to solidify its meaning.

The formula for r begins by standardizing the observations. Suppose, for example, that x is height in centimeters and y is weight in kilograms and that we have height and weight measurements for n people. Then $\bar{x}$ and s_x are the mean and standard deviation of the n heights, both in centimeters. The value

$$\frac{x_i - \bar{x}}{s_x}$$

is the standardized height of the ith person, from Chapter 3. The standardized height says how many standard deviations above or below the mean a person's height lies. Standardized values have no units—in this example, they are no longer measured in centimeters. Standardize the weights also. The correlation r is an average of the products of the standardized height and the standardized weight for all the individuals. Just as in the case of the standard deviation s, the "average" here divides by one fewer than the number of individuals.

Apply Your Knowledge

4.10 Coral Reefs. Exercise 4.2 discusses a study in which scientists examined data on mean sea surface temperatures (in degrees Celsius) and mean coral growth (in millimeters per year) over a several-year period at locations in the Red Sea. Here are the data:[7] CORAL

Sea surface temperature	29.68	29.87	30.16	30.22	30.48	30.65	30.90
Growth	2.63	2.58	2.68	2.60	2.48	2.38	2.26

(a) Make a scatterplot. Which is the explanatory variable? The plot shows a negative linear pattern.

(b) Find the correlation r step-by-step. You may wish to round off to two decimal places in each step. First find the mean and standard deviation of each variable. Then find the seven standardized values for each variable. Finally, use the formula for r. Explain how your value for r matches your graph in (a).

(c) Enter these data into your calculator or software, and use the correlation function to find r. Check that you get the same result as in (b), up to roundoff error.

Facts about correlation

The formula for correlation helps us see that r is positive when there is a positive association between the variables. Height and weight, for example, have a positive association. People who are above average in height tend to also be above average in weight. Both the standardized height and the standardized weight are positive. People who are below average in height tend to also have below-average weight. Then both standardized height and standardized weight are negative. In both cases, the products in the formula for r are mostly positive, and so r is positive. In the same way, we can see that r is negative when the association between x and y is negative. More detailed study of the formula gives more detailed properties of r. Here is what you need to know to interpret correlation.

1. *Correlation makes no distinction between explanatory and response variables.* It makes no difference which variable you call x and which you call y in calculating the correlation.

2. Because *r* uses the standardized values of the observations, *r does not change when we change the units of measurement of x, y, or both.* Measuring height in inches rather than centimeters and weight in pounds rather than kilograms does not change the correlation between height and weight. The correlation *r* itself has no unit of measurement; it is just a number.

3. *Positive r indicates positive association between the variables, and negative r indicates negative association.*

4. *The correlation r is always a number between* −1 *and* 1. Values of *r* near 0 indicate a very weak linear relationship. The strength of the linear relationship increases as *r* moves away from 0 toward either −1 or 1. Values of *r* close to −1 or 1 indicate that the points in a scatterplot lie close to a straight line. The extreme values *r* = −1 and *r* = 1 occur only in the case of a perfect linear relationship, when the points lie exactly along a straight line.

EXAMPLE 4.6	From Scatterplot to Correlation

The scatterplots in Figure 4.5 illustrate how values of *r* closer to 1 or −1 correspond to stronger linear relationships. To make the meaning of *r* clearer, the standard deviations of both variables in these plots are equal, and the horizontal and vertical scales are the same. In general, it is not so easy to guess the value of *r* from the appearance of a scatterplot. Remember that changing the plotting scales in a scatterplot may mislead our eyes, but it does not change the correlation.

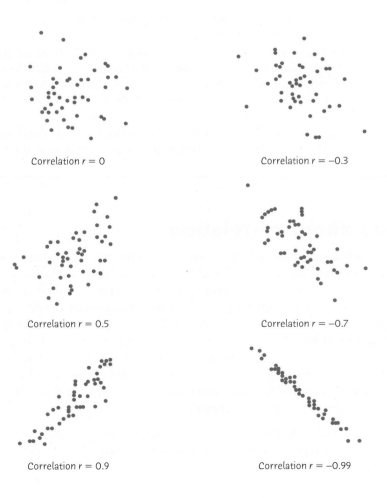

FIGURE 4.5

How correlation measures the strength of a linear relationship, for Example 4.6. Patterns closer to a straight line have correlations closer to 1 or −1.

The scatterplots in Figure 4.6 show four sets of real data. The patterns are less regular than those in Figure 4.5, but they also illustrate how correlation measures the strength of linear relationships.[8]

(a) This repeats the manatee plot in Figure 4.2. There is a strong positive linear relationship, $r = 0.953$.

(b) Here are the number of named tropical storms each year between 1984 and 2012 plotted against the number predicted before the start of hurricane season by William Gray of Colorado State University. There is a moderate linear relationship, $r = 0.637$.

STORMS

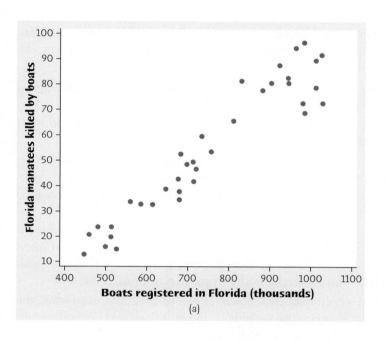

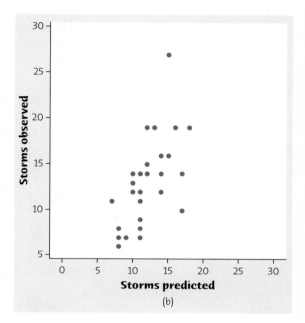

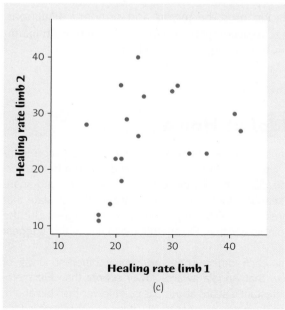

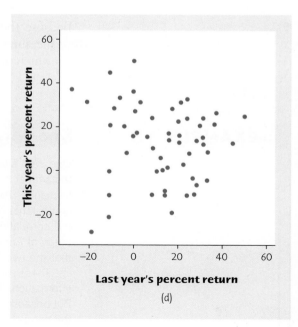

FIGURE 4.6

How correlation measures the strength of a linear relationship, for Example 4.6. Four sets of real data with (a) $r = 0.953$, (b) $r = 0.637$, (c) $r = 0.358$, and (d) $r = -0.081$.

(c) These data come from an experiment that studied how quickly cuts in the limbs of newts heal. Each point represents the healing rate in micrometers (millionths of a meter) per hour for the two front limbs of the same newt. This relationship is weaker than those in (a) and (b), with $r = 0.358$.

(d) Does last year's stock market performance help predict how stocks will do this year? No. The correlation between last year's percent return and this year's percent return over 56 years is only $r = -0.081$. The scatterplot shows a cloud of points with no visible linear pattern. ■

Describing the relationship between two variables is a more complex task than describing the distribution of one variable. Here are some more facts about correlation, cautions to keep in mind when you use r:

1. ⚠ *Correlation requires that both variables be quantitative, so that it makes sense to do the arithmetic indicated by the formula for r.* We cannot calculate a correlation between the incomes of a group of people and what city they live in because city is a categorical variable.

2. ⚠ Correlation measures the strength of only the *linear* relationship between two variables. *Correlation does not describe curved relationships between variables, no matter how strong they are.* Exercise 4.13 (page 111) illustrates this important fact.

3. ⚠ *Like the mean and standard deviation, the correlation is not resistant: r is strongly affected by a few outlying observations.* Use r with caution when outliers appear in the scatterplot. Reporting the correlation both with the outliers included and with the outliers removed is informative.

4. ⚠ *Correlation is not a complete summary of two-variable data,* even when the relationship between the variables is linear. You should give the means and standard deviations of both x and y along with the correlation.

Because the formula for correlation uses the means and standard deviations, these measures are the proper choice to accompany a correlation. Here is an example in which understanding requires both means and correlation.

EXAMPLE 4.7   **Scoring *American Idol* at Home**

One Web site recommends that fans of the television show *American Idol* score contestants at home on a scale from 1 to 10, with higher scores indicating a better performance. Two friends, Angela and Elizabeth, decide to follow this advice and score contestants over the course of a season. How well do they agree? We calculate that the correlation between their scores is $r = 0.9$, suggesting that they agree. But the mean of Angela's scores is 0.8 point lower than Elizabeth's mean. Does this suggest that the two friends disagree?

These facts do not contradict each other. They are simply different kinds of information. The mean scores show that Angela awards lower scores than Elizabeth. But because Angela gives *every* contestant a score about 0.8 point lower than Elizabeth, the correlation remains high. Adding the same number to all values of either x or y does not change the correlation. Angela and Elizabeth actually score consistently because they agree on which performances are better. The high r shows their agreement. ■

Of course, even giving means, standard deviations, and the correlation for state SAT scores and percent taking will not point out the clusters in Figure 4.1 (page 100). Numerical summaries complement plots of data, but they don't replace them.

Apply Your Knowledge

4.11 Changing the Units. The healing rates plotted in Figure 4.6(c) (page 109) are measured in micrometers (millionths of a meter) per hour. The correlation between healing rates for the two front limbs of newts is $r = 0.358$. If the measurements were made in inches per day, would the correlation change? Explain your answer.

4.12 Changing the Correlation. Use your calculator, software, or the *Two-Variable Statistical Calculator* applet to demonstrate how outliers can affect correlation.

   (a) What is the correlation between lean body mass and metabolic rate for the 12 women in Exercise 4.4? ▮▮ᵢₗᵢ METAB

   (b) Make a scatterplot of the data with two new points added. Point A: mass 65 kilograms, metabolic rate 1761 calories. Point B: mass 35 kilograms, metabolic rate 1400 calories. Find two new correlations: one for the original data plus Point A, and another for the original data plus Point B. ▮▮ᵢₗᵢ METAB3

   (c) By looking at your plot, explain why adding Point A makes the correlation stronger (closer to 1) and adding Point B makes the correlation weaker (closer to 0).

4.13 Strong Association But No Correlation. The gas mileage of an automobile first increases and then decreases as the speed increases. Suppose this relationship is very regular, as shown by the following data on speed (miles per hour) and mileage (miles per gallon): ▮▮ᵢₗᵢ MPG

Speed	30	40	50	60	70
Mileage	24	28	30	28	24

Make a scatterplot of mileage versus speed. Show that the correlation between speed and mileage is $r = 0$. Explain why the correlation is 0 even though there is a strong relationship between speed and mileage.

CHAPTER 4 SUMMARY

Chapter Specifics

■ To study relationships between variables, we must measure the variables on the same group of individuals.

■ If we think that a variable x may explain or even cause changes in another variable y, we call x an **explanatory variable** and y a **response variable.**

■ A **scatterplot** displays the relationship between two quantitative variables measured on the same individuals. Mark values of one variable on the horizontal axis (x axis) and values of the other variable on the vertical axis (y axis). Plot each individual's data as a point on the graph. Always plot the explanatory variable, if there is one, on the x axis of a scatterplot.

■ Plot points with different colors or symbols to see the effect of a categorical variable in a scatterplot.

- In examining a scatterplot, look for an overall pattern showing the **direction, form,** and **strength** of the relationship and then for **outliers** or other deviations from this pattern.

- **Direction:** If the relationship has a clear direction, we speak of either **positive association** (high values of the two variables tend to occur together) or **negative association** (high values of one variable tend to occur with low values of the other variable).

- **Form: Linear relationships,** where the points show a straight-line pattern, are an important form of relationship between two variables. Curved relationships and **clusters** are other forms to watch for.

- **Strength:** The **strength** of a relationship is determined by how close the points in the scatterplot lie to a simple form such as a line.

- The **correlation** r measures the direction and strength of the linear association between two quantitative variables x and y. Although you can calculate a correlation for any scatterplot, r measures only straight-line relationships.

- Correlation indicates the direction of a linear relationship by its sign: $r > 0$ for a positive association and $r < 0$ for a negative association. Correlation always satisfies $-1 \leq r \leq 1$ and indicates the strength of a relationship by how close it is to -1 or 1. Perfect correlation, $r = \pm 1$, occurs only when the points on a scatterplot lie exactly on a straight line.

- Correlation ignores the distinction between explanatory and response variables. The value of r is not affected by changes in the unit of measurement of either variable. Correlation is not resistant, so outliers can greatly change the value of r.

Link It

In Chapters 1 to 3, we focused on exploring features of a single variable. In this chapter we continued our study of exploratory data analysis but for the purpose of examining relationships *between* variables. A useful tool for exploring the relationship between two variables is the scatterplot. When the relationship is linear, correlation is a numerical measure of the strength of the linear relationship.

It is tempting to assume that the patterns we observe in our data hold for values of our variables that we have not observed—in other words, that additional data would continue to conform to these patterns. The process of identifying underlying patterns would seem to assume that this is the case. But is this assumption justified? Parts II to V of the book answer this question.

CHECK YOUR SKILLS

4.14 You have data for many years on the average price of a barrel of oil and the average retail price of a gallon of unleaded regular gasoline. When you make a scatterplot, the explanatory variable on the *x* axis

(a) is the price of oil.
(b) is the price of gasoline.
(c) can be either oil price or gasoline price.

4.15 In a scatterplot of the average price of a barrel of oil and the average retail price of a gallon of gasoline, you expect to see

(a) a positive association.
(b) very little association.
(c) a negative association.

4.16 Figure 4.7 is a scatterplot of school GPA against IQ test scores for 15 seventh-grade students. There is one low outlier in the plot. The IQ and GPA scores for this student are

(a) IQ = 0.5, GPA = 103.
(b) IQ = 103, GPA = 0.5.
(c) IQ = 103, GPA = 7.6.

(a) a strong straight-line pattern.
(b) a cloud of points with no visible pattern.
(c) no straight-line pattern, but there might be a strong pattern of another form.

4.20 The points on a scatterplot lie very close to a straight line. The correlation between *x* and *y* is close to

(a) −1.   (b) 1.   (c) either −1 or 1, we can't say which.

4.21 If men always married women who were two years younger than themselves, the correlation between the ages of husband and wife would be

(a) 1.   (b) −1.   (c) Can't tell without seeing the data.

4.22 For a biology project, you measure the weight in grams and the tail length in millimeters of a group of mice. The correlation is $r = 0.7$. If you had measured tail length in centimeters instead of millimeters, what would be the correlation? (There are 10 millimeters in a centimeter.)

(a) 0.7/10 = 0.07   (b) 0.7   (c) (0.7)(10) = 7

4.23 Because elderly people may have difficulty standing to have their heights measured, a study looked at predicting

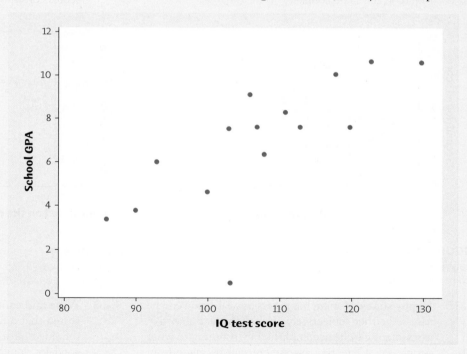

FIGURE 4.7
Scatterplot of school GPA against IQ test scores for seventh-grade students, for Exercises 4.16 and 4.17.

4.17 If we leave out the low outlier, the correlation for the remaining 14 points in Figure 4.7 is closest to

(a) 0.9.   (b) −0.9   (c) 0.1.

4.18 What are all the values that a correlation *r* can possibly take?

(a) $r \geq 0$ (b) $0 \leq r \leq 1$ (c) $-1 \leq r \leq 1$

4.19 If the correlation between two variables is close to 0, you can conclude that a scatterplot would show

overall height from height to the knee. Here are data (in centimeters) for six elderly men: KNEEHT

Knee height x	57.7	47.4	43.5	44.8	55.2	54.6
Height y	192.1	153.3	146.4	162.7	169.1	177.8

Use your calculator or software: The correlation between knee height and overall height is about

(a) $r = 0.08$.   (b) $r = 0.89$.   (c) $r = 0.74$.

CHAPTER 4 EXERCISES

4.24 **Scores at the Masters.** The Masters is one of the four major golf tournaments. Figure 4.8 is a scatterplot of the scores for the first two rounds of the 2012 Masters for all the golfers entered. Only the 60 golfers with the lowest two-round total advance to the final two rounds (unless several people are tied for 60th place, in which case all those tied for 60th place advance). The plot has a grid pattern because golf scores must be whole numbers.[9] 🖥📊 MASTR12

(a) Read the graph: What was the lowest score in the first round of play? How many golfers had this low score? What were their scores in the second round?
(b) Read the graph: Ben Crenshaw had the highest score in the second round. What was this score? What was Crenshaw's score in the first round?
(c) Is the correlation between first-round scores and second-round scores closest to $r = 0.3$, $r = 0.7$, or $r = 0.9$? Explain your choice. Does the graph suggest that knowing a professional golfer's score for one round is much help in predicting his score for another round on the same course?

system of health surveys. Lower scores indicate a greater degree of happiness. To objectively assess happiness, the investigators computed a mean well-being score (called the compensating-differentials score) for each state, based on objective measures that have been found to be related to happiness or well-being. The states were then ranked according to this score (Rank 1 being the happiest). Figure 4.9 is a scatterplot of mean BRFSS scores (response) against the rank based on the compensating-differentials (explanatory).[10] 🖥📊 HAPPY

(a) Is there an overall positive association or an overall negative association between mean BRFSS score and rank based on the compensating-differentials method?
(b) Does the overall association indicate agreement or disagreement between the mean subjective BRFSS score and the ranking based on objective data used in the compensating-differentials method?
(c) Are there any outliers? If so, what are the BRFSS scores corresponding to these outliers?

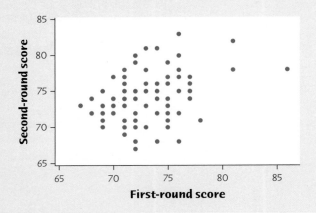

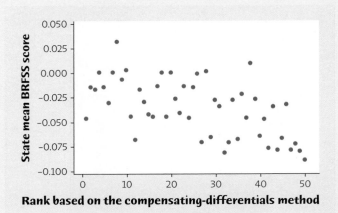

FIGURE 4.8
Scatterplot of the scores in the first two rounds of the 2012 Masters Tournament, for Exercise 4.24.

FIGURE 4.9
Scatterplot of mean BRFSS score in each state against each state's well-being rank, for Exercise 4.25.

4.25 **Happy states.** Human happiness or well-being can be assessed either subjectively or objectively. Subjective assessment can be accomplished by listening to what people say. Objective assessment can be made from data related to well-being such as income, climate, availability of entertainment, housing prices, lack of traffic congestion, and so on. Do subjective and objective assessments agree? To study this, investigators made both subjective and objective assessments of happiness for each of the 50 states. The subjective measurement was the mean score on a life-satisfaction question found on the Behavioral Risk Factor Surveillance System (BRFSS), which is a state-based

4.26 **Wine and cancer in women.** Some studies have suggested that a nightly glass of wine may not only take the edge off a day but also improve health. Is wine good for your health? A study of nearly 1.3 million middle-aged British women examined wine consumption and the risk of breast cancer. The researchers were interested in how risk changed as wine consumption increased. Risk is based on breast cancer rates in drinkers relative to breast cancer rates in nondrinkers in the study, with higher values indicating greater risk. In particular, a value greater than 1 indicates a greater breast cancer rate than that of nondrinkers. Wine intake is the mean wine intake, in grams per day, of all women in the study who drank approximately the same amount of wine per week. Here are the data (for drinkers only):[11] 📊 CANCER

Wine intake x (grams per day)	2.5	8.5	15.5	26.5
Relative risk y	1.00	1.08	1.15	1.22

(a) Make a scatterplot of these data. Based on the scatterplot, do you expect the correlation to be positive or negative? Near ± 1 or not?

(b) Find the correlation r between wine intake and relative risk. Do the data show that women who consume more wine tend to have higher relative risks of breast cancer?

4.27 Ebola and gorillas. The deadly Ebola virus is a threat to both people and gorillas in Central Africa. An outbreak in 2002 and 2003 killed 91 of the 95 gorillas in seven home ranges in the Congo. To study the spread of the virus, measure "distance" by the number of home ranges separating a group of gorillas from the first group infected. Here are data on distance and time in number of days until deaths began in each later group:[12]

📊 EBOLA

Distance	1	3	4	4	4	5
Time	4	21	33	41	43	46

(a) Make a scatterplot. Which is the explanatory variable? What kind of pattern does your plot show?

(b) Find the correlation r between distance and time.

(c) If time in days were replaced by time in number of weeks until death began in each later group (fractions allowed so that 4 days becomes 4/7 weeks), would the correlation between distance and time change? Explain your answer.

4.28 Sparrowhawk colonies. One of nature's patterns connects the percent of adult birds in a colony that return from the previous year and the number of new adults that join the colony. Here are data for 13 colonies of sparrowhawks:[13]

📊 SPARROW

age fotostock/Superstock

Percent return	74	66	81	52	73	62	52	45	62	46	60	46	38
New adults	5	6	8	11	12	15	16	17	18	18	19	20	20

(a) Plot the count of new adults (response) against the percent of returning birds (explanatory).

Describe the direction and form of the relationship. Is the correlation r an appropriate measure of the strength of this relationship? If so, find r.

(b) For short-lived birds, the association between these variables is positive: changes in weather and food supply drive the populations of new and returning birds up or down together. For long-lived territorial birds, on the other hand, the association is negative because returning birds claim their territories in the colony and don't leave room for new recruits. Which type of species is the sparrowhawk?

4.29 Our brains don't like losses. Most people dislike losses more than they like gains. In money terms, people are about as sensitive to a loss of $10 as to a gain of $20. To discover what parts of the brain are active in decisions about gain and loss, psychologists presented subjects with a series of gambles with different odds and different amounts of winnings and losses. From a subject's choices, they constructed a measure of "behavioral loss aversion." Higher scores show greater sensitivity to losses. Observing brain activity while subjects made their decisions pointed to specific brain regions. Here are data for 16 subjects on behavioral loss aversion and "neural loss aversion," a measure of activity in one region of the brain:[14] 📊 LOSSES

Neural	−50.0	−39.1	−25.9	−26.7	−28.6	−19.8	−17.6	5.5
Behavioral	0.08	0.81	0.01	0.12	0.68	0.11	0.36	0.34
Neural	2.6	20.7	12.1	15.5	28.8	41.7	55.3	155.2
Behavioral	0.53	0.68	0.99	1.04	0.66	0.86	1.29	1.94

(a) Make a scatterplot that shows how behavior responds to brain activity.

(b) Describe the overall pattern of the data. There is one clear outlier. What is the behavioral score associated with this outlier?

(c) Find the correlation r between neural and behavioral loss aversion both with and without the outlier. Does the outlier have a strong influence on the value of r? By looking at your plot, explain why adding the outlier to the other data points causes r to increase.

4.30 Sulfur, the ocean, and the sun. Sulfur in the atmosphere affects climate by influencing formation of clouds. The main natural source of sulfur is dimethyl sulfide (DMS) produced by small organisms in the upper layers of the oceans. DMS production is in turn influenced by the amount of energy the upper ocean receives from sunlight. Here are monthly data on solar radiation dose (SRD, in watts per square meter) and surface DMS concentration (in nanomolars) for a region in the Mediterranean:[15] 📊 SULFUR

SRD	12.55	12.91	14.34	19.72	21.52	22.41	37.65	48.41
DMS	0.796	0.692	1.744	1.062	0.682	1.517	0.736	0.720
SRD	74.41	94.14	109.38	157.79	262.67	268.96	289.23	
DMS	1.820	1.099	2.692	5.134	8.038	7.280	8.872	

(a) Make a scatterplot that shows how DMS responds to SRD.

(b) Describe the overall pattern of the data. Find the correlation r between DMS and SRD. Because SRD changes with the seasons of the year, the close relationship between SRD and DMS helps explain other seasonal patterns.

4.31 Alcohol and cancer in women. Exercise 4.26 discusses a study of the relationship between wine consumption and the risk of breast cancer in women. The researchers were also interested in how risk changed as consumption of alcoholic beverages other than wine increased. Intake of alcoholic beverages other than wine is the mean intake, in grams per day, of all women in the study who drank approximately the same amount of alcohol other than wine per week. Here are the data for both women who drank wine and women who drank alcoholic beverages other than wine: 📊📶 CANCER

Wine intake (grams per day)	2.5	8.5	15.5	26.5
Relative risk	1.00	1.08	1.15	1.22
Alcohol intake (grams per day)	2.0	7.0	13.0	24.0
Relative risk	0.96	1.06	1.11	1.20

(a) Make a scatterplot of the relative risk versus intake, using separate symbols for the two types of drinks.

(b) What does your plot show about the pattern of risk? What does it show about the effect of type of drink on risk?

4.32 Feed the birds. Canaries provide more food to their babies when the babies beg more intensely. Researchers wondered if begging was the main factor determining how much food baby canaries receive, or if parents also take into account whether the babies are theirs or not. To investigate, researchers conducted an experiment allowing canary parents to raise two broods: one of their own and one fostered from a different pair of parents. If begging determines how much food babies receive, then differences in the "begging intensities" of the broods should be strongly associated with differences in the amount of food the broods receive. The researchers decided to use the relative growth rates (the growth rate of the foster babies relative to that of the natural babies, with values greater than 1 indicating that the foster babies grew more rapidly than the natural babies) as a measure of the difference in the amount of food received. They recorded the difference in begging intensities (the begging intensity of the foster babies minus that of the natural babies) and relative growth rates. Here are data from the experiment:[16] 📊📶 CANARY

Difference in begging intensity	-14.0	-12.5	-12.0	-8.0	-8.0	-6.5	-5.5
Relative growth rate	0.85	1.00	1.33	0.85	0.90	1.15	1.00
Difference in begging intensity	-3.5	-3.0	-2.0	-1.5	-1.5	0.0	0.0
Relative growth rate	1.30	1.33	1.03	0.95	1.15	1.13	1.00
Difference in begging intensity	2.00	2.00	3.00	4.50	7.00	8.00	8.50
Relative growth rate	1.07	1.14	1.00	0.83	1.15	0.93	0.70

(a) Make a scatterplot that shows how relative growth rate responds to the difference in begging intensity.

(b) Describe the overall pattern of the relationship. Is it linear? Is there a positive or negative association, or neither? Find the correlation r. Is r a helpful description of this relationship?

(c) If begging intensity is the main factor determining food received, with higher intensity leading to more food, one would expect the relative growth rate to increase as the difference in begging intensity increases. However, if both begging intensity and a preference for their own babies determine the amount of food received (and hence the relative growth rate), we might expect growth rate to increase initially as begging intensity increases but then to level off (or even decrease) as the parents begin to ignore increases in begging by the foster babies. Which of these theories do the data appear to support? Explain your answer.

4.33 Good weather and tipping. Favorable weather has been shown to be associated with increased tipping. Will just the belief that future weather will be favorable lead  to higher tips? Researchers gave 60 index cards to a waitress at an Italian restaurant in New Jersey. Before delivering the bill to each customer, the waitress randomly selected a card and wrote on the bill the same message that was printed on the index card. Twenty of the cards had the message "The weather is supposed to be really good tomorrow. I hope you enjoy the day!" Another 20 cards contained the message "The weather is supposed to be not so good tomorrow. I hope you enjoy the day anyway!" The remaining 20 cards were blank, indicating that the waitress was not supposed to write any message. Choosing a card at random ensured that there was a random assignment of the diners to the three experimental conditions. Here are the percentage tips for the three messages:[17] 📊 TIPPING

WEATHER REPORT	PERCENTAGE TIP									
Good	20.8	18.7	19.9	20.6	22.0	23.4	22.8	24.9	22.2	20.3
	24.9	22.3	27.0	20.4	22.2	24.0	21.2	22.1	22.0	22.7
Bad	18.0	19.0	19.2	18.8	18.4	19.0	18.5	16.1	16.8	14.0
	17.0	13.6	17.5	19.9	20.2	18.8	18.0	23.2	18.2	19.4
None	19.9	16.0	15.0	20.1	19.3	19.2	18.0	19.2	21.2	18.8
	18.5	19.3	19.3	19.4	10.8	19.1	19.7	19.8	21.3	20.6

(a) Make a plot of percentage tip against the weather report on the bill (space the three weather reports equally on the horizontal axis). Which weather report appears to lead to the best tip?

(b) Does it make sense to speak of a positive or negative association between weather report and percentage tip? Why? Is correlation r a helpful description of the relationship? Why?

4.34 Thinking about correlation. Exercise 4.26 presents data on wine intake and the relative risk of breast cancer in women.

(a) If wine intake is measured in ounces per day rather than grams per day, how would the correlation change? (There are 0.035 ounces in a gram.)

(b) How would r change if all the relative risks were 0.25 less than the values given in the table? Does the correlation tell us that among women who drink, those who drink more wine tend to have a greater relative risk of cancer than women who don't drink at all?

(c) If drinking an additional gram of wine each day raised the relative risk of breast cancer by exactly 0.01, what would be the correlation between wine intake and relative risk of breast cancer? (*Hint:* Draw a scatterplot for several values of wine intake.)

4.35 The effect of changing units. Changing the units of measurement can dramatically alter the appearance of a scatterplot. Return to the data on knee height and overall height in Exercise 4.23: 📊 KNEEHT2

Knee height x	57.7	47.4	43.5	44.8	55.2	54.6
Height y	192.1	153.3	146.4	162.7	169.1	177.8

Both heights are measured in centimeters. A mad scientist decides to measure knee height in millimeters and height in meters. The same data in these units are

Knee height x	577	474	435	448	552	546
Height y	1.921	1.533	1.464	1.627	1.691	1.778

(a) Make a plot with the x axis extending from 0 to 600 and the y axis from 0 to 250. Plot the original data on these axes. Then plot the new data using a different color or symbol. The two plots look very different.

(b) Nonetheless, the correlation is exactly the same for the two sets of measurements. Why do you know that this is true without doing any calculations? Find the two correlations to verify that they are the same.

4.36 Statistics for investing. Investment reports now often include correlations. Following a table of correlations among mutual funds, a report adds: "Two funds can have perfect correlation, yet different levels of risk. For example, Fund A and Fund B may be perfectly correlated, yet Fund A moves 20% whenever Fund B moves 10%." Write a brief explanation, for someone who knows no statistics, of how this can happen. Include a sketch to illustrate your explanation.

4.37 Statistics for investing. A mutual funds company's newsletter says, "A well diversified portfolio includes assets with low correlations." The newsletter includes a table of correlations between the returns on various classes of investments. For example, the correlation between municipal bonds and large-cap stocks is 0.50, and the correlation between municipal bonds and small-cap stocks is 0.21.

(a) Rachel invests heavily in municipal bonds. She wants to diversify by adding an investment whose returns do not closely follow the returns on her bonds. Should she choose large-cap stocks or small-cap stocks for this purpose? Explain your answer.

(b) If Rachel wants an investment that tends to increase when the return on her bonds drops, what kind of correlation should she look for?

4.38 Teaching and research. A college newspaper interviews a psychologist about student ratings of the teaching of faculty members. The psychologist says, "The evidence indicates that the correlation between the research productivity and teaching rating of faculty members is close to zero." The paper reports this as "Professor McDaniel said that good researchers tend to be poor teachers, and vice versa." Explain why the paper's report is wrong. Write a statement in plain language (don't use the word *correlation*) to explain the psychologist's meaning.

4.39 **Sloppy writing about correlation.** Each of the following statements contains a blunder. Explain in each case what is wrong.

(a) "There is a high correlation between the sex of American workers and their income."

(b) "We found a high correlation ($r = 1.09$) between students' ratings of faculty teaching and ratings made by other faculty members."

(c) "The correlation between height and weight of the subjects was $r = 0.63$ centimeter."

4.40 **Correlation is not resistant.** Go to the *Correlation and Regression* applet. Click on the scatterplot to create a group of 10 points in the lower-left corner of the scatterplot with a strong straight-line pattern (correlation about 0.9).

(a) Add one point at the upper right that is in line with the first 10. How does the correlation change?

(b) Drag this last point down until it is opposite the group of 10 points. How small can you make the correlation? Can you make the correlation negative? You see that a single outlier can greatly strengthen or weaken a correlation. Always plot your data to check for outlying points.

4.41 **Match the correlation.** You are going to use the *Correlation and Regression* applet to make scatterplots with 10 points that have correlation close to 0.7. The lesson is that many patterns can have the same correlation. Always plot your data before you trust a correlation.

(a) Click on the scatterplot to add the first 2 points. What is the value of the correlation? Why does it have this value?

(b) Make a lower-left to upper-right pattern of 10 points with correlation about $r = 0.7$. (You can drag points up or down to adjust r after you have 10 points.) Make a rough sketch of your scatterplot.

(c) Make another scatterplot with 9 points in a vertical stack at the left of the plot. Add 1 point far to the right, and move it until the correlation is close to 0.7. Make a rough sketch of your scatterplot.

(d) Make yet another scatterplot with 10 points in a curved pattern that starts at the lower left, rises to the right, then falls again at the far right. Adjust the points up or down until you have a quite smooth curve with correlation close to 0.7. Make a rough sketch of this scatterplot also.

*The following exercises ask you to answer questions from data without having the details outlined for you. The exercise statements give you the **State** step*

of the four-step process. In your work, follow the **Plan, Solve,** *and* **Conclude** *steps of the process, described on page 60.*

4.42 **Brighter sunlight?** The brightness of sunlight at the earth's surface changes over time, depending on whether the earth's atmosphere is more or less clear. Sunlight dimmed between 1960 and 1990. After 1990, air pollution dropped in industrial countries. Did sunlight brighten? Here are annual averages computed by researchers from data from Ny Alesund, Spitsbergen, Norway, averaging over only clear days each year. (Other locations show similar trends.) The response variable is solar radiation in watts per square meter.[18] LIGHT

Year	1993	1994	1995	1996	1997	1998	1999	2000
Sun	116.0	120.0	123.0	123.5	125.5	125.0	129.0	128.0

4.43 **Will women outrun men?** Does the physiology of women make them better suited than men to long-distance running? Will women eventually outperform men in long-distance races? Researchers examined data on world record times (in seconds) for men and women in the marathon. Here are data for women:[19] RUNNING

Year	1926	1964	1967	1970	1971	1974	1975
Time	13,222.0	11,973.0	11,246.0	10,973.0	9990.0	9834.5	9499.0

Year	1977	1980	1981	1982	1983	1985	
Time	9287.5	9027.0	8806.0	8771.0	8563.0	8466.0	

Here are data for men:

Year	1908	1909	1913	1920	1925	1935	1947
Time	10,518.4	9751.0	9366.6	9155.8	8941.8	8802.0	8739.0

Year	1952	1953	1954	1958	1960	1963	1964
Time	8442.2	8314.8	8259.4	8117.0	8116.2	8068.0	7931.2

Year	1965	1967	1969	1981	1984	1985	1988
Time	7920.0	7776.4	7713.6	7698.0	7685.0	7632.0	7610.0

(a) What do the data show about women's and men's times in the marathon? (Start by plotting both sets of data on the same plot, using two different plotting symbols.)

(b) Based on these data, researchers (in 1992) predicted that women would outrun men in the marathon in 1998. How do you think they arrived at this date? Was their prediction accurate? (You may want to look on the Web; try doing a Google search on "women's world record marathon times.")

4.44 **Toucan's beak.** The toco toucan, the largest member of the toucan family, possesses the largest beak relative to body size of all birds. This exaggerated feature has received various interpretations, such as being a refined adaptation for feeding. However, the large surface area may also be an important mechanism for radiating heat (and hence cooling the bird) as

© Kevin Schafer/Alamy

outdoor temperature increases. Here are data for beak heat loss, as a percent of total body heat loss, at various temperatures in degrees Celsius:[20] TOUCAN

Temperature (°C)	15	16	17	18	19	20	21	22
Percent heat loss from beak	32	34	35	33	37	46	55	51
Temperature (°C)	23	24	25	26	27	28	29	30
Percent heat loss from beak	43	52	45	53	58	60	62	62

Investigate the relationship between outdoor temperature and beak heat loss, as a percentage of total body heat loss.

4.45 **Does social rejection hurt?** We often describe our emotional reaction to social rejection as "pain." Does social rejection cause activity in areas of the brain that are known to be activated by physical pain? If it does, we really do experience social and physical pain in similar ways. Psychologists first included and then deliberately excluded individuals from a social activity while they measured changes in brain activity. After each activity, the subjects filled out questionnaires that assessed how excluded they felt. Here are data for 13 subjects:[21] REJECT

SUBJECT	SOCIAL DISTRESS	BRAIN ACTIVITY
1	1.26	−0.055
2	1.85	−0.040
3	1.10	−0.026
4	2.50	−0.017
5	2.17	−0.017
6	2.67	0.017
7	2.01	0.021
8	2.18	0.025
9	2.58	0.027
10	2.75	0.033
11	2.75	0.064
12	3.33	0.077
13	3.65	0.124

The explanatory variable is "social distress" measured by each subject's questionnaire score after exclusion relative to the score after inclusion. (So values greater than 1 show the degree of distress caused by exclusion.) The response variable is change in activity in a region of the brain that is activated by physical pain. Negative values show a decrease in activity, suggesting less distress. Discuss what the data show.

4.46 **Bushmeat.** African peoples often eat "bushmeat," the meat of wild animals. Bushmeat is widely traded in Africa, but its consumption threatens the survival of some animals in the wild. Bushmeat is often not the first choice of consumers— they eat bushmeat when other sources of protein are in short supply. Researchers looked at declines in 41 species of mammals in nature reserves in Ghana and at catches of fish (the primary source of animal protein) in the same region. The data appear in Table 4.2.[22] Fish supply is measured in kilograms per person. The other variable is the percent change in the total "biomass" (weight in tons) for the 41 animal species in six nature reserves. Most of the yearly percent changes in wildlife mass are negative because most years saw fewer wild animals in West Africa. Discuss how the data support the idea that more animals are killed for bushmeat when the fish supply is low. BUSHMT

TABLE 4.2 FISH SUPPLY AND WILDLIFE DECLINE IN WEST AFRICA

YEAR	FISH SUPPLY (kg PER PERSON)	BIOMASS CHANGE (PERCENT)	YEAR	FISH SUPPLY (kg PER PERSON)	BIOMASS CHANGE (PERCENT)
1971	34.7	2.9	1985	21.3	−5.5
1972	39.3	3.1	1986	24.3	−0.7
1973	32.4	−1.2	1987	27.4	−5.1
1974	31.8	−1.1	1988	24.5	−7.1
1975	32.8	−3.3	1989	25.2	−4.2
1976	38.4	3.7	1990	25.9	0.9
1977	33.2	1.9	1991	23.0	−6.1
1978	29.7	−0.3	1992	27.1	−4.1
1979	25.0	−5.9	1993	23.4	−4.8
1980	21.8	−7.9	1994	18.9	−11.3
1981	20.8	−5.5	1995	19.6	−9.3
1982	19.7	−7.2	1996	25.3	−10.7
1983	20.8	−4.1	1997	22.0	−1.8
1984	21.1	−8.6	1998	21.0	−7.4

 Exploring the Web

4.47 **Drive for show, putt for dough.** A popular saying in golf is "You drive for show but you putt for dough." The point is that hitting the golf ball a long way with a driver looks impressive, but putting well is more important for the final score and hence the amount of money you win. You can find this season's Professional Golfers Association (PGA) Tour statistics at the PGA Tour Web site: www.pgatour.com/stats.html (click on "View All" under any category displayed to see the statistics for all golfers). You can also find these statistics at the ESPN Web site: espn.go.com/golf/statistics/_/year. Look at the most recent putting, driving, and money earnings data for the current season on the PGA Tour.

(a) Make a scatterplot of earnings and putting average. Use earnings as the response variable. Describe the direction, form, and strength of the relationship in the plot. Are there any outliers?

(b) Make a scatterplot of earnings and driving distance. Use earnings as the response variable. Describe the direction, form, and strength of the relationship in the plot. Are there any outliers?

(c) Do your plots support the maxim "You drive for show but you putt for dough"?

4.48 **Olympic medals.** Go to the *Chance News* Web site at www.causeweb.org/wiki/chance/index.php/Chance_News_61#Predicting_medal_counts and read the article "Predicting Medal Counts." Next, search the Web and locate the Winter Olympics medal counts for 2010 and 2014 (I found Winter Olympics medal counts on Wikipedia). Make a scatterplot that is similar to the one in the *Chance News* article but that uses the 2010 medal counts to predict the 2014 medal counts. How does your plot compare with the plot in the *Chance News* article?

Overview

As mentioned in the overview to Chapter 4, linear regression models are one of the research tools most widely used by the social sciences, life sciences, physical sciences, and engineering. The simple linear regression model is the most basic kind of regression model and is typically the introductory statistics student's first exposure to constructing a model that can be used to make a prediction or to explore a relationship between two variables.

The main idea in simple linear regression is to exploit the relationship between two quantitative variables to gain information about one variable (the *response* variable) if information is known about the other (the *explanatory* variable). The process requires first discovering the relationship between the variables, then, knowing this is a roughly linear relationship, constructing a model (in this case, a straight-line model) relating the response variable to the explanatory variable. As emphasized in the previous chapter, the distinction between the explanatory and response variables is critical.

Three distinct issues are investigated in this chapter. First, the question of how to fit a simple linear regression line is considered. Here, students need to understand that the least-squares regression line is the line that uniquely minimizes the sum of the squared prediction errors. Second, the question of how to measure the regression line's goodness of fit is considered. We often use the squared correlation (often called the coefficient of determination) to measure this, but this is, in fact, an incomplete summary. We should also add the regression line to the original scatterplot and consider any potential undue influence from outliers. Third, we consider the question of how to use the regression line to make a prediction of the response variable, given a known value of the explanatory variable. In addition to these three main points, we use residual plots as a diagnostic tool for regression—residual plots may highlight a problem that was not clear in the original scatterplot.

Regression models have limitations, as discussed at the end of the chapter. For example, the dangers of extrapolation and the effects of lurking variables are discussed in detail, and students should be made aware of these cautions. Students should also recognize that association does not imply causation.

LEARNING OBJECTIVES**

- Understand that regression requires an explanatory variable and a response variable and that both variables must be quantitative. Correctly identifying which variable is the explanatory variable and which is the response variable is important—switching these will result in different regression lines.

- Use software or a calculator to find the least-squares regression line of a response variable y on an explanatory variable x from data.

- Explain what the slope b and intercept a mean in the equation $\hat{y} = a + bx$ of a regression line. For particular examples, these explanations should be in context.

- Draw a graph of a regression line when you are given its equation. (Note: Exercise 5.28 is a particularly good example, although students tend to get confused when asked to do this exercise on their own. Thus, you might want to work through this with them.)

- Use a regression line to predict y for a given x. Recognize extrapolation, and be aware of its dangers.

- Find the slope and intercept of the least-squares regression line, given the means and standard deviations of x and y and their correlation.

- Use r^2, the square of the correlation, to describe how much of the variation in one variable can be accounted for by a straight-line relationship with another variable.

- Recognize outliers and potentially influential observations from a scatterplot with the regression line drawn on it.

- Calculate the residuals and plot them against the explanatory variable x. Recognize that a residual plot magnifies the pattern of the scatterplot of y versus x.

- Understand that both r and the least-squares regression line can be strongly influenced by a few extreme observations.

- Recognize possible lurking variables that may explain the observed association between two variables, x and y.

- Understand that even a strong correlation does not mean that there is a cause-and-effect relationship between x and y.

- Give plausible explanations for an observed association between two variables: direct cause and effect, the influence of lurking variables, or both.

**These learning outcomes appear later for the students in Chapter 7: Part I Review.

Teaching Suggestions and Additional Examples/Activities for the Classroom

1. Identify "Best Fit" by Eye

Students should understand what it means for a line to fit the data in a scatterplot well. Consider revisiting an example used in your discussion of correlation, in Chapter 4. Pick an example for which there is a strong linear relationship between response and explanatory variables. Most of the students would agree on a line that "best" fits the points in the scatterplot, but this isn't always the case. Try plotting the points (2,4), (3,3), and (1,2) on a scatterplot. It isn't at all obvious which line (or even its direction) fits the points best. Now discuss with students the idea that we usually want to study how Y is affected by X, or even predict Y from X. Illustrate the notion of a prediction error (the residual), and make clear that the prediction errors are represented by vertical deviations of points from the fitted regression line. In this light, the "best" line (in the sense of minimizing the sum of the squared prediction errors) becomes clear. Discuss the need to square prediction errors, reminding students that we squared deviations from the mean back in Chapter 2, when we studied standard deviation. To help your students visualize this idea, have them work with the *Correlation and Regression* applet.

2. Emphasize Interpretation in Context

Students should be able to interpret the slope and intercept of the least-squares regression line in the context of the problem under consideration. Take an example in which the slope of the fitted regression line is negative, and ask them to interpret the slope. A discussion of the intercept often invites a discussion of extrapolation because the intercept is often far from the observed values of the explanatory variable (the data from which the fitted line was constructed).

3. Deconstruct the Slope

Students should also understand the relationship between correlation and prediction. The least-squares regression line associates a change of $r \cdot s_y$ in the predicted value of Y for every increase by s_x of X. For example, if the correlation between "height" and "weight" is 0.7, then a person two standard deviations above average in height would be predicted to be $(0.7)(2) = 1.4$ standard deviations above average in weight. This relationship is explicitly where the word "regression" comes from—sons of tall fathers are likely to be tall, but not as tall as their fathers. (Note: both Pearson and Galton worked with father–son height data, and you can find the data on the Internet. It might also be interesting to read the *Chance* article "Mommy's Baby, Daddy's Maybe: A Closer Look at Regression to the Mean" (http://chance.amstat.org/2013/09/1-pagano/). Although some of the article may be beyond the scope of the course, it's interesting to show students that these data are still used and discussed.)

4. Discuss the Impact of Averages on Regression

You might consider revisiting the notion of ecological correlation in the context of regression. Average scores by state for the three parts of the SAT are available for the 2013 testing year at http://www.commonwealthfoundation.org/policyblog/detail/sat-scores-by-state-2013. Would the results of this regression describe individuals? As discussed in Chapter 4, regression lines fit to data involving averages are not reliable, just as correlations computed in these settings are not very meaningful.

5. Examine the Influence of Outliers

Depending on the backgrounds of your students, you might examine how regression lines are affected by outliers. The *Correlation and Regression* applet allows students to manipulate data values and to see the impact those data values have on fitted regression lines.

6. Don't Extrapolate

Use stark examples to illustrate the pitfalls of extrapolation. Giving a plant water is good for the plant's growth up to a point—if we assume that more water will make the plant grow, we will be saddened to see the plant drown. In a study of the amount of chemical additive to chicken feed, chickens are seen to grow larger with more additive. But, clearly, chickens cannot continue to get larger. Plots of gas prices over a short period of time may exhibit a relationship quite different from the same gas prices plotted over a longer period of time (to access such plots, visit GasBuddy.com at http://www.gasbuddy.com/gb_retail_price_chart.aspx?time=24). Essentially, if you didn't experience like values of the explanatory variable in your data, then you can't know what would happen outside the range of the data. Certainly, everyone knows of the horrors of Hurricane Katrina. Weather forecasters are forced to extrapolate beyond the range of their data; a forecast map for Katrina four days before she struck New Orleans (http://www.weather.com/outlook/weather-news/news/articles/trop-katrina-aug26-diary_2010-08-17) illustrates the dangers; essentially the map shows the hurricane striking "somewhere" on the Gulf Coast.

7. Discuss the Issue of Lurking Variables

To address lurking variables, consider the following: ice cream sales and murder rates (or incidences of the flu) during a year (lurking variable: temperature); number of stork pairs and number of births in Europe (lurking variable: land mass of the countries; see "Storks Deliver Babies," *Teaching Statistics*, 22(2), Summer 2000 for more details).

8. Emphasize that Association Does Not Imply Causation

It is also important to remind students that association does not imply causation. At the same time, it is good to show students the criteria needed to establish causation when an experiment cannot be performed (Example 5.10).

Other Resources (LaunchPad)

StatClips
Regression Introduction and Motivation

Snapshots Video
Intro to Regression

EESEE Case Studies
Blood Alcohol Content
Will Women Outrun Men?
Brain Size and Intelligence

Applet
Correlation and Regression

Masakazu Watanabe/Aflo/Getty Images

Regression

Linear (straight-line) relationships between two quantitative variables are easy to understand and quite common. In Chapter 4, we found linear relationships in settings as varied as Florida manatee deaths, the risk of cancer, and predicting tropical storms. Correlation measures the direction and strength of these linear relationships. When a scatterplot shows a linear relationship, we would like to summarize the overall pattern by drawing a line on the scatterplot.

Regression lines

A *regression line* summarizes the relationship between two variables, but only in a specific setting: one of the variables helps explain or predict the other. That is, regression describes a relationship between an explanatory variable and a response variable.

Regression Line

A **regression line** is a straight line that describes how a response variable y changes as an explanatory variable x changes. We often use a regression line to predict the value of y for a given value of x.

EXAMPLE 5.1   Does Fidgeting Keep You Slim?

FATGAIN

Why is it that some people find it easy to stay slim? Here, following our four-step process (page 60), is an account of a study that sheds some light on gaining weight.

STATE: Some people don't gain weight even when they overeat. Perhaps fidgeting and other "nonexercise activity" (NEA) explains why. Some people may spontane-ously increase nonexercise activity when fed more. Researchers deliberately overfed 16 healthy young adults for 8 weeks. They measured fat gain (in kilograms) and, as an explanatory variable, change in energy use (in calories) from activity other than deliberate exercise—fidgeting, daily living, and the like. Change in energy use was energy use measured the last day of the 8-week period minus energy use measured the day before the overfeeding began. Here are the data:[1]

NEA change (cal)	−94	−57	−29	135	143	151	245	355
Fat gain (kg)	4.2	3.0	3.7	2.7	3.2	3.6	2.4	1.3
NEA change (cal)	392	473	486	535	571	580	620	690
Fat gain (kg)	3.8	1.7	1.6	2.2	1.0	0.4	2.3	1.1

Do people with larger increases in NEA tend to gain less fat?

PLAN: Make a scatterplot of the data, and examine the pattern. If it is linear, use correlation to measure its strength and draw a regression line on the scatterplot to predict fat gain from change in NEA.

SOLVE: Figure 5.1 is a scatterplot of these data. The plot shows a mod-erately strong negative linear association with no outliers. The correlation is

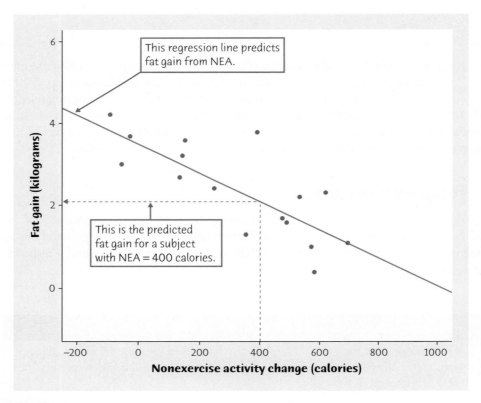

FIGURE 5.1
Fat gain after 8 weeks of overeating, plotted against increase in nonexercise activity over the same period, for Example 5.1.

$r = -0.7786$. The line on the plot is a regression line for predicting fat gain from change in NEA.

CONCLUDE: People with larger increases in NEA do indeed gain less fat. To add to this conclusion, we must study regression lines in more detail.

We can, however, already use the regression line to predict fat gain from NEA. Suppose that an individual's NEA increases by 400 calories when she overeats. Go "up and over" on the graph in Figure 5.1. From 400 calories on the *x* axis, go up to the regression line and then over to the *y* axis. The graph shows that the predicted gain in fat is a bit more than 2 kilograms. ■

Many calculators and software programs will give you the equation of a regression line from keyed-in data. Understanding and using the line are more important than the details of where the equation comes from.

REGRESSION TOWARD THE MEAN

To "regress" means to go backward. Why are statistical methods for predicting a response from an explanatory variable called "regression"? Sir Francis Galton (1822–1911), who was the first to apply regression to biological and psychological data, looked at examples such as the heights of children versus the heights of their parents. He found that taller-than-average parents tended to have children who were also taller than average but not as tall as their parents. Galton called this fact "regression toward the mean," and the name came to be applied to the statistical method.

Review of Straight Lines

Suppose that *y* is a response variable (plotted on the vertical axis) and *x* is an explanatory variable (plotted on the horizontal axis). A straight line relating *y* to *x* has an equation of the form

$$y = a + bx$$

In this equation, *b* is the **slope**, the amount by which *y* changes when *x* increases by one unit. The number *a* is the **intercept**, the value of *y* when *x* = 0.

EXAMPLE 5.2   Using a Regression Line

Any straight line describing the NEA data has the form

fat gain = *a* + (*b* × NEA change)

The line in Figure 5.1 is the regression line with the equation

fat gain = 3.505 − 0.00344 × NEA change

Be sure you understand the role of the two numbers in this equation:

■ The slope *b* = −0.00344 tells us that, on average, fat gained goes down by 0.00344 kilogram for each added calorie of NEA change. The slope of a regression line is the *rate of change* in the response, on average, as the explanatory variable changes.

■ The intercept, *a* = 3.505 kilograms, is the estimated fat gain if NEA does not change when a person overeats.

The equation of the regression line makes it easy to predict fat gain. If a person's NEA change increases by 400 calories when she overeats, substitute *x* = 400 in the equation. The predicted fat gain is

fat gain = 3.505 − (0.00344 × 400) = 2.13 kilograms

This is a bit more than 2 kilograms, as we estimated directly from the plot in Example 5.1.

To **plot the line** on the scatterplot, use the equation to find the predicted *y* for two values of *x*, one near each end of the range of *x* in the data. Plot each *y* above its *x*-value, and draw the line through the two points. ■

plotting a line

The slope of a regression line is an important numerical description of the relationship between the two variables. Although we need the value of the intercept to draw the line, this value is statistically meaningful only when, as in Example 5.2, the explanatory variable can actually take values close to zero. The slope $b = -0.00344$ in Example 5.2 is small. This does *not* mean that change in NEA has little effect on fat gain. The size of the slope depends on the units in which we measure the two variables. In this example, the slope is the change in fat gain in kilograms when NEA increases by one calorie. There are 1000 grams in a kilogram. If we measured fat gain in grams, the slope would be 1000 times larger, $b = 3.44$. *You can't say how important a relationship is by looking at the size of the slope of the regression line.*

Apply Your Knowledge

5.1 **City Mileage, Highway Mileage.** We expect a car's highway gas mileage to be related to its city gas mileage. Data for all 1137 vehicles in the government's *2013 Fuel Economy Guide* give the regression line

$$\text{highway mpg} = 6.785 + (1.033 \times \text{city mpg})$$

for predicting highway mileage from city mileage.

(a) What is the slope of this line? Say in words what the numerical value of the slope tells you.

(b) What is the intercept? Explain why the value of the intercept is not statistically meaningful.

(c) Find the predicted highway mileage for a car that gets 16 miles per gallon in the city. Do the same for a car with city mileage of 28 mpg.

(d) Draw a graph of the regression line for city mileages between 10 and 50 mpg. (Be sure to show the scales for the x and y axes.)

5.2 **What's the Line?** You use the same bar of soap to shower each morning. The bar weighs 80 grams when it is new. Its weight goes down by 5 grams per day on the average. What is the equation of the regression line for predicting weight from days of use?

The least-squares regression line

In most cases, no line will pass exactly through all the points in a scatterplot. Different people will draw different lines by eye. We need a way to draw a regression line that doesn't depend on our guess of where the line should go. Because we use the line to predict y from x, the prediction errors we make are errors in y, the vertical direction in the scatterplot. *A good regression line makes the vertical distances of the points from the line as small as possible.*

Figure 5.2 illustrates the idea. This plot shows three of the points from Figure 5.1, along with the line, on an expanded scale. The line passes above one of the points and below two of them. The three prediction errors appear as vertical line segments.

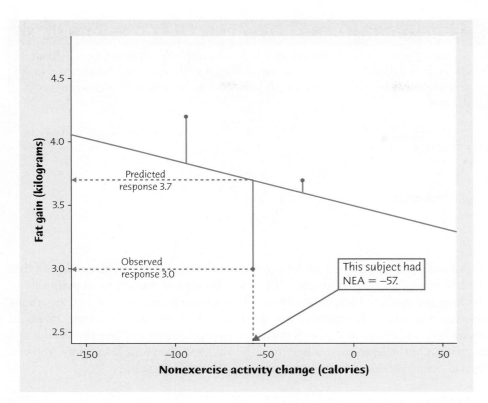

FIGURE 5.2
The least-squares idea. For each observation, find the vertical distance of each point on the scatterplot from a regression line. The least-squares regression line makes the sum of the squares of these distances as small as possible.

For example, one subject had $x = -57$, a decrease of 57 calories in NEA. The line predicts a fat gain of 3.7 kilograms, but the actual fat gain for this subject was 3.0 kilograms. The prediction error is

$$\text{error} = \text{observed response} - \text{predicted response}$$
$$= 3.0 - 3.7 = -0.7 \text{ kilogram}$$

There are many ways to make the collection of vertical distances "as small as possible." The most common is the *least-squares* method.

Least-Squares Regression Line

The **least-squares regression line** of y on x is the line that makes the sum of the squares of the vertical distances of the data points from the line as small as possible.

One reason for the popularity of the least-squares regression line is that the problem of finding the line has a simple answer. We can give the equation for the least-squares line in terms of the means and standard deviations of the two variables and the correlation between them.

Equation of the Least-Squares Regression Line

We have data on an explanatory variable x and a response variable y for n individuals. From the data, calculate the means $\bar{x}$ and $\bar{y}$ and the standard deviations s_x and s_y of the two variables and their correlation r. The least-squares regression line is the line

$$\hat{y} = a + bx$$

with **slope**

$$b = r\frac{s_y}{s_x}$$

and **intercept**

$$a = \bar{y} - b\bar{x}$$

We write $\hat{y}$ (read "y hat") in the equation of the regression line to emphasize that the line gives a *predicted* response $\hat{y}$ for any x. Because of the scatter of points about the line, the predicted response will usually not be exactly the same as the actually *observed* response y. In practice, you don't need to calculate the means, standard deviations, and correlation first. Software or your calculator will give the slope b and intercept a of the least-squares line from the values of the variables x and y. You can then concentrate on understanding and using the regression line.

Using technology

Least-squares regression is one of the most common statistical procedures. Any technology you use for statistical calculations will give you the least-squares line and related information. Figure 5.3 displays the regression output for the data of

Texas Instruments Graphing Calculator

```
LinReg
 y=a+bx
 a=3.505122916
 b=-.003441487
 r²=.6061492049
 r=-.7785558457
```

Minitab

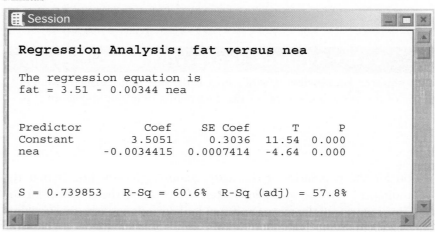

FIGURE 5.3

Least-squares regression for the nonexercise activity data: output from a graphing calculator, two statistical programs, and a spreadsheet program.

(Continued)

CrunchIt!

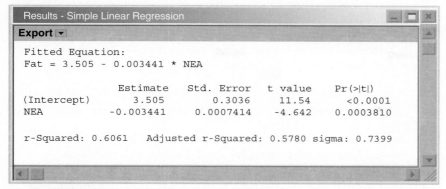

FIGURE 5.3
(Continued)

Microsoft Excel

	A	B	C	D	E	F
1	SUMMARY OUTPUT					
2						
3	*Regression statistics*					
4	Multiple R	0.778555846				
5	R Square	0.606149205				
6	Adjusted R Square	0.578017005				
7	Standard Error	0.739852874				
8	Observations	16				
9						
10		*Coefficients*	*Standard Error*	*t Stat*	*P-value*	
11	Intercept	3.505122916	0.303616403	11.54458	1.53E-08	
12	nea	-0.003441487	0.00074141	-4.64182	0.000381	
13						

Output / nea data

Examples 5.1 and 5.2 from a graphing calculator, two statistical programs, and a spreadsheet program. Each output records the slope and intercept of the least-squares line. The software also provides information that we do not yet need, although we will use much of it later. (In fact, we left out part of the Minitab and Excel outputs.) Be sure that you can locate the slope and intercept on all four outputs. *Once you understand the statistical ideas, you can read and work with almost any software output.*

Apply Your Knowledge

5.3   Coral Reefs. Exercises 4.2 and 4.10 discuss a study in which scientists examined data on mean sea surface temperatures (in degrees Celsius) and mean coral growth (in millimeters per year) over a several-year period at locations in the Red Sea. Here are the data:[2] CORAL

Sea surface temperature	29.68	29.87	30.16	30.22	30.48	30.65	30.90
Growth	2.63	2.58	2.68	2.6	2.48	2.38	2.26

(a) Use your calculator to find the mean and standard deviation of both sea surface temperature x and growth y and the correlation r between x and y. Use these basic measures to find the equation of the least-squares line for predicting y from x.

(b) Enter the data into your software or calculator, and use the regression function to find the least-squares line. The result should agree with your work in (a) up to roundoff error.

5.4   **Do Heavier People Burn More Energy?** We have data on the lean body mass and resting metabolic rate for 12 women who are subjects in a study of dieting. Lean body mass, given in kilograms, is a person's weight leaving out all fat. Metabolic rate, in calories burned per 24 hours, is the rate at which the body consumes energy. METAB

| Mass | 36.1 | 54.6 | 48.5 | 42.0 | 50.6 | 42.0 | 40.3 | 33.1 | 42.4 | 34.5 | 51.1 | 41.2 |
| Rate | 995 | 1425 | 1396 | 1418 | 1502 | 1256 | 1189 | 913 | 1124 | 1052 | 1347 | 1204 |

(a) Make a scatterplot that shows how metabolic rate depends on body mass. There is a quite strong linear relationship, with correlation $r = 0.876$.

(b) Find the least-squares regression line for predicting metabolic rate from body mass. Add this line to your scatterplot.

(c) Explain in words what the slope of the regression line tells us.

(d) Another woman has a lean body mass of 45 kilograms. What is her predicted metabolic rate?

Facts about least-squares regression

One reason for the popularity of least-squares regression lines is that they have many convenient properties. Here are some facts about least-squares regression lines.

Fact 1. The distinction between explanatory and response variables is essential in regression. Least-squares regression makes the distances of the data points from the line small only in the y direction. If we reverse the roles of the two variables, we get a different least-squares regression line.

EXAMPLE 5.3	**Predicting Fat Gain, Predicting Change in NEA**

Figure 5.4 repeats the scatterplot of the NEA data in Figure 5.1, but with *two* least-squares regression lines. The solid line is the regression line for predicting fat gain from change in NEA. This is the line that appeared in Figure 5.1.

We might also use the data on these 16 subjects to predict the change in NEA for another subject from that subject's fat gain when overfed for 8 weeks. Now the roles of the variables are reversed: fat gain is the explanatory variable, and change in NEA is the response variable. The dashed line in Figure 5.4 is the least-squares line for predicting NEA change from fat gain. The two regression lines are not the same. *In the regression setting, you must know clearly which variable is explanatory.* ■

Fact 2. There is a close connection between correlation and the slope of the least-squares line. The slope is

$$b = r \frac{s_y}{s_x}$$

You see that **the slope and the correlation always have the same sign.** For example, if a scatterplot shows a positive association, then both b and r are positive. The formula for the slope b says more: along the regression line, **a change of one standard deviation in x corresponds to a change of r standard deviations in y.**

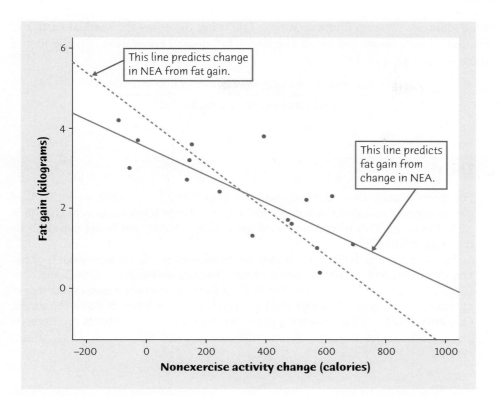

FIGURE 5.4
Two least-squares regression lines
for the nonexercise activity data, for
Example 5.3. The solid line predicts fat
gain from change in nonexercise activity.
The dashed line predicts change in
nonexercise activity from fat gain.

When the variables are perfectly correlated ($r = 1$ or $r = -1$), the change in the predicted response $\hat{y}$ is the same (in standard deviation units) as the change in x. Otherwise, because $-1 \leq r \leq 1$, the change in $\hat{y}$ (in standard deviation units) is less than the change in x. As the correlation grows less strong, the prediction $\hat{y}$ moves less in response to changes in x.

Fact 3. The least-squares regression line always passes through the point $(\bar{x}, \bar{y})$ on the graph of y against x. This is a consequence of the equation of the least-squares regression line (box on page 126). In Exercise 5.48 we ask you to confirm this.

Fact 4. The correlation r describes the strength of a straight-line relationship. In the regression setting, this description takes a specific form: **the square of the correlation, r^2, is the fraction of the variation in the values of y that is explained by the least-squares regression of y on x.**

The idea is that when there is a linear relationship, some of the variation in y is accounted for by the fact that as x changes, y changes along with it. Look again at Figure 5.1 (page 122), the scatterplot of the NEA data. The variation in y appears as the spread of fat gains from 0.4 to 4.2 kg. Some of this variation is explained by the fact that x (change in NEA) varies from a loss of 94 calories to a gain of 690 calories. As x changes from -94 to 690, y changes along the line. You would predict a smaller fat gain for a subject whose NEA increased by 600 calories than for someone with 0 change in NEA. But the straight-line tie of y to x doesn't explain *all* of the variation in y. The remaining variation appears as the scatter of points above and below the line.

Although we won't do the algebra, it is possible to break the variation in the observed values of y into two parts. One part measures the variation in $\hat{y}$ along the least-squares regression line as x varies. The other measures the vertical scatter of the data points above and below the line. The squared correlation r^2 is the first of these as a fraction of the whole:

$$r^2 = \frac{\text{variation in } \hat{y} \text{ along the regression line as } x \text{ varies}}{\text{total variation in observed values of } y}$$

EXAMPLE 5.4   **Using r^2**

For the NEA data, $r = -0.7786$ and $r^2 = (-0.7786)^2 = 0.6062$. About 61% of the variation in fat gained is accounted for by the linear relationship with change in NEA. The other 39% is individual variation among subjects that is not explained by the linear relationship.

Figure 4.2 (page 103) shows a stronger linear relationship between boat registrations in Florida and manatees killed by boats. The correlation is $r = 0.953$ and $r^2 = (0.953)^2 = 0.908$. Slightly more than 90% of the year-to-year variation in number of manatees killed by boats is explained by regression on number of boats registered. Only about 10% is variation among years with similar numbers of boats registered. ■

You can find a regression line for any relationship between two quantitative variables, but the usefulness of the line for prediction depends on the strength of the linear relationship. So r^2 is almost as important as the equation of the line in reporting a regression. All the outputs in Figure 5.3 (pages 126–127) include r^2, either in decimal form or as a percent. When you see a correlation, square it to get a better feel for the strength of the association. Perfect correlation ($r = -1$ or $r = 1$) means the points lie exactly on a line. Then $r^2 = 1$, and all the variation in one variable is accounted for by the linear relationship with the other variable. If $r = -0.7$ or $r = 0.7$, $r^2 = 0.49$ and about half the variation is accounted for by the linear relationship. In the r^2 scale, correlation ± 0.7 is about halfway between 0 and ± 1.

Facts 2, 3, and 4 are special properties of least-squares regression. They are not true for other methods of fitting a line to data.

Apply Your Knowledge

5.5   How Useful Is Regression? Figure 4.8 (page 114) displays the relationship between golfers' scores on the first and second rounds of the 2012 Masters Tournament. The correlation is $r = 0.296$. Exercise 4.30 (page 115) gives data on solar radiation (SRD) and concentration of dimethyl sulfide (DMS) over a region of the Mediterranean. The correlation is $r = 0.969$. Explain in simple language why knowing only these correlations enables you to say that prediction of DMS from SRD by a regression line will be much more accurate than prediction of a golfer's second-round score from his first-round score.

5.6   Feed the Birds. Exercise 4.32 (page 116) gives data from a study in which canary parents cared for both their own babies and those of other parents. Investigators looked at how the growth rate of the foster babies relative to the growth rate of the natural babies changed as the begging intensity for food by the foster babies increased over the begging intensity of the natural

babies. If begging intensity is the main factor determining food received, with higher intensity leading to more food, one would expect the relative growth rate to increase as the difference in begging intensity increases. However, if both begging intensity and a preference for their own babies determine the amount of food received (and hence the relative growth rate), we might expect growth rate to increase initially as begging intensity increases but then to level off (or even decrease) as the parents begin to ignore further increases in begging by the foster babies. 📊 CANARY

(a) Make a scatterplot of the data. Find the least-squares regression line for predicting relative growth rate of the foster brood from the difference in begging intensity between the foster brood and the actual babies of the parents, and add this line to your plot. Should we *not* use the regression line for prediction in this setting?

(b) What is r^2? What does this value say about the success of the regression line in predicting relative growth rate?

Residuals

One of the first principles of data analysis is to look for an overall pattern and also for striking deviations from the pattern. A regression line describes the overall pattern of a linear relationship between an explanatory variable and a response variable. We see deviations from this pattern by looking at the scatter of the data points about the regression line. The vertical distances from the points to the least-squares regression line are as small as possible, in the sense that they have the smallest possible sum of squares. Because they represent "leftover" variation in the response after fitting the regression line, these distances are called *residuals*.

Residuals

A **residual** is the difference between an observed value of the response variable and the value predicted by the regression line. That is, a residual is the prediction error that remains after we have chosen the regression line:

$$\text{residual} = \text{observed } y - \text{predicted } y$$
$$= y - \hat{y}$$

EXAMPLE 5.5   I Feel Your Pain

"Empathy" means being able to understand what others feel. To see how the brain expresses empathy, researchers recruited 16 couples in their mid-twenties who were married or had been dating for at least two years. They zapped the man's hand with an electrode while the woman watched and measured the activity in several parts of the woman's brain that would respond to her own pain. Brain activity was recorded as a fraction of the activity observed when the woman herself was zapped with the electrode. The women also completed a psychological test that measures empathy. Will women who score higher in empathy respond more

EMPATHY

strongly when their partner has a painful experience? Here are data for one brain region:[3]

Subject	1	2	3	4	5	6	7	8
Empathy score	38	53	41	55	56	61	62	48
Brain activity	−0.120	0.392	0.005	0.369	0.016	0.415	0.107	0.506
Subject	9	10	11	12	13	14	15	16
Empathy score	43	47	56	65	19	61	32	105
Brain activity	0.153	0.745	0.255	0.574	0.210	0.722	0.358	0.779

Figure 5.5 is a scatterplot, with empathy score as the explanatory variable x and brain activity as the response variable y. The plot shows a positive association. That is, women who score higher in empathy do indeed react more strongly to their partner's pain. The overall pattern is moderately linear, with correlation $r = 0.515$.

The line on the plot is the least-squares regression line of brain activity on empathy score. Its equation is

$$\hat{y} = -0.0578 + 0.00761x$$

For Subject 1, with empathy score 38, we predict

$$\hat{y} = -0.0578 + (0.00761)(38) = 0.231$$

This subject's actual brain activity level was −0.120. The residual is

$$\text{residual} = \text{observed } y - \text{predicted } y$$
$$= -0.120 - 0.231 = -0.351$$

The residual is negative because the data point lies below the regression line. The dashed line segment in Figure 5.5 shows the size of the residual. ■

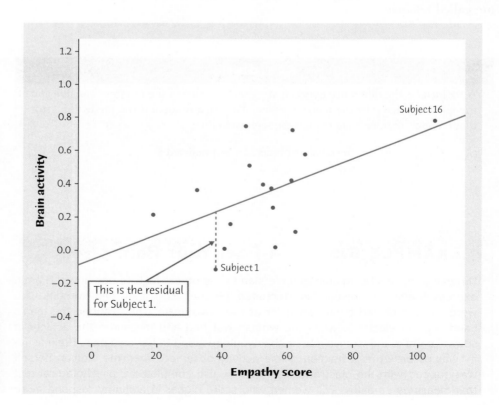

FIGURE 5.5

Scatterplot of activity in a region of the brain that responds to pain versus score on a test of empathy, for Example 5.5. Brain activity is measured as the subject watches her partner experience pain. The line is the least-squares regression line.

There is a residual for each data point. Finding the residuals is a bit unpleasant because you must first find the predicted response for every *x*. Software or a graphing calculator gives you the residuals all at once. Here are the 16 residuals for the empathy study data, from software:

```
residuals:
 -0.3515  0.0463  -0.2494  0.0080  -0.3526  0.0084  -0.3072  0.1983
 -0.1166  0.4449  -0.1136  0.1369   0.1231  0.3154   0.1721  0.0374
```

Because the residuals show how far the data fall from our regression line, examining the residuals helps us assess how well the line describes the data. Although residuals can be calculated from any curve or line fitted to the data, the residuals from the least-squares line have a special property: **the mean of the least-squares residuals is always zero.**

Compare the scatterplot in Figure 5.5 with the *residual plot* for the same data in Figure 5.6. The horizontal line at zero in Figure 5.6 helps orient us. This "residual = 0" line corresponds to the regression line in Figure 5.5.

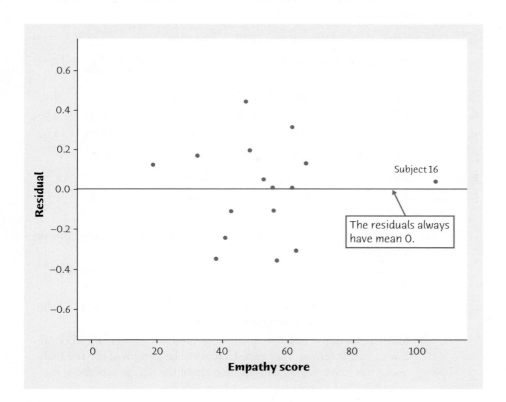

FIGURE 5.6
Residual plot for the data shown in Figure 5.5. The horizontal line at zero residual corresponds to the regression line in Figure 5.5.

Residual Plots

A **residual plot** is a scatterplot of the regression residuals against the explanatory variable. Residual plots help us assess how well a regression line fits the data.

A residual plot in effect turns the regression line horizontal. It magnifies the deviations of the points from the line and makes it easier to see unusual observations and patterns.

Apply Your Knowledge

5.7 **Residuals by Hand.** In Exercise 5.3 (page 127) you found the equation of the least-squares line for predicting coral growth y from mean sea surface temperature x.

(a) Use the equation to obtain the 7 residuals step-by-step. That is, find the prediction $\hat{y}$ for each observation and then find the residual $y - \hat{y}$.

(b) Check that (up to roundoff error) the residuals add to 0.

(c) The residuals are the part of the response y left over after the straight-line tie between y and x is removed. Show that the correlation between the residuals and x is 0 (up to roundoff error). That this correlation is always 0 is another special property of least-squares regression.

5.8 **Does Fast Driving Waste Fuel?** Exercise 4.8 (page 103) gives data on the fuel consumption y of a car at various speeds x. Fuel consumption is measured in liters of gasoline per 100 kilometers driven, and speed is measured in kilometers per hour. Software tells us that the equation of the least-squares regression line is

$$\hat{y} = 11.058 - 0.01466x$$

Using this equation we can add the residuals to the original data: ▦ FASTDR2

Speed	10	20	30	40	50	60	70	80
Fuel	21.00	13.00	10.00	8.00	7.00	5.90	6.30	6.95
Residual	10.09	2.24	−0.62	−2.47	−3.33	−4.28	−3.73	−2.94

Speed	90	100	110	120	130	140	150
Fuel	7.57	8.27	9.03	9.87	10.79	11.77	12.83
Residual	−2.17	−1.32	−0.42	0.57	1.64	2.76	3.97

(a) Make a scatterplot of the observations, and draw the regression line on your plot.

(b) Would you use the regression line to predict y from x? Explain your answer.

(c) Verify the value of the first residual, for x = 10. Verify that the residuals have sum zero (up to roundoff error).

(d) Make a plot of the residuals against the values of x. Draw a horizontal line at height zero on your plot. How does the pattern of the residuals about this line compare with the pattern of the data points about the regression line in your scatterplot from (a)?

Influential observations

Figures 5.5 and 5.6 show one unusual observation. Subject 16 is an outlier in the x direction, with empathy score 40 points higher than any other subject. Because of its extreme position on the empathy scale, this point has a strong influence on the correlation. Dropping Subject 16 reduces the correlation from $r = 0.515$ to $r = 0.331$. You can see that this point extends the linear pattern in Figure 5.5 and so increases the correlation. We say that Subject 16 is *influential* for calculating the correlation.

Influential Observations

An observation is **influential** for a statistical calculation if removing it would markedly change the result of the calculation.

The result of a statistical calculation may be of little practical use if it depends strongly on a few influential observations.

Points that are outliers in either the x or the y direction of a scatterplot are often influential for the correlation. Points that are outliers in the x direction are often influential for the least-squares regression line.

If an observation is influential, it can be informative to report statistical calculations with both the influential observation included and the influential observation removed. This provides readers with the ability to assess the effect of the influential observation.

EXAMPLE 5.6   An Influential Observation?

Subject 16 in Example 5.5 is influential for the correlation between empathy score and brain activity because removing it reduces r from 0.515 to 0.331. Calculating that $r = 0.515$ is not a very useful description of the data, because the value depends so strongly on just one of the 16 subjects.

Is this observation also influential for the least-squares line? Figure 5.7 shows that it is not. The regression line calculated without Subject 16 (dashed) differs little from the line that uses all the observations (solid). The reason that the outlier has little influence on the regression line is that it lies close to the dashed regression line calculated from the other observations. ■

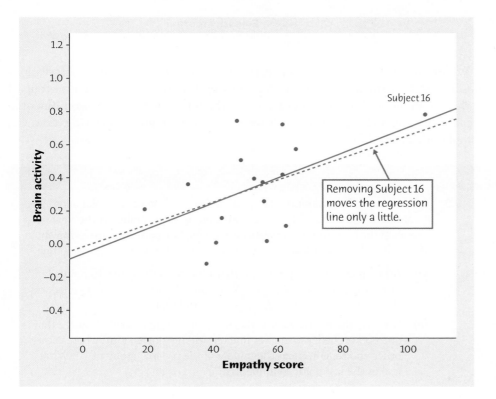

FIGURE 5.7
Subject 16 is an outlier in the x direction. The outlier is not influential for least-squares regression because removing it moves the regression line only a little.

To see why points that are outliers in the *x* direction are often influential for regression, let's try an experiment. Suppose that Subject 16's point in the scatterplot moves straight down. What happens to the regression line? Figure 5.8 gives the answer. The dashed line is the regression line with the outlier in its new, lower position. Because there are no other points with similar *x*-values, the line chases the outlier. The *Correlation and Regression* applet allows you to try this experiment yourself— see Exercise 5.9. *An outlier in x pulls the least-squares line toward itself. If the outlier does not lie close to the line calculated from the other observations, it will be influential.*

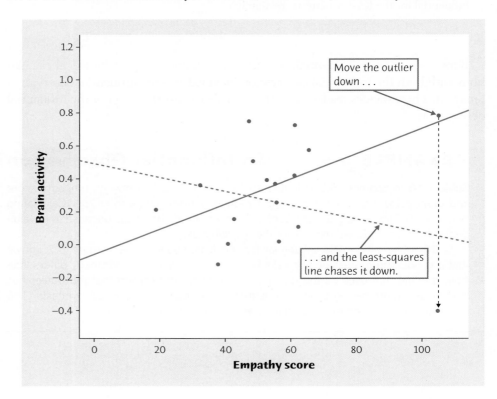

FIGURE 5.8

An outlier in the *x* direction pulls the least-squares line to itself because there are no other observations with similar values of *x* to hold the line in place. When the outlier moves down, the regression line chases it down. The original regression line is solid, and the final position of the regression line is dashed.

We did not need the distinction between outliers and influential observations in Chapter 2. A single high salary that pulls up the mean salary $\bar{x}$ for a group of workers is an outlier because it lies far above the other salaries. It is also influential because the mean changes when it is removed. In the regression setting, however, not all outliers are influential.

Apply Your Knowledge

5.9   Influence in Regression. The *Correlation and Regression* applet allows you to animate Figure 5.8. Click to create a group of 10 points in the lower-left corner of the scatterplot with a strong straight-line pattern (correlation about 0.9). Click the "Show least-squares line" box to display the regression line.

(a)   Add 1 point at the upper right that is far from the other 10 points but exactly on the regression line. Why does this outlier have no effect on the line even though it changes the correlation?

(b)   Now use the mouse to drag this last point straight down. You see that one end of the least-squares line chases this single point, whereas the other end remains near the middle of the original group of 10. What makes the last point so influential?

5.10 Do Heavier People Burn More Energy? Return to the data of Exercise 5.4 (page 128) on body mass and metabolic rate. We will use these data to illustrate influence. METAB2

(a) Make a scatterplot of the data that is suitable for predicting metabolic rate from body mass, with two new points added. Point A: mass 42 kilograms, metabolic rate 1500 calories. Point B: mass 70 kilograms, metabolic rate 1400 calories. In which direction is each of these points an outlier?

(b) Add three least-squares regression lines to your plot: for the original 12 women, for the original women plus Point A, and for the original women plus Point B. Which new point is more influential for the regression line? Explain in simple language why each new point moves the line in the way your graph shows.

5.11 Outsourcing by Airlines. Exercise 4.5 (page 101) gives data for 11 airlines on the percent of major maintenance outsourced and the percent of flight delays blamed on the airline. AIRLINE

(a) Make a scatterplot with outsourcing percent as *x* and delay percent as *y*. Would you consider Hawaiian Airlines to be influential?

(b) Find the correlation *r* with and without Hawaiian Airlines. How influential is the outlier for correlation?

(c) Find the least-squares line for predicting *y* from *x* with and without Hawaiian Airlines. Draw both lines on your scatterplot. Use both lines to predict the percent of delays blamed on an airline that has outsourced 78.6% of its major maintenance. How influential is the outlier for the least-squares line?

Cautions about correlation and regression

Correlation and regression are powerful tools for describing the relationship between two variables. When you use these tools, you must be aware of their limitations. You already know that

- *Correlation and regression lines describe only linear relationships.* You can do the calculations for any relationship between two quantitative variables, but the results are useful only if the scatterplot shows a linear pattern.

- *Correlation and least-squares regression lines are not resistant.* Always plot your data and look for observations that may be influential.

Here are three more things to keep in mind when you use correlation and regression.

Beware ecological correlation There is a large positive correlation between *average* income and number of years of education. The correlation is smaller if we compare the incomes of *individuals* with number of years of education. The correlation based on average income ignores the large variation in the incomes of individuals having the same amount of education. The variation from individual to individual increases the scatter in a scatterplot, reducing the correlation. The correlation between average income and education overstates the strength of the relation between the incomes of individuals and number of years of education. *Correlations based on averages can be misleading if they are interpreted to be about individuals.*

Ecological Correlation

A correlation based on averages rather than on individuals is called an **ecological correlation.**

Beware extrapolation Suppose that you have data on a child's growth between 3 and 8 years of age. You find a strong linear relationship between age x and height y. If you fit a regression line to these data and use it to predict height at age 25 years, you will predict that the child will be 8 feet tall. Growth slows down and then stops at maturity, so extending the straight line to adult ages is foolish. *Few relationships are linear for all values of x. Don't make predictions far outside the range of x that actually appears in your data.*

Extrapolation

Extrapolation is the use of a regression line for prediction far outside the range of values of the explanatory variable x that you used to obtain the line. Such predictions are often not accurate.

DO LEFT-HANDERS DIE EARLY?

Yes, said a study of 1000 deaths in California. Left-handed people died at an average age of 66 years; right-handers at 75 years of age. Should left-handed people fear an early death? No—the lurking variable has struck again. Older people grew up in an era when many natural left-handers were forced to use their right hands. So right-handers are more common among older people, and left-handers are more common among the young. When we look at deaths, the left-handers who die are younger on the average because left-handers in general are younger. Mystery solved.

Beware the lurking variable Another caution is even more important: *the relationship between two variables can often be understood only by taking other variables into account. Lurking variables can make a correlation or regression misleading.*

Lurking Variable

A **lurking variable** is a variable that is not among the explanatory or response variables in a study and yet may influence the interpretation of relationships among those variables.

You should always think about possible lurking variables before you draw conclusions based on correlation or regression.

EXAMPLE 5.7   Magic Mozart?

The Kalamazoo (Michigan) Symphony once advertised a "Mozart for Minors" program with this statement: "Question: Which students scored 51 points higher in verbal skills and 39 points higher in math? Answer: Students who had experience in music."[4]

We could as well answer "Students who played soccer." Why? Children with prosperous and well-educated parents are more likely than poorer children to have experience with music and also to play soccer. They are also likely to attend good schools, get good health care, and be encouraged to study hard. These advantages lead to high test scores. Family background is a lurking variable that explains why test scores are related to experience with music. ■

Apply Your Knowledge

5.12 One More Inch, Three More Pounds. Data on the *average* weight of men who are between 5 feet 2 inches and 6 feet 4 inches tall (rounded to the nearest inch) show a very high positive correlation. Would the correlation be greater, smaller, or about the same if you calculated the correlation between the weights of individual men and their heights (rounded to the nearest inch)? Explain your answer.

5.13 The Endangered Manatee. Table 4.1 (page 102) gives 36 years of data on boats registered in Florida and manatees killed by boats. Figure 4.2 (page 103) shows a strong positive linear relationship. The correlation is $r = 0.953$.

PhotoDisc

(a) Find the equation of the least-squares line for predicting manatees killed from thousands of boats registered. Because the linear pattern is so strong, we expect predictions from this line to be quite accurate—but only if conditions in Florida remain similar to those of the past 36 years. ▦ MANATEE

(b) In 2012, experts predicted that the number of boats registered in Florida would be 900,000 in 2013. What would you predict the number of manatees killed by boats to be if there are 900,000 boats registered? Explain why we can trust this prediction.

(c) Predict manatee deaths if there were *no* boats registered in Florida. Explain why the predicted count of deaths is impossible. (We use $x = 0$ to find the intercept of the regression line, but unless the explanatory variable x actually takes values near 0, prediction for $x = 0$ is an example of extrapolation.)

5.14 Is Math the Key to Success in College? A College Board study of 15,941 high school graduates found a strong correlation between how much math minority students took in high school and their later success in college. News articles quoted the head of the College Board as saying that "math is the gatekeeper for success in college."[5] Maybe so, but we should also think about lurking variables. What might lead minority students to take more or fewer high school math courses? Would these same factors influence success in college?

Association does not imply causation

Thinking about lurking variables leads to the most important caution about correlation and regression. When we study the relationship between two variables, we often hope to show that changes in the explanatory variable *cause* changes in the response variable. *A strong association between two variables is not enough to draw conclusions about cause and effect.* Sometimes an observed association really does reflect cause and effect. A household that heats with natural gas uses more gas in colder months because cold weather requires burning more gas to stay warm. In other cases, an association is explained by lurking variables, and the conclusion that x causes y is either wrong or not proved.

EXAMPLE 5.8   Does Having More Cars Make You Live Longer?

A serious study once found that people with two cars live longer than people who own only one car.[6] Owning three cars is even better, and so on. There is a substantial positive correlation between number of cars x and length of life y.

The basic meaning of causation is that by changing x we can bring about a change in y. Could we lengthen our lives by buying more cars? No. The study used number of cars as a quick indicator of affluence. Well-off people tend to have more cars. They also tend to live longer, probably because they are better educated, take better care of themselves, and get better medical care. The cars have nothing to do with it. There is no cause-and-effect tie between number of cars and length of life. ■

 THE SUPER BOWL EFFECT

The Super Bowl is the most-watched TV broadcast in the United States. Data show that on Super Bowl Sunday we consume 3 times as many potato chips as on an average day, and 17 times as much beer. What's more, the number of fatal traffic accidents goes up in the hours after the game ends. Could that be celebration? Or catching up with tasks left undone? Or maybe it's the beer.

Correlations such as that in Example 5.8 are sometimes called "nonsense correlations." The correlation is real. What is nonsense is the conclusion that changing one of the variables causes changes in the other. A lurking variable—such as personal affluence in Example 5.8—that influences both x and y can create a high correlation even though there is no direct connection between x and y.

Association Does Not Imply Causation

An association between an explanatory variable x and a response variable y, even if it is very strong, is not by itself good evidence that changes in x actually cause changes in y.

EXAMPLE 5.9   Overweight Mothers, Overweight Daughters

Overweight parents tend to have overweight children. The results of a study of Mexican American girls aged 9 to 12 years are typical. The investigators measured body mass index (BMI), a measure of weight relative to height, for both the girls and their mothers. People with high BMI are overweight. The correlation between the BMI of daughters and the BMI of their mothers was $r = 0.506$.[7]

Body type is in part determined by heredity. Daughters inherit half their genes from their mothers. There is therefore a direct cause-and-effect link between the BMI of mothers and daughters. But perhaps mothers who are overweight also set an example of little exercise, poor eating habits, and lots of television viewing. Their daughters may pick up these habits, so the influence of heredity is mixed up with influences from the girls' environment. Both contribute to the mother–daughter correlation. ■

 The lesson of Example 5.9 is more subtle than just "association does not imply causation." *Even when direct causation is present, it may not be the whole explanation for a correlation.* You must still worry about lurking variables. Careful statistical studies try to anticipate lurking variables and measure them. The mother–daughter study did measure TV viewing, exercise, and diet. Elaborate statistical analysis can remove the effects of these variables to come closer to the direct effect of mother's BMI on daughter's BMI. This remains a second-best approach to *experiment*   causation. The best way to get good evidence that x causes y is to do an **experiment** in which we change x and keep lurking variables under control. We will discuss experiments in Chapter 9.

When experiments cannot be done, explaining an observed association can be difficult and controversial. Many of the sharpest disputes in which statistics plays a role involve questions of causation that cannot be settled by experiment. Do gun control laws reduce violent crime? Does using cell phones cause brain tumors? Has increased free trade widened the gap between the incomes of more educated and less educated American workers? All these questions have become public issues. All concern associations among variables. And all have this in common: they try to pinpoint cause and effect in a setting involving complex relations among many interacting variables.

EXAMPLE 5.10   Does Smoking Cause Lung Cancer?

Despite the difficulties, it is sometimes possible to build a strong case for causation in the absence of experiments. The evidence that smoking causes lung cancer is about as strong as nonexperimental evidence can be.

Doctors had long observed that most lung cancer patients were smokers. Comparison of smokers and "similar" nonsmokers showed a very strong association between smoking and death from lung cancer. Could the association be explained by lurking variables? Might there be, for example, a genetic factor that predisposes people both to nicotine addiction and to lung cancer? Smoking and lung cancer would then be positively associated even if smoking had no direct effect on the lungs. How were these objections overcome? ■

Let's answer this question in general terms: what are the criteria for establishing causation when we cannot do an experiment?

■ *The association is strong.* The association between smoking and lung cancer is very strong.

■ *The association is consistent.* Many studies of different kinds of people in many countries link smoking to lung cancer. That reduces the chance that a lurking variable specific to one group or one study explains the association.

■ *Higher doses are associated with stronger responses.* People who smoke more cigarettes per day or who smoke over a longer period get lung cancer more often. People who stop smoking reduce their risk.

■ *The alleged cause precedes the effect in time.* Lung cancer develops after years of smoking. The number of men dying of lung cancer rose as smoking became more common, with a lag of about 30 years. Lung cancer kills more men than any other form of cancer. Lung cancer was rare among women until women began to smoke. Lung cancer in women rose along with smoking, again with a lag of about 30 years, and has now passed breast cancer as the leading cause of cancer death among women.

■ *The alleged cause is plausible.* Experiments with animals show that tars from cigarette smoke do cause cancer.

Medical authorities do not hesitate to say that smoking causes lung cancer. The U.S. Surgeon General has long stated that cigarette smoking is "the largest avoidable cause of death and disability in the United States."[8] The evidence for causation is overwhelming—but it is not as strong as the evidence provided by well-designed experiments.

Apply Your Knowledge

5.15 Another Reason Not to Smoke? A stop-smoking booklet says, "Children of mothers who smoked during pregnancy scored nine points lower on intelligence tests at ages three and four than children of nonsmokers." Suggest some lurking variables that may help explain the association between smoking during pregnancy and children's later test scores. The association by itself is not good evidence that mothers' smoking *causes* lower scores.

5.16 Education and Income. There is a strong positive association between workers' education and their income. For example, the U.S. Census Bureau reported in 2011 that the mean income of young adults (ages 25 to 34) who worked full-time increased from $18,993 for those with less than a ninth-grade education, to $28,272 for high school graduates, to $47,757 for holders of a bachelor's degree, and on up for yet more education. In part, this association reflects causation—education helps people qualify for better jobs. Suggest several lurking variables that also contribute. (Ask yourself what kinds of people tend to get more education.)

5.17 To Earn More, Get Married? Data show that men who are married, and also divorced or widowed men, earn quite a bit more than men the same age who have never been married. This does not mean that a man can raise his income by getting married because men who have never been married are different from married men in many ways other than marital status. Suggest several lurking variables that might help explain the association between marital status and income.

CHAPTER 5 SUMMARY

Chapter Specifics

■ A **regression line** is a straight line that describes how a response variable y changes as an explanatory variable x changes. You can use a regression line to **predict** the value of y for any value of x by substituting this x into the equation of the line.

■ The **slope** b of a regression line $\hat{y} = a + bx$ is the rate at which the predicted response $\hat{y}$ changes along the line as the explanatory variable x changes. Specifically, b is the change in $\hat{y}$ when x increases by 1.

■ The **intercept** a of a regression line $\hat{y} = a + bx$ is the predicted response $\hat{y}$ when the explanatory variable $x = 0$. This prediction is of no statistical interest unless x can actually take values near 0.

■ The most common method of fitting a line to a scatterplot is least squares. The **least-squares regression line** is the straight line $\hat{y} = a + bx$ that minimizes the sum of the squares of the vertical distances of the observed points from the line.

■ The least-squares regression line of y on x is the line with slope $b = rs_y/s_x$ and intercept $a = \bar{y} - b\bar{x}$. This line always passes through the point $(\bar{x}, \bar{y})$.

■ **Correlation and regression** are closely connected. The correlation r is the slope of the least-squares regression line when we measure both x and y in standardized units. The **square of the correlation** r^2 is the fraction of the variation in one variable that is explained by least-squares regression on the other variable.

■ Correlation and regression must be **interpreted with caution. Plot the data** to be sure the relationship is roughly linear and to detect outliers and influential observations. A plot of the **residuals** makes these effects easier to see.

- Look for **influential observations,** individual points that substantially change the correlation or the regression line. Outliers in the x direction are often influential for the regression line.

- Be aware of **ecological correlation,** the tendency for correlations based on averages to be stronger than correlations based on individuals. Be careful not to misinterpret correlations based on averages as applying to individuals.

- Avoid **extrapolation,** the use of a regression line for prediction for values of the explanatory variable far outside the range of the data from which the line was calculated.

- **Lurking variables** may explain the relationship between the explanatory and response variables. Correlation and regression can be misleading if you ignore important lurking variables.

- Most of all, be careful not to conclude that there is a cause-and-effect relationship between two variables just because they are strongly associated. **High correlation does not imply causation.** The best evidence that an association is due to causation comes from an **experiment** in which the explanatory variable is directly changed and other influences on the response are controlled.

Link It

In this chapter we continued our study of exploring relationships between two variables that we began in Chapter 4. We use the least-squares regression line to describe the straight-line relationship between two variables when such a pattern is seen in a scatterplot. The equation of the least-squares regression line is a numerical summary that makes precise the notion of "straight-line relationship."

Even if the least-squares regression line is a good description of the relationship between the observed values of two variables, we must exercise caution in how we interpret this relationship. Such interpretations rest on the assumption that the relationship is valid in some broader sense. We will explore this more carefully later in this book, but in this chapter we content ourselves with providing some cautions. Association, as indicated by a large correlation, does not imply that there is an underlying cause-and-effect relation between the response and explanatory variables. There may, in fact, be a lurking variable that influences the interpretation of any relation between the response and explanatory variables. Correlations based on averages of measurements tend to be higher than correlations based on the individual observations used to compute the averages. Be careful not to misinterpret correlations based on averages as applying to individuals. Finally, be careful not to use the least-squares regression line to make predictions outside the range of values of the explanatory variable that you used to obtain the line.

In this chapter we studied relationships between two quantitative variables. In the next chapter, we explore relationships between two qualitative variables.

CHECK YOUR SKILLS

5.18 Figure 5.9 is a scatterplot of school GPA against IQ test scores for 15 seventh-grade students. The line is the least-squares regression line for predicting school GPA from IQ score. If another child in this class has IQ score 110, you predict the school GPA to be close to

(a) 2.   (b) 7.5.   (c) 11.

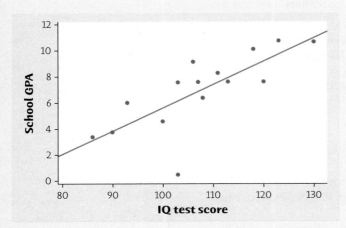

FIGURE 5.9
Scatterplot of IQ test scores and school GPA for 15 seventh-grade students, for Exercises 5.18 and 5.19.

5.19 The slope of the line in Figure 5.9 is closest to

(a) −11.   (b) 0.2.   (c) 2.0.

5.20 The points on a scatterplot lie close to the line whose equation is $y = 4 - 3x$. The slope of this line is

(a) 4.   (b) 3.   (c) −3.

5.21 Fred keeps his savings in his mattress. He began with $1000 from his mother and adds $100 each year. His total savings y after x years are given by the equation

(a) $y = 1000 + 100x$.
(b) $y = 100 + 1000x$.
(c) $y = 1000 + x$.

5.22 Smokers don't live as long (on the average) as non-smokers, and heavy smokers don't live as long as light smokers. You regress the age at death of a group of male smokers on the number of packs per day they smoked. The slope of your regression line

(a) will be greater than 0.
(b) will be less than 0.
(c) Can't tell without seeing the data.

5.23 An owner of a home in the Midwest installed solar panels to reduce heating costs. After installing the solar panels, he measured the amount of natural gas used y (in cubic feet) to heat the home and outside temperature x (in degree-days, where a day's degree-days are the number of degrees its average temperature falls below 65°F) over a 23-month period. He then computed the least-squares regression line for predicting y from x and found it to be[9]

$$\hat{y} = 85 + 16x$$

How much, on average, does gas used increase for each additional degree-day?

(a) 23 cubic feet   (b) 85 cubic feet
(c) 16 cubic feet

5.24 According to the regression line in Exercise 5.23, the predicted amount of gas used when the outside temperature is 20 degree-days is about

(a) 405 cubic feet.
(b) 320 cubic feet.
(c) 105 cubic feet.

5.25 By looking at the equation of the least-squares regression line in Exercise 5.23, you can see that the correlation between amount of gas used and degree-days is

(a) greater than zero.
(b) less than zero.
(c) Can't tell without seeing the data.

5.26 The software used to compute the least-squares regression line in Exercise 5.23 says that $r^2 = 0.98$. This suggests that

(a) although degree-days and gas used are correlated, degree-days does not predict gas used very accurately.
(b) gas used increases by $\sqrt{0.98} = 0.99$ cubic feet for each additional degree-day.
(c) prediction of gas used from degree-days will be quite accurate.

5.27 Because elderly people may have difficulty standing to have their heights measured, a study looked at predicting overall height from height to the knee. Here are data (in centimeters) for six elderly men: KNEEHT

Knee height x	57.7	47.4	43.5	44.8	55.2	54.6
Height y	192.1	153.3	146.4	162.7	169.1	177.8

Use your calculator or software: what is the equation of the least-squares regression line for predicting overall height from knee height?

(a) $\hat{y} = 42.9 + 2.5x$
(b) $\hat{y} = -3.4 + 0.3x$
(c) $\hat{y} = 2.5 + 42.9x$

CHAPTER 5 EXERCISES

5.28 **Penguins diving.** A study of king penguins looked for a relationship between how deep the penguins dive to seek food and how long they stay underwater.[10]

© Paul A. Souders/CORBIS

For all but the shallowest dives, there is a linear relationship that is different for different penguins. The study report gives a scatterplot for one penguin titled "The relation of dive duration (DD) to depth (D)." Duration DD is measured in minutes, and depth D is in meters. The report then says, "The regression equation for this bird is: DD = 2.69 + 0.0138D."

(a) What is the slope of the regression line? Explain in specific language what this slope says about this penguin's dives.

(b) According to the regression line, how long does a typical dive to a depth of 200 meters last?

(c) The dives varied from 40 meters to 300 meters in depth. Plot the regression line from D = 40 to D = 300.

5.29 **The price of diamond rings.** A newspaper advertisement in the *Straits Times* of Singapore contained pictures of diamond rings and listed their prices, diamond weight (in carats), and gold purity. Based on data for only the 20-carat gold ladies' rings in the advertisement, the least-squares regression line for predicting price (in Singapore dollars) from the weight of the diamond (in carats) is[11]

Price = 259.63 + 3721.02 Carats

(a) What does the slope of this line say about the relationship between price and number of carats?

(b) What is the predicted price when number of carats = 0? How would you interpret this price?

5.30 **Does social rejection hurt?** Exercise 4.45 (page 119) gives data from a study that shows that social exclusion causes "real pain." That is, activity in an area of the brain that responds to physical pain goes up as distress from social exclusion goes up. A scatterplot shows a moderately strong linear relationship. Figure 5.10 shows Minitab regression output for these data. [icon] REJECT

(a) What is the equation of the least-squares regression line for predicting brain activity from social distress score? Use the equation to predict brain activity for a social distress score of 2.0.

(b) What percent of the variation in brain activity among these subjects is explained by the straight-line relationship with social distress score?

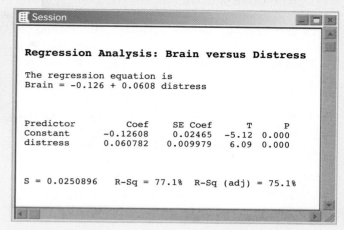

FIGURE 5.10
Minitab regression output for a study of the effects of social rejection on brain activity, for Exercise 5.30.

(c) Use the information in Figure 5.10 to find the correlation r between social distress score and brain activity. How do you know whether the sign of r is + or −?

5.31 **Toucan's beak.** Exercise 4.44 (page 118) gives data on beak heat loss, as a percent of total body heat loss from all sources, at various temperatures. The data show that beak heat loss is higher at higher temperatures and that the relationship is roughly linear. Figure 5.11 shows Minitab regression output for these data. [icon] TOUCAN

(a) What is the equation of the least-squares regression line for predicting beak heat loss, as a percent of total body heat loss from all sources, from temperature? Use the equation to predict beak heat loss, as a percent of total body heat loss from all sources, at a temperature of 25 degrees Celsius.

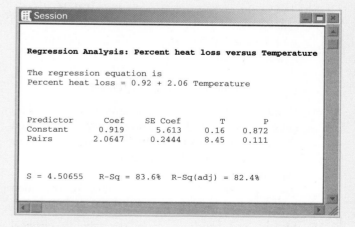

FIGURE 5.11
Minitab regression output for a study of how temperature affects beak heat loss in toucans, for Exercise 5.31.

(b) What percent of the variation in beak heat loss is explained by the straight-line relationship with temperature?

(c) Use the information in Figure 5.11 to find the correlation r between beak heat loss and temperature. How do you know whether the sign of r is $+$ or $-$?

5.32   **Husbands and wives.** The mean height of American women in their twenties is about 64.3 inches, and the standard deviation is about 2.7 inches. The mean height of men the same age is about 69.9 inches, with standard deviation about 3.1 inches. Suppose that the correlation between the heights of husbands and wives is about $r = 0.5$.

(a) What are the slope and intercept of the regression line of the husband's height on the wife's height in young couples?

(b) Draw a graph of this regression line for heights of wives between 56 and 72 inches. Predict the height of the husband of a woman who is 67 inches tall, and plot the wife's height and predicted husband's height on your graph.

(c) You don't expect this prediction for a single couple to be very accurate. Why not?

5.33   **What's my grade?** In Professor Krugman's economics course the correlation between the students' total scores prior to the final examination and their final-examination scores is $r = 0.5$. The pre-exam totals for all students in the course have mean 280 and standard deviation 40. The final-exam scores have mean 75 and standard deviation 8. Professor Krugman has lost Julie's final exam but knows that her total before the exam was 300. He decides to predict her final-exam score from her pre-exam total.

(a) What is the slope of the least-squares regression line of final-exam scores on pre-exam total scores in this course? What is the intercept?

(b) Use the regression line to predict Julie's final-exam score.

(c) Julie doesn't think this method accurately predicts how well she did on the final exam. Use r^2 to argue that her actual score could have been much higher (or much lower) than the predicted value.

5.34   **Going to class.** A study of class attendance and grades among first-year students at a state university showed that, in general, students who attended a higher percent of their classes earned higher grades. Class attendance explained 16% of the variation in grade index among the students. What is the numerical value of the correlation between percent of classes attended and grade index?

5.35   **Sisters and brothers.** How strongly do physical characteristics of sisters and brothers correlate? Here are data on the heights (in inches) of 12 adult pairs:[12] **BROSIS**

Brother	71	68	66	67	70	71	70	73	72	65	66	70
Sister	69	64	65	63	65	62	65	64	66	59	62	64

(a) Use your calculator or software to find the correlation and the equation of the least-squares line for predicting sister's height from brother's height. Make a scatterplot of the data, and add the regression line to your plot.

(b) Damien is 70 inches tall. Predict the height of his sister Tonya. Based on the scatterplot and the correlation r, do you expect your prediction to be very accurate? Why?

5.36   **Keeping water clean.** Keeping water supplies clean requires regular measurement of levels of pollutants. The measurements are indirect—a typical analysis involves forming a dye by a chemical reaction with the dissolved pollutant, then passing light through the solution and measuring its "absorbence." To calibrate such measurements, the laboratory measures known standard solutions and uses regression to relate absorbence and pollutant concentration. This is usually done every day. Here is one series of data on the absorbence for different levels of nitrates. Nitrates are measured in milligrams per liter of water.[13] **NITRATE**

Nitrates	50	50	100	200	400	800	1200	1600	2000	2000
Absorbence	7.0	7.5	12.8	24.0	47.0	93.0	138.0	183.0	230.0	226.0

(a) Chemical theory says that these data should lie on a straight line. If the correlation is not at least 0.997, something went wrong, and the calibration procedure is repeated. Plot the data and find the correlation. Must the calibration be done again?

(b) The calibration process sets nitrate level and measures absorbence. The linear relationship that results is used to estimate the nitrate level in water from a measurement of absorbence. What is the equation of the line used to estimate nitrate level? What does the slope of this line say about the relationship between nitrate level and absorbence? What is the estimated nitrate level in a water specimen with absorbence 40?

(c) Do you expect estimates of nitrate level from absorbence to be quite accurate? Why?

5.37   **Sparrowhawk colonies.** One of nature's patterns connects the percent of adult birds in a colony that return from the previous year and the number of new adults that join the colony. Here are data for 13 colonies of sparrowhawks:[14] **SPARROW**

Percent return x	74	66	81	52	73	62	52	45	62	46	60	46	38
New adults y	5	6	8	11	12	15	16	17	18	18	19	20	20

You saw in Exercise 4.28 that there is a moderately strong linear relationship, with correlation $r = -0.748$.

(a) Find the least-squares regression line for predicting y from x. Make a scatterplot and draw your line on the plot.

(b) Explain in words what the slope of the regression line tells us.

(c) An ecologist uses the line, based on 13 colonies, to predict how many new birds will join another colony, to which 60% of the adults from the previous year return. What is the prediction?

5.38 **Our brains don't like losses.** Exercise 4.29 (page 115) describes an experiment that showed a linear relationship between how sensitive people are to monetary losses ("behavioral loss aversion") and activity in one part of their brains ("neural loss aversion"). **LOSSES**

(a) Make a scatterplot with neural loss aversion as x and behavioral loss aversion as y. One point is a high outlier in both the x and y directions.

(b) Find the least-squares line for predicting y from x, *leaving out the outlier*, and add the line to your plot.

(c) The outlier lies very close to your regression line. Looking at the plot, you now expect that adding the outlier will increase the correlation but will have little effect on the least-squares line. Explain why.

(d) Find the correlation and the equation of the least-squares line with and without the outlier. Your results verify the expectations from (c).

5.39 **Always plot your data!** Table 5.1 presents four sets of data prepared by the statistician Frank Anscombe to illustrate the dangers of calculating without first plotting the data.[15] **ANSCOMA**

(a) Without making scatterplots, find the correlation and the least-squares regression line for all four data sets. What do you notice? Use the regression line to predict y for x = 10. **ANSCOMB**

(b) Make a scatterplot for each of the data sets, and add the regression line to each plot. **ANSCOMC**

(c) In which of the four cases would you be willing to use the regression line to describe the dependence of y on x? Explain your answer in each case. **ANSCOMD**

5.40 **Managing diabetes.** People with diabetes must manage their blood sugar levels carefully. They measure their fasting plasma glucose (FPG) several times a day with a glucose meter. Another measurement, made at regular medical checkups, is called HbA. This is roughly the percent of red blood cells that have a glucose molecule attached. It measures average exposure to glucose over a period of several months.

© Glow Wellness/Alamy

Table 5.2 (page 148) gives data on both HbA and FPG for 18 diabetics five months after they had completed a diabetes education class.[16] **DIABET**

(a) Make a scatterplot with HbA as the explanatory variable. There is a positive linear relationship, but it is surprisingly weak.

(b) Subject 15 is an outlier in the y direction. Subject 18 is an outlier in the x direction. Find the correlation for all 18 subjects, for all except Subject 15, and for all except Subject 18. Are either or both of these subjects influential for the correlation? Explain in simple language why r changes in opposite directions when we remove each of these points.

5.41 **The effect of changing units.** The equation of a regression line, unlike the correlation, depends on the units we use to measure the explanatory and response variables. Here are data on knee height

TABLE 5.1	FOUR DATA SETS FOR EXPLORING CORRELATION AND REGRESSION									
DATA SET A										
x	10	8	13	9	11	14	6	4	12	7	5
y	8.04	6.95	7.58	8.81	8.33	9.96	7.24	4.26	10.84	4.82	5.68
DATA SET B										
x	10	8	13	9	11	14	6	4	12	7	5
y	9.14	8.14	8.74	8.77	9.26	8.10	6.13	3.10	9.13	7.26	4.74
DATA SET C										
x	10	8	13	9	11	14	6	4	12	7	5
y	7.46	6.77	12.74	7.11	7.81	8.84	6.08	5.39	8.15	6.42	5.73
DATA SET D										
x	8	8	8	8	8	8	8	8	8	8	19
y	6.58	5.76	7.71	8.84	8.47	7.04	5.25	5.56	7.91	6.89	12.50

TABLE 5.2   TWO MEASURES OF GLUCOSE LEVEL IN DIABETICS

SUBJECT	HbA (%)	FPG (mg/ml)	SUBJECT	HbA (%)	FPG (mg/ml)	SUBJECT	HbA (%)	FPG (mg/ml)
1	6.1	141	7	7.5	96	13	10.6	103
2	6.3	158	8	7.7	78	14	10.7	172
3	6.4	112	9	7.9	148	15	10.7	359
4	6.8	153	10	8.7	172	16	11.2	145
5	7.0	134	11	9.4	200	17	13.7	147
6	7.1	95	12	10.4	271	18	19.3	255

and overall height (in centimeters) for six elderly men: **KNEETH2**

Knee height x	57.7	47.4	43.5	44.8	55.2	54.6
Height y	192.1	153.3	146.4	162.7	169.1	177.8

(a) Find the equation of the regression line for predicting overall height in centimeters from knee height in centimeters.

(b) A mad scientist decides to measure knee height in millimeters and height in meters. The same data in these units are

Knee height x	577	474	435	448	552	546
Height y	1.921	1.533	1.464	1.627	1.691	1.778

Find the equation of the regression line for predicting overall height in meters from knee height in millimeters.

(c) Use both lines to predict the overall height of a man whose knee height is 50 centimeters, which is the same as 500 millimeters. Use the fact that there are 100 centimeters in a meter to show that the two predictions are the same (up to roundoff error).

5.42 **Managing diabetes, continued.** Add three regression lines for predicting FPG from HbA to your scatterplot from Exercise 5.40: for all 18 subjects, for all except Subject 15, and for all except Subject 18. Is either Subject 15 or Subject 18 strongly influential for the least-squares line? Explain in simple language what features of the scatterplot explain the degree of influence. **DIABET**

5.43 **Are you happy?** Exercise 4.25 (page 114) discusses a study in which the mean BRFSS life-satisfaction score of individuals in each state was compared with the mean of an objective measure of well-being (based on the "compensating-differentials method") for each state. Suppose that instead of the means for the states, the BRFSS life-satisfaction scores for individuals were compared with the corresponding measure of well-being (based on the compensating-differentials method) for these individuals. Would you expect the correlation between the mean state scores on these two measures to be lower, about the same, or higher than the correlation between the scores of individuals on these two measures? Explain your answer.

5.44 **Do artificial sweeteners cause weight gain?** People who use artificial sweeteners in place of sugar tend to be heavier than people who use sugar. Does this mean that artificial sweeteners cause weight gain? Give a more plausible explanation for this association.

5.45 **Learning online.** Many colleges offer online versions of courses that are also taught in the classroom. It often happens that the students who enroll in the online version do better than the classroom students on the course exams. This does not show that online instruction is more effective than classroom teaching because the people who sign up for online courses are often quite different from the classroom students. Suggest some differences between online and classroom students that might explain why online students do better.

5.46 **Grade inflation and the SAT.** The effect of a lurking variable can be surprising when individuals are divided into groups. In recent years, the mean SAT score of all high school seniors has increased. But the mean SAT score has decreased for students at each level of high school grades (A, B, C, and so on). Explain how grade inflation in high school (the lurking variable) can account for this pattern.

5.47 **Workers' incomes.** Here is another example of the group effect cautioned about in the previous exercise. Explain how, as a nation's population grows older, median income can go down for workers in each age group, yet still go up for all workers.

5.48 **Some regression math.** Use the equation of the least-squares regression line (box on page 126) to show that the regression line for predicting y from x always passes through the point $(\bar{x}, \bar{y})$. That is, when $x = \bar{x}$, the equation gives $\hat{y} = \bar{y}$.

5.49 **Regression to the mean.** Figure 4.8 (page 114) displays the relationship between golfers' scores on the first and second rounds of the 2012 Masters

Tournament. The least-squares line for predicting second-round scores from first-round scores has equation $\hat{y} = 51.42 + 0.314x$. Find the predicted second-round scores for a player who shot 80 in the first round and for a player who shot 70. The mean second-round score for all players was 74.43. So a player who does well in the first round is predicted to do less well, but still better than average, in the second round. And a player who does poorly in the first is predicted to do better, but still worse than average, in the second.

regression to the mean   (*Comment:* This is **regression to the mean.** If you select individuals with extreme scores on some measure, they tend to have less extreme scores when measured again. That's because their extreme position is partly merit and partly luck. The luck will be different next time. Regression to the mean contributes to lots of "effects." The rookie of the year often doesn't do as well the next year; the best player in an orchestral audition may play less well once hired than the runners-up; a student who feels she needs coaching after taking the SAT often does better on the next try without coaching.)

5.50 **Regression to the mean.** We expect that students who do well on the midterm exam in a course will usually also do well on the final exam. Gary Smith of Pomona College looked at the exam scores of all 346 students who took his statistics class over a 10-year period.[17] The least-squares line for predicting final exam score from midterm-exam score was $\hat{y} = 46.6 + 0.41x$. (Both exams have a 100-point scale.)

Octavio scores 10 points above the class mean on the midterm. How many points above the class mean do you predict that he will score on the final? (*Hint:* Use the fact that the least-squares line passes through the point $(\bar{x}, \bar{y})$ and the fact that Octavio's midterm score is $\bar{x} + 10$.) This is another example of regression to the mean: students who do well on the midterm will, in general, do less well, but still above average, on the final.

5.51 **Is regression useful?** In Exercise 4.41 (page 118) you used the *Correlation and Regression* applet to create three scatterplots having correlation about $r = 0.7$ between the horizontal variable x and the vertical variable y. Create three similar scatterplots again, and click the "Show least-squares line" box to display the regression lines. Correlation $r = 0.7$ is considered reasonably strong in many areas of work. Because there is a reasonably strong correlation, we might use a regression line to predict y from x. In which of your three scatterplots does it make sense to use a straight line for prediction?

5.52 **Guessing a regression line.** In the *Correlation and Regression* applet, click on the scatterplot to create a group of 15 to 20 points from lower left to upper right with a clear positive straight-line pattern (correlation around 0.7). Click the "Draw line" button and use the mouse (right-click and drag) to draw a line through the middle of the cloud of points from lower left to upper right. Note the "thermometer" above the plot. The red portion is the sum of the squared vertical distances from the points in the plot to the least-squares line. The green portion is the "extra" sum of squares for your line—it shows by how much your line misses the smallest possible sum of squares.

(a) You drew a line by eye through the middle of the pattern. Yet the right-hand part of the bar is probably almost entirely green. What does that tell you?

(b) Now click the "Show least-squares line" box. Is the slope of the least-squares line smaller (the new line is less steep) or larger (line is steeper) than that of your line? If you repeat this exercise several times, you will consistently get the same result. The least-squares line minimizes the *vertical* distances of the points from the line. It is *not* the line through the "middle" of the cloud of points. This is one reason why it is hard to draw a good regression line by eye.

The following exercises ask you to answer questions from data without having the details outlined for you. The exercise statements give you the State step of the four-step process. In your work, follow the Plan, Solve, and Conclude steps of the process, described on page 60.

5.53 **Beavers and beetles.** Do beavers benefit beetles? Researchers laid out 23 circular plots, each 4 meters in diameter, in an area where beavers were cutting down cottonwood trees. In each plot, they counted the number of stumps from trees cut by beavers and the number of clusters of beetle larvae. Ecologists think that the new sprouts from stumps are more tender than other cottonwood growth, so that beetles prefer them. If so, more stumps should produce more beetle larvae. Here are the data:[18] **BEAVERS**

Stumps	2	2	1	3	3	4	3	1	2	5	1	3
Beetle larvae	10	30	12	24	36	40	43	11	27	56	18	40
Stumps	2	1	2	2	1	1	4	1	2	1	4	
Beetle larvae	25	8	21	14	16	6	54	9	13	14	50	

Analyze these data to see if they support the "beavers benefit beetles" idea.

5.54 **A computer game.** A multimedia statistics learning system includes a test of skill in using the computer's mouse. The software displays a circle at a random location on the computer screen. The subject clicks

TABLE 5.3   REACTION TIMES (IN MILLISECONDS) IN A COMPUTER GAME

TIME	DISTANCE	HAND	TIME	DISTANCE	HAND
115	190.70	right	240	190.70	left
96	138.52	right	190	138.52	left
110	165.08	right	170	165.08	left
100	126.19	right	125	126.19	left
111	163.19	right	315	163.19	left
101	305.66	right	240	305.66	left
111	176.15	right	141	176.15	left
106	162.78	right	210	162.78	left
96	147.87	right	200	147.87	left
96	271.46	right	401	271.46	left
95	40.25	right	320	40.25	left
96	24.76	right	113	24.76	left
96	104.80	right	176	104.80	left
106	136.80	right	211	136.80	left
100	308.60	right	238	308.60	left
113	279.80	right	316	279.80	left
123	125.51	right	176	125.51	left
111	329.80	right	173	329.80	left
95	51.66	right	210	51.66	left
108	201.95	right	170	201.95	left

in the circle with the mouse as quickly as possible. A new circle appears as soon as the subject clicks in the old one. Table 5.3 gives data for one subject's trials, 20 with each hand. Distance is the distance from the cursor location to the center of the new circle, in units whose actual size depends on the size of the screen. Time is the time required to click in the new circle, in milliseconds.[19] We suspect that time depends on distance. We also suspect that performance will not be the same with the right and left hands. Analyze the data with a view to predicting performance separately for the two hands.

COMGAME

5.55   Predicting tropical storms. William Gray heads the Tropical Meteorology Project at Colorado State University (well away from the hurricane belt). His forecasts before each year's hurricane season attract lots of attention. Here are data on the number of named Atlantic tropical storms predicted by Dr. Gray and the actual number of storms for the years 1984 to 2012:[20]

STORMS2

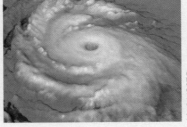

NASA/GSFC

YEAR	FORECAST	ACTUAL	YEAR	FORECAST	ACTUAL
1984	10	12	1999	14	12
1985	11	11	2000	12	14
1986	8	6	2001	12	15
1987	8	7	2002	11	12
1988	11	12	2003	14	16
1989	7	11	2004	14	14
1990	11	14	2005	15	27
1991	8	8	2006	17	10
1992	8	6	2007	17	14
1993	11	8	2008	15	16
1994	9	7	2009	11	9
1995	12	19	2010	18	19
1996	10	13	2011	16	19
1997	11	7	2012	13	19
1998	10	14			

Analyze these data. How accurate are Dr. Gray's forecasts? How many tropical storms would you expect in a year when his preseason forecast calls for 16 storms? What is the effect of the disastrous 2005 season on your answers?

TABLE 5.4 ARCTIC RIVER DISCHARGE (CUBIC KILOMETERS), 1936 TO 2010

YEAR	DISCHARGE	YEAR	DISCHARGE	YEAR	DISCHARGE	YEAR	DISCHARGE
1936	1721	1955	1656	1974	2000	1993	1845
1937	1713	1956	1721	1975	1928	1994	1902
1938	1860	1957	1762	1976	1653	1995	1842
1939	1739	1958	1936	1977	1698	1996	1849
1940	1615	1959	1906	1978	2008	1997	2007
1941	1838	1960	1736	1979	1970	1998	1903
1942	1762	1961	1970	1980	1758	1999	1970
1943	1709	1962	1849	1981	1774	2000	1905
1944	1921	1963	1774	1982	1728	2001	1890
1945	1581	1964	1606	1983	1920	2002	2085
1946	1834	1965	1735	1984	1823	2003	1780
1947	1890	1966	1883	1985	1822	2004	1900
1948	1898	1967	1642	1986	1860	2005	1930
1949	1958	1968	1713	1987	1732	2006	1910
1950	1830	1969	1742	1988	1906	2007	2270
1951	1864	1970	1751	1989	1932	2008	2078
1952	1829	1971	1879	1990	1861	2009	1900
1953	1652	1972	1736	1991	1801	2010	1813
1954	1589	1973	1861	1992	1793		

5.56 Great Arctic rivers. One effect of global warming is to increase the flow of water into the Arctic Ocean from rivers. Such an increase may have major effects on the world's climate. Six rivers (Yenisey, Lena, Ob, Pechora, Kolyma, and Severnaya Dvina) drain two-thirds of the Arctic in Europe and Asia. Several of these are among the largest rivers on earth. Table 5.4 presents the total discharge from these rivers each year from 1936 to 2010.[21] Discharge is measured in cubic kilometers of water. Analyze these data to uncover the nature and strength of the trend in total discharge over time. ARCTIC

5.57 Will women outrun men? Does the physiology of women make them better suited than men to long-distance running? Will women eventually outperform men in long-distance races? Researchers examined data on world record times (in seconds) for men and women in the marathon. Based on these data, researchers (in 1992) attempted to predict when women would outrun men in the marathon. Here are data for women:[22] ▦ RUNNING

Year	1926	1964	1967	1970	1971	1974	1975
Time	13,222.0	11,973.0	11,246.0	10,973.0	9990.0	9834.5	9499.0

Year	1977	1980	1981	1982	1983	1985
Time	9287.5	9027.0	8806.0	8771.0	8563.0	8466.0

Here are data for men:

Year	1908	1909	1913	1920	1925	1935	1947
Time	10,518.4	9751.0	9366.6	9155.8	8941.8	8802.0	8739.0

Year	1952	1953	1954	1958	1960	1963	1964
Time	8442.2	8314.8	8259.4	8117.0	8116.2	8068.0	7931.2

Year	1965	1967	1969	1981	1984	1985	1988
Time	7920.0	7776.4	7713.6	7698.0	7685.0	7632.0	7610.0

Analyze these data using least-squares regression to estimate when men and women's record times will be equal. How reliable is your estimate?

Exploring the Web

5.58 Association and causation. Find an example of a study in which the issue of association and causation is present. This can be either an example in which association is confused with causation or an example in which the association is not confused with causation. Summarize the study and its conclusions in your own words. Be sure to include either a copy of the actual article or at least the Web source, title, and where the article was published. The *Chance News* Web site at www.causeweb.org/wiki/chance/index.php/Main_Page is a good place to look for examples.

5.59 Predicting batting averages. Go to www.mlb.com/and find the batting averages for a diverse set of 30 players for both the 2012 and 2013 seasons. You can click on the "Stats" tab to find the results for the current season as well as historical data. You should select only players who played in at least 50 games both seasons. Make a scatterplot of the batting averages using the 2012 season average as the explanatory variable and the 2013 season average as the response. Is it reasonable to fit a straight line to these data? If so, find the least-squares regression line for predicting batting average in 2013 from that in 2012 based on your sample of 30 players. In 2012, the major league leader in batting was Buster Posey, who had a batting average of .344. What does your least-squares regression line predict for the 2012 batting average of someone who hit .336 in 2012? Is the 2013 predicted batting average higher or lower than .336?

5.60 Predicting the federal budget. Go to the Congressional Budget Office Web site, www.cbo.gov/topics/. Click on the "topics and subtopics" button and select "Budget and Economic Outlook." Then click on "apply." What is the current prediction for the federal budget in five years' time? Is a surplus or a deficit predicted? Do you think the prediction is accurate? Why or why not?

Two-Way Tables

Overview

Some would argue that this chapter can be skipped without compromising the continuity of course material and that skills learned in this chapter are not essential to success in later chapters. However, there are at least four arguments that spending time on this material is a good investment. First, this material deals with categorical data, which is central to the work of many social scientists. Second, there is a direct connection between this material and that in Chapter 18 ("Comparing Two Proportions"). Third, understanding that conditional and marginal distributions each convey different (and sometimes conflicting) information about relationships between the two variables is part of what constitutes basic statistical literacy—we want students to think carefully about numbers they encounter in the media; consideration of these distributions can ease the discussion of conditional probability to come in Chapter 12 (if you plan to cover that material). Finally, some of the ideas described in this chapter parallel the upcoming discussions of experimental design and sampling in Chapters 8 and 9. For example, the notion of a "lurking variable" arises in discussions of Simpson's paradox. Still, if you prefer to wait on Chapter 6 until later in the course (perhaps pairing it with Chapter 24, "Two Categorical Variables: The Chi-Square Test"), that is fine.

> ### LEARNING OBJECTIVES**
>
> - From a two-way table of counts, find the marginal distributions of both variables by obtaining the row sums and column sums.
>
> - Express any distribution in percents by dividing the category counts by their total.
>
> - Describe the relationship between two categorical variables by computing and comparing percents. Often this involves comparing the conditional distributions of one variable for the different categories of the other variable.
>
> - Recognize Simpson's paradox and be able to explain it.
>
> **These learning outcomes appear later for the students in Chapter 7: "Part I Review."

This chapter is concerned with revealing relationships between two categorical variables, as opposed to that of Chapters 4 and 5 on regression, which dealt with relating two quantitative variables. Perhaps the most interesting revelation for students is the notion that a marginal distribution may reveal a different relationship than the conditional distribution. It is probably sufficient to spend one or two class periods on this material, all of which can be described using one compelling example.

When teaching this chapter, it is important to recognize that students sometimes have trouble determining conditional probabilities, both when they are required and how to calculate them. We have found it helpful to ask students to focus on "what we know first." For example, the following three questions are all different: (1) What proportion of registered voters are male Republicans? (2) What proportion of registered voters who are male are Republican? (3) What proportion of registered voters who are Republican are male? Question #1 is not a conditional probability—here we are being asked about those registered voters who are male <u>and</u> Republican. Questions #2 and #3 are both conditional—in Question #2 we first know that the registered voter is male, while in Question #3 we first know that the registered voter is Republican. Wording of these questions can be subtle, so it is important to have your students think about what the question is asking. You might point out that "key words" such as "if," "when," and "given" signal a conditional question.

Teaching Suggestions and Additional Examples/Activities for the Classroom

1. Motivate Ideas with an Interesting Example

Consider using one or two interesting examples to describe all the material in this chapter. A well-known and interesting example is provided by considering admission rates for men and women applicants to six academic departments at Berkeley for Fall 1973 (P. J. Bickel, E. A. Hammel, and J. W. O'Connell (1975). "Sex Bias in Graduate Admissions: Data from Berkeley." *Science* 187(4175): 398–404).

You can use these data to illustrate bar graphs (again), compute the marginal distribution of sex, and compute the marginal distribution of admission status. You can then examine the conditional distributions of admission, given sex. Finally, you can use these data to illustrate Simpson's paradox:

When we examine the table "collapsed" on department, men are admitted at a higher rate than women:

	Number of Applicants	% Admitted
Men	8442	44%
Women	4321	35%

However, when we look at each department individually, this isn't the case. In fact, in most departments, women have a higher admission rate than men.

The "lurking variable," or explanation, is that men and women choose to apply to different kinds of departments—most men applied to departments that had higher admission rates (for both sexes), whereas most women applied to departments with lower admission rates (for both sexes).

Department	Men		Women	
	Applicants	% Admitted	Applicants	% Admitted
A	825	62%	108	82%
B	560	63%	25	68%
C	325	37%	593	34%
D	417	33%	375	35%
E	191	28%	393	24%
F	272	6%	341	7%

2. Discuss Lurking Variables and Simpson's Paradox

Also consider using the data provided in Example 6.4 ("Do Medical Helicopters Save Lives?"). In this example, "accident severity" is the lurking variable—victims of more serious accidents tend to be transported by helicopter and are also more likely to die than victims of less serious accidents.

3. Encourage Visualization of Simpson's Paradox

If you would like to help your students visualize Simpson's paradox, the applet developed by Scheiter and Symanzik might be helpful ("An Applet for the Investigation of Simpson's Paradox," *Journal of Statistics Education*, 21(1), 2013; available online at http://www.amstat.org/publications/jse/v21n1/schneiter.pdf).

Other Resources (LaunchPad)

EESEE Case Studies
    Surviving the Titanic
    On Time Flights
    Does Smoking Improve Survival?
    Cancer and Power Lines

© Purestock/age fotostock

Two-Way Tables*

In this chapter we cover...

- Marginal distributions
- Conditional distributions
- Simpson's paradox

We have concentrated on relationships in which at least the response variable is quantitative. Now we will describe relationships between two categorical variables. Some variables—such as sex, race, and occupation—are categorical by nature. Other categorical variables are created by grouping values of a quantitative variable into classes. Published data often appear in grouped form to save space. To analyze categorical data, we use the *counts* or *percents* of individuals that fall into various categories.

EXAMPLE 6.1	A Job Outside the Home

A sample survey of adults (aged 18 and over) asked, "If you were free to do either, would you prefer to have a job outside the home, or would you prefer to stay home and take care of the house and family?" Table 6.1 shows the responses.[1] This is a **two-way table** because it describes two categorical variables. One is the sex and education level of the respondent. The other is the preferred lifestyle (a job outside the home, stay home, or no preference). Sex and education is the **row variable** because each row in the table describes a combination of sex and education level. Because the education level has a natural order from "No college" to "College," the rows are also in this order for women and men. Preferred lifestyle is the **column variable** because each column describes one choice. The entries in the table are the counts of individuals in each sex-and-education-level-by-preferred-lifestyle class. ■

two-way table

row and column variables

*This material is important in statistics, but it is needed later in this book only for Chapter 24. You may omit it if you do not plan to read Chapter 24 or delay reading it until you reach Chapter 24.

	PREFERRED LIFESTYLE			
SEX AND EDUCATION	**JOB OUTSIDE HOME**	**STAY HOME AND CARE FOR HOUSE, FAMILY**	**NO PREFERENCE**	**TOTAL**
Women no college	81	104	10	195
Women college	173	115	15	303
Men no college	92	32	2	126
Men college	299	81	8	388
Total	645	332	35	1012

TABLE 6.1   ADULTS BY PREFERRED LIFESTYLE AND SEX AND EDUCATION

Marginal distributions

How can we best grasp the information contained in Table 6.1? First, *look at the distribution of each variable separately.* The distribution of a categorical variable says how often each outcome occurred. The "Total" column at the right of the table contains the totals for each of the rows. These row totals give the distribution of sex and education level in the entire group of 1012 adults: 195 were women with no college education, 303 were women with a college education, and so on.

If the row and column totals are missing, the first thing to do in studying a two-way table is to calculate them. The distributions of sex and education alone and

marginal distribution   preferred lifestyle alone are called **marginal distributions** because they appear at the right and bottom margins of the two-way table.

Percents are often more informative than counts. We can display the marginal distribution of sex and education in percents by dividing each row total by the table total and converting to a percent.

EXAMPLE 6.2   **Calculating a Marginal Distribution**

The percent of these adults who were women with no college is

$$\frac{\text{women with no college total}}{\text{table total}} = \frac{195}{1012} = 0.193 = 19.3\%$$

Do three more such calculations to obtain the marginal distribution of sex and education in percents for each group. Here is the complete distribution:

Response	Percent
Women no college	$\frac{195}{1012} = 19.3\%$
Women college	$\frac{303}{1012} = 29.9\%$
Men no college	$\frac{126}{1012} = 12.5\%$
Men college	$\frac{388}{1012} = 38.3\%$

It seems that more women and more men have attended college than not attended college. The total is 100% because everyone belongs to one of the four sex and education classes. ■

Each marginal distribution from a two-way table is a distribution for a single categorical variable. As we saw in Chapter 1, we can use a bar graph or a pie chart to display such a distribution. Figure 6.1 is a bar graph of the distribution of sex and education among adults.

In working with two-way tables, you must calculate lots of percents. Here's a tip to help you decide what fraction gives the percent you want. Ask, "What group represents the total of which I want a percent?" The count for that group is the denominator of the fraction that leads to the percent. In Example 6.2, we want a percent "of adults," so the count of adults (the table total) is the denominator.

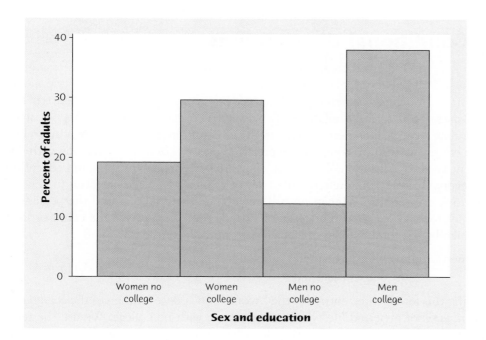

FIGURE 6.1

Bar graph of the distribution of sex and education of adults who participated in the survey. This is one of the marginal distributions for Table 6.1.

Apply Your Knowledge

6.1 Video Gaming and Grades. The popularity of computer, video, online, and virtual reality games has raised concerns about their ability to negatively affect youth. The data in this exercise are based on a recent survey of 14- to 18-year-olds in Connecticut high schools. Here are the grade distributions of boys who have and have not played video games.[2] ▦|▥ **GAMING**

	Grade Average		
	A's and B's	C's	D's and F's
Played games	736	450	193
Never played games	205	144	80

(a) How many people does this table describe? How many of these have played video games?

(b) Give the marginal distribution of the grades. What percent of the boys represented in the table received a grade of C or lower?

6.2 Ages of College Students. Here is a two-way table of U.S. Census Bureau data describing the age and sex of all American students enrolled in college. The table entries are counts in thousands of students.[3] ▦|▥ **AGES**

Keith Bedford/The New York Times/Redux

Age Group	Female	Male
18 to 24 years	6432	5640
25 to 34 years	2450	1843
35 years or older	2124	1069

(a)   How many college undergraduates are there?

(b)   Find the marginal distribution of age group. What percent of undergraduates are in the 18- to 24-year-old college age group?

Conditional distributions

 Table 6.1 contains much more information than the two marginal distributions of sex and education alone and preferred lifestyle alone. *Marginal distributions tell us nothing about the relationship between two variables.* To describe a relationship between two categorical variables, we must calculate some well-chosen percents from the counts given in the body of the table.

Let's say we want to compare the preferred lifestyles of women and men with different levels of education. To do this, compare percents for each sex and education category. To study the preferred lifestyles of women who have not been to college, we look only at the "Women no college" row in Table 6.1. To find the percent *of women with no college* who prefer a job outside the home, divide the count of such women by the total number of women with no college (the row total):

$$\frac{\text{women with no college who prefer a job outside the home}}{\text{row total}} = \frac{81}{195} = 0.415 = 41.5\%$$

Doing this for all three entries in the "Women no college" row gives the *conditional distribution* of preferred lifestyles among women with no college. We use the term *conditional* because this distribution describes only adults who satisfy the condition that they are women with no college.

 SMILING FACES
Women smile more than men. The same data that produce this fact allow us to link smiling to other variables in two-way tables. For example, as the second variable add whether or not the person thinks they are being observed. If yes, that's when women smile more. If no, there's no difference between women and men. Next, take the second variable to be the person's social role (for example, is he or she the boss in an office?). Within each role, there is very little difference in smiling between women and men.

Marginal and Conditional Distributions

The **marginal distribution** of one of the categorical variables in a two-way table of counts is the distribution of values of that variable among all individuals described by the table.

A **conditional distribution** of a variable is the distribution of values of that variable among only individuals who have a given value of the other variable. There is a separate conditional distribution for each value of the other variable.

EXAMPLE 6.3   Comparing Different Sex and Education Groups

STATE: How do women and men with different levels of education differ in their responses to the question "If you were free to do either, would you prefer to have a job outside the home, or would you prefer to stay home and take care of the house and family?"

PLAN: Make a two-way table of response by sex and education category. Find the four conditional distributions of response for each sex and education category alone. Compare these four distributions.

SOLVE: Table 6.1 is the two-way table we need. Look first at just the "Women no college" row to find the conditional distribution for women with no college, then at just the "Women college" row to find the conditional distribution for women who have been to college, and so on. Here are the calculations and the four conditional distributions:

Response	Job Outside Home	Stay Home	No Preference
Women no college	$\frac{81}{195} = 41.5\%$	$\frac{104}{195} = 53.3\%$	$\frac{10}{195} = 5.1\%$
Women college	$\frac{173}{303} = 57.1\%$	$\frac{115}{303} = 38.0\%$	$\frac{15}{303} = 5.0\%$
Men no college	$\frac{92}{126} = 73.0\%$	$\frac{32}{126} = 25.4\%$	$\frac{2}{126} = 1.6\%$
Men college	$\frac{299}{388} = 77.1\%$	$\frac{81}{388} = 20.9\%$	$\frac{8}{388} = 2.1\%$

The percents in each row should be 100% because for each sex and education category, everyone holds one of the three lifestyle preferences. In fact, the percents add to 99.9%, 100.0%, or 100.1% because we round each one to the nearest tenth. This is **roundoff error**.

roundoff error

Each set of percents adds to 100% because everyone holds one of the three opinions.

CONCLUDE: Men are more likely to prefer a job outside the home than are women, whether they went to college or not. Adults who attended college are more likely to prefer a job outside the home, whether they are women or men. ■

Software will do these calculations for you. Most programs allow you to choose which conditional distributions you want to compare. The output in Figure 6.2 presents the four conditional distributions of preferred lifestyle, for each sex and education category, and also the marginal distribution of opinion for all the adults. The distributions agree (up to roundoff) with the results in Examples 6.2 and 6.3.

FIGURE 6.2

Minitab and CrunchIt! output for the two-way table of adults by preferred lifestyle and sex and education. Each entry in the Minitab output includes the percent of its row total. The "Men college," "Men no college," "Women college," and "Women no college" rows give the conditional distributions of responses for each sex and education category, and the "All" row shows the marginal distribution of responses for all these adults. Notice that Minitab orders variables in the table alphabetically. Each entry in the CrunchIt! output includes the percent of its row total, the percent of its column total, and the percent of the entire table total. The second entry in each cell gives the conditional distribution of responses for the different sex and education categories. The third entry in each cell for the preferred lifestyle columns gives the conditional distributions of responses for each category of preferred living. The "All" row and column show the corresponding marginal distribution of responses for all these adults.

```
Session                                                    _ □ ×

 Rows: Sex and education  Columns: Preferred lifestyle
                         Job
                      outside        No
                        home  preference  Stay home       All

 Men college             299         8         81         388
                       77.06      2.06      20.88      100.00

 Men no college          92         2         32         126
                       73.02      1.59      25.40      100.00

 Women college          173        15        115         303
                       57.10      4.95      37.95      100.00

 Women no college        81        10        104         195
                       41.54      5.13      53.33      100.00

 All                    645        35        332        1012
                       63.74      3.46      32.81      100.00

 Cell Contents:       Count
                      % of Row
```

(Continued)

FIGURE 6.2
(Continued)

	Stay home	Job outside home	No preference	All
Women no college	104	81	10	195
	53.33	41.54	5.128	100
	31.33	12.56	28.57	19.27
	10.28	8.004	0.9881	19.27
Men no college	32	92	2	126
	25.40	73.02	1.587	100
	9.639	14.26	5.714	12.45
	3.162	9.091	0.1976	12.45
Men college	81	299	8	388
	20.88	77.06	2.062	100
	24.40	46.36	22.86	38.34
	8.004	29.55	0.7905	38.34
Women college	115	173	15	303
	37.95	57.10	4.950	100
	34.64	26.82	42.86	29.94
	11.36	17.09	1.482	29.94
All	332	645	35	1012
	3281	63.74	3.458	100
	100	100	100	100
	3281	63.74	3.458	100

Count   % of Row   % of Col   % of Total

```
Chi-squared statistic: 83.10
df:                    6
P-value:               <0.0001
```

Remember that there are two sets of conditional distributions for any two-way table. Example 6.3 looked at the conditional distributions of preferred lifestyles for the four sex and education categories. We could also examine the three conditional distributions of sex and education, one for each of the three preferred lifestyles, by looking separately at the three columns in Table 6.1 (page 154). Figure 6.3 makes this comparison in a bar graph. Each bar is divided (segmented) into four parts, represented by four colors. The upper portion of each bar represents the percent of women with no college among adults who prefer each lifestyle. The other portions represent the percents of each of the other sex and education categories. Each bar has a height of 100% because each bar represents all the adults in each different group of people. Bar graphs like that in Figure 6.3 in which each bar is divided into parts, each part representing a different

segmented bar graphs   category, are sometimes called **segmented bar graphs.**

No single graph (such as a scatterplot) portrays the form of the relationship between categorical variables. No single numerical measure (such as the correlation) summarizes the strength of the association. Bar graphs are flexible enough to be helpful, but you must think about what comparisons you want to display. For numerical measures, we rely on well-chosen percents. You must decide

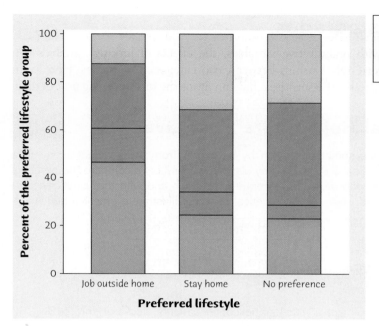

FIGURE 6.3
Bar graph comparing the percents of women and no college (orange), women and college (blue), men and no college (green), and men and college (red) among those who prefer each lifestyle.

which percents you need. Here is a hint: *if there is an explanatory-response relationship, compare the conditional distributions of the response variable for the separate values of the explanatory variable.* If you think that sex and education influences adults' preferred lifestyle, compare the conditional distributions of preferred lifestyle for each sex and education category, as in Example 6.3.

Apply Your Knowledge

6.3 Video Gaming and Grades. Exercise 6.1 (page 155) gives data on the grade distribution of boys who have and have not played video games. To see the relationship between grades and game-playing experience, find the conditional distributions of grades (the response variable) for players and nonplayers. What do you conclude? **GAMING**

6.4 Ages of College Students. Exercise 6.2 (page 155) gives U.S. Census Bureau data describing the age and sex of all American students enrolled in college. We suspect that the percent of women is higher among students in the 25- to 34-year old age group than in the 18- to 24-year age group. Do the data support this suspicion? Follow the four-step process as illustrated in Example 6.3. **AGES**

6.5 Marginal Distributions Aren't the Whole Story. Here are the row and column totals for a two-way table with two rows and two columns:

a	b	50
c	d	50
60	40	100

Make up *two different* sets of counts a, b, c, and d for the body of the table that give these same totals. This shows that the relationship between two variables cannot be obtained from the two individual distributions of the variables.

Simpson's paradox

As is the case with quantitative variables, the effects of lurking variables can change or even reverse relationships between two categorical variables. Here is an example that demonstrates the surprises that can await the unsuspecting user of data.

EXAMPLE 6.4	Do Medical Helicopters Save Lives?

© Ashley Cooper/CORBIS

Accident victims are sometimes taken by helicopter from the accident scene to a hospital. Helicopters save time. Do they also save lives? Let's compare the percents of accident victims who die with helicopter evacuation and with the usual transport to a hospital by road. Here are hypothetical data that illustrate a practical difficulty.[4]

	Helicopter	Road
Victim died	64	260
Victim survived	136	840
Total	200	1100

We see that 32% (64 out of 200) of helicopter patients died, but only 24% (260 out of 1100) of the others did. That seems discouraging.

The explanation is that the helicopter is sent mostly to serious accidents, so that the victims transported by helicopter are more often seriously injured. They are more likely to die with or without helicopter evacuation. Here are the same data broken down by the seriousness of the accident:

Serious Accidents				Less Serious Accidents		
	Helicopter	Road			Helicopter	Road
Died	48	60		Died	16	200
Survived	52	40		Survived	84	800
Total	100	100		Total	100	1000

Inspect these tables to convince yourself that they describe the same 1300 accident victims as the original two-way table. For example, 200 (100 + 100) were moved by helicopter, and 64 (48 + 16) of these died.

Among victims of serious accidents, the helicopter saves 52% (52 out of 100) compared with 40% for road transport. If we look only at less serious accidents, 84% of those transported by helicopter survive, versus 80% of those transported by road. Both groups of victims have a higher survival rate when evacuated by helicopter. ■

How can it happen that the helicopter does better for both groups of victims but worse when all victims are lumped together? Examining the data makes the explanation clear. Half the helicopter transport patients are from serious accidents, compared with only 100 of the 1100 road transport patients. So the helicopter carries patients who are more likely to die. The seriousness of the accident was a lurking variable that, until we uncovered it, hid the true relationship between survival and mode of transport to a hospital. Example 6.4 illustrates *Simpson's paradox.*

Simpson's Paradox

An association or comparison that holds for all of several groups can reverse direction when the data are combined to form a single group. This reversal is called **Simpson's paradox.**

The lurking variable in Simpson's paradox is categorical. That is, it breaks the individuals into groups, as when accident victims are classified as injured in a "serious accident" or a "less serious accident." Simpson's paradox is just an extreme form of the fact that observed associations can be misleading when there are lurking variables.

Apply Your Knowledge

6.6 Field Goal Shooting. Here are data on field goal shooting for two members of the Kent State University 2002–2003 women's basketball team:[5]

BSKTBLL

	Jamie Rubis		Lindsay Shearer	
	Made	Missed	Made	Missed
Two-pointers	119	115	86	84
Three-pointers	36	61	5	16

(a) What percent of all field goal attempts did Jamie Rubis make? What percent of all field goal attempts did Lindsay Shearer make?

(b) Now find the percent of all two-point field goals and all three-point field goals that Jamie made. Do the same for Lindsay.

(c) Lindsay had a lower percent for *both* types of field goals, but had a better overall percent. That sounds impossible. Explain carefully, referring to the data, how this can happen.

6.7 Bias in the Jury Pool? The New Zealand Department of Justice did a study of the composition of juries in court cases. Of interest was whether Maori, the indigenous people of New Zealand, were adequately represented in jury pools. Here are the results for two districts, Rotorua and Nelson, in New Zealand (similar results were found in all districts):[6] **JURY**

Mike Powell/Getty Images

Rotorua	Maori	Non-Maori
In jury pool	79	258
Not in jury pool	8810	23,751
Total	8889	24,009

Nelson	Maori	Non-Maori
In jury pool	1	56
Not in jury pool	1328	32,602
Total	1329	32,658

(a) Compare percents to show that the percent of all Maori in the jury pool in each district is less than the percent of non-Maori in the jury pool.

(b) Combine the data into a single two-way table of outcome ("in jury pool" or "not in jury pool") by ethnicity (Maori or non-Maori). The original study only reported such an overall rate. Which ethnic group has a higher percent of its people in the jury pool?

(c) Explain from the data, in language that a reporter can understand, how Maori can have a higher percent overall even though non-Maori have higher percents for both districts.

CHAPTER 6 SUMMARY

Chapter Specifics

■ A **two-way table** of counts organizes data about two categorical variables. Values of the **row variable** label the rows that run across the table, and values of the **column variable** label the columns that run down the table. Two-way tables are often used to summarize large amounts of information by grouping outcomes into categories.

- The **row totals** and **column totals** in a two-way table give the **marginal distributions** of the two individual variables. It is clearer to present these distributions as percents of the table total. Marginal distributions tell us nothing about the relationship between the variables.

- There are two sets of **conditional distributions** for a two-way table: the distributions of the row variable for each fixed value of the column variable and the distributions of the column variable for each fixed value of the row variable. Comparing one set of conditional distributions is one way to describe the association between the row and the column variables.

- To find the **conditional distribution** of the row variable for one specific value of the column variable, look only at that one column in the table. Find each entry in the column as a percent of the column total.

- **Bar graphs** are a flexible means of presenting categorical data. There is no single best way to describe an association between two categorical variables.

- A comparison between two variables that holds for each individual value of a third variable can be changed or even reversed when the data for all values of the third variable are combined. This is **Simpson's paradox.** Simpson's paradox is an example of the effect of lurking variables on an observed association.

Link It

In Chapters 4 and 5 we considered relationships between two quantitative variables. In this chapter we use two-way tables to describe relationships between two *categorical* variables. To explore relationships between two categorical variables, we examine their conditional distributions. Changes in the pattern of the conditional distribution of one variable as the value of the other varies provides information about the relationship between the variables. No change in this pattern suggests that there is no relationship.

As in Chapters 4 and 5, we must be careful not to assume that the patterns we observe would continue to hold for additional data or in a broader setting. Simpson's paradox is an example of how such an assumption could mislead us.

An important question is whether a pattern we observe while exploring a given set of data holds for values of our variables that we have not observed—in other words, that additional data would continue to conform to these patterns. We will begin to answer this question in Part II. And in Chapter 24 we will learn the answer to this question for two-way tables.

CHECK YOUR SKILLS

The Pew Internet and American Life Project interviewed several hundred teens (ages 12 to 17). One question asked was, "How often do you take your cell phone to school?" Here is a two-way table of the responses by how permissive the school is with regard to cell phone use:[7]

FREQUENCY	FORBID	ALLOW IN SCHOOL BUT NOT IN CLASS	ALLOW IN CLASS
Never	25	19	4
Less often	14	31	6
At least several times per week	14	23	8
Every day	97	314	57

Exercises 6.8 to 6.16 are based on this table. PHONE

6.8 How many individuals are described by this table?

(a) 468   (b) 612   (c) Need more information

6.9 How many teens from schools that forbid cell phones were among the respondents?

(a) 48   (b) 150   (c) Need more information

6.10 The percent of teens from schools that forbid cell phones among the respondents was

(a) about 8%.
(b) about 25%.
(c) about 48%.

6.11 Your percent from the previous exercise is part of

(a) the marginal distribution of school permissiveness.
(b) the marginal distribution of the frequency at which a teen brought a cell phone to school.
(c) the conditional distribution of the frequency at which a teen brought a cell phone to school among schools with a given level of permissiveness.

6.12 What percent of teens from schools that forbid cell phones brought their cell phone to school every day?

(a) about 16%   (b) about 21%   (c) about 65%

6.13 Your percent from the previous exercise is part of

(a) the marginal distribution of the frequency at which a teen brought a cell phone to school.
(b) the conditional distribution of school permissiveness among those who brought a cell phone to school every day.

(c) the conditional distribution of the frequency that a teen brought a cell phone to school among schools that forbid cell phones.

6.14 What percent of those who brought their cell phone to school every day were from schools that forbid cell phones?

(a) about 16%   (b) about 21%   (c) about 65%

6.15 Your percent from the previous exercise is part of

(a) the marginal distribution of the frequency at which a teen brought a cell phone to school.
(b) the conditional distribution of school permissiveness among those who brought a cell phone to school every day.
(c) the conditional distribution of the frequency at which a teen brought a cell phone to school among schools with a given level of permissiveness.

6.16 A bar graph showing the conditional distribution of the frequency at which a teen brought a cell phone to school among schools with a given level of permissiveness would have

(a) 3 bars.   (b) 4 bars.   (c) 12 bars.

6.17 A college looks at the grade point average (GPA) of its full-time and part-time students. Grades in science courses are generally lower than grades in other courses. There are few science majors among part-time students but many science majors among full-time students. The college finds that full-time students who are science majors have higher GPAs than part-time students who are science majors. Full-time students who are not science majors also have higher GPAs than part-time students who are not science majors. Yet part-time students as a group have higher GPAs than full-time students. This finding is

(a) not possible: if both science and other majors who are full-time have higher GPAs than those who are part-time, then all full-time students together must have higher GPAs than all part-time students together.
(b) an example of Simpson's paradox: full-time students do better in both kinds of courses but worse overall because they take more science courses.
(c) due to comparing two conditional distributions that should not be compared.

6.18 Is astrology scientific? The University of Chicago's General Social Survey (GSS) is the nation's most important social science sample survey. The GSS asked a random sample of adults their opinion about whether astrology is very scientific, sort of scientific, or not at all scientific. Here is a two-way table of counts for people in the sample who had three levels of higher education degrees:[8] ASTRLGY

	DEGREE HELD		
	JUNIOR COLLEGE	BACHELOR	GRADUATE
Not at all scientific	87	198	111
Very or sort of scientific	43	57	28

Find the two conditional distributions of degree held, one for those who hold the opinion that astrology is not at all scientific and one for those who say astrology is very or sort of scientific. Based on your calculations, describe with a graph and in words the differences between those who say astrology is not at all scientific and those who say it is very or sort of scientific.

6.19 Weight-lifting injuries. Resistance training is a popular form of conditioning aimed at enhancing sports performance and is widely used among high school, college, and professional athletes, although its use for younger athletes is controversial. A random sample of 4111 patients between the ages of 8 and 30 admitted to U.S. emergency rooms with the injury code "weight-lifting" was obtained. These injuries were classified as "accidental" if caused by dropped weight or improper equipment use. The patients were also classified into the four age categories 8 to 13 years, 14 to 18, 19 to 22, and 23 to 30. Here is a two-way table of the results:[9] LIFTING

AGE	ACCIDENTAL	NOT ACCIDENTAL
8–13	295	102
14–18	655	916
19–22	239	533
23–30	363	1008

Compare the distributions of ages for accidental and nonaccidental injuries. Use percents and draw a bar graph. What do you conclude?

Marital status and job level. *We sometimes hear that getting married is good for your career. Table 6.2 presents data from one of the studies behind this generalization. To avoid gender effects, the investigators looked only at men. The data describe the marital status and the job level of all 8235 male managers and professionals employed by a large manufacturing firm.[10] The firm assigns each position a grade that reflects the value of that particular job to the company.*

TABLE 6.2 MARITAL STATUS AND JOB LEVEL

JOB GRADE	MARITAL STATUS				TOTAL
	SINGLE	MARRIED	DIVORCED	WIDOWED	
1	58	874	15	8	955
2	222	3927	70	20	4239
3	50	2396	34	10	2490
4	7	533	7	4	551
Total	337	7730	126	42	8235

The authors of the study grouped the many job grades into quarters. Grade 1 contains jobs in the lowest quarter of the job grades, and Grade 4 contains those in the highest quarter. Exercises 6.20 to 6.24 are based on these data.

6.20 Marginal distributions. Give (in percents) the two marginal distributions, for marital status and for job grade. Do each of your two sets of percents add to exactly 100%? If not, why not? STATUS

6.21 Percents. What percent of single men hold Grade 1 jobs? What percent of Grade 1 jobs are held by single men? STATUS

6.22 Conditional distribution. Give (in percents) the conditional distribution of job grade among single men. Should your percents add to 100% (up to roundoff error)? STATUS

6.23 Marital status and job grade. One way to see the relationship is to look at who holds Grade 1 jobs. STATUS

(a) There are 874 married men with Grade 1 jobs, and only 58 single men with such jobs. Explain why these counts by themselves don't describe the relationship between marital status and job grade.

(b) Find the percent of men in each marital status group who have Grade 1 jobs. Then find the percent in each marital group who have Grade 4 jobs. What do these percents say about the relationship?

6.24 Association is not causation. The data in Table 6.2 show that single men are more likely to hold lower-grade jobs than are married men. We should not conclude that single men can help their career by getting married. What lurking variables might help explain the association between marital status and job grade? STATUS

6.25 Race and the death penalty. Whether a convicted murderer gets the death penalty seems to be influenced by the race of the victim. Several researchers studied this issue in the 1970s and 1980s, resulting in several landmark, oft-cited, and controversial papers. Here are data on 326 cases in which the defendant was convicted of murder from one of these studies:[11] DISCRIM

WHITE DEFENDANT			BLACK DEFENDANT		
	WHITE VICTIM	BLACK VICTIM		WHITE VICTIM	BLACK VICTIM
Death	19	0	Death	11	6
Not	132	9	Not	52	97

(a) Use these data to make a two-way table of defendant's race (white or black) versus death penalty (Yes or No).

(b) Show that Simpson's paradox holds: a higher percent of white defendants are sentenced to death overall, but for both black and white victims a higher percent of black defendants are sentenced to death.

(c) Use the data to explain why the paradox holds in language that a judge could understand.

6.26 **Obesity and health.** To estimate the health risks of obesity, we might compare how long obese and nonobese people live. Smoking is a lurking variable that may reduce the gap between the two groups because smoking tends to both reduce weight and lead to earlier death. So if we ignore smoking, we may underestimate the health risks of obesity. Illustrate Simpson's paradox by a simplified version of this situation: make up two-way tables of obese (Yes or No) by early death (Yes or No) separately for smokers and nonsmokers such that

■ Obese smokers and obese nonsmokers are both more likely to die earlier than those who are not obese.

■ But when smokers and nonsmokers are combined into a two-way table of obese by early death, persons who are not obese are more likely to die earlier because more of them are smokers.

The following exercises ask you to answer questions from data without having the details outlined for you. The exercise statements give you the State step of the four-step process. In your work, follow the Plan, Solve, and Conclude steps of the process as illustrated in Example 6.3 (page 156).

6.27 **Smoking cessation.** A large randomized trial was conducted to assess the efficacy of Chantix for smoking cessation compared with bupropion (more commonly known as Wellbutrin or Zyban) and a placebo. Chantix is different from most other quit-smoking products in that it targets nicotine receptors in the brain, attaches to them,

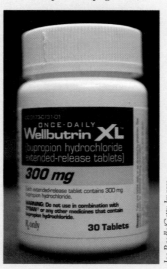

Joe Raedle/Getty Images

and blocks nicotine from reaching them, whereas bupropion is an antidepressant often used to help people stop smoking. Generally healthy smokers who smoked at least 10 cigarettes per day were assigned at random to take Chantix ($n = 352$), bupropion ($n = 329$), or a placebo ($n = 344$). The response measure is continuous cessation from smoking for Weeks 9 through 12 of the study. Here is a two-way table of the results:[12] SMOKE

	TREATMENT		
	CHANTIX	BUPROPION	PLACEBO
No smoking in Weeks 9–12	155	97	61
Smoked in Weeks 9–12	197	232	283

How does whether a subject smoked in Weeks 9 to 12 depend on the treatment received?

6.28 **Animal testing.** "It is right to use animals for medical testing if it might save human lives." The General Social Survey asked 1152 adults to react to this statement. Here is the two-way table of their responses: ANTEST

RESPONSE	MALE	FEMALE
Strongly agree	76	59
Agree	270	247
Neither agree nor disagree	87	139
Disagree	61	123
Strongly disagree	22	68

How do the distributions of opinion differ between men and women?

6.29 **College degrees.** "Colleges and universities across the country are grappling with the case of the mysteriously vanishing male." So said an article in the *Washington Post*. Here are data on the numbers of degrees earned in 2012–2013, as projected by the National Center for Education Statistics. The table entries are counts of degrees in thousands.[13] DEGREES

DEGREE	FEMALE	MALE
Associate's	519	304
Bachelor's	989	731
Master's	418	266
Professional	49	50
Doctor's	40	35

Briefly contrast the counts and distributions of men and women in earning degrees. Are men "vanishing" from colleges and universities across the country?

6.30 **Complications of bariatric surgery.** Bariatric surgery, or weight-loss surgery, includes a variety of procedures performed on people who are obese. Weight loss is achieved by reducing the size of the stomach with an implanted medical device (gastric banding), by removing a portion of the stomach (sleeve gastrectomy), or by resecting and rerouting the small intestines to a small stomach pouch (gastric bypass surgery). Because there can be complications using any of these methods, the National Institute of Health recommends bariatric surgery for obese people with a body mass index (BMI) of at least 40, and for people with BMI 35 and serious coexisting medical conditions such as diabetes. Serious complications include potentially life-threatening, permanently disabling, and fatal outcomes. Here is a two-way table for data collected in Michigan over several years giving counts of non-life-threatening complications, serious complications, and no complications for these three types of surgeries:[14] BARI

| | TYPE OF COMPLICATION | | | |
	NON-LIFE-THREATENING	SERIOUS	NONE	TOTAL
Gastric banding	81	46	5253	5380
Sleeve gastrectomy	31	19	804	854
Gastric bypass	606	325	8110	9041

What do the data say about differences in complications for the three types of surgeries?

6.31 **Smokers rate their health.** The University of Michigan Health and Retirement Study (HRS) surveys more than 22,000 Americans over the age of 50 every two years. A subsample of the HRS participated in a 2009 Internet-based survey that collected information on a number of topical areas, including health (physical and mental health behaviors), psychosocial items, economics (income, assets, expectations, and consumption), and retirement.[15] Two of the questions asked were "Would you say your health is excellent, very good, good, fair, or poor?" and "Do you smoke cigarettes now?" The two-way table summarizes the answers on these two questions. SMRATE

| | CURRENT SMOKER | |
HEALTH	YES	NO
Excellent	25	484
Very good	115	1557
Good	145	1309
Fair	90	545
Poor	29	11

What do the data say about differences in self-evaluation of health for current smokers and nonsmokers?

6.32 **Python eggs.** How is the hatching of water python eggs influenced by the temperature of the snake's nest? Researchers placed 104 newly laid eggs in a hot environment, 56 in a neutral environment, and 27 in a cold environment. Hot duplicates the warmth provided by the mother python. Neutral and cold are cooler, as when the mother is absent. The results: 75 of the hot eggs hatched, along with 38 of the neutral eggs and 16 of the cold eggs.[16]

(a) Make a two-way table of "environment temperature" against "hatched or not."
(b) The researchers anticipated that eggs would hatch less well at cooler temperatures. Do the data support that anticipation?

6.33 A random sample of 871 students between the ages of 20 and 24 at a large midwestern university completed a survey including questions about their sleep quality, moods, academic performance, physical health and psychoactive drug use. Sleep quality was measured using the Pittsburgh Sleep Quality Index (PSQI), with students scoring less than or equal to 5 on the index classified as optimal sleepers, those scoring a 6 or 7 classified as borderline, and those scoring over 7 classified as poor sleepers. The following table looks at the relationship between sleep quality classification and the use of over-the-counter (OTC) or prescription (Rx) stimulant medication more than once a month to help keep awake.[17] SLEEPQ

| USE OF OTC/RX MEDS TO WAKE > 1X/MONTH | SLEEP QUALITY ON PSQI INDEX | | |
	OPTIMAL	BORDERLINE	POOR
Yes	37	53	84
No	266	186	245

What do the data say about differences in sleep quality for those that use over-the-counter or prescription stimulant medication medication more than once a month to keep awake and those that don't?

Exploring the Web

6.34   Promoting women. In academics, faculty typically start as assistant professors, are promoted to associate professor (and gain tenure), and finally reach the rank of full professor. Some have argued that women have a harder time gaining promotion to associate and full professor than do men. Do data support this argument? Search the Web to find the number of faculty by rank and gender at some university. Do you see a pattern that suggests that the proportion of women decreases as rank increases? We found several sources of data by doing a Google search on "faculty head count by rank and gender." In addition to discussing the pattern you find, provide the data, the name of the school, and the source of the data.

6.35   Accidental deaths and age. Accidental deaths are shocking and tragic. Do the ways in which people die by accident change with age? Look at the most recent *Statistical Abstract of the United States* (www.census.gov/compendia/statab/) and make a two-way table that provides the counts of deaths due to accidents from various causes for three different age groups. What do you conclude?

6.36   Simpson's paradox. Find an example of Simpson's paradox and discuss how your example illustrates the paradox. Two examples that we found (thanks to Patricia Humphrey at Georgia Southern University) are www.nytimes. com/2006/07/15/education/15report.html and online.wsj. com/article/SB125970744553071829.html.

Overview

This chapter serves as an overview of the material covered thus far in the text, which covers exploratory data analysis methods (graphical displays and numerical summaries). Part I of the text is a set of chapters that provide us with simple tools for exploring data to see what the data are telling us. Students should, at this time, consider that many of the methods in Chapters 1–6 are typically combined to paint an informative sketch of the data, possibly even pointing out questions of further interest. Some students in an introductory course will tend to view chapters as disconnected bits of information to learn, and this can lead to a failure to grasp the larger picture. To help prevent this, have your students think of this chapter as an attempt to bring together all the concepts they have learned so far. The Test Yourself section will help students review the basic ideas and skills from Chapters 1–6, whereas the problems in the Supplementary Exercises section of the chapter provide example problems useful for in-class review or student study. Notice that many of the exercises are more thorough and require the use of several approaches simultaneously, rather than just one or two.

> **LEARNING OUTCOMES:**
> Refer to Learning Outcomes for Chapters 1–6, as well as the Summary provided in Chapter 7.

Many of the concepts and techniques learned in Part I will be used throughout the text. It is always worth foreshadowing topics the students may encounter repeatedly. Students have seen and will see again the importance of the roles variables play in a study and also the dangers and other effects that lurking variables may have on results. Finally, students are reminded that the key steps in any data analysis are to State, Plan, Solve, and Conclude. This four-step process will be used in future chapters as well.

Teaching Suggestions and Additional Examples/Activities for the Classroom

Consider working one or two comprehensive problems that bring together the most important concepts of the preceding chapters. Some suggested problems include the following:

■ Exercises 7.40 and 7.41: These exercises provide the opportunity to review basic graphical and numerical methods for exploring data in the context of an interesting data set and point out that different graphical displays tell us different things.

■ Exercise 7.46: Along with providing the opportunity to review graphical displays (side-by-side boxplots, in particular) and describing distributions, the exercise invites a discussion of some experimental design issues, as well as how comparisons might be made when there's a good deal of variation, as opposed to when there isn't much variation. This problem also allows you to review the four-step process while discussing these issues.

■ Exercise 7.50: A scatterplot will reveal that linear regression is an appropriate technique for investigating the relationship between "wildebeest abundance" and "percent burned." Students may not quickly recognize the role of linear regression here. Be sure to have students compute correlation (which is negative) given r^2 obtained from the regression output. This exercise provides another opportunity for your students to work through the four-step process.

Exploring Data:
Part I Review

CHAPTER 7

**In this chapter
we cover...**

- Part I Summary
- Test Yourself
- Supplementary Exercises

Data analysis is the art of describing data using graphs and numerical summaries. The purpose of exploratory data analysis is to help us see and understand the most important features of a set of data. Chapter 1 commented on graphs to display distributions: pie charts and bar graphs for categorical variables, histograms and stemplots for quantitative variables. In addition, time plots show how a quantitative variable changes over time. Chapter 2 presented numerical tools for describing the center and spread of the distribution of one variable. Chapter 3 discussed density curves for describing the overall pattern of a distribution, with emphasis on the Normal distributions.

The first STATISTICS IN SUMMARY figure on the next page organizes the big ideas for exploring a quantitative variable. Plot your data, then describe their center and spread using either the mean and standard deviation or the five-number summary. The last step, which makes sense only for some data, is to summarize the data in compact form by using a Normal curve as a description of the overall pattern. The question marks at the last two stages remind us that the usefulness of numerical summaries and Normal distributions depends on what we find when we examine graphs of our data. No short summary does justice to irregular shapes or to data with several distinct clusters.

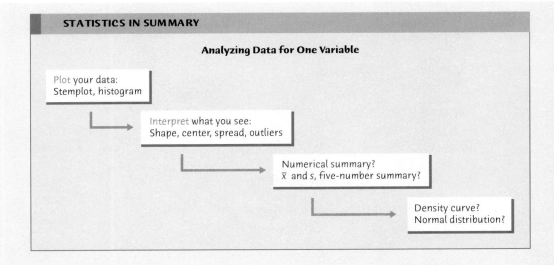

Chapters 4 and 5 applied the same ideas to relationships between two quantitative variables. The second STATISTICS IN SUMMARY figure retraces the big ideas, with details that fit the new setting. Always begin by making graphs of your data. In the case of a scatterplot, we have learned a numerical summary only for data that show a roughly linear pattern on the scatterplot. The summary is then the means and standard deviations of the two variables and their correlation. A regression line drawn on the plot gives a compact description of the overall pattern that we can use for prediction. Once again there are question marks at the last two stages to remind us that correlation and regression describe only straight-line relationships. Chapter 6 shows how to understand relationships between two categorical variables; comparing well-chosen percents is the key.

You can organize your work in any open-ended data analysis setting by following the four-step STATE, PLAN, SOLVE, and CONCLUDE process first introduced in Chapter 2 (page 60). After we have mastered the extra background needed for statistical inference, this process will also guide practical work on inference later in the book.

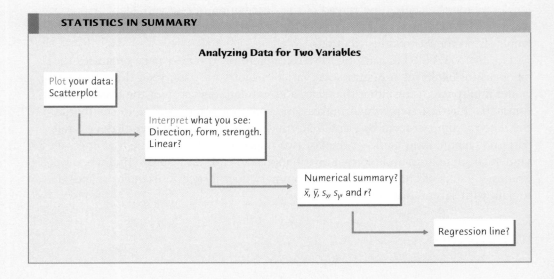

Part I Summary

Here are the most important skills you should have acquired from reading Chapters 1 to 6.

A. Data

1. Identify the individuals and variables in a set of data.
2. Identify each variable as categorical or quantitative. Identify the units in which each quantitative variable is measured.
3. Identify the explanatory and response variables in situations where one variable explains or influences another.

B. Displaying Distributions

1. Recognize when a pie chart can and cannot be used.
2. Make a bar graph to display the distribution of a categorical variable or, in general, to compare related quantities.
3. Interpret pie charts and bar graphs.
4. Make a histogram of the distribution of a quantitative variable.
5. Make a stemplot of the distribution of a small set of observations. Round leaves or split stems as needed to make an effective stemplot.
6. Make a time plot of a quantitative variable over time. Recognize patterns such as trends and cycles in time plots.

C. Describing Distributions (Quantitative Variable)

1. Look for the overall pattern and for major deviations from the pattern.
2. Assess from a histogram or stemplot whether the shape of a distribution is roughly symmetric, distinctly skewed, or neither. Assess whether the distribution has one or more major peaks.
3. Describe the overall pattern by giving numerical measures of center and spread in addition to a verbal description of shape.
4. Decide which measures of center and spread/variability are more appropriate: the mean and standard deviation (especially for symmetric distributions) or the five-number summary (especially for skewed distributions).
5. Recognize outliers, and give plausible explanations for them.

D. Numerical Summaries of Distributions

1. Find the median M and the quartiles Q_1 and Q_3 for a set of observations.
2. Find the five-number summary and draw a boxplot; assess center, spread, symmetry, and skewness from a boxplot.
3. Find the mean $\bar{x}$ and the standard deviation s for a set of observations.
4. Understand that the median is more resistant than the mean. Recognize that skewness in a distribution moves the mean away from the median toward the long tail.
5. Know the basic properties of the standard deviation: $s \geq 0$ always; $s = 0$ only when all observations are identical and increases as the spread increases; s has the same units as the original measurements; s is pulled strongly up by outliers or skewness.

E. Density Curves and Normal Distributions

1. Know that areas under a density curve represent proportions of all observations and that the total area under a density curve is 1.

2. Approximately locate the median (equal-areas point) and the mean (balance point) on a density curve.

3. Know that the mean and median both lie at the center of a symmetric density curve and that the mean moves farther toward the long tail of a skewed density curve.

4. Recognize the shape of Normal curves and estimate by eye both the mean and standard deviation from such a curve.

5. Use the 68–95–99.7 rule and symmetry to state what percent of the observations from a Normal distribution fall between two points when both points lie at the mean or one, two, or three standard deviations on either side of the mean.

6. Find the standardized value (z-score) of an observation. Interpret z-scores and understand that any Normal distribution becomes the standard Normal $N(0, 1)$ distribution when standardized.

7. Given that a variable has a Normal distribution with a stated mean μ and standard deviation σ, calculate the proportion of values above a stated number, below a stated number, or between two stated numbers.

8. Given that a variable has a Normal distribution with a stated mean μ and standard deviation σ, calculate the point having a stated proportion of all values above it or below it.

F. Scatterplots and Correlation

1. Make a scatterplot to display the relationship between two quantitative variables measured on the same subjects. Place the explanatory variable (if any) on the horizontal scale of the plot.

2. Add a categorical variable to a scatterplot by using a different plotting symbol or color.

3. Describe the direction, form, and strength of the overall pattern of a scatterplot. In particular, recognize positive or negative association and linear (straight-line) patterns. Recognize outliers in a scatterplot.

4. Judge whether it is appropriate to use correlation to describe the relationship between two quantitative variables. Find the correlation r.

5. Know the basic properties of correlation: r measures the direction and strength of only straight-line relationships; r is always a number between -1 and 1; $r = \pm 1$ only for perfect straight-line relationships; r moves away from 0 toward ± 1 as the straight-line relationship gets stronger.

G. Regression Lines

1. Understand that regression requires an explanatory variable and a response variable. Correctly identifying which variable is the explanatory variable and which is the response variable is important. Switching these will result in different regression lines. Use a calculator or software to find the least-squares regression line of a response variable y on an explanatory variable x from data.

2. Explain what the slope b and the intercept a mean in the equation $\hat{y} = a + bx$ of a regression line.

3. Draw a graph of a regression line when you are given its equation.

4. Use a regression line to predict y for a given x. Recognize extrapolation and be aware of its dangers.

5. Find the slope and intercept of the least-squares regression line from the means and standard deviations of x and y and their correlation.

6. Use r^2, the square of the correlation, to describe how much of the variation in one variable can be accounted for by a straight-line relationship with another variable.

7. Recognize outliers and potentially influential observations from a scatter-plot with the regression line drawn on it.

8. Calculate the residuals and plot them against the explanatory variable x. Recognize that a residual plot magnifies the pattern of the scatterplot of y versus x.

H. Cautions about Correlation and Regression

1. Understand that both r and the least-squares regression line can be strongly influenced by a few extreme observations.

2. Recognize possible lurking variables that may explain the observed association between two variables x and y.

3. Understand that even a strong correlation does not mean that there is a cause-and-effect relationship between x and y.

4. Give plausible explanations for an observed association between two variables: direct cause and effect, the influence of lurking variables, or both.

I. Categorical Data (optional)

1. From a two-way table of counts, find the marginal distributions of both variables by obtaining the row sums and column sums.

2. Express any distribution in percents by dividing the category counts by their total.

3. Describe the relationship between two categorical variables by computing and comparing percents. Often this involves comparing the conditional distributions of one variable for the different categories of the other variable.

4. Recognize Simpson's paradox and be able to explain it.

 DRIVING IN CANADA

Canada is a civilized and restrained nation, at least in the eyes of Americans. A survey sponsored by the Canada Safety Council suggests that driving in Canada may be more adventurous than expected. Of the Canadian drivers surveyed, 88% admitted to aggressive driving in the past year, and 76% said that sleep-deprived drivers were common on Canadian roads. What really alarms us is the name of the survey: the Nerves of Steel Aggressive Driving Study.

Test Yourself

The following questions include multiple-choice, calculations, and short-answer questions. They will help you review the basic ideas and skills presented in Chapters 1 to 6.

7.1 As part of a database on new births at a hospital, some variables recorded are the age of the mother, marital status of the mother (single, married, divorced, other), weight of the baby, and sex of the baby. Of these variables

(a) age, marital status, and weight are quantitative variables.
(b) age and weight are categorical variables.
(c) sex and marital status are categorical variables.
(d) sex, marital status, and age are categorical variables.

7.2 You are interested in obtaining information about the performance of students in your statistics class and seeing how this performance is affected by several factors such as sex. To do this, you are going to give a questionnaire to all students in the class. Give two questions for which the response is categorical and two questions for which the response is quantitative. For the categorical variables, give the possible values, and for the quantitative variables give the unit of measurement.

Pocket change. *In a statistics class with 136 students, the professor records how much money each student has in his or her possession during the first class of the semester. Figure 7.1 gives the histogram of the data collected. Use this histogram to help answer Questions 7.3 to 7.5.*

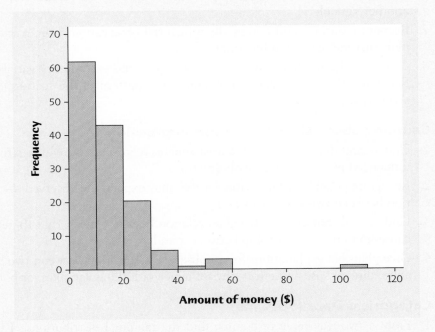

FIGURE 7.1
Histogram of the distribution of the amount of money carried by each student on the first day of class, for Questions 7.3 to 7.5.

7.3   The *number* of students with under $10 in their possession is closest to

(a) 40.   (b) 50.   (c) 60.   (d) 70.

7.4   The histogram

(a) is skewed right.   (b) has an outlier.
(c) is asymmetric.   (d) is all of the above.

7.5   The *percent* of students with $20 or more in their possession is

(a) about 10%.   (b) about 20%.   (c) about 30%.   (d) over 40%.

7.6   A reporter wishes to portray baseball players as overpaid. Which measure of center should he report as the average salary of major league players?

(a) The mean
(b) The median
(c) Either the mean or the median. It doesn't matter because they will be equal.
(d) Neither the mean nor the median. Both will be much lower than the actual average salary.

Weeds among the corn. Velvetleaf is a particularly annoying weed in corn fields. It produces lots of seeds, and the seeds wait in the soil for years until conditions are right. How many seeds do velvetleaf plants produce? Here are counts from 28 plants that came up in a corn field when no herbicide was used. The counts have been ordered to simplify your calculations.[1] Use the following counts to answer Questions 7.7 to 7.9.

130	164	218	363	721	863	880	1050	1051	1110
1136	1531	1623	1711	1716	1809	1911	1961	2008	2101
2114	2137	2228	2450	2504	2819	5973	8015		

7.7 What is the median number of seeds for the 28 plants?

    (a) 1711 seeds   (b) 1713.5 seeds   (c) 1716 seeds   (d) 1809 seeds

7.8 What is the first quartile for these data?

    (a) 880 seeds   (b) 965 seeds   (c) 1050 seeds   (d) 1711 seeds

7.9 What is the interquartile range for these data?

    (a) 880 seeds   (b) 965 seeds   (c) 1160.5 seeds   (d) 2125.5 seeds

7.10 Which of the following is likely to have a mean that is smaller than the median?

    (a) The salaries of all National Football League players
    (b) The scores of students (out of 100 points) on a very easy exam in which most students score perfectly, but a few do very poorly
    (c) The prices of homes in a large city
    (d) The scores of students (out of 100 points) on a very difficult exam in which most students score poorly, but a few do very well

7.11 For a biology project, you measure the tail length in centimeters and weight in grams of 12 mice of the same variety. What units of measurement do each of the following have?

    (a) The mean length of the tails
    (b) The first quartile of the tail lengths
    (c) The standard deviation of the tail lengths
    (d) The variance of the weights

Employment times. *A sample of 40 employees from the local Honda plant was obtained, and the length of time (in months) worked was recorded for each employee. A stemplot of these data follows. Use the stemplot to answer Questions 7.12 and 7.13. In the stemplot 5|2 represents 52 months.*

```
5 | 2 2 3 3 4 5 7 8 9 9
6 | 0 0 0 2 3 4 4 4 5 6 7 7 8 8 8 9
7 | 3 4 5 5 6 6 7 7 7 8 8 9 9
8 |
9 | 8
```

7.12 What would be a better way to represent this data set?

    (a) Display the data in a time plot
    (b) Split the stems
    (c) Use a pie chart
    (d) Use a histogram with class width equal to 10

7.13 The *percent* of employees in the sample who have worked at the plant for less than 5 years is

    (a) approximately zero.   (b) 10%.   (c) 15%.   (d) 25%.

Search engine use. *In 2012, approximately 91% of Internet users used a search engine to find information on the web. Search engine users were asked "How often do you use a search engine to find information online?". The bar graph in Figure 7.2 gives the percentages for the most frequent use categories.[2] Use the bar graph to help answer Questions 7.15 and 7.16.*

7.14 Approximately what percent of search engine users used a search engine 1–2 days per week?

    (a) 5%   (b) 10%   (c) 15%   (d) 20%

7.15 The total of the percents of the bars in the graph add to 92%.

    (a) The percent of search engine users who use a search engine less than once every few weeks is greater than 10%.

FIGURE 7.2

Bar graph of the distribution of search engine use for the most frequent use categories in 2012, for Questions 7.15 and 7.16.

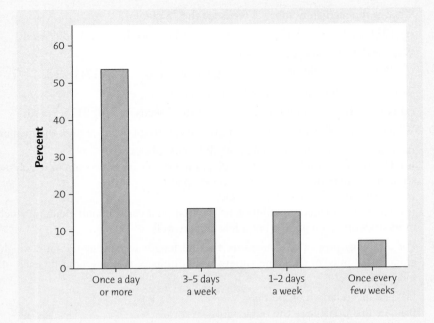

(b) A pie chart could be drawn for the categories given in the bar graph.

(c) The percent of search engine users who use a search engine *at least* 3–5 days a week is about 70%.

(d) None of the above.

7.16 Reports on a student's ACT, SAT, or MCAT usually give the percentile as well as the actual score. The percentile is just the cumulative proportion stated as a percent: the percent of all scores that were lower than this one. In 2012, the total MCAT scores were close to Normal with mean 25.2 and standard deviation 6.4.[3] William scored 32. What was his percentile?

7.17 The length of the thorax in a population of male fruit flies is approximately Normal with mean 0.800 millimeters (mm) and standard deviation 0.078 mm. Use the 68–95–99.7 rule to answer the following questions.

(a) What range of thorax lengths covers almost all (99.7%) of this distribution?

(b) What percent of male fruit flies have a thorax length exceeding 0.878 mm?

7.18 A professor knows from past experience that the time for students to complete a quiz has an $N(19, 3)$ distribution.

(a) If he allows 20 minutes for the quiz, what percent of the students will not complete the quiz?

(b) Suppose that he wants to allow sufficient time so that 95% of the students will complete the quiz in the allotted time. How much time should he allow for the quiz?

7.19 The Aleppo pine and the Torrey pine are widely planted as ornamental trees in Southern California. Here are the lengths (centimeters) of 15 Aleppo pine needles:[4]

ALEPPO

10.2   7.2   7.6   9.3   12.1   10.9   9.4   11.3   8.5   8.5   12.8   8.7   9.0   9.0   9.4

(a) Find the five-number summary for the distribution of Aleppo pine needles.

Figure 7.3 gives a boxplot for the distribution of the lengths (centimeters) of 18 Torrey pine needles. Use this information to help answer the remainder of this question.

(b) The median of the distribution of Torrey pine needles is closest to which of the following values?

$$24 \quad 25 \quad 27 \quad 30$$

(c) Twenty-five percent of the Torrey pine needles exceed what value?

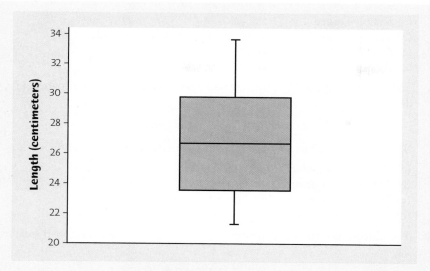

FIGURE 7.3
Boxplot for the distribution of the lengths (centimeters) of 18 Torrey pine needles, for Question 7.19.

(d) Given only the length of a needle, do you think you could say which pine species it comes from? Explain briefly.

Soap in the shower. *From Rex Boggs in Australia comes an unusual data set: before showering in the morning, he weighed the bar of soap in his shower stall. The weight goes down as the soap is used. The data appear in the following table (weights in grams). Notice that Mr. Boggs forgot to weigh the soap on some days. Questions 7.20 to 7.23 are based on the soap data set.*

DAY	WEIGHT	DAY	WEIGHT	DAY	WEIGHT
1	124	8	84	16	27
2	121	9	78	18	16
5	103	10	71	19	12
6	96	12	58	20	8
7	90	13	50	21	6

7.20 Figure 7.4 is a scatterplot of the weight of the bar of soap against day. How would you describe the overall pattern?

BEER IN SOUTH DAKOTA

Take a break from doing exercises to apply your math to beer cans in South Dakota. A newspaper there reported that every year an average of 650 beer cans per mile are tossed onto the state's highways. South Dakota has about 83,000 miles of roads. How many beer cans is that in all? The U.S. Census Bureau says that there are about 810,000 people in South Dakota. How many beer cans does each man, woman, and child in the state toss on the road each year? That's pretty impressive. Maybe the paper got its numbers wrong.

SOAP

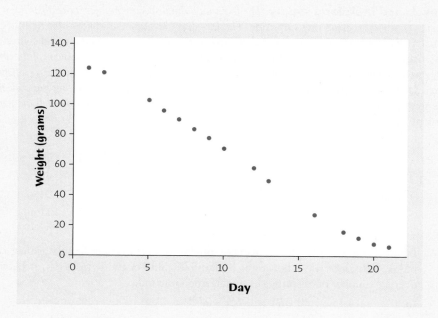

FIGURE 7.4
Scatterplot of the weight of a bar of soap against day, for Question 7.20.

(a) Sharply curved.

(b) There are two distinct clusters that are widely separated.

(c) A very weak positive association.

(d) A strong negative association.

7.21 The equation of the least-squares regression line for predicting soap weight from day is

$$\text{weight} = 133.2 - 6.3 \times \text{day}$$

What does this tell us about the rate at which the soap lost weight?

(a) The soap lost about −6.3 grams per day.

(b) The soap lost about 6.3 grams per day.

(c) The soap lost about −133.2 grams per day.

(d) The soap lost about 133.2 grams per day.

7.22 The equation of the least-squares regression line for predicting soap weight from day is

$$\text{weight} = 133.2 - 6.3 \times \text{day}$$

Mr. Boggs did not measure the weight of the soap on Day 4. Use the regression equation to predict that weight.

(a) 108 grams   (b) 126.9 grams   (c) 157.3 grams   (d) 526.5 grams

7.23 The equation of the least-squares regression line for predicting soap weight from day is

$$\text{weight} = 133.2 - 6.3 \times \text{day}$$

I use the regression equation to predict the weight of the soap on Day 30. I conclude that

(a) the soap will last at least a month.

(b) the prediction is not sensible because the prediction is far outside the range of values of the response variable.

(c) the prediction is not sensible because 30 days is far outside the range of values of the explanatory variable.

(d) the prediction is not sensible because of the outlier present in the data.

Squirrels and their food supply. *The fact that animal species produce more offspring when their supply of food goes up isn't surprising. The fact that some animals appear able to anticipate unusual food abundance is surprising. Red squirrels eat seeds from pine cones, a food source that occasionally has very large crops (called seed masting). Following are data on an index of the abundance of pine cones (larger values indicate greater abundance) and average number of offspring per female over 16 years.[5] What makes these data interesting is that the offspring are conceived in the spring, before the cones mature in the fall to feed the new young squirrels through the winter. Questions 7.24 to 7.26 are based on these data.*

Cone index x	0.00	2.02	0.25	3.22	4.68	0.31	3.37	3.09
Offspring y	1.49	1.10	1.29	2.71	4.07	1.29	3.36	2.41
Cone index x	2.44	4.81	1.88	0.31	1.61	1.88	0.91	1.04
Offspring y	1.97	3.41	1.49	2.02	3.34	2.41	2.15	2.12

7.24 Figure 7.5 is a scatterplot of average number of offspring per female against cone index. Which of the following is a plausible value of the correlation, *r*, between average number of offspring per female and cone index?

(a) 0   (b) 1   (c) 0.75   (d) −0.75

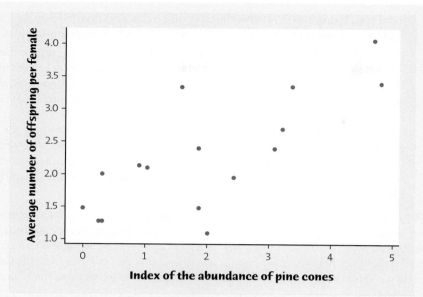

FIGURE 7.5
Scatterplot of the average number of offspring per female against cone index, for Question 7.24.

7.25 The equation of the least-squares regression line for predicting average number of offspring per female from cone index is

$$\text{offspring} = 1.41 + 0.44 \times \text{cone index}$$

What does the intercept of 1.41 tell us?

(a) The average number of offspring per female is 1.41.
(b) The predicted number of offspring per female is 1.41.
(c) The predicted number of offspring per female when the cone index is 0 is 1.41.
(d) All of the above.

7.26 The equation of the least-squares regression line for predicting average number of offspring per female from cone index is

$$\text{offspring} = 1.41 + 0.44 \times \text{cone index}$$

Use this to predict the average number of offspring per female for a year with a cone index of 0.25.

(a) 1.52   (b) 1.29   (c) 0.44   (d) 0.11

7.27 How well do people remember their past diet? Data are available for 91 people who were asked about their diet when they were 18 years old. Researchers asked them at about age 55 to describe their eating habits at age 18. For each subject, the researchers calculated the correlation between actual intakes of many foods at age 18 and the intakes the subjects now remember. The median of the 91 correlations was $r = 0.217$.[6] Which of the following conclusions is consistent with this correlation?

(a) We conclude that subjects remember approximately 21.7% of their food intakes at age 18.
(b) We conclude that subjects remember approximately $r^2 = 0.217^2 = 0.047$ of their food intakes at age 18.
(c) We conclude that food intake at age 55 is about 21.7% that of food intake at age 18.
(d) We conclude that memory of food intake in the distant past is fair to poor.

7.28 Joe's retirement plan invests in stocks through an "index fund" that follows the behavior of the stock market as a whole, as measured by the Standard & Poor's (S&P) 500 stock index. Joe wants to buy a mutual fund that does not track the index closely. He reads that monthly returns from Fidelity Technology Fund have correlation $r = 0.77$ with the S&P 500 index and that Fidelity Real Estate Fund has correlation $r = 0.37$ with the index. Which of the following is correct?

(a) The Fidelity Technology Fund has a closer relationship to returns from the stock market as a whole and also has higher returns than the Fidelity Real Estate Fund.

(b) The Fidelity Technology Fund has a closer relationship to returns from the stock market as a whole, but we cannot say that it has higher returns than the Fidelity Real Estate Fund.

(c) The Fidelity Real Estate Fund has a closer relationship to returns from the stock market as a whole and also has higher returns than the Fidelity Technology Fund.

(d) The Fidelity Real Estate Fund has a closer relationship to returns from the stock market as a whole, but we cannot say that it has higher returns than the Fidelity Technology Fund.

Monkey calls. *The usual way to study the brain's response to sounds is to have subjects listen to "pure tones." The response to recognizable sounds may differ. To compare responses, researchers anesthetized macaque monkeys. They fed pure tones and also monkey calls directly to their brains by inserting electrodes. Response to the stimulus was measured by the firing rate (electrical spikes per second) of neurons in various areas of the brain. Table 7.1 contains the responses for 37 neurons.[7]*

TABLE 7.1 NEURON RESPONSE (ELECTRICAL FIRING RATE PER SECOND) TO PURE TONES AND MONKEY CALLS

NEURON	TONE	CALL	NEURON	TONE	CALL	NEURON	TONE	CALL	NEURON	TONE	CALL
1	474	500	11	19	66	21	74	62	31	56	201
2	256	138	12	20	54	22	72	112	32	47	279
3	241	485	13	35	103	23	20	193	33	46	62
4	226	338	14	145	42	24	21	129	34	41	84
5	185	194	15	141	241	25	26	135	35	26	203
6	174	159	16	129	194	26	71	134	36	28	192
7	176	341	17	113	123	27	68	65	37	31	70
8	168	85	18	112	182	28	59	182			
9	161	303	19	102	141	29	59	97			
10	150	208	20	100	118	30	57	318			

Figure 7.6 is a scatterplot of monkey call response against pure-tone response (explanatory variable). Questions 7.29 and 7.30 refer to these data and the scatterplot.

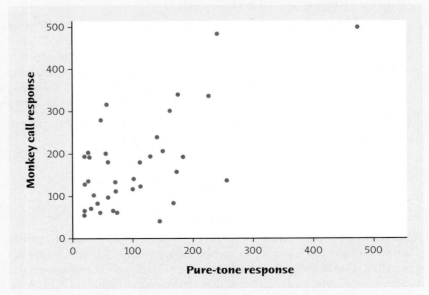

FIGURE 7.6

Scatterplot of monkey tone response against pure-tone response, for Questions 7.29 and 7.30.

7.29 We might expect some neurons to have strong responses to any stimulus and others to have consistently weak responses. There would then be a strong relationship between tone response and call response. From the scatterplot of monkey call response against pure-tone response in Figure 7.6 what would you estimate the correlation r to be?

MONKEY

(a) −0.6   (b) −0.1   (c) 0.1   (d) 0.6

7.30 Which of the following statements about the scatterplot in Figure 7.6 is correct?

(a) There is moderate evidence that pure-tone response causes monkey call response.
(b) There is moderate evidence that monkey call response causes pure-tone response.
(c) There are one or two outliers, and at least one of these may also be influential.
(d) None of the above.

Catalog shopping (optional). *What is the most important reason that students buy from catalogs? The answer may differ for different groups of students. Here are counts for samples of American and East Asian students at a large midwestern university.*[8] *Use these counts to answer Questions 7.31 to 7.33.*

REASON	AMERICAN	ASIAN
Save time	29	10
Easy	28	11
Low price	17	34
Live far from stores	11	4
No pressure to buy	10	3
Other	20	7
Total	115	69

7.31 What percent of all students say that the most important reason to buy from a catalog is to save time?

SHOP

(a) 74%   (b) 25%   (c) 21%   (d) 14%

7.32 What percent of East Asian students say that the most important reason to buy from a catalog is low price?

SHOP

(a) 67%   (b) 49%   (c) 28%   (d) 18%

7.33 What are the most important differences between the two groups of students?

SHOP

(a) The most important reasons for American students to buy from a catalog are to save time and because it is easy, whereas for East Asian students it is low price.
(b) American students appear to be almost three times more likely to live far from stores than East Asian students.
(c) East Asian students are twice as likely to purchase from a catalog because of low price than American students.
(d) All of the above.

Investment strategies. *One reason to invest abroad is that markets in different countries don't move in step. When American stocks go down, foreign stocks may go up. So an investor who holds both bears less risk. That's the theory. But then we read in a magazine article that the correlation between changes in American and European stock prices rose from 0.4 in the mid-1990s to 0.8 in 2000.*[9] *Questions 7.34 and 7.35 refer to this article.*

7.34 Explain to an investor who knows no statistics why the fact stated in this article reduces the protection provided by buying European stocks.

7.35 The same article that claims that the correlation between changes in stock prices in Europe and the United States is 0.8 goes on to say: "Crudely, that means that movements on Wall Street can explain 80% of price movements in Europe."

(a) Is this true? Explain your answer.
(b) What is the correct percent explained if $r = 0.8$?

INTROSP

7.36 Researchers wished to determine whether individual differences in introspective ability are reflected in the anatomy of brain regions responsible for this function. They measured introspective ability (using a score on a test of introspective ability, with larger values indicating greater introspective ability) and gray-matter volume in milliliters in the anterior prefrontal cortex of the brain of 29 subjects. Here are the data:

Introspective ability	0.55	0.58	0.59	0.59	0.59	0.61	0.62	0.63	0.63	0.63
Volume	59	62	43	63	83	61	55	57	57	67
Introspective ability	0.63	0.64	0.65	0.65	0.65	0.65	0.65	0.66	0.66	0.67
Volume	72	62	58	62	65	70	75	60	63	71
Introspective ability	0.67	0.67	0.68	0.69	0.70	0.70	0.71	0.72	0.75	
Volume	71	80	68	72	66	73	61	80	75	

The researchers wished to determine the equation of the least-squares regression line for predicting introspective ability (y) from gray-matter volume (x). To do this they calculated the following summary statistics:

$$\bar{x} = 65.90, \; s_x = 8.69$$
$$\bar{y} = 0.649, \; s_y = 0.045$$
$$r = 0.448$$

(a) Use this information to calculate the equation of the least-squares regression line.
(b) Based on the least-squares regression line, what would you predict introspective ability to be for someone with gray-matter volume 65?
(c) Based on the least-squares regression line, what would you predict introspective ability to be for someone with gray-matter volume 30? How reliable do you think this prediction is? Explain your answer.

THINFAT

7.37 Animals and people that take in more energy than they expend will get fatter. Here are data on 12 rhesus monkeys: 6 lean monkeys (4% to 9% body fat) and 6 obese monkeys (13% to 44% body fat). The data report the energy expended in 24 hours (kilojoules per minute) and the lean body mass (kilograms, leaving out fat) for each monkey.[10]

LEAN		OBESE	
MASS	ENERGY	MASS	ENERGY
6.6	1.17	7.9	0.93
7.8	1.02	9.4	1.39
8.9	1.46	10.7	1.19
9.8	1.68	12.2	1.49
9.7	1.06	12.1	1.29
9.3	1.16	10.8	1.31

(a) Compute the mean lean body mass of the lean monkeys.
(b) Compute the mean lean body mass of the obese monkeys.
(c) The goal of the study is to compare the energy expended in 24 hours by the lean monkeys with that of the obese monkeys. However, animals with higher lean mass usually expend more energy. Based on your calculations in parts (a) and (b),

would it make sense to simply compute the mean energy expended by lean and obese monkeys and compare the means? Explain.

(d) To investigate how energy expended is related to body mass, make a scatterplot of energy versus mass, using different plot symbols for lean and obese monkeys.

(e) What do the trends in your scatterplot suggest about the monkeys?

7.38 The number of adult Americans who smoke continues to drop. Here are estimates of the percents of adults (aged 18 and over) who were smokers in the years between 1965 and 2009:[11]

SMOKERS

Year x	1965	1974	1979	1983	1987	1990	1993	1997	2000	2002	2006	2009
Smokers y	41.9	37.0	33.3	31.9	28.6	25.3	24.8	24.6	23.1	22.5	20.8	20.6

(a) Make a scatterplot of these data.

(b) Describe the direction, form, and strength of the relationship between percent of smokers and year. Are there any outliers?

(c) Here are the means and standard deviations for both variables and the correlation between percent of smokers and year:

$$\bar{x} = 1990.4, s_x = 13.4$$
$$\bar{y} = 27.9, s_y = 6.8$$
$$r = -0.98$$

Use this information to find the least-squares regression line for predicting percent of smokers from year and add the line to your plot.

(d) According to your regression line, how much did smoking decline per year during this period, on the average?

(e) What percent of the observed variation in percent of adults who smoke can be explained by linear change over time?

(f) One of the government's national health objectives was to reduce smoking to no more than 12% of adults by 2010. Use your regression line to predict the percent of adults who smoked in 2010. In 2009 did the regression line suggest that this health objective would be met?

(g) Use your regression line to predict the percent of adults who will smoke in 2050. Why is your result impossible? Why was it foolish to use the regression line for this prediction?

7.39 **(Optional)** People who get angry easily tend to have more heart disease. That's the conclusion of a study that followed a random sample of 12,986 people from three locations for about four years. All subjects were free of heart disease at the beginning of the study. The subjects took the Spielberger Trait Anger Scale test, which measures how prone a person is to sudden anger. Here are data for the 8474 people in the sample who had normal blood pressure. CHD stands for "coronary heart disease." This includes people who had heart attacks and those who needed medical treatment for heart disease.[12]

ANGER

	LOW ANGER	MODERATE ANGER	HIGH ANGER	TOTAL
CHD	53	110	27	190
No CHD	3057	4621	606	8284
Total	3110	4731	633	8474

(a) What percent of all 8474 people with normal blood pressure had CHD?

(b) What percent of all 8474 people were classified as having high anger?

(c) What percent of those classified as having high anger had CHD?

(d) What percent of those with no CHD were classified as having moderate anger?

(e) Do these data provide any evidence that as anger score increases, the percent who suffer CHD increases? Explain.

Supplementary Exercises

*Supplementary exercises apply the skills you have learned in ways that require more thought or more elaborate use of technology. Some of these exercises ask you to follow the **Plan**, **Solve**, and **Conclude** steps of the four-step process introduced on page 60.*

7.40 **The Mississippi River.** Table 7.2 gives the volume of water discharged by the Mississippi River into the Gulf of Mexico for each year from 1954 to 2001.[13] The units are cubic kilometers of water—the Mississippi is a big river. DATA MISSIP

AP Photo/The Gleaner, Mike Lawrence

(a) Make a graph of the distribution of water volume. Describe the overall shape of the distribution and any outliers.

(b) Based on the shape of the distribution, do you expect the mean to be close to the median, clearly less than the median, or clearly greater than the median? Why? Find the mean and the median to check your answer.

(c) Based on the shape of the distribution, does it seem reasonable to use $\bar{x}$ and s to describe the center and spread of this distribution? Why? Find $\bar{x}$ and s if you think they are a good choice. Otherwise, find the five-number summary.

7.41 **More on the Mississippi River.** The data in Table 7.2 are a time series. Make a time plot that shows how the volume of water in the Mississippi changed between 1954 and 2001. What does the time plot reveal that the histogram from the previous exercise does not? It is a good idea to always make a time plot of time series data because a histogram cannot show changes over time. DATA MISSIP

Falling through the ice. *The Nenana Ice Classic is an annual contest to guess the exact time in the spring thaw when a tripod erected on the frozen Tanana River near Nenana, Alaska, will fall through the ice. The 2012 jackpot prize was $350,000. The contest has been run since 1917. Table 7.3 gives simplified*

TABLE 7.2   YEARLY DISCHARGE (CUBIC KILOMETERS OF WATER) OF THE MISSISSIPPI RIVER

YEAR	DISCHARGE	YEAR	DISCHARGE	YEAR	DISCHARGE
1954	290	1970	540	1986	600
1955	420	1971	480	1987	450
1956	390	1972	600	1988	420
1957	610	1973	880	1989	630
1958	550	1974	710	1990	680
1959	440	1975	670	1991	700
1960	470	1976	420	1992	510
1961	600	1977	430	1993	900
1962	550	1978	560	1994	640
1963	360	1979	800	1995	590
1964	390	1980	500	1996	670
1965	500	1981	420	1997	680
1966	410	1982	640	1998	690
1967	460	1983	770	1999	580
1968	510	1984	710	2000	390
1969	560	1985	680	2001	580

TABLE 7.3   DAYS FROM APRIL 20 FOR THE TANANA RIVER TRIPOD TO FALL

YEAR	DAY	YEAR	DAY	YEAR	DAY	YEAR	DAY
1917	11	1941	14	1965	18	1989	12
1918	22	1942	11	1966	19	1990	5
1919	14	1943	9	1967	15	1991	12
1920	22	1944	15	1968	19	1992	25
1921	22	1945	27	1969	9	1993	4
1922	23	1946	16	1970	15	1994	10
1923	20	1947	14	1971	19	1995	7
1924	22	1948	24	1972	21	1996	16
1925	16	1949	25	1973	15	1997	11
1926	7	1950	17	1974	17	1998	1
1927	23	1951	11	1975	21	1999	10
1928	17	1952	23	1976	13	2000	12
1929	16	1953	10	1977	17	2001	19
1930	19	1954	17	1978	11	2002	18
1931	21	1955	20	1979	11	2003	10
1932	12	1956	12	1980	10	2004	5
1933	19	1957	16	1981	11	2005	9
1934	11	1958	10	1982	21	2006	13
1935	26	1959	19	1983	10	2007	8
1936	11	1960	13	1984	20	2008	16
1937	23	1961	16	1985	23	2009	12
1938	17	1962	23	1986	19	2010	10
1939	10	1963	16	1987	16	2011	14
1940	1	1964	31	1988	8	2012	3

data that record only the date on which the tripod fell each year. The earliest date so far is April 20. To make the data easier to use, the table gives the date each year in days starting with April 20. That is, April 20 is 1, April 21 is 2, and so on. Exercises 7.42 to 7.44 concern these data.[14]

7.42 When does the ice break up? We have 96 years of data on the date of ice breakup on the Tanana River. Describe the distribution of the breakup date with both a graph or graphs and appropriate numerical summaries. What is the median date (month and day) for ice breakup? TANANA

7.43 Global warming? Because of the high stakes, the falling of the tripod has been carefully observed for many years. If the date the tripod falls has been getting earlier, that may be evidence for the effects of global warming. TANANA

(a) Make a time plot of the date the tripod falls against year.
(b) There is a great deal of year-to-year variation. Fitting a regression line to the data may help us see the trend. Fit the least-squares line, and add it to your time plot. What do you conclude?
(c) There is much variation about the line. Give a numerical description of how much of the year-to-year variation in ice breakup time is accounted for by the time trend represented by the regression line. (This simple example is typical of more complex evidence for the effects of global warming: large year-to-year variation requires many years of data to see a trend.)

7.44 More on global warming. Side-by-side boxplots offer a different look at the data. Group the data into periods of roughly equal length: 1917 to 1940, 1941 to 1964, 1965 to 1988, and 1989 to 2012. Make boxplots to compare ice breakup dates in these four time periods. Write a brief description of what the plots show. TANANA

7.45 Teacher's salaries. The Organization for Economic Cooperation and Development (OECD) began in 1961 to stimulate economic progress and world trade. It originally consisted of European countries, the United States and Canada, but has now grown to include 34 countries spanning the globe. The data *TEACHSL* on the Web site, gives the average starting salaries in 2012 of primary public school teachers (PPP, US$) for the member nations.[15] TEACHSL

(a) Make a stemplot or a histogram to display the distribution of teacher's salaries in the OECD countries.

(b) There is one high outlier. What country is this? What is the overall shape of the distribution if you ignore the outlier?
(c) Based on your work in (b), give a numerical summary of the center and spread of the distribution, omitting the outlier.
(d) Some Americans complain about teachers being overpaid. Where does the United States ($36,858) stand in this international comparison?

7.46 Cicadas as fertilizer? Every 17 years, swarms of cicadas emerge from the ground in the eastern United States, live for about six weeks, then die.

© Alastair Shay; Papilio/CORBIS

(There are several "broods," so we experience cicada eruptions more often than every 17 years.) There are so many cicadas that their dead bodies can serve as fertilizer and increase plant growth. In an experiment, a researcher added 10 cicadas under some plants in a natural plot of American bellflowers in a forest, leaving other plants undisturbed. One of the response variables was the size of seeds produced by the plants. Here are data (seed mass in milligrams) for 39 cicada plants and 33 undisturbed (control) plants:[16] CICADA

CICADA PLANTS				CONTROL PLANTS			
0.237	0.277	0.241	0.142	0.212	0.188	0.263	0.253
0.109	0.209	0.238	0.277	0.261	0.265	0.135	0.170
0.261	0.227	0.171	0.235	0.203	0.241	0.257	0.155
0.276	0.234	0.255	0.296	0.215	0.285	0.198	0.266
0.239	0.266	0.296	0.217	0.178	0.244	0.190	0.212
0.238	0.210	0.295	0.193	0.290	0.253	0.249	0.253
0.218	0.263	0.305	0.257	0.268	0.190	0.196	0.220
0.351	0.245	0.226	0.276	0.246	0.145	0.247	0.140
0.317	0.310	0.223	0.229	0.241			
0.192	0.201	0.211					

Describe and compare the two distributions. Do the data support the idea that dead cicadas can serve as fertilizer?

7.47 A big-toe problem. Hallux abducto valgus (call it HAV) is a deformation of the big toe that is not common in youth and often requires surgery. Doctors used X-rays to measure the angle (in degrees) of deformity in 38 consecutive patients under the age of 21 who came to a medical center for surgery to correct HAV.[17] The angle is a measure of the seriousness of the deformity. The data appear in Table 7.4 as "HAV angle." Describe the distribution of the angle of deformity among young patients needing surgery for this condition.   BIGTOE

TABLE 7.4   ANGLE OF DEFORMITY (DEGREES) FOR TWO TYPES OF FOOT DEFORMITY

HAV ANGLE	MA ANGLE	HAV ANGLE	MA ANGLE	HAV ANGLE	MA ANGLE
28	18	21	15	16	10
32	16	17	16	30	12
25	22	16	10	30	10
34	17	21	7	20	10
38	33	23	11	50	12
26	10	14	15	25	25
25	18	32	12	26	30
18	13	25	16	28	22
30	19	21	16	31	24
26	10	22	18	38	20
28	17	20	10	32	37
13	14	18	15	21	23
20	20	26	16		

7.48   Prey attract predators. Here is one way in which nature regulates the size of animal populations: high population density attracts predators, who remove a higher proportion of the population than when the density of the prey is low. One study looked at kelp perch and their common predator, the kelp bass. The researcher set up four large circular pens on sandy ocean bottom in Southern California. He chose young perch at random from a large group and placed 10, 20, 40, and 60 perch in the four pens. Then he dropped the nets protecting the pens, allowing bass to swarm in, and counted the perch left after 2 hours. Here are data on the proportions of perch eaten in four repetitions of this setup:[18]   PREY

PERCH	PROPORTION KILLED			
10	0.0	0.1	0.3	0.3
20	0.2	0.3	0.3	0.6
40	0.075	0.3	0.6	0.725
60	0.517	0.55	0.7	0.817

Do the data support the principle that "more prey attract more predators, who drive down the number of prey"?

7.49   Predicting foot problems. Metatarsus adductus (call it MA) is a turning in of the front part of the foot that is common in adolescents and usually corrects itself. Table 7.4 gives the severity of MA ("MA angle"). Doctors speculate that the severity of MA can help predict the severity of HAV. Describe the relationship between MA and HAV. Do you think the data confirm the doctors' speculation? Why or why not?   BIGTOE

7.50   Change in the Serengeti. Long-term records from the Serengeti National Park in Tanzania show interesting ecological relationships. When wildebeest are more abundant, they graze the grass more heavily, so there are fewer fires and more trees grow. Lions feed more successfully when there are more trees, so the lion population increases. Here are data on one part of this cycle, wildebeest abundance (in thousands of animals) and the percent of the grass area that burned in the same year:[19]   SERENG

Gallo Images-Anthony Bannister/Getty Images

WILDEBEEST (1000s)	PERCENT BURNED	WILDEBEEST (1000s)	PERCENT BURNED
396	56	622	60
476	50	600	56
698	25	902	45
1049	16	1440	21
1178	7	1147	32
1200	5	1173	31
1302	7	1178	24
360	88	1253	24
444	88	1249	53
524	75		

To what extent do these data support the claim that more wildebeest reduce the percent of grasslands that burn? How rapidly does burned area decrease as the number of wildebeest increases? Include a graph and suitable calculations.

7.51 **Casting aluminum.** In casting metal parts, molten metal flows through a "gate" into a die that shapes the part. The gate velocity (the speed at which metal is forced through the gate) plays a critical role in die casting. A firm that casts cylindrical aluminum pistons examined 12 types formed from the same alloy. How does the piston wall thickness (inches) influence the gate velocity (feet per second) chosen by the skilled workers who do the casting? If there is a clear pattern, it can be used to direct new workers or to automate the process. Analyze these data and report your findings.[20] ALUM

THICKNESS	VELOCITY	THICKNESS	VELOCITY
0.248	123.8	0.628	326.2
0.359	223.9	0.697	302.4
0.366	180.9	0.697	145.2
0.400	104.8	0.752	263.1
0.524	228.6	0.806	302.4
0.552	223.8	0.821	302.4

7.52 **How are schools doing? (optional)** The non-profit group Public Agenda conducted telephone interviews with parents of high school children. Interviewers chose equal numbers of black, Hispanic, and non-Hispanic white parents at random. One question asked was, "Are the high schools in your state doing an excellent, good, fair, or poor job, or don't you know enough to say?" Here are the survey results:[21] SCHOOLS

OPINION	BLACK PARENTS	HISPANIC PARENTS	WHITE PARENTS
Excellent	12	34	22
Good	69	55	81
Fair	75	61	60
Poor	24	24	24
Don't know	22	28	14
Total	202	202	201

Write a brief analysis of these results that focuses on the relationship between parent group and opinions about schools.

7.53 **Influence: Hot sector funds?** Investment advertisements always warn that "past performance does not guarantee future results." Here is an example that shows why you should pay attention to this warning.

Stocks fell sharply in 2002, then rose sharply in 2003. The following table gives the percent returns from 23 Fidelity Investments "sector funds" in these two years. Sector funds invest in narrow segments of the stock market. They often rise and fall faster than the market as a whole. MFUNDS

2002 RETURN	2003 RETURN	2002 RETURN	2003 RETURN	2002 RETURN	2003 RETURN
−17.1	23.9	−0.7	36.9	−37.8	59.4
−6.7	14.1	−5.6	27.5	−11.5	22.9
−21.1	41.8	−26.9	26.1	−0.7	36.9
−12.8	43.9	−42.0	62.7	64.3	32.1
−18.9	31.1	−47.8	68.1	−9.6	28.7
−7.7	32.3	−50.5	71.9	−11.7	29.5
−17.2	36.5	−49.5	57.0	−2.3	19.1
−11.4	30.6	−23.4	35.0		

(a) Make a scatterplot of 2003 return (response) against 2002 return (explanatory). The funds with the best performance in 2002 tend to have the worst performance in 2003. Fidelity Gold Fund, the only fund with a positive return in both years, is an extreme outlier.

(b) To demonstrate that correlation is not resistant, find r for all 23 funds and then find r for the 22 funds other than Gold. Explain from Gold's position in your plot why omitting this point makes r more negative.

(c) Find the equations of two least-squares lines for predicting 2003 return from 2002 return, one for all 23 funds and one omitting Fidelity Gold Fund. Add both lines to your scatterplot. Starting with the least-squares idea, explain why adding Fidelity Gold Fund to the other 22 funds moves the line in the direction that your graph shows.

7.54 **Influence: Monkey calls.** Table 7.1 (page 180) contains data on the response of 37 monkey neurons to pure tones and to monkey calls. Figure 7.6 (page 180) is a scatterplot of these data. MONKEY

(a) Find the least-squares line for predicting a neuron's call response from its pure tone response. Add the line to your scatterplot. Mark on your plot the point (call it A) with the largest residual (either positive or negative) and also the point (call it B) that is an outlier in the x direction.

(b) How influential are each of these points for the correlation r?

(c) How influential are each of these points for the regression line?

7.55 Influence: Bushmeat. Table 4.2 (page 120) gives data on fish catches in a region of West Africa and the percent change in the biomass (total weight) of 41 animals in nature reserves. It appears that years with smaller fish catches see greater declines in animals, probably because local people turn to "bushmeat" when other sources of protein are not available. The next year (1999) had a fish catch of 23.0 kilograms per person and animal biomass change of −22.9%. 📊 BUSHMT

(a) Make a scatterplot that shows how change in animal biomass depends on fish catch. Be sure to include the additional data point. Describe the overall pattern.

The added point is a low outlier in the y direction. (b) Find the correlation between fish catch and change in animal biomass both with and without the outlier. The outlier is influential for correlation. Explain from your plot why adding the outlier makes the correlation smaller.

(c) Find the least-squares line for predicting change in animal biomass from fish catch both with and without the additional data point for 1999. Add both lines to your scatterplot from (a). The outlier is not influential for the least-squares line. Explain from your plot why this is true.

Producing Data

T he purpose of statistics is to gain understanding from data. We can seek understanding in different ways, depending on the circumstances. We have studied one approach to data, *exploratory data analysis*, in some detail. Now we begin the move from data analysis toward *statistical inference*. Both types of reasoning are essential to effective work with data. Here is a brief sketch of the differences between them:

EXPLORATORY DATA ANALYSIS	STATISTICAL INFERENCE
Purpose is unrestricted exploration of the data, searching for interesting patterns.	Purpose is to answer specific questions, posed before the data were produced.
Conclusions apply only to the individuals and circumstances for which we have data in hand.	Conclusions apply to a larger group of individuals or a broader class of circumstances.
Conclusions are informal, based on what we see in the data.	Conclusions are formal, backed by a statement of our confidence in them.

Our journey toward inference begins in Chapters 8 and 9, which describe statistical designs for *producing data* by samples and experiments. The important lesson of Part II is that the quality of the inferences we make from data depends heavily on how the data are produced.

Producing Data: Sampling

Overview

The quality of any decision, estimate, or judgment about a population based on a sample—any statistical inference—depends on the sample being representative of the population. Chapters 8 and 9 are all about collecting data sensibly—making the sample more likely to represent the population. There are two essential ways in which data are collected: surveys and random samples (Chapter 8) and both observational studies and well-designed experiments (Chapter 9). Note that surveys often come from random samples and lead to observational studies. There is a good chance that a number of your students will one day take entire courses devoted to the topics in these chapters, and introductory courses often fail to devote enough attention to these important concepts. There's no overcoming bad data—"garbage in, garbage out." Another quote that comes to mind is "You can't fix by analysis what you bungled by design" (Richard Light, Judith Singer, and John Willett; accessed www.causeweb.org, 3/4/14).

> ### LEARNING OUTCOMES**
> - Identify the population in a sampling situation.
> - Recognize bias due to voluntary response samples and other inferior sampling methods.
> - Use software or Table B of random digits to select a simple random sample (SRS) from a population.
> - Recognize the presence of undercoverage and nonresponse as sources of error in a sample survey. Recognize the effect of the wording of questions on the responses.
> - Use random digits to select a stratified random sample from a population when the strata are identified.
> - Recognize when stratified random sampling would provide an advantage over simple random sampling.
>
> **These learning outcomes appear later for the students in Chapter 10: Part II Review.

Why do statisticians sample? One primary reason is that collecting data takes time and effort. If the population of interest is very large, it is impossible to reach all members; similarly, in a quality control setting, it would be impossible to determine, for example, the average lifetime of every product a company makes—if they did, they would have nothing left to sell; we must rely on testing a sample of the products.

The sampling methods presented here start with the simplest and easiest and build to more complicated, but often more useful, methods. Convenience samples and voluntary response samples are both straightforward and tempting because they are easy to obtain, but they are among the least favorable sample methods and generally produce useless data, as they are unlikely to provide a representative sample of the target population. The simple random sample (SRS) is the first proper random sampling design students will see, and it is the easiest to understand. Stratified random samples are more complicated, but are often more efficient than simple random samples; these samples are relatively easy to describe, but their analysis is beyond the scope of this text.

Students should be constantly reminded of some of the cautions associated with sampling. Confounding variables, which often arise in observational studies, are defined here and discussed again in future chapters. Even samples that are selected with good methodology will suffer from random (sampling) error—if you repeat the survey, results will vary by random chance. However, samples selected with poor methodology may also suffer from nonsampling, systematic error—types of bias due to undercoverage and nonresponse, for example. And sample surveys are subject to additional issues such as response bias and wording effects.

Technology is rapidly changing the landscape of sample surveys. With more and more mobile phones and fewer landline phones, polling companies must be careful to ensure that their samples represent the populations of interest. Web surveys are very tempting, but are problematic—we must consider who visits the site and whether a person can respond more than once. In some cases, people who were not the intended audience of the poll responded en masse, and the poll "backfired"—take, for example, the unexpected results of a 2004 American Family Association online poll asking its audience about gay marriage (see http://archive.wired.com/culture/lifestyle/news/2004/01/61982 for a nice write-up of the issue).

Teaching Suggestions and Additional Examples/Activities for the Classroom

Note: If you prefer to begin your course with generating data, you can move Chapters 8 and 9 to where they best fit your curriculum.

1. Involve Students in a Discussion of Sampling Methods

Start a class session by asking students how they might estimate the proportion of all students at the school who regularly engage in aerobic exercises. Hopefully, students will make all kinds of suggestions—using the students enrolled in the class, for example, or surveying people at the campus gym or student union. Use their suggestions to point out some of the problems that surveyors encounter—this class is not representative of all students, for example; students at the campus gym are more likely to do aerobic exercise because they are at the gym; students who might be encountered at the student union may have more free time than those elsewhere on campus (or are merely free at the time of the survey).

2. Emphasize the Importance of Proper Sampling Techniques

Provide students with a record of some of the Gallup Poll's final predictions (all based on samples) for the U.S. presidential election results dating back to 1936. You can find this data online (http://www.gallup.com/poll/110548/gallup-presidential-election-trialheat-trends-19362004.aspx). This site excludes results for the 2012 election. For that election, you can show students polls with two different outcomes: among registered voters, Gallup found 49% for Obama and 46% for Romney (http://www.gallup.com/poll/150743/Obama-Romney.aspx), whereas among likely voters, Gallup found 49% for Romney and 48% for Obama (http://www.gallup.com/poll/157817/election-2012-likely-voters-trial-heat-obama-romney.aspx). Final results for the 2012 presidential election had Obama with 51.0% and Romney with 47.2% (http://www.fec.gov/pubrec/fe2012/2012presgeresults.pdf). Gallup's samples taken prior to 1948 were nonrandom, and they systematically favored Republican candidates, as demonstrated by the actual election returns compared with Gallup's final predictions. However, once Gallup started using random samples (beginning 1948), this systematic error (a selection bias) disappeared, and the magnitude of error diminished as well.

It's most important for students to recognize the difference between random (sampling) error and nonrandom (nonsampling) error and that larger samples will help with the former but not with the latter. With the preceding example concerning aerobic exercise, a larger sample of students standing in front of the campus gym (or relaxing at the student union) will not improve our estimate.

3. Discuss Stratified Sampling

The use of stratified sampling can be illustrated with a simple example. If you're teaching a high-school class, ask what proportion of students has a driver's license—class rank provides a natural stratification variable. If you're teaching a college class, ask what the average income of all college employees is—type of employee (administrator, faculty, staff, graduate student, football coach, etc.) provides a natural stratification variable; another possibility that always elicits avid interest is to ask about parking on campus; commuter students differ from those who live in residence halls, and both groups typically differ from faculty and staff.

4. Emphasize the Difference Between Voluntary Response and Nonresponse

Students often confuse voluntary response with nonresponse. Try to clarify that in cases of voluntary response, the decision about whether to participate is entirely up to the respondent (they "self-select" into the poll); cases of nonresponse arise when randomly selected individuals refuse to cooperate.

Other Resources (LaunchPad)

StatClips
> *Data and Sampling*
> *Introduction to Statistics: Populations, Parameters, Samples, and Statistics*

Snapshots Videos
> *Sampling*

EESEE Case Studies
> *Strollers at the Zoo*
> *Nurses and Glove Use* (example of stratified sampling)
> *Columbus' 1993 Election Poll*
> *Survey Response by Physicians*
> *Survey Response by Sports Fans*
> *Puerto Rican vs. U.S. Consumers*

Applet
> *Simple Random Sample*

Frank Fell/Getty Images

Producing Data: Sampling

Statistics, the science of data, provides ideas and tools that we can use in many settings. Sometimes we have data that describe a group of individuals and want to learn what the data say. That's the job of exploratory data analysis. Sometimes we have specific questions but no data to answer them. To get sound answers, we must *produce data* in a way that is designed to answer our questions.

Suppose our question is, "What percent of college students think that people should not obey laws that violate their personal values?" To answer the question, we interview undergraduate college students. We can't afford to ask all students, so we put the question to a *sample* chosen to represent the entire student *population*. How shall we choose a sample that truly represents the opinions of the entire population? Statistical designs for choosing samples are the topic of this chapter. We will see that

- A sound statistical design is necessary if we are to trust data from a sample for drawing sound conclusions about the population.

- In sampling from large human populations, however, "practical problems" can overwhelm even sound designs.

- The impact of technology (particularly cell phones and the Internet) is making it harder to produce trustworthy national data by sampling.

Population versus sample

A political scientist wants to know what percent of college-age adults consider themselves conservatives. An automaker hires a market research firm to learn what percent of adults aged 18 to 35 recall seeing television advertisements for a new gas-electric hybrid car. Government economists inquire about average household income. In all these cases, we want to gather information about a large group of individuals. Time, cost, and inconvenience forbid contacting every individual. So we gather information about only part of the group to draw conclusions about the whole.

Population, Sample, Sampling Design

The **population** in a statistical study is the entire group of individuals about which we want information.

A **sample** is a part of the population from which we actually collect information. We use a sample to draw conclusions about the entire population.

A **sampling design** describes exactly how to choose a sample from the population.

Pay careful attention to the details of the definitions of "population" and "sample." Look at Exercise 8.1 on the next page right now to check your understanding.

We often draw conclusions about a whole on the basis of a sample. Everyone has tasted a sample of ice cream and ordered a cone on the basis of that taste. But ice cream is uniform, so that the single taste represents the whole. Choosing a repre-sentative sample from a large and varied population is not so easy. The first step in *sample survey* planning a **sample survey** is to say exactly *what population* we want to describe. The second step is to say exactly *what we want to measure,* that is, to give exact defini-tions of our variables. These preliminary steps can be complicated, as the following example illustrates.

EXAMPLE 8.1   The Current Population Survey

The most important government sample survey in the United States is the monthly Current Population Survey (CPS) conducted by the Bureau of the Census for the Bureau of Labor Statistics. The CPS contacts about 60,000 households each month. It produces the monthly unemployment rate and much other economic and social information. (See Figure 8.1.) To measure unemployment, we must first specify the population we want to describe. Which age groups will we include? Will we include illegal immigrants or people in prisons? The CPS defines its population as all U.S. residents (legal or not) 16 years of age and over who are civilians and are not in an institution such as a prison. The unemployment rate announced in the news refers to this specific population.

The second question is harder: what does it mean to be "unemployed"? Someone who is not looking for work—for example, a full-time student—should not be called unemployed just because she is not working for pay. If you are chosen for the CPS sample, the interviewer first asks whether you are available to work and whether you actually looked for work in the past four weeks. If not, you are neither employed nor unemployed—you are not in the labor force. So discouraged workers who haven't looked for a job in four weeks are excluded from the count.

If you are in the labor force, the interviewer goes on to ask about employment. If you did any work for pay or in your own business during the week of the survey,

U.S. Bureau of Labor Statistics

FIGURE 8.1

The home page of the Current Population Survey at the Bureau of Labor Statistics.

you are employed. If you worked at least 15 hours in a family business without pay, you are employed. You are also employed if you have a job but didn't work because of vacation, being on strike, or other good reason. An unemployment rate of 6.7% means that 6.7% of the sample was unemployed, using the exact CPS definitions of both "labor force" and "unemployed." ■

The final step in planning a sample survey is the sampling design. We will now introduce basic statistical designs for sampling.

Apply Your Knowledge

8.1 **Sampling Students.** A political scientist wants to know how college students feel about the Social Security system. She obtains a list of the 3456 undergraduates at her college and mails a questionnaire to 250 students selected at random. Only 104 questionnaires are returned.

(a) What is the population in this study? Be careful: what group does she *want information about?*

(b) What is the sample? Be careful: from what group does she *actually obtain information?*

The important message in this problem is that the sample can redefine the population about which information is obtained.

8.2 **Student Archaeologists.** An archaeological dig turns up large numbers of pottery shards, broken stone implements, and other artifacts. Students working on the project classify each artifact and assign it a number. The counts in different categories are important for understanding the site, so the project

director chooses 2% of the artifacts at random and checks the students' work. What are the population and the sample here?

8.3   **Customer Satisfaction.** A department store mails a customer satisfaction survey to people who make credit card purchases at the store. This month, 45,000 people made credit card purchases. Surveys are mailed to 1000 of these people, chosen at random, and 137 people return the survey form.

(a) What is the population of interest for this survey?

(b) What is the sample? From what group is information actually obtained?

How to sample badly

How can we choose a sample that we can trust to represent the population? A sampling design is a specific method for choosing a sample from the population. The easiest—but not the best—design just chooses individuals close at hand. If we are interested in finding out how many people have jobs, for example, we might go to a shopping mall and ask people passing by if they are employed. A sample selected by taking the members of the population that are easiest to reach is called a **convenience sample.** Convenience samples often produce unrepresentative data.

convenience sample

| EXAMPLE 8.2 | Sampling at the Mall |

A sample of mall shoppers is fast and cheap. But people at shopping malls tend to be more prosperous than typical Americans. They are also more likely to be teenagers or retired. Moreover, unless interviewers are carefully trained, they tend to question well-dressed, respectable-looking people and avoid poorly dressed or tough-looking individuals. The type of people at the mall will also vary by time of day and day of week. In short, mall interviews will not contact a sample that is representative of the entire population. ■

Interviews at shopping malls will almost surely overrepresent middle-class and retired people and underrepresent the poor. This will happen almost every time we take such a sample. That is, it is a systematic error caused by a bad sampling design, not just bad luck on one sample. This is *bias:* the outcomes of mall surveys will repeatedly miss the truth about the population in the same ways.

Bias

The design of a statistical study is **biased** if it systematically favors certain outcomes.

| EXAMPLE 8.3 | Online Polls |

Former CNN evening commentator Lou Dobbs doesn't like illegal immigration. One of his broadcasts in 2007 was largely devoted to attacking a proposal by the governor of New York State to offer driver's licenses to illegal immigrants as a public safety measure. During the show, Mr. Dobbs invited his viewers to go to loudobbs.com to vote on the question "Would you be more or less likely to vote for a presidential candidate who supports giving drivers' licenses to illegal aliens?" We aren't surprised that 97% of the 7350 people who voted by the end of the broadcast said, "Less likely." ■

The loudobbs.com poll was biased because people chose whether or not to participate (see Exercise 8.34, page 209). Most who voted were viewers of Lou Dobbs's program who had just heard him denounce the governor's idea. *People who take the trouble to respond to an open invitation are usually not representative of any clearly defined population.* That's true of the people who bother to respond to write-in, call-in, or online polls in general. Polls like these are examples of *voluntary response sampling.*

Voluntary Response Sample

A **voluntary response sample** consists of people who choose themselves by responding to a broad appeal. Voluntary response samples are biased because people with strong opinions are most likely to respond.

Apply Your Knowledge

8.4   Sampling on Campus. You see a student standing in front of the student center, now and then stopping other students to ask them questions. She says that she is collecting student opinions for a class assignment. Explain why this sampling method is almost certainly biased.

8.5   More Sampling on Campus. You would like to start a club for psychology majors on campus, and you are interested in finding out what proportion of psychology majors would join. The dues would be $35 and used to pay for speakers to come to campus. You ask five psychology majors from your senior psychology honors seminar whether they would be interested in joining this club and find that four of the five students questioned are interested. Is this sampling method biased, and if so, what is the likely direction of bias?

Simple random samples

In a voluntary response sample, people choose whether to respond. In a convenience sample, the interviewer makes the choice. In both cases, personal choice produces bias. The statistician's remedy is to allow impersonal chance to choose the sample. A sample chosen by chance rules out both favoritism by the sampler and self-selection by respondents. Choosing a sample by chance attacks bias by giving all individuals an equal chance to be chosen. Rich and poor, young and old, liberal and conservative, all have the same chance to be in the sample.

The simplest way to use chance to select a sample is to place names in a hat (the population) and draw out a handful (the sample). This is the idea of *simple random sampling.* Although the idea of drawing names from a hat is a good way to conceptualize a simple random sample, it is generally *not* a good method for obtaining a simple random sample. Writing names on slips of paper can lead to bias if the slips of paper are not well mixed or there is a tendency to select them from, say, the top or bottom. Drawing names from a hat would be particularly difficult to implement if the population size is large, possibly requiring thousands of slips of paper.

Simple Random Sample

A **simple random sample (SRS)** of size n consists of n individuals from the population chosen in such a way that every set of n individuals has an equal chance to be the sample actually selected.

"Kept as they are now." Can we trust the opinions of this sample to fairly represent the opinions of all adults? Here's part of the statement by the *Times* on how the poll was conducted:

> The latest New York Times/CBS News poll is based on telephone interviews conducted January 11 through January 15 with 1110 adults throughout the United States.
>
> The sample of land line telephone exchanges called was randomly selected by a computer from a complete list of more than 72,000 active residential exchanges across the country. The exchanges were chosen so as to ensure that each region of the country was represented in proportion to its population.
>
> Within each exchange, random digits were added to form a complete telephone number, thus permitting access to listed and unlisted numbers alike. Within each household, one adult was designated by a random procedure to be the respondent for the survey.
>
> To increase coverage, this landline sample was supplemented by respondents reached through random dialing of cell phone numbers.[1]

random digit dialing

The selection of landline phone numbers is a good description of a common method for choosing national samples, called **random digit dialing.** Although the information about the conduct of the poll provides the reader with some basic details, data collection and analysis of a national survey is a great deal more complex than this short description suggests. We'll come back to random digit dialing and its problems later (see Exercise 8.16, page 206), but this statement on the conduct of the poll does contain important information. We know the size of the sample, when the poll was taken, and the comforting word *random* appears four times. ■

Apply Your Knowledge

8.6 Apartment Living. You are planning a report on apartment living in a college town. You decide to select three apartment complexes at random for in-depth interviews with residents. Use the *Simple Random Sample* applet, other software, or Table B to select a simple random sample of four of the following apartment complexes. If you use Table B, start at line 122.

Ashley Oaks	Country View	Mayfair Village
Bay Pointe	Country Villa	Nobb Hill
Beau Jardin	Crestview	Pemberly Courts
Bluffs	Del-Lynn	Peppermill
Brandon Place	Fairington	Pheasant Run
Briarwood	Fairway Knolls	River Walk
Brownstone	Fowler	Sagamore Ridge
Burberry Place	Franklin Park	Salem Courthouse
Cambridge	Georgetown	Village Square
Chauncey Village	Greenacres	Waterford Court

8.7 Minority Managers. A firm wants to understand the attitudes of its minority managers toward its system for assessing management performance. Following is a list of all the firm's managers who are members of minority groups. Use the *Simple Random Sample* applet, other software, or Table B at line 134 to choose five to be interviewed in detail about the performance appraisal system.

Adelaja	Draguljic	Huo	Modur
Ahmadiani	Fernandez	Ippolito	Rettiganti
Barnes	Fox	Jiang	Rodriguez
Bonds	Gao	Jung	Sanchez
Burke	Gemayel	Mani	Sgambellone
Deis	Gupta	Mazzeo	Yajima
Ding	Hernandez		

8.8 **Sampling Gravestones.** The local genealogical society in Coles County, Illinois, has compiled records on all 55,914 gravestones in cemeteries in the county for the years 1825 to 1985. Historians plan to use these records to learn about African Americans in Coles County's history. They first choose an SRS of 395 records to check their accuracy by visiting the actual gravestones.[2]

© The Photo Works

(a) How would you label the 55,914 records?

(b) Use software or Table B, starting at line 120, to choose the first 5 records for the SRS.

Inference about the population

The purpose of a sample is to give us information about a larger population. The process of drawing conclusions about a population on the basis of sample data is called **inference** because we *infer* information about the population from what we *know* about the sample.

inference

Inference from convenience samples or voluntary response samples would be misleading because these methods of choosing a sample are biased. We are almost certain that the sample does *not* fairly represent the population. *The first reason to rely on random sampling is to eliminate bias in selecting samples from the list of available individuals.*

Nonetheless, it is unlikely that results from a random sample are exactly the same as for the entire population. Sample results, like the unemployment rate obtained from the monthly Current Population Survey, are only estimates of the truth about the population. If we select two samples at random from the same population, we will almost certainly draw different individuals. So the sample results will differ somewhat, just by chance. Properly designed samples avoid systematic bias, but their results are rarely exactly correct, and they vary from sample to sample.

Why can we trust random samples? The big idea is that the results of random sampling don't change haphazardly from sample to sample. Because we deliberately use chance, the results obey the laws of probability that govern chance behavior. These laws allow us to say how likely it is that sample results are close to the truth about the population. *The second reason to use random sampling is that the laws of probability allow trustworthy inference about the population.* Results from random samples come with a margin of error that sets bounds on the size of the likely error. How to do this is part of the technique of statistical inference. We will describe the reasoning in Chapter 16 and present details throughout the rest of the book.

One point is worth making now: *larger random samples give more accurate results than smaller random samples.* By taking a very large sample, you can be confident that the sample result is very close to the truth about the population. The Current Population Survey contacts about 60,000 households, so it estimates the national unemployment rate very accurately. Opinion polls that contact 1000 or 1500 people give less accurate results. Of course, only samples chosen by chance carry this guarantee. Lou Dobbs's online sample tells us little about overall American public opinion even though 7350 people clicked a response.

Apply Your Knowledge

8.9 **Ask More People.** Just before a presidential election, a national opinion-polling firm increases the size of its weekly sample from the usual 1500 people to 4000 people. Why do you think the firm does this?

8.10 Sampling Pentecostals. Pentecostals are among the fastest-growing Christian groups in many countries. In 2006, the Pew Forum on Religion and Public Life surveyed Pentecostal Christians in 10 countries and compared their opinions with those of the general population. In South Korea, random samples by Gallup Korea had margins of error (we will give more detail in later chapters) of ±4% for the general public and ±9% for Pentecostals.[3] What do you think explains the fact that estimates for Pentecostals were less precise?

Other sampling designs

Random sampling, the use of chance to select the sample, is the essential principle of statistical sampling. Designs for random sampling from large populations spread out over a wide area are usually more complex than an SRS. For example, it is common to sample important groups within the population separately, then combine these samples. This is the idea of a *stratified random sample.*

GOLFING AT RANDOM

Random drawings give everyone the same chance to be chosen, so they offer a fair way to decide who gets a scarce good—like a round of golf. Lots of golfers want to play the famous Old Course at St. Andrews, Scotland. Some can reserve in advance, at considerable expense. Most must hope that chance favors them in the daily random drawing for tee times. At the height of the summer season, only 1 in 6 wins the right to pay $250 for a round.

Stratified Random Sample

To select a **stratified random sample,** first classify the population into groups of similar individuals, called **strata.** Then choose a separate SRS in each stratum and combine these SRSs to form the full sample.

Choose the strata based on facts known before the sample is taken. For example, a population of election districts might be divided into urban, suburban, and rural strata. A stratified design can produce more precise information than an SRS of the same size by taking advantage of the fact that individuals in the same stratum are similar to one another.

EXAMPLE 8.6   **Seat Belt Use in Hawaii**

Ryan McVay/Photo Disc/Getty Images

Each state conducts an annual survey of seat belt use by drivers, following guidelines set by the federal government. The guidelines require random sampling. Seat belt use is observed at randomly chosen road locations at random times during daylight hours. The locations are not an SRS of all locations in the state but rather a stratified sample using the state's counties as strata.

In Hawaii, the counties are the islands that make up the state's territory. The seat belt survey sample consists of 135 road locations in the four most populated islands: 66 in Oahu, 24 in Maui, 23 in Hawaii, and 22 in Kauai. The sample sizes on the islands are proportional to the amount of road traffic.[4] ■

multistage sample

Most large-scale sample surveys use **multistage samples.** For example, the opinion poll described in Example 8.5 has three stages: choose a random sample of telephone exchanges (stratified by region of the country), then an SRS of household telephone numbers within each exchange, then a random adult in each household.

Analysis of data from sampling designs more complex than an SRS takes us beyond basic statistics. But the SRS is the building block of more elaborate designs, and analysis of other designs differs more in complexity of detail than in fundamental concepts.

Apply Your Knowledge

8.11 Sampling Metro Chicago. Cook County, Illinois, has the second-largest population of any county in the United States (after Los Angeles County, California). Cook County has 30 suburban townships and an additional 8 townships that make up the city of Chicago. The suburban townships are

Barrington	Elk Grove	Maine	Orland	Riverside
Berwyn	Evanston	New Trier	Palatine	Schaumburg
Bloom	Hanover	Niles	Palos	Stickney
Bremen	Lemont	Northfield	Proviso	Thornton
Calumet	Leyden	Norwood Park	Rich	Wheeling
Cicero	Lyons	Oak Park	River Forest	Worth

The Chicago townships are

Hyde Park	Lake	North Chicago	South Chicago
Jefferson	Lake View	Rogers Park	West Chicago

Because city and suburban areas may differ, the first stage of a multistage sample chooses a stratified sample of 5 suburban townships and 3 of the more heavily populated Chicago townships. Use software, the simple random sample applet, or Table B to choose this sample. (If you use Table B, assign labels in alphabetical order and start at line 105 for the suburbs and at line 115 for Chicago.)

8.12 Academic Dishonesty. A study of academic dishonesty among college students used a two-stage sampling design. The first stage chose a sample of 30 colleges and universities. Then the study authors mailed questionnaires to a stratified sample of 200 seniors, 100 juniors, and 100 sophomores at each school.[5] One of the schools chosen has 1127 freshmen, 989 sophomores, 943 juniors, and 895 seniors. You have alphabetical lists of the students in each class. Explain how you would assign labels for stratified sampling. Then use software or Table B, starting at line 122, to select the first 5 students in the sample from each stratum. After selecting 5 students for a stratum, continue to select the students for the next stratum.

Cautions about sample surveys

Random selection eliminates bias in the choice of a sample from a list of the population. When the population consists of human beings, however, accurate information from a sample requires more than a good sampling design.

To begin, we need an accurate and complete list of the population. Because such a list is rarely available, most samples suffer from some degree of *undercoverage*. A sample survey of households, for example, will miss not only homeless people but also prison inmates and students in dormitories. An opinion poll conducted by calling landline telephone numbers will miss households that have only cell phones as well as households without a phone. The results of national sample surveys therefore have some bias if the people not covered differ from the rest of the population.

A more serious source of bias in most sample surveys is *nonresponse*, which occurs when a selected individual cannot be contacted or refuses to cooperate. Nonresponse to sample surveys often exceeds 50%, even with careful planning and several callbacks. Because nonresponse is higher in urban areas, most sample surveys substitute other people in the same area to avoid favoring rural areas in the final sample. If the people contacted differ from those who are rarely at home or who refuse to answer questions, some bias remains.

Undercoverage and Nonresponse

Undercoverage occurs when some groups in the population are left out of the process of choosing the sample.

Nonresponse occurs when an individual chosen for the sample can't be contacted or refuses to participate.

EXAMPLE 8.7   ## How Bad Is Nonresponse?

The U.S. Census Bureau's American Community Survey (ACS) has the lowest nonresponse rate of any poll we know: in 2011 only about 1.1% of the households in the sample refused to respond; the overall nonresponse rate, including "never at home" and other causes, was just 2.4%.[6] This monthly survey of about 250,000 households replaces the "long form" that in the past was sent to some households in the every-ten-years national census. Participation in the ACS is mandatory, and the U.S. Census Bureau follows up by telephone and then in person if a household fails to return the mail questionnaire.

The University of Chicago's General Social Survey (GSS) is the nation's most important social science survey. (See Figure 8.3.) The GSS contacts its sample in person, and it is run by a university. Its response rates are about 70%, among the highest in the world.

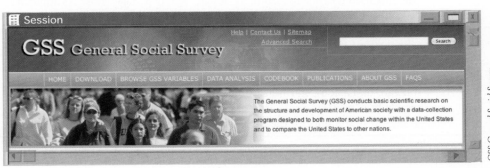

FIGURE 8.3
The home page of the General Social Survey at the University of Chicago's National Opinion Research Center. The GSS has tracked opinions about a wide variety of issues since 1972.

What about opinion polls by news media and opinion-polling firms? We don't know the rates of nonresponse for many of these surveys because they don't say, which in itself is a bad sign. In a February 2013 survey on social media use, the Pew Research Center provided a full disposition of the sampled phone numbers. Here are the details. Initially, 33,785 landline and 19,800 cell phones were dialed, with 12,335 of the landlines and 11,504 of the cell numbers corresponding to working numbers. Of these 23,839 working numbers, they were able to contact about 75%, as a large portion of the calls went to voice mail. Among those contacted, about 17% cooperated. To sum it up, the 23,839 working numbers dialed resulted in a final sample of 2,989, giving a response rate of about 12.5%.[7] ■

---

response bias   In addition, the behavior of the respondent or of the interviewer can cause **response bias** in sample results. People know that they should take the trouble to vote, for example, so many who didn't vote in the last election will tell an interviewer that they did. The race or sex of the interviewer can influence responses to questions about race relations or attitudes toward feminism. Answers to questions that ask respondents to recall past events are often inaccurate because of faulty memory. For example, many people "telescope" events in the past, bringing them forward in memory to more recent time periods. "Have you visited a dentist in the last 6 months?" will often draw a "Yes" from someone who last visited a dentist 8 months ago.[8] Careful training of interviewers and careful supervision to avoid variation among the interviewers can reduce response bias. Good interviewing technique is another aspect of a well-done sample survey.

The **wording of questions** is the most important influence on the answers given to a sample survey. Confusing or leading questions can introduce strong bias, and changes in wording can greatly change a survey's outcome. Even the order in which questions are asked matters. Here are some examples.[9]

wording effects

EXAMPLE 8.8   What Was That Question?

How do Americans feel about illegal immigrants? "Should illegal immigrants be prosecuted and deported for being in the United States illegally, or shouldn't they?" Asked this question in an opinion poll, 69% favored deportation. But when the very same sample was asked whether illegal immigrants who have worked in the United States for two years "should be given a chance to keep their jobs and eventually apply for legal status," 62% said that they should. Different questions give quite different impressions of attitudes toward illegal immigrants.

What about government help for the poor? Only 13% think we are spending too much on "assistance to the poor," but 44% think we are spending too much on "welfare." ■

EXAMPLE 8.9   Are You Happy?

Ask a sample of college students these two questions:

"How happy are you with your life in general?" (Answers on a scale of 1 to 5)

"How many dates did you have last month?"

The correlation between answers is $r = -0.012$ when asked in this order. It appears that dating has little to do with happiness. Reverse the order of the questions, however, and $r = -0.66$. Asking a question that brings dating to mind makes dating success a big factor in happiness. ■

Don't trust the results of a sample survey until you have read the exact questions asked. The amount of nonresponse and the date of the survey are also important. Good statistical design is a part, but only a part, of a trustworthy survey.

Apply Your Knowledge

8.13 A Survey of 100,000 Physicians. In 2010, the Physicians Foundation conducted a survey of physicians' attitude about health care reform, calling the report "a survey of 100,000 physicians." The survey was sent to 100,000 randomly selected physicians practicing in the United States, 40,000 via post-office mail and 60,000 via email. A total of 2,379 completed surveys were received.[10]

(a) State carefully what population is sampled in this survey and what is the sample size. Could you draw conclusions from this study about all physicians practicing in the United States?

(b)   What is the rate of nonresponse for this survey? How might this affect the credibility of the survey results?

(c)   Why is it misleading to call the report "a survey of 100,000 physicians?"

8.14 **Gays in the Military.** In 2010, a Quinnipiac University Poll and a CNN Poll each asked a nationwide sample about their views on openly gay men and women serving in the military.[11] Here are the two questions:

> Question A: *Federal law currently prohibits openly gay men and women from serving in the military. Do you think this law should be repealed or not?*
>
> Question B: *Do you think people who are openly gay or homosexual should or should not be allowed to serve in the U.S. military?*

One of these questions had 78% responding "should," and the other question had only 57% responding "should." Which wording is slanted toward a more negative response on gays in the military? Why?

The impact of technology

A few national sample surveys, including the General Social Survey and the government's American Community Survey and Current Population Survey, interview some or all of their subjects in person. This is expensive and time-consuming, so most national surveys contact subjects by telephone using the random digit dialing (RDD) method described in Example 8.5 (page 197). Technology, especially the spread of cell phones, is making traditional RDD methods outdated.

First, *call screening* is now common. A large majority of American households have answering machines, voice mail, or caller ID, and many use these methods to screen their calls. Calls from polling organizations are rarely returned.

More seriously, the number of *cell-phone-only households* is increasing rapidly. By mid-2007, 14% of American households had a cell phone but no landline phone, by the end of 2009 that number had increased to almost 25%, and in 2012 the number was almost 36%. It's clear from these numbers that RDD reaching only landline numbers is in trouble. Can surveys just add cell phone numbers? Not easily. Federal regulations prohibit automated dialing to cell phones, which rules out computerized RDD sampling and requires hand dialing of cell phone numbers, which is expensive. A cell phone can be anywhere, and many people keep their cell number despite moving, so stratifying by location becomes difficult. And a cell phone user may be driving or otherwise unable to talk safely.

People who screen calls and people who have only a cell phone tend to be younger than the general population. In 2012, just over 60% of adults aged 25 to 29 years lived in households with no landline phone. So RDD surveys using only landlines may be biased (see Exercise 8.16, page 206). Careful surveys weight their responses to reduce bias. For example, if a sample contains too few young adults, the responses of the young adults who do respond are given extra weight. But with response rates steadily dropping and cell-phone-only use steadily growing, the future of RDD landline telephone surveys is not promising. Many polling organizations now include a minimum quota of cell phone users in their samples to help adjust for bias[12] (see Example 8.5, page 197 and Exercise 8.44, page 211).

One alternative is to use *Web surveys* rather than telephone surveys, an increasingly popular survey method. Web surveys have several advantages over more traditional survey methods. It is possible to collect large amounts of survey data at lower costs than traditional methods allow. Anyone can put survey questions on

DO NOT CALL!

People who do sample surveys hate telemarketing. We all get so many unwanted sales pitches by phone that many people hang up before learning that the caller is conducting a survey rather than selling vinyl siding. You can eliminate calls from commercial telemarketers by placing your phone number on the National Do Not Call Registry. Just go to www.donotcall.gov to sign up.

dedicated sites offering free services; thus large-scale data collection is available to almost every person with access to the Internet. Furthermore, Web surveys allow one to deliver multimedia survey content to respondents, opening up new realms of survey possibilities that would be extremely difficult to implement using traditional methods. Some argue that eventually Web surveys will replace traditional survey methods.

Although Web surveys are easy to do, they are not easy to do well. Three major problems are voluntary response, undercoverage, and nonresponse. Voluntary response appears in several forms in online surveys. Example 8.3 (page 194) is a survey that invited individuals to a particular Web site to participate in a poll. Other Web surveys solicit participation through announcements in news groups, email invitations, and banner ads on high-traffic sites.[13]

EXAMPLE 8.10   More to Makeup Than Meets the Eye

Dougal Waters/Getty Images

The Renfrew Center Foundation, a nonprofit charitable organization, reports that 44% of women have negative feelings when they don't wear makeup.[14] How did they come up with this number? It is based on an online survey of 1292 women 18 years of age and older conducted for them by Harris Interactive from December 20–22, 2011. This is a more sophisticated example of voluntary response. It occurs when the polling organization, in this case Harris Interactive, maintains a panel of volunteers. The panel is recruited to fill out questionnaires on the Internet, often in exchange for points redeemable for cash and gifts. Panelists join by going to the polling company Web site and filling out some personal information later used to select them for specific surveys. Harris Interactive uses an online research panel of over 6 million volunteers worldwide, from which it selected the sample for the Renfrew Center Foundation poll.

In their report on online panels,[15] the American Association of Public Opinion Research states, "Researchers should avoid nonprobability online panels when one of the research objectives is to accurately estimate population values. From a total survey error perspective, the principal source of error in estimates from these types of sample sources is a combination of the lack of Internet access in roughly one in three U.S. households and the self-selection bias inherent in the panel recruitment processes." They also state that "The reporting of a margin of sampling error associated with an opt-in or self-identified sample is misleading." In the methodology section of the poll, the Renfew Center Foundation does state, "This online survey is not based on a probability sample, and therefore, no estimate of theoretical sampling error can be calculated," although the issue of potential bias in the estimate is not addressed. ■

Undercoverage is a serious problem for even careful Web surveys because about a quarter of Americans lack Internet access, and only about 60% have broadband access. People without Internet access are more likely to be poor, elderly, minority, or rural than the overall population, so the potential for bias in a Web survey is clear. There is no easy way to choose a random sample even from people with Web access because there is no technology that generates personal email addresses at random in the way that RDD generates residential telephone numbers. Even if such technology existed, etiquette and regulations aimed at spammers would prevent mass emailing. For the present, Web surveys work well only for restricted populations, for example, surveying students at your university using the school's list of student email addresses. Here is an example of a successful Web survey.

| EXAMPLE 8.11 | **Doctors and Placebos** |

A placebo is a dummy treatment such as a pill that has no direct effect on a patient but may bring about a response because patients expect it to. Do academic physicians who maintain private practices sometimes give their patients placebos? A Web survey of doctors in internal medicine departments at Chicago-area medical schools was possible because almost all the doctors had listed email addresses.

An email was sent to each doctor explaining the purpose of the study, promising anonymity, and giving an individual Web link for response. In all, 231 of 443 doctors responded. The response rate was helped by the fact that the email came from a team at a medical school. Result: 45% said they sometimes used placebos in their clinical practice.[16] ■

Apply Your Knowledge

8.15 NPR Facebook Survey. In 2010, National Public Radio (NPR) conducted a survey of preferences and habits of their Facebook fans by recruiting respondents through messages posted on their Facebook page. The survey was conducted online and deployed July 12–19. A total of 40,043 respondents began the survey with 33,304 completing all questions. It was found that people accessed NPR on the radio, at NPR.org, through iPhone apps, and several other platforms. Asked about time spent with NPR, about 20% of respondents indicated that they spent more than three hours per day, including radio listening.

(a) Here is what NPR says about the survey methodology: "Respondents were self-selected and the resulting sample is non-random—therefore a margin of error cannot be calculated, and the survey results cannot be projected to any population other than the sample itself."[17] Why can't inference about any population be made?

(b) Suppose that people who spent more time with NPR were more likely to respond to the survey. Do you think the true percentage of NPR's Facebook fans who spend more than three hours with NPR is higher or lower than the 20% found from the survey? Explain why.

8.16 More on Random Digit Dialing. By the beginning of 2012 about 35% of adults lived in households with a cell phone and no landline phone, and among adults aged 25 to 29 this number was about 60%.

(a) Write a survey question for which the opinions of adults with landline phones only are likely to differ from the opinions of adults with cell phones only. Give the direction of the difference of opinion.

(b) For the survey question in (a), suppose a survey was conducted using random digit dialing of landline phones only. Would the results be biased? What would be the direction of bias?

(c) Most surveys now supplement the landline sample contacted by RDD with a second sample of respondents reached through random dialing of cell phone numbers. The landline respondents are weighted to take account of household size and number of telephone lines into the residence, whereas the cell phone respondents are weighted according to whether they were reachable only by cell phone or also by landline. Explain why it is important to include both a landline sample and a cell phone sample. Why is the number of telephone lines into the residence important? (Hint: How does the number of telephone lines into the residence affect the chance of the household being included in the RDD sample?)

CHAPTER 8 SUMMARY

Chapter Specifics

- A **sample survey** selects a **sample** from the **population** of all individuals about which we desire information. We base conclusions about the population on data from the sample. It is important to specify exactly what population you are interested in and what variables you will measure.

- The **design** of a sample describes the method used to select the sample from the population. **Random sampling** designs use chance to select a sample.

- The basic random sampling design is a **simple random sample (SRS).** An SRS gives every possible sample of a given size the same chance to be chosen.

- Choose an SRS by labeling the members of the population and using **random digits** to select the sample. Software can automate this process.

- To choose a **stratified random sample,** classify the population into **strata,** groups of individuals that are similar in some way that is important to the response. Then choose a separate SRS from each stratum.

- Failure to use random sampling often results in **bias,** or systematic errors in the way the sample represents the population. **Voluntary response samples,** in which the respondents choose themselves, are particularly prone to large bias.

- In human populations, even random samples can suffer from bias due to **undercoverage** or **nonresponse,** from **response bias,** or from misleading results due to **poorly worded questions.** Sample surveys must deal expertly with these potential problems in addition to using a random sampling design.

- Most national sample surveys are carried out by telephone, using **random digit dialing** to choose residential telephone numbers at random. Because call screening is increasing nonresponse to such surveys, and the rise of cell-phone-only households is increasing undercoverage, many surveys include a minimum quota of cell phone users in their samples to help adjust for bias. **Web surveys** are becoming more frequent, but many suffer from volunteer response, undercoverage, and nonresponse.

Link It

The methods of Chapters 1 to 6 can be used to describe data regardless of how the data were obtained. However, if we want to reason from data to give answers to specific questions or to draw conclusions about the larger population, then the method that was used to collect the data is important. Sampling is one way to collect data, but it does not guarantee that we can draw meaningful conclusions. Biased sampling methods, such as convenience sampling and voluntary response samples, produce data that can be misleading, resulting in incorrect conclusions. Simple random sampling avoids bias and produces data that can lead to valid conclusions regarding the population. Even with perfect sampling methods, there is still sample-to-sample variation; we will begin our study of the connection between sampling variation and drawing conclusions in Chapter 15.

Even when we take a simple random sample, our conclusions can be weakened by undercoverage, nonresponse, and poor wording of questions. Careful attention must be given to all aspects of the sampling process to ensure that the conclusions we make are valid. In some cases, more complex designs are required, such as stratified sampling or multistage sampling. But the use of impersonal chance to select the sample remains a key ingredient in the sampling process. And issues such as undercoverage and nonresponse still remain for these more complex designs.

The next chapter is about statistical designs for experiments, a quite different way to produce data than sample surveys.

8.17 An online store contacts 1000 customers from its list of customers who have purchased something from them in the last year. In all, 696 of the 1000 say that they are very satisfied with the store's Web site. The population in this setting is

(a) all customers who have purchased something in the last year.
(b) the 1000 customers contacted.
(c) the 696 customers who were very satisfied with the store's Web site.

8.18 A state representative wants to know how voters in his district feel about enacting a statewide smoking ban in all enclosed public places, including bars and restaurants, as well as several other current statewide issues. He mails a questionnaire addressing these issues to an SRS of 800 voters in his district. Of the 800 questionnaires mailed, 152 were returned. The sample is

(a) the 800 voters receiving the questionnaire.
(b) the 152 voters returning the questionnaire.
(c) all voters in his district.

8.19 In the survey for the previous exercise, there are 8741 registered voters in his district. You label the voters 0001 to 8741 in alphabetical order. Using line 123 of Table B to select the sample, the first 5 voters in your sample would be

(a) 5458, 0815, 7271, 2560, 2755.
(b) 5458, 0815, 0727, 1025, 6027.
(c) 5458, 8150, 7271, 2560, 2755.

8.20 The Web site www.twiigs.com allows you to vote on polls that interest you or to post one of your own. Once you have found a poll of interest, you just click on "Vote," and your response becomes part of the sample. One of the questions in July 2010 was, "How many times have you been pulled over by the police?" Of the 780 people responding, 70% said "1–5 times." You can conclude that

(a) about 70% of Americans have been pulled over by the police "1–5 times."
(b) the poll uses voluntary response, so the results tell us little about the population of all adults.
(c) more people still need to vote on the question, as a larger sample is required to reduce bias.

8.21 Archaeologists plan to examine a sample of two-meter-square plots near an ancient Greek city for artifacts visible in the ground. They choose separate samples of plots from floodplain, coast, foothills, and high hills. What kind of sample is this?

(a) A simple random sample
(b) A stratified random sample
(c) A voluntary response sample

8.22 You must choose an SRS of 10 of the 440 retail outlets in New York that sell your company's products. How would you label this population to select a simple random sample?

(a) 001, 002, 003, . . . , 439, 440
(b) 000, 001, 002, . . . , 439, 440
(c) 1, 2, . . . , 439, 440

8.23 You are using the table of random digits to choose a simple random sample of 6 students from a class of 30 students. You label the students 01 to 30 in alphabetical order and then select a simple random sample. Which of the following is a possible sample that could be obtained?

(a) 45, 74, 04, 18, 07, 65
(b) 04, 18, 07, 13, 02, 07
(c) 04, 18, 07, 13, 02, 05

8.24 A sample of households in a community is selected at random from the telephone directory. In this community, 4% of households have no telephone, 10% have only cell phones, and another 25% have unlisted telephone numbers. The sample will certainly suffer from

(a) nonresponse.
(b) undercoverage.
(c) false responses.

8.25 The Pew Research Center survey asked a random sample of 1500 adults, "Do you think the use of marijuana should be made legal, or not?" In the entire sample, 41% said, "Yes, legal." But only 24% of the Republicans in the sample said, "Yes, legal." Which of these two sample percents will be more accurate as an estimate of the truth about the population?

(a) The result for Republicans is more accurate because it is easier to estimate a proportion for a smaller group.
(b) The result for the entire sample is more accurate because it comes from a larger sample.
(c) Both are equally accurate because both come from the same sample.

CHAPTER 8 EXERCISES

In all exercises asking for an SRS, you may use software, the Simple Random Sample applet, or Table B.

8.26 **Immigration reform priorities.** A Gallup Poll asked, "If you had to choose, what should be the main focus of the U.S. government in dealing with the issue of illegal immigration—developing a plan for halting the flow of illegal immigrants into the U.S. (or) developing a plan to deal with immigrants who are currently in the U.S. illegally?" Gallup's report said, "Results are based on telephone interviews conducted June 7–11,

2012, with a random sample of 1,004 adults, aged 18 and older, living in the continental U.S."[18] What is the population for this sample survey? What is the sample?

8.27 Sampling stuffed envelopes. A large retailer prepares its customers' monthly credit card bills using an automatic machine that folds the bills, stuffs them into envelopes, and seals the envelopes for mailing. Are the envelopes completely sealed? Inspectors choose 40 envelopes from the 1000 stuffed each hour for visual inspection. What is the population for this sample survey? What is the sample?

8.28 Do you trust the Internet? You want to ask a sample of college students the question, "How much do you trust information about health that you find on the Internet—a great deal, somewhat, not much, or not at all?" You try out this and other questions on a pilot group of 8 students chosen from your class. The class members are

Adams	Devore	Guo	Newberg	Shoepf
Aeffner	Ding	Heaton	Paulsen	Spagnola
Barnes	Drake	Huling	Payton	Terry
Bower	Eckstein	Kahler	Prince	Vore
Burke	Fassnacht	Kessis	Pulak	Wallace
Cao	Fullmer	Lu	Rabin	Wanner
Cisse	Gandhi	Mattos	Roberts	Zhang

Choose an SRS of 8 students. If you use Table B, start at line 131.

8.29 Sampling telephone area codes. The United States currently has approximately 287 Numbering Plan Areas (NPAs) in service, corresponding to geographic regions. Each NPA is identified by a three-digit code, commonly called an area code. (More are created regularly.)[19] You want to choose an SRS of 20 of these area codes for a study of available telephone numbers. Label the codes 001 to 287, and use the *Simple Random Sample* applet or other software to choose your sample. (If you use Table B, start at line 135 and choose only the first 5 area codes in the sample.)

8.30 Sampling the forest. To gather data on a 1200-acre pine forest in Louisiana, the U.S. Forest Service laid a grid of 1410 equally spaced circular plots over a map of the forest. A ground survey visited a sample of 10% of these plots.[20]

(a) How would you label the plots?
(b) Choose the first 5 plots in an SRS of 141 plots. (If you use Table B, start at line 105.)

8.31 Paying taxes. In April 2013, a Gallup Poll asked two questions about the amount one pays in federal income taxes.[21] Here are the two questions:

> Question A: *Do you regard the income tax which you will have to pay this year as fair?*
>
> Question B: *Do you consider the amount of federal income tax you have to pay as too high, about right, or too low?*

One of these questions drew 55% saying the amount was fair or about right; the other, only 45%. Which wording produced the higher percentage? Why?

8.32 Random digits. Which of the following statements are true of a table of random digits, and which are false? Briefly explain your answers.

(a) There are exactly four 0s in each row of 40 digits.
(b) Each pair of digits has chance 1/100 of being 00.
(c) The digits 0000 can never appear as a group, because this pattern is not random.

8.33 Movie viewing. An opinion poll calls 2000 randomly chosen residential telephone numbers and asks to speak with an adult member of the household. The interviewer asks, "How many movies have you watched in a movie theater in the past 12 months?"

(a) What population do you think the poll has in mind?
(b) In all, 831 people respond. What is the rate (percent) of nonresponse?
(c) What source of response error is likely for the question asked?

8.34 Online polls. Example 8.3 (page 194) reports an online poll in which 97% of the respondents opposed issuing driver's licenses to illegal immigrants. National random samples taken at the same time showed about 70% of the respondents opposed to such licenses. Explain briefly to someone who knows no statistics why the random samples report public opinion more reliably than the online poll.

8.35 Nonresponse. Academic sample surveys, unlike commercial polls, often discuss nonresponse. A survey of drivers began by randomly sampling all listed residential telephone numbers in the United States. Of 45,956 calls to these numbers, 5029 were completed.[22] What was the rate of nonresponse for this sample? (Only one call was made to each number. Nonresponse would be lower if more calls were made.)

8.36 Running red lights. The sample described in the previous exercise produced a list of 5024 licensed drivers. The investigators then chose an SRS of 880 of these drivers to answer questions about their driving habits.

(a) How would you assign labels to the 5024 drivers? Choose the first 5 drivers in the sample. If you use Table B, start at line 104.
(b) One question asked was, "Recalling the last ten traffic lights you drove through, how many of them were red when you entered the intersections?" Of the 880 respondents, 171 admitted that at least one light had been red. A practical problem with this survey is that people may not give truthful answers. What is the likely direction of the bias: do you think more or fewer than 171 of the 880 respondents really ran a red light? Why?

8.37 **Seat belt use.** A study in El Paso, Texas, looked at seat belt use by drivers. Drivers were observed at randomly chosen convenience stores. After they left their cars, they were invited to answer questions that included questions about seat belt use. In all, 75% said they always used seat belts, yet only 61.5% were wearing seat belts when they pulled into the store parking lots.[23] Explain the reason for the bias observed in responses to the survey. Do you expect bias in the same direction in most surveys about seat belt use?

8.38 **Sampling at a party.** At a large block party there are 40 men and 30 women. You want to ask opinions about how to improve the next party. You choose at random 4 of the men and separately choose at random 3 of the women to interview.

(a) What is the probability that any of the 40 men is in your random sample of 4 men to be interviewed? What is the probability that any of the 30 women is in your random sample of 3 women to be interviewed?

(b) If you have done the calculations correctly in part (a), the probability of any person at the party being interviewed is the same. Why is your sample of 7 men and women not an SRS of people from the party?

8.39 **Ring-no-answer.** A common form of nonresponse in telephone surveys is "ring-no-answer." That is, a call is made to an active number but no one answers. The Italian National Statistical Institute looked at nonresponse to a government survey of households in Italy during the periods January 1 to Easter and July 1 to August 31. All calls were made between 7 and 10 P.M., but 21.4% gave "ring-no-answer" in one period versus 41.5% "ring-no-answer" in the other period.[24] Which period do you think had the higher rate of no answers? Why? Explain why a high rate of nonresponse makes sample results less reliable.

8.40 **Sampling pharmacists.** All pharmacists in the Canadian province of Ontario are required to be members of the Ontario College of Pharmacists. In 2012, there were 12,829 members of the college divided geographically into 6 voting districts. The number of members in each district follow:[25]

District	1	2	3	4	5	6
Membership	1944	1566	4122	2660	1859	678

Suppose they are interested in obtaining members' views on the 2012 Expanded Scope of Practice Regulations, which authorizes pharmacists to provide additional services including prescribing drug products for smoking cessation and administering the publicly funded influenza vaccine. To be sure that the opinions of all districts are represented, you choose a stratified random sample of 5 pharmacists from each district. Explain how you will assign labels within each district,

and then give the labels of the pharmacists from Districts 1 and 2 in your sample. If you use Table B, start at line 122 for District 1 and at line 131 for District 2. Why should you not start at the same line in Table B to obtain your samples for Districts 1 and 2?

8.41 **Sampling Amazon forests.** Stratified samples are widely used to study large areas of forest. Based on satellite images, a forest area in the Amazon basin is divided into 14 types. Foresters studied the four most commercially valuable types: alluvial climax forests of quality levels 1, 2, and 3 and mature secondary forest. They divided the area of each type into large parcels, chose parcels of each type at random, and counted tree species in a 20- by 25-meter rectangle randomly placed within each parcel selected. Here is some detail:

© Age Fotostock/SuperStock

FOREST TYPE	TOTAL PARCELS	SAMPLE SIZE
Climax 1	36	4
Climax 2	72	7
Climax 3	31	3
Secondary	42	4

Choose the stratified sample of 18 parcels. Be sure to explain how you assigned labels to parcels. If you use Table B, start at line 102.

8.42 **Canadian health care survey.** The Tenth Annual Health Care in Canada Survey is a survey of the opinions of the Canadian public and health care providers on a variety of health care issues, including quality of health care, access to health care, health and the environment, and so forth. A description of the survey follows:

The 10th edition of the Health Care in Canada Survey was conducted by POLLARA Research between October 3rd and November 8th, 2007. Results for the survey are based on telephone interviews with nationally representative samples of 1,223 members of the Canadian public, 202 doctors, 201 nurses, 202 pharmacists and 201 health managers. Public results are considered to be accurate within ±2.8%, while the margin of error for results for doctors, nurses, pharmacists and managers is ±6.9%.[26]

(a) Why is the accuracy greater for the public than for health care providers and managers?

(b) Why do you think they sampled the public as well as health care providers and managers?

8.43 Systematic random samples. *Systematic random samples* go through a list of the population at fixed intervals from a randomly chosen starting point. For example, a study of dating among college students chose a systematic sample of 200 single male students at a university as follows.[27] Start with a list of all 9000 single male students. Because 9000/200 = 45, choose one of the first 45 names on the list at random and then every 45th name after that. For example, if the first name chosen is at position 23, the systematic sample consists of the names at positions, 23, 68, 113, 158, and so on up to 8978.

(a) Choose a systematic random sample of 5 names from a list of 200. If you use Table B, enter the table at line 120.

(b) Like an SRS, a systematic sample gives all individuals the same chance to be chosen. Explain why this is true, then explain carefully why a systematic sample is nonetheless *not* an SRS.

8.44 Why random digit dialing is common. The list of individuals from which a sample is actually selected is called the *sampling frame*. Ideally, the frame should list every individual in the population, but in practice this is often difficult. A frame that leaves out part of the population is a common source of undercoverage.

(a) Suppose that a sample of households in a community is selected at random from the telephone directory. What households are omitted from this frame? What types of people do you think are likely to live in these households? These people will probably be underrepresented in the sample.

(b) It is usual in telephone surveys to use random digit dialing equipment that selects the last four digits of a telephone number at random after being given the exchange (the first three digits), as described in Example 8.5 (page 197). Which of the households you mentioned in your answer to (a) will be included in the sampling frame by random digit dialing?

8.45 Concerns about global warming on the rise. After years of skepticism about the dangers of global warming, 57% of Americans now believe that the increase in the earth's temperature over the last century are due more to the effects of pollution from human activities than to natural changes in the environment not due to human activities. According to the survey methods section, "Results for this Gallup poll are based on telephone interviews conducted March 7–10, 2013, with a random sample of 1,022 adults, aged 18 and older, living in all 50 U.S. states and the District of Columbia. Interviews are conducted with respondents on landline telephones and cellular phones. Landline telephone numbers are chosen at random among listed telephone numbers. Cell phone numbers are selected using random-digit-dial methods. Landline respondents are chosen at random within each household on the basis of which member had the most recent birthday."[28]

(a) The survey wants the opinion of an individual adult, but a landline phone reaches a household in which several adults may live. In that case, the survey interviewed the adult with the most recent birthday. Why is this preferable to simply interviewing the person who answers the phone?

(b) What is the population that this survey wants to describe? Why do you think it is important to include both landline and cellular phones in your sample?

8.46 Wording survey questions. Comment on each of the following as a potential sample survey question. Is the question sufficiently clear? Is it slanted toward a desired response?

(a) "In light of skyrocketing gasoline prices, we should consider opening up a very small amount of Alaskan wilderness for oil exploration as a way of reducing our dependence on foreign oil. Do you agree or disagree?"

(b) "Do you agree that a national system of health insurance should be favored because it would provide health insurance for everyone and would reduce administrative costs?"

(c) "In view of the negative externalities in parent labor force participation and pediatric evidence associating increased group size with morbidity of children in day care, do you support government subsidies for day care programs?"

8.47 Your own bad questions. Write your own examples of bad sample survey questions.

(a) Write a biased question designed to get one answer rather than another.

(b) Write a question to which many people may not give truthful answers.

8.48 The Canadian census. The Canadian government's decision to eliminate the mandatory long-form version of the census and to move these questions to an optional survey has many concerned. Many members of the business community and economists stressed the importance of the census data for crafting public policy. The minister of industry was given the task of defending the government's decision. In response to an argument that making the long form of the census voluntary would skew the data by eliminating the statistical randomness of the survey, the minister replied: "Wrong. Statisticians can ensure validity with a larger sample size."[29] Is the minister correct? If not, explain in simple terms the error in his statement.

Exploring the Web

8.49   **Poor survey designs.** The Web site for the American Association for Public Opinion Research discusses several issues about polls. This information can be found at www.aapor.org. Under *Resources & Education*, click on the link *Poll & Survey FAQ*. Click on the link *What Information Should Survey Researchers Disclose* and read Section A for recommendations about the information that should be available about how the poll was conducted. After reading Section A, go back to the previous Web page and click on the link *What Is a Random Sample?* and then the link *Bad Samples* for some examples of flawed samples.

(a) You are going to design a survey at your university. Give a question of interest and two examples of bad ways to collect your sample, along with the likely direction of bias that would result. Explain your answers.

(b) How would you modify your examples in (a) to produce a better sample? What are some difficulties you might encounter when collecting your sample?

8.50   **Find a survey.** The Web site for the Pew Research Center for the People and the Press is www.people-press.org. Go to the Web site and read one of the featured surveys. What information under Section A of *What Information Should Survey Researchers Disclose* from the previous exercise is included in the featured survey you have read? You may find the concluding section entitled *About the Survey* and some of the links at the end of the featured survey helpful for answering this question.

Producing Data: Experiments

Overview

Proper design of experiments is central to research, especially in the life sciences, physical sciences, and engineering (in many social science settings—economics, psychology, and sociology, for example—observational studies are more typical sources of data). Experimental design is concerned with efficiently collecting data, so that the maximum amount of information about a problem of interest can be obtained, given certain constraints (e.g., a budget). In this chapter, we focus on the most basic principles of experimentation.

Although observational studies may be used to determine if a relationship exists between two variables, it is only through formal experimentation that a cause and effect relationship can be shown. The experiments discussed in this chapter are taken from the life sciences, agriculture, medicine, and many other fields.

The roles of subjects, factor(s), treatment(s), and the response variable are key in any experimental design, so students should be reminded throughout the chapter of the importance and roles of these concepts. The simplest designs, randomized comparative experiments and completely randomized designs, are introduced first; then more advanced and thus more complicated designs, such as matched pair and block designs, are covered. For a nice explanation on why randomized comparative experiments are designed to provide good evidence of cause and effect, have your students read the section titled "The logic of randomized comparative experiments." Note that the idea of statistical significance is briefly introduced as "an observed effect so large that it would rarely occur by chance." You might mention to students that statistical significance will resurface when we study inference.

Teaching Suggestions and Additional Examples/ Activities for the Classroom

Students should understand that we have a kind of "golden rule" for experimentation—groups being compared should be alike in all ways except for the variable being controlled and studied. This point can be dramatically illustrated with two examples.

1. Emphasize the Importance of a Well-Designed Experiment

The first example compares the poorly designed National Foundation for Infantile Paralysis (NFIP) experiment to the well-designed Public Health Service (PHS) experiment, both

---

LEARNING OUTCOMES**

■ Distinguish between an observational study and an experiment.

■ Recognize bias due to confounding of explanatory variables with lurking variables in either an observational study or an experiment.

■ Recognize that experiments are superior to observational studies in establishing causal relationships between the response and explanatory variables because random assignment to groups maximizes the chance that the groups being compared are alike in all ways, except for the imposed difference in explanatory variable level.

■ Identify the factors (explanatory variables), treatments, response variables, and individuals or subjects in an experiment.

■ Outline the design of a completely randomized experiment using a diagram like that in Figure 9.3. The diagram in a specific case should show the sizes of the groups, the specific treatments, and the response variable.

■ Use software or Table B of random digits to carry out the random assignment of subjects to groups in a completely randomized experiment.

■ Recognize the placebo effect. Recognize when blinding or double-blinding should be used.

■ Explain why a randomized comparative experiment can give good evidence for cause-and-effect relationships.

■ Define statistical significance.

■ Understand the role of blocking in reducing the effect of an extraneous source of variation in a problem. If possible, demonstrate that blocking in an experiment is analogous to "controlling" for a known lurking variable (stratifying) in an observational study.

■ Understand that a matched pairs design is a kind of randomized block experiment in which we block over subjects to reduce the impact of subject variation.

■ Be able to outline a block design and a matched pairs experiment.

**These learning outcomes appear later for the students in Chapter 10: Part II Review.

designed to assess the effectiveness of Jonas Salk's polio vaccine in 1954. Both experiments involved more than 1 million children, but the NFIP design resulted in a treatment group different in many ways from the control group. In particular, the treatment group consisted only of "volunteers," who tended to be more susceptible to polio for socioeconomic reasons. The PHS design was a classic randomized, double-blind experiment. If you discuss the problems with the NFIP design, many students will suggest modifications that lead to the PHS design, and so they may effectively discover this basic design for themselves. For more details, see "A Calculated Risk: The Salk polio Vaccine Field Trials of 1954" (M. Meldrum, *British Medical Journal*, 317(7167): 1233–1236, 1998; http://www.ncbi.nlm.nih.gov/pmc/articles/PMC1114166/).

2. Emphasize the Importance of Randomization

Another example involves the drug Ribavirin for treating HIV positive patients. Initial results were promising ("Ribavarin: a clinical overview," *European Journal of Epidemiology*, 2(1), 1–14, 1986; http://www.ncbi.nlm.nih.gov/pubmed/3021519); however, concerns soon surfaced regarding the randomization methods used (e.g., "Ribavirin: A role in HIV infection?" *Journal of Acquired Immune Deficiency Syndromes*, 3:893–895; http://journals.lww.com/jaids/abstract/1990/09000/ribavirin—a_role_in_hiv_infection_.7.aspx). These flaws included a proportionately larger number of symptomatic patients "randomized" to the placebo treatment (this was among the first potential drugs to combat the disease) and the fact that patients in the placebo group were "more immunocompromised as measured by every parameter of immune function." Subsequent (properly randomized) trials found no evidence for the efficacy of the drug ("Multicenter clinical trial of oral ribavirin in symptomatic HIV-infected patients. The Ribavirin ARC Study Group," *Journal of Acquired Immune Deficiency Syndromes*, 6(1): 32–41, 1993; http://www.ncbi.nlm.nih.gov/pubmed/8417172). Even though the drug has since been found useful for other diseases, the controversy at the time caused the initial manufacturer of Ribavirin to go bankrupt.

3. Discuss When (and Why) Blocking Can Be Useful

A good example for discussing the use of blocking follows: Osteoporosis results from a depletion of calcium in bones. It typically afflicts women more often and more severely than men. In studying the effectiveness of a calcium supplement designed to treat osteoporosis, consider a set of 10 subjects—6 women and 4 men. A completely randomized experiment might lead to the treatment group consisting of proportionally more or fewer women. In addition, there is a lot of variation between the sexes—possibly more than within each sex. Therefore, blocking by sex is natural. Simply design a completely randomized experiment for the women and one for the men. Compare women on the treatment to women on the placebo, and compare men on the treatment to men on the placebo. It is worth noting that in the "old days," all drug trials were done on men only. It has since been shown and accepted (see, for example, "Sex-Based Differences in Drug Activity," *American Family Physician*, 80(11), 1254–1258, 2009; http://www.aafp.org/afp/2009/1201/p1254.html) that women react differently (as do racial and ethnic groups that are not Caucasian).

4. Motivate Matched Pairs Designs

A simple example of a matched pairs design is provided in a study comparing two suntan lotions. The subjects vary a lot in their skin sensitivities, so we might eliminate this extraneous source of variation by blocking on subjects—use each lotion on each subject, randomizing the side of the body each lotion is applied to.

Other Resources (LaunchPad)

StatClips
> *Types of Studies*

Snapshots Videos
> *Types of Studies*
> *Experimental Design*

EESEE Case Studies
> *Blinded Knee Doctors*
> *Surgery in a Blanket*
> *Cancer and Power Lines*
> *Radar Detectors and Speeding*
> *Survey Response by Sports Fans*
> *Nurses and Glove Use*

Applet
> *Simple Random Sample*

Producing Data: Experiments

A sample survey aims to gather information about a population without disturbing the population in the process. Sample surveys are one kind of *observational study*. Other observational studies observe the behavior of animals in the wild or the interactions between teacher and students in the classroom. This chapter is about statistical designs for *experiments*, a quite different way to produce data.

Observation versus experiment

In contrast to observational studies, experiments don't just observe individuals or ask them questions. They actively impose some treatment to observe the response. Experiments can answer questions such as "Does aspirin reduce the chance of a heart attack?" and "Do a majority of college students prefer Pepsi to Coke when they taste both without knowing which they are drinking?"

Observation Versus Experiment

An **observational study** observes individuals and measures variables of interest but does not attempt to influence the responses. The purpose of an observational study is to describe some group or situation.

An **experiment,** on the other hand, deliberately imposes some treatment on individuals to observe their responses. The purpose of an experiment is to study whether the treatment causes a change in the response.

An observational study, even one based on a statistical sample, is a poor way to gauge the effect of a treatment. To see the response to a change, we must actually impose the change. *When our goal is to understand cause and effect, experiments are the only source for fully convincing data.* For this reason, the distinction between observation and experiment is one of the most important in statistics.

EXAMPLE 9.1   Drink a Little, but Not a Lot

Many observational studies show that people who drink a moderate amount of alcohol have less heart disease than people who drink no alcohol or who drink heavily.[1] ("Moderate" means one or two drinks a day for men and one drink a day for women.) Is this *association* good reason to think that moderate drinking actually *causes* less heart disease? People who choose to drink in moderation are, as a group, different from both heavy drinkers and abstainers. They are more likely to maintain a healthy weight, get enough sleep, and exercise regularly. Moderate drinkers may be healthier because of these healthy habits rather than because of the effect of alcohol on health.

It is easy to imagine an experiment that would settle the issue of whether moderate drinking really *causes* reduced heart disease. Choose half of a large group of adults at random to be the "treatment" group. The remaining half becomes the "control" group. Require the treatment group to have one alcoholic drink every day. Require the control group to abstain from alcohol. Follow both groups for a decade. This experiment isolates the effect of alcohol. Of course, it isn't practical or ethical to carry out such an experiment. ■

The point of Example 9.1 is the contrast between observing people who choose for themselves what to drink and an experiment that requires some people to drink and others to abstain. When we simply observe people's drinking choices, the effect of moderate drinking is *confounded* with (mixed up with) the characteristics of people who choose to drink in moderation. These characteristics are lurking variables (see page 138) that make it hard to see the true relationship between the explanatory and response variables. Figure 9.1 shows the confounding in picture form.

Confounding

Two variables (explanatory variables or lurking variables) are **confounded** when their effects on a response variable cannot be distinguished from each other.

 Observational studies of the effect of one variable on another often fail because the explanatory variable is confounded with lurking variables. Well-designed experiments take steps to prevent confounding.

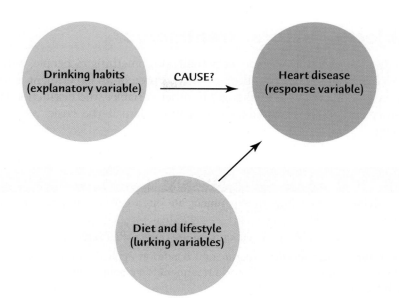

FIGURE 9.1
Confounding: We can't distinguish the effects of drinking habits from the effects of overall diet and lifestyle.

Apply Your Knowledge

9.1   Cell Phones and Brain Cancer. A study of cell phones and the risk of brain cancer looked at a group of 469 people who have brain cancer. The investigators matched each cancer patient with a person of the same sex, age, and race who did not have brain cancer, then asked about use of cell phones. Result: "Our data suggest that use of handheld cellular telephones is not associated with risk of brain cancer."[2] Is this an observational study or an experiment? Why? What are the explanatory and response variables?

9.2   The Font Matters! In general, when trying to change your behavior, if the effort required is perceived as high, this will be an impediment to change, whether it is modifying your diet or your study habits. Divide 40 students into two groups of 20. The first group reads instructions for an exercise program printed in an easy-to-read font (Arial, 12 point), and the second group reads identical instructions in a difficult-to-read font (Brush, 12 point). Each subject estimates how many minutes the program would take and also uses a 7-point rating scale to report whether they are likely to include the exercise program as part of their daily routine (7 = very likely). The researchers hypothesized that those reading about the exercise program in the more difficult-to-read font would estimate that the program would take longer and they would be less likely to make the exercise program part of their regular routine.[3] Is this an experiment? Why or why not? What are the explanatory and response variables?

9.3   Quitting Smoking and Risk for Type 2 Diabetes. Researchers studied a group of 10,892 middle-aged adults over a period of nine years. They found that smokers who quit had a higher risk for diabetes within three years of quitting than either nonsmokers or continuing smokers.[4] Does this show that stopping smoking causes the short-term risk for type 2 diabetes to increase? (Weight gain has been shown to be a major risk factor for developing type 2 diabetes and is often a side effect of quitting smoking.) Based on this research, should you tell a middle-aged adult who smokes that stopping smoking can *cause* diabetes and advise him or her to continue smoking? Carefully explain your answers to both questions.

Subjects, factors, treatments

A study is an experiment when we actually do something to people, animals, or objects to observe the response. Because the purpose of an experiment is to reveal the response of one variable to changes in other variables, the distinction between explanatory and response variables is essential. Here is the basic vocabulary of experiments.

> ### Subjects, Factors, Treatments
>
> The **individuals** studied in an experiment are often called **subjects,** particularly when they are people.
>
> The explanatory variables in an experiment are often called **factors.**
>
> A **treatment** is any specific experimental condition applied to the subjects. If an experiment has more than one factor, a treatment is a combination of specific values of each factor.

EXAMPLE 9.2   Foster Care Versus Orphanages

Do abandoned children placed in foster homes do better than similar children placed in an institution? The Bucharest Early Intervention Project found that the answer is a clear "Yes." The *subjects* were 136 young children abandoned at birth and living in orphanages in Bucharest, Romania. Half of the children, chosen at random, were placed in foster homes. The other half remained in the orphanages. The experiment compared these two *treatments*. There is a single *factor,* type of care, with two values, foster and institutional care. When there is only one factor, the levels or values of the factor correspond to the treatments. The *response variables* included measures of mental and physical development.[5] (Foster care was not easily available in Romania at the time and so was paid for by the study. See Exercise 15 on page 246 in the Data Ethics essay for ethical questions concerning this experiment.) ■

EXAMPLE 9.3   Effects of TV Advertising

What are the effects of repeated exposure to an advertising message? The answer may depend both on the length of the ad and on how often it is repeated. An experiment investigated this question using undergraduate students as *subjects*. All subjects viewed a 40-minute television program that included ads for a digital camera. Some subjects saw a 30-second commercial; others, a 90-second version. The same commercial was shown either 1, 3, or 5 times during the program.

This experiment has 2 *factors:* length of the commercial, with 2 values, and repetitions, with 3 values. The 6 combinations of 1 value of each factor form 6 *treatments*. Figure 9.2 shows the layout of the treatments. After viewing, all the subjects answered questions about their recall of the ad, their attitude toward the camera, and their intention to purchase it. These are the *response variables*. ■

Examples 9.2 and 9.3 illustrate the advantages of experiments over observational studies. In an experiment, we can study the effects of the specific treatments we are interested in. By assigning subjects to treatments, we can avoid confounding.

**Factor B
Repetitions**

	1 time	3 times	5 times
Factor A **Length** 30 seconds	1	2	3
90 seconds	4	5	6

> Subjects assigned to Treatment 3 see a 30-second ad five times during the program.

FIGURE 9.2

The treatments in the experimental design of Example 9.3. Combinations of values of the two factors form six treatments.

For example, observational studies of the effects of foster homes versus institutions on the development of children have often been biased because healthier or more alert children tend to be placed in homes. The random assignment in Example 9.2 eliminates bias in placing the children. Moreover, we can control the environment of the subjects to hold constant factors that are of no interest to us, such as the specific product advertised in Example 9.3.

Another advantage of experiments is that we can study the combined effects of several factors simultaneously. The interaction of several factors can produce effects that could not be predicted from looking at the effect of each factor alone. Perhaps longer commercials increase interest in a product, and more commercials also increase interest, but if we both make a commercial longer and show it more often, viewers get annoyed, and their interest in the product drops. The two-factor experiment in Example 9.3 will help us find out.

Apply Your Knowledge

For the experiments described in Exercises 9.4 and 9.5, identify the subjects, the factors, the treatments, and the response variables.

9.4 Ginkgo Extract and the Postlunch Dip. The postlunch dip is the drop in mental alertness after a midday meal. Does an extract of the leaves of the ginkgo tree reduce the postlunch dip? Assign healthy people aged 18 to 40 to take either ginkgo extract or a placebo pill. After lunch, ask them to read seven pages of random letters and place an X over every *e*. Count the number of misses.

9.5 Reactions to Simulated News Reports. A sample of University of Colorado students each viewed one of two simulated news reports about a terrorist bombing against the United States by a fictitious country. One report showed the bombing attack on a military target and the other on a cultural/educational site. In addition, before viewing the news report, each student read one of two "primes." The first was a prime for *forgiveness* based on the biblical saying "Love thy enemy," and the second was a *retaliatory* prime based on the biblical saying "An eye for an eye, and a tooth for a tooth." After viewing the news report, the students were asked to rate on a scale of 1 to 12 what the U.S. reaction should be, with the lowest score (1) corresponding to the United States sending a special ambassador to the country and the highest score (12) corresponding to an all-out nuclear attack against the country.[6]

9.6 Ripening Mangoes. The mango is considered the "king of fruits" in many parts of the world. Mangoes are generally harvested at the mature green stage

Howard Bjornson/Getty Images

and ripen up during the marketing process of transport, storage, and so on. During this process, about 30% of the fruit is wasted. Because of this, the impact of harvest stage and storage conditions on the postharvest quality are of interest. In this experiment, the fruit was harvested at 80, 95, or 110 days after the fruit setting (the transition from flower to fruit) and then stored at temperatures of 20, 30, or 40 degrees centigrade. For each harvest time and storage temperature, a random sample of mangoes was selected, and the time to ripening was measured.[7]

(a) What are the factors, the treatments, and the response variables? Use a diagram like Figure 9.2 to display the factors and treatments.

(b) For simplicity, the experimenter thought about selecting nine mango trees and randomly assigning one tree to each of the nine treatments. All the mangoes on a tree would then receive the same treatment. Do you think this would be a good way to assign the mangoes to the treatments? Explain your reasoning in simple English.

How to experiment badly

Experiments are the preferred method for examining the effect of one variable on another. By imposing the specific treatment of interest and controlling other influences, we can pin down cause and effect. Statistical designs are often essential for effective experiments. To see why, let's look at an example in which an experiment suffers from confounding just as observational studies do.

EXAMPLE 9.4　An Uncontrolled Experiment

A college regularly offers a review course to prepare candidates for the Graduate Management Admission Test (GMAT), which is required by most graduate business schools. This year, it offers only an online version of the course. The average GMAT score of students in the online course is 10% higher than the longtime average for those who took the classroom review course. Is the online course more effective?

This experiment has a very simple design. A group of subjects (the students) were exposed to a treatment (the online course), and the outcome (GMAT scores) was observed. Here is the design:

Subjects ⟶ Online course ⟶ GMAT scores

A closer look at the GMAT review course showed that the students in the online review course were quite different from the students who in past years took the classroom course. In particular, they were older and more likely to be employed. An online course appeals to these mature people, but we can't compare their performance with that of the undergraduates who previously dominated the course. The online course might even be less effective than the classroom version. The effect of online versus in-class instruction is confounded with the effect of lurking variables. As a result of confounding, the experiment is biased in favor of the online course.

Would the situation have been different if both the online and the classroom courses had been given this year? If students still chose the course they wanted, with older students tending to sign up for the online course and younger students tending to sign up for the classroom course, then the effect of type of course would still be confounded with the lurking variable age. The solution will be described in the next section. ▪

Most laboratory experiments use a design like that in Example 9.4:

$$\text{Subjects} \longrightarrow \text{Treatment} \longrightarrow \text{Measure response}$$

In the controlled environment of the laboratory, simple designs often work well. Field experiments and experiments with living subjects are exposed to more variable conditions and deal with more variable subjects. *Outside the laboratory, uncontrolled experiments often yield worthless results because of confounding with lurking variables.*

Apply Your Knowledge

9.7   Reducing Unemployment. Will cash bonuses speed the return to work of unemployed people? A state department of labor notes that last year 41% of people who filed claims for unemployment insurance found a new job within 15 weeks. As an experiment, the state offers $500 to people filing unemployment claims if they find a job within 15 weeks. The percent who do so increases to 53%. Suggest some conditions that might make it easier or harder to find a job this year as opposed to last year. Confounding with these lurking variables makes it impossible to say whether the $500 bonus really caused the increase.

Randomized comparative experiments

The remedy for the confounding in Example 9.4 is to be sure that we do a *comparative experiment* in which some students are taught in the classroom and other, similar students take the course online. In Example 9.4, the classroom group is called a **control group**. A control group receives either a "standard" treatment or in some cases a "sham" treatment and provides a basis for comparison with the other treatment groups. Although in many experiments one of the treatments is a control treatment, a comparative experiment does not require this. In Example 9.4, a comparative experiment could simply compare two newly developed online courses without including a control or classroom treatment. Most well-designed experiments compare two or more treatments. Part of the design of an experiment is a description of the factors (explanatory variables) and the layout of the treatments, with comparison as the leading principle.

control group

However, as discussed at the end of Example 9.4, comparison alone isn't enough to produce results we can trust. If the treatments are given to groups that differ markedly when the experiment begins, bias will result. If we allow students to select online or classroom instruction, students who are older and employed are likely to sign up for the online course. Personal choice will bias our results in the same way that volunteers bias the results of online opinion polls. The solution to the problem of bias in sampling is random selection, and the same is true in experiments. The subjects assigned to any treatment should be chosen at random from the available subjects.

Randomized Comparative Experiment

An experiment that uses both comparison of two or more treatments and random assignment of subjects to treatments is a **randomized comparative experiment.**

| EXAMPLE 9.5 | Classroom Versus Online |

The college decides to compare the progress of 25 on-campus students taught in the classroom with that of 25 students taught the same material online. Select the students who will be taught online by taking a simple random sample size of 25 from the 50 available subjects. The remaining 25 students form the control group. They will receive classroom instruction. The result is a randomized comparative experiment with two groups. Figure 9.3 outlines the design in graphical form.

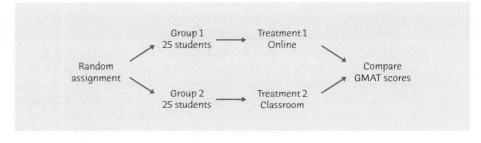

FIGURE 9.3
Outline of a randomized comparative experiment to compare online and classroom instruction, for Example 9.5.

The selection procedure is exactly the same as it is for sampling. **Label:** Label the 50 students 01 to 50. **Table:** Go to the table of random digits and read successive two-digit groups. The first 25 labels encountered select the online group. As usual, ignore repeated labels and groups of digits not used as labels. For example, if you begin at line 125 in Table B, the first 5 students chosen are those labeled 21, 49, 37, 18, and 44. Software such as the *Simple Random Sample* applet makes it particularly easy to choose treatment groups at random. ■

The design in Example 9.5 is *comparative* because it compares two treatments (the two instructional settings). It is *randomized* because the subjects are assigned to the treatments by chance. This "flowchart" outline in Figure 9.3 presents all the essentials: randomization, the sizes of the groups and which treatment they receive, and the response variable. There are, as we will see later, statistical reasons for generally using treatment groups that are about equal in size. We call designs like that in Figure 9.3 *completely randomized*.

Completely Randomized Design

In a **completely randomized** experimental design, all the subjects are allocated at random among all the treatments.

Completely randomized designs can compare any number of treatments. Here is an example that compares three treatments.

| EXAMPLE 9.6 | Conserving Energy |

Many utility companies have introduced programs to encourage energy conservation among their customers. An electric company considers placing small digital displays in households to show current electricity use and what the cost would be if this use continued for a month. Will the displays reduce electricity use? Would cheaper methods work almost as well? The company decides to conduct an experiment.

One cheaper approach is to give customers a chart and information about monitoring their electricity use from their outside meter. The experiment compares these two approaches (display, chart) and also a control. The control group of customers receives information about energy conservation but no help in monitoring electricity use. The response variable is total electricity used in a year. The company finds 60 single-family residences in the same city willing to participate, so it assigns 20 residences at random to each of the three treatments. Figure 9.4 outlines the design.

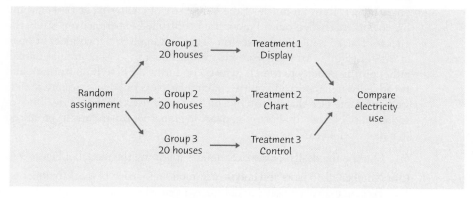

FIGURE 9.4

Outline of a completely randomized design comparing three energy-saving programs, for Example 9.6.

To use the *Simple Random Sample* applet, set the population labels as 1 to 60 and the sample size to 20 and click "Reset" and "Sample." The 20 households chosen receive the displays. The "Population hopper" now contains the 40 remaining households, in scrambled order. Click "Sample" again to choose 20 of these to receive charts. The 20 households remaining in the "Population hopper" form the control group.

To use Table B, label the 60 households 01 to 60. Enter the table to select an SRS of 20 to receive the displays. Continue in Table B, selecting 20 more to receive charts. The remaining 20 form the control group. ■

Examples 9.5 and 9.6 describe completely randomized designs that compare values of a single factor. In Example 9.5, the factor is the type of instruction. In Example 9.6, it is the method used to encourage energy conservation. Completely randomized designs can have more than one factor. The advertising experiment of Example 9.3 has two factors: the length and the number of repetitions of a television commercial. Their combinations form the six treatments outlined in Figure 9.2. A completely randomized design assigns subjects at random to these six treatments. Once the layout of treatments is set, the randomization needed for a completely randomized design is tedious but straightforward.

Apply Your Knowledge

9.8 Adolescent Obesity. Adolescent obesity is a serious health risk affecting more than 5 million young people in the United States alone. Laparoscopic adjustable gastric banding has the potential to provide a safe and effective treatment. Fifty adolescents between 14 and 18 years old with a body mass index higher than 35 were recruited from the Melbourne, Australia, community for the study. Twenty-five were randomly selected to undergo gastric banding, and the remaining 25 were assigned to a supervised lifestyle intervention program involving diet, exercise, and behavior modification. All subjects were followed for two years, and their weight loss was recorded.[8]

(a) Outline the design of this experiment, following the model of Figure 9.3. What is the response variable?

(b) Carry out the random assignment of 25 adolescents to the gastric-banding group, using the *Simple Random Sample* applet, other software, or Table B, starting at line 130.

Suttle, K. B., M. A. Thomsen, and M. E. Power. "Species Interactions Reverse Grassland Responses to Changing Climate." Science 315.5812 (2007): 640-42.

9.9   More Rain for California? The changing climate will probably bring more rain to California, but we don't know whether the additional rain will come during the winter wet season or extend into the long dry season in spring and summer. Kenwyn Suttle of the University of California at Berkeley and his coworkers carried out a randomized controlled experiment to study the effects of more rain in either season. They randomly assigned plots of open grassland to three treatments: added water equal to 20% of annual rainfall either during January to March (winter) or during April to June (spring), and no added water (control). Thirty-six circular plots of area 70 square meters were available (see the photo), of which 18 were used for this study. One response variable was total plant biomass, in grams per square meter, produced in a plot over a year.[9]

(a)   Outline the design of the experiment, following the model of Figure 9.4.

(b)   Number all 36 plots and choose 6 at random for each of the 3 treatments. Be sure to explain how you did the random selection.

9.10   Effects of TV Advertising. Figure 9.2 (page 217) displays the 6 treatments for the two-factor experiment on TV advertising described in Example 9.3. The 24 students named here will serve as subjects. Outline the design and randomly assign the subjects to the 6 treatments, an equal number of subjects to each treatment. If you use Table B, start at line 132.

Abramson	Biery	Cohen	Greenberg	Linder	Stanley
Anthony	Blake	Cote	Kessis	Minor	Tory
Austen	Brower	Delp	Koster	Schwartz	Truitt
Baker	Carroll	Disbro	Kruger	Shi	Walsh

The logic of randomized comparative experiments

Randomized comparative experiments are designed to give good evidence that differences in the treatments actually *cause* the differences we see in the response. The logic is as follows:

- Random assignment of subjects forms groups that should be similar in all respects before the treatments are applied. Exercise 9.50 (page 235) uses the *Simple Random Sample* applet to demonstrate this.

- A comparative experiment with randomization ensures that influences other than the experimental treatments operate equally on all groups.

- Therefore, differences in average response must be due either to the treatments or to the play of chance in the random assignment of subjects to the treatments.

That "either-or" deserves more thought. In Example 9.5, we cannot say that *any* difference between the average GMAT scores of students enrolled online and in the classroom must be caused by a difference in the effectiveness of the two types of instruction. There would be some difference even if both groups received the same instruction, because of variation among students in background and study habits. Chance assigns students to one group or the other, and this creates a chance difference between the groups. We would not trust an experiment with just one student in each group, for example. The results would depend too much on which group got lucky and received the stronger student. If we assign many subjects to each group, however, the effects of chance will average out, and there will be little difference in the average responses in the two groups unless the treatments themselves cause a difference. "Use enough subjects to reduce chance variation" is the third big idea of statistical design of experiments.

Principles of Experimental Design

The basic principles of statistical design of experiments are

1. **Control** the effects of lurking variables on the response, most simply by comparing two or more treatments.
2. **Randomize**—use chance to assign subjects to treatments.
3. **Use enough subjects** in each group to reduce chance variation in the results.

We hope to see a difference in the responses so large that it is unlikely to happen just because of chance variation. We can use the laws of probability, which describe chance behavior, to learn if the treatment effects are larger than we would expect to see if only chance were operating. If they are, we call them *statistically significant*.

Statistical Significance

An observed effect so large that it would rarely occur by chance is called **statistically significant**.

If we observe statistically significant differences among the groups in a randomized comparative experiment, we have good evidence that the treatments actually caused these differences. You will often see the phrase "statistically significant" in reports of investigations in many fields of study. The great advantage of randomized comparative experiments is that they can produce data that give good evidence for a cause-and-effect relationship between the explanatory and response variables. We know that in general a strong association does not imply causation. A statistically significant association in data from a well-designed experiment *does* imply causation.

Apply Your Knowledge

9.11 Prayer and Meditation. You read in a magazine that "nonphysical treatments such as meditation and prayer have been shown to be effective in controlled scientific studies for such ailments as high blood pressure, insomnia, ulcers, and asthma." Explain in simple language what the article means by "controlled scientific studies." Why can such studies in principle provide good evidence that, for example, meditation is an effective treatment for high blood pressure?

9.12 Conserving Energy. Example 9.6 (page 220) describes an experiment to learn whether providing households with digital displays or charts will reduce their electricity consumption. An executive of the electric company objects to including a control group. He says: "It would be simpler to just compare electricity use last year (before the display or chart was provided) with consumption in the same period this year. If households use less electricity this year, the display or chart must be working." Explain clearly why this design is inferior to that in Example 9.6.

9.13 Healthy Diet and Cataracts. The relationship between healthy diet and prevalence of cataracts was assessed using a sample of 1808 participants from the Women's Health Initiative Observational Study. Having a high Healthy Eating Index score was the strongest predictor of a reduced risk of cataracts, among modifiable behaviors considered. The Healthy Eating Index score was created by the U.S. Department of Agriculture and measures how well a person's diet conforms to recommended healthy eating patterns. The report concludes: "These data add to the body of evidence suggesting that eating foods rich in a variety of vitamins and minerals may contribute to postponing the occurrence of the most common type of cataract in the United States."[10]

(a) Explain why this is an observational study rather than an experiment.

(b) Although the result was statistically significant, the authors did not use strong language in stating their conclusions, using words such as "suggesting" and "may." Do you think that their language is appropriate given the nature of the study? Why?

Cautions about experimentation

The logic of a randomized comparative experiment depends on our ability to treat all the subjects identically in every way except for the actual treatments being compared. Good experiments therefore require careful attention to details to ensure that all subjects really are treated identically.

placebo

If some subjects in a medical experiment take a pill each day and a control group takes no pill, the subjects are not treated identically. Many medical experiments are therefore "placebo controlled." A study of the effects of taking vitamin E on heart disease is typical. All the subjects receive the same medical attention during the several years of the experiment. All take a pill every day, vitamin E in the treatment group and a placebo in the control group. A **placebo** is a dummy treatment. Many patients respond favorably to any treatment, even a placebo, perhaps because they trust the doctor. The response to a dummy treatment is called the *placebo effect*. If the control group did not take any pills, the effect of vitamin E in the treatment group would be confounded with the placebo effect, the effect of simply taking pills.

SCRATCH MY FURRY EARS

Rats and rabbits, specially bred to be uniform in their inherited characteristics, are the subjects in many experiments. Animals, like people, are quite sensitive to how they are treated. This can create opportunities for hidden bias. For example, human affection can change the cholesterol level of rabbits. Choose some rabbits at random and regularly remove them from their cages to have their heads scratched by friendly people. Leave other rabbits unloved. All the rabbits eat the same diet, but the rabbits that receive affection have lower cholesterol.

In addition, such studies are usually *double-blind*. The subjects don't know whether they are taking vitamin E or a placebo. Neither do the medical personnel who work with them. The double-blind method avoids unconscious bias by, for example, a doctor who is convinced that a vitamin must be better than a placebo. In many medical studies, only the statistician who does the randomization knows which treatment each patient is receiving.

Double-Blind Experiments

In a **double-blind** experiment, neither the subjects nor the people who interact with them know which treatment each subject is receiving.

lack of realism

Placebo controls and the double-blind method are more ways to eliminate possible confounding. But even well-designed experiments often face another problem: **lack of realism.** Practical constraints may mean that the subjects or treatments or setting of an experiment don't realistically duplicate the conditions we really want to study. Here are two examples.

EXAMPLE 9.7   Response to Advertising

The study of television advertising in Example 9.3 (page 216) showed a 40-minute video to students who knew an experiment was going on. We can't be sure that the results apply to everyday television viewers. Many behavioral science experiments use as subjects students or other volunteers who know they are subjects in an experiment. That's not a realistic setting. ■

EXAMPLE 9.8   Center Brake Lights

Do those high center brake lights, required on all cars sold in the United States since 1986, really reduce rear-end collisions? Randomized comparative experiments with fleets of rental and business cars, done before the lights were required, showed that the third brake light reduced rear-end collisions by as much as 50%. Alas, requiring the third light in all cars led to only a 5% drop.

What happened? Most cars did not have the extra brake light when the experiments were carried out, so it caught the eye of following drivers. Now that almost all cars have the third light, they no longer capture attention. ■

© Corbis

Lack of realism can limit our ability to apply the conclusions of an experiment to the settings of greatest interest. Most experimenters want to generalize their conclusions to some setting wider than that of the actual experiment. *Statistical analysis of an experiment cannot tell us how far the results will generalize.* Nonetheless, the randomized comparative experiment, because of its ability to give convincing evidence for causation, is one of the most important ideas in statistics.

Apply Your Knowledge

9.14 Testosterone for Older Men. As men age, their testosterone levels gradually decrease. This may cause a reduction in lean body mass, an increase in fat, and other undesirable changes. Do testosterone supplements reverse some of these effects? A study in the Netherlands assigned 237 men aged 60 to 80 with low or low-normal testosterone levels to either a testosterone supplement or a placebo. The report in the *Journal of the American Medical Association* described the study as a "double-blind, randomized, placebo-controlled trial."[11] Explain each of these terms to someone who knows no statistics.

9.15 Does Meditation Reduce Anxiety? An experiment that claimed to show that meditation reduces anxiety proceeded as follows. The experimenter interviewed the subjects and rated their level of anxiety. Then the subjects were randomly assigned to two groups. The experimenter taught one group how to meditate, and they meditated daily for a month. The other group was simply told to relax more. At the end of the month, the experimenter interviewed all the subjects again and rated their anxiety level. The meditation group now had less anxiety. Psychologists said that the results were suspect because the ratings were not blind. Explain what this means and how lack of blindness could bias the reported results.

Matched pairs and other block designs

Completely randomized designs are the simplest statistical designs for experiments. They illustrate clearly the principles of control, randomization, and adequate number of subjects. However, completely randomized designs are often inferior to more elaborate statistical designs. In particular, matching the subjects in various ways can produce more precise results than simple randomization.

matched pairs design

One common design that combines matching with randomization is the **matched pairs design.** A matched pairs design compares just two treatments. Choose pairs of subjects that are as closely matched as possible. Use chance to decide which subject in a pair gets the first treatment. The other subject in that pair gets the other treatment. That is, the random assignment of subjects to treatments is done within each matched pair, not for all subjects at once. Sometimes each "pair" in a matched pairs design consists of just one subject, who gets both treatments one after the other. Each subject serves as his or her own control. The *order* of the treatments can influence the subject's response, so we randomize the order for each subject.

| EXAMPLE 9.9 | Cell Phones and Driving |

© Tom Grill/Corbis

Does talking on a hands-free cell phone distract drivers? Undergraduate students "drove" in a high-fidelity driving simulator equipped with a hands-free cell phone. The car ahead brakes: how quickly does the subject react? Let's compare two designs for this experiment. There are 40 student subjects available.

In design 1, the *completely randomized design,* all 40 subjects are assigned at random, 20 to simply drive and the other 20 to talk on the cell phone while driving. In design 2, the *matched pairs design* that was actually used, all subjects drive both with and without using the cell phone. The two drives are on separate days to reduce carryover effects. The *order* of the two treatments is assigned at random: 20 subjects are chosen to drive first with the phone, and the remaining 20 drive first without the phone.[12]

Some subjects naturally react faster than others. The completely randomized design relies on chance to distribute the faster subjects roughly evenly between the two groups. The matched pairs design compares each subject's reaction time with and without the cell phone. This makes it easier to see the effects of using the phone. ■

Matched pairs designs use the principles of comparison of treatments and randomization. However, the randomization is not complete—we do not randomly assign all the subjects at once to the two treatments. Instead, we randomize only within each matched pair. This allows matching to reduce the effect of variation among the subjects. Matched pairs are one kind of *block design*, with each pair forming a *block.*

Block Design

A **block** is a group of individuals that are known before the experiment to be similar in some way that is expected to affect the response to the treatments.

In a **block design,** the random assignment of individuals to treatments is carried out separately within each block.

A block design combines the idea of creating equivalent treatment groups by matching with the principle of forming treatment groups at random. Blocks are another form of *control.* They control the effects of some outside variables by bringing those variables into the experiment to form the blocks. Here are some typical examples of block designs.

EXAMPLE 9.10   Men, Women, and Advertising

Women and men respond differently to advertising. An experiment to compare the effectiveness of three advertisements for the same product will want to look separately at the reactions of men and women, as well as assess the overall response to the ads.

A *completely randomized design* considers all subjects, both men and women, as a single pool. The randomization assigns subjects to three treatment groups without regard to their sex. This ignores the differences between men and women. A *block design* considers women and men separately. Randomly assign the women to three groups, one to view each advertisement. Then separately assign the men at random to three groups. Figure 9.5 outlines this improved design. ■

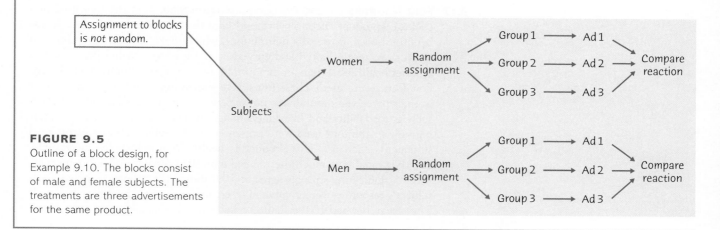

FIGURE 9.5
Outline of a block design, for Example 9.10. The blocks consist of male and female subjects. The treatments are three advertisements for the same product.

EXAMPLE 9.11   Comparing Welfare Policies

A social policy experiment will assess the effect on family income of several proposed new welfare systems and compare them with the present welfare system. Because the future income of a family is strongly related to its present income, the families who agree to participate are divided into blocks of similar income levels. The families in each block are then allocated at random among the welfare systems. ■

A block design allows us to draw separate conclusions about each block, for example, about men and women in Example 9.10. Blocking also allows more precise overall conclusions because the systematic differences between men and women can be removed when we study the overall effects of the three advertisements. The idea of blocking is an important additional principle of statistical design of experiments. A wise experimenter will form blocks based on the most important unavoidable sources of variability among the subjects. Randomization will then average out the effects of the remaining variation and allow an unbiased comparison of the treatments.

Like the design of samples, the design of complex experiments is a job for experts. Now that we have seen a bit of what is involved, we will concentrate for the most part on completely randomized experiments.

9.16 Comparing Breathing Frequencies in Swimming. Researchers from the United Kingdom studied the effect of two breathing frequencies on both performance times and several physiological parameters in front crawl swimming.[13] The breathing frequencies were one breath every second stroke (B2) and one breath every fourth stroke (B4). Subjects were 10 male collegiate swimmers. Each subject swam 200 meters, once with breathing frequency B2 and once on a different day with breathing frequency B4.

(a) Describe the design of this matched pairs experiment, including the randomization required by this design.

(b) Could this experiment be conducted using a completely randomized design? How would the design differ from the matched pairs experiment?

(c) Are there any problems with having swimmers choose their own breathing frequency and then swim 200 meters using their selected frequency?

9.17 Font Naturalness and Perceived Healthiness. Can the font used in a packaged product affect our perception of the product's healthiness? It was hypothesized that use of a natural font, which looks more handwritten and tends to be more slanted and curved, would lead to a higher perception of product healthiness than an unnatural font. Two fonts, Impact and Sketchflow Print, were used. These fonts were shown in a previous study to differ in their perceived naturalness, but otherwise were rated similarly on factors such as readability and likeability. Images of two identical packages differing only in the font used were presented to the subjects. Participants read statements such as "This product is healthy"/natural/wholesome/organic, and they rated how much they agreed with the statement on a 7 point scale with "1" indicating strongly agree and "7" indicating strongly disagree. Each subject's responses were combined to create a perceived healthiness score.[14] The researcher had 100 students available to serve as subjects.

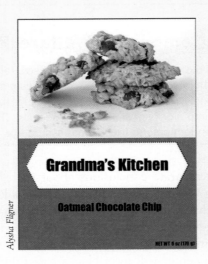 

(a) What are the response and the explanatory variables?

(b) Describe the design of a completely randomized experiment to learn the effect of font naturalness on perceived healthiness.

(c) Describe the design of a matched pairs experiment using the same 100 subjects.

9.18 Technology for Teaching Statistics. The Brigham Young University statistics department is performing randomized comparative experiments to compare teaching methods. Response variables include students' final-exam scores and a measure of their attitude toward statistics. One study compares two levels of technology for large lectures: standard (overhead projectors and chalk) and multimedia. The individuals in the study are the eight lectures in a basic statistics course. There are four instructors, each of whom teaches two lectures. Because the lecturers differ, their lectures form four blocks.[15] Suppose the lectures and lecturers are as follows:

Lecture	Lecturer	Lecture	Lecturer
1	Hilton	5	Tolley
2	Christensen	6	Hilton
3	Hadfield	7	Tolley
4	Hadfield	8	Christensen

Outline a block design and do the randomization that your design requires.

CHAPTER 9 SUMMARY

Chapter Specifics

- We can produce data intended to answer specific questions by **observational studies** or **experiments.** Sample surveys that select a part of a population of interest to represent the whole are one type of observational study. **Experiments,** unlike observational studies, actively impose some treatment on the subjects of the experiment.

- Variables are **confounded** when their effects on a response can't be distinguished from each other. Observational studies and uncontrolled experiments often fail to show that changes in an explanatory variable actually cause changes in a response variable because the explanatory variable is confounded with lurking variables.

- In an experiment, we impose one or more **treatments** on individuals, often called **subjects.** Each treatment is a combination of values of the explanatory variables, which we call **factors.**

- The **design** of an experiment describes the choice of treatments and the manner in which the subjects are assigned to the treatments. The basic principles of statistical design of experiments are **control** and **randomization** to combat bias and **using enough subjects** to reduce chance variation.

- The simplest form of control is **comparison.** Experiments should compare two or more treatments to avoid confounding of the effect of a treatment with other influences, such as lurking variables.

- **Randomization** uses chance to assign subjects to the treatments. Randomization creates treatment groups that are similar (except for chance variation) before the treatments are applied. Randomization and comparison together prevent **bias,** or systematic favoritism, in experiments.

- You can carry out randomization by using software or by giving numerical labels to the subjects and using a **table of random digits** to choose treatment groups.

- Applying each treatment to many subjects reduces the role of chance variation and makes the experiment more sensitive to differences among the treatments.

■ Good experiments also require attention to detail as well as good statistical design. Many behavioral and medical experiments are **double-blind.** Some give a **placebo** to a control group. **Lack of realism** in an experiment can prevent us from generalizing its results.

■ In addition to comparison, a second form of control is to restrict randomization by forming **blocks** of individuals that are similar in some way that is important to the response. Randomization is then carried out separately within each block.

■ **Matched pairs** are a common form of blocking for comparing just two treatments. In some matched pairs designs, each subject receives both treatments in a random order. In others, the subjects are matched in pairs as closely as possible, and each subject in a pair receives one of the treatments.

Link It

Observational studies and experiments are two methods for producing data. Observational studies are useful when the conclusion involves describing a group or situation without disturbing the scene we observe. Sample surveys, discussed in Chapter 8, are an important type of observational study in which we draw conclusions about a population by observing only a part of the population (the sample). In contrast, experiments are used when the situation calls for a conclusion about whether a treatment *causes* a change in a response. The distinction between observational studies and experiments will be important when stating your conclusions in later chapters.

Only well-designed experiments provide a sound basis for concluding cause-and-effect relationships. In a simple comparative experiment, two treatments are imposed on two groups of individuals. Reaching the conclusion that the difference between the groups is caused by the treatments, rather than lurking variables, requires that the two groups of individuals be similar at the outset. A randomized comparative experiment is used to create groups that are similar. If there is a sufficiently large difference between the groups after imposing the treatments, we can say that the results are statistically significant and conclude that the differences in the response were *caused* by the treatments. In later chapters, the specific statistical procedures for reaching these conclusions will be described.

As with sampling, when conducting an experiment, attention to detail is important because our conclusions can be weakened by several factors. A lack of blinding can result in the expectations of the researcher influencing the results, while the placebo effect can confound the comparison between a treatment and a control group. In many instances, a more complex design is required to overcome difficulties and can produce more precise results.

CHECK YOUR SKILLS

9.19 A large representative random sample of 6906 U.S. adults collected over 20 years showed that "parents reported higher levels of life satisfaction than nonparents," with the observed difference in life satisfaction between the two groups being statistically significant.[16] This is an example of

(a) an observational study.
(b) a randomized comparative experiment.
(c) a single blind experiment because the respondents in the sample were aware of whether or not they were parents.

9.20 In the study described in the previous exercise we can conclude that

(a) having children leads to higher levels of life satisfaction. We can reach this conclusion because we have a representative sample.
(b) having children leads to higher levels of life satisfaction. We can reach this conclusion because we have both a large and a representative sample.
(c) parents tend to have higher satisfaction in their lives than nonparents.

9.21 What electrical changes occur in muscles as they get tired? Student subjects hold their arms above their shoulders until they have to drop them. Meanwhile, the electrical activity in their arm muscles is measured. This is

(a) an observational study.
(b) an uncontrolled experiment.
(c) a randomized comparative experiment.

9.22 Do violence and sex in television programs help sell products in advertisements? Subjects were randomly assigned to watch one of four types of TV shows: (1) neither sex nor violence in the content code, (2) violence but no sex in the content code, (3) sex but no violence in the content code, and (4) both sex and violence in the content code. For each TV show, the original advertisements were replaced with the same set of 12 advertisements. Subjects were not told the purpose of the study but were instead told that the researchers were studying attitudes toward TV shows. After viewing the show, subjects received a surprise memory test to check their recall of the products advertised.[17] This experiment has

(a) 4 factors, the four TV shows being compared.
(b) 12 factors, the advertisements being shown.
(c) 2 factors, with/without violent content and with/without sexual content.

9.23 In the experiment of the previous exercise, the 336 subjects are labeled 001 to 336. Labels are selected at random by software, with the first 84 selected assigned to view TV show 1, the next 84 to view TV show 2, and the next 84 to view TV show 3. The 84 remaining subjects view TV show 4. This is a

(a) matched pairs design because subjects are matched to the TV shows.
(b) completely randomized design.
(c) block design with TV shows representing the four blocks.

9.24 In the experiment described in Exercise 9.22,

(a) it would have been better to have subjects choose the type of TV show they preferred to view to improve their recall and reduce confounding.
(b) the score on the memory test of their recall of advertisements is the response.
(c) the experimenters should have used different advertisements for each type of TV show to reduce confounding.

9.25 A medical experiment compares an antidepression medicine with a placebo for relief of chronic headaches. There are 36 headache patients available to serve as subjects. To choose 18 patients to receive the medicine, you would

(a) assign labels 01 to 36 and use Table B or a random number generator to choose 18.
(b) assign labels 01 to 18, because only 18 need to be chosen.
(c) assign the first 18 who signed up to get the medicine.

9.26 The Community Intervention Trial for Smoking Cessation asked whether a community-wide advertising campaign would reduce smoking. The researchers located 11 pairs of communities, each pair similar in location, size, economic status, and so on. One community in each pair participated in the advertising campaign, and the other did not. This is

(a) an observational study.
(b) a matched pairs experiment.
(c) a completely randomized experiment.

9.27 To decide which community in each pair in the previous exercise should get the advertising campaign, it is best to

(a) toss a coin.
(b) choose the community that will help pay for the campaign.
(c) choose the community with a mayor who will participate.

9.28 A marketing class designs two videos advertising an expensive Mercedes sports car. They test the videos by asking fellow students to view both (in random order) and say which makes them more likely to buy the car.

Mercedes should be reluctant to agree that the video favored in this study will sell more cars because

(a) the study used a matched pairs design instead of a completely randomized design.

(b) results from students may not generalize to the older and richer customers who might buy a Mercedes.

(c) this is an observational study, not an experiment.

CHAPTER 9 EXERCISES

In all exercises that require randomization, you may use software, the Simple Random Sample applet, or Table B. See Example 9.6 (page 220) for directions on using the applet for more than two treatment groups.

9.29 Red meat and mortality. Many studies have found an association between red meat consumption and an increased risk of chronic diseases. What is the relationship between red meat consumption and mortality? A large study followed 120,000 men and women who were free of coronary heart disease and cancer at the beginning of the study. Participants were asked detailed questions about their eating habits every four years, and the study spanned almost 30 years. It was found that the risk of dying at an early age—from heart disease, cancer, or any other cause, rises with the amount of red meat that they consumed.[18]

(a) Is this an observational study or an experiment? What are the explanatory and response variables?
(b) The authors noted that "Men and women with higher intake of red meat were less likely to be physically active and were more likely to be current smokers, to drink alcohol, and to have a higher body mass index." Explain carefully why differences in these variables make it more difficult to conclude that higher intake of red meat explains the increased death rate. What are the variables physical activity, smoking status, drinking behavior, and body mass index called?
(c) Suggest at least one lurking variable related to diet that may be confounded with higher intake of red meat. Explain why you chose these variables.

9.30 Neighborhood's effect on grades. To study the effect of neighborhood on academic performance, one thousand families were given federal housing vouchers to move out of their low-income neighborhoods. No improvement in the academic performance of the children in the families was found one year after the move. Explain clearly why the lack of improvement in academic performance after one year does not necessarily mean that neighborhood does not affect academic performance. In particular, identify some lurking variables whose effect on academic performance may be confounded with the effect of neighborhood. Use a figure like Figure 9.1 (page 215) to illustrate your explanation.

9.31 Reducing nonresponse. How can we reduce the rate of refusals in telephone surveys? Most people who answer at all listen to the interviewer's introductory remarks and then decide whether to continue. One study made telephone calls to randomly selected households to ask opinions about the next election. In some calls, the interviewer gave her name, in others she identified the university she was representing, and in still others she identified both herself and the university. The study recorded what percent of each group of interviews was completed. Is this an observational study or an experiment? Why? What are the explanatory and response variables?

9.32 Samples versus experiments. Give an example of a question about college students, their behavior, or their opinions that would best be answered by

(a) a sample survey.
(b) an experiment.

9.33 Observation versus experiment. An *LA Times* article reported that the artery walls of people living within 100 meters of a highway thicken more than twice as fast as the average person's.[19] Researchers used ultrasound to measure the carotid artery wall thickness of 1483 people living near freeways in the Los Angeles area. The artery wall thickness among those living within 100 meters of a highway increased by 5.5 micrometers (roughly 1/20th the thickness of a human hair) each year during the three-year study, which is more than twice the progression observed in participants who did not live within this distance of a highway.

(a) The study compared artery thickening of subjects in the study that were living within 100 meters of a highway with those that were not. Without reading any further details of this study, how do you know this was an observational study?
(b) Suggest some variables that might differ between the subjects in the study living within 100 meters of a highway and those that were further away. Are any of these possible confounding variables? Explain. (Think about whether living very close to a highway is a desirable neighborhood).
(c) Could this study be conducted as a randomized comparative experiment? What would be the difficulties?

9.34 Attitudes toward homeless people. Negative attitudes toward poor people are common. Are attitudes more negative when a person is homeless? To find out, a description of a poor person is read to subjects. There are two versions of this description. One begins

Jim is a 30-year-old single man. He is currently living in a small single-room apartment.

The other description begins

Jim is a 30-year-old single man. He is currently homeless and lives in a shelter for homeless people.

Otherwise, the descriptions are the same. After reading the description, you ask subjects what they believe about Jim and what they think should be done to help him. The subjects are 544 adults interviewed by telephone.[20] Outline the design of this experiment.

9.35 **Getting teachers to come to school.** Elementary schools in rural India are usually small, with a single teacher. The teachers often fail to show up for work. Here is an idea for improving attendance: give the teacher a digital camera with a tamper-proof time and date stamp and ask a student to take a photo of the teacher and class at the beginning and end of the day. Offer the teacher better pay for good attendance—verified by the photos. Will this work? A randomized comparative experiment started with 120 rural schools in Rajasthan and assigned 60 to this treatment and 60 to a control group. Random checks for teacher attendance showed that 21% of teachers in the treatment group were absent, as opposed to 42% in the control group.[21]

(a) Outline the design of this experiment.
(b) Label the schools, and choose the first 10 schools for the treatment group. If you use Table B, start at line 108.

9.36 **Marijuana and work.** How does smoking marijuana affect willingness to work? Canadian researchers persuaded young adult men who used marijuana to live for 98 days in a "planned environment." The men earned money by weaving belts. They used their earnings to pay for meals and other consumption and could keep any money left over. One group smoked two potent marijuana cigarettes every evening. The other group smoked two weak marijuana cigarettes. All subjects could buy more cigarettes but were given strong or weak cigarettes depending on their group. Did the weak and strong groups differ in work output and earnings?[22]

(a) Outline the design of this experiment.
(b) Here are the names of the 30 subjects. Use software or Table B at line 120 to carry out the randomization your design requires.

Abel	DeVore	Kennedy	Reichert	Stout
Aeffner	Fleming	Lamone	Riddle	Williams
Birkel	Fritz	Mani	Sawant	Wilson
Bower	Giriunas	Mattos	Scannell	Worbis
Burke	Glosup	Molnar	Sheldon	Zaccai
Deis	Heaton	Newlen	Simmons	Zelaski

(c) Do you think this can be run as a double-blind experiment? Explain.

9.37 **The benefits of red wine.** Some people think that red wine protects moderate drinkers from heart disease better than other alcoholic beverages. This calls for a randomized comparative experiment. The subjects were healthy men aged 35 to 65. They were randomly assigned to drink red wine (9 subjects), drink white wine (9 subjects), drink white wine and also take polyphenols from red wine (6 subjects), take polyphenols alone (9 subjects), or drink vodka and lemonade (6 subjects).[23] Outline the design of the experiment, and randomly assign the 39 subjects to the 5 groups. If you use Table B, start at line 107.

9.38 **Can low-fat food labels lead to obesity?** What are the effects of low-fat food labels on food consumption? Do people eat more of a snack food when the food is labeled as low-fat? The answer may depend both on whether the snack food is labeled low-fat and whether the label includes serving-size information. An experiment investigated this question using university staff, graduate students, and undergraduate students at a large university as subjects. Subjects were asked to evaluate a pilot episode for an upcoming TV show in a theater on campus and were given a cold 24-ounce bottle of water and a bag of granola from a respected campus restaurant called The Spice Box. They were told to enjoy as much or as little of the granola as they wanted. Depending on the condition randomly assigned to the subjects, the granola was labeled as either "Regular Rocky Mountain Granola" or "Low-Fat Rocky Mountain Granola." Below this, the label indicated "Contains 1 Serving," or "Contains 2 Servings," or it provided no serving-size information.[24] Twenty subjects are assigned to each treatment, and their granola bags were weighed at the end of the session to determine how much granola was eaten.

(a) What are the factors and the treatments? How many subjects does the experiment require?
(b) Outline a completely randomized design for this experiment. (You need not actually do the randomization.)

9.39 **Relieving headaches.** Doctors identify "chronic tension-type headaches" as headaches that occur almost daily for at least six months. Can antidepressant medications or stress management training reduce the number and severity of these headaches? Are both together more effective than either alone?

(a) Use a diagram like Figure 9.2 (page 217) to display the treatments in a design with two factors: "medication, yes or no" and "stress management, yes or no." Then outline the design of a completely randomized experiment to compare these treatments.
(b) The headache sufferers named here have agreed to participate in the study. Randomly assign the

subjects to the treatments. If you use the *Simple Random Sample* applet or other software, assign all the subjects. If you use Table B, start at line 125 and assign subjects to only the first treatment group.

Abbott	Decker	Herrera	Lucero	Richter
Abdalla	Devlin	Hersch	Masters	Riley
Alawi	Engel	Hurwitz	Morgan	Samuels
Broden	Fuentes	Irwin	Nelson	Smith
Chai	Garrett	Jiang	Nho	Suarez
Chuang	Gill	Kelley	Ortiz	Upasani
Cordoba	Glover	Kim	Ramdas	Wilson
Custer	Hammond	Landers	Reed	Xiang

Treating sinus infections. *Sinus infections are common, and doctors commonly treat them with antibiotics. Another treatment is to spray a steroid solution into the nose. A well-designed clinical trial found that these treatments, alone or in combination, do not reduce the severity or the length of sinus infections.*[25] *Exercises 9.40 to 9.42 concern this trial.*

9.40 Experimental design. The clinical trial was a completely randomized experiment that assigned 240 patients at random among four treatments as follows:

	ANTIBIOTIC PILL	PLACEBO PILL
Steroid spray	53	64
Placebo spray	60	63

(a) Outline the design of the experiment.
(b) How will you label the 240 subjects?
(c) Explain briefly how you would do the random assignment of patients to treatments. Assign the first five patients who will receive the first treatment.

9.41 Describing the design. The report of this study in the *Journal of the American Medical Association* describes it as a "double-blind, randomized, placebo-controlled factorial trial." "Factorial" means that the treatments are formed from more than one factor. What are the factors? What do "double-blind" and "placebo-controlled" mean?

9.42 Checking the randomization. If the random assignment of patients to treatments did a good job of eliminating bias, possible lurking variables such as smoking history, asthma, and hay fever should be similar in all four groups. After recording and comparing many such variables, the investigators said that "all showed no significant difference between groups." Explain to someone who knows no statistics what "no significant difference" means. Does it mean that the presence of all these variables was exactly the same in all four treatment groups?

9.43 Frappuccino light? Here's the opening of a Starbucks press release: "Starbucks Corp. on Monday said it would roll out a line of blended coffee drinks intended to tap into the growing popularity of reduced-calorie and reduced-fat menu choices for Americans." You wonder if Starbucks customers like the new "Mocha Frappuccino Light" as well as the regular Mocha Frappuccino coffee.

(a) Describe a matched pairs design to answer this question. Be sure to include proper blinding of your subjects. What is your response variable going to be?
(b) You have 20 regular Starbucks customers on hand. Use the *Simple Random Sample* applet or Table B at line 141 to do the randomization that your design requires.

9.44 Growing trees faster. The concentration of carbon dioxide (CO_2) in the atmosphere is increasing rapidly due to our use of fossil fuels. Because green plants use CO_2 to fuel photosynthesis, more CO_2 may cause trees to grow faster. An elaborate apparatus allows researchers to pipe extra CO_2 to a 30-meter circle of forest. We want to compare the growth in base area of trees in treated and untreated areas to see if extra CO_2 does in fact increase growth. We can afford to treat three circular areas.[26]

(a) Describe the design of a completely randomized experiment using six well-separated 30-meter circular areas in a pine forest. Sketch the circles, and carry out the randomization your design calls for.
(b) Areas within the forest may differ in soil fertility. Describe a matched pairs design using three pairs of circles that will reduce the extra variation due to different fertility. Sketch the circles and carry out the randomization your design calls for.

9.45 Athletes taking oxygen. We often see players on the sidelines of a football game inhaling oxygen. Their coaches think this will speed their recovery. We might measure recovery from intense exertion as follows: Have a football player run 100 yards three times in quick succession. Then allow three minutes to rest before running 100 yards again. Time the final run. Because players vary greatly in speed, you plan a matched pairs experiment using 25 football players as subjects. Discuss the design of such an experiment to investigate the effect of inhaling oxygen during the rest period.

AP Photo/Wade Payne

9.46 **Protecting ultramarathon runners.** An ultramarathon, as you might guess, is a footrace longer than the 26.2 miles of a marathon. Runners commonly develop respiratory infections after an ultramarathon. Will taking 600 milligrams of vitamin C daily reduce these infections? Researchers randomly assigned ultramarathon runners to receive either vitamin C or a placebo. Separately, they also randomly assigned these treatments to a group of nonrunners the same age as the runners. All subjects were watched for 14 days after the big race to see if infections developed.[27]

(a) What is the name for this experimental design?

(b) Use a diagram to outline the design.

9.47 **Wine, beer, or spirits?** There is good evidence that moderate alcohol use improves health. Some people think that red wine is better for your health than other alcoholic drinks. You have recruited 300 adults aged 45 to 65 who are willing to follow your orders about alcohol consumption over the next five years. You want to compare the effects on heart disease of moderate drinking of just red wine, just beer, or just spirits. Outline the design of a completely randomized experiment to do this. (No such experiment has been done because subjects aren't willing to have their drinking regulated for years.)

9.48 **Wine, beer, or spirits?** Women as a group develop heart disease much later than men. We can improve the completely randomized design of Exercise 9.47 by using women and men as blocks. Your 300 subjects include 120 women and 180 men. Outline a block design for comparing wine, beer, and spirits. Be sure to say how many subjects you will put in each group in your design.

9.49 **Quick randomizing.** Here's a quick and easy way to randomize. You have 100 subjects, 50 women and 50 men. Toss a coin. If it's heads, assign all the men to the treatment group and all the women to the control group. If the coin comes up tails, assign all the women to treatment and all the men to control. This gives every individual subject a 50–50 chance of being assigned to treatment or control. Why isn't this a good way to randomly assign subjects to treatment groups?

9.50 **Do antioxidants prevent cancer?** People who eat lots of fruits and vegetables have lower rates of colon cancer than those who eat little of these foods. Fruits and vegetables are rich in "antioxidants" such as vitamins A, C, and E. Will taking antioxidants help prevent colon cancer? A medical experiment studied this question with 864 people who were at risk of colon cancer. The subjects were divided into four groups: daily beta-carotene, daily vitamins C and E, all three vitamins every day, or daily placebo. After four years, the researchers were surprised to find no significant difference in colon cancer among the groups.[28]

(a) What are the explanatory and response variables in this experiment?

(b) Outline the design of the experiment. Use your judgment in choosing the group sizes.

(c) The study was double-blind. What does this mean?

(d) What does "no significant difference" mean in describing the outcome of the study?

(e) Suggest some lurking variables that could explain why people who eat lots of fruits and vegetables have lower rates of colon cancer. The experiment suggests that these variables, rather than the antioxidants, may be responsible for the observed benefits of fruits and vegetables.

9.51 **An herb for depression?** Does the herb Saint-John's-wort relieve major depression? Here are some excerpts from the report of a study of this issue.[29] The study concluded that the herb is no more effective than a placebo.

(a) "Design: Randomized, double-blind, placebo-controlled clinical trial. . . ." A clinical trial is a medical experiment using actual patients as subjects. Explain the meaning of each of the other terms in this description.

(b) "Participants . . . were randomly assigned to receive either Saint-John's-wort extract ($n = 98$) or placebo ($n = 102$). . . . The primary outcome measure was the rate of change in the Hamilton Rating Scale for Depression over the treatment period." Based on this information, use a diagram to outline the design of this clinical trial.

© Organics Image Library/Alamy

9.52 **Randomization avoids bias.** Suppose that the 25 even-numbered students among the 50 students available for the comparison of classroom and online instruction (Example 9.5, page 220) are older, employed students. We hope that randomization will distribute these students roughly equally between the classroom and online groups. Use the *Simple Random Sample* applet to take 20 samples of size 25 from the 50 students.

These 25 students will be the classroom instruction group. (Be sure to click "Reset" after each sample.) Record the counts of even-numbered students in each of your 20 samples.

(a) How many older students would you expect to see in the classroom instruction group?

(b) You see that there is considerable chance variation in the number of older (even-numbered) students assigned to the classroom group. Draw a stem-and-leaf plot of the number of older students assigned to the classroom group. Do you see any systematic bias in favor of one or the other group being assigned the older students? Larger samples from a larger population will, on the average, do an even better job of creating two similar groups.

 Exploring the Web

9.53 **Smoking cessation.** Go to the *New England Journal of Medicine* Web site, www.nejm.org, and find the article "A Randomized, Controlled Trial of Financial Incentives for Smoking Cessation" by Volpp et al. in the February 12, 2009, issue. Under the "ISSUES" link, you need to go to the "Browse full index" link and then to the February 12, 2009, issue. You can then download the pdf of the article for free. Was this a comparative study? Was randomization used? How many subjects took part? There were 22 subjects in the control group and 64 in the incentive group who were still not smoking six months after they stopped. What were the percents in each group? This difference is statistically significant. Explain in simple language what this means.

9.54 **Find an experiment.** You can find the latest medical research in the *Journal of the American Medical Association* at www.jama.ama-assn.org and the *New England Journal of Medicine* at www.nejm.org. Many of the articles describe randomized comparative experiments and use the language of statistical significance when giving conclusions. Look through the abstracts and find an experiment of interest to you. If your institution has a subscription to these journals, you should be able to view the entire article. Otherwise, use the information in the abstract to answer as many of these questions as you can. What was the purpose of the experiment? How many factors were in the experiment, and what were the levels of the factors? What response(s) were measured? How many subjects were assigned to each of the treatments, and was randomization used? Was it a double-blind experiment? What were the conclusions, and were the results statistically significant?

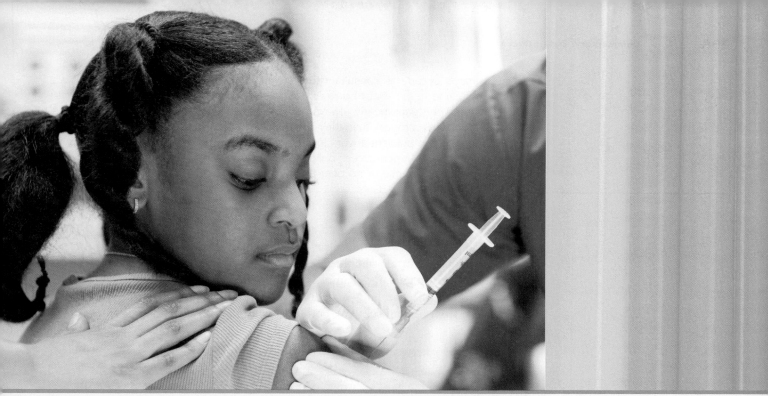

Commentary: Data Ethics*

In this commentary we cover...

- Institutional review boards
- Informed consent
- Confidentiality
- Clinical trials
- Behavioral and social science experiments

The production and use of data, like all human endeavors, raise ethical questions. We won't discuss the telemarketer who begins a telephone sales pitch with "I'm conducting a survey." Such deception is clearly unethical. It enrages legitimate survey organizations, which find the public less willing to talk with them. Neither will we discuss those few researchers who, in the pursuit of professional advancement, publish fake data. There is no ethical question here—faking data to advance your career is just wrong. It will end your career when uncovered. But just how honest must researchers be about real, unfaked data? Here is an example that suggests the answer is "More honest than they often are."

EXAMPLE 1   The Whole Truth?

Papers reporting scientific research are supposed to be short, with no extra baggage. Brevity, however, can allow researchers to avoid complete honesty about their data. Did they choose their subjects in a biased way? Did they report data on only some of their subjects? Did they try several statistical analyses and report only the ones that looked best? The statistician John Bailar screened more than 4000 medical papers in

*This short essay concerns a very important topic, but the material is not needed to read the rest of the book.

more than a decade as consultant to the *New England Journal of Medicine*. He says, "When it came to the statistical review, it was often clear that critical information was lacking, and the gaps nearly always had the practical effect of making the authors' conclusions look stronger than they should have."[1] The situation is no doubt worse in fields that screen published work less carefully. This problem continues to grow with the proliferation of open-access online journals that "will print seemingly anything for a fee" and provide little or no peer review.[2] ■

The most complex issues of data ethics arise when we collect data from people. The ethical difficulties are more severe for experiments that impose some treatment on people than for sample surveys that simply gather information. Trials of new medical treatments, for example, can do harm as well as good to their subjects. Here are some basic standards of data ethics that must be obeyed by all studies that gather data from human subjects, both observational studies and experiments.

Basic Data Ethics for Human Subjects

All planned studies must be reviewed in advance by an **institutional review board** charged with protecting the safety and well-being of the subjects.

All individuals who are subjects in a study must give their **informed consent** before data are collected.

All individual data must be kept **confidential.** Only statistical summaries for groups of subjects may be made public.

If subjects are children then their consent is needed in addition to that of the parents or guardians.

Many journals have a formal requirement of explicitly addressing human subjects issues if the study is classified as human subjects research. Also, the law requires that studies carried out or funded by the federal government obey these principles.[3] But neither the law nor the consensus of experts is completely clear about the details of their application.

Institutional review boards

The purpose of an institutional review board is not to decide whether a proposed study will produce valuable information or whether it is statistically sound. The board's purpose is, in the words of one university's board, "to protect the rights and welfare of human subjects (including patients) recruited to participate in research activities." The board reviews the plan of the study and can require changes. It reviews the consent form to ensure that subjects are informed about the nature of the study and about any potential risks. Once research begins, the board monitors the study's progress at least once a year.

The most pressing issue concerning institutional review boards is whether their workload has become so large that their effectiveness in protecting subjects drops. When the government temporarily stopped human subject research at Duke University Medical Center in 1999 due to inadequate protection of subjects, more than 2000 studies were going on. That's a lot of review work. There are shorter review procedures for projects that involve only minimal risks to subjects, such as most sample surveys. When a board is overloaded, there is a temptation to put more proposals in the minimal-risk category to speed the work.

INSTITUTIONAL REVIEW BOARD (IRB)

Home

Education and Training

Federalwide Assurance (FWA)

Definitions

Policy Manual

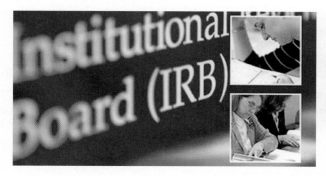

Director, Mayo Clinic Office of Human Research Protection

William J Tremaine, M.D.

QUALITY
DEDICATION
COMMITMENT

MAYO CLINIC IS HIGHLY RANKED BY EIGHT OF THE BEST-KNOWN ASSESSMENT ORGANIZATIONS. READ THE FULL STORY.

OVERVIEW

The Mayo Clinic Institutional Review Board (IRB) reviews all human subject research conducted at Mayo Clinic Florida (MCF), Mayo Clinic Rochester (MCR), or Mayo Clinic Arizona (MCA) and research conducted at other facilities under the direction of MCF, MCR, or MCA staff. A guarantee that all human subject research at Mayo will be reviewed by the IRB has been given to the U.S. Department of Health and Human Services (HHS) in a Federalwide Assurance (FWA00005001).

Read more ▶

Mission

The primary mission of Mayo Clinic's IRB is to ensure the protection of rights, privacy and welfare of all human participants in research programs conducted by Mayo Clinic and associated faculty, professional staff, and students. Coexistent with participant protection is the goal of providing quality service to enhance the conduct of research. To achieve this goal, the IRB has the authority to review, approve, modify or disapprove research protocols submitted by faculty, staff and student investigators. The IRB review process is guided by federal rules and regulations, and is based on the Protection of Human Subject Code of Federal Regulations, the Belmont Report and provisions of 45CFR46 – Protection of Human Subjects requiring institutions receiving federal funds to have all research involving human participants be approved by an IRB.

Full Accreditation

Related Resources

▢ Food and Drug Administration (FDA)

▢ Guidance for Institutional Review Boards and Clinical Investigators (FDA)

▢ Office for Human Research Protections (OHRP)

▢ National Institutes of Health (NIH)

Courtesy of Mayo Clinic

The Web page of the Mayo Clinic's institutional review board. It begins by describing the job of such boards.

Informed consent

Both words in the phrase "informed consent" are important, and both can be controversial. Subjects must be *informed* in advance about the nature of a study and any risk of harm it may bring. In the case of a sample survey, physical harm is not possible. The subjects should be told what kinds of questions the survey will ask and about how much of their time it will take. Experimenters must tell subjects the nature and purpose of the study and outline possible risks. Subjects must then *consent* in writing.

EXAMPLE 2   Who Can Consent?

© Tim Pannell/Corbis

Are there some subjects who can't give informed consent? It was once common, for example, to test new vaccines on prison inmates who gave their consent in return for good-behavior credit. Now we worry that prisoners are not really free to refuse, and the law forbids almost all medical research in prisons.

Children can't give fully informed consent, so the usual procedure is to ask their parents. A study of new ways to teach reading is about to start at a local elementary school, so the study team sends consent forms home to parents. Many parents don't return the forms. Can their children take part in the study because the parents did not say "No," or should we allow only children whose parents returned the form and said "Yes"?

What about research into new medical treatments for people with mental disorders? What about studies of new ways to help emergency room patients who may be unconscious? In most cases, there is not time to get the consent of the family. Does the principle of informed consent bar realistic trials of new treatments for unconscious patients?

These are questions without clear answers. Reasonable people differ strongly on all of them. There is nothing simple about informed consent.[4] ■

The difficulties of informed consent do not vanish even for capable subjects. Some researchers, especially in medical trials, regard consent as a barrier to getting patients to participate in research. They may not explain all possible risks; they may not point out that there are other therapies that might be better than those being studied; they may be too optimistic in talking with patients, even when the consent form has all the right details. On the other hand, mentioning every possible risk leads to very long consent forms that really are barriers. "They are like rental car contracts," one lawyer said. Some subjects don't read forms that run five or six printed pages. Others are frightened by the large number of possible (but unlikely) disasters that might happen and so refuse to participate. Of course, unlikely disasters sometimes happen. When they do, lawsuits follow, and the consent forms become yet longer and more detailed.

Confidentiality

Ethical problems do not disappear once a study has been cleared by the review board, has obtained consent from its subjects, and has actually collected data about the subjects. It is important to protect the subjects' privacy. Privacy refers to a person's interest in controlling the access of others to himself or herself, including information about himself or herself. One way this is done is by keeping all data about individuals confidential. Confidentiality refers to the agreement between the investigator and participant in how data will be managed and used. The report of an opinion poll may say what percent of the 1200 respondents felt that legal immigration should be reduced. It may not report what *you* said about this or any other issue. However the investigator who collected the data will know what you said about this or other issues in the poll.

Confidentiality is not the same as **anonymity.** Anonymity means that subjects are anonymous—their names are not known even to the director of the study. Anonymity provides a high degree of privacy, but anonymity is rare in statistical studies. Even where it is possible (mainly in surveys conducted by mail), anonymity prevents any follow-up to improve nonresponse or inform subjects of results.

anonymity

Any breach of confidentiality is a serious violation of data ethics. The best practice is to separate the identity of the subjects from the rest of the data at once. Sample surveys, for example, use the identification only to check on who did or did not respond. In an era of advanced technology, however, it is no longer enough to be sure that each individual set of data protects people's privacy. The government, for example, maintains a vast amount of information about citizens in many

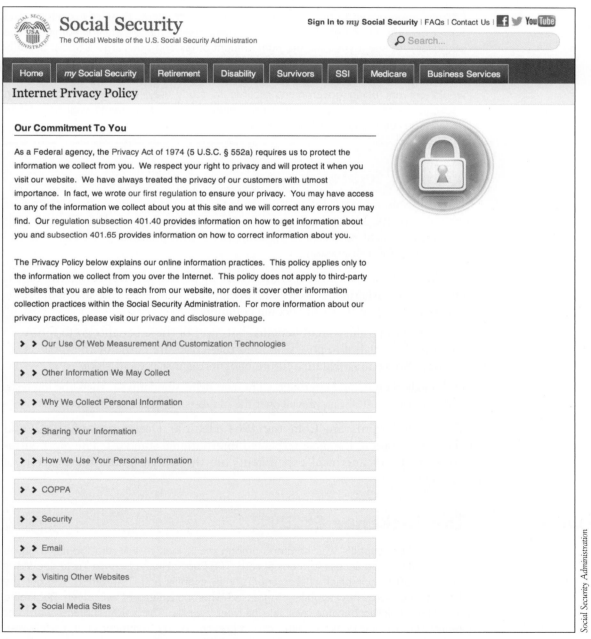

The privacy policy of the government's Social Security Administration Web site.

separate databases—census responses, tax returns, Social Security information, data from surveys such as the Current Population Survey, and so on. Many of these databases can be searched by computers for statistical studies. A clever computer search of several databases might be able, by combining information, to identify you and learn a great deal about you, even if your name and other identification have been removed from the data available for search. A colleague from Germany once remarked that "female full professor of statistics with a PhD from the United States" was enough to identify her among all the 83 million residents of Germany. Privacy and confidentiality of data are hot issues among statisticians in the computer age. Computer hacking and thefts of laptops containing data add to the difficulties.

EXAMPLE 3   **Uncle Sam Knows**

Citizens are required to give information to the government. Think of tax returns and Social Security contributions. The government needs these data for administrative purposes—to see if you paid the right amount of tax and how large a Social Security benefit you are owed when you retire. Some people feel that individuals should be able to forbid any other use of their data, even with all identification removed. This would prevent using government records to study, say, the ages, incomes, and household sizes of Social Security recipients. Such a study could well be vital to debates on reforming Social Security. ■

Clinical trials

Clinical trials are experiments that study the effectiveness of medical treatments on actual patients. Medical treatments can harm as well as heal, so clinical trials spotlight the ethical problems of experiments with human subjects. Here are the starting points for a discussion:

■ Randomized comparative experiments are by far the best way to see the true effects of new treatments. Without them, risky treatments that are no more effective than placebos will become common.

■ Clinical trials produce great benefits, but most of these benefits go to future patients. The trials also pose risks, and these risks are borne by the subjects of the trial. So we must balance future benefits against present risks.

■ Both medical ethics and international human rights standards say that "the interests of the subject must always prevail over the interests of science and society."

The quoted words are from the 1964 Helsinki Declaration of the World Medical Association, the most respected international standard. The most outrageous examples of unethical experiments are those that ignore the interests of the subjects.

EXAMPLE 4   **The Tuskegee Study**

In the 1930s, syphilis was common among black men in the rural South, a group that had almost no access to medical care. The Public Health Service Tuskegee study recruited 399 poor black sharecroppers with syphilis and 201 others without the disease to observe how syphilis progressed when no treatment was given. Beginning in 1943, penicillin became available to treat syphilis. The study subjects were not treated. In fact, the Public Health Service prevented any treatment until word leaked out and forced an end to the study in the 1970s.

The Tuskegee study is an extreme example of investigators following their own interests and ignoring the well-being of their subjects. A 1996 review said, "It has come to symbolize racism in medicine, ethical misconduct in human research, paternalism by physicians, and government abuse of vulnerable people." In 1997, President Clinton formally apologized to the surviving participants in a White House ceremony.[5] ■

Because "the interests of the subject must always prevail," medical treatments can be tested in clinical trials only when there is reason to hope that they will help the patients who are subjects in the trials. Future benefits aren't enough to justify experiments with human subjects. Of course, if there is already strong evidence that a treatment works and is safe, it is unethical *not* to give it. Here are the words of Dr. Charles Hennekens of the Harvard Medical School, who directed the large clinical trial that showed that aspirin reduces the risk of heart attacks:

> There's a delicate balance between when to do or not do a randomized trial. On the one hand, there must be sufficient belief in the agent's potential to justify exposing half the subjects to it. On the other hand, there must be sufficient doubt about its efficacy to justify withholding it from the other half of subjects who might be assigned to placebos.[6]

Why is it ethical to give a control group of patients a placebo? Well, we know that placebos often work. Moreover, placebos have no harmful side effects. So in the state of balanced doubt described by Dr. Hennekens, the placebo group may be getting a better treatment than the drug group. If we *knew* which treatment was better, we would give it to everyone. When we don't know, it is ethical to try both and compare them.[7]

Behavioral and social science experiments

When we move from medicine to the behavioral and social sciences, the direct risks to experimental subjects are less acute, but so are the possible benefits to the subjects. Consider, for example, the experiments conducted by psychologists in their study of human behavior.

EXAMPLE 5   Psychologists in the Men's Room

Psychologists observe that people have a "personal space" and are uneasy if others come too close to them. We don't like strangers to sit at our table in a coffee shop if other tables are available, and we see people move apart in elevators if there is room to do so. Americans tend to require more personal space than people in most other cultures. Can violations of personal space have physical, as well as emotional, effects?

Investigators set up shop in a men's public restroom. They blocked off urinals to force men walking in to use either a urinal next to an experimenter (treatment group) or a urinal separated from the experimenter (control group). Another experimenter, using a periscope from a toilet stall, measured how long the subject took to start urinating and how long he continued.[8] ■

MEN

David Pollack/CORBIS

This personal space experiment illustrates the difficulties facing those who plan and review behavioral studies.

■ There is no risk of harm to the subjects, although they would certainly object to being watched through a periscope. What should we protect subjects from when physical harm is unlikely? Possible emotional harm? Undignified situations? Invasion of privacy?

■ What about informed consent? The subjects did not even know they were participating in an experiment. Many behavioral experiments rely on hiding the true purpose of the study. The subjects would change their behavior if told in advance what the investigators were looking for. Subjects are asked to consent on the basis of vague information. They receive full information only after the experiment.

The "Ethical Principles" of the American Psychological Association require consent unless a study merely observes behavior in a public place. They allow deception only when it is necessary to the study, does not hide information that might influence a subject's willingness to participate, and is explained to subjects as soon as possible. The personal space study (from the 1970s) does not meet current ethical standards.

We see that the basic requirement for informed consent is understood differently in medicine and psychology. Here is an example of another setting with yet another interpretation of what is ethical. The subjects get no information and give no consent. They don't even know that an experiment may be sending them to jail for the night.

EXAMPLE 6	**Reducing Domestic Violence**

How should police respond to domestic violence calls? In the past, the usual practice was to remove the offender and order him to stay out of the household overnight. Police were reluctant to make arrests because the victims rarely pressed charges. Women's groups argued that arresting offenders would help prevent future violence even if no charges were filed. Is there evidence that arrest will reduce future offenses? That's a question that experiments have tried to answer.

A typical domestic violence experiment compares two treatments: arrest the suspect and hold him overnight, or warn the suspect and release him. When police officers reach the scene of a domestic violence call, they calm the participants and investigate. Weapons or death threats require an arrest. If the facts permit an arrest but do not require it, an officer radios headquarters for instructions. The person on duty opens the next envelope in a file prepared in advance by a statistician. The envelopes contain the treatments in random order. The police either arrest the suspect or warn and release him, depending on the contents of the envelope. The researchers then watch police records and visit the victim to see if the domestic violence reoccurs.

Such experiments show that arresting domestic violence suspects does reduce their future violent behavior.[9] As a result of this evidence, arrest has become the common police response to domestic violence. ■

The domestic violence experiments shed light on an important issue of public policy. Because there is no informed consent, the ethical rules that govern clinical trials and most social science studies would forbid these experiments. They were cleared by review boards because, in the words of one domestic violence researcher, "These people became subjects by committing acts that allow the police to arrest them. You don't need consent to arrest someone."

DISCUSSION EXERCISES

Most of these exercises pose issues for discussion. There are no right or wrong answers, but there are more and less thoughtful answers.

1. **Minimal risk?** You are a member of your college's institutional review board. You must decide whether several research proposals qualify for less rigorous review because they involve only minimal risk to subjects. Federal regulations say that "minimal risk" means the risks are no greater than "those ordinarily encountered in daily life or during the performance of routine physical or psychological examinations or tests." That's vague. Which of these do you think qualifies as "minimal risk"?

   (a) Take hair and nail clippings in a nondisfiguring manner.
   (b) Draw a drop of blood by pricking a finger to measure blood sugar.
   (c) Draw blood from the arm for a full set of blood tests.
   (d) Insert a tube that remains in the arm so that blood can be drawn regularly.
   (e) Take extra specimens from a subject who is undergoing an invasive clinical procedure such as a bronchoscopy (a procedure in which a physician views the inside of the airways for diagnostic and therapeutic purposes using an instrument that is inserted into the airways, usually through the nose or mouth).

2. **Who reviews?** Government regulations require that institutional review boards consist of at least five people, including at least one scientist, one nonscientist, and one person from outside the institution. Most boards are larger, but many contain just one outsider.

   (a) Why should review boards contain people who are not scientists?
   (b) Do you think that one outside member is enough? How would you choose that member? (For example, would you prefer a medical doctor? A member of the clergy? An activist for patients' rights?)

3. **Informed consent.** A researcher suspects that people with ultraliberal political beliefs tend to be more prone to depression. She prepares a questionnaire that measures depression and also asks many political questions. Write a description of the purpose of this research to be read by subjects to obtain their informed consent. You must balance the conflicting goals of not deceiving the subjects about what the questionnaire will tell about them and of not biasing the sample by scaring off people with ultraliberal political views.

4. **Is consent needed?** In which of the following circumstances would you allow collecting personal information without the subjects' consent?

   (a) A government agency takes a random sample of income tax returns to obtain information on the marital status and average income of people who identify themselves as clergy. Only the marital status and income are recorded from the returns, not the names.

   (b) A social psychologist attends public meetings of a religious group to study the behavior patterns of members.
   (c) A social psychologist pretends to be converted to membership in a religious group and attends private meetings to study the behavior patterns of members.

5. **Studying your blood.** Long ago, doctors drew a blood specimen from you as part of treating minor anemia. Unknown to you, the sample was stored. Now researchers plan to use stored samples from you and many other people to look for genetic factors that may influence anemia. It is no longer possible to ask your consent. Modern technology can read your entire genetic makeup from the blood sample.

   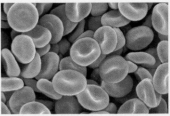
   Susumu Nishinaga/Science Source

   (a) Do you think it violates the principle of informed consent to use your blood sample if your name is on it but you were not told that it might be saved and studied later?
   (b) Suppose that your identity is not attached. The blood sample is known only to come from (say) "a 20-year-old white female being treated for anemia." Is it now okay to use the sample for research?
   (c) Perhaps we should use biological materials such as blood samples only from patients who have agreed to allow the material to be stored for later use in research. It isn't possible to say in advance what kind of research, so this falls short of the usual standard for informed consent. Is it nonetheless acceptable, given complete confidentiality and the fact that using the sample can't physically harm the patient?

6. **Anonymous? Confidential?** One of the most important nongovernment surveys in the United States is the National Opinion Research Center's General Social Survey. The GSS regularly monitors public opinion on a wide variety of political and social issues. Interviews are conducted in person in the subject's home. Are a subject's responses to GSS questions anonymous, confidential, or both? Explain your answer.

7. **Anonymous or confidential?** The University of Wisconsin at Madison, like many universities, offers free screening for HIV, the virus that causes AIDS. The announcement at the University Health Services Web site says that for persons who seek testing one option is the following. "A code is used instead of a name. The person tested receives a copy of the report for their own information, but only the code identifies the report as theirs. The report does not go into a medical record." Does this practice offer anonymity or just confidentiality? Explain your answer.

8. **Political polls.** Suppose the presidential election campaign is in full swing, and the candidates have hired polling organizations to take sample surveys to find out what the voters think about the issues. What information should the pollsters be required to give out?

(a) What does the standard of informed consent require the pollsters to tell potential respondents?

(b) The standards accepted by polling organizations also require giving respondents the name and address of the organization that carries out the poll. Why do you think this is required?

(c) The polling organization usually has a professional name such as "Samples Incorporated," so respondents don't know that the poll is being paid for by a political party or candidate. Would revealing the sponsor to respondents bias the poll? Should the sponsor always be announced whenever poll results are made public?

9. **Making poll results public.** Some people think that the law should require that all political poll results be made public. Otherwise, the possessors of poll results can use the information to their own advantage. They can act on the information, release only selected parts of it, or time the release for best effect. A candidate's organization replies that they are paying for the poll to gain information for their own use, not to amuse the public. Do you favor requiring complete disclosure of political poll results? What about other private surveys, such as market research surveys of consumer tastes?

10. **Student subjects.** Students taking Psychology 001 are required to serve as experimental subjects. Students in Psychology 002 are not required to serve, but they are given extra credit if they do so. Students in Psychology 003 are required either to sign up as subjects or to write a term paper. Serving as an experimental subject may be educational, but current ethical standards frown on using "dependent subjects" such as prisoners or charity medical patients. Students are certainly somewhat dependent on their teachers. Do you object to any of these course policies? If so, which ones, and why?

11. **The Willowbrook hepatitis studies.** In the 1960s, children entering the Willowbrook State School, an institution for the mentally retarded, were deliberately infected with hepatitis. The researchers argued that almost all children in the institution quickly became infected anyway. The studies showed for the first time that two strains of hepatitis existed. This finding contributed to the development of effective vaccines. Despite these valuable results, the Willowbrook studies are now considered an example of unethical research. Explain why, according to current ethical standards, useful results are not enough to allow a study.

12. **Unequal benefits.** Researchers on aging proposed to investigate the effect of supplemental health services on the quality of life of older people. Eligible patients on the rolls of a large medical clinic were to be randomly assigned to treatment and control groups. The treatment group would be offered hearing aids, dentures, transportation, and other services not available without charge to the control group. The review board felt that providing these services to some but not other persons in the same institution raised ethical questions. Do you agree?

13. **How many have HIV?** Researchers from Yale, working with medical teams in Tanzania, wanted to know how common infection with HIV, the virus that causes AIDS, is among pregnant women in that African country. To do this, they planned to test blood samples drawn from pregnant women.

Yale's institutional review board insisted that the researchers get the informed consent of each woman and tell her the results of the test. This is the usual procedure in developed nations. The Tanzanian government did not want to tell the women why blood was drawn or tell them the test results. The government feared panic if many people turned out to have an incurable disease for which the country's medical system could not provide care. The study was canceled. Do you think that Yale was right to apply its usual standards for protecting subjects?

14. **AIDS trials in Africa.** The drug programs that treat AIDS in rich countries are very expensive, so some African nations cannot afford to give them to large numbers of people. Yet AIDS is more common in parts of Africa than anywhere else. "Short-course" drug programs that are much less expensive might help, for example, in preventing infected pregnant women from passing the infection to their unborn children. Is it ethical to compare a short-course program with a placebo in a clinical trial? Some say "No": this is a double standard, because in rich countries the full drug program would be the control treatment. Others say "Yes": the intent is to find treatments that are practical in Africa, and the trial does not withhold any treatment that subjects would otherwise receive. What do you think?

15. **Abandoned children in Romania.** The study described in Example 9.2 (page 216) randomly assigned abandoned children in Romanian orphanages to move to foster homes or to remain in an orphanage. All the children would otherwise have remained in an orphanage. The foster care was paid for by the study. There was no informed consent because the children had been abandoned and had no adult to speak for them. The experiment was considered ethical because "people who cannot consent can be protected by enrolling them only in minimal-risk research, whose risks do not exceed those of everyday life," and because the study "aimed to produce results that would primarily benefit abandoned, institutionalized children."[10] Do you agree?

16. **Asking teens about sex.** The Centers for Disease Control and Prevention, in a survey of teenagers, asked the subjects if they had ever had sexual intercourse. Males who said "Yes" were then asked, "That very first time that you had sexual intercourse with a female, how old were you?" and "Please tell me the name or initials of your first sexual partner so that I can refer to her during the interview." Should consent of parents be required to ask minors about sex, drugs, and other such issues, or is consent of the minors themselves enough? Give reasons for your opinion.

17. **Deceiving subjects.** Students sign up to be subjects in a psychology experiment. When they arrive, they are placed in a room and assigned a task. During the task, the subject hears a loud thud from an adjacent room and then a piercing cry for help. Some subjects are placed in a room by themselves. Others are placed in a room with "confederates" who have been instructed by the researcher to look up on hearing the cry, then return to their task. The treatments being compared are whether the subject is alone in the room or in the room with confederates. Will the subject ignore the cry for help?

   The students had agreed to take part in an unspecified study, and the true nature of the experiment is explained to them afterward. Do you think this study is ethically okay?

18. **Deceiving subjects.** A psychologist conducts the following experiment: he measures the attitude of subjects toward cheating, then has them take a mathematics skills exam in which the subjects are tempted to cheat. Subjects are told that high scores will receive a $100 gift certificate and that the purpose of the experiment is to see if rewards affect performance. The exam is computer-based and multiple choice. Subjects are left alone in a room with a computer on which the exam is available and are told that they are to click on the answer they believe is correct. However, when subjects click on an answer, a small pop-up window appears with the correct answer indicated. When the pop-up window is closed, it is possible to change the answer selected. The computer records—unknown to the subjects—whether or not they change their answers after closing the pop-up window. After completing the exam, attitude toward cheating is retested.

   Subjects who cheat tend to change their attitudes to find cheating more acceptable. Those who resist the temptation to cheat tend to condemn cheating more strongly on the second test of attitude. These results confirm the psychologist's theory.

   This experiment tempts subjects to cheat. The subjects are led to believe that they can cheat secretly when in fact they are observed. Is this experiment ethically objectionable? Explain your position.

Overview

This chapter serves as an overview of the data production methods learned in Chapters 8 and 9. As in Chapter 7, the Test Yourself section will help students review the basic ideas and skills from Chapters 8 and 9, whereas the problems in the Supplementary Exercises section of the chapter provide example problems useful for in-class review or student study.

It is important that students recognize the differences among sampling objects (for quality control as an example), sample surveys, observational studies, and experiments. Sometimes students think that observational studies cannot have a control group, but they can. Consider two similar famous cases that have made great books and movies based on case-control studies: *A Civil Action* (starring John Travolta) and *Erin Brockovich* (starring Julia Roberts). Both involved observation of "cancer clusters" in the towns of Woburn, Massachusetts, and Hinkley, California, respectively. Comparison of residents in locations where the cancers developed with residents in other nearby areas yielded the conclusion that contaminated water might be the reason. Brockovich (and the lawyers she worked for) successfully sued Pacific Gas and Electric, winning $333 million for affected individuals. In Woburn, the sum awarded the plaintiffs was much smaller (about $1 million), but the Environmental Protection Agency eventually forced the responsible companies to pay for the largest chemical cleanup (to that date) in the Northeastern United States (approximately $68 million).

Case-control studies (including the two studies mentioned above) and cohort studies are popular methods in medical journals and are both observational, not experimental, in nature. The main point you will want to convey to your students is that observational studies are often used in place of designed experiments and can look like designed experiments (e.g., have control groups), but they require great care in interpreting results. In fact, it is not until an effect shows up repeatedly in many different observational studies that one should consider it statistically significant. See, for example, *Interpreting observational studies: why empirical calibration is needed to correct p-values* (Schuemie, Ryan, DuMouchel, Suchard, and Madigan, *Statistics in Medicine* 33(2), 209–218, 2014, http://onlinelibrary.wiley.com/doi/10.1002/sim.5925/full). The article by Schuemie et al. is one publication of the Observational Health Data Sciences and Informatics collaborative, which was formed to find "open-source solutions that bring out the value of observational health data through large-scale analytics" (http://ohdsi.org/).

Teaching Suggestions and Additional Examples/Activities for the Classroom

Consider working one or two comprehensive problems that bring together the most important concepts of the preceding chapters. Some suggested problems include the following:

■ Exercises 10.3–10.5: These exercises provide the opportunity to review sample, population, and bias.

■ Exercise 10.9: This exercise gives students the chance to think about lurking variables that cannot be controlled in an observational study.

■ Exercise 10.22: This exercise provides students with the chance to think about how the wording of questions may result in bias.

■ Exercise 10.25: Students often confuse treatments and factors in an experiment. This exercise gives them an opportunity to solidify the difference, as well as to consider how an effective randomization can be accomplished.

■ Exercise 10.26: This exercise allows students to design an experiment with individual subjects and with matched pairs. You might add a question about why the matched pairs design is superior.

© *Cultura RM/Alamy*

Producing Data: Part II Review

In this chapter we cover...

- Part II Summary
- Test Yourself
- Supplementary Exercises

I n Part I of this book, you mastered **data analysis,** the use of graphs and numerical summaries to organize and explore any set of data. Part II has introduced designs for data production. Part III will discuss basic probability, which provides the basis for statistical inference. Parts IV and V will deal in detail with statistical inference.

Designs for producing data are essential if the data are intended to represent some wider population or process. Figures 10.1 and 10.2 display the big ideas visually. Random sampling and randomized comparative experiments are perhaps the most important statistical inventions of the 20th century. Both were slow to gain acceptance, and you will still see many voluntary response samples and uncontrolled experiments. You should now understand good designs for producing data and also why bad designs often produce data that are worthless for inference. The deliberate use of chance in producing data is a central idea in statistics. It not only reduces bias but also allows us to use **probability,** the mathematics of chance, as the basis for inference.

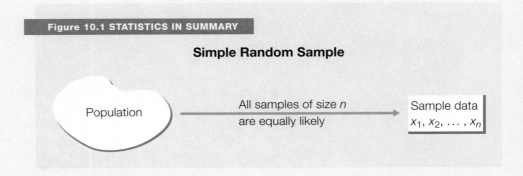

Figure 10.1 STATISTICS IN SUMMARY

Simple Random Sample

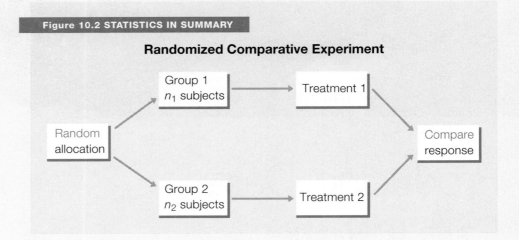

Figure 10.2 STATISTICS IN SUMMARY

Randomized Comparative Experiment

Part II Summary

Here are the most important skills you should have acquired from reading Chapters 8 and 9.

A. Sampling

1. Identify the population in a sampling situation.
2. Recognize bias due to voluntary response samples and other inferior sampling methods.
3. Use software or Table B of random digits to select a simple random sample (SRS) from a population.
4. Recognize the presence of undercoverage and nonresponse as sources of error in a sample survey. Recognize the effect of the wording of questions on the responses.
5. Use random digits to select a stratified random sample from a population when the strata are identified.

B. Experiments

1. Recognize whether a study is an observational study or an experiment.
2. Recognize bias due to confounding of explanatory variables with lurking variables in either an observational study or an experiment.

3. Identify the factors (explanatory variables), treatments, response variables, and individuals or subjects in an experiment.

4. Outline the design of a completely randomized experiment using a diagram like that in Figure 9.3 (page 220). The diagram in a specific case should show the sizes of the groups, the specific treatments, and the response variable.

5. Use software or Table B of random digits to carry out the random assignment of subjects to groups in a completely randomized experiment.

6. Recognize the placebo effect. Recognize when the double-blind technique should be used.

7. Explain why randomized comparative experiments can give good evidence for cause-and-effect relationships.

 ICING THE KICKER
The football team lines up for what they hope will be the winning field goal . . . and the other team calls time-out. "Make the kicker think about it" is their motto. Does "icing the kicker" really work? That is, does the probability of making a field goal go down when the kicker must wait around during the time-out? This isn't a simple question. A detailed statistical study considered the distance, the weather, the kicker's skill, and so on. The conclusion is cheering to coaches: yes, icing the kicker does reduce the probability of success.

Test Yourself

The following questions include both multiple-choice and short-answer questions and calculations. They will help you review the basic ideas and skills presented in Chapters 8 and 9.

Elephants and bees. *Elephants sometimes damage crops in Africa. It turns out that elephants dislike bees. They recognize beehives in areas where they are common and avoid them. Can this be used to keep elephants away from trees? A group in Kenya placed active beehives in some trees, empty beehives in others, whereas others received no beehives.[1] Will elephant damage be less in trees with hives? Will even empty hives keep elephants away? Use this information to answer Questions 10.1 and 10.2.*

© David Paynter/age fotostock

10.1 This experiment has

(a) two factors, beehives present or absent.
(b) matched pairs.
(c) three treatments.
(d) stratification by beehive.

10.2 The response in this experiment is

(a) the type of crop.
(b) the presence or absence of bees.
(c) the presence or absence of hives.
(d) elephant damage.

American Community Survey. *Each month the U.S. Census Bureau's American Community Survey mails survey forms to 250,000 households asking questions about demographic, social, economic, and housing characteristics such as mortgage and utility costs. Telephone calls are made to households that don't return the form. In one month, responses were obtained from 240,000 of the households contacted. Use this information to answer Questions 10.3 to 10.5.*

10.3 The sample is

(a) the 250,000 households initially contacted.
(b) the 240,000 households that responded.
(c) the 10,000 households that did not respond.
(d) all U.S. households.

10.4 The population of interest is

(a) all households with mortgages.
(b) the 250,000 households contacted.
(c) only U.S. households with phones.
(d) all U.S. households.

10.5 A source of bias in this survey is

(a) voluntary response.

(b) nonresponse.

(c) the fact that the survey was not double-blind.

(d) only U.S. households were contacted.

10.6 The Nurses' Health Study has interviewed a sample of more than 100,000 female registered nurses every two years since 1976. The study finds that "light-to-moderate drinkers had a significantly lower risk of death" than either nondrinkers or heavy drinkers. The Nurses' Health Study is

(a) an observational study.

(b) an experiment.

(c) Can't tell without more information.

10.7 At a local health club, a researcher samples 75 people whose primary exercise is cardiovascular and 75 people whose primary exercise is strength training. The researcher's objective is to assess the effect of type of exercise on cholesterol. Each subject reported to a clinic to have his or her cholesterol measured. The subjects were unaware of the purpose of the study, and the technician measuring the cholesterol was not aware of the subjects' type of exercise. This is

(a) an observational study.

(b) an experiment, but not a double-blind experiment.

(c) a double-blind experiment.

(d) a matched pairs experiment.

10.8 A university's financial aid office wants to know how much it can expect students to earn from summer employment. This information will be used to set the level of financial aid. The population contains 3478 students who have completed at least one year of study but have not yet graduated. The university will send a questionnaire to an SRS of 100 of these students, drawn from an alphabetized list.

(a) Describe how you will label the students to select the sample.

(b) Use Table B, beginning at line 105, to select the first five students in the sample.

(c) What is the response variable in this study?

10.9 A common definition of "binge drinking" is five or more drinks at one setting for men, and four or more for women. An observational study finds that students who binge have lower average GPA than those who don't. Suggest two lurking variables that may be confounded with binge drinking, and be sure to give a reason why you have chosen each of these variables. The possibility of confounding means that we can't conclude that binge drinking *causes* lower GPA.

10.10 The evidence linking chocolate to chronic headaches is inconsistent. In one study, 64 women with chronic headaches ate a restricted diet for two weeks. They then ate candy bars containing either chocolate or carob, prepared to taste the same, and reported whether they had a headache in the next 12 hours.[2]

(a) Outline the design of this experiment.

(b) Use Table B, beginning at line 110, to choose the first five members of the chocolate group.

10.11 In 2000, when the federal budget showed a large surplus, the Pew Research Center asked random samples of adults two questions about using the remaining surplus. Both questions stated that Social Security would be "fixed."

> Question A: *Should the money be used for a tax cut, or should it be used to fund new government programs?*

> Question B: *Should the money be used for a tax cut, or should it be spent on programs for education, the environment, health care, crime-fighting, and military defense?*

One of these questions drew 60% favoring a tax cut. The other drew only 22%. Which wording pulls respondents toward a tax cut? Why?

Snacking and movies. *In a study of human development, investigators showed two movies that were different types to a group of children. Crackers were available in a bowl at each movie, and the investigators compared the number of crackers eaten by children watching each movie. One movie was shown at 8 A.M. (right after the children had breakfast) and the other at 11 A.M. (right before the children had lunch). It was found that during the movie shown at 11 A.M., more crackers were eaten than during the movie shown at 8 A.M. The investigators concluded that the different types of movies had different effects on appetite. Use this information to answer Questions 10.12 and 10.13.*

10.12 The results cannot be trusted because

(a) the study was not double-blind. Neither the investigators nor the children should have been aware of which movie was being shown.
(b) the investigators were biased. They knew beforehand what the study would show.
(c) the investigators should have used several bowls of crackers randomly placed in the room.
(d) the time each movie was shown is a confounding variable.

10.13 The treatment in this experiment is

(a) the number of crackers eaten.
(b) the different types of movies.
(c) the time each movie was shown.
(d) the type of cracker.

10.14 The Web site of the PBS television program *NOVA Science Now* invites viewers to vote on issues such as re-creating the virus responsible for the deadly flu epidemic of 1918. This online poll is unusual in offering detailed arguments for both sides. Of the 790 viewers who read the arguments and voted, 64% said that re-creating the virus was justified.[3] Explain to someone who knows no statistics why these 790 responses probably don't represent the opinions of all American adults.

10.15 A study attempts to determine whether a football filled with helium travels farther when kicked than one filled with air. Each subject kicks twice, once with a football filled with helium and once with a football filled with air. The order of the type of football kicked is randomized. This is an example of

(a) a matched pairs experiment.
(b) a randomized controlled experiment.
(c) a stratified experiment.
(d) the placebo effect.

A student survey. *To assess the opinion of students about campus safety at The Ohio State University, a reporter for the student newspaper interviews 15 students she meets walking on the campus late at night who are willing to give their opinion. Use this information to answer Questions 10.16 to 10.18.*

Justin Sullivan/Getty Images

10.16 The sample is

(a) all those students walking on campus late at night.
(b) all students at universities with safety issues.
(c) the 15 students interviewed.
(d) all students approached by the reporter.

10.17 The population is

(a) all students at The Ohio State University.
(b) all students at The Ohio State University with safety issues.
(c) all students at The Ohio State University who walk on campus late at night.
(d) all university students in the United States.

10.18 The sample obtained is

(a) a simple random sample of students who feel safe.
(b) a stratified random sample of students who feel safe.
(c) a probability sample of students with night classes.
(d) probably biased.

Supplementary Exercises

Supplementary exercises apply the skills you have learned in ways that require more thought or more elaborate use of technology.

10.19 Sampling students. You want to investigate the attitudes of students at your school toward the school's policy on sexual harassment. You have a grant that will pay the costs of contacting about 500 students.

(a) Specify the exact population for your study. For example, will you include part-time students?

(b) Describe your sample design. Will you use a stratified sample?

(c) Briefly discuss the practical difficulties that you anticipate. For example, how will you contact the students in your sample?

10.20 The placebo effect. A survey of physicians found that some doctors give a placebo to a patient who complains of pain for which the physician can find no cause. If the patient's pain improves, these doctors conclude that it had no physical basis. The medical school researchers who conducted the survey claimed that these doctors do not understand the placebo effect. Why?

10.21 Informed consent. The requirement that human subjects give their informed consent to participate in an experiment can greatly reduce the number of available subjects. For example, a study of new teaching methods asks the consent of parents for their children to be taught by either a new method or the standard method. Many parents do not return the forms, so their children must continue to follow the standard curriculum. Why is it not correct to consider these children as part of the control group along with children who are randomly assigned to the standard method?

10.22 Fixing health care. The cost of health care and health insurance is the biggest health concern among Americans, even ahead of cancer and other diseases. Changing to a national government health insurance system is controversial. An opinion poll will give different results depending on the wording of the question asked. For each of the following claims, say whether including it in the question would *increase* or *decrease* the percent of a poll sample who support a government health insurance system.

(a) A national system would mean that everybody has health insurance.

(b) A national system would probably require an increase in taxes.

(c) Eliminating private insurance companies and their profits would reduce insurance costs.

(d) A national system would limit the medical treatments available to contain costs.

10.23 Let's go polling. Use Google or your favorite search engine to search the Web for "Web polling software." Choose one of the sites that offer software that allows you to conduct your own online opinion polls. (At the time of writing, www.twiigs.com and www.micropoll.com were two options, but things change quickly on the Web.)

(a) Choose a site, and give the name of the site that you are using.

(b) Briefly describe two attractive features that the software offers. (For example, you would like to list the answer choices in random order, so that the same choice is not always in the first position.)

(c) Despite these features in (b), all such polls share a fatal weakness. What is this?

10.24 Market research. Stores advertise price reductions to attract customers. What type of price cut is most attractive? Market researchers prepared ads for athletic shoes announcing different levels of discounts (20%, 40%, or 60%). The student subjects who read the ads were also given "inside information" about the fraction of shoes on sale (50% or 100%). Each subject then rated the attractiveness of the sale on a scale of 1 to 7.[4]

(a) There are two factors. Make a sketch like Figure 9.2 (page 217) that displays the treatments formed by all combinations of levels of the factors.

(b) Outline a completely randomized design using 60 student subjects. Use software or Table B at line 111 to choose the subjects for the first treatment.

10.25 Making french fries. Few people want to eat discolored french fries. Potatoes are kept refrigerated before being cut for french fries to prevent spoiling and preserve flavor. But immediate processing of cold potatoes causes discoloring due to complex chemical reactions. The potatoes must therefore be brought to

© Zave Smith/age fotostock

room temperature before processing. Design an experiment in which tasters will rate the color and flavor of french fries prepared from several groups of potatoes. The potatoes will be freshly picked or stored for a month at room temperature or stored for a month refrigerated. They will then be sliced and cooked either immediately or after an hour at room temperature.

(a) What are the factors and their levels, the treatments, and the response variables?

(b) Describe and outline the design of this experiment.

(c) It is efficient to have each taster rate fries from all treatments. How will you use randomization in presenting fries to the tasters?

10.26 How long did I work? A psychologist wants to know if the difficulty of a task influences our estimate of how long we spend working at it. She designs two sets of mazes that subjects can work through on a computer. One set has easy mazes, and the other has hard mazes. Subjects work until told to stop (after six minutes, but subjects do not know this). They are then asked to estimate how long they worked. The psychologist has 30 students available to serve as subjects.

(a) Describe the design of a completely randomized experiment to learn the effect of difficulty on estimated time.

(b) Describe the design of a matched pairs experiment using the same 30 subjects.

10.27 Alcohol and heart attacks. Many studies have found that people who drink alcohol in moderation have lower risk of heart attacks than either nondrinkers or heavy drinkers. Does alcohol consumption also improve survival after a heart attack? One study followed 1913 people who were hospitalized after severe heart attacks. In the year before their heart attacks, 47% of these people did not drink, 36% drank moderately, and 17% drank heavily. After four years, fewer of the moderate drinkers had died.[5]

(a) Is this an observational study or an experiment? Why? What are the explanatory and response variables?

(b) Suggest some lurking variables that may be confounded with the drinking habits of the subjects. The possible confounding makes it difficult to conclude that drinking habits explain death rates.

Chris Clinton/Getty Images

Probability

A rmed with designs for producing trustworthy data, we continue our journey toward *statistical inference.* Exploratory data analysis allows us to examine the data obtained from sampling or experiments, but simply describing or looking for patterns in the data at hand is often not the primary goal. Usually data are used to answer specific questions, posed before the data are collected. If the sample has been selected using the principles presented in Chapter 8, the sample can tell us about important aspects of the population from which it was obtained. In a comparative experiment, the data can indicate how strong the evidence is that our treatment would be superior to the placebo for a broader class of circumstances.

Generalizing the results of sampling or experiments to a larger group of individuals or a broader class of circumstances is one goal of statistical inference. The conclusions of inference use the language of *probability*, the mathematics of chance. Chapter 11 presents the ideas we need, and the optional Chapters 12 and 13 add more detail. With this language of probability, we will be prepared to understand the big ideas of inference in Part IV.

PART III

Introducing Probability

Overview

An important concept to convey to students in a statistics course is the difference between deductive and inductive reasoning. We are generalizing to the whole from the part. Such generalizations can never be certain, and we will use probability to express the uncertainty in this generalization. Probabilistic reasoning is deductive: we make statements about a sample (part of the population) based on information about the whole population. Statistical reasoning is inductive in nature: we make statements about a population based on information in a sample and include a description of our belief (or confidence) that we have a correct conclusion, so we're generalizing to the whole from the part. These are opposite problems, and understanding the probabilistic statements we make in the practice of statistics requires understanding something about probability. In an introductory course such as this, students don't need to grasp much of probability to understand its use in statistical reasoning.

Without question, the most important concept for beginning students to grasp is that the probability of an event represents a long-run proportion of times the event would occur if the random experiment is repeated endlessly. Later, when we interpret a *P*-value or a confidence interval, this notion will be central, as all these are interpreted in terms of what we might expect to see if we could repeatedly draw more samples and recompute the test or interval. Other concepts, such as formulas for computing probabilities, are arguably less important, but some of them (such as the probability of a complementary event or the union of two disjoint events) will be used in the future

> ### LEARNING OUTCOMES**
>
> ■ Recognize that some phenomena are random. Probability describes the long-run regularity of random phenomena.
>
> ■ Understand that the probability of an event is the proportion of times the event occurs in very many repetitions of a random phenomenon. Use the idea of probability as a long-run proportion to think about probability.
>
> ■ Use basic probability rules to detect illegitimate assignments of probability: Any probability must be a number between 0 and 1, and the total probability assigned to all possible outcomes must be 1.
>
> ■ Use basic probability rules to find the probability of events that are formed from other events. The probability that an event does not occur is 1 minus its probability. If two events are disjoint, the probability that one or the other occurs is the sum of their individual probabilities.
>
> ■ Find probabilities in a finite probability model by adding the probabilities of their outcomes. Find probabilities in a continuous probability model as areas under a density curve.
>
> ■ Use the notation of random variables to make compact statements about random outcomes, such as $P(X \le 4) = 0.3$. Be able to interpret such statements.
>
> ---
> **These learning outcomes appear later for the students in Chapter 14: Part III Review.

also (e.g., a two-tailed *P*-value is calculated using the union of two disjoint parts of a distribution curve). In addition, it is useful to understand the concept of a random variable.

One of the central ideas of probability is that chance behavior is unpredictable in the short run but has regular and predictable patterns over the long run. So, when a fair coin is flipped twice, we may not observe one head and one tail; however, we know that if the coin is flipped continually, the proportion of heads will converge to 0.5. Have your students use the *Probability* applet to visualize the long-run behavior of flipping a coin.

Teaching Suggestions and Additional Examples/Activities for the Classroom

Note: If you plan on covering both Chapters 11 and 12 of the text, you might choose to combine the topics in such a way that you introduce the general addition rule and general multiplication rule before the addition rule for disjoint events and the multiplication rule for independent events, respectively. I tell my students that it's always safe to use a general rule, but that they cannot assume that the specific rule holds. This is why I teach the two chapters together.

1. Explain Why We Cannot Assume Two Events Are Disjoint

An example that makes clear the point that considering nondisjoint events as disjoint may result in an illegitimate probability will serve to make the distinction obvious. One such simple example (these probabilities are approximately correct for any college) would be P(female student) = 0.55, P(underclass) = 0.59. If we simply add the probabilities, we have 1.14; students will immediately understand there are numerous female underclass individuals who were included twice. It is helpful to have the students tell you that we should not double-count these individuals.

2. Use Manipulatives to Demonstrate Random Phenomena

Use a very familiar object, such as a (presumably) fair six-sided die, to demonstrate that occurrences of random phenomena are unpredictable in the short term. Discuss the notion of a sample space. Students will understand that the probability of observing any one of the single values in the sample space on a single role of the die is 1/6. This example can be used throughout the chapter as concepts are covered. Events can be defined, such as rolling an even number or rolling a number less than 4, and then probabilities can be determined for these events.

3. Examine Short-Term Probabilities to Begin to Detect a Long-Run Pattern

To develop the notion of probability as a long-run proportion of times an event occurs, ask students what they think the probability of rolling an even number is. Have each student roll one fair six-sided die (say) 10 times, and take a look at the rolling proportion of times an even number appears. Students will sense that it is converging, especially if you're generating effectively 200 (or more) die rolls in total. The main thing is to get them to start thinking about the observed frequency of even rolls as a number that is moving as the number of trials changes.

4. Use Technology to Simulate Many Trials

Now use the *Probability* applet or a computer simulation to demonstrate a simulation involving thousands of trials. Because students have computed these sample fractions themselves for smaller numbers of trials, they'll be able to understand what is happening in the simulation. Use a graph to illustrate how the proportion of even rolls is converging to 0.5 as the number of rolls increases. Note that the *Probability* applet will produce such a graph for the students.

5. Turn to Applied Examples for Further Practice

Now move to a more applied example. You can use the probability model for Benford's law from Example 11.7 (page 268). You can ask students for the probability that the first digit is even. Then you could ask them to use the complement rule to find the probability that the first digit is odd. Ask the students to compute $P(X \leq 3)$ and $P(X < 3)$. If you do, make sure to ask why the two probabilities are different.

Other Resources (LaunchPad)

StatClips
> *Probability Overview and Motivation*
> *Probability Distributions*

Snapshots Videos
> *Probability*

EESEE Case Studies
> *On-time Flights*
> *Surviving the Titanic*
> *Psychic Probability*
> *Alcoholism in Twins*

Applet
> *Probability*
> *Normal Curve*

© Eyebyte/Alamy

Introducing Probability

In this chapter we cover...

- The idea of probability
- The search for randomness*
- Probability models
- Probability rules
- Finite and discrete probability models
- Continuous probability models
- Random variables
- Personal probability*

Why is probability, the mathematics of chance behavior, needed to understand statistics, the science of data? Let's look at a typical sample survey.

EXAMPLE 11.1   Do You Own a Gun?

What proportion of all adults own a gun? We don't know, but we do have results from the Gallup Poll. Gallup took a random sample of 1005 adults. The poll found that 342 of the people in the sample own a gun. The proportion who own a gun is

$$\text{sample proportion} = \frac{342}{1005} = 0.34 \text{ (that is, 34\%)}$$

If the sample was a simple random sample of all adults[1] then all adults had the same chance to be among the chosen 1005. It would be reasonable to use this 34% as an estimate of the unknown proportion in the population. It's a *fact* that 34% of the sample claimed to own a gun—we know because Gallup asked them. We don't know what percent of all adults own a gun, but we *estimate* that about 34% did. This is a basic move in statistics: use a result from a sample to estimate something about a population. ■

259

What if Gallup took a second random sample of 1005 adults? The new sample would have different people in it. It is almost certain that there would not be exactly 342 positive responses. That is, Gallup's estimate of the proportion of adults who own a gun will vary from sample to sample. Could it happen that one random sample finds that 34% of adults own a gun and a second random sample finds that 56% own a gun? *Random samples eliminate bias from the act of choosing a sample, but they can still be wrong because of the variability that results when we choose at random.* If the variation when we take repeated samples from the same population is too great, we can't trust the results of any one sample.

This is where we need facts about probability to make progress in statistics. Because Gallup uses chance to choose its samples, the laws of probability govern the behavior of the samples. Gallup says that the probability is 0.95 that an estimate from one of their samples comes within ±4 percentage points of the truth about the population of all adults. The first step toward understanding this statement is to understand what "probability 0.95" means. Our purpose in this chapter is to understand the language of probability, but without going into the mathematics of probability theory.

The idea of probability

To understand why we can trust random samples and randomized comparative experiments, we must look closely at chance behavior. The big fact that emerges is this: **chance behavior is unpredictable in the short run but has a regular and predictable pattern in the long run.**

Toss a coin, or choose a random sample. The result can't be predicted in advance because the result will vary when you toss the coin or choose the sample repeatedly. But there is still a regular pattern in the results, a pattern that emerges clearly only after many repetitions. This remarkable fact is the basis for the idea of probability.

EXAMPLE 11.2 **Coin Tossing**

When you toss a coin, there are only two possible outcomes, heads or tails. Figure 11.1 shows the results of tossing a coin 5000 times twice. For each number of tosses from 1 to 5000, we have plotted the proportion of those tosses that gave a head. Trial A (solid red line) begins tail, head, tail, tail. You can see that the proportion of heads for Trial A starts at 0 on the first toss, rises to 0.5 when the second toss gives a head, then falls to 0.33 and 0.25 as we get two more tails. Trial B, on the other hand, starts with five straight heads, so the proportion of heads is 1 until the sixth toss.

The proportion of tosses that produce heads is quite variable at first. Trial A starts low, and Trial B starts high. As we make more and more tosses, however, the proportion of heads for both trials gets close to 0.5 and stays there. If we made yet a third trial at tossing the coin a great many times, the proportion of heads would again settle down to 0.5 in the long run. This is the intuitive idea of probability. Probability 0.5 means "occurs half the time in a very large number of trials." The probability 0.5 appears as a horizontal line on the graph. ■

We might suspect that a coin has probability 0.5 of coming up heads just because the coin has two sides. But we can't be sure. In fact, spinning a penny on a flat surface, rather than tossing the coin, gives heads probability about 0.45 rather than 0.5.[2] The idea of probability is empirical. That is, it is based on observation rather than theorizing. Probability describes what happens in very many trials, and we must actually observe many trials to pin down a probability. In the case of tossing a coin, some diligent people have in fact made thousands of tosses.

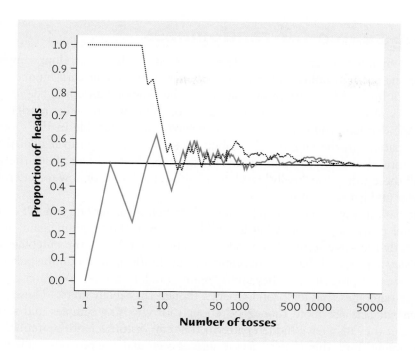

FIGURE 11.1
The proportion of tosses of a coin that give a head changes as we make more tosses. Eventually, however, the proportion approaches 0.5, the probability of a head. This figure shows the results of two trials of 5000 tosses each.

EXAMPLE 11.3 Some Coin Tossers

The French naturalist Count Buffon (1707–1788) tossed a coin 4040 times. Result: 2048 heads, or proportion 2048/4040 = 0.5069 for heads.

Around 1900, the English statistician Karl Pearson heroically tossed a coin 24,000 times. Result: 12,012 heads, a proportion of 0.5005.

While imprisoned by the Germans during World War II, the South African mathematician John Kerrich tossed a coin 10,000 times. Result: 5067 heads, a proportion of 0.5067. ■

Randomness and Probability

We call a phenomenon **random** if individual outcomes are uncertain but there is nonetheless a regular distribution of outcomes in a large number of repetitions.

The **probability** of any outcome of a random phenomenon is the proportion of times the outcome would occur in a very long series of repetitions.

The best way to understand randomness is to observe random behavior, as in Figure 11.1. You can do this with physical devices like coins, but computer simulations (imitations) of random behavior allow faster exploration. The *Probability* applet is a computer simulation that animates Figure 11.1. It allows you to choose the probability of a head and simulate any number of tosses of a coin with that probability. Experience shows that the proportion of heads gradually settles down close to the probability. Equally important, it also shows that *the proportion in a small or moderate number of tosses can be far from the probability. Probability describes only what happens in the long run.* Of course, we can never observe a probability exactly. We could always continue tossing the coin, for example. Mathematical probability is an idealization based on imagining what would happen in an indefinitely long series of trials.

DOES GOD PLAY DICE?

Few things in the world are truly random in the sense that no amount of information will allow us to predict the outcome. But according to the branch of physics called quantum mechanics, randomness does rule events inside individual atoms. Although Albert Einstein helped quantum theory get started, he always insisted that nature must have some fixed reality, not just probabilities. "I shall never believe that God plays dice with the world," said the great scientist. A century after Einstein's first work on quantum theory, it appears that he was wrong.

The search for randomness*

Random numbers are valuable. They are used to choose random samples, to shuffle the cards in online poker games, to encrypt our credit card numbers when we buy online, and as part of simulations of the flow of traffic and the spread of epidemics. Where does randomness come from, and how can we get random numbers? We defined randomness by how it behaves: unpredictable in the short run, regular pattern in the long run. Probability describes the long-run regular pattern. That many things are random in this sense is an observed fact about the world. Not all these things are "really" random. Here's a quick tour of how to find random behavior and get random numbers.

The easiest way to get random numbers is from a *computer program*. Of course, a computer program just does what it is told to do. Run the program again, and you get exactly the same result. The random numbers in Table B, the outcomes of the *Probability* applet, and the random numbers that shuffle cards for online poker come from computer programs, so they aren't "really" random. Clever computer programs produce outcomes that look random even though they really aren't. These *pseudo-random numbers* are more than good enough for choosing samples and shuffling cards. But they may have hidden patterns that can distort scientific simulations.

You might think that *physical devices such as coins and dice* produce really random outcomes. But a tossed coin obeys the laws of physics. If we knew all the inputs of the toss (forces, angles, and so on), then we could say in advance whether the outcome will be heads or tails. The outcome of a toss is predictable rather than random. Why do the results of tossing a coin *look* random? The outcomes are extremely sensitive to the inputs, so that very small changes in the forces you apply when you toss a coin change the outcome from heads to tails and back again. In practice, the outcomes are not predictable. Probability is a lot more useful than physics for describing coin tosses.

We call a phenomenon with "small changes in, big changes out" behavior *chaotic*. If we can feed chaotic behavior into a computer, we can do better than pseudorandom numbers. Coins and dice are awkward, but you can go to the Web site www.random.org to get random numbers from radio noise in the atmosphere, a chaotic phenomenon that is easy to feed to a computer.

Is anything really random? As far as current science can say, behavior inside atoms really is random—that is, there isn't any way to predict behavior in advance no matter how much information we have. It was this "really, truly random" idea that Einstein disliked as he watched the new science of quantum mechanics emerge. You can go to the HotBits Web site www.fourmilab.ch/hotbits to get really, truly random numbers generated from the radioactive decay of atoms.

Apply Your Knowledge

© Cut and Deal Ltd./Alamy

11.1   Texas Hold 'em. In the popular Texas hold 'em variety of poker, players make their best five-card poker hand by combining the two cards they are dealt with three of five cards available to all players. You read in a book on poker that if you hold a pair (two cards of the same rank) in your hand, the probability of getting four of a kind is 2/245. Explain carefully what this means. In particular, explain why it does *not* mean that if you play 245 such hands, exactly 2 will be four of a kind.

11.2   Probability Says . . . Probability is a measure of how likely an event is to occur. Match one of the probabilities that follow with each statement of

*This short discussion is optional.

likelihood given. (The probability is usually a more exact measure of likelihood than is the verbal statement.)

$$0 \quad 0.01 \quad 0.45 \quad 0.50 \quad 0.55 \quad 0.99 \quad 1$$

(a) This event is impossible. It can never occur.

(b) This event is certain. It will occur on every trial.

(c) This event is very likely, but it will not occur once in a while in a long sequence of trials.

(d) This event will occur slightly less often than not.

11.3 Random Digits. The table of random digits (Table B) was produced by a random mechanism that gives each digit probability 0.1 of being a 0.

(a) What proportion of the first 200 digits (those in the first five lines) in the table are 0s? This proportion is an estimate, based on 200 repetitions, of the true probability, which we know is 0.1.

(b) The *Probability* applet can imitate random digits. Set the probability of heads in the applet to 0.1. Check "Show true probability" to show this value on the graph. A head stands for a 0 in the random digit table, and a tail stands for any other digit. Simulate 200 digits (keep clicking "Toss" to get 40 at a time—don't click "Reset"). If you kept going forever, presumably you would get 10% heads. What was the result of your 200 tosses?

11.4 The Long Run but Not the Short Run. Our intuition about chance behavior is not very accurate. In particular, we tend to expect that the long-run pattern described by probability will show up in the short run as well. For example, we tend to think that tossing a coin 20 times will give close to 10 heads.

(a) Set the probability of heads in the *Probability* applet to 0.5 and the number of tosses to 12. Click "Toss" to simulate 20 tosses of a balanced coin. What was the proportion of heads?

(b) Click "Reset" and toss again. The simulation is fast, so do it 25 times and keep a record of the proportion of heads in each set of 20 tosses. Make a stemplot of your results. You see that the result of tossing a coin 20 times is quite variable and need not be very close to the probability 0.5 of heads.

Probability models

Gamblers have known for centuries that the fall of coins, cards, and dice displays clear patterns in the long run. The idea of probability rests on the observed fact that the average result of many thousands of chance outcomes can be known with near certainty. How can we give a mathematical description of long-run regularity?

To see how to proceed, think first about a very simple random phenomenon, tossing a coin once. When we toss a coin, we cannot know the outcome in advance. What *do* we know? We are willing to say that the outcome will be either heads or tails. We believe that each of these outcomes has probability 1/2. This description of coin tossing has two parts:

■ A list of possible outcomes

■ A probability for each outcome

Such a description is the basis for all *probability models*. Here is the basic vocabulary we use.

Probability Models

The **sample space** *S* of a random phenomenon is the set of all possible outcomes.

An **event** is an outcome or a set of outcomes of a random phenomenon. That is, an event is a subset of the sample space.

A **probability model** is a mathematical description of a random phenomenon consisting of two parts: a sample space *S* and a way of assigning probabilities to events.

A sample space *S* can be very simple or very complex. When we toss a coin once, there are only two outcomes, heads and tails. The sample space is $S = \{H, T\}$. When Gallup draws a random sample of 1005 adults, the sample space contains all possible choices of 1005 of the 237 million adults in the United States. This *S* is extremely large. Each member of *S* is a possible sample, so *S* is the collection or "space" of all possible samples. This explains the term *sample space*.

EXAMPLE 11.4   ## Rolling Dice

Rolling two dice is a common way to lose money in casinos. There are 36 possible outcomes when we roll two dice and record the up-faces in order (first die, second die). Figure 11.2 displays these outcomes. They make up the sample space *S*. "Roll a 5" is an event, call it *A*, that contains four of these 36 outcomes:

$$A = \{ \boxdot\ \vdots \qquad \because\ \vdots \qquad \because\ \because \qquad \vdots\ \boxdot \}$$

How can we assign probabilities to this sample space? We can find the actual probabilities for two specific dice only by actually tossing the dice many times, and even then only approximately. So we will give a probability model that assumes ideal, perfectly balanced dice. This model will be quite accurate for carefully made casino dice and less accurate for the cheap dice that come with a board game.

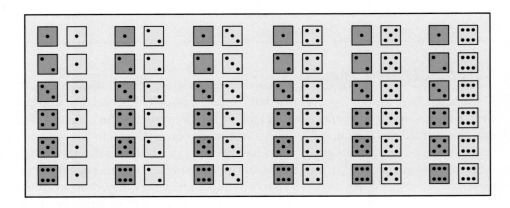

FIGURE 11.2

The 36 possible outcomes in rolling two dice. If the dice are carefully made, all these outcomes have the same probability.

If the dice are perfectly balanced, all 36 outcomes in Figure 11.2 will be *equally likely*. That is, each of the 36 outcomes will come up on one thirty-sixth of all rolls in the long run. So each outcome has probability 1/36. There are 4 outcomes in the event *A* ("roll a 5"), so this event has probability 4/36. In this way we can assign a probability to any event. So we have a complete probability model. ■

EXAMPLE 11.5   Rolling Dice and Counting the Spots

Gamblers care only about the total number of spots on the up-faces of the dice. The sample space for rolling two dice and counting the spots is

$$S = \{2, 3, 4, 5, 6, 7, 8, 9, 10, 11, 12\}$$

⚠ **CAUTION** Comparing this S with Figure 11.2 reminds us that *we can change S by changing the detailed description of the random phenomenon we are describing.*

What are the probabilities for this new sample space? The 11 possible outcomes are *not* equally likely because there are six ways to roll a 7 and only one way to roll a 2 or a 12. That's the key: each outcome in Figure 11.2 has probability 1/36. So "roll a 7" has probability 6/36 because this event contains 6 of the 36 outcomes. Similarly, "roll a 2" has probability 1/36, and "roll a 5" (4 outcomes from Figure 11.2) has probability 4/36. Here is the complete probability model:

Total spots	2	3	4	5	6	7	8	9	10	11	12
Probability	1/36	2/36	3/36	4/36	5/36	6/36	5/36	4/36	3/36	2/36	1/36

■

Apply Your Knowledge

11.5 Sample Space. Choose a student at random from a large statistics class. Describe a sample space S for each of the following. (In some cases you may have some freedom in specifying S.)

(a)   Does the student live on campus or off campus?

(b)   What is the student's age in years?

(c)   Ask how much money in coins (not bills) the student is carrying.

(d)   Record the student's letter grade at the end of the course.

11.6 Role-Playing Games. Computer games in which the players take the roles of characters are very popular. They go back to earlier tabletop games such as Dungeons & Dragons. These games use many different types of dice. A four-sided die has faces with one of the numbers 1, 2, 3, or 4 appearing at the bottom of each visible face. This number is your score.

(a)   What is the sample space for rolling a four-sided die twice (numbers on first and second rolls)? Follow the example of Figure 11.2.

(b)   What is the assignment of probabilities to outcomes in this sample space? Assume that the die is perfectly balanced, and follow the method of Example 11.4.

11.7 Role-Playing Games. The intelligence of a character in a game is determined by rolling the four-sided die twice and adding 1 to the sum of the numbers. Start with your work in the previous exercise to give a probability model (sample space and probabilities of outcomes) for the character's intelligence. Follow the method of Example 11.5.

Slpix/Dreamstime.com

Probability rules

In Examples 11.4 and 11.5 we found probabilities for tossing dice. As random phenomena go, dice are pretty simple. Even so, we had to assume idealized, perfectly balanced dice. In most situations, it isn't easy to give a "correct" probability model. We can make progress by listing some facts that must be true for *any* assignment of probabilities. These facts follow from the idea of probability as "the long-run proportion of repetitions on which an event occurs."

1. **Any probability is a number between 0 and 1.** Any proportion is a number between 0 and 1, so any probability is also a number between 0 and 1. An event with probability 0 never occurs, and an event with probability 1 occurs in every trial. An event with probability 0.5 occurs in half the trials in the long run.

2. **All possible outcomes together must have probability 1.** Because some outcome must occur on every trial, the sum of the probabilities for all possible outcomes must be exactly 1.

3. **If two events have no outcomes in common, the probability that one or the other occurs is the sum of their individual probabilities.** If one event occurs in 40% of all trials, a different event occurs in 25% of all trials, and the two can never occur together, then one or the other occurs on 65% of all trials because 40% + 25% = 65%.

4. **The probability that an event does not occur is 1 minus the probability that the event does occur.** If an event occurs in (say) 70% of all trials, it fails to occur in the other 30%. The probability that an event occurs and the probability that it does not occur always add to 100%, or 1.

We can use mathematical notation to state Facts 1 to 4 more concisely. Capital letters near the beginning of the alphabet denote events. If A is any event, we write its probability as $P(A)$. Here are our probability facts in formal language. As you apply these rules, remember that they are just another form of intuitively true facts about long-run proportions.

Probability Rules

Rule 1. The probability $P(A)$ of any event A satisfies $0 \leq P(A) \leq 1$.

Rule 2. If S is the sample space in a probability model, then $P(S) = 1$.

Rule 3. Two events A and B are **disjoint** if they have no outcomes in common and so can never occur together. If A and B are disjoint,

$$P(A \text{ or } B) = P(A) + P(B)$$

This is the **addition rule for disjoint events.**

Rule 4. For any event A,

$$P(A \text{ does not occur}) = 1 - P(A)$$

The addition rule extends to more than two events that are disjoint in the sense that no two have any outcomes in common. If events A, B, and C are disjoint, the probability that one of these events occurs is $P(A) + P(B) + P(C)$.

EQUALLY LIKELY?

A game of bridge begins by dealing all 52 cards in the deck to the four players, 13 to each. If the deck is well shuffled, all the immense number of possible hands will be equally likely. But don't expect the hands that appear in newspaper bridge columns to reflect the equally likely probability model. Writers on bridge choose "interesting" hands, especially those that lead to high bids that are rare in actual play.

EXAMPLE 11.6	**Using the Probability Rules**

We already used the addition rule for disjoint events, without calling it by that name, to find the probabilities in Example 11.5. The event "roll a 5" contains the four disjoint outcomes displayed in Example 11.4, so the addition rule (Rule 3) says that its probability is

$$P(\text{roll a 5}) = P\left(\boxed{\cdot}\ \boxed{\because}\right) + P\left(\boxed{\because}\ \boxed{\cdot}\right) + P\left(\boxed{\because}\ \boxed{\cdot}\right) + P\left(\boxed{\because}\ \boxed{\cdot}\right)$$

$$= \frac{1}{36} + \frac{1}{36} + \frac{1}{36} + \frac{1}{36}$$

$$= \frac{4}{36} = 0.111$$

Check that the probabilities in Example 11.5, found using the addition rule, are all between 0 and 1 and add to exactly 1. That is, this probability model obeys Rules 1 and 2.

What is the probability of rolling anything other than a 5? By Rule 4,

$$P(\text{roll does not give a } 5) = 1 - P(\text{roll a } 5)$$
$$= 1 - 0.111 = 0.889$$

Our model assigns probabilities to individual outcomes. To find the probability of an event, just add the probabilities of the outcomes that make up the event. For example:

$$P(\text{outcome is odd}) = P(3) + P(5) + P(7) + P(9) + P(11)$$
$$= \frac{2}{36} + \frac{4}{36} + \frac{6}{36} + \frac{4}{36} + \frac{2}{36}$$
$$= \frac{18}{36} = \frac{1}{2} \blacksquare$$

© Image Source/Alamy

Apply Your Knowledge

11.8 Who Takes the GMAT? In many settings, the "rules of probability" are just basic facts about percents. The Graduate Management Admission Test (GMAT) Web site provides the following information about the undergraduate majors of those who took the test in 2011–2012: 54% majored in business or commerce; 16% majored in engineering; 16% majored in the social sciences; 6% majored in the sciences; 5% majored in the humanities; and 3% listed some major other than the preceding.[3]

(a) What percent of those who took the test in 2011–2012 majored in either engineering or the sciences? Which rule of probability did you use to find the answer?

(b) What percent of those who took the test in 2011–2012 majored in something other than business or commerce? Which rule of probability did you use to find the answer?

11.9 Overweight? Although the rules of probability are just basic facts about percents or proportions, we need to be able to use the language of events and their probabilities. Choose an American adult at random. Define two events:

A = the person chosen is obese
B = the person chosen is overweight, but not obese

According to the National Center for Health Statistics, $P(A) = 0.34$ and $P(B) = 0.33$.

(a) Explain why events A and B are disjoint.

(b) Say in plain language what the event "A or B" is. What is $P(A \text{ or } B)$?

(c) If C is the event that the person chosen has normal weight or less, what is $P(C)$?

11.10 Languages in Canada. Canada has two official languages, English and French. Choose a Canadian at random and ask, "What is your mother tongue?" Here is the distribution of responses, combining many separate languages from the province of Quebec:[4]

Language	English	French	Italian	Other
Probability	0.08	0.80	0.02	?

(a) What probability should replace "?" in the distribution?

(b) What is the probability that a Canadian's mother tongue is not English?

(c) What is the probability that a Canadian's mother tongue is a language other than English or French?

Finite and discrete probability models

Examples 11.4, 11.5, and 11.6 illustrate one way to assign probabilities to events: assign a probability to every individual outcome, then add these probabilities to find the probability of any event. This idea works well when there are only a finite (fixed and limited) number of outcomes.

Finite Probability Model

A probability model with a finite sample space is called **finite**.

To assign probabilities in a finite model, list the probabilities of all the individual outcomes. These probabilities must be numbers between 0 and 1 that add to exactly 1. The probability of any event is the sum of the probabilities of the outcomes making up the event.

Finite probability models are sometimes called **discrete** probability models. However, discrete probability models include finite sample spaces as well as sample spaces that are infinite and equivalent to the set of all positive integers. An example of a discrete but not finite sample space would be the sample space for the number of free-throw attempts until a basketball player makes her first free throw. This could occur on her first attempt, her second attempt, her third attempt, and so on. Assigning probabilities to individual outcomes in an infinite discrete sample space is more complicated than for a finite sample space. In this book we will often refer to finite probability models as discrete, and in practice statisticians often refer to finite probability models as discrete.

EXAMPLE 11.7   Benford's Law

Faked numbers in tax returns, invoices, or expense account claims often display patterns that aren't present in legitimate records. Some patterns, such as too many round numbers, are obvious and easily avoided by a clever crook. Others are more subtle. It is a striking fact that the first digits of numbers in legitimate records often follow a model known as Benford's law.[5] Call the first digit of a randomly chosen record X for short. Benford's law gives this probability model for X (note that a first digit can't be 0):

First digit X	1	2	3	4	5	6	7	8	9
Probability	0.301	0.176	0.125	0.097	0.079	0.067	0.058	0.051	0.046

Check that the probabilities of the outcomes sum to exactly 1. This is therefore a legitimate finite (or discrete) probability model. With these probabilities, investigators can detect fraud by comparing the first digits in records such as invoices paid by a business.

The probability that a first digit is equal to or greater than 6 is

$$P(X \geq 6) = P(X = 6) + P(X = 7) + P(X = 8) + P(X = 9)$$
$$= 0.067 + 0.058 + 0.051 + 0.046 = 0.222$$

This is less than the probability that a record has first digit 1,

$$P(X = 1) = 0.301$$

Fraudulent records tend to have too few 1s and too many higher first digits.

 Note that the probability that a first digit is greater than or equal to 6 is not the same as the probability that a first digit is strictly greater than 6. The latter probability is

$$P(X > 6) = 0.058 + 0.051 + 0.046 = 0.155$$

The outcome $X = 6$ is included in "greater than or equal to" and is not included in "strictly greater than." ■

Apply Your Knowledge

11.11 Rolling a Die. Figure 11.3 displays several finite probability models for rolling a die. We can learn which model is actually *accurate* for a particular die only by rolling the die many times. However, some of the models are not *legitimate*. That is, they do not obey the rules. Which are legitimate and which are not? In the case of the illegitimate models, explain what is wrong.

			Probability		
Outcome		Model 1	Model 2	Model 3	Model 4
⚀		1/7	1/3	1/3	1
⚁		1/7	1/6	1/6	1
⚂		1/7	1/6	1/6	2
⚃		1/7	0	1/6	1
⚄		1/7	1/6	1/6	1
⚅		1/7	1/6	1/6	2

FIGURE 11.3

Four assignments of probabilities to the six faces of a die, for Exercise 11.11.

11.12 Benford's Law. The first digit of a randomly chosen expense account claim follows Benford's law (Example 11.7). Consider the events

$$A = \{\text{first digit is 4 or greater}\}$$
$$B = \{\text{first digit is even}\}$$

(a) What outcomes make up the event A? What is $P(A)$?

(b) What outcomes make up the event B? What is $P(B)$?

(c) What outcomes make up the event "A or B"? What is $P(A \text{ or } B)$? Why is this probability not equal to $P(A) + P(B)$?

11.13 Weighty Behavior. Choose an adult in the United States at random and ask, "How many days per week do you lift weights?" Call the response X for short. Based on a large sample survey, here is a probability model for the answer you will get:[6]

Days	0	1	2	3	4	5	6	7
Probability	0.73	0.06	0.06	0.06	0.04	0.02	0.01	0.02

(a) Verify that this is a legitimate finite probability model.

(b) Describe the event $X < 4$ in words. What is $P(X < 4)$?

(c) Express the event "lifted weights at least once" in terms of X. What is the probability of this event?

Continuous probability models

When we use the table of random digits to select a digit between 0 and 9, the finite probability model assigns probability 1/10 to each of the 10 possible outcomes. Suppose that we want to choose a number at random between 0 and 1, allowing *any* number between 0 and 1 as the outcome. Software random number generators will do this. For example, here is the result of asking software to produce 5 random numbers between 0 and 1:

0.2893511   0.3213787   0.5816462   0.9787920   0.4475373

The sample space is now an entire interval of numbers:

$$S = \{\text{all numbers between 0 and 1}\}$$

Call the outcome of the random number generator Y for short. How can we assign probabilities to such events as $\{0.3 \le Y \le 0.7\}$? As in the case of selecting a random digit, we would like all possible outcomes to be equally likely. But we cannot assign probabilities to each individual value of Y and then add them, because there is an infinite continuum of possible values. In fact, we cannot even make a list of the individual values of Y. For example, what is the next largest value of Y after 0.3?

We use a new way of assigning probabilities directly to events—as *areas under a density curve*. Any density curve has area exactly 1 underneath it, corresponding to total probability 1. We met density curves as models for data in Chapter 3 (page 71).

Continuous Probability Model

A **continuous probability model** assigns probabilities as areas under a density curve. The area under the curve and above any range of values is the probability of an outcome in that range.

EXAMPLE 11.8   Random Numbers

The random number generator will spread its output uniformly across the entire interval from 0 to 1 as we allow it to generate a long sequence of numbers. Figure 11.4 is a histogram of 10,000 random numbers. They are quite uniform, but not exactly so. The bar heights would all be exactly equal (1000 numbers for each bar) if the 10,000 numbers were exactly uniform. In fact, the counts vary from a low of 978 to a high of 1060.

As in Chapter 3, we have adjusted the histogram scale so that the total area of the bars is exactly 1. Now we can add the density curve that describes the distribution of

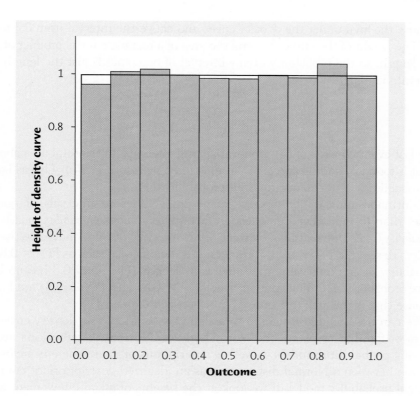

perfectly random numbers. This density curve also appears in Figure 11.4. It has height 1 over the interval from 0 to 1. This is the density curve of a **uniform distribution.** It is the continuous probability model for the results of generating very many random numbers. Like the probability models for perfectly balanced coins and dice, the density curve is an idealized description of the outcomes of a perfectly uniform random number generator. It is a good approximation for software outcomes, but even 10,000 tries isn't enough for actual outcomes to look exactly like the idealized model. ■

uniform distribution

The uniform density curve has height 1 over the interval from 0 to 1. The area under the curve is 1, and the probability of any event is the area under the curve and above the event in question. Figure 11.5 illustrates finding probabilities as areas under the density curve. The probability that the random number generator produces a number between 0.3 and 0.7 is

$$P(0.3 \leq Y \leq 0.7) = 0.4$$

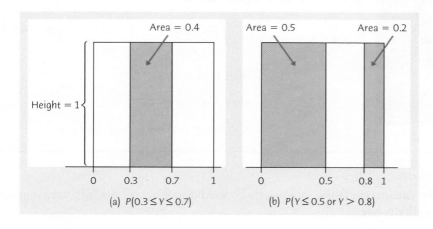

FIGURE 11.5
Probability as area under a density curve. The uniform density curve spreads probability evenly between 0 and 1.

because the area under the density curve and above the interval from 0.3 to 0.7 is 0.4. The height of the curve is 1, and the area of a rectangle is the product of height and length, so the probability of any interval of outcomes is just the length of the interval. Similarly,

$$P(Y \leq 0.5) = 0.5$$
$$P(Y > 0.8) = 0.2$$
$$P(Y \leq 0.5 \text{ or } Y > 0.8) = 0.7$$

The last event consists of two nonoverlapping intervals, so the total area above the event is found by adding two areas, as illustrated by Figure 11.5(b). This assignment of probabilities obeys all our rules for probability.

Continuous probability models assign probabilities to intervals of outcomes rather than to individual outcomes. In fact, *all continuous probability models assign probability 0 to every individual outcome.* Only intervals of values have positive probability. To see that this is true, consider a specific outcome such as $P(Y = 0.8)$. The probability of any interval is the same as its length. The point 0.8 has no length, so its probability is 0. Put another way, $P(Y > 0.8)$ and $P(Y \geq 0.8)$ are both 0.2 because that is the area in Figure 11.5(b) between 0.8 and 1.

We can use any density curve to assign probabilities. The density curves that are most familiar to us are the Normal curves. **Normal distributions are continuous probability models** as well as descriptions of data. There is a close connection between a Normal distribution as an idealized description for data and a Normal probability model. If we look at the heights of all young women, we find that they closely follow the Normal distribution with mean $\mu = 64.3$ inches and standard deviation $\sigma = 2.7$ inches. This is a distribution for a large set of data. Now choose one young woman at random. Call her height X. If we repeat the random choice very many times, the distribution of values of X is the same Normal distribution that describes the heights of all young women.

EXAMPLE 11.9	**The Heights of Young Women**

What is the probability that a randomly chosen young woman has height between 68 and 70 inches? The height X of the woman we choose has the $N(64.2, 2.8)$ distribution. We want $P(68 \leq X \leq 70)$. This is the area under the Normal curve in Figure 11.6. Software or the *Normal Curve* applet will give us the answer at once: $P(68 \leq X \leq 70) = 0.0679$.

We can also find the probability by standardizing and using Table A, the table of standard Normal probabilities. We will reserve capital Z for a standard Normal variable.

$$P(68 \leq X \leq 70) = P\left(\frac{68 - 64.2}{2.8} \leq \frac{X - 64.2}{2.8} \leq \frac{70 - 64.2}{2.8}\right)$$

$$= P(1.36 \leq Z \leq 2.07)$$

$$= P(Z \leq 2.07) - P(Z \leq 1.36)$$

$$= 0.9808 - 0.9131 = 0.0677 \ ■$$

Getty Images/Blend Images

The calculation is the same as those we did in Chapter 3. Only the language of probability is new.

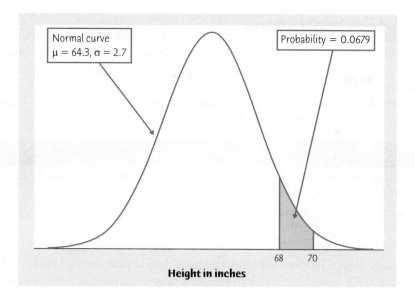

FIGURE 11.6
The probability in Example 11.9 as an area under a Normal curve.

Apply Your Knowledge

11.14 Random Numbers. Let Y be a random number between 0 and 1 produced by the idealized random number generator described in Example 11.8 and Figure 11.4 (pages 270–71). Find the following probabilities:

(a) $P(Y \leq 0.6)$ (b) $P(Y < 0.6)$ (c) $P(0.4 \leq Y \leq 0.8)$

11.15 Adding Random Numbers. Generate two random numbers between 0 and 1 and take X to be their sum. The sum X can take any value between 0 and 2. The density curve of X is the triangle shown in Figure 11.7.

(a) Verify by geometry that the area under this curve is 1.

(b) What is the probability that X is less than 1? (Sketch the density curve, shade the area that represents the probability, then find that area. Do this for (c) also.)

(c) What is the probability that X is less than 0.5?

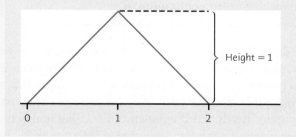

FIGURE 11.7
The density curve for the sum of two random numbers, for Exercise 11.15. This density curve spreads probability between 0 and 2.

11.16 The Medical College Admission Test. The Normal distribution with mean $\mu = 25.0$ to 25.2 and standard deviation $\sigma = 6.4$ is a good description of the total score on the Medical College Admission Test (MCAT). This is a continuous probability model for the score of a randomly chosen student. Call the score of a randomly chosen student X for short.

(a) Write the event "the student chosen has a score of 35 or higher" in terms of X.

(b) Find the probability of this event.

Random variables

Examples 11.7 to 11.9 use a shorthand notation that is often convenient. In Example 11.9, we let X stand for the result of choosing a woman at random and measuring her height. We know that X would take a different value if we made another random choice. Because its value changes from one random choice to another, we call the height X a *random variable*.

Random Variable

A **random variable** is a variable whose value is a numerical outcome of a random phenomenon.

The **probability distribution** of a random variable X tells us what values X can take and how to assign probabilities to those values.

We usually denote random variables by capital letters near the end of the alphabet, such as X or Y. Of course, the random variables of greatest interest to us are outcomes such as the mean $\bar{x}$ of a random sample, for which we will keep the familiar notation. There are two main types of random variables, corresponding to two types of probability models: *discrete* and *continuous*. Notice that neither a finite sample space nor a sample space consisting of all positive integers would be considered continuous, because neither sample space is an interval of all possible numbers (to an arbitrary number of decimal places) between two given values. Thus, we classify random variables as either discrete or continuous, rather than as either finite or continuous.

EXAMPLE 11.10	**Discrete and Continuous Random Variables**
	The first digit X in Example 11.7 is a random variable whose possible values are the whole numbers {1, 2, 3, 4, 5, 6, 7, 8, 9}. The distribution of X assigns a probability to each of these outcomes. Random variables that have a finite list of possible outcomes are called **discrete**.
discrete random variable	
	Compare the output Y of the random number generator in Example 11.8. The values of Y fill the entire interval of numbers between 0 and 1. The probability distribution of Y is given by its density curve, shown in Figure 11.4. Random variables that can take on any value in an interval, with probabilities given as areas under a density curve, are called **continuous**. ▪
continuous random variable	

We have defined a random variable to be a variable whose value is a *numerical* outcome of a random phenomenon. However, there are situations in which the outcomes of a random phenomenon are not numerical. In Example 11.2, the possible outcomes of a coin toss are heads and tails. We can represent these outcomes by numbers, letting 1 represent heads and 0 represent tails. Using numbers to represent the nonnumerical outcomes of a random phenomenon is a common practice in statistics. For mathematical purposes it is convenient to restrict random variables to take on numerical values, even if this means representing nonnumerical outcomes by numbers.

Apply Your Knowledge

11.17 Grades in an Economics Course. Indiana University posts the grade distributions for its courses online.[7] Students in Economics 201 in the fall 2009 semester received 9% As, 8% A−s, 10% B+s, 14% Bs, 13% B−s, 10% C+s, 12% Cs, 4% C−s, 4% D+s, 8% Ds, and 8% Fs. Choose an Economics

201 student at random. To "choose at random" means to give every student the same chance to be chosen. The student's grade on a four-point scale (with A = 4, A− = 3.7, B+ = 3.3, B = 3.0, B− = 2.7, C+ = 2.3, C = 2.0, C− = 1.7, D+ = 1.3, D = 1.0, and F = 0.0) is a discrete random variable X with this probability distribution:

Value of X	0.0	1.0	1.3	1.7	2.0	2.3	2.7	3.0	3.3	3.7	4.0
Probability	0.08	0.08	0.04	0.04	0.12	0.10	0.13	0.14	0.10	0.08	0.09

(a) Say in words what the meaning of $P(X \geq 3.0)$ is. What is this probability?

(b) Write the event "the student got a grade poorer than B−" in terms of values of the random variable X. What is the probability of this event?

11.18 Running a Mile. A study of 12,000 able-bodied male students at the University of Illinois found that their times for the mile run were approximately Normal with mean 7.11 minutes and standard deviation 0.74 minute.[8] Choose a student at random from this group and call his time for the mile Y.

(a) Say in words what the meaning of $P(Y \geq 8)$ is. What is this probability?

(b) Write the event "the student could run a mile in less than 6 minutes" in terms of values of the random variable Y. What is the probability of this event?

Personal probability*

We began our discussion of probability with one idea: the probability of an outcome of a random phenomenon is the proportion of times that outcome would occur in a very long series of repetitions. This idea ties probability to actual outcomes. It allows us, for example, to estimate probabilities by simulating random phenomena. Yet we often meet another, quite different, idea of probability.

 WHAT ARE THE ODDS?

Gamblers often express chance in terms of *odds* rather than probability. Odds of A to B against an outcome means that the probability of that outcome is B/(A + B). So "odds of 5 to 1" is another way of saying "probability 1/6." A probability is always between 0 and 1, but odds range from 0 to infinity. Although odds are mainly used in gambling, they give us a way to make very small probabilities clearer. "Odds of 999 to 1" may be easier to understand than "probability 0.001."

EXAMPLE 11.11   Joe and the Chicago Cubs

Joe sits staring into his beer as his favorite baseball team, the Chicago Cubs, loses another game. The Cubbies have some good young players, so let's ask Joe, "What's the chance that the Cubs will go to the World Series next year?" Joe brightens up. "Oh, about 10%," he says.

Does Joe assign probability 0.10 to the Cubs appearing in the World Series? The outcome of next year's pennant race is certainly unpredictable, but we can't reasonably ask what would happen in many repetitions. Next year's baseball season will happen only once and will differ from all other seasons in players, weather, and many other ways. If probability measures "what would happen if we did this many times," Joe's 0.10 is not a probability. Probability is based on data about many repetitions of the same random phenomenon. Joe is giving us something else, his personal judgment. ■

*This short section is optional.

Although Joe's 0.10 isn't a probability in our usual sense, it gives useful information about Joe's opinion. More seriously, a company asking, "How likely is it that building this plant will pay off within five years?" can't employ an idea of probability based on many repetitions of the same thing. The opinions of company officers and advisers are nonetheless useful information, and these opinions can be expressed in the language of probability. These are *personal probabilities*.

Personal Probability

A **personal probability** of an outcome is a number between 0 and 1 that expresses an individual's judgment of how likely the outcome is.

Rachel's opinion about the Cubs may differ from Joe's, and the opinions of several company officers about the new plant may differ. Personal probabilities are indeed personal: they vary from person to person. Moreover, if two people assign different personal probabilities to an event, it may be difficult or impossible to determine who is more correct. If we say, "In the long run, this coin will come up heads 60% of the time," we can find out if we are right by actually tossing the coin several thousand times. If Joe says, "I think the Cubs have a 10% chance of going to the World Series next year," that's just Joe's opinion. Why think of personal probabilities as probabilities? Because *any set of personal probabilities that makes sense obeys the same basic Rules 1 to 4 that describe any legitimate assignment of probabilities to events.* If Joe thinks there's a 10% chance that the Cubs will go to the World Series, he must also think that there's a 90% chance that they won't go. There is just one set of rules of probability, even though we now have two interpretations of what probability means.

Apply Your Knowledge

11.19 Will You Have an Accident? The probability that a randomly chosen driver will be involved in an accident in the next year is about 0.2. This is based on the proportion of millions of drivers who have accidents. "Accident" includes things like crumpling a fender in your own driveway, not just highway accidents.

(a) What do you think is your own probability of being in an accident in the next year? This is a personal probability.

(b) Give some reasons why your personal probability might be a more accurate prediction of your "true chance" of having an accident than the probability for a random driver.

(c) Almost everyone says their personal probability is lower than the random driver probability. Why do you think this is true?

11.20 Winning the ACC Tournament. The annual Atlantic Coast Conference men's basketball tournament has temporarily taken Joe's mind off the Chicago Cubs. He says to himself, "I think that Florida State has probability 0.1 of winning. North Carolina's probability is twice Florida State's, and Duke's probability is three times Florida State's."

(a) What are Joe's personal probabilities for North Carolina and Duke?

(b) What is Joe's personal probability that one of the 12 teams other than Florida State, North Carolina, and Duke will win the tournament?

CHAPTER 11 SUMMARY

Chapter Specifics

- A **random phenomenon** has outcomes that we cannot predict but that nonetheless have a regular distribution in very many repetitions.

- The **probability** of an event is the proportion of times the event occurs in many repeated trials of a random phenomenon.

- A **probability model** for a random phenomenon consists of a sample space S and an assignment of probabilities P.

- The **sample space S** is the set of all possible outcomes of the random phenomenon. Sets of outcomes are called **events.** P assigns a number $P(A)$ to an event A as its probability.

- Any assignment of probability must obey the rules that state the basic properties of probability:

  1. $0 \leq P(A) \leq 1$ for any event A.
  2. $P(S) = 1$.
  3. **Addition rule for disjoint events:** Events A and B are **disjoint** if they have no outcomes in common. If A and B are disjoint, then $P(A \text{ or } B) = P(A) + P(B)$.
  4. For any event A, $P(A \text{ does not occur}) = 1 - P(A)$.

- When a sample space S contains finitely many possible values, a **finite probability model** assigns each of these values a probability between 0 and 1 such that the sum of all the probabilities is exactly 1. The probability of any event is the sum of the probabilities of all the values that make up the event. Finite probability models are also referred to as discrete probability models.

- A sample space can contain all values in some interval of numbers. A **continuous probability model** assigns probabilities as areas under a **density curve.** The probability of any event is the area under the curve above the values that make up the event.

- A **random variable** is a variable taking numerical values determined by the outcome of a random phenomenon. The **probability distribution** of a random variable X tells us what the possible values of X are and how probabilities are assigned to those values.

- A random variable X and its distribution can be **discrete** or **continuous.** The distribution of a **discrete random variable** with finitely many possible values gives the probability of each value. A **continuous random variable** takes all values in some interval of numbers. A density curve describes the probability distribution of a continuous random variable.

Link It

This chapter begins our study of probability. The important fact is that random phenomena are unpredictable in the short run but have a regular and predictable behavior in the long run. Probability rules and probability models provide the tools for describing and predicting the long-run behavior of random phenomena.

Probability helps us understand why we can trust random samples and randomized comparative experiments, the subjects of Chapters 8 and 9. It is the key to generalizing what we learn from data produced by random samples and randomized comparative experiments to some wider universe or population. How we use probability to do this will be the topic of the remainder of this book.

CHECK YOUR SKILLS

11.21 You read in a book on poker that the probability of being dealt two pairs in a five-card poker hand is 1/20. This means that

(a) if you deal thousands of poker hands, the fraction of them that contain two pairs will be very close to 1/20.
(b) if you deal 20 poker hands, exactly 1 of them will contain two pairs.
(c) if you deal 10,000 poker hands, exactly 500 of them will contain two pairs.

11.22 A basketball player shoots 5 free throws during a game. The sample space for counting the number she makes is

(a) S = any number between 0 and 1.
(b) S = whole numbers 0 to 5.
(c) S = all sequences of 5 hits or misses, like HMMHH.

Here is the probability model for the political affiliation of a randomly chosen adult in the United States.[9] Exercises 11.23 to 11.26 use this information.

Political affiliation	Republican	Independent	Democrat	Other
Probability	0.26	0.40	0.33	?

11.23 This probability model is

(a) continuous.   (b) finite.   (c) equally likely.

11.24 The probability that a randomly chosen American adult's political affiliation is "Other" must be

(a) any number between 0 and 1.
(b) 0.01.   (c) 0.1.

11.25 What is the probability that a randomly chosen American adult is a member of one of the two major political parties (Republicans and Democrats)?

(a) 0.41   (b) 0.40   (c) 0.59

11.26 What is the probability that a randomly chosen American adult is not a Republican?

(a) 0.26   (b) 0.74   (c) 0.01

11.27 In a table of random digits such as Table B, each digit is equally likely to be any of 0, 1, 2, 3, 4, 5, 6, 7, 8, or 9. What is the probability that a digit in the table is a 7?

(a) 1/9   (b) 1/10   (c) 9/10

11.28 In a table of random digits such as Table B, each digit is equally likely to be any of 0, 1, 2, 3, 4, 5, 6, 7, 8, or 9. What is the probability that a digit in the table is 7 or greater?

(a) 7/10   (b) 4/10   (c) 3/10

11.29 Choose an American household at random, and let the random variable X be the number of cars (including SUVs and light trucks) they own. Here is the probability model if we ignore the few households that own more than 6 cars:[10]

Number of cars X	0	1	2	3	4	5	6
Probability	0.09	0.33	0.36	0.14	0.05	0.02	0.01

A housing company builds houses with two-car garages. What percent of households have more cars than the garage can hold?

(a) 14%   (b) 22%   (c) 42%

11.30 Choose a common fruit fly *Drosophila melanogaster* at random. Call the length of the thorax (where the wings and legs attach) Y. The random variable Y has the Normal distribution with mean $\mu = 0.800$ millimeter (mm) and standard deviation $\sigma = 0.078$ mm. The probability $P(Y > 1)$ that the fly you choose has a thorax more than 1 mm long is about

(a) 0.995.   (b) 0.5.   (c) 0.005.

CHAPTER 11 EXERCISES

11.31 **Sample space.** In each of the following situations, describe a sample space S for the random phenomenon.

(a) A basketball player shoots four free throws. You record the sequence of hits and misses.

(b) A basketball player shoots four free throws. You record the number of baskets she makes.

Darrell Walker/HWMS/Icon SMI/Newscom

11.32 **Probability models?** In each of the following situations, state whether or not the given assignment of probabilities to individual outcomes is legitimate, that is, satisfies the rules of probability. Remember, a legitimate model need not be a practically reasonable model. If the assignment of probabilities is not legitimate, give specific reasons for your answer.

(a) Roll a six-sided die, and record the count of spots on the up-face:

$P(1) = 0$ $P(2) = 1/6$ $P(3) = 1/3$
$P(4) = 1/3$ $P(5) = 1/6$ $P(6) = 0$

(b) Deal a card from a shuffled deck:

P(clubs) = 12/52 P(diamonds) = 12/52
P(hearts) = 12/52 P(spades) = 16/52

(c) Choose a college student at random and record sex and enrollment status:

P(female full-time) = 0.56 P(male full-time) = 0.44
P(female part-time) = 0.24 P(male part-time) = 0.17

11.33 Education among young adults. Choose a young adult (aged 25 to 29) at random. The probability is 0.13 that the person chosen did not complete high school, 0.31 that the person has a high school diploma but no further education, and 0.29 that the person has at least a bachelor's degree.

(a) What must be the probability that a randomly chosen young adult has some education beyond high school but does not have a bachelor's degree?
(b) What is the probability that a randomly chosen young adult has at least a high school education?

11.34 Land in Canada. Canada's national statistics agency, Statistics Canada, says that the land area of Canada is 9,094,000 square kilometers. Of this land, 4,176,000 square kilometers are forested. Choose a square kilometer of land in Canada at random.

(a) What is the probability that the area you choose is forested?
(b) What is the probability that it is not forested?

11.35 Foreign-language study. Choose a student in a U.S. public high school at random and ask if he or she is studying a language other than English. Here is the distribution of results:

Language	Spanish	French	German	All others	None
Probability	0.30	0.08	0.02	0.03	0.57

(a) Explain why this is a legitimate probability model.
(b) What is the probability that a randomly chosen student is studying a language other than English?
(c) What is the probability that a randomly chosen student is studying French, German, or Spanish?

11.36 Car colors. Choose a new car or light truck at random and note its color. Here are the probabilities of the most popular colors for vehicles sold globally in 2012:[11]

Color	White	Black	Silver	Gray	Red	Blue	Beige, brown
Probability	0.23	0.21	0.18	0.14	0.08	0.06	0.06

(a) What is the probability that the vehicle you choose has any color other than those listed?
(b) What is the probability that a randomly chosen vehicle is neither white nor silver?

11.37 Drawing cards. You are about to draw a card at random (that is, all choices have the same probability)

from a set of 7 cards. Although you can't see the cards, here they are:

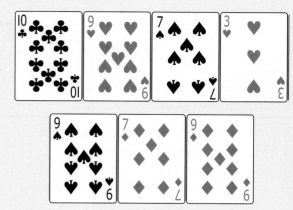

(a) What is the probability that you draw a 9?
(b) What is the probability that you draw a red 9?
(c) What is the probability that you do not draw a 7?

11.38 Loaded dice. There are many ways to produce crooked dice. To *load* a die so that 6 comes up too often and 1 (which is opposite 6) comes up too seldom, add a bit of lead to the filling of the spot on the 1 face. If a die is loaded so that 6 comes up with probability 0.2 and the probabilities of the 2, 3, 4, and 5 faces are not affected, what is the assignment of probabilities to the six faces?

11.39 A door prize. A party host gives a door prize to one guest chosen at random. There are 48 men and 42 women at the party. What is the probability that the prize goes to a woman? Explain how you arrived at your answer.

11.40 Race and ethnicity. The U.S. Census Bureau allows each person to choose from a long list of races. That is, in the eyes of the U.S. Census Bureau, you belong to whatever race you say you belong to. "Hispanic/Latino" is a separate category; Hispanics may be of any race. If we choose a resident of the United States at random, the U.S. Census Bureau gives these probabilities:[12]

	HISPANIC	NOT HISPANIC
Asian	0.002	0.050
Black	0.008	0.126
White	0.151	0.649
Other	0.005	0.009

(a) Verify that this is a legitimate assignment of probabilities.
(b) What is the probability that a randomly chosen American is Hispanic?
(c) Non-Hispanic whites are the historical majority in the United States. What is the probability that a randomly chosen American is not a member of this group?

Choose at random a person aged 15 to 44 years. Ask their age and who they live with (alone, with spouse, with other persons). Here is the probability model for 12 possible answers:[13]

	AGE IN YEARS			
	15–19	**20–24**	**25–34**	**35–44**
Alone	0.001	0.010	0.032	0.028
With spouse	0.001	0.021	0.151	0.208
With others (not a spouse)	0.168	0.139	0.149	0.092

Exercises 11.41 to 11.43 use this probability model.

11.41 Living arrangements.

(a) Why is this a legitimate finite probability model?
(b) What is the probability that the person chosen is a 15- to 19-year-old who lives with others (not a spouse)?
(c) What is the probability that the person is 15 to 19 years old?
(d) What is the probability that the person chosen lives with others (not a spouse)?

11.42 Living arrangements, continued.

(a) List the outcomes that make up the event

$A = $ {The person chosen is *either* 15 to 19 years old *or* lives with others (not a spouse), or both}

(b) What is $P(A)$? Explain carefully why $P(A)$ is not the sum of the probabilities you found in parts (b) and (c) of the previous exercise.

11.43 Living arrangements, continued.

(a) What is the probability that the person chosen is 20 years old or older?
(b) What is the probability that the person chosen does not live alone?

11.44 Spelling errors. Spell-checking software catches "nonword errors" that result in a string of letters that is not a word, as when "the" is typed as "teh." When undergraduates are asked to type a 250-word essay (without spell-checking), the number X of nonword errors has the following distribution:

Value of X	0	1	2	3	4
Probability	0.1	0.2	0.3	0.3	0.1

(a) Is the random variable X discrete or continuous? Why?
(b) Write the event "at least one nonword error" in terms of X. What is the probability of this event?
(c) Describe the event $X \leq 2$ in words. What is its probability? What is the probability that $X < 2$?

11.45 First digits again. A crook who never heard of Benford's law might choose the first digits of his faked invoices so that all of 1, 2, 3, 4, 5, 6, 7, 8, and 9 are equally likely. Call the first digit of a randomly chosen fake invoice W for short.

(a) Write the probability distribution for the random variable W.
(b) Find $P(W \geq 6)$ and compare your result with the Benford's law probability from Example 11.7.

11.46 Who gets interviewed? Abby, Deborah, Mei-Ling, Sam, and Roberto are students in a small seminar course. Their professor decides to choose two of them to interview about the course. To avoid unfairness, the choice will be made by drawing two names from a hat. (This is an SRS of size 2.)

(a) Write down all possible choices of two of the five names. This is the sample space.
(b) The random drawing makes all choices equally likely. What is the probability of each choice?
(c) What is the probability that Mei-Ling is chosen?
(d) Abby, Deborah, and Mei-Ling liked the course. Sam and Roberto did not like the course. What is the probability that both people selected liked the course?

11.47 Birth order. A couple plans to have three children. There are 8 possible arrangements of girls and boys. For example, GGB means the first two children are girls and the third child is a boy. All 8 arrangements are (approximately) equally likely.

© Picture Press/Alamy

(a) Write down all 8 arrangements of the sexes of three children. What is the probability of any one of these arrangements?
(b) Let X be the number of girls the couple has. What is the probability that $X = 2$?
(c) Starting from your work in (a), find the distribution of X. That is, what values can X take, and what are the probabilities for each value?

11.48 Unusual dice. Nonstandard dice can produce interesting distributions of outcomes. You have two balanced, six-sided dice. One is a standard die, with faces having 1, 2, 3, 4, 5, and 6 spots. The other die has three faces with 0 spots and three faces with 6 spots. Find the probability distribution for the total number of spots Y on the up-faces when you roll these two dice. (*Hint:* Start with a picture like Figure 11.2 (page 264) for the possible up-faces. Label the three 0 faces on the second die 0a, 0b, 0c in your picture, and similarly distinguish the three 6 faces.)

11.49 **Random numbers.** Many random number generators allow users to specify the range of the random numbers to be produced. Suppose you specify that the random number Y can take any value between 0 and 2. Then the density curve of the outcomes has constant height between 0 and 2, and height 0 elsewhere.

(a) Is the random variable Y discrete or continuous? Why?

(b) What is the height of the density curve between 0 and 2? Draw a graph of the density curve.

(c) Use your graph from (b) and the fact that probability is area under the curve to find $P(Y \leq 1)$.

11.50 **More random numbers.** Find these probabilities as areas under the density curve you sketched in Exercise 11.49.

(a) $P(0.5 < Y < 1.3)$

(b) $P(Y \geq 0.8)$

11.51 **Survey accuracy.** A sample survey contacted an SRS of 2854 registered voters shortly before the 2012 presidential election and asked respondents whom they planned to vote for. Election results show that 51% of registered voters voted for Barack Obama. We will see later that in this situation the proportion of the sample who planned to vote for Barack Obama (call this proportion V) has approximately the Normal distribution with mean $\mu = 0.51$ and standard deviation $\sigma = 0.009$.

(a) If the respondents answer truthfully, what is $P(0.49 \leq V \leq 0.53)$? This is the probability that the sample proportion V estimates the population proportion 0.51 within plus or minus 0.02.

(b) In fact, 49% of the respondents said they planned to vote for Barack Obama ($V = 0.49$). If respondents answer truthfully, what is $P(V \leq 0.49)$?

11.52 **Friends.** How many close friends do you have? Suppose that the number of close friends adults claim to have varies from person to person with mean $\mu = 9$ and standard deviation $\sigma = 2.5$. An opinion poll asks this question of an SRS of 1100 adults. We will see later in Chapter 19 that in this situation the sample mean response $\bar{x}$ has approximately the Normal distribution with mean 9 and standard deviation 0.075. What is $P(8.9 \leq \bar{x} \leq 9.1)$, the probability that the sample result $\bar{x}$ estimates the population truth $\mu = 9$ to within ± 0.1?

11.53 **Playing Pick 4.** The Pick 4 games in many state lotteries announce a four-digit winning number each day. Each of the 10,000 possible numbers 0000 to 9999 has the same chance of winning. You win if your choice matches the winning digits. Suppose your chosen number is 5974.

(a) What is the probability that the winning number matches your number exactly?

(b) What is the probability that the winning number has the same digits as your number *in any order*?

11.54 **Nickels falling over.** You may feel that it is obvious that the probability of a head in tossing a coin is about 1/2 because the coin has two faces. Such opinions are not always correct. Stand a nickel on edge on a hard, flat surface. Pound the surface with your hand so that the nickel falls over. What is the probability that it falls with heads upward? Make at least 50 trials to estimate the probability of a head.

11.55 **What probability doesn't say.** The idea of probability is that the *proportion* of heads in many tosses of a balanced coin eventually gets close to 0.5. But does the actual *count* of heads get close to one-half the number of tosses? Let's find out. Set the "Probability of heads" in the *Probability* applet to 0.5 and the number of tosses to 40. You can extend the number of tosses by clicking "Toss" again to get 40 more. Don't click "Reset" during this exercise.

(a) After 40 tosses, what is the proportion of heads? What is the count of heads? What is the difference between the count of heads and 20 (one-half the number of tosses)?

(b) Keep going to 120 tosses. Again record the proportion and count of heads and the difference between the count and 60 (half the number of tosses).

(c) Keep going. Stop at 240 tosses and again at 480 tosses to record the same facts. Although it may take a long time, the laws of probability say that the proportion of heads will always get close to 0.5 and also that the difference between the count of heads and half the number of tosses will always grow without limit.

11.56 **LeBron's free throws.** The basketball player LeBron James makes about three-quarters of his free throws over an entire season. Use the *Probability* applet or statistical software to simulate 100 free throws shot by a player who has probability 0.75 of making each shot. (In most software, the key phrase to look for is "Bernoulli trials." This is the technical term for independent trials with Yes/No outcomes. Our outcomes here are "Hit" and "Miss.")

(a) What percent of the 100 shots did he hit?

(b) Examine the sequence of hits and misses. How long was the longest run of shots made? Of shots missed? (Sequences of random outcomes often show runs longer than our intuition thinks likely.)

11.57 **Simulating an opinion poll.** A 2012 opinion poll showed that about 38% of the American public have very little or no confidence in big business. Suppose that this is exactly true. Choosing a person at random then has probability 0.38 of getting one who has very little or no confidence in big business. Use the *Probability* applet or statistical software to simulate choosing many people at random. (In most software, the key phrase to look for is "Bernoulli trials." This is the technical term for independent trials with Yes/No outcomes. Our outcomes here are "Favorable" or not.)

(a) Simulate drawing 50 people, then 100 people, then 400 people. What proportion have very little or no confidence in big business in each case? We expect (but because of chance variation we can't be sure) that the proportion will be closer to 0.38 in longer runs of trials.

(b) Simulate drawing 50 people 10 times and record the percents in each sample who have very little or no confidence in big business. Then simulate drawing 400 people 10 times and again record the 10 percents. Which set of 10 results is less variable? We expect the results of samples of size 400 to be more predictable (less variable) than the results of samples of size 50. That is "long-run regularity" showing itself.

 Exploring the Web

11.58 Super Bowl odds. Oddsmakers often list the odds for certain sporting events on the Web. For example, one can find the current odds of winning the next Super Bowl for each NFL team. We found a list of such odds at `www.vegas.com/gaming/futures/superbowl.html`. When an oddsmaker says the odds are A to B of winning, he or she means that the probability of winning is $B/(A + B)$. For example, when we checked the Web site listed, the odds that the Denver Broncos would win Super Bowl XLVII were 5 to 2. This corresponds to a probability of winning of $2/(5 + 2) = 2/7$.

On the Web, find the current odds, according to an oddsmaker, of winning the Super Bowl for each NFL team. Convert these odds to probabilities. Do these probabilities satisfy Rules 1 and 2 given in this chapter? If they don't, can you think of a reason why?

Overview

The material in this chapter is often omitted from introductory courses, especially when students in the course have limited background in mathematics. Omitting this material will not prevent students from understanding the more important methodological or applied material concerning statistical inference, and doing so may preserve valuable time in an otherwise time-pressed course. However, some concepts in this chapter will enhance understanding of material in this class or future classes at a deeper level. For example, the concept of independence is based on conditional probability—with independence, knowing that something *has* occurred does not change the probability that something else of interest will also occur. Independence (of trials) is also central to the binomial distribution (if you cover Chapter 13). More important, it is the independence of outcomes in a random sample that provides part of the foundation of inference.

Many of the concepts of this chapter can be illustrated with simple examples familiar to students. Examples that make use of a deck of cards, dice, or coins present a visual tool for students; situations where the entire sample space can physically be drawn are often intuitive. Events can be defined, and probabilities can be determined, using the different concepts of the chapter as they are encountered. Another useful tool introduced in the chapter is the Venn diagram; these appear throughout the chapter. In addition to using Venn diagrams to illustrate certain concepts in class, suggest to students that they use them on homework and exams to help with the calculations of probabilities for complex events.

> **LEARNING OUTCOMES****
> - Use Venn diagrams to picture relationships among several events.
> - Use the general addition rule to find probabilities that involve overlapping events.
> - Understand the idea of independence. Judge when it is reasonable to assume independence as part of a probability model.
> - Use the multiplication rule for independent events to find the probability that all of several independent events occur.
> - Use the multiplication rule for independent events in combination with other probability rules to find the probabilities of complex events.
> - Understand the idea of conditional probability. Find conditional probabilities for individuals chosen at random from a table of counts of possible outcomes.
> - Use the general multiplication rule to find $P(A$ and $B)$ from $P(A)$ and the conditional probability $P(B|A)$.
> - Use tree diagrams to organize several-stage probability models.
>
> ****These learning outcomes appear later for the students in Chapter 14: Part III Review.

Teaching Suggestions and Additional Examples/Activities for the Classroom

If this material is covered in class, students will benefit most if they are able to "discover" or construct the probability rules for themselves. Rules 1–3, which are boxed on page 284 of the text, form the three axioms of probability—they are assumed to be true and can, in fact, be used to prove all results in probability.

1. Discuss Probability Rules 1–3

Consider starting class with a brief discussion of what Rules 1–3 mean, and how they are intuitively appealing. Use a simple device, such as a fair six-sided die, to motivate them. For example, Rule 3 can be motivated by asking students to compute the probability that a fair die roll results in a "low" (1, 2) or "high" (5, 6) number.

2. Motivate the General Addition Rule with a Venn Diagram

After Rules 1–3 have been discussed, let students construct rules for themselves. Use a Venn diagram to motivate the general addition rule (page 288); however, care must be taken with nondisjoint events in attaching probabilities to the different regions. For example, if P(female) = 0.55 in the population of your college, some of those students are under-class and some are upperclass; P(female and underclass) + P(female and upperclass) = 0.55. If P(underclass) = 0.58 and P(female and underclass) = 0.30, then P(female or underclass) = P(female) + P(underclass) − P(female and underclass) = 0.55 + 0.58 − 0.30 = 0.83. Too often, students will forget to subtract the probability for the overlap region. If you have a strong class, consider asking them to create a formula for P(A or B or C). In doing so, they'll be discovering the inclusion–exclusion principle, and many will enjoy viewing the challenge as a puzzle.

3. Address Common Misconceptions with the General Addition Rule

Students tend to have problems with the concept of "A or B," assuming that the "or" we use is exclusive (A or B, but not both). The "or" in statistics is an inclusive or (A or B or both). Students also tend to have the problem of assuming independence. For example, students assume that P(A or B) = P(A) + P(B) − P(A)P(B). This is only true if the events A and B are independent. Otherwise, we must be told the probability of the overlap of the two events to properly calculate P(A or B) = P(A) + P(B) − P(A and B).

4. Compare and Contrast Conditional Independence

Conditional probability can be motivated by asking students to compute the probability that a fair die roll is "low," given that it is "even." Of course, these are independent. With a simple example like this, students can construct a formula for the conditional probability of event A, given that event B has occurred, and they can define the concept of independence formally in these terms. The general multiplication rule follows from the definition of conditional probability, whereas the multiplication rule for independent events follows from the definition of independence.

5. Encourage Students to Think Carefully about the Question Being Asked

The concepts in this chapter are not difficult to grasp, but students typically struggle to represent events symbolically. For example, the question, "*What proportion of green cars have an automatic transmission?*" calls for a conditional probability— P(*automatic transmission* | *green*). Developing student ability to express events with probability notation is a worthwhile objective throughout the chapter. This is also a good time to go back to the supplementary material for Chapter 6 and remind your students that, although wording of conditional probability questions can be subtle, "key words" such as "if," "when," and "given" may signal a conditional question.

Other Resources (LaunchPad)

StatClips
   Probability Basic Rules
EESEE Case Studies
   Surviving the Titanic

Getty Images/Lonely Planet Images

General Rules of Probability*

CHAPTER 12

Probability models can describe the flow of traffic through a highway system, a telephone interchange, or a computer processor; the genetic makeup of populations; the energy states of subatomic particles; the spread of epidemics or rumors; and the rate of return on risky investments. Although we are interested in probability mainly because it is the foundation for statistical inference, the mathematics of chance is important in many fields of study. Our introduction to probability in Chapter 11 concentrated on basic ideas and facts. Now we look at some further details. With more probability at our command, we can model more complex random phenomena.

Although we won't emphasize the math, everything in this chapter (and much more) follows from the four rules we met in Chapter 11. Here they are again.

---

*This more advanced chapter introduces some of the mathematics of probability. The material is not needed to read the rest of the book.

Probability Rules

Rule 1. For any event A, $0 \leq P(A) \leq 1$.

Rule 2. If S is the sample space, $P(S) = 1$.

Rule 3. Addition rule for disjoint events: If A and B are **disjoint** events,

$$P(A \text{ or } B) = P(A) + P(B)$$

Rule 4. For any event A,

$$P(A \text{ does not occur}) = 1 - P(A)$$

Independence and the multiplication rule

Rule 3, the addition rule for disjoint events, describes the probability that *one or the other* of two events A and B occurs in the special situation when A and B cannot occur together. Now we will describe the probability that *both* events A and B occur, again only in a special situation.

You may find it helpful to draw a picture to display relations among several events. A picture like Figure 12.1 that shows the sample space S as a rectangular area and events as areas within S is called a **Venn diagram.** The events A and B in Figure 12.1 are disjoint because they do not overlap. The Venn diagram in Figure 12.2 illustrates two events that are not disjoint. The event $\{A \text{ and } B\}$ appears as the overlapping area that is common to both A and B. Can we find the probability $P(A \text{ and } B)$ that both events occur if we know the individual probabilities $P(A)$ and $P(B)$?

Venn diagram

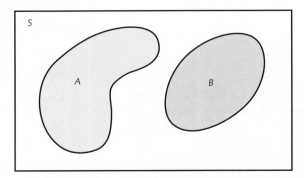

FIGURE 12.1

Venn diagram showing disjoint events A and B.

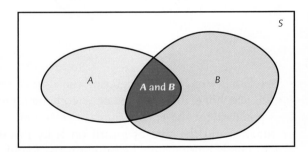

FIGURE 12.2

Venn diagram showing events A and B that are not disjoint. The event $\{A \text{ and } B\}$ consists of outcomes common to A and B.

EXAMPLE 12.1   ## Can You Taste PTC?

That molecule in the diagram is PTC, a substance with an unusual property: 70% of people find that it has a bitter taste, and the other 30% can't taste it at all. The difference is genetic, depending on a single gene. Ask two people chosen at random to taste PTC. We are interested in the events

$$A = \{\text{first person can taste PTC}\}$$
$$B = \{\text{second person can taste PTC}\}$$

We know that $P(A) = 0.7$ and $P(B) = 0.7$. What is the probability $P(A \text{ and } B)$ that both can taste PTC?

We can think our way to the answer. The first person chosen can taste PTC in 70% of all samples and then the second person can taste it in 70% of those samples. We will get two tasters in 70% of 70% of all samples. That's $P(A \text{ and } B) = 0.7 \times 0.7 = 0.49$. ■

The argument in Example 12.1 works because knowing that the first person can taste PTC tells us nothing about the second person. The probability is still 0.7 that the second person can taste PTC whether or not the first person can. We say that the events "first person can taste PTC" and "second person can taste PTC" are **independent.** Now we have another rule of probability.

independent events

Multiplication Rule for Independent Events

Two events A and B are **independent** if knowing that one occurs does not change the probability that the other occurs. If A and B are independent,

$$P(A \text{ and } B) = P(A)P(B)$$

EXAMPLE 12.2   Independent or Not?

To use this multiplication rule, we must decide whether events are independent. Here are some examples to help you recognize when you can assume events are independent.

In Example 12.1, we think that the ability of one randomly chosen person to taste PTC tells us nothing about whether or not a second person, also randomly chosen, can taste PTC. That's independence. But if the two people are members of the same family, the fact that ability to taste PTC is inherited warns us that they are not independent.

Independence is clearly recognized in artificial settings such as games of chance. Because a coin has no memory and most coin tossers cannot influence the fall of the coin, it is safe to assume that successive coin tosses are independent, so that the probability of three heads in succession is $0.5 \times 0.5 \times 0.5 = 0.125$.

On the other hand, the colors of successive cards dealt from the same deck are not independent. A standard 52-card deck contains 26 red and 26 black cards. For the first card dealt from a shuffled deck, the probability of a red card is $26/52 = 0.50$. Once we see that the first card is red, we know that there are only 25 reds among the remaining 51 cards. The probability that the second card is red is therefore only $25/51 = 0.49$. Knowing the outcome of the first deal changes the probabilities for the second. ■

 CONDEMNED BY INDEPENDENCE

Assuming independence when it isn't true can lead to disaster. Several mothers in England were convicted of murder simply because two of their children had died in their cribs with no visible cause. An "expert witness" for the prosecution said that the probability of an unexplained crib death in a nonsmoking middle-class family is 1/8500. He then multiplied 1/8500 by 1/8500 to claim that there is only a 1 in 73 million chance that two children in the same family could have died naturally. This is nonsense: it assumes that crib deaths are independent, and data suggest that they are not. Some common genetic or environmental cause, not murder, probably explains the deaths.

The multiplication rule extends to collections of more than two events, provided that all are independent. Independence of events A, B, and C means that no information about any one or any two can change the probability of the remaining events. Independence is often assumed in setting up a probability model when the events we are describing seem to have no connection.

If two events A and B are independent, the event that A does not occur is also independent of B, and so on. For example, choose two people at random and ask if they can taste PTC. Because 70% can taste PTC and 30% cannot, the probability that the first person is a taster and the second is not is $(0.7)(0.3) = 0.21$.

EXAMPLE 12.3   Surviving?

© Corbis

During World War II, the British found that the probability that a bomber was lost through enemy action on a mission over occupied Europe was 0.05. The probability that the bomber returned safely from a mission was therefore 0.95. It is reasonable to assume that missions were independent. Take A_i to be the event that a bomber survived its ith mission. The probability of surviving 2 missions is

$$P(A_1 \text{ and } A_2) = P(A_1)P(A_2)$$
$$= (0.95)(0.95) = 0.9025$$

The multiplication rule also applies to more than two independent events, so the probability of surviving 3 missions is

$$P(A_1 \text{ and } A_2 \text{ and } A_3) = P(A_1)P(A_2)P(A_3)$$
$$= (0.95)(0.95)(0.95) = 0.8574$$

In 1941, the tour of duty for an airman was established as 30 missions. The probability of surviving 30 missions is only

$$P(A_1 \text{ and } A_2 \text{ and } \dots \text{ and } A_{30}) = P(A_1)P(A_2)\cdots P(A_{30})$$
$$= (0.95)(0.95)\cdots(0.95)$$
$$= (0.95)^{30} = 0.2146$$

The probability of surviving two tours of duty was much smaller. ■

Here is another example of using the multiplication rule for independent events to compute probabilities.

EXAMPLE 12.4   Rapid HIV Testing

STATE: Many people who come to clinics to be tested for HIV, the virus that causes AIDS, don't come back to learn the test results. Clinics now use "rapid HIV tests" that give a result while the client waits. In a clinic in Malawi, for example, use of rapid tests increased the percent of clients who learned their test results from 69% to 99.7%.

The trade-off for fast results is that rapid tests are less accurate than slower laboratory tests. Applied to people who have no HIV antibodies, one rapid test has probability about 0.004 of producing a false positive (that is, of falsely indicating that antibodies are present).[1] If a clinic tests 200 people who are free of HIV antibodies, what is the chance that at least one false positive will occur?

PLAN: It is reasonable to assume that the test results for different individuals are independent. We have 200 independent events, each with probability 0.004. What is the probability that at least one of these events occurs?

SOLVE: "At least one" combines many outcomes. It is much easier to use Rule 4 (page 284), which says that

$$P(\text{at least one positive}) = 1 - P(\text{no positives})$$

and find $P(\text{no positives})$ first.

The probability of a negative result for any one person is $1 - 0.004 = 0.996$. To find the probability that all 200 people tested have negative results, use the multiplication rule:

$$P(\text{no positives}) = P(\text{all 200 negative})$$
$$= (0.996)(0.996)\cdots(0.996)$$
$$= 0.996^{200} = 0.4486$$

The probability we want is therefore

$$P(\text{at least one positive}) = 1 - 0.4486 = 0.5514$$

CONCLUDE: The probability is greater than 1/2 that at least one of the 200 people will test positive for HIV, even though no one has the virus. ■

The multiplication rule $P(A \text{ and } B) = P(A)P(B)$ *holds if A and B are independent but not otherwise. The addition rule* $P(A \text{ or } B) = P(A) + P(B)$ *holds if A and B are disjoint but not otherwise.* Resist the temptation to use these simple rules when the circumstances that justify them are not present. *You must also be careful not to confuse disjointness and independence.* If A and B are disjoint, then the fact that A occurs tells us that B cannot occur—look again at Figure 12.1. So disjoint events are not independent. Unlike disjointness, we cannot picture independence in a Venn diagram because it involves the probabilities of the events rather than just the outcomes that make up the events.

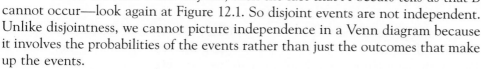

Apply Your Knowledge

12.1 Older College Students. Government data show that 8% of adults are full-time college students and that 30% of adults are age 55 or older. Nonetheless, we can't conclude that because $(0.08)(0.30) = 0.024$, about 2.4% of adults are college students 55 or older. Why not?

12.2 Common Names. The U.S. Census Bureau says that the 10 most common names in the United States are (in order) Smith, Johnson, Williams, Brown, Jones, Miller, Davis, Garcia, Rodriguez, and Wilson. These names account for 9.6% of all U.S. residents. Out of curiosity, you look at the authors of the textbooks for your current courses. There are 9 authors in all. Would you be surprised if none of the names of these authors were among the 10 most common? (Assume that authors' names are independent and follow the same probability distribution as the names of all residents.)

12.3 Lost Internet Sites. Internet sites often vanish or move, so that references to them can't be followed. In fact, 13% of Internet sites referenced in major scientific journals are lost within two years after publication.[2] If a paper contains seven Internet references, what is the probability that all seven are still good two years later? What specific assumptions did you make to calculate this probability?

The general addition rule

We know that if A and B are disjoint events, then $P(A \text{ or } B) = P(A) + P(B)$. If events A and B are *not* disjoint, they can occur together. The probability that one or the other occurs is then *less* than the sum of their probabilities. As Figure 12.3 illustrates, outcomes common to both are counted twice when we add probabilities, so we must subtract this probability once. Here is the addition rule for any two events, disjoint or not.

FIGURE 12.3

The general addition rule: for any events A and B, $P(A \text{ or } B) = P(A) + P(B) - P(A \text{ and } B)$.

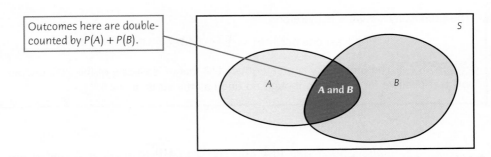

Outcomes here are double-counted by $P(A) + P(B)$.

Addition Rule for Any Two Events

For any two events A and B,

$$P(A \text{ or } B) = P(A) + P(B) - P(A \text{ and } B)$$

If A and B are disjoint, the event $\{A \text{ and } B\}$ that both occur contains no outcomes and therefore has probability 0. Because the general addition rule includes Rule 3, the addition rule for disjoint events, we can always use the general addition rule to find $\{A \text{ or } B\}$.

EXAMPLE 12.5   Motor Vehicle Sales

Motor vehicles sold in the United States (ignoring heavy trucks) are classified as either cars or light trucks and as either domestic or imported. "Light trucks" include SUVs and minivans. "Domestic" means made in Canada, Mexico, or the United States, so that a Toyota made in Canada counts as domestic.

In 2012, 78.3% of the new vehicles sold to individuals were domestic, 49.5% were light trucks, and 42.1% were domestic light trucks.[3] Choose a vehicle sale at random. Then

$$P(\text{domestic or light truck}) = P(\text{domestic}) + P(\text{light truck}) - P(\text{domestic light truck})$$
$$= 0.783 + 0.495 - 0.421 = 0.857$$

That is, 85.7% of vehicles sold were either domestic or light trucks (or both). A vehicle is an imported car if it is *neither* domestic nor a light truck. So

$$P(\text{imported car}) = 1 - 0.857 = 0.143 \quad ■$$

Venn diagrams clarify events and their probabilities because you can just think of adding and subtracting areas. Figure 12.4 shows all the events formed from "domestic"

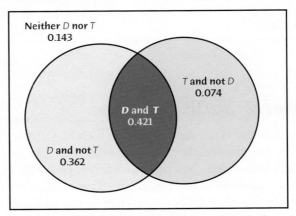

FIGURE 12.4

Venn diagram and probabilities for motor vehicle sales, for Example 12.5.

$D = $ vehicle is domestic $T = $ vehicle is a light truck

and "truck" in Example 12.5. The four probabilities that appear in the figure add to 1 because they refer to four disjoint events that make up the entire sample space. All of these probabilities come from the information in Example 12.5. For example, the probability that a randomly chosen vehicle sale is a domestic car ("D and not T" in the figure) is

$$P(\text{domestic car}) = P(\text{domestic}) - P(\text{domestic light truck})$$
$$= 0.783 - 0.421 = 0.362$$

Apply Your Knowledge

12.4 College Degrees. Of all college degrees awarded in the United States, 50% are bachelor's degrees, 59% are earned by women, and 29% are bachelor's degrees earned by women. Make a Venn diagram and use it to answer these questions.

Barry Austin Photography/
Getty Images

(a)   What percent of all degrees are earned by men?

(b)   What percent of all degrees are bachelor's degrees earned by men?

(c)   Are the events earning a bachelor's degree and being a man independent? Why?

12.5 Distance Learning. A study of the students taking distance learning courses at a university finds that they are mostly older students not living in the university town. Choose a distance learning student at random. Let A be the event that the student is 25 years old or older and B the event that the student is local. The study finds that $P(A) = 0.7$, $P(B) = 0.25$, and $P(A \text{ and } B) = 0.05$.

(a)   Make a Venn diagram similar to Figure 12.4 showing the events {A and B}, {A and not B}, {B and not A}, and {neither A nor B}.

(b)   Describe each of these events in words.

(c)   Find the probabilities of all four events, and add the probabilities to your Venn diagram.

Conditional probability

The probability we assign to an event can change if we know that some other event has occurred. This idea is the key to many applications of probability.

EXAMPLE 12.6   Trucks Among Imported Motor Vehicles

Figure 12.4, based on the information in Example 12.5, gives the following probabilities for a randomly chosen light motor vehicle sold at retail in the United States:

	Domestic	Imported	Total
Light truck	0.421	0.074	0.495
Car	0.362	0.143	0.505
Total	0.783	0.217	1

The four probabilities in the body of the table add to 1 because they describe all vehicles sold. We obtain the "Total" row and column from these probabilities by the addition rule. For example, the probability that a randomly chosen vehicle is a light truck is

$$P(\text{truck}) = P(\text{truck and domestic}) + P(\text{truck and imported})$$
$$= 0.421 + 0.074 = 0.495$$

Suppose we are told that the vehicle chosen is imported. That is, it is one of the 21.7% in the "Imported" column of the table. The probability that a vehicle is a light truck, *given the information that it is imported,* is the proportion of trucks in the "Imported" column,

$$P(\text{truck} \mid \text{imported}) = \frac{0.074}{0.217} = 0.341$$

conditional probability

This is a **conditional probability.** You can read the bar | as "given the information that." ■

Although 49.5% of all vehicles sold are trucks, only 34.1% of imported vehicles are trucks. It's common sense that knowing that one event (the vehicle is imported) occurs often changes the probability of another event (the vehicle is a truck). The example also shows how we should define conditional probability. The idea of a conditional probability $P(B \mid A)$ of one event B given that another event A occurs is the proportion *of all occurrences of* A for which B also occurs.

> ### Conditional Probability
>
> When $P(A) > 0$, the **conditional probability** of B given A is
>
> $$P(B \mid A) = \frac{P(A \text{ and } B)}{P(A)}$$

 The conditional probability $P(B \mid A)$ makes no sense if the event A can never occur, so we require that $P(A) > 0$ whenever we talk about $P(B \mid A)$. *Be sure to keep in mind the distinct roles of the events A and B in $P(B \mid A)$.* Event A represents the information we are given, and B is the event whose probability we are calculating. Here is an example that emphasizes this distinction.

EXAMPLE 12.7   Imports Among Trucks

What is the conditional probability that a randomly chosen vehicle is imported, *given the information that it is a truck?* Using the definition of conditional probability,

$$P(\text{imported} \mid \text{truck}) = \frac{P(\text{imported and truck})}{P(\text{truck})}$$

$$= \frac{0.074}{0.495} = 0.149$$

Only 14.9% of trucks sold are imports. ■

Be careful not to confuse the two different conditional probabilities

$$P(\text{truck} \mid \text{imported}) = 0.341$$
$$P(\text{imported} \mid \text{truck}) = 0.149$$

The first answers the question "What proportion of imports are trucks?" The second answers "What proportion of trucks are imports?"

Apply Your Knowledge

12.6 College Degrees. In the setting of Exercise 12.4, what is the conditional probability that a degree is earned by a woman, given that it is a bachelor's degree?

12.7 Distance Learning. In the setting of Exercise 12.5, what is the conditional probability that a student is local, given that he or she is less than 25 years old?

12.8 Computer Games. Here is the distribution of computer games sold by type of game:[4]

Game Type	Probability
Strategy	0.354
Role playing	0.139
Family entertainment	0.127
Shooters	0.109
Children's	0.057
Other	0.214

What is the conditional probability that a computer game is a role-playing game, given that it is not a strategy game?

The general multiplication rule

The definition of conditional probability reminds us that in principle all probabilities, including conditional probabilities, can be found from the assignment of probabilities to events that describe a random phenomenon. More often, however, conditional probabilities are part of the information given to us in a probability model. The definition of conditional probability then turns into a rule for finding the probability that both of two events occur.

Multiplication Rule for Any Two Events

The probability that both of two events A and B happen together can be found by

$$P(A \text{ and } B) = P(A)P(B \mid A)$$

Here $P(B \mid A)$ is the conditional probability that B occurs, given the information that A occurs.

In words, this rule says that for both of two events to occur, first one must occur and then, given that the first event has occurred, the second must occur. This is just common sense expressed in the language of probability, as the following example illustrates.

WINNING THE LOTTERY TWICE

In 1986, Evelyn Marie Adams won the New Jersey lottery for the second time, adding $1.5 million to her previous $3.9 million jackpot. The *New York Times* claimed that the odds of one person winning the big prize twice were 1 in 17 trillion. Nonsense, said two statisticians in a letter to the *Times*. The chance that Evelyn Marie Adams would win twice is indeed tiny, but it is almost certain that *someone* among the millions of lottery players would win two jackpots. Sure enough, Robert Humphries won his second Pennsylvania lottery jackpot ($6.8 million total) in 1988, and more recently Ernest Pullen of St. Louis won the Missouri lottery in June of 2010 and then won again in September of 2010 ($3 million total). When commenting on the double win, Ernest Pullen said he considers himself to be a "lucky guy."

EXAMPLE 12.8   Teens with Online Profiles

The Pew Internet and American Life Project finds that 93% of teenagers (ages 12 to 17) use the Internet, and that 55% of online teens have posted a profile on a social networking site.[5] What percent of teens are online *and* have posted a profile?

Use the multiplication rule:

$$P(\text{online}) = 0.93$$
$$P(\text{profile} \mid \text{online}) = 0.55$$
$$P(\text{online and have profile}) = P(\text{online}) \times P(\text{profile} \mid \text{online})$$
$$= (0.93)(0.55) = 0.5115$$

That is, about 51% of all teens use the Internet and have a profile on a social networking site.

You should think your way through this: if 93% of teens are online and 55% *of these* have posted a profile, then 55% of 93% are both online and have a profile. ■

We can extend the multiplication rule to find the probability that all of several events occur. The key is to condition each event on the occurrence of *all* of the preceding events. So for any three events A, B, and C,

$$P(A \text{ and } B \text{ and } C) = P(A)P(B \mid A)P(C \mid \text{both } A \text{ and } B)$$

Here is an example of the extended multiplication rule.

EXAMPLE 12.9   Fundraising by Telephone

STATE: A charity raises funds by calling a list of prospective donors to ask for pledges. It is able to talk with 40% of the names on its list. Of those the charity reaches, 30% make a pledge. But only half of those who pledge actually make a contribution. What percent of the donor list contributes?

PLAN: Express the information we are given in terms of events and their probabilities:

If A = {the charity reaches a prospect}	then	$P(A) = 0.4$
If B = {the prospect makes a pledge}	then	$P(B \mid A) = 0.3$
If C = {the prospect makes a contribution}	then	$P(C \mid \text{both } A \text{ and } B) = 0.5$

We want to find $P(A \text{ and } B \text{ and } C)$.

SOLVE: Use the multiplication rule:

$$P(A \text{ and } B \text{ and } C) = P(A)P(B \mid A)P(C \mid \text{both } A \text{ and } B)$$
$$= 0.4 \times 0.3 \times 0.5 = 0.06$$

CONCLUDE: Only 6% of the prospective donors make a contribution. ■

As Example 12.9 illustrates, formulating a problem in the language of probability is often the key to success in applying probability ideas.

Apply Your Knowledge

12.9   At the Gym. Suppose that 10% of adults belong to health clubs, and 40% of these health club members go to the club at least twice a week. What percent of all adults go to a health club at least twice a week? Write the information given in terms of probabilities and use the general multiplication rule.

12.10   Teens Online. We saw in Example 12.8 that 93% of teenagers are online and that 55% of online teens have posted a profile on a social networking site. Of online teens with a profile, 76% have placed comments on a friend's blog. What percent of all teens are online, have a profile, and comment on a friend's blog? Define events and probabilities, and follow the pattern of Example 12.9 (see page 292).

12.11 The Probability of a Flush. A poker player holds a flush when all 5 cards in the hand belong to the same suit (clubs, diamonds, hearts, or spades). We will find the probability of a flush when 5 cards are dealt. Remember that a deck contains 52 cards, 13 of each suit, and that when the deck is well shuffled, each card dealt is equally likely to be any of those that remain in the deck.

(a) Concentrate on spades. What is the probability that the first card dealt is a spade? What is the conditional probability that the second card is a spade, given that the first is a spade? (*Hint:* How many cards remain? How many of these are spades?)

(b) Continue to count the remaining cards to find the conditional probabilities of a spade on the third, the fourth, and the fifth card, given in each case that all previous cards are spades.

(c) The probability of being dealt 5 spades is the product of the 5 probabilities you have found. Why? What is this probability?

(d) The probability of being dealt 5 hearts or 5 diamonds or 5 clubs is the same as the probability of being dealt 5 spades. What is the probability of being dealt a flush?

Independence again

The conditional probability $P(B|A)$ is generally not equal to the unconditional probability $P(B)$. That's because the occurrence of event A generally gives us some additional information about whether or not event B occurs. If knowing that A occurs gives no additional information about B, then A and B are independent events. The precise definition of independence is expressed in terms of conditional probability.

Independent Events

Two events A and B that both have positive probability are **independent** if

$$P(B|A) = P(B)$$

We now see that the multiplication rule for independent events, $P(A \text{ and } B) = P(A)P(B)$, is a special case of the general multiplication rule, $P(A \text{ and } B) = P(A)P(B|A)$, just as the addition rule for disjoint events is a special case of the general addition rule. We rarely use the definition of independence because most often independence is part of the information given to us in a probability model.

Apply Your Knowledge

12.12 Independent? The Clemson University Fact Book and Data Center for 2011 shows that 102 of the university's 256 assistant professors were women, along with 79 of the 258 associate professors and 73 of the 349 full professors.

(a) What is the probability that a randomly chosen Clemson professor is a woman?

(b) What is the conditional probability that a randomly chosen professor is a woman, given that the person chosen is a full professor?

(c) Are the rank and sex of Clemson professors independent? How do you know?

Tree diagrams

Probability models often have several stages, with probabilities at each stage conditional on the outcomes of earlier stages. These models require us to combine several of the basic rules into a more elaborate calculation. Here is an example.

EXAMPLE 12.10   Who Visits YouTube?

STATE: Video sharing sites, led by YouTube, are popular destinations on the Internet. Let's look only at adult Internet users, aged 18 and over. About 27% of adult Internet users are 18 to 29 years old, another 45% are 30 to 49 years old, and the remaining 28% are 50 and over. The Pew Internet and American Life Project finds that 70% of Internet users aged 18 to 29 have visited a video sharing site, along with 51% of those aged 30 to 49 and 26% of those 50 or older. What percent of all adult Internet users visit video sharing sites?

PLAN: To use the tools of probability, restate all these percents as probabilities. If we choose an online adult at random,

$$P(\text{aged 18 to 29}) = 0.27$$
$$P(\text{aged 30 to 49}) = 0.45$$
$$P(\text{aged 50 and older}) = 0.28$$

These three probabilities add to 1 because all adult Internet users are in one of the three age groups. The percents of each group who visit video sharing sites are *conditional* probabilities:

$$P(\text{video yes} \mid \text{aged 18 to 29}) = 0.70$$
$$P(\text{video yes} \mid \text{aged 30 to 49}) = 0.51$$
$$P(\text{video yes} \mid \text{aged 50 and older}) = 0.26$$

We want to find the unconditional probability $P(\text{video yes})$.

tree diagram

SOLVE: The **tree diagram** in Figure 12.5 organizes this information. Each segment in the tree is one stage of the problem. Each complete branch shows a path through the two stages. The probability written on each segment is the conditional probability of an Internet user following that segment, given that he or she has reached the node from which it branches.

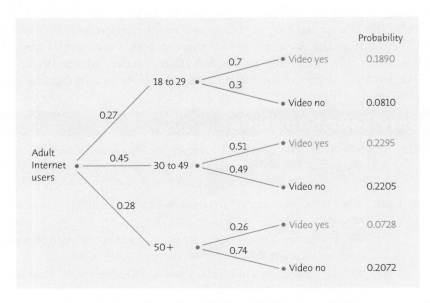

FIGURE 12.5

Tree diagram for use of the Internet and video sharing sites such as YouTube, for Example 12.10. The three disjoint paths to the outcome that an adult Internet user visits video sharing sites are colored red.

Starting at the left, an Internet user falls into one of the three age groups. The probabilities of these groups mark the leftmost segments in the tree. Look at age 18 to 29, the top branch. The two segments going out from the "18 to 29" branch point carry the conditional probabilities

$$P(\text{video yes} \mid \text{aged 18 to 29}) = 0.70$$
$$P(\text{video no} \mid \text{aged 18 to 29}) = 0.30$$

The full tree shows the probabilities for all three age groups.

Now use the multiplication rule. The probability that a randomly chosen Internet user is an 18- to 29-year-old who visits video sharing sites is

$$P(\text{aged 18 to 29 and video yes}) = P(\text{aged 18 to 29})\,P(\text{video yes} \mid \text{aged 18 to 29})$$
$$= (0.27)(0.70) = 0.1890$$

This probability appears at the end of the topmost branch. The multiplication rule says that the probability of any complete branch in the tree is the product of the probabilities of the segments in that branch.

There are three disjoint paths to "video yes," one for each of the three age groups. These paths are colored red in Figure 12.5. Because the three paths are disjoint, the probability that an adult Internet user visits video sharing sites is the sum of their probabilities:

$$P(\text{video yes}) = (0.27)(0.70) + (0.45)(0.51) + (0.28)(0.26)$$
$$= 0.1890 + 0.2295 + 0.0728 = 0.4913$$

CONCLUDE: About 49% of all adult Internet users have visited a video sharing site. ■

It takes longer to explain a tree diagram than it does to use it. Once you have understood a problem well enough to draw the tree, the rest is easy. Here is another question about video sharing sites that the tree diagram helps us answer.

EXAMPLE 12.11   Young Adults at Video Sharing Sites

STATE: What percent of adult Internet users who visit video sharing sites are aged 18 to 29?

PLAN: In probability language, we want the conditional probability $P(\text{aged 18 to 29} \mid \text{video yes})$. Use the tree diagram and the definition of conditional probability:

$$P(\text{aged 18 to 29} \mid \text{video yes}) = \frac{P(\text{aged 18 to 29 and video yes})}{P(\text{video yes})}$$

SOLVE: Look again at the tree diagram in Figure 12.5. $P(\text{video yes})$ is the sum of the three red probabilities, as in Example 12.10. $P(\text{aged 18 to 29 and video yes})$ is the result of following just the top branch in the tree diagram. So

$$P(\text{aged 18 to 29} \mid \text{video yes}) = \frac{P(\text{aged 18 to 29 and video yes})}{P(\text{video yes})}$$
$$= \frac{0.1890}{0.4913} = 0.3847$$

CONCLUDE: About 38% of adults who visit video sharing sites are between 18 and 29 years old. Compare this conditional probability with the original information (unconditional) that 27% of adult Internet users are between 18 and 29 years old. Knowing that a person visits video sharing sites increases the probability that he or she is young. ■

Examples 12.10 and 12.11 illustrate a common setting for tree diagrams. Some outcome (such as visiting video sharing sites) has several sources (such as the three age groups). Starting from

■ the probability of each source, and

■ the conditional probability of the outcome given each source

the tree diagram leads to the overall probability of the outcome. Example 12.10 does this. You can then use the probability of the outcome and the definition of conditional probability to find the conditional probability of one of the sources, given that the outcome occurred. Example 12.11 shows how.

Apply Your Knowledge

12.13 PSA Screening for Prostate Cancer. The prostate-specific antigen (PSA) test is a simple blood test to screen for prostate cancer. It has been used in men over 50 as a routine part of a physical exam with levels above 4 ng/ml indicating possible prostate cancer. The test result is not always correct, sometimes indicating prostate cancer when it is not present and often missing prostate cancer that is present. Here are the approximate probabilities of a positive (above 4 ng/ml) and negative test result when cancer is present or absent:[6]

	Test Result	
	Positive	Negative
Cancer present	0.21	0.79
Cancer absent	0.06	0.94

In a large study of prostate cancer screening, it was found that about 6.3% of the population has prostate cancer.

(a) Draw a tree diagram for selecting a person from this population (outcomes: cancer present or absent) and testing his blood (outcomes: test positive or negative).

(b) What is the probability that the test is positive for a randomly chosen person from this population?

12.14 Eye Color, Hair Color, and Freckles. A large study of children of Caucasian descent in Germany looked at the effect of eye color, hair color, and freckles on the reported extent of burning from sun exposure.[7] The population's distribution of hair color, eye color, and freckles was as shown in the tree diagram of Figure 12.6 (page 297). Find the following probabilities and describe them in plain English:

(a) P(blue eyes | red hair) and P(blue eyes and red hair).

(b) P(freckles | red hair and blue eyes) and P(freckles and red hair and blue eyes).

12.15 False PSA Positives. Continue your work from Exercise 12.13. The probabilities given in the table of Exercise 12.13 are properties of the PSA test. The false-positive rate is a property of the PSA test as applied to a population.

(a) Suppose that 6.3% of the population has prostate cancer. What is the probability that a person does not have cancer, given that the PSA test is positive? This is the false-positive rate.

(b) Treatment for prostate cancer can have serious side effects including incontinence and impotence, and many men diagnosed with prostate cancer would eventually die of another cause if the cancer were left untreated. In fact, it has been found in some large clinical trials that screening did not reduce overall mortality.[8] In October 2011, the U.S. Preventive Services Task Force (USPSTF) released a draft report in which they recommended against using the PSA test to screen for prostate cancer in the general population. Based on the facts given about prostate cancer and the calculations that you have done, explain simply the reasoning behind the USPSTF recommendation.

FIGURE 12.6

Tree diagram of the features of children of Caucasian descent in Germany, for Exercise 12.14. The three stages are hair color, eye color, and whether or not the child has freckles.

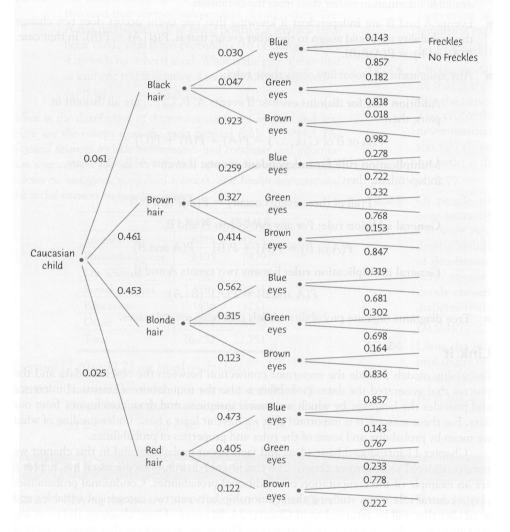

POLITICALLY CORRECT

In 1950, the Soviet mathematician B. V. Gnedenko (1912–1995) wrote *The Theory of Probability*, a text that was popular around the world. The introduction contains a mystifying paragraph that begins, "We note that the entire development of probability theory shows evidence of how its concepts and ideas were crystallized in a severe struggle between materialistic and idealistic conceptions." It turns out that "materialistic" is jargon for "Marxist-Leninist." It was good for the health of Soviet scientists in the Stalin era to add such statements to their books.

12.16 Eye Color, Hair Color, and Freckles, Continued. Continue your work from Exercise 12.14 using the tree diagram of Figure 12.6. Find the following probabilities and describe them in plain English:

(a) P(red hair), P(blue eyes), and P(freckles).

(b) P(freckles and red hair) and P(freckles | red hair).

(c) What can you say about the events freckles and red hair in this population?

slot machines were like this: you pull the lever to spin three wheels; each wheel has 20 symbols, all equally likely to show when the wheel stops spinning; the three wheels are independent of each other. Suppose that the middle wheel has 9 cherries among its 20 symbols, and the left and right wheels have 1 cherry each.

(a) You win the jackpot if all three wheels show cherries. What is the probability of winning the jackpot?

(b) There are three ways that the three wheels can show two cherries and one symbol other than a cherry. Find the probability of each of these ways.

(c) What is the probability that the wheels stop with exactly two cherries showing among them?

12.30 A whale of a time. Hacksaw's Boats of St. Lucia takes tourists on a daily dolphin/whale watch cruise. Their brochure claims an 80% chance of sighting

© Mark Conlin/Alamy

a whale or a dolphin, and you can assume that sightings from day to day are independent.

(a) If you take the dolphin/whale watch cruise on two consecutive days, what is the probability that you see a whale or a dolphin on both days?

(b) If you take the dolphin/whale watch cruise on two consecutive days, what is the probability you will see a dolphin or a whale on at least one day? (Hint: First compute the probability that this event does not occur.)

(c) If you want to have a 99% probability of seeing a whale or a dolphin at least once, what is the minimum number of days that you will need to take the cruise?

12.31 Tendon surgery. You have torn a tendon and are facing surgery to repair it. The surgeon explains the risks to you: infection occurs in 3% of such operations, the repair fails in 14%, and both infection and failure occur together in 1%. What percent of these operations succeed and are free from infection? Follow the four-step process in your answer.

12.32 A whale of a time, continued. Hacksaw's Boats of St. Lucia takes tourists on a daily dolphin/whale watch cruise. Their brochure claims an 80% chance of sighting a whale or a dolphin. Suppose there is a 75% chance of seeing dolphins and a 15% chance of seeing both dolphins and whales. Make a Venn diagram. Then answer these questions.

(a) What is the probability of seeing a whale on the cruise?

(b) What is the probability of seeing a whale but not a dolphin?

(c) Are seeing a whale and seeing a dolphin independent events?

12.33 Tendon surgery, continued. You have torn a tendon and are facing surgery to repair it. The surgeon explains the risks to you: infection occurs in 3% of such operations, the repair fails in 14%, and both infection and failure occur together in 1%. What is the probability of infection given that the repair is successful? Follow the four-step process in your answer.

12.34 Screening job applicants. A company retains a psychologist to assess whether job applicants are suited for assembly-line work. The psychologist classifies applicants as one of A (well suited), B (marginal), or C (not suited). The company is concerned about the event D that an employee leaves the company within a year of being hired. Data on all people hired in the past five years give these probabilities:

$$P(A) = 0.4 \qquad P(B) = 0.3 \qquad P(C) = 0.3$$
$$P(A \text{ and } D) = 0.1 \quad P(B \text{ and } D) = 0.1 \quad P(C \text{ and } D) = 0.2$$

Sketch a Venn diagram of the events A, B, C, and D and mark on your diagram the probabilities of all combinations of psychological assessment and leaving (or not) within a year. What is $P(D)$, the probability that an employee leaves within a year?

12.35 Cancer-detecting dogs. Research has shown that specific biochemical markers are found exclusively in the breath of patients with lung cancer. However, no lab test can currently distinguish the breath of lung cancer patients from that of other subjects. Could dogs be trained to identify these markers in specimens of human breath, as they can be to detect illegal substances or to follow a person's scent? An experiment trained dogs to distinguish breath specimens of lung cancer patients from breath specimens of control individuals by using a food-reward training method. After the training was complete, the dogs were tested on new breath specimens without any reward or clue using a double-blind, completely randomized design. Here are the results for a random sample of 1286 breath specimen:[10]

| | BREATH SPECIMEN FROM A | | |
DOG TEST RESULT	CONTROL SUBJECT	CANCER SUBJECT	TOTAL
Negative	708	10	718
Positive	4	564	568
Total	712	574	1286

(a) The sensitivity of a diagnostic test is its ability to correctly give a positive result when a person tested has the disease, or P(positive test | disease). Find the sensitivity of the dog cancer-detection test for lung cancer.

(b) The specificity of a diagnostic test is the conditional probability that the subject tested doesn't have the disease, given that the test has come up negative. Find the specificity of the dog cancer-detection test for lung cancer.

12.36 **Income tax returns.** Here is the distribution of the adjusted gross income (in thousands of dollars) reported on individual federal income tax returns in 2010:[11]

Income	<15	15–29	30–74	75–199	≥200
Probability	0.263	0.216	0.310	0.181	0.030

(a) What is the probability that a randomly chosen return shows an adjusted gross income of $30,000 or more?

(b) Given that a return shows an income of at least $30,000, what is the conditional probability that the income is at least $75,000?

12.37 **Thomas's pizza.** You work at Thomas's pizza shop. You have the following information about the 7 pizzas in the oven: 3 of the 7 have thick crust, and of these one has only sausage and 2 have only mushrooms; the remaining 4 pizzas have regular crust, and of these 2 have only sausage and 2 have only mushrooms. Choose a pizza at random from the oven.

(a) Are the events {getting a thick crust pizza} and {getting a pizza with mushrooms} independent? Explain.

(b) You add an eighth pizza to the oven. This pizza has thick crust with only cheese. Now are the events {getting a thick crust pizza} and {getting a pizza with mushrooms} independent? Explain.

12.38 **A probability teaser.** Suppose (as is roughly correct) that each child born is equally likely to be a boy or a girl and that the sexes of successive children are independent. If we let BG mean that the older child is a boy and the younger child is a girl, then each of the combinations BB, BG, GB, GG has probability 0.25. Ashley and Brianna each have two children.

(a) You know that at least one of Ashley's children is a boy. What is the conditional probability that she has two boys?

(b) You know that Brianna's older child is a boy. What is the conditional probability that she has two boys?

12.39 **College degrees.** A striking trend in higher education is that more women than men reach each level of attainment. The National Center for Educational Statistics provides projections for the number of degrees earned, classified by level and by the sex of the degree recipient. Here are the projected number of earned degrees (in thousands) in the United States for the 2015–2016 academic year:[12]

(a) If you choose a degree recipient at random, what is the probability that the person you choose is a man?

(b) What is the conditional probability that you choose a man, given that the person chosen received a master's?

(c) Are the events "choose a man" and "choose a master's degree recipient" independent? How do you know?

12.40 **College degrees.** Exercise 12.39 gives the projected counts (in thousands) of earned degrees in the United States in the 2015–2016 academic year. Use these data to answer the following questions.

(a) What is the probability that a randomly chosen degree recipient is a woman?

(b) What is the conditional probability that the person chosen received an associate's degree, given that she is a woman?

(c) Use the general multiplication rule to find the probability of choosing a female associate's degree recipient. Check your result by finding this probability directly from the table of counts.

12.41 **Deer and pine seedlings.** As suburban gardeners know, deer will eat almost anything green. In a study of pine seedlings at an environmental center in Ohio, researchers noted how deer damage varied with how much of the seedling was covered by thorny undergrowth:[13]

	DEER DAMAGE	
THORNY COVER	YES	NO
None	60	151
<1/3	76	158
1/3 to 2/3	44	177
>2/3	29	176

Peter Skinner/Science Source

	ASSOCIATE'S	BACHELOR'S	MASTER'S	PROFESSIONAL	DOCTORATE	TOTAL
Female	556	1034	450	54	45	2139
Male	311	737	282	53	38	1421
Total	867	1771	732	107	83	3560

(a) What is the probability that a randomly selected seedling was damaged by deer?

(b) What are the conditional probabilities that a randomly selected seedling was damaged, given each level of cover?

(c) Does knowing about the amount of thorny cover on a seedling change the probability of deer damage? If so, cover and damage are not independent.

12.42   Deer and pine seedlings. In the setting of Exercise 12.41, what percent of the trees that were not damaged by deer were more than two-thirds covered by thorny plants?

12.43   Deer and pine seedlings. In the setting of Exercise 12.41, what percent of the trees that were damaged by deer were less than one-third covered by thorny plants?

Julie is graduating from college. She has studied biology, chemistry, and computing and hopes to use her science background in crime investigation. Late one night she thinks about some jobs for which she has applied. Let A, B, and C be the events that Julie is offered a job by

A = the Connecticut Office of the Chief Medical Examiner

B = the New Jersey Division of Criminal Justice

C = the Federal Disaster Mortuary Operations Response Team

Julie writes down her personal probabilities for being offered these jobs:

$P(A) = 0.5$ $P(B) = 0.4$ $P(C) = 0.2$

$P(A \text{ and } B) = 0.1$ $P(A \text{ and } C) = 0.05$ $P(B \text{ and } C) = 0.05$

$P(A \text{ and } B \text{ and } C) = 0$

Make a Venn diagram of the events A, B, and C. As in Figure 12.4 (see page 288), mark the probabilities of every intersection involving these events. Use this diagram for Exercises 12.44 to 12.46.

12.44   Will Julie get a job offer? What is the probability that Julie is not offered any of the three jobs?

12.45   Will Julie get just these offers? What is the probability that Julie is offered the Connecticut job, but not the New Jersey or federal job?

12.46   Julie's conditional probabilities. If Julie is offered the federal job, what is the conditional probability that she is also offered the New Jersey job? If Julie is offered the New Jersey job, what is the conditional probability that she is also offered the federal job?

12.47   The geometric distributions. You are rolling a pair of balanced dice in a board game. Rolls are independent. You land in a danger zone that requires you to roll doubles (both faces show the same number of spots) before you are allowed to play again. How long will you wait to play again?

(a) What is the probability of rolling doubles on a single toss of the dice? (If you need review, the possible outcomes appear in Figure 11.2 (page 264). All 36 outcomes are equally likely.)

(b) What is the probability that you do not roll doubles on the first toss, but you do on the second toss?

(c) What is the probability that the first two tosses are not doubles and the third toss is doubles? This is the probability that the first doubles occurs on the third toss.

(d) Now you see the pattern. What is the probability that the first doubles occurs on the fourth toss? On the fifth toss? Give the general result: what is the probability that the first doubles occurs on the kth toss?

(e) What is the probability that you get to go again within 3 turns?

(*Comment:* The distribution of the number of trials to the first success is called a *geometric distribution*. In this problem you have found geometric distribution probabilities when the probability of a success on each trial is 1/6. The same idea works for any probability of success.)

12.48   Winning at tennis. A player serving in tennis has two chances to get a serve into play. If the first serve is out, the player serves again. If the second serve is also out, the player loses the point. Here are probabilities based on four years of the Wimbledon Championship:[14]

$$P(\text{1st serve in}) = 0.59$$
$$P(\text{win point} \mid \text{1st serve in}) = 0.73$$
$$P(\text{2nd serve in} \mid \text{1st serve out}) = 0.86$$
$$P(\text{win point} \mid \text{1st serve out and 2nd serve in}) = 0.59$$

Make a tree diagram for the results of the two serves and the outcome (win or lose) of the point. (The branches in your tree have different numbers of stages depending on the outcome of the first serve.) What is the probability that the serving player wins the point?

12.49   Peanut allergies among children. About 2% of children in the United States are allergic to peanuts.[15] Choose 3 children at random and let the random variable X be the number in this sample who are allergic to peanuts. The possible values X can take are 0, 1, 2, and 3. Make a three-stage tree diagram of the outcomes (allergic or not allergic) for the 3 individuals and use it to find the probability distribution of X.

12.50   Winning at tennis, continued. Based on your work in Exercise 12.48, in what percent of points won by the server was the first serve in? (Write this as a conditional probability and use the definition of conditional probability.)

12.51   Peanut allergies among children, continued. Continue your work from Exercise 12.49. What is the conditional probability that exactly 2 of the children will be allergic to peanuts, given that at least 1 of the 3 children suffers from this allergy?

12.52 **Lactose intolerance.** Lactose intolerance causes difficulty digesting dairy products that contain lactose (milk sugar). It is particularly common among people of African and Asian ancestry. In the United States (ignoring other groups and people who consider themselves to belong to more than one race), 82% of the population is white, 14% is black, and 4% is Asian. Moreover, 15% of whites, 70% of blacks, and 90% of Asians are lactose intolerant.[16]

(a) What percent of the entire population is lactose intolerant?

(b) What percent of people who are lactose intolerant are Asian?

12.53 **Fundraising by telephone.** Tree diagrams can organize problems having more than two stages. Figure 12.7 shows probabilities for a charity calling potential donors by telephone.[17] Each person called is either a recent donor, a past donor, or a new prospect. At the next stage, the person called either does or does not pledge to contribute, with conditional probabilities that depend on the donor class the person belongs to. Finally, those who make a pledge either do or don't actually make a contribution.

(a) What percent of calls result in a contribution?

(b) What percent of those who contribute are recent donors?

DNA Forensics. *When a suspect's DNA is compared to a sample of DNA collected at a crime scene, the comparison is made between certain sections of the DNA called loci. Each locus has two alleles (gene forms), one inherited from our mother and the other from our father. Suppose there are two alleles for a particular locus called A and B. These alleles can be present at the locus in three combinations. A person could have both alleles at the locus be A, one allele be A and the other B, or both alleles be B, giving the three combinations (A and A), (A and B), and (B and B). Here's*

how the math works. If the proportion of the population with allele A as one of their alleles at the locus is a, and the proportion of the population with allele B as one of their alleles at the locus is b, then the proportion of the population with the three combinations of these allele types at the locus follows:

ALLELES AT THE LOCUS	POPULATION PROPORTION WITH ALLELE COMBINATION
A and A	a^2
A and B	$2ab$
B and B	b^2

Use this information in Exercises 12.54 and 12.55. The numbers used in the exercises are from the FBI database.[18]

12.54 Suppose the locus D21S11 has two alleles called 29 and 31. The proportion of the Caucasian population with allele 29 is 0.181 and with allele 31 is 0.071. What proportion of the Caucasian population has the combination (29, 31) at the locus D21S11? What proportion has the combination (29, 29)?

12.55 Suppose the locus D3S1358 has two alleles called 16 and 17. The proportion of the Caucasian population with allele 16 is 0.232 and with allele 17 is 0.212. What proportion of the Caucasian population has the combination (16, 17) at the locus D3S1358?

One important fact regarding the loci evaluated in such forensic tests is that the allele combinations at each locus have been shown to be independent. Use this information in Exercise 12.56.

12.56 What proportion of the Caucasian population has the combination (29, 31) at the loci D21S11 and combination (16, 17) at the loci D3S1358? As we specify the alleles present at more loci, what will happen to the proportion of the Caucasian population that matches the allele combinations at all the loci?

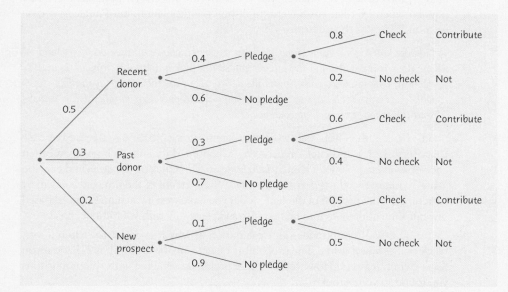

FIGURE 12.7

Tree diagram for fundraising by telephone, for Exercise 12.53. The three stages are the type of prospect called, whether or not the person makes a pledge, and whether or not a person who pledges actually makes a contribution.

A defendant in Ohio was indicted on December 17, 2009, on the charges of Aggravated Burglary and Assault (State of Ohio vs. Myers, Case No. 09 CR 666). In this case, a hair found at the crime scene was tested at six loci and demonstrated a specific combination of alleles found in a proportion of about 1 in 1.6 million individuals in the population. Comparison of the DNA profile found on the hair to a database of convicted felons revealed a match between the allelic profile found on the hair and an individual in the database (the defendant). Defense attorneys in the case requested that the state perform additional DNA testing because several previously untested loci were available to test. The results of this testing revealed that the defendant did not match at some of these newly tested markers, indicating that the DNA from the hair was not the defendant's and the charges were dropped. Use this information in Exercise 12.57.

12.57 If the DNA profile (or combination of alleles) found on the hair is possessed by 1 in 1.6 million individuals, and the database of convicted felons contains 4.5 million individuals, approximately how many individuals in the database would demonstrate a match between their DNA and that found on the hair?

 Exploring the Web

12.58 SAT Mathematics scores. The Web site `http://research.college-board.org/programs/sat/data` presents data for high school seniors who participated in the SAT Program from both the current year, as well as previous years. Under *Data*, click on the link for *College Bound Seniors* for the most recent year given. In the window that opens, click on the link for *Total Group Report* for this year. The Total Group Profile Report presents data for high school graduates who participated in the SAT program that year. Students are counted only once, no matter how often they are tested, and only their latest scores are summarized. Suppose a high school graduate is selected at random. Use the information in the tables available to answer the following questions.

(a) What is the probability that the selected student is female?

(b) What is the probability that the selected student scores 600 or over on the Mathematics section of the SAT?

(c) What is the conditional probability that the selected student scores 600 or over on the Mathematics section given that the student is male?

(d) What is the conditional probability that the selected student scores 600 or over on the Mathematics section given that the student is female?

(e) Are scoring over 600 on the Mathematics section and sex independent? If not, explain in a simple sentence the nature of the dependence.

12.59 Let's make a deal. The Monty Hall Problem is an example of a simple probability problem with an answer that is counterintuitive. The problem was made popular when Marilyn vos Savant published the problem in her *Parade Magazine* column.[19] Here is the question:

Suppose you're on a game show, and you're given a choice of three doors: Behind one door is a car; behind the others, goats. You pick a door, say number 1, and the host, who knows what's behind the doors, opens another door, say number. 3, which has a goat. He says to you, "Do you want to pick door number 2?" Is it to your advantage to switch your choice of doors?

(a) Go to the Web site `www.nytimes.com/2008/04/08/science/08monty.html` and click on *Let's play*. Choose a door and click on it, and then click on *Continue*. Now click on *Switch*. Did you win? Click on *Try again* and play the game 30 times, switching doors each time. What percent of the time did you win by switching? What percent of the time would you have won if you didn't switch? Based on your simulation does it seem better to switch or stay with your initial choice?

(b) The Web site provides an explanation of why it is better to switch doors. Click on *See how it works*. The probability of winning if you switch is 2/3 and if you don't switch it is 1/3. How well do these probabilities agree with the proportions you found in your simulation?

Binomial Distributions CHAPTER 13

Overview

In practice, we're often interested in the number of *successes* in a fixed number of independent trials with constant probability of success p. For example, we may be interested in the chance that no more than 4 of 25 randomly selected parts are defective. The probability distribution describing this count is a binomial distribution.

Emphasize to your students that there are four components to the binomial setting:

1. There are a fixed number n of observations.
2. The n observations are all independent.
3. Each observation falls into just one of two categories ("success" or "failure").
4. The probability of success, p, is assumed to be the same for each observation.

If you emphasize these components, your students will be able to connect them to inference for proportions, beginning in Chapter 15. This should also help in differentiating proportions from means in inference.

The material in this chapter is considered optional, but there are some compelling reasons to consider covering it. First, the mean and standard deviation of a binomial random variable will be seen within the mean and standard error of the sample proportion when we discuss inference for a population proportion, p (beginning in Chapter 15). This chapter concludes with the Normal approximation to binomial distributions, which sets us up for study of the sampling distribution for a proportion in Chapter 15. Second, the structure of the binomial distribution is intuitively appealing, at least for students with a reasonable mathematical background.

> ### LEARNING OUTCOMES**
>
> ■ Recognize the binomial setting: a fixed number n of independent success–failure trials with the same probability p of success on each trial.
>
> ■ Recognize and use the binomial distribution of the count of successes in a binomial setting.
>
> ■ Use the binomial probability formula to find the probability of events involving the count X of successes in a binomial setting for small values of n.
>
> ■ Find the mean and standard deviation of a binomial count X.
>
> ■ Recognize when you can use the Normal approximation to a binomial distribution. Use the Normal approximation to calculate probabilities that concern a binomial count X.
>
> ---
> **These learning outcomes appear later for the students in Chapter 14: Part III Review.

Teaching Suggestions and Additional Examples/Activities for the Classroom

If you are covering this chapter, you may have opted to cover at least part of Chapter 12, on general probability rules. These simple rules can be used to construct the formula for a binomial probability.

1. Motivate the Formula for the Binomial

Start with an example for which the success probability is *not* 0.5. For example, if you plant 3 seeds and each will germinate independently with probability 0.8, what is the probability that exactly 2 of them will germinate? You might begin by asking your students if the probability is $(0.8)^2(0.2) = 0.128$. Ask them if this probability includes all possible ways that exactly 2 of the 3 seeds will germinate. Once they recognize that it does not, you can motivate the formula for the binomial distribution with a tree diagram. The number of branches in the tree diagram that agree in the number of successes (2 in the preceding example) is the combination appearing at the beginning of the formula. You can apply both the addition rule for disjoint events and the multiplication rule for independent events in deriving the probability in question.

2. Use Several Examples to Conceptualize the Binomial Distribution Formula

Students will likely require several examples before the binomial distribution's structure begins to appeal intuitively. Focus on the structure—there are three components: (1) probability of success raised to the power of the number of successes, (2) probability of failure raised to the power of the number of failures, and (3) number of branches in the tree diagram leading to the desired number of successes (or number of ways to get k successes in n trials). Emphasize that in n trials, something must happen on each trial (either a success or a failure), so the sum of the exponents must be n.

3. Discuss When and How the Normal Can Be Used to Approximate the Binomial

When you introduce your students to the Normal approximation to the binomial, you should return to the ideas of the Normal distribution from Chapter 3. Make sure to have your students check the conditions for using the Normal approximation to the binomial: np and $n(1 - p)$ must *both* be at least 10; the reason for the restriction can be visually demonstrated using the *Normal Approximation to the Binomial* applet. This will set your students up for checking conditions when we get to the chapters on inference.

4. Remind Students What Standardizing Does and to Check for Reasonability

When we standardize in this chapter, the z-score still tells us how many standard deviations away from the mean the observed result is, and we can make a quick judgment about how unusual our sample was. You can remind your students of the 68–95–99.7 rule as a way to check for reasonability of their answers.

Other Resources (LaunchPad)
EESEE Case Studies
Psychic Probability
Is Caffeine Dependence Real?
Applets
Normal Approximation to Binomial Distributions
Probability

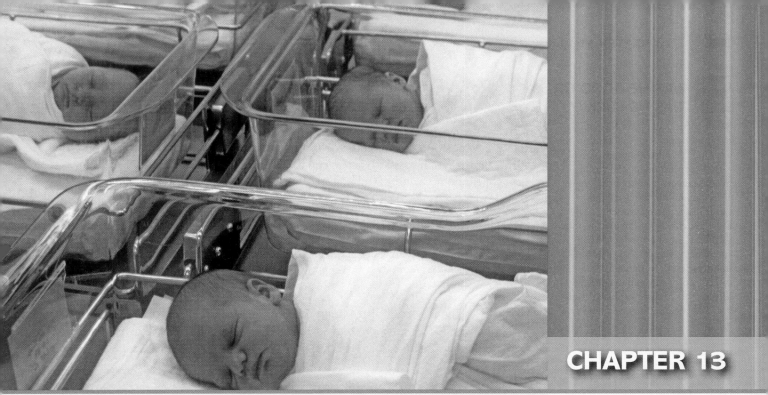

© Randy Duchaine/Alamy

Binomial Distributions*

In this chapter we cover...

- The binomial setting and binomial distributions
- Binomial distributions in statistical sampling
- Binomial probabilities
- Using technology
- Binomial mean and standard deviation
- The Normal approximation to binomial distributions

A basketball player shoots 5 free throws. How many does she make? A sample survey dials 1200 residential phone numbers at random. How many live people answer the phone? You plant 10 dogwood trees. How many live through the winter? In all these situations, we want a probability model for a *count* of successful outcomes.

The binomial setting and binomial distributions

The distribution of a count depends on how the data are produced. Here is a common situation.

The Binomial Setting

1. There are a fixed number *n* of observations.
2. The *n* observations are all **independent.** That is, knowing the result of one observation does not change the probabilities we assign to other observations.
3. Each observation falls into one of just two categories, which for convenience we call "success" and "failure."
4. The probability of a success, call it *p*, is the same for each observation.

*This more advanced chapter concerns a special topic in probability. The material is not needed to read the rest of the book.

When a baseball player hits .300, everyone applauds. A .300 hitter gets a hit in 30% of times at bat. Could a .300 year just be luck? Typical major leaguers bat about 500 times a season and hit about .260. A hitter's successive tries seem to be independent, so we have a binomial setting. From this model, we can calculate or simulate the probability of hitting .300. It is about 0.025. Out of 100 run-of-the-mill major league hitters, two or three each year will bat .300 because they were lucky.

Think of tossing a coin n times as an example of the binomial setting. Each toss gives either heads or tails. Knowing the outcome of one toss doesn't change the probability of a head on any other toss, so the tosses are independent. If we call heads a success, then p is the probability of a head and remains the same as long as we toss the same coin. For tossing a coin, p is close to 0.5. If we spin the coin on a flat surface rather than toss it, p is not equal to 0.5. The number of heads we count is a discrete random variable X. The distribution of X is called a *binomial distribution*.

Binomial Distribution

The count X of successes in the binomial setting has the **binomial distribution** with parameters n and p. The parameter n is the number of observations, and p is the probability of a success on any one observation. The possible values of X are the whole numbers from 0 to n.

The binomial distributions are an important class of discrete probability models. *Pay attention to the binomial setting, because not all counts have binomial distributions.*

EXAMPLE 13.1   Blood Types

Genetics says that children receive genes from their parents independently. Each child of a particular pair of parents has probability 0.25 of having type O blood. If these parents have 5 children, the number who have type O blood is the count X of successes in 5 independent observations with probability 0.25 of a success on each observation. So X has the binomial distribution with $n = 5$ and $p = 0.25$. ■

EXAMPLE 13.2   Counting Boys

Here is set of genetic examples that require more thought.

Choose two births at random from the last year's births at a large hospital and count the number of boys (0, 1, or 2). The sexes of children born to different mothers are surely independent. The probability that a randomly chosen birth in Canada and the United States is a boy is about 0.52. (Why it is not 0.5 is something of a mystery.) So the count of boys has a binomial distribution with $n = 2$ and $p = 0.52$.

Next, observe successive births at a large hospital, and let X be the number of births until the first boy is born. Births are independent, and each has probability 0.52 of being a boy. Yet X is *not* binomial because there is no fixed number of observations. "Count observations until the first success" is a different setting than "count the number of successes in a fixed number of observations."

Finally, choose at random a family with exactly two children and count the number of boys. Careful study of such families shows that the count of boys is *not* binomial: the probability of exactly 1 boy is too high.[1] Families are less likely to have a third child if the first two are a boy and a girl, so when we look at families that stopped at two children, "one of each" is more common than if we look at randomly chosen births. The sexes of successive children in two-child families are *not independent* because the parents' choices interfere with the genetics. ■

Binomial distributions in statistical sampling

The binomial distributions are important in statistics when we wish to make inferences about the proportion p of "successes" in a population. Here is a typical example.

EXAMPLE 13.3   Choosing an SRS of Tomatoes

Physical damage to tomatoes, which can occur throughout the distribution system from field to consumer, has a major impact on market loss of fresh tomatoes. At the packinghouse, a distributor inspects an SRS of 10 tomatoes from a shipment of 10,000 tomatoes. Suppose that (unknown to the distributor) 11% of tomatoes in the shipment can be considered unmarketable due to physical damage, most generally bruising.[2] Count the number X of unmarketable tomatoes in the sample.

This is not quite a binomial setting. Removing one tomato changes the proportion of unmarketable tomatoes remaining in the shipment. So the probability that the second tomato chosen is unmarketable changes when we know whether the first is unmarketable or not. But removing one tomato from a shipment of 10,000 changes the makeup of the remaining 9999 tomatoes very little. In practice, the distribution of X is very close to the binomial distribution with $n = 10$ and $p = 0.11$. ■

Example 13.3 shows how we can use the binomial distributions in the statistical setting of selecting an SRS. When the population is much larger than the sample, a count of successes in an SRS of size n has approximately the binomial distribution with n equal to the sample size and p equal to the proportion of successes in the population.

Sampling Distribution of a Count

Choose an SRS of size n from a population with proportion p of successes. When the population is much larger than the sample, the count X of successes in the sample has approximately the binomial distribution with parameters n and p.

Apply Your Knowledge

In each of Exercises 13.1 to 13.3, X is a count. Does X have a binomial distribution? Give your reasons in each case.

13.1 Random Digit Dialing. When an opinion poll calls residential telephone numbers at random, only 20% of the calls reach a live person. You watch the random dialing machine make 15 calls. X is the number that reach a live person.

13.2 Random Digit Dialing. When an opinion poll calls residential telephone numbers at random, only 20% of the calls reach a live person. You watch the random dialing machine make calls. X is the number of calls until the first live person answers.

13.3 Boxes of Tiles. Boxes of six-inch slate flooring tile contain 40 tiles per box. The count X is the number of cracked tiles in a box. You have noticed that most boxes contain no cracked tiles, but if there are cracked tiles in a box, then there are usually several.

13.4 Smoking in Canada. Statistics Canada reports that 19.9% of Canadians aged 12 and older smoke either daily or occasionally.[3] If you take an SRS of 1500 Canadians aged 12 and over, what is the approximate distribution of the number in your sample who smoke either daily or occasionally?

Binomial probabilities

We can find a formula for the probability that a binomial random variable takes any value by adding probabilities for the different ways of getting exactly that many successes in n observations. Here is an example that illustrates the idea.

EXAMPLE 13.4 Inheriting Blood Type

The blood types of successive children born to the same parents are independent and have fixed probabilities that depend on the genetic makeup of the parents. Each child born to a particular set of parents has probability 0.25 of having blood type O. If these parents have 5 children, what is the probability that exactly 2 of them have type O blood?

The count of children with type O blood is a binomial random variable X with $n = 5$ tries and probability $p = 0.25$ of a success on each try. We want $P(X = 2)$. ■

Because the method doesn't depend on the specific example, let's use "S" for success and "F" for failure for short, with a success representing type O blood. Do the work in two steps.

Step 1 Find the probability that a specific 2 of the 5 tries, say the first and the third, give successes. This is the outcome SFSFF. Because tries are independent, the multiplication rule for independent events applies. The probability we want is

$$P(\text{SFSFF}) = P(S)P(F)P(S)P(F)P(F)$$
$$= (0.25)(0.75)(0.25)(0.75)(0.75)$$
$$= (0.25)^2(0.75)^3$$

Step 2 Observe that *any one arrangement* of 2 Ss and 3 Fs has this same probability. This is true because we multiply together 0.25 twice and 0.75 three times whenever we have 2 Ss and 3 Fs. The probability that $X = 2$ is the probability of getting 2 Ss and 3 Fs in any arrangement whatsoever. Here are all the possible arrangements:

SSFFF	SFSFF	SFFSF	SFFFS	FSSFF
FSFSF	FSFFS	FFSSF	FFSFS	FFFSS

There are 10 of them, all with the same probability. The overall probability of 2 successes is therefore

$$P(X = 2) = 10(0.25)^2(0.75)^3 = 0.2637$$

The pattern of this calculation works for any binomial probability. To use it, we must count the number of arrangements of k successes in n observations. We use the following fact to do the counting without actually listing all the arrangements.

WHAT LOOKS RANDOM?

Toss a coin six times and record heads (H) or tails (T) on each toss. Which of these outcomes is more probable: HTHTTH or TTTHHH? Almost everyone says that HTHTTH is more probable because TTTHHH does not "look random." In fact, both are equally probable. That heads has probability 0.5 says that about half of a very long sequence of tosses will be heads. It doesn't say that heads and tails must come close to alternating in the short run. The coin doesn't know what past outcomes were, and it can't try to create a balanced sequence.

Binomial Coefficient

The number of ways of arranging k successes among n observations is given by the **binomial coefficient**

$$\binom{n}{k} = \frac{n!}{k!(n-k)!}$$

for $k = 0, 1, 2, \ldots, n$.

The formula for binomial coefficients uses the **factorial** notation. For any positive whole number n, its factorial $n!$ is

factorial

$$n! = n \times (n-1) \times (n-2) \times \cdots \times 3 \times 2 \times 1$$

In addition, we define $0! = 1$.

The larger of the two factorials in the denominator of a binomial coefficient will cancel much of the $n!$ in the numerator. For example, the binomial coefficient we need for Example 13.4 is

$$
\begin{aligned}
\binom{5}{2} &= \frac{5!}{2!3!} \\
&= \frac{(5)(4)(3)(2)(1)}{(2)(1) \times (3)(2)(1)} \\
&= \frac{(5)(4)}{(2)(1)} = \frac{20}{2} = 10
\end{aligned}
$$

⚠ **CAUTION** *The binomial coefficient $\binom{5}{2}$ is not related to the fraction $\frac{5}{2}$. A helpful way to remember its meaning is to read it as "5 choose 2."* Binomial coefficients have many uses, but we are interested in them only as an aid to finding binomial probabilities. The binomial coefficient $\binom{n}{k}$ counts the number of different ways in which k successes can be arranged among n observations. The binomial probability $P(X = k)$ is this count multiplied by the probability of any one specific arrangement of the k successes. Here is the result we seek.

Binomial Probability

If X has the binomial distribution with n observations and probability p of success on each observation, the possible values of X are $0, 1, 2, \ldots, n$. If k is any one of these values,

$$P(X = k) = \binom{n}{k} p^k (1-p)^{n-k}$$

EXAMPLE 13.5   Inspecting Tomatoes

The number X of unmarketable tomatoes in Example 13.3 has approximately the binomial distribution with $n = 10$ and $p = 0.11$.

The probability that the sample contains no more than 1 unmarketable tomato is

$$
\begin{aligned}
P(X \le 1) &= P(X = 1) + P(X = 0) \\
&= \binom{10}{1}(0.11)^1(0.89)^9 + \binom{10}{0}(0.11)^0(0.89)^{10} \\
&= \frac{10!}{1!9!}(0.11)(0.3504) + \frac{10!}{0!10!}(1)(0.3118) \\
&= (10)(0.11)(0.3504) + (1)(1)(0.3118) \\
&= 0.3854 + 0.3118 = 0.6972
\end{aligned}
$$

This calculation uses the facts that $0! = 1$ and that $a^0 = 1$ for any number a other than 0. We see that about 70% of all samples will contain no more than 1 unmarketable tomato. In fact, about 31% of the samples will contain no unmarketable tomatoes. A sample of size 10 cannot be trusted to alert the distributor to the presence of unacceptable tomatoes in the shipment.

The complement rule described in Chapter 11 can make the computation of certain binomial probabilities simpler. For example, the probability that the sample contains at least 1 unmarketable tomato is

$$P(X \geq 1) = P(X = 1) + P(X = 2) + \cdots + P(X = 10)$$
$$= 1 - P(X = 0)$$
$$= 1 - 0.3118 = 0.6882$$

When computing binomial probabilities by hand, it is useful to keep the complement rule in mind. ■

Using technology

The binomial probability formula is awkward to use unless the number of observations n is quite small. You can find tables of binomial probabilities $P(X = k)$ and cumulative probabilities $P(X \leq k)$ for selected values of n and p, but the most efficient way to do binomial calculations is to use technology.

Figure 13.1 shows output for the calculation in Example 13.5 from a graphing calculator, two statistical programs, and a spreadsheet program. We asked all four to give cumulative probabilities. The calculator, Minitab, and CrunchIt! have menu entries for binomial cumulative probabilities. Excel has no menu entry, but the worksheet function BINOMDIST is available. All the outputs agree with the result 0.6972 of Example 13.5.

Texas Instruments Graphing Calculator

```
binomcdf(10,0.11,1)
                .6972
```

Minitab

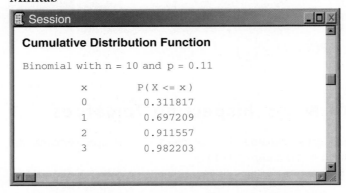

CrunchIt!

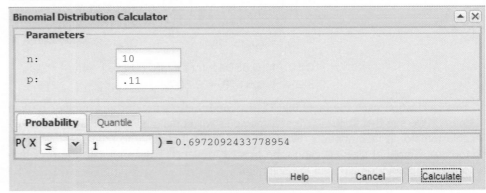

FIGURE 13.1

The binomial probability $P(X \leq 1)$ for Example 13.5: output from a graphing calculator, two statistical programs, and a spreadsheet program.

Microsoft Excel

FIGURE 13.1
(Continued)

```
Microsoft Excel - Book1                                    _ □ X
   A1            ▼      fx =  =BINOMDIST(1,10,0.11,
              A              B      C      D      E
  1               0.697209
  2
   Sheet1  Sheet2  Sheet3
```

Apply Your Knowledge

13.5 Proofreading. Typing errors in a text are either nonword errors (as when "the" is typed as "teh") or word errors that result in a real but incorrect word. Spell-checking software will catch nonword errors but not word errors. Human proofreaders catch 70% of word errors. You ask a fellow student to proofread an essay in which you have deliberately made 10 word errors.

(a) If the student matches the usual 70% rate, what is the distribution of the number of errors caught? What is the distribution of the number of errors missed?

(b) Missing 3 or more out of 10 errors seems a poor performance. What is the probability that a proofreader who catches 70% of word errors misses exactly 3 out of 10? If you use software, also find the probability of missing 3 or more out of 10.

13.6 Random Digit Dialing. When an opinion poll calls residential telephone numbers at random, only 20% of the calls reach a live person. You watch the random digit dialing machine make 15 calls.

(a) What is the probability that exactly 3 calls reach a person?

(b) What is the probability that at most 3 calls reach a person?

(c) What is the probability that at least 3 calls reach a person?

(d) What is the probability that fewer than 3 calls reach a person?

(e) What is the probability that more than 3 calls reach a person?

13.7 Google Does Binomial. Point your Web browser to www.google.com. Instead of searching the Web or looking for images, you can request a calculation in the Search box.

(a) Enter 5 choose 2 and click Search. What does Google return?

(b) You see that Google calculates the binomial coefficient "5 choose 2." What are the values of the binomial coefficients for "500 choose 2" and "500 choose 100"? We expect that there are more ways to choose 100 than to choose 2, but how many more may be a surprise. That e+107 in Google's answer means a 1 followed by 107 zeros.

(c) Google also does binomial probabilities. Enter (10 choose 1)*0.11 *0.89^9 to find $P(X = 1)$ in Example 13.5 (page 309). What is Google's answer with all its decimal places?

Binomial mean and standard deviation

If a count X has the binomial distribution based on n observations with probability p of success, what is its mean μ? That is, in very many repetitions of the binomial setting, what will be the average count of successes? We can guess the answer. If a basketball player makes 80% of her free throws, the mean number made in 10 tries should be 80% of 10, or 8. In general, the mean of a binomial distribution should be $\mu = np$. Here are the facts.

Binomial Mean and Standard Deviation

If a count X has the binomial distribution with number of observations n and probability of success p, the **mean** and **standard deviation** of X are

$$\mu = np$$
$$\sigma = \sqrt{np(1-p)}$$

 Remember that these short formulas are good only for binomial distributions. They can't be used for other distributions.

EXAMPLE 13.6   Inspecting Tomatoes

 RANDOMNESS TURNS SILVER TO BRONZE

After many charges of favoritism by judges, the rules for scoring international figure skating competitions changed in 2004. The big change is that 12 judges score all performances, then scores from 3 judges chosen at random are dropped for each part of the program. So there are $\binom{12}{9} = 220$ possible panels of 9 judges for (say) the "Free Skate," and these panels will have slightly different scores. Result: at the 2006 World Figure Skating Championships, the Russian pair Maria Petrova and Alexei Tikhonov received the bronze medal when the consensus of all 12 judges would have given them the silver medal. Perhaps the system needs another change.

Continuing Example 13.5, the count X of unmarketable tomatoes is binomial with $n = 10$ and $p = 0.11$. The histogram in Figure 13.2 displays this probability distribution. (Because probabilities are long-run proportions, using probabilities as the heights of the bars shows what the distribution of X would be in very many repetitions.) The distribution is strongly right skewed. Although X can take any whole-number value from 0 to 10, the probabilities of values larger than 5 are so small that they do not appear in the histogram.

The mean and standard deviation of the binomial distribution in Figure 13.2 are

$$\mu = np$$
$$= (10)(0.11) = 1.1$$
$$\sigma = \sqrt{np(1-p)}$$
$$= \sqrt{(10)(0.11)(0.89)} = \sqrt{0.979} = 0.9894$$

The mean is marked on the probability histogram in Figure 13.2. ■

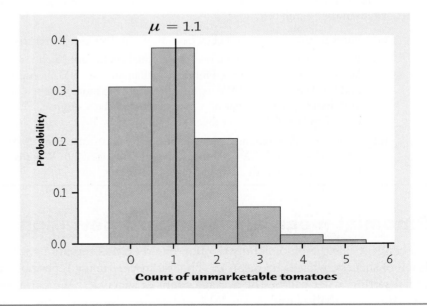

FIGURE 13.2

Probability histogram for the binomial distribution with $n = 10$ and $p = 0.11$.

Apply Your Knowledge

13.8 Random Digit Dialing. When an opinion poll calls residential telephone numbers at random, only 20% of the calls reach a live person. You watch the random digit dialing machine make 15 calls.

(a) What is the mean number of calls that reach a person?

(b) What is the standard deviation σ of the count of calls that reach a person?

(c) If calls are made to New York City rather than nationally, the probability that a call reaches a person is only $p = 0.08$. How does this new p affect the standard deviation? What would be the standard deviation if $p = 0.01$? What does your work show about the behavior of the standard deviation of a binomial distribution as the probability of a success gets closer to 0?

13.9 Proofreading. Return to the proofreading setting of Exercise 13.5 (page 311).

(a) If X is the number of word errors missed, what is the distribution of X? If Y is the number of word errors caught, what is the distribution of Y?

(b) What is the mean number of errors caught? What is the mean number of errors missed? The mean counts of successes and of failures always add to n, the number of observations.

(c) What is the standard deviation of the number of errors caught? What is the standard deviation of the number of errors missed? The standard deviations of the count of successes and the count of failures are always the same.

The Normal approximation to binomial distributions

It isn't practical to use the formula for binomial probabilities when the number of observations n is large. (Look at part (b) of Exercise 13.7 (page 311) to see why.) Software or a graphing calculator will handle many problems that are beyond the reach of hand calculation. If technology does not rescue you, there is another alternative: *as the number of observations n gets larger, the binomial distribution gets close to a Normal distribution.* When n is large, we can use Normal probability calculations to approximate binomial probabilities. Here are the facts.

Normal Approximation for Binomial Distributions

Suppose that a count X has the binomial distribution with n observations and success probability p. When n is large, the distribution of X is approximately Normal, $N(np, \sqrt{np(1-p)})$.

As a rule of thumb, we will use the Normal approximation when n is so large that $np \geq 10$ and $n(1-p) \geq 10$.

The Normal approximation is easy to remember because it says to act as if X is Normal with exactly the same mean and standard deviation as the binomial. The accuracy of the Normal approximation improves as the sample size n increases. It is most accurate for any fixed n when p is close to $1/2$ and least accurate when p is near 0 or 1. This is why the rule of thumb in the box depends on p as well as n.

EXAMPLE 13.7   Tracking Your Health

Sample surveys show that more people are tracking changes in their health on paper, spreadsheet, mobile device, or just "in their heads." A survey asked a nationwide random sample of 3014 adults "Now thinking about your health overall, do you currently keep track of your own weight, diet, or exercise routine, or is this not something you currently do?"[4] The population that the poll wants to draw conclusions about is all U.S. residents aged 18 and over. Suppose that in fact 62% of all adult U.S. residents would say "Yes" if asked this question. What is the probability that 1900 or more of the sample say "Yes"? ■

Because there are about 235 million adults in the United States, the responses of 3014 randomly chosen adults are very close to independent. So the number in our sample who are tracking their weight, diet, or exercise routine is a random variable X having the binomial distribution with $n = 3014$ and $p = 0.62$. To find the probability $P(X \geq 1900)$ that at least 1900 of the people in the sample are tracking their weight, diet, or exercise routine, we must add the binomial probabilities of all outcomes from $X = 1900$ to $X = 3014$. Figure 13.3 is a probability histogram of this binomial distribution, from Minitab. As the Normal approximation suggests, the shape of the distribution looks Normal. The probability we want is the sum of the heights of the shaded bars. Here are three ways to find this probability.

FIGURE 13.3
Probability histogram for the binomial distribution with $n = 3014$, $p = 0.62$. The bars at and above 1900 are shaded to highlight the probability of getting at least 1900 successes. The shape of this binomial probability distribution closely resembles a Normal curve.

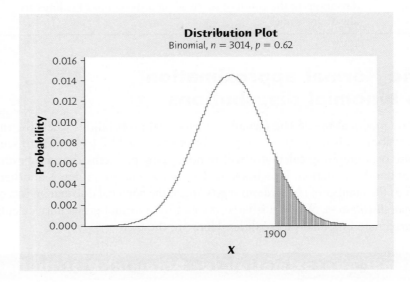

1. Use technology. Statistical software can find the exact binomial probability. In most cases, software finds cumulative probabilities $P(X \leq x)$. So start by writing

$$P(X \geq 1900) = 1 - P(X \leq 1899)$$

Here is Minitab's answer for $P(X \leq 1899)$:

```
Binomial with n = 3014 and p = 0.62

    X     P(X<=x)
  1899    0.876384
```

The probability we want is $1 - 0.876384 = 0.123616$, correct to 6 decimal places.

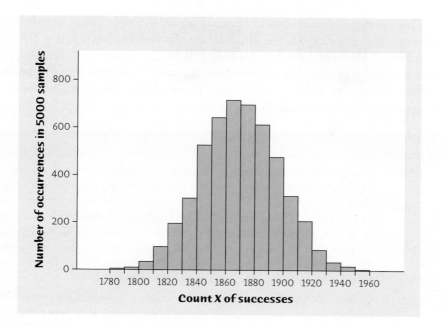

FIGURE 13.4
Histogram of 5000 simulated binomial counts ($n = 3014, p = 0.62$).

2. Simulate a large number of samples. Figure 13.4 displays a histogram of the counts X from 5000 samples of size 3014 when the truth about the population is $p = 0.62$. The simulated distribution, like the exact distribution in Figure 13.3, looks Normal. Because 649 of these 5000 samples have X at least 1900, the probability estimated from the simulation is

$$P(X \geq 1900) = \frac{649}{5000} = 0.1298$$

This estimate misses the true probability by about 0.006. The law of large numbers says that the results of such simulations always get closer to the true probability as we simulate more and more samples.

3. Both of the previous methods require software. We can avoid the need for software by using the Normal approximation.

EXAMPLE 13.8   Normal Calculation of a Binomial Probability

Act as though the count X in Example 13.7 has the Normal distribution with the same mean and standard deviation as the binomial distribution:

$$\mu = np = (3014)(0.62) = 1868.68$$
$$\sigma = \sqrt{np(1 - p)} = \sqrt{(3014)(0.62)(0.38)} = 26.648$$

Standardizing X gives a standard Normal variable Z. The probability we want is

$$P(X \geq 1900) = P\left(\frac{X - 1868.68}{26.648} \geq \frac{1900 - 1868.68}{26.648}\right)$$
$$= P(Z \geq 1.18)$$
$$= 1 - 0.8810 = 0.1190$$

The Normal approximation 0.1190 misses the true probability calculated in Example 13.7 by about 0.005. ■

The *Normal Approximation to Binomial* applet shows in visual form how well the Normal approximation fits the binomial distribution for any n and p. You can slide n and watch the approximation get better. Whether or not the Normal approximation is satisfactory depends on how accurate your calculations need to be. For most statistical purposes, great accuracy is not required. Our rule of thumb for use of the Normal approximation reflects this judgment.

Apply Your Knowledge

13.10 Using Benford's Law. According to Benford's law (Example 11.7, page 268) the probability that the first digit of the amount of a randomly chosen invoice is a 1 or a 2 is 0.477. You examine 90 invoices from a vendor and find that 29 have first digits 1 or 2. If Benford's law holds, the count of 1s and 2s will have the binomial distribution with $n = 90$ and $p = 0.477$. Too few 1s and 2s suggests fraud. What is the approximate probability of 29 or fewer 1s and 2s if the invoices follow Benford's law? Do you suspect that the invoice amounts are not genuine?

13.11 College Admissions. A small liberal arts college in Ohio would like to have an entering class of 475 students next year. Past experience shows that about 31% of the students admitted will decide to attend. The college is planning to admit 1520 students. Suppose that students make their decisions independently and that the probability is 0.31 that a randomly chosen student will accept the offer of admission.

  (a) What are the mean and standard deviation of the number of students who accept the admissions offer from this college?

  (b) Use the Normal approximation: what is the approximate probability that the college gets more students than they want? Be sure to check that you can safely use the approximation.

  (c) Use software to compute the exact probability that the college gets more students than they want. How good is the approximation in part (b)?

13.12 Checking for Survey Errors. One way of checking the effect of undercoverage, nonresponse, and other sources of error in a sample survey is to compare the sample with known facts about the population. About 24% of the Canadian population over 15 are first generation, that is, they were born outside Canada. The number X of first-generation Canadians in random samples of 1000 persons over 15 should therefore vary with the binomial ($n = 1000$, $p = 0.24$) distribution.

  (a) What are the mean and standard deviation of X?

  (b) Use the Normal approximation to find the probability that the sample will contain between 210 and 270 first-generation Canadians. Be sure to check that you can safely use the approximation.

CHAPTER 13 SUMMARY

Chapter Specifics

■ A count X of successes has a **binomial distribution** in the **binomial setting:** there are n observations; the observations are independent of each other; each observation results in a success or a failure; each observation has the same probability p of a success.

■ The binomial distribution with n observations and probability p of success gives a good approximation to the sampling distribution of the count of successes in an SRS of size n from a large population containing proportion p of successes.

- If X has the binomial distribution with parameters n and p, the possible values of X are the whole numbers $0, 1, 2, \ldots, n$. The **binomial probability** that X takes any of these values is

$$P(X = k) = \binom{n}{k} p^k (1 - p)^{n-k}$$

Binomial probabilities in practice are best found using software.

- The **binomial coefficient**

$$\binom{n}{k} = \frac{n!}{k!(n - k)!}$$

counts the number of ways k successes can be arranged among n observations. Here the **factorial $n!$** is

$$n! = n \times (n - 1) \times (n - 2) \times \cdots \times 3 \times 2 \times 1$$

for positive whole numbers n, and $0! = 1$.

- The **mean** and **standard deviation** of a binomial count X are

$$\mu = np$$
$$\sigma = \sqrt{np(1 - p)}$$

- The **Normal approximation** to the binomial distribution says that if X is a count having the binomial distribution with parameters n and p, then when n is large, X is approximately $N(np, \sqrt{np(1 - p)})$. Use this approximation only when $np \geq 10$ and $n(1 - p) \geq 10$.

Link It

The binomial distribution is used to compute probabilities for the count of successes among n observations that are produced under the binomial setting. An important situation for which the binomial setting can be used is when we choose an SRS from a population with a proportion p of successes. When the success probability p is known, probabilities associated with the number of successes among n observations can be computed using either the binomial formula, or the Normal approximation when the sample size is large enough.

An important application of the binomial distribution is in making inferences about the *proportion* of some outcome in a population. This is described in Chapters 16 and 17 where our data is collected under the binomial setting, but the proportion with the given outcome in the population is not known. For example, we may be interested in learning about the proportion of young adults (aged 19 to 25) that still live at home with their parents based on a sample from the population of young adults. When we are interested in making inferences about an unknown proportion in a population, we generally work with the *proportion of successes in the sample* rather than the count of successes in the sample. When using the proportion of successes in the sample to answer questions and draw conclusions about this unknown proportion in the population, the statistical methods are still related to the binomial distribution and the Normal approximation to the binomial.

CHECK YOUR SKILLS

13.13 Larry reads that 1 out of 4 eggs contains salmonella bacteria. So he never uses more than 3 eggs in cooking. If eggs do or don't contain salmonella independent of each other, the number of contaminated eggs when Larry uses 3 chosen at random has the distribution

(a) binomial with $n = 4$ and $p = 1/4$.
(b) binomial with $n = 3$ and $p = 1/4$.
(c) binomial with $n = 3$ and $p = 1/3$.

13.14 In the previous exercise, the probability that at least 1 of Larry's 3 eggs contains salmonella is about

(a) 0.68.   (b) 0.58.   (c) 0.30.

13.15 In a group of 10 college students, 4 are business majors. You choose 3 of the 10 students at random and ask their major. The distribution of the number of business majors you choose is

(a) binomial with $n = 10$ and $p = 0.4$.
(b) binomial with $n = 3$ and $p = 0.4$.
(c) not binomial.

In a test for ESP (extrasensory perception), a subject is told that cards the experimenter can see but he cannot contain either a star, a circle, a wave, or a square. As the experimenter looks at each of 5 cards in turn, the subject names the shape on the card. A subject who is just guessing has probability 0.25 of guessing correctly on each card. Exercises 13.16 to 13.18 use this information.

13.16 If the subject guesses 2 shapes correctly and 3 incorrectly, in how many ways can you arrange the sequence of correct and incorrect guesses?

(a) $\binom{3}{2} = 3$ (b) $\binom{5}{2} = 20$ (c) $\binom{5}{2} = 10$

13.17 Assume the subject's guesses are independent of each other. The probability that the subject guesses the shape correctly on the first and last card, but incorrectly on the other three cards is about

(a) 0.264.   (b) 0.026.   (c) 0.009.

13.18 Assume the subject's guesses are independent of each other. The probability that the subject guesses the shape correctly on 2 out of the 5 cards is about

(a) 0.264.   (b) 0.026.   (c) 0.009.

Each entry in a table of random digits like Table B has probability 0.1 of being any of the 10 digits 0 to 9, and digits are independent of each other. Exercises 13.19 to 13.21 use this setting.

13.19 The probability of an entry being either an 8 or a 9 is

(a) 0.1.   (b) 0.2.   (c) 0.4.

13.20 Each line in Table B has 40 digits. The number of times an 8 or a 9 occurs in *two lines* of the table is

(a) binomial with $n = 80$ and $p = 0.2$.
(b) binomial with $n = 80$ and $p = 0.1$.
(c) binomial with $n = 40$ and $p = 0.2$.

13.21 The mean number of times an 8 or a 9 occurs in *two lines* of the table is

(a) 16.   (b) 12.8.   (c) 8.

CHAPTER 13 EXERCISES

13.22 **Binomial setting?** In each of the following situations, is it reasonable to use a binomial distribution for the random variable X? Give reasons for your answer in each case.

(a) An auto manufacturer chooses one car from each hour's production for a detailed quality inspection. One variable recorded is the count X of finish defects (dimples, ripples, etc.) in the car's paint.
(b) The pool of potential jurors for a murder case contains 100 persons chosen at random from the adult residents of a large city. Each person in the pool is asked whether he or she opposes the death penalty; X is the number who say "Yes."
(c) Joe buys a ticket in his state's Pick 3 lottery game every week; X is the number of times in a year that he wins a prize.

13.23 **Binomial setting?** A binomial distribution will be approximately correct as a model for one of these two

sports settings and not for the other. Explain why by briefly discussing both settings.

(a) A National

© David Bergman/Corbis

Football League kicker has made 80% of his field goal attempts in the past. This season he attempts 20 field goals. The attempts differ widely in distance, angle, wind, and so on.
(b) A National Basketball Association player has made 80% of his free-throw attempts in the past. This season he takes 150 free throws. Basketball free throws are always attempted from 15 feet away from the basket with no interference from other players.

13.24 Antibiotic resistance. Antibiotic resistance occurs when disease-causing microbes become resistant to antibiotic drug therapy. Because this resistance is typically genetic and transferred to the next generations of microbes, it is a very serious public health problem. According to the CDC, approximately 30% of gonorrhea cases tested in 2011 were resistant to at least one of the three major antibiotics commonly used to treat gonorrhea.[5] A physician treated 10 cases of gonorrhea during one week of 2011.

(a) What is the distribution of the cases resistant to at least one of the three major antibiotics?

(b) What is the probability that exactly 1 out of the 10 cases was resistant to at least one of the three major antibiotics? What is that probability for exactly 2 out of 10?

(c) Also find the probability that 1 or more out of the 10 were resistant to at least one of the three major antibiotics. (Hint: It is easier to first find the probability that exactly 0 of the 10 cases were resistant.)

13.25 Random stock prices. A believer in the random walk theory of stock markets thinks that an index of stock prices has probability 0.65 of increasing in any year. Moreover, the change in the index in any given year is not influenced by whether it rose or fell in earlier years. Let X be the number of years among the next 5 years in which the index rises.

(a) X has a binomial distribution. What are n and p?

(b) What are the possible values that X can take?

(c) Find the probability of each value of X. Draw a probability histogram for the distribution of X. (See Figure 13.2 (page 312) for an example of a probability histogram.)

(d) What are the mean and standard deviation of this distribution? Mark the location of the mean on your histogram.

13.26 Roulette—betting on red. A roulette wheel has 38 slots, numbered 0, 00, and 1 to 36. The slots 0 and 00 are colored green, 18 of the others are red, and 18 are black. The dealer spins the wheel and at the same time rolls a small ball along the wheel in the opposite direction. The wheel is carefully balanced so that the ball is equally likely to land in any slot when the wheel slows. Gamblers can bet on various combinations of numbers and colors.

(a) If you bet on "red," you win if the ball lands in a red slot. What is the probability of winning with a bet on red in a single play of roulette?

(b) You decide to play roulette 4 times, each time betting on red. What is the distribution of X, the number of times you win?

(c) If you bet the same amount on each play and win on exactly 2 of the 4 plays, then you will "break even." What is the probability that you will break even?

(d) If you win on *fewer than* 2 of the 4 plays, then you will lose money. What is the probability that you will lose money?

13.27 The birth control shot. The birth control shot is one of the most effective methods of birth control available, and it works best when you get the shot regularly, every 12 weeks. Under ideal conditions, only 1% of women taking the shot become pregnant within one year. In typical use, however, 3% become pregnant.[6] Choose at random 20 women using the shot as their method of birth control. We count the number who become pregnant in the next year.

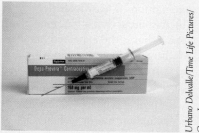

Urbano Delvalle/Time Life Pictures/ Getty Images

(a) Explain why this is a binomial setting.

(b) What is the probability that at least one of the women becomes pregnant under ideal conditions? What is the probability in typical use?

13.28 Roulette, continued. You decide to play roulette 200 times, each time betting the same amount on red. You will lose money if you win on fewer than 100 of the plays. Based on the information in Exercise 13.26, what is the probability that you will lose money? (Check that the Normal approximation is permissible and use it to find this probability. If your software allows, find the exact binomial probability and compare the two results.) In general, if you bet the same amount on red every time, you will lose money if you win on fewer than half of the plays. What do you think happens to the probability of making money the longer you continue to play?

13.29 The birth control shot, continued. A study of the effectiveness of the birth control shot interviews a random sample of 600 women who are using the shot as their method of birth control.

(a) Based on the information about typical use in Exercise 13.27, what is the probability that at least 20 of these women become pregnant in the next year? (Check that the Normal approximation is permissible, and use it to find this probability. If your software allows, find the exact binomial probability, and compare the two results.)

(b) We can't use the Normal approximation to the binomial distributions to find this probability under ideal conditions as described in Exercise 13.27. Why not?

13.30 **Hitting the fairway.** One statistic used to assess professional golfers is driving accuracy, the percent of drives that land in the fairway. In 2012, driving accuracy for PGA Tour professionals ranged from about 45% to about 75%. Rory McIlroy, the highest money winner on the PGA tour in 2012, only hits the fairway about 57% of the time.[7] McIlroy is also one of the longest drivers on the tour, and increased distance is generally associated with decreased accuracy.

(a) McIlroy hits 14 drives in a round. What assumptions must you make to use a binomial distribution for the count X of fairways he hits? Which of these assumptions is least realistic?

(b) Assuming that a binomial distribution can be used, what is the expected number of fairways that McIlroy hits in a round in which he hits 14 drives?

13.31 **Genetics.** According to genetic theory, the blossom color in the second generation of a certain cross of sweet peas should be red or white in a 3:1 ratio. That is, each plant has probability 3/4 of having red blossoms, and the blossom colors of separate plants are independent.

© blickwinkel/Alamy

(a) What is the probability that exactly 3 out of 4 of these plants have red blossoms?

(b) What is the mean number of red-blossomed plants when 60 plants of this type are grown from seeds?

(c) What is the probability of obtaining at least 45 red-blossomed plants when 60 plants are grown from seeds? Use the Normal approximation. If your software allows, find the exact binomial probability and compare the two results.

13.32 **False positives in testing for HIV.** A rapid test for the presence in the blood of antibodies to HIV, the virus that causes AIDS, gives a positive result with probability about 0.004 when a person who is free of HIV antibodies is tested. A clinic tests 1000 people who are all free of HIV antibodies.

(a) What is the distribution of the number of positive tests?

(b) What is the mean number of positive tests?

(c) You cannot safely use the Normal approximation for this distribution. Explain why.

13.33 **Chevrolet sales in 2012.** Chevrolet, the world's fastest-growing major automotive brand, sold more than 4.95 million vehicles in 2012, setting a global sales record and driving General Motors Co. global sales to more than 9.2 million vehicles, a 2.9% gain compared with 2011. The five best-selling Chevrolet nameplates in the first nine months of 2012 were Silverado pickups with approximately 298,000 sold, Cruze compact cars with 181,000, Malibu with 179,000, Equinox SUVs with 167,000, and Impalas with 140,000.[8] They want to undertake a survey of buyers among these five nameplates to ask them about satisfaction with their purchases.

(a) What proportion of the top five nameplates sold in the first nine months of 2012 were Impalas?

(b) If they plan to survey a total of 1000 customers, what is the expected number and standard deviation of the number of Impala buyers in the sample?

(c) What is the probability they will get more than 150 Impala buyers in their sample?

13.34 **Retention rates in a weight loss program.** Americans spend over $30 billion dollars on a variety of weight loss products and services. In a study of retention rates of those using the Platinum Program at Jenny Craig in May, 2001–May, 2002, it was found that about 25% of those who began the program, dropped out in the first four weeks.[9] Assume we have a random sample of 300 people beginning the program.

(a) What is the mean number of people who would drop out of the Platinum Program in the first four weeks in a sample of this size? What is the standard deviation?

(b) What is the approximate probability that at least 210 people in the sample will still be in the Platinum Program after the first four weeks?

13.35 **Multiple-choice tests.** Here is a simple probability model for multiple-choice tests. Suppose each student has probability p of correctly answering a question chosen at random from a universe of possible questions. (A strong student has a higher p than a weak student.) Answers to different questions are independent.

(a) Stacey is a good student for whom $p = 0.75$. Use the Normal approximation to find the probability that Stacey scores between 70% and 80% on a 100-question test.

(b) If the test contains 250 questions, what is the probability that Stacey will score between 70% and 80%? You see that Stacey's score on the longer test is more likely to be close to her "true score."

13.36 **Is this coin balanced?** While he was a prisoner of war during World War II, John Kerrich tossed a coin 10,000 times. He got 5067 heads. If the coin is perfectly balanced, the probability of a head is 0.5. Is there reason to think that Kerrich's coin was not balanced? To answer this question, find the probability that tossing a balanced coin 10,000 times would give a count of heads at least this far from 5000 (that is, at least 5067 heads or no more than 4933 heads.)

13.37 **Binomial variation.** Never forget that probability describes only what happens in the long run. Example 13.5 (page 309) concerns the count of unmarketable tomatoes in inspection samples of size 10. The count has the binomial distribution with $n = 10$ and $p = 0.11$. The *Probability* applet simulates inspecting an SRS of 10 tomatoes if you set the probability of heads to 0.11, toss 10 times, and let each head stand for a unmarketable tomato.

(a) The mean number of unmarketable tomatoes in a sample is 1.1. Click "Toss" and "Reset" repeatedly to simulate 20 samples. How many unmarketable tomatoes did you find in each sample? How close to the mean 1.1 is the average number of unmarketable tomatoes in these samples?

(b) Example 13.5 shows that the probability of exactly 1 unmarketable tomato is 0.3854. How close to the probability is the proportion of the 20 samples that have exactly 1 unmarketable tomato?

Whooping cough. *Whooping cough (pertussis) is a highly contagious bacterial infection that was a major cause of childhood deaths before the development of vaccines. About 80% of unvaccinated children who are exposed to whooping cough will develop the infection, as opposed to only about 5% of vaccinated children. Exercises 13.38 to 13.41 are based on this information.*

13.38 **Vaccination at work.** A group of 20 children at a nursery school are exposed to whooping cough by playing with an infected child.

(a) If all 20 have been vaccinated, what is the mean number of new infections? What is the probability that no more than 2 of the 20 children develop infections?

(b) If none of the 20 have been vaccinated, what is the mean number of new infections? What is the probability that 18 or more of the 20 children develop infections?

13.39 **A whooping cough outbreak.** In 2007, Bob Jones University in Greenville, SC, ended its fall semester a week early because of a whooping cough outbreak; 158 students were isolated and another 1200

given antibiotics as a precaution.[10] Authorities react strongly to whooping cough outbreaks because the disease is so contagious. Because the effect of childhood vaccination often wears off by late adolescence, treat the Bob Jones students as if they were unvaccinated. It appears that about 1400 students were exposed. What is the probability that at least 75% of these students develop infections if not treated? (Fortunately, whooping cough is much less serious after infancy.)

13.40 **A mixed group: means.** A group of 20 children at a nursery school are exposed to whooping cough by playing with an infected child. Of these children 17 have been vaccinated and 3 have not.

(a) What is the distribution of the number of new infections among the 17 vaccinated children? What is the mean number of new infections?

(b) What is the distribution of the number of new infections among the 3 unvaccinated children? What is the mean number of new infections?

(c) Add your means from parts (a) and (b). This is the mean number of new infections among all 20 exposed children.

13.41 **A mixed group: probabilities.** We would like to find the probability that exactly 2 of the 20 exposed children in the previous exercise develop whooping cough.

(a) One way to get 2 infections is to get 1 among the 17 vaccinated children and 1 among the 3 unvaccinated children. Find the probability of exactly 1 infection among the 17 vaccinated children. Find the probability of exactly 1 infection among the 3 unvaccinated children. These events are independent: what is the probability of exactly 1 infection in each group?

(b) Write down all the ways in which 2 infections can be divided between the two groups of children. Follow the pattern of part (a) to find the probability of each of these possibilities. Add all of your results (including the result of part (a)) to obtain the probability of exactly 2 infections among the 20 children.

13.42 **Estimating π from random numbers.** Kenyon College student Eric Newman used basic geometry to evaluate software random number generators as part of a summer research project. He generated 2000 independent random points (X, Y) in the unit square. (That is, X and Y are independent random numbers between 0 and 1, each having the density function illustrated in Figure 11.4 (page 271). The probability that (X, Y) falls in any region within the unit square is the area of the region.)[11]

(a) Sketch the unit square, the region of possible values for the point (X, Y).

(b) The set of points (X, Y) where $X^2 + Y^2 < 1$ describes a circle of radius 1. Add this circle to your sketch in part (a), and label the intersection of the two regions A.

(c) Let T be the total number of the 2000 points that fall into the region A. T follows a binomial distribution. Identify n and p. (Hint: Recall that the area of a circle is πr^2.)

(d) What are the mean and standard deviation of T?

(e) Explain how Eric used a random number generator and the facts given here to estimate π.

13.43 **The continuity correction.** One reason why the Normal approximation may fail to give accurate estimates of binomial probabilities is that the binomial distributions are discrete and the Normal distributions are continuous. That is, counts take only whole number values, but Normal variables can take any value. We can improve the Normal approximation by treating each whole number count as if it occupied the interval from 0.5 below the number to 0.5 above the number. For example, approximate a binomial probability $P(X \geq 10)$ by finding the Normal probability $P(X \geq 9.5)$. Be careful: binomial $P(X > 10)$ is approximated by Normal $P(X \geq 10.5)$.

We saw in Exercise 13.30 that Rory McIlroy hits the fairway in 57% of his drives. We will assume that his drives are independent and that each has probability 0.57 of hitting the fairway. Suppose McIlroy drives 24 times. The exact binomial probability that he hits 17 or more fairways is 0.121.

(a) Show that this setting satisfies the rule of thumb for use of the Normal approximation (just barely).

(b) What is the Normal approximation to $P(X \geq 17)$?

(c) What is the Normal approximation using the continuity correction? That's a lot closer to the true binomial probability.

 Exploring the Web

13.44 **MCAT physical science score.** Go to the Web site www.aamc.org/students/applying/mcat/admissionsadvisors/mcat_stats, which reports the distribution of the total scaled MCAT composite score as well as the scores on the individual areas of assessment for recent years. The areas of assessment include three multiple-choice portions scored on a 15-point scale, which are combined to give the overall composite score.

(a) Open the PDF with the percentage and scaled score tables for the most recent year provided. Find the percentage of students who obtained a physical science section score of 9 or greater.

(b) A survey organization is planning on contacting an SRS of 300 of the examinees from the most recent year to see how they prepared for the MCAT exam. What is the distribution of the number of examinees in the sample that had a physical science section score of 9 or greater?

(c) Use the Normal distribution to approximate the probability that at least half of the examinees in the sample had a physical science section score of 9 or greater.

(d) Use software or a calculator to compute the exact probability that at least half of the examinees in the sample had a physical science section score of 9 or greater. How do the exact probability and the Normal approximation to this probability compare? Which of the two answers would you report to the survey organization? Why?

13.45 **Binomial calculators.** A number of Web sites will do exact binomial probability calculations for you.

(a) Find a Web site with a binomial calculator and give its URL.

(b) In Example 13.7 (page 314), the number of people in the sample who track their health is a binomial random variable X with $n = 3014$ and $p = .62$. Use the binomial calculator to compute the probability that 1900 or more of the sample agree.

Overview

Part III addresses aspects of probability and forms the bridge from data production and description to statistical inference. The chapters you chose to present to your students and the order in which you taught them should inform your use of the Part III Review in terms of the exercises you might assign to your students. The exercises associated with Chapters 12 and 13 are highlighted as optional in the Part III Review.

> **LEARNING OUTCOMES:**
> Refer to Learning Outcomes for Chapters 11–13, as well as the Summary provided in this chapter.

Students should understand the basics of probability (covered in Chapter 11) so that probabilities they encounter in the rest of the course make sense to them. A basic understanding of probability can help your students understand the probabilities discussed in the chapters on sampling distributions (Chapters 15 and 19), what we mean by confidence (the long-run success rate of confidence intervals), and P-values in hypothesis testing.

If you chose to go beyond Chapter 11 with your students, you will find that this review chapter provides the students with ample opportunity to practice probability concepts without knowing the specific chapter. Watch that your students do not slip into some of the common traps addressed in the instructor material for Chapters 11 and 12 (e.g., assuming that two events are independent or disjoint).

Teaching Suggestions and Additional Examples/Activities for the Classroom

Because the course material is compartmentalized into chapters, students focus on one specific topic at a time. This chapter provides them the opportunity to learn to distinguish between the kinds of problems they have already encountered. Consider working through some comprehensive problems that bring together the most important concepts of the preceding chapters. Some suggested problems include the following:

■ Exercises 14.1 and 14.26: Each of these problems focuses on a specific concept related to probability.

■ Exercises 14.27 and 14.28: These two problems allow students to review basic probability.

■ If you covered Chapters 12 and 13, Exercises 14.22 and 14.23 provide a distinction between simply using the multiplication rule for independent events and a binomial situation. Exercise 14.25 uses both the binomial distribution and its Normal approximation and can be a nice lead-in to inference for binomial proportions, as studied in the next section.

Chris Clinton/Getty Images

Probability: Part III Review

In Part I of this book, you mastered **data analysis,** the use of graphs and numerical summaries to organize and explore any set of data. Part II introduced designs for **data production,** and Part III **probability,** the mathematics that provides the basis for statistical inference. Parts IV and V will deal in detail with **statistical inference.**

The deliberate use of chance in producing data is a central idea in statistics. It not only reduces bias but also allows us to use probability, the mathematics of chance, as the basis for inference. Fortunately, we need only some basic facts about probability to understand statistical inference.

Part III Summary

Here are the most important skills you should have acquired from reading Chapters 11 to 13.

A. Probability

1. Recognize that some phenomena are random. Probability describes the long-run regularity of random phenomena.

2. Understand that the probability of an event is the proportion of times the event occurs in very many repetitions of a random phenomenon. Use the idea of probability as long-run proportion to think about probability.

3. Use basic probability rules to detect illegitimate assignments of probability: any probability must be a number between 0 and 1, and the total probability assigned to all possible outcomes must be 1.

4. Use basic probability rules to find the probabilities of events that are formed from other events. The probability that an event does not occur is 1 minus its probability. If two events are disjoint, the probability that one or the other occurs is the sum of their individual probabilities.

5. Find probabilities in a finite probability model by adding the probabilities of their outcomes. Find probabilities in a continuous probability model as areas under a density curve.

6. Use the notation of random variables to make compact statements about random outcomes, such as $P(X \leq 4) = 0.3$. Be able to interpret such statements.

B. General Rules of Probability (not required for later chapters)

1. Use Venn diagrams to picture relationships among several events.

2. Use the general addition rule to find probabilities that involve overlapping events.

3. Understand the idea of independence. Judge when it is reasonable to assume independence as part of a probability model.

4. Use the multiplication rule for independent events to find the probability that all of several independent events occur.

5. Use the multiplication rule for independent events in combination with other probability rules to find the probabilities of complex events.

6. Understand the idea of conditional probability. Find conditional probabilities for individuals chosen at random from a table of counts of possible outcomes.

7. Use the general multiplication rule to find $P(A \text{ and } B)$ from $P(A)$ and the conditional probability $P(B \mid A)$.

8. Use tree diagrams to organize several-stage probability models.

C. Binomial Distributions (not required for later chapters)

1. Recognize the binomial setting: a fixed number n of independent success–failure trials with the same probability p of success on each trial.

2. Recognize and use the binomial distribution of the count of successes in a binomial setting.

3. Use the binomial probability formula to find probabilities of events involving the count X of successes in a binomial setting for small values of n.

4. Find the mean and standard deviation of a binomial count X.

5. Recognize when you can use the Normal approximation to a binomial distribution. Use the Normal approximation to calculate probabilities that concern a binomial count X.

Test Yourself

The following questions include both multiple-choice and short-answer questions and calculations. They will help you review the basic ideas and skills presented in Chapters 11 to 13.

14.1 You read online that the chance of winning the Powerball lottery is one in 175 million. This means

(a) if 175 million lottery tickets are sold, exactly one of them will be the winning ticket.

(b) if 175 million lottery tickets are sold, at least one of them will be the winning ticket.

(c) if someone wins the lottery, at least 175 million tickets must have been sold.

(d) none of the above.

14.2 A randomly chosen subject arrives for a study of exercise and fitness. Describe a sample space for each of the following. (In some cases, you may have some freedom in your choice of S.)

(a) The subject is either female or male.

(b) After 10 minutes on an exercise bicycle, you ask the subject to rate his or her effort on the Rate of Perceived Exertion (RPE) scale. RPE ranges in whole-number steps from 6 (no exertion at all) to 20 (maximal exertion).

(c) You measure VO_2, the maximum volume of oxygen consumed per minute during exercise. VO_2 is generally between 2.5 and 6.1 liters per minute.

(d) You measure the maximum heart rate (beats per minute).

Ascent Xmedia/Getty Images

Internet search engines. *Internet search sites compete for users because they sell advertising space on their sites and can charge more if they are heavily used. Choose an Internet search attempt at random. Here is the probability distribution for the site the search uses:*[1]

SITE	GOOGLE	YAHOO	MSN	ASK.COM	OTHERS
Probability	0.66	0.21	0.07	0.04	?

Use this information to answer Questions 14.3 and 14.4.

14.3 What is the probability that a search attempt is made at a site other than the leading four?

(a) 0.02

(b) 0.34

(c) 0.98

(d) Cannot be determined from the information given.

14.4 What is the probability that a search attempt is directed to a site other than Google?

(a) 0.02

(b) 0.34

(c) 0.98

(d) Cannot be determined from the information given.

How many in the house? *In government data, a household consists of all occupants of a dwelling unit. Here is the distribution of household size in the United States:*

NUMBER OF PERSONS	1	2	3	4	5	6	7
Probability	0.26	0.33	0.16	0.15	0.07	0.02	0.01

Choose an American household at random, and let the random variable Y be the number of persons living in the household. Use this information to answer Questions 14.5 to 14.7.

14.5 Express "more than one person lives in this household" in terms of Y. What is the probability of this event?

14.6   What is $P(2 < Y \leq 4)$?

14.7   What is $P(Y \neq 2)$?

   (a) 0.26   (b) 0.33   (c) 0.41   (d) 0.67

How many children? *How many children do women give birth to during their childbearing years? Choose at random an American woman who is past childbearing:*[2]

NUMBER OF CHILDREN	0	1	2	3	4	5
Probability	0.193	0.174	0.344	0.181	0.074	0.034

(The few women with 6 or more children are included in the "5 children" group.) Use this information to answer Questions 14.8 to 14.11.

14.8   Check that this distribution satisfies the two requirements for a legitimate finite probability model.

14.9   Describe in words the event $P(X \leq 2)$. What is the probability of this event?

14.10   What is $P(X < 2)$?

   (a) 0.289   (b) 0.344   (c) 0.367   (d) 0.711

14.11   Write the event "a woman gives birth to 3 or more children" in terms of values of X. What is the probability of this event?

Random number generators. *Many random number generators allow users to specify the range of the random numbers to be produced. Suppose you specify that the random number Y can take any value between 0 and 5. The density curve of the outcome has a constant height between 0 and 5, and height 0 elsewhere. Use this information to answer Questions 14.12 to 14.14.*

14.12   The random variable Y is

   (a) discrete.
   (b) continuous, but not Normal.
   (c) continuous and Normal.
   (d) none of the above.

14.13   The height of the density curve between 0 and 5 is

   (a) 0.2.
   (b) 1.
   (c) 5.
   (d) none of the above.

14.14   Draw a graph of the density curve and find $P(1 \leq Y \leq 3)$.

14.15   **(Optional)** Byron claims that the probability that Florida and Alabama will play in the Southeastern Conference (SEC) championship football game this year is 78%. The number 78% is

   (a) the proportion of times Florida and Alabama have played in the championship game in the past.
   (b) Byron's personal probability that Florida and Alabama will play in the SEC championship football game this year.
   (c) the area under a Normal density curve.
   (d) all of the above.

14.16   The Wechsler Adult Intelligence Scale (WAIS) is a common "IQ test" for adults. The distribution of WAIS scores for persons over 16 years of age is approximately Normal with mean 100 and standard deviation 15. What is the probability that a randomly chosen individual has a WAIS score of 105 or higher?

   (a) 0.0005   (b) 0.3707   (c) 0.4400   (d) 0.6293

14.17 **(Optional)** *Accidents, suicide, and murder are the leading causes of death for young adults. Here are the counts of violent deaths in a recent year among people 20 to 24 years of age:*

	FEMALE	MALE
Accidents	1818	6457
Homicide	457	2870
Suicide	345	2152

(a) Choose a violent death in this age group at random. What is the probability that the victim was male?

(b) Find the conditional probability that the victim was male, given that the death was accidental.

Parallel systems (optional). *A system has two components that operate in parallel, as shown in Figure 14.1. Because the components operate in parallel, at least one of the components must function properly if the system is to function properly. The probabilities of failure for Components 1 and 2 during one period of operation are 0.20 and 0.03, respectively. Let F denote the event that Component 1 fails during one period of operation and G denote the event that Component 2 fails during one period of operation. The component failures are independent. Use this information to answer Questions 14.18 and 14.19.*

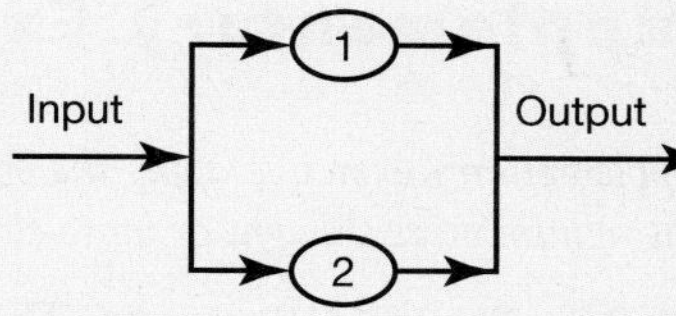

FIGURE 14.1

Parallel systems for Exercise 14.18 and 14.19.

14.18 The event corresponding to the system failing during one period of operation is

(a) F and G.

(b) F or G.

(c) not F or not G.

(d) not F and not G.

14.19 The probability that the system functions properly during one period of operation is closest to

(a) 0.994.   (b) 0.970.   (c) 0.940.   (d) 0.776.

14.20 **(Optional)** A survey of college students finds that 35% like country music, 25% like gospel music, and 15% like both. The proportion of students that like country music but not gospel music is

(a) 15%.   (b) 20%.   (c) 25%.   (d) 40%.

14.21 **(Optional)** Alysha is the star player on her middle school basketball team. If she makes 2 free throws and misses 3 free throws during a game, in how many ways can you arrange the sequence of hits and misses?

(a) $\binom{3}{2} = 3$ (b) $\binom{5}{2} = 20$ (c) $\binom{5}{3} = 10$

14.22 **(Optional)** Alysha makes 40% of her free throws. She takes 5 free throws in a game. If the shots are independent of each other, the probability that she misses the first two shots but makes the other three is about

(a) 0.230.   (b) 0.115.   (c) 0.023.

14.23 **(Optional)** Alysha makes 40% of her free throws. She takes 5 free throws in a game. If the shots are independent of each other, the probability that she makes 3 out of the 5 shots is about

(a) 0.230.   (b) 0.115.   (c) 0.023.

14.24 **(Optional)** Opinion polls find that 63% of American teens say their parents put at least some pressure on them to get into a good college.[3] If you take an SRS of 1000 teens, the approximate distribution of the number in your sample who say that they feel at least some pressure from their parents to get into a good college is

(a) $N(0.63, 15.27)$.   (c) $N(630, 15.27)$.

(b) $N(0.63, 233.1)$.   (d) $N(630, 233.1)$.

14.25 **(Optional)** What kinds of Web sites do males aged 18 to 34 visit? About 50% of male Internet users in this age group visit an auction site such as eBay at least once a month.[4]

(a) If we interview a random sample of 12 male Internet users aged 18 to 34, what is the probability that exactly 8 of the 12 have visited an auction site in the past month?

(b) Suppose we had interviewed a random sample of 500 men aged 18 to 34. What is the probability that at least 235 of the men in the sample visit an online auction site at least once a month? (Check that the Normal approximation is permissible and use it to find this probability.)

Supplementary Exercises

Supplementary exercises apply the skills you have learned in ways that require more thought or more elaborate use of technology.

14.26 **The addition rule.** The addition rule for probabilities, $P(A \text{ or } B) = P(A) + P(B)$, is not always true. Give (in words) an example of real-world events A and B for which this rule is not true.

14.27 **Comparing wine tasters.** Two wine tasters rate each wine they taste on a scale of 1 to 5. From data on their ratings of a large number of wines, we obtain the following probabilities for both tasters' ratings of a randomly chosen wine:

	TASTER 2				
TASTER 1	**1**	**2**	**3**	**4**	**5**
1	0.03	0.02	0.01	0.00	0.00
2	0.02	0.08	0.05	0.02	0.01
3	0.01	0.05	0.25	0.05	0.01
4	0.00	0.02	0.05	0.20	0.02
5	0.00	0.01	0.01	0.02	0.06

(a) Why is this a legitimate finite probability model?

(b) What is the probability that the tasters agree when rating a wine?

(c) What is the probability that Taster 1 rates a wine higher than Taster 2? What is the probability that Taster 2 rates a wine higher than Taster 1?

14.28 **A 14-sided die.** An ancient Korean drinking game involves a 14-sided die. The players roll the die in turn and must submit to whatever humiliation is written on the up-face: something like "Keep still when tickled on face." Six of the 14 faces are squares. Let's call them A, B, C, D, E, and F for short. The other eight faces are triangles, which we will call 1, 2, 3, 4, 5, 6, 7, and 8. Each of the squares is equally likely. Each of the triangles is also equally likely, but the triangle probability differs from the square probability. The probability of getting a square is 0.72. Give the probability model for the 14 possible outcomes.

David Moore

14.29 **Testing for ESP (optional).** In a test for ESP (extrasensory perception), a subject is told that cards the experimenter can see but he cannot contain either a star, a circle, a wave, or a square. As the experimenter looks at each of 20 cards in turn, the subject names the shape on the card. A subject who is just guessing has probability 0.25 of guessing correctly on each card.

(a) The count of correct guesses in 20 cards has a binomial distribution. What are n and p?

(b) What is the mean number of correct guesses in many repetitions of the experiment?

(c) What is the probability of exactly 5 correct guesses?

14.30 **Internet use in Canada (optional).** Surveys find that in 2009, 80% of Canadians aged 16 and older used the Internet for personal reasons.[5] In the survey, an "Internet user" is defined as someone who used the Internet for personal reasons from any location in the 12 months preceding the survey. If you take an SRS of 1500 Canadians aged 16 and over, what is the approximate distribution of the number in your sample who have used the Internet for personal reasons?

14.31 Testing for AIDS (optional). Enzyme immunoassay tests are used to screen blood specimens for the presence of antibodies to HIV, the virus that causes AIDS. Antibodies indicate the presence of the virus. The test is quite accurate but is not always correct. Here are approximate probabilities of positive and negative test results when the blood tested does and does not actually contain antibodies to HIV:[6]

| | **TEST RESULT** | |
	POSITIVE	**NEGATIVE**
Antibodies present	0.9985	0.0015
Antibodies absent	0.0060	0.9940

Suppose that 1% of a large population carries antibodies to HIV in their blood.

(a) Draw a tree diagram for selecting a person from this population (outcomes: antibodies present or absent) and testing his or her blood (outcomes: test positive or negative).

(b) What is the probability that the test is positive for a randomly chosen person from this population?

14.32 Testing for AIDS, continued (optional). Continue your work from the previous exercise. What is the probability that a person has the antibody, given that the test is positive? (Your result illustrates a fact that is important when considering proposals for widespread testing for HIV, illegal drugs, or agents of biological warfare: if the condition being tested is uncommon in the population, most positives will be false positives.)

14.33 What type of high school did you attend? (optional). Choose a college freshman at random and ask what type of high school they attended. Here is the distribution of results:[7]

TYPE	**REGULAR PUBLIC**	**PUBLIC CHARTER**	**PUBLIC MAGNET**
Probability	0.781	0.018	0.031

TYPE	**PRIVATE RELIGIOUS**	**PRIVATE INDEPENDENT**	**HOMESCHOOL**
Probability	0.105	0.059	0.006

What is the conditional probability that a college freshman was homeschooled, given that he or she did not attend a regular public high school?

© Rosemary Harris/Alamy in the Dyers' Souk, Marrakech, Morocco

Inference about Variables

Armed with designs for producing trustworthy data, data analysis to examine the data, and the language of probability, we are now prepared to understand the big ideas of inference. In the remaining chapters of this book, you will meet many of the most commonly used statistical procedures. We have grouped these procedures into two classes, corresponding to our division of data analysis into exploring variables and distributions and exploring relationships. Part IV concerns inference about the distribution of a single variable and inference for comparing the distribution of two variables. Part V deals with inference for relationships among variables.

We begin Part IV with inference for a categorical response variable, making use of counts and proportions of outcomes. Chapters 15, 16, and 17 introduce the basic ideas of inference in the context of a single proportion. These ideas are the foundation for the discussion of inference that occupies the rest of the book. Chapter 18 applies these basic ideas to inference for comparing two proportions.

In Chapters 19, 20, and 21, we discuss inference for a quantitative response variable, focusing on inference for means. Chapter 22 discusses some important issues that arise when using inference in practice. Chapter 23 reviews this part of the text.

The four-step process for approaching a statistical problem can guide much of your work in these chapters. The statement of an exercise usually does the "State" step for you, leaving the "Plan," "Solve," and "Conclude" steps for you to complete. It is helpful to first summarize the *State* step in your own words to organize your thinking. Many examples and exercises in these chapters involve both carrying out inference and thinking about inference in practice. Remember that any inference method is useful only under certain conditions, and that you must judge these conditions before rushing to inference.

Sampling Distribution for a Proportion

Overview

Sampling distributions form the underpinning of statistical inference, so this chapter discusses concepts crucial to the success of introductory students. Unfortunately, the notion of a sampling distribution is often difficult for beginning students to grasp. On the surface, it is easy to see that different samples of the same size may yield different values for a sample statistic (that we have sampling variability), but the idea that the statistic has its own distribution, with its own mean and standard deviation, is a more abstract concept.

In *Statistics in Practice*, inference for proportions is taught before inference for means. Arguments can be (and have been) made to teach proportions before means (and vice versa).

If we choose to teach inference for means first, we are able to rely on what students know about the Normal distribution if we assume that we know the population standard deviation σ. We build on that foundation by introducing the sampling distribution for a sample mean, which is Normal (by assumption or through using the central limit theorem). The inference for means based on this assumption is easier for our students to learn because they are already familiar with the standard Normal distribution. It would be a challenge for us to introduce the sampling distribution for the sample mean, admit that we will rarely know σ, and plunge immediately into using the Student's t distribution, as the students then have to work in a new, more difficult setting, *and* with a new distribution.

For those instructors who wish to remain in the realm of the comfortable standard Normal distribution but do not want to address the sampling distribution for the sample mean when σ is known (an artificial setting), presenting the sampling distribution for the sample proportion first is a natural solution. However, care must be taken with the theory—we do not want to overwhelm our students with theory, but we need to acknowledge that there is a difference in teaching inference for proportions before inference for means. At the same time, students may be more familiar with proportions than with means due to the prevalence of surveys in the news media, so presenting proportions first benefits those students. There is not one good answer to "should means or proportions come first?" For arguments for and against, see Chance, B. L., and Rossman, A. J. "Sequencing Topics in Introductory Statistics: A Debate on What to Teach When," *The American Statistician*, 2001, 55 (2), pp. 140–144.

LEARNING OUTCOMES**

- Identify parameters and statistics in a statistical study.

- Recognize the fact of sampling variability: a statistic will take different values when you repeat a sample or experiment.

- Interpret a sampling distribution as describing the values taken by a statistic in *all* possible repetitions of a sample or experiment under the same conditions (i.e., the sample you observe and all other possible samples of the same size).

- Interpret the sampling distribution of a statistic as describing the probabilities of its possible values.

- Recognize when a problem involves the proportion $\hat{p}$ of a sample. Understand that $\hat{p}$ estimates a proportion p in the population and that the population proportion p is fixed, but (almost always) unknown.

  - Note: Although a proportion may change over time (e.g., attitudes toward same-sex marriage), the proportion is considered fixed for each sample. This is the reason that most polls are done over a short time frame.

- Use the law of large numbers to describe the behavior of $\hat{p}$ as the size of the sample increases.

- Find the mean and standard deviation of a sample proportion $\hat{p}$ from an SRS of size n when the population proportion p is known.

- Understand that $\hat{p}$ is an unbiased estimator of p and that the variability of $\hat{p}$ about its mean p gets smaller as the sample size increases.

- Understand and check the conditions for which $\hat{p}$ has approximately a Normal distribution. Use this Normal distribution to calculate probabilities that concern $\hat{p}$.

**These learning outcomes appear later for the students in Chapter 23: Part IV Review.

Often, a researcher is interested in estimating a population proportion. This is a parameter that we label p. For example, p could denote the proportion of all U.S. adults who approve of the president's job performance; or it could be the proportion of all parts manufactured by a factory that are defective. In all such problems, the researcher is observing qualitative data ("yes" versus "no," or "success" versus "fail" outcomes). Here, we will use the sample proportion, $\hat{p}$, to estimate the population proportion, p.

The chapter begins by reminding students about the definitions of parameter and statistic and introduces the sample proportion $\hat{p}$. The law of large numbers for proportions is introduced as a bridge between the idea that samples vary and the idea of a sampling distribution. In examining the sampling distribution of a sample proportion, students are reminded of the 68–95–99.7 rule, thus allowing them to connect the behavior of a sample proportion back to the Normal distribution they learned about in Chapter 3.

Teaching Suggestions and Additional Examples/Activities for the Classroom

1. Discuss Sampling Variability

When you first discuss the notion of a sampling distribution, ask students to consider a survey of (say) 100 students. Imagine having all the students in the class survey 100 randomly selected students at the school, and each using their sample to estimate the proportion of students who play intramural sports based on their sample. Students should easily recognize that (1) their sample proportion would differ if they were to take another sample, and (2) the differences between their sample proportions will tend to decrease if each surveys 200 (or 500) students instead of 100 because more of the population has been included in each sample.

2. Use an Applet to Help Students Visualize a Sampling Distribution

Once students grasp that the value of a statistic changes from sample to sample, they are ready to explore the notion of a sampling distribution. This is easily done using simulation—students should simulate for themselves, using the *Sampling Distribution of a Proportion* applet—the distribution of a sample proportion for different values of n and p. With this applet, students can simulate individual sample proportions (and see those sample proportions cluster around p), or generate 10,000 different samples and observe the approximate sampling distribution.

3. Use an Applet to Address the Law of Large Numbers for Proportions

Working with the law of large numbers for proportions is equivalent to observing the long-run behavior of a proportion. The *Probability* applet can simulate sampling from a population with any proportion of successes. You can have your students work with the applet as is (i.e., talk about the probability that a fair coin comes up heads) or use a proportion that may be of interest to them (e.g., find results from a Gallup poll and use the observed proportion from a survey—this will be similar to what the students are asked to do in Exercise 15.5). The *Probability: Spinning a Roulette Wheel* applet is also applicable here; this applet allows you to simulate the proportion of "successes" for several types of "bets."

Other Resources (LaunchPad)

Applet
Probability
Probability: Spinning a Roulette Wheel
Sampling Distribution of a Proportion

© Blend Images/Alamy

Sampling Distribution for a Proportion

In this chapter we cover...

- Parameters and statistics
- Statistical estimation using the sample proportion
- Sampling distributions
- The sampling distribution of $\hat{p}$

What can be said about the emotional health of college freshmen? The Freshman Survey administered nationally by the Higher Education Research Institute (HERI) at UCLA's Graduate School of Education & Information Studies sampled 201,818 college freshmen in 2010.[1] Only 51.9% of students surveyed reported that their emotional health was "above average," the lowest value since the survey began in 1985. The 51.9% describes the sample but we use it to *estimate* the percentage of all freshman who would report their emotional health as "above average." This is an example of statistical inference: we use information from a sample to infer something about a wider population.

Because the results of random samples and randomized comparative experiments include an element of chance, we can't guarantee that our inferences are correct. What we can guarantee is that our methods usually give correct answers. The reasoning of statistical inference rests on asking, "How often would this method give a correct answer if we used it very many times?" If our data come from random sampling or randomized comparative experiments, the laws of probability answer the question "What would happen if we did this many times?" This chapter presents some facts about probability that help answer this question.

Parameters and statistics

Our study of inference begins with populations that have some outcome of interest. For convenience, call the outcome we are looking for a "success." In the college student survey, the population is all college freshmen, and a "success" is a student who would report that their emotional health was "above average."

When using sample data to draw conclusions about a wider population, we must take care to keep straight whether a number describes a sample or a population. Here is the vocabulary we use.

WHO IS A SMOKER?

When estimating a proportion p, be sure you know what counts as a "success." The news says that 20% of adolescents smoke. Shocking. It turns out that this is the percent who smoked at least once in the past month. If we say that a smoker is someone who smoked on at least 20 of the past 30 days and smoked at least half a pack on those days, fewer than 4% of adolescents qualify.

Parameter, Statistic

A **parameter** is a number that describes the population. In statistical practice, the value of a parameter is not known because we cannot examine the entire population.

A **statistic** is a number that can be computed from the sample data without making use of any unknown parameters. In practice, we often use a statistic to estimate an unknown parameter.

EXAMPLE 15.1 Risky Behavior in the Age of AIDS

How common is behavior that puts people at risk of AIDS? In the early 1990s, the landmark National AIDS Behavioral Surveys used random dialing of telephone numbers to contact a sample of 2673 adult heterosexuals. Of these, 170 had more than one sexual partner in the past year.[2] The population that the survey wants to draw conclusions about is all heterosexual adults. ■

The *parameter* of interest is p, the proportion of all heterosexual adults who had more than one sexual partner in the last year. We don't know the value of this parameter. The *statistic* that estimates the proportion of all adult heterosexuals having multiple sexual partners in the last year is the **sample proportion**

sample proportion

$$\hat{p} = \frac{\text{number of successes in the sample}}{\text{total number of individuals in the sample}}$$
$$= \frac{170}{2673} = 0.0636$$

Read the sample proportion $\hat{p}$ as "p-hat." The proportion 0.0636 is a statistic because it describes this one sample. A different sample of 2673 heterosexual adults would almost certainly lead to a value of $\hat{p}$ that differs from 0.0636.

Remember **s**tatistics come from **s**amples, and **p**arameters come from **p**opulations. As long as we were just doing data analysis, searching for patterns or summarizing features of our data, the distinction between population and sample was not as important. Now, as we begin to understand what our data (sample) tell us about a population, it is essential. The notation we use must reflect this distinction. This is why we write p for the **proportion in a population** and $\hat{p}$ for the **proportion in a sample.** The population proportion p is a fixed parameter that is unknown when we use a sample for inference. The sample proportion $\hat{p}$ is a statistic that would almost certainly take a different value if we chose another sample from the same population.

Apply Your Knowledge

15.1 Genetic Engineering. Here's a new idea for treating advanced melanoma, the most serious kind of skin cancer. Genetically engineer white blood cells to better recognize and destroy cancer cells, then infuse these cells into patients. The subjects in a small initial study of this approach were 11 patients whose melanoma had not responded to existing treatments. One outcome of this experiment was measured by a test for the presence of cells that trigger an immune response in the body and so may help fight cancer. The mean counts of active cells per 100,000 cells for the 11 subjects were **3.8** before infusion and **160.2** after infusion. Is each of the boldface numbers a parameter or a statistic?

15.2 Florida Voters. Florida played a key role in the 2000 and 2004 presidential elections. Voter registration records in October 2012 show that **41%** of Florida voters are registered as Democrats and **35%** as Republicans. (Most of the others did not choose a party.) To test a random-digit dialing device that you plan to use to poll voters for the next election, you use it to call 250 randomly chosen residential telephones in Florida. Of the registered voters contacted, **34%** are registered Democrats. Is each of the boldface numbers a parameter or a statistic?

15.3 Human Growth Hormone. Researchers surveyed 250 American male weight lifters, ranging in age from 18 to 40, to learn about the behaviors of all American male weight lifters in this age range. They found that **12%** of them had used human growth hormone (HGH), which has been banned in sports for more than 20 years now. The median usage time for those who reported HGH use was **23** weeks. Is each of the boldface numbers a parameter or a statistic?

© Rubberball/age fotostock

Statistical estimation using the sample proportion

Statistical inference uses sample data to draw conclusions about the entire population. Because good samples are chosen with random sampling, statistics such as $\hat{p}$ computed from these samples are random variables. We can describe the behavior of a sample statistic by a probability model that answers the question "What would happen if we took random samples of the same size many times?" Here is an example that will lead us toward the probability ideas most important for statistical inference.

EXAMPLE 15.2   More Ohio Uninsured Drivers on the Road!

The proportion of uninsured drivers varies widely between states as do the enforcement policies. Maine and Massachusetts have the lowest rates in the nation, and both require drivers to show proof of insurance when they register a vehicle. Ohio is a passive enforcement state: although it requires drivers to indicate that they have insurance when they register a vehicle, no proof of insurance is required.[3]

The proportion p of Ohio drivers without insurance is a parameter describing the population of Ohio drivers. To estimate p, we take a simple random sample of 150 Ohio drivers and find that 27 do not have insurance. The sample proportion of these subjects without insurance is $\hat{p} = 27/150 = 0.18$. It seems reasonable to use the sample result $\hat{p} = 0.18$ to estimate the unknown p. An SRS should fairly represent the population, so the proportion $\hat{p}$ of the sample should be somewhere near the proportion p of the population. Of course, we don't expect $\hat{p}$ to be exactly equal to p. We realize that if we choose another SRS, the sample will probably produce a different $\hat{p}$. ■

Blend Images/SuperStock

If $\hat{p}$ is rarely exactly right and varies from sample to sample, why is it nonetheless a reasonable estimate of the population proportion p? Here is one answer: *if we continue to take larger and larger samples, the statistic $\hat{p}$ is guaranteed to get closer and closer to the parameter p.* We have the comfort of knowing that if we can afford to continue sampling more subjects, eventually we will estimate the proportion of uninsured Ohio drivers very accurately. This is a special case of a more general mathematical result called the *law of large numbers* that we will encounter in Chapter 19.

Law of Large Numbers for Proportions

Draw observations at random from any population with population proportion p. As the number of observations drawn increases, the sample proportion $\hat{p}$ gets closer and closer to the population proportion p.

You should recognize the similarity between the law of large numbers for proportions and the idea of probability (page 260). The idea of probability states that in the long run, the *proportion* of outcomes taking any value gets close to the probability of that value. Suppose the value of the outcome corresponds to obtaining a success when sampling a single individual from the population. Then the probability of the value of the outcome is p, the proportion of successes in the population. The fact that $\hat{p}$ gets closer and closer to the population proportion p is just the idea of probability rephrased in our new terminology. Because of the importance of this concept, here is an example similar to Example 11.2 (page 260), but in the context of inference.

EXAMPLE 15.3	**The Law of Large Numbers for Proportions in Action**

Extensive studies conducted throughout the year show that the proportion of uninsured drivers in Ohio is very close to 16%.[4] Because of this, we will take $p = 0.16$ as the value of the population proportion, or the true value of the parameter. Figure 15.1 shows how the sample sample proportion $\hat{p}$ of an SRS drawn from the population of drivers changes as we add more subjects to our sample.

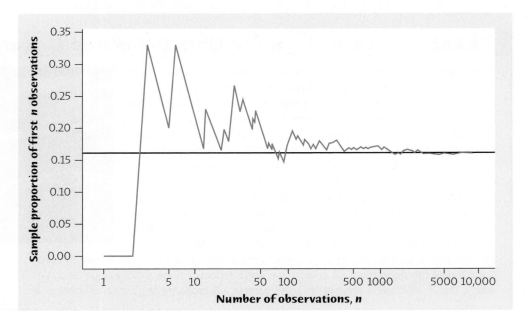

FIGURE 15.1

The law of large numbers for proportions in action: as we take more observations, the sample proportion $\hat{p}$ always approaches the population proportion p.

The first subject in the sample had insurance, so the sample proportion after sampling one subject is zero, and the line in Figure 15.1 starts there. The second subject also had insurance, and the sample proportion after sampling two subjects remains at zero as in Figure 15.1. The third subject did not have insurance, so after sampling three subjects the sample proportion increases to

$$\hat{p} = \frac{1}{3} = 0.33,$$

the third point on the graph. At first, the graph shows that the sample proportion changes considerably as we take more observations. Eventually, however, the sample proportion gets close to the population proportion $p = 0.16$ and settles down to that value.

If we started over, again choosing people at random from the population, we would get a different path from left to right in Figure 15.1. The law of large numbers says that whatever path we get will always settle down to 0.16 as we draw more and more people. ■

We know that if we continue to sample subjects, eventually we will estimate the population proportion very accurately. In practice, we take a sample of a fixed size. The key idea that will allow us to calculate the accuracy of an estimate for a fixed sample size is presented in the next section.

Apply Your Knowledge

15.4 **Prayer Among the Millennials.** The Millennial generation (so called because they were born after 1980 and began to come of age around the year 2000) are less religiously active than older Americans. One of the questions in the General Social Survey in 2012 was "How often does the respondent pray?" Among the 377 respondents in the survey between 18 and 30 years of age, 225 prayed at least once a week.[5]

   (a) Describe the population and explain in words what the parameter p is.

   (b) Give the numerical value of the statistic $\hat{p}$ that estimates p.

15.5 **The Law of Large Numbers for Proportions Made Visible.** You are going to simulate sampling from a population with proportion of successes $p = 0.16$ as in Example 15.3. The law of large numbers says that the proportion $\hat{p}$ from a sample gets closer and closer to $p = 0.16$ as the sample increases in size.

   (a) The *Probability* applet can simulate sampling from a population with any proportion of successes. Set the probability of heads to 0.16 and the number of tosses to 50 using the sliders. Then check "Show true probability" to show this value on the graph. Now click "Toss." You see that the graph displays at each point the proportion of successes up to the last one. This is exactly like Figure 15.1. What is the value of $\hat{p}$ after 50 tosses? This will be to the right of "heads =" in the window.

   (b) Click "Toss" again to generate 50 more observations—don't click "Reset." Repeat this one more time so that there are now a total of 150 observations. What changes do you notice after clicking "Toss" each time in the graph? What portion of your final graph is similar to Figure 15.1? Is there any portion of the graph that is different? Sketch (or print out) the final graph.

Sampling distributions

The law of large numbers assures us that if we measure enough subjects, the statistic $\hat{p}$ will eventually get very close to the unknown parameter p. But what can we say about estimating p by $\hat{p}$ from a sample of 150 subjects in Example 15.2? To answer this question, put this one sample in the context of all such samples by asking, "What would happen if we took many samples of 150 subjects from this population?" Here's how to answer this question:

■ Take a large number of samples of size 150 from the population.

■ Calculate the sample proportion $\hat{p}$ for each sample.

■ Make a histogram of the values of $\hat{p}$.

■ Examine the shape, center, and spread/variability of the distribution displayed in the histogram.

In practice it is too expensive to take many samples from a large population such as all Ohio drivers or all adults in the United States. But we can imitate taking many samples by using software? Using software to imitate chance behavior is called ***simulation.***

simulation

EXAMPLE 15.4   What Would Happen in Many Samples?

Extensive studies in Ohio have found that the proportion of uninsured drivers is approximately 16%. This is the *population proportion.*

Figure 15.2 illustrates the process of choosing many samples and finding the sample proportion of uninsured drivers $\hat{p}$ for each one. Follow the flow of the figure from the population at the left, to choosing an SRS and finding the $\hat{p}$ for this sample, to collecting together the $\hat{p}$'s from many samples. The first sample has $\hat{p} = 0.120$. The second sample contains a different 150 people, with $\hat{p} = 0.147$, and so on. The histogram at the right of the figure shows the distribution of the values of $\hat{p}$ from 2000 separate SRSs of size 150. This histogram displays the *sampling distribution* of the statistic $\hat{p}$. ■

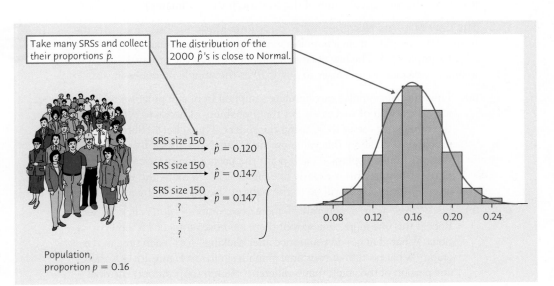

FIGURE 15.2

The idea of a sampling distribution: take many samples from the same population, collect the $\hat{p}$'s from all the samples, and display the distribution of the $\hat{p}$'s. The histogram shows the results of 2000 samples.

Sampling Distribution

The **sampling distribution** of a statistic is the distribution of values taken by the statistic in *all* possible samples of the same size from the same population.

Strictly speaking, the sampling distribution is the ideal pattern that would emerge if we looked at all possible samples of size 150 from our population. A histogram obtained from a fixed number of trials, like the 2000 trials in Figure 15.2, is only an approximation to the sampling distribution. One of the uses of probability theory in statistics is to obtain sampling distributions without simulation. The interpretation of a sampling distribution is the same, however, whether we obtain it by simulation or by the mathematics of probability.

We can use the tools of data analysis to describe any distribution. Let's apply those tools to Figure 15.2. What can we say about the shape, center, and spread/variability of this distribution?

■ **Shape:** It looks Normal! Detailed examination confirms that the distribution of $\hat{p}$ from many samples is very close to Normal.

■ **Center:** The mean of the 2000 $\hat{p}$'s is 0.1602. That is, the distribution is centered very close to the population proportion $p = 0.16$.

■ **Spread:** The standard deviation of the 2000 $\hat{p}$'s is 0.0292. The relationship between the standard deviation, the population proportion, and the sample size is described in the next section.

Although these results describe just one simulation of a sampling distribution, they reflect facts that are true whenever we use random sampling.

Apply Your Knowledge

15.6 **Generating a Sampling Distribution.** Let's illustrate the idea of a sampling distribution in the case of a very small sample from a very small population. The population is the sex of 10 students in a class:

Student	0	1	2	3	4	5	6	7	8	9
Score	F	F	M	M	F	M	M	M	F	M

The parameter of interest is the proportion of females p in this population. The sample is an SRS of size $n = 4$ drawn from the population. Because the students are labeled 0 to 9, a single random digit from Table B chooses one student for the sample.

(a) Find the proportion of females among the 10 students in the population. This is the population proportion p.

(b) Use technology or the first digits in row 116 of Table B to draw an SRS of size 4 from this population. What is the sex of the four students in your sample? What is the proportion of females $\hat{p}$? This statistic is an estimate of p.

(c) Repeat this process 9 more times, using technology or the first digits in rows 117 to 125 of Table B. Make a histogram of the 10 values of $\hat{p}$. You are simulating the sampling distribution of $\hat{p}$, with 10 simulations. Is the center of your histogram close to p?

The sampling distribution of $\hat{p}$

Figure 15.2 suggests that when we choose many SRSs from a population, the sampling distribution of the sample proportions is centered at the population proportion p. The spread/variability depends on both the population proportion p and the sample size n. Here are the facts we need to get started.[6]

Mean and Standard Deviation of a Sample Proportion

Draw an SRS of size n from a large population that contains proportion p of successes. Let $\hat{p}$ be the **sample proportion** of successes,

$$\hat{p} = \frac{\text{number of successes in the sample}}{n}$$

Then:

- The **mean** of the sampling distribution is p.
- The **standard deviation** of the sampling distribution is

$$\sqrt{\frac{p(1-p)}{n}}$$

These results about the mean and the standard deviation of the sampling distribution of $\hat{p}$ have important implications for statistical inference:

- The mean of the statistic $\hat{p}$ is always equal to the mean p of the population. That is, the sampling distribution of $\hat{p}$ is centered at p. In repeated sampling, $\hat{p}$ will sometimes be above the true value of the parameter p and sometimes below, but there is no systematic tendency to overestimate or underestimate the parameter. This makes the idea of lack of bias in the sense of "no favoritism" more precise. Because the mean of $\hat{p}$ is equal to p, we say that the statistic $\hat{p}$ is an **unbiased estimator** of the parameter p.

- An unbiased estimator is "correct on the average" in many samples. How close the estimator falls to the parameter in most samples is determined by the spread/variability of the sampling distribution. The sample proportion $\hat{p}$ from samples of size n has standard deviation $\sqrt{(p(1-p)/n)}$. For any population proportion p, the standard deviation of the distribution of $\hat{p}$ gets smaller as we take larger samples. **The results of large samples are less variable than the results of small samples.**

The upshot of all this is that we can trust the sample proportion from a large random sample to estimate the population proportion accurately. If the sample size n is large, the standard deviation of $\hat{p}$ is small, and almost all samples will give values of $\hat{p}$ that lie very close to the true parameter p. *However, the standard deviation of the sampling distribution gets smaller only at the rate $\sqrt{n}$. To cut the standard deviation of $\hat{p}$ in half, we must take four times as many observations, not just twice as many.* So very precise estimates (estimates with very small standard deviation) may be expensive, time consuming, and impractical.

unbiased estimator

SAMPLE SIZE MATTERS

The new thing in baseball is using statistics to evaluate players, with new measures of performance to help decide which players are worth the high salaries they demand. This challenges traditional subjective evaluation of young players and the usefulness of traditional measures such as batting average. But success has led many major league teams to hire statisticians. The statisticians say that sample size matters in baseball also: the 162-game regular season is long enough for the better teams to come out on top, but 5-game and 7-game play-off series are so short that luck has a lot to say about who wins.

We have described the center and spread of the sampling distribution of a sample proportion $\hat{p}$, but not its shape. As the sample size gets larger, the shape of the sampling distribution approaches a Normal distribution for any value of the population proportion.

Sampling Distribution of a Sample Proportion

As the sample size increases, the sampling distribution of $\hat{p}$ becomes **approximately Normal.** That is, for large n, $\hat{p}$ has approximately the $N(p, \sqrt{p(1-p)/n})$ distribution. As a rule of thumb, use this approximation when the sample size n is so large that both np and $n(1-p)$ are 10 or more.

The accuracy of the Normal approximation improves as the sample size n increases. It is most accurate for any fixed n when p is close to 1/2 and least accurate when p is near 0 or 1. This is why the rule of thumb in the box depends on p as well as n. Here is an example on using the sampling distribution.

EXAMPLE 15.5   The 68–95–99.7 Rule and $\hat{p}$

Although over 50% of American adults believe the maxim that breakfast is the most important meal of the day, only about 30% eat breakfast daily.[7] A cereal manufacturer plans to select an SRS of 1000 American adults. What does the 68–95–99.7 rule tell us about $\hat{p}$, the proportion in the sample who eat breakfast every day?

First verify that the rule of thumb to use the Normal approximation is satisfied. Both $np = (1000)(0.3) = 300$ and $n(1-p) = (1000)(0.7) = 700$ are 10 or more. The sampling distribution for $\hat{p}$ based on an SRS of 1000 is approximately Normal with mean $p = 0.3$ and standard deviation $\sqrt{p(1-p)/n} = \sqrt{0.3(0.7)/1000} = 0.014$. The 68 part of the 68–95–99.7 rule says that 68% of samples of size 1000 should have $\hat{p}$ within one standard deviation of the mean, or between

$$(0.3 - 0.014) = 0.286 \text{ and } (0.3 + 0.014) = 0.314.$$

Similarly, 95% of samples of size 1000 should have $\hat{p}$ between

$$(0.3 - (2)(0.014)) = 0.272 \text{ and } (0.3 + (2)(0.014)) = 0.328$$

and almost all samples should have $\hat{p}$ within 3 standard deviations of the mean, or between 0.258 and 0.342. ■

In Example 15.2 we estimated the proportion of uninsured drivers in Ohio with an SRS of 150 drivers. How close is this estimate to the true proportion? Here is another example showing how the sampling distribution can be used to begin to answer this question.

EXAMPLE 15.6   How Close Is the Estimate?

STATE: We are conducting a poll to estimate the proportion of Ohio drivers who are uninsured. It is desired to have our sample estimate be within 2% of the population proportion with high probability, say at least 95%. A simple random sample of 150 Ohio drivers will be taken. Is this sample size sufficient to obtain the desired accuracy in our estimate? Or do we need to take a larger sample?

PLAN: The true proportion of uninsured drivers in the state is approximately 16%. For our estimate based on an SRS of 150 drivers to be within 2% of the population proportion requires that $\hat{p}$ be between 14% and 18%. What is the probability that $\hat{p}$ based on an SRS of 150 drivers is between 14% and 18%?

SOLVE: First verify that the rule of thumb to use the Normal approximation is satisfied. Both $np = (150)(0.16) = 24$ and $n(1 - p) = (150)(0.84) = 126$ are 10 or more. The sampling distribution for a proportion says that the sample proportion $\hat{p}$ of uninsured drivers based on a sample of 150 from a population in which 16% of all drivers are uninsured has approximately the Normal distribution with mean equal to the population mean $p = 0.16$ and standard deviation

$$\sqrt{\frac{p(1 - p)}{n}} = \sqrt{\frac{0.16(1 - 0.16)}{150}} = 0.03$$

The distribution of $\hat{p}$ is therefore approximately $N(0.16, 0.03)$.

Using this Normal distribution, the probability we want is

$$P(0.14 < \hat{p} < 0.18) = 0.4950$$

Software gives this probability immediately, or you can standardize and use Table A. For example,

$$P(0.14 < \hat{p} < 0.18) = P\left(\frac{0.14 - 0.16}{0.03} < \frac{\hat{p} - 0.16}{0.03} < \frac{0.18 - 0.16}{0.03}\right)$$

$$= P(-0.67 < Z < 0.67) = 0.7486 - 0.2514 = 0.4972$$

with the usual roundoff error.

CONCLUDE: If you sample 150 drivers, the probability that your sample proportion is within 2% of the population proportion is only about 50%. To increase this probability to at least 95%, it is necessary to take a larger sample. ■

Example 15.6 raises two important issues. The first is "How do we determine the necessary sample size so that our estimate is within a certain percentage of the true value with a high probability?" The second issue is more subtle. In Example 15.6, the value of the population proportion p was used in our calculation of the accuracy of the estimate. In practice, the value of p is unknown or there would be no need to estimate it. These issues will be addressed in the next chapter, which presents confidence intervals for a population proportion, our first methodology for statistical inference.

Apply Your Knowledge

15.7 Lead-Based Paint. The U.S. Environmental Protection Agency defines lead-based paint as any paint that contains more than 0.5% lead by weight (or about 1 milligram per square centimeter of painted surface). This is the "Action Level" at which the EPA recommends removal of lead paint if it is deteriorating and chipping. A government survey plans to estimate the proportion of schools in your state that exceed the Action Level. The researchers will report the proportion $\hat{p}$ from their sample as an estimate of the population proportion p.

(a) Explain to someone who knows no statistics what it means to say that $\hat{p}$ is an "unbiased" estimator of p.

(b) The sample result $\hat{p}$ is an unbiased estimator of the population proportion p no matter what size SRS the study uses. Explain to someone who knows no statistics why a large sample gives more trustworthy results than a small sample.

15.8 Larger Sample, More Accurate Estimate. About 90% of young adult Internet users (ages 18 to 29) use social-networking sites.[8]

(a) Suppose a sample survey contacts an SRS of 1500 young adult Internet users and calculates the proportion $\hat{p}$ in this sample who use social-networking sites. What is the approximate distribution of $\hat{p}$? What is the probability that $\hat{p}$ is between 87% and 93%? This is the probability that $\hat{p}$ estimates p within 3%.

(b) If the sample size were 6000 rather than 1500, what would be the approximate distribution of $\hat{p}$? Now what is the probability that $\hat{p}$ falls within 3% of p? The larger sample size is much more likely to give an accurate estimate of p.

15.9 Teen Binge Drinking. In 2012, about 24% of high school seniors reported binge drinking (defined as 5 or more drinks in a row in the past 2 weeks), a substantial drop since the late 1990s.[9] An SRS of 500 high school seniors is to be taken.

(a) What is the standard deviation of $\hat{p}$, the proportion of high school seniors in the sample who would report binge drinking?

(b) How large a sample is required to reduce the standard deviation of $\hat{p}$ to 0.01?

CHAPTER 15 SUMMARY

Chapter Specifics

- A **parameter** in a statistical problem is a number that describes a population, such as the **population proportion p.** To estimate an unknown parameter, use a **statistic** calculated from a sample, such as the **sample proportion $\hat{p}$.**

- The **law of large numbers for proportions** states that the actual observed sample proportion $\hat{p}$ must approach the population proportion p as the number of observations increases.

- The **sampling distribution** of a statistic describes the values of the statistic in all possible samples of the same size from the same population.

- When the sample is an SRS from the population, the **mean** of the sampling distribution of the sample proportion $\hat{p}$ is the same as the population proportion p. That is, $\hat{p}$ is an **unbiased estimator** of p.

- The **standard deviation** of the sampling distribution of $\hat{p}$ is $\sqrt{p(1-p)/n}$ for an SRS of size n when the population proportion is p. The standard deviation of the sampling distribution gets smaller only at the rate $\sqrt{n}$.

- Choose an SRS of size n from a population with population proportion p. When n is large, **the sampling distribution of $\hat{p}$ is approximately Normal.** We can use the $N(p, \sqrt{p(1-p)/n})$ distribution to calculate approximate probabilities for events involving $\hat{p}$, provided both np and $n(1-p)$ are 10 or more.

Link It

As we mentioned in Chapter 11, probability is the tool we will use to generalize from data produced by random samples and randomized comparative experiments to some wider population. In this chapter we begin to formalize this process. We use a statistic to estimate an unknown parameter. We use the sampling distribution to summarize the behavior of a statistic in all possible random samples of the same size from a population.

More specifically, in this chapter we begin to think about how a sample proportion, $\hat{p}$, can provide information about a population proportion, p. When the sample proportion is computed from an SRS drawn from a large population, its sampling distribution has properties that help us understand how the sample proportion can be used to draw conclusions about a population proportion. The law of large numbers tells us that a sample proportion computed from a *random* sample from some population gets closer and closer to the population proportion as the sample size increases. The approximate Normality of the sampling distribution of the sample proportion for "large" SRSs allows us to make probability statements about possible values of the sample proportion. In the next two chapters, we will develop specific methods for drawing conclusions about a population proportion based on a sample proportion computed from an SRS. These methods will use the tools developed in this chapter.

CHECK YOUR SKILLS

15.10 The Bureau of Labor Statistics announces that last month it interviewed all members of the labor force in a sample of 60,000 households; **7.8%** of the people interviewed were unemployed. The boldface number is a

(a) sampling distribution.
(b) statistic.
(c) parameter.

15.11 A study of voting chose 663 registered Canadian voters at random shortly after the 2011 elections. Of these, 69% said they had voted in the election. Election records show that only **61.4%** of registered voters voted in the election, up from 59.1% in 2008. The boldface number is a

(a) sampling distribution.
(b) statistic.
(c) parameter.

15.12 A 2011 study finds that in a random sample of 3000 American adults aged 18 to 34, 2140 owned an MP3 player such as an iPod. The sample proportion $\hat{p}$ who own an MP3 player is

(a) 71.   (b) 0.67.   (c) 0.71.

15.13 In the previous exercise, the sample proportion is an *unbiased estimator* of the proportion p of all American adults aged 18 to 34 who own an MP3 player. Unbiasedness in this situation means that

(a) in many samples from this population, the mean of the many values of $\hat{p}$ will be equal to p.
(b) as we take larger and larger samples from this population, $\hat{p}$ will get closer and closer to p.

(c) in many samples from this population, the many values of $\hat{p}$ will have a distribution that is close to Normal.

15.14 The proportion of drivers who use seat belts depends on things like age, sex, and ethnicity. As part of a broader study, investigators observed a random sample of 117 female Hispanic drivers in Boston. Suppose that in fact 60% of all female Hispanic drivers in the Boston area wear seat belts. In repeated samples, the sample proportion $\hat{p}$ would follow approximately a Normal distribution with mean

(a) 70.2.   (b) 0.6.   (c) 0.4.

15.15 The standard deviation of the distribution of $\hat{p}$ in the previous exercise is about

(a) 0.002.   (b) 0.045.   (c) 0.24.

15.16 Low birthweight is the second-leading cause of infant mortality in the United States. A newborn baby has low birth weight if it weighs less than 2500 grams, with approximately 8% of infants born at a low birth-weight. A sample of 500 babies is selected. The approximate distribution of $\hat{p}$, the proportion of low birth weight infants in the sample, is

(a) $N(8, 0.074)$.
(b) $N(0.8, 0.012)$.
(c) $N(0.08, 0.012)$.

15.17 In the previous exercise, the probability that more than 45 infants in the sample are born at a low birth-weight is approximately

(a) 0.82.   (b) 0.20.   (c) 0.18.

CHAPTER 15 EXERCISES

15.18 **Staph infections.** A study investigated ways to prevent staph infections in surgery patients. In a first step, the researchers examined the nasal secretions of a random sample of 6771 patients admitted to various hospitals for surgery. They found that 1251 of these patients tested positive for *Staphylococcus aureus*, a bacterium responsible for most staph infections.[10]

(a) Describe the population and explain in words what the parameter p is.
(b) Give the numerical value of the statistic $\hat{p}$ that estimates p.

15.19 **Texting.** In 2012, about 85% of American adults owned a cell phone, and the percent of cell phone owners who used their cell phone to send or receive text messages was at an all-time high of 80%.[11] A polling firm contacts an SRS of 1200 people chosen from the population of cell phone owners.

(a) What are the mean and standard deviation of $\hat{p}$, the proportion of cell phone owners in the sample who use their cell phone to text?
(b) What is the approximate probability that $\hat{p}$ is between 78% and 82%? This is the probability that $\hat{p}$ estimates p within 2%.

15.20 **Statistics anxiety.** What can teachers do to alleviate statistics anxiety in their students? To explore this question, statistics anxiety for students in two classes was compared. In one class, the instructor lectured in a formal manner, including dressing formally. In the other, the instructor was less formal, dressed informally, was more personal, used humor, and called on students by their first names. Anxiety was measured using a questionnaire. Higher scores indicate a greater level of anxiety. The mean anxiety score for students in the formal lecture class was **25.40**; in the informal class the mean was **20.41.** For each of the boldface numbers, indicate whether it is a parameter or a statistic. Explain your answer.

15.21 Texting, continued. In 2012, about 85% of American adults owned a cell phone with the percent of cell phone owners who used their cell phone to send or receive text messages at an all-time high of 80%. A polling firm plans to contact an SRS of people chosen from the population of cell phone owners. You want the proportion $\hat{p}$ of cell phone owners in the sample who use their cell phone to text, to estimate p with an error of no more than 2% in either direction. That is, you want $\hat{p}$ to be between 78% and 82%.

(a) What standard deviation must $\hat{p}$ have so that 99.7% of all samples give a $\hat{p}$ within 2% of p? (Hint: Using the 68–95–99.7 rule, how many standard deviations does 2% need to equal for the interval between 78% and 82% to contain 99.7% of all sample values of $\hat{p}$? How large must the standard deviation of $\hat{p}$ be to achieve this?)

(b) How large an SRS do you need to reduce the standard deviation of $\hat{p}$ to the value you found in part (a)? This is the number of observations that you need so that the sample proportion $\hat{p}$ will estimate p with an error of no more than 2% in either direction. We will return to this type of sample size calculation in the next chapter.

15.22 Roulette. A roulette wheel has 38 slots, of which 18 are black, 18 are red, and 2 are green. When the wheel is spun, the ball is equally likely to come to rest in any of the slots. One of the simplest wagers chooses red or black. A bet of $1 on red returns $2 if the ball lands in a red slot. Otherwise, the player loses his dollar. When gamblers bet on red or black, they need to win more than 50% of their wagers to have a winning day. Because of the two green slots, the probability of winning when a gambler bets on red is 18/38 = 47.4%. Explain what the law of large numbers for proportions tells us about the chance of having a winning day if a gambler makes very many bets on red.

© WidStock/Alamy

15.23 Sampling voters. Your state representative is planning on supporting a bill legalizing same-sex marriage in your state and wants to estimate the proportion of voters in her district who support this bill. She mails a survey to an SRS of 1300 registered voters in her district, and 800 surveys are returned. In her district, 54% of the registered voters are female.

(a) What is the population for this survey?

(b) What is the sample for this survey?

(c) Treating the 800 respondents as an SRS from her district, what is the probability that the proportion of females among the respondents differs from the proportion of registered female voters in her district by more than 5%?

15.24 Sampling voters, continued. In the previous exercise, suppose that in the sample of 800 surveys returned to your state representative, 528 are from female voters.

(a) What is the sample proportion $\hat{p}$ of females that responded to the survey? Using your results from Exercise 15.23, if the 800 returned surveys were an SRS, do you feel that the difference between the proportion of female voters that responded to the survey and the population proportion of 54% can be easily explained by chance variation? If not, what is an alternative explanation for the high proportion of female respondents?

(b) Among the 800 surveys that are returned, 72% support the bill. If women are more likely to support the bill legalizing same-sex marriage than men, what would be the direction of the bias in the estimate? Explain your answer.

15.25 Online shopping. Over a typical quarter of the year, about 75 million people, or one-third of the total Internet population, buy goods online to be shipped to them. Of those, approximately 67% select Economy Ground shipping, which takes 5–7 days.[12] A survey of 1200 online shoppers is to be taken. What is the probability that the proportion of shoppers in the sample who select Economy Ground shipping for their order is larger than 70%? Use the four-step process to guide your work.

15.26 Too close to call! In the 2012 Presidential election, Florida was one of the swing states in which no single candidate or party had overwhelming support. In a poll completed November 1, just one week before the election, the Tampa Bay Times found **51%** in favor of Romney, 45% in favor of Obama and 4% undecided. On November 3, the opening line in the article of the *Tampa Bay Times* that reported the results of this poll said "Florida continues to look good for Mitt Romney."[13] The poll was based on a sample of 800 likely voters. On election day, Romney received **49.1%** of the votes in Florida and lost the state. What happened?

(a) For each of the boldface numbers, indicate whether it is a parameter or a statistic. Explain your answer.

(b) Suppose 49.1% of all voters supported Mitt Romney at the time of the poll. What are the mean and standard deviation of $\hat{p}$, the proportion of voters in a SRS of 800 voters who would support Romney?

(c) Suppose 49.1% of all voters supported Mitt Romney at the time of the poll. What is the probability that at least 51% of the voters in an SRS of 800 voters would support Romney?

Polls conducted by NBC News and Public Policy Polling in the same week showed Obama in the lead in Florida. In a very close election, be careful not to overinterpret the results from a single poll. Unfortunately chance variation doesn't make an interesting headline!

AP Photo/Nikolas Giakoumidis

15.27 Sex and social networking. Although women tend to visit social-networking sites somewhat more frequently than men, the recent social media phenomenon Pinterest is an extreme example with 83% of the users being female in the United States. The usage is so one-sided that *Time* declared "Men are from Google+, Women are from Pinterest."[14] An SRS of 500 Pinterest users is to be taken.

(a) What are the mean and standard deviation of $\hat{p}$, the proportion of Pinterest users in the sample that are female?

(b) What is the approximate probability that $\hat{p}$ is between 80% and 86%? This is the probability that $\hat{p}$ estimates p within 3%.

15.28 Greeks lead the world in pessimism. Greeks bucked a global trend in which people in most countries expect their lives in five years to be better than their current lives. Even among those countries with much lower current life ratings, greater optimism was found because people cannot fathom their lives getting worse. But in Greece, only 25% expect their lives to be better in five years than currently, possibly reflecting the government's debt crisis and the political and economic uncertainty in the country.[15] In an SRS of 500 Greek citizens, what is the probability that less than 30% of the sample expect their lives in five years to be better than their current lives?

15.29 Sex and social networking, continued. Although women tend to visit social-networking sites more frequently than men, the recent social media phenomenon Pinterest is dominated by women, with 83% of the users being female in the United States. The usage is so one-sided that *Time* declared "Men are from Google+, Women are from Pinterest." An SRS of Pinterest users is to be taken, and you want your sample proportion $\hat{p}$ to estimate p with an error of no more than 3% in either direction, or $\hat{p}$ to be between 80% and 86%.

(a) What standard deviation must $\hat{p}$ have so that 95% of all samples give a $\hat{p}$ within 3% of p?

(b) How large an SRS do you need to reduce the standard deviation of $\hat{p}$ to the value you found in part (a)?

15.30 Car colors on campus. Approximately 20% of cars in the United States are white. You take an SRS of 300 cars parked in student lots on your campus and find that 42 are white. For this problem, you can assume that you have obtained an SRS of student cars on campus.

(a) What is the proportion, $\hat{p}$, of white cars in your sample?

(b) If 20% of student cars on campus are white, what proportion of samples would give a value of $\hat{p}$ as small as the value computed in (a)?

(c) There are two explanations for your sample result. The first is that you obtained a sample with an unusually low percentage of white cars. What is an alternative explanation?

Exploring the Web

15.31 Health care access/coverage. The Behavioral Risk Factor Surveillance System (BRFSS) is an ongoing data collection program designed to measure behavioral risk factors for the adult population (18 years of age or older) living in households. Data are collected from a random sample of adults (one per household) through a telephone survey. Go to the Web site `apps.nccd.cdc.gov/BRFSS/`, and, using the most recent year, under "Category" go to "Health Care Access/Coverage." Under the topic "Adults aged 18–64 who have any kind of health care coverage," you will find the percent with coverage in each state.

(a) Which state has the highest percent of coverage, and what is the reported value? Which state has the lowest percent, and what is its value? Are the reported percents statistics or parameters?

(b) Choose a state of interest to you, and click on the link. In the table that opens, there is a line for n, and the entries are the numbers who answered "Yes" and "No." Find the percent in the sample who answered "Yes." Notice that it is different from the percent reported in the table. The table estimates are weighted to try to reduce bias. If it is determined that certain portions of the population are underrepresented in the sample, then that portion of the sample receives more weight when computing the estimate of the percent. The assumptions for an SRS are rarely met in practice, and more complicated methods are often necessary to estimate proportions.

15.32 **More on weighting of estimates.** The Web site for the American Association for Public Opinion Research discusses several issues about polls. Read the discussion of weighting at www.aapor.org/Weighting1.htm. The necessity of weighting arises in Exercises 15.23 and 15.24 (page 346). Here are the important points from these exercises.

A state representative wants to estimate the proportion of voters in her district who support a bill legalizing same-sex marriage. In her district, it is known that 54% of the registered voters are female. She mails a survey to an SRS of 1300 registered voters, and 800 surveys are returned. Among the 800 surveys returned, over 65% were from women, and the women tended to support the bill more than the men. Because the women are both overrepresented in the sample compared with the population and they also have stronger support for the bill, the proportion in the sample that support the bill will tend to be larger than in the population, producing bias in the estimate.

(a) The online discussion of weighting gives three uses of weighting to adjust poll results. Which of these three uses is relevant in this situation?

Here is a breakdown of the respondents.

	SUPPORT THE BILL		
	YES	NO	TOTAL
Females	422	106	528
Males	154	118	272
Total	576	224	800

(b) What is the proportion of females in the sample, $\hat{p}_F$, that support the bill and the proportion of males in the sample, $\hat{p}_M$, that support the bill?

(c) If the sample proportions $\hat{p}_F$ and $\hat{p}_M$ were equal to the true proportions of females and males that support the bill, what would be the true proportion of all registered voters in the population who support the bill? This is a weighted estimator. (Hint: You will need to use the fact that 54% of the population is female and 46% is male in your computation.)

(d) The weighted estimator is smaller than the unweighted estimator $\hat{p} = 576/800 = 0.72$. Why is this the correct direction to adjust the estimator?

Confidence Intervals: The Basics CHAPTER 16

Overview

This chapter begins our coverage of statistical inference—the heart of the course. We have spent a lot of time studying descriptive statistics, data collection (experimental design and sampling), and probability along with the sampling distribution for proportions. The new ideas of inference may feel somewhat inverted to students; where we previously used information about the entire population to characterize likely samples, now we will make inferences about the population using information in a sample.

Because a sample statistic is practically never exactly the same as the parameter, confidence intervals are used to estimate the unknown population parameters by providing a set of a plausible values and attaching a level of certainty/confidence to that set. A strong understanding of the sampling distribution for proportions is the key to understanding both construction and properties of confidence intervals.

In constructing a confidence interval for a population parameter, we invoke the sampling distribution of an estimator of the parameter. The sampling distribution of the statistic (the estimator) yields a margin of error associated with the point estimate. The margin of error is a function of the estimator's standard error (or standard deviation).

By studying confidence intervals before hypothesis tests, students can get a sense of what we mean about a reasonable (plausible) range of values—we go out from the sample statistic by a specified number of standard errors in either direction. This is helpful because students are already familiar with the 68–95–99.7 rule, so extending the idea to other probabilities (here, levels of confidence) should be a natural extension. Another benefit of studying confidence intervals before hypothesis tests is that we can use confidence intervals to "set up" hypothesis tests by asking what values might be *unreasonable* for the parameter based on our data. In hypothesis testing, we go on to ask about a specific value for the parameter and how unusual our sample is, assuming the value we specify is indeed the value of the unknown parameter.

An important first step in many studies is to determine the sample size needed to meet certain predetermined criteria. One method of determining such a sample size is presented in this chapter. It is important to talk with your students about why we choose $\hat{p}$ to be 0.5 or to be a value based on prior results; 0.5 is always "safe," but may cause unneeded work if the population proportion is very different from that value

> ## LEARNING OUTCOMES**
>
> - State in nontechnical language what is meant by "95% confidence" or other statements of confidence in statistical reports.
>
> - Know the four-step process (page 357) for any confidence interval.
>
> - Understand how the margin of error of a confidence interval changes with the sample size and the level of confidence C.
>
> - Find the sample size required to obtain a confidence interval of specified margin of error m when the confidence level and other information are given.
>
> - Identify sources of error in a study that are *not* included in the margin of error of a confidence interval, such as undercoverage or nonresponse.
>
> - Verify that you can use either the large-sample (or, optionally, the plus four) z procedure in a particular setting. Check the study design and the guidelines for sample size.
>
> - Use the large-sample z procedure to give a confidence interval for a population proportion p. Understand that the true confidence level may be substantially less than you ask for unless the sample is very large and the true p is not close to 0 or 1.
>
> - (Optional) Use the plus four modification of the z procedure to give a confidence interval for p that is accurate even for small samples and for any value of p.
>
> **These learning outcomes appear later for the students in Chapter 23: Part IV Review.

Teaching Suggestions and Additional Examples/Activities for the Classroom

1. Motivate Confidence Interval Construction

Open with a discussion of how the law of large numbers for proportions describes the sampling distribution of the sample proportion. You can use this to motivate the construction of a confidence interval in the following manner:

- Draw a Normal curve describing the sampling distribution of the sample proportion, marking off the population proportion, p, and the two values $p - 1.96\sqrt{\dfrac{p(1-p)}{n}}$ and $p + 1.96\sqrt{\dfrac{p(1-p)}{n}}$.

- Make sure that students acknowledge that under repeated sampling, approximately 95% of observed sample proportions will fall between these two values, or

$$P\left(p - 1.96\sqrt{\frac{p(1-p)}{n}} \leq \hat{p} \leq p + 1.96\sqrt{\frac{p(1-p)}{n}}\right) = 0.95.$$

- Use the graph you've created to demonstrate the important point: When $\hat{p}$ falls between $p - 1.96\sqrt{\dfrac{p(1-p)}{n}}$ and $p + 1.96\sqrt{\dfrac{p(1-p)}{n}}$, the constructed interval $\left(\hat{p} - 1.96\sqrt{\dfrac{p(1-p)}{n}}, \hat{p} + 1.96\sqrt{\dfrac{p(1-p)}{n}}\right)$ will capture the true parameter p.

- Move the sample proportion $\hat{p}$ around on the graph, having students visualize the interval constructed until they recognize the connection between the random location of $\hat{p}$ and the interval's success or failure in capturing p.

Be sure students understand that confidence refers to faith in the process of random sampling and only represents a probability before a sample is taken. For example, if we repeatedly sample randomly, each time constructing a (say) 90% confidence interval for p, then in the long run 90% of the intervals will capture p. Conduct an in-class demonstration of this interpretation. It might be helpful for you to have the students work with the *Confidence Interval for Proportions* applet.

2. Discuss Impact of Sample Size and Confidence Level

Discuss the impact of sample size and confidence level on the size of the margin of error. This can easily be seen using the applet and may naturally lead to the question of sample size determination. Students will be surprised at the somewhat counterintuitive notion that a required sample size does not depend on the size of a population (as long as the population is much larger than the sample—a topic beyond the scope of this text).

Consider bringing regular-sized bags of milk chocolate M&M's candies to class. Students can use their own samples of M&M's to estimate the proportion of candies that are blue (at the time of this writing, that percentage is 24%). Make sure that students have samples that are large enough to satisfy the conditions for constructing confidence intervals for proportions (we suggest a sample size of 50 M&M's). Find the percentage of confidence intervals constructed by your students that contain the true proportion of blue M&M's and compare the percentage of "good" intervals to the confidence level. Make sure to discuss with your students that we cannot expect the percentage to be exactly the same as the confidence level, as the size of the class will certainly not be large enough to emulate a long-run event. Again, you can use the *Confidence Interval for Proportions* applet to have the students "see" what would happen in the long run.

3. Encourage Students to Determine If the Data Are Categorical or Quantitative

Students who get in the habit of considering for each problem whether the researcher described collected quantitative or qualitative data can help themselves organize the material effectively. If you push your students to develop this habit by asking each time you work an example problem in class whether the data collected are quantitative or qualitative, you will help students to overcome this difficulty. (This will be especially important as you move from inference on proportions to inference on means.) Similarly, this is the time to encourage your students to make sure that they check necessary conditions for the procedure of interest and communicate their results in context. By doing so, you will help the students develop the ability to think about the process of doing statistics.

4. Relate Course Material to Surveys from the Media (and Revisit Sampling Issues)

Students are often interested in political or social surveys. Because we want our students to be savvy consumers of media, this chapter provides a good opportunity to relate course material to survey results as encountered in the media. The Gallup Organization (www.gallup.com) provides current, regularly updated survey results on far-ranging issues affecting society. Importantly, you can obtain details about these surveys—how many people were surveyed, the reported margin of error, and so on. Have students examine the details of such a survey and compute the margin of error, as reported. Organizations such as Gallup *always*, from custom, compute their margins of error assuming 95% confidence. In recent years, the typical margins of error have changed from $\pm 3\%$ to $\pm 4\%$. The margins of error have increased because Gallup has been using more complicated sampling schemes (e.g., including results from both landline phones and mobile phones).

Here there is the opportunity to revisit issues related to sampling. All of statistical inference is based on an assumption of random sampling or random assignment of subjects to experimental groups. Remind students that when they encounter results from a survey, they should question how the sample was selected. For example, "phone-in" surveys on the local news are invariably unreliable because they are not based on a random sample, but rather on a voluntary response survey. Computing a standard error does not address issues of bias. Ask students whether it would be reasonable to estimate the proportion of students at your school that have cheated on an exam or assignment, using your class as a sample. Again, the sample isn't random and probably does not represent all students at the school with respect to cheating habits. Moreover, it is difficult to solicit honest answers to difficult questions (response bias). You might mention the concept of "randomized response" surveys, which attempt to deal with this problem. A discussion of randomized response surveys is provided on the appropriate Wikipedia page (http://en.wikipedia.org/wiki/Randomized_response).

5. Clarify When/If Your Students Should Use the "Plus Four" Approach

Note that "plus four" confidence intervals for a proportion are optional in this chapter. If you choose to use this interval, make sure that your students know why you would like them to use this approach. Also point out to your students that we have different conditions for the plus-four confidence interval than for the (Wald) confidence interval that is used elsewhere in the chapter.

Other Resources (LaunchPad)

StatClips

Confidence Intervals: Intervals for Proportions (Note that there is some mention of inference for means in this StatClip.)

Snapshots Videos
> *Inference for One Proportion* (Note that there is some mention of inference for means in this video.)

EESEE Case Studies
> *Columbus' 1993 Election Poll*
> *Radar Detectors and Speeding*
> *Seasonal Weevil Migration*

Applet
> *Confidence Interval for Proportions*

Confidence Intervals: The Basics

A fter we have selected a sample, we know the responses of the individuals in the sample. The usual reason for taking a sample is not to learn about the individuals in the sample but to *infer* from the sample data some conclusion about the wider population that the sample represents.

Statistical Inference

Statistical inference provides methods for drawing conclusions about a population from sample data.

Because a different sample might lead to different conclusions about the population, we can't be certain that our conclusions are correct. Statistical inference uses the language of probability to say how trustworthy our conclusions are. This chapter introduces one of the two most common types of inference, *confidence intervals* for estimating the value of a population parameter. The next chapter discusses the other common type of inference, *tests of significance* for assessing the evidence for a claim about a population. Both types of inference are based on the sampling distributions of statistics. That is, both use probability to say what would happen if we applied the inference method many times.

This chapter presents the basic reasoning of statistical inference. To make the reasoning as clear as possible, we start with the simple setting of inference for a single population proportion.

In Chapter 15 we learned that the sampling distribution of a sample proportion $\hat{p}$, computed from an SRS selected from a population in which the true proportion is p, has approximately the $N(p, \sqrt{p(1-p)/n})$ distribution for large n. In addition, in Chapter 15 we recommend that we use this Normal approximation when the sample size n is so large that both np and $n(1-p)$ are 10 or more. In practice we don't know p, but for large samples we know that $\hat{p}$ will be close to p. So for purposes of inference we recommend that we use this Normal approximation when both $n\hat{p}$ and $n(1-\hat{p})$ are 10 or more.[1] These comments provide the basis for inference about a proportion.

Conditions for Inference About a Proportion

1. We have an SRS of size n from the population of interest. There is no nonresponse or other practical difficulty.

2. We don't know the population proportion p.

3. The sample size n is sufficiently large so that both $n\hat{p}$ and $n(1-\hat{p})$ are 10 or more. In other words, when the number of successes and the number of failures is at least 10.

In some cases you will see this Normal approximation used when the sample is not actually an SRS but there is reason to believe the sample can be regarded as though it were an SRS. But if the sample was not selected by some formal random sampling method and the argument for regarding it as an SRS rests only on "expert opinion," we recommend that you treat subsequent inferences with caution.

We may also wish to do inference about a proportion based on data generated by an experiment. We will meet an example (an experiment seeking to detect the ability to sense the future) in the next chapter. For data generated by an experiment, we will require that the experiment is properly randomized and that the number of subjects is sufficiently large (both $n\hat{p}$ and $n(1-\hat{p})$ are 10 or more). Inference will tell us what would happen if we repeated the experiment many times.

The reasoning of statistical estimation

The "boomerang generation" is a term applied to the current generation of young adults. They are so named because of the frequency with which they choose to live with their parents after a brief period of living on their own, thus boomeranging back to their place of origin.

RANGES ARE FOR STATISTICS?

Many people like to think that statistical estimates are exact. The Nobel Prize–winning economist Daniel McFadden tells a story of his time on the Council of Economic Advisers. Presented with a range of forecasts for economic growth, President Lyndon Johnson replied: "Ranges are for cattle; give me one number."

EXAMPLE 16.1   The Boomerang Generation

A Pew Research Center survey of 808 adults nationwide conducted December 6–19, 2011, found that 39% of 18- to 34-year-olds lived with their parents or had moved back in temporarily because of the economy in the past few years. On the basis of this sample, we want to estimate (at the time of the survey) the proportion p in the population of all 18- to 34-year-olds in the United States who lived with their parents or had moved back in temporarily because of the economy in the past few years.

Are the conditions for inference met? The Pew Research Center sample was not an SRS but is described as a "nationally representative sample." We proceed with some caution, but will treat the Pew

Research Center sample as an SRS from the population of all 18- to 34-year-olds in the United States.

The number of "successes" in the sample is $n\hat{p} = 0.39 \times 808 = 315$, and the number of "failures" is $n(1 - \hat{p}) = 0.61 \times 808 = 493$. Both are much larger than 10, so we regard the sample size as sufficiently large for using the Normal approximation. ■

Assuming our conditions for inference are met, here is the reasoning of statistical estimation in a nutshell:

1. To estimate the unknown population proportion p of all 18- to 34-year-olds in the United States who lived with their parents or had moved back in temporarily because of the economy in the past few years, use the proportion $\hat{p} = 0.39$ of the random sample. We don't expect $\hat{p}$ to be exactly equal to p, so we want to say how accurate this estimate is.

2. Because our conditions for inference are met, we know that the sampling distribution of $\hat{p}$ has approximately the Normal distribution with mean p and standard deviation $\sqrt{p(1 - p)/n}$.

3. The 95 part of the 68–95–99.7 rule for Normal distributions says that $\hat{p}$ and p are within 2 standard deviations, $-2\sqrt{p(1 - p)/n}$ and $+2\sqrt{p(1 - p)/n}$, of each other in approximately 95% of all samples of size 808. We can put this another way: approximately 95% of all samples of size 808 give an outcome $\hat{p}$ such that the population proportion p is captured by the interval $\hat{p} - 2\sqrt{p(1 - p)/n}$ and $\hat{p} + 2\sqrt{p(1 - p)/n}$. So if we estimate that p lies somewhere in the interval from $\hat{p} - 2\sqrt{p(1 - p)/n}$ to $\hat{p} + 2\sqrt{p(1 - p)/n}$, we'll be right for approximately 95% of all possible samples of size 808.

4. Unfortunately, we need to know p to compute $\sqrt{p(1 - p)/n}$ in this interval. But if we knew p we wouldn't need to do inference for it! What can we do? Well, the standard deviation of the statistic $\hat{p}$ does depend on the parameter p, but it doesn't change a lot when p changes. For values of p near $\hat{p} = 0.39$ and sample sizes of 808 we find the following:

Value of p:	0.35	0.37	0.39	0.41	0.43
Standard deviation:	0.0168	0.0170	0.0172	0.0173	0.0174

We see that if we guess a value of p reasonably close to the true value, the standard deviation found from the guessed value will be about right (all the values listed round to 0.017). We know that, when we take a large random sample, the statistic $\hat{p}$ is almost always close to the parameter p. So we will use $\hat{p}$ as the guessed value of the unknown p. Now we have an interval that we can calculate from the sample data.[2]

So if we estimate that p lies somewhere in the interval from $\hat{p} - 2\sqrt{\hat{p}(1 - \hat{p})/n}$ to $\hat{p} + 2\sqrt{\hat{p}(1 - \hat{p})/n}$, we'll capture the population proportion p in approximately 95% of all possible samples of size 808. Here,

$$\sqrt{\hat{p}(1 - \hat{p})/n} = \sqrt{0.39(1 - 0.39)/808} = 0.017 \text{ (rounded off)}$$

So for this particular sample, this interval is

$$\hat{p} - 0.034 = 0.39 - 0.034 = 0.356$$

to

$$\hat{p} + 0.034 = 0.39 + 0.034 = 0.424$$

5. Because we got the interval 0.356 to 0.424 from a method that captures the population proportion for 95% of all possible samples, we say that we are 95% *confident* that the proportion p of all 18- to 34-year-olds in the United States who lived with their parents or had moved back in temporarily because of the economy in the past few years is some value in that interval, no lower than 0.356 and no higher than 0.424.

The big idea is that the sampling distribution of $\hat{p}$ allows us to determine how close to p the sample proportion $\hat{p}$ is likely to be for large n. Statistical estimation just turns that information around to say how close to $\hat{p}$ the unknown population proportion p is likely to be. We call the interval of numbers between the values

$$\hat{p} \pm 2\sqrt{\hat{p}(1 - \hat{p})/n}$$

a large-sample 95% *confidence interval* for p.

Apply Your Knowledge

© Daniel Sicolo/Design Pics/Corbis

16.1 Moral Values. Americans have historically favored promoting traditional values. However, a 2012 Gallup Poll found that 52% of those surveyed think that government should not favor any particular set of values. These results were based on a sample of 1017 adults aged 18 and older, living in the United States.[3] Consider the Gallup sample as an SRS.

(a) Is the sample size $n = 1017$ large enough for $\hat{p}$ to have a sampling distribution that is approximately a Normal distribution?

(b) What is the large-sample 95% confidence interval for the population proportion p based on this one sample?

16.2 Belief in Santa Claus. A poll conducted by GfK Roper Public Affairs & Corporate Communications asked a sample of 1000 adults in the United States, "As a child, did you ever believe in Santa Claus, or not?"[4] Of those surveyed, 84% said they had believed as a child. Consider the sample as an SRS. We want to estimate the proportion p of all adults in the United States who would answer that they had believed to the question "As a child, did you ever believe in Santa Claus, or not?" Verify that the condition for the sample size to be large enough is met, and give a large-sample 95% confidence interval for p based on this sample.

Margin of error and confidence level

The large-sample 95% confidence interval for the proportion p of all 18- to 34-year-olds in the United States who lived with their parents or moved back in temporarily because of the economy in the past few years, based on the Pew Research Center sample, is $\hat{p} \pm 2\sqrt{\hat{p}(1 - \hat{p})/n}$. Once we have the sample results in hand, we know that for this sample $\hat{p} = 0.39$, so that our confidence interval is 0.39 ± 0.034. Most confidence intervals have a form similar to this,

estimate ± margin of error

margin of error

The estimate ($\hat{p} = 0.39$ in our example) is our guess for the value of the unknown parameter. The **margin of error** (± 0.034 in our example) shows how accurate we believe our guess is, based on the variability of the estimate. We have a 95% confidence interval because the interval $\hat{p} \pm 2\sqrt{\hat{p}(1 - \hat{p})/n}$ catches the unknown parameter in 95% of all possible samples.

Confidence Interval

A **level *C* confidence interval** for a parameter has two parts:

- An interval calculated from the data, usually of the form

$$\text{estimate} \pm \text{margin of error}$$

- **A confidence level** C, which gives the probability that the interval will capture the true parameter value in repeated samples. That is, the confidence level is the success rate for the method.

Users can choose the confidence level, usually 90% or higher, because we usually want to be quite sure of our conclusions. The most common confidence level is 95%.

Interpreting a Confidence Level

The confidence level is the success rate of the method that produces the interval. We don't know whether the 95% confidence interval from a particular sample is one of the 95% that capture p or one of the unlucky 5% that miss.

To say that we are **95% confident** that the unknown p lies between 0.356 and 0.424 is shorthand for **"We got these numbers using a method that gives correct results 95% of the time."**

EXAMPLE 16.2   Statistical Estimation in Pictures

Figures 16.1 and 16.2 illustrate the behavior of confidence intervals. Study these figures carefully. If you understand what they tell us, you have mastered one of the big ideas of statistics.

Figure 16.1 illustrates the behavior of the interval $\hat{p} \pm 2\sqrt{\hat{p}(1 - \hat{p})/n}$ for the proportion of all 18- to 34-year-olds in the United States who lived with their parents or moved back in temporarily because of the economy in the past few years. Starting with the population, imagine taking many SRSs of 808 18- to 34-year-olds. The first sample has $\hat{p} = 0.39$, the second has $\hat{p} = 0.41$, the third has $\hat{p} = 0.38$, and so on. For each of these values of $\hat{p}$ we find that, to three decimal places, $\sqrt{\hat{p}(1 - \hat{p})/n} = 0.017$. The sample proportion varies from sample to sample, but when we use the formula

$$\hat{p} \pm 2\sqrt{\hat{p}(1 - \hat{p})/n} = \hat{p} \pm 0.034$$

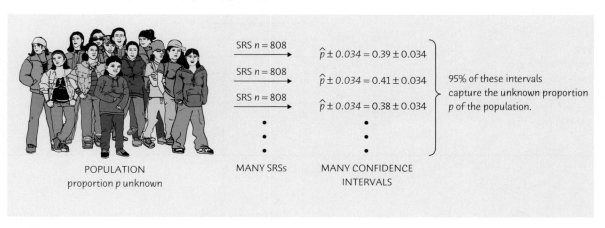

POPULATION
proportion *p* unknown

MANY SRSs

MANY CONFIDENCE INTERVALS

SRS *n* = 808 → $\hat{p} \pm 0.034 = 0.39 \pm 0.034$

SRS *n* = 808 → $\hat{p} \pm 0.034 = 0.41 \pm 0.034$

SRS *n* = 808 → $\hat{p} \pm 0.034 = 0.38 \pm 0.034$

95% of these intervals capture the unknown proportion *p* of the population.

FIGURE 16.1

To say that $\hat{p} \pm 2\sqrt{\hat{p}(1 - \hat{p})/n}$ is a 95% confidence interval for the population proportion *p* is to say that, in repeated samples, 95% of these intervals capture *p*.

to get an interval based on each sample, approximately *95% of these intervals capture the unknown population proportion p for samples of size 808.*

Figure 16.2 illustrates the idea of a 95% confidence interval in a different form. It shows the result of drawing many SRSs from the same population with $p = 0.5$ and calculating a 95% confidence interval from each sample. The center of each interval is at $\hat{p}$ and therefore varies from sample to sample. The sampling distribution of $\hat{p}$ appears at the top of the figure to show the long-term pattern of this variation. The population proportion p is at the center of the sampling distribution. The 95% confidence intervals from 25 SRSs appear underneath. The center $\hat{p}$ of each interval is marked by a dot. The arrows on either side of the dot span the confidence interval. All except 1 of these 25 intervals capture the true value of p. If we take a very large number of samples, 95% of the confidence intervals will contain p. ■

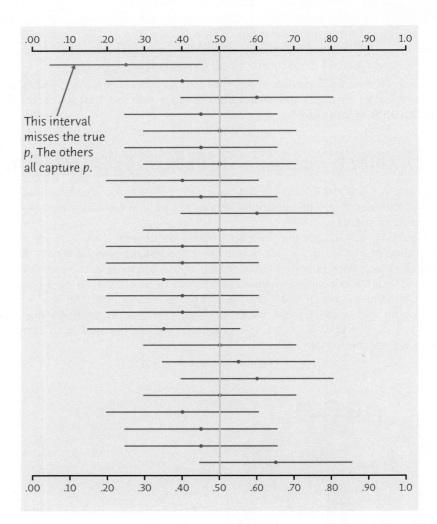

FIGURE 16.2

Twenty-five samples from the same population gave these 95% confidence intervals. In the long run, 95% of all samples give an interval that contains the population proportion $p = 0.5$.

The *Confidence Intervals for Proportions* applet animates Figure 16.2. You can use the applet to watch confidence intervals from one sample after another capture or fail to capture the true parameter.

Apply Your Knowledge

16.3 Confidence Intervals in Action. The idea of an 80% confidence interval is that in 80% of all samples the method produces an interval that captures the true parameter value. That's not high enough confidence for practical use, but 80% hits and 20% misses make it easy to see how a confidence interval behaves in repeated samples from the same population. Go to the *Confidence Intervals for Proportions* applet.

(a) Set the confidence level to 80%. Click "Sample" to choose an SRS of size $n = 20$, and calculate the confidence interval. Do this 10 times to simulate 10 SRSs with their 10 confidence intervals. How many of the 10 intervals captured the true proportion p? How many missed?

(b) You see that we can't predict whether the next sample will hit or miss. The confidence level, however, tells us what percent will hit in the long run. Reset the applet and click "Sample 25" twice to get the confidence intervals from 50 SRSs. How many hit?

(c) Click "Sample 25" repeatedly, and write down the number of hits each time. What was the percent of hits among 100, 200, 300, 400, 500, 600, 700, 800, and 1000 SRSs? Even 1000 samples is not truly "the long run," but we expect the percent of hits in 1000 samples to be fairly close to the confidence level, 80%.

16.4 Losing Weight. A Gallup Poll in November 2012 found that 54% of the people in the sample said they want to lose weight. Gallup announced, "For results based on the total sample of national adults, one can say with 95% confidence that the maximum margin of sampling error is ±4 percentage points."

(a) What is the 95% confidence interval for the percent of all adults who want to lose weight?

(b) What does it mean to have 95% confidence in this interval?

Confidence intervals for a population proportion

In the setting of Example 16.1 (page 350) we outlined the reasoning that leads to a 95% confidence interval for the unknown proportion p of a population. Now we will reduce the reasoning to a formula.

To find a 95% confidence interval for the proportion of all 18- to 34-year-olds in the United States who lived with their parents or moved back in temporarily because of the economy in the past few years, we first caught the central 95% of the Normal sampling distribution by going out two standard deviations (or more precisely, $2\sqrt{\hat{p}(1-\hat{p})/n}$) in both directions from the mean of the sampling distribution. To find a level C confidence interval, we first catch the central area C under the Normal sampling distribution. Because all Normal distributions are the same in the standard scale, we can obtain everything we need from the standard Normal curve.

Figure 16.3 shows how the central area C under a standard Normal curve is marked off by two points $-z^*$ and z^*. Numbers like z^* that mark off specified areas are called **critical values** of the standard Normal distribution. Values of z^* for many

critical value

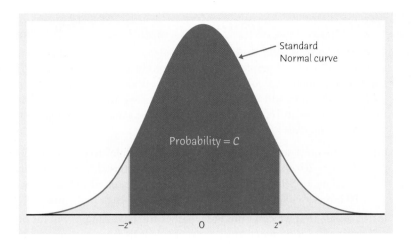

choices of C appear at the bottom of Table C in the back of the book, in the row labeled z^*. Here are the entries for the most common confidence levels:

Confidence level C	90%	95%	99%
Critical value z^*	1.645	1.960	2.576

You see that for $C = 95\%$ the table gives $z^* = 1.960$. This is a bit more precise than the approximate value $z^* = 2$ based on the 68–95–99.7 rule. You can of course use software to find critical values z^*, as well as the entire confidence interval.

Figure 16.3 shows that there is area C under the standard Normal curve between $-z^*$ and z^*. So *any* Normal curve has area C within z^* standard deviations on either side of its mean. The Normal sampling distribution of $\hat{p}$ has approximately area C within $z^*\sqrt{p(1-p)/n}$ on either side of the population proportion p because it has mean p and standard deviation $\sqrt{p(1-p)/n}$. If we start at $\hat{p}$ and go out $z^*\sqrt{p(1-p)/n}$ in both directions, we get an interval that contains the population mean p in a proportion C of all samples. As before, because we do not know p, we replace it with $\hat{p}$ in $\sqrt{p(1-p)/n}$. The resulting interval is

$$\text{from } \hat{p} - z^*\sqrt{\hat{p}(1-\hat{p})/n} \text{ to } \hat{p} + z^*\sqrt{\hat{p}(1-\hat{p})/n}$$

or

$$\hat{p} \pm z^*\sqrt{\hat{p}(1-\hat{p})/n}$$

This is our approximate level C confidence interval for p when n is large.

Large-Sample Confidence Interval for a Population Proportion

Draw an SRS of size n from a population having unknown proportion p with some characteristic. A level C **confidence interval for p** is

$$\hat{p} \pm z^*\sqrt{\hat{p}(1-\hat{p})/n}$$

provided both $n\hat{p}$ and $n(1-\hat{p})$ are 10 or more. The critical value z^* is illustrated in Figure 16.3 and found at the bottom of Table C.

In the large-sample confidence interval for a population proportion, the margin of error is $z^*\sqrt{\hat{p}(1 - \hat{p})/n}$. The term $\sqrt{\hat{p}(1 - \hat{p})/n}$ is known as the **standard error** for the sample proportion $\hat{p}$, and the margin of error has the form $z^* \times$ standard error.

standard error

The steps in finding a confidence interval mirror the overall four-step process for organizing statistical problems.

Confidence Intervals: The Four-Step Process

STATE: What is the practical question that requires estimating a parameter?

PLAN: Identify the parameter, choose a level of confidence, and select the type of confidence interval that fits your situation.

SOLVE: Carry out the work in two phases:

1. **Check the conditions** for the interval you plan to use.
2. Calculate the **confidence interval.**

CONCLUDE: Return to the practical question to describe your results in the context of the problem.

EXAMPLE 16.3   Risky Behavior in the Age of AIDS

STATE: How common is behavior that puts people at risk for AIDS? In the early 1990s, the landmark National AIDS Behavioral Surveys interviewed a random sample of 2673 adult heterosexuals. Of these, 170 had more than one sexual partner in the year prior to the survey.[5] That is,

$$\hat{p} = \frac{170}{2673} = 0.0636$$

Based on these data, what can we say about the proportion of all adult heterosexuals who had multiple partners in the year prior to the survey?

PLAN: We will estimate the proportion of all adult heterosexuals who have multiple partners by giving a large-sample 95% confidence interval. The confidence interval just introduced fits this situation.

SOLVE: We should start by checking the conditions for inference.

■ The sampling design was a complex stratified sample, and the survey used inference procedures for that design. The overall effect is close to an SRS, however.

■ The sample is large enough: $n\hat{p} = 170$ and $n(1 - \hat{p}) = 2503$ are both much larger than 10.

The sample size condition is easily satisfied. The condition that the sample be an SRS is only approximately met, so we proceed with some caution.

A 95% confidence interval for the proportion p of all adult heterosexuals with multiple partners uses the standard Normal critical value $z^* = 1.96$. The confidence interval is

$$\hat{p} \pm z^*\sqrt{\frac{\hat{p}(1 - \hat{p})}{n}} = 0.0636 \pm 1.960\sqrt{\frac{(0.0636)(0.9364)}{2673}}$$
$$= 0.0636 \pm 0.0093$$
$$= 0.0543 \text{ to } 0.0729$$

CONCLUDE: We are 95% confident that the percent of adult heterosexuals who had more than one sexual partner in the year prior to the survey lies between about 5.4% and 7.3%. ■

As usual, the practical problems of a large sample survey weaken our confidence in the AIDS survey's conclusions. Only people in households with landline telephones could be reached. Although at the time of the survey about 89% of American households had landline telephones, as the number of cellphone-only users increases, using a sample of households with landline phones is becoming less acceptable for surveys of the general population (see page 204). In addition, some groups at high risk for AIDS, such as people who inject illegal drugs, often don't live in settled households and were therefore underrepresented in the sample. About 30% of the people reached refused to cooperate. A nonresponse rate of 30% is not unusual in large sample surveys, but it may cause some bias if those who refuse to participate differ systematically from those who cooperate. The survey used statistical methods that adjust for unequal response rates in different groups. Finally, some respondents may not have told the truth when asked about their sexual behavior. The survey team tried to make respondents feel comfortable. For example, Hispanic women were interviewed only by Hispanic women, and Spanish speakers were interviewed by Spanish speakers with the same regional accent (Cuban, Mexican, or Puerto Rican). Nonetheless, the survey report says that some bias is probably present:

> It is more likely that the present figures are underestimates; some respondents may underreport their numbers of sexual partners and intravenous drug use because of embarrassment and fear of reprisal, or they may forget or not know details of their own or of their partner's HIV risk and their antibody testing history.[6]

Reading the report of a large study like the National AIDS Behavioral Surveys reminds us that statistics in practice involves much more than formulas for inference.

Apply Your Knowledge

16.5 Find a Critical Value. The critical value z^* for confidence level 85% is not in Table C. Use software or Table A of standard Normal probabilities to find z^*. Include in your answer a sketch like Figure 16.3 (page 356) with C = 0.85 and your critical value z^* marked on the axis.

16.6 Canadian Attitudes Toward Guns. Canada has much stronger gun control laws than the United States, and Canadians support gun control more strongly than do Americans. A sample survey asked a random sample of 1505 adult Canadians, "Do you agree or disagree that all firearms should be registered?" Of the 1505 people in the sample, 1288 answered either "Agree strongly" or "Agree somewhat."[7]

(a) The survey dialed residential telephone numbers at random in all ten Canadian provinces (omitting the sparsely populated northern territories). Based on what you know about sample surveys, what is likely to be the biggest weakness in this survey?

(b) Nonetheless, act as if we have an SRS from adults in the Canadian provinces. Give a 95% confidence interval for the proportion who support registration of all firearms. Be sure to verify that the sample is large enough.

16.7 Weight-Lifting Injuries. Resistance training is a popular form of conditioning aimed at enhancing sports performance and is widely used among high school, college, and professional athletes, although its use for younger athletes is controversial. Researchers obtained a random sample of 4111 patients between the ages of 8 and 30 who were admitted to U.S. emergency rooms with injuries classified by the Consumer Product Safety Commission code "weightlifting." These injuries were further classified as "accidental" if caused by dropped weights or improper equipment use. Of the 4111 weight-lifting injuries, 1552 were classified as accidental.[8] Give a 90% confidence interval for the proportion of weight-lifting injuries in this age group that were accidental. Follow the four-step process as illustrated in Example 16.3 (page 357).

16.8 No Confidence Interval. In the National AIDS Behavioral Surveys sample of 2673 adult heterosexuals, 0.2% (that's 0.002 as a decimal fraction) had both received a blood transfusion and had a sexual partner from a group at high risk for AIDS. Explain why we can't use the large-sample confidence interval to estimate the proportion p in the population who share these two risk factors.

Sports Illustrated/Getty Images

How confidence intervals behave

The large-sample confidence interval

$$\hat{p} \pm z^* \sqrt{\hat{p}(1 - \hat{p})/n}$$

for a population proportion illustrates several important properties that are shared by all confidence intervals in common use. The user chooses the confidence level, and the margin of error follows from this choice. We would like high confidence and also a small margin of error. High confidence says that our method almost always gives correct answers. A small margin of error says that we have pinned down the parameter quite precisely. The factors that influence the margin of error of the confidence interval for a population proportion are typical of most confidence intervals.

How do we get a small margin of error? The margin of error for the confidence interval is

$$\text{margin of error} = z^* \sqrt{\hat{p}(1 - \hat{p})/n} = \frac{z^* \sqrt{\hat{p}(1 - \hat{p})}}{\sqrt{n}}$$

This expression has z^* in the numerator and $\sqrt{n}$ in the denominator. Therefore, the margin of error gets smaller when

- z^* gets smaller. Smaller z^* is the same as lower confidence level C (look again at Figure 16.3 on page 356). *There is a trade-off between the confidence level and the margin of error. To obtain a smaller margin of error from the same data, you must be willing to accept lower confidence.*

- n gets larger. Increasing the sample size n reduces the margin of error for any confidence level. Larger samples thus allow more precise estimates. However, *because n appears under a square root sign, we must take four times as many observations to cut the margin of error in half.*

EXAMPLE 16.4   Changing the Margin of Error

In Example 16.3, the National AIDS Behavioral Surveys interviewed a random sample of 2673 adult heterosexuals. Of these, the proportion who said they had more than one sexual partner in the year prior to the survey was 0.0636. The 99% confidence interval for the proportion p of all adult heterosexuals with multiple partners in the year prior to the survey is

$$\hat{p} \pm z^* \sqrt{\frac{\hat{p}(1 - \hat{p})}{n}} = 0.0636 \pm 2.576 \sqrt{\frac{(0.0636)(0.9364)}{2673}}$$

$$= 0.0636 \pm 0.0122$$

$$= 0.0514 \text{ to } 0.0758$$

The 90% confidence interval based on the same data replaces the 99% critical value $z^* = 2.576$ by the 90% critical value $z^* = 1.645$. This interval is

$$\hat{p} \pm z^* \sqrt{\frac{\hat{p}(1 - \hat{p})}{n}} = 0.0636 \pm 1.645 \sqrt{\frac{(0.0636)(0.9364)}{2673}}$$

$$= 0.0636 \pm 0.0078$$

$$= 0.0558 \text{ to } 0.0714$$

Lower confidence results in a smaller margin of error, ± 0.0078 in place of ± 0.0122. We saw that the margin of error for the 95% confidence, ± 0.0093, is between these two values. Figure 16.4 compares these three confidence intervals.

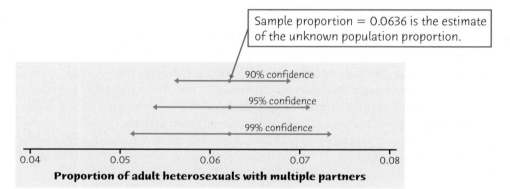

FIGURE 16.4
The lengths of three confidence intervals for Example 16.4. All three are centered at the estimate $\hat{p} = 0.0636$. When the data and the sample size remain the same, higher confidence results in a larger margin of error.

If we had a sample of only 1336 adult heterosexuals, you could check that the margin of error for 99% confidence increases from ± 0.0122 to ± 0.0172. Cutting the sample size in half does *not* double the margin of error because the sample size n appears under a square root sign. ■

Apply Your Knowledge

16.9 Confidence Level and Margin of Error. Example 16.1 (page 350) described a Pew Research Center survey of 808 18- to 34-year-olds in the United States in which $\hat{p} = 0.39$ lived with their parents or had moved back in temporarily because of the economy in the past few years. We treated these data as an SRS from the population of all 18- to 34-year-olds in the United States.

(a) Give three confidence intervals for the proportion p of all 18- to 34-year-olds in the United States who lived with their parents or had moved back in temporarily because of the economy in the past few years, using 90%, 95%, and 99% confidence.

(b) What are the margins of error for 90%, 95%, and 99% confidence? How does increasing the confidence level change the margin of error of a confidence interval when the sample size and population standard deviation remain the same?

16.10 Sample Size and Margin of Error. Example 16.1 (page 350) described a Pew Research Center survey of 808 18- to 34-year-olds in the United States in which $\hat{p} = 0.39$ lived with their parents or had moved back in temporarily because of the economy in the past few years. We treated these data as an SRS from the population of all 18- to 34-year-olds in the United States.

(a) Suppose we had an SRS of just 100 18- to 34-year-olds. What would be the margin of error for 95% confidence?

(b) Find the margins of error for 95% confidence based on SRSs of 400 18- to 34-year-olds and 1600 18- to 34-year-olds.

(c) Compare the three margins of error. How does increasing the sample size change the margin of error of a confidence interval when the confidence level and population standard deviation remain the same?

16.11 Belief in Santa Claus. In Exercise 16.2 (page 352) we saw that in an SRS of 1000 adults in the United States, 84% said they had believed as a child in response to the question, "As a child, did you ever believe in Santa Claus, or not?" We computed a 95% confidence interval for the p of all adults in the United States who would answer that they had believed as a child in response to the question "As a child, did you every believe in Santa Claus, or not?"

(a) Find a 90% confidence interval for p based on this sample.

(b) What is the margin of error for 90%? How does decreasing the confidence level change the margin of error of a confidence interval when the sample size and population standard deviation remain the same?

(c) Suppose we had an SRS of just 100 adults in the United States. What would be the margin of error for 95% confidence? (Verify that a sample of size 100 is sufficiently large so that we can use the large-sample confidence interval.)

(d) How does decreasing the sample size change the margin of error of a confidence interval when the confidence level remains the same?

Choosing the sample size

NEW YORK, NEW YORK

In planning a study, we may want to choose a sample size that will allow us to estimate the parameter within a given margin of error. The margin of error in the large-sample confidence interval for p is

$$m = z^* \sqrt{\frac{\hat{p}(1 - \hat{p})}{n}}$$

Here z^* is the standard Normal critical value for the level of confidence we want. Because the margin of error involves the sample proportion of successes $\hat{p}$, we need to guess this value when choosing n. Call our guess p^*. Here are two ways to get p^*:

1. Use a guess p^* based on a pilot study or on past experience with similar studies. You can do several calculations to cover the range of values of $\hat{p}$ you might get.

2. Use $p^* = 0.5$ as the guess. The margin of error m is largest when $\hat{p} = 0.5$, so this guess is conservative in the sense that if we get any other $\hat{p}$ when we do our study, we will get a margin of error smaller than planned.

Once you have a guess p^*, the recipe for the margin of error can be solved to give the sample size n needed. Here is the result for the large-sample confidence interval.

New York City, they say, is bigger, richer, faster, ruder. Maybe there's something to that. The sample survey firm Zogby International says that as a national average it takes 5 telephone calls to reach a live person. When calling to New York, it takes 12 calls. Survey firms assign their best interviewers to make calls to New York and often pay them bonuses to cope with the stress.

> ### Sample Size for Desired Margin of Error
>
> The level C confidence interval for a population proportion p will have margin of error approximately equal to a specified value m when the sample size is
>
> $$n = \left(\frac{z^*}{m}\right)^2 p^*(1 - p^*)$$
>
> where p^* is a guessed value for the sample proportion. The margin of error will always be less than or equal to m if you take the guess p^* to be 0.5.

Which method for finding the guess p^* should you use? The n you get doesn't change much when you change p^* as long as p^* is not too far from 0.5. You can use the conservative guess $p^* = 0.5$ if you expect the true $\hat{p}$ to be roughly between 0.3 and 0.7. If the true $\hat{p}$ is close to 0 or 1, using $p^* = 0.5$ as your guess will give a sample much larger than you need. Try to use a better guess from a pilot study when you suspect that $\hat{p}$ will be less than 0.3 or greater than 0.7.

EXAMPLE 16.5   Planning a Poll

STATE: Gloria Chavez and Ronald Flynn are the candidates for mayor in a large city. You are planning a sample survey to determine what percent of the voters intend to vote for Chavez. You will contact an SRS of registered voters in the city. You want to estimate the proportion p of Chavez voters with 95% confidence and a margin of error no greater than 3%, or 0.03. How large a sample do you need?

PLAN: Find the sample size n needed for margin of error $m = 0.03$ and 95% confidence. The winner's share in all but the most lopsided elections is between 30% and 70% of the vote. You can use the guess $p^* = 0.5$.

SOLVE: The sample size you need is

$$n = \left(\frac{1.96}{0.03}\right)^2 (0.5)(1 - 0.5) = 1067.1$$

Round the result up to $n = 1068$. (Rounding down would give a margin of error slightly greater than 0.03.)

CONCLUDE: An SRS of 1068 registered voters is adequate for margin of error ±3%.

If you want a 2.5% margin of error rather than 3%, then (after rounding up)

$$n = \left(\frac{1.96}{0.025}\right)^2 (0.5)(1 - 0.5) = 1537$$

For a 2% margin of error the sample size you need is

$$n = \left(\frac{1.96}{0.02}\right)^2 (0.5)(1 - 0.5) = 2401$$

As usual, smaller margins of error call for larger samples. ■

Apply Your Knowledge

16.12 Canadians and Doctor-Assisted Suicide. A Gallup Poll asked a sample of Canadian adults if they thought the law should allow doctors to end the life of a patient who is in great pain and near death if the patient makes a request in writing. The poll included 270 people in Quebec, 221 of whom agreed that doctor-assisted suicide should be allowed.[9]

(a) What is the margin of error of the large-sample 95% confidence interval for the proportion of all Quebec adults who would allow doctor-assisted suicide?

(b) How large a sample is needed to get the common ±3 percentage point margin of error? Use the previous sample as a pilot study to get p^*.

16.13 **Can You Taste PTC?** PTC is a substance that has a strong bitter taste for some people and is tasteless for others. The ability to taste PTC is inherited. About 75% of Italians can taste PTC, for example. You want to estimate the proportion of Americans with at least one Italian grandparent who can taste PTC. Starting with the 75% estimate for Italians, how large a sample must you collect to estimate the proportion of PTC tasters within ±0.04 with 90% confidence?

Cautions about confidence intervals

The most important caution about confidence intervals in general is a consequence of the use of a sampling distribution. A sampling distribution shows how a statistic such as $\hat{p}$ varies in repeated random sampling. This variation causes *random sampling error* because the statistic misses the true parameter by a random amount. No other source of variation or bias in the sample data influences the sampling distribution. So *the margin of error in a confidence interval ignores everything except the sample-to-sample variation due to choosing the sample randomly.*

The Margin of Error Doesn't Cover All Errors

The margin of error in a confidence interval covers only random sampling errors.

Practical difficulties such as undercoverage and nonresponse are often more serious than random sampling error. The margin of error does not take such difficulties into account.

Recall from Chapter 8 that national opinion polls often have response rates less than 50% and that even small changes in the wording of questions can strongly influence results. In such cases, the announced margin of error is probably unrealistically small. And of course there is no way to assign a meaningful margin of error to results from voluntary response or convenience samples because there is no random selection. Look carefully at the details of a study before you trust a confidence interval.

Apply Your Knowledge

16.14 **Are You Overweight?** A Gallup Poll asked a national random sample of 494 adult women to state their current weight. The proportion of women who said they were "very" or "somewhat" overweight was 0.42. We will treat these data as an SRS.

(a) Give a 95% confidence interval for the proportion of adult women who would say they were "very" or "somewhat" overweight based on these data. Verify that the sample is large enough for us to use the large-sample confidence interval.

(b) Do you trust the interval you computed in part (a) as a 95% confidence interval for the proportion of all U.S. adult women who would say they were "very" or "somewhat" overweight? Why or why not?

16.15 **Is Your Food Safe?** "Do you feel confident or not confident that the food available at most grocery stores is safe to eat?" When a Gallup Poll asked this question, 82% of the sample said they were confident.[10] Gallup announced the poll's margin of error for 95% confidence as ±3 percentage points. Which of the following sources of error are included in this margin of error?

(a) Gallup dialed landline telephone numbers at random and so missed all people without landline phones, including people whose only phone is a cell phone.

(b) Some people whose numbers were chosen never answered the phone in several calls or answered but refused to participate in the poll.

(c) There is chance variation in the random selection of telephone numbers.

Plus four confidence intervals for a proportion*

The large-sample confidence interval $\hat{p} \pm z^* \sqrt{\hat{p}(1 - \hat{p})/n}$ for a sample proportion p is easy to calculate. It is also easy to understand because it is based directly on the approximately Normal distribution of $\hat{p}$. However, confidence levels from this interval can be inaccurate, particularly with smaller samples. The actual confidence level is usually *less* than the confidence level you asked for in choosing the critical value z^*. That's bad. What is worse, accuracy does not consistently get better as the sample size n increases. There are "lucky" and "unlucky" combinations of the sample size n and the true population proportion p.

Fortunately, there is a simple modification that is almost magically effective in improving the accuracy of the confidence interval. We call it the "plus-four" method because all you need to do is *add four imaginary observations, two successes and two failures*. With the added observations, the **plus four estimate** of p is

plus four estimate

$$\tilde{p} = \frac{\text{number of successes in the sample} + 2}{n + 4}$$

The formula for the confidence interval is exactly as before, with the new sample size and number of successes.[11] You do not need software that offers the plus four interval—just enter the new sample size (actual size + 4) and number of successes (actual number + 2) into the large-sample procedure.

PLUS FOUR CONFIDENCE INTERVAL FOR A PROPORTION

Draw an SRS of size n from a large population that contains an unknown proportion p of successes. To get the **plus four confidence interval for p,** add four imaginary observations, two successes and two failures. Then use the large-sample confidence interval with the new sample size ($n + 4$) and number of successes (actual number + 2).

Use this interval when the confidence level is at least 90% and the sample size n is at least 10, with any counts of successes and failures.

*Plus four confidence intervals are more accurate than the large-sample confidence interval we have discussed so far, but this more advanced material is not needed to read the rest of the book.

EXAMPLE 16.6   Cocaine Traces in Spanish Currency

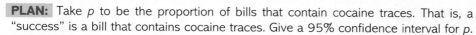

STATE: Cocaine users commonly snort the powder up the nose through a rolled-up paper currency bill. Spain has a high rate of cocaine use, so it's not surprising that euro paper currency in Spain often contains traces of cocaine. Researchers collected 20 euro bills in each of several Spanish cities. In Madrid, 17 out of 20 contained traces of cocaine.[12] The researchers noted that we can't tell whether the bills had been used to snort cocaine or had been contaminated in currency-sorting machines. Estimate the proportion of all euro bills in Madrid that have traces of cocaine.

PLAN: Take p to be the proportion of bills that contain cocaine traces. That is, a "success" is a bill that contains cocaine traces. Give a 95% confidence interval for p.

SOLVE: It is not clear how the bills in the sample were selected, so we don't know if we have an SRS. We will act as thought we have an SRS, but we proceed with caution. The conditions for use of the large-sample interval are not met because there are only 3 failures. To apply the plus-four method, add two successes and two failures to the original data. The plus four estimate of p is

$$\tilde{p} = \frac{17 + 2}{20 + 4} = \frac{19}{24} = 0.7917$$

We calculate the plus four confidence interval in the same way as we do the large-sample interval, but we base it on 19 successes in 24 observations. Here it is:

$$\tilde{p} \pm z^* \sqrt{\frac{\tilde{p}(1 - \tilde{p})}{n + 4}} = 0.7917 \pm 1.960 \sqrt{\frac{(0.7917)(0.2083)}{24}}$$
$$= 0.7917 \pm 0.1625$$
$$= 0.6292 \text{ to } 0.9542$$

CONCLUDE: Assuming the sample can be regarded as an SRS, we estimate with 95% confidence that between about 63% and 95% of all euro bills in Madrid contain traces of cocaine. ■

For comparison, the ordinary sample proportion is

$$\hat{p} = \frac{17}{20} = 0.85$$

The plus four estimate $\tilde{p} = 0.7917$ in Example 16.6 is farther away from 1 than $\hat{p} = 0.85$. The plus four estimate gains its added accuracy by always moving toward 0.5 and away from 1 or 0, whichever is closer. This is particularly helpful when the sample contains only a few successes or a few failures. The numerical difference between a large-sample interval and the corresponding plus four interval is often small. Remember that the confidence level is the probability that the interval will catch the true population proportion *in very many uses*. Small differences every time add up to accurate confidence levels from plus four versus inaccurate levels from the large-sample interval.

How much more accurate is the plus four interval? Computer studies have asked how large n must be to guarantee that the actual probability that a 95% confidence interval covers the true parameter value is at least 0.94 for all samples of size n or larger. If $p = 0.1$, for example, the answer is $n = 646$ for the large-sample interval and $n = 11$ for the plus four interval.[13] The consensus of computational and theoretical studies is that plus four is very much better than the large-sample interval for many combinations of n and p. (If you use software such as Minitab, you may find an "exact method" on the menu. Despite the appealing name, this method is often less accurate than plus four.)

Apply Your Knowledge

rtyreel/Getty Images

16.16 Black Raspberries and Cancer. Sample surveys usually contact large samples, so we can use the large-sample confidence interval if the sample design is close to an SRS. Scientific studies often use smaller samples that require the plus-four method. For example, familial adenomatous polyposis (FAP) is a rare inherited disease characterized by the development of an extreme number of polyps early in life and by colon cancer in virtually 100% of patients before the age of 40. A group of 14 people suffering from FAP and being treated at the Cleveland Clinic drank black raspberry powder in a slurry of water every day for nine months. The number of polyps was reduced in 11 out of 14 of these patients.[14]

(a) Why can't we use the large-sample confidence interval for the proportion p of patients suffering from FAP who will have the number of polyps reduced after 9 months of treatment? Can we use the plus-four method?

(b) The plus-four method adds four observations: two successes and two failures. What are the sample size and the number of successes after you do this? What is the plus four estimate $\tilde{p}$ of p?

(c) Give the plus four 90% confidence interval for the proportion of patients suffering from FAP who will have the number of polyps reduced after nine months of treatment.

16.17 Computer/Internet-Based Crime. With over 50% of adults spending more than an hour a day on the Internet, the number experiencing computer- or Internet-based crime continues to rise. A survey in 2010 of a random sample of 1025 adults, aged 18 and older, reached by random digit dialing found 113 adults in the sample who said that they or a household member was a victim of a computer or Internet crime on their home computer in the past year.[15]

(a) Give the 95% large-sample confidence interval for the proportion p of all households that have experienced computer or Internet crime during the year before the survey was conducted. Be sure to verify that the sample size is large enough to use the large-sample confidence interval.

(b) Give the plus four 95% confidence interval for p. If you express the two intervals in percents, rounded to the nearest tenth of a percent, how do they differ? (The plus four interval always pulls the results away from 0% or 100%, whichever is closer. Even though the condition for using the large-sample interval is met, the plus four interval is more trustworthy.)

16.18 Cocaine Traces in Spanish Currency, Continued. The plus-four method is particularly useful when there are *no* successes or *no* failures in the data. The study of Spanish currency described in Example 16.6 found that in Seville, all 20 of a sample of 20 euro bills had cocaine traces.

(a) What is the sample proportion $\hat{p}$ of contaminated bills? What would the large-sample 95% confidence interval give for p (ignoring the fact that the sample size condition is not met)? It's not plausible that *every* bill in Seville has cocaine traces, as this interval says.

(b) Find the plus four estimate $\tilde{p}$ and the plus four 95% confidence interval for p. These results are more reasonable.

CHAPTER 16 SUMMARY

Chapter Specifics

■ A **confidence interval** uses sample data to estimate an unknown population parameter with an indication of how accurate the estimate is and of how confident we are that the result is correct.

■ Any confidence interval has two parts: an interval calculated from the data and a confidence level C. The **confidence interval** often has the form

$$\text{estimate} \pm \text{margin of error}$$

■ The **confidence level** is the success rate of the method that produces the interval. That is, C is the probability that the method will give a correct answer. If you use 95% confidence intervals often, in the long run 95% of your intervals will contain the true parameter value. You do not know whether or not a 95% confidence interval calculated from a particular set of data contains the true parameter value.

■ A level C **confidence interval for p** when the sample size is large is

$$\hat{p} \pm z^* \sqrt{\frac{\hat{p}(1 - \hat{p})}{n}}$$

where z^* is the critical value for the standard Normal curve with area C between $-z^*$ and z^*.

■ Other things being equal, the margin of error of a confidence interval gets smaller as
   ■ the confidence level C decreases,
   ■ the sample size n increases.

■ The **sample size** needed to obtain a confidence interval with approximate margin of error m for a population proportion is

$$n = \left(\frac{z^*}{m}\right)^2 p^*(1 - p^*)$$

where p^* is a guessed value for the sample proportion $\hat{p}$, and z^* is the standard Normal critical point for the level of confidence you want. If you use $p^* = 0.5$ in this formula, the margin of error of the interval will be less than or equal to m no matter what the value of $\hat{p}$ is.

■ A specific confidence interval is correct only under specific conditions. The most important conditions concern the method used to produce the data. Other factors such as the shape of the population distribution may also be important.

■ Whenever you use statistical inference, you are acting as if your data are a random sample or come from a randomized comparative experiment.

■ The margin of error in a confidence interval accounts for only the chance variation due to random sampling. In practice, errors due to nonresponse or undercoverage are often more serious.

■ **(Optional material)** To get a more accurate confidence interval, add four imaginary observations, two successes and two failures, to your sample. Then use the same formula for the confidence interval. This is the **plus four confidence interval.** Use this interval in practice for confidence levels 90% or higher and sample size n at least 10.

■ **(Optional material)** The true confidence level of the large-sample interval can be substantially less than the planned level C for many combinations of n and p unless the sample is very large. Use of the plus four interval avoids these inaccuracies.

Link It

The reason we collect data is not to learn about the individuals that we observed but to infer from the data to some wider population that the individuals represent. Chapters 8 and 9 tell us that the way we produce the data (sampling, experimental design) affects whether we have a good basis for generalizing to some wider population. Chapters 11, 12, and 13 discuss probability, the formal mathematical tool that determines the nature of the inferences we make. In addition, Chapter 15 discusses sampling distributions, which tell us how repeated SRSs behave and hence what a statistic (in particular, a sample proportion) computed from our sample is likely to tell us about the corresponding parameter of the population (in particular, a population proportion) from which the sample was selected.

In this chapter we discussed the basic reasoning of statistical estimation, with emphasis on estimating a population proportion. To an estimate of the population proportion we attach a margin of error and a confidence level. The result is a confidence interval. The sampling distribution of the sample proportion, discussed in Chapter 15, provides the mathematical basis for constructing confidence intervals and understanding their properties. Although we apply the reasoning of statistical estimation in a simple setting, we will use the same logic in future chapters to construct confidence intervals for population parameters in more complicated settings.

When making inferences about a proportion in a population, an important assumption for the methods of this chapter is that the data are an SRS from the population. This is the most difficult assumption to guarantee because of the many difficulties associated with obtaining an SRS. These difficulties were described in Chapter 8. With nonresponse rates in many surveys that are over 80%, the final sample may not be representative of the population, even when researchers initially selected an SRS. Because of this, most surveys of large populations need to use more complicated sampling schemes as well as modify the estimates to adjust for problems such as nonresponse.

CHECK YOUR SKILLS

16.19 To give a 99.9% confidence interval for a population proportion p, you would use the critical value

(a) $z^* = 1.960$.   (b) $z^* = 2.576$.   (c) $z^* = 3.291$.

Use the following information for Exercises 16.20 through 16.22. The proportion of drivers who use seat belts depends on things like age, sex, and ethnicity. As part of a broader study, investigators observed a random sample of 117 female Hispanic drivers in Boston. Seventy of those in the sample were observed wearing a seat belt.

16.20 A large-sample 95% confidence interval for the proportion of all female Hispanic drivers in the Boston area who wear seat belts is

(a) 0.60 ± 0.05.
(b) 70 ± 0.007.
(c) 0.60 ± 0.09.

16.21 You want a large-sample 99% confidence interval for the proportion of all female Hispanic drivers in the Boston area who wear seat belts. The margin of error for this interval will be

(a) smaller than the margin of error for 95% confidence.
(b) greater than the margin of error for 95% confidence.
(c) about the same as the margin of error for 95% confidence.

16.22 Suppose we had observed a random sample of 468 female Hispanic drivers in Boston and found that 280 of those in the sample were observed wearing a seat belt. A large-sample 95% confidence interval for the proportion of all female Hispanic drivers in the Boston area who wear seat belts based on this sample would be

(a) larger than the interval in Exercise 16.20.
(b) smaller than the interval in Exercise 16.20.
(c) either larger or smaller than the interval in Exercise 16.20 because the sample is random.

Use the following information for Exercises 16.23 through 16.26. A 2011 NBC News survey found that 80% of a sample of 4500 American teens said they owned an MP3 player such as an iPod. Assume that the sample was an SRS.

16.23 Based on the sample, the large-sample 90% confidence interval for the proportion of all American teens who own an MP3 player is

(a) 0.80 ± 0.0060.
(b) 0.80 ± 0.0098.
(c) 0.80 ± 0.0117.

16.24 In the previous exercise, suppose we computed a large-sample 80% confidence interval for the proportion of all American teens who own an MP3 player.

(a) This 80% confidence interval would have a smaller margin of error than the 90% confidence interval.

(b) This 80% confidence interval would have a larger margin of error than the 90% confidence interval.
(c) This 80% confidence interval could have either a smaller or a larger margin of error than the 90% confidence interval. This varies from sample to sample.

16.25 Suppose NBC News took an SRS of 450 teens rather than 4500, but again found that 80% of those sampled said they owned an MP3 player such as an iPod. Compared with an SRS of 4500 teens, the margin of error for a large-sample 90% confidence interval for the proportion of all American teens who own an MP3 player is

(a) smaller.
(b) larger.
(c) either smaller or larger, but we can't say which.

16.26 How many American teens must be interviewed to estimate the proportion who own MP3 players within ± 0.02 with 99% confidence using the large-sample confidence interval? Use 0.5 as the conservative guess for p.

(a) $n = 1692$ (b) $n = 2401$ (c) $n = 4148$

16.27 An opinion poll asks an SRS of 40 college seniors how likely they are to seek a job out of state. In all, 8 say "very likely." The large-sample 95% confidence interval for estimating the proportion of all college seniors who think they are very likely to seek a job out of state is

(a) 0.20 ± 0.06.
(b) 0.20 ± 0.12.
(c) inappropriate because the large sample method cannot be used here.

16.28 A February 2012 Gallup Poll found that 53% of American adults have a very favorable opinion of Canada. The poll's margin of error for 95% confidence was 4%. This means that

(a) the poll used a method that gets an answer within 4% of the truth about the population 95% of the time.
(b) we can be sure that the percent of all American adults who have a very favorable opinion of Canada is between 49% and 57%.
(c) if Gallup takes another poll using the same method, the results of the second poll will lie between 49% and 57%.

16.29 An opinion poll asks an SRS of 100 college seniors how they view their job prospects. In all, 53 say "Good." A large-sample 95% confidence interval for estimating the proportion of all college seniors who think their job prospects are good is

(a) 0.53 ± 0.082.
(b) 0.53 ± 0.098.
(c) 0.53 ± 0.050.

16.30 The sample survey in Exercise 16.29 actually called 130 seniors, but 30 of the seniors refused to answer. This nonresponse could cause the survey result to be in error. The error due to nonresponse

(a) is in addition to the margin of error found in Exercise 16.29.

(b) is included in the margin of error found in Exercise 16.29.

(c) can be ignored because it isn't random.

CHAPTER 16 EXERCISES

16.31 **Should we ban texting while driving?** A February 2012 Harris Poll found that in a sample of 2211 U.S. adults over age 18, 2012 said they would support a ban on texting while driving.

(a) Suppose the people interviewed were an SRS of U.S. adults over age 18. Verify that the sample-size conditions are met and give a 95% confidence interval for the percent of all U.S. adults over age 18 who would support a ban on texting while driving.

(b) Harris states that the respondents for this survey were selected from among those who have agreed to participate in Harris Interactive surveys. Based on what you know about national sample surveys, what is likely to be the biggest weakness in the survey?

16.32 **Reporting cheating.** Students are reluctant to report cheating by other students. A student project put this question to an SRS of 172 undergraduates at a large university: "You witness two students cheating on a quiz. Do you go to the professor?" Only 19 answered "Yes."[16] Verify that the sample size conditions are met, and give a 95% confidence interval for the proportion of all undergraduates at this university who would report cheating.

16.33 **Would you do it over again?** A September 2012 Gallup Poll asked a sample of smokers, "If you had to do it over again, would you start smoking or not?" Of the 166 people in the sample, 17 said "Yes."

Give a 95% confidence interval for the proportion of smokers who would start smoking if they could do it over again. Interpret your interval in the context of the poll.

16.34 **Gallup announces a margin of error.** Exercise 16.33 describes a Gallup Poll survey of smokers in which 17 of a sample of 166 smokers said that they would start smoking if they could do it over again. Gallup announces a maximum margin of error of ±10 percentage points. Opinion polls announce the margin of error for 95% confidence.

(a) What is the actual margin of error (in percent) for the large-sample confidence interval from this sample?

(b) The margin of error is largest when $\hat{p} = 0.5$. What would the margin of error (in percent) be if the sample had resulted in $\hat{p} = 0.5$?

(c) Why do you think that Gallup announces a maximum ±10% margin of error?

16.35 **Explaining confidence.** A student reads in a 2012 Gallup poll that a 95% confidence interval for the proportion of adult American women who worry some or all of the time about their weight is 0.55 ± 0.06. Asked to explain the meaning of this interval, the student says, "There is a 95% probability that the proportion all adult American women who worry some or all of the time about their weight is between 0.49 and 0.61." Is the student right? Explain your answer.

16.36 **Explaining confidence.** You ask another student to explain the confidence interval described in the previous exercise. The student answers, "We can be 95% confident that future samples of adult American women will say that the proportion who worry some or all of the time about their weight is between 0.49 and 0.61." Is this explanation correct? Explain your answer.

16.37 **Explaining confidence.** Here is an explanation from the Associated Press concerning one of its opinion polls. Explain briefly but clearly in what way this explanation is incorrect.

> For a poll of 1,600 adults, the variation due to sampling error is no more than three percentage points either way. The error margin is said to be valid at the 95 percent confidence level. This means that, if the same questions were repeated in 20 polls, the results of at least 19 surveys would be within three percentage points of the results of this survey.

16.38 **Testing the waters.** In August 2010, *The Columbus Dispatch* took water samples at 20 Ohio State Park swimming areas and tested for fecal coliform, which are bacteria found in human and animal feces. Experts warn that the tests are a snapshot of the quality of the water at the time they were taken, and levels can change as weather and other conditions vary. An unsafe level of fecal coliform means that there's a higher chance that disease-causing bacteria are present and more risk that a swimmer will become ill. Of the 20 swimming areas tested, 13 were found to have unsafe levels of fecal coliform according to state standards. Assume that the swimming areas tested represent a random sample of swimming areas throughout the state.[17] Show that the conditions for the large-sample confidence interval discussed in this chapter are not met.

16.39 **Internet searches and cell phones.** Pew Internet and American Life Project asked a random sample of 2485 cell phone users whether they had used their cell phone to look up health or medical information. Of these, 422 said "Yes."[18]

(a) Pew dialed cell phone telephone numbers at random in the continental United States in an attempt to contact a random sample of adults. Based on what you know about national sample surveys, what is likely to be the biggest weakness in the survey?

(b) Act as if the sample is an SRS. Verify that the sample size conditions are met and give a large-sample 90% confidence interval for the proportion p of all cell phone users who have used their cell phone to look up health or medical information.

(c) Three out of the five most popular health-related searches on cell phones have to do with sex: "pregnancy," "herpes," and "STDs" (sexually transmitted diseases). Sex-related queries don't even show up on Google and Yahoo's lists of the top five health searches on computers. What do you think explains the difference in the topics of health-related searches on cell phones versus computers? When drawing conclusions from a sample, you must always be careful to consider the relevant population.

16.40 **Running red lights.** A random digit dialing telephone survey of 880 drivers asked, "Recalling the last ten traffic lights you drove through, how many of them were red when you entered the intersections?" Of the 880 respondents, 171 admitted that at least one light had been red.[19]

(a) Verify that the sample size conditions are met and give a 95% confidence interval for the proportion of all drivers who ran one or more of the last ten red lights they encountered.

(b) Nonresponse is a practical problem for this survey—only 21.6% of calls that reached a live person were completed. Another practical problem is that people may not give truthful answers. What is the likely direction of the bias: do you think more or fewer than 171 of the 880 respondents really ran a red light? Why?

16.41 **Customer satisfaction.** An automobile manufacturer would like to know what proportion of its customers are not satisfied with the service provided by the local dealer. The customer relations department will survey a random sample of customers and compute a 99% confidence interval for the proportion who are not satisfied.

(a) Past studies suggest that this proportion will be about 0.2. Find the sample size needed if the margin of error of the confidence interval is to be about 0.015.

(b) When the sample of the size found in (a) is actually contacted, 10% of the sample say they are not satisfied. What is the margin of error of the 99% confidence interval?

In responding to Exercises 16.42 to 16.45, follow the **Plan, Solve,** *and* **Conclude** *steps of the four-step process.*

16.42 **College-educated parents.** The National Assessment of Educational Progress (NAEP) includes a "long-term trend" study that tracks reading and mathematics skills over time and obtains demographic information. In the 2008 study, a random sample of 9600 17-year-old students was selected.[20] The NAEP sample used a multistage design, but the overall effect is quite similar to an SRS of 17-year-olds who are still in school.

(a) In the sample, 46% of students had at least one parent who was a college graduate. Estimate with 99% confidence the proportion of all 17-year-old students in 2008 who had at least one parent who was a college graduate.

(b) The sample does not include 17-year-olds who dropped out of school, so your estimate is valid only for students. Do you think the proportion of all 17-year-olds with at least one parent who was a college graduate would be higher or lower than 46%? Explain.

16.43 **Opinions about evolution.** A sample survey funded by the National Science Foundation asked a random sample of American adults about biological evolution.[21] One question asked subjects to answer "True," "False," or "Not sure" to the statement "Human beings, as we know them today, developed from earlier species of animals." Of the 1484 respondents, 594 said "True." What can you say with 95% confidence about the percent of all American adults who think that humans developed from earlier species of animals?

16.44 **Order in choice.** Does the order in which wine is presented make a difference? Several choices of wine are presented one at a time, and the subject is then asked to choose his or her preferred wine at the end of the sequence. In one study, subjects were asked to taste two wine samples in sequence. Both samples given to a subject were the *same* wine, although the subjects were expecting to taste two different samples of a particular variety. Of the 32 subjects in the study, 22 selected the wine presented first when presented with two identical wine samples.[22]

(a) Give a 95% confidence interval for the proportion of subjects who would select the first choice presented.

(b) The subjects were recruited in Ontario, Canada, via advertisements to participate in a study of "attitudes and values towards wine." What assumption are you making about these subjects?

16.45 **Chick-fil-A gets it right.** Which fast-food chain fills orders most accurately at the drive-thru window? The *Quick Service Restaurant* (QSR) magazine drive-thru study involved a total of 7594 visits to restaurants in the 25 largest fast-food chains in all 50 states. All visits occurred during the lunch hours of 11:00 A.M. to 2:30 P.M. or during the dinner hours of 4:00 to 7:00 P.M. During each visit, the researcher ordered a main item, a side item, and a drink. One item was left off each order; for example, a field researcher could order a burger with no pickles. After receiving the order, all food and drink items were checked for complete accuracy. Any food or drink item received that was not exactly as ordered resulted in the order being classified as inaccurate. Also included in the measurement of accuracy were condiments asked for, napkins, straws, and correct change. Any errors in these resulted in the order being classified as inaccurate. Chick-fil-A had the fewest inaccuracies, with only 14 of 196 orders classified as inaccurate.[23] What proportion of orders are filled *accurately* by Chick-fil-A? (Use 95% confidence.)

© Julie Dermansky/Julie Dermansky/
Corbis

16.46 **Order in choice: planning a study.** How large a sample would be needed to obtain margin of error ±0.05 in the study of choice order for tasting wine? Use the $\hat{p}$ from Exercise 16.44 as your guess for the unknown p.

16.47 **Why are larger samples better?** Statisticians prefer large samples. Describe briefly the effect of increasing the size of a sample on the margin of error of a 95% confidence interval.

The following exercises concern the optional material on the plus-four method.

16.48 **Testing the waters.** In Exercise 16.38 (page 370) you showed that the conditions for the large-sample confidence interval for a population proportion are not met. Use the plus-four method to give a 90% confidence interval for the percent of Ohio State Park swimming areas that have unsafe levels of fecal coliform.

16.49   **Shrubs that survive fires.** Some shrubs have the useful ability to resprout from their roots after their tops are destroyed. Fire is a particular threat to shrubs in dry climates, as it can injure the roots as well as destroy the aboveground material. One study of resprouting took place in a dry area of Mexico.[24] The investigators clipped the tops of samples of several species of shrubs. In some cases, they also applied a propane torch to the stumps to simulate a fire. Of 12 specimens of the shrub *Krameria cytisoides,* 5 resprouted after fire. Estimate with 90% confidence the proportion of all shrubs of this species that will resprout after fire.

16.50 **Prayer among the Millennials.** The Millennial generation (so called because they were born after 1980 and began to come of age around the year 2000) are less religiously active than older Americans. One of the questions in the General Social Survey in 2010 was "How often does the respondent pray?" Among the 411 respondents in the survey between 18 and 30 years of age, 277 prayed at least once a week.[25] Assume that the sample is an SRS.

(a) Verify that the sample size conditions are met for the large-sample confidence interval. What is the large-sample 99% confidence interval for the proportion p of all adults between 18 and 30 years of age who pray at least once a week?

(b) Give the plus four 99% confidence interval for p. If you express the two intervals in percents and round to the nearest tenth of a percent, how do they differ? (As always, the plus-four method pulls results away from 0% or 100%, whichever is closer. Although the condition for the large-sample interval is met, the plus four interval is more trustworthy.)

 Exploring the Web

16.51 **A statistics glossary.** An editorial was published in the *Journal of the National Cancer Institute*, Vol. 101, No. 23 (December 2, 2009) that announced some online resources for journalists, including a statistics glossary. The glossary can be found at `www.oxfordjournals.org/our_journals/jnc/resource/ statistics%20glossary.pdf`. Read the definition of a confidence interval. Is this an accurate definition? Explain your answer.

16.52 **Getting around No Child Left Behind.** The PBS Web site has an interesting article from 2007 discussing how school districts were getting around certain requirements of the No Child Left Behind law. You can find the article at `http://www. pbs.org/newshour/bb/education-july-dec07-nclb_08-14/`. What does the article have to say about the use of confidence intervals in reporting results about the percent of students passing proficiency tests?

Overview

In this chapter, students encounter a second type of formal statistical inference: tests of significance. With confidence intervals, we were interested in determining a set of plausible values of a parameter—values consistent with the sample observed. With tests of significance, the focus is on a particular value of the parameter, with an interest in whether the particular value of interest is plausible or may be rejected on seeing the sample. Hence, there is a conceptual duality between confidence intervals and hypothesis tests.

Many students will find that the terminology and symbols encountered in this chapter are almost overwhelming. However, there are only a handful of crucial ideas here, the most important being the concept of a *P*-value. The *P*-value is a measure of how unusual our sample results are, assuming the null hypothesis is true. Common challenges for students include (1) correct specification of hypotheses (including letting the research question and not the data dictate the alternative hypothesis); (2) calculating the *P*-value based on the alternative hypothesis (one-tailed to the left, one-tailed to the right, two-tailed); (3) understanding that smaller *P*-values indicate stronger evidence against the null hypothesis; and (4) understanding that the decision is rejecting the null hypothesis or not rejecting it—failure to reject the null hypothesis *does not* imply that the null hypothesis is true.

Teaching Suggestions and Additional Examples/Activities for the Classroom

1. Carry Over Examples from Chapter 15

Many of the teaching suggestions from Chapter 16 can be used in this chapter as well. For example, if you had your students calculate confidence intervals for the proportion of blue milk chocolate M&M's, you can use the same data to have your students test the hypotheses $H_0: p = 0.24$ versus $H_a: p \neq 0.24$ (at the time of this writing the proportion of blue M&M's is 0.24).

2. Contrast Standard Errors for Hypothesis Tests and Confidence Intervals

Students often confuse the standard error of $\hat{p}$ for confidence intervals and significance tests. Many will, for example, use $\hat{p}$ instead of p_0 in computing a test statistic. Though this might not always result in a large computational error, it does constitute a substantial conceptual error. Emphasize to students that a null hypothesis is also a statement about the standard error of $\hat{p}$, whereas for confidence intervals the standard error of $\hat{p}$ is estimated from the data only (because we don't know what p is). You can return to Chapter 15 to discuss the tie between the mean and standard deviation for a sample proportion if we know p. In hypothesis testing, if we assume the null hypothesis is true, the mean of the

LEARNING OUTCOMES**

- State the null and alternative hypotheses in a testing situation in terms of the parameter of interest.
- Explain in nontechnical language the meaning of the *P*-value when you are given the numerical value for *P* for a test.
- Know the four-step process (page 383) for any significance test.
- Assess statistical significance at standard levels α, by comparing *P* with α (or, optionally, by comparing the value of the test statistic with standard critical values).
- Recognize that significance testing does not measure the size or importance of an effect. Explain why a small effect can be significant in a large sample and why a large effect can fail to be significant in a small sample.
- Recognize that any inference procedure acts as if the data were properly produced. In this (single sample) case, the data should be an SRS from the population.
- Verify the conditions under which we can safely use the large-sample *z* procedures in a particular setting. Check the study design and the guidelines for sample size.
- Use the *z* statistic to carry out a test of significance for the hypothesis $H_0: p = p_0$ about a population proportion *p* against either a one-sided or a two-sided alternative. Use software or Table A to find the *P*-value, or Table C to get an approximate value.

**These learning outcomes appear later for the students in Chapter 23: Part IV Review.

sampling distribution is p_0 and the standard error is $\sqrt{\dfrac{p_0(1-p_0)}{n}}$. This might also be a good time to foreshadow that this relationship between the mean and standard deviation will not hold when we turn our attention to means.

Other Resources (LaunchPad)

Snapshots Videos
Inference for One Proportion

EESEE Case Studies
Is Caffeine Dependence Real?
Anecdotes of Significance Testing

Applets
Reasoning of a Statistical Test
Statistical Significance for One Proportion
P-Value for a Test of One Proportion

Main: Niels Hariot/Shutterstock; Left: Sofia Andreevna/Shutterstock; Right: Franck Boston/Shutterstock

Tests of Significance: The Basics

Confidence intervals are one of the two most common types of statistical inference. Use a confidence interval when your goal is to estimate a population parameter. The second common type of inference, called *tests of significance*, has a *different goal:* to assess the evidence provided by data about some claim concerning a population. Here is the reasoning of statistical tests in a nutshell.

In this chapter we cover...

- The reasoning of tests of significance
- Stating hypotheses
- *P*-value and statistical significance
- Large sample tests for a population proportion
- Cautions about significance tests
- Significance from a table*

EXAMPLE 17.1   I'm a Good Free-Throw Shooter

I claim that I make 75% of my basketball free throws. To test my claim, you ask me to shoot 20 free throws. I make only 8 of the 20. "Aha!" you say. "Someone who makes 75% of his free throws would almost never make only 8 out of 20. So I don't believe your claim."

Your reasoning is based on asking what would happen if my claim were true and we repeated the sample of 20 free throws many times—I would almost never make as few as 8. This outcome is so unlikely that it gives strong evidence that my claim is not true.

You can say how strong the evidence against my claim is by giving the probability that I would make as few as 8 out of 20 free throws if I really make 75% in the long run. This probability is 0.0009. I would make as few as 8 of 20 only 9 times in 10,000 tries in the long run if my claim to make 75% were true. The small probability convinces you that my claim is false. ∎

The *Reasoning of a Statistical Test* applet animates Example 17.1. You can ask a player to shoot free throws until the data do (or don't) convince you that he makes fewer than 75%. Significance tests use an elaborate vocabulary, but the basic idea is simple: *an outcome that would rarely happen if a claim were true is good evidence that the claim is not true.*

The reasoning of tests of significance

The reasoning of statistical tests, like that of confidence intervals, is based on asking what would happen if we repeated the sample or experiment many times. We will act as if the conditions listed on page 350 are true: we have a perfect SRS from a population (or we have conducted a properly randomized experiment) with a sufficiently large sample size (both the number of successes and the number of failures are at least 10) so that the sampling distribution of the sample proportion $\hat{p}$ is approximately Normal.

Here is an example we will explore.

EXAMPLE 17.2   **Can We Feel the Future?**

Precognition is the conscious awareness of a future event that could not otherwise be anticipated through basic reasoning. Undergraduate volunteers at Cornell University[1] were tested to see if an unconscious need to avoid negative experience could stimulate precognition. Participants were shown a picture and its mirror image side by side on a computer screen and were asked to indicate which image they liked better. The computer then randomly selected one of the pictures to be the target. If the participant indicated a preference for the target-to-be (which would occur through precognition), the computer flashed a "positive" picture three times on the screen. A "positive" picture is one that that has been previously identified as evoking positive emotions. If the participant indicated a preference for the nontarget, the computer subliminally flashed a highly arousing "negative" picture (one evoking negative emotions). The researcher was interested in whether, over the course of the experiment, precognition would be strengthened by an unconscious desire to avoid the negative emotions evoked by the negative pictures.

The researcher reported that in 5400 trials, participants selected the target-to-be 2790 times. The proportion of successful selections is $\hat{p} = 2790/5400 = 0.517$. If participants were just guessing, we would expect the proportion of successful selections to be 0.5. The observed proportion of successes, 0.517, is only slightly larger than 0.5, but perhaps it indicates some small increase in precognition. Do these data provide good evidence that precognition can be strengthened by an unconscious desire to avoid negative emotions? In particular, would an outcome as large as $\hat{p} = 0.517$ rarely occur if subjects are just guessing? If not, the data are not good evidence of precognition. ◼

The reasoning is the same as in Example 17.1. We make a claim and ask if the data give evidence *against* it. Evidence that there *is* precognition occurs when the results cannot be explained by guessing, so the claim we test is that subjects are just guessing (there *is not* precognition). In that case, the proportion of times participants will select the target-to-be would be $p = 0.5$.

◼ If the claim that $p = 0.5$ is true, the sampling distribution of $\hat{p}$ in 5400 trials is approximately Normal with mean $p = 0.5$ and standard deviation

$$\sqrt{\frac{p(1-p)}{n}} = \sqrt{\frac{0.5(1-0.5)}{5400}} = 0.0068$$

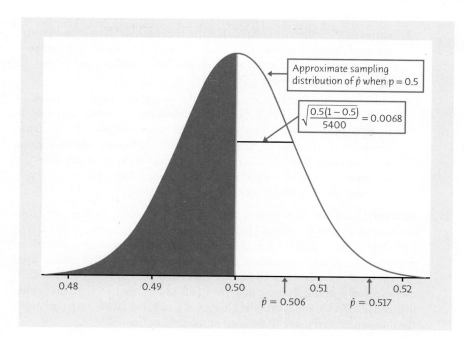

FIGURE 17.1
If subjects do not have precognition, the proportion of successful selections of the target-to-be $\hat{p}$ in 5400 trials will have this sampling distribution. A proportion of successes $\hat{p} = 0.506$ could easily happen just by chance. A proportion of successes $\hat{p} = 0.517$ is far out on the Normal curve and is good evidence against the assumption that subjects are simply guessing.

Here $np = 5400(0.5) = 2700$ and $n(1 - p) = 5400(1 - 0.5) = 2700$, and both are much larger than 10. Thus, it is safe to use this approximation. This is just like the calculations we did in Chapter 15 (see Example 15.5 on page 341) and Chapter 16 (see Example 16.1 on page 350). Figure 17.1 shows this sampling distribution. We can judge whether any observed $\hat{p}$ is surprising by locating it on this distribution.

■ If in 5400 trials participants select the target-to-be 2730 times, the proportion of successful selections is $\hat{p} = 2730/5400 = 0.506$. It is clear from Figure 17.1 that a $\hat{p}$ this large could easily occur just by chance when subjects are simply guessing and the true proportion is $p = 0.5$. That in 5400 trials, the proportion of times subjects select the target to be is $\hat{p} = 0.506$ is not evidence for precognition.

■ The experiment at Cornell University produced $\hat{p} = 0.517$. That's far out on the Normal curve in Figure 17.1—so far out that *an observed value this large would rarely occur just by chance if the true p were 0.5.* This observed value is good evidence that the true p is in fact greater than 0.5, that is, that the subjects are selecting the target-to-be more frequently than can be explained by chance. One explanation might be that precognition was stimulated by an avoidance of negative emotions.

Apply Your Knowledge

17.1 More Feeling the Future. The study described in Example 17.2 also identified, in advance, volunteers who were sensitive to negative emotions. For these volunteers, an experiment similar to that in Example 17.2 was conducted. The researcher reported that in 1800 trials, participants selected the target-to-be 963 times.

(a) We seek evidence *against* the claim that $p = 0.5$. What is the sampling distribution of the sample proportion $\hat{p}$ of times the target-to-be is correctly identified in 1800 trials if the claim is true? Draw the density curve of this distribution. (Sketch a Normal curve, then mark on the axis the values of the mean and 1, 2, and 3 standard deviations on either side of the mean.)

(b)   Calculate $\hat{p}$ for the data observed by the researcher. Mark this point on the axis of your sketch. Suppose that instead, the researcher found participants selected the target-to-be 918 times. Calculate $\hat{p}$ and mark this point on your sketch. Using your sketch, explain in simple language why one result is good evidence that the proportion of participants sensitive to negative emotions is good evidence that subjects are not guessing and why the other outcome is not.

17.2   Stricter Laws for Gun Sales. A December 27, 2012, Gallup poll (conducted shortly after the Sandy Hook Elementary School shootings) found that $\hat{p} = 0.58$ of those polled favored strengthening the laws covering the sale of firearms. The results were based on a random sample of 1038 adults, aged 18 years and older, living in all 50 U.S. states and the District of Columbia. Is there reason to think that the majority of all adults, aged 18 years and older, living in all 50 U.S. states and the District of Columbia favored strengthening the laws covering the sale of firearms at the time of the poll?

(a)   We seek evidence *against* the claim that $p = 0.5$, where p is the actual proportion of all adults, aged 18 years and older, living in all 50 U.S. states and the District of Columbia who favored strengthening the laws covering the sale of firearms at the time of the poll. What is the sampling distribution of the proportion $\hat{p}$ in many random samples of size 1038 if the claim is true? Verify that the conditions for inference hold and make a sketch of the Normal curve for this distribution. (Draw a Normal curve, then mark on the axis the values of the mean and 1, 2, and 3 standard deviations on either side of the mean.)

(b)   Mark the sample proportion $\hat{p} = 0.58$ on the axis of your sketch. Suppose the sample proportion had been $\hat{p} = 0.51$ (the value Gallup found in a similar 2008 poll). Mark this value on the axis as well. Explain in simple language why one result is good evidence that the true proportion is greater than 0.5 and why the other result gives no reason to doubt that 0.5 is correct.

Stating hypotheses

A statistical test starts with a careful statement of the claims we want to compare. In Example 17.2, we saw that the precognition data are not plausible if participants are just guessing. Because the reasoning of tests looks for evidence *against* a claim, we start with the claim we seek evidence against, such as "just guessing."

Null and Alternative Hypotheses

The claim tested by a statistical test is called the **null hypothesis.** The test is designed to assess the strength of the evidence *against* the null hypothesis. Usually the null hypothesis is a statement of "no effect" or "no difference."

The claim about the population that we are trying to find evidence *for* is the **alternative hypothesis.** The alternative hypothesis is **one sided** if it states that a parameter is *larger than* or *smaller than* the null hypothesis value. It is **two sided** if it states that the parameter is *different from* the null value (it could be either smaller or larger).

We abbreviate the null hypothesis as H_0 and the alternative hypothesis as H_a. *Hypotheses always refer to a population, not to a particular outcome. Be sure to state H_0 and H_a in terms of population parameters.* Because H_a expresses the effect that we hope to find evidence *for*, it is sometimes easier to begin by stating H_a and then set up H_0 as the statement that the hoped-for effect is not present.

In Example 17.2, we are seeking evidence *for* precognition. The null hypothesis says "just guessing" on the average in a large sequence of trials. The alternative hypothesis says the proportion of trials in which the target-to-be is correctly identified is greater than what would occur if participants are just guessing because this is what would happen if subjects have precognition. So the hypotheses are

$$H_0: p = 0.5$$
$$H_a: p > 0.5$$

The alternative hypothesis is *one sided* because we are interested only in whether the proportion is greater than what would happen if subjects were just guessing.

EXAMPLE 17.3   Order in Choice

Does the order in which wine is presented make a difference? In one Canadian study, subjects were asked to taste two wine samples in sequence and then indicate which was preferred. Both samples given to a subject were the *same* wine, although the subjects were expecting to taste two different samples of a particular variety.[2]

The parameter of interest is the proportion p of subjects in the population of all adult wine drinkers who would select the wine presented first. The null hypothesis is that order makes no difference and thus that half the time the wine presented first will be selected, that is,

$$H_0: p = 0.5$$

The authors of the study wanted to know if order makes a difference, but did not specify whether first or second wine would be preferred. The alternative hypothesis is therefore *two sided:*

$$H_a: p \neq 0.5 \; \blacksquare$$

 HONEST HYPOTHESES? Chinese and Japanese, for whom the number 4 is unlucky, die more often on the fourth day of the month than on other days. The authors of a study did a statistical test of the claim that the fourth day has more deaths than other days and found good evidence in favor of this claim. Can we trust this? Not if the authors looked at all days, picked the one with the most deaths, then made "this day is different" the claim to be tested. A critic raised that issue, and the authors replied: "No, we had day 4 in mind in advance, so our test was legitimate."

*The hypotheses should express the hopes or suspicions we have **before** we see the data. It is cheating to first look at the data and then frame hypotheses to fit what the data show.* For example, the data for the study in Example 17.3 showed that participants preferred the first wine, but this should not influence the choice of H_a. If you do not have a specific direction firmly in mind in advance, use a two-sided alternative.

Apply Your Knowledge

17.3 More Feeling the Future. State the null and alternative hypotheses for the study of precognition in subjects who are sensitive to negative stimuli described in Exercise 17.1. (Is the alternative hypothesis one sided or two sided?)

17.4 Stricter Laws for Gun Sales. State the null and alternative hypotheses for the Gallup poll concerning stricter laws for the sale of firearms described in Exercise 17.2. (Is the alternative hypothesis one sided or two sided?)

17.5 Grading a Teaching Assistant. The examinations in a large statistics class are scaled after grading so that 40% of the grades are A or B. The professor thinks that one teaching assistant is a poor teacher and suspects that his students have a lower proportion of As and Bs than the class as a whole. The TA's students this semester can be considered a sample from the population of all students in the course, so the professor compares the proportion of As and Bs with 0.40. State the hypotheses H_0 and H_a.

17.6 College Enrollment of Women. According to the Bureau of Labor Statistics, the proportion of American women who graduated from high school in 2011 and went on to college was 0.723. You wonder whether the proportion of female graduates who enrolled in college from your school district in your county in 2011 is different from the national average. You obtain graduation information from an SRS of 40 female 2011 graduates and find that $\hat{p} = 0.625$. What are your null and alternative hypotheses?

17.7 Stating Hypotheses. In planning a study of the proportion of high school students who have experienced cyberbullying, a researcher states the hypotheses as

$$H_0: \hat{p} = 0.5$$
$$H_a: \hat{p} > 0.5$$

What's wrong with this?

P-value and statistical significance

The idea of stating a null hypothesis that we want to find evidence *against* seems odd at first. It may help to think of a criminal trial. The defendant is "innocent until proven guilty." That is, the null hypothesis is innocence, and the prosecution must try to provide convincing evidence against this hypothesis. That's exactly how statistical tests work, though in statistics we deal with evidence provided by data and use a probability to say how strong the evidence is.

The probability that measures the strength of the evidence against a null hypothesis is called a *P-value*. Statistical tests generally work like this:

Test Statistic and P-Value

A **test statistic** calculated from the sample data measures how far the data diverge from what we would expect if the null hypothesis H_0 were true. Large values of the statistic show that the data are not consistent with H_0.

The probability, computed assuming that H_0 is true, that the test statistic would take a value as extreme or more extreme than that actually observed is called the **P-value** of the test. The smaller the P-value, the stronger the evidence against H_0 provided by the data.

Small P-values are evidence against H_0 because they say that the observed result would be unlikely to occur if H_0 were true. Large P-values fail to give evidence against H_0. Statistical software will give you the P-value of a test when you enter your null and alternative hypotheses and your data. So your most important task is to understand what a P-value says.

EXAMPLE 17.4   Can We Feel the Future: One-Sided P-Value

The study of precognition in Example 17.2 tests the hypotheses

$$H_0: p = 0.5$$
$$H_a: p > 0.5$$

Because the alternative hypothesis says that $p > 0.5$, values of $\hat{p}$ greater than 0.5 favor H_a over H_0. The test statistic compares the observed $\hat{p}$ with the hypothesized value $p = 0.5$. For now, let's concentrate on the P-value.

For the experiment presented in Example 17.2 (page 374), the proportion of trials in which participants selected the target-to-be is $\hat{p} = 0.517$. *The P-value is the probability of getting a $\hat{p}$ this large when participants are just guessing, and so the proportion of times participants would select the target to be is really $p = 0.5$.*

The shaded area in Figure 17.2 shows the P-value when $\hat{p} = 0.517$. The Normal curve is the sampling distribution of $\hat{p}$ when the null hypothesis $H_0: p = 0.5$ is true. A Normal probability calculation (which you will find in Exercise 17.8 on page 381) shows that the P-value is $P(\hat{p} \geq 0.517) = 0.006$.

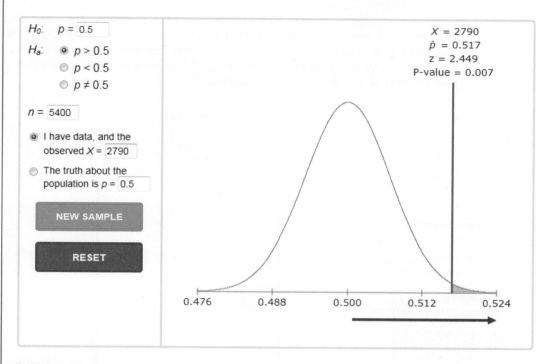

H_0: $p = \boxed{0.5}$

H_a: ⦿ $p > 0.5$
    ○ $p < 0.5$
    ○ $p \neq 0.5$

$n = \boxed{5400}$

⦿ I have data, and the observed $X = \boxed{2790}$

○ The truth about the population is $p = \boxed{0.5}$

[NEW SAMPLE]

[RESET]

$X = 2790$
$\hat{p} = 0.517$
$z = 2.449$
P-value = 0.007

0.476 0.488 0.500 0.512 0.524

FIGURE 17.2

The one-sided P-value for the proportion of successful selections $\hat{p} = 0.517$ in Example 17.4. The figure shows both the input and the output for the *P-Value of a Test of Significance* applet. Note that the P-value is the shaded area under the curve, not the unshaded area.

A value as large as $\hat{p} = 0.517$ would occur just by chance in 0.6% of all samples when $H_0: p = 0.5$ is true. We would rarely observe a proportion of correct selections as large or larger than 0.517 if H_0 were true. This small P-value provides strong evidence against H_0 and in favor of the alternative $H_a: p > 0.5$. ■

So observing $\hat{p} = 0.517$ is strong evidence against H_0. Figure 17.2 is actually the output of the *P-Value of a Test of Significance* applet, along with the information we entered into the applet. This applet automates the work of finding P-values under the "simple conditions" for inference about a proportion.

The alternative hypothesis sets the direction that counts as evidence against H_0. In Example 17.4, only values larger than 0.5 count because the alternative is one sided on the high side. If the alternative is two sided, both directions count.

EXAMPLE 17.5   **Order in Choice: Two-Sided *P*-Value**

The study of order on choice in Example 17.3 (page 377) requires that we test

$$H_0: p = 0.5$$
$$H_a: p \neq 0.5$$

where p is the proportion of all adult wine drinkers who would select the wine presented first.

There were 32 subjects in the study. This is a randomized experiment and if the null hypothesis is true both the expected number of successes $(32(0.5) = 16)$ and the expected number of failures $(32(1 - 0.5) = 16)$ are at least 10. So our conditions for inference are satisfied.

Twenty-two subjects in the study selected the wine presented first when presented with two identical samples. These data give $\hat{p} = 22/32 = 0.6875$. That is, these subjects tend to prefer the wine presented first. Because the alternative is two sided, the *P*-value is the probability of getting a $\hat{p}$ at least as far from $p = 0.5$ *in either direction* as the observed $\hat{p} = 0.6875$.

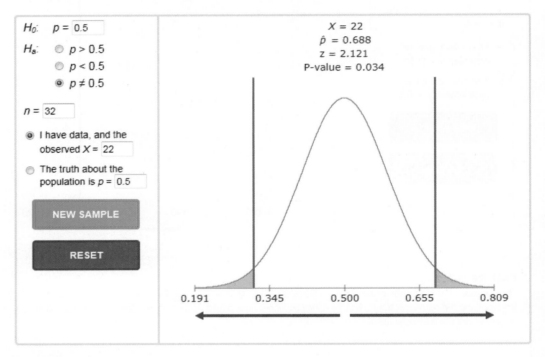

FIGURE 17.3

The two-sided *P*-value for Example 17.5. The figure shows both the input and the output for the *P-Value of a Test of Significance* applet. Note that the *P*-value is the shaded area under the curve, not the unshaded area.

Enter the information for this example into the *P-Value of a Test of Significance* applet and click "Show P." Figure 17.3 shows the applet output as well as the information we entered. The *P*-value is the sum of the two shaded areas under the Normal curve. It is $P = 0.0340$. Values as far from 0.5 as $\hat{p} = 0.6875$ (in either direction) would happen 3.4% of the time when the true population proportion is $p = 0.5$. An outcome that would occur this rarely when H_0 is true is reasonably good evidence against H_0. ▪

The conclusion of Example 17.5 is that we have evidence against H_0. What would we conclude if 19 of the 32 subjects selected the wine presented first? In this case, $\hat{p} = 19/32 = 0.5938$. The applet tells us that the *P*-value now is $P = 0.2888$.

Values as far from 0.5 as $\hat{p} = 0.5938$ (in either direction) would happen 29% of the time when the true population proportion is $p = 0.5$. An outcome that would occur so often when H_0 is true is not good evidence against H_0. However, the conclusion in this case is *not* that H_0 is true. The study looked for evidence against H_0: $p = 0.5$ and failed to find strong evidence. That is all we can say.

Tests of significance assess the evidence against H_0. If the evidence is strong, we can confidently reject H_0 in favor of the alternative. *Failing to find evidence against H_0 means only that the data are not inconsistent with H_0, not that we have clear evidence that H_0 is true.* Only data that are inconsistent with H_0 provide evidence against H_0.

In Example 17.4, we decided that P-value $P = 0.006$ was strong evidence against the null hypothesis. In Example 17.5, a P-value of $P = 0.0340$ was also reasonably strong evidence against the null hypothesis. However, if only 19 subjects had selected the wine presented first in Example 17.5, the P-value of $P = 0.2888$ does not give convincing evidence. There is no rule for how small a P-value we should require to reject H_0—it's a matter of judgment and depends on the specific circumstances.

Nonetheless, we can compare a P-value with some fixed values that are in common use as standards for evidence against H_0. The most common fixed values are 0.05 and 0.01. If $P \leq 0.05$, there is no more than 1 chance in 20 that a sample would give evidence this strong just by chance when H_0 is actually true. If $P \leq 0.01$, we have a result that in the long run would happen no more than once per 100 samples if H_0 were true. These fixed standards for P-values are called **significance levels.** We use α, the Greek letter alpha, to stand for a significance level.

significance level

Statistical Significance

If the P-value is as small or smaller than α, we say that the data are **statistically significant at level α.** The quantity α is called the **significance level** or the **level of significance.**

"Significant" in the statistical sense does not mean "important." It means simply "not likely to happen just by chance." In Example 17.4 (page 378) the P-value is quite small, but the proportion of trials in which participants selected the target-to-be is $\hat{p} = 0.517$. Would you consider this an impressive increase over $p = 0.5$? The significance level α makes "not likely" more exact. Significance at level 0.01 is often expressed by the statement. "The results were significant ($P < 0.01$)." Here P stands for the P-value. The actual P-value is more informative than a statement of significance because it allows us to assess significance at any level we choose. For example, a result with $P = 0.0340$ is significant at the $\alpha = 0.05$ level but is not significant at the $\alpha = 0.01$ level.

 **SIGNIFICANCE STRIKES DOWN A NEW DRUG**

The pharmaceutical company Pfizer spent $1 billion developing a new cholesterol-fighting drug. The final test for its effectiveness was a clinical trial with 15,000 subjects. To enforce double-blindness, only an independent group of experts saw the data during the trial. Three years into the trial, the monitors declared that there was a statistically significant excess of deaths and of heart problems in the group assigned to the new drug. Pfizer ended the trial. There went $1 billion.

Apply Your Knowledge

17.8 Can We Feel the Future: Find the P-Value. The P-value in Example 17.4 (page 378) is the probability (taking the null hypothesis $p = 0.5$ to be true) that $\hat{p}$ takes a value at least as large as 0.517.

(a) What is the sampling distribution of $\hat{p}$ when $p = 0.5$? This distribution appears in Figure 17.2 (page 379).

(b) Do a Normal probability calculation to find the P-value. Your result should agree with Example 17.4 up to roundoff error.

17.9 **Order in Choice: Find the *P*-Value.** The *P*-value in Example 17.5 (page 380) is the probability (taking the null hypothesis $p = 0.5$ to be true) that $\hat{p}$ takes a value at least as far from 0.5 as 0.6875.

    (a) What is the sampling distribution of $\hat{p}$ when $p = 0.5$? This distribution is shown in Figure 17.3 (page 380).

    (b) Do a Normal probability calculation to find the *P*-value. Your result should agree with Example 17.5 up to roundoff error.

17.10 **Do Cash Incentives Improve Learning?** A high-school teacher in a low-income urban school in Worcester, Massachusetts, used cash incentives to encourage student learning in his AP statistics class.[3] In 2010, 15 of the 61 students enrolled in his class scored a 5 on the AP statistics exam. Worldwide, the proportion of students who scored a 5 in 2010 was 0.15. Is this evidence that the proportion of students who would score a 5 on the AP statistics exam when taught by the teacher in Worcester using cash incentives is higher than the worldwide proportion of 0.15? The *P*-value is $P = 0.0180$.

    (a) Explain to someone who knows no statistics why these results mean that there is reasonably good reason to think that the proportion of students who would score a 5 on the AP statistics exam when taught by the teacher in Worcester using cash incentives is higher than the worldwide proportion of 0.15. Include an explanation of what $P = 0.0180$ means.

    (b) Are the conditions for inference listed on page 350 satisfied here? Explain your answer.

    (c) Does this study provide evidence that cash incentives *cause* an increase in the proportion of 5s on the AP statistics exam? Explain your answer.

17.11 **More Feeling the Future.** Exercise 17.1 (page 375) describes a further study of precognition involving subjects who were identified as sensitive to negative stimuli. You stated the null and alternative hypotheses in Exercise 17.3 (page 377).

    (a) In 1800 trials, participants selected the target-to-be 963 times. Compute $\hat{p}$ and enter this value, along with the other required information, into the *P-Value of a Test of Significance* applet. What is the *P*-value? Is this outcome statistically significant at the $\alpha = 0.05$ level? At the $\alpha = 0.01$ level?

    (b) Suppose instead that the researcher found that participants selected the target-to-be 918 times in 1800 trials. Use the applet to find the *P*-value for this outcome. Is it statistically significant at the $\alpha = 0.05$ level? At the $\alpha = 0.01$ level?

    (c) Explain briefly why these *P*-values tell us that one outcome is strong evidence against the null hypothesis and that the other outcome is not.

17.12 **Stricter Laws for Gun Sales.** Exercise 17.2 (page 376) describes a 2012 Gallup poll in which 1038 adults were asked whether they favored strengthening the laws covering the sale of firearms. We are interested in whether this was evidence that the majority of all adults, aged 18 years and older, living in all 50 U.S. states and the District of Columbia favored strengthening the laws covering the sale of firearms. You stated the null and alternative hypotheses in Exercise 17.4 (page 377).

(a) The poll found that the proportion of the sample favoring strengthening the laws covering the sale of firearms was $\hat{p} = 0.58$. Enter this $\hat{p}$, along with the other required information, into the *P-Value of a Test of Significance* applet. What is the *P*-value? Is this outcome statistically significant at the $\alpha = 0.05$ level? At the $\alpha = 0.01$ level?

(b) A poll that asked the same question in 2008 found that the proportion of the sample favoring strengthening the laws covering the sale of firearms was $\hat{p} = 0.51$. Assume the same sample size as the 2012 poll and use the applet to find the *P*-value for this outcome. Is it statistically significant at the $\alpha = 0.05$ level? At the $\alpha = 0.01$ level?

(c) Explain briefly why these *P*-values tell us that one outcome is strong evidence against the null hypothesis and that the other outcome is not.

Large sample tests for a population proportion

We have used tests for hypotheses about a population proportion p, under the "simple conditions," to introduce tests of significance. The big idea is the reasoning of a test: *data that would rarely occur if the null hypothesis H_0 were true provide evidence that H_0 is not true.* The *P*-value gives us a probability to measure "would rarely occur." In practice, the steps in carrying out a significance test mirror the overall four-step process for organizing realistic statistical problems.

Tests of Significance: The Four-Step Process

STATE: What is the practical question that requires a statistical test?

PLAN: Identify the parameter, state null and alternative hypotheses, and choose the type of test that fits your situation.

SOLVE: Carry out the test in three phases:

1. **Check the conditions** for the test you plan to use.

2. Calculate the **test statistic.**

3. Find the **P-value.**

CONCLUDE: Return to the practical question to describe your results in this setting.

Once you have stated your question, formulated hypotheses, and checked the conditions for your test, you or your software can find the test statistic and *P*-value by following a rule. Here is the rule for the test we have used in our examples.

Significance Tests for a Proportion

Draw an SRS of size n from a large population that contains an unknown proportion p of successes. To **test the hypothesis H_0: $p = p_0$,** compute the z statistic

$$z = \frac{\hat{p} - p_0}{\sqrt{\dfrac{p_0(1 - p_0)}{n}}}$$

In terms of a variable Z having the standard Normal distribution, the approximate P-value for a test of H_0 against

$H_a: p > p_0$ is $P(Z \geq z)$

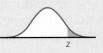

$H_a: p < p_0$ is $P(Z \leq z)$

$H_a: p \neq p_0$ is $2P(Z \geq |z|)$

Use this test when the sample size n is so large that both the expected number of successes and failures are at least 10 when H_0 is true. Specifically, check that both np_0 and $n(1 - p_0)$ are 10 or more.[4]

As promised, the test statistic z measures how far the observed sample proportion $\hat{p}$ deviates from the hypothesized population value p_0. The measurement is in the familiar standard scale obtained by dividing by the standard deviation of $\hat{p}$ when H_0 is true. So we have a common scale for all z tests, and the 68–95–99.7 rule helps us see at once if $\hat{p}$ is far from p_0. The pictures that illustrate the P-value look just like the curves in Figures 17.2 and 17.3 except that they are in the standard scale.

EXAMPLE 17.6   Are Boys More Likely?

STATE: We hear that newborn babies are more likely to be boys than girls, presumably to compensate for higher mortality among boys in early life. Is this true? A random sample collected in 2005 found 13,173 boys among 25,468 firstborn children.[5] The sample proportion of boys was

$$\hat{p} = \frac{13{,}173}{25{,}468} = 0.5172$$

Boys do make up more than half of the sample, but of course we don't expect a perfect 50–50 split in a random sample. Is this sample evidence that boys are more common than girls in the entire population?

PLAN: Take p to be the proportion of boys among all firstborn children of American mothers. (Biology says that this should be the same as the proportion among all children, but the survey data concern first births.) We want to test the hypotheses

$$H_0: p = 0.5$$
$$H_a: p > 0.5$$

SOLVE: The conditions for inference require that we have a random sample and that $np_0 = (25{,}468)(0.5) = 12{,}734$ and $n(1 - p_0) = (25{,}468)(0.5) = 12{,}734$

are both at least 10. Because the conditions for inference are met, we can go on to find the *z* test statistic:

$$z = \frac{\hat{p} - p_0}{\sqrt{\dfrac{p_0(1 - p_0)}{n}}}$$

$$= \frac{0.5172 - 0.5}{\sqrt{\dfrac{(0.5)(0.5)}{25{,}468}}} = 5.49$$

Using the *P-Value of a Test of Significance* applet, we find that the *P*-value is $P < 0.001$.

CONCLUDE: There is very strong evidence that more than half of firstborns are boys ($P < 0.001$). ∎

In this chapter we sometimes act as if the conditions stated on page 350 are true. In practice, you must verify these conditions.

Apply Your Knowledge

17.13 Spinning Euros. All euros have a national image on the "heads" side and a common design on the "tails" side. Spinning a coin, unlike tossing it, may not give heads and tails equal probabilities. Polish students spun the Belgian euro 250 times, with its portly king, Albert, displayed on the heads side. The result was 140 heads.[6] How significant is this evidence against equal probabilities? Follow the four-step process as illustrated in Example 17.6.

17.14 Vote for the Best Face? We often judge other people by their faces. It appears that some people judge candidates for elected office by their faces. In a 2005 study, psychologists showed head-and-shoulders photos of the two main candidates in 32 races for the U.S. Senate to many subjects (dropping subjects who recognized one of the candidates) to see which candidate was rated "more competent" based on nothing but the photos. On election day, the candidates whose faces looked more competent won 22 of the 32 contests.[7] If faces don't influence voting, half of all races in the long run should be won by the candidate with the better face. Is there evidence that the candidate with the better face wins more than half the time? Follow the four-step process as illustrated in Example 17.6.

© Photos 12/Alamy

17.15 No Test. Explain whether we can use the *z* test for a proportion in these situations:

(a) You toss a coin 10 times to test the hypothesis $H_0: p = 0.5$ that the coin is balanced.

(b) A local candidate contacts an SRS of 900 of the registered voters in his district to see if there is evidence that more than half support the bill he is sponsoring.

(c) A college president says, "99% of the alumni support my firing of Coach Boggs." You contact an SRS of 200 of the college's 15,000 living alumni to test the hypothesis $H_0: p = 0.99$.

Cautions about significance tests

Significance tests are widely used in most areas of statistical work. New pharmaceutical products require significant evidence of effectiveness and safety. Courts inquire about statistical significance in hearing class action discrimination cases. Marketers want to know whether a new package design will significantly increase sales. Medical researchers want to know whether a new therapy performs significantly better than the existing therapy. In all these uses, statistical significance is valued because it points to an effect that is unlikely to occur simply by chance. Here are some points to keep in mind when you use or interpret significance tests.

How small a *P* is convincing? The purpose of a test of significance is to describe the degree of evidence provided by the sample against the null hypothesis. The *P*-value does this. But how small a *P*-value is convincing evidence against the null hypothesis? This depends mainly on two circumstances:

- *How plausible is* H_0? If H_0 represents an assumption that the people you must convince have believed for years, strong evidence (small *P*) will be needed to persuade them.

- *What are the consequences of rejecting* H_0? If rejecting H_0 in favor of H_a means making an expensive changeover from one type of product packaging to another, you need strong evidence that the new packaging will boost sales.

These criteria are a bit subjective. Different people will often insist on different levels of significance. Giving the *P*-value allows each of us to decide individually if the evidence is sufficiently strong.

Users of statistics have often emphasized standard levels of significance such as 10%, 5%, and 1%. For example, courts have tended to accept 5% as a standard in discrimination cases.[8] This emphasis reflects the time when tables of critical values rather than software dominated statistical practice. The 5% level ($\alpha = 0.05$) is

 particularly common. *There is no sharp border between "significant" and "not significant," only increasingly strong evidence as the P-value decreases. There is no practical distinction between the P-values 0.049 and 0.051. It makes no sense to treat P ≤ 0.05 as a universal rule for what is significant.*

Significance depends on the alternative hypothesis You may have noticed that the *P*-value for a one-sided test is one-half the *P*-value for the two-sided test of the same null hypothesis based on the same data. The two-sided *P*-value combines two equal areas, one in each tail of a Normal curve. The one-sided *P*-value is just one of these areas, in the direction specified by the alternative hypothesis. It makes sense that the evidence against H_0 is stronger when the alternative is one sided because it is based on the data *plus* information about the direction of possible deviations from H_0. If you lack this added information prior to observing the data, always use a two-sided alternative hypothesis.

Significance depends on sample size A sample survey conducted in 2001 by researchers at the Harvard School of Public Health showed that significantly fewer students are heavy drinkers at colleges that ban alcohol on campus. "Significantly fewer" is not enough information to decide whether there is an important difference in drinking behavior at schools that ban alcohol. *How important an effect is depends on the size of the effect as well as on its statistical significance.* If the number of heavy drinkers is only 1% less at colleges that ban alcohol than at other colleges, this is not an important effect even if it is statistically significant. In fact, the sample survey found that 38% of students at colleges that ban alcohol are "heavy episodic drinkers" compared with 48% at other colleges.[9] That difference is large enough to be important. (Of course, this observational study doesn't *prove* that an alcohol ban directly reduces drinking; it may be that colleges that ban alcohol attract more students who don't want to drink heavily.)

Such examples remind us to always look at the size of an effect (like 38% versus 48%) as well as its significance. They also raise a question: can a tiny effect really be highly significant? Yes. The behavior of the z test statistic is typical. The statistic is

$$z = \frac{\hat{p} - p_0}{\sqrt{\dfrac{p_0(1 - p_0)}{n}}}$$

The numerator measures how far the sample proportion deviates from the hypothesized mean p_0. Larger values of the numerator give stronger evidence against H_0: $p = p_0$. The denominator is the standard deviation of $\hat{p}$ when H_0: $p = p_0$ is true. It measures how much random variation we expect. There is less variation when the number of observations n is large. So z gets larger (more significant) when the estimated effect $\hat{p} - p_0$ gets larger *or* when the number of observations n gets larger. Significance depends both on the size of the effect we observe *and* on the size of the sample. Understanding this fact is essential to understanding significance tests. In Example 17.4 (page 378), the increase in the proportion of times subjects selected the target-to-be as compared with guessing is $\hat{p} - p_0 = 0.017$. This is not a substantial increase. However, the sample size is 5400, and $z = 2.50$.

SHOULD TESTS BE BANNED?

Significance tests don't tell us how large or how important an effect is. Research in psychology has emphasized tests, so much so that some think their weaknesses should ban them from use. The American Psychological Association asked a group of experts. They said: "Use anything that sheds light on your study. Use more data analysis and confidence intervals." But: "The task force does not support any action that could be interpreted as banning the use of null hypothesis significance testing or *P*-values in psychological research and publication."

Sample Size Affects Statistical Significance

Because large random samples have small chance variation, very small population effects can be highly significant if the sample is large.

Because small random samples have a lot of chance variation, even large population effects can fail to be significant if the sample is small.

Statistical significance does not tell us whether an effect is large enough to be important. That is, **statistical significance is not the same thing as practical importance.**

Keep in mind that "statistical significance" means "the sample showed an effect larger than would often occur just by chance." The extent of chance variation changes with the size of the sample, so the size of the sample does matter. Exercise 17.19 (page 389) demonstrates in detail how increasing the sample size drives down the *P*-value. Here is another example.

EXAMPLE 17.7   It's Significant. Or Not. So What?

We are testing the hypothesis of no correlation between two variables. With 1000 observations, an observed correlation of only $r = 0.08$ is significant evidence at the 1% level that the correlation in the population is not zero but positive. *The small P-value does not mean that there is a strong association, only that there is strong evidence of some association.* The true population correlation is probably quite close to the observed sample value, $r = 0.08$. We might well conclude that for practical purposes we can ignore the association between these variables, even though we are confident (at the 1% level) that the correlation is positive.

On the other hand, if we have only 10 observations, a correlation of $r = 0.5$ is not significantly greater than zero even at the 5% level. Small samples vary so much that a large r is needed if we are to be confident that we aren't just seeing chance variation at work. So a small sample will often fall short of significance, even if the true population correlation is quite large. ■

Beware of multiple analyses Statistical significance ought to mean that you have found an effect that you were looking for. The reasoning behind statistical significance works well if you decide what effect you are seeking, design a study to search for it, and use a test of significance to weigh the evidence you get. In other settings, significance may have little meaning.

EXAMPLE 17.8   **Cell Phones and Brain Cancer**

Might the radiation from cell phones be harmful to users? Many studies have found little or no connection between using cell phones and various illnesses. Here is part of a news account of one study published in 1999:

> A hospital study that compared brain cancer patients and a similar group without brain cancer found no statistically significant association between cell phone use and a group of brain cancers known as gliomas. But when 20 types of glioma were considered separately an association was found between phone use and one rare form. Puzzlingly, however, this risk appeared to decrease rather than increase with greater mobile phone use.[10]

Think for a moment. Suppose that the 20 null hypotheses (no association) for these 20 significance tests are all true. Then each test has a 5% chance of being significant at the 5% level. That's what $\alpha = 0.05$ means: results this extreme occur 5% of the time just by chance when the null hypothesis is true. Because 5% is 1/20, we expect about 1 of 20 tests to give a significant result just by chance. That's what the study observed. ■

Running one test and reaching the 5% level of significance is reasonably good evidence that you have found something. Running 20 tests and reaching that level only once is not. The caution about multiple analyses applies to confidence intervals as well. A single 95% confidence interval has probability 0.95 of capturing the true parameter each time you use it. The probability that all of 20 confidence intervals will capture their parameters is much less than 95%. If you think that multiple tests or intervals may have discovered an important effect, you need to gather new data to do inference about that specific effect.

Apply Your Knowledge

17.16 Is It Significant? In the absence of special preparation, according to the data from the College Board Web site SAT Mathematics (SATM) scores in 2012 varied Normally with mean 514. Thus, we expect half of the students who take the SATM to have a score greater than 514. One hundred students go through a rigorous training program designed to raise their SATM scores by improving their mathematics skills. Either by hand, by statistical software, or by using the *P-Value of a Test of Significance* applet, carry out a test of

$$H_0: p = 0.5$$
$$H_a: p > 0.5$$

in each of the following situations:

(a) Fifty-eight of the students score greater than 514. Is this result significant at the 5% level?

(b) Fifty-nine of the students score greater than 514. Is this result significant at the 5% level?

The difference between the two outcomes in (a) and (b) is of no practical importance. Beware attempts to treat $\alpha = 0.05$ as sacred.

17.17 Global Warming and the Media. A March 2012 Gallup poll asked American adults whether they thought the news media exaggerate the seriousness of global warming. The proportion who responded "No" was 0.58. Is this good evidence that the majority of American adults do not believe the news media exaggerate the seriousness of global warming? The answer depends on the size of the sample.

Let p denote the true proportion of American adults who do not believe the news media exaggerate the seriousness of global warming. Either by hand, statistical software, or using the *P-Value of a Test of Significance* applet, carry out three tests of

$$H_0: p = 0.5$$
$$H_a: p > 0.5$$

Use $\hat{p} = 0.58$ in all three tests. But use three different sample sizes, $n = 25$, $n = 100$, and $n = 400$.

(a) What are the *P*-values for the three tests? *The P-value of the same result $\hat{p} = 0.58$ gets smaller (more significant) as the sample size increases.*

(b) For each test, sketch the Normal curve for the sampling distribution of $\hat{p}$ when H_0 is true. This curve has mean 0.5 and standard deviation $0.5/\sqrt{n}$. Mark the observed $\hat{p} = 0.58$ on each curve. (If you use the applet, you can just copy the curves displayed by the applet.) *The same result $\hat{p} = 0.58$ gets more extreme on the sampling distribution as the sample size increases.*

17.18 Confidence Intervals Help. Give a 95% confidence interval for the true proportion p for each sample size in the previous exercise. The intervals, unlike the *P*-values, give a clear picture of what values of the true proportion are plausible for each sample.

17.19 Searching for ESP. A researcher looking for evidence of extrasensory perception (ESP) tests 500 subjects. Four of these subjects do significantly better ($P < 0.01$) than random guessing.

(a) You can't conclude that these four people have ESP. Why not?

(b) What should the researcher now do to test whether any of these four subjects have ESP?

Significance from a table*

Statistics in practice uses technology (graphing calculator or software) to get *P*-values quickly and accurately. In the absence of suitable technology, you can get approximate *P*-values quickly by comparing the value of your test statistic with critical values from a table. For the z statistic, the table is Table C, the same table we used for confidence intervals.

*This material can be skipped if you use software to compute *P*-values.

Look at the bottom row of critical values in Table C, labeled z^*. At the top of the table, you see the confidence level C for each z^*. At the bottom of the table, you see both the one-sided and two-sided P-values for each z^*. Values of a test statistic z that are farther out than a z^* (in the direction given by the alternative hypothesis) are significant at the level that matches z^*.

Significance from a Table of Critical Values

To find the approximate P-value for any z statistic, compare z (ignoring its sign) with the critical values z^* at the bottom of Table C. If z falls between two values of z^*, the P-value falls between the two corresponding values of P in the "One-sided P" or the "Two-sided P" row of Table C.

EXAMPLE 17.9   Is it significant?

z^*	2.326	2.576
One-sided P	0.01	0.005

The z statistic for a one-sided test is $z = 2.50$. How significant is this result? Compare $z = 2.50$ with the z^* row in Table C.

It lies between $z^* = 2.326$ and $z^* = 2.576$. So the P-value lies between the corresponding entries in the "One-sided P" row, which are $P = 0.01$ and $P = 0.005$. This z *is* significant at the $\alpha = 0.01$ level and *is not* significant at the $\alpha = 0.005$ level.

Figure 17.4 illustrates the situation. The shaded area under the Normal curve is the P-value for $z = 2.50$. You can see that P falls between the areas to the right of the two critical values, for $P = 0.01$ and $P = 0.005$.

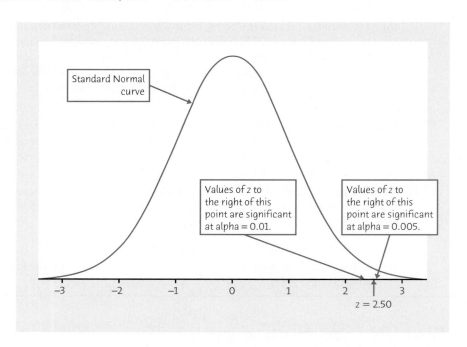

FIGURE 17.4

Is it significant? The test statistic value $z = 2.50$ falls between the critical values required for significance at the $\alpha = 0.01$ and $\alpha = 0.005$ levels. So the test *is* significant at $\alpha = 0.01$ and *is not* significant at $\alpha = 0.005$.

z^*	2.054	2.326
Two-sided P	0.04	0.02

The z statistic in Example 17.5 (page 380) turns out to be $z = 2.12$. The alternative hypothesis is two sided. Compare $z = 2.12$ with the z^* row in Table C.

It lies between $z^* = 2.054$ and $z^* = 2.326$. So the P-value lies between the matching entries in the "Two-sided P" row, $P = 0.04$ and $P = 0.02$. This is enough to conclude that the data provide reasonably good evidence against the null hypothesis. ■

Apply Your Knowledge

17.20 Significance from a Table. A test of $H_0: p = 0.5$ against $H_a: p > 0.5$ has test statistic $z = 1.876$. Is this test significant at the 5% level ($\alpha = 0.05$)? Is it significant at the 1% level ($\alpha = 0.01$)?

17.21 Significance from a Table. A test of $H_0: p = 0.5$ against $H_a: p \neq 0.5$ has test statistic $z = 1.876$. Is this test significant at the 5% level ($\alpha = 0.05$)? Is it significant at the 1% level ($\alpha = 0.01$)?

17.22 Testing a Random Number Generator. A random number generator is supposed to produce random numbers that are uniformly distributed on the interval from 0 to 1. If this is true, the numbers generated come from a population in which the proportion that are less than 0.5 is $p = 0.5$. A command to generate 100 random numbers gives outcomes for which the proportion that are less than 0.5 is $\hat{p} = 0.43$. We want to test

$$H_0: p = 0.5$$
$$H_a: p \neq 0.5$$

(a) Calculate the value of the z test statistic.

(b) Use Table C: is z significant at the 5% level ($\alpha = 0.05$)?

(c) Use Table C: is z significant at the 1% level ($\alpha = 0.01$)?

(d) Between which two Normal critical values z^* in the bottom row of Table C does z lie? Between what two numbers does the P-value lie? Does the test give good evidence against the null hypothesis?

CHAPTER 17 SUMMARY

Chapter Specifics

- A **test of significance** assesses the evidence provided by data against a **null hypothesis** H_0 in favor of an **alternative hypothesis** H_a.

- Hypotheses are always stated in terms of population parameters. Usually H_0 is a statement that no effect is present, and H_a says that a parameter differs from its null value in a specific direction (**one-sided alternative**) or in either direction (**two-sided alternative**).

- The essential reasoning of a significance test is as follows. Suppose for the sake of argument that the null hypothesis is true. If we repeated our data production many times, would we often get data as inconsistent with H_0 as the data we actually have? Data that would rarely occur if H_0 were true provide evidence against H_0.

- A test is based on a **test statistic** that measures how far the sample outcome is from the value stated by H_0.

- The **P-value** of a test is the probability, computed supposing H_0 to be true, that the test statistic will take a value at least as extreme as that actually observed. Small P-values indicate strong evidence against H_0. To calculate a P-value we must know the sampling distribution of the test statistic when H_0 is true.

- If the P-value is as small or smaller than a specified value α, the data are **statistically significant** at **significance level** α.

- **Significance tests for $H_0: p = p_0$** are based on the z statistic

$$z = \frac{\hat{p} - p_0}{\sqrt{\dfrac{p_0(1 - p_0)}{n}}}$$

with P-values calculated from the standard Normal distribution. Use this test in practice when $np_0 \geq 10$ and $n(1 - p_0) \geq 10$.

- There is no universal rule for how small a *P*-value in a test of significance is convincing evidence against the null hypothesis. Beware of placing too much weight on traditional significance levels such as $\alpha = 0.05$.

- Very small effects can be highly significant (small *P*) when a test is based on a large sample. A statistically significant effect need not be practically important. Plot the data to display the effect you are seeking, and use confidence intervals to estimate the actual values of parameters.

- On the other hand, lack of significance does not imply that H_0 is true. Even a large effect can fail to be significant when a test is based on a small sample.

- Many tests run at once will probably produce some significant results by chance alone, even if all the null hypotheses are true.

Link It

In this chapter we discuss tests of significance, the second type of statistical inference. The mathematics of probability, in particular the sampling distributions discussed in Chapter 15, provides the formal basis for a test of significance. The sampling distribution allows us to assess "probabilistically" the strength of evidence against a null hypothesis, either through a level of significance or a *P*-value. The goal of hypothesis testing, which is used to assess the evidence provided by data about some claim concerning a population, is different from the goal of confidence interval estimation, which is used to estimate a population parameter.

In this chapter we also provide some cautions about tests of significance. Some of these echo statements made in Chapters 8 and 9. Some are based on the behavior of significance tests. These cautions will help you evaluate studies that report the results of a test of significance.

Although we apply the reasoning of tests of significance for a population proportion based on a sufficiently large simple random sample or completely randomized experimental design with a sufficiently large number of subjects, we will use the same logic in future chapters to construct tests of significance for population parameters in other settings.

CHECK YOUR SKILLS

17.23 You use software to carry out a test of significance. The program tells you that the P-value is $P = 0.031$. You conclude

(a) that the probability, computed assuming that H_0 is true, that the test statistic would take a value as extreme or more extreme than that actually observed is 0.031.

(b) that the probability, computed assuming that H_0 is true, that the test statistic would take a value as extreme or less extreme than that actually observed is 0.031.

(c) that the probability, computed assuming that H_0 is false, that the test statistic would take a value as extreme or more extreme than that actually observed is 0.031.

17.24 You use software to carry out a test of significance. The program tells you that the P-value is $P = 0.031$. This result is

(a) not significant at either $\alpha = 0.05$ or $\alpha = 0.01$.

(b) significant at $\alpha = 0.05$ but not at $\alpha = 0.01$.

(c) significant at $\alpha = 0.01$ but not at $\alpha = 0.05$.

17.25 Experiments on learning in animals sometimes measure how long it takes mice to find their way through a maze. Only half of all mice complete one particular maze in less than 18 seconds. A researcher thinks that a loud noise will cause the mice to complete the maze faster. She measures the proportion of 40 mice that complete the maze in less than 18 seconds with a noise as stimulus. The proportion of mice that completed the maze in less than 18 seconds is $\hat{p} = 0.7$. The null hypothesis for the significance test is

Garry Gay/Getty Images

(a) $H_0: p < 0.5$.

(b) $H_0: p = 0.5$.

(c) $H_0: p > 0.5$.

17.26 The alternative hypothesis for the test in Exercise 17.25 is

(a) $H_a: p < 0.5$.

(b) $H_a: p \neq 0.5$.

(c) $H_a: p > 0.5$.

17.27 The z statistic for a one-sided test is $z = 2.433$. This test is

(a) not significant at either $\alpha = 0.05$ or $\alpha = 0.01$.

(b) significant at $\alpha = 0.05$ but not at $\alpha = 0.01$.

(c) significant at both $\alpha = 0.05$ and $\alpha = 0.01$.

17.28 A June 2012 newspaper article stated that more than half of all American adults aged 18 years and older

have very little or no confidence in Congress. This is based on a June 2012 Gallup poll in which 52% of a random sample of 1004 American adults aged 18 years and older indicated little or no confidence in Congress. The z statistic for testing $H_0: p = 0.5$, where p is the proportion of all American adults aged 18 years and older who have very little or no confidence in Congress, based on the Gallup poll result is

(a) $z = 1.27$.

(b) $z = -0.90$.

(c) $z = -1.27$.

17.29 You read an article about an experiment in which the researcher conducted a test of significance. The article tells you that the P-value is $P = 0.19$. This means that

(a) the probability that the null hypothesis is true is 0.19.

(b) the value of the test statistic is not particularly large.

(c) neither of the above is true.

17.30 You are testing $H_0: p = 0.5$ against $H_0: p \neq 0.5$ based on an SRS of 100 observations. What values of the z statistic are statistically significant at the $\alpha = 0.005$ level?

(a) All values for which $z > 2.576$

(b) All values for which $z > 2.807$

(c) All values for which $|z| > 2.807$

17.31 You are testing $H_0: p = 0.5$ against $H_0: p > 0.5$ based on an SRS of 100 observations. What values of the z statistic are statistically significant at the $\alpha = 0.005$ level?

(a) All values for which $z > 2.576$

(b) All values for which $z > 2.807$

(c) All values for which $|z| > 2.807$

17.32 An opinion poll asks an SRS of 100 college seniors how they view their job prospects. In all, 53 say "Good." Does this poll give reason to conclude that more than half of all seniors think their job prospects are good? The hypotheses for a test to answer this question are

(a) $H_0: p = 0.5$, $H_a: p > 0.5$.

(b) $H_0: p > 0.5$, $H_a: p = 0.5$.

(c) $H_0: p = 0.5$, $H_a: p \neq 0.5$.

17.33 The value of the z statistic for the test of the previous exercise is

(a) $z = 12$. (b) $z = 6$. (c) $z = 0.6$.

17.34 The most important condition for sound conclusions from statistical inference is usually

(a) that the data can be thought of as a random sample from the population of interest or that the data come from a completely randomized experiment.

(b) that the sample size is very large.

(c) that the data contain no outliers.

17.35 You turn your Web browser to the online Harris Interactive Poll. Based on 2163 responses, a 2010 poll reports that 43% of U.S. adults said they would like to be richer, 21% said thinner, 14% said smarter, and 12% said younger. Nearly 1 in 10 (9%) said they would not want to choose any of the given options.[11] You should refuse to test hypotheses about the proportion of all U.S. adults who would not want to choose any of the given options based on this sample because

(a) the observed proportion is too small.

(b) this is not an SRS.

(c) there were more than two choices available to those surveyed.

17.36 A news story reports that: "A randomized double blind experiment with 14 women found a significant proportion reported that their mood improved after taking a mood supplement for three weeks ($P = 0.016$)." You should be skeptical about the significant improvement because

(a) the results are not practically important.

(b) the P-value is not small enough to be convincing.

(c) the sample size is not sufficiently large.

17.37 Vigorous exercise helps people live several years longer (on the average). Whether mild activities like slow walking extend life is not clear. Suppose that the added life expectancy from regular slow walking is just 2 months. A statistical test is more likely to find a significant increase in mean life if

(a) it is based on a very large random sample.

(b) it is based on a very small random sample.

(c) The size of the sample doesn't have any effect on the significance of the test.

CHAPTER 17 EXERCISES

In all exercises that call for P-values, give the actual value if you use software or the P-value applet. Otherwise, use Table C to give values between which P must fall.

17.38 **Is this what P means?** When asked to explain the meaning of "the P-value was $P = 0.03$," a student says, "This means there is only probability 0.03 that the null hypothesis is true." Explain what $P = 0.03$ really means in a way that makes it clear that the student's explanation is wrong.

17.39 **How to show that you are rich.** Every society has its own marks of wealth and prestige. In ancient China, it appears that owning pigs was such a mark. Evidence comes from examining burial sites. The skulls of sacrificed pigs tend to appear along with expensive ornaments, which suggests that the pigs, like the ornaments, signal the wealth and prestige of the person buried. A 1994 study of burials from around 3500 B.C. concluded that "there are striking differences in grave goods between burials with pig skulls and burials without them. . . . A test indicates that the two samples of total artifacts are significantly different at the 0.01 level."[12] Explain clearly why "significantly different at the 0.01 level" gives good reason to think that there really is a systematic difference between burials that contain pig skulls and those that lack them.

17.40 **Alleviating test anxiety.** Research suggests that pressure to perform well can reduce performance on exams. Are there effective strategies to deal with pressure? In an experiment conducted in 2011, researchers had students take a test on mathematical skills. The same students were asked to take a second test on the same skills, but now each student was paired with a partner and only if both improved their scores would they receive a monetary reward for participating in the experiment. They were also told that their performance would be videotaped and watched by teachers and students. To help them cope with the pressure, ten minutes before the second exam they were asked to write as candidly as possible about their thoughts and feelings regarding the exam. "Students who expressed their thoughts before the high-pressure test showed a significant 5% math accuracy improvement from the pretest to posttest" ($P < 0.03$).[13] A colleague who knows no statistics says that an increase of 5% isn't a lot—maybe it's just an accident due to natural variation among the students. Explain in simple language how "$P < 0.03$" answers this objection.

17.41 **Treating Parkinson's disease.** A randomized comparative experiment conducted in 2010 compared the effects of two types of deep-brain stimulation (pallidal stimulation and subthalamic stimulation) on change in motor function, as blindly assessed on the Unified Parkinson's Disease Rating Scale, part III (UPDRS-III). The abstract of the study said: "Mean changes in the primary outcome did not differ significantly between the two study groups ($P = 0.50$)."[14] The P-value refers to a null hypothesis of "no change" in measurements between pallidal stimulation and subthalamic stimulation. Explain clearly why this *P*-value provides no evidence of change.

17.42 **5% versus 1%.** Sketch the standard Normal curve for the z test statistic and mark off areas under the curve to show why a value of z that is significant at the 1% level in a one-sided test is always significant at the 5% level. If z is significant at the 5% level, what can you say about its significance at the 1% level?

17.43 **The wrong alternative.** A graduate student is comparing final-exam test scores of male and female students in an introductory physics class. She starts with no expectations about which sex will have a higher proportion of final-exam test scores above 90 (the score needed for a grade of A). After seeing that men did better than women on the first quiz, she tests a one-sided alternative about the proportion of final-exam scores above 90,

$$H_0: p_M = p_F$$
$$H_a: p_M > p_F$$

She finds $z = 1.9$ with one-sided *P*-value $P = 0.0287$.

(a) Explain why she should have used the two-sided alternative hypothesis.

(b) What is the correct *P*-value for $z = 1.9$?

17.44 **The wrong P.** The report of a 2001 study of seat belt use by drivers says, "Hispanic drivers were not significantly more likely than White/non-Hispanic drivers to overreport safety belt use (27.4 vs. 21.1%, respectively; $z = 1.33$, $P > 1.0$)."[15] How do you know that the *P*-value given is incorrect? What is the correct one-sided *P*-value for test statistic $z = 1.33$?

*Exercises 17.45 to 17.49 ask you to answer questions from data. The exercise statements give you the **State** step of the four-step process. In your work, follow the **Plan, Solve,** and **Conclude** steps, illustrated in Example 17.6 (page 384) for a test of significance.*

17.45 **Low-fat or low-carb?** Recent research, including a 2012 article in the *Journal of the American Medical Association*,[16] has suggested that a low-carb diet boosts overall energy expenditure, meaning that low-carb dieters are burning more calories per day than their low-fat counterparts. Has this affected public opinion regarding whether a low-fat or low-carb diet is better? An August 2012 Gallup poll asked a random sample of 1014 adults, aged 18 and older, living in all 50 U.S. states and the District of Columbia "from a health perspective, which of the following do you think is more beneficial for the average American—a diet low in fat, or a diet low in carbohydrates?" Thirty percent of those sampled said "a diet low in carbohydrates." Gallup polls in previous years found that slightly less than one-quarter of respondents said "a diet low in carbohydrates." Does the 2012 poll provide evidence that more than one-quarter of all adults, aged 18 and older, living in all 50 U.S. states and the District of Columbia believed that a diet low in carbohydrates is more beneficial than a diet low in fat at the time of the poll?

17.46 **Downloading music.** A husband and wife, Ted and Suzanne, share a digital music player that has a feature that randomly selects which song to play. A total of 3476 songs have been loaded into the player, some by Ted and the rest by Suzanne. They are interested in determining whether they have each loaded a different proportion of songs into the player. Suppose that when the player was in the random-selection mode, 22 of the first 30 songs selected were songs loaded by Suzanne. Let p denote the proportion of songs that were loaded by Suzanne. State the null and alternative hypotheses to be tested. How strong is the evidence that Ted and Suzanne have each loaded a different proportion of songs into the player?

17.47 **Opinions about evolution.** A 2006 sample survey funded by the National Science Foundation asked a random sample of American adults about biological evolution.[17] One question asked subjects to answer "True," "False," or "Not sure" to the statement "Human beings, as we know them today, developed from earlier species of animals." Of the 1484 respondents, 594 said "True." Does this sample give good evidence to support the claim "Fewer than half of American adults think that humans developed from earlier species of animals" at the time the survey was conducted?

17.48 **Gastric bypass surgery.** How effective is gastric bypass surgery in maintaining weight loss in extremely obese people? A Utah-based study conducted between 2000 and 2011 found that 76% of 418 subjects who had received gastric bypass surgery maintained at least a 20% weight loss 6 years after surgery.[18] Do these data provide evidence that more than three-quarters of those receiving gastric bypass surgery will maintain at least a 20% weight loss 6 years after surgery?

17.49 **Final election poll.** Gallup's final presidential election poll was conducted November 1–4, 2012. The poll included a random sample of 2551 adults, aged 18 and over, living in all 50 U.S. states and the District of Columbia, who were identified as likely to vote. Forty-eight percent of these likely voters said they would vote for Barack Obama. Does this provide evidence that, among all likely voters at the time the poll was conducted, less than half supported Barack Obama?

Exploring the Web

17.50   **Significance in journals.** Choose a major journal in your field of study. Use a Web search engine to find its Web site—just search on the journal's name. Find a paper that uses a phrase like "significant ($P = 0.01$)" and summarize the findings in the paper.

17.51   **A statistics glossary.** An editorial was published in the *Journal of the National Cancer Institute,* Vol. 101, No. 23 (December 2, 2009) that announced some online resources for journalists, including a statistics glossary. The glossary can be found at `www.oxfordjournals.org/our_journals/jnc/resource/statistics%20glossary.pdf`. Read the definition of a *P*-value. Is this an accurate definition? Explain your answer.

Overview

In Chapters 16 and 17 we discussed confidence intervals and hypothesis tests for one proportion. Here we turn to the comparison of two population proportions using two independent samples. If the two population proportions are p_1 and p_2, we compare them by estimating their difference, $p_1 - p_2$. In this chapter we develop confidence intervals and significance tests for this difference.

We begin by considering the sampling distribution of the difference in sample proportions, $\hat{p}_1 - \hat{p}_2$. The theory presented in Chapter 15 is extended to the difference between two population proportions. The sampling distribution of $\hat{p}_1 - \hat{p}_2$ is approximately Normal if each sample size is large enough. "Large enough" differs for confidence intervals and hypothesis tests—for confidence intervals, the number of successes and failures must each be at least 10 in both samples; for hypothesis tests, the number of successes and failures must each be at least 5 in both samples. Note that some texts use the expected number of successes and failures (assuming the null hypothesis is true) for hypothesis testing.

Hopefully, students will begin to feel a sense of redundancy in studying this material. The ideas explored here are, in essence, repackaged and reformulated for the particular setting in question—if the students understand the concepts involved in and the process of working with confidence intervals and hypothesis tests for one proportion, that will help them extend the ideas to inference for the difference between two proportions.

LEARNING OUTCOMES**

- Recognize the problem setting as involving two-sample z procedures.

- Verify that you can safely use the large-sample z procedures in a particular setting. Check the study design and the guidelines for sample sizes.

- Use the large-sample z procedure to give a confidence interval for the difference $p_1 - p_2$ between proportions in two populations based on independent samples from the populations.

- Use a z statistic to test the hypothesis H_0: $p_1 = p_2$ that the proportions in two distinct populations are equal. Use software or Table A to find the P-value or Table C to get an approximate value.

- (Optional) When the conditions for the large-sample confidence interval are not met, understand when to use the plus four modification of the z procedure to give a confidence interval for $p_1 - p_2$ that is accurate, even for very small samples and for any values of p_1 and p_2.

**These learning outcomes appear later for the students in Chapter 23: Part IV Review.

Teaching Suggestions and Additional Examples/Activities for the Classroom

1. Emphasize the Structure of Confidence Intervals and Make Connections

The methods studied in this chapter are natural extensions of those studied earlier. To many students, formulas such as the standard error of $\hat{p}_1 - \hat{p}_2$ will look new, but the overall process of constructing a confidence interval or conducting a test of significance remains the same. For example, emphasize the structure of confidence intervals as identical to those seen earlier—and this theme will continue in future chapters, when we study inference for means. For all confidence intervals we see in this course, the basic structure is *estimate* $\pm$ (#)SE$_{estimate}$, where "#" denotes the number of standard errors we need to go out on either side of the estimate and depends on both the confidence level and the sampling distribution.

2. Motivate the Need for a Two-Sample Confidence Interval

Consider asking students whether the proportion of in-state students at your school who have an academic scholarship differs from that for out-of-state students. For illustration, you can use your class for data, mentioning that this will formally invalidate statistical inference because the sample isn't likely to be representative. Use the sample of in-state students or the sample of out-of-state students to review material from the last chapter—we can construct two confidence intervals, one for the proportion of in-state students who have an academic scholarship and one for the proportion of out-of-state students have an academic scholarship, for example. This leads to the question of comparison. You can ask your students if it is okay to compare the two confidence intervals or whether you should use one confidence interval to examine the difference in proportions.

Once you have explained the construction of a confidence interval for $p_1 - p_2$, work your class example concerning academic scholarships for in-state and out-of-state students. Is "0" contained in the confidence interval?

3. Explain Why and How a Common Proportion Arises in Hypothesis Testing

When constructing significance tests for comparing two population proportions, the null hypothesis of no difference imposes pooling of the two samples to obtain an estimate of the common proportion.

4. Use Real-World Examples

A nice example relevant to college students concerns a large study comparing men and women college students and "binge drinking" habits, described in the paper "College Binge Drinking in the 1990s: A Continuing Problem," by Wechsler et al. A copy of the paper and the data can be obtained from the Harvard School of Public Health Web site (https://galileo.seas.harvard.edu/images/material/2800/1140/Wechsler_College BingeDrinkinginthe1990_s_AContinuingProblem_ResultsfromtheHarvardSchoolofPublic Health1999CollegeAlcoholStudy.pdf). The data suggest that a far greater proportion of male college students than female college students binge drink. Another great example is Physicians' Health Study I, which concluded low-dose aspirin decreases the risk of a first myocardial infarction. Information on the Physicians' Health Study can be found at http://phs.bwh.harvard.edu/.

Other Resources (LaunchPad)

Snapshots Videos
Comparing Two Proportions

EESEE Case Studies
How Many Poets Wrote Cleanness?

Universal Images Group via Getty Images

Comparing Two Proportions

In this chapter
we cover...

- Two-sample problems: proportions

- The sampling distribution of a difference between proportions

- Large-sample confidence intervals for comparing proportions

- Using technology

- Significance tests for comparing proportions

- Plus four confidence intervals for comparing proportions*

Comparing two populations or two treatments is one of the most common situations encountered in statistical practice. We call such situations *two-sample problems.*

Two-Sample Problems

- The goal of inference is to compare the responses to two treatments or to compare the characteristics of two populations.
- We have a separate sample from each treatment or each population.

Two-sample problems: proportions

A two-sample problem can arise from a randomized comparative experiment that randomly divides subjects into two groups and exposes each group to a different treatment. Comparing random samples separately selected from two populations is also a two-sample problem. In this chapter we consider two-sample problems in which the measurement on the individual can be categorized as either a success or a failure. Our goal is to compare the *proportions* of successes in the two populations. Here are some questions that you will answer in the exercises of this chapter:

EXAMPLE 18.1   Two-Sample Problems for Proportions

■ Does the proportion of males that have used the Internet to obtain information about a specific disease or medical problem in the last year differ from the proportion of females that have done so? A sample of males and a sample of females are obtained, and the proportions in each sample that have used the Internet to obtain information about a specific disease or medical problem in the last year are to be compared (see Exercise 18.1, page 402).

■ Does hand washing with alcohol-based hand sanitizers reduce the risk of infection for the common cold? A randomized experiment assigned 100 subjects to a treatment group that followed a regimen of hand washing with alcohol-based sanitizers and 100 subjects to a control group that used routine hand washing without alcohol-based sanitizers. After 10 weeks compare the proportions in the two groups who were infected with the common cold virus over the study period (see Exercise 18.35, page 413). ■

COMPUTER-ASSISTED INTERVIEWING

The days of the interviewer with a clipboard are past. Interviewers now read questions from a computer screen and use the keyboard to enter responses. The computer skips irrelevant items—once a woman says that she has no children, further questions about her children never appear. The computer can even present questions in random order to avoid bias due to always following the same order. Software keeps records of who has responded and prepares a file of data from the responses. The tedious process of transferring responses from paper to computer, once a source of errors, has disappeared.

Although we will use notation similar to that used in our study of proportions, we need to make a small change because there are now two populations instead of one. We refer to the groups we want to compare as Population 1 and Population 2. We have a separate SRS from each population or responses from two treatments in a randomized comparative experiment. A subscript shows which group a parameter or statistic describes. Here is our notation:

Population	Population Proportion	Sample Size	Sample Proportion
1	p_1	n_1	$\hat{p}_1$
2	p_2	n_2	$\hat{p}_2$

We compare the populations by doing inference about the difference $p_1 - p_2$ between the population proportions. The statistic that estimates this difference is the difference between the two sample proportions, $\hat{p}_1 - \hat{p}_2$.

EXAMPLE 18.2   Interracial Dating

STATE: "Would you date a person of a different race?" Researchers answered this question for black males and females by collecting data from the Internet dating site `Match.com`. When people post profiles on the site, they indicate which races they are willing to date. A random sample of 100 black males and a random sample of 100 black females were selected from the dating site, with 75 of the black males indicating their willingness to date white females and 56 of the black females indicating their willingness to date white males.[1] Is this good evidence that different proportions of black males and females on this Internet dating site would be willing to date someone who is white? How large is the difference between the proportions of black males and females who would be willing to date someone who is white?

PLAN: Take black males to be Population 1 and black females to be Population 2. The population proportions who are willing to date whites are p_1 for black males and p_2 for black females. We want to test the hypotheses

$$H_0: p_1 = p_2 \quad (\text{the same as } H_0: p_1 - p_2 = 0)$$
$$H_a: p_1 \neq p_2 \quad (\text{the same as } H_a: p_1 - p_2 \neq 0)$$

We also want to give a confidence interval for the difference $p_1 - p_2$.

aaremar/Shutterstock

SOLVE: Inference about population proportions is based on the sample proportions

$$\hat{p}_1 = \frac{75}{100} = 0.75 \quad \text{(men)}$$

$$\hat{p}_2 = \frac{56}{100} = 0.56 \quad \text{(women)}$$

We see that 75% of black males but only 56% of black females would be willing to date someone who is white. Because the sample sizes are of moderate size and the sample proportions are quite different, we expect that a test will be highly significant (in fact, $P = 0.0046$). So we concentrate on the confidence interval. To estimate $p_1 - p_2$, start from the difference between sample proportions

$$\hat{p}_1 - \hat{p}_2 = 0.75 - 0.56 = 0.19$$

To complete the "Solve" step, we must know how this difference behaves. ■

The sampling distribution of a difference between proportions

To use $\hat{p}_1 - \hat{p}_2$ for inference, we must know its sampling distribution. Here are the facts we need:

- When the samples are large, the distribution of $\hat{p}_1 - \hat{p}_2$ is **approximately Normal.**

- The **mean** of the sampling distribution is $p_1 - p_2$. That is, the difference between sample proportions is an unbiased estimator of the difference between population proportions.

- The **standard deviation** of the distribution is

$$\sqrt{\frac{p_1(1 - p_1)}{n_1} + \frac{p_2(1 - p_2)}{n_2}}$$

Figure 18.1 displays the distribution of $\hat{p}_1 - \hat{p}_2$. The standard deviation of $\hat{p}_1 - \hat{p}_2$ involves the unknown parameters p_1 and p_2. Just as in Chapters 16 and 17, we must replace these by estimates to do inference. And just as in these chapters, we do this a bit differently for confidence intervals and for hypothesis tests.

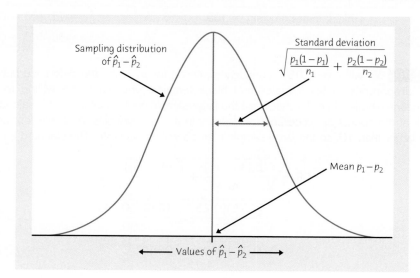

Sampling distribution of $\hat{p}_1 - \hat{p}_2$

Standard deviation
$$\sqrt{\frac{p_1(1 - p_1)}{n_1} + \frac{p_2(1 - p_2)}{n_2}}$$

Mean $p_1 - p_2$

← Values of $\hat{p}_1 - \hat{p}_2$ →

FIGURE 18.1

Select independent SRSs from two populations having proportions of successes p_1 and p_2. The proportions of successes in the two samples are $\hat{p}_1$ and $\hat{p}_2$. When the samples are large, the sampling distribution of the difference $\hat{p}_1 - \hat{p}_2$ is approximately Normal.

Large-sample confidence intervals for comparing proportions

standard error

To obtain a confidence interval, replace the population proportions p_1 and p_2 in the standard deviation by the sample proportions. The result is the **standard error** of the statistic $\hat{p}_1 - \hat{p}_2$:

$$SE = \sqrt{\frac{\hat{p}_1(1 - \hat{p}_1)}{n_1} + \frac{\hat{p}_2(1 - \hat{p}_2)}{n_2}}$$

The confidence interval has the same form we met in Chapter 16:

$$estimate \pm z^*SE_{estimate}$$

Large-Sample Confidence Interval for Comparing Two Proportions

Draw an SRS of size n_1 from a large population having proportion p_1 of successes and draw an independent SRS of size n_2 from another large population having proportion p_2 of successes. When n_1 and n_2 are large, an approximate level C **confidence interval for $p_1 - p_2$** is

$$(\hat{p}_1 - \hat{p}_2) \pm z^*SE$$

In this formula the standard error SE of $\hat{p}_1 - \hat{p}_2$ is

$$SE = \sqrt{\frac{\hat{p}_1(1 - \hat{p}_1)}{n_1} + \frac{\hat{p}_2(1 - \hat{p}_2)}{n_2}}$$

and z^* is the critical value for the standard Normal density curve with area C between $-z^*$ and z^*.

Use this interval only when the numbers of successes and failures are each 10 or more in both samples.

EXAMPLE 18.3   Interracial Dating, Continued

We can now complete Example 18.2. Here is a summary of the basic information:

Population	Population Description	Sample Size	Number of Successes	Sample Proportion
1	black males	$n_1 = 100$	75	$\hat{p}_1 = 75/100 = 0.75$
2	black females	$n_2 = 100$	56	$\hat{p}_2 = 56/100 = 0.56$

SOLVE: We will give a 95% confidence interval for $p_1 - p_2$, the difference between the proportions of black males and black females who would be willing to date someone who is white. To check that the large-sample confidence interval is safe to use, look at the counts of successes and failures in the two samples. All these four counts are larger than 10, so the large-sample method will be accurate. The standard error is

$$SE = \sqrt{\frac{\hat{p}_1(1 - \hat{p}_1)}{n_1} + \frac{\hat{p}_2(1 - \hat{p}_2)}{n_2}}$$

$$= \sqrt{\frac{(0.75)(0.25)}{100} + \frac{(0.56)(0.44)}{100}}$$

$$= \sqrt{0.004339} = 0.0659$$

The 95% confidence interval is

$$(\hat{p}_1 - \hat{p}_2) \pm z^*SE = (0.75 - 0.56) \pm (1.960)(0.0659)$$
$$= 0.19 \pm 0.13$$
$$= 0.06 \text{ to } 0.32$$

CONCLUDE: We are 95% confident that the percent of black males willing to date whites is between 6 and 32 percentage points higher than the percent of black females who are willing to date white men on comparable Internet dating sites. Even with sample sizes of 100 in each group, the resulting confidence interval 0.06 to 0.32 is quite wide. As with a single proportion, fairly large sample sizes are required to obtain narrow confidence intervals. In a similar study by the authors, it was found that white men were more willing than white women to date someone who is black. ■

Using technology

Figure 18.2 displays software output for Example 18.3 from a graphing calculator and two statistical software programs. As usual, you can understand the output even without knowledge of the program that produced it. Minitab gives the test as

Graphing Calculator

Minitab

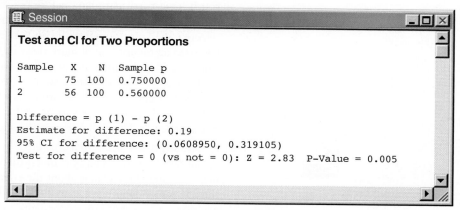

CrunchIt!

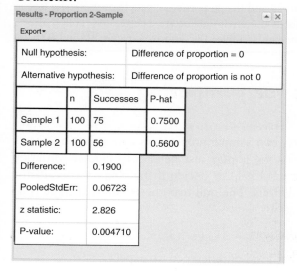

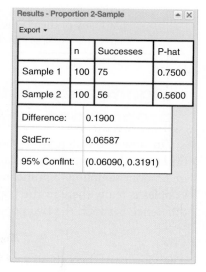

FIGURE 18.2

Output from a graphing calculator, Minitab, and CrunchIt! for the 95% confidence interval of Example 18.3.

well as the confidence interval, confirming that the difference between men and women is highly significant. In CrunchIt!, the test and the confidence interval must be requested using separate commands resulting in the two outputs in the figure. Excel spreadsheet output is not shown because Excel lacks menu items for inference about proportions. You must use the spreadsheet's formula capability to program the confidence interval or test statistic and then to find the *P*-value of a test.

Apply Your Knowledge

18.1 Health Online. In 2012, a random sample of 2392 Internet users found approximately 55% had used the Internet to obtain health information about a specific disease or medical problem in the last year. Of the 1084 men in the survey, 520 used the Internet to learn about a specific disease or medical problem, whereas 811 of the 1308 women in the survey had done so.[2] Give a 95% confidence interval for the difference between the proportions of male and female Internet users who used the Internet to obtain health information about a specific disease or medical problem in the last year. Follow the four-step process as illustrated in Examples 18.2 and 18.3 (pages 398 and 400).

18.2 Marijuana Use in High School. The Youth Risk Behavior Surveillance System (YRBSS) monitors health risk behaviors among U.S. high school students, which include tobacco use, alcohol and drug use, inadequate physical activity, unhealthy diet and risky sexual behavior. Approximately 8% of youth have tried marijuana before reaching age 13 years. How does marijuana use change over the high school years? In 2011, the survey randomly selected 3642 ninth-graders and 3606 twelfth-graders and asked them if they had used marijuana one or more times during the previous 30 days.[3] Of these students, 656 ninth-graders and 1010 twelfth-graders said yes. Give a 99% confidence interval for the difference between the proportions of all ninth- and twelfth-grade high school students who smoked marijuana at least once in the 30 days prior to the survey. Follow the four-step process as illustrated in Examples 18.2 and 18.3 (pages 398 and 400).

18.3 A Question on the Environment. In 2010, respondents to the General Social Survey were asked how much they agreed or disagreed with the statement: "Many of the claims about the environment are exaggerated."[4] Among the 251 respondents aged 18 to 29, 75 said either "Agree" or "Strongly agree," whereas among the 376 respondents over age 60, 174 gave one of these responses. Give a 95% confidence interval for the difference between the proportion of people aged 18 to 29 and the proportion of people over 60 who feel that many of the claims about the environment are exaggerated. Follow the four-step process as illustrated in Examples 18.2 and 18.3 (pages 398 and 400).

THE COOKIE STRIKES

How many different people clicked on your business Web site last month? Technology tries to help: when someone visits your site, a little piece of code called a cookie is left on their computer. When the same person clicks again, the cookie says not to count them as a "unique visitor" because this isn't their first visit. But lots of Web users delete cookies, either by hand or automatically with software. These people get counted again when they visit your site again. That's bias: your counts of unique visitors are systematically too high. One study found that unique-visitor counts were as much as 50% too high.

Significance tests for comparing proportions

An observed difference between two sample proportions can reflect an actual difference between the populations, or it may just be due to chance variation in random sampling. Significance tests help us decide if the effect we see in the samples is really there in the populations. The null hypothesis says that there is no difference between the two populations:

$$H_0: p_1 = p_2 \text{ (the same as } H_0: p_1 - p_2 = 0)$$

The alternative hypothesis says what kind of difference we expect.

EXAMPLE 18.4   False Memories

STATE: The political event depicted in Figure 18.3 was remembered by about 31% of those surveyed, despite the fact that it never occurred. In 2010, *Slate*, a current affairs and culture magazine, surveyed over 5000 readers regarding their perspective on several past political events and their memories of these events. Unbeknownst to those surveyed, one of the events shown to each participant was fabricated, making this the largest false memory study conducted to date.[5] The hypothesis of interest was whether political preferences guided the formation of false memories, as false memories have been shown to be more easily implanted in memory when they are congruent with a person's preexisting attitudes. Figure 18.3 was viewed by 616 participants who categorized themselves as progressive and 49 participants who categorized themselves as conservative. The event was falsely remembered as having occurred by 212 of the progressives surveyed and by 7 of the conservatives. How strong is the evidence that a larger proportion of progressives have a false memory of this event than conservatives?

PLAN: Call the population proportions p_1 for progressives and p_2 for conservatives. Because the image of a conservative president relaxing at his ranch during a major crisis is more congruent with the preexisting attitude of progressives, our hypothesis gives a direction for the difference before looking at the data, so we have the one-sided alternative:

$$H_0: p_1 = p_2$$
$$H_a: p_1 > p_2$$

SOLVE: Consider those who classify themselves as progressives and conservatives as separate SRSs of progressive and conservative *Slate* readers. The sample proportions who falsely remembered the event in Figure 18.3 are

$$\hat{p}_1 = \frac{212}{616} = 0.344 \quad \text{(progressives)}$$

$$\hat{p}_2 = \frac{7}{49} = 0.143 \quad \text{(conservatives)}$$

That is, 34% of those classifying themselves as progressive falsely remembered the event in Figure 18.3, but only 14% of conservatives falsely remembered it. Is this apparent difference statistically significant? To continue the solution, we must learn the proper test. ■

FIGURE 18.3
September 1, 2005: As parts of New Orleans lie underwater in the wake of Hurricane Katrina, President Bush entertains Houston Astros pitcher Roger Clemens at his ranch in Crawford, Texas.

Photo illustration by Holly Allen (Slate Magazine)

To do a hypothesis test, standardize the difference between the sample proportions $\hat{p}_1 - \hat{p}_2$ to get a z statistic. If H_0 is true, both samples come from populations in which the same unknown proportion p have a false memory of the event depicted in Figure 18.3. We take advantage of this by combining the two samples to estimate this single p instead of estimating p_1 and p_2 separately. Call this the **pooled sample proportion.** It is

pooled sample proportion

$$\hat{p} = \frac{\text{number of successes in both samples combined}}{\text{number of individuals in both samples combined}}$$

Use $\hat{p}$ in place of both $\hat{p}_1$ and $\hat{p}_2$ in the expression for the standard error SE of $\hat{p}_1 - \hat{p}_2$ to get a z statistic that has the standard Normal distribution when H_0 is true. Here is the test.

Significance Test for Comparing Two Proportions

Draw an SRS of size n_1 from a large population having proportion p_1 of successes and draw an independent SRS of size n_2 from another large population having proportion p_2 of successes. **To test the hypothesis H_0: $p_1 = p_2$,** first find the pooled proportion $\hat{p}$ of successes in both samples combined. Then compute the z statistic

$$z = \frac{\hat{p}_1 - \hat{p}_2}{\sqrt{\hat{p}(1 - \hat{p})\left(\dfrac{1}{n_1} + \dfrac{1}{n_2}\right)}}$$

In terms of a variable Z having the standard Normal distribution, the P-value for a test of H_0 against

H_a: $p_1 > p_2$ is $P(Z \geq z)$

H_a: $p_1 < p_2$ is $P(Z \leq z)$

H_a: $p_1 \neq p_2$ is $2P(Z \geq |z|)$
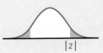

Use this test when the counts of successes and failures are each 5 or more in both samples.[6]

EXAMPLE 18.5   False Memories, Continued

SOLVE: The data come from an SRS, and the counts of successes and failures are all larger than 5. The pooled proportion of progressives and conservatives who falsely remembered this event is

$$\hat{p} = \frac{\text{number "falsely remembered event" among progressives and conservatives combined}}{\text{number of progressives and conservatives combined}}$$

$$= \frac{212 + 7}{616 + 49}$$

$$= \frac{219}{665} = 0.329$$

The z test statistic is

$$z = \frac{\hat{p}_1 - \hat{p}_2}{\sqrt{\hat{p}(1 - \hat{p})\left(\dfrac{1}{n_1} + \dfrac{1}{n_2}\right)}}$$

$$= \frac{0.344 - 0.143}{\sqrt{(0.329)(0.671)\left(\dfrac{1}{616} + \dfrac{1}{49}\right)}}$$

$$= \frac{0.201}{0.0697} = 2.88$$

The one-sided *P*-value is the area under the standard Normal curve greater than 2.88. Figure 18.4 shows this area. Software tells us that *P* = 0.00199.

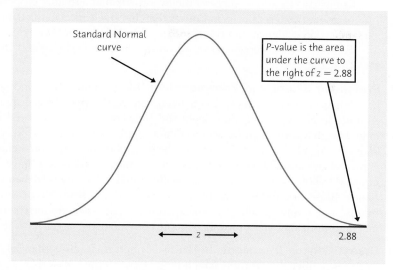

Standard Normal curve

P-value is the area under the curve to the right of *z* = 2.88

z

2.88

FIGURE 18.4
The *P*-value for the one-sided test of Example 18.5.

z^*	2.807	3.091
One-sided *P*	.0025	.001

Without software, you can compare *z* = 2.88 with the bottom row of Table C (standard Normal critical values) to approximate *P*.

It lies between the critical values 2.807 and 3.091 for one-sided *P*-values 0.0025 and 0.001.

CONCLUDE: There is strong evidence (*P* < 0.0025) that, among *Slate* readers, progressives are more likely than conservatives to have a false memory of Bush vacationing with Roger Clemens in the wake of the Katrina hurricane. In a second fabricated event by the authors, it was found that conservatives were more likely than progressives to falsely remember President Obama shaking hands with Iranian President Ahmadinejad at a United Nations conference. ■

The sample survey in this example selected a single sample of *Slate* readers, not two separate samples of progressives and conservatives. To get two samples, we divided the single sample by political orientation. This means that we did not know the two sample sizes n_1 and n_2 until after the data were in hand. The two-sample z procedures for comparing proportions are valid in such situations. This is an important fact about these methods.

Apply Your Knowledge

18.4 Seat Belt Use. The proportion of drivers who use seat belts depends on things like age (young people are more likely to go unbelted) and sex (women are more likely to use belts). It also depends on local law. In New York City, police can stop a driver who is not belted. In Boston at the time of the survey, police could cite a driver for not wearing a seat belt only if the driver had been stopped for some other violation. Here are data from observing random samples of female Hispanic drivers in these two cities in 2002:[7]

City	Drivers	Belted
New York	220	183
Boston	117	68

(a) Is this an experiment or an observational study? Why?

(b) Comparing local laws suggests the hypothesis that a smaller proportion of drivers wear seat belts in Boston than in New York. Do the data give good evidence that this is true for female Hispanic drivers? Follow the four-step process as illustrated in Examples 18.4 and 18.5 (pages 403–404).

18.5 Protecting Skiers and Snowboarders. Most alpine skiers and snowboarders do not use helmets. Do helmets reduce the risk of head injuries? A study in Norway compared skiers and snowboarders who suffered head injuries with a control group who were not injured. Of 578 injured subjects, 96 had worn a helmet. Of the 2992 in the control group, 656 wore helmets.[8] Is helmet use less common among skiers and snowboarders who have head injuries? Follow the four-step process as illustrated in Examples 18.4 and 18.5 (pages 403–404). (Note that this is an observational study that compares injured and uninjured subjects. An experiment that assigned subjects to helmet and no-helmet groups would be more convincing.)

18.6 Breast Cancer Treatment. In sentinel lymph node dissection (SLND), surgeons remove two or three lymph nodes close to the breast that are most likely to contain cancer cells. If these "sentinel" lymph nodes are free of tumors, SLND alone is the accepted management for patients. When the sentinel lymph nodes contain metastases, axillary lymph node dissection (ALND; that is the removal of further nodes) remains the standard of care, although its contribution to survival is controversial. ALND carries the additional risk of complications such as seroma, infection, and lymphedema. In one study, patients with sentinel metastases identified by SLND were randomized to undergo ALND or no further treatment (SLND alone). Here are the five-year disease-free survival numbers for the two groups:[9]

Group	Sample Size	Disease-Free After Five Years
ALND	420	345
SLND alone	436	366

How strong is the evidence that the proportions of disease-free patients after five years differ for patients who underwent ALND or only SLND? Follow the four-step process as illustrated in Examples 18.4 and 18.5 (pages 403–404).

Plus four confidence intervals for comparing proportions*

Like the large-sample confidence interval for a single proportion p, the large-sample interval for $p_1 - p_2$ generally has a true confidence level less than the level you asked for. The inaccuracy is not as serious as in the one-sample case, at least if our guidelines for use are followed. Once again, adding imaginary observations greatly improves the accuracy.[10]

*This section is optional.

Plus Four Confidence Interval for Comparing Two Proportions

Draw independent SRSs from two large populations with population proportions of successes p_1 and p_2. To get the **plus four confidence interval for the difference $p_1 - p_2$**, add four imaginary observations, one success, and one failure in each of the two samples. Then use the large-sample confidence interval with the new sample sizes (actual sample sizes + 2) and counts of successes (actual counts + 1).

Use this interval when the sample size is at least 5 in each group, with any counts of successes and failures.

If your software does not offer the plus four method, just enter the new plus four sample sizes and success counts into the large-sample procedure.

EXAMPLE 18.6   Abecedarian Early Childhood Education Program: Adult Outcomes

STATE: The Abecedarian Project was a randomized controlled study to assess the effects of intensive early childhood education on children who were at high risk based on several sociodemographic indicators. The project randomly assigned children to a treatment group, which was provided with early educational activities before kindergarten, and the remainder to a control group. A recent follow-up study interviewed subjects at age 30 and compared the college graduation rates (earning a 4-year degree).[11] Here are the data for two groups:

Population	Population Description	Sample Size	Number of Successes	Sample Proportion
1	treatment	$n_1 = 52$	12	$\hat{p}_1 = 12/52 = 0.2308$
2	control	$n_2 = 49$	3	$\hat{p}_2 = 3/49 = 0.0612$

How much does early childhood education increase the proportion earning a 4-year degree by age 30?

PLAN: Give a 90% confidence interval for the difference between population proportions, $p_1 - p_2$.

SOLVE: The conditions for the large-sample interval are not met because there are only 3 successes in the control group. However, we can use the plus four method because the sample sizes for the treatment and the control group are both at least 5. Add four imaginary observations. The new data summary is

Population	Population Description	Sample Size	Number of Successes	Plus Four Sample Proportion
1	treatment	$n_1 + 2 = 54$	$12 + 1 = 13$	$\tilde{p}_1 = 13/54 = 0.2407$
2	control	$n_2 + 2 = 51$	$3 + 1 = 4$	$\tilde{p}_2 = 4/51 = 0.0784$

The standard error based on the new facts is

$$SE = \sqrt{\frac{\tilde{p}_1(1 - \tilde{p}_1)}{n_1 + 2} + \frac{\tilde{p}_2(1 - \tilde{p}_2)}{n_2 + 2}}$$

$$= \sqrt{\frac{(0.2407)(0.7583)}{54} + \frac{(0.0784)(0.9216)}{51}}$$

$$= \sqrt{0.00480} = 0.0693$$

The plus four 90% confidence interval is

$$(\tilde{p}_1 - \tilde{p}_2) \pm z^*\text{SE} = (0.2407 - 0.0784) \pm (1.645)(0.0693)$$
$$= 0.1623 \pm 0.1140$$
$$= 0.048 \text{ to } 0.276$$

CONCLUDE: We are 90% confident that the early childhood education program increases the proportion of high-risk children who earn a four-year degree by age 30 by between 4.8% and 27.6%. ■

The plus four interval may be conservative (that is, the true confidence level may be *higher* than you asked for) for very small samples and population p's close to 0 or 1. It is generally much more accurate than the large-sample interval when the samples are small. Nevertheless, the plus four interval in Example 18.6 cannot save us from the fact that sample sizes around 50 will produce a wide confidence interval when comparing two proportions.

Apply Your Knowledge

©Jupiterimages/Comstock Images/Alamy

18.7 **In-line Skaters.** A study of injuries to in-line skaters used data from the National Electronic Injury Surveillance System, which collects data from a random sample of hospital emergency rooms. The researchers interviewed 161 people who came to emergency rooms with injuries from in-line skating. Wrist injuries (mostly fractures) were the most common.[12]

(a) The interviews found that 53 people were wearing wrist guards, and 6 of these had wrist injuries. Of the 108 who did not wear wrist guards, 45 had wrist injuries. Why should we not use the large-sample confidence interval for these data?

(b) The plus four method adds one success and one failure in each sample. What are the sample sizes and counts of successes after you do this?

(c) Give the plus four 95% confidence interval for the difference between the two population proportions of wrist injuries. State carefully what populations your inference compares.

18.8 **Shrubs That Withstand Fire.** Fire is a serious threat to shrubs in dry climates. Some shrubs can resprout from their roots after their tops are destroyed. One study of resprouting took place in a dry area of Mexico.[13] The investigators first clipped the tops of all the shrubs in the study. For the treatment, they then applied a propane torch to the stumps of the shrubs to simulate a fire, and for the control the stumps were left alone. The study included 24 shrubs of which 12 were randomly assigned to the treatment and the remaining 12 to the control. A shrub is a success if it resprouts. For the shrub *Xerospirea hartwegiana*, all 12 shrubs in the control group resprouted, whereas only 8 in the treatment group resprouted. How much does burning reduce the proportion of shrubs of this species that resprout? Give the 95% plus four confidence interval for the amount by which burning reduces the proportion of shrubs of this species that resprout. The plus four method is particularly helpful when, as here, a count of either successes or failures is zero. Follow the four-step process as illustrated in Example 18.6.

CHAPTER 18 SUMMARY

Chapter Specifics

■ The data in a **two-sample problem** are two independent SRSs, each drawn from a separate population.

- Tests and confidence intervals to compare the proportions p_1 and p_2 of successes in the two populations are based on the difference $\hat{p}_1 - \hat{p}_2$ between the sample proportions of successes in the two SRSs.

- When the sample sizes n_1 and n_2 are large, the sampling distribution of $\hat{p}_1 - \hat{p}_2$ is close to Normal with mean $p_1 - p_2$.

- The level C **large-sample confidence interval for $p_1 - p_2$** is

$$(\hat{p}_1 - \hat{p}_2) \pm z^*\text{SE}$$

where the **standard error** of $\hat{p}_1 - \hat{p}_2$ is

$$\text{SE} = \sqrt{\frac{\hat{p}_1(1 - \hat{p}_1)}{n_1} + \frac{\hat{p}_2(1 - \hat{p}_2)}{n_2}}$$

and z^* is a standard Normal critical value.

- The true confidence level of the large-sample interval can be substantially less than the planned level C. Use this interval only if the counts of successes and failures in both samples are 10 or greater.

- **Significance tests for H_0: $p_1 = p_2$** use the **pooled sample proportion**

$$\hat{p} = \frac{\text{number of successes in both samples combined}}{\text{number of individuals in both samples combined}}$$

and the z statistic

$$z = \frac{\hat{p}_1 - \hat{p}_2}{\sqrt{\hat{p}(1 - \hat{p})\left(\dfrac{1}{n_1} + \dfrac{1}{n_2}\right)}}$$

P-values come from the standard Normal distribution. Use this test when there are 5 or more successes and 5 or more failures in both samples.

- **(Optional material)** When the conditions for the large-sample confidence interval are not met, to get a more accurate confidence interval, add four imaginary observations, one success and one failure in each sample. Then use the same formula for the confidence interval. This is the **plus four confidence interval.** You can use it whenever both samples have 5 or more observations.

Link It

Most studies compare two or more treatments rather than investigating a single treatment. These can be observational studies or comparative experiments, and the differences in the type of conclusions that can be reached were described in Chapter 9. This chapter considers the case where the response classifies individuals into two categories such as black men and women being willing to date someone who is white, or not. The comparison is then a comparison of this proportion for two groups, such as the proportion of black males who are willing to date someone who is white versus the proportion of black females who are willing to date someone who is white. Because the methods of this chapter are approximate, it is important to always check the conditions required for the approximations to work well, whether you are using the large-sample confidence interval, the plus four method, or the z test.

After the appropriate statistical method has been applied, care must be taken when stating the conclusion. The issues described in Chapter 9 are still important. For example, a lack of blinding can result in the expectations of the researcher influencing the results, whereas confounding can mix up the comparison of the two groups with other factors.

When there are two treatments and the response is a quantitative measurement, the comparison between the treatments is often based on a comparison of the treatment means. These methods will be described in Chapter 21.

In 2010, respondents to the General Social Survey were asked: "There are always some people whose ideas are considered bad or dangerous by other people. For instance, somebody who is against churches and religion. If such a person wanted to make a speech in your (city/town/community) against churches and religion, should he be allowed to speak, or not?"[14]

Among the 407 respondents who considered themselves Democrats, 288 said, "Allow," whereas among the 293 respondents who considered themselves Republicans, 227 said, "Allow." Is there a difference between the proportion of Democrats and the proportion of Republicans that would allow an antireligionist to speak? Exercises 18.9 to 18.13 are based on these results.

18.9 Take p_D and p_R to be the proportions of all Democrats and Republicans who would allow an antireligionist to speak. The hypotheses to be tested are

(a) $H_0: p_D = p_R$ versus $H_a: p_D \neq p_R$.
(b) $H_0: p_D = p_R$ versus $H_a: p_D > p_R$.
(c) $H_0: p_D = p_R$ versus $H_a: p_D < p_R$.

18.10 The sample proportions of Democrats and Republicans who would allow an antireligionist to speak are

(a) $\hat{p}_D = 0.708$ and $\hat{p}_R = 0.775$.
(b) $\hat{p}_D = 0.775$ and $\hat{p}_R = 0.708$.
(c) $\hat{p}_D = 0.775$ and $\hat{p}_R = 0.736$.

18.11 The pooled sample proportion of respondents who would allow an antireligionist to speak is

(a) $\hat{p} = 0.725$.   (b) $\hat{p} = 0.736$.   (c) $\hat{p} = 0.740$.

18.12 The z test for comparing the proportions of Democrats and Republicans who would allow an antireligionist to speak has

(a) $P < 0.01$.
(b) $0.01 < P < 0.05$.
(c) $P > 0.05$.

18.13 The 90% large-sample confidence interval for the difference $p_D - p_R$ in the proportions of Democrats and Republicans who would allow an antireligionist to speak is about

(a) -0.067 ± 0.055.
(b) -0.067 ± -0.065.
(c) -0.067 ± 0.086.

18.14 In an experiment to learn if substance M can help restore memory, the brains of 20 rats were treated to damage their memories. The rats were trained to run a maze. After a day, 10 rats were given M, and 7 of them succeeded in the maze; only 2 of the 10 control rats were successful. The z test for "no difference" against "a higher proportion of the M group succeeds" has

(a) $z = 2.25, P < 0.02$.
(b) $z = 2.60, P < 0.005$.
(c) $z = 2.25, 0.02 < P < 0.04$.

18.15 The z test in the previous exercise

(a) may be inaccurate because the populations are too small.
(b) may be inaccurate because some counts of successes and failures are too small.
(c) is reasonably accurate because the conditions for inference are met.

18.16 The plus four 90% confidence interval for the difference between the proportion of rats that succeed when given M and the proportion that succeed without it is

(a) 0.455 ± 0.312.
(b) 0.417 ± 0.304.
(c) 0.417 ± 0.185.

CHAPTER 18 EXERCISES

When using the large-sample methods of this chapter, be sure to check that the guidelines for its use are met, and to state your conclusions in context.

18.17 **Truthfulness in online profiles.** Many teens have posted profiles on a social-networking Web site. A sample survey in 2007 asked a random sample of teens with online profiles if they included false information in their profiles. Of 170 younger teens (aged 12 to 14), 117 said "Yes." Of 317 older teens (aged 15 to 17), 152 said "Yes."[15]

(a) Do these samples satisfy the guidelines for the large-sample confidence interval?
(b) Give a 95% confidence interval for the difference between the proportions of younger and older teens who include false information in their online profiles.

18.18 **Effects of an appetite suppressant.** Subjects with preexisting cardiovascular symptoms who were receiving subitramine, an appetite suppressant, were found to be at increased risk of cardiovascular events while taking the drug. The study included 9804 overweight or obese subjects with preexisting cardiovascular disease and/or type 2 diabetes. The subjects were randomly assigned to subitramine (4906 subjects) or a placebo (4898 subjects) in a double-blind fashion. The primary outcome measured was the occurrence of any of the following events: nonfatal myocardial infarction or stroke, resuscitation after cardiac arrest, or cardiovascular death. The primary outcome was observed in 561 subjects in the subitramine group and 490 subjects in the placebo group.[16]

(a) Find the proportion of subjects experiencing the primary outcome for both the subitramine and placebo groups.

(b) Can we safely use the large-sample confidence interval for comparing the proportions of subitramine and placebo subjects who experienced the primary outcome? Explain.

(c) Give a 95% confidence interval for the difference between the proportions of subitramine and placebo subjects who experienced the primary outcome.

18.19 **(Optional) Genetically altered mice.** Genetic influences on cancer can be studied by manipulating the genetic makeup of mice. One of the processes that turn genes on or off (so to speak) in particular locations is called "DNA methylation." Do low levels of this process help cause tumors? Compare mice altered to have low levels with normal mice. Of 33 mice with lowered levels of DNA methylation, 23 developed tumors. None of the control group of 18 normal mice developed tumors in the same time period.[17]

Mark Harmel/Getty Images

(a) Explain why we cannot safely use either the large-sample confidence interval or the test for comparing the proportions of normal and altered mice that develop tumors.

(b) The plus four method adds two observations, a success and a failure, to each sample. What are the sample sizes and the numbers of mice with tumors after you do this?

(c) Give a 99% confidence interval for the difference in the proportions of the two populations that develop tumors.

18.20 **Effects of an appetite suppressant, continued.** Exercise 18.18 describes a study to determine if subjects with preexisting cardiovascular symptoms were at an increased risk of cardiovascular events while taking subitramine. Do the data give good reason to think that there is a difference between the proportions of treatment and placebo subjects who experienced the primary outcome? (Note that subitramine is no longer available in the United States due to manufacturer's concerns over increased risk of heart attack or stroke.)

(a) State hypotheses, find the test statistic, and use either software or the bottom row of Table C for the P-value. Be sure to state your conclusion.

(b) Explain simply why it was important to have a placebo group in this study.

Adolescence, music, and algebra. *Research has suggested that musicians process music in the same cortical regions in which adolescents process algebra. When taking introductory algebra, will students who were enrolled in formal instrumental or choral music instruction during middle school outperform those who experienced neither of these modes of musical instruction? The sample consisted of 6026 ninth-grade students in Maryland who had completed introductory algebra. Of these, 3239 students had received formal instrumental or choral instruction during all three years of middle school, whereas the remaining students had not. Of those receiving formal musical instruction, 2818 received a passing grade on the Maryland Algebra/Data Analysis High School Assessment (HSA). In contrast, 2091 of the 2787 students not receiving musical instruction received a passing grade.[18] Exercises 18.21 to 18.23 are based on this study.*

18.21 **Does music make a difference?**

(a) Is there a significant difference in the proportions of students with and without musical instruction who receive a passing grade on the Maryland HSA? State hypotheses, find the test statistic, and use software or the bottom row of Table C to get a P-value.

(b) Is this an observational study or an experiment? Why?

(c) In view of your answer in (b), carefully state your conclusions about the relationship between music instruction and success in algebra.

18.22 **How many students pass?** Give a 95% confidence interval for the proportion of ninth-grade students who receive a passing grade on the HSA.

18.23 **How big a difference?** Give a 95% confidence interval for the difference between the proportions of students passing the HSA who have received or not received formal musical instruction in middle school.

18.24 **The design matters.** Due to concerns about the safety of bariatric weight loss surgery, in 2006 the Centers for Medicare and Medicaid Services (CMS) restricted coverage of bariatric surgery to hospitals designated as Centers for Excellence. Did the CMS restriction improve the outcomes of bariatric surgery for Medicare patients? Among the 1847 Medicare patients in the study having surgery in the 18 months preceding the restriction, 270 experienced overall complications, whereas in the 1639 patients having bariatric surgery in the 18 months following the restriction, 170 experienced overall complications.[19]

(a) What are the sample proportions of Medicare patients who experienced overall complications from bariatric surgery before and after the CMS restriction on coverage? How strong is the evidence that the proportions of overall complications are different before and after the CMS restriction? Use an appropriate hypothesis test to answer this question.

(b) Is this an observational study or an experiment? Can we conclude that the CMS restriction has reduced the proportion of overall complications?

(c) Improved outcomes may be due to several factors, including the use of lower-risk bariatric procedures, increased surgeon experience, or healthier patients receiving the surgery. What types of variables are these, and how do they affect the type of conclusions that you can make?

(d) In a second study,[20] a control group consisting of non-Medicare patients obtained before and after the restrictions on coverage was obtained. When compared with this control group, no significant evidence was found that the CMS restriction reduced the proportion of overall complications for the Medicare group. Explain in simple language how comparison with a control group could help reduce the effects of some of the variables described in (c)?

18.25 **Significant does not mean important.** Never forget that even small effects can be statistically significant if the samples are large. To illustrate this fact, consider a sample of 148 small businesses. During a three-year period, 15 of the 106 headed by men and 7 of the 42 headed by women failed.[21]

(a) Find the proportions of failures for businesses headed by women and businesses headed by men. These sample proportions are quite close to each other. Give the P-value for the z test of the hypothesis that the same proportion of women's and men's businesses fail. (Use the two-sided alternative.) The test is very far from being significant.

(b) Now suppose that the same sample proportions came from a sample 30 times as large. That is, 210 out of 1260 businesses headed by women and 450 out of 3180 businesses headed by men fail. Verify that the proportions of failures are exactly the same as in (a). Repeat the z test for the new data, and show that it is now significant at the $\alpha = 0.05$ level.

(c) It is wise to use a confidence interval to estimate the size of an effect rather than just giving a P-value. Give 95% confidence intervals for the difference between the proportions of women's and men's businesses that fail for the settings of both (a)

and (b). What is the effect of larger samples on the confidence interval?

*In responding to Exercises 18.26 to 18.35, follow the **Plan, Solve,** and **Conclude** steps of the four-step process.*

18.26 **Demographics of social-networking sites: sex.** A Pew Internet survey in 2012 examined several demographic variables of users of social-networking sites including sex, age, race, education and income. Of the 1802 Internet users included in the survey, 846 were men and 525 of these used social networking sites. Among the 956 women in the survey, 679 of the women used social-networking sites.[22] Is there good evidence that the proportion of Internet users who use social-networking sites is different for men and women?

18.27 **Demographics of social-networking sites: race.** The survey in the previous exercise also looked at possible differences in the proportions of white and Hispanic Internet users who used social-networking sites. They found that 866 of the 1332 white Internet users and 111 of the 154 Hispanics used social-networking sites. Is there evidence of a difference between the proportions of white and Hispanic Internet users who use social-networking sites?

18.28 **More on social-networking and sex.** Continue your work from Exercise 18.26. Estimate the difference between the proportions of male and female Internet users who use social-networking sites. (Use 90% confidence.)

18.29 **Smoking cessation.** Chantix is different from most other quit-smoking products in that it targets nicotine receptors in the brain, attaches to them, and blocks nicotine from reaching them. As part of a larger randomized controlled trial, generally healthy smokers who smoked at least 10 cigarettes per day were assigned at random to take Chantix or a placebo. The study was double-blind, with the response measure being continuous absence from smoking for Weeks 9 through 12 of the study. Of the 352 subjects taking Chantix, 155 abstained from smoking during Weeks 9 through 12, whereas 61 of the 344 subjects taking the placebo abstained during this same time period.[23] Give a 99% confidence interval for the difference (treatment minus placebo) in the proportions of smokers who abstained from smoking during Weeks 9 through 12.

18.30 **The Gold Coast.** A historian examining British colonial records for the Gold Coast in Africa suspects that the death rate was higher among African miners than among European miners. In the year 1936, there were 223 deaths among 33,809 African miners and 7 deaths among 1541

European miners on the Gold Coast.[24] (The Gold Coast became the independent nation of Ghana in 1957.) Consider this year as a random sample from the colonial era in West Africa. Is there good evidence that the proportion of African miners who died was higher than the proportion of European miners who died?

18.31 **I refuse!** Do our emotions influence economic decisions? One way to examine the issue is to have subjects play an "ultimatum game" against other people and against a computer. Your partner (person or computer) gets $10, on the condition that it be shared with you. The partner makes you an offer. If you refuse, neither of you gets anything. So it's to your advantage to accept even the unfair offer of $2 out of the $10. Some people get mad and refuse unfair offers. Here are data on the responses of 76 subjects randomly assigned to receive an offer of $2 from either a person they were introduced to or a computer:[25]

	ACCEPT	REJECT
Human offers	20	18
Computer offers	32	6

We suspect that emotion will lead to offers from another person being rejected more often than offers from an impersonal computer. Do a test to assess the evidence for this conjecture.

18.32 **Did the random assignment work?** A large clinical trial of the effect of diet on breast cancer assigned women at random to either a normal diet or a low-fat diet. To check that the random assignment did produce comparable groups, we can compare the two groups at the start of the study. Asked if there is a family history of breast cancer: 3396 of the 19,541 women in the low-fat group and 4929 of the 29,294 women in the control group said "Yes."[26] If the random assignment worked well, there should *not* be a significant difference in the proportions with a family history of breast cancer. How significant is the observed difference?

18.33 **(Optional) Lyme disease.** Lyme disease is spread in the northeastern United States by infected ticks. The ticks are infected mainly by feeding on mice, so more mice result in more infected ticks. The mouse population in turn rises and falls with the abundance of acorns, their favored food. Experimenters studied two similar forest areas in a year when the acorn crop failed. They added hundreds of thousands of acorns to one area to imitate an abundant acorn crop, while leaving the other area untouched. The next spring, 54 of the 72 mice

trapped in the first area were in breeding condition, versus 10 of the 17 mice trapped in the second area.[27] Estimate the difference between the proportions of mice ready to breed in good acorn years and bad acorn years. (Use 90% confidence. Be sure to justify your choice of confidence interval.)

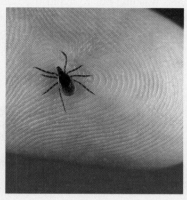

Scott Camazine/Science Source

18.34 **Abecedarian early childhood education program: Adult outcomes.** The Abecedarian Project is a randomized controlled study to assess the effects of intensive early childhood education on children who were at high risk based on several sociodemographic indicators.[28] The project randomly assigned children to a treatment group that was provided with early educational activities before kindergarten and the remainder to a control group. A recent follow-up study interviewed subjects at age 30 and evaluated educational, economic, and social-emotional outcomes to learn if the positive effects of the program continued into adulthood. The follow-up study included 52 individuals from the treatment group and 49 from the control group. Out of these, 39 from the treatment group and 26 from the control group were considered "consistently" employed (working 30+ hours per week in at least 18 of the 24 months prior to the interview). Does the study provide significant evidence that children who had early childhood education have a higher proportion of consistent employment than those who did not? How large is the difference between the proportions in the two populations that are consistently employed? Do inference to answer both questions. Be sure to explain exactly what inference you choose to do.

18.35 **Hand sanitizers.** Hand disinfection is frequently recommended for prevention of transmission of the rhinovirus that causes the common cold. In particular, hand lotion containing 2% citric acid and 2% malic acid in 70% ethanol (HL+) has been found to have both immediate and persistent ability to inactivate rhinovirus (RV) on the hands in an experimental setting. Is hand disinfection effective in reducing the risk of infection in a natural setting? A total of 212 volunteers were assigned at random to either the HL+ group, which used the hand lotion every three hours or after hand washing, and a control group, which was asked to use routine hand washing but to avoid the use of alcohol-based hand sanitizers. Here are the data on the numbers of subjects with and without RV infection in the two groups over the 10-week study period:[29]

	RV INFECTION	
	YES	NO
HL+	49	67
Control group	49	47

(a) Is this an experiment or an observational study? Why?

(b) Do the data give good evidence that hand sanitizers reduce the chance of an RV infection?

 Exploring the Web

18.36 **Hearing loss in adolescents.** Go to the *Journal of the American Medical Association* Web site, `http://jama.ama-assn.org/content/by/year`, and find the article "Change in Prevalence of Hearing Loss in US Adolescents" by Shargorodsky et al. in the August 18, 2010, issue. If you cannot get the full text of the article, use the information in the abstract plus the information given here to answer the questions. NHANES III is the earlier sample, and NHANES 2005–2006 is the more recent sample.

(a) Is this an observational study or an experiment?

(b) How many people were in the earlier sample, and how many were in the later sample?

(c) If you do not have access to Table 2 of the full article, here are the facts that you will need: in the earlier study 480 people experienced some hearing loss, whereas in the later study 333 people experienced some hearing loss. Is there evidence of an increase in hearing loss for children aged 12 to 19 in the later study? State hypotheses, find the test statistic, and use either software or Table A to compute the P-value. Although the article used a more sophisticated analysis, your P-value should be quite close to the P-value of 0.02 reported in the abstract.

18.37 **Compare two surveys.** Go to the Web site `www.pollingreport.com`, which contains the results of surveys conducted by several survey organizations. Choose a topic of interest to you, and then, to see if attitudes have changed over time, find two surveys that were conducted at two different times but that ask the same question. For example, you might choose the topic of abortion and compare the percents of people who feel abortion should always be legal at points in time separated by several years. State hypotheses to check for a difference over time, find the test statistic, and use either software or Table A to compute the P-value. What is your conclusion in context?

Overview

In Chapter 15, we introduced the law of large numbers and the sampling distribution for proportions. Both of these results are applications of the more general results presented in this chapter, which parallel the coverage of Chapter 15, but in the context of means rather than proportions. We begin with the law of large numbers in its usual form, which says that $\bar{x}$ gets closer to the population mean as we take larger and larger samples. The sampling distribution of $\bar{x}$ is centered at the population mean with variance $\dfrac{\sigma^2}{n}$; if the population is Normal, the sampling distribution of $\bar{x}$ is also Normal.

More generally, the approximate sampling distribution for $\bar{x}$ is based on the central limit theorem, which states the sampling distribution of $\bar{x}$ will be approximately Normal, no matter what population the sample is taken from, provided the sample size is large enough. This should not be confused with the fact that if the population has a Normal distribution, the sample mean has a Normal distribution for all sample sizes (a distinction we will return to in the following chapters). The variability in the sampling distribution of $\bar{x}$ decreases as the sample size increases, making the sample mean $\bar{x}$ a more precise estimator of the population mean μ.

Teaching Suggestions and Additional Examples/ Activities for the Classroom

1. Remind Students about Sampling Variability

It is important for you to remind your students about sampling variability. As with proportions, this is probably best done using simulation—students should simulate for themselves, using the *Central Limit Theorem* applet described later. Consider having students explore the sampling distribution of the sample mean using this applet.

2. Show Students the Central Limit Theorem in Action

At some point, you'll want to allow students to "discover" the central limit theorem, constructing this idea themselves, rather than presenting it to them. Remember that sampling distributions are difficult for beginning students to understand, and the central limit theorem's formality is especially difficult. However, your students have experience with sampling distributions from Chapter 15, so the formality of the central limit theorem should not be problematic. You can demonstrate the central limit theorem by providing them with the results of a series of simulations you've done before class—start with a heavily skewed population, then examine the sampling distribution of the sample mean for increasing sam-

> ### LEARNING OUTCOMES**
> ■ Identify parameters and statistics in a statistical study.
> ■ Recognize the fact of sampling variability: a statistic will take different values when you repeat a sample or experiment.
> ■ Interpret a sampling distribution as describing the values taken by a statistic in *all* possible repetitions of a sample or experiment under the same conditions (i.e., the sample you observe and all other possible samples).
> ■ Interpret the sampling distribution of a statistic as describing the probabilities of its possible values.
> ■ Recognize when a problem involves the mean $\bar{x}$ of sample. Understand that $\bar{x}$ estimates the mean μ of the population from which the sample is drawn.
> ■ Use the law of large numbers to describe the behavior of $\bar{x}$ as the size of the sample increases.
> ■ Find the mean and standard deviation of the sample mean $\bar{x}$ from an SRS of size n when the mean μ and standard deviation σ of the population are known.
> ■ Understand that $\bar{x}$ is an unbiased estimator of μ and that the variability of μ about its mean μ gets smaller as the sample size increases.
> ■ Understand that $\bar{x}$ has approximately a Normal distribution when the sample is large (central limit theorem). Use this Normal distribution to calculate probabilities that concern $\bar{x}$.
>
> **These learning outcomes appear later for the students in Chapter 23: Part IV Review.

ple sizes. The objective here is that students recognize for themselves that the sampling distributions become more symmetric as the sample size increases, and that when the sample is large enough, it is approximately Normal. If students discover this for themselves, they will have no trouble generalizing this to other populations. They should also easily grasp that if the population is fairly symmetric, the sampling distribution of the sample mean will converge to Normal more quickly. You might also want to have your students experiment with David Lane's sampling distribution applet (see Exercise 19.33 on page 433) or the *Central Limit Theorem* applet on the companion Web site.

The central limit theorem and the law of large numbers play key roles in industries such as casino gambling and insurance. You can discuss how small house advantages in games of chance translate to large "guaranteed" profits for casinos (or insurance companies) if enough games are played. What's more, the distribution of individual wager outcomes is irrelevant—if enough games are played, the sample mean profit per wager for a casino in a given night is Normally distributed, with very small standard deviation. (You might remind your students that this is the reason they will not become rich at a casino!)

3. Emphasize the Roles of Sample Size and Population Variability in Sampling Distributions

Have students work some example problems computing probabilities for a sample mean in a simple setting. For example, if the mean contents of potato chip bags is 14.1 ounces with standard deviation 0.2 ounces (better still, have students suggest a reasonable value for standard deviation), what is the probability that a random sample of 100 bags will average less than the labeled 14.0 ounces in contents? Work out said problem, then (importantly) have them consider the following questions: (1) What will happen to their answer if the sample size increases or decreases? (2) What would happen if the population standard deviation increases or decreases? and (3) Does it matter whether the population distribution of individual bag contents is Normal? Why? Would it matter if the sample were only 10 bags? These sorts of questions get students to think about the roles of sample size and population variability in the sampling distribution of the sample mean.

Other Resources (LaunchPad)

StatClips
Sampling Distributions: Overview and Motivation, Parts I and II
Central Limit Theorem, Parts I, II, and III

Snapshots Videos
Sampling Distributions and CLT

Applets
Law of Large Numbers
Central Limit Theorem

age fotostock/SuperStock

CHAPTER 19

Sampling Distribution for a Mean

In this chapter we cover...

- Estimation of a mean and the law of large numbers
- The sampling distribution of $\bar{x}$
- The central limit theorem
- The connection to proportions*

What is the average income of American households? Each March, the government's Current Population Survey asks detailed questions about income. The 96,659 households contacted in March 2012 had a mean "total money income" of \$69,677 in 2011.[1] The mean income of the sample of 96,659 households contacted by the Current Population Survey is a *statistic* because it describes this one Current Population Survey sample. The population that the poll wants to draw conclusions about is all 121 million U.S. households, and the *parameter* of interest is the mean income of all these households. This is another example of statistical inference in which we want to use information from a sample to infer something about a wider population.

The reasoning of statistical inference remains the same and rests on asking, "How often would this method give a correct answer if I used it very many times?" As with proportions, our two methods of inference are confidence intervals and hypothesis tests. This chapter presents the results on the sampling distribution of the mean, which are necessary to develop confidence intervals and hypothesis tests.

Estimation of a mean and the law of large numbers

population mean μ
population standard deviation σ

sample mean x̄
sample standard deviation s

In our study of inference for means, we need our notation to distinguish between parameters and statistics. We write μ (the Greek letter mu) for the **mean of a population** and σ (the Greek letter sigma) for the **standard deviation of a population.** These are the two parameters that we will use throughout this chapter. The **mean of the sample** is the familiar $\bar{x}$, the average of the observations in the sample. The **standard deviation of the sample** is denoted by s, the standard deviation of the observations in the sample. The sample mean $\bar{x}$ and sample standard deviation s from a sample or an experiment are estimates of the mean μ and standard deviation σ of the underlying population. Here is an example that we will use to motivate many of the ideas in this chapter.

EXAMPLE 19.1	Does This Wine Smell Bad?

Jupiterimages/Getty Images

Sulfur compounds such as dimethyl sulfide (DMS) are sometimes present in wine. DMS causes "off-odors" in wine, so winemakers want to know the odor threshold, the lowest concentration of DMS that the human nose can detect. Different people have different thresholds, so we start by asking about the mean threshold μ in the population of all adults. The number μ is a parameter that describes this population.

To estimate μ, we present tasters with both natural wine and the same wine spiked with DMS at different concentrations to find the lowest concentration at which they identify the spiked wine. Here are the odor thresholds (measured in micrograms of DMS per liter of wine) for 10 randomly chosen subjects:

$$28 \quad 40 \quad 28 \quad 33 \quad 20 \quad 31 \quad 29 \quad 27 \quad 17 \quad 21$$

Because an SRS should fairly represent the population, it seems reasonable to use the mean threshold for these subjects, $\bar{x} = 27.4$, to estimate the unknown μ. ∎

When we use a sample statistic to estimate a parameter, an important property of the statistic is that if we keep on taking larger and larger samples, the statistic is guaranteed to get closer and closer to the parameter. In Example 19.1, we hope that if we can afford to keep on measuring more subjects, eventually we will estimate the mean odor threshold of all adults very accurately. This important property is true when we use the sample mean to estimate the population mean and is a consequence of a remarkable fact called the *law of large numbers*. It is remarkable because it holds for *any* population, not just for some special class such as Normal distributions.

Law of Large Numbers

Draw observations at random from any population with finite mean μ. As the number of observations drawn increases, the mean $\bar{x}$ of the observed values gets closer and closer to the mean μ of the population.

The law of large numbers can be proved mathematically starting from the basic laws of probability, and we have seen an application of this law when studying proportions. Just as the sample proportion gets close to the population proportion in the long run, the *sample average* gets close to the population mean. Here is an example of how sample means approach the population mean.

EXAMPLE 19.2   The Law of Large Numbers in Action

In fact, the distribution of odor thresholds among all adults has mean 25. The mean $\mu = 25$ is the true value of the parameter we seek to estimate. Figure 19.1 shows how the sample mean $\bar{x}$ of an SRS drawn from this population changes as we add more subjects to our sample.

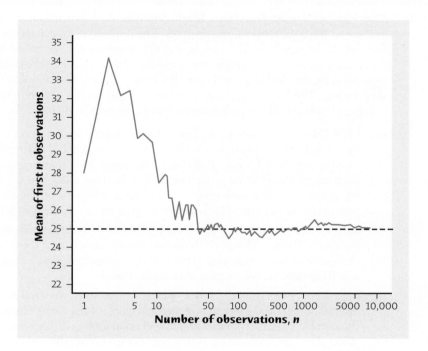

FIGURE 19.1

The law of large numbers in action: as we take more observations, the sample mean $\bar{x}$ always approaches the mean μ of the population.

The first subject in Example 19.1 had threshold 28, so the line in Figure 19.1 starts there. The mean for the first two subjects is

$$\bar{x} = \frac{28 + 40}{2} = 34$$

This is the second point on the graph. At first, the graph shows that the mean of the sample changes as we take more observations. Eventually, however, the mean of the observations gets close to the population mean $\mu = 25$ and settles down at that value.

If we started over, again choosing people at random from the population, we would get a different path from left to right in Figure 19.1. The law of large numbers says that whatever path we get will always settle down at 25 as we draw more and more people. ■

The *Law of Large Numbers* applet animates Figure 19.1 in a different setting. You can use the applet to watch $\bar{x}$ change as you average more observations until it eventually settles down at the mean μ.

The law of large numbers is the foundation of such business enterprises as gambling casinos and insurance companies. The winnings (or losses) of a gambler on a few plays are uncertain—that's why some people find gambling exciting. In

Figure 19.1, the mean of even 100 observations is not yet very close to μ. It is only *in the long run* that the mean outcome is predictable. The casino plays tens of thousands of times. So the casino, unlike individual gamblers, can count on the long-run regularity described by the law of large numbers. The average winnings of the casino on tens of thousands of plays will be very close to the mean of the distribution of winnings. Needless to say, this mean guarantees the casino a profit. That's why gambling can be a business.

Apply Your Knowledge

19.1 **The Law of Large Numbers Made Visible.** Roll two balanced dice, and count the total spots on the up-faces. The probability model appears in Example 11.5 (page 265). You can see that this distribution is symmetric with 7 as its center, so it's no surprise that the mean is $\mu = 7$. This is the population mean for the idealized population that contains the results of rolling two dice forever. The law of large numbers says that the average $\bar{x}$ from a finite number of rolls gets closer and closer to 7 as we do more and more rolls.

(a) Click "More dice" once in the *Law of Large Numbers* applet to get two dice. Click "Show mean" to see the mean 7 on the graph. Leaving the number of rolls at 1, click "Roll dice" three times. How many spots did each roll produce? What is the average for the three rolls? You see that the graph displays at each point the average number of spots for all rolls up to the last one. This is similar to what we saw in Figure 19.1.

(b) Set the number of rolls to 100, and click "Roll dice." The applet rolls the two dice 100 times. The graph shows how the average count of spots changes as we make more rolls. That is, the graph shows $\bar{x}$ as we continue to roll the dice. Sketch (or print out) the final graph.

(c) Repeat your work from (b). Click "Reset" to start over, then roll two dice 100 times. Make a sketch of the final graph of the mean $\bar{x}$ against the number of rolls. Your two graphs will often look very different. What they have in common is that the average eventually gets close to the population mean $\mu = 7$. The law of large numbers says that this will *always* happen if you keep on rolling the dice.

19.2 **What's the Mean?** Suppose you roll five balanced dice. We wonder what the mean number of spots on the up-faces of the five dice is. The law of large numbers says that we can find out by experience: roll five dice many times, and the average number of spots will eventually approach the true mean. Set up the *Law of Large Numbers* applet to roll five dice. Don't click "Show mean" yet. Roll the dice until you are confident you know the mean quite closely, then click "Show mean" to verify your discovery. What is the mean? Make a rough sketch of the path the averages $\bar{x}$ followed as you kept adding more rolls.

19.3 **Insurance.** The idea of insurance is that we all face risks that are unlikely but carry high cost. Think of a fire or flood destroying your apartment. Insurance spreads the risk: we all pay a small amount, and the insurance policy pays a large amount to those few of us whose apartments are damaged. An insurance company looks at the records for millions of apartment owners and sees that the mean loss from apartment damage in a year is $\mu = 75$ per person. (Most of us have no loss, but a few lose most of their possessions. The $75 is the average loss.) The company plans to sell insurance for $75 plus enough to cover its costs and profit. Explain clearly why it would be unwise to sell only 12 policies. Then explain why selling thousands of such policies is a safe business.

The sampling distribution of $\bar{x}$

Although the law of large numbers tells us we can estimate the mean odor threshold very precisely if we sample many subjects, we want to know what can be said about estimating μ by $\bar{x}$ from a sample of 10 subjects as in Example 19.1 (page 416). The sampling distribution of $\bar{x}$ addresses this by answering the question, "What would happen if we took many samples of 10 subjects from this population?" We begin with a simulation in which software is used to imitate the process of taking many samples.

EXAMPLE 19.3   Simulating the Sampling Distribution of $\bar{x}$

Extensive studies have found that the DMS odor threshold of adults follows roughly a Normal distribution with mean $\mu = 25$ micrograms per liter and standard deviation $\sigma = 7$ micrograms per liter. We call this the *population distribution* of odor threshold.

Figure 19.2 illustrates the process of simulating many samples and finding the sample mean threshold $\bar{x}$ for each one. Follow the flow of the figure from the population at the left, to simulating an SRS and finding the $\bar{x}$ for this sample, to collecting together the $\bar{x}$'s from many samples. The first sample has $\bar{x} = 27.40$. The second sample contains a different 10 people, with $\bar{x} = 24.28$, and so on. The histogram at the right of the figure shows the distribution of the values of $\bar{x}$ from 1000 simulated SRSs of size 10. The simulated distribution obtained from a fixed number of trials, like the 1000 trials in Figure 19.2, is only an approximation to the true sampling distribution. The solid Normal curve superimposed over the histogram in Figure 19.2 can be proven mathematically to be the exact sampling distribution of $\bar{x}$ for a sample of size 10 from this population. Although we would use this exact distribution in practice, thinking about the simulation process helps us understand what the sampling distribution represents. ■

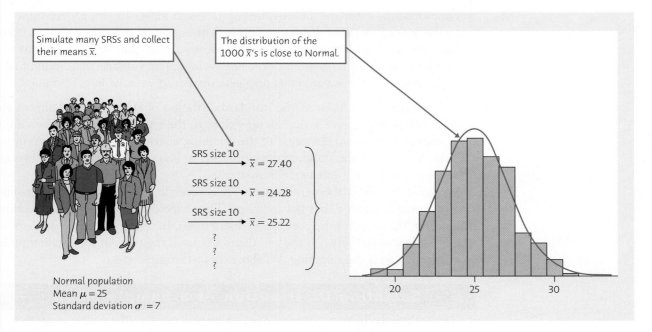

FIGURE 19.2

Simulating the sampling distribution of the sample mean: simulate many samples from the population in Example 19.3, collect the $\bar{x}$'s from all the samples, and display the distribution of the $\bar{x}$'s. The histogram shows the results of 1000 samples with the true sampling distribution superimposed on the histogram.

Population Distribution, Sampling Distribution

The **population distribution** of a variable is the distribution of values of the variable among all the individuals in the population.

The **sampling distribution** of a statistic is the distribution of values taken by the statistic in all possible samples of the same size from the same population.

 Be careful: The population distribution describes the *individuals* that make up the population. A sampling distribution describes how a *statistic* varies in many samples from the population. Here are two important facts about the sampling distribution of $\bar{x}$.

Mean and Standard Deviation of a Sample Mean[2]

Suppose that $\bar{x}$ is the mean of an SRS of size n drawn from a large population with mean μ and standard deviation σ. Then the sampling distribution of $\bar{x}$ has **mean μ** and **standard deviation $\sigma/\sqrt{n}$**.

These facts about the mean and the standard deviation of the sampling distribution of $\bar{x}$ are true for *any* population, not just for some special class such as Normal distributions. They have important consequences for making inferences about the population mean:

■ The mean of the statistic $\bar{x}$ is always equal to the mean μ of the population. Although $\bar{x}$ will sometimes fall above the true value of the parameter μ and sometimes below, there is no systematic tendency to overestimate or underestimate the parameter. We say that the statistic $\bar{x}$ is an *unbiased estimator* of the parameter μ.

■ How close the estimator falls to the parameter in most samples is determined by the spread/variability of the sampling distribution. If individual observations have standard deviation σ, then sample means $\bar{x}$ from samples of size n have standard deviation $\sigma/\sqrt{n}$. Not only are averages less variable than individual observations, but this variability also gets smaller as we take larger samples.

If the sample size n is large, the standard deviation of $\bar{x}$ is small, and almost all samples will give values of $\bar{x}$ that lie very close to the true parameter μ. As with a proportion, the standard deviation of the sampling distribution gets smaller only at the rate $\sqrt{n}$. To cut the standard deviation of $\bar{x}$ in half, we must take four times as many observations, not just twice as many.

We have described the center and spread/variability of the sampling distribution of a sample mean $\bar{x}$, but not its shape. The shape of the sampling distribution depends on the shape of the population distribution. In one important case there is a simple relationship between the two distributions: if the population distribution is Normal, then so is the sampling distribution of the sample mean.

Sampling Distribution of a Sample Mean

If individual observations have the $N(\mu, \sigma)$ distribution, then the sample mean $\bar{x}$ of an SRS of size n has the $N(\mu, \sigma/\sqrt{n})$ distribution.

Notice that if the population distribution is Normal, then the sampling distribution of the sample mean is Normal *regardless of the sample size n*. Because the population of DMS thresholds is Normal, we were able to draw a Normal curve as the sampling distribution of $\bar{x}$ in Figure 19.2.

EXAMPLE 19.4   Population Distribution, Sampling Distribution

If we measure the DMS odor thresholds of individual adults, the values follow the Normal distribution with mean $\mu = 25$ micrograms per liter and standard deviation $\sigma = 7$ micrograms per liter. This is the population distribution of odor threshold.

Take many SRSs of size 10 from this population and find the sample mean $\bar{x}$ for each sample, as in Figure 19.2. The sampling distribution describes how the values of $\bar{x}$ vary among samples. That sampling distribution is also Normal, with mean $\mu = 25$ and standard deviation

$$\frac{\sigma}{\sqrt{n}} = \frac{7}{\sqrt{10}} = 2.2136$$

Figure 19.3 contrasts these two Normal distributions. Both are centered at the population mean, but sample means are much less variable than individual observations.

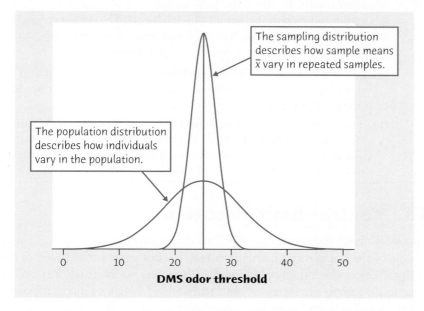

The sampling distribution describes how sample means $\bar{x}$ vary in repeated samples.

The population distribution describes how individuals vary in the population.

DMS odor threshold

FIGURE 19.3

The distribution of single observations (the population distribution) compared with the sampling distribution of the means $\bar{x}$ of 10 observations. Both have the same mean, but averages are less variable than individual observations.

The smaller variation of sample means shows up in probability calculations. You can show (using software or standardizing and using Table A) that about 52% of all adults have odor thresholds between 20 and 30. But almost 98% of means of samples of size 10 lie in this range. ■

Apply Your Knowledge

19.4 Sampling Distribution Versus Population Distribution. During World War II, 12,000 able-bodied male undergraduates at the University of Illinois participated in required physical training. Each student ran a timed mile. Their times followed the Normal distribution with mean 7.11 minutes and standard deviation 0.74 minute. An SRS of 100 of these students has mean time $\bar{x} = 7.15$ minutes. A second SRS of size 100 has mean $\bar{x} = 6.97$ minutes. After many SRSs, the many values of the sample mean $\bar{x}$ follow the Normal distribution with mean 7.11 minutes and standard deviation 0.074 minute.

(a) What is the population? What values does the population distribution describe? What is this distribution?

(b) What values does the sampling distribution of $\bar{x}$ describe? What is the sampling distribution?

19.5 **Larger Sample, More Accurate Estimate.** Suppose that in fact the blood cholesterol level of all men aged 20 to 34 follows the Normal distribution with mean $\mu = 186$ milligrams per deciliter (mg/dl) and standard deviation $\sigma = 41$ mg/dl.

(a) Choose an SRS of 100 men from this population. What is the sampling distribution of $\bar{x}$? What is the probability that $\bar{x}$ takes a value between 183 and 189 mg/dl? This is the probability that $\bar{x}$ estimates μ within ± 3 mg/dl.

(b) Choose an SRS of 1000 men from this population. Now what is the probability that $\bar{x}$ falls within ± 3 mg/dl of μ? The larger sample is much more likely to give an accurate estimate of μ.

19.6 **Measurements in the Lab.** Juan makes a measurement in a chemistry laboratory and records the result in his lab report. The standard deviation of students' lab measurements is $\sigma = 10$ milligrams. Juan repeats the measurement 4 times and records the mean $\bar{x}$ of his 4 measurements.

(a) What is the standard deviation of Juan's mean result? (That is, if Juan kept on making 4 measurements and averaging them, what would be the standard deviation of all his $\bar{x}$'s?)

(b) How many times must Juan repeat the measurement to reduce the standard deviation of $\bar{x}$ to 2? Explain to someone who knows no statistics the advantage of reporting the average of several measurements rather than the result of a single measurement.

The central limit theorem

The facts about the mean and standard deviation of $\bar{x}$ are true no matter what the shape of the population distribution may be. But what is the shape of the sampling distribution when the population distribution is not Normal? *It is a remarkable fact that as the sample size increases, the distribution of $\bar{x}$ changes shape: it looks less like that of the population and more like a Normal distribution.* When the sample is large enough, the distribution of $\bar{x}$ is very close to Normal. This is true no matter what shape the population distribution has, as long as the population has a finite standard deviation σ. This famous fact of probability theory is called the *central limit theorem*. It is much more useful than the fact that the distribution of $\bar{x}$ is exactly Normal if the population is exactly Normal.

WHAT WAS THAT PROBABILITY AGAIN?

Wall Street uses fancy mathematics to predict the probabilities that fancy investments will go wrong. The probabilities are always too low—sometimes because something was assumed to be Normal but was not. Probability predictions in other areas also go wrong. In mid-September 2007, the New York Mets had probability 0.998 of making the National League playoffs, or so an elaborate calculation said. Then the Mets lost 12 of their final 17 games, the Phillies won 13 of their final 17, and the Mets were out. Maybe someday?

Central Limit Theorem

Draw an SRS of size n from any population with mean μ and finite standard deviation σ. The **central limit theorem** says that when n is large, the sampling distribution of the sample mean $\bar{x}$ is approximately Normal:

$$\bar{x} \text{ is approximately } N\left(\mu, \frac{\sigma}{\sqrt{n}}\right)$$

The central limit theorem allows us to use Normal probability calculations to answer questions about sample means from many observations even when the population distribution is not Normal.

More general versions of the central limit theorem say that the distribution of any sum or average of many small random quantities is close to Normal. This is true even if the quantities are correlated with each other or have different distributions, provided certain technical conditions are satisfied. The central limit theorem suggests why the Normal distributions are common models for observed data. Any variable that is a sum of many small influences will have approximately a Normal distribution.

How large a sample size n is needed for $\bar{x}$ to be close to Normal depends on the population distribution. More observations are required if the shape of the population distribution is far from Normal. Here are two examples in which the population is far from Normal.

EXAMPLE 19.5   The Central Limit Theorem in Action

In March 2012, the Current Population Survey contacted 96,659 households. Figure 19.4(a) is a histogram of the earnings of the 73,366 households that had earned income greater than zero in 2011.[3] As we expect, the distribution of earned incomes is strongly skewed to the right and very spread out. The right tail of the distribution is even longer than the histogram shows because there are too few high incomes for their bars to be visible on this scale. In fact, we cut off the earnings scale at $400,000 to save space—a few households earned even more than $400,000. The mean earnings for these 73,366 households was $72,728.

Regard these 73,366 households as a population with mean $\mu = \$72,728$. Take an SRS of 100 households. The mean earnings in this sample is $\bar{x} = \$83,418$. That's higher than the mean of the population. Take another SRS of size 100. The mean for this sample is $\bar{x} = \$56,978$. That's less than the mean of the population. *What would happen if we did this many times?* Figure 19.4(b) is a histogram of the mean earnings for 500 samples, each of size 100. The scales in Figures 19.4(a) and 19.4(b) are the same, for easy comparison. Although the distribution of individual

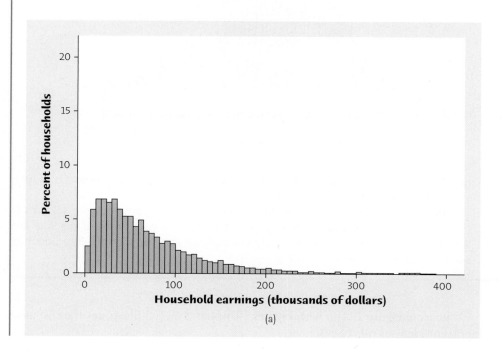

(a)

FIGURE 19.4

The central limit theorem in action, for Example 19.5. (a) The distribution of earned income in a population of 73,366 households. (b) The distribution of the mean earnings for 500 SRSs of 100 households each from this population. (c) The distribution of the sample means in more detail: the shape is close to Normal.

FIGURE 19.4
(Continued)

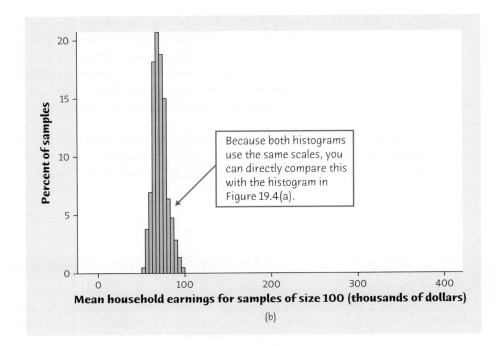

Because both histograms use the same scales, you can directly compare this with the histogram in Figure 19.4(a).

(b)

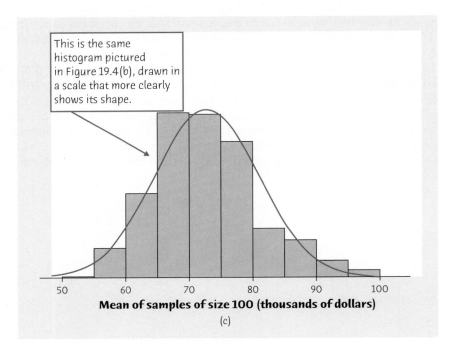

This is the same histogram pictured in Figure 19.4(b), drawn in a scale that more clearly shows its shape.

(c)

earnings is skewed and very spread out, the distribution of sample means is roughly symmetric and much less spread out.

Figure 19.4(c) zooms in on the center part of the histogram in Figure 19.4(b) to more clearly show its shape. Although $n = 100$ is not a very large sample size and the population distribution is extremely skewed, we can see that the distribution of sample means is close to Normal. ■

Comparing Figure 19.4(a) with Figures 19.4(b) and 19.4(c) illustrates the two most important ideas of this chapter.

Thinking About Sample Means

Means of random samples are **less variable** than individual observations.

Means of random samples are **more Normal** than individual observations.

EXAMPLE 19.6   The Central Limit Theorem in Action

Exponential distributions are used as models for the lifetime in service of electronic components and for the time required to serve a customer or repair a machine. Figure 19.5(a) shows the exponential population distribution, that is, the density curve of a single observation. This distribution is strongly right-skewed, and the most probable outcomes are near 0. The mean μ of this distribution is 1, and its standard deviation σ is also 1.

Using mathematics, we can derive the theoretical sampling distribution of $\bar{x}$ when sampling from an exponential distribution. Figures 19.5(b), (c), and (d) are the theoretical density curves of the sample means of 2, 10, and 25 observations from this population. As n increases, the shape becomes more Normal. The mean remains at $\mu = 1$, and the standard deviation decreases, taking the value $1/\sqrt{n}$. The density curve for 10 observations is still somewhat skewed to the right but already resembles a Normal curve having $\mu = 1$ and $\sigma = 1/\sqrt{10} = 0.32$. The density curve for $n = 25$ is yet more Normal. The contrast between the shapes of the population distribution and of the distribution of the mean of 10 or 25 observations is striking. ■

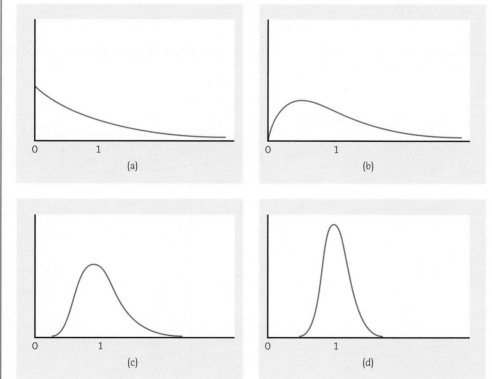

FIGURE 19.5

The central limit theorem in action. The distribution of sample means $\bar{x}$ from a strongly non-Normal population becomes more Normal as the sample size increases. (a) The distribution of 1 observation. (b) The distribution of $\bar{x}$ for 2 observations. (c) The distribution of $\bar{x}$ for 10 observations. (d) The distribution of $\bar{x}$ for 25 observations.

The *Central Limit Theorem* applet allows you to watch the central limit theorem in action. The Applet simulates the sampling distribution of $\bar{x}$ for several population distributions and you can see how the sampling distribution changes shape as the sample size increases.

Let's use Normal calculations based on the central limit theorem to answer a question about the very non-Normal distribution in Figure 19.5(a).

| EXAMPLE 19.7 | **Maintaining Air Conditioners** |

STATE: The time (in hours) that a technician requires to perform preventive maintenance on an air-conditioning unit is governed by the exponential distribution whose density curve appears in Figure 19.5(a). The exponential distribution arises in many engineering and industrial problems, such as time until failure of a machine or time until a success. The mean time is $\mu = 1$ hour, and the standard deviation is $\sigma = 1$ hour. Your company has a contract to maintain 70 of these units in an apartment building. You must schedule technicians' time for a visit to this building. Is it safe to budget an average of 1.1 hours for each unit? Or should you budget an average of 1.25 hours?

PLAN: We believe that the manufacturing and distribution process associated with this type of air-conditioning unit is such that variation from one unit to the next is random. Thus, we treat these 70 air conditioners as an SRS from all units of this type. What is the probability that the average maintenance time for 70 units exceeds 1.1 hours? That the average time exceeds 1.25 hours?

SOLVE: The central limit theorem says that the sample mean time $\bar{x}$ spent working on 70 units has approximately the Normal distribution with mean equal to the population mean $\mu = 1$ hour and standard deviation

$$\frac{\sigma}{\sqrt{70}} = \frac{1}{\sqrt{70}} = 0.12 \text{ hour}$$

The distribution of $\bar{x}$ is therefore approximately $N(1, 0.12)$. This Normal curve is the solid curve in Figure 19.6.

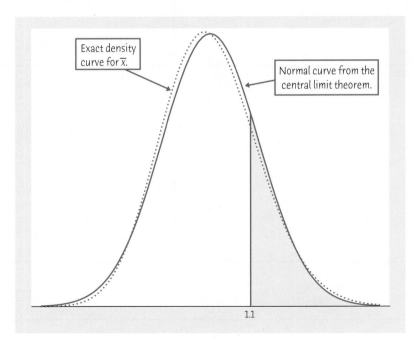

FIGURE 19.6
The exact distribution (dotted) and the Normal approximation from the central limit theorem (solid) for the average time needed to maintain an air conditioner. The probability we want is the area to the right of 1.1.

Using this Normal distribution, the probabilities we want are

$$P(\bar{x} > 1.10 \text{ hours}) = 0.2014$$
$$P(\bar{x} > 1.25 \text{ hours}) = 0.0182$$

Software gives these probabilities immediately, or you can standardize and use Table A. For example,

$$P(\bar{x} > 1.10) = P\left(\frac{\bar{x} - 1}{0.12}\right) > P\left(\frac{1.10 - 1}{0.12}\right)$$
$$= P(Z > 0.83) = 1 - 0.7967 = 0.2033$$

with the usual roundoff error. Don't forget to use standard deviation 0.12 in your software or when you standardize $\bar{x}$.

CONCLUDE: If you budget 1.1 hours per unit, there is a 20% chance that the technicians will not complete the work in the building within the budgeted time. This chance drops to about 2% if you budget 1.25 hours. You should therefore budget 1.25 hours per unit. ■

Using more mathematics, we can start with the exponential distribution and find the actual density curve of $\bar{x}$ for 70 observations. This is the dotted curve in Figure 19.6. You can see that the solid Normal curve is a good approximation. The exactly correct probability for 1.1 hours is an area to the right of 1.1 under the dotted density curve. It is 0.1977. The central limit theorem Normal approximation 0.2014 is off by only about 0.004.

Apply Your Knowledge

19.7 What Does the Central Limit Theorem Say? Asked what the central limit theorem says, a student replies, "As you take larger and larger samples from a population, the histogram of the sample values looks more and more Normal." Is the student right? Explain your answer.

19.8 Detecting Gypsy Moths. The gypsy moth is a serious threat to oak and aspen trees. A state agriculture department places traps throughout the state to detect the moths. When traps are checked periodically, the mean number of moths trapped is only 0.5, but some traps have several moths. The distribution of moth counts is finite and strongly skewed, with standard deviation 0.7.

© Bruce Coleman Inc./Alamy

   (a) What are the mean and standard deviation of the average number of moths $\bar{x}$ in 50 traps?

   (b) Use the central limit theorem to find the probability that the average number of moths in 50 traps is greater than 0.6.

19.9 More on Insurance. An insurance company knows that in the entire population of millions of apartment owners, the mean annual loss from damage is $\mu = \$75$ and the standard deviation of the loss is $\sigma = \$300$. The distribution of losses is strongly right-skewed: most policies have $0 loss, but a few have large losses. If the company sells 10,000 policies, can it safely base its rates on the assumption that its average loss will be no greater than $85? Follow the four-step process as illustrated in Example 19.7.

The connection to proportions*

In Chapter 15 we learned about using the sample proportion $\hat{p}$ to estimate the population proportion p. The results of that chapter are very similar to what we have just learned about means. There is a simple reason for this: a proportion is

*This short section is optional.

related to a mean for a population that contains only 1's and 0's. Here is the connection:

EXAMPLE 19.8   **Thinking of a Proportion as a Mean**

Let's illustrate the connection with a very small population consisting of the sex of 10 students in a class, where we will call a female a success. We can think of the students in the class as having an F or M associated with them to indicate their sex, or we can associate a numeric variable with each student, which is a 1 if the student is a female and a 0 if the student is a male. This is done in the following table.

Student	0	1	2	3	4	5	6	7	8	9
Sex	F	F	M	M	F	M	M	M	F	M
0 or 1	1	1	0	0	1	0	0	0	1	0

The parameter of interest is p, the proportion of females in this population. The numerical value of this parameter is $p = 0.4$ because 4 of the 10 students are female. If we define a numeric variable, which is a 1 if the student is a female and a 0 if the student is a male and compute the population mean μ for this variable, we find that $\mu = 0.4$ as well. The reason for this is that the sum of the 1's and 0's in the last line of the table is just the number of females in the population. When we divide this sum by the population size of 10 to get the population mean, the answer is the same as the proportion of females in the population.

Now suppose we sample four students at random and obtain the sample with students numbered 4, 2, 9, and 6. The sample proportion of females is $\hat{p} = 0.25$ because there is one female in the sample. If we take the sample mean of the 1's and 0's associated with students numbered 4, 2, 9, and 6, we get $\bar{x} = 0.25$. The sample proportion $\hat{p}$ is just the sample mean when we define the population variable as 1 if the student is a female and 0 if the student is a male. ■

The upshot of this is that when our interest is in a proportion, we can always think of the population as consisting of 1's for successes and 0's for failures. Because the law of large numbers says that the sample mean gets closer to the population mean for *any* population, a consequence of the law of large numbers is that the sample proportion gets closer to the population proportion. This is the law of large numbers for proportions on page 336. To see how the sampling distribution of $\hat{p}$ follows from the central limit theorem requires one additional fact.

The Population Standard Deviation for Proportions

Suppose we have a population consisting of two outcomes, where we assign the value 1 to the outcome we call a success and 0 to the outcome we call a failure. Then the population standard deviation is $\sigma = \sqrt{p(1 - p)}$, where p is the proportion of 1's, or successes, in the population.

The central limit theorem says that for *any* population,

$$\bar{x} \text{ is approximately } N\left(\mu, \frac{\sigma}{\sqrt{n}}\right)$$

If we have a population with two values, a 1 for a success and a 0 for a failure, then we know that $\bar{x}$ is the same as $\hat{p}$ and μ is the same as p. Making these substitutions in the general result for the central theorem, and replacing σ by $\sqrt{p(1-p)}$, it follows that

$$\hat{p} \text{ is approximately } N\left(p, \frac{\sqrt{p(1-p)}}{\sqrt{n}}\right)$$

This is the result for the sampling distribution of a proportion on page 420. We have shown that this result is just an application of the central limit theorem.

CHAPTER 19 SUMMARY

Chapter Specifics

- The **law of large numbers** states that the actual observed mean outcome $\bar{x}$ must approach the mean μ of the population as the number of observations increases.

- The **population distribution** of a variable describes the values of the variable for all individuals in a population.

- The **sampling distribution** of a statistic describes the values of the statistic in all possible samples of the same size from the same population.

- When the sample is an SRS from the population, the **mean** of the sampling distribution of the sample mean $\bar{x}$ is the same as the population mean μ. That is, $\bar{x}$ is an **unbiased estimator** of μ.

- The **standard deviation** of the sampling distribution of $\bar{x}$ is $\sigma/\sqrt{n}$ for an SRS of size n if the population has standard deviation σ. That is, averages are less variable than individual observations.

- When the sample is an SRS from a population that has a Normal distribution, the sample mean $\bar{x}$ also has a Normal distribution.

- Choose an SRS of size n from any population with mean μ and finite standard deviation σ. The **central limit theorem** states that when n is large, the sampling distribution of $\bar{x}$ is approximately Normal. That is, averages are more Normal than individual observations. We can use the $N(\mu, \sigma/\sqrt{n})$ distribution to calculate approximate probabilities for events involving $\bar{x}$.

Link It

In Chapters 15 through 18, we discussed inference for proportions. We began with the sampling distribution of $\hat{p}$. The properties of the sampling distribution helped us understand how the sample proportion could be used to draw conclusions about a population proportion. The ideas of confidence intervals and hypothesis tests were then introduced in the context of a single proportion. Both of these methods of inference make use of the sampling distribution of $\hat{p}$. We then extended this methodology to the comparison of two proportions. Our study of means will parallel our study of proportions.

   In this chapter we began to think about how a sample mean, $\bar{x}$, can provide information about a population mean, μ. The law of large numbers tells us that a sample mean computed from a *random* sample from some population gets closer and closer to the population mean as the sample size increases. The central limit theorem describes the sampling distribution of the sample mean for "large" SRSs and allows us to make probability statements about possible values of the sample mean. In Chapter 20, we will develop specific methods for obtaining confidence intervals and hypothesis tests about a population mean. These methods will use the tools developed in this chapter. Chapter 21 expands on these ideas and develops statistical methods for the comparison of two means.

CHECK YOUR SKILLS

19.10 Annual returns on the more than 5000 common stocks available to investors vary a lot. In a recent year, the mean return was 8.3%, and the standard deviation of returns was 28.5%. The law of large numbers says that

(a) you can get an average return higher than the mean 8.3% by investing in a large number of stocks.
(b) as you invest in more and more stocks chosen at random, your average return on these stocks gets ever closer to 8.3%.
(c) if you invest in a large number of stocks chosen at random, your average return will have approximately a Normal distribution.

19.11 Scores on the Critical Reading part of the SAT exam in a recent year were roughly Normal with mean 501 and standard deviation 112. You choose an SRS of 100 students and average their SAT Critical Reading scores. If you do this many times, the mean of the average scores you get will be close to

(a) 501.
(b) $501/100 = 5.01$.
(c) $501/\sqrt{100} = 50.1$.

19.12 Scores on the Critical Reading part of the SAT exam in a recent year were roughly Normal with mean 501 and standard deviation 112. You choose an SRS of 100 students and average their SAT Critical Reading scores. If you do this many times, the standard deviation of the average scores you get will be close to

(a) 112.
(b) $112/100 = 1.12$.
(c) $112/\sqrt{100} = 11.2$.

19.13 A newborn baby has extremely low birth weight (ELBW) if it weighs less than 1000 grams. A study of the health of such children in later years examined a random sample of 219 children. Their mean weight at birth was $\bar{x} = 810$ grams. This sample mean is an *unbiased estimator* of the mean weight μ in the population of all ELBW babies. This means that

(a) in many samples from this population, the mean of the many values of $\bar{x}$ will be equal to μ.
(b) as we take larger and larger samples from this population, $\bar{x}$ will get closer and closer to μ.
(c) in many samples from this population, the many values of $\bar{x}$ will have a distribution that is close to Normal.

19.14 The number of hours a battery lasts before failing varies from battery to battery. The distribution of failure times follows an exponential distribution, which is strongly skewed to the right (the exponential density curve is shown in Figure 19.5(a) on page 425). The central limit theorem says that

(a) as we look at more and more batteries, their average failure time gets ever closer to the mean μ for all batteries of this type.
(b) the average failure time of a large number of batteries has a distribution of the same shape (strongly skewed) as the distribution for individual batteries.
(c) the average failure time of a large number of batteries has a distribution that is close to Normal.

19.15 The length of human pregnancies from conception to birth varies according to a distribution that is approximately Normal with mean 266 days and standard deviation 16 days. The probability that the average pregnancy length for 6 randomly chosen women exceeds 270 days is about

(a) 0.40.   (b) 0.27.   (c) 0.07.

CHAPTER 19 EXERCISES

19.16 **Roulette.** A roulette wheel has 38 slots, of which 18 are black, 18 are red, and 2 are green. When the wheel is spun, the ball is equally likely to come to rest in any of the slots. One of the simplest wagers chooses red or black. A bet of $1 on red returns $2 if the ball lands in a red slot. Otherwise, the player loses his dollar. When gamblers bet on red or black, the two green slots belong to the house. Because the probability of winning $2 is 18/38, the mean payoff from a $1 bet is twice 18/38, or 94.7 cents. Explain what the law of large numbers tells us about what will happen if a gambler makes very many bets on red.

19.17 **Monsoon rains.** The summer monsoon rains in India follow approximately a Normal distribution with mean 852 millimeters (mm) of rainfall and standard deviation 82 mm. Rainfall is to be recorded each year for a decade and the mean rainfall $\bar{x}$ computed. What are the mean and standard deviation of $\bar{x}$, the mean rainfall per year?

19.18 **The Medical College Admission Test.** Almost all medical schools in the United States require students to take the Medical College Admission Test (MCAT). To estimate the mean score μ of those who took the MCAT on your campus, you will obtain the scores of an SRS of students. The scores follow a Normal distribution, and from published information you know that the standard deviation is 6.4. Suppose that (unknown to you) the mean score of those taking the MCAT on your campus is 25.0.

(a) If you choose one student at random, what is the probability that the student's score is between 20 and 30?

(b) You sample 25 students. What are the shape, center and variability of the sampling distribution of their average score $\bar{x}$?

(c) What is the probability that the mean score of your sample is between 20 and 30?

19.19 Glucose testing. Shelia's doctor is concerned that she may suffer from gestational diabetes (high blood glucose levels during pregnancy). There is variation both in the actual glucose level and in the blood test that measures the level. A patient is classified as having gestational diabetes if the glucose level is above 140 milligrams per deciliter (mg/dl) one hour after having a sugary drink. Shelia's measured glucose level one hour after the sugary drink varies according to the Normal distribution with $\mu = 122$ mg/dl and $\sigma = 12$ mg/dl.

(a) If a single glucose measurement is made, what is the probability that Shelia is diagnosed as having gestational diabetes?

(b) If measurements are made on 4 separate days and the mean result is compared with the criterion 140 mg/dl, what is the probability that Shelia is diagnosed as having gestational diabetes?

19.20 Daily activity. It appears that people who are mildly obese are less active than leaner people. One study looked at the average number of minutes per day that people spend standing or walking.[4] Among mildly obese people, the mean number of minutes of daily activity (standing or walking) is approximately Normally distributed with mean 373 minutes and standard deviation 67 minutes. The mean number of minutes of daily activity for lean people is approximately Normally distributed with mean 526 minutes and standard deviation 107 minutes. A researcher records the minutes of activity for an SRS of 5 mildly obese people and an SRS of 5 lean people.

(a) What is the probability that the mean number of minutes of daily activity of the 5 mildly obese people exceeds 420 minutes?

(b) What is the probability that the mean number of minutes of daily activity of the 5 lean people exceeds 420 minutes?

19.21 Glucose testing, continued. Shelia's measured glucose level one hour after having a sugary drink varies according to the Normal distribution with $\mu = 122$ mg/dl and $\sigma = 12$ mg/dl. What is the level L such that there is probability only 0.05 that the mean glucose level of 4 test results falls above L? (*Hint:* This requires a backward Normal calculation. See page 87 in Chapter 3 if you need to review.)

19.22 Pollutants in auto exhausts. Light vehicles sold in the United States must emit an average of no more than 0.07 grams per mile (g/mi) of nitrogen oxides (NOX). NOX emissions for one car model vary Normally with mean 0.05 g/mi and standard deviation 0.01 g/mi.

(a) What is the probability that a single car of this model emits more than 0.07 g/mi of NOX?

(b) A company has 25 cars of this model in its fleet. What is the probability that the average NOX level $\bar{x}$ of these cars is above 0.07 g/mi?

19.23 Runners. In a study of exercise, a large group of male runners walk on a treadmill for 6 minutes. After this exercise, their heart rates vary with mean 8.8 beats per five seconds and standard deviation 1.0 beats per five seconds. This distribution takes only whole-number values, so it is certainly not Normal.

Bruce Laurance/Getty Images

(a) Let $\bar{x}$ be the mean number of beats per five seconds after measuring heart rate for 12 five-second intervals (a minute). What is the approximate distribution of $\bar{x}$ according to the central limit theorem?

(b) What is the approximate probability that $\bar{x}$ is less than 8?

(c) What is the approximate probability that the heart rate of a runner is less than 100 beats per minute? (*Hint:* Restate this event in terms of $\bar{x}$.)

19.24 Pollutants in auto exhausts, continued. The level of nitrogen oxides (NOX) in the exhaust of cars of a particular model varies Normally with mean 0.05 g/mi and standard deviation 0.01 g/mi. A company has 25 cars of this model in its fleet. What is the level L such that the probability that the average NOX level $\bar{x}$ for the fleet is greater than L is only 0.01? (*Hint:* This requires a backward Normal calculation. See page 87 in Chapter 3 if you need to review.)

19.25 **Returns on stocks.** Andrew plans to retire in 40 years. He plans to invest part of his retirement funds in stocks, so he seeks out information on past returns. He learns that from 1960 to 2009, the annual returns on U.S. common stocks had mean 10.8% and standard deviation 17.1%.[5] The distribution of annual returns on common stocks is roughly symmetric, so the mean return over even a moderate number of years is close to Normal. What is the probability (assuming that the past pattern of variation continues) that the mean annual return on common stocks over the next 40 years will exceed 10%? What is the probability that the mean return will be less than 5%? Follow the four-step process as illustrated in Example 19.7 (page 426).

19.26 **Airline passengers get heavier.** In response to the increasing weight of airline passengers, the Federal Aviation Administration (FAA) in 2003 told airlines to assume that passengers average 190 pounds in the summer, including clothing and carry-on baggage. But passengers vary, and the FAA did not specify a standard deviation. A reasonable standard deviation is 35 pounds. Weights are not Normally distributed, especially when the population includes both men and women, but they are not very non-Normal. A commuter plane carries 22 passengers. What is the approximate probability that the total weight of the passengers exceeds 4500 pounds? Use the four-step process to guide your work. (*Hint:* To apply the central limit theorem, restate the problem in terms of the mean weight.)

© *Jeff Greenberg/The Image Works*

19.27 **Sampling students.** To estimate the mean score μ of those who took the Medical College Admission Test on your campus, you will obtain the scores of an SRS of students. From published information you know that the scores are approximately Normal with standard deviation about 6.4. How large an SRS

must you take to reduce the standard deviation of the sample mean score to 1?

19.28 **Sampling students, continued.** To estimate the mean score μ of those who took the Medical College Admission Test on your campus, you will obtain the scores of an SRS of students. From published information you know that the scores are approximately Normal with standard deviation about 6.4. You want your sample mean $\bar{x}$ to estimate μ with an error of no more than 1 point in either direction.

(a) What standard deviation must $\bar{x}$ have so that 99.7% of all samples give an $\bar{x}$ within 1 point of μ? (Hint: Use the 68–95–99.7 rule.)

(b) How large an SRS do you need to reduce the standard deviation of $\bar{x}$ to the value you found in part (a)?

19.29 **Playing the numbers.** The numbers racket is a well-entrenched illegal gambling operation in most large cities. One version works as follows: you choose one of the 1000 three-digit numbers 000 to 999 and pay your local numbers runner a dollar to enter your bet. Each day, one three-digit number is chosen at random and pays off $600. The mean payoff for the population of thousands of bets is $\mu = 60$ cents. Joe makes one bet every day for many years. Explain what the law of large numbers says about Joe's results as he keeps on betting.

19.30 **Playing the numbers: A gambler gets chance outcomes.** The law of large numbers tells us what happens in the long run. Like many games of chance, the numbers racket has outcomes so variable—one three-digit number wins $600 and all others win nothing—that gamblers never reach "the long run." Even after many bets, their average winnings may not be close to the mean. For the numbers racket, the mean payout for single bets is $0.60 (60 cents), and the standard deviation of payouts is about $18.96. If Joe plays 350 days a year for 40 years, he makes 14,000 bets.

(a) What are the mean and standard deviation of the average payout $\bar{x}$ that Joe receives from his 14,000 bets?

(b) The central limit theorem says that his average payout is approximately Normal with the mean and standard deviation you found in part (a). What is the approximate probability that Joe's average payout per bet is between $0.50 and $0.70? You see that Joe's average may not be very close to the mean $0.60 even after 14,000 bets.

19.31 **Playing the numbers: The house has a business.** Unlike Joe (see the previous exercise) the operators of the numbers racket can rely on the law of large numbers. It is said that the New York City mobster Casper Holstein took as many as 25,000 bets per day in the Prohibition era. That's 150,000 bets in a week if he takes Sunday off. Casper's mean winnings per bet are $0.40 (he pays out 60 cents of each dollar bet to people like Joe and keeps the other 40 cents.) His standard deviation for single bets is about $18.96, the same as Joe's.

NY Daily News via Getty Images

(a) What are the mean and standard deviation of Casper's average winnings $\bar{x}$ on his 150,000 bets?
(b) According to the central limit theorem, what is the approximate probability that Casper's average winnings per bet are between $0.30 and $0.50? After only a week, Casper can be pretty

confident that his winnings will be quite close to $0.40 per bet.

19.32 Can we trust the central limit theorem? The central limit theorem says that "when n is large" we can act as if the distribution of a sample mean $\bar{x}$ is close to Normal. How large a sample we need depends on how far the population distribution is from being Normal. Example 19.7 (page 426) shows that we can trust this Normal approximation for quite moderate sample sizes even when the population has a strongly skewed continuous distribution.

The central limit theorem requires much larger samples for Joe's bets with his local numbers racket. The population of individual bets has a finite distribution with only two possible outcomes: $600 (probability 0.001) and $0 (probability 0.999). This distribution has mean $\mu = 0.6$ and standard deviation about $\sigma = 18.96$. With more math and good software, we can find exact probabilities for Joe's average winnings.

(a) If Joe makes 14,000 bets, the exact probability $P(0.5 \le \bar{x} \le 0.7) = 0.4961$. How accurate was your Normal approximation from part (b) of Exercise 19.30?
(b) If Joe makes only 3500 bets, $P(0.5 \le \bar{x} \le 0.7) = 0.4048$. How accurate is the Normal approximation for this probability?
(c) If Joe and his buddies make 150,000 bets, $P(0.5 \le \bar{x} \le 0.7) = 0.9629$. How accurate is the Normal approximation?

 Exploring the Web

19.33 Sampling distributions in action. The sampling distribution Applet by David Lane at `www.onlinestatbook.com/stat_sim/sampling_dist/index.html` allows you to explore various aspects of sampling distributions. You will use the Applet to investigate the effect of both the population distribution and the sample size on the sampling distribution of the mean.

(a) When you begin the Applet, four histograms are shown. The histogram at the top will be a Normal distribution with mean 16 and standard deviation 5. The third histogram will be set to simulate the sampling distribution of $\bar{x}$ for a sample size of 5. Change the sample size to 2, and highlight the box *Fit normal*. In the boxes to the right of the bottom histogram, use the pull-down menu to set the statistic to the mean and the sample size to 16, and highlight the box *Fit normal*. Finally, in the box next to the second histogram, click on 10000, and the Applet will simulate 1000 samples of size 2 and then 10000 samples of size 16, both from the $N(16, 5)$ distribution. For each sample size, a histogram of the 10000 values of $\bar{x}$ will be drawn. Because we are sampling from a Normal distribution, we know the sampling distribution of $\bar{x}$ is Normal. What are the mean and standard deviation of the sampling distribution of $\bar{x}$

for each of the two sample sizes? The information to the left of each sampling distribution histogram gives the mean and standard deviation of the 10000 values of $\bar{x}$ that were simulated. How do they compare with the theoretical values?

(b) To the right of the top histogram, click on *Clear lower 3*, and change the distribution of the population (parent) distribution to *Skewed*. The skewed distribution has a mean of 8.08 and a standard deviation of 6.22. Keep the samples sizes at $n = 2$ and $n = 16$. Again, simulate 10000 values of $\bar{x}$ from this skewed distribution and examine the histograms of the values of $\bar{x}$ for samples of size 2 and samples of size 16. For which sample size is the sampling distribution closer to a Normal? Why is that? What can you say about the comparison of the mean and standard deviation of the sampling distribution of $\bar{x}$ for samples of size 2 and 16 from this skewed distribution?

(c) To the right of the top histogram, click on *Clear lower 3*, and change the distribution of the parent distribution to *Custom*. Use the mouse to draw a distribution with two peaks, and answer the questions from part (b) about the sampling distribution of $\bar{x}$ for the distribution you have created.

19.34   **Work the law of large numbers.** Read the online article at `ezinearticles.com/?Work-The-Law-of-Large-Numbers-But-Remember-It-Only-Takes-One-to-Succeed!&id=932026`. Does this article accurately describe the law of large numbers? Explain your answer.

Inference about a Population Mean   CHAPTER 20

Overview

In this chapter, students are exposed to realistic applications of statistical inference about a population mean. We can go directly to the *t* procedures for confidence intervals and hypothesis tests for a population mean because students have already worked with confidence intervals and hypothesis tests for a population proportion (Chapters 16 and 17) and have seen the sampling distribution for a sample mean in Chapter 19. The new idea in Chapter 20 is that we no longer use *z* procedures to do inference—students are introduced to *t* distributions. The authors state that if we knew the population standard deviation σ, we could use the *z* procedures. Because we almost always do not, we use *t* procedures. The explanation on text pages 437–439 is quite nice.

The *t* distribution is similar to the standard Normal distribution in that it has mean 0 and is symmetric and bell shaped. However, the *t* distribution is more variable than the standard Normal (heavier tails), and this variability depends on the sample size through *degrees of freedom*. One upshot of this is that use of the *t* distribution will result in wider confidence intervals and lower power in hypothesis testing than if we knew the population standard deviation σ and could use *z* procedures. However, when σ is unknown, our inference should reflect this additional uncertainty.

Most students will find that the procedures of this chapter are easily learned because they are similar to those initially taught in Chapters 16 and 17. For this reason, this chapter serves as an opportunity to review previous concepts, even as new material is covered. Even the matched pairs procedure covered at the end of the chapter should not be taught as a new topic, but rather as an extension of what students have already seen. This material serves as a prelude to the material to be covered in Chapter 21, concerning inference for the difference between two means.

The *t* procedures are based on an assumption that the population being sampled from is Normal, at least for small sample sizes. (See "Conditions for Inference about a Mean" on page 436.) However, the chapter ends with a discussion of the robustness of *t* procedures (beginning on page 450), emphasizing that they perform well even if data is non-Normal, unless extreme outliers or skewness are present. It is this robustness that makes the *t* procedures so applicable.

Teaching Suggestions and Additional Examples/Activities for the Classroom

1. Help Students Differentiate Between Working with Means and Proportions

At this stage of the course, many students will begin to become confused about when different procedures should be applied. Some will look for rules to apply, such as "use *z* procedures for proportions and *t* procedures for means." Focusing on rules like this may make life easier for weaker students in the short run, but it will do little to develop any understanding of how these procedures are related, or why they are appropriate. You should work hard not to focus on rules like this during class.

As you work examples using t procedures, you should work to make sure students can articulate what is being assumed to apply the procedure. When we use the z procedures for inference for proportions, it is because we use the Normal approximation to the binomial distribution, provided certain conditions are satisfied. (Note: it is okay if you do not cover Chapter 13 with your students. As long as you know which theory is being used where, that will help your students.) When we use t procedures, we assume the sample size is large (and use the central limit theorem to ensure an approximately Normal sampling distribution) or that the population we're sampling from is Normal or nearly Normal, or at least that it is not heavily skewed and lacks outliers. Students should phrase these in the context of the problem.

Consider starting with a discussion similar to that of Example 20.1 on page 437. This will lead naturally into the motivation for using the t procedures. Your students should be able to combine their knowledge of the process of confidence intervals and hypothesis tests for proportions with this new distribution for inference for means in a way that makes sense.

2. Use Class Data to Determine If Data Can Be Considered Normal

Work an example using class data, if the class is small enough to manage it. For example, if you ask your students to count their pulse rate over a minute, use of a t procedure is probably appropriate. If you ask students how many pairs of shoes they own, it may not be appropriate because there are likely to be one or two very large outliers. You can inspect the data to check this and then work the example. Make sure to take this opportunity to remind the students that they are certainly not a random sample from the population of students at your school and to remind students that proper sampling is paramount. These two examples provide you with the opportunity to say more: in both examples, the measurements are integers, so neither distribution can actually be Normal. The issue is whether either distribution looks approximately Normal (or enough so that it is safe to use t procedures).

3. Use a Class Project to Construct Confidence Intervals and Address Sampling Issues

If you have time, consider the following more involved class project: Take a large collection of pennies to class—say 1000 of them. Shake them, and then have each student "randomly" select (say) 10 or 15. Have each student construct a 99% confidence interval for the average age of pennies in the bucket. More interestingly, if you shake the bucket enough, you'll notice that most confidence intervals will significantly underestimate the average age of pennies. This is because newer pennies (minted after 1983) are lighter, made from zinc rather than copper. Shaking the pennies causes heavier (older) pennies to lie at the bottom of the bucket and be less likely to be selected by students, who will tend to simply select 10 conveniently located pennies.

Other Resources (LaunchPad)

StatClips
One-Sample Hypothesis Testing
Confidence Intervals: Interpretation

Snapshots Videos
Statistical Inference
Inference for One Mean
Confidence Intervals
Hypothesis Tests

© Tim Tadder/Corbis

Inference about a Population Mean

This chapter describes confidence intervals and significance tests for the mean μ of a population. We develop these procedures in a manner similar to the one we followed for a population proportion p in Chapters 16 and 17. The sampling distribution of a sample proportion $\hat{p}$ was the basis for inference for a population proportion. In this chapter, the sampling distribution for a sample mean $\bar{x}$, discussed in Chapter 18, forms the basis for inference for a population mean.

Conditions for inference about a mean

Confidence intervals and tests of significance for the mean μ of a Normal population are based on the sample mean $\bar{x}$. Confidence intervals and P-values involve probabilities calculated from the sampling distribution of $\bar{x}$. Here are the conditions needed for realistic inference about a population mean.

Conditions for Inference about a Mean

▪ We can regard our data as a **simple random sample** (SRS) from the population. This condition is very important.

▪ Observations from the population have a **Normal distribution** with mean μ and standard deviation σ. In practice, it is enough that the distribution be symmetric and single peaked unless the sample is very small. Both μ and σ are unknown parameters.

▪ For observations from a population with moderate skewness and no outliers, the sample size should be at least 15.

▪ For observations from a population with strong skewness and no outliers, the sample size should be at least 40.

The last two conditions concerning the sample size will be discussed in more detail at the end of this chapter.

There is another condition that applies to all the inference methods in this book: *the population must be much larger than the sample, say at least 20 times as large.*[1] All our examples and exercises satisfy this condition. Practical settings in which the sample is a large part of the population are rather special, and we will not discuss them.

When the conditions for inference are satisfied, we learned in Chapter 19 that the sample mean $\bar{x}$ has the Normal distribution (or has approximately the Normal distribution) with mean μ and standard deviation $\sigma/\sqrt{n}$. In Chapter 19 we assumed that we knew the population standard deviation σ and thus were able to compute probabilities for a sample mean. In practice we do not know σ. Because we don't know σ, we estimate it by the sample standard deviation s. We then estimate the standard deviation of $\bar{x}$ by $s/\sqrt{n}$. This quantity is called the *standard error* of the sample mean $\bar{x}$.

Standard Error

When the standard deviation of a statistic is estimated from data, the result is called the **standard error** of the statistic. The standard error of the sample mean $\bar{x}$ is $s/\sqrt{n}$.

Apply Your Knowledge

20.1 Travel Time to Work. A study of commuting times reports the travel times to work of a random sample of 1000 employed adults. The mean is $\bar{x} = 49.2$ minutes, and the standard deviation is $s = 63.9$ minutes. What is the standard error of the mean?

20.2 Comparing Breathing Frequencies in Swimming. Researchers from the United Kingdom studied the effect of two breathing frequencies on performance times and on several physiological parameters in front crawl swimming. The breathing frequencies were one breath every second stroke (B2) and one breath every fourth stroke (B4). Subjects were 10 male collegiate swimmers. Each subject swam 200 meters using each breathing frequency: once with breathing frequency B2 and once on a different day with breathing frequency B4. A paper states that the results are expressed as mean plus or minus the standard deviation.[2] One result reported in the paper states that the immediate postexercise heart rate for subjects when using breathing frequency B2 was 163 ± 15 beats per minute. What are $\bar{x}$ and the standard error of the mean for these subjects?

The *t* distributions

Body mass index (BMI) is used to screen for possible weight problems. It is calculated as weight divided by the square of height, measuring weight in kilograms and height in meters. Many online BMI calculators allow you to enter weight in pounds and height in inches. Adults with BMI less than 18.5 are considered underweight, and those with BMI greater than 25 may be overweight. For data about BMI, we turn to the National Health and Nutrition Examination Survey (NHANES), a continuing government sample survey that monitors the health of the American population.

EXAMPLE 20.1   Body Mass Index of Young Women

An NHANES report gives data for 654 women aged 20 to 29 years.[3] The mean BMI of these 654 women was $\bar{x} = 26.8$ and the sample standard deviation $s = 7.5$. On the basis of this sample, we want to estimate the mean BMI μ in the population of all 20.6 million women in this age group.

We will treat the NHANES sample as an SRS from a Normal population with unknown standard deviation σ.

1. To estimate the unknown population mean BMI μ, use the mean $\bar{x} = 26.8$ of the random sample. We don't expect $\bar{x}$ to be exactly equal to μ, so we want to say how accurate this estimate is.

2. We know the sampling distribution of $\bar{x}$. In repeated samples, $\bar{x}$ has the Normal distribution with mean μ and standard deviation $\sigma/\sqrt{n}$. So the average BMI $\bar{x}$ of an SRS of 654 young women has standard deviation

$$\frac{\sigma}{\sqrt{n}} = \frac{\sigma}{\sqrt{654}}$$

3. From Chapter 16 (page 356) any Normal curve has area C within z^* standard deviations on either side of the mean, where z^* can be found at the bottom of Table C, for many choices of C. The sampling distribution of $\bar{x}$ therefore has area C within $z^*\dfrac{\sigma}{\sqrt{654}}$.

4. By reasoning similar to that in Chapter 16, the interval is

$$\text{from} \quad \bar{x} - z^*\frac{\sigma}{\sqrt{654}} \quad \text{to} \quad \bar{x} + z^*\frac{\sigma}{\sqrt{654}}$$

or

$$\bar{x} \pm z^*\frac{\sigma}{\sqrt{654}}$$

This is our level C confidence interval for μ. The problem is that we do not know σ. ■

If we knew the value of σ, we would base probability calculations for $\bar{x}$ on the z statistic

$$z = \frac{\bar{x} - \mu}{\sigma/\sqrt{n}}$$

As we learned in Chapter 19, this z statistic has the standard Normal distribution $N(0,1)$. In practice, we don't know σ, so we substitute the standard error $s/\sqrt{n}$ of $\bar{x}$ for its standard deviation $\sigma/\sqrt{n}$. The statistic that results does not have a Normal distribution. It has a distribution that is new to us, called a *t distribution*.

The One-Sample *t* Statistic and the *t* Distributions

Draw an SRS of size n from a large population that has the Normal distribution with mean μ and standard deviation σ. The **one-sample *t* statistic**

$$t = \frac{\bar{x} - \mu}{s/\sqrt{n}}$$

has the ***t* distribution** with $n - 1$ degrees of freedom.

degrees of freedom

The t statistic has the same interpretation as any standardized statistic: it says how far $\bar{x}$ is from its mean μ in standard deviation units. There is a different t distribution for each sample size. We specify a particular t distribution by giving its **degrees of freedom.** The degrees of freedom for the one-sample t statistic come from the sample standard deviation s in the denominator of t. We saw in Chapter 2 (page 56) that s has $n - 1$ degrees of freedom. There are other t statistics with different degrees of freedom, some of which we will meet later. We will write the t distribution with $n - 1$ degrees of freedom as $t(n - 1)$ for short.

Figure 20.1 compares the density curves of the standard Normal distribution and the t distributions with 2 and 9 degrees of freedom. The figure illustrates these facts about the t distributions:

■ The density curves of the t distributions are similar in shape to the standard Normal curve. They are symmetric about 0, single peaked, and bell shaped.

■ The variability of the t distributions is a bit greater than that of the standard Normal distribution. The t distributions in Figure 20.1 have more probability in the tails and less in the center than does the standard Normal. This is true because substituting the estimate s for the fixed parameter σ introduces more variation into the statistic.

■ As the degrees of freedom increase, the t density curve approaches the $N(0, 1)$ curve ever more closely. This happens because s estimates σ more accurately as the sample size increases. So using s in place of σ causes little extra variation when the sample is large.

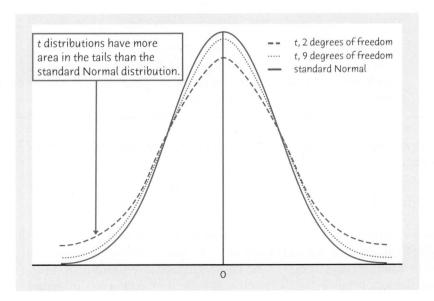

FIGURE 20.1

Density curves for the t distributions with 2 and 9 degrees of freedom and for the standard Normal distribution. All are symmetric with center 0. The t distributions are somewhat more spread out.

Table C in the back of the book gives critical values for the *t* distributions. Each row in the table contains critical values for the *t* distribution whose degrees of freedom appear at the left of the row. For convenience, we label the table entries both by the confidence level C (in percent) required for confidence intervals and by the one-sided and two-sided *P*-values for each critical value. You have already used the standard Normal critical values in the *z** row at the bottom of Table C. By looking down any column, you can check that the *t* critical values approach the Normal values as the degrees of freedom increase. If you use statistical software, you don't need Table C.

EXAMPLE 20.2   *t* Critical Values

Figure 20.1 shows the density curve for the *t* distribution with 9 degrees of freedom. What point on this distribution has probability 0.05 to its right? In Table C, look in the df = 9 row above one-sided *P*-value .05, and you will find that this critical value is *t** = 1.833. To use software, enter the degrees of freedom and the probability you want to the *left,* 0.95 in this case. Here is Minitab's output:

```
Student's t distribution with 9 DF
P( X <= x )        x
    0.95        1.83311 ■
```

Apply Your Knowledge

20.3 Critical Values. Use software or Table C to find

(a) the critical value for a one-sided test with level $\alpha = 0.05$ based on the $t(4)$ distribution.

(b) the critical value for a 98% confidence interval based on the $t(26)$ distribution.

20.4 More Critical Values. You have an SRS of size 30 and calculate the one-sample *t* statistic. What is the critical value *t** such that

(a) *t* has probability 0.025 to the right of *t**?

(b) *t* has probability 0.75 to the left of *t**?

The one-sample *t* confidence interval

In Example 20.1 (page 437) we argued that the interval

$$\bar{x} \pm z^* \frac{\sigma}{\sqrt{n}}$$

would be a level C confidence interval for μ if we knew σ, where z^* is the critical value for the standard Normal distribution corresponding to confidence level C. With unknown σ, we replace the standard deviation $\sigma/\sqrt{n}$ of $\bar{x}$ by its standard error $s/\sqrt{n}$. With this change, critical values now come from the *t* distribution with $n - 1$ degrees of freedom rather than the standard Normal distribution. The confidence interval and test that result are *one-sample t procedures.*

BETTER STATISTICS, BETTER BEER

The *t* distribution and the *t* inference procedures were invented by William S. Gosset (1876–1937). Gosset worked for the Guinness brewery, and his goal in life was to make better beer. He used his new *t* procedures to find the best varieties of barley and hops. Gosset's statistical work helped him become head brewer, a more interesting title than professor of statistics. Because Gosset published under the pen name "Student" you will often see the *t* distribution called "Student's *t*" in his honor.

The One-Sample *t* Confidence Interval

Draw an SRS of size n from a large population having unknown mean μ. A level C **confidence interval for μ** is

$$\bar{x} \pm t^* \frac{s}{\sqrt{n}}$$

where t^* is the critical value for the $t(n - 1)$ density curve with area C between $-t^*$ and t^*. This interval is exact when the population distribution is Normal and is approximately correct for $n \geq 15$ if there are no outliers and the population distribution is moderately skewed, and for $n \geq 40$ if there are no outliers and the population is strongly skewed.

EXAMPLE 20.3

TIP2

© Ariel Skelley/age fotostock

Good Weather, Good Tips?

STATE: Does the expectation of good weather lead to more generous behavior? Psychologists studied the size of the tip in a restaurant when a message indicating that the next day's weather would be good was written on the bill. Here are tips from 20 patrons, measured in percent of the total bill:[4]

| 20.8 | 18.7 | 19.9 | 20.6 | 21.9 | 23.4 | 22.8 | 24.9 | 22.2 | 20.3 |
| 24.9 | 22.3 | 27.0 | 20.4 | 22.2 | 24.0 | 21.1 | 22.1 | 22.0 | 22.7 |

This is one of three sets of measurements made, the others being tips received when the message on the bill said that the next day's weather would not be good or there was no message on the bill. We want to estimate the mean tip for comparison with tips under the other conditions.

PLAN: We will estimate the mean percentage tip μ for all patrons of this restaurant when they receive a message on their bill indicating that the next day's weather will be good by giving a 95% confidence interval.

SOLVE: We must first check the conditions for inference.

- **SRS:** We don't have an actual SRS from the population of all patrons of this restaurant. Scientists often act as if subjects are SRSs if there is nothing special about how the subjects were obtained. But it is always better to have an actual SRS because otherwise we can never be sure that hidden biases aren't present. This study was actually a randomized comparative experiment in which these 20 patrons were assigned at random from a larger group of patrons to get one of the treatments being compared. As a consequence, we are willing to regard these patrons as an SRS from all patrons of this restaurant.

- **Normal distribution:** The psychologists expect from past experience that measurements like this on patrons of the same restaurant under the same conditions will follow approximately a Normal distribution. We can't look at the population, but we can examine the sample. The stemplot in Figure 20.2 does not suggest any strong departures from Normality. The sample size is at least 15, so we are comfortable using the *t* interval.

We can proceed to calculation. For these data,

$$\bar{x} = 22.21 \quad \text{and} \quad s = 1.963$$

```
18 | 7
19 | 9
20 | 3 4 6 8
21 | 1 9
22 | 0 1 2 2 3 7 8
23 | 4
24 | 0 9 9
25 |
26 |
27 | 0
```

FIGURE 20.2

Stemplot of the percentage tips, for Example 20.3.

The degrees of freedom are $n - 1 = 19$. From Table C we find that for 95% confidence $t^* = 2.093$. The confidence interval is

$$\bar{x} \pm t^* \frac{s}{\sqrt{n}} = 22.21 \pm 2.093 \frac{1.963}{\sqrt{20}}$$
$$= 22.21 \pm 0.92$$
$$= 21.29 \text{ to } 23.13\%$$

CONCLUDE: We are 95% confident that the mean percentage tip for all patrons of this restaurant when their bill contains a message that the next day's weather will be good is between 21.29 and 23.13. ■

The one-sample *t* confidence interval has the form

$$\text{estimate} \pm t^* \text{SE}_{\text{estimate}}$$

where "SE" stands for "standard error." We will meet a number of confidence intervals that have this common form. In Example 20.3, the estimate is the sample mean $\bar{x}$, and its standard error is

$$\text{SE}_{\bar{x}} = \frac{s}{\sqrt{n}}$$
$$= \frac{1.963}{\sqrt{20}} = 0.439$$

Software will find $\bar{x}$, s, $\text{SE}_{\bar{x}}$, and the confidence interval from the data. Figure 20.5 (page 446) displays typical software output for Example 20.3.

Apply Your Knowledge

20.5 Critical Values. What critical value t^* from Table C (or from software) would you use for a confidence interval for the mean of the population in each of the following situations?

(a) A 95% confidence interval based on $n = 12$ observations.

(b) A 99% confidence interval from an SRS of 18 observations.

(c) A 90% confidence interval from a sample of size 6.

20.6 How Much Will I Bet? Our decisions depend on how the options are presented to us. Here's an experiment that illustrates this phenomenon. Tell 20 subjects that they have been given $50 but can't keep it all. Then present them with a long series of choices between bets they can make with the $50. Scattered among these choices in random order are 64 choices that ask the subject to choose between betting a fixed amount and an all-or-nothing gamble. The odds for all the bets are the same, but in 32 of the choices, the fixed option reads "Keep $20" and in the other 32 choices the fixed option reads "Lose $30." These two fixed options lead to exactly the same outcome, but people are more likely to choose the fixed option that says they lose money. Here are the percent differences ("Number of times chose 'Lose $30'" minus "Number of times chose 'Keep $20'" divided by the number of trials on which

the 20 subjects chose the fixed-option gamble rather than the all-or-nothing bet).[5] GAMB1

| 37.5 | 30.8 | 6.2 | 17.6 | 14.3 | 8.3 | 16.7 | 20.0 | 10.5 | 21.7 |
| 30.8 | 27.3 | 22.7 | 38.5 | 8.3 | 10.5 | 8.3 | 10.5 | 25.0 | 7.7 |

(a) Make a stemplot. Is there any sign of a major deviation from Normality?

(b) All 20 subjects gambled a fixed amount more often when faced with a sure loss than when faced with a sure win. Give a 95% confidence interval for the mean percent increase in gambling a fixed amount when faced with a sure loss. What must we assume about the 20 subjects for this interval to be valid?

20.7 Ancient Air. The composition of the earth's atmosphere may have changed over time. To try to discover the nature of the atmosphere long ago, we can examine the gas in bubbles inside ancient amber. Amber is tree resin that has hardened and been trapped in rocks. The gas in bubbles within amber should be a sample of the atmosphere at the time the amber was formed. Measurements on specimens of amber from the late Cretaceous era (75 to 95 million years ago) give these percents of nitrogen:[6] 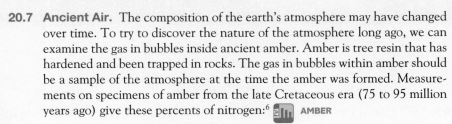 AMBER

| 63.4 | 65.0 | 64.4 | 63.3 | 54.8 | 64.5 | 60.8 | 49.1 | 51.0 |

Assume (this is not yet agreed on by experts) that these observations are an SRS from the late Cretaceous atmosphere. Use a 90% confidence interval to estimate the mean percent of nitrogen in ancient air. Follow the four-step process as illustrated in Example 20.3 (page 440). (Our present-day atmosphere is about 78.1% nitrogen). Be sure to check the conditions for inference.

The one-sample *t* test

Testing hypotheses about the mean μ of a Normal population follows the same reasoning as for testing hypotheses about a population proportion that we met earlier (see page 383). But now we state the null and alternative hypotheses in terms of the parameter μ and use the one-sample *t* statistic

$$t = \frac{\bar{x} - \mu}{s/\sqrt{n}}$$

to find the *P*-value. Here is the rule for the test we use.

The One-Sample *t* Test

Draw an SRS of size *n* from a large population having unknown mean μ. To **test the hypothesis** $H_0: \mu = \mu_0$, compute the **one-sample *t* statistic**

$$t = \frac{\bar{x} - \mu_0}{s/\sqrt{n}}$$

In terms of a variable *T* having the $t(n-1)$ distribution, the *P*-value for a test of H_0 against

$H_a: \mu > \mu_0$ is $P(T \geq t)$

$$H_a: \mu < \mu_0 \quad \text{is} \quad P(T \leq t)$$

$$H_a: \mu \neq \mu_0 \quad \text{is} \quad 2P(T \geq |t|)$$

These *P*-values are exact if the population distribution is Normal and are approximately correct for $n \geq 15$ if there are no outliers and the population distribution is moderately skewed, and for $n \geq 40$ if there are no outliers and the population is strongly skewed.

EXAMPLE 20.4　　Water Quality

We follow the four-step process for a significance test, outlined on page 383.

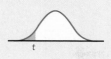

STATE: To investigate water quality, on August 8, 2011, the *Columbus Dispatch* took water samples at 20 Ohio State Park swimming areas. Those samples were taken to laboratories and tested for fecal coliform, which are bacteria found in human and animal feces. An unsafe level of fecal coliform means there's a higher chance that disease-causing bacteria are present and more risk that a swimmer will become ill. Ohio considers it unsafe if a 100-milliliter sample (about 3.3 ounces) of water contains more than 400 coliform bacteria. Here are the fecal coliform levels found by the laboratories:[7]

WQUAL

160	40	2800	80	2000	2000	1500	400	150	500
3000	2200	15	80	2000	2000	2600	600	1000	1500

Are these data good evidence that, on average, the fecal coliform levels in these swimming areas were unsafe?

PLAN: Experts caution that the tests are a snapshot of the quality of the water at the time they were taken. Fecal coliform levels can change as weather and other conditions change. So we ask the question in terms of the mean fecal coliform level μ for all these swimming areas. The null hypothesis is "level is not unsafe," and the alternative hypothesis is "level is unsafe."

$$H_0: \mu = 400$$
$$H_a: \mu > 400$$

SOLVE: First check the conditions for inference. We are willing to regard these particular 20 samples as an SRS from a large population of possible samples. Figure 20.3 is a histogram of the data. We can't accurately judge Normality from 20 observations; there are no outliers but the data are somewhat skewed. The sample size is larger than 15, so it is safe to use the one-sample *t* test, but *P*-values for the test may be only approximately accurate.

The basic statistics are

$$\bar{x} = 1231.3 \quad \text{and} \quad s = 1037.5$$

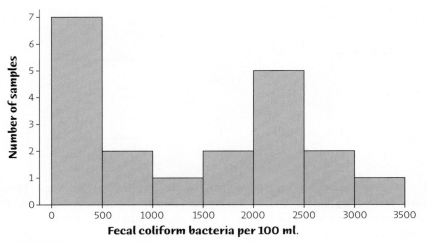

FIGURE 20.3

Histogram of the fecal coliform level in Example 20.4.

The one-sample t statistic is

$$t = \frac{\bar{x} - \mu_0}{s/\sqrt{n}} = \frac{1231.3 - 400}{1037.5/\sqrt{20}}$$
$$= 3.583$$

The P-value for $t = 3.583$ is the area to the right of 3.583 under the t distribution curve with degrees of freedom $n - 1 = 19$. Figure 20.4 shows this area. Software (see Figure 20.6) tells us that $P = 0.001$.

Without software, we can pin P between two values by using Table C. Search the $df = 19$ row of Table C for entries that bracket $t = 3.583$.

The observed t lies between the critical values for one-sided P-values 0.001 and 0.0005 as shown in the margin.

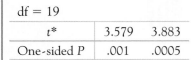

$df = 19$		
t^*	3.579	3.883
One-sided P	.001	.0005

CONCLUDE: There is quite strong evidence ($P < 0.001$) that, on average, fecal coliform levels in these Ohio State Park swimming areas are unsafe. ■

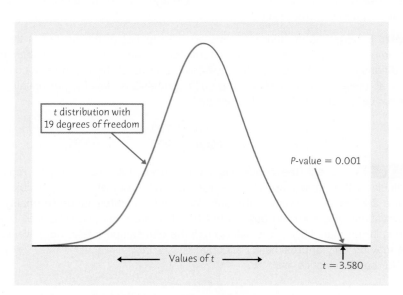

FIGURE 20.4

The P-value for the one-sided t test in Example 20.4.

Apply Your Knowledge

20.8 **Is It Significant?** The one-sample t statistic for testing

$$H_0: \mu = 0$$
$$H_a: \mu > 0$$

from a sample of $n = 20$ observations has the value $t = 1.84$.

(a) What are the degrees of freedom for this statistic?

(b) Give the two critical values t^* from Table C that bracket t. What are the one-sided P-values for these two entries?

(c) Is the value $t = 1.84$ significant at the 5% level? Is it significant at the 1% level?

(d) (Optional) If you have access to suitable technology, give the exact one-sided P-value for $t = 1.84$.

20.9 **Is It Significant?** The one-sample t statistic from an SRS of $n = 15$ observations from a Normal population for the two-sided test of

$$H_0: \mu = 64$$
$$H_a: \mu \neq 64$$

has the value $t = 2.12$.

(a) What are the degrees of freedom for t?

(b) Locate the two critical values t^* from Table C that bracket t. What are the two-sided P-values for these two entries?

(c) Is the value $t = 2.12$ statistically significant at the 10% level? At the 5% level?

(d) (Optional) If you have access to suitable technology, give the exact two-sided P-value for $t = 2.12$.

20.10 **Ancient Air, Continued.** Do the data of Exercise 20.7 (page 442) give good reason to think that the percent of nitrogen in the air during the Cretaceous era was different from the present 78.1%? Carry out a test of significance, following the four-step process as illustrated in Example 20.4 (page 440). Be sure to check the conditions for inference if you did not do Exercise 20.7. AMBER

Using technology

Any technology suitable for statistics will implement the one-sample t procedures. As usual, you can read and use almost any output now that you know what to look for. Figure 20.5 displays output for the 95% confidence interval of Example 20.3 (page 440) from a graphing calculator, a statistical program, a spreadsheet program, and the CrunchIt! software package. The calculator, Minitab, and CrunchIt! outputs are straightforward. All three give the estimate $\bar{x}$ and the confidence interval plus a clearly labeled selection of other information. The confidence interval agrees with our hand calculation in Example 20.3. In general, software results are more accurate because of the rounding in hand calculations. Excel gives several descriptive measures but does not give the confidence interval. The entry labeled "Confidence Level (95.0%)" is the margin of error. You can use this together with $\bar{x}$ to get the interval using either a calculator or the spreadsheet's formula capability.

Figure 20.6 displays output for the t test in Example 20.4. The graphing calculator, Minitab, and CrunchIt! give the sample mean $\bar{x}$, the t statistic, and its P-value.

Texas Instruments Graphing Calculator

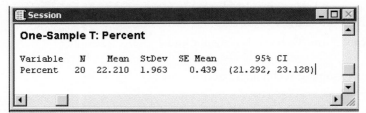

```
TInterval
 (21.292,23.128)
 x̄=22.2100
 Sx=1.9625
 n=20.0000
```

Minitab

```
Session                                      _ □ ×

One-Sample T: Percent

Variable   N    Mean  StDev  SE Mean      95% CI
Percent   20  22.210  1.963    0.439  (21.292, 23.128)
```

Excel

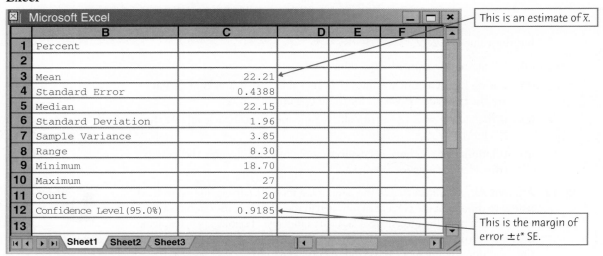

	B	C	D	E	F
1	Percent				
2					
3	Mean	22.21			
4	Standard Error	0.4388			
5	Median	22.15			
6	Standard Deviation	1.96			
7	Sample Variance	3.85			
8	Range	8.30			
9	Minimum	18.70			
10	Maximum	27			
11	Count	20			
12	Confidence Level(95.0%)	0.9185			
13					

This is an estimate of x̄.

This is the margin of error ±*t** SE.

CrunchIt!

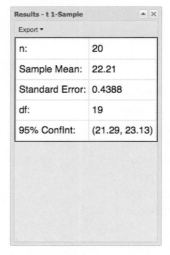

Results - t 1-Sample	▲ ✕
Export ▾	
n:	20
Sample Mean:	22.21
Standard Error:	0.4388
df:	19
95% ConfInt:	(21.29, 23.13)

FIGURE 20.5

The *t* confidence interval for Example 20.3: output from a graphing calculator, two statistical programs, and a spreadsheet program.

Texas Instruments Graphing Calculator

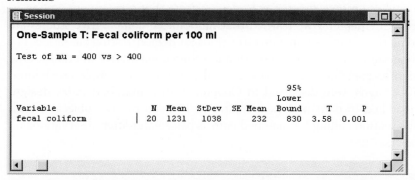

```
T-Test
 μ>400.0000
 t=3.5803
 P=.0010
 x̄=1231.2500
 Sx=1037.5374
 n=20.0000
```

Minitab

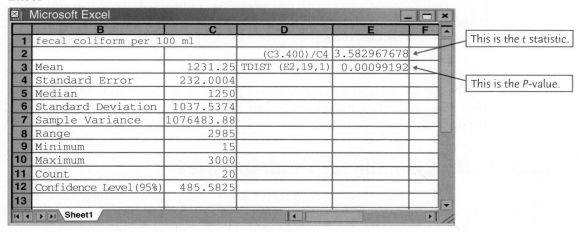

Session

One-Sample T: Fecal coliform per 100 ml

Test of mu = 400 vs > 400

Variable	N	Mean	StDev	SE Mean	95% Lower Bound	T	P
fecal coliform	20	1231	1038	232	830	3.58	0.001

Excel

Microsoft Excel

	B	C	D	E	F
1	fecal coliform per 100 ml				
2			(C3.400)/C4	3.582967678	
3	Mean	1231.25	TDIST (E2,19,1)	0.00099192	
4	Standard Error	232.0004			
5	Median	1250			
6	Standard Deviation	1037.5374			
7	Sample Variance	1076483.88			
8	Range	2985			
9	Minimum	15			
10	Maximum	3000			
11	Count	20			
12	Confidence Level(95%)	485.5825			
13					

Sheet1

This is the *t* statistic.

This is the *P*-value.

CrunchIt!

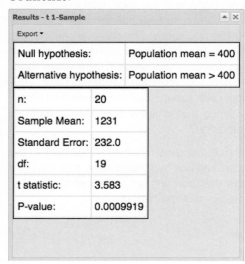

Results - t 1-Sample

Export ▾

Null hypothesis:	Population mean = 400
Alternative hypothesis:	Population mean > 400

n:	20
Sample Mean:	1231
Standard Error:	232.0
df:	19
t statistic:	3.583
P-value:	0.0009919

FIGURE 20.6

The *t* test for Example 20.4: output from a graphing calculator, two statistical programs, and a spreadsheet program.

Accurate *P*-values are the biggest advantage of software for the *t* procedures. Excel is, as usual, more awkward than software designed for statistics. It lacks a one-sample *t* test menu selection but does have a function named TDIST for tail areas under *t* density curves. The Excel output shows functions for the *t* statistic and its *P*-value to the right of the main display, along with their values $t = 3.582967678$ and $P = 0.00099192$.

Matched pairs *t* procedures

Often the goal of an investigation is to demonstrate that a treatment causes an observed effect. In Chapter 9 we learned that randomized comparative studies are more convincing than single-sample investigations for demonstrating causation. For that reason, one-sample inference is less common than comparative inference. One common design to compare two treatments makes use of one-sample procedures. *matched pairs design* Matched pairs designs were discussed in Chapter 9. In a **matched pairs design,** subjects are matched in pairs and each treatment is given to one subject in each pair. Another situation calling for matched pairs is before-and-after observations on the same subjects.

Matched Pairs *t* Procedures

To compare the responses to the two treatments in a matched pairs design, find the difference between the responses within each pair. Then apply the one-sample *t* procedures to these differences.

The parameter μ in a matched pairs *t* procedure is the mean difference in the responses to the two treatments within matched pairs of subjects in the entire population.

EXAMPLE 20.5   ## Do Chimpanzees Collaborate?

CHIMPS

Manoj Shah/Getty Images

STATE: Humans often collaborate to solve problems. Will chimpanzees recruit another chimp when solving a problem requires collaboration? Researchers presented chimpanzee subjects with food outside their cage that they could bring within reach by pulling two ropes, one attached to each end of the food tray. If a chimp pulled only one rope, the rope came loose and the food was lost. Another chimp in an adjacent cage was available as a partner to help pull the ropes, but only if the subject chimp opened a door joining the two cages. The same 8 chimpanzee subjects faced this problem in two versions: the two ropes were close enough together that one chimp could pull both (no collaboration needed) or the two ropes were too far apart for one chimp to pull both (collaboration needed). Table 20.1 shows how often in 24 trials for each version, each subject opened the door to recruit another chimp as a partner.[8] Is there evidence that chimpanzees recruit partners more often when a problem requires collaboration?

PLAN: Take μ to be the mean difference (collaboration required minus not) in the number of times a subject recruited a partner. The null hypothesis says that the need for collaboration has no effect, and H_a says that partners are recruited more often when the problem requires collaboration. So we test the hypotheses

$$H_0: \mu = 0$$
$$H_a: \mu > 0$$

TABLE 20.1 TRIALS (OUT OF 24) ON WHICH CHIMPANZEES RECRUITED A PARTNER

| CHIMPANZEE | COLLABORATION NEEDED | | DIFFERENCE |
	YES	NO	
Namuiska	16	0	16
Kalema	16	1	15
Okech	23	5	18
Baluku	19	3	16
Umugenzi	15	4	11
Indi	20	9	11
Bili	24	16	8
Asega	24	20	4

SOLVE: The subjects are "semi-free-ranging chimpanzees at Ngamba Island Chimpanzee Sanctuary in Uganda." We are willing to regard them as an SRS from their species. To analyze the data, we examine the difference in the number of times a chimp recruited a partner, so subtract the "no collaboration needed" count from the "collaboration needed" count for each subject. The 8 differences form a single sample from a population with unknown mean μ. They appear in the "Difference" column in Table 20.1. All the chimpanzees recruited a partner more often when the ropes were too far apart to be pulled by one chimp.

The stemplot in Figure 20.7 creates the impression of a left-skew. This is a bit misleading, as the *dotplot* in the bottom part of Figure 20.7 shows. A dotplot simply places the observations on an axis, stacking observations that have the same value. It gives a good picture of distributions with only whole-number values. We know that observations that can take only whole-number values cannot come from a Normal population. In practice, researchers are willing to treat such observations as coming from a Normal population if there are more than just a few possible values and the distribution appears approximately Normal. Of course, we can't assess approximate Normality from just 8 observations, but there are no signs of major departures from Normality. The researchers used the matched pairs *t* test.

The 8 differences have

$$\bar{x} = 12.375 \quad \text{and} \quad s = 4.749$$

The one-sample *t* statistic is therefore

$$t = \frac{\bar{x} - 0}{s/\sqrt{n}} = \frac{12.375 - 0}{4.749/\sqrt{8}}$$
$$= 7.37$$

Find the *P*-value from the *t*(7) distribution. (Remember that the degrees of freedom are 1 less than the sample size.) Table C shows that 7.37 is greater than the critical value for one-sided *P* = 0.0005, as the display in the margin indicates. The *P*-value is therefore less than 0.0005. Software says that *P* = 0.000077.

CONCLUDE: The data give very strong evidence (*P* < 0.0005) that chimpanzees recruit a collaborator more often when faced with a problem that requires a collaborator to solve. That is, chimpanzees recognize when collaboration is necessary, a skill that they share with humans. ■

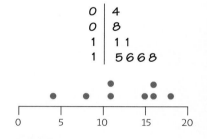

```
0 | 4
0 | 8
1 | 1 1
1 | 5 6 6 8
```

FIGURE 20.7
Stemplot and dotplot of the differences, for Example 20.5.

df = 7		
t*	4.785	5.408
One-sided *P*	.001	.0005

Example 20.5 illustrates how to turn matched pairs data into single-sample data by taking differences within each pair. We are making inferences about a single population, the population of all differences within matched pairs. *It is incorrect to ignore the matching and analyze the data as if we had two samples of chimpanzees, one facing ropes close together and the other facing ropes far apart.* Inference procedures for comparing two samples assume that the samples are selected independent of each other. This condition does not hold when the same subjects are measured twice. The proper analysis depends on the design used to produce the data.

Apply Your Knowledge

Many exercises from this point on ask you to give the P-value of a t test. If you have suitable technology, give the exact P-value. Otherwise, use Table C to give two values between which P lies.

20.11 **The Brain Responds to Sound.** The usual way to study the brain's response to sounds is to have subjects listen to "pure tones." The response to recognizable sounds may differ. To compare responses, researchers anesthetized macaque monkeys. They fed pure tones and also monkey calls directly to their brains by inserting electrodes. Response to the stimulus was measured by the firing rate (electrical spikes per second) of neurons in various areas of the brain. Table 20.2 contains the responses for 37 neurons.[9] You may assume the 37 responses are independent. Researchers suspected that the response to monkey calls would be stronger than the response to a pure tone. Do the data support this idea? Complete the "Plan," "Solve," and "Conclude" steps of the four-step process, following the model of Example 20.5. 📊 BRESPN

TABLE 20.2	NEURON RESPONSE TO TONES AND MONKEY CALLS						
TONE	CALL	TONE	CALL	TONE	CALL	TONE	CALL
474	500	145	42	71	134	35	103
256	138	141	241	68	65	31	70
241	485	129	194	59	182	28	192
226	338	113	123	59	97	26	203
185	194	112	182	57	318	26	135
174	159	102	141	56	201	21	129
176	341	100	118	47	279	20	193
168	85	74	62	46	62	20	54
161	303	72	112	41	84	19	66
150	208						

20.12 **The Brain Responds, Continued.** How much more strongly do monkey brains respond to monkey calls than to pure tones? Give a 99% confidence interval to answer this question. 📊 BRESPN

Robustness of *t* procedures

The *t* confidence interval and test are exactly correct when the distribution of the population is exactly Normal. No real data are exactly Normal. The usefulness of the *t* procedures in practice therefore depends on how strongly they are affected by lack of Normality.

Robust Procedures

A confidence interval or significance test is called **robust** if the confidence level or *P*-value does not change very much when the conditions for use of the procedure are violated.

The condition that the population is Normal rules out outliers, so the presence of outliers shows that this condition is not fulfilled. The *t* procedures are not robust against outliers unless the sample is large because $\bar{x}$ and *s* are not resistant to outliers.

Fortunately, the *t* procedures are quite robust against non-Normality of the population except when outliers or strong skewness are present. (Skewness is more serious than other kinds of non-Normality.) As the size of the sample increases, the central limit theorem ensures that the distribution of the sample mean $\bar{x}$ becomes more nearly Normal and that the *t* distribution becomes more accurate for critical values and *P*-values of the *t* procedures.

Always make a plot to check for skewness and outliers before you use the *t* procedures for small samples. For most purposes, you can safely use the one-sample *t* procedures when $n \geq 15$ unless an outlier or quite strong skewness is present. Here are practical guidelines for inference on a single mean.[10]

CATCHING CHEATERS

A certification test for surgeons asks 277 multiple-choice questions. Smith and Jones have 193 common right answers and 53 identical wrong choices. The computer flags their 246 identical answers as evidence of possible cheating. They sue. The court wants to know how unlikely it is that exams this similar would occur just by chance. That is, the court wants a *P*-value. Statisticians offer several *P*-values based on different models for the exam-taking process. They all say that results this similar would almost never happen just by chance. Smith and Jones fail the exam.

Using the *t* Procedures

- Except in the case of small samples, the condition that the data are an SRS from the population of interest is more important than the condition that the population distribution is Normal.

- *Sample size less than 15:* Use *t* procedures if the data appear close to Normal (roughly symmetric, single peak, no outliers). If the data are clearly skewed or if outliers are present, do not use *t*.

- *Sample size at least 15:* The *t* procedures can be used except in the presence of outliers or strong skewness.

- *Large samples:* The *t* procedures can be used even for clearly skewed distributions when the sample is large, roughly $n \geq 40$.

These guidelines are the same as those in the conditions for inference about a mean on page 436.

EXAMPLE 20.6   Can We Use *t*?

Figure 20.8 shows plots of several data sets. For which of these can we safely use the *t* procedures?[11]

- Figure 20.8(a) is a histogram of the percent of each state's adult residents who are college graduates. *We have data on the entire population of 50 states, so inference is not needed.* We can calculate the exact mean for the population. There is no uncertainty due to having only a sample from the population, and no need for a confidence interval or test. *If these data were an SRS from a larger population, t inference would be safe despite the mild skewness because n = 50.*

- ■ Figure 20.8(b) is a stemplot of the force required to pull apart 20 pieces of Douglas fir. *The data are strongly skewed to the left with possible low outliers, so we cannot trust the t procedures for n = 20.*

- ■ Figure 20.8(c) is a stemplot of the lengths of 23 specimens of the red variety of the tropical flower *Heliconia. The data are mildly skewed to the right, and there are no outliers. We can use the t distributions for such data.*

- ■ Figure 20.8(d) is a histogram of the heights of the students in a college class. *This distribution is quite symmetric and appears close to Normal. We can use the t procedures for any sample size.* ■

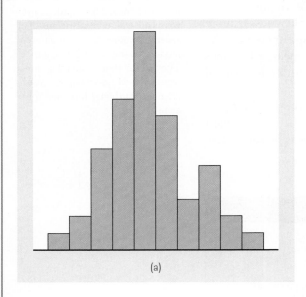

(a)

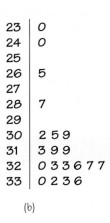

```
23 | 0
24 | 0
25 |
26 | 5
27 |
28 | 7
29 |
30 | 2 5 9
31 | 3 9 9
32 | 0 3 3 6 7 7
33 | 0 2 3 6
```

(b)

```
37 | 4 8 9
38 | 0 0 1 1 2 2 8 9
39 | 2 6 8
40 | 6 7
41 | 5 7 9 9
42 | 0 2
43 | 1
```

(c)

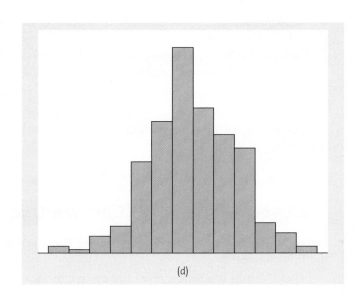

(d)

FIGURE 20.8

Can we use *t* procedures for these data? (a) Percent of adult college graduates in the 50 states. *No,* this is an entire population, not a sample. (b) Force required to pull apart 20 pieces of Douglas fir. *No,* there are just 20 observations and strong skewness. (c) Lengths of 23 tropical flowers of the same variety. *Yes,* the sample is large enough to overcome the mild skewness. (d) Heights of female students. *Yes, for any size sample,* because the distribution is close to Normal.

Apply Your Knowledge

20.13 Diamonds. A group of earth scientists studied the small diamonds found in a nodule of rock carried up to the earth's surface in surrounding rock. This is an opportunity to examine a sample from a single population of diamonds formed in a single event deep in the earth.[12] Table 20.3 presents data on the nitrogen content (parts per million) and the abundance of carbon-13 in these diamonds. (Carbon has several isotopes, forms with different numbers of neutrons in the nuclei of their atoms. Carbon-12 makes up almost 99% of natural carbon. The abundance of carbon-13 is measured by the ratio of carbon-13 to carbon-12, in parts per thousand more or less than a standard. The minus signs in the data mean that the ratio is smaller in these diamonds than in standard carbon.) DMNDS

© Eric Nathan/Alamy

TABLE 20.3 NITROGEN AND CARBON-13 IN A SAMPLE OF DIAMONDS

DIAMOND	NITROGEN (PPM)	CARBON-13 RATIO	DIAMOND	NITROGEN (PPM)	CARBON-13 RATIO
1	487	−2.78	13	273	−2.73
2	1430	−1.39	14	94	−2.33
3	60	−4.26	15	69	−3.83
4	244	−1.19	16	262	−2.04
5	196	−2.12	17	120	−2.82
6	274	−2.87	18	302	−0.84
7	41	−3.68	19	75	−3.57
8	54	−3.29	20	242	−2.42
9	473	−3.79	21	115	−3.89
10	30	−4.06	22	65	−3.87
11	98	−1.83	23	311	−1.58
12	41	−4.03	24	61	−3.97

We would like to estimate the mean abundance of both nitrogen and carbon-13 in the population of diamonds represented by this sample. Examine the data for nitrogen. Can we use a t confidence interval for mean nitrogen? Explain your answer. Give a 90% confidence interval if you think the result can be trusted.

20.14 Diamonds, Continued. Examine the data in Table 20.3 on abundance of carbon-13. Can we use a t confidence interval for mean carbon-13? Explain your answer. Give a 90% confidence interval if you think the result can be trusted. DMNDS

CHAPTER 20 SUMMARY

Chapter Specifics

■ Tests and confidence intervals for the mean μ of a Normal population are based on the sample mean $\bar{x}$ of an SRS. Because of the central limit theorem, the resulting procedures are approximately correct for other population distributions when the sample is large.

■ The standardized sample mean is

$$\frac{\bar{x} - \mu}{\sigma/\sqrt{n}}$$

■ If we knew σ, we would use the standard Normal distribution to calculate probabilities. In practice, we do not know σ. Replace the standard deviation $\sigma/\sqrt{n}$ of $\bar{x}$ by the **standard error** $s/\sqrt{n}$ to get the **one-sample t statistic**

$$t = \frac{\bar{x} - \mu}{s/\sqrt{n}}$$

■ The t statistic has the **t distribution** with $n-1$ degrees of freedom.

■ There is a t distribution for every positive **degree of freedom.** All are symmetric distributions similar in shape to the standard Normal distribution. The t distribution approaches the $N(0, 1)$ distribution as the degrees of freedom increase.

■ A level C **confidence interval for the mean** μ of a Normal population is

$$\bar{x} \pm t^* \frac{s}{\sqrt{n}}$$

■ The **critical value** t^* is chosen so that the t curve with $n-1$ degrees of freedom has area C between $-t^*$ and t^*.

■ **Significance tests** for H_0: $\mu = \mu_0$ are based on the t statistic. Use P-values or fixed significance levels from the $t(n-1)$ distribution.

■ Use these one-sample procedures to analyze **matched pairs** data by first taking the difference within each matched pair to produce a single sample.

■ The t procedures are quite **robust** when the population is non-Normal, especially for larger sample sizes. The t procedures are useful for non-Normal data when $n \geq 15$ unless the data show outliers or strong skewness. When $n \geq 40$ the t procedures can be used even for clearly skewed distributions.

Link It

In Chapters 16 and 17 we began our study of inference. We focused on inference for a population proportion based on a sample from a Normal population. In this chapter we continue our study of inference but focus on a population mean based on a sample from a Normal population. The basic ideas of Chapters 16 and 17 still apply, but now we use the t distribution rather than the Normal distribution to find confidence intervals and calculate P-values. Unfortunately, the mathematics associated with the t distribution is more complicated than that associated with the Normal distribution. We must rely on approximations from tables or statistical software to find confidence intervals and calculate P-values. As we continue our study of statistical inference, statistical software will be invaluable.

As we saw in Chapter 9, statistical studies comparing two or more groups are preferable to one-sample procedures if we want to demonstrate that a treatment causes an observed response. As a first step toward developing methods for comparing means from two populations, we considered matched pairs studies (also discussed in Chapter 9) and saw that by looking at differences, we could use our one-sample t procedures to compare two means. In the next chapter, we consider inference for comparing the means of two populations when we have independent samples from the two populations. This will expand our tools for doing statistical inference in settings that we encounter in practice.

CHECK YOUR SKILLS

20.15 We prefer the t procedures to the z procedures for inference about a population mean because

(a) z can be used only for large samples.
(b) z requires that you know the population standard deviation σ.
(c) z requires that you can regard your data as an SRS from the population.

20.16 You are testing $H_0: \mu = 10$ against $H_a: \mu < 10$ based on an SRS of 16 observations from a Normal population. The data give $\bar{x} = 8$ and $s = 4$. The value of the t statistic is

(a) -0.5.   (b) -2.   (c) -8.

20.17 You are testing $H_0: \mu = 100$ against $H_a: \mu < 100$ based on an SRS of 25 observations from a Normal population. The t statistic is $t = -2.6$. The degrees of freedom for the t statistic are

(a) 26.   (b) 25.   (c) 24.

20.18 The P-value for the statistic in the previous exercise

(a) falls between 0.02 and 0.04.
(b) falls between 0.01 and 0.02.
(c) is less than 0.01.

20.19 You have an SRS of 12 observations from a Normally distributed population. What critical value would you use to obtain a 98% confidence interval for the mean μ of the population?

(a) 2.718   (b) 2.681   (c) 2.650

20.20 You are testing $H_0: \mu = 0$ against $H_a: \mu \neq 0$ based on an SRS of 12 observations from a Normal population. What values of the t statistic are statistically significant at the $\alpha = 0.005$ level?

(a) $t > 3.497$
(b) $t < -3.497$ or $t > 3.497$
(c) $t < -3.428$ or $t > 3.428$

20.21 Data on the blood cholesterol levels of 10 rats (milligrams per deciliter of blood) give $\bar{x} = 85$ and $s = 12$. A 99% confidence interval for the mean blood cholesterol of rats is

(a) 76.4 to 93.6.
(b) 73.0 to 97.0.
(c) 72.7 to 97.3.

20.22 Which of the following would cause the most worry about the validity of the confidence interval you calculated in the previous exercise?

(a) There is a clear outlier in the data.
(b) A stemplot of the data shows a mild right-skew.
(c) You do not know the population standard deviation σ.

20.23 Which of these settings does *not* allow use of a matched pairs t procedure?

(a) You interview both the husband and the wife in 64 married couples and ask each about their ideal number of children.
(b) You interview a sample of 64 unmarried male students and another sample of 64 unmarried female students and ask each about their ideal number of children.
(c) You interview 64 female students in their freshman year and again in their senior year and ask each about their ideal number of children.

20.24 Because the t procedures are robust, the most important condition for their safe use is that

(a) the population standard deviation σ is known.
(b) the population distribution is exactly Normal.
(c) the data can be regarded as an SRS from the population.

CHAPTER 20 EXERCISES

20.25 **Read carefully.** You read in the report of a psychology experiment: "Separate analyses for our two groups of 12 participants revealed no overall placebo effect for our student group (mean = 0.08, SD = 0.37, $t(11) = 0.49$) and a significant effect for our nonstudent group (mean = 0.35, SD = 0.37, $t(11) = 3.25$, $p < 0.01$)."[13] The null hypothesis is that the mean effect is zero. What are the correct values of the two t statistics based on the means and standard deviations? Compare each correct t-value with the critical values in Table C. What can you say about the two-sided P-value in each case?

20.26 **Body mass index of young women.** In Example 20.1 (page 437) we discussed finding a confidence interval for the mean body mass index (BMI) of women aged 20 to 29 years, based on a national random sample of 654 such women. The sample data had mean BMI $\bar{x} = 26.8$ and standard deviation $s = 7.42$. What is the 95% t confidence interval for the mean BMI of all young women?

20.27 **Reading scores in Atlanta.** The Trial Urban District Assessment (TUDA) is a government-sponsored study of student achievement in large urban school districts. TUDA gives a reading test scored from 0 to 500. A score of 243 is a "basic" reading level, and a score of 281 is "proficient." Scores for a random sample of 3000 eighth-graders in Atlanta had $\bar{x} = 250$ with standard error 1.0.[14]

(a) We don't have the 3000 individual scores, but use of the t procedures is surely safe. Why?

(b) Give a 99% confidence interval for the mean score of all Atlanta eighth graders. (Be careful: the report gives the standard error of $\bar{x}$, not the standard deviation s.)

(c) Urban children often perform below the basic level. Is there good evidence that the mean for all Atlanta eighth-graders is less than the basic level?

20.28 Color and cognition. In a randomized comparative experiment on the effect of color on the performance of a cognitive task, researchers randomly divided 69 subjects (27 males and 42 females ranging in age from 17 to 25 years) into three groups. Participants were asked to solve a series of 6 anagrams. One group was presented with the anagrams on a blue screen, one group saw them on a red screen, and one group had a neutral screen. The time, in seconds, taken to solve the anagrams was recorded. The paper reporting the study gives $\bar{x} = 11.58$ and $s = 4.37$ for the times of the 23 members of the neutral group.[15]

(a) Give a 95% confidence interval for the mean time in the population from which the subjects were recruited.

(b) What conditions for the population and the study design are required by the procedure you used in (a)? Which of these conditions are important for the validity of the procedure in this case?

20.29 The placebo effect. The placebo effect is particularly strong in patients with Parkinson's disease. To understand the workings of the placebo effect, scientists measure activity at a key point in the brain when patients receive a placebo that they think is an active drug and also when no treatment is given.[16] The same 6 patients are measured both with and without the placebo, at different times.

(a) Explain why the proper procedure to compare the mean response to placebo with control (no treatment) is a matched pairs t test.

(b) The 6 differences (treatment minus control) had $\bar{x} = -0.326$ and $s = 0.181$. Is there significant evidence of a difference between treatment and control? Are the conditions for inference about a mean satisfied?

20.30 The conductivity of fibrous-glass board. How well materials conduct heat matters when designing houses, for example. Conductivity is measured in terms of watts of heat power transmitted per square meter of surface per degree Celsius of temperature difference on the two sides of the material. The National Institute of Standards and Technology (NIST) provides data on properties of materials. Here are 9 NIST measurements of the heat conductivity of a particular type of fibrous-glass board:[17] **HEATCON**

> 0.0339 0.0337 0.0334 0.0334 0.0333 0.0333
> 0.0333 0.0332 0.0330

(a) We can consider this an SRS of all specimens of fibrous-glass board of this type. Make a stemplot. Is there any sign of major deviation from Normality?

(b) Give a 95% confidence interval for the mean conductivity.

(c) Is there significant evidence at the 5% level that the mean conductivity of this type of fibrous-glass board is not 0.0330?

20.31 Exhaust from school buses. In a study of exhaust emissions from school buses, the pollution intake by passengers was determined for a sample of 9 school buses used in the Southern California Air Basin. The pollution intake is the amount of exhaust emissions, in grams per person, that would be breathed in while traveling on the bus during its usual 18-mile trip on congested freeways from South Central LA to a magnet school in West LA. (As a reference, the average intake of motor emissions of carbon monoxide in the LA area is estimated to be about 0.000046 grams per person.) Here are the amounts for the 9 buses when driven with the windows open:[18] **EMIT**

> 1.15 0.33 0.40 0.33 1.35 0.38 0.25 0.40 0.35

(a) Make a stemplot. Are there outliers or strong skewness that would forbid use of the t procedures?

(b) A good way to judge the effect of outliers is to do your analysis twice, once with the outliers and a second time without them. Give two 90% confidence intervals, one with all the data and one with the outliers removed, for the mean pollution intake among all school buses used in the Southern California Air Basin that travel the route investigated in the study.

(c) Compare the two intervals in part (b). What is the most important effect of removing the outliers?

20.32 A big toe problem. Hallux abducto valgus (call it HAV) is a deformation of the big toe that often requires surgery. Doctors used X-rays to measure the angle (in degrees) of deformity in 38 consecutive patients under the age of 21 who came to a medical center for surgery to correct HAV. The

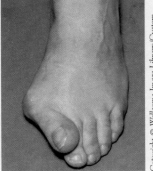

angle is a measure of the seriousness of the deformity. Here are the data:[19] ▦ᵢₗᵢ BIGTOE

```
28 32 25 34 38 26 25 18 30 26 28 13 20
21 17 16 21 23 14 32 25 21 22 20 18 26
16 30 30 20 50 25 26 28 31 38 32 21
```

It is reasonable to regard these patients as a random sample of young patients who require HAV surgery. Carry out the "Solve" and "Conclude" steps for a 95% confidence interval for the mean HAV angle in the population of all such patients.

20.33 **An outlier's effect.** Our bodies have a natural electrical field that is known to help wounds heal. Does changing the field strength slow healing? A series of experiments with newts investigated this question. In one experiment, the two hind limbs of 12 newts were assigned at random to either experimental or control groups. This is a matched pairs design. The electrical field in the experimental limbs was reduced to zero by applying a voltage. The control limbs were left alone. Here are the rates at which new cells closed a razor cut in each limb, in micrometers per hour:[20] ▦ᵢₗᵢ NEWTS

Newt	1	2	3	4	5	6	7	8	9	10	11	12
Control limb	36	41	39	42	44	39	39	56	33	20	49	30
Experimental limb	28	31	27	33	33	38	45	25	28	33	47	23

(a) Make a stemplot of the differences between limbs of the same newt (control limb minus experimental limb). There is a high outlier.

(b) A good way to judge the effect of an outlier is to do your analysis twice, once with the outlier and a second time without it. Carry out two t tests to see if the mean healing rate is significantly lower in the experimental limbs, one test including all 12 newts and another that omits the outlier. What are the test statistics and their P-values? Does the outlier have a strong influence on your conclusion?

20.34 **An outlier's effect.** A good way to judge the effect of an outlier is to do your analysis twice, once with the outlier and a second time without it. The data in Exercise 20.32 follow a Normal distribution quite closely except for one patient with HAV angle 50 degrees, a high outlier. ▦ᵢₗᵢ BIGTOE

(a) Find the 95% confidence interval for the population mean based on the 37 patients who remain after you drop the outlier.

(b) Compare your interval in (a) with your interval from Exercise 20.32. What is the most important effect of removing the outlier?

20.35 **Men of few words?** Researchers claim that women speak significantly more words per day than men. One estimate is that a woman uses about 20,000 words per day, whereas a man uses about 7,000. To investigate such claims, one study used a special device to record the conversations of male and female university students over a four-day period. From these recordings, the daily word count of the 20 men in the study was determined. Here are their daily word counts:[21] ▦ᵢₗᵢ TALKING

28,408	10,084	15,931	21,688	37,786
10,575	12,880	11,071	17,799	13,182
8,918	6,495	8,153	7,015	4,429
10,054	3,998	12,639	10,974	5,255

(a) Examine the data. Is it reasonable to use the t procedures? (Assume these men are an SRS of all male students at this university).

(b) If your conclusion in part (a) is "Yes," do the data give convincing evidence that the mean number of words per day of men at this university differs from 7,000?

20.36 **Genetic engineering for cancer treatment.** Here's a new idea for treating advanced melanoma, the most serious kind of skin cancer. Genetically engineer white blood cells to better recognize and destroy cancer cells, then infuse these cells into patients. The subjects in a small initial study were 11 patients whose melanoma had not responded to existing treatments. One question was how rapidly the new cells would multiply after infusion, as measured by the doubling time in days. Here are the doubling times:[22] ▦ᵢₗᵢ CNCRTRT

```
1.4  1.0  1.3  1.0  1.3  2.0  0.6  0.8  0.7  0.9  1.9
```

(a) Examine the data. Is it reasonable to use the t procedures?

(b) Give a 90% confidence interval for the mean doubling time. Are you willing to use this interval to make an inference about the mean doubling time in a population of similar patients?

20.37 **Genetic engineering for cancer treatment, continued.** Another outcome in the cancer experiment described in Exercise 20.36 is measured by a test for the presence of cells that trigger an immune response in the body and so may help fight cancer. Here are data for the 11 subjects: counts of active cells per 100,000 cells before and after infusion of the modified cells. The difference (after infusion minus before infusion) is the response variable. ▦ᵢₗᵢ MORECAN

Before	14	0	1	0	0	0	0	20	1	6	0
After	41	7	1	215	20	700	13	530	35	92	108
Difference	27	7	0	215	20	700	13	510	34	86	108

(a) Examine the data. Is it reasonable to use the t procedures?

(b) If your conclusion in part (a) is "Yes," do the data give convincing evidence that the count of active cells is higher after treatment?

20.38 **Kicking a helium-filled football.** Does a football filled with helium travel farther than one filled with ordinary air? To test this, the *Columbus Dispatch* conducted a study. Two identical footballs, one filled with helium and one filled with ordinary air, were used. A casual observer was unable to detect a difference in the two footballs. A novice kicker was used to punt the footballs. A trial consisted of kicking both footballs in a random order. The kicker did not know which football (the helium-filled or the air-filled football) he was kicking. The distance of each punt was recorded. Then another trial was conducted. A total of 39 trials were run. Here are the data for the 39 trials, in yards that the footballs traveled. The difference (helium minus air) is the response variable.[23] FTBALL

Helium	25	16	25	14	23	29	25	26	22	26
Air	25	23	18	16	35	15	26	24	24	28
Difference	0	−7	7	−2	−12	14	−1	2	−2	−2
Helium	12	28	28	31	22	29	23	26	35	24
Air	25	19	27	25	34	26	20	22	33	29
Difference	−13	9	1	6	−12	3	3	4	2	−5
Helium	31	34	39	32	14	28	30	27	33	11
Air	31	27	22	29	28	29	22	31	25	20
Difference	0	7	17	3	−14	−1	8	−4	8	−9
Helium	26	32	30	29	30	29	29	30	26	
Air	27	26	28	32	28	25	31	28	28	
Difference	−1	6	2	−3	2	4	−2	2	−2	

(a) Examine the data. Is it reasonable to use the *t* procedures?

(b) If your conclusion in part (a) is "Yes," do the data give convincing evidence that the helium-filled football travels farther than the air-filled football?

20.39 **Growing trees faster.** The concentration of carbon dioxide (CO_2) in the atmosphere is increasing rapidly due to our use of fossil fuels. Because plants use CO_2 to fuel photosynthesis, more CO_2 may cause trees and other plants to grow faster. An elaborate apparatus allows researchers to pipe extra CO_2 to a 30-meter circle of forest. They selected two nearby circles in each of three parts of a pine forest and randomly chose one of each pair to receive extra CO_2. The response variable is the mean increase in base area for 30 to 40 trees in a circle during a growing season. We measure this in percent increase per year. Here are one year's data:[24] TREES

PAIR	CONTROL PLOT	TREATED PLOT
1	9.752	10.587
2	7.263	9.244
3	5.742	8.675

(a) State the null and alternative hypotheses. Explain clearly why the investigators used a one-sided alternative.

(b) Carry out a test and report your conclusion in simple language.

(c) The investigators used the test you just carried out. Any use of the *t* procedures with samples this size is risky. Why?

20.40 **Fungus in the air.** The air in poultry-processing plants often contains fungus spores. Inadequate ventilation can affect the health of the workers. The problem is most serious during the summer. To measure the presence of spores, air samples are pumped to an agar plate, and "colony-forming units (CFUs)" are counted after an incubation period. Here are data from two locations in a plant that processes 37,000 turkeys per day, taken on four days in the summer. The units are CFUs per cubic meter of air.[25] FUNGUS

	DAY 1	DAY 2	DAY 3	DAY 4
Kill room	3175	2526	1763	1090
Processing	529	141	362	224

(a) Explain carefully why these are matched pairs data.

(b) The spore count is clearly higher in the kill room. Give sample means and a 90% confidence interval to estimate how much higher. Be sure to state your conclusion in plain English.

(c) You will often see the *t* procedures used for data like these. You should regard the results as only rough approximations. Why?

20.41 **Weeds among the corn.** Velvetleaf is a particularly annoying weed in cornfields. It produces lots of seeds, and the seeds wait in the soil for years until conditions are right. How many seeds do velvetleaf plants produce? Here are counts from 28 plants that came up in a cornfield when no herbicide was used:[26] WEEDS

2450	2504	2114	1110	2137	8015	1623	1531	2008	1716
721	863	1136	2819	1911	2101	1051	218	1711	164
2228	363	5973	1050	1961	1809	130	880		

We would like to give a confidence interval for the mean number of seeds produced by velvetleaf plants. Alas, the *t* interval can't be safely used for these data. Why not?

20.42 **Sweetening colas.** Cola makers test new recipes for loss of sweetness during storage. Trained tasters rate the sweetness before and after storage. Here are the sweetness losses (sweetness before storage minus sweetness after storage) found by 10 tasters for one new cola recipe: ▦ COLA

2.0 0.4 0.7 2.0 −0.4 2.2 −1.3 1.2 1.1 2.3

Take the data from these 10 carefully trained tasters as an SRS from a large population of all trained tasters.

(a) Use these data to see if there is good evidence that the cola lost sweetness.

(b) It is not uncommon to see the *t* procedures used for data like these. However, you should regard the results as only rough approximations. Why?

20.43 **How much oil?** How much oil wells in a given field will ultimately produce is key information in deciding whether to drill more wells. Here are the estimated total amounts of oil recovered from 64 wells in the Devonian Richmond Dolomite area of the Michigan basin, in thousands of barrels:[27] ▦ OIL

21.7	53.2	46.4	42.7	50.4	97.7	103.1	51.9
43.4	69.5	156.5	34.6	37.9	12.9	2.5	31.4
79.5	26.9	18.5	14.7	32.9	196	24.9	118.2
82.2	35.1	47.6	54.2	63.1	69.8	57.4	65.6
56.4	49.4	44.9	34.6	92.2	37.0	58.8	21.3
36.6	64.9	14.8	17.6	29.1	61.4	38.6	32.5
12.0	28.3	204.9	44.5	10.3	37.7	33.7	81.1
12.1	20.1	30.5	7.1	10.1	18.0	3.0	2.0

Take these wells to be an SRS of wells in this area.

(a) Give a 95% *t* confidence interval for the mean amount of oil recovered from all wells in this area.

(b) Make a graph of the data. The distribution is very skewed, with several high outliers. A computer-intensive method that gives accurate confidence intervals without assuming any specific shape for the distribution gives a 95% confidence interval of 40.28 to 60.32. How does the *t* interval compare with this? Should the *t* procedures be used with these data?

20.44 **E. coli in swimming areas.** To investigate water quality, the *Columbus Dispatch* took water samples at 16 Ohio State Park swimming areas in central Ohio. Those samples were taken to laboratories and tested for *E. coli*, which are bacteria that can cause serious gastrointestinal problems. If a 100-milliliter sample (about 3.3 ounces) of water

contains more than 130 *E. coli* bacteria, it is considered unsafe. Here are the *E. coli* levels found by the laboratories:[28] ▦ ECOLI

291.0	190.4	47.0	86.0	44.0	18.9	1.0	50.0
10.9	45.7	28.5	8.6	9.6	16.0	34.0	18.9

Take these water samples to be an SRS of the water in all swimming areas in central Ohio.

(a) Are these data good evidence that on average the *E. coli* levels in these swimming areas were unsafe?

(b) Make a graph of the data. The distribution is very skewed. Another method that gives *P*-values without assuming any specific shape for the distribution gives a *P*-value of 0.9997 for the question in part (a). How does the one-sample *t* test compare with this? Should the *t* procedures be used with these data?

The following exercises ask you to answer questions from data without having the details outlined for you. The four-step process is illustrated in Examples 20.3, 20.4, and 20.5. The exercise statements give you the State step. Follow the **Plan, Solve,** *and* **Conclude** *steps in your work.*

20.45 **Natural weed control?** Fortunately, we aren't really interested in the number of seeds velvetleaf plants produce (see Exercise 20.41). The velvetleaf seed beetle feeds on the seeds and might be a natural weed control. Here are the total seeds, seeds infected by the beetle, and percent of seeds infected for 28 velvetleaf plants: ▦ WDCTRL

Seeds	2450	2504	2114	1110	2137	8015	1623	1531	2008	1716
Infected	135	101	76	24	121	189	31	44	73	12
Percent	5.5	4.0	3.6	2.2	5.7	2.4	1.9	2.9	3.6	0.7
Seeds	721	863	1136	2819	1911	2101	1051	218	1711	164
Infected	27	40	41	79	82	85	42	0	64	7
Percent	3.7	4.6	3.6	2.8	4.3	4.0	4.0	0.0	3.7	4.3
Seeds	2228	363	5973	1050	1961	1809	130	880		
Infected	156	31	240	91	137	92	5	23		
Percent	7.0	8.5	4.0	8.7	7.0	5.1	3.8	2.6		

Do a complete analysis of the percent of seeds infected by the beetle. Include a 90% confidence interval for the mean percent infected in the population of all velvetleaf plants. Do you think that the beetle is very helpful in controlling the weed?

20.46 **Recruiting T cells.** There is evidence that cytotoxic T lymphocytes (T cells) participate in controlling tumor growth and that they can be harnessed to use the body's immune system to treat cancer. One study investigated the use of a T cell-engaging antibody, blinatumomab, to recruit T cells to control tumor growth. The data below are T cell counts (1000 per microliter) at baseline (beginning of the study) and after 20 days on blinatumomab for 6 subjects in the study.[29] The difference (after 20 days minus baseline) is the response variable. TCELLS

Baseline	0.04	0.02	0.00	0.02	0.38	0.33
After 20 days	0.28	0.47	1.30	0.25	1.22	0.44
Difference	0.24	0.45	1.30	0.23	0.84	0.11

Do the data give convincing evidence that the mean count of T cells is higher after 20 days on blinatumomab?

20.47 Recruiting T cells, continued. Give a 95% confidence interval for the mean difference in T cell counts (after 20 days minus baseline) in the previous exercise. TCELLS

20.48 Mutual funds performance. Mutual funds often compare their performance with a benchmark provided by an "index" that describes the performance of the class of assets in which the fund invests. For example, the Vanguard International Growth Fund benchmarks its performance against the EAFE (Europe, Australasia, Far East) index. Table 20.4 gives annual returns (percent) for the fund and the index. Does the fund's performance differ significantly from that of its benchmark? MFUND

(a) Explain clearly why the matched pairs t test is the proper choice to answer this question.

(b) Do a complete analysis that answers the question posed.

TABLE 20.4 A MUTUAL FUND VERSUS ITS BENCHMARK INDEX

YEAR	FUND RETURN (%)	INDEX RETURN (%)	YEAR	FUND RETURN (%)	INDEX RETURN (%)
1984	−1.02	7.38	1998	16.93	20.00
1985	56.94	56.16	1999	26.34	26.96
1986	56.71	69.44	2000	−8.60	−14.17
1987	12.48	24.63	2001	−18.92	−21.44
1988	11.61	28.27	2002	−17.79	−15.94
1989	24.76	10.54	2003	34.45	38.59
1990	−12.05	−23.45	2004	18.95	20.25
1991	4.74	12.13	2005	15.00	13.54
1992	−5.79	−12.17	2006	25.92	26.34
1993	44.74	32.56	2007	15.98	11.17
1994	0.76	7.78	2008	−44.94	−43.38
1995	14.89	11.21	2009	41.63	31.78
1996	14.65	6.05	2010	15.66	8.13
1997	4.12	1.78			

20.49 Right versus left. The design of controls and instruments affects how easily people can use them. Timothy Sturm investigated this effect in a course project, asking 25 right-handed students to turn a knob (with their right hands) that moved an indicator by screw action. There were two identical instruments, one with a right-hand thread (the knob turns clockwise) and the other with a left-hand thread (the knob turns counterclockwise). Table 20.5 gives the times in seconds each subject took to move the indicator a fixed distance.[30] RTLFT

(a) Each of the 25 students used both instruments. Explain briefly how you would use randomization in arranging the experiment.

(b) The project hoped to show that right-handed people find right-hand threads easier to use. Do an analysis that leads to a conclusion about this issue.

TABLE 20.5 PERFORMANCE TIMES (SECONDS) USING RIGHT-HAND AND LEFT-HAND THREADS

SUBJECT	RIGHT THREAD	LEFT THREAD	SUBJECT	RIGHT THREAD	LEFT THREAD
1	113	137	14	107	87
2	105	105	15	118	166
3	130	133	16	103	146
4	101	108	17	111	123
5	138	115	18	104	135
6	118	170	19	111	112
7	87	103	20	89	93
8	116	145	21	78	76
9	75	78	22	100	116
10	96	107	23	89	78
11	122	84	24	85	101
12	103	148	25	88	123
13	116	147			

20.50 Comparing two drugs. Makers of generic drugs must show that they do not differ significantly from the "reference" drugs that they imitate. One aspect in which drugs might differ is their extent of absorption in the blood. Table 20.6 gives data taken from 20 healthy nonsmoking male subjects for one pair of drugs.[31] This is a matched pairs design. Numbers 1 to 20 were assigned at random to the subjects. Subjects 1 to 10 received the generic drug first, followed by the reference drug. Subjects 11 to 20 received the reference drug first, followed by the generic drug. In all cases, a washout period separated the two drugs so that the first had disappeared from the blood before the subject took the second. By randomizing the order, we eliminate the order in which the drugs were administered from being confounded with the difference in the absorption in the blood. Do the drugs differ significantly in the amount absorbed in the blood? DRUGS

TABLE 20.6 ABSORPTION EXTENT FOR TWO VERSIONS OF A DRUG

SUBJECT	REFERENCE DRUG	GENERIC DRUG
15	4108	1755
3	2526	1138
9	2779	1613
13	3852	2254
12	1833	1310
8	2463	2120
18	2059	1851
20	1709	1878
17	1829	1682
2	2594	2613
4	2344	2738
16	1864	2302
6	1022	1284
10	2256	3052
5	938	1287
7	1339	1930
14	1262	1964
11	1438	2549
1	1735	3340
19	1020	3050

20.51 **Practical significance?** Give a 90% confidence interval for the mean time advantage of right-hand over left-hand threads in the setting of Exercise 20.49. Do you think that the time saved would be of practical importance if the task were performed many times—for example, by an assembly-line worker? To help answer this question, find the mean time for right-hand threads as a percent of the mean time for left-hand threads. ▦ DRUGS

20.52 **Bad weather, bad tips?** As part of the study of tipping in a restaurant that we met in Example 20.3 (page 440) the psychologists also studied the size of the tip in a restaurant when a message indicating that the next day's weather would be bad was written on the bill. Here are tips from 20 patrons, measured in percent of the total bill:[32] ▦ TIP3

```
18.0  19.1  19.2  18.8  18.4  19.0  18.5  16.1  16.8  14.0
17.0  13.6  17.5  20.0  20.2  18.8  18.0  23.2  18.2  19.4
```

Do the data give convincing evidence that the mean percentage tip for all patrons of this restaurant when their bill contains a message that the next day's weather will be bad is less than 20%? (20% is an often-recommended size for restaurant tips.)

20.53 **Tests from confidence intervals.** A confidence interval for the population mean μ tells us which values of μ are plausible (those inside the interval) and which values are not plausible (those outside the interval) at the chosen level of confidence. You can use this idea to carry out a test of any null hypothesis H_0: $\mu = \mu_0$ starting with a confidence interval: *reject H_0 if μ_0 is outside the interval and fail to reject if μ_0 is inside the interval.*

The alternative hypothesis is always two-sided, H_a: $\mu \neq \mu_0$, because the confidence interval extends in both directions from $\bar{x}$. A 95% confidence interval leads to a test at the 5% significance level because the interval is wrong 5% of the time. In general, confidence level C leads to a test at significance level $\alpha = 1 - C$.

(a) In Example 20.3, psychologists found the mean percentage tip to be $\bar{x} = 22.1$ for 20 patrons. The standard deviation of the percentage tip was $s = 1.963$. Give a 90% confidence interval for the mean percentage tip μ of all patrons of this restaurant when their bill contains a message that the next day's weather will be good.

(b) The hypothesized value $\mu_0 = 22$ falls *inside* this confidence interval. Carry out the t test for H_0: $\mu = 22$ against the two-sided alternative. Show that the test is *not significant* at the 10% level.

(c) The hypothesized value $\mu_0 = 21$ falls *outside* this confidence interval. Carry out the t test for H_0: $\mu = 21$ against the two-sided alternative. Show that the test *is significant* at the 10% level.

20.54 **Tests from confidence intervals.** A 95% confidence interval for a population mean is 30.7 ± 3.2. Use the method described in the previous exercise to answer these questions.

(a) With a two-sided alternative, can you reject the null hypothesis that $\mu = 33$ at the 5% ($\alpha = 0.05$) significance level? Why?

(b) With a two-sided alternative, can you reject the null hypothesis that $\mu = 34$ at the 5% significance level? Why?

20.55 **Tests from confidence intervals.** The previous two exercises might suggest that you can also use a confidence interval for a population proportion p to carry out a test of any null hypothesis H_0: $p = p_0$ starting with a confidence interval: *reject H_0 if p_0 is outside the interval and fail to reject if p_0 is inside the interval* when the alternative hypothesis is two-sided. Unfortunately, this is only approximately correct. To see this, you will work out an example. You will need to use the methods for inference about a population proportion discussed in Chapters 16 and 17.

(a) You wish to determine the proportion p of students at a large state university who are from in-state. You take an SRS of 33 students and find that 22 are from in-state. Use the method discussed in

Chapter 17 (page 383) to test for H_0: $p = 0.5$ against the two-sided alternative at the 5% ($\alpha = 0.05$) significance level. What do you conclude?

(b) Use the data in (a) to find a 95% confidence interval for p. If you use this interval to test for H_0: $p = 0.5$ against the two-sided alternative at the 5% ($\alpha = 0.05$) significance level (rejecting H_0 if 0.5 is outside the interval), what do you conclude?

Notice that it is only when the hypothesized value of p_0 is very close to the boundary of the confidence interval for p that the test based on the confidence interval may disagree with the test discussed in Chapter 17.

Exploring the Web

20.56   **A matched pairs study.** Find an example of a matched pairs study on the Web. The *Journal of the American Medical Association* (jama.ama-assn.org), *Science Magazine* (www.sciencemag.org), the *Canadian Medical Association Journal* (www.cmaj.ca), the *Journal of Statistics Education* (www.amstat.org/publications/jse), or perhaps the *Journal of Quantitative Analysis in Sports* (www.degruyter.com/view/j/jqas) are possible sources. To help locate an article, look through the abstracts of articles. Once you find a suitable article, read it, and then briefly describe the study (including why it is a matched pairs study) and its conclusions. If *P*-values, means, standard deviations, and so on are reported, be sure to include them in your summary. Also, be sure to give the reference (either the Web link or the journal, issue, year, title of the paper, authors, and page numbers).

20.57   **An improper use of a *t* procedure.** Search the Web for an example of an improper use of a *t* procedure. You might try using Google to do a search on "improper use of a t-test." Summarize the study in which the *t* procedure was used and discuss how it was used improperly. Be sure to provide a link to your example or an appropriate reference.

20.58   **How big a sample size do you need?** In practice, people use software rather than tables such as Table C to construct confidence intervals and calculate *P*-values. If you don't have access to statistical software, but do have access to the Internet, you can find applets that compute *t* probabilities. Professor R. Webster West, Department of Statistics, North Carolina State University, has created such an applet. The link to the applet is www.stat.tamu.edu/~west/applets/tdemo.html.

If you examine Table C you will notice that critical values of the *t* distribution get closer and closer to the corresponding critical values of the Normal distribution as the number of degrees of freedom increase. You can see this by comparing the *z* critical values at the bottom of Table C with the *t* critical values in the corresponding column.

Use Professor West's applet to determine how large a sample size (or how many degrees of freedom) is needed for the critical value of the *t* distribution to be within 0.001 of the corresponding critical value of the Normal distribution for a 95%, and 99% confidence interval for a population mean.

Comparing Two Means

CHAPTER 21

Overview

The subject of comparing two population means, covered in this chapter, represents another version of a keystone topic in the course because a key application for statistical inference is the problem of comparing two treatments. Researchers are often concerned with problems such as which of two drugs is more effective (e.g., which provides relief in a shorter period of time), or whether a new teaching method is superior to an old teaching method in teaching children to read (e.g., which method results in better test scores).

The standard way to compare the means of two population means, μ_1 and μ_2, is through estimation of their difference, $\mu_1 - \mu_2$. For example, if this difference is estimated to be a large positive number, the conclusion may be reached that μ_1 is larger than μ_2. Similarly, if the difference is estimated to be near zero, the conclusion can be made that the populations may have nearly the same mean. Here, the words "large" and "near" are determined, as always, relative to the standard error of the estimator, $\bar{x}_1 - \bar{x}_2$.

There are two options for computing degrees of freedom when using a two-sample t procedure. When technology is used, software can obtain accurate critical values using degrees of freedom obtained via an advanced method. Details of this initial option for the t distribution are addressed in an optional section later in the chapter. When technology is not used, the more conservative, but simpler, option of basing degrees of freedom on the size of the smaller sample is suggested. The chapter ends with a few advanced topics presented in optional sections.

Hopefully, students will begin to feel a sense of redundancy in studying this material. The ideas explored here are, in essence, repackaged and reformulated for the particular setting in question—this material does for two population means what the material in Chapter 18 did for two population proportions.

LEARNING OUTCOMES**

- Verify that the two-sample t procedures are appropriate in a particular setting. Check the study design and the distribution of the data and take advantage of robustness against lack of Normality.

- Give a confidence interval for the difference between two means. Use software if you have it. Use the two-sample t statistic with conservative degrees of freedom and Table C if you do not have statistical software.

- Test the hypothesis that two populations have equal means against either a one-sided or two-sided alternative. Use software if you have it. Use the two-sample t test with conservative degrees of freedom and Table C if you do not have statistical software.

- Know that procedures for comparing the standard deviations of two Normal populations are available, but that these procedures are risky because they are not at all robust against non-Normal distributions.

**These learning outcomes appear later for the students in Chapter 23: Part IV Review.

Teaching Suggestions and Additional Examples/Activities for the Classroom

1. Make Ties Between New Methods and Methods Students Already Know

The methods studied in this chapter are natural extensions of those studied earlier. To many students, formulas such the standard error of $\bar{x}_1 - \bar{x}_2$ will look new, but the overall process of constructing a confidence interval or conducting a test of significance remains the same. Remind students of how we compared two population proportions by estimating the difference $p_1 - p_2$, and how the statistic $\hat{p}_1 - \hat{p}_2$ served as an estimator. Review the basic construction of a confidence interval for $p_1 - p_2$, and how the standard error of $\hat{p}_1 - \hat{p}_2$ is constructed. This should lead comfortably to the construction of a standard error for $\bar{x}_1 - \bar{x}_2$, especially if you have also reviewed confidence intervals for a single mean.

2. Focus on the Difference Between Two-Sample Problems and Matched-Pairs Problems

Students also often struggle to distinguish between two-sample problems covered in this chapter and matched-pairs problems covered in the previous chapter. In matched-pairs data, we cannot rearrange the data in one of the "rows" without loss of information (ruining the data). In a real two-sample problem, the samples are independent, so order within a sample does not matter. Make sure that your students understand that they cannot just rely on the sample size(s) to determine whether the data are paired or not—independent samples may have equal sample sizes, but this does not mean that we have paired data.

3. Use Real Data from Your Class for Examples

You can illustrate the computations with a fun in-class activity involving the students. Construct a confidence interval for the difference in mean height of women and men. Be sure to discuss whether it is reasonable to infer to the population of all students at the school (or nationwide) based on the sample of students in your class.

4. Touch On Pooled Procedures If You Plan to Teach ANOVA

You might choose to spend time covering material in one of the optional sections at the end of the chapter. One reason to at least touch on the pooled procedures is if you plan to cover ANOVA in Chapter 26. ANOVA has as an assumption that the population standard deviations are equal and is an extension of the pooled t test.

5. (Optional) Let Students Calculate Degrees of Freedom for Option 1

A strong group of students can deal with details of the t distribution's degrees of freedom computation, especially if armed with a calculator that will work the details. Any of the Texas Instruments TI-84+, TI-89, or TI-NSpire series will do this, as will any of a number of scientific calculators with statistics functionality. You can demonstrate that the fast, simple degrees of freedom formula for hand-calculations is (sometimes absurdly) more conservative than the more accurate version presented here.

Other Resources (LaunchPad)

Snapshots Videos
 Comparing Two Means

EESEE Case Studies
 Is Friday the 13th Unhealthy?
 Leave Survey at the Beep
 Sleeping Patterns in Ducks
 Keeping Balance as You Age

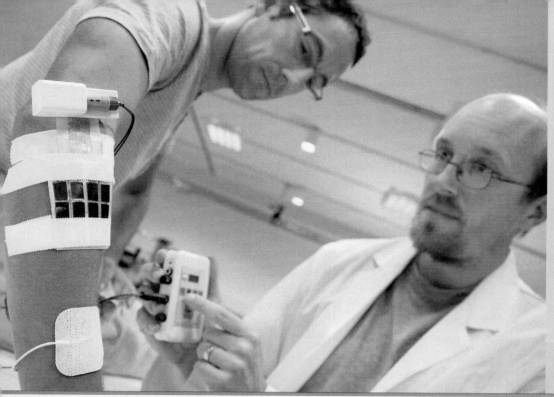

© Javier Larrea/age fotostock

Comparing Two Means

CHAPTER 21

As we discussed in Chapter 18, comparing two populations or two treatments is one of the most common situations encountered in statistical practice. In Chapter 18 we studied two-sample problems when the goal is to compare proportions. Now we study two-sample problems when the goal is to compare means.

Two-sample problems: means

We use random samples separately selected from two populations to compare the population means. Unlike the matched pairs designs studied in Chapter 20, there is no matching of the individuals in the two samples, and the two samples can be of different sizes. Inference procedures for means for two-sample data differ from those for matched pairs. Here are some typical two-sample problems for means:

EXAMPLE 21.1   Two-Sample Problems for Means

- Does regular physical therapy help lower-back pain? A randomized experiment assigned patients with lower-back pain to two groups: 142 received an examination and advice from a physical therapist; another 144 received regular physical therapy for up to five weeks. After a year, the change in their level of pain (measured on a scale of 0 to 30, with 0 meaning no pain) was assessed by a doctor who did not know which treatment the patients had received.

463

- A psychologist develops a test that measures social insight. He compares the social insight of female college students with that of male college students by giving the test to a sample of female students and a separate sample of male students.

- A bank wants to know which of two incentive plans will most increase the use of its credit cards. It offers each incentive to a random sample of credit card customers and compares the amounts charged during the following six months. ■

Apply Your Knowledge

DRIVING WHILE FASTING

Muslims fast from sunrise to sunset during the month of Ramadan. Does this affect the rate of traffic accidents? Fasting can improve alertness, reducing accidents. Or it can cause dehydration, increasing accidents. Data from Turkey show a statistically significant increase, starting two weeks into Ramadan. Ah, but because Ramadan follows a lunar calendar, it cycles through the year. Perhaps accidents go down during a winter Ramadan (alertness) but go up during a summer Ramadan (longer fast and dehydration). Ask the statisticians this question, and get their favorite answer: we need more data.

Which Data Design? *Each situation described in Exercises 21.1 to 21.4 requires inference about a mean or means. Identify each as involving (1) a single sample, (2) matched pairs, or (3) two independent samples. The procedures of Chapter 20 apply to designs (1) and (2). We are about to learn procedures for (3).*

21.1 Managing the Finances. Choose 50 engaged couples who have not been previously married. Interview the man and woman separately about how their joint finances will be handled after marriage. Compare the views of men and women.

21.2 Does Peer Discussion Promote Learning? Undergraduate students in a biology class were randomly divided into two groups. In the first group, students worked alone on a multiple-choice exam on material recently covered in the class. In the second group, students worked in teams of four on the same multiple-choice exam. The teams were encouraged to discuss the questions among themselves before answering the questions. All exams were graded and returned to the students. Then all students worked alone on another multiple-choice exam on the same material as the first exam. Compare the mean scores of the two groups on the second exam.

21.3 Chemical Analysis. To check a new analytical method, a chemist obtains a reference specimen of known concentration from the National Institute of Standards and Technology. She then makes 20 measurements of the concentration of this specimen with the new method and checks for bias by comparing the mean result with the known concentration.

21.4 Chemical Analysis, Continued. Another chemist is checking the same new method. He has no reference specimen, but a familiar analytic method is available. He wants to know if the new and old methods agree. He takes a specimen of unknown concentration and measures the concentration 10 times with the new method and 10 times with the old method.

Comparing two population means

Comparing two populations or the responses to two treatments starts with data analysis: make boxplots, stemplots (for small samples), or histograms (for larger samples) and compare the shapes, centers, and spreads of the two samples. The most common goal of inference is to compare the average or typical responses in the two populations. When data analysis suggests that both population distributions are symmetric, and especially when they are at least approximately Normal, we want to compare the population means. Here are the conditions for inference about means.

Conditions for Inference Comparing Two Means

■ We have **two SRSs,** from two distinct populations. The samples are **independent.** That is, one sample has no influence on the other. Matching violates independence, for example. We measure the same response variable for both samples.

■ Both populations are **Normally distributed.** The means and standard deviations of the populations are unknown. In practice, it is enough that the distributions have similar shapes and that the data have no strong outliers.

■ For observations from a population that is not Normally distributed and with moderate skewness and no outliers, the sum of the two sample sizes should be at least 15.

■ For observations from a population that is not Normally distributed and with strong skewness and no outliers, the sum of the two sample sizes should be at least 40.

The last two conditions concerning the sample sizes will be discussed in more detail later in this chapter.

Call the variable we measure x_1 in the first population and x_2 in the second because the variable may have different distributions in the two populations. Here is how we describe the two populations:

Population	Variable	Mean	Standard Deviation
1	x_1	μ_1	σ_1
2	x_2	μ_2	σ_2

There are four unknown parameters: the two means and the two standard deviations. The subscripts remind us which population a parameter describes. We want to compare the two population means, either by giving a confidence interval for their difference $\mu_1 - \mu_2$ or by testing the hypothesis of no difference, H_0: $\mu_1 = \mu_2$.

We use the sample means and standard deviations to estimate the unknown parameters. Again, subscripts remind us which sample a statistic comes from. Here is how we describe the samples:

Population	Sample Size	Sample Mean	Sample Standard Deviation
1	n_1	$\bar{x}_1$	s_1
2	n_2	$\bar{x}_2$	s_2

To do inference about the difference $\mu_1 - \mu_2$ between the means of the two populations, we start from the difference $\bar{x}_1 - \bar{x}_2$ between the means of the two samples.

EXAMPLE 21.2   Daily Activity and Obesity

STATE: People gain weight when they take in more energy from food than they expend. James Levine and his collaborators at the Mayo Clinic investigated the link between obesity and energy spent on daily activity.[1]

Choose 20 healthy volunteers who don't exercise. Deliberately choose 10 who are lean and 10 who are mildly obese but still healthy. Attach sensors that monitor the subjects' every move for 10 days. Table 21.1 presents data on the time (in minutes per day) that the subjects spent standing or walking, sitting, and lying

ACTIVE

TABLE 21.1 TIME (MINUTES PER DAY) SPENT IN THREE DIFFERENT POSTURES BY LEAN AND OBESE SUBJECTS

GROUP	SUBJECT	STAND/WALK	SIT	LIE
Lean	1	511.100	370.300	555.500
Lean	2	607.925	374.512	450.650
Lean	3	319.212	582.138	537.362
Lean	4	584.644	357.144	489.269
Lean	5	578.869	348.994	514.081
Lean	6	543.388	385.312	506.500
Lean	7	677.188	268.188	467.700
Lean	8	555.656	322.219	567.006
Lean	9	374.831	537.031	531.431
Lean	10	504.700	528.838	396.962
Obese	11	260.244	646.281	521.044
Obese	12	464.756	456.644	514.931
Obese	13	367.138	578.662	563.300
Obese	14	413.667	463.333	532.208
Obese	15	347.375	567.556	504.931
Obese	16	416.531	567.556	448.856
Obese	17	358.650	621.262	460.550
Obese	18	267.344	646.181	509.981
Obese	19	410.631	572.769	448.706
Obese	20	426.356	591.369	412.919

down. Do lean and obese people differ in the average time they spend standing and walking?

PLAN: Examine the data and carry out a test of hypotheses. We suspect in advance that lean subjects (Group 1) are more active than obese subjects (Group 2), so we test the hypotheses

$$H_0: \mu_1 = \mu_2$$
$$H_a: \mu_1 > \mu_2$$

SOLVE (first steps): Are the conditions for inference met? The subjects are volunteers, so they are not SRSs from all lean and mildly obese adults. The study tried to recruit comparable groups: all worked in sedentary jobs, none smoked or were taking medication, and so on. Setting clear standards like these helps make up for the fact that we can't reasonably get SRSs for so invasive a study. The subjects were not told that they were chosen from a larger group of volunteers because they did not exercise and were either lean or mildly obese. Because their willingness to volunteer isn't related to the purpose of the experiment, we will treat them as two independent SRSs.

A back-to-back stemplot (Figure 21.1) displays the data in detail. To make the plot, we rounded the data to the nearest 10 minutes and used 100s as stems and 10s as leaves. The distributions are a bit irregular, as we expect with just 10 observations. There are no clear departures from Normality such as extreme outliers or skewness. The lean subjects as a group spent much more time standing

```
   Lean        Obese
           2 | 6 7
        72 | 3 | 5 6 7
           4 | 1 1 2 3 6
   886410 | 5 |
        81 | 6 |
```

FIGURE 21.1

Back-to-back stemplot of the times spent walking or standing, for Example 21.2.

and walking than did the obese subjects. Calculating the group means confirms this:

Group	*n*	Mean $\bar{x}$	Standard Deviation *s*
Group 1 (lean)	10	525.751	107.121
Group 2 (obese)	10	373.269	67.498

The observed difference in mean time per day spent standing or walking is

$$\bar{x}_1 - \bar{x}_2 = 525.751 - 373.269 = 152.482 \text{ minutes}$$

To complete the "Solve" step, we must learn the details of inference for comparing two means. ■

Two-sample *t* procedures

To assess the significance of the observed difference between the means of our two samples, we follow a familiar path. Whether an observed difference is surprising depends on the spread/variability of the observations as well as on the two means. Widely different means can arise just by chance if the individual observations vary a great deal. To take variation into account, we would like to standardize the observed difference $\bar{x}_1 - \bar{x}_2$ by subtracting its mean, $\mu_1 - \mu_2$, and dividing the result by its standard deviation. This standard deviation of the difference in sample means is

$$\sqrt{\frac{\sigma_1^2}{n_1} + \frac{\sigma_2^2}{n_2}}$$

This standard deviation gets larger as either population gets more variable, that is, as σ_1 or σ_2 increases. It gets smaller as the sample sizes n_1 and n_2 increase.

Because we don't know the population standard deviations, we estimate them by the sample standard deviations from our two samples. The result is the **standard error,** or estimated standard deviation, of the difference in sample means:

standard error

$$\sqrt{\frac{s_1^2}{n_1} + \frac{s_2^2}{n_2}}$$

When we standardize the estimate by subtracting its mean, $\mu_1 - \mu_2$, and dividing the result by its standard error, the result is the **two-sample *t* statistic:**

two-sample t statistic

$$t = \frac{(\bar{x}_1 - \bar{x}_2) - (\mu_1 - \mu_2)}{\sqrt{\dfrac{s_1^2}{n_1} + \dfrac{s_2^2}{n_2}}}$$

The two-sample statistic *t* has the same interpretation as any *z* or *t* statistic: it says how far $\bar{x}_1 - \bar{x}_2$ is from $\mu_1 - \mu_2$ in standard deviation units. For purposes of testing the hypotheses with null hypothesis

$$H_0: \mu_1 = \mu_2$$

$\mu_1 - \mu_2 = 0$ and the two-sample *t* statistic simplifies to

$$t = \frac{\bar{x}_1 - \bar{x}_2}{\sqrt{\dfrac{s_1^2}{n_1} + \dfrac{s_2^2}{n_2}}}$$

The two-sample *t* statistic has approximately a *t* distribution. It does not have exactly a *t* distribution even if the populations are both exactly Normal. In practice, however, the approximation is very accurate. There are two practical options for using the two-sample t procedures:

Option 1. Use software. Software uses the statistic t with accurate critical values from the approximating t distribution. The degrees of freedom are calculated from the data by a somewhat messy formula. Moreover, the degrees of freedom may not be a whole number.

Option 2. Without software, use the statistic t with critical values from the t distribution with *degrees of freedom equal to the smaller of $n_1 - 1$ and $n_2 - 1$.* These procedures are always conservative for any two Normal populations. The confidence interval has a margin of error *as large as or larger than* is needed for the desired confidence level. The significance test gives a P-value *equal to or greater than* the true P-value.

The two options are exactly the same except for the degrees of freedom used for t critical values and P-values. As the sample sizes increase, confidence levels and P-values from Option 2 become more accurate. The gap between what Option 2 reports and the truth is quite small unless the sample sizes are both small and unequal.[2]

The Two-Sample t Procedures

Draw an SRS of size n_1 from a large Normal population with unknown mean μ_1, and draw an independent SRS of size n_2 from another large Normal population with unknown mean μ_2. A level C **confidence interval for $\mu_1 - \mu_2$** is given by

$$(\bar{x}_1 - \bar{x}_2) \pm t^* \sqrt{\frac{s_1^2}{n_1} + \frac{s_2^2}{n_2}}$$

Here t^* is the critical value for confidence level C for the t distribution with degrees of freedom from either Option 1 (software) or Option 2 (the smaller of $n_1 - 1$ and $n_2 - 1$).

To **test the hypothesis $H_0: \mu_1 = \mu_2$,** calculate the **two-sample t statistic**

$$t = \frac{\bar{x}_1 - \bar{x}_2}{\sqrt{\dfrac{s_1^2}{n_1} + \dfrac{s_2^2}{n_2}}}$$

Find P-values from the t distribution with degrees of freedom from either Option 1 (software) or Option 2 (the smaller of $n_1 - 1$ and $n_2 - 1$).

EXAMPLE 21.3   Daily Activity and Obesity

We can now complete Example 21.2.

SOLVE (inference): The two-sample t statistic comparing the average minutes spent standing and walking in Group 1 (lean) and Group 2 (obese) is

$$t = \frac{\bar{x}_1 - \bar{x}_2}{\sqrt{\dfrac{s_1^2}{n_1} + \dfrac{s_2^2}{n_2}}}$$

$$= \frac{525.751 - 373.269}{\sqrt{\dfrac{107.121^2}{10} + \dfrac{67.498^2}{10}}}$$

$$= \frac{152.482}{40.039} = 3.808$$

Software (Option 1) gives a one-sided *P*-value $P = 0.0008$ based on df = 15.174.

Without software, use the conservative Option 2. Because $n_1 - 1 = 9$ and $n_2 - 1 = 9$, there are 9 degrees of freedom. Because H_a is one-sided, the *P*-value is the area to the right of $t = 3.808$ under the $t(9)$ curve. Figure 21.2 illustrates this *P*-value. Table C shows that $t = 3.808$ lies between the critical values t^* for 0.0025 and 0.001. So $0.001 < P < 0.0025$. Option 2 gives a larger (more conservative) *P*-value than Option 1. As usual, the practical conclusion is the same for both versions of the test.

df = 9		
t^*	3.690	4.297
One-sided *P*	.0025	.001

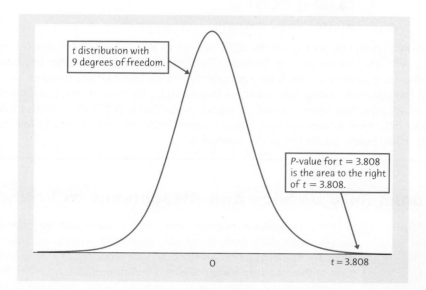

FIGURE 21.2
Using the conservative Option 2, the *P*-value in Example 21.3 comes from the *t* distribution with 9 degrees of freedom.

CONCLUDE: There is very strong evidence ($P = 0.0008$) that, on average, lean people spend more time walking and standing than do moderately obese people. ■

Does lack of daily activity *cause* obesity? This is an observational study, and that affects our ability to draw cause-and-effect conclusions. It may be that some people are naturally more active and are therefore less likely to gain weight. Or it may be that people who gain weight reduce their activity level. The study went on to enroll most of the obese subjects in a weight-reduction program and most of the lean subjects in a supervised program of overeating. After 8 weeks, the obese subjects had lost weight (mean 8 kilograms) and the lean subjects had gained weight (mean 4 kilograms). But both groups kept their original allocation of time to the different postures. This suggests that time allocation is biological and influences weight, rather than the other way around. The authors remarked: "It should be emphasized that this was a pilot study and that the results need to be confirmed in larger studies."

EXAMPLE 21.4   How Much More Active Are Lean People?

PLAN: Give a 90% confidence interval for $\mu_1 - \mu_2$, the difference in average daily minutes spent standing and walking between lean and mildly obese adults.

SOLVE and CONCLUDE: As in Example 21.3, the conservative Option 2 uses 9 degrees of freedom. Table C shows that the $t(9)$ critical value is $t^* = 1.833$. We are 90% confident that $\mu_1 - \mu_2$ lies in the interval

$$(\bar{x}_1 - \bar{x}_2) \pm t^* \sqrt{\frac{s_1^2}{n_1} + \frac{s_2^2}{n_2}}$$

$$= (525.751 - 373.269) \pm 1.833 \sqrt{\frac{107.121^2}{10} + \frac{67.498^2}{10}}$$

$$= 152.482 \pm 73.391$$

$$= 79.09 \text{ to } 225.87 \text{ minutes}$$

Software using Option 1 gives the 90% interval as 82.35 to 222.62 minutes, based on t with 15.174 degrees of freedom. The Option 2 interval is wider because this method is conservative. Both intervals are quite wide because the samples are small and the variation among individuals, as measured by the two sample standard deviations, is large. Whichever interval we report, we are (at least) 90% confident that the mean difference in average daily minutes spent standing and walking between lean and mildly obese adults lies in this interval. ■

EXAMPLE 21.5   Community Service and Attachment to Friends

Hero Images/Getty Images

STATE: Do college students who have volunteered for community service work differ from those who have not? A study obtained data from 57 students who had done service work and 17 who had not. One of the response variables was a measure of attachment to friends (roughly, secure relationships), measured by the Inventory of Parent and Peer Attachment. In particular, the response is a score based on the responses to 25 questions. Here are the results:[3]

Group	Condition	n	$\bar{x}$	s
1	Service	57	105.32	14.68
2	No service	17	96.82	14.26

PLAN: The investigator had no specific direction for the difference in mind before looking at the data, so the alternative is two-sided. We will test the hypotheses

$$H_0: \mu_1 = \mu_2$$
$$H_a: \mu_1 \neq \mu_2$$

SOLVE: The investigator says that the individual scores, examined separately in the two samples, appear roughly Normal. There is a serious problem with the more important condition that the two samples can be regarded as SRSs from two student populations. We will discuss that after we illustrate the calculations.

The two-sample t statistic is

$$t = \frac{\bar{x}_1 - \bar{x}_2}{\sqrt{\dfrac{s_1^2}{n_1} + \dfrac{s_2^2}{n_2}}}$$

$$= \frac{105.32 - 96.82}{\sqrt{\dfrac{14.68^2}{57} + \dfrac{14.26^2}{17}}}$$

$$= \frac{8.5}{3.9677} = 2.142$$

Software (Option 1) says that the two-sided *P*-value is $P = 0.0414$.

Without software, use Option 2 to find a conservative *P*-value. There are 16 degrees of freedom, the smaller of

$$n_1 - 1 = 57 - 1 = 56 \quad \text{and} \quad n_2 - 1 = 17 - 1 = 16$$

Figure 21.3 illustrates the *P*-value. Find it by comparing $t = 2.142$ with the two-sided critical values for the $t(16)$ distribution. Table C shows that the *P*-value is between 0.05 and 0.04.

df = 16		
t^*	2.120	2.235
Two-sided *P*	.05	.04

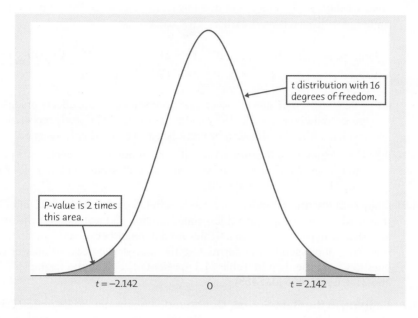

FIGURE 21.3
The *P*-value in Example 21.5. Because the alternative is two-sided, the *P*-value is double the area to the left of $t = -2.142$.

CONCLUDE: The data give moderately strong evidence ($P < 0.05$) that students who have engaged in community service are, on the average, more attached to their friends. ■

Is the *t* test in Example 21.5 justified? The student subjects were "enrolled in a course on U.S. Diversity at a large midwestern university." Unless this course is required of all students, the subjects cannot be considered a random sample even from this campus. Students were placed in the two groups on the basis of a questionnaire, 39 in the "no service" group and 71 in the "service" group. The data were gathered from a follow-up survey two years later; 17 of the 39 "no service" students responded (44%), compared with 80% response (57 of 71) in the "service" group. Nonresponse is confounded with group: students who had done community service were much more likely to respond. Finally, 75% of the "service" respondents were women, compared with 47% of the "no service" respondents. Sex, which can strongly affect attachment, is badly confounded with the presence or absence of community service. The data are so far from meeting the SRS condition for inference that the *t* test is meaningless. Difficulties like these are common in social science research, where confounding variables have stronger effects than is usual when biological or physical variables are measured. This researcher honestly disclosed the weaknesses in data production but left it to readers to decide whether to trust her inferences.

Apply Your Knowledge

Digital Vision/Getty Images

In exercises that call for two-sample t procedures, use Option 1 if you have technology that implements that method. Otherwise, use Option 2 (degrees of freedom the smaller of $n_1 - 1$ and $n_2 - 1$).

21.5 Logging in the Rain Forest. "Conservationists have despaired over destruction of tropical rain forest by logging, clearing, and burning." These words begin a report on a statistical study of the effects of logging in Borneo.[4] Here are data on the number of tree species in 12 unlogged forest plots and 9 similar plots logged 8 years earlier: **LOG2**

Unlogged	22	18	22	20	15	21	13	13	19	13	19	15
Logged	17	4	18	14	18	15	15	10	12			

(a) The study report says, "Loggers were unaware that the effects of logging would be assessed." Why is this important? The study report also explains why the plots can be considered to be randomly assigned.

(b) Does logging significantly reduce the mean number of species in a plot after 8 years? Follow the four-step process as illustrated in Examples 21.2 and 21.3 (pages 465 and 468).

21.6 Daily Activity and Obesity. We can conclude from Examples 21.2 and 21.3 that mildly obese people spend less time standing and walking (on the average) than lean people. Is there a significant difference between the mean times the two groups spend lying down? Use the four-step process to answer this question from the data in Table 21.1 (page 466). Follow the Examples 21.2 and 21.3 (pages 465 and 468).

21.7 Logging in the Rain Forest, Continued. Use the data in Exercise 21.5 to give a 99% confidence interval for the difference in mean number of species between unlogged and logged plots. **LOG2**

Using technology

Software should use Option 1 for the degrees of freedom to give accurate confidence intervals and *P*-values. Unfortunately, there is variation in how well software implements Option 1. Figure 21.4 displays output from a graphing calculator, two statistical programs, and a spreadsheet program for the test of Example 21.3. All four claim to use Option 1. The two-sample *t* statistic is exactly as in Example 21.3, $t = 3.808$. You can find this in all four outputs (though Minitab rounds to 3.81; Excel and the graphing calculator give additional decimal places). The different technologies use different methods to find the *P*-value for $t = 3.808$.

■ CrunchIt! and the calculator get Option 1 completely right. The accurate approximation uses the *t* distribution with approximately 15.174 (CrunchIt! rounds this to 15.17) degrees of freedom. The *P*-value is $P = 0.0008$.

■ Minitab uses Option 1, but it *truncates* the exact degrees of freedom to the next smaller whole number to get critical values and *P*-values. In this example, the exact df = 15.174 is truncated to df = 15, so that Minitab's results are slightly conservative. That is, Minitab's *P*-value (rounded to $P = 0.001$ in the output) is slightly larger than the full Option 1 *P*-value.

■ Excel *rounds* the exact degrees of freedom to the nearest whole number, so that df = 15.174 becomes df = 15. Excel's method agrees with Minitab's in this

Texas Instruments Graphing Calculator

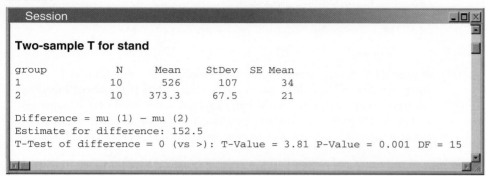

```
2-SampTTest
 μ1>μ2
 t=3.808375604
 P=8.4101904E-4
 df=15.17355038
 x̄1=525.7513
↓x̄2=373.2692
```

Minitab

```
Session                                                    _□×

Two-sample T for stand

group          N      Mean    StDev   SE Mean
1             10       526      107        34
2             10     373.3     67.5        21

Difference = mu (1) − mu (2)
Estimate for difference: 152.5
T-Test of difference = 0 (vs >): T-Value = 3.81 P-Value = 0.001 DF = 15
```

Microsoft Excel

```
Excel                                              _ □ ×
         A                    B              C
1  t-Test: Two-Sample Assuming Unequal Variances
2
3                            Lean           Obese
4  Mean                   525.7513       373.2692
5  Variance             11474.8903    4556.019849
6  Observations               10             10
7  Hypothesized Mean           0
8  df                         15
9  t Stat               3.808375604
10 P(T<=t) one-tail      0.000856818
11 t Critical one-tail   1.753051038
12 P(T<=t) two-tail      0.001713635
13 P(T<t) two-tail       2.131450856
14
  Sheet4
```

CrunchIt!

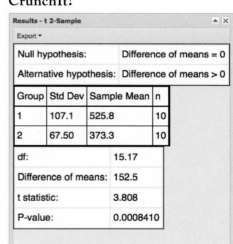

Results - t 2-Sample				
Export ▾				
Null hypothesis:		Difference of means = 0		
Alternative hypothesis:		Difference of means > 0		
Group	Std Dev	Sample Mean		n
1	107.1	525.8		10
2	67.50	373.3		10
df:		15.17		
Difference of means:		152.5		
t statistic:		3.808		
P-value:		0.0008410		

FIGURE 21.4

The two-sample *t* procedures applied to the data on activity and obesity: output from a graphing calculator, two statistical programs, and a spreadsheet program.

example. But when rounding moves the degrees of freedom up to the next higher whole number, Excel's *P*-values are slightly smaller than is correct. This is misleading, another illustration of the fact that Excel is not always as careful as other statistical software.

Excel's label for the test, "Two-Sample Assuming Unequal Variances," is seriously misleading. *The two-sample t procedures we have described work whether or not the two populations have the same variance.* There is an old-fashioned special procedure that works only when the two variances are equal. We discuss this method in an optional section on page 478, but you should never use it.

Although different calculators and software give slightly different *P*-values, in practice you can just accept what your technology says. The small differences in *P* don't affect the conclusion. Even "between 0.001 and 0.0025" from Option 2 (Example 21.3 (page 468)) is close enough for practical purposes.

META-ANALYSIS

Small samples have large margins of error. Large samples are expensive. Often we can find several studies of the same issue; if we could combine their results, we would have a large sample with a small margin of error. That is the idea of "meta-analysis." Of course, we can't just lump the studies together because of differences in design and quality. Statisticians have more sophisticated ways of combining the results. Meta-analysis has been applied to issues ranging from the effect of secondhand smoke to whether coaching improves SAT scores.

Apply Your Knowledge

21.8 Perception of Life Expectancy. Do women and men differ in how they perceive their life expectancy? A researcher asked a sample of men and women to indicate their life expectancy. This was compared with values from actuarial tables, and the relative percent difference was computed (perceived life expectancy minus life expectancy from actuarial tables was divided by life expectancy from actuarial tables and converted to a percent). Here are the relative percent differences for all men and women over the age of 70 in the sample:[5] 📊 LIFEEXP

Men	−28	−23	−20	−19	−14	−13	
Women	−20	−19	−15	−12	−10	−8	−5

Figure 21.5 shows output for the two-sample t test using Option 1. (This output is from CrunchIt! software, which does Option 1 without rounding or truncating the degrees of freedom.) Do men and women over 70 years old differ in their perceptions of life expectancy? Using the output in Figure 21.5, write a summary in a sentence or two, including t, df, P, and a conclusion.

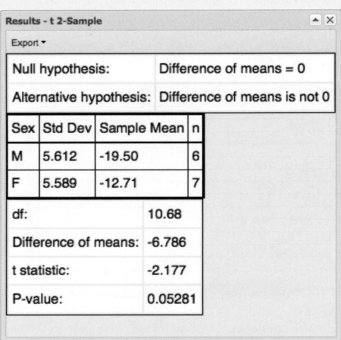

FIGURE 21.5
Two-sample t output from CrunchIt! for Exercise 21.8.

Robustness again

The two-sample t procedures are more robust than the one-sample t methods, particularly when the distributions are not symmetric. When the sizes of the two samples are equal and the two populations being compared have distributions with similar shapes, probability values from the t table are quite accurate for a broad range of distributions when the sample sizes are as small as $n_1 = n_2 = 5$.[6] When the two population distributions have different shapes, larger samples are needed.

As a guide to practice, adapt the guidelines given on page 451 for the use of one-sample t procedures to two-sample procedures by replacing "sample size" with the "sum of the sample sizes," $n_1 + n_2$. These guidelines are the same as those in the conditions for inference comparing two means on page 465, and err on the side of safety, *especially when the two samples are of equal size. In planning a two-sample study, choose equal sample sizes whenever possible. The two-sample t procedures are most robust against non-Normality in this case, and the conservative Option 2 probability values are most accurate.*

Apply Your Knowledge

21.9 Do Good Smells Bring Good Business? Businesses know that customers often respond to background music. Do they also respond to odors? One study of this question took place in a small pizza restaurant in France on two Saturday evenings in May. On one of these evenings, a relaxing lavender odor was spread through the restaurant. On the other evening, no scent was used. Table 21.2 gives the time (in minutes) that two samples of 30 customers spent in the restaurant and the amount they spent (in euros).[7] The two evenings were comparable in many ways (weather, customer count, and so on), so we

TABLE 21.2	TIME (MINUTES) AND SPENDING (EUROS) BY RESTAURANT CUSTOMERS		
NO ODOR		**LAVENDER**	
MINUTES	**EUROS SPENT**	**MINUTES**	**EUROS SPENT**
103	15.9	92	21.9
68	18.5	126	18.5
79	15.9	114	22.3
106	18.5	106	21.9
72	18.5	89	18.5
121	21.9	137	24.9
92	15.9	93	18.5
84	15.9	76	22.5
72	15.9	98	21.5
92	15.9	108	21.9
85	15.9	124	21.5
69	18.5	105	18.5
73	18.5	129	25.5
87	18.5	103	18.5
109	20.5	107	18.5
115	18.5	109	21.9
91	18.5	94	18.5
84	15.9	105	18.5
76	15.9	102	24.9
96	15.9	108	21.9
107	18.5	95	25.9
98	18.5	121	21.9
92	15.9	109	18.5
107	18.5	104	18.5
93	15.9	116	22.8
118	18.5	88	18.5
87	15.9	109	21.9
101	25.5	97	20.7
75	12.9	101	21.9
86	15.9	106	22.5

are willing to regard the data as independent SRSs from spring Saturday evenings at this restaurant. The authors say, "Therefore at this stage it would be impossible to generalize the results to other restaurants." ODORS2

(a) Does a lavender odor encourage customers to stay longer in the restaurant? Examine the time data and explain why they are suitable for two-sample t procedures. Use the two-sample t test to answer the question posed.

(b) Does a lavender odor encourage customers to spend more while in the restaurant? Examine the spending data. In what ways do these data deviate from Normality? With 30 observations, the t procedures are nonetheless reasonably accurate. Use the two-sample t test to answer the question posed.

21.10 Compressing Soil. Farmers know that driving heavy equipment on wet soil compresses the soil and injures future crops. Here are data on the "penetrability" of the same type of soil at two levels of compression.[8] Penetrability is a measure of how much resistance plant roots will meet when they try to grow through the soil. SOILCM

Compressed Soil									
2.86	2.68	2.92	2.82	2.76	2.81	2.78	3.08	2.94	2.86
3.08	2.82	2.78	2.98	3.00	2.78	2.96	2.90	3.18	3.16

Intermediate Soil									
3.14	3.38	3.10	3.40	3.38	3.14	3.18	3.26	2.96	3.02
3.54	3.36	3.18	3.12	3.86	2.92	3.46	3.44	3.62	4.26

(a) Make stemplots to investigate the shape of the distributions. The penetrabilities for intermediate soil are skewed to the right and have a high outlier. Returning to the source of the data shows that the outlying sample had unusually low soil density, so that it belongs in the "loose soil" class. We are justified in removing the outlier.

(b) We suspect that the penetrability of compressed soil is less than that of intermediate soil. Do the data (with the outlier removed) support this suspicion?

21.11 Weeds Among the Corn. Lamb's-quarter is a common weed that interferes with the growth of corn. An agriculture researcher planted corn at the same rate in 16 small plots of ground, then weeded the plots by hand to allow a fixed number of lamb's-quarter plants to grow in each meter of corn row. No other weeds were allowed to grow. Here are the yields of corn (bushels per acre) for only the experimental plots controlled to have 1 weed per meter of row and 9 weeds per meter of row:[9] WDCORN

1 weed/meter	166.2	157.3	166.7	161.1
9 weeds/meter	162.8	142.4	162.8	162.4

Explain carefully why a two-sample t confidence interval for the difference in mean yields may not be accurate.

21.12 Compressing Soil, Continued. Use the data in Exercise 21.10, omitting the outlier, to give a 90% confidence interval for the decrease in penetrability of compressed soil relative to intermediate soil. SOILCM2

Details of the *t* approximation*

The exact distribution of the two-sample *t* statistic is not a *t* distribution. Moreover, the distribution changes as the unknown population standard deviations σ_1 and σ_2 change. However, an excellent approximation is available. We call this Option 1 for *t* procedures.

Approximate Distribution of the Two-Sample *t* Statistic

The distribution of the two-sample *t* statistic is very close to the *t* distribution with degrees of freedom df given by

$$df = \frac{\left(\dfrac{s_1^2}{n_1} + \dfrac{s_2^2}{n_2}\right)^2}{\dfrac{1}{n_1 - 1}\left(\dfrac{s_1^2}{n_1}\right)^2 + \dfrac{1}{n_2 - 1}\left(\dfrac{s_2^2}{n_2}\right)^2}$$

This approximation is accurate when both sample sizes n_1 and n_2 are 5 or larger.

EXAMPLE 21.6 Daily Activity and Obesity

In the experiment of Examples 21.2 and 21.3 (pages 465 and 468), the data on minutes per day spent standing and walking give

ACTIVE

Group	n	$\bar{x}$	s
Group 1 (lean)	10	525.751	107.121
Group 2 (obese)	10	373.269	67.498

The two-sample *t* test statistic calculated from these values is $t = 3.808$.

The one-sided *P*-value is the area to the right of 3.808 under a *t* density curve, as in Figure 21.2 (page 469). The conservative Option 2 uses the *t* distribution with 9 degrees of freedom. Option 1 finds a very accurate *P*-value by using the *t* distribution with degrees of freedom df given by

$$df = \frac{\left(\dfrac{107.121^2}{10} + \dfrac{67.498^2}{10}\right)^2}{\dfrac{1}{9}\left(\dfrac{107.121^2}{10}\right)^2 + \dfrac{1}{9}\left(\dfrac{67.498^2}{10}\right)^2}$$

$$= \frac{2{,}569{,}894}{169{,}367.2} = 15.1735$$

These degrees of freedom appear in the graphing calculator output in Figure 21.4 (page 473). Because the formula is messy and roundoff errors are likely, we don't recommend calculating df by hand. ■

The degrees of freedom df is generally not a whole number. It is always at least as large as the smaller of $n_1 - 1$ and $n_2 - 1$. The larger degrees of freedom that result from Option 1 give slightly shorter confidence intervals and slightly smaller *P*-values than the conservative Option 2 produces. There is a *t* distribution for any

---

*This section can be omitted unless you are using software and wish to understand what the software does.

positive degrees of freedom, even though Table C contains entries only for whole-number degrees of freedom.

The difference between the t procedures using Options 1 and 2 is rarely of practical importance. That is why we recommend the simpler, conservative Option 2 for inference without software. With software, the more accurate Option 1 procedures are painless.

Apply Your Knowledge

21.13 Students' Self-Concept. A study of the self-concept of seventh-grade students asked if male and female students differ in mean score on the Piers-Harris Children's Self-Concept Scale.[10] Software that uses Option 1 gives these summary results:

```
Sex      n    Mean  Std dev  Std err        t    df       P
F       31 55.5161  12.6961   2.2803  -0.8276  62.8  0.4110
M       47 57.9149  12.2649   1.7890
```

Starting from the sample means and standard deviations, verify each of these entries: the standard errors of the means, the degrees of freedom for two-sample t, and the value of t.

21.14 Perception of Life Expectancy. Figure 21.5 (page 474) gives output for the life expectancy data in Exercise 21.8 (page 474) from software that does Option 1 with the correct degrees of freedom. What are $\bar{x}_i$ and s_i for the two samples? Starting from these values, find the t test statistic and its degrees of freedom. Your work should agree with Figure 21.5.

21.15 Students' Self-Concept, Continued. Write a sentence or two summarizing the comparison of female and male students in Exercise 21.13, as if you were preparing a report for publication. Use the output in Exercise 21.13.

 IS MONEY THE ROOT OF ALL EVIL?

That may go too far, but Kathleen Vohs and her coworkers show that even thinking about money has strong effects. Exercise 21.46 describes a small part of their work. What does Professor Vohs say about the consequences of having money? "Money makes people feel self-sufficient and behave accordingly." With money, you can achieve your goals with less help from others. You feel less dependent on others and more willing to work toward your own goals. Maybe that's good. You also prefer to be less involved with others, so that self-sufficiency is a barrier to close relationships with others. Maybe that's not good. Scientists don't tell us what's good or not good, just that money increases our sense of self-sufficiency.

Avoid the pooled two-sample t procedures*

Most software and graphing calculators, including all four illustrated in Figure 21.4, offer a choice of two-sample t statistics. One is often labeled for "unequal" variances; the other for "equal" variances. The "unequal" variance procedure is our two-sample t. *This test is valid whether or not the population variances are equal.* The other choice is a special version of the two-sample t statistic that assumes the two populations have the same variance. This procedure averages (the statistical term is "pools") the two sample variances to estimate the common population variance. The resulting statistic is called the *pooled two-sample t statistic.* It is equal to our t statistic if the two sample sizes are the same, but not otherwise. We could choose to use the pooled t for tests and confidence intervals.

The pooled t statistic has exactly the t distribution with $n_1 + n_2 - 2$ degrees of freedom *if the two population variances really are equal and the population distributions are exactly Normal.* The pooled t was in common use before software made it easy to use Option 1 for our two-sample t statistic. Of course, in the real world distributions are not exactly Normal, and population variances are not exactly

 equal. In practice, the Option 1 two-sample t procedures are almost always more accurate than the pooled procedures. Our advice: *Never use the pooled t procedures if you have software that will implement Option 1.*

*This short section offers advice on what not to do. This material is not needed to read the rest of the book.

Avoid inference about standard deviations*

Two basic features of a distribution are its center and spread/variability. In a Normal population, we measure center by the mean and spread by the standard deviation. We use the t procedures for inference about population means for Normal populations, and we know that t procedures are widely useful for non-Normal populations as well. It is natural to turn next to inference about the standard deviations of Normal populations. Our advice here is short and clear: don't do it without expert advice.

 There are methods for inference about the standard deviations of Normal populations. The most common such method is the "F test" for comparing the standard deviations of two Normal populations. You will find this test in the menus of most statistical software. *Unlike the t procedures for means, the F test for standard deviations is extremely sensitive to non-Normal distributions.* This lack of robustness does not improve in large samples. It is difficult in practice to tell whether a significant test result is evidence of unequal population spreads/variability or simply a sign that the populations are not Normal. Because this test is of little use in practice, we don't give its details.

The deeper difficulty underlying the very poor robustness of Normal population procedures for inference about spread/variability already appeared in our work on describing data. The standard deviation is a natural measure of spread/variability for Normal distributions but not for distributions in general. In fact, because skewed distributions have unequal tails, no single numerical measure does a good job of describing the spread/variability of a skewed distribution. In summary, the standard deviation is not always a useful parameter, and even when it is (for symmetric distributions), the results of inference about the standard deviation are not trustworthy. Consequently, *we do not recommend trying to do inference about population standard deviations in basic statistical practice.*[11]

CHAPTER 21 SUMMARY

Chapter Specifics

■ The data in a **two-sample problem** are two independent SRSs, each drawn from a separate population.

■ Tests and confidence intervals for the difference between the means μ_1 and μ_2 of two Normal populations start from the difference $\bar{x}_1 - \bar{x}_2$ between the two sample means. Because of the central limit theorem, the resulting procedures are approximately correct for other population distributions when the sample sizes are large.

■ Draw independent SRSs of sizes n_1 and n_2 from two Normal populations with parameters μ_1, σ_1 and μ_2, σ_2. The **two-sample t statistic** is

$$t = \frac{(\bar{x}_1 - \bar{x}_2) - (\mu_1 - \mu_2)}{\sqrt{\dfrac{s_1^2}{n_1} + \dfrac{s_2^2}{n_2}}}$$

■ The statistic t has approximately a t distribution.

■ There are two choices for the **degrees of freedom** of the two-sample t statistic. Option 1: software produces accurate probability values using degrees of freedom calculated from the data. Option 2: for conservative inference procedures, use degrees of freedom equal to the smaller of $n_1 - 1$ and $n_2 - 1$.

■ The **confidence interval for $\mu_1 - \mu_2$** is

$$(\bar{x}_1 - \bar{x}_2) \pm t^* \sqrt{\frac{s_1^2}{n_1} + \frac{s_2^2}{n_2}}$$

■ The critical value t^* from Option 1 gives a confidence level very close to the desired level C. Option 2 produces a margin of error at least as wide as is needed for the desired level C.

■ **Significance tests for H_0: $\mu_1 = \mu_2$** are based on

$$t = \frac{\bar{x}_1 - \bar{x}_2}{\sqrt{\dfrac{s_1^2}{n_1} + \dfrac{s_2^2}{n_2}}}$$

■ P-values calculated from Option 1 are very accurate. Option 2 P-values are always at least as large as the true P.

■ The two-sample t procedures are quite **robust** against departures from Normality. Guidelines for practical use are similar to those for one-sample t procedures. Equal sample sizes are recommended.

■ Procedures for inference about the standard deviations of Normal populations are very sensitive to departures from Normality. Avoid inference about standard deviations unless you have expert advice.

Link It

In Chapter 20 we studied inference for the mean of a Normal population using procedures based on the t distribution. In practice, the most common use of t procedures for a single population mean is with matched pairs data because most research studies make comparisons between two or more populations. Thus, it is useful to have procedures for making comparisons between two or more populations.

In this chapter, we discuss t procedures for comparing the means of two Normal populations when we have independent samples from these two populations. These procedures can also be used to compare the means of two populations that are not Normal provided we have sufficiently large samples from these populations. These t procedures are quite common in practice, and one encounters them in many research papers. Researchers typically report the means and standard deviations for the two samples and, in the case of tests of hypotheses, the value of the t statistic and the corresponding P-value.

In Chapters 16, 17, 18, 20, and this chapter we have made a point of emphasizing the conditions that are needed for the types of inference discussed. However, it is not always easy to determine if these conditions are met. In the next chapter we take a more careful look at determining when we can use our methods for inference in practice.

CHECK YOUR SKILLS

21.16 The 2009 National Assessment of Educational Progress (NAEP) gave a mathematics test to a random sample of eighth-graders in Texas. The mean score was 287 out of 500. To give a confidence interval for the mean score of all Texas eighth-graders, you would use

(a) a two-sample t interval.
(b) a matched pairs t interval.
(c) a one-sample t interval.

21.17 In the 2009 NAEP sample of Texas eighth-graders, the mean mathematics scores were 286 for female students and 287 for male students. To see if this difference is statistically significant, you would use

(a) a two-sample t test.
(b) a matched pairs t test.
(c) a one-sample t test.

21.18 A new method for measuring blood pressure in laboratory mice using radiotelemetry has been proposed. How does this compare with the existing method? You apply both methods to a sample of 16 mice and use

(a) a two-sample t test.
(b) a matched pairs t test.
(c) a one-sample t test.

21.19 One major reason that the two-sample t procedures are widely used is that they are quite *robust*. This means that

(a) t procedures do not require that we know the standard deviations of the populations.
(b) confidence levels and P-values from the t procedures are quite accurate even if the population distribution is not exactly Normal.
(c) t procedures compare population means, a comparison that answers many practical questions.

21.20 A study of the effect of exposure to color (red or blue) on the ability to solve puzzles used 42 subjects. Half the subjects (21) were asked to solve a series of puzzles while in a red-colored environment. The other half were asked to solve the same series of puzzles while in a blue-colored environment. The time taken to solve the puzzles was recorded for each subject. To compare the mean times for the two groups of subjects using the two-sample t procedures with the conservative Option 2, the correct degrees of freedom is

(a) 41. (b) 40. (c) 20.

21.21 The 21 subjects in the red-colored environment had a mean time for solving the puzzles of 9.64 seconds with standard deviation 3.43; the 21 subjects in the blue-colored environment had a mean time of 15.84 seconds with standard deviation 8.65. The two-sample t statistic for comparing the population means has value

(a) 1.50. (b) 3.05. (c) 6.2.

21.22 A study of the use of social media asked a sample of 488 American adults under the age of 40 and a sample of 421 American adults aged 40 or over about their use of social media. Based on their answers, each subject was assigned a social media usage score on a scale of 0 to 25. Higher scores indicate greater usage. The subjects were chosen by random digit dialing of telephone numbers. Are the conditions for two-sample t inference satisfied?

(a) Maybe: the SRS condition is OK, but we need to look at the data to check Normality.
(b) No: scores in a range between 0 and 25 can't be Normal.
(c) Yes: the SRS condition is OK and large sample sizes make the Normality condition unnecessary.

21.23 We suspect that younger adults use social media more than adults aged 40 or over. To see if this is true, test these hypotheses for the mean social media usage scores of all adults under 40 and all adults 40 and over:

(a) $H_0: \mu_{<40} = \mu_{\geq 40}$ versus $H_a: \mu_{<40} > \mu_{\geq 40}$
(b) $H_0: \mu_{<40} = \mu_{\geq 40}$ versus $H_a: \mu_{<40} \neq \mu_{\geq 40}$
(c) $H_0: \mu_{<40} = \mu_{\geq 40}$ versus $H_a: \mu_{<40} < \mu_{\geq 40}$

21.24 The two-sample t statistic for the social media use study ("under 40" mean minus "40 and over" mean) is $t = 3.18$. The P-value for testing the hypotheses from the previous exercise satisfies

(a) $0.001 < P < 0.005$.
(b) $0.0005 < P < 0.001$.
(c) $0.001 < P < 0.002$.

CHAPTER 21 EXERCISES

Exercises 21.25 to 21.34 are based on summary statistics rather than raw data. This information is typically all that is presented in published reports. You can perform inference procedures by hand from the summaries. Use the conservative Option 2 (degrees of freedom the smaller of $n_1 - 1$ and $n_2 - 1$) for two-sample t confidence intervals and P-values. You must

trust that the authors understood the conditions for inference and verified that they apply. This isn't always true.

21.25 **Do women talk more than men?** Equip male and female students with a small device that secretly records sound for a random 30 seconds during each 12.5-minute period over two days. Count the words each subject speaks during each recording period, and from this, estimate how many words per day each subject speaks. The published report includes a table summarizing six such studies.[12] Here are two of the six:

| STUDY | SAMPLE SIZE | | ESTIMATED AVERAGE NUMBER (SD) OF WORDS SPOKEN PER DAY | |
	WOMEN	MEN	WOMEN	MEN
1	56	56	16,177 (7520)	16,569 (9108)
2	27	20	16,496 (7914)	12,867 (8343)

Readers are supposed to understand that, for example, the 56 women in the first study had $\bar{x} = 16,177$ and $s = 7520$. It is commonly thought that women talk more than men. Does either of the two samples support this idea? For each study:

(a) State hypotheses in terms of the population means for men (μ_M) and women (μ_F).
(b) Find the two-sample t statistic.
(c) What degrees of freedom does Option 2 use to get a conservative P-value?
(d) Compare your value of t with the critical values in Table C. What can you say about the P-value of the test?
(e) What do you conclude from the results of these two studies?

21.26 Alcohol and zoning out. Healthy men aged 21 to 35 were randomly assigned to one of two groups: half received 0.82 grams of alcohol per kilogram of body weight; half received a placebo. Participants were then given 30 minutes to read up to 34 pages of Tolstoy's *War and Peace* (beginning at Chapter 1, with each page containing approximately 22 lines of text). Every two to four minutes participants were prompted to indicate whether they were "zoning out." The proportion of times participants indicated they were zoning out was recorded for each subject. The following table summarizes data on the proportion of episodes of zoning out.[13] (The study report gave the standard error of the mean $s/\sqrt{n}$, abbreviated as SEM, rather than the standard deviation s.)

GROUP	n	$\bar{x}$	SEM
Alcohol	25	0.25	0.05
Placebo	25	0.12	0.03

(a) What are the two sample standard deviations?
(b) What degrees of freedom does the conservative Option 2 use for two-sample t procedures for these samples?
(c) Using Option 2, give a 90% confidence interval for the mean difference between the two groups.

21.27 Stress and weight in rats. In a study of the effects of stress on behavior in rats, 71 rats were randomly assigned to either a stressful environment or a control (nonstressful) environment. All rats were provided the same amount of food. After 21 days, the change in weight (in grams) was determined for each rat. The following table summarizes data on weight gain.[14] (The study report gave the standard error of the mean $s/\sqrt{n}$, abbreviated as SEM, rather than the standard deviation s.)

Olivier Voisin/Science Source

GROUP	n	$\bar{x}$	SEM
Stress	20	26	3
No stress	51	32	2

(a) What are the standard deviations for the two groups?
(b) What degrees of freedom does the conservative Option 2 use for two-sample t procedures for these data?
(c) Test the null hypothesis of no difference between the two group means against the two-sided alternative. Use the degrees of freedom from part (b).

21.28 Is Montessori preschool beneficial? Do education programs for preschool children that follow the Montessori method perform better than other programs? A study compared 5-year-old children in Milwaukee, Wisconsin, who had been enrolled in preschool programs from the age of 3.[15]

(a) Explain why comparing children whose parents chose a Montessori school with children of other parents would not show whether Montessori schools perform better than other programs.
(b) In fact, all the children in the study applied to the Montessori school. The school district assigned students to Montessori or other preschools by a random lottery. In all, 54 children were assigned to the Montessori school and 112 to other schools at age 3. When the children were 5, parents of 30 of the Montessori children and 25 of the others could be located and agreed to and subsequently participated in testing. This information reveals a possible source of bias in the comparison of outcomes. Explain why.
(c) One of the many response variables was score on a test of ability to apply basic mathematics to solve problems. Here are summaries for the children who took this test:

GROUP	n	$\bar{x}$	s
Montessori	30	19	3.11
Control	25	17	4.19

Is there evidence of a difference in the population mean scores? (The researchers used two-sided alternative hypotheses.)

21.29 **Ginkgo extract and the postlunch dip.** The postlunch dip is the drop in mental alertness after a midday meal. Does an extract of the leaves of the ginkgo tree reduce the postlunch dip? Assign healthy people aged 18 to 40 to take either ginkgo extract or a placebo pill. After lunch, ask them to read seven pages of random letters and place an X over every *e*. Count the number of misses per line read.[16]

© Emilio Ereza/Alamy

(a) What is a placebo, and why was one group given a placebo?

(b) What is the double-blind method, and why should it be used in this experiment?

(c) Here are summaries of performance after 13 weeks of either ginkgo extract or placebo:

GROUP	GROUP SIZE	MEAN	STD. DEV.
Ginkgo	21	0.06383	0.01462
Placebo	18	0.05342	0.01549

Is there a significant difference between the two groups? What do these data show about the effect of ginkgo extract?

21.30 **Illusory pattern perceptions.** When one experiences a lack of control in one's life, does one compensate by seeking structure elsewhere? Assign 36 undergraduate students to one of two conditions. All are asked to identify a concept associated with a series of "grainy" pictures that contain an embedded image and are presented to them sequentially. During the task they can ask questions to help them determine the associated concept. Half of the subjects (the lack-of-control group) receive feedback that is random and noncontingent on their questions. The other half receive useful feedback (the in-control group). After attempting to complete the task, all subjects are presented with 12 grainy pictures that are similar to those used in the task but lack an embedded image. All are asked whether they perceive an image in the pictures, and the number of pictures identified as containing an image is counted for each subject. Here are the summary statistics:[17]

GROUP	GROUP SIZE	MEAN	STD. DEV.
Lack-of-control	18	5.16	3.5
In-control	18	3.47	2.0

(a) What degrees of freedom would you use in the conservative two-sample *t* procedures to compare the lack-of-control and in-control groups?

(b) What is the two-sample *t* test statistic for comparing the mean number of pictures identified as having an image for the two groups?

(c) Test the null hypothesis of no difference between the two population means against the two-sided alternative. Use your statistic from part (b) with degrees of freedom from part (a).

21.31 **Asperger's syndrome and the ability to mentalize.** Asperger's syndrome is a form of autism. People with this syndrome have difficulty interacting and communicating socially. Some researchers believe that behind these difficulties is an inability to "mentalize," that is, to automatically attribute mental states to the self and others. To test this, researchers had 36 subjects (19 with Asperger's syndrome and 17 without Asperger's syndrome) watch a play in which an actor is known to have a false belief (mental state) and his actions are different than would be the case if he did not have this false belief. In particular, the eye movements of subjects were tracked to see if subjects anticipated the direction the actor would move based on his false belief rather than the direction he would move if he did not have the false belief. A score reflecting the bias toward looking in the correct direction (the direction the actor moved based on his false belief) was recorded for each subject. Higher scores indicate the subject looked in this "correct" direction more often. (The study report gave the standard error of the mean $s/\sqrt{n}$, abbreviated as SEM, rather than the standard deviation s.) Here are the summary statistics:[18]

GROUP	GROUP SIZE	MEAN	SEM
Asperger's	19	−0.001	0.15
Non-Asperger's	17	0.42	0.17

Is there evidence of a difference in mean scores between those with Asperger's syndrome and those without it?

21.32 Coaching and SAT scores. Coaching companies claim that their courses can raise the SAT scores of high school students. Of course, students who retake the SAT without paying for coaching generally raise their scores, too. A random sample of students who took the SAT twice found 427 who were coached and 2733 who were uncoached.[19] Starting with their Verbal scores on the first and second tries, we have these summary statistics:

	n	TRY 1 $\bar{x}$	TRY 1 s	TRY 2 $\bar{x}$	TRY 2 s	GAIN $\bar{x}$	GAIN s
Coached	427	500	92	529	97	29	59
Uncoached	2733	506	101	527	101	21	52

Let's first ask if students who are coached increased their scores significantly.

(a) You could use the information on the Coached line to carry out either a two-sample t test comparing Try 1 with Try 2 for coached students or a matched pairs t test using Gain. Which is the correct test? Why?

(b) Carry out the proper test. What do you conclude?

(c) Give a 99% confidence interval for the mean gain of all students who are coached.

21.33 Coaching and SAT scores, continued. What we really want to know is whether coached students improve more than uncoached students, and whether any advantage is large enough to be worth paying for. Use the information in the previous exercise to answer these questions:

(a) Is there good evidence that coached students gained more on the average than uncoached students?

(b) How much more do coached students gain on the average? Give a 99% confidence interval.

(c) Based on your work, what is your opinion: do you think coaching courses are worth paying for?

21.34 Coaching and SAT scores: critique. The data you used in the previous two problems came from a random sample of students who took the SAT twice. The response rate was 63%, which is pretty good for nongovernment surveys, so let's accept that the respondents do represent all students who took the exam twice. Nonetheless, we can't be sure that coaching actually *caused* the coached students to gain more than the uncoached students. Explain briefly but clearly why this is so.

Exercises 21.35 to 21.42 include the actual data. To apply the two-sample t procedures, use Option 1 if you have technology that implements that method. Otherwise, use Option 2.

21.35 Improving your tips. Researchers gave 40 index cards to a waitress at an Italian restaurant in New Jersey. Before delivering the bill to each customer, the waitress randomly selected a card and wrote on the bill the same message that was printed on the index card. Twenty of the cards had the message "The weather is supposed to be really good tomorrow. I hope you enjoy the day!" Another 20 cards contained the message "The weather is supposed to be not so good tomorrow. I hope you enjoy the day anyway!" After the customers left, the waitress recorded the amount of the tip (percent of bill) before taxes. Here are the tips from those receiving the good-weather message:[20] TIP4

20.8 18.7 19.9 20.6 21.9 23.4 22.8 24.9 22.2 20.3
24.9 22.3 27.0 20.5 22.2 24.0 21.2 22.1 22.0 22.7

The tips from the 20 customers who received the bad weather message are

18.0 19.1 19.2 18.8 18.4 19.0 18.5 16.1 16.8 14.0
17.0 13.6 17.5 20.0 20.2 18.8 18.0 23.2 18.2 19.4

(a) Make stemplots or histograms of both sets of data. Because the distributions are reasonably symmetric with no extreme outliers, the t procedures will work well.

(b) Is there good evidence that the two different messages produce different percent tips? State hypotheses, carry out a two-sample t test, and report your conclusions.

21.36 Do good smells bring good business? In Exercise 21.9 (page 475) you examined the effects of a lavender odor on customer behavior in a small restaurant. Lavender is a relaxing odor. The researchers also looked at the effects of lemon, a stimulating odor. The design of the study is described in Exercise 21.9. Here are the times in minutes that customers spent in the restaurant when no odor was present: ODORS3

103 68 79 106 72 121 92 84 72 92
85 69 73 87 109 115 91 84 76 96
107 98 92 107 93 118 87 101 75 86

When a lemon odor was present, customers lingered for these times:

78 104 74 75 112 88 105 97 101 89
88 73 94 63 83 108 91 88 83 106
108 60 96 94 56 90 113 97

(a) Examine both samples. Does it appear that use of two-sample t procedures is justified? Do the sample means suggest that a lemon odor changes the average length of stay?

(b) Does a lemon odor influence the length of time customers stay in the restaurant? State hypotheses, carry out a t test, and report your conclusions.

21.37 Improving your tips, continued. Use the data in Exercise 21.35 to give a 95% confidence interval for the difference between the mean percent tips for the two different messages. TIP4

21.38 **How strong are durable press fabrics?** "Durable press" cotton fabrics are treated to improve their recovery from wrinkles after washing. Unfortunately, the treatment also reduces the strength of the fabric. A study compared the breaking strength of fabrics treated by two commercial durable press processes. Five swatches of the same fabric were assigned at random to each process. Here are the data, in pounds of pull needed to tear the fabric:[21] **FABRICS**

Permafresh	29.9	30.7	30.0	29.5	27.6
Hylite	28.8	23.9	27.0	22.1	24.2

Is there good evidence that the two processes result in different mean breaking strengths?

(a) Do the sample means suggest that one of the processes is superior in breaking strength?

(b) Make stemplots for both samples. The Permafresh sample contains a mild outlier. With just 5 observations per group, we worry that this outlier will affect our conclusions.

(c) Test the hypothesis $H_0: \mu_1 = \mu_2$ against the two-sided alternative twice: once using all the data and again without the outlier in the Permafresh sample. Do the two tests lead to similar conclusions? Can we safely conclude that one treatment has significantly higher mean breaking strength than the other?

21.39 **Reducing wrinkles.** Of course, the reason for durable press treatment is to reduce wrinkling. "Wrinkle recovery angle" measures how well a fabric recovers from wrinkles. Higher is better. Here are data on the wrinkle recovery angle (in degrees) for the same fabric swatches discussed in the previous exercise: **WRINK**

Permafresh	136	135	132	137	134
Hylite	143	141	146	141	145

Is there a significant difference in wrinkle resistance?

(a) Do the sample means suggest that one process has better wrinkle resistance?

(b) Make stemplots for both samples. There are no obvious deviations from Normality.

(c) Test the hypothesis $H_0: \mu_1 = \mu_2$ against the two-sided alternative. What do you conclude from part (a) and from the result of your test?

21.40 **How much stronger?** Continue your work from Exercise 21.38. A fabric manufacturer wants to know how large an advantage in strength fabrics treated by the Permafresh method have over fabrics treated by the Hylite process. Give a 90% confidence interval for the difference in mean breaking strengths. (Use all five fabric swatches.) **FABRICS**

21.41 **How much less wrinkling?** In Exercise 21.39, you found that the Hylite process results in significantly greater wrinkle resistance than the Permafresh process. How large is the difference in mean wrinkle recovery angle? Give a 90% confidence interval. **WRINK**

21.42 **Do women talk more than men? Another study.** Exercise 21.25 (page 481) described a series of six studies investigating the number of words women and men speak per day. Exercise 21.25 gives results from two of these studies. Here are the results from another of these studies. The estimated numbers of words spoken per day for 27 women are **TALK2**

15,357	13,618	9,783	26,451	12,151	8,391
19,763	25,246	8,427	6,998	24,876	6,272
10,047	15,569	39,681	23,079	24,814	19,287
10,351	8,866	10,827	12,584	12,764	19,086
26,852	17,639	16,616			

The estimated numbers of words spoken per day for 20 men are

28,408	10,084	15,931	21,688	37,786	10,575
12,880	11,071	17,799	13,182	8,918	6,495
8,153	7,015	4,429	10,054	3,998	12,639
10,974	5,255				

Does this study provide good evidence that women talk more than men, on average?

(a) Make stemplots for both samples. Are there any obvious deviations from Normality? Despite these deviations from Normality, it is safe to use the t procedures. Explain.

(b) Test the hypothesis $H_0: \mu_1 = \mu_2$ against the one-sided alternative that the mean number of words per day for women (μ_1) is greater than the mean number of words per day for men (μ_2). What do you conclude?

Do birds learn to time their breeding? *Blue titmice eat caterpillars. The birds would like lots of caterpillars around when they have young to feed, but they breed earlier than peak caterpillar season. Do the birds time when they breed based on the previous year's caterpillar supply? Researchers randomly assigned 7 pairs of birds to have the natural caterpillar supply supplemented while feeding their young and another 6 pairs to serve as a control group relying on natural food supply. The next year, they measured how many days after the caterpillar peak the birds produced their nestlings.[22] Exercises 21.43 to 21.45 are based on this experiment.*

© Hugh Clark/Frank Lane Picture Agency/CORBIS

21.43 **Did the randomization produce similar groups?** The first thing to do is to compare the two groups in the first year. The only difference should be the chance effect of the random assignment. The study report says: "In the experimental year, the degree of synchronization did not differ between food-supplemented and control females." For this comparison, the report gives $t = -1.05$. What type of t statistic (paired or two-sample) is this? What are the degrees of freedom for this statistic? Show that this t leads to the quoted conclusion.

21.44 **Did the treatment have an effect?** The investigators expected the control group to adjust their breeding date the next year, whereas the well-fed supplemented group had no reason to change. The report continues: "but in the following year food-supplemented females were more out of synchrony with the caterpillar peak than the controls." Here are the data (days behind the caterpillar peak): **BREED**

Control	4.6	2.3	7.7	6.0	4.6	−1.2	
Supplemented	15.5	11.3	5.4	16.5	11.3	11.4	7.7

Carry out a t test and show that it leads to the quoted conclusion.

21.45 **Year-to-year comparison.** Rather than comparing the two groups in each year, we could compare the behavior of each group in the first and second years. The study report says: "Our main prediction was that females receiving additional food in the nestling period should not change laying date the next year, whereas controls, which (in our area) breed too late in their first year, were expected to advance their laying date in the second year."

Comparing days behind the caterpillar peak in Years 1 and 2 gave $t = 0.63$ for the control group and $t = -2.63$ for the supplemented group. Are these paired or two-sample t statistics? What are the degrees of freedom for each t? Show that these t-values do *not* agree with the prediction.

*The remaining exercises ask you to answer questions from data without having the details outlined for you. The exercise statements give you the **State** step of the four-step process. Follow the **Plan**, **Solve**, and **Conclude** steps as illustrated in Examples 21.2 and 21.3 (pages 465 and 468) for tests and Example 21.4 (page 469) for confidence intervals. Remember that examining the data and discussing the conditions for inference are part of the **Solve** step.*

21.46 **Thinking about money changes behavior.** Kathleen Vohs of the University of Minnesota and her coworkers carried out several randomized comparative experiments on the effects of thinking about money. Here's part of one such experiment.[23] Ask student subjects to unscramble 30 sets of five words to make a meaningful phrase from four of the five words. The control group

unscrambled phrases like "cold it desk outside is" into "it is cold outside." The treatment group unscrambled phrases that lead to thinking about money, turning "high a salary desk paying" into "a high-paying salary." Then each subject worked a hard puzzle, knowing that he or she could ask for help. Here are the times in seconds until subjects asked for help. For the treatment group: **MNYTHNK**

609	444	242	199	174	55	251	466	443
531	135	241	476	482	362	69	160	

For the control group:

118	272	413	291	140	104	55	189	126
400	92	64	88	142	141	373	156	

The researchers suspected that money is connected with self-sufficiency, so that the treatment group will ask for help less quickly on the average. Do the data support this idea?

21.47 **Adolescent obesity.** Adolescent obesity is a serious health risk affecting more than 5 million young people in the United States alone. Laparoscopic adjustable gastric banding has the potential to provide a safe and effective treatment. Fifty adolescents between 14 and 18 years old with a body mass index higher than 35 were recruited from the Melbourne, Australia, community for the study.[24] Twenty-five were randomly selected to undergo gastric banding, and the remaining twenty-five were assigned to a supervised lifestyle intervention program involving diet, exercise, and behavior modification. All subjects were followed for two years. Here are the weight losses in kilograms for the subjects who completed the study. In the gastric banding group: **ADOBESE**

35.6	81.4	57.6	32.8	31.0	37.6	36.5	19.4
−5.4	27.9	49.0	64.8	39.0	43.0	33.9	32.3
29.7	20.2	15.2	41.7	53.4	13.4	24.8	22.0

In the lifestyle intervention group:

6.0	2.0	−3.0	20.6	11.6	15.5	−17.0	1.4	4.0
−4.6	15.8	34.6	6.0	−3.1	−4.3	−16.7	−1.8	−12.8

Is there good evidence that gastric banding is superior to the lifestyle intervention program?

21.48 **Active versus traditional learning.** Can active learning improve knowledge retention? Two undergraduate calculus-based engineering statistics courses were taught in different academic quarters, with one employing active-learning methods and another using traditional learning methods. The traditional class was taught lecture-style with relatively little in-class interaction between peers and with the instructor. The active-learning course integrated four group projects into the curriculum, with in-class time devoted to group work on the projects and fewer homework assignments. To assess knowledge retention, two five-question versions of a test were created. They had similar but not identical questions covering core statistics topics, worth a total of 18 possible points. All students in both sections were randomly given one version of the test as part of their final exam. Then, eight months later, a volunteer subset of the original students were given the version that they had not taken previously. To encourage students to take the second version of the exam, a ten-dollar gift card to the university bookstore was given to each participant. The change in the score from the first version to the second (the score on the second version minus the score on the first version) is used to measure a student's long-term ability to retain the course material. ACTVLRN

Here are the changes in exam scores for the 15 students in the active group:[25]

```
0  5  7  8  0  3  6  2  5  1
3  2  4  3  5
```

The changes in exam scores for the 23 students in the traditional group are

```
7  0  8  2  4  3  1  2  5  8
5  6  3  12  1  6  3  6  7  7
5  6  2
```

Is there good evidence that active learning is superior to traditional lecturing?

21.49 **Each day I am getting better in math.** A "subliminal" message is below our threshold of awareness but may nonetheless influence us. Can subliminal messages help students learn math? A group of students who had failed the mathematics part of the City University of New York Skills Assessment Test agreed to participate in a study to find out.

All received a daily subliminal message, flashed on a screen too rapidly to be consciously read. The treatment group of 10 students (chosen at random) was exposed to "Each day I am getting better in math." The control group of 8 students was exposed to a neutral message, "People are walking on the street." All students participated in a summer program designed to raise their math skills, and all took the assessment test again at the end of the program. Table 21.3 gives data on the subjects' scores before and after the program.[26] Is there good evidence that the treatment brought about a greater improvement in math scores than the neutral message? How large is the mean difference in gains between treatment and control? (Use 90% confidence.) SUBLIM

TABLE 21.3 MATHEMATICS SKILLS SCORES BEFORE AND AFTER A SUBLIMINAL MESSAGE

TREATMENT GROUP		CONTROL GROUP	
BEFORE	AFTER	BEFORE	AFTER
18	24	18	29
18	25	24	29
21	33	20	24
18	29	18	26
18	33	24	38
20	36	22	27
23	34	15	22
23	36	19	31
21	34		
17	27		

21.50 **Active versus traditional learning, continued.**

(a) Use the data in Exercise 21.48 to give a 90% confidence interval for the difference in the mean change in score for students in the active and traditional classes. ACTVLRN

(b) Give a 90% confidence interval for the mean change in score of students in the active-learning class.

21.51 **Tropical flowers.** Different varieties of the tropical flower *Heliconia* are fertilized by different species of hummingbirds. Over time, the lengths of the flowers and the forms of the hummingbirds' beaks have evolved to match each other. Here are data on the lengths in millimeters of two color varieties of the same species of flower on the island of Dominica:[27]

Art Wolfe/Getty Images

H. caribaea RED

41.90	42.01	41.93	43.09	41.47	41.69	39.78	40.57
39.63	42.18	40.66	37.87	39.16	37.40	38.20	38.07
38.10	37.97	38.79	38.23	38.87	37.78	38.01	

H. caribaea YELLOW

36.78	37.02	36.52	36.11	36.03	35.45	38.13	37.1
35.17	36.82	36.66	35.68	36.03	34.57	34.63	

Is there good evidence that the mean lengths of the two varieties differ? Estimate the difference between the population means. (Use 95% confidence.)

📊 FLOWERS

21.52 Student drinking. A professor asked her sophomore students, "How many drinks do you typically have per session? (A drink is defined as one 12-oz beer, one 4-oz glass of wine, or one 1-oz shot of liquor.)" Some of the students didn't drink. Table 21.4 gives the responses of the female and male students who did drink.[28] It is likely that some of the students exaggerated a bit. The sample is all students in one large sophomore-level class. The class is popular, so we are tentatively willing to regard its members as an SRS of sophomore students at this college. Do a complete analysis that reports on DRINKS

(a) the drinking behavior claimed by sophomore women.
(b) the drinking behavior claimed by sophomore men.
(c) a comparison of the behavior of women and men.

TABLE 21.4 DRINKS PER SESSION CLAIMED BY FEMALE AND MALE STUDENTS

FEMALE STUDENTS												
2.5	9	1	3.5	2.5	3	1	3	3	3	3	2.5	2.5
5	3.5	5	1	2	1	7	3	7	4	4	6.5	4
3	6	5	3	8	6	6	3	6	8	3	4	7
4	5	3.5	4	2	1	5	5	3	3	6	4	2
7	7	7	3.5	3	2.5	10	5	4	9	8	1	6
2	5	2.5	3	4.5	9	5	4	4	3	4	6	7
4	5	1	5	3	4	10	7	3	4	4	4	4
2	1	2.5	2.5									

MALE STUDENTS												
7	7.5	8	15	3	4	1	5	11	4.5	6	4	10
16	4	8	5	9	7	7	3	5	6.5	1	12	4
6	8	8	4.5	10.5	8	6	10	1	9	8	7	8
15	3	10	7	4	6	5	2	10	7	9	5	8
7	3	7	6	4	5	2	5	5.5	9	10	10	4
8	4	2	4	12.5	3	15	2	6	3	4	3	10
6	4.5	5										

🌐 **Exploring the Web**

21.53   **A two-sample t test example.** Find an example of a two-sample t test on the Web. The *Journal of the American Medical Association* (jama.ama-assn.org), *Science Magazine* (www.sciencemag.org), the *Canadian Medical Association Journal* (www.cmaj.ca), the *Journal of Statistics Education* (www.amstat.org/publications/jse), or perhaps the *Journal of Quantitative Analysis in Sports* (www.degruyter.com/view/j/jqas) are possible sources. To help locate an article, look through the abstracts of articles. Once you find a suitable article, read it and then briefly describe the study (including why it is a two-sample study) and its conclusions. If P-values, means, standard deviations, t statistics, and so on are reported, be sure to include them in your summary. Also, be sure to give the reference (either the Web link or the journal, issue, year, title of the paper, authors, and page numbers).

21.54   **Antibiotics after surgery.** If your college has online access to the *Archives of Otolaryngology—Head and Neck Surgery*, read the article "Duration-Related Efficacy of Postoperative Antibiotics Following Pediatric Tonsillectomy: A Prospective, Randomized, Placebo-Controlled Trial" (available online at archotol.ama-assn.org/cgi/content/full/135/10/984). The authors appear to use two-sample t procedures in the paper. After reading the article, read the (brief) discussion on the *Chance* Web site, test.causeweb.org/wiki/chance/index.php/Chance_News_57. What are some criticisms or concerns expressed about the study in the *Chance* article?

Overview

The previous chapters provided a basic introduction to the two primary vehicles for inference. In Chapter 16 we learned about confidence intervals for a population proportion, and in Chapter 17 we learned about hypothesis tests for a population proportion. Chapter 18 extended the ideas from Chapters 16 and 17 to investigate inference on the difference between two population proportions. In Chapters 20 and 21, we learned about confidence intervals and hypothesis tests for one population mean, matched pairs, and the difference between two population means, These methods are based on the sampling distributions of the respective parameter estimators: $\hat{p}$ for p; $\hat{p}_1 - \hat{p}_2$ for $p_1 - p_2$; $\bar{x}$ for μ; and $\bar{x}_1 - \bar{x}_2$ for $\mu_1 - \mu_2$. In this chapter, we focus on developing a sense for when it is valid to apply these methods and learn about the kinds of practical problems that can threaten the validity of inferences made using them.

If the researcher's objective is to make inferences about the population(s), it is important to consider how the data were collected. All methods for statistical inference are based on an assumption that the sample(s) selected is(are) random, or that the data come from a randomized comparative experiment. If the method of selecting a sample induces bias, then the integrity of inference is threatened. The margin of error associated with a confidence interval does not account for this kind of error.

In significance testing, a statistically significant result may not be practically significant, whereas lack of statistical significance does not imply that the null hypothesis is true (nor does it imply that the results are not practically important). In interpreting a *P*-value, we should reflect on whether it is best to require less evidence to reject the null hypothesis (select a larger level of significance), or best to require more evidence (select a lower level of significance). The selection of the level of significance depends on the consequences of an incorrect decision. The chapter ends with two optional sections: a discussion of the power of a test of hypotheses—the probability that a false null hypothesis will be rejected; and the meaning of Type I and Type II errors in selecting a level of significance. Note that you can omit the discussion of power and talk about Type I and II errors if you desire; these ideas help to drive home what the α level of a test and confidence level of an interval really imply in terms of making decisions from data. In an imperfect world (where complete information is not feasible), it is always possible that an incorrect decision will be made.

LEARNING OUTCOMES**

- Recognize that any inference procedure acts as if the data were properly produced. For a single sample, the data should be an SRS from the population. In a two-sample problem, the data can arise from a randomized comparative experiment or a comparison of SRSs from two populations. More complicated sampling and experimental methods will have more complicated calculations (beyond the scope of this text), but the ideas are transferable.

- Recognize that there is no simple rule for deciding when you can act as if a sample is an SRS. In particular, pay attention to the following:

  - Recognize that practical problems such as nonresponse in samples or dropouts from an experiment can hinder inference even from a well-designed study.

  - Recognize that different methods are needed for different designs.

  - Recognize that there is no cure for fundamental flaws like voluntary response surveys or uncontrolled experiments.

- (Optional) Recognize the relationship among significance level, effect size, and power.

- (Optional) Use technology to calculate the power of a significance test.

- (Optional) Define and differentiate between Type I and Type II errors. Be able to explain their consequences in context.

- (Optional) Calculate the probabilities of Type I and Type II errors.

_____

**These learning outcomes appear later for the students in Chapter 23: Part IV Review.

Teaching Suggestions and Additional Examples/Activities for the Classroom

1. Discuss Power, Statistical Significance, and Practical Importance

Start a discussion by telling your class that you'll use them to construct a confidence interval for the (population) average number of math classes taken by all students at the school. This is a convenience sample and will clearly provide a biased estimate of the average number of math classes taken. The problem, of course, is that you haven't selected a random sample of students.

Now suppose that you obtain a list of the number of math courses taken by every student at your school. Why would it not be reasonable to construct a confidence interval for the average number of math classes taken by all students at the school?

Spend some time working through an example demonstrating that a large sample size will lead to rejection of the null hypothesis if the population mean is even slightly different from the value stated under the null hypothesis. This is really a discussion of power—but, more important, it illustrates that statistical significance does not imply practical importance.

2. Discuss the Consequences of Selecting a Significance Level

Selecting a level of significance is a topic many instructors don't emphasize—but this discussion provides an opportunity to emphasize where subjectivity plays a role in inference. Consider the two examples: H_0: Defendant is innocent, and H_0: Airport security is adequate. In the first example, a Type I error would be seen by most as more problematic than a Type II error because many legal systems are based on the idea that sentencing an innocent person to prison is worse than letting a guilty person go free. In the second example, a Type II error is more catastrophic than a Type I error—if we decide airport security is adequate when it is not, terrible things could happen, whereas if we decide airport security is not adequate when it really is, there may be additional inconveniences for the passengers, but the result will be a safer traveling experience. We would want to select a smaller level of significance (perhaps 0.01 or 0.02) in the first example and a larger level of significance (perhaps 0.08 or 0.10) in the second example. Students should understand that raising the level of significance makes it "easier" to reject the null hypothesis, which is sometimes desirable.

Other Resources (LaunchPad)

Snapshots Videos
Interpreting Inference

EESEE Case Studies
Blinded Knee Doctors
Visibility of Highway Signs
Passive Smoking and Respiratory Health

Applets
Statistical Power
Statistical Significance

Inference in Practice

In this chapter we cover...

■ Conditions for inference in practice

■ Planning studies: the power of a statistical test*

■ Assessing the performance of a significance test: Type I and Type II errors*

There is a saying among statisticians that "mathematical theorems are true; statistical methods are effective when used with judgment." That the one-sample t statistic has the t distribution with $n - 1$ degrees of freedom for an SRS of size n from a Normal population when the null hypothesis is true is a mathematical theorem. Effective use of statistical methods requires more than knowing such facts. It requires even more than understanding the underlying reasoning. In Chapters 16 and 17 we provided cautions about the use of confidence intervals and significance tests in practice. This chapter continues the process of helping you develop the judgment needed to use statistics in practice.

Conditions for inference in practice

CAUTION

Any confidence interval or significance test can be trusted only under specific conditions. It's up to you to understand these conditions and judge whether they fit your situation. With that in mind, let's look back at the conditions for inference about a proportion (page 350) and for inference about a mean (page 436).

Two conditions are hard to escape: for inference about both proportions and means, we must have an SRS; for inference about a mean, we must also have some knowledge of the shape of the population distribution for all but large samples. In fact, they represent the kinds of conditions needed if we are to trust almost any statistical inference. As you plan inference, you should always ask, "Where did the

489

DON'T TOUCH THE PLANTS

We know that confounding can distort inference. We don't always recognize how easy it is to confound data. Consider the innocent scientist who visits plants in the field once a week to measure their size. To measure the plants, he has to touch them. A study of six plant species found that one touch a week significantly increased leaf damage by insects in two species and significantly decreased damage in another species.

data come from?" and you must often also ask, "What is the shape of the population distribution?" This is the point where knowing mathematical facts gives way to the need for judgment.

Where did the data come from? *The most important requirement for any inference procedure is that the data come from a process to which the laws of probability apply.* Inference is most reliable when the data come from a random sample or a randomized comparative experiment. Random samples use chance to choose respondents. Randomized comparative experiments use chance to assign subjects to treatments. The deliberate use of chance ensures that the laws of probability apply to the outcomes, and this in turn ensures that statistical inference makes sense.

Where the Data Come from Matters

When you use statistical inference, you are acting as if your data are a random sample or come from a randomized comparative experiment.

⚠️ **CAUTION**   *If your data don't come from a random sample or a randomized comparative experiment, your conclusions may be challenged.* To answer the challenge, you must usually rely on subject-matter knowledge, not on statistics. It is common to apply statistical inference to data that are not produced by random selection. When you see such a study, ask whether the data can be trusted as a basis for the conclusions of the study.

EXAMPLE 22.1   The Psychologist and the Sociologist

A psychologist is interested in how our visual perception can be fooled by optical illusions. Her subjects are students in Psychology 101 at her university. Most psychologists would agree that it's safe to treat the students as an SRS of all people with normal vision. There is nothing special about being a student that changes visual perception.

A sociologist at the same university uses students in Sociology 101 to examine attitudes toward poor people and antipoverty programs. Students as a group are younger than the adult population as a whole. Even among young people, students as a group come from more prosperous and better-educated homes. Even among students, this university isn't typical of all campuses. Even on this campus, students in a sociology course may have opinions that are quite different from those of engineering students. The sociologist can't reasonably act as if these students are a random sample from any interesting population. ■

Our first examples of inference, using the *t* procedures, act as if the data are an SRS from the population of interest. Let's look back at the examples in Chapter 20.

EXAMPLE 22.2   Is It Really an SRS?

The NHANES survey that produced the BMI data for Example 20.1 used a complex multistage sample design, so it's a bit oversimplified to treat the BMI data as coming from an SRS from the population of young women.[1] Although the overall effect of the NHANES sample is close to an SRS, professional statisticians would use more complex inference procedures to match the more complex design of the sample.

The 20 patrons in the tipping study in Example 20.3 were chosen from those eating at a particular restaurant to receive one of several treatments being compared in a randomized comparative experiment. Recall that each treatment group in a completely randomized experiment is an SRS of the available subjects. Researchers sometimes act as if the available subjects are an SRS from some population if there is nothing special about where the subjects came from. In some cases, researchers collect demographic data on subjects to help justify the assumption that the subjects are a representative sample from some population. We are willing to regard the subjects as an SRS from the population of patrons of this particular restaurant, but perhaps this needs to be explored further.

The water quality test in Example 20.4 uses water samples from 20 Ohio State Park swimming areas. These samples were taken at a randomly selected location in each swimming area. We are willing to treat these samples as an SRS from the population of all possible samples from these 20 swimming areas on the date the samples were taken.

The subjects in Example 20.5 are "semi free-ranging chimpanzees at Ngamba Island Chimpanzee Sanctuary in Uganda." Researchers are willing to regard the chimpanzees in the sanctuary as representative of their species and the chimpanzees used in the study as representative of all chimpanzees in the sanctuary. Thus, we are willing to regard them as an SRS from their species. However, this conclusion relies strongly on the researchers' subject-matter knowledge that the subjects are representative of their species. ■

These examples are typical. Three are situations in which common practice is to act as if the sample were an SRS from some population. In the remaining example procedures that assume we have an SRS are used for a quick

 analysis from a more complex random sample. *There is no simple rule for deciding when you can act as if a sample is an SRS. Pay attention to these cautions:*

■ *Practical problems such as nonresponse in samples or dropouts from an experiment can hinder inference even from a well-designed study.* The NHANES survey has about an 80% response rate. This is much higher than opinion polls and most other national surveys, so by realistic standards NHANES data are quite trustworthy. (NHANES uses advanced methods to try to correct for nonresponse, but these methods work a lot better when response is high to start with.)

■ *Different methods are needed for different designs.* The *t* procedures aren't correct for random sampling designs more complex than an SRS. Later chapters give methods for some other designs, but we won't discuss inference for really complex designs like that used by NHANES. Always be sure that you (or your statistical consultant) know how to carry out the inference your design calls for.

■ *There is no cure for fundamental flaws like voluntary response surveys or uncontrolled experiments.* Look back at the bad examples in Chapters 8 and 9 and steel yourself to just ignore data from such studies.

What is the shape of the population distribution? Most statistical inference procedures require some conditions on the shape of the population distribution. Many of the most basic methods of inference are designed for Normal populations. That's the case for the one-sample *t* procedures. When we have an SRS of size *n* from a Normal population, the one-sample *t* statistic has the

REALLY WRONG NUMBERS

By now you know that "statistics" that don't come from properly designed studies are often dubious and sometimes just made up. It's rare to find wrong numbers that anyone can see are wrong, but it does happen. A German physicist claimed that 2006 was the first year since 1441 with more than one Friday the 13th. Sorry: Friday the 13th occurred in February and August of 2004, which is a bit more recent than 1441. Also, there were three Friday the 13ths in 2012 (January, April, and July).

t distribution with *n* − 1 degrees of freedom. Fortunately, because of the robustness of *t* procedures, this condition is less essential than where the data come from.

This is true because the *t* procedures and many other procedures designed for Normal distributions are based on Normality of the sample mean $\bar{x}$, not Normality of individual observations. The central limit theorem tells us that $\bar{x}$ is more Normal than the individual observations and that $\bar{x}$ becomes more Normal as the size of the sample increases. In practice, the *t* procedures are reasonably accurate for any roughly symmetric distribution for samples of even moderate size. If the sample is large, $\bar{x}$ will be close to Normal even if individual measurements are strongly skewed, as Figures 19.4 (page 423) and 19.5 (page 425) illustrate. Later chapters give practical guidelines for specific inference procedures.

There is one important exception to the principle that the shape of the population is less critical than how the data were produced. Outliers can distort the results of inference. *Any inference procedure based on sample statistics like the sample mean $\bar{x}$ that are not resistant to outliers can be strongly influenced by a few extreme observations.*

We rarely know the shape of the population distribution. In practice we rely on previous studies and on data analysis. Sometimes long experience suggests that our data are likely to come from a roughly Normal distribution, or not. For example, heights of people of the same sex and similar ages are close to Normal, but weights are not. Always explore your data before doing inference. When the data are chosen at random from a population, the shape of the data distribution mirrors the shape of the population distribution. Make a stemplot or histogram of your data and look to see whether the shape is roughly Normal. Remember that small samples have a lot of chance variation, so that Normality is hard to judge from just a few observations. Always look for outliers and try to correct them or justify their removal before performing the *t* procedures or other inference based on statistics like $\bar{x}$ that are not resistant.

When outliers are present or the data suggest that the population is strongly non-Normal, consider alternative methods that don't require Normality and are not sensitive to outliers. Some of these methods appear in Chapter 27 (available online).

Apply Your Knowledge

22.1 **Rate the Lecture.** A professor is interested in how the 500 students in his class will rate today's lecture. He selects the first 20 students on his class list, reads the names at the beginning of the lecture, and asks them to go online to the course Web site and rate the lecture on a scale of 0 to 5. Which of the following is the most important reason why a confidence interval for the mean rating by all his students based on these data is of little use? Comment briefly on each reason to explain your answer.

(a) The number of students selected is small, so the margin of error will be large.

(b) Most of the students selected will not respond.

(c) The students selected can't be considered a random sample from the population of all students in the course.

22.2 **Running Red Lights.** A survey of licensed drivers inquired about running red lights. One question asked, "Of every ten motorists who run a red light, about how many do you think will be caught?" The mean result for 880 respondents was $\bar{x} = 1.92$ and the standard deviation was $s = 1.83$.[2]

(a) Give a 95% confidence interval for the mean opinion in the population of all licensed drivers.

(b) The distribution of responses is skewed to the right rather than Normal. This will not strongly affect the t confidence interval for this sample. Why not?

(c) The 880 respondents are an SRS from completed calls among 45,956 calls to randomly chosen residential telephone numbers listed in telephone directories. Only 5029 of the calls were completed. This information gives two reasons to suspect that the sample may not represent all licensed drivers. What are these reasons?

22.3 Sampling Shoppers. A marketing consultant observes 50 consecutive shoppers at a department store the Friday after Thanksgiving, recording how much each shopper spends in the store. Suggest some reasons why it may be risky to act as if 50 consecutive shoppers at this particular time are an SRS of all shoppers at this store.

Planning studies: the power of a statistical test*

How large a sample should we take when we plan to carry out a test of significance? We know that if our sample is too small, even large effects in the population will often fail to give statistically significant results. Here are the questions we must answer to decide how many observations we need:

Significance level. How much protection do we want against getting a significant result from our sample when there really is no effect in the population?

Effect size. How large an effect in the population is important in practice?

Power. How confident do we want to be that our study will detect an effect of the size we think is important?

The three boldface terms are statistical shorthand for three pieces of information. *Power* is a new idea.

EXAMPLE 22.3   Sweetening Colas: Planning a Study

Diet colas use artificial sweeteners to avoid sugar. These sweeteners gradually lose their sweetness over time. Manufacturers therefore test new colas for loss of sweetness before marketing them. Trained tasters sip the cola along with drinks of standard sweetness and score the cola on a "sweetness score" of 1 to 10. The cola is then stored for a month at high temperature to imitate the effect of four months' storage at room temperature. Each taster scores the cola again after storage. This is a matched pairs experiment. Our data are the differences (score before storage minus score after storage) in the tasters' scores. The bigger these differences, the bigger the loss of sweetness.

From experience, suppose we know that sweetness loss scores vary from taster to taster according to a Normal distribution with

*Power calculations are important in planning studies, but this more advanced material is not needed to read the rest of the book.

standard deviation about $\sigma = 1$. To see if the taste test gives reason to think that the cola does lose sweetness, we will test

$$H_0: \mu = 0$$
$$H_a: \mu > 0$$

Will 10 tasters be sufficient for this study, or should we use more?

Significance level. Requiring significance at the 5% level is enough protection against declaring there is a loss in sweetness when in fact there is no change if we could look at the entire population. This means that when there is no change in sweetness in the population, 1 out of 20 samples of tasters will wrongly find a significant loss.

Effect size. A mean sweetness loss of 0.8 point on the 10-point scale will be noticed by consumers and so is important in practice.

Power. We want to be 90% confident that our test will detect a mean loss of 0.8 point in the population of all tasters. We agreed to use significance at the 5% level as our standard for detecting an effect. So we want probability at least 0.9 that a test at the $\alpha = 0.05$ level will reject the null hypothesis $H_0: \mu = 0$ when the true population mean is $\mu = 0.8$. ∎

The probability that the test successfully detects a sweetness loss of the specified size is the *power* of the test. You can think of tests with high power as being highly sensitive to deviations from the null hypothesis. In Example 22.3, we decided that we want power 90% when the truth about the population is that $\mu = 0.8$, and we are willing to assume we know the population standard deviation is $\sigma = 1$.

Power

The **power** of a test against a specific alternative is the probability that the test will reject H_0 at a chosen significance level α when the specified alternative value of the parameter is true.

For most statistical tests, calculating power is a job for comprehensive statistical software. Calculating power for tests of hypotheses about a population mean when the population is Normal and the population standard deviation is known is relatively easy, but we will nonetheless skip the details. The two following examples illustrate two approaches: an applet that shows the meaning of power and statistical software.

EXAMPLE 22.4   Finding Power: Use an Applet

Finding the power of the test for a population mean for a Normal population with known standard deviation is less challenging than most other power calculations because it requires only a Normal distribution probability calculation. The *Statistical Power* applet does this and illustrates the calculation with Normal curves. Enter the information from Example 22.3 into the applet: hypotheses, significance level $\alpha = 0.05$, alternative value $\mu = 0.8$, standard deviation $\sigma = 1$, and sample size $n = 10$. Click "Update." The applet output appears in Figure 22.1.

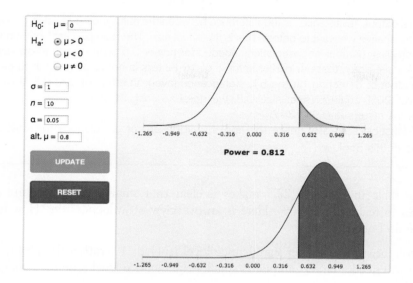

The power of the test against the specific alternative $\mu = 0.8$ is 0.812. That is, the test will reject H_0 about 81% of the time when this alternative is true. So 10 tasters are too few to give power 90%. ■

The two Normal curves in Figure 22.1 show the sampling distribution of $\bar{x}$ under the null hypothesis $\mu = 0$ (top) and also under the specific alternative $\mu = 0.8$ (bottom) when $\sigma = 1$. The curves have the same shape because σ does not change. The top curve is centered at $\mu = 0$ and the bottom curve at $\mu = 0.8$. The shaded region at the right of the top curve has area 0.05. It marks off values of $\bar{x}$ that are statistically significant at the $\alpha = 0.05$ level. The lower curve shows the probability of these same values when $\mu = 0.8$. This area is the power, 0.812.

The applet will find the power for any given sample size. It's more helpful in practice to turn the process around and learn what sample size we need to achieve a given power. Statistical software will do this but usually doesn't show the helpful Normal curves that are part of the applet's output.

EXAMPLE 22.5   Finding Power: Use Software

Some software packages (for example, SAS, JMP, Minitab, and R) will calculate power. We asked Minitab to find the number of observations needed for the one-sided test to have power 0.9 against several specific alternatives at the 5% significance level when the population standard deviation is $\sigma = 1$. Here is the table that results:

Difference	Sample Size	Target Power	Actual Power
0.1	857	0.9	0.900184
0.2	215	0.9	0.901079
0.3	96	0.9	0.902259
0.4	54	0.9	0.902259
0.5	35	0.9	0.905440
0.6	24	0.9	0.902259
0.7	18	0.9	0.907414
0.8	14	0.9	0.911247
0.9	11	0.9	0.909895
1.0	9	0.9	0.912315

FISH, FISHERMEN, AND POWER

Are the stocks of cod in the ocean off eastern Canada declining? Studies over many years failed to find significant evidence of a decline. These studies had low power—that is, they might fail to find a decline even if one were present. When it became clear that the cod were vanishing, quotas on fishing ravaged the economy in parts of Canada. If the earlier studies had had high power, they would likely have detected the decline. Quick action might have reduced the economic and environmental costs.

In this output, "Difference" is the difference between the null hypothesis value $\mu = 0$ and the alternative we want to detect. This is the effect size. The "Sample Size" column shows the smallest number of observations needed for power 0.9 against each effect size.

We see again that our earlier sample of 10 tasters is not large enough to be 90% confident of detecting (at the 5% significance level) an effect of size 0.8. If we want power 90% against effect size 0.8, we need at least 14 tasters. The actual power with 14 tasters is 0.911247.

Statistical software, unlike the applet, will do power calculations for most of the tests in this book. ■

The table in Example 22.5 makes it clear that smaller effects require larger samples to reach power 90%. Here is an overview of influences on "How large a sample do I need?"

- If you insist on a smaller significance level (such as 1% rather than 5%), you will need a larger sample. A smaller significance level requires stronger evidence to reject the null hypothesis.

- If you insist on higher power (such as 99% rather than 90%), you will need a larger sample. Higher power gives a better chance of detecting an effect when it is really there.

- At any significance level and desired power, a two-sided alternative requires a larger sample than a one-sided alternative.

- At any significance level and desired power, detecting a small effect requires a larger sample than detecting a large effect.

Planning a serious statistical study always requires an answer to the question, "How large a sample do I need?" If you intend to test the hypothesis $H_0: \mu = \mu_0$ about the mean μ of a population, you need at least a rough idea of the size of the population standard deviation σ and of how big a deviation $\mu - \mu_0$ of the population mean from its hypothesized value you want to be able to detect. More elaborate settings, such as comparing the mean effects of several treatments, require more elaborate advance information. You can leave the details to experts, but you should understand the idea of power and the factors that influence how large a sample you need.

To calculate the power of a test, we act as if we are interested in a fixed level of significance such as $\alpha = 0.05$. That's essential to do a power calculation, but remember that in practice we think in terms of P-values rather than a fixed level α. To effectively plan a statistical test we must find the power for several significance levels and for a range of sample sizes and effect sizes to get a full picture of how the test will behave.

Apply Your Knowledge

22.4 What Is Power? You manufacture and sell an iron rod whose electrical conductivity is supposed to be 10.1. You plan to make 6 measurements of the conductivity of a rod you plan to sell. You know that the standard deviation of your measurements is $\sigma = 0.1$. If the rod meets specifications, the mean of many measurements will be 10.1. You will therefore test

$$H_0: \mu = 10.1$$
$$H_a: \mu \neq 10.1$$

If the true conductivity is 10.15, the rod is not suitable for its intended use. You learn that the power of your test at the 5% significance level against the alternative $\mu = 10.15$ is 0.24.

(a) Explain in simple language what "power = 0.24" means.

(b) Explain why the test you plan will not adequately protect you against selling an iron rod with conductivity 10.15.

22.5 Thinking about Power. Answer these questions in the setting of the previous exercise about measuring the conductivity of an iron rod.

(a) You could get higher power against the same alternative with the same α by changing the number of measurements you make. Should you make more measurements or fewer to increase power?

(b) If you decide to use $\alpha = 0.10$ in place of $\alpha = 0.05$, with no other changes in the test, will the power increase or decrease?

(c) If you shift your interest to the alternative $\mu = 10.2$ with no other changes, will the power increase or decrease?

22.6 How Power Behaves. In the setting of Exercise 22.4, use the *Statistical Power* applet to find the power in each of the following circumstances. Be sure to set the applet to the two-sided alternative.

(a) Standard deviation $\sigma = 0.1$, significance level $\alpha = 0.05$, alternative $\mu = 10.15$, and sample sizes $n = 6$, $n = 12$, and $n = 24$. How does increasing the sample size with no other changes affect the power?

(b) Standard deviation $\sigma = 0.1$, significance level $\alpha = 0.05$, sample size $n = 6$, and alternatives $\mu = 10.15$, $\mu = 10.2$, and $\mu = 10.25$. How do alternatives more distant from the hypothesis (larger effect sizes) affect the power?

(c) Standard deviation $\sigma = 0.1$, sample size $n = 6$, alternative $\mu = 10.15$, and significance levels $\alpha = 0.05$, $\alpha = 0.10$, and $\alpha = 0.25$. (Click the $+$ and $-$ buttons to change α.) How does increasing the desired significance level affect the power?

22.7 How Power Behaves. Another approach to improving the unsatisfactory power of the test in Exercise 22.4 is to improve the measurement process. That is, use a measurement process that is less variable. Use the *Statistical Power* applet to find the power of the test in Exercise 22.4 in each of these circumstances: significance level $\alpha = 0.05$, alternative $\mu = 10.15$, sample size $n = 6$, and $\sigma = 0.1$, $\sigma = 0.05$, and $\sigma = 0.025$. How does decreasing the variability of the population of measurements affect the power?

Assessing the performance of a significance test: Type I and Type II errors*

We can assess the performance of a test by giving two probabilities: the significance level α and the power for an alternative that we want to be able to detect. The significance level of a test is the probability of reaching the *wrong* conclusion when the null hypothesis is true. The power for a specific alternative is the probability of reaching the *right* conclusion when that alternative is true. We can just as well describe the test by giving the probabilities of being *wrong* under both conditions.

---

*Calculating Type I and Type II errors is important in assessing the performance of a significance test, but this more advanced material is not needed to read the rest of the book.

Type I and Type II Errors

If we reject H_0 when in fact H_0 is true, this is a **Type I error**.

If we fail to reject H_0 when in fact H_a is true, this is a **Type II error**.

The **significance level α** of any fixed level test is the probability of a Type I error.

The **power** of a test against any alternative is 1 minus the probability of a Type II error for that alternative.

The possibilities are summed up in Figure 22.2. If H_0 is true, our conclusion is correct if we fail to reject H_0 and is a Type I error if we reject H_0. If H_a is true, our conclusion is either correct or a Type II error. Only one error is possible at one time.

		Truth about the population	
		H_0 true	H_a true
Conclusion based on sample	Reject H_0	Type I error	Correct conclusion
	Fail to reject H_0	Correct conclusion	Type II error

FIGURE 22.2

The two types of error in testing hypotheses.

EXAMPLE 22.6   Calculating Error Probabilities

Because the probabilities of the two types of error are just a rewording of significance level and power, we can see from Figure 22.1 what the error probabilities are for the test in Example 22.4 (page 494).

$$P(\text{Type I error}) = P(\text{reject } H_0 \text{ when in fact } \mu = 0)$$
$$= \text{significance level } \alpha = 0.05$$
$$P(\text{Type II error}) = P(\text{fail to reject } H_0 \text{ when in fact } \mu = 0.8)$$
$$= 1 - \text{power} = 1 - 0.812 = 0.188$$

The two Normal curves in Figure 22.1 (page 495) are used to find the probabilities of a Type I error (top curve, $\mu = 0$) and of a Type II error (bottom curve, $\mu = 0.8$). ▪

Apply Your Knowledge

22.8 **Two Types of Error.** Your company markets a computerized medical diagnostic program used to evaluate thousands of people. The program scans the results of routine medical tests (pulse rate, blood tests, etc.) and refers the case to a doctor if there is evidence of a medical problem. The program makes a decision about each person.

(a) What are the two hypotheses and the two types of error that the program can make? Describe the two types of error in terms of "false-positive" and "false-negative" test results.

(b) The program can be adjusted to decrease one error probability, at the cost of an increase in the other error probability. Which error probability would you choose to make smaller, and why? (This is a matter of judgment. There is no single correct answer.)

CHAPTER 22 SUMMARY

Chapter Specifics

- A specific confidence interval or test is correct only under specific conditions. The most important conditions concern the method used to produce the data. Other factors such as the shape of the population distribution may also be important.

- Whenever you use statistical inference, you are acting as if your data are a random sample or come from a randomized comparative experiment.

- Always do data analysis before inference to detect outliers or other problems that would make inference untrustworthy.

- When you plan a statistical study, plan the inference as well. In particular, ask what sample size you need for successful inference.

- The **power** of a significance test measures its ability to detect an alternative hypothesis. The power against a specific alternative is the probability that the test will reject H_0 at a particular level α when that alternative is true (Optional).

- Increasing the size of the sample increases the power of a significance test. You can use statistical software to find the sample size needed to achieve a desired power (Optional).

- If we reject H_0 when in fact H_0 is true, this is a **Type I error.** If we fail to reject H_0 when in fact H_a is true, this is a **Type II error (Optional).**

- The probability of a Type I error is the significance level α of any fixed level test. The probability of a Type II error against any alternative is 1 minus the power of the test for that alternative (Optional).

Link It

In Chapters 16, 17, 18, 20, and 21 we discussed methods for making inferences about population proportions and means. For these methods to be valid, certain conditions were required. In this chapter we discuss more carefully the conditions under which our procedures for inference do and do not hold. Where the data come from is crucial. For inferences about means, even if the population distribution is not Normal, our procedures are approximately correct if we have a moderate sample size and the shape of the population distribution is roughly symmetric. If the sample size is large, our procedures are approximately correct even if the population distribution is strongly skewed. However, any procedure based on sample statistics like the sample mean that are not resistant to outliers can be influenced by a few extreme observations. This is a caution that we first encountered in Chapter 4. Methods for data analysis introduced in Part I are useful for assessing the shape of the population distribution and hence for deciding if the conditions for inference about a mean are met. Where the data come from and the shape of the population distribution will continue to be important conditions for the methods for inference about relationships in Part V.

CHECK YOUR SKILLS

22.9 The most important condition for sound conclusions from statistical inference is usually

(a) that the data can be thought of as a random sample from the population of interest.
(b) that the population distribution is exactly Normal.
(c) that the data contain no outliers.

22.10 The coach of a college men's soccer team records the resting heart rates of the 27 team members. You should not trust a confidence interval for the mean resting heart rate of all male students at this college based on these data because

(a) with only 27 observations, the margin of error will be large.
(b) heart rates may not have a Normal distribution.
(c) the members of the soccer team can't be considered a random sample of all students.

22.11 You turn your Web browser to the online Harris Interactive Poll. Based on 2163 responses, the poll reports that 43% of U.S. adults said they would like to be richer, 21% said thinner, 14% said smarter, and 12% said younger. Nearly 1 in 10 (9%) said they would not want to choose any of the given options.[3] You should refuse to calculate a 95% confidence interval for the proportion of all U.S. adults who would like to be richer based on this sample because

(a) the poll was taken a week ago.
(b) inference from a voluntary response sample can't be trusted.
(c) the sample is too large.

22.12 Many sample surveys use well-designed random samples, but half or more of the original sample can't be contacted or refuse to take part. Any errors due to this nonresponse

(a) have no effect on the accuracy of confidence intervals.
(b) are included in the announced margin of error.
(c) are in addition to the random variation accounted for by the announced margin of error.

22.13 A writer in a medical journal says: "An uncontrolled experiment in 37 women found a significantly improved mean clinical symptom score after treatment. Methodologic flaws make it difficult to interpret the results of this study." The writer is skeptical about the significant improvement because

(a) there is no control group, so the improvement might be due to the placebo effect or to the fact that many medical conditions improve over time.
(b) the P-value given was $P = 0.048$, which is too large to be convincing.
(c) the response variable might not have an exactly Normal distribution in the population.

22.14 A medical experiment compared the herb echinacea with a placebo for preventing colds. One response variable was "volume of nasal secretions" (if you have a cold, you blow your nose a lot). Take the average volume of nasal secretions in people without colds to be $\mu = 1$. An increase to $\mu = 3$ indicates a cold. The significance level of a test of $H_0: \mu = 1$ versus $H_a: \mu > 1$ is defined as

(a) the probability that the test rejects H_0 when $\mu = 1$ is true.
(b) the probability that the test rejects H_0 when $\mu = 3$ is true.
(c) the probability that the test fails to reject H_0 when $\mu = 3$ is true.

22.15 (Optional) The power of the test in the previous exercise against the specific alternative $\mu = 3$ is defined as

(a) the probability that the test rejects H_0 when $\mu = 1$ is true.
(b) the probability that the test rejects H_0 when $\mu = 3$ is true.
(c) the probability that the test fails to reject H_0 when $\mu = 3$ is true.

22.16 (Optional) The power of a test is important in practice because power

(a) describes how well the test performs when the null hypothesis is actually true.
(b) describes how sensitive the test is to violations of conditions such as Normal population distribution.
(c) describes how well the test performs when the null hypothesis is actually not true.

CHAPTER 22 EXERCISES

22.17 **Color blindness in Africa.** An anthropologist claims that color blindness is less common in societies that live by hunting and gathering than in settled agricultural societies. He tests a number of adults in two populations in Africa, one of each type. The proportion of color-blind people is significantly lower ($P < 0.05$) in the hunter-gatherer population. What additional information would you want to help you decide whether you believe the anthropologist's claim?

22.18 **Sampling at the mall.** A market researcher chooses at random from women entering a large suburban shopping mall. One outcome of the study is a 95% confidence interval for the mean of "the highest price you would pay for a pair of jeans."

(a) Explain why this confidence interval does not give useful information about the population of all women.

(b) Explain why it may give useful information about the population of women who shop at large suburban malls.

22.19 **An outlier strikes.** You have data on an SRS of recent graduates from your college that shows how long each student took to complete a bachelor's degree. The data contain one high outlier. Will this outlier have a greater effect on a confidence interval for mean completion time if your sample is small or if it is large? Why?

22.20 **Can we trust this interval?** Here are data on the percent change in the total mass (in tons) of wildlife in several West African game preserves in the years 1971 to 1999:[4] **WLDMASS**

1971	1972	1973	1974	1975	1976	1977	1978	1979	1980
2.9	3.1	−1.2	−1.1	−3.3	3.7	1.9	−0.3	−5.9	−7.9

1981	1982	1983	1984	1985	1986	1987	1988	1989	1990
−5.5	−7.2	−4.1	−8.6	−5.5	−0.7	−5.1	−7.1	−4.2	0.9

1991	1992	1993	1994	1995	1996	1997	1998	1999
−6.1	−4.1	−4.8	−11.3	−9.3	−10.7	−1.8	−7.4	−22.9

Software gives the 95% confidence interval for the mean annual percent change as −6.66% to −2.55%. There are several reasons why we might not trust this interval.

(a) Examine the distribution of the data. What feature of the distribution throws doubt on the validity of statistical inference?

(b) Plot the percents against year. What trend do you see in this time series? Explain why a trend over time casts doubt on the condition that years 1971 to 1999 can be treated as an SRS from a larger population of years.

22.21 **When to use pacemakers.** A medical panel prepared guidelines for when cardiac pacemakers should be implanted in patients with heart problems. The panel reviewed a large number of medical studies to judge the strength of the evidence supporting each recommendation. For each recommendation, they ranked the evidence as level A (strongest), B, or C (weakest). Here, in scrambled order, are the panel's descriptions of the three levels of evidence.[5] Which is A, which B, and which C? Explain your ranking.

Evidence was ranked as level ——— when data were derived from a limited number of trials involving comparatively small numbers of patients or from well-designed data analysis of nonrandomized studies or observational data registries.

Evidence was ranked as level ——— if the data were derived from multiple randomized clinical trials involving a large number of individuals.

Evidence was ranked as level ——— when consensus of expert opinion was the primary source of recommendation.

The following exercises concern the optional material on the power of a test.

22.22 **The first child has higher IQ.** Does the birth order of a family's children influence their IQ scores? A careful study of 241,310 Norwegian 18- and 19-year-olds found that firstborn children scored 2.3 points higher on the average than second children in the same family. This difference was highly significant ($P < 0.001$). A commentator said, "One puzzle highlighted by these latest findings is why certain other within-family studies have failed to show equally consistent results. Some of these previous null findings, which have all been obtained in much smaller samples, may be explained by inadequate statistical power."[6] Explain in simple language why tests having low power often fail to give evidence against a null hypothesis even when the hypothesis is really false.

22.23 **How valium works.** Valium is a common antidepressant and sedative. A study investigated how valium works by comparing its effect on sleep in 7 genetically modified mice and 8 normal control mice. There was no significant difference between the two groups. The authors say that this lack of significance "is related to the large inter-individual variability that is also reflected in the low power (20%) of the test."[7]

(a) Explain exactly what power 20% against a specific alternative means.

(b) Explain in simple language why tests having low power often fail to give evidence against a null hypothesis even when the null hypothesis is really false.

(c) What fact about this experiment most likely explains the low power?

22.24 **Dialysis.** An article in the *New England Journal of Medicine* describes a randomized controlled trial that compared early versus late initiation of dialysis on the survival of adults with progressive chronic kidney disease. The experiment found no significant difference between early and late initiation of dialysis. According to the article, the study was designed to have power 80%, with a two-sided Type I error of 0.05, to detect a clinically important difference of approximately 10 percentage points in the absolute risk of death.[8]

(a) What fixed significance level was used in calculating the power?

(b) Explain to someone who knows no statistics why power 80% means that the experiment would probably have been significant if there was a difference between early and late initiation of dialysis.

22.25 **Power.** Software can generate samples from (almost) exactly Normal distributions. Here is a random sample of size 5 from the Normal distribution with mean 8 and standard deviation 2:

$$4.47 \quad 5.51 \quad 8.10 \quad 11.63 \quad 7.91$$

Although we know the true value of μ, suppose we pretend that we do not and we test the hypotheses

$$H_0: \mu = 6$$
$$H_a: \mu \neq 6$$

at the $\alpha = 0.05$ significance level. Enter the information from this example into the *Statistical Power* applet. (Don't forget that the alternative hypothesis is two-sided.) What is the power of the test against the alternative $\mu = 8$ (the actual population mean)?

RSAMPLE

22.26 **Finding power by hand.** Even though software is used in practice to calculate power, doing the work by hand builds your understanding. Return to the test in Example 22.6 (page 495). There are $n = 10$ observations from a population with standard deviation $\sigma = 1$ and unknown mean μ. We will test

$$H_0: \mu = 0$$
$$H_a: \mu > 0$$

with fixed significance level $\alpha = 0.05$. Find the power against the alternative $\mu = 0.8$ by following these steps.

(a) The z test statistic is

$$z = \frac{\bar{x} - \mu_0}{\sigma/\sqrt{n}} = \frac{\bar{x} - 0}{1/\sqrt{10}} = 3.162\bar{x}$$

(Remember that you won't know the numerical value of $\bar{x}$ until you have data.) What values of z lead to rejecting H_0 at the 5% significance level?

(b) Starting from your result in (a), what values of $\bar{x}$ lead to rejecting H_0? The area above these values is shaded under the top curve in Figure 22.1 (page 495).

(c) The power is the probability that you observe any of these values of $\bar{x}$ when $\mu = 0.8$. This is the shaded area under the bottom curve in Figure 22.1. What is this probability?

22.27 **Finding power by hand: two-sided test.** The previous exercise shows how to calculate the power of a one-sided z test. Power calculations for two-sided tests follow the same outline. We will find the power

of a test based on 6 measurements of the conductivity of an iron rod, reported in Exercise 22.4 (page 496). The hypotheses are

$$H_0: \mu = 10.1$$
$$H_a: \mu \neq 10.1$$

The population of all measurements is Normal with standard deviation $\sigma = 0.1$, and the alternative we hope to be able to detect is $\mu = 10.15$. (If you used the *Statistical Power* applet for Exercise 22.6, the two Normal curves for $n = 6$ illustrate parts (a) and (b) below.)

(a) Write the z test statistic in terms of the sample mean $\bar{x}$. For what values of z does this two-sided test reject H_0 at the 5% significance level?

(b) Restate your result from part (a): what values of $\bar{x}$ lead to rejection of H_0?

(c) Now suppose that $\mu = 10.15$. What is the probability of observing an $\bar{x}$ that leads to rejection of H_0? This is the power of the test.

22.28 **Error probabilities.** You read that a statistical test at significance level $\alpha = 0.01$ has power 0.78. What are the probabilities of Type I and Type II errors for this test?

22.29 **Power.** You read that a statistical test at the $\alpha = 0.05$ level has probability 0.14 of making a Type II error when a specific alternative is true. What is the power of the test against this alternative?

22.30 **Find the error probabilities.** You have an SRS of size $n = 16$ from a Normal distribution with $\sigma = 1$. You wish to test

$$H_0: \mu = 0$$
$$H_a: \mu > 0$$

You decide to reject H_0 if $\bar{x} > 0$ and to accept H_0 otherwise.

(a) Find the probability of a Type I error. That is, find the probability that the test rejects H_0 when in fact $\mu = 0$.

(b) Find the probability of a Type II error when $\mu = 0.2$. This is the probability that the test accepts H_0 when in fact $\mu = 0.2$.

(c) Find the probability of a Type II error when $\mu = 0.5$.

22.31 **Two types of error.** Go to the *Statistical Significance* applet. This applet carries out tests at a fixed significance level. When you arrive, the applet is set for the cola-tasting test of Example 22.3 (page 493). That is, the hypotheses are

$$H_0: \mu = 0$$
$$H_a: \mu > 0$$

We have an SRS of size 10 from a Normal population with standard deviation $\sigma = 1$, and we will do a test at level $\alpha = 0.05$. At the left side of the screen, a button allows you to choose the true value of the mean μ and then to generate samples from a population with that mean.

(a) Set $\mu = 0$, so that the null hypothesis is true. Each time you click the button, a new sample appears. If the sample $\bar{x}$ lands in the colored region, that sample rejects H_0 at the 5% level. Click 100 times rapidly, keeping track of how many samples

reject H_0. Use your results to estimate the probability of a Type I error. If you kept clicking forever, what probability would you get?

(b) Now set $\mu = 0.8$. Example 22.3 (page 493) shows that the test has power 0.812 against this alternative. Click 100 times rapidly, keeping track of how many samples fail to reject H_0. Use your results to estimate the probability of a Type II error. If you kept clicking forever, what probability would you get?

 ## Exploring the Web

22.32 **The American Psychological Association and inference.** The report of the American Psychological Association's Task Force on Statistical Inference is an excellent brief introduction to wise use of inference. The report appeared in the journal *American Psychologist* in 1999. You can find a copy of the initial report and a longer follow-up report on the Web in the list of "TFSI Publications/Links" from this journal at `www.apa.org/science/leadership/bsa/statistical/index.aspx`. Skim the follow-up report. What do the authors say about using a convenience sample rather than a random sample? What do the authors say about assumptions in the analysis of data?

Inference about Variables: Part IV Review

Overview

Part IV has provided students with a first look at the tools needed for statistical inference. Students should understand and be comfortable with the material covered in Chapter 15 (sampling distribution for a proportion) before moving on to Chapters 16–18 (inference for proportions) and Chapter 19 (sampling distribution for a mean) before moving on to Chapters 20–22 (inference for means). If students do not understand the basics of sampling distributions, confidence intervals, and tests of significance, they will struggle with the material to come. This review chapter provides an opportunity to examine problems that assess all of these topics.

LEARNING OUTCOMES
Refer to Learning Outcomes for Chapters 15–22, as well as the Summary provided in Chapter 23.

Part IV also introduced students to how inferences are conducted in practice. They have seen one-sample and two-sample inferences for the most common settings—estimating population means and proportions in both one- and two-sample cases. Hopefully, they have also realized that the basic concepts driving these methods are the same across these settings. If this is the case, then groundwork has been laid for understanding other inferential statistical methods, in the rest of the course and possibly in future statistics courses.

The "Part IV Review" gives a nice summary of the connections among the parts of the text to this point. The discussion in the "Part IV Summary" of the chapter provides a flow-chart students can use to organize the different kinds of problems they have encountered recently. This is particularly useful because students often struggle to recognize the kinds of formulas and methods that should be applied to different problems.

This chapter also provides a good reminder of the importance of the process of doing inference: (1) choose the correct inference procedure, (2) check the appropriate conditions for the procedure, (3) carry out the details (i.e., do the calculations), and (4) state your conclusion in the context of the problem.

Teaching Suggestions and Additional Examples/Activities for the Classroom

Consider taking some class time to have students solve a handful of problems selected from the "Test Yourself" questions provided with the chapter. Rather than worrying about the calculations, focus on having students recognize the methods appropriate for solving the problems (Exercises 23.51 through 23.55 specifically address this) and on stating and checking conditions required to use them. Alternatively, consider working some of the following problems:

- Exercises 23.25–23.27: This problem set provides a review of the sampling distribution for a sample mean and the central limit theorem.

- Exercises 23.28–23.30: This problem set provides a review of the sampling distribution for a sample proportion.

- Exercises 23.31–23.36: This problem set allows the students to examine inference for proportions and means in the same basic setting and to see that the question asked determines the methods. Recall that earlier we suggested reinforcing this idea by repeatedly asking students about the type of data that was collected and the number of samples (and whether "two" samples are independent).

■ (Optional) Exercise 23.61: This problem allows students to review confidence intervals for a population proportion using the plus-four method.

■ Exercise 23.62: This problem allows students to review significance tests for a population mean.

■ Exercise 23.63: This problem allows students to review significance tests for comparing two population means.

■ Exercise 23.69: This problem allows students to review significance tests for comparing two population means. Also, this problem examines the impact of an outlier.

■ Exercises 23.73 and 23.74: These problems present significance tests for comparing two population means and allow students to practice interpreting a P-value.

■ Exercise 23.76: This problem allows students to review confidence intervals for the population mean.

■ Exercise 23.77: In this problem, students are reminded of the duality between confidence intervals and two-sided significance tests.

© Rosemary Harris/Alamy

Inference about Variables: Part IV Review

In this chapter we cover...

- Part IV Summary
- Test Yourself
- Supplementary Exercises

The procedures of Chapters 15 to 22 include the most common of all statistical inference methods. Now that you have mastered important ideas and practical methods for inference, it's time to review the big ideas of statistics in outline form. Here is a summary of how the ideas of Parts I and II of this book lead up to Part IV. The outline contains some important warnings: look for the Caution icon.

1. **Data Production**
   - Data basics:

     Individuals (subjects).
     Variables: categorical versus quantitative, units of measurement, explanatory versus response.
     Purpose of study.
   - Data production basics:

     Observation versus experiment.
     Simple random samples.
     Completely randomized experiments.
   - Beware: really bad data production (e.g., voluntary response, confounding) can make interpretation impossible.

Beware: weaknesses in data production (for example, sampling students at only one campus) can make generalizing conclusions difficult.

2. **Data Analysis**

- Plot your data. Look for an overall pattern and striking deviations.
- Add numerical descriptions based on what you see.
- Beware: averages and other simple descriptions can miss the real story.
- One quantitative variable:
    Graphs: stemplot, histogram, boxplot.
    Pattern: distribution shape, center, spread/variability. Outliers?
    Density curves (such as Normal curves) to describe overall pattern.
    Numerical descriptions: five-number summary or $\bar{x}$ and s.

- Relationships between two quantitative variables:
    Graph: scatterplot.
    Pattern: relationship form, direction, strength. Outliers? Influential observations?
    Numerical description for linear relationships: correlation, regression line.

    Beware the lurking variable: correlation does not imply causation.
- Beware the effects of outliers and influential observations.

3. **The Reasoning of Inference**

- Inference uses data to infer conclusions about a wider population.

- When you do inference, you are acting as if your data come from random samples or randomized comparative experiments. Beware: if they don't, you may have "garbage in, garbage out."
- Always examine your data before doing inference. Inference often requires a regular pattern, such as roughly Normal with no strong outliers.
- Key idea: "What would happen if we did this many times?"
- Confidence intervals: estimate a population parameter.
    95% confidence: I used a method that captures the true parameter 95% of the time in repeated use.
    Beware: the margin of error of a confidence interval does not include the effects of practical errors such as undercoverage and nonresponse.

- Significance tests: assess evidence against H_0 in favor of H_a.
    P-value: If H_0 were true, how often would I get an outcome favoring the alternative this strongly? Smaller P = stronger evidence against H_0.
    Statistical significance at the 5% level, $P < 0.05$, means that an outcome this extreme would occur less than 5% of the time if H_0 were true.
    Beware: $P < 0.05$ is not sacred.

    Beware: statistical significance is not the same as practical importance. Large samples can make small effects significant. Small samples can fail to declare large effects significant.
    Always try to estimate the size of an effect (for example, with a confidence interval), not just its significance.

4. Methods of Inference

- Choose the right inference procedure.
- Check the conditions for the procedure.
- Carry out the details.
- State your conclusion in the context of the problem.

Part IV of this book introduces parts 3 and 4 of this outline. In addition to understanding the important ideas of inference, to actually do inference you must be able to choose the right procedure, including checking the conditions for the procedure, and carry out the details. The Statistics in Summary flowchart on the next page offers a brief guide. It is important to do some of the review exercises because now, for the first time, you must decide which of several inference procedures to use. Learning to recognize problem settings to choose the right type of inference is a key step in advancing your mastery of statistics. This is the "Plan" step in the four-step process, in which you translate the real-world problem from the "State" step into a specific inference procedure.

The flowchart organizes one way of planning inference problems. Let's go through it from left to right.

1. *Do you want to test a claim or estimate an unknown quantity?* That is, will you need a test of significance or a confidence interval?

2. *Are your data a single sample representing one population or two samples chosen to compare two populations or responses to two treatments in an experiment?* Remember that to work with *matched pairs* data, you form one sample from the differences within pairs.

3. *Is the response variable quantitative or categorical?* Quantitative variables take numerical values with some unit of measurement such as inches or grams. The most common inference questions about quantitative variables concern *mean* responses. If the response variable is categorical, inference most often concerns the *proportion* of some category (call it a "success") among the responses.

The flowchart leads you to a specific test or confidence interval, indicated by a formula at the end of each path. The formula is just an aid to guide you toward the "Solve" and "Conclude" steps. You (or your technology) will use the formula as part of the "Solve" step, but don't forget that you must do more.

- *Are the conditions for this procedure met?* Can you act as if the data come from a random sample or randomized comparative experiment? Does data analysis show extreme outliers or strong skewness that forbid use of inference based on Normality? Do you have enough observations for your intended procedure?

- *Do your data come from an experiment or from an observational study?* The details of inference methods are the same for both. But the design of the study determines what conclusions you can reach because experiments give much better evidence that an effect uncovered by inference can be explained by direct causation.

You may ask, as you study the Statistics in Summary flowchart, "What if I have an experiment comparing four treatments, or samples from three populations?" The flowchart allows only one or two, not three or four or more. Be patient: methods for comparing more than two means or proportions, as well as some other settings for inference, appear in Part V.

 HOW MANY WAS THAT?

Good causes often breed bad statistics. An advocacy group claims, without much evidence, that 150,000 Americans suffer from the eating disorder anorexia nervosa. Soon someone misunderstands and says that 150,000 people *die* from anorexia nervosa each year. This wild number gets repeated in countless books and articles. It really is a wild number: only about 55,000 women aged 15 to 44 (the main group affected) die of *all causes* each year.

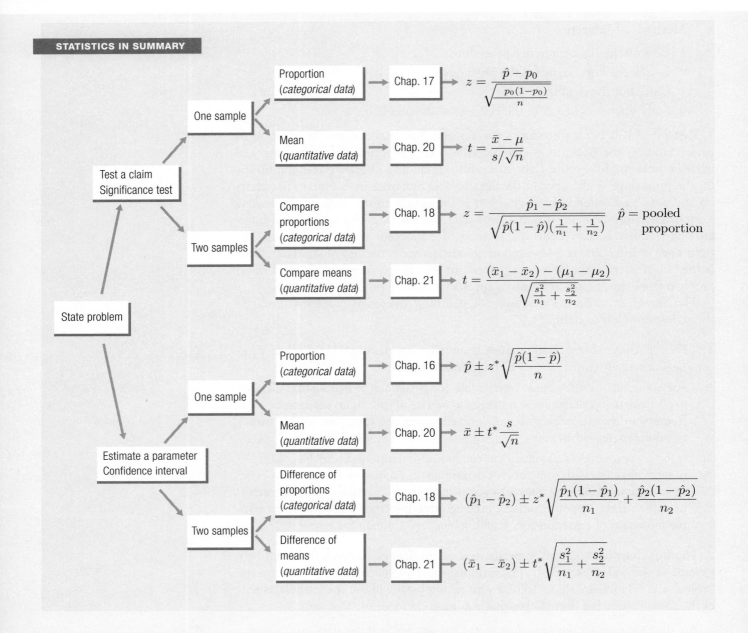

Part IV Summary

Here are the most important skills you should have acquired from reading Chapters 15 to 22.

A. Sampling Distributions

1. Identify parameters and statistics in a statistical study.

2. Recognize the fact of sampling variability: a statistic will take different values when you repeat a sample or experiment.

3. Interpret a sampling distribution as describing the values taken by a statistic in *all* possible repetitions of a sample or experiment under the same conditions.

4. Interpret the sampling distribution of a statistic as describing the probabilities of its possible values.

B. The Sampling Distribution of a Sample Proportion

1. Recognize when a problem involves the proportion $\hat{p}$ of a sample. Understand that $\hat{p}$ estimates a proportion p in the population.

2. Use the law of large numbers to describe the behavior of $\hat{p}$ as the size of the sample increases.

3. Find the mean and standard deviation of a sample proportion $\hat{p}$ from an SRS of size n when the population proportion p is known.

4. Understand that $\hat{p}$ is an unbiased estimator of p and that the variability of $\hat{p}$ about its mean p gets smaller as the sample size increases.

5. Understand and check the conditions for which $\hat{p}$ has approximately a Normal distribution. Use this Normal distribution to calculate probabilities that concern $\hat{p}$.

C. The Sampling Distribution of a Sample Mean

1. Recognize when a problem involves the mean $\bar{x}$ of a sample. Understand that $\bar{x}$ estimates the mean μ of the population from which the sample is drawn.

2. Use the law of large numbers to describe the behavior of $\bar{x}$ as the size of the sample increases.

3. Find the mean and standard deviation of a sample mean $\bar{x}$ from an SRS of size n when the mean μ and standard deviation σ of the population are known.

4. Understand that $\bar{x}$ is an unbiased estimator of μ and that the variability of $\bar{x}$ about its mean μ gets smaller as the sample size increases.

5. Understand that $\bar{x}$ has approximately a Normal distribution when the sample is large (central limit theorem). Use this Normal distribution to calculate probabilities that concern $\bar{x}$.

D. Confidence Intervals

1. State in nontechnical language what is meant by "95% confidence" or other statements of confidence in statistical reports.

2. Know the four-step process (page 357) for any confidence interval. This process will be used more extensively in later chapters.

3. Understand how the margin of error of a confidence interval changes with the sample size and the level of confidence C.

4. Find the sample size required to obtain a confidence interval of specified margin of error m when the confidence level and other information are given.

5. Identify sources of error in a study that are *not* included in the margin of error of a confidence interval, such as undercoverage or nonresponse.

E. Significance Tests

1. State the null and alternative hypotheses in a testing situation in terms of the parameter of interest.

2. Explain in nontechnical language the meaning of the P-value when you are given the numerical value of P for a test.

3. Know the four-step process (page 383) for any significance test.

4. Assess statistical significance at standard levels α, either by comparing P with α or by comparing the value of the test statistic with standard critical values.

5. Recognize that significance testing does not measure the size or importance of an effect. Explain why a small effect can be significant in a large sample and why a large effect can fail to be significant in a small sample.

6. Recognize that any inference procedure acts as if the data were properly produced. For a single sample, the data should be an SRS from the population. In a two-sample problem, the data can arise from a randomized comparative experiment or a comparison of SRSs from two populations.

F. Recognition

1. Recognize when a problem requires inference about population proportions (usually a categorical response variable) or population means (quantitative response variable).

2. Recognize from the design of a study whether one-sample, matched pairs, or two-sample procedures are needed.

3. Based on recognizing the problem setting, choose among the one- and two-sample z procedures for proportions and the one- and two-sample t procedures for means.

G. Inference about One Proportion

1. Verify the conditions under which you can safely use the large-sample z procedures in a particular setting. Check the study design and the guidelines for sample size.

2. Use the large-sample z procedure to give a confidence interval for a population proportion p.

3. Use the z statistic to carry out a test of significance for the hypothesis $H_0: p = p_0$ about a population proportion p against either a one-sided or a two-sided alternative. Use software or Table A to find the P-value, or Table C to get an approximate value.

4. (Optional) When the conditions for the large sample confidence are not met, understand when to use the plus four modification of the z procedure to give a confidence interval for p that is accurate even for small samples and for any value of p.

H. Comparing Two Proportions

1. Verify the conditions under which you can safely use the large-sample z procedures in a particular setting. Check the study design and the guidelines for sample sizes.

2. Use the large-sample z procedure to give a confidence interval for the difference $p_1 - p_2$ between proportions in two populations based on independent samples from the populations.

3. Use a z statistic to test the hypothesis $H_0: p_1 = p_2$ that proportions in two distinct populations are equal. Use software or Table A to find the P-value or Table C to get an approximate value.

4. (Optional) When the conditions for the large-sample confidence interval are not met, understand when to use the plus four modification of the z procedure to give a confidence interval for $p_1 - p_2$ that is accurate, even for very small samples and for any values of p_1 and p_2.

I. Inference about One Mean

1. Verify that the t procedures are appropriate in a particular setting. Check the study design and the distribution of the data and take advantage of robustness against lack of Normality.

2. Recognize when poor study design, outliers, or a small sample from a skewed distribution make the t procedures risky.

3. Use the one-sample t procedure to obtain a confidence interval at a stated level of confidence for the mean μ of a population.

4. Carry out a one-sample t test for the hypothesis that a population mean μ has a specified value against either a one-sided or a two-sided alternative. Use software to find the P-value or Table C to get an approximate value.

5. Recognize matched pairs data and use the t procedures to obtain confidence intervals and to perform tests of significance for such data.

J. Comparing Two Means

1. Verify that the two-sample t procedures are appropriate in a particular setting. Check the study design and the distribution of the data and take advantage of robustness against lack of Normality.

2. Give a confidence interval for the difference between two means. Use software if you have it. Use the two-sample t statistic with conservative degrees of freedom and Table C if you do not have statistical software.

3. Test the hypothesis that two populations have equal means against either a one-sided or a two-sided alternative. Use software if you have it. Use the two-sample t test with conservative degrees of freedom and Table C if you do not have statistical software.

4. Know that procedures for comparing the standard deviations of two Normal populations are available, but that these procedures are risky because they are not at all robust against non-Normal distributions.

Test Yourself

The following questions include both multiple-choice and short-answer questions and calculations. They will help you review the basic ideas and skills presented in Chapters 15 to 22.

Calcium and blood pressure. *In a randomized comparative experiment on the effect of dietary calcium on blood pressure, researchers divided 54 healthy white males at random into two groups. One group received calcium; the other, a placebo. At the beginning of the study, the researchers measured many variables on the subjects. The paper reporting the study gives $\bar{x} = 114.9$ and $s = 9.3$ for the seated systolic blood pressure of the 27 members of the placebo group. Use this information to answer Questions 23.1 and 23.2.*

23.1 A 95% confidence interval for the mean blood pressure in the population from which the subjects were recruited is

(a) 113.1 to 116.7.
(b) 111.8 to 118.0.
(c) 111.2 to 118.6.
(d) 109.9 to 119.9.

23.2 What conditions for the population and the study design are required by the procedure you used to construct your confidence interval? Which of these conditions are important for the validity of the procedure in this case?

Does nature heal better? *Our bodies have a natural electrical field that is known to help wounds heal. Does changing the field strength slow healing? A series of experiments with newts investigated this question. The following data are the healing rates of cuts (micrometers per hour)* *in a matched pairs experiment. The pairs are the two hind limbs of the same newt, with the body's natural field in one limb (control) and half the natural value in the other limb (experimental).*[1]

Newt	1	2	3	4	5	6	7	8	9	10	11	12	13	14
Control	25	13	44	45	57	42	50	36	35	38	43	31	26	48
Experimental	24	23	47	42	26	46	38	33	28	28	21	27	25	45
Difference (Control − Experimental)	1	−10	−3	3	31	−4	12	3	7	10	22	4	1	3

The mean and standard deviation of the differences are 5.71 and 10.56 micrometers per hour, respectively. Use this information to answer Questions 23.3 and 23.4.

23.3   Is there good evidence that changing the electrical field from its natural level slows healing? The *P*-value for your test is

(a) less than 0.01.
(b) between 0.01 and 0.05.
(c) between 0.05 and 0.10.
(d) greater than 0.10.

23.4   Give a 99% confidence interval for the difference in healing rates (control minus experimental).

Game players. *A government survey randomly selected 6889 female high school students and 7028 male high school students.*[2] *Of these students, 1020 females and 1926 males played video or computer games for three or more hours a day. Use this information to answer Questions 23.5 to 23.8.*

23.5   The estimate of the proportion of males who played video or computer games for three or more hours a day is

(a) 0.148.   (c) 0.231.
(b) 0.212.   (d) 0.274.

23.6   The estimate of the proportion of females who played video or computer games for three or more hours a day is

(a) 0.148.   (c) 0.231.
(b) 0.212.   (d) 0.274.

23.7   The sampling distribution for the difference in the sample proportions has standard error

(a) 0.0026.   (c) 0.0053.
(b) 0.0043.   (d) 0.0068.

23.8   A 99% confidence interval for the difference in proportions of male and female high school students who play video or computer games for at least three hours a day is

(a) 0.115 to 0.137.   (c) 0.108 to 0.144.
(b) 0.113 to 0.139.   (d) 0.106 to 0.146.

23.9   Wikipedia. A sample survey of 1497 adult Internet users found that 36% consult the online collaborative encyclopedia Wikipedia.[3]

(a) Give the standard error SE of $\hat{p}$, the proportion of all adult Internet users who refer to Wikipedia.

(b) Give a 95% confidence interval for the proportion of all adult Internet users who refer to Wikipedia.

An IQ test. *The Wechsler Adult Intelligence Scale (WAIS) is a common "IQ test" for adults. The distribution of WAIS scores for persons over 16 years of age is approximately Normal with mean 100 and standard deviation 15. Use this information to answer Questions 23.10 to 23.13.*

23.10   What is the probability that a randomly chosen individual has a WAIS score of 105 or higher?

    (a) 0.0005   (c) 0.4400

    (b) 0.3707   (d) 0.6293

23.11   What are the mean and standard deviation of the average WAIS score $\bar{x}$ for an SRS of 60 people?

    (a) mean = 13.56, standard deviation = 15.

    (b) mean = 100, standard deviation = 15.

    (c) mean = 100, standard deviation = 1.94.

    (d) mean = 100, standard deviation = 0.25.

23.12   What is the probability that the average WAIS score of an SRS of 60 people is 105 or higher?

    (a) 0.0049

    (b) 0.3607

    (c) 0.9738

    (d) None of the above.

23.13   Would your answers to any of Questions 23.10, 23.11, or 23.12 be affected if the distribution of WAIS scores in the adult population were distinctly non-Normal? Explain.

Men and muscle. *Ask young men to estimate their own degree of body muscle by choosing from a set of 100 photos. Then ask them to choose what they think women prefer. The researchers know the actual degree of muscle, measured as kilograms per square meter of fat-free mass, for each of the photos. They can therefore measure the difference between what a subject thinks women prefer and the subject's own self-image. Call this difference the "muscle gap." Here are summary statistics for the muscle gap from two samples, one of American and European young men and the other of Chinese young men from Taiwan:*[4]

GROUP	*n*	$\bar{x}$	*s*
American/European	200	2.35	2.5
Chinese	55	1.20	3.2

Use this information to answer Questions 23.14 to 23.17.

23.14   A 95% confidence interval for the mean size of the muscle gap for all American and European young men is

    (a) 2.35 ± 0.18.   (c) 2.35 ± 4.95.

    (b) 2.35 ± 0.35.   (d) 1.15 ± 0.93.

23.15   A 95% confidence interval for the mean size of the muscle gap for all Chinese young men is

    (a) 1.20 ± 0.43.   (c) 1.20 ± 6.40.

    (b) 1.20 ± 0.87.   (d) 1.15 ± 0.93.

23.16   Is there a significant difference between the mean sizes of the muscle gap for American/European men and Chinese men? The value of the *t* statistic for testing the null hypothesis of no difference in the mean sizes of the muscle gap is

    (a) 0.47.   (c) 2.13.

    (b) 1.15.   (d) 2.47.

23.17   Is there a significant difference between the mean sizes of the muscle gap for American/European men and Chinese men? The degrees of freedom using the conservative Option 2 (page 468) for the *t* statistic for testing the null hypothesis of no difference in the mean sizes of the muscle gap is

    (a) 54.   (c) 199.

    (b) 126.5.   (d) 253.

 HOW MANY MILES PER GALLON?

As gasoline prices rise, more people pay attention to the government's gas mileage ratings of their vehicles. Until recently these ratings overstated the miles per gallon we can expect in real-world driving. The ratings assumed a top speed of 60 miles per hour, slow acceleration, and no air conditioning. That doesn't resemble what we see around us on the highway. Maybe it doesn't resemble the way we ourselves drive. Starting with 2008 models, the ratings assume higher speeds (80 miles per hour tops), faster acceleration, and air conditioning in warm weather. Mileage ratings of the same vehicle dropped by about 12% in the city and 8% on the highway.

Michael Lustbader/Science Source

23.18 **Butterflies mating.** Here's how butterflies mate: a male passes to a female a packet of sperm called a spermatophore. Females may mate several times. Will they remate sooner if the first spermatophore they receive is small? Among 20 females who received a large spermatophore (greater than 25 milligrams), the mean time to the next mating was 5.15 days, with standard deviation 0.18 day. For 21 females who received a small spermatophore (about 7 milligrams), the mean was 4.33 days, and the standard deviation was 0.31 day.[5] Is the observed difference in means statistically significant? Test using the conservative Option 2 for the degrees of freedom. The P-value is

(a) less than 0.01.
(b) between 0.01 and 0.05.
(c) between 0.05 and 0.10.
(d) greater than 0.10.

Mouse endurance. *A study of the inheritance of speed and endurance in mice found a trade-off between these two characteristics, both of which help mice survive. To test endurance, mice were made to swim in a bucket with a weight attached to their tails. (The mice were rescued when exhausted.) Here are data on endurance in minutes for female and male mice:*[6]

GROUP	n	MEAN	STANDARD DEVIATION
Female	162	11.4	26.09
Male	135	6.7	6.69

Use this information to answer Questions 23.19 to 23.22.

23.19 Both sets of endurance data are skewed to the right. Why are t procedures none-theless reasonably accurate for these data?

23.20 A 90% confidence interval for the mean endurance of female mice swimming is

(a) 9.35 to 13.45. (c) 7.34 to 15.46.
(b) 8.01 to 14.79. (d) 7.14 to 15.66.

23.21 A 90% confidence interval for the mean difference (female minus male) in endurance times is

(a) 8.00 to 14.80. (c) 5.75 to 7.65.
(b) 1.17 to 8.23. (d) 2.25 to 7.15.

23.22 Do the data show that female mice have significantly higher endurance on the average than male mice?

23.23 **(Optional) Prereaders in kindergarten.** A school has two kindergarten classes. There are 21 children in Ms. Toodle's kindergarten class. Of these, 17 are "prereaders"—children on the verge of reading. There are 19 children in Mr. Grimace's kindergarten class. Of these, 13 are prereaders. Using the plus four confidence interval method, a 90% confidence interval for the difference in proportions of children in these classes who are prereaders is 0.104 to 0.336. Which of the following statements is correct?

(a) This confidence interval is not reliable because the samples are so small.
(b) This confidence interval is of no use because it does not contain 0, the value of no difference between classes.
(c) This confidence interval is reasonable because the sample sizes are both at least 5.
(d) This confidence interval is not reliable because these samples cannot be viewed as simple random samples taken from a larger population.

23.24 **State of the economy.** If we want to estimate p, the population proportion of likely voters who believe the state of the economy is the most urgent national concern, with 99% confidence and a margin of error no greater than 2%, how many likely voters need to be surveyed? Assume that you have no idea of the value of p.

(a) 2401   (c) 4148
(b) 3484   (d) 8256

Reaction times. *The time that people require to react to a stimulus usually has a right-skewed distribution, as lack of attention or tiredness causes some lengthy reaction times. Reaction times for children with attention-deficit/hyperactivity disorder (ADHD) are more skewed, as their condition causes more frequent lack of attention. In one study, children with ADHD were asked to press the spacebar on a computer keyboard when any letter other than X appeared on the screen. With 2 seconds between letters, the mean reaction time was 445 milliseconds (ms), and the standard deviation was 82 ms.[7] Take these values to be the population μ and σ for ADHD children. Use this information to answer Questions 23.25 to 23.27.*

23.25 What are the mean and standard deviation of the mean reaction time $\bar{x}$ for a randomly chosen group of 15 ADHD children? For a group of 150 such children?

23.26 The distribution of reaction time is strongly skewed. Explain briefly why we hesitate to regard $\bar{x}$ as Normally distributed for 15 children but are willing to use a Normal distribution for the mean reaction time of 150 children.

23.27 What is the approximate probability that the mean reaction time in a group of 150 ADHD children is greater than 450 ms?

Favoritism for college athletes? *Sports Illustrated surveyed a random sample of 757 Division I college athletes in 36 sports. One question asked was "Have you ever received preferential treatment from a professor because of your status as an athlete?" Of the athletes polled, 225 said "Yes." Use this information to answer Questions 23.28 to 23.30.*

23.28 The sample proportion of athletes who have received preferential treatment from a professor is

(a) 0.160.   (c) 0.703.
(b) 0.297.   (d) 0.840.

23.29 The standard error SE of $\hat{p}$, the proportion of athletes who have received preferential treatment from a professor, is

(a) 0.003.   (c) 0.209.
(b) 0.017.   (d) 0.457.

23.30 A 90% confidence interval for the proportion of athletes who have received preferential treatment from a professor is

(a) 0.276 to 0.320.   (c) 0.265 to 0.331.
(b) 0.269 to 0.325.   (d) 0.255 to 0.342.

Very-low-birth-weight babies. *Starting in the 1970s, medical technology allowed babies with very low birth weight (VLBW, less than 1500 grams, about 3.3 pounds) to survive without major handicaps. It was noticed that these children nonetheless had difficulties in school and as adults. A long-term study has followed 242 VLBW babies to age 20 years, along with a control group of 233 babies from the same population who had normal birth weight.[8] At age 20, 179 of the VLBW group and 193 of the control group had graduated from high school. Use this information to answer Questions 23.31 to 23.36.*

23.31 This is an example of

(a) an observational study.
(b) a nonrandomized experiment.
(c) a randomized controlled study.
(d) a matched pairs experiment.

23.32 Take p_{VLBW} and p_{control} to be the proportions of all VLBW and normal-birth-weight (control) babies who would graduate from high school. The hypotheses to be tested are

(a) H_0: $p_{\text{VLBW}} = p_{\text{control}}$ versus H_a: $p_{\text{VLBW}} \neq p_{\text{control}}$.
(b) H_0: $p_{\text{VLBW}} = p_{\text{control}}$ versus H_a: $p_{\text{VLBW}} > p_{\text{control}}$.
(c) H_0: $p_{\text{VLBW}} = p_{\text{control}}$ versus H_a: $p_{\text{VLBW}} < p_{\text{control}}$.
(d) H_0: $p_{\text{VLBW}} > p_{\text{control}}$ versus H_a: $p_{\text{VLBW}} = p_{\text{control}}$.

23.33 The pooled sample proportion of babies who would graduate from high school is

(a) $\hat{p} = 0.74$.
(b) $\hat{p} = 0.78$.
(c) $\hat{p} = 0.81$.
(d) $\hat{p} = 0.83$.

23.34 The numerical value of the z test for comparing the proportions of all VLBW and normal-birth-weight (control) babies who would graduate from high school is

(a) $z = -1.65$.
(b) $z = -2.34$.
(c) $z = -2.77$.
(d) $z = -3.14$.

23.35 IQ scores were available for 113 men in the VLBW group and for 106 men in the control group. The mean IQ for the 113 men in the VLBW group was 87.6, and the standard deviation was 15.1. The 106 men in the control group had mean IQ 94.7, with standard deviation 14.9. Is there good evidence that mean IQ is lower among VLBW men than among controls from similar backgrounds? To test this with a two-sample t test, the test statistic would be

(a) $t = -1.72$.
(b) $t = -3.50$.
(c) $t = -5.00$.
(d) $t = -7.10$.

23.36 Of the 126 women in the VLBW group, 38 said they had used illegal drugs; 54 of the 124 control group women had done so. The IQ scores for the VLBW women who had used illegal drugs had mean 86.2 (standard deviation 13.4), and the normal-birth-weight controls who had used illegal drugs had mean IQ 89.8 (standard deviation 14.0). Is there a statistically significant difference between the two groups in mean IQ? The P-value for this test is

(a) less than 0.01.
(b) between 0.01 and 0.05.
(c) between 0.05 and 0.10.
(d) greater than 0.10.

Binge drinking. *According to the National Institute on Alcohol Abuse and Alcoholism (NIAAA) and the National Institutes of Health (NIH), 41% of college students nationwide engage in "binge-drinking" behavior: having 5 or more drinks on one occasion during the past two weeks. A college president wonders if the proportion of students enrolled at her college who binge drink is actually lower than the national proportion. In a commissioned study, 348 students are selected randomly from a list of all students enrolled at the college. Of these, 132 admit to having engaged in binge drinking. Use this information to answer Questions 23.37 to 23.40.*

23.37 Based on the results of the commissioned study, a 95% confidence interval for the proportion of all students at this college who engage in binge drinking is

(a) 0.328 to 0.430.   (c) 0.341 to 0.420.
(b) 0.338 to 0.423.   (d) 0.343 to 0.418.

23.38 The college president is interested in testing her belief that the proportion of students at her college who engage in binge drinking is lower than the national proportion of 0.41. If p is the proportion of students at her college who engage in binge drinking, she should test the hypotheses

(a) H_0: $\hat{p} = 0.41$ versus H_a: $\hat{p} < 0.41$.
(b) H_0: $\hat{p} = 0.41$ versus H_a: $\hat{p} > 0.41$.
(c) H_0: $p = 0.41$ versus H_a: $p < 0.41$.
(d) H_0: $p = 0.41$ versus H_a: $p > 0.41$.

23.39 The college president is interested in testing her belief that the proportion of students at her college who engage in binge drinking is lower than the national proportion of 0.41. The P-value for this test is

(a) between 0.15 and 0.20.
(b) between 0.10 and 0.15.
(c) between 0.05 and 0.10.
(d) below 0.05.

23.40 Which of the following conclusions is reasonable, based on the P-value computed in the previous exercise?

(a) There is little evidence to support a conclusion that the proportion of students at this particular college who binge drink is lower than the national proportion of 0.41.
(b) There is moderate but not strong evidence that the proportion of binge-drinking students at this college is lower than the national proportion of 0.41.
(c) There is strong evidence that the proportion of students at this college who binge drink is lower than the national proportion of 0.41.
(d) We can't reach any reasonable conclusion because the assumptions necessary for a significance test for a proportion are not met in this case.

23.41 **Americans with bachelor's degrees.** About 30% of Americans over 25 years of age have a bachelor's degree. A polling firm contacts an SRS of 1500 people chosen from the population of Americans over 25 years of age.

(a) What are the mean and standard deviation of $\hat{p}$, the proportion of Americans over 25 in the sample who have college degrees?
(b) What is the approximate probability that $\hat{p}$ is greater than 35%?

23.42 In the setting of the previous problem,

(a) How large an SRS do you need for the standard deviation of $\hat{p}$ to be less than 0.01?
(b) How large an SRS would you need to estimate the population proportion to within 0.02 with 95% confidence if you did not have this poll to use as an estimate?

Listening to rap. *The Black Youth Project of the University of Chicago interviewed random samples of black, Hispanic, and white young people aged 15 to 25. We can consider this a stratified sample or three separate random samples of 634 blacks, 314 Hispanics, and 567 whites. The survey found that 58% of black youth listen to rap music every day, compared with 45% of Hispanics and 23% of whites. But attitudes were quite similar in the three groups. For example, 72% of blacks, 72% of Hispanics, and 68% of whites agreed that "rap music videos contain too many references to sex."[9] Questions 23.43 to 23.45 are based on this study.*

23.43 Give a 90% confidence interval for the proportion of all black young people who listen to rap every day.

23.44 Give a 90% confidence interval for the difference between the proportions of all Hispanic and all white young people who listen to rap every day.

23.45 Is there a significant difference between the proportions of black and white young people who think that rap videos contain too much sex? State hypotheses, check appropriate conditions, and find the test statistic, and use either software or the bottom row of Table C for the P-value. Be sure to state your conclusion.

© Henry Horenstein/CORBIS

Cholesterol in dogs. *High levels of cholesterol in the blood are not healthy in either humans or dogs. Because a diet rich in saturated fats raises the cholesterol level, it is plausible that dogs owned as pets have higher cholesterol levels than dogs owned by a veterinary research clinic. "Normal" levels of cholesterol based on the clinic's dogs would then be misleading. A clinic compared healthy dogs it owned with healthy pets brought to the clinic to be neutered. The summary statistics for blood cholesterol levels (milligrams per deciliter of blood) appear in the following table.*[10]

GROUP	n	$\bar{x}$	s
Pets	26	193	68
Clinic	23	174	44

Questions 23.46 to 23.50 are based on this study.

23.46 A 95% confidence interval for the mean cholesterol level in pets is

(a) 179.7 to 206.3.

(b) 176.8 to 209.2.

(c) 165.5 to 220.5.

(d) 159.6 to 226.48.

23.47 Is there strong evidence that pets have a higher mean cholesterol level than clinic dogs? To test this with a two-sample t test, the values of the t statistic and its degrees of freedom using conservative Option 2 are

(a) $t = 1.17$, df $= 22$.

(b) $t = 1.17$, df $= 47$.

(c) $t = 8.92$, df $= 22$.

(d) $t = 8.92$, df $= 47$.

23.48 A 95% confidence interval for the difference in mean cholesterol levels between pets and clinic dogs is (use conservative Option 2 for the degrees of freedom)

(a) -26.1 to 64.1.

(b) -14.6 to 52.6.

(c) -8.7 to 46.7.

(d) 2.8 to 35.2.

23.49 What conditions must be satisfied to justify the procedures you used in Exercise 23.46? In Exercise 23.47? In Exercise 23.48?

23.50 Assuming that the cholesterol measurements have no outliers and are not strongly skewed, what is the chief threat to the validity of the results of this study?

Choosing an inference procedure. *In each of Questions 23.51 to 23.56, say which type of inference procedure from the Statistics in Summary flowchart (page 508) you would use, or explain why none of these procedures fits the problem. You do not need to carry out any procedures.*

23.51 Driving too fast. How seriously do people view speeding in comparison with other annoying behaviors? A large random sample of adults was asked to rate a number of behaviors on a scale of 1 (no problem at all) to 5 (very severe problem). Do speeding drivers get a higher average rating than noisy neighbors?

23.52 Preventing drowning. Drowning in bathtubs is a major cause of death in children less than 5 years old. A random sample of parents was asked many questions related to bathtub safety. Overall, 85% of the sample said they used baby bathtubs for infants. Estimate the percent of all parents of young children who use baby bathtubs.

23.53 Acid rain? You have data on rainwater collected at 16 locations in the Adirondack Mountains of New York State. One measurement is the acidity of the water, measured by pH on a scale of 0 to 14 (the pH of distilled water is 7.0). Estimate the average acidity of rainwater in the Adirondacks.

23.54 Athletes' salaries. Looking online, you find the salaries of the 29 players on the roster of the Chicago Cubs as of opening day of the 2013 baseball season. The team total was $104.3 million, fourteenth highest in the major leagues. Estimate the average salary of the Cubs players.

23.55 Looking back on love. How do young adults look back on adolescent romance? Investigators interviewed 40 couples in their midtwenties. The female and male partners were interviewed separately. Each was asked about his or her current relationship and also about a romantic relationship that lasted at least two months when they were aged 15 or 16. One response variable was a measure on a numerical scale of how much the attractiveness of the adolescent partner mattered. You want to compare the men and women on this measure.

23.56 Preventing AIDS through education. The Multisite HIV Prevention Trial was a randomized comparative experiment to compare the effects of twice-weekly small-group AIDS discussion sessions (the treatment) with a single one-hour session (the control). Compare the effects of treatment and control on each of the following response variables:

(a) A subject does or does not use condoms six months after the education sessions.
(b) The number of unprotected intercourse acts by a subject between four and eight months after the sessions.
(c) A subject is or is not infected with a sexually transmitted disease six months after the sessions.

Supplementary Exercises

*Supplementary exercises apply the skills you have learned in ways that require more thought or more use of technology. Some of these exercises start from actual data rather than from data summaries. Many of these exercises ask you to follow the **Plan, Solve,** and **Conclude** steps of the four-step process. Remember that the **Solve** step includes checking the conditions for the inference you plan.*

23.57 Do you have confidence? A report of a survey distributed to randomly selected e-mail addresses at a large university says: "We have collected 427 responses from our sample of 2,100 as of April 30, 2004. This number of responses is large enough to achieve a 95% confidence interval with ±5% margin of sampling error in generalizing the results to our study population."[11] Why would you be reluctant to trust a confidence interval based on these data?

23.58 Pain from a rubber hand. People who have had limbs amputated sometimes feel sensations from the limb that is no longer there. To study this effect, psychologists asked subjects to place their right arm on a table. They then put a rubber arm and hand next to the real arm, with a high partition arranged so that the subject could see only the rubber arm. After a few minutes during which the real and fake hand were both tapped by an experimenter, the subjects felt the taps coming from the location of the rubber hand they could see, not from the real hand they couldn't see. Now the experiment begins: bend back a finger of the fake hand in a way that would cause pain, while merely lifting a real finger. Do electrical measurements show a response to pain? Because there would be some response from the surprise of being touched, a control treatment delayed the touch to the real hand to separate surprise from "pain." Here are summary data for 16 undergraduate students who were subject to both stimuli:[12]

STIMULUS	$\bar{x}$	s
Treatment	0.39	0.28
Control	0.18	0.20

(a) Which t procedures are correct for comparing the mean response to treatment and control: one-sample, matched pairs, or two-sample?
(b) The data summary given is not enough information to carry out the correct t procedures. Explain why not.

23.59 Monkeys and music. Humans generally prefer music to silence. What about monkeys? Allow a tamarin monkey to enter a V-shaped cage with food in both arms of the V. After the monkey eats the food, which arm will it prefer? The monkey's location determines what it hears, a lullaby played by a flute in one arm and silence in the other. Each of 4 monkeys was tested 6 times, on different days and with the music arm alternating between left and right (in case a monkey prefers one direction). The monkeys chose silence for about 65% of their time in the cage. The researchers reported a one-sample t test for the mean percent of time spent in the music arm, H_0: μ = 50% against the two-sided alternative, t = −5.26, df = 23, P < 0.0001.[13]

Although the result is interesting, the statistical analysis is not correct. The degrees of freedom df = 23 show that the researchers assumed that they had 24 independent observations. Explain why the results of the 24 trials are not independent.

23.60   **(Optional) Drug-detecting rats?** Dogs are big and expensive. Rats are small and cheap. Might rats be trained to replace dogs in sniffing out illegal drugs? A first study of this idea trained rats to rear up on their hind legs when they smelled simulated cocaine. To see how well rats performed after training, they were let loose on a surface with many cups sunk in it, one of which contained simulated cocaine. Four out of six trained rats succeeded in 80 out of 80 trials.[14] How should we estimate the long-term success rate p of a rat that succeeds in every one of 80 trials?

(a) What is the rat's sample proportion $\hat{p}$? What is the large-sample 95% confidence interval for p? It's not plausible that the rat will *always* be successful, as this interval says.

(b) Find the plus four estimate $\tilde{p}$ and the plus four 95% confidence interval for p. These results are more reasonable.

23.61   **(Optional) A new vaccine.** In 2006, the pharmaceutical company Merck released a vaccine named Gardasil for human papilloma virus, the most common cause of cervical cancer in young women. The Merck Web site gives results from "four placebo-controlled, double-blind, randomized clinical studies" with women 16 to 26 years of age, as follows:[15]

	n	CERVICAL CANCER	n	GENITAL WARTS
Gardasil	8487	0	7897	1
Placebo	8460	32	7899	91

(a) Give a 99% confidence interval for the difference in the proportions of young women who develop cervical cancer with and without the vaccine.

(b) Do the same for the proportions who develop genital warts.

(c) What do you conclude about the overall effectiveness of the vaccine?

23.62   **Starting to talk.** At what age do infants speak their first word of English? Here are data on 20 children (ages in months):[16]

| 15 | 26 | 10 | 9 | 15 | 20 | 18 | 11 | 8 | 20 |
| 7 | 9 | 10 | 11 | 11 | 10 | 12 | 17 | 11 | 10 |

(In fact, the sample contained one more child, who began to speak at 42 months. Child development experts consider this abnormally late, so the investigators dropped the outlier to get a sample of "normal" children. We are willing to treat these data as an SRS.) Is there good evidence that the mean age at first word among all normal children is greater than one year? 1STWORD

23.63    **Fertilizing a tropical plant.** Bromeliads are tropical flowering plants. Many are epiphytes that attach to trees and obtain moisture and nutrients from air and rain. Their leaf bases form cups that collect water and are home to the larvae of many insects. In an experiment in Costa Rica, Jacqueline Ngai and Diane Srivastava studied whether added nitrogen increases the productivity of bromeliad plants. Bromeliads were randomly assigned to nitrogen or control groups. Here are data on the number of new leaves produced over a 7-month period:[17] FERT

| Control | 11 | 13 | 16 | 15 | 15 | 11 | 12 |
| Nitrogen | 15 | 14 | 15 | 16 | 17 | 18 | 17 | 13 |

Is there evidence that adding nitrogen increases the mean number of new leaves formed?

23.64   **Starting to talk, continued.** Use the data in Exercise 23.62 to give a 90% confidence interval for the mean age at which children speak their first word. 1STWORD

23.65   **Dyeing fabrics.** Different fabrics respond differently when dyed. This matters to clothing manufacturers, who want the color of the fabric to be just right. A researcher dyed fabrics made of cotton and of ramie with the same "procion blue" dye applied in the same way. Then she used a colorimeter to measure the lightness of the color on a scale in which black is 0 and white is 100. Here are the data for 8 pieces of each fabric:[18] FBCDYE

| Cotton | 48.82 | 48.88 | 48.98 | 49.04 | 48.68 | 49.34 | 48.75 | 49.12 |
| Ramie | 41.72 | 41.83 | 42.05 | 41.44 | 41.27 | 42.27 | 41.12 | 41.49 |

Is there a significant difference between the fabrics? Which fabric is darker when dyed in this way?

23.66   **More on dyeing fabrics.** The color of a fabric depends on the dye used and also on how the dye is applied. This matters to clothing manufacturers, who

G Newport/Science Source

want the color of the fabric to be just right. The study discussed in the previous exercise went on to dye fabric made of ramie with the same procion blue dye applied in two different ways. Here are the lightness scores for 8 pieces of identical fabric dyed in each way: FBCDYE2

| Method B | 40.98 | 40.88 | 41.30 | 41.28 | 41.66 | 41.50 | 41.39 | 41.27 |
| Method C | 42.30 | 42.20 | 42.65 | 42.43 | 42.50 | 42.28 | 43.13 | 42.45 |

(a) This is a randomized comparative experiment. Outline the design.

(b) A clothing manufacturer wants to know which method gives the darker color (lower lightness score). Use sample means to answer this question. Is the difference between the two sample means statistically significant? Can you tell from just the *P*-value whether the difference is large enough to be important in practice?

23.67 Do parents matter? A professor asked her sophomore students, "Does either of your parents allow you to drink alcohol around him or her?" and "How many drinks do you typically have per session? (A drink is defined as one 12 oz beer, one 4 oz glass of wine, or one 1 oz shot of liquor.)" Table 23.1 contains the responses of the female students who are not abstainers.[19] The sample is all students in one large sophomore-level class. The class is popular, so we are tentatively willing to regard its members as an SRS of sophomore students at this college. Does the behavior of parents make a significant difference in how many drinks students have on the average? FMDRINK

23.68 Parents' behavior. We wonder what proportion of female students have at least one parent who allows them to drink around him or her. Table 23.1 contains information about a sample of 94 students. Use this sample to give a 95% confidence interval for this proportion. FMDRINK

23.69 Diabetic mice. The body's natural electrical field helps wounds heal. If diabetes changes this field, that might explain why people with diabetes heal slowly. A study of this idea compared normal mice and mice bred to spontaneously develop diabetes. The investigators attached sensors to the right hip and front feet of the mice and measured the difference in electrical potential (millivolts) between these locations. Here are the data:[20] MICE

DIABETIC MICE					
14.70	13.60	7.40	1.05	10.55	16.40
10.00	22.60	15.20	19.60	17.25	18.40
9.80	11.70	14.85	14.45	18.25	10.15
10.85	10.30	10.45	8.55	8.85	19.20

NORMAL MICE				
13.80	9.10	4.95	7.70	9.40
7.20	10.00	14.55	13.30	6.65
9.50	10.40	7.75	8.70	8.85
8.40	8.55	12.60		

(a) Make a stemplot of each sample of potentials. There is a low outlier in the diabetic group. Does it appear that potentials in the two groups differ in a systematic way?

(b) Is there significant evidence of a difference in mean potentials between the two groups?

(c) Repeat your inference without the outlier. Does the outlier affect your conclusion?

23.70 Keeping crackers from breaking. We don't like to find broken crackers when we open the package. How can makers reduce breaking? One idea is to microwave the crackers for 30 seconds right after baking them. Analyze the following results from two experiments intended to examine this idea.[21] Does microwaving significantly improve indicators of future breaking? How large is the improvement?

TABLE 23.1	DRINKS PER SESSION BY FEMALE STUDENTS											
PARENT ALLOWS STUDENT TO DRINK												
2.5	1	2.5	3	1	3	3	3	2.5	2.5	3.5	5	2
7	7	6.5	4	8	6	6	3	6	3	4	7	5
3.5	2	1	5	3	3	6	4	2	7	5	8	1
6	5	2.5	3	4.5	9	5	4	4	3	4	6	4
5	1	5	3	10	7	4	4	4	4	2	2.5	2.5
PARENT DOES NOT ALLOW STUDENT TO DRINK												
9	3.5	3	5	1	1	3	4	4	3	6	5	3
8	4	4	5	7	7	3.5	3	10	4	9	2	7
4	3	1										

What do you conclude about the idea of microwaving crackers?

(a) The experimenter randomly assigned 65 newly baked crackers to be microwaved and another 65 to a control group that is not microwaved. Fourteen days after baking, 3 of the 65 microwaved crackers and 57 of the 65 crackers in the control group showed visible checking, which is the starting point for breaks.

(b) The experimenter randomly assigned 20 crackers to be microwaved and another 20 to a control group. After 14 days, he broke the crackers. Here are summaries of the pressure needed to break them, in pounds per square inch:

	MICROWAVE	CONTROL
Mean	139.6	77.0
Standard deviation	33.6	22.6

23.71 Falling through the ice. Table 7.3 (page 184) gives the dates on which a wooden tripod fell through the ice of the Tanana River in Alaska, thus deciding the winner of the Nenana Ice Classic contest, for the years 1917 to 2012. Give a 95% confidence interval for the mean date on which the tripod falls through the ice. After calculating the interval in the scale used in the table (days from April 20, which is Day 1), translate your result into calendar dates and hours within the dates. (Each hour is 1/24, or 0.042, of a day.) TANANA

23.72 A case for the Supreme Court. In 1986, a Texas jury found a black man guilty of murder. The prosecutors had used "peremptory challenges" to remove 10 of the 11 blacks and 4 of the 31 whites in the pool from which the jury was chosen.[22] The law says that there must be a plausible reason (that is, a reason other than race) for different treatment of blacks and whites in the jury pool. When the case reached the Supreme Court 17 years later, the Court said that "happenstance is unlikely to produce this disparity." Explain why the methods we know can't be safely used to do the inference that lies behind the Court's finding that chance is unlikely to produce so large a black–white difference.

23.73 Mouse genes. A study of genetic influences on diabetes compared normal mice with similar mice genetically altered to remove a gene called *aP2*. Mice of both types were allowed to become obese by eating a high-fat diet. The researchers then measured the levels of insulin and glucose in their blood plasma. Here are some excerpts from their findings.[23] The normal mice are called "wild-type," and the altered mice are called "aP2$^{-/-}$."

Each value is the mean $\pm$ SEM of measurements on at least 10 mice. Mean values of each plasma component are compared between aP2$^{-/-}$ mice and wild-type controls by Student's t test (*P $<$ 0.05 and **P $<$ 0.005).

PARAMETER	WILD TYPE	aP2$^{-/-}$
Insulin (ng/ml)	5.9 $\pm$ 0.9	0.75 $\pm$ 0.2**
Glucose (mg/dl)	230 $\pm$ 25	150 $\pm$ 17*

Despite much greater circulating amounts of insulin, the wild-type mice had higher blood glucose than the aP2$^{-/-}$ animals. These results indicate that the absence of aP2 interferes with the development of dietary obesity-induced insulin resistance.

Other biologists are supposed to understand the statistics reported so tersely.

(a) What does "SEM" mean? What is the expression for SEM based on n, $\bar{x}$, and s from a sample?

(b) Which of the tests we have studied did the researchers apply?

(c) Explain to a biologist who knows no statistics what $P < 0.05$ and $P < 0.005$ mean. Which is stronger evidence of a difference between the two types of mice?

23.74 Mouse genes, continued. The report quoted in the previous exercise says only that the sample sizes were "at least 10." Suppose that the results are based on exactly 10 mice of each type. Use the values in the table to find $\bar{x}$ and s for the insulin concentrations in the two types of mice. Carry out a test to assess the significance of the difference in mean insulin concentration. Does your P-value confirm the claim in the report that $P < 0.005$?

23.75 Normal body temperature? Here are the daily average body temperatures (degrees Fahrenheit) for 20 healthy adults:[24] BDYTEMP

98.74 98.83 96.80 98.12 97.89 98.09 97.87
97.42 97.30 97.84 100.27 97.90 99.64 97.88
98.54 98.33 97.87 97.48 98.92 98.33

(a) Make a stemplot of the data. The distribution is roughly symmetric and single-peaked. There is one mild outlier. We expect the distribution of the sample mean $\bar{x}$ to be close to Normal.

(b) Do these data give evidence that the mean body temperature for all healthy adults is not equal to the traditional 98.6 degrees? Follow the four-step process for significance tests (page 383).

23.76 Normal body temperature, continued. Use the data in Exercise 23.75 to estimate mean body temperature with 90% confidence. Follow the four-step process for confidence intervals (page 357). 📊 BDYTEMP

23.77 Tests from confidence intervals. You read in a U.S. Census Bureau report that a 99% confidence interval for the mean income in 2005 of American households headed by a college-educated person at least 25 years old was $100,272 ± $1651. (The median income of these households was lower, $77,179.) Based on this interval, can you reject the null hypothesis that the mean income in this group is $95,000? What is the alternative hypothesis of the test? What is its significance level?

23.78 (Optional) Low power? It appears that eating oat bran lowers cholesterol slightly. At a time when oat bran was something of a fad, a paper in the *New England Journal of Medicine* found that it had no significant effect on cholesterol.[25] The paper reported a study with just 20 subjects. Letters to the journal denounced publication of a negative finding from a study with very low power. Explain why lack of significance in a study with low power gives no reason to accept the null hypothesis that oat bran has no effect.

23.79 (Optional) Type I and Type II errors. Exercise 23.75 asks for a significance test of the null hypothesis that the mean body temperature for all healthy adults is 98.6 degrees against the alternative hypothesis that the mean is different from 98.6 degrees.

© Galina Mikhalishina/Alamy

Inference about Relationships

Statistical inference offers more methods than anyone can know well, as a glance at the offerings of any large statistical software package demonstrates. In an introductory text, we must be selective. Parts I to IV have laid a foundation for understanding statistics:

- The nature and purpose of data analysis

- The central ideas of designs for data production

- The reasoning behind confidence intervals and significance tests

- Experience applying these ideas in practice

Each of the three chapters of Part V offers an introduction to a more advanced topic in statistical inference. You may choose to read any or all of them, in any order.

What makes a statistical method "more advanced"? More complex data, for one thing. In Part IV, we looked only at methods for inference about a single population parameter and for comparing two parameters. All the chapters in Part V present methods for studying relationships between two variables. In Chapter 24, both variables are categorical, with data given as a two-way table of counts of outcomes. Chapter 25 considers inference in the setting of regressing a response variable on an explanatory variable. This is an important type of relationship between two quantitative variables. In Chapter 26 we meet methods for comparing the mean response in more than two groups. Here, the explanatory variable (group) is categorical, and the response variable is quantitative. These chapters together bring our knowledge of inference to the same point that our study of data analysis reached in Chapters 1 to 7.

With greater complexity comes greater reliance on technology. In these final three chapters you will be interpreting the output of statistical software or using software yourself more often. With effort, you can do the calculations needed in Chapter 24 with a basic calculator. In Chapters 25 and 26, the pain of doing calculations by hand is too great and the contribution to learning too small. Fortunately, you can grasp the ideas without step-by-step arithmetic.

Another aspect of "more advanced" methods is new concepts and ideas. This is where we draw the line in deciding what statistical topics we can master in a first course. Part V builds elaborate methods on the foundation we have laid without introducing fundamentally new concepts. You can see that statistical practice does need additional big ideas by reading the sections on "the problem of multiple comparisons" in Chapters 24 and 26. But the ideas you already know place you among the world's statistical sophisticates.

525

Two Categorical Variables: The Chi-Square Test

Overview

In Chapter 6, we discussed two-way tables and some associated topics, such as bar graphs and Simpson's paradox. That discussion provided students with an early exposure to categorical data. In this chapter we use statistical inferential methods to analyze data in two-way tables. On the surface, the formulas used (to construct the test statistic, for example) will be unfamiliar to students. However, like all test statistics, it measures the "distance" between observed data and expected values under a null hypothesis, and interpreting the P-value remains essentially unchanged.

The chapter begins with a discussion of two-way tables and the problem of multiple comparisons. Under the null hypothesis of no relationship between the two categorical variables, we can calculate expected counts for cells. The chi-square test statistic measured the "distance" between expected counts and observed counts across the table. As with other inference procedures, there is a discussion of assumptions and conditions required for the test's validity. There is also an optional discussion of chi-square test statistic's use as a test for goodness of fit, which is particularly useful in modeling.

This chapter provides opportunity for the instructor to establish connections to topics possibly included in the course to date. The expected cell counts under the model of independence, for example, are easily explained as coming from the multiplication rule for independent events, possibly studied in Chapter 12. In an $r \times 2$ table, with one dimension (columns, without loss of generality) viewed as success or failure, the number of observations in each cell has the binomial distribution (Chapter 13) with number of trials equal to row total. In fact, in a more general $r \times c$ table, individual cell counts have the more general *multinomial* distribution. In the 2×2 case, the chi-square test statistic, when computed, is equal to the square of the z-test statistic for comparing two proportions, studied in Chapter 18. As a result, the test for independence studied here yields exactly the same P-value in the two-sided test $H_0: p_1 = p_2$ versus $H_a: p_1 \neq p_2$. More generally, it can be viewed as good for students to see examples of test statistic constructions that don't follow the same formulaic mold as the t or z statistics examined earlier. If the students have a good foundation about the process of statistical inference, the extension to the chi-square tests should make more sense to them.

LEARNING OUTCOMES

- Understand that the data for a chi-square test must be presented as a two-way table of counts of outcomes.

- Recognize a two-way table can occur from a single SRS with each individual classified according to two categorical variables, or from independent SRSs from two or more populations with each individual classified according to one categorical variable.

- Use percents to describe the relationship between the row and column variables, starting from the counts in a two-way table.

- Locate the chi-square statistic, its P-value, and other useful facts (row or column percents, expected counts, terms of chi-square) in output from your software or calculator.

- Explain what null hypothesis the chi-square statistic tests in a specific two-way table. Understand the difference between testing for independence or homogeneity in a two-way table.

- Calculate the expected count for any cell from the observed counts in a two-way table. Check whether you can safely use the chi-square test.

- Calculate the term of the chi-square statistic for any cell, as well as the overall statistic.

- Give the degrees of freedom of a chi-square statistic. Make a quick assessment of the significance of the statistic by comparing the observed value with the degrees of freedom.

- Use the chi-square critical values in Table D to approximate the P-value of a chi-square test.

- If the test is significant, compare percents, compare observed with expected cell counts, or look for the largest terms of the chi-square statistic to see what deviations from the null hypothesis are most important.

Teaching Suggestions and Additional Examples/Activities for the Classroom

1. Introduce Chi-Square Methods with a Simple Example

Introduce the topics in this chapter with a simple example involving data collected from your students. For example, you can construct a 2×2 table of state status (in-state, out-of-state) versus academic scholarship recipient. Discuss with your students whether in-state students are more or less likely than out-of-state students to have an academic scholarship, and compute the corresponding proportions we would compare. This should remind students of material covered in Chapter 18 concerning two sample proportions. The difference is that you are not taking two independent samples of fixed sample sizes, but the number of in-state and out-of-state sampled is, in fact, random. Introduce and discuss the hypothesis of independence at this point.

2. Connect the Calculation of Expected Counts to General Probability Rules

If you covered Chapter 12 on general probability rules, you might use the general multiplication rule for independent events to motivate the formula for expected cell counts: If P(Female and Sports Team Membership) $= P$(Female) $\times P$(Sports Team Membership), so the expected cell count for females on a sports team is (Number of Students) $\times P$(Female) $\times P$(Sports Team Membership). We estimate P(Female) by $\dfrac{\text{Number of Females}}{\text{Number of Students}}$, and P(Sports Team Membership) by $\dfrac{\text{Number on Sports Team}}{\text{Number of Students}}$. It follows then that the expected cell count for females on a sports team reduces to $\dfrac{\text{Number of Females} \times \text{Number on Sports Team}}{\text{Number of Students}}$.

Ask students to compute expected cell counts for the two-way table constructed from your class data, and then ask them to reflect on the difference between expected cell counts and observed cell counts. If you covered Chapter 18, you can remind students that you would be able to use the methods of that chapter to compare two proportions. But, what if we are comparing sports team membership across class ranks (freshman, sophomore, junior, senior)? There would be six different comparisons to make. This introduces the problem of multiple comparisons, which motivates the need for a more general approach.

Other, more interesting data sets can be used. Consider using the Berkeley admissions data described in Chapter 6 (AIE material). Alternatively, use the data on Titanic survival rates, provided in the EESEE Case Studies.

3. Motivate Goodness of Fit with a Fun Activity

If the optional section on goodness of fit at the end of the chapter is covered, a fun activity is to perform an analysis on the color of milk chocolate M&M's candies. As of this writing, the percentages of colors in milk chocolate M&M's candies are 24% blue, 20% orange, 16% green, 14% yellow, 13% red, and 13% brown. Provide each student with a sample of milk chocolate M&M's, and allow them to construct an appropriate cross-classified table for analyzing the distribution of colors, based on their particular sample. Each student can then look for evidence that the distribution of colors differs from that stated by the company. (Note: The M&M Mars company has reiterated many times that the candies are mixed according to the stated proportions; any "deviations" are due solely to packaging equipment.) Make sure to have your students check the conditions needed for the goodness of fit test.

Other Resources (LaunchPad)

EESEE Case Studies
Alcoholism in Twins
Baldness and Heart Attacks
Mud Wrestling
Instructor Behaviors

© Axel Dupeux/Corbis

Two Categorical Variables: The Chi-Square Test

In this chapter we cover...

- Two-way tables
- The problem of multiple comparisons
- Expected counts in two-way tables
- The chi-square test statistic
- Cell counts required for the chi-square test
- Using technology
- Uses of the chi-square test: independence and homogeneity
- The chi-square distributions
- The chi-square test for goodness of fit*

The two-sample z procedures of Chapter 18 allow us to compare the proportions of successes in two groups, either two populations or two treatment groups in an experiment. In the initial example of this chapter, we investigate where young people live. Suppose we want to compare the proportions of men and women who live with their parents. This is a question about the relationship between two categorical variables, sex (female or male) and "Where do you live?" (with parents or not), which can be answered using the two-sample z procedures of Chapter 18. In fact, the data include three more outcomes for "Where do you live?": in another person's home, in your own place, and in group quarters such as a dormitory. When there are more than two outcomes, or when we want to compare more than two groups, we need a new statistical test. The new test addresses a general question: *is there a relationship between two categorical variables?*

Two-way tables

We saw in Chapter 6 that we can present data on two categorical variables in a **two-way table** of counts. That's our starting point. Let's begin our exploration of where college-age young people live.

two-way table

EXAMPLE 24.1   Where Do Young People Live?

LIVING

cell

A sample survey asked a random sample of young adults, "Where do you live now? That is, where do you stay most often?" Table 24.1 is a two-way table of all 2984 people in the sample (both men and women) classified by their age and by where they lived.[1] Living arrangement is a categorical variable. Even though age is quantitative, the two-way table treats age as dividing young adults into four categories. Table 24.1 gives the counts for all 20 combinations of age and living arrangement. Each of the 20 counts occupies a **cell** of the table. ■

TABLE 24.1 YOUNG ADULTS BY AGE AND LIVING ARRANGEMENT

LIVING ARRANGEMENT	AGE (YEARS)				TOTAL
	19	20	21	22	
Parents' home	324	378	337	318	1357
Another person's home	37	47	40	38	162
Your own place	116	279	372	487	1254
Group quarters	58	60	49	25	192
Other	5	2	3	9	19
Total	540	766	801	877	2984

As usual, we prepare for inference by first doing data analysis. Because we think that age helps explain where young people live, find the percents of people in each age group who have each living arrangement. The percents appear in Table 24.2. Each column adds to 100% (up to roundoff error) because we are looking at each age group separately. In the language of Chapter 6 (page 156), Table 24.2 shows the four *conditional distributions* of living arrangements given a specific age.

Figure 24.1 is a bar graph comparing the four conditional distributions. The graph shows a strong relationship between age and living arrangement. As young adults age from 19 to 22, the percent living with their parents drops and the percent living in their own place rises. The percent living in group quarters also declines with age as college students move out of dormitories. Are these differences among the four age groups large enough to be statistically significant?

TABLE 24.2 PERCENTS OF EACH AGE GROUP WHO HAVE EACH LIVING ARRANGEMENT (READ DOWN COLUMNS)

	AGE			
	19	20	21	22
Parents' home	60.0%	49.3%	42.1%	36.3%
Another person's home	6.9%	6.1%	5.0%	4.3%
Your own place	21.5%	36.4%	46.4%	55.5%
Group quarters	10.7%	7.8%	6.1%	2.9%
Other	0.9%	0.3%	0.4%	1.0%
Total	100.0%	99.9%	100.0%	100.0%

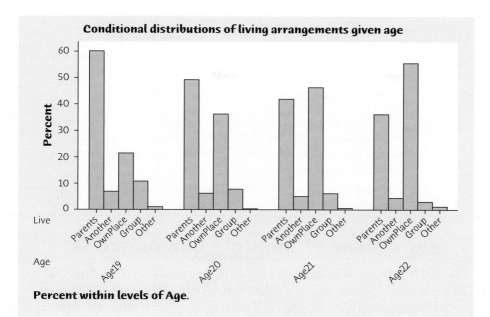

Conditional distributions of living arrangements given age

Percent within levels of Age.

FIGURE 24.1

Bar graph comparing the four conditional distributions of living arrangements given age, for Example 24.1.

Apply Your Knowledge

24.1 Facebook at Penn State. Pennsylvania State University has its main campus in University Park and more than 20 smaller "commonwealth campuses" around the state. The Penn State Division of Student Affairs polled a random sample of undergraduates about their use of online social networking. (The response rate was only about 20%, which casts some doubt on the usefulness of the data.) Facebook was the most popular site, with more than 80% of students having an account. Here is a comparison of Facebook use by undergraduates at the University Park and commonwealth campuses:[2] FACEBOOK

	University Park	Commonwealth
Do not use Facebook	68	248
Several times a month or less	55	76
At least once a week	215	157
At least once a day	640	394

(a) What percent of University Park students fall in each Facebook category? What percent of commonwealth campus students fall in each category? Each column should add to 100% (up to roundoff error). These are the conditional distributions of Facebook use given campus setting.

(b) Make a bar graph that compares the two conditional distributions. What are the most important differences in Facebook use between the two campus settings?

24.2 Video Gaming and Grades. The popularity of computer, video, online, and virtual reality games have raised concerns about their ability to negatively impact youth, although the existing literature has been inconsistent. The data in this exercise are based on a recent survey of 14- to 18-year olds in Connecticut high schools. Here are the grade distributions of boys that have and have not played video games.[3] GAMING

| | Grade Average | | |
	A's and B's	C's	D's and F's
Played games	736	450	193
Never played games	205	144	80

(a) It appears that boys who have played video games have better grades than those that have never played video games. Give percents to back up this claim. Make a bar graph that compares your percents for boys that have and have not played video games.

(b) Association does not prove causation. Explain why you can't conclude from this study that playing video games improves grades for boys.

HE STARTED IT!
A study of deaths in bar fights showed that in 90% of the cases, the person who died started the fight. You shouldn't believe this. If you killed someone in a fight, what would you say when the police ask you who started the fight? After all, dead men tell no tales.

The problem of multiple comparisons

The null hypothesis in Example 24.1 is that in the population of all American young adults there is *no difference* among the four conditional distributions of living arrangements for people aged 19, 20, 21, and 22. If the null hypothesis is true, the differences in the sample are just accidents due to random selection of the sample. Put more generally, the null hypothesis is that there is *no relationship* between two categorical variables,

H_0: there is no relationship between age and living arrangement for the population of all American young adults

The alternative hypothesis says that there *is* a relationship but does not specify any particular kind of relationship,

H_a: there is some relationship between age and living arrangement for the population of all American young adults

Any difference among the four distributions of living arrangements in the population of all young adults means that the null hypothesis is false and the alternative hypothesis is true. The alternative hypothesis is not one-sided or two-sided. We might call it "many-sided" because it allows any kind of difference.

With only the methods we already know, we might start by comparing the proportions of people aged 19 and 22 who live with their parents. We could similarly compare other pairs of proportions, ending up with many tests and many *P*-values. This is a bad idea. The *P*-values belong to each test separately, not to the collection of all the tests together. Think of the distinction between the probability that a basketball player makes a free throw and the probability that she makes all of her free throws in a game. *When we do many individual tests or confidence intervals, the individual P-values and confidence levels don't tell us how confident we can be in all of the inferences taken together.*

Because of this, it's cheating to pick out one large difference from Table 24.2 and then test its significance as if it were the only comparison we had in mind. For example, the percents of people aged 19 and 22 who live with their parents are significantly different ($z = 8.92$, $P < 0.001$) if we make just this one comparison. But we could also pick a comparison that is not significant; for example, the proportions of people aged 21 and 22 who live in another person's home do not differ significantly ($z = 0.64$, $P = 0.522$). Individual comparisons can't tell us whether the four distributions, each with five outcomes, are significantly different.

The problem of how to do many comparisons at once with an overall measure of confidence in all our conclusions is common in statistics. This is the problem of *multiple comparisons* **multiple comparisons.** Statistical methods for dealing with multiple comparisons usually have two steps:

1. An *overall test* to see if there is good evidence of *any* differences among the parameters that we want to compare.

2. A detailed *follow-up analysis* to decide which of the parameters differ and to estimate how large the differences are.

The overall test, though more complex than the tests we met earlier, is reasonably straightforward. The follow-up analysis can be quite elaborate. We will concentrate on the overall test and use data analysis to describe in detail the nature of the differences.

Apply Your Knowledge

24.3 Facebook at Penn State. In the setting of Exercise 24.1 (see page 529), we might do several significance tests to compare University Park with the commonwealth campuses. **FACEBOOK**

(a) Is there a significant difference between the proportions of students in the two locations who do not use Facebook? Give the *P*-value.

(b) Is there a significant difference between the proportions of students in the two locations who are in the "at least once a week" category? Give the *P*-value.

(c) Explain clearly why *P*-values for individual outcomes like these can't tell us whether the two distributions for all four outcomes in the two locations differ significantly.

24.4 Is Astrology Scientific? The University of Chicago's General Social Survey (GSS) is the nation's most important social science sample survey. The GSS asked a random sample of adults their opinion about whether astrology is very scientific, sort of scientific, or not at all scientific. Here is a two-way table of counts for people in the sample who had three levels of higher education degrees:[4] **ASTRLGY**

	Degree Held		
	Junior College	Bachelor	Graduate
Not at all scientific	44	122	71
Very or sort of scientific	31	62	27

(a) Give three 95% confidence intervals, for the percents of people with each degree who think that astrology is not at all scientific.

(b) Explain clearly why we are *not* 95% confident that *all three* of these intervals capture their respective population proportions.

Expected counts in two-way tables

Our general null hypothesis H_0 is that there is *no relationship* between the two categorical variables that label the rows and columns of a two-way table. To test H_0, we compare the observed counts in the table with the *expected counts*, the counts we would expect—except for random variation—if H_0 were true. If the observed counts are far from the expected counts, that is evidence against H_0. It is easy to find the expected counts.

Expected Counts

The **expected count** in any cell of a two-way table when H_0 is true is

$$\text{expected count} = \frac{\text{row total} \times \text{column total}}{\text{table total}}$$

EXAMPLE 24.2	**Where Young People Live: Expected Counts**

LIVING

Let's find the expected counts for the study of where young people live. Look back at the two-way table of counts, Table 24.1 (page 528). That table includes the row and column totals. The expected count of 19-year-olds who live in their parents' home is

$$\frac{\text{row 1 total} \times \text{column 1 total}}{\text{table total}} = \frac{(1357)(540)}{2984} = 245.57$$

The expected count of 22-year-olds who live with their parents is

$$\frac{\text{row 1 total} \times \text{column 4 total}}{\text{table total}} = \frac{(1357)(877)}{2984} = 398.82$$

The actual counts are 324 and 318, respectively. More younger people and fewer older people live with their parents than we would expect if there were no relationship between age and living arrangement. Table 24.3 shows all 20 expected counts.

TABLE 24.3	YOUNG ADULTS BY AGE AND LIVING ARRANGEMENT: EXPECTED CELL COUNTS				
	AGE				
	19	**20**	**21**	**22**	**TOTAL**
Parents' home	245.57	348.35	364.26	398.82	1357
Another person's home	29.32	41.59	43.49	47.61	162
Your own place	226.93	321.90	336.61	368.55	1254
Group quarters	34.75	49.29	51.54	56.43	192
Other	3.44	4.88	5.10	5.58	19
Total	540	766	801	877	2984

As this table shows, *the expected counts have exactly the same row and column totals (up to roundoff error) as the observed counts.* That's a good way to check your work. Comparing the actual counts (Table 24.1) and the expected counts (Table 24.3) shows in what ways the data diverge from the null hypothesis. ▪

Why the formula works Where does the formula for an expected count come from? Think of a basketball player who makes 70% of her free throws in the long run. If she shoots 10 free throws in a game, we expect her to make 70% of them, or 7 of the 10. Of course, she won't make exactly 7 every time she shoots 10 free throws in a game. There is chance variation from game to game. But in the long run, 7 of 10 is what we expect. In more formal language, if we have n independent tries and the probability of a success on each try is p, we expect np successes.

Now go back to the count of 19-year-olds living in their parents' home. The proportion of all 2984 subjects who live with their parents is

$$\frac{\text{count of successes}}{\text{table total}} = \frac{\text{row 1 total}}{\text{table total}} = \frac{1357}{2984}$$

Think of this as p, the overall proportion of successes. If H_0 is true, we expect (except for random variation) this same proportion of successes in all four age groups. So the expected count of successes among the 540 nineteen-year-olds is

$$np = (540)\left(\frac{1357}{2984}\right) = 245.57$$

That's the formula in the Expected Counts box.

Apply Your Knowledge

24.5 Facebook at Penn State. The two-way table in Exercise 24.1 (page 529) displays data on use of Facebook by two groups of Penn State students. It's clear that nonusers are much more frequent at the commonwealth campuses. Let's look just at students who have Facebook accounts: ☷⁞ᵢᵢₗ **FACEBOOK**

Use Facebook	University Park	Commonwealth
Several times a month or less	55	76
At least once a week	215	157
At least once a day	640	394
Total Facebook users	910	627

The null hypothesis is that there is no relationship between campus and Facebook use.

(a) If this hypothesis is true, what are the expected counts for Facebook use among commonwealth campus students? This is one column of the two-way table of expected counts. Find the column total and verify that it agrees with the column total for the observed counts.

(b) Commonwealth campus students as a group are older and more likely to be married and employed than University Park students. What does comparing the observed and expected counts in this column show about Facebook use by these students?

24.6 Video Gaming and Grades. Exercise 24.2 (page 529) describes a comparison of the grade distribution of a sample of 14- to 18-year old boys in Connecticut who do and don't play video games. The null hypothesis "no relationship" says that in the population of all 14- to 18-year old boys in Connecticut, the proportions who have each grade average are the same for those who play and don't play video games. ☷⁞ᵢᵢₗ **GAMING**

(a) Find the expected cell counts if this hypothesis is true, and display them in a two-way table. Add the row and column totals to your table, and check that they agree with the totals for the observed counts.

(b) Are there any large deviations between the observed counts and the expected counts? What kind of relationship between the two variables do these deviations point to?

The chi-square test statistic

To test whether the observed differences among the four distributions of living arrangements given age are statistically significant, we compare the observed and expected counts. The test statistic that makes the comparison is the *chi-square statistic.*

Chi-Square Statistic

The **chi-square statistic** is a measure of how far the observed counts in a two-way table are from the expected counts if H_0 were true. The formula for the statistic is

$$\chi^2 = \sum \frac{(\text{observed count} - \text{expected count})^2}{\text{expected count}}$$

The sum is over all cells in the table.

As you might guess, the symbol χ in the box is the Greek letter chi. The chi-square statistic is a sum of terms, one for each cell in the table.

EXAMPLE 24.3	**Where Young People Live: The Test Statistic**

LIVING

In the study of where young people live, 324 19-year-olds lived with their parents. The expected count for this cell is 245.57. So the term of the chi-square statistic from this cell is

$$\frac{(\text{observed count} - \text{expected count})^2}{\text{expected count}} = \frac{(324 - 245.57)^2}{245.57}$$

$$= \frac{6151.26}{245.57} = 25.05$$

The chi-square statistic χ^2 is the sum of 20 terms like this one. Here they are, arranged to match the layout of the two-way table:

$$\chi^2 = 25.05 + 2.53 + 2.04 + 16.38$$
$$+ \ 2.01 + 0.71 + 0.28 + \ 1.94$$
$$+ 54.23 + 5.72 + 3.72 + 38.07$$
$$+ 15.56 + 2.33 + 0.13 + 17.51$$
$$+ \ 0.71 + 1.70 + 0.87 + \ 2.09$$
$$= \ 193.58$$

To find the value $\chi^2 = 193.58$, we had to calculate the 20 expected cell counts in Table 24.3 and then the 20 terms of the sum. Each term in the sum represents the "contribution" to the chi-square statistic for one of the cells in the table. *Moreover,* *even rounding each term to two decimal places as we have done can still create roundoff error in the sum. Because of this, software is very handy in finding χ^2.* ■

Think of χ^2 as a measure of the distance of the observed counts from the expected counts if H_0 were true. Like any distance, it is always zero or positive, and it is zero only when the observed counts are exactly equal to the expected counts. Large values of χ^2 are evidence against H_0 because they say that the observed counts are far from what we would expect if H_0 were true. *Although the alternative hypothesis H_a is* *many-sided, the chi-square test is one-sided because any violation of H_0 tends to* produce a large value of χ^2. Small values of χ^2 are not evidence against H_0.

Cell counts required for the chi-square test

The chi-square test, like the z procedures for comparing two proportions, is an approximate method that becomes more accurate as the counts in the cells of the table get larger. We must therefore check that the counts are large enough to allow us to trust the P-value. Fortunately, the chi-square approximation is accurate for quite modest counts. Here is a practical guideline.[5]

Cell Counts Required for the Chi-Square Test
You can safely use the chi-square test with critical values from the chi-square distribution when no more than 20% of the expected counts are less than 5, and all individual expected counts are 1 or greater. In particular, all four expected counts in a 2 × 2 table should be 5 or greater.

Note that the guideline uses *expected* cell counts. The expected counts for the living arrangements study of Example 24.1 appear in Table 24.3. Only 2 of the 20 expected counts (that's 10%) are less than 5, and all are greater than 1, so the data meet the guideline for safe use of chi-square.

Using technology

Calculating the expected counts and then the chi-square statistic by hand is time consuming. As usual, software saves time and always gets the arithmetic right. Figure 24.2 shows output for the chi-square test for the living arrangements data from a graphing calculator and two statistical programs.

Texas Instruments Graphing Calculator

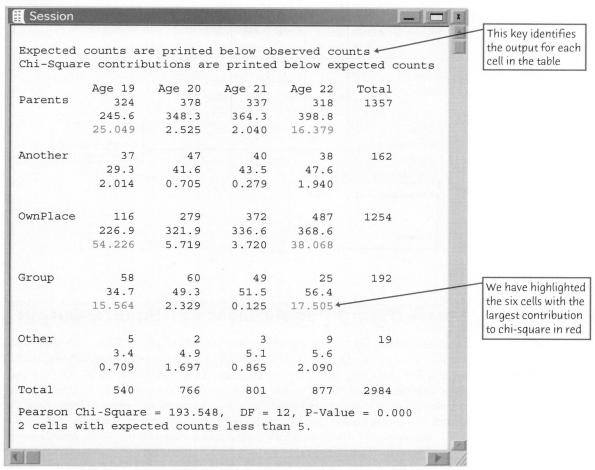

```
X²-Test
 X²=193.5482798
 P=6.981157E-35
 df=12
```

```
round([B],2)
[[245.57 348.35…
 [29.32  41.59 …
 [226.93 321.9 …
 [34.75  49.29 …
 [3.44   4.88  …
```

Minitab

Expected counts are printed below observed counts   *This key identifies the output for each cell in the table*
Chi-Square contributions are printed below expected counts

	Age 19	Age 20	Age 21	Age 22	Total
Parents	324	378	337	318	1357
	245.6	348.3	364.3	398.8	
	25.049	2.525	2.040	16.379	
Another	37	47	40	38	162
	29.3	41.6	43.5	47.6	
	2.014	0.705	0.279	1.940	
OwnPlace	116	279	372	487	1254
	226.9	321.9	336.6	368.6	
	54.226	5.719	3.720	38.068	
Group	58	60	49	25	192
	34.7	49.3	51.5	56.4	
	15.564	2.329	0.125	17.505	
Other	5	2	3	9	19
	3.4	4.9	5.1	5.6	
	0.709	1.697	0.865	2.090	
Total	540	766	801	877	2984

We have highlighted the six cells with the largest contribution to chi-square in red

Pearson Chi-Square = 193.548, DF = 12, P-Value = 0.000
2 cells with expected counts less than 5.

FIGURE 24.2

Output from a graphing calculator, Minitab, and CrunchIt! for the two-way table in the study of where young people live, for Example 24.4.

(Continued)

FIGURE 24.2
(Continued)

CrunchIt!

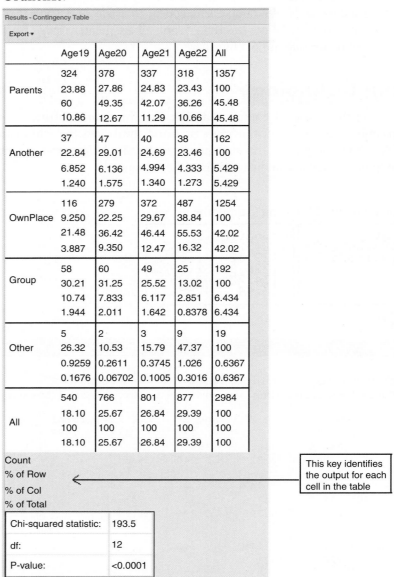

This key identifies the output for each cell in the table

EXAMPLE 24.4	Where Young People Live: Chi-Square Output

LIVING

All three outputs tell us that the chi-square statistic is $\chi^2 = 193.5$, with very small P-value. Minitab reports $P = 0.000$, rounded to three decimal places, while CrunchIt! reports $P < 0.0001$. The graphing calculator says that $P = 6.98 \times 10^{-35}$, a very small number indeed. This P-value comes from an approximation to the sampling distribution of χ^2. The approximation is less accurate far out in the tail than near the center of the distribution, so you should not take 6.98×10^{-35} literally. Just read it as "P is very small." The sample gives very strong evidence that living arrangements differ among the four age groups.

Statistical software generally offers additional information on request. We asked Minitab to show the observed counts and expected counts and also the term in the chi-square statistic for each cell, called the "chi-square contribution." The top left cell has expected count 245.57 and contributes 25.049 to the chi-square statistic, as we calculated earlier. (Roundoff errors are smaller with software than in hand calculation.) The graphing calculator also displays the observed and expected cell counts on request. We told the calculator to display the expected counts rounded to two decimal places. To see the remaining columns of expected counts on the calculator's small screen, you must scroll to the right. CrunchIt! displays the counts, the row and column percents associated with each count, and the overall percent of the total associated with the count. Depending on the problem, only one of these percents will generally be of interest.

What about the Excel spreadsheet program? Excel is in general a poor choice for statistics. It is particularly awkward for chi-square because its Data Analysis tool pack omits this common test. You will need add-in modules to use Excel effectively for chi-square. ■

The chi-square test is an overall test for detecting relationships between two categorical variables. If the test is significant, it is important to look at the data to learn the nature of the relationship. We have three ways to look at the living arrangements data:

■ **Compare selected percents:** which living arrangements occur in quite different percents of the four age groups? This is the method we learned in Chapter 6.

■ **Compare observed and expected cell counts:** which cells have more or fewer observations than we would expect if H_0 were true?

■ **Look at the terms of the chi-square statistic:** which cells contribute the most to the value of χ^2?

EXAMPLE 24.5   *Where Young People Live: Conclusion*

There is very strong evidence ($\chi^2 = 193.55$, $P < 0.001$) that living arrangements of young people are not the same for ages 19, 20, 21, and 22. Comparing selected percents—specifically, the four conditional distributions of living arrangements for each age in Table 24.2 and Figure 24.1—shows how young people become more independent as they grow older.

The additional information provided by programs like Minitab shows what differences among the age groups explain the large value of the chi-square statistic. Look at the 20 terms in the chi-square statistic in the Minitab output and compare the observed and expected counts in the cells that contribute most to chi-square. Just 6 of the 20 cells, those highlighted in red in Figure 24.2, contribute 166.79 of the total chi-square $\chi^2 = 193.55$. These 6 highlighted cells occur in pairs:

■ 54.226 and 38.068: fewer 19-year-olds than expected and more 22-year-olds than expected live in their own place.

■ 25.049 and 16.379: more 19-year-olds than expected and fewer 22-year-olds than expected live in their parents' home.

■ 15.564 and 17.505: more 19-year-olds than expected and fewer 22-year-olds than expected live in group quarters.

These three trends display the increase in independent living between age 19 and age 22. ■

Apply Your Knowledge

24.7 **Facebook at Penn State.** Figure 24.3 displays Minitab output for how frequently students at the University Park and commonwealth campuses of Penn State University who have Facebook accounts make use of their accounts. The output includes the two-way table of observed counts, the expected counts, and each cell's contribution to the chi-square statistic. **USERS**

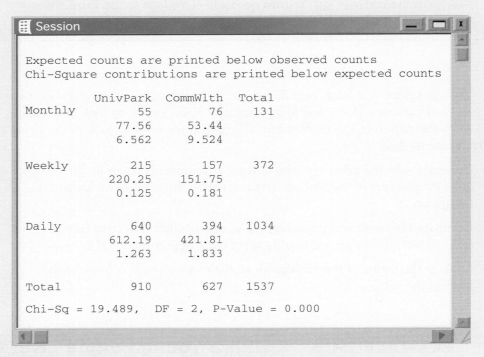

```
┌─────────────────────────────────────────────────────────────────┐
│ ▦ Session                                          ─  □  X        │
├─────────────────────────────────────────────────────────────────┤
│                                                                   │
│ Expected counts are printed below observed counts                 │
│ Chi-Square contributions are printed below expected counts        │
│                                                                   │
│           UnivPark  CommWlth   Total                              │
│ Monthly        55        76     131                               │
│             77.56     53.44                                       │
│             6.562     9.524                                       │
│                                                                   │
│ Weekly        215       157     372                               │
│            220.25    151.75                                       │
│             0.125     0.181                                       │
│                                                                   │
│ Daily         640       394    1034                               │
│            612.19    421.81                                       │
│             1.263     1.833                                       │
│                                                                   │
│ Total         910       627    1537                               │
│ Chi-Sq = 19.489,  DF = 2, P-Value = 0.000                         │
│                                                                   │
└─────────────────────────────────────────────────────────────────┘
```

FIGURE 24.3
Minitab output for the two-way table of Facebook use by Penn State campus.

(a) Verify from the output that the data meet the cell count requirement for use of chi-square.

(b) What hypotheses does chi-square test? What are the test statistic and its P-value?

(c) Which cells contribute the most to χ^2? Compare the observed and expected counts in these cells and comment on the most important differences in Facebook use between students at the two locations.

24.8 **Video Gaming and Grades.** Your data analysis in Exercise 24.2 (page 529) found that boys who have played video games tend to have higher grades than those who have not. Figure 24.4 gives Minitab output for the two-way table in Exercise 24.2. **GAMING**

(a) Verify from the output that the data meet the cell count requirement for use of chi-square.

(b) What are the chi-square statistic and its P-value? Explain in simple language what it means to reject H_0 in this setting.

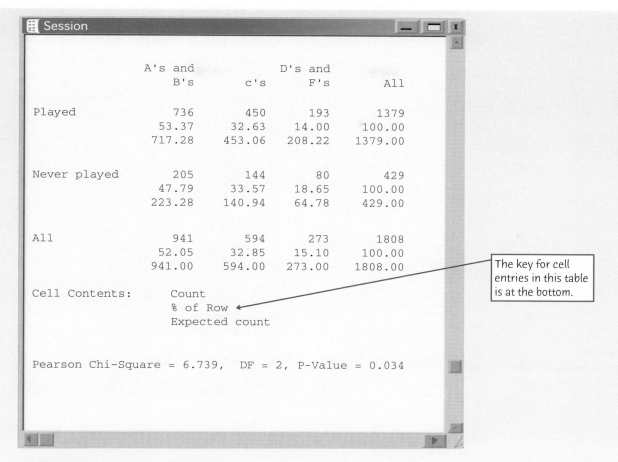

```
 Session                                        _  □  X

                A's and              D's and
                  B's        c's       F's        All

 Played            736        450       193       1379
                 53.37      32.63     14.00     100.00
                717.28     453.06    208.22    1379.00


 Never played      205        144        80        429
                 47.79      33.57     18.65     100.00
                223.28     140.94     64.78     429.00


 All               941        594       273       1808
                 52.05      32.85     15.10     100.00
                941.00     594.00    273.00    1808.00


 Cell Contents:        Count
                       % of Row
                       Expected count


 Pearson Chi-Square = 6.739,  DF = 2, P-Value = 0.034
```

The key for cell entries in this table is at the bottom.

FIGURE 24.4
Minitab output for the study of video gaming and grades.

(c) Give an overall conclusion that refers to row percents to describe the nature of the relationship between playing video games and grades.

24.9 Is Astrology Scientific? The General Social Survey asked a random sample of adults about their education and about their view of astrology as scientific or not. Here are the data for people with three levels of higher education degrees: ASTRLGY

	Degree Held		
	Junior College	**Bachelor**	**Graduate**
Not at all scientific	44	122	71
Very or sort of scientific	31	62	27

Figure 24.5 gives Minitab chi-square output for these data. Follow the *Plan, Solve,* and *Conclude* steps of the four-step process in using the information in the output to describe how people with these levels of education differ in their opinions about astrology. Be sure that your *Solve* step includes data analysis and checking conditions for inference as well as a formal test.

```
  Session                                                    _  □  X

                 Junior
                 college   Bachelor   Graduate       All

  Not Science         44        122         71       237
                   58.67      66.30      72.45     66.39
                   49.79     122.15      65.06    237.00
                  0.67329    0.00019    0.54255         *

  Science             31         62         27       120
                   41.33      33.70      27.55     33.61
                   25.21      61.85      32.94    210.00
                  1.32975    0.00037    1.07153         *

  All                 75        184         98       357
                  100.00     100.00     100.00    100.00
                   75.00     184.00      98.00    357.00
                       *          *          *         *


  Cell Contents:          Count
                          % of Column
                          Expected count
                          Contribution to Chi-square

  Pearson Chi-Square = 3.618,   DF = 2,  P-Value = 0.16
```

FIGURE 24.5

Minitab output for the two-way table of opinion about astrology by degree held, for Exercise 24.9.

Uses of the chi-square test: independence and homogeneity

Two-way tables can arise in several ways. Most commonly, the subjects in a single sample are classified by two categorical variables. For example, we classified young adults by their age group and where they lived. The question of whether or not there is a relationship between these two classification variables can be stated in terms of the independence of these variables, using the definition of independence from our study of probability in Chapter 12. Recall that for two events, A and B, to be independent, we must have $P(A) = P(A \mid B)$. In this setting, independence of age and living arrangement would imply that the conditional probability that a randomly selected young adult lives at home given their age would be the same for all ages. That is, knowing a young adult's age gives us no information about the probability of living at home. The test we have been using to this point is generally referred to as the **chi-square test for independence,** as thus far all the examples and exercises have been questions about whether two classification variables are independent or not.

chi-square test for independence

The next example illustrates a different setting for a two-way table, in which we compare separate samples from two or more populations, or from two or more treatments in a randomized controlled experiment. In the setting of a randomized controlled experiment, we think of the different treatment groups as the separate populations. "Which population" is now one of the variables for the two-way table. For each sample, we classify individuals according to one variable, and we are interested in whether or not the probabilities of being classified in each category of this variable are the same for each population. Although our calculations for the chi-square test are unchanged, the method of collecting the data is different. This use of the chi-square test is referred to as the **chi-square test for homogeneity** because we are interested in whether or not the populations from which the samples are selected are homogeneous (the same) with respect to the single classification variable.

chi-square test for homogeneity

EXAMPLE 24.6   Are Cell-Only Telephone Users Different?

AB/Getty Images

STATE: Random digit dialing (RDD) telephone surveys do not call cell phone numbers. We know that cell-only users tend to be younger (see Exercise 24.31, page 552), which results in younger adults being underrepresented in an RDD sample. Survey organizations can compensate for this by weighting the younger adults in the RDD sample more heavily, but this assumes that the young adults accessible by landline are similar on the survey issue to their cell-only peers, who are excluded. There is growing evidence that this is not always the case. In young adults between 18 and 25 years of age, Pew interviewed separate random samples of cell-only and landline telephone users and compared them on demographic, lifestyle, and attitudinal issues.[6] Here are the results on the living situations for the two groups:

	Landline Sample	Cell-Only Sample
Live with parents	165	25
Rent	95	74
Own	46	13
Live in a dorm	7	15
Other/refused	16	3
Total	329	130

PLAN: Carry out a chi-square test for

H_0: homogeneity; that is, the distribution of living situation is the same in both populations

H_a: lack of homogeneity; that is, the living situation distribution for the cell-only population differs from that for landline users

Compare column percents or observed versus expected cell counts or terms of chi-square to see the nature of the differences in the distributions of living situations in the two populations.

SOLVE: The Minitab output in Figure 24.6 includes the column percents. These give the distribution of living situation for each type of telephone use. Cell-only users are less likely to live with parents (19.23% versus 50.15% of landline users), more likely to rent (56.92% versus 28.88% of landline users), and more likely to live in a dorm (11.54% versus 2.13% of landline users). There is little difference in the percentages for the other two categories. To see if the differences are significant, first check the guidelines for use of chi-square. The samples can be assumed to be SRSs. The Minitab output shows 3 of the 10 cells have expected counts less than 5, although one of these is 4.86. The other two cells are greater than 1 but less than 5, so the condition for use of chi-square (page 534) is close enough to being satisfied for use. The chi-square test shows a highly significant difference between the distributions of living situations of the two groups of young adults ($\chi^2 = 61.63$, $P = 0.000$). Comparing observed and expected cell counts again shows that cell-only young adults are less likely to live with parents than would be expected if the null hypothesis of homogeneity were true, and more likely to rent or live in a dorm. The contributions to chi-square for the other two categories are quite small.

FIGURE 24.6

Minitab output for the two-way table of living situation and telephone use.

```
┌─────────────────────────────────────────────────────────────┐
│ ▦ Session                                      ─  □  X        │
├─────────────────────────────────────────────────────────────┤
│                                                           ▲   │
│                   Landline   Cell-only                    ▓   │
│                    sample     sample       All            ▓   │
│                                                               │
│   Live with parents    165         25          190           │
│                      50.15      19.23        41.39           │
│                     136.19      53.81       190.00           │
│                       6.096     15.427            *          │
│                                                               │
│   Rent                  95         74          169           │
│                      28.88      56.92        36.82           │
│                     121.14      47.86       169.00           │
│                       5.639     14.270            *          │
│                                                               │
│   Own                   46         13           59           │
│                      13.98      10.00        12.85           │
│                      42.29      16.71        59.00           │
│                       0.326      0.824            *          │
│                                                               │
│   Live in a dorm         7         15           22           │
│                       2.13      11.54         4.79           │
│                      15.77       6.23        22.00           │
│                       4.876     12.341            *          │
│                                                               │
│   Other/refused         16          3           19           │
│                       4.86       2.31         4.14           │
│                      13.62       5.38        19.00           │
│                       0.416      1.054            *          │
│                                                               │
│   All                  329        130          459           │
│                     100.00     100.00       100.00           │
│                     329.00     130.00       459.00           │
│                          *          *            *           │
│                                                               │
│   Cell Contents:       Count                                 │
│                        % of Column                           │
│                        Expected count                        │
│                        Contribution to Chi-square            │
│                                                               │
│   Pearson Chi-Square = 61.269,  DF = 4, P-Value = 0.000      │
│                                                           ▼   │
├─────────────────────────────────────────────────────────────┤
│ ◄ ▮                                             ►             │
└─────────────────────────────────────────────────────────────┘
```

CONCLUDE: The association between living situation and type of telephone service is statistically significant. Young adults accessible by landline phone are more likely to live with their parents and less likely to rent or live in a dorm. The Pew study also found young adults with landlines tend to be more likely to attend religious services at least once a week and less likely to report drinking alcohol in the past seven days or to say that it is okay for people to smoke marijuana. They also tend to be less technology savvy using email, texting, and visiting social networking sites less frequently. Because of the rapid increase in cell-only users plus the differences in a number of dimensions between cell-only users and their landline peers, most survey organizations now routinely include a cell-only sample when conducting a survey.[7] ■

One of the most useful properties of chi-square is that it tests the null hypothesis "the row and column variables are not related to each other" whenever this hypothesis makes sense for a two-way table. It makes sense when we are comparing a categorical response in two or more samples, as when we compared people who have only a cell phone with people who have a landline phone. This is the chi-square test for homogeneity. The hypothesis also makes sense when we have data on two categorical variables for the individuals in a single sample, as when we examined age group and living arrangement for a sample of young adults. This is the chi-square test for independence. Statistical significance has the same intuitive meaning in both settings: "A relationship this strong is not likely to happen just by chance."

Uses of the Chi-Square Test

Use the chi-square test to test the null hypothesis

H_0: there is no relationship between two categorical variables

when you have a two-way table from one of these situations:

■ A single SRS, with each individual classified according to both of two categorical variables. In this case, the null hypothesis of no relationship says that the two categorical variables are independent, and the test is called the chi-square test of independence.

■ Independent SRSs from two or more populations, with each individual classified according to one categorical variable. (The other variable says which sample the individual comes from.) In this case, the null hypothesis of no relationship says the populations are homogeneous, and the test is called the chi-square test of homogeneity.

MORE CHI-SQUARE TESTS

There are other chi-square tests for hypotheses more specific than "no relationship." A sociologist places people in classes by social status, waits ten years, then classifies the same people again. The row and column variables are the classes at the two times. She might test the hypothesis that there has been no change in the overall distribution of social status in the group. Or she might ask if moves up in status are balanced by matching moves down. These and other null hypotheses can be tested by variations of the chi-square test.

Apply Your Knowledge

24.10 Do You Use Cocaine? Sample surveys on sensitive issues can give different results depending on how the question is asked. A University of Wisconsin study divided 2400 respondents into 3 groups at random. All were asked if they had ever used cocaine. One group of 800 was interviewed by phone; 21% said they had used cocaine. Another 800 people were asked the question in a one-on-one personal interview; 25% said "Yes." The remaining 800 were allowed to make an anonymous written response; 28% said "Yes."[8] Are there statistically significant differences among these proportions?

(a) Convert the information given into a two-way table of counts.

(b) Should you use a chi-square test of independence or homogeneity?

(c) Do a complete analysis of these data, following the four-step process.

24.11 Who's More Informed about Politics? In 2012, the General Social Survey asked a random sample of adults, "Compared to most people, how informed are you about politics?" Here are the data classified by their responses to this question and their age group:

	Not at All	A Little	Somewhat	Very	Extremely
Age 20–29	8	29	28	13	0
Age 30–39	15	28	55	23	9
Age 40–49	2	25	49	26	14
Age 50 or older	21	59	120	79	23

(a) Should you use a chi-square test of independence or homogeneity? Explain.

(b) Do a complete analysis of these data, following the four-step process.

The chi-square distributions

Software usually finds P-values for us. The P-value for a chi-square test comes from comparing the value of the chi-square statistic with critical values for a *chi-square distribution*.

The Chi-Square Distributions

The **chi-square distributions** are a family of distributions that take only positive values and are skewed to the right. A specific chi-square distribution is specified by giving its **degrees of freedom.**

The chi-square test for a two-way table with r rows and c columns uses critical values from the chi-square distribution with $(r - 1)(c - 1)$ degrees of freedom. The P-value is the area under the density curve of this chi-square distribution to the right of the value of test statistic.

Figure 24.7 shows the density curves for three members of the chi-square family of distributions. As the degrees of freedom increase, the density curves become less skewed, and larger values become more probable. Table D in the back of the book gives critical values for chi-square distributions. You can use Table D if you do not have software that gives you P-values for a chi-square test.

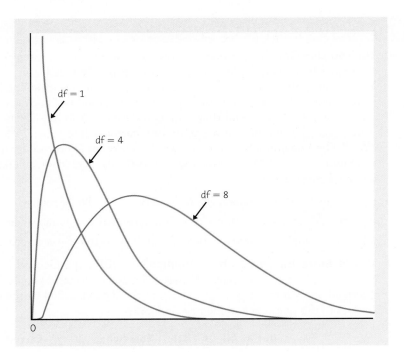

FIGURE 24.7

Density curves for the chi-square distributions with 1, 4, and 8 degrees of freedom. Chi-square distributions take only positive values and are right-skewed.

EXAMPLE 24.7   Using the Chi-Square Table

The two-way table of 5 outcomes by 4 age groups for the living arrangements study (Table 24.1) study has 5 rows and 4 columns. That is, $r = 5$ and $c = 4$. The chi-square statistic therefore has degrees of freedom

$$(r - 1)(c - 1) = (5 - 1)(4 - 1) = (4)(3) = 12$$

Both outputs in Figure 24.2 give 12 as the degrees of freedom.

The observed value of the chi-square statistic is $\chi^2 = 193.55$. Look in the df = 12 row of Table D. The value $\chi^2 = 193.55$ falls above the largest critical value in the table, for $P = 0.0005$. Remember that the chi-square test is always one-sided. So the P-value of $\chi^2 = 193.55$ is less than 0.0005. ∎

df = 12		
p	.001	.0005
x^*	32.91	34.82

We know that all z and t statistics measure the size of an effect in the standard scale centered at zero. We can roughly assess the size of any z or t statistic by the 68–95–99.7 rule, though this is exact only for z. The chi-square statistic does not have any such natural interpretation. But here is a helpful fact: *the mean of any chi-square distribution is equal to its degrees of freedom.* In Example 24.7, χ^2 would have mean 12 if the null hypothesis were true. The observed value $\chi^2 = 193.55$ is so much larger than 12 that we suspect it is significant even before we look at Table D.

Apply Your Knowledge

24.12 Facebook at Penn State. The Minitab output in Figure 24.3 (see page 538) gives the degrees of freedom for a table of Facebook use by students at two campus locations as DF = 2.

(a) Show that this is correct for a table with 3 rows and 2 columns.

(b) Minitab gives the chi-square statistic as Chi-Sq = 19.489. Where does this value fall when compared with critical values of the chi-square distribution with 2 degrees of freedom in Table D? How does Minitab's result P-Value = 0.000 compare with the P-value from the table?

(c) The table included only students who have Facebook accounts. The original table in Exercise 24.1 had 4 rows and 2 columns. What is the proper degrees of freedom for that table?

24.13 Video Gaming and Grades. The Minitab output in Figure 24.4 (see page 539) gives 2 degrees of freedom for the table in Exercise 24.2.

(a) Verify that this is correct.

(b) The computer gives the value of the chi-square statistic as $\chi^2 = 6.739$. Between what two entries in Table D does this value lie? Verify that Minitab's P-value does fall between the tail probabilities p for these two entries.

(c) What is the mean value of the statistic χ^2 if the null hypothesis is true? How does the observed value of χ^2 compare with this mean?

The chi-square test for goodness of fit*

The most common and most important use of the chi-square statistic is to test the hypothesis that there is *no relationship between two categorical variables.* A variation of the statistic can be used to test a different kind of null hypothesis: that *a categorical variable has a specified distribution.* Here is an example that illustrates this use of chi-square.

*This special topic is optional.

EXAMPLE 24.8   Never on Sunday?

BIRTH140

Births are not evenly distributed across the days of the week. Fewer babies are born on Saturday and Sunday than on other days, probably because doctors find weekend births inconvenient.

A random sample of 140 births from local records shows this distribution across the days of the week:

Day	Sun.	Mon.	Tue.	Wed.	Thu.	Fri.	Sat.
Births	13	23	24	20	27	18	15

Sure enough, the two smallest counts of births are on Saturday and Sunday. Do these data give significant evidence that local births are not equally likely on all days of the week? ■

The chi-square test answers the question of Example 24.8 by comparing observed counts with expected counts under the null hypothesis. The null hypothesis for births says that they *are* evenly distributed. To state the hypotheses carefully, write the discrete probability distribution for days of birth:

Day	Sun.	Mon.	Tue.	Wed.	Thu.	Fri.	Sat.
Probability	p_1	p_2	p_3	p_4	p_5	p_6	p_7

The null hypothesis says that the probabilities are the same on all days. In that case, all 7 probabilities must be 1/7. So the null hypothesis is

$$H_0: p_1 = p_2 = p_3 = p_4 = p_5 = p_6 = p_7 = \frac{1}{7}$$

The alternative hypothesis says that days are *not* all equally probable:

$$H_a: \text{not all } p_i = \frac{1}{7}$$

As usual in chi-square tests, H_a is a "many-sided" hypothesis that simply says that H_0 is not true. The chi-square statistic is also as usual:

$$\chi^2 = \sum \frac{(\text{observed count} - \text{expected count})^2}{\text{expected count}}$$

The expected count for an outcome with probability p is np, as we saw in the discussion following Example 24.2. Under the null hypothesis, all the probabilities p_i are the same, so all 7 expected counts are equal to

$$np_i = 140 \times \frac{1}{7} = 20$$

These expected counts easily satisfy our guidelines for using chi-square. The chi-square statistic is

$$\chi^2 = \sum \frac{(\text{observed count} - 20)^2}{20}$$
$$= \frac{(13 - 20)^2}{20} + \frac{(23 - 20)^2}{20} + \cdots + \frac{(15 - 20)^2}{20}$$
$$= 7.6$$

This new use of χ^2 requires a different degrees of freedom. To find the P-value, compare χ^2 with critical values from the chi-square distribution with degrees of freedom one less than the number of values the birth day can take. That's $7 - 1 = 6$ degrees of freedom. From Table D, we see that $\chi^2 = 7.6$ is smaller than the smallest entry in the df = 6 row, which is the critical value for tail area 0.25.

The P-value is therefore greater than 0.25 (software gives the more exact value $P = 0.269$). These 140 births don't give convincing evidence that births are not equally likely on all days of the week.

The chi-square test applied to the hypothesis that a categorical variable has a specified distribution is called the test for *goodness of fit*. The idea is that the test assesses whether the observed counts "fit" the distribution. The chi-square statistic is the same as for the two-way table test, but the expected counts and degrees of freedom are different. Here are the details.

df = 6		
p	.25	.20
x^*	7.84	8.56

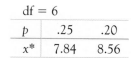 **CHI-SQUARE IN THE CASINO**

Gambling devices such as slot machines and roulette wheels are supposed to have a fixed and known distribution of outcomes. Here's a job for the chi-square test of goodness of fit: state gambling regulators use it to verify that casino devices are honest. How much deviation a casino can get away with depends on the state. Nevada cracks down if chi-square is significant at the 5% level. Mississippi gives more leeway, acting only when the 1% level is reached.

The Chi-Square Test for Goodness of Fit

A categorical variable has k possible outcomes, with probabilities $p_1, p_2, p_3, \ldots, p_k$. That is, p_i is the probability of the ith outcome. We have n independent observations from this categorical variable.

To test the null hypothesis that the probabilities have specified values

$$H_0: p_1 = p_{10}, p_2 = p_{20}, \ldots, p_k = p_{k0}$$

find the **expected count** for the ith possible outcome as np_{i0} and use the **chi-square statistic**

$$\chi^2 = \sum \frac{(\text{observed count} - \text{expected count})^2}{\text{expected count}}$$

The sum is over all the possible outcomes.

The P-value is the area to the right of χ^2 under the density curve of the chi-square distribution with $k - 1$ degrees of freedom.

In Example 24.8, the outcomes are days of the week, with $k = 7$. The null hypothesis says that the probability of a birth on the ith day is $p_{i0} = 1/7$ for all days. We observe $n = 140$ births and count how many fall on each day. These are the counts used in the chi-square statistic.

Apply Your Knowledge

24.14 Saving Birds from Windows. Many birds are injured or killed by flying into windows. It appears that birds don't see windows. Can tilting windows down so that they reflect earth rather than sky reduce bird strikes? Place six windows at the edge of a woods: two vertical, two tilted 20 degrees, and two tilted 40 degrees. During the next four months, there were 53 bird strikes, 31 on the vertical windows, 14 on the 20-degree windows, and 8 on the 40-degree windows.[9] If the tilt has no effect, we expect strikes on windows with all three tilts to have equal probability. Test this null hypothesis. What do you conclude?

24.15 More on Birth Days. Births really are not evenly distributed across the days of the week. The data in Example 24.8 failed to reject this null hypothesis because of random variation in a quite small number of births. Here are data on 700 births in the same locale: **BIRTH700**

Day	Sun.	Mon.	Tue.	Wed.	Thu.	Fri.	Sat.
Births	84	110	124	104	94	112	72

(a) The null hypothesis is that all days are equally probable. What are the probabilities specified by this null hypothesis? What are the expected counts for each day in 700 births?

(b) Calculate the chi-square statistic for goodness of fit.

(c) What are the degrees of freedom for this statistic? Do these 700 births give significant evidence that births are not equally probable on all days of the week?

24.16 Police Harassment? Police may use minor violations such as not wearing a seat belt to stop motorists for other reasons. A large study in Michigan first studied the population of drivers not wearing seat belts during daylight hours by observation at more than 400 locations around the state. Here is the population distribution of seat belt violators by age group:[10]

Age group	16 to 29	30 to 59	60 or older
Proportion	0.328	0.594	0.078

The researchers then looked at court records and called a random sample of 803 drivers who had actually been cited by police for not wearing a seat belt. Here are the counts:

Age group	16 to 29	30 to 59	60 or older
Count	401	382	20

Does the age distribution of people cited differ significantly from the distribution of ages of all seat belt violators? Which age groups have the largest contributions to chi-square? Are these age groups cited more or less frequently than is justified? (The study found that males, blacks, and younger drivers were all overcited.)

24.17 Order in Choice. Does the order in which wine is presented make a difference? Several choices of wine are presented one at a time and in sequence, and the subject is then asked to choose their preferred wine at the end of the sequence. In this study, subjects were asked to taste four wine samples in sequence. All four samples given to a subject were the *same* wine, although subjects were expecting to taste four different samples of a particular variety.[11] There were 33 subjects in the study, and the positions in the sequence they selected for their preferred wine were

Position	1	2	3	4
Count	15	5	2	11

(a) What percent of the subjects chose each position?

(b) If the subjects were equally likely to select each position, what are the expected counts for each position?

(c) Does the chi-square test for goodness of fit give good evidence that the subjects were not equally likely to choose each position?

(State hypotheses, check the guidelines for using chi-square, give the test statistic and its P-value, and state your conclusion.)

(d) The *primacy* effect is a tendency for subjects to choose the first wine tasted, whereas the *recency* effect is a tendency for subjects to choose the most recent wine tasted. Are either of these effects present in this data?

24.18 **What's Your Sign?** For reasons known only to social scientists, the General Social Survey (GSS) regularly asks its subjects their astrological sign. Here are the counts of responses for the 2012 GSS: SIGNS

Sign	Aries	Taurus	Gemini	Cancer	Leo	Virgo
Count	145	162	163	148	186	161
Sign	Libra	Scorpio	Sagittarius	Capricorn	Aquarius	Pisces
Count	178	147	145	151	170	157

If births are spread uniformly across the year, we expect all 12 signs to be equally likely. Are they? Follow the four-step process in your answer.

CHAPTER 24 SUMMARY

Chapter Specifics

■ The **chi-square test** for a two-way table tests the null hypothesis H_0 that there is no relationship between the row variable and the column variable. The alternative hypothesis H_a says that there is some relationship but does not say what kind. When the two-way table results from classifying individuals in a single SRS according to two categorical variables, no relationship says the two categorical variables are independent, and the test is called the **chi-square test of independence.** If the two-way table is formed from independent SRSs from two or more populations with each individual classified according to one categorical variable, no relationship says the populations are the same and the test is called the **chi-square test of homogeneity.**

■ The test compares the observed counts of observations in the cells of the table with the counts that would be expected if H_0 were true. The **expected count** in any cell is

$$\text{expected count} = \frac{\text{row total} \times \text{column total}}{\text{table total}}$$

■ The **chi-square statistic** is

$$\chi^2 = \sum \frac{(\text{observed count} - \text{expected count})^2}{\text{expected count}}$$

■ The chi-square test compares the value of the statistic χ^2 with critical values from the **chi-square distribution** with $(r - 1)(c - 1)$ **degrees of freedom.** Large values of χ^2 are evidence against H_0, so the P-value is the area under the chi-square density curve to the right of χ^2.

■ The chi-square distribution is an approximation to the distribution of the statistic χ^2. You can safely use this approximation when all expected cell counts are at least 1 and no more than 20% are less than 5.

■ If the chi-square test finds a statistically significant relationship between the row and column variables in a two-way table, do data analysis to describe the nature of the relationship. You can do this by comparing well-chosen percents, comparing the observed counts with the expected counts, and looking for the largest **terms of the chi-square statistic.**

STATISTICS IN SUMMARY

Here are the most important skills you should have acquired from reading this chapter.

A. Two-Way Tables

1. Understand that the data for a chi-square test must be presented as a two-way table of counts of outcomes.

2. Recognize a two-way table can occur from a single SRS with each individual classified according to two categorical variables, or from independent SRSs from two or more populations with each individual classified according to one categorical variable.

3. Use percents to describe the relationship between the row and column variables, starting from the counts in a two-way table.

B. Interpreting Chi-Square Tests

1. Locate the chi-square statistic, its *P*-value, and other useful facts (row or column percents, expected counts, terms of chi-square) in output from your software or calculator.

2. Use the expected counts to check whether you can safely use the chi-square test.

3. Explain what null hypothesis the chi-square statistic tests in a specific two-way table. Understand the difference between testing for independence or homogeneity in a two-way table.

4. If the test is significant, compare percents, compare observed with expected cell counts, or look for the largest terms of the chi-square statistic to see what deviations from the null hypothesis are most important.

C. Doing Chi-Square Tests by Hand

1. Calculate the expected count for any cell from the observed counts in a two-way table. Check whether you can safely use the chi-square test.

2. Calculate the term of the chi-square statistic for any cell, as well as the overall statistic.

3. Give the degrees of freedom of a chi-square statistic. Make a quick assessment of the significance of the statistic by comparing the observed value with the degrees of freedom.

4. Use the chi-square critical values in Table D to approximate the *P*-value of a chi-square test.

Link It

The final portion of the text studies relationships between variables. In this chapter, the case of two-way tables is considered and a formal test for answering the question "Is there a relationship between the row and column variables?" is developed. As with procedures described in earlier chapters, we must first consider how the data were produced, as this plays an important role in the conclusions we can reach. In addition, we should begin with data analysis rather than a formal test. In the case of two-way tables, this typically involves looking at percents, both numerically and graphically, to first understand the nature of the relationship. When considering the relationship between the age of young adults and living arrangements in Example 24.1, we can see from our data analysis that as young adults age from 19 to 22, the percent living with their parents drops as the percent living in their own place rises.

Even though there appears to be a clear relationship between age and living arrangements in Example 24.1, we must still answer the question whether the observed differences are large enough to be statistically significant. The chi-square test can be used for this, but it is an approximate procedure, and the conditions for cell sizes need to be checked before applying the test. Unlike some of the simpler procedures in earlier chapters, if the differences are statistically significant, then we have only determined from the chi-square test that there is evidence of a relationship, not the nature of the relationship. Although there are formal statistical procedures to further investigate the nature of the relationship, at this point we need to be satisfied describing the relationship between the two categorical variables using our data analysis tools.

CHECK YOUR SKILLS

Resistance training is a popular form of conditioning aimed at enhancing sports performance and is widely used among high school, college, and professional athletes, although its use for younger athletes is controversial. A random sample of 4111 patients between the ages of 8 to 30 admitted to U.S. emergency rooms with the injury code "weightlifting" were obtained. These injuries were classified as "accidental"  *if caused by dropped weight or improper equipment use. The patients were also classified into the four age categories "8–13," "14–18," "19–22," and "23–30." Here is a two-way table of the results:*[12]

	ACCIDENTAL	NOT ACCIDENTAL
8–13	295	102
14–18	655	916
19–22	239	533
23–30	363	1008

24.19 The number of "accidental" injuries in the sample is

(a) 1552.   (b) 2559.   (c) 4111.

24.20 The percent of 14- to 18-year-olds in the sample whose injuries were classified as "accidental" is about

(a) 42.2%.   (b) 41.7%.   (c) 74.3%.

24.21 The percent of 14- to 18-year-olds in the sample whose injuries were classified as "accidental" is

(a) higher than the percent for 23- to 30-year-olds.
(b) about the same as the percent for 23- to 30-year-olds.
(c) lower than the percent for 23- to 30-year-olds.

24.22 For this two-way table, we should do a chi-square test of

(a) homogeneity.
(b) independence.
(c) goodness of fit.

24.23 The expected count of 14- to 18-year-olds whose injuries were classified as "accidental" is about

(a) 593.09.   (b) 655.   (c) 977.91.

24.24 The term in the chi-square statistic for the cell of 14- to 18-year-olds whose injuries were classified as "accidental" is about

(a) 593.09.   (b) 3.919.   (c) 6.463.

24.25 The degrees of freedom for the chi-square test for this two-way table are

(a) 3.   (b) 4.   (c) 8.

24.26 The null hypothesis for the chi-square test for this two-way table is

(a) the proportion of "Accidental" and "Not accidental" injuries are the same.
(b) there is no difference in the conditional probability of an "accidental" injury for each of the four age groups.
(c) "accidental" injuries are more likely for the younger age groups.

24.27 The alternative hypothesis for the chi-square test for this two-way table is

(a) the proportion of "Accidental" and "Not accidental" injuries are different.
(b) the conditional probabilities of an "accidental" injury for each of the four age groups are not the same.
(c) "accidental" injuries are more likely for the younger age groups.

24.28 Software gives chi-square statistic $\chi^2 = 325.459$ for this table. From the table of critical values, we can say that the P-value is

(a) between 0.0025 and 0.001.
(b) between 0.001 and 0.0005.
(c) less than 0.0005.

24.29 We can use our results to make inference about the relationship between age and type of weightlifting injury in the population of patients between 18 and 30 admitted to emergency rooms because

(a) the sample is large, 4111 weightlifting injuries in all.
(b) the sample is close to an SRS of all weightlifting injuries for this age group.
(c) there are only two types of weightlifting injury.

CHAPTER 24   EXERCISES

If you have access to software or a graphing calculator, use it to speed your analysis of the data in these exercises. Exercises 24.30 to 24.35 are suitable for hand calculation if necessary.

24.30 **Smoking cessation.** A large randomized trial was conducted to assess the efficacy of Chantix® for smoking cessation compared with bupropion (more

commonly known as Wellbutrin® or Zyban®) and a placebo. Chantix® is different from most other quit-smoking products in that it targets nicotine receptors in the brain, attaches to them, and blocks nicotine from reaching them, whereas bupropion is an antidepressant often used to help people stop smoking. Generally healthy smokers who smoked at least 10 cigarettes per day were assigned at random to take Chantix® ($n = 352$), bupropion ($n = 329$), or a placebo ($n = 344$). The study was double blind, with the response measure being continuous absence from smoking for weeks 9 through 12 of the study. Here is a two-way table of the results:[13] ░ SMOKE

	TREATMENT		
	CHANTIX®	**BUPROPION**	**PLACEBO**
No smoking in weeks 9–12	155	97	61
Smoked in weeks 9–12	197	232	283

(a) Give a 95% confidence interval for the difference between the proportions of smokers in the bupropion and placebo groups who did not smoke in weeks 9 through 12 of the study.

(b) What proportion of each of the three groups in the sample did not smoke in weeks 9 through 12 of the study? Are there statistically significant differences among these proportions? State hypotheses and give a test statistic and its P-value.

(c) In part (b), did you use a chi-square test for independence or homogeneity? Explain.

24.31 **Cell-only versus landline users.** We suspect that people who rely entirely on cell phones will as a group be younger than those who have a landline telephone. Do data confirm this guess? The Pew Research Center interviewed separate random samples of cell-only and landline telephone users and broke down the samples by age group:[14] ░ CELLAGE

	LANDLINE SAMPLE	**CELL-ONLY SAMPLE**
Age 18–29	335	374
Age 30–49	1242	347
Age 50–64	1625	146
Age 65 or older	1481	36
Total	4683	903

(a) Should you use a chi-square test of independence or homogeneity?

(b) Do a complete analysis of these data, following the four-step process.

24.32 **College students need better sleep!** A random sample of 871 students between the ages of 20 and 24 at a large midwestern university completed a survey including questions about their sleep quality, moods, academic performance, physical health, and psychoactive drug use. Sleep quality was measured using the Pittsburgh Sleep Quality Index (PSQI), with students scoring less than or equal to five on the index classified as optimal sleepers, those scoring a 6 or 7 classified as borderline, and those scoring over 7 classified as poor sleepers. The following table

looks at the relationship between sleep quality classification and the use of over-the-counter (OTC) or prescription (Rx) stimulant medication more than once a month to help keep awake.[15] ░ SLEEPQ

USE OF OTC/Rx MEDS TO WAKE > 1X/MONTH	SLEEP QUALITY ON PSQI INDEX		
	OPTIMAL	**BORDERLINE**	**POOR**
Yes	37	53	84
No	266	186	245

(a) Regarding this survey as approximately an SRS of college students, give a 95% confidence interval for the proportion of college students who use over-the-counter or prescription stimulant medication more than once a month to keep awake.

(b) Find the conditional distributions of sleep quality for students who use over-the-counter or prescription stimulant medication more than once a month to keep awake and those that don't. Make a graph that compares the two conditional distributions. Use your work to describe the overall relationship between students who use and don't use medications to keep awake and their sleep quality.

(c) Do students who use over-the-counter or prescription stimulant medication more than once a month to keep awake differ significantly in their quality of sleep than those who don't? State hypotheses, give the chi-square statistic and its P-value, and state your conclusion.

24.33 **Did the randomization work?** After randomly assigning subjects to treatments in a randomized comparative experiment, we can compare the treatment groups to see how well the randomization worked. We hope to find no significant differences among the groups. A study of how to provide premature infants with a substance essential to their development assigned infants at random to receive one of four types of supplement, called PBM, NLCP, PL-LCP, and TG-LCP.[16]

(a) The subjects were 77 premature infants. Outline the design of the experiment if 20 are assigned to the PBM group and 19 to each of the other treatments.

(b) The random assignment resulted in 9 females in the TG-LCP group and 11 females in each of the other groups. Make a two-way table of group by sex, and do a chi-square test to see if there are significant differences among the groups. What do you find?

24.34 **More on video gaming.** The data for comparing two sample proportions can be presented in a two-way table containing the counts of successes and failures in both samples, with two rows and two columns. In Exercise 24.2 (see page 529), a survey of the consequences of video gaming on 14- to 18-year-olds is described. Here is another question from the survey. The question was about aggressive behavior as evidenced by getting into serious fights, and the comparison was between girls that have and have not played video games. Here are the data:

	SERIOUS FIGHTS	
	YES	NO
Played games	36	55
Never played games	578	1436

(a) Is there evidence that the proportions of all 14- to 18-year-old girls who played or have never played video games and have gotten into serious fights differ? Find the two sample proportions, the z statistic, and its P-value.

(b) Is there evidence that the proportions for 14- to 18-year-old girls who have or have not gotten into serious fights differs between those who have played or have never played video games? Find the chi-square statistic χ^2 and its P-value.

(c) Show that (up to roundoff error) your χ^2 is the same as z^2. The two P-values are also the same. These facts are always true, so you will often see chi-square for 2×2 tables used to compare two proportions.

(d) Suppose we were interested in finding out if the data gave good evidence that video gaming was associated with *increased* aggression in girls as evidenced by getting into serious fights. Can we use the z test for this hypothesis? What about the χ^2 test? What is the important difference between these two procedures?

24.35 Unhappy rats and tumors. Some people think that the attitude of cancer patients can influence the progress of their disease. We can't experiment with humans, but here is a rat experiment on this theme. Inject 60 rats with tumor cells, and then divide them at random into two groups of 30. All the rats receive electric shocks, but rats in Group 1 can end the shock by pressing a lever. (Rats learn this sort of thing quickly.) The rats in Group 2 cannot control the shocks, which presumably makes them feel helpless and unhappy. We suspect that the rats in Group 1 will develop fewer tumors. The results: 11 of the Group 1 rats and 22 of the Group 2 rats developed tumors.[17]

(a) Make a two-way table of tumors by group. State the null and alternative hypotheses for this investigation.

(b) Although we have a two-way table, the chi-square test can't test a one-sided alternative. Carry out the z test and report your conclusion.

24.36 I think I'll be rich by age 30. A sample survey asked young adults (aged 19 to 25), "What do you think are the chances you will have much more than a middle-class income at age 30?" The Minitab output in Figure 24.8 shows the two-way table and related information, omitting a few subjects who refused to respond or who said they were already rich.[18] RICHBY30

```
Session                                        _  □  X

                                  Female      Male

A: Almost no chance                   96        98
                                    95.2      98.8
                                  0.0076    0.0073

B: Some chance but probably not      426       286
                                   349.2     362.8
                                 16.8842   16.2525

C: A 50 50 chance                    696       720
                                   694.5     721.5
                                  0.0032    0.0031

D: A good chance                     663       758
                                   697.0     724.0
                                  1.6543    1.5924

E: Almost certain                    486       597
                                   531.2     551.8
                                  3.8424    3.6986

Cell Contents:        Count
                      Expected count
                      Contribution to Chi-square

Pearson Chi-Square = 43.946,  DF = 4, P-Value = 0.000
```

FIGURE 24.8
Minitab output for the sample survey responses of Exercise 24.36.

(a) Should you use a chi-square test of independence or homogeneity?

(b) Use the output as the basis for a discussion of the relationship between sex and a person's assessment of their chances of being rich by age 30.

24.37 Sexy magazine ads? Look at full-page ads in magazines with a young adult readership. Classify ads that show a model as "not sexual" or "sexual" depending on how the model is dressed (or not dressed). Here are data on 1509 ads in magazines aimed at young men, at young women, or at young adults in general:[19]

	READERS		
	MEN	WOMEN	GENERAL
Sexual	105	225	66
Not sexual	514	351	248

Figure 24.9 displays Minitab chi-square output. Use the information in the output to describe the relationship between the target audience and the sexual content of ads in magazines for young adults. SEXYADS

Mistakes in using the chi-square test are unusually common. Exercises 24.38 to 24.41 illustrate several kinds of mistakes.

24.38 Sorry, no chi-square. An experimenter hid a toy from a dog behind either Screen A or Screen B. In the first phase the toy was always hidden behind Screen A, whereas in the second phase the toy was always hidden behind Screen B. Will the dog continue to look behind Screen A in the second phase? This was tried

under three conditions. In the Social-Communicative condition, the experimenter communicated with the dog by establishing eye contact and addressing the dog while hiding the toy; in the Noncommunicative condition the toy was hidden without communication; in the Nonsocial the toy was dragged by a string to be hidden without any interaction from the experimenter. There were 12 dogs assigned at random to each condition, and the dog had up to three trials to find the toy hidden behind Screen B in phase 2. An error occurred if the dog continued to search behind Screen A, with the number of errors ranging from zero if the dog found the toy behind Screen B on the initial trial up to three if the dog never correctly chose Screen B. Here are the data:[20] HIDETOY

	NUMBER OF ERRORS			
	0	1	2	3
Social-Communicative	0	3	3	6
Noncommunicative	5	3	1	3
Nonsocial	8	2	2	0

(a) The data do show a difference in the number of errors for the different conditions. Show this by comparing suitable percents.

(b) The researchers used a more complicated but exact procedure rather than chi-square to assess significance for these data. Why can't the chi-square test be trusted in this case?

```
Session                                              _ □ X

              Men      Women    General      All

Sexual         105        225        66        396
             16.96      39.06     21.02      26.24
             162.4      151.2      82.4      396.0
            20.312     36.074     3.265         *

Not sexual     514        351       248       1113
             83.04      60.94     78.98      73.76
             456.6      424.8     231.6     1113.0
             7.227     12.835     1.162         *

All            619        576       314       1509
            100.00     100.00    100.00     100.00
             619.0      576.0     314.0     1509.0
                 *          *         *         *


Cell Contents:        Count
                      % of Column
                      Expected count
                      Contribution to Chi-square

Pearson Chi-Square = 80.874,  DF = 2, P-Value = 0.00
```

FIGURE 24.9
Minitab output for a study of ads in magazines, Exercise 24.37.

(c) If you use software, does the chi-square output for these data warn you against using the test?

24.39 Sorry, no chi-square. How do U.S. residents who travel overseas for leisure differ from those who travel for business? Here is the breakdown by occupation:[21]

	LEISURE TRAVELERS	BUSINESS TRAVELERS
Professional/technical	36%	39%
Manager/executive	23%	48%
Retired	14%	3%
Student	7%	3%
Other	20%	7%
Total	100%	100%

Explain why we don't have enough information to use the chi-square test to learn whether these two distributions differ significantly.

24.40 Sorry, no chi-square. Socially isolated individuals have been shown to be at greater risk for the development of a variety of illnesses and increased mortality. It is a more serious problem among the elderly as decreasing economic resources, mobility impairment, and death of contemporaries combine to limit social contact. In a study of 6500 men and women in the English Longitudinal Study of Aging, individuals were classified according to social isolation, with 5269 classified as low/average and the remaining 1231 classified as high social isolation. The following table shows the incidence of several limiting long-standing illnesses for the two groups.[22] Explain why it is not correct to use a chi-square test on this table to compare the "Low/average isolation" and "High isolation" groups. Note that to use the chi-square test in a two-way table, each individual must fall into *one* cell of the table.

	LOW/AVERAGE ISOLATION	HIGH ISOLATION
Cancer	168	36
Diabetes	332	89
Coronary heart disease	193	43
Chronic lung disease	65	34
Stroke	134	43
Arthritis	1778	477
Mobility impairment	2940	769
Depression	960	359

24.41 Sorry, no chi-square. Does eating chocolate trigger headaches? To find out, women with chronic headaches followed the same diet except for eating chocolate bars and carob bars that looked and tasted the same. Each subject ate both chocolate and carob bars in random order with at least three days

between. Each woman then reported whether or not she had a headache within 12 hours of eating the bar. Here is a two-way table of the results for the 64 subjects:[23]

	NO HEADACHE	HEADACHE
Chocolate	53	11
Carob (placebo)	38	26

The researchers carried out a chi-square test on this table to see if the two types of bar differ in triggering headaches. Explain why this test is incorrect. (*Hint:* There are 64 subjects. How many observations appear in the two-way table?)

*The remaining exercises concern larger tables that require software for easy analysis. In many cases, you should follow the **Plan, Solve,** and **Conclude** steps of the four-step process in your answers.*

24.42 Smokers rate their health. The University of Michigan Health and Retirement Study (HRS) surveys more than 22,000 Americans over the age of 50 every two years. A subsample of the Health and Retirement Study (HRS) participated in the 2009 Internet-based survey that collected information on a number of topical areas, including health (physical and mental, health behaviors); psychosocial items; economics (income, assets, expectations, and consumption); and retirement.[24] Two of the questions asked on the Internet survey were, "Would you say your health is excellent, very good, good, fair, or poor?" and "Do you smoke cigarettes now?" The two-way table summarizes the answers on these two questions. SMRATE

	CURRENT SMOKER	
	YES	NO
Excellent	25	484
Very good	115	1557
Good	145	1309
Fair	90	545
Poor	29	11

(a) Regarding the HRS Internet survey as approximately an SRS of Americans over the age of 50, give a 99% confidence interval for the proportion of Americans over the age of 50 who are current smokers.
(b) Compare the conditional distributions of self-evaluation of health for current smokers and nonsmokers using both a table and a graph. What are the most important differences?
(c) Carry out the chi-square test for the hypothesis of no difference between the self-evaluation of health for current smokers and nonsmokers. What would be the mean of the test statistic if the null hypothesis were true? The value of the statistic is so

far above this mean that you can see at once that it must be highly significant. What is the approximate *P*-value?

(d) Look at the terms of the chi-square statistic and compare observed and expected counts in the cells that contribute the most to chi-square. Based on this and your findings in part (b), write a short comparison of the differences in self-evaluation of health for current smokers and nonsmokers.

24.43 **Who goes to religious services?** The General Social Survey (GSS) asked this question: "Have you attended religious services in the last week?" Here are the responses for those whose highest degree was high school or above:  ▪||▪ SERVICES

	HIGHEST DEGREE HELD			
	HIGH SCHOOL	JUNIOR COLLEGE	BACHELOR	GRADUATE
Attended services	400	62	146	76
Did not attend services	880	101	232	105

(a) Carry out the chi-square test for the hypothesis of no relationship between the highest degree attained and attendance at religious services in the last week. What do you conclude?

(b) Make a 2 × 3 table by omitting the column corresponding to those whose highest degree was high school. Carry out the chi-square test for the hypothesis of no relationship between the type of advanced degree attained and attendance at religious services in the last week. What do you conclude?

(c) Make a 2 × 2 table by combining the counts in the three columns that have a highest degree beyond high school, so that you are comparing adults whose highest degree was high school against those whose highest degree was beyond high school. Carry out the chi-square test for the hypothesis of no relationship between attaining a degree beyond high school and attendance at religious services for this 2 × 2 table. What do you conclude?

(d) Using the results from these three chi-square tests, write a short report explaining the relationship between attendance at religious services in the last week and the highest degree attained. As part of your report, you should give the percentages who attended religious services for each of the four highest degrees.

24.44 **Condom usage among high school students.** The Centers for Disease Control developed the Youth Risk Behavior Surveillance System (YRBSS) to monitor six categories of priority health-risk behaviors among youth—behaviors that contribute to unintentional injuries and violence; tobacco use; alcohol and other drug use; sexual behaviors that contribute to unintended pregnancy and sexually transmitted diseases; unhealthy dietary behaviors; and physical inactivity. A multistage sample design is used to produce a representative samples of students in grades 9–12, who then fill out a questionnaire on these behaviors. The following data are for the question, "Did Not Use A Condom During Last Sexual Intercourse?" The two-way table of grade and condom usage only included students who were currently sexually active. Here are the results:[25] ▪||▪ CONDOMS

	CONDOM USED	
	YES	NO
9th grade	281	463
10th grade	388	669
11th grade	597	938
12th grade	727	936

Describe the most important differences between condom usage and grade. Is there a significant overall difference between the proportion who used condoms in the different grades?

24.45 **How are schools doing?** The nonprofit group Public Agenda conducted telephone interviews with a stratified sample of parents of high school children. There were 202 black parents, 202 Hispanic parents, and 201 white parents. One question asked was, "Are the high schools in your state doing an excellent, good, fair, or poor job, or don't you know enough to say?" Here are the survey results:[26] ▪||▪ HIGHSCHL

	BLACK PARENTS	HISPANIC PARENTS	WHITE PARENTS
Excellent	12	34	22
Good	69	55	81
Fair	75	61	60
Poor	24	24	24
Don't know	22	28	14
Total	202	202	201

Are the differences in the distributions of responses for the three groups of parents statistically significant? What departures from the null hypothesis "no relationship between group and response" contribute most to the value of the chi-square statistic? Write a brief conclusion based on your analysis.

24.46 **Complications of bariatric surgery.** Bariatric surgery, or weight-loss surgery, includes a variety of procedures performed on people who are obese. Weight loss is achieved by reducing the size of the stomach with an implanted medical device (gastric banding), through removal of a portion of the stomach (sleeve gastrectomy), or by resecting and rerouting the small intestines to a small stomach pouch (gastric bypass surgery). Because there can be complications using any of these methods, the National Institute of Health recommends bariatric surgery for obese people with a body mass index (BMI) of at least 40, and for people with BMI 35 and serious coexisting medical conditions such as diabetes. Serious complications include potentially life-threatening, permanently disabling, and fatal outcomes. Here is a two-way table for data collected in Michigan over several years giving counts of non-life-threatening complications, serious complications, and no complications for these three types of surgeries:[27] **BARI**

	TYPE OF COMPLICATION			
	NON-LIFE-THREATENING	SERIOUS	NONE	TOTAL
Gastric banding	81	46	5253	5380
Sleeve gastrectomy	31	19	804	854
Gastric bypass	606	325	8110	9041

(a) Is this study an experiment? Explain your answer.
(b) Is there a significant difference in the distribution of type of complication for the three types of surgery? Which surgeries have the greatest chance of complications? Can we conclude that the surgery is more dangerous, or could there be other factors associated with the increased risk?

24.47 **Market research.** Before bringing a new product to market, firms carry out extensive studies to learn how consumers react to the product and how best to advertise its advantages. Here are data from a study of a new laundry detergent.[28] The subjects are people who don't currently use the established brand that the new product will compete with. Give subjects free samples of both detergents. After they have tried both for a while, ask which they prefer. The answers may depend on other facts about how people do laundry. **LAUNDRY**

	LAUNDRY PRACTICES			
	SOFT WATER, WARM WASH	SOFT WATER, HOT WASH	HARD WATER, WARM WASH	HARD WATER, HOT WASH
Prefer standard product	53	27	42	30
Prefer new product	63	29	68	42

How do laundry practices (water hardness and wash temperature) influence the choice of detergent? In which settings does the new detergent do best? Are the differences between the detergents statistically significant?

Support for political parties. *Political parties want to know what groups of people support them. The General Social Survey (GSS) asked its 2012 sample, "Generally speaking, do you usually think of yourself as a Republican, Democrat, Independent, or what?" The GSS is essentially an SRS of American adults. Here is a large two-way table breaking down the responses by the highest degree the subject held:*

	NONE	HIGH SCHOOL	JR. COLLEGE	BACHELOR	GRADUATE
Strong Democrat	47	161	35	60	53
Not strong Democrat	43	176	23	63	38
Independent, near Democrat	37	117	10	43	28
Independent	101	186	26	36	24
Independent, near Republican	20	76	16	29	16
Not strong Republican	18	130	16	64	22
Strong Republican	14	101	19	43	15
Other party	5	22	5	14	8

Exercises 24.48 to 24.50 are based on this table.

24.48 **Other parties.** Give a 95% confidence interval for the proportion of adults who are "Independent."

24.49 **Party support in brief.** Make a 2 × 5 table by combining the counts in the three rows that mention Democrat and in the three rows that mention Republican and ignoring strict independents and supporters of other parties. We might think of this table as comparing all adults who lean Democrat and all adults who lean Republican. How does support for the two major parties differ among adults with different levels of education? **COMBPART**

24.50 **Party support in full.** Use the full table to analyze the differences in political party support among levels of education. The sample is so large that the differences are bound to be highly significant, but give the chi-square statistic and its *P*-value nonetheless. The main challenge is in seeing what the data say. Does the full table yield any insights not found in the compressed table you analyzed in the previous exercise? **FULLPART**

Exploring the Web

24.51   **Make your own table.** The Behavioral Risk Factor Surveillance System (BRFSS) is an ongoing data collection program designed to measure behavioral risk factors for the adult population (18 years of age or older) living in households. Data are collected from a random sample of adults (one per household) through a telephone survey. Go to the Web site `apps.nccd.cdc.gov/BRFSS/` and under BRFSS contents click on *Web Enabled Analysis Tool* (WEAT) and then click on *Cross Tabulation Analysis*. After selecting a year, a window will open that will allow you to produce two-way tables.

(a) Choose a state of interest to you and two variables for the two-way table. For example, you could choose Connecticut and look at the relationship between a demographic variable such as education level and a variable such as health care coverage. Once you have chosen your state and two variables, click on *Run Report* on the bottom of the page. A two-way table will appear in a new window.

(b) Is there a relationship between the two variables you have selected? If the relationship is statistically significant, describe the relationship in a brief report using percentages from the table and an appropriate graph.

24.52   **What do the voters think?** The American National Election Studies (ANES) is the leading academically run national survey of voters in the United States, conducted before and after every presidential election. SDA (Survey Documentation and Analysis) is a set of programs that allows you to analyze survey data and includes the ANES survey as part of its archive. Go to the Web site `sda.berkeley.edu/` and click on *Archive*. Go to the 2012 ANES survey.

(a) Open the pre-election survey data. Under Liberal/Conservative choose the variable "liberal/conservative self-placement on a 7 point scale." Use this as your row variable. Under Energy and Climate Change, choose the variable "Anthropogenic climate change", or climate change caused or produced by humans. Use this as your column variable. In the details for the table, set *Weight* to none and for *N of Cases to Display*, make sure the unweighted box is checked. For *Percentaging*, choose row percents. Now click on "run the table."

(b) To analyze the data, make a 3 × 3 table by combining the rows for extremely liberal and liberal, slightly liberal, middle of the road, slightly conservative, and conservative and extremely conservative. Carry out a formal test to determine if there is a relationship between these two variables, and then describe the relationship in a brief report using percentages from the table or an appropriate graph.

(c) Select two other variables of interest to you, and analyze the relationship between these two variables. If there is a more recent survey than 2012, you should use it.

Inference for Regression

Overview

This chapter demonstrates the use of statistical inference in the setting of simple linear regression and correlation (see Chapters 4 and 5 for an introduction to these concepts). Earlier in the text, students learned about the least-squares regression line $\hat{y} = a + bx$, but this was done in a descriptive manner (without the context of a population and sample). Here, we consider the sample as having been selected from a population. We propose that the population of bivariate pairs is described by a population regression line, $\mu_y = \alpha + \beta x$, and that the least-squares regression line fit earlier is an estimate of this population line.

Conditions necessary for inference are presented to students early in the chapter, when the formal model is presented. We discuss the regression standard error in this context. A computational formula is presented, but requires tedious calculations, so the use of technology is emphasized. Students then encounter their first formal inference regarding regression, which is conducting a test of hypothesis on the slope parameter. The hypothesis $H_0: \beta = 0$ is the claim that no linear relationship exists between the explanatory and response variables. Inference is based on an underlying t distribution. The test statistic here is equivalent to that for testing H_0: Population Correlation $= 0$. Confidence intervals for the regression slope, β, are also discussed. Finally, the discussion of inference concludes with confidence intervals for the mean response and prediction intervals for an individual response are explained. The chapter ends with another discussion of checking conditions for inference, including (1) the relationship is linear in nature, (2) the response varies Normally about the population regression line, (3) observations are independent, and (4) the standard deviation of the responses is the same for all values of x. Note that many statistical packages will easily provide the diagnostics for these four conditions.

LEARNING OUTCOMES

- Make a scatterplot to show the relationship between an explanatory and a response variable.

- Use software or a calculator to find the correlation and the equation of the least-squares regression line.

- Recognize the regression setting: a straight-line relationship between an explanatory variable x and a response variable y.

- Recognize which type of inference you need in a particular regression setting.

- Explain in any specific regression setting the meaning of the slope β of the true regression line in context.

- Understand software output for regression. Find in the output the slope and intercept of the least-squares line, their standard errors, and the regression standard error.

- Use that information to carry out tests of $H_0: \beta = 0$ and calculate confidence intervals for β.

- Explain the distinction between a confidence interval for the mean response and a prediction interval for an individual response.

- If software gives output for prediction, use that output to give either confidence or prediction intervals.

- Check the conditions for regression inference:
  - Make a stemplot or histogram of the residuals and look for strong departures from Normality.
  - Make a residual plot and look for departures from a linear pattern or unequal variability about the "residual = 0" line.
  - Ask whether the study design suggests that observations are independent.

Teaching Suggestions and Additional Examples/Activities for the Classroom

Note: This chapter provides an opportunity to review material from Chapters 4 and 5 on correlation and simple linear regression, and doing so provides a natural introduction for this material on regression inference.

1. Use Real Data to Refresh the Ideas of Correlation and Regression

If you visit the Web site for a fast-food restaurant, you can download nutritional information for entrée menu items that provide interesting data for exploration. For example, at the

McDonald's® Web site (http://www.mcdonalds.com/us/en/food.html), you might down-load Calorie Content and Fat Content for all major entrée items (sandwiches, wraps, Mighty Wings, etc.), then present the data with a corresponding scatterplot (using "fat content" as the explanatory variable). There is a very strong linear relationship between these variables.

Use an example (such as the one described above) to review all the major ideas of Chapters 4 and 5—correlation, construction and meaning of a least-squares regression line, prediction of the response variable given a value of the explanatory variable, extrapolation (ask if it would be reasonable, for example, to predict the calories in a sandwich like the new Bacon Clubhouse Burger at 40 grams of fat), and the meaning of r^2. After you've done this, you can begin to introduce the topics of Chapter 25, concerning inference for the regression model.

2. Explain the Difference Between the Population and Least-Squares Regression Lines

Start with a discussion of the difference between the population regression line (which would fit all values in the population) and the least-squares regression line (which fits our sample). Students often struggle to grasp the idea that the regression line we computed in Chapter 5 was really an estimate of some unknown population regression line. Ask students to visualize the population, then imagine taking samples of n bivariate observations repeatedly, each time fitting a regression line. The key is to get them to understand that the sample regression line would vary from sample to sample (this is a good time to remind them that they are familiar with sampling variability). The slope of the regression lines would change, as would the predictions resulting from the lines. Notice that the notation we use for the population regression line involves Greek symbols, consistent with notation for parameters.

3. Connect Regression Standard Error to Students' Prior Knowledge

Another concept students sometimes struggle to understand is the regression standard error, s. Once you have explained the conditions for regression inference at the beginning of the chapter, use the Normality of y variables at a fixed value of x to provide a visual explanation of this value's meaning. For example, according to the 68–95–99.7 rule, we would expect virtually all fast-food menu items with 30 grams of fat to have between $\hat{y} - 3s$ and $\hat{y} + 3s$ calories. It is far less important for students to know the formula for s than it is to understand its meaning.

4. Tie Inference for Regression to Prior Inference Methods

When discussing the confidence interval for the regression slope, focus should be on how the interval construction parallels all the other confidence intervals we have seen. Remind students of some of these intervals and how any margin of error is (some number) $\times$ (standard error). When you illustrate the construction of a confidence interval for the regression slope, use the "repeated samples" interpretation of the interval to remind students that we're envisioning repeating the regression study—the estimated slopes would vary from sample to sample. Likewise, when you illustrate the t test for regression slope (test for linear relationship), interpret the P-value in a similar manner: If in the population there was really no linear relationship between X and Y, then an observed slope as far from zero as the one we sampled would occur with this probability.

5. Emphasize the Difference Between Confidence Intervals and Prediction Intervals

Students often struggle to understand the difference between confidence intervals for the mean response and prediction intervals for a single response. Ask students the following question: In which situation am I better able to come up with a more precise estimate: (1) predict the weight of the very next 6-foot-tall person I encounter; or (2) estimate the typical weight of all people 6 feet tall? Given their experience to date, students should easily see that in (2) the resulting interval would be narrower, reflecting greater precision. Most software packages provide both types of intervals, and little is gained by having students grind through the rather tedious computations by hand.

6. Use a Culminating Example to Examine All Aspects of Inference for Regression

Work through one very thorough example (such as the one relating fat and calories in fast-food menu items), including a discussion of checking conditions for inference. Provide the students with a histogram of the residuals (to check for Normality), a residual plot (to check for patterns and/or nonconstant variance), and a plot of the residuals over time (to check for independence when time might be a lurking variable) as part of the output.

Other Resources (LaunchPad)

StatClips
Regression: Assumptions
Regression: Inference

Snapshots Videos
Regression Inference

EESEE Case Studies
Weighing Trucks in Motion
Counting Calories

Applets
Two-Variable Statistical Calculator
Correlation and Regression

Barbara Peacock/Getty Images

Inference for Regression

CHAPTER 25

W hen a scatterplot shows a linear relationship between a quantitative explanatory variable x and a quantitative response variable y, we can use the least-squares line fitted to the data to predict y for a given value of x. When the data are a sample from a larger population, we need statistical inference to answer questions like these about the population:

- Is there really a linear relationship between x and y in the population, or might the pattern we see in the scatterplot plausibly arise just by chance?

- What is the slope (rate of change) that relates y to x in the population, including a margin of error for our estimate of the slope?

- If we use the least-squares line to predict y for a given value of x, how accurate is our prediction (again, with a margin of error)?

This chapter shows you how to answer these questions. Here is an example we will explore.

EXAMPLE 25.1   Crying and IQ

STATE: Infants who cry easily may be more easily stimulated than others. This may be a sign of higher IQ. Child development researchers explored the relationship between the crying of infants four to ten days old and their later IQ test scores. A snap of a rubber band on the sole of the foot caused the infants to cry. The research-ers recorded the crying and measured its intensity by the number of peaks in the

CRYING

559

most active 20 seconds. They later measured the children's IQ at age three years using the Stanford-Binet IQ test. Table 25.1 contains data on 38 infants.[1] Do children with higher crying counts tend to have higher IQs?

TABLE 25.1 INFANTS' CRYING (NUMBER OF PEAKS) AND IQ SCORES							
CRYING	IQ	CRYING	IQ	CRYING	IQ	CRYING	IQ
10	87	20	90	17	94	12	94
12	97	16	100	19	103	12	103
9	103	23	103	13	104	14	106
16	106	27	108	18	109	10	109
18	109	15	112	18	112	23	113
15	114	21	114	16	118	9	119
12	119	12	120	19	120	16	124
20	132	15	133	22	135	31	135
16	136	17	141	30	155	22	157
33	159	13	162				

PLAN: Make a scatterplot. If the relationship appears linear, use correlation and regression to describe it. Finally, ask whether there is a *statistically significant* linear relationship between crying and IQ.

SOLVE (first steps): Chapters 4 and 5 introduced the data analysis that must come before inference. The first steps we take are a review of this data analysis. Figure 25.1 is a **scatterplot** of the crying data. Plot the explanatory variable (count of crying peaks) horizontally and the response variable (IQ) vertically. Look for the form, direction, and strength of the relationship as well as for outliers or other deviations. There is a moderately strong positive linear relationship, with no extreme outliers or potentially influential observations.

scatterplot

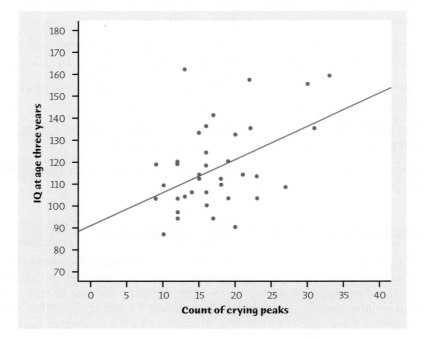

FIGURE 25.1

Scatterplot of the IQ score of infants at age three years against the intensity of their crying soon after birth, with the least-squares regression line, for Example 25.1.

Because the scatterplot shows a roughly linear (straight-line) pattern, the **correlation** describes the direction and strength of the relationship. The correlation between crying and IQ is $r = 0.455$. We are interested in predicting the response from information about the explanatory variable. So we find the **least-squares regression line** for predicting IQ from crying. The equation of the regression line is

$$\hat{y} = a + bx$$
$$= 91.27 + 1.493x$$

correlation

least-squares line

CONCLUDE (first steps): Children who cry more vigorously do tend to have higher IQs. Because $r^2 = 0.207$, only about 21% of the variation in IQ scores is explained by crying intensity. Prediction of IQ will not be very accurate. It is nonetheless impressive that behavior soon after birth can even partly predict IQ three years later. Is this observed relationship statistically significant? We must now develop tools for inference in the regression setting. ■

Conditions for regression inference

We can fit a regression line to *any* data relating two quantitative variables, though the results are useful only if the scatterplot shows a linear pattern. Statistical inference requires more detailed conditions. Because the conclusions of inference always concern some *population,* the conditions describe the population and how the data are produced from it. The slope b and intercept a of the least-squares line are *statistics.* That is, we calculated them from the sample data. These statistics would take somewhat different values if we repeated the study with different infants. To do inference, think of a and b as estimates of unknown *parameters* that describe the population of all infants.

Conditions for Regression Inference

We have n observations on an explanatory variable x and a response variable y. Our goal is to study or predict the behavior of y for given values of x.

■ For any fixed value of x, the response y varies according to a **Normal distribution.** Repeated responses y are **independent** of each other.

■ The mean response μ_y has a **straight-line relationship** with x given by a **population regression line**

$$\mu_y = \alpha + \beta x$$

The slope β and intercept α are unknown parameters.

■ The **standard deviation** of y (call it σ) is the same for all values of x. The value of σ is unknown.

There are thus three population parameters that we must estimate from the data: α, β, and σ.

These conditions say that in the population there is an "on the average" straight-line relationship between y and x. The population regression line $\mu_y = \alpha + \beta x$ says that the *mean* response μ_y moves along a straight line as the explanatory variable x changes. We can't observe the population regression line. The values of y that we do observe vary about their means according to a Normal distribution. If we hold x

fixed and take many observations on y, the Normal pattern will eventually appear in a stemplot or histogram. In practice, we observe y for many different values of x, so that we see an overall linear pattern formed by points scattered about the population line. The standard deviation σ determines whether the points fall close to the population regression line (small σ) or are widely scattered (large σ).

Figure 25.2 shows the conditions for regression inference in picture form. The line in the figure is the population regression line. The mean of the response y moves along this line as the explanatory variable x takes different values. The Normal curves show how y will vary when x is held fixed at different values. All the curves have the same σ, so the variability of y is the same for all values of x. You should check the conditions for inference when you do inference about regression. We will see later how to do that.

FIGURE 25.2
The nature of regression data when the conditions for inference are met. The line is the population regression line, which shows how the mean response μ_y changes as the explanatory variable x changes. For any fixed value of x, the observed response y varies according to a Normal distribution having mean μ_y and standard deviation σ.

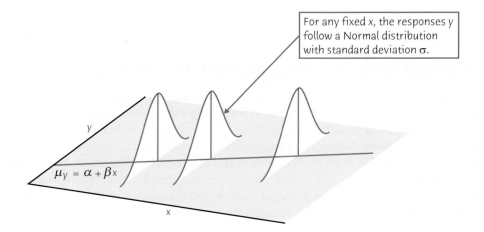

For any fixed x, the responses y follow a Normal distribution with standard deviation σ.

$$\mu_y = \alpha + \beta x$$

Estimating the parameters

The first step in inference is to estimate the unknown parameters α, β, and σ.

Estimating the Population Regression Line

When the conditions for regression are met and we calculate the least-squares line $\hat{y} = a + bx$, the slope b of the least-squares line is an unbiased estimator of the population slope β, and the intercept a of the least-squares line is an unbiased estimator of the population intercept α.

EXAMPLE 25.2   **Crying and IQ: Slope and Intercept**

The data in Figure 25.1 satisfy the condition of scatter about an invisible population regression line reasonably well. The least-squares line is $\hat{y} = 91.27 + 1.493x$. The slope is particularly important. *A slope is a rate of change.* The population slope β says how much higher average IQ is for children with one more peak in their crying measurement. Because $b = 1.493$ estimates the unknown β, we estimate that, on the average, IQ is about 1.5 points higher for each added crying peak.

We need the intercept $a = 91.27$ to draw the line, but it has no statistical meaning in this example. No child had fewer than 9 crying peaks, so we have no data near $x = 0$. We suspect that all normal children would cry when snapped with a rubber band, so that we will never observe $x = 0$. ■

The remaining parameter is the standard deviation σ, which describes the variability of the response y about the population regression line. The least-squares line estimates the population regression line. So the **residuals** estimate how much y varies about the population line. Recall that the residuals are the vertical deviations of the data points from the least-squares line:

$$\text{residual} = \text{observed } y - \text{predicted } y$$
$$= y - \hat{y}$$

There are n residuals, one for each data point. Because σ is the standard deviation of responses about the population regression line, we estimate it by a sample standard deviation of the residuals. We call this sample standard deviation the *regression standard error* to emphasize that it is estimated from data. The residuals from a least-squares line always have mean zero. That simplifies their standard error.

residuals

Regression Standard Error

The **regression standard error** is

$$s = \sqrt{\frac{1}{n-2}\sum \text{residual}^2}$$
$$= \sqrt{\frac{1}{n-2}\sum (y - \hat{y})^2}$$

Use s to estimate the standard deviation σ of responses about the mean given by the population regression line.

Because we use the regression standard error so often, we just call it s. The quantity $\sum(y - \hat{y})^2$ is the sum of the squared deviations of the data points from the line. We average the squared deviations by dividing by $n - 2$, the number of data points less 2. It turns out that if we know $n - 2$ of the n residuals, the other two are determined. That is, $n - 2$ are the **degrees of freedom** of s. We first met the idea of degrees of freedom in the case of the ordinary sample standard deviation of n observations, which has $n - 1$ degrees of freedom. Now we observe two variables rather than one, and the proper degrees of freedom are $n - 2$ rather than $n - 1$.

Calculating s is unpleasant. You must find the predicted response for each x in your data set, then the residuals, and then s. In practice you will use software that does this arithmetic instantly. Nonetheless, here is an example to help you understand the standard error s.

THE JINX!
Athletes are often jinxed. We read of "the rookie of the year jinx," the "cover of *Sports Illustrated* jinx," and many others. That is, athletes who are recognized for an outstanding performance often fail to do as well in the future. No, nature isn't retaliating against them. It's just random variation about their long-term mean performance. They were recognized because they randomly varied above their typical performance, and in the future they return to the mean or randomly vary down from it. If they randomly vary down, they can hope for a "comeback" award the next year.

degrees of freedom

EXAMPLE 25.3   Crying and IQ: Residuals and Standard Error

CRYINGR

Table 25.1 shows that the first infant studied had 10 crying peaks and a later IQ of 87. The predicted IQ for $x = 10$ is

$$\hat{y} = 91.27 + 1.493x$$
$$= 91.27 + 1.493(10) = 106.2$$

The residual for this observation is

$$\text{residual} = y - \hat{y}$$
$$= 87 - 106.2 = -19.2$$

That is, the observed IQ for this infant lies 19.2 points below the least-squares line on the scatterplot.

Repeat this calculation 37 more times, once for each subject. The 38 residuals are

−19.20	−31.13	−22.65	−15.18	−12.18	−15.15	−16.63	−6.18
−1.70	−22.60	−6.68	−6.17	−9.15	−23.58	−9.14	2.80
−9.14	−1.66	−6.14	−12.60	0.34	−8.62	2.85	14.30
9.82	10.82	0.37	8.85	10.87	19.34	10.89	−2.55
20.85	24.35	18.94	32.89	18.47	51.32		

Check the calculations by verifying that the sum of the residuals is zero. It is 0.04, not quite zero, because of roundoff error. Another reason to use software in regression is that roundoff errors in hand calculation can accumulate to make the results inaccurate.

The variance about the line is

$$s^2 = \frac{1}{n-2} \sum \text{residual}^2$$

$$= \frac{1}{38-2}\left[(-19.20)^2 + (-31.13)^2 + \cdots + (51.32)^2\right]$$

$$= \frac{1}{36}(11{,}023.3) = 306.20$$

Finally, the regression standard error is

$$s = \sqrt{306.20} = 17.50 \quad\blacksquare$$

We will study several kinds of inference in the regression setting. The regression standard error s is the key measure of the variability of the responses in regression. It is part of the standard error of all the statistics we will use for inference.

Apply Your Knowledge

© Glowimages RM/age fotostock

25.1 Wine and Cancer in Women. Some studies have suggested that a nightly glass of wine may not only take the edge off a day but also improve health. Is wine good for your health? A study of nearly 1.3 million middle-aged British women examined wine consumption and the relative risk of breast cancer. The relative risk is the proportion of those in the study who drank a given amount of wine and who developed breast cancer divided by the proportion of nondrinkers in the study who developed breast cancer. For example, if 10% of the women in the study who drank 10 grams of wine developed breast cancer and 9% of nondrinkers in the study developed breast cancer, the relative risk of breast cancer for women drinking 10 grams of wine per day would be 10%/9% = 1.11. A relative risk greater than 1 indicates a greater proportion of drinkers in the study developed breast cancer than nondrinkers. Wine intake is the mean wine intake, in grams per day, of all women in the study who drank some wine but less than or equal to 2 drinks per week; who drank between 3 and 6 drinks per week; who drank between 7 and 14 drinks per week; and who drank 15 or more drinks per week. Here are the data (for drinkers only):[2] ▥ CANCER

Wine intake (grams per day) (x)	2.5	8.5	15.5	26.5
Relative risk (y)	1.00	1.08	1.15	1.22

(a) Examine the data. Make a scatterplot with wine intake as the explanatory variable, and find the correlation. There is a strong linear relationship.

(b) Explain in words what the slope β of the population regression line would tell us if we knew it (note that these data represent averages over large numbers of women and are an example of an ecological correlation (see page 137), and one must be careful not to interpret the data as applying to individuals). Based on the data, what are the estimates of β and the intercept α of the population regression line?

(c) Calculate by hand the residuals for the four data points. Check that their sum is 0 (up to roundoff error). Use the residuals to estimate the standard deviation σ that measures variation in the responses (relative risk) about the means given by the population regression line. You have now estimated all three parameters.

Using technology

Basic "two-variable statistics" calculators will find the slope b and intercept a of the least-squares line from keyed-in data. Inference about regression requires in addition the regression standard error s. At this point, software or a graphing calculator that includes procedures for regression inference becomes almost essential for practical work.

Figure 25.3 shows regression output for the data of Table 25.1 from a graphing calculator, two statistical programs, and a spreadsheet program. When we entered the data into the programs, we called the explanatory variable "Crycount." The software outputs use that label. The graphing calculator just uses "x" and "y" to label the explanatory and response variables. You can locate the basic information in all the outputs. The regression slope is $b = 1.4929$, and the regression intercept is $a = 91.268$. The equation of the least-squares line is therefore (after rounding) just as given in Example 25.1. The regression standard error is $s = 17.4987$, and the squared correlation is $r^2 = 0.207$. Both of these results reflect the rather wide scatter of the points in Figure 25.1 about the least-squares line.

Each output contains other information, some of which we will need shortly and some of which we don't need. In fact, we left out some output to save space. Once you know what to look for, you can find what you want in almost any output and ignore what doesn't interest you.

Texas Instruments Graphing Calculator

FIGURE 25.3

Regression of IQ on crying peaks: output from a graphing calculator, two statistical programs, and a spreadsheet program.

FIGURE 25.3
(Continued)

Minitab

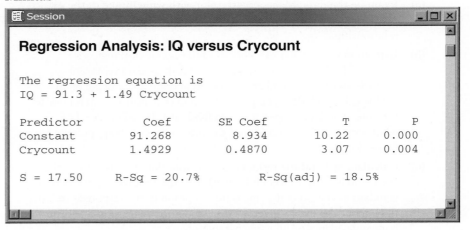

Session

Regression Analysis: IQ versus Crycount

The regression equation is
IQ = 91.3 + 1.49 Crycount

Predictor	Coef	SE Coef	T	P
Constant	91.268	8.934	10.22	0.000
Crycount	1.4929	0.4870	3.07	0.004

S = 17.50   R-Sq = 20.7%   R-Sq(adj) = 18.5%

Excel

Microsoft Excel - Book1

	A	B	C	D	E	F	G
1	SUMMARY OUTPUT						
2							
3	*Regression statistics*						
4	Multiple R	0.4550					
5	R Square	0.2070					
6	Adjusted R Square	0.1850					
7	Standard Error	17.4987					
8	Observations	38					
9							
10		Coefficients	Standard Error	t Stat	P-value	Lower 95%	Upper 95%
11	Intercept	91.2683	8.9342	10.2156	3.5E-12	73.1489	109.3877
12	Crycount	1.4929	0.4870	3.0655	0.004105	0.5052	2.4806
13							

Sheet4 / Sheet1 / Sheet2 / Sheet3 /

CrunchIt!

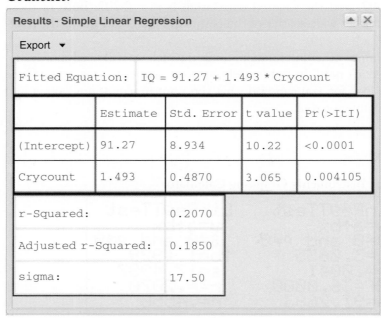

Results - Simple Linear Regression

Export ▾

Fitted Equation: IQ = 91.27 + 1.493 * Crycount

	Estimate	Std. Error	t value	Pr(>ItI)
(Intercept)	91.27	8.934	10.22	<0.0001
Crycount	1.493	0.4870	3.065	0.004105

r-Squared:	0.2070
Adjusted r-Squared:	0.1850
sigma:	17.50

Apply Your Knowledge

25.2 Introspection and Gray Matter. The ability to introspect about self-performance is key to human subjective experience. Accurate introspection requires discriminating correct decisions from incorrect ones, a capacity that varies substantially across individuals. Are individual differences in introspective ability reflected in the anatomy of brain regions responsible for this function? The following data are a measure of introspective ability (labeled Aroc and based on the performance of subjects on a task) and a measure of gray-matter volume (Brodmann area) in the anterior prefrontal cortex of the brain of 29 subjects.[3] INTROSP

Aroc	0.55	0.58	0.59	0.59	0.59	0.61	0.62	0.63	0.63	0.63
Volume	59	62	43	63	83	61	55	57	57	67
Aroc	0.63	0.64	0.65	0.65	0.65	0.65	0.65	0.66	0.66	0.67
Volume	72	62	58	62	65	70	75	60	63	71
Aroc	0.67	0.67	0.68	0.69	0.70	0.70	0.71	0.72	0.75	
Volume	71	80	68	72	66	73	61	80	75	

We want to predict Aroc from volume. Figure 25.4 shows Minitab regression output for these data.

(a) Make a scatterplot suitable for predicting Aroc from volume. What is the squared correlation r^2?

(b) For regression inference, we must estimate the three parameters α, β, and σ. From the output, what are the estimates of these parameters?

(c) What is the equation of the least-squares regression line of Aroc on volume? Add this line to your plot. We will continue the analysis of these data in later exercises.

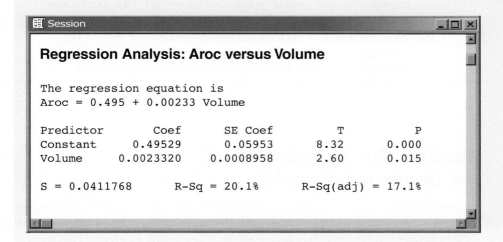

FIGURE 25.4
Minitab output for the introspective ability data, for Exercise 25.2.

25.3 Great Arctic Rivers. One effect of global warming is to increase the flow of water into the Arctic Ocean from rivers. Such an increase may have major effects on the world's climate. Six rivers (Yenisey, Lena, Ob, Pechora, Kolyma, and Severnaya Dvina) drain two-thirds of the Arctic in Europe and Asia. Several of these are among the largest rivers on earth. Table 25.2 presents the total discharge (amount of water flowing from these rivers) each year from 1936 to 2010.[4] Discharge is measured in cubic kilometers of water. Use software to analyze these data. ARCTIC

TABLE 25.2 ARCTIC RIVER DISCHARGE (CUBIC KILOMETERS), 1936 TO 2010

YEAR	DISCHARGE	YEAR	DISCHARGE	YEAR	DISCHARGE	YEAR	DISCHARGE
1936	1721	1955	1656	1974	2000	1993	1845
1937	1713	1956	1721	1975	1928	1994	1902
1938	1860	1957	1762	1976	1653	1995	1842
1939	1739	1958	1936	1977	1698	1996	1849
1940	1615	1959	1906	1978	2008	1997	2007
1941	1838	1960	1736	1979	1970	1998	1903
1942	1762	1961	1970	1980	1758	1999	1970
1943	1709	1962	1849	1981	1774	2000	1905
1944	1921	1963	1774	1982	1728	2001	1890
1945	1581	1964	1606	1983	1920	2002	2085
1946	1834	1965	1735	1984	1823	2003	1780
1947	1890	1966	1883	1985	1822	2004	1900
1948	1898	1967	1642	1986	1860	2005	1930
1949	1958	1968	1713	1987	1732	2006	1910
1950	1830	1969	1742	1988	1906	2007	2270
1951	1864	1970	1751	1989	1932	2008	2078
1952	1829	1971	1879	1990	1861	2009	1900
1953	1652	1972	1736	1991	1801	2010	1813
1954	1589	1973	1861	1992	1793		

(a) Make a scatterplot of river discharge against time. Is there a clear increasing trend? Calculate r^2 and briefly interpret its value. There is considerable year-to-year variation, so we wonder if the trend is statistically significant.

(b) As a first step, find the least-squares line and draw it on your plot. Then find the regression standard error s, which measures scatter about this line. We will continue the analysis in later exercises.

Testing the hypothesis of no linear relationship

Example 25.1 asked, "Do children with higher crying counts tend to have higher IQ?" Data analysis supports this conjecture. But is the positive association statistically significant? That is, is it too strong to often occur just by chance? To answer this question, test hypotheses about the slope β of the population regression line:

$$H_0: \beta = 0$$
$$H_a: \beta > 0$$

A regression line with slope 0 is horizontal. That is, the mean of y does not change at all when x changes. So H_0 says that there is *no linear relationship* between x and y in the population. Put another way, H_0 says that *linear regression of y on x is of no value for predicting y.*

The test statistic is just the standardized version of the least-squares slope b, using the hypothesized value $\beta = 0$ for the mean of b. It is another t statistic. Here are the details.

Significance Test for Regression Slope

To **test the hypothesis** $H_0: \beta = 0$, compute the t statistic

$$t = \frac{b}{\mathrm{SE}_b}$$

In this formula, the standard error of the least-squares slope b is

$$\mathrm{SE}_b = \frac{s}{\sqrt{\Sigma(x - \bar{x})^2}}$$

The sum runs over all observations on the explanatory variable x. In terms of a random variable T having the $t(n - 2)$ distribution, the P-value for a test of H_0 against

$H_a: \beta > 0$ is $P(T \geq t)$

$H_a: \beta < 0$ is $P(T \leq t)$

$H_a: \beta \neq 0$ is $2P(T \geq |t|)$

As advertised, the standard error of b is a multiple of the regression standard error s. The degrees of freedom $n - 2$ are the degrees of freedom of s. Although we give the formula for this standard error, you should not try to calculate it by hand. Regression software gives the standard error SE_b along with b itself.

EXAMPLE 25.4   Crying and IQ: Is the Relationship Significant?

The hypothesis $H_0: \beta = 0$ says that crying has no straight-line relationship with IQ. We conjecture that there is a positive relationship, so we use the one-sided alternative $H_a: \beta > 0$.

Figure 25.1 (page 560) shows that there is a positive relationship, and from Figure 25.3 we see $b = 1.4929$ and $\mathrm{SE}_b = 0.4870$. Thus,

$$t = \frac{b}{\mathrm{SE}_b} = \frac{1.4929}{0.4870} = 3.07$$

so it is not surprising that all the outputs in Figure 25.3 give $t = 3.07$ with two-sided P-value 0.004. The P-value for the one-sided test is half of this, $P = 0.002$. There is very strong evidence that IQ increases as the intensity of crying increases. ■

Apply Your Knowledge

25.4 Wine and Cancer in Women. Exercise 25.1 (page 564) gives data on daily wine consumption and the relative risk of breast cancer in women. Software tells us that the least-squares slope is $b = 0.009012$ with standard error $SE_b = 0.001112$.

(a) What is the t statistic for testing $H_0: \beta = 0$?

(b) How many degrees of freedom does t have? Use Table C to approximate the P-value of t against the one-sided alternative $H_a: \beta > 0$. What do you conclude?

25.5 Great Arctic Rivers: Testing. The most important question we ask of the data in Table 25.2 is this: is the increasing trend visible in your plot (Exercise 25.3, page 567) statistically significant? If so, changes in the Arctic may already be affecting the earth's climate. Use software to answer this question. Give a test statistic, its P-value, and the conclusion you draw from the test. ARCTIC

25.6 Does Fast Driving Waste Fuel? Exercise 4.8 (page 103) gives data on the fuel consumption of a small car at various speeds from 10 to 150 kilometers per hour. Is there significant evidence of straight-line dependence between speed and fuel use? Make a scatterplot and use it to explain the result of your test. FASTDR

Testing lack of correlation

The least-squares slope b is closely related to the correlation r between the explanatory and response variables x and y. In the same way, the slope β of the population regression line is closely related to the correlation between x and y in the population. In particular, the slope is 0 exactly when the correlation is 0.

Testing the null hypothesis $H_0: \beta = 0$ is therefore exactly the same as testing that there is *no correlation* between x and y in the population from which we drew our data. You can use the test for zero slope to test the hypothesis of zero correlation between any two quantitative variables. That's a useful trick.

Because correlation also makes sense when there is no explanatory-response distinction, it is handy to be able to test correlation without doing regression. Table E in the back of the book gives critical values of the sample correlation r under the null hypothesis that the correlation is 0 in the population. Use this table when both variables have at least approximately Normal distributions or when the sample size is large.

EXAMPLE 25.5 Testing Lack of Correlation

AP Photo/David J. Phillip

Figure 25.5 displays two scatterplots that we will use to illustrate testing lack of correlation and also to illustrate once again the need for formal statistical tests. On the left are data from an experiment on the healing of cuts in the limbs of newts. The data are the healing rates (micrometers per hour) for the two front limbs of 18 newts. The right-hand scatterplot shows the first- and second-round scores for the 94 golfers in the 2012 Masters Tournament. (There are fewer than 94 points because of duplicate scores.)

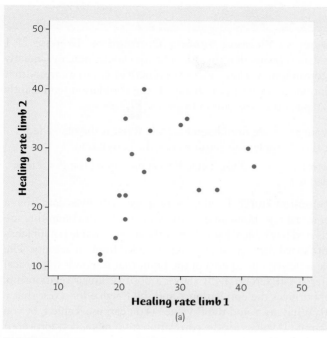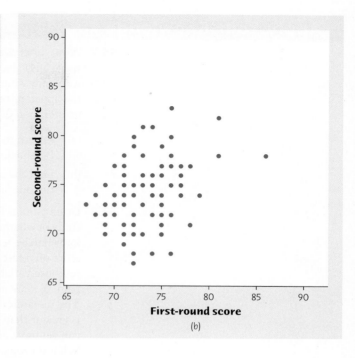

FIGURE 25.5

Two scatterplots for inference about the population correlation, for Example 25.5. (a) Healing rates for the two front limbs of 18 newts. (b) Scores on the first two rounds of the 2012 Masters Tournament.

We will test the hypotheses

$$H_0: \text{population correlation} = 0$$
$$H_a: \text{population correlation} \neq 0$$

for both sets of data. (The Masters scores are all whole numbers, but with $n = 94$ the robustness of t procedures allows their use.) Software gives

```
newts     r = 0.3581   t = 1.5342   P = 0.1445
masters   r = 0.2955   t = 2.97     P = 0.0038
```

The two-sided P-values for the t statistic for testing slope 0 are also the two-sided P-values for testing correlation 0.

Without software, compare the correlation $r = 0.3581$ for newts with the critical values in the $n = 18$ row of Table E. It falls between the table entries for one-tail probabilities 0.05 and 0.10, so the two-sided P-value lies between 0.10 and 0.20. For the Masters data, use the $n = 80$ row of Table E. (There is no table entry for sample size $n = 94$, so we use the next smaller sample size.) The two-sided P-value lies between 0.0025 to 0.005. ■

The evidence for nonzero correlation is strong for the Masters scores ($t = 2.97$, $P = 0.0038$) but not for newts ($t = 1.5$, $P = 0.14$). Yet the correlation for the newts is slightly larger than that for the Masters, and the scatterplots suggest similar linear relationships for both. What happened? The larger sample size for the Masters data is largely responsible. The same r will have a smaller P-value for $n = 94$ than for $n = 18$. Our eyeball impression, even aided by calculating r, can't assess significance. We need the P-value from a formal test to guide us.

Apply Your Knowledge

25.7 Wine and Cancer in Women: Testing Correlation. Exercise 25.1 (page 564) gives data showing that the risk of breast cancer increases linearly with daily wine consumption. There are only 4 observations, so we worry that the apparent relationship may be just chance. Is the correlation significantly greater than 0? Answer this question in two ways. **CANCER**

(a) Return to your t statistic from Exercise 25.4. What is the one-sided P-value for this t? Apply your result to test the correlation.

(b) Find the correlation r, and use Table E to approximate the P-value of the one-sided test.

25.8 Does Social Rejection Hurt? Exercise 4.45 (page 119) gives data from a study of whether social rejection causes activity in areas of the brain that are known to be activated by physical pain. The explanatory variable is a subject's score on a test of "social distress" after being excluded from an activity. The response variable is activity in an area of the brain that responds to physical pain. Your scatterplot (Exercise 4.45) shows a positive linear relationship. The research report gives the correlation r and the P-value for a test that r is greater than 0. What are r and the P-value? (You can use Table E or you can get more accurate P-values for the correlation from regression software.) What do you conclude about the relationship? **REJECT**

Confidence intervals for the regression slope

The slope β of the population regression line is usually the most important parameter in a regression problem. The slope is the rate of change of the mean response as the explanatory variable increases. We often want to estimate β. The slope b of the least-squares line is an unbiased estimator of β. A confidence interval is more useful because it shows how accurate the estimate b is likely to be. The confidence interval for β has the familiar form

$$\text{estimate} \pm t^* \text{SE}_{\text{estimate}}$$

Because b is our estimate, the confidence interval is $b \pm t^* \text{SE}_b$. Here are the details.

Confidence Interval for Regression Slope

A level C **confidence interval for the slope β** of the population regression line is

$$b \pm t^* \text{SE}_b$$

Here t^* is the critical value for the $t(n - 2)$ density curve with area C between $-t^*$ and t^*. The formula for SE_b appears in the box on page 569.

EXAMPLE 25.6   Crying and IQ: Estimating the Slope

All the software outputs in Figure 25.3 (page 566) give the slope $b = 1.4929$ (or $b = 1.493$), and three also give the standard error $\text{SE}_b = 0.4870$. The outputs giving both use a similar arrangement, a table in which each regression coefficient is followed by its standard error. Excel also gives the lower and upper endpoints of the 95% confidence interval for the population slope β, 0.505 and 2.481.

Once we know b and SE_b, it is easy to find the confidence interval. There are 38 data points, so the degrees of freedom are $n - 2 = 36$. Because Table C does not have a row for df $= 36$, we must use either software or the next smaller degrees of freedom in the table, df $= 30$. To use software, enter 36 degrees of freedom. For 95% confidence, enter the cumulative proportion 0.975 that corresponds to upper-tail area 0.025. Minitab gives

```
Student's t distribution with 36 DF
P( X <= x )          x
  0.975        2.02809
```

The 95% confidence interval for the population slope β is

$$b \pm t^*SE_b = 1.4929 \pm (2.02809)(0.4870)$$
$$= 1.4929 \pm 0.9877$$
$$= 0.505 \text{ to } 2.481$$

This agrees with Excel's result. We are 95% confident that mean IQ increases by between about 0.5 and 2.5 points for each additional peak in crying. ■

You can find a confidence interval for the intercept α of the population regression line in the same way, using a and SE_a from the "Constant" line of the Minitab output or the "Intercept" line in Excel and CrunchIt!. We rarely need to estimate α.

Apply Your Knowledge

25.9   Wine and Cancer in Women: Estimating Slope. Exercise 25.1 gives data on wine consumption and the risk of breast cancer. Software tells us that the least-squares slope is $b = 0.009012$ with standard error $SE_b = 0.001112$. Because there are only 4 observations, the observed slope b may not be an accurate estimate of the population slope β. Give a 90% confidence interval for β.

25.10   Introspection and Gray Matter: Estimating Slope. Exercise 25.2 (page 567) gives data on introspective ability and gray-matter volume of the brains of subjects. We want a 95% confidence interval for the slope of the population regression line. Starting from the information in the Minitab output in Figure 25.4, find this interval. Say in words what the slope of the population regression line tells us about the relationship between Aroc and gray-matter volume.

25.11   Great Arctic Rivers: Estimating Slope. Use the data in Table 25.2 (page 568) to give a 90% confidence interval for the slope of the population regression of Arctic river discharge on year. Does this interval convince you that discharge is actually increasing over time? Explain your answer.

ARCTIC

Inference about prediction

One of the most common reasons to fit a line to data is to predict the response to a particular value of the explanatory variable. This is another setting for regression inference: we want not simply a prediction, but a prediction with a margin of error that describes how accurate the prediction is likely to be.

EXAMPLE 25.7   Beer and Blood Alcohol

STATE: The EESEE story "Blood Alcohol Content" describes a study in which 16 student volunteers at the Ohio State University drank a randomly assigned number of cans of beer. Thirty minutes later, a police officer measured their blood alcohol content (BAC) in grams of alcohol per deciliter of blood. Here are the data:[5]

Student	1	2	3	4	5	6	7	8
Beers	5	2	9	8	3	7	3	5
BAC	0.10	0.03	0.19	0.12	0.04	0.095	0.07	0.06
Student	9	10	11	12	13	14	15	16
Beers	3	5	4	6	5	7	1	4
BAC	0.02	0.05	0.07	0.10	0.085	0.09	0.01	0.05

The students were equally divided between men and women and differed in weight and usual drinking habits. Because of this variation, many students don't believe that number of drinks predicts blood alcohol well. Steve thinks he can drive legally 30 minutes after he finishes drinking 5 beers. The legal limit for driving is BAC 0.08 in all states. We want to predict Steve's blood alcohol content, using no information except that he drinks 5 beers.

PLAN: Regress BAC on number of beers. Use the regression line to predict Steve's BAC. Give a margin of error that allows us to have 95% confidence in our prediction.

SOLVE: The scatterplot in Figure 25.6 and the regression output in Figure 25.7 show that student opinion is wrong: number of beers predicts blood alcohol content quite well. In fact, $r^2 = 0.80$, so that number of beers explains 80% of the observed variation in BAC. To predict Steve's BAC after 5 beers, use the equation of the regression line:

$$\hat{y} = -0.0127 + 0.0180x$$
$$= -0.0127 + 0.0180(5) = 0.077$$

That's dangerously close to the legal limit 0.08. What about 95% confidence? The "Predicted Values" part of the output in Figure 25.7 shows *two* 95% intervals. Which should we use? ■

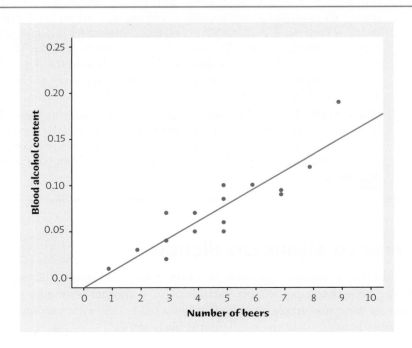

FIGURE 25.6

Scatterplot of students' blood alcohol content against the number of cans of beer consumed, with the least-squares regression line, for Example 25.7.

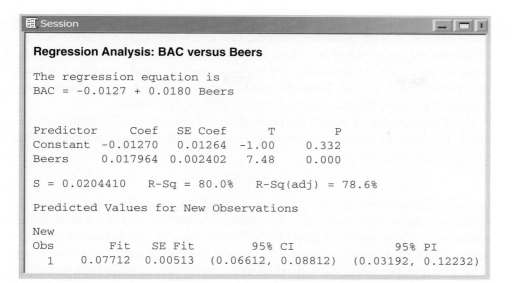

 To decide which interval to use, you must answer this question: do you want to predict the *mean* BAC for *all students* who drink 5 beers, or do you want to predict the BAC of *one individual student* who drinks 5 beers? *Both of these predictions may be interesting, but they are two different problems.* The actual prediction is the same, $\hat{y} = 0.077$. But the margin of error is different for the two kinds of prediction. Individual students who drink 5 beers don't all have the same BAC. So we need a larger margin of error to pin down Steve's result than to estimate the mean BAC for all students who have 5 beers.

Write the given value of the explanatory variable x as x^*. In Example 25.7, $x^* = 5$. The distinction between predicting a single outcome and predicting the mean of all outcomes when $x = x^*$ determines what margin of error is correct. To emphasize the distinction, we use different terms for the two intervals.

- To estimate the *mean* response, we use a *confidence interval.* It is an ordinary confidence interval for the mean response when x has the value x^*, which is $\mu_y = \alpha + \beta x^*$. This is a parameter, a fixed number whose value we don't know.

- To estimate an *individual* response y, we use a **prediction interval.** A prediction interval estimates a single random response y rather than a parameter like μ_y. The response y is not a fixed number. If we took more observations with $x = x^*$, we would get different responses.

 IS REGRESSION GARBAGE?

No—but garbage can be the setting for regression. The U.S. Census Bureau once asked if weighing a neighborhood's garbage would help count its people. So 63 households had their garbage sorted and weighed. It turned out that pounds of plastic in the trash gave the best garbage prediction of the number of people in a neighborhood. The margin of error for a 95% prediction interval in a neighborhood of about 100 households, based on five weeks' worth of garbage, was about ± 2.5 people. Alas, that is not accurate enough to help the U.S. Census Bureau.

prediction interval

EXAMPLE 25.8   Beer and Blood Alcohol: Conclusion

Steve is one individual, so we must use the prediction interval. The output in Figure 25.7 helpfully labels the confidence interval as "95% CI" and the prediction interval as "95% PI." We are 95% confident that Steve's BAC after 5 beers will lie between 0.032 and 0.122. The upper part of that range will get him arrested if he drives. The 95% confidence interval for the mean BAC of all students who drink 5 beers is much narrower, 0.066 to 0.088. ■

The meaning of a prediction interval is very much like the meaning of a confidence interval. A 95% prediction interval, like a 95% confidence interval, is right 95% of the time in repeated use. "Repeated use" now means that we take an observation on y for each of the n values of x in the original data and then take one more observation y with $x = x^*$. Form the prediction interval from the n observations, then see if it covers the one more y. It will in 95% of all repetitions.

The interpretation of prediction intervals is a minor point. The main point is that it is harder to predict one response than to predict a mean response. Both intervals have the usual form

$$\hat{y} \pm t^*SE$$

but the prediction interval is wider than the confidence interval because individuals are more variable than averages. As a further illustration, consider a professional athlete such as the basketball player Kobe Bryant. You can find his career statistics online. During his career as a starter, his season scoring *average* ranges from about 20 to 35 points per game, but individual game performances range from just a few points to a career high of 81 points. Individual game performances of professional athletes are more variable than season average performances. So the margin of error for predictions of individual game performances will be larger than for predictions of season averages. You will rarely need to know the details because software automates the calculation, but here they are.

Confidence and Prediction Intervals for Regression Response

A level C **confidence interval for the mean response** μ_y when x takes the value x^* is

$$\hat{y} \pm t^*SE_{\hat{\mu}}$$

The standard error $SE_{\hat{\mu}}$ is

$$SE_{\hat{\mu}} = s\sqrt{\frac{1}{n} + \frac{(x^* - \bar{x})^2}{\Sigma(x - \bar{x})^2}}$$

A level C **prediction interval for a single observation** y when x takes the value x^* is

$$\hat{y} \pm t^*SE_{\hat{y}}$$

The standard error for prediction $SE_{\hat{y}}$ is

$$SE_{\hat{y}} = s\sqrt{1 + \frac{1}{n} + \frac{(x^* - \bar{x})^2}{\Sigma(x - \bar{x})^2}}$$

In both intervals, t^* is the critical value for the $t(n - 2)$ density curve with area C between $-t^*$ and t^*.

There are two standard errors: $SE_{\hat{\mu}}$ for estimating the mean response μ_y and $SE_{\hat{y}}$ for predicting an individual response y. The only difference between the two standard errors is the extra 1 under the square root sign in the standard error for prediction. The extra 1 makes the prediction interval wider. Both standard errors are multiples of the regression standard error s. The degrees of freedom are again $n - 2$, the degrees of freedom of s.

Apply Your Knowledge

25.12 Wine and Cancer in Women: Prediction. Exercise 25.1 gives data on wine consumption and the risk of breast cancer. For a new group of women who drink an average of 10 grams of wine per day, predict their relative risk of breast cancer.

(a) Figure 25.8 is part of the output from Minitab for prediction when $x^* = 10.0$. Which interval in the output is the proper 95% interval for predicting the relative risk?

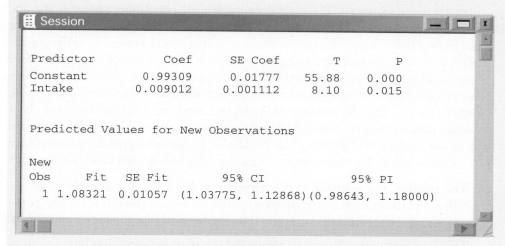

```
Session                                                    _ □ X

Predictor            Coef      SE Coef        T        P
Constant          0.99309     0.01777     55.88     0.000
Intake           0.009012    0.001112      8.10     0.015


Predicted Values for New Observations

New
Obs     Fit    SE Fit        95% CI                 95% PI
 1   1.08321  0.01057   (1.03775, 1.12868)(0.98643, 1.18000)
```

FIGURE 25.8
Partial Minitab output for regressing relative risk of breast cancer on mean daily intake of wine, for Exercise 25.12.

(b) Minitab gives only one of the two standard errors used in prediction. It is $SE_{\hat{\mu}}$, the standard error for estimating the mean response. Use this fact along with the output to give a 90% confidence interval for the mean relative risk of breast cancer in all women who drink an average of 10 grams of wine per day.

25.13 Introspection and Gray Matter: Prediction. Analysis of the data in Exercise 25.2 (page 567) shows that the relationship of gray-matter volume and introspective ability, as measured by Aroc, is roughly linear. We might want to predict the mean introspective ability of a person with a gray matter volume of 65. Here is the Minitab output for prediction when $x^* = 65$:

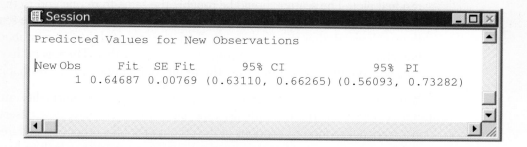

```
Session                                                  _ □ X
Predicted Values for New Observations

New Obs     Fit   SE Fit       95% CI               95% PI
     1   0.64687 0.00769  (0.63110, 0.66265) (0.56093, 0.73282)
```

(a) Use the regression line from Figure 25.4 (page 567) to verify that "Fit" is the predicted value for $x^* = 65$. (Start with the results in the "Coef" column of Figure 25.4 to reduce roundoff error.)

(b) What is the 95% interval we want?

Checking the conditions for inference

You can fit a least-squares line to any set of explanatory-response data when both variables are quantitative. If the scatterplot doesn't show a roughly linear pattern, the fitted line may be almost useless. But it is still the line that fits the data best in the least-squares sense. To use regression inference, however, the data must satisfy additional conditions. *Before you can trust the results of inference, you must check the conditions for inference one by one.* There are ways to deal with violations of any of the conditions. If you see a clear violation, get expert advice.

Although the conditions for regression inference are a bit elaborate, it is not hard to check for major violations. The conditions involve the population regression line and the deviations of responses from this line. We can't observe the population line, but the least-squares line estimates it, and the residuals estimate the deviations from the population line. *You can check all the conditions for regression inference by looking at graphs of the residuals.* This is what we recommend in practice (and a failure to do so can sometimes lead to unjustified conclusions), and most regression software will calculate and save the residuals for you. Start by making a stemplot or histogram of the residuals and also a *residual plot,* **residual plot,** a plot of the residuals against the explanatory variable x, with a horizontal line at the "residual = 0" position. The "residual = 0" line represents the position of the least-squares line in the scatterplot of y against x. Let's look at each condition in turn.

- **The relationship is linear in the population.** Look for curved patterns or other departures from a straight-line overall pattern in the residual plot. You can also use the original scatterplot, but the residual plot magnifies any effects.

- **The response varies Normally about the population regression line.** Because different y-values usually come from different x-values, the responses themselves need not be Normal. It is the deviations from the population line—estimated by the residuals—that must be Normal. Check for clear skewness or other major departures from Normality in your stemplot or histogram of the residuals.

- **Observations are independent.** In particular, repeated observations on the same individual are not allowed. You should not use ordinary regression to make inferences about the growth of a single child over time, for example. Signs of dependence in the residual plot are a bit subtle, so we usually rely on common sense.

- **The standard deviation of the responses is the same for all values of x.** Look at the scatter of the residuals above and below the "residual = 0" line in the residual plot. The scatter should be roughly the same from one end to the other. You will sometimes find that, as the response y gets larger, so does the scatter of the residuals. Rather than remaining fixed, the standard deviation σ about the line changes with x as the mean response changes with x. There is no fixed σ for s to estimate. You cannot trust the results of inference when this happens.

You will always see some irregularity when you look for Normality and fixed standard deviation in the residuals, especially when you have few observations. Don't overreact to minor violations of the conditions. Like other t procedures, inference for regression is (with one exception) not very sensitive to lack of Normality, especially when we have many observations. Do beware of influential observations, which can greatly affect the results of inference.

The exception is the prediction interval for a single response y. This interval relies on Normality of individual observations, not just on the approximate Normality of statistics like the slope a and intercept b of the least-squares line. The statistics a

and *b* become more Normal as we take more observations. This contributes to the robustness of regression inference, but it isn't enough for the prediction interval. We will not study methods that carefully check Normality of the residuals, so *you should regard prediction intervals as rough approximations.*

EXAMPLE 25.9   Climate Change Chases Fish North

ANGFISH

STATE: As the climate grows warmer, we expect many animal species to move toward the poles in an attempt to maintain their preferred temperature range. Do data on fish in the North Sea confirm this expectation? Table 25.3 gives data for 25 years on mean winter temperatures at the bottom of the North Sea (degrees Celsius) and the center of the distribution of anglerfish in degrees of north latitude.[6]

TABLE 25.3 WINTER TEMPERATURE (°C) AND ANGLERFISH LATITUDE, 1977 TO 2001

YEAR	TEMP.	LATITUDE	YEAR	TEMP.	LATITUDE	YEAR	TEMP.	LATITUDE
1977	6.26	57.20	1986	6.52	57.72	1994	7.02	58.71
1978	6.26	57.96	1987	6.68	57.83	1995	7.09	58.07
1979	6.27	57.65	1988	6.76	57.87	1996	7.13	58.49
1980	6.31	57.59	1989	6.78	57.48	1997	7.15	58.28
1981	6.34	58.01	1990	6.89	58.13	1998	7.29	58.49
1982	6.32	59.06	1991	6.90	58.52	1999	7.34	58.01
1983	6.37	56.85	1992	6.93	58.48	2000	7.57	58.57
1984	6.39	56.87	1993	6.98	57.89	2001	7.65	58.90
1985	6.42	57.43						

© *Jurgen Freund/naturepl.com*

PLAN: Regress latitude on temperature. Look for a positive linear relationship, and assess its significance. Be sure to check the conditions for regression inference.

SOLVE: The scatterplot in Figure 25.9 shows a clear positive linear relationship. The solid line in the plot is the least-squares regression line of the center of the fish

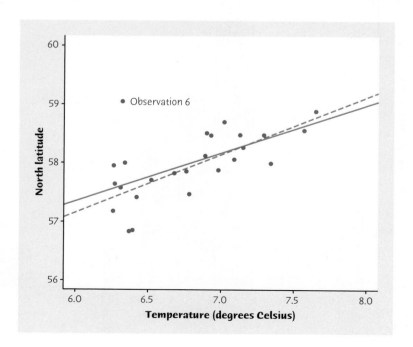

FIGURE 25.9
Scatterplot of the latitude of the center of the distribution of anglerfish in the North Sea against mean winter temperature at the bottom of the sea, for Example 25.9. The two regression lines are for the data with (solid) and without (dashed) Observation 6.

distribution (north latitude) on winter ocean temperature. Software shows that the slope is $b = 0.818$. That is, each degree of ocean warming moves the fish about 0.8 degree of latitude farther north. The t statistic for testing $H_0: \beta = 0$ is $t = 3.6287$ with one-sided P-value $P = 0.0007$. There is very strong evidence that the population slope is positive, $\beta > 0$.

CONCLUDE: The data give highly significant evidence that anglerfish have moved north as the ocean has grown warmer. Before relying on this conclusion, we must check the conditions for inference. ■

The software that did the regression calculations also finds the 25 residuals. In the same order as the observations in Example 25.9, they are

```
-0.3731   0.3869   0.0687  -0.0240   0.3714   1.4378  -0.8131
-0.8095  -0.2740  -0.0658  -0.0867  -0.1121  -0.5185   0.0415
 0.4234   0.3588  -0.2721   0.5152  -0.1821   0.2052  -0.0211
 0.0743  -0.4466  -0.0747   0.1899
```

Begin by making two graphs of the residuals. Figure 25.10 is a histogram of the residuals. Figure 25.11 is the residual plot, a plot of the residuals against the explanatory variable, sea-bottom temperature. The "residual = 0" line marks the position of the regression line. Patterns in residual plots are often easier to see if you use a wider vertical scale than your software's default plot, and we suggest you do so if possible. Both graphs show that Observation 6 is a high outlier. Let's check the conditions for regression inference.

■ **Linear relationship.** The scatterplot in Figure 25.9 and the residual plot in Figure 25.11 both show a linear relationship except for the outlier.

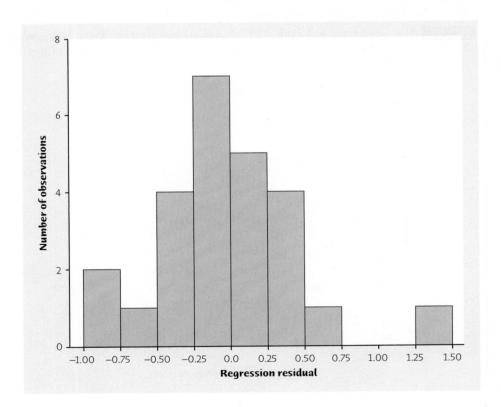

FIGURE 25.10

Histogram of the residuals from the regression of latitude on temperature in Example 25.9.

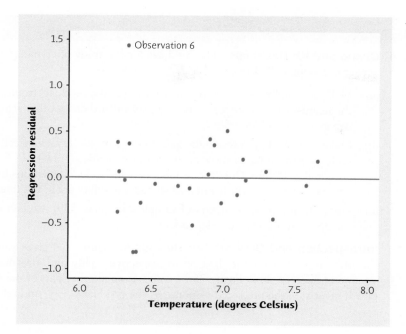

FIGURE 25.11
Residual plot for the regression of latitude on temperature in Example 25.9.

- **Normal residuals.** The histogram in Figure 25.10 is roughly symmetric and single peaked. There are no important departures from Normality except for the outlier.

- **Independent observations.** The observations were taken a year apart, so we are willing to regard them as close to independent. The residual plot shows no obvious pattern of dependence, such as runs of points all above or all below the line.

- **Constant standard deviation.** Again excepting the outlier, the residual plot shows no unusual variation in the scatter of the residuals above and below the line as x varies.

The outlier is the only serious violation of the conditions for inference. How influential is the outlier? The dashed line in Figure 25.9 is the regression line without Observation 6. Because there are several other observations with similar values of temperature, dropping Observation 6 does not move the regression line very much. *Even though the outlier is not very influential for the regression line, it influences regression inference because of its effect on the regression standard error.* The standard error is $s = 0.4734$ with Observation 6 and $s = 0.3622$ without it. When we omit the outlier, the t statistic changes from $t = 3.6287$ to $t = 5.5599$, and the one-sided P-value changes from $P = 0.0007$ to $P < 0.00001$. Fortunately, the outlier does not affect the conclusion we drew from the data. Dropping Observation 6 makes the test for the population slope *more* significant and *increases* the percent of variation in fish location explained by ocean temperature.

One more caution about inference in this example: as usual in an observational study, the possibility of lurking variables makes us hesitant to conclude that rising temperature is *causing* anglerfish to move north. Ocean temperature was steadily rising during these years. The effect on fish latitude of any lurking variable that increased over time—perhaps increased commercial fishing—is confounded with the effect of temperature.

Apply Your Knowledge

25.14 Crying and IQ: Residuals. The residuals for the study of crying and IQ appear in Example 25.3 (page 563). CRYINGR

(a) Make a stemplot to display the distribution of the residuals (round to the nearest whole number). Are there strong outliers or other signs of departures from Normality?

(b) Make a residual plot, residuals against crying peaks. Try a vertical scale of −60 to 60 to show patterns more clearly. Draw the "residual = 0" line. Does the residual plot show clear deviations from a linear pattern or clearly unequal spread/variability about the line?

(c) Using the information given in Example 25.1 (page 559), explain why the 38 observations are independent.

25.15 Introspection and Gray Matter: Residuals. Figure 25.4 gives part of the Minitab output for the data on introspective ability and gray-matter volume in Exercise 25.2. Figure 25.12 comes from another part of the output. It gives x, y, the predicted response $\hat{y}$, the residual $y - \hat{y}$, and related

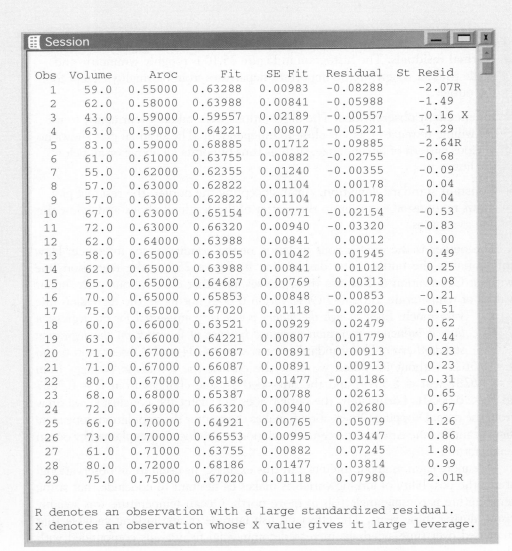

Obs	Volume	Aroc	Fit	SE Fit	Residual	St Resid
1	59.0	0.55000	0.63288	0.00983	−0.08288	−2.07R
2	62.0	0.58000	0.63988	0.00841	−0.05988	−1.49
3	43.0	0.59000	0.59557	0.02189	−0.00557	−0.16 X
4	63.0	0.59000	0.64221	0.00807	−0.05221	−1.29
5	83.0	0.59000	0.68885	0.01712	−0.09885	−2.64R
6	61.0	0.61000	0.63755	0.00882	−0.02755	−0.68
7	55.0	0.62000	0.62355	0.01240	−0.00355	−0.09
8	57.0	0.63000	0.62822	0.01104	0.00178	0.04
9	57.0	0.63000	0.62822	0.01104	0.00178	0.04
10	67.0	0.63000	0.65154	0.00771	−0.02154	−0.53
11	72.0	0.63000	0.66320	0.00940	−0.03320	−0.83
12	62.0	0.64000	0.63988	0.00841	0.00012	0.00
13	58.0	0.65000	0.63055	0.01042	0.01945	0.49
14	62.0	0.65000	0.63988	0.00841	0.01012	0.25
15	65.0	0.65000	0.64687	0.00769	0.00313	0.08
16	70.0	0.65000	0.65853	0.00848	−0.00853	−0.21
17	75.0	0.65000	0.67020	0.01118	−0.02020	−0.51
18	60.0	0.66000	0.63521	0.00929	0.02479	0.62
19	63.0	0.66000	0.64221	0.00807	0.01779	0.44
20	71.0	0.67000	0.66087	0.00891	0.00913	0.23
21	71.0	0.67000	0.66087	0.00891	0.00913	0.23
22	80.0	0.67000	0.68186	0.01477	−0.01186	−0.31
23	68.0	0.68000	0.65387	0.00788	0.02613	0.65
24	72.0	0.69000	0.66320	0.00940	0.02680	0.67
25	66.0	0.70000	0.64921	0.00765	0.05079	1.26
26	73.0	0.70000	0.66553	0.00995	0.03447	0.86
27	61.0	0.71000	0.63755	0.00882	0.07245	1.80
28	80.0	0.72000	0.68186	0.01477	0.03814	0.99
29	75.0	0.75000	0.67020	0.01118	0.07980	2.01R

R denotes an observation with a large standardized residual.
X denotes an observation whose X value gives it large leverage.

FIGURE 25.12

Residuals from Minitab for Exercise 25.15. The table gives the predicted value ("Fit") and the residual for each observation.

quantities for each of the 29 observations. Most statistical software provides similar output. Examine the conditions for regression inference one by one. This example illustrates mild violations of the conditions that did not prevent the researchers from doing inference. [DATA] INTROSPR

(a) **Linear relationship.** Your scatterplot and r^2 from Exercise 25.2 show that the relationship is roughly linear. Plot the residuals against volume. Are any deviations from a straight line apparent?

(b) **Normal variation about the line.** Make a stemplot of the residuals (round to the third decimal place and don't forget that −0.00 and 0.00 are separate stems). With only 29 observations, a small amount of skew is not disturbing. Minitab suggests that Observations 1, 5, and 29 may be outliers. Does your plot confirm or refute this suggestion?

(c) **Independent observations.** The data come from 29 different subjects who were each measured separately. Can you conclude the observations are independent?

(d) **Spread/variability about the line stays the same.** Is there any evidence that the spread/variability may be larger at one end?

CHAPTER 25 SUMMARY

Chapter Specifics

■ **Least-squares regression** fits a straight line to data to predict a response variable y from an explanatory variable x. Inference about regression requires more conditions.

■ The **conditions for regression inference** say that there is a **population regression line** $\mu_y = \alpha + \beta x$ that describes how the mean response varies as x changes. The observed response y for any x has a Normal distribution with mean given by the population regression line and with the same standard deviation σ for any value of x. Observations on y are independent.

■ The **parameters to be estimated** are the intercept α and the slope β of the population regression line and also the standard deviation σ. The slope a and intercept b of the least-squares line estimate α and β. Use the **regression standard error s** to estimate σ.

■ The regression standard error s has $n - 2$ **degrees of freedom.** All t procedures for regression inference have $n - 2$ degrees of freedom.

■ To test **the hypothesis that the slope is zero in the population,** use the t statistic $t = b/\text{SE}_b$. This null hypothesis says that straight-line dependence on x has no value for predicting y. In practice, use software to find the slope b of the least-squares line, its standard error SE_b, and the t statistic.

■ The t test for regression slope is also a test for **the hypothesis that the population correlation between x and y is zero.** To do this test without software, use the sample correlation r and Table E.

■ **Confidence intervals for the slope** of the population regression line have the form $b \pm t^*\text{SE}_b$.

■ **Confidence intervals for the mean response** when x has value x^* have the form $\hat{y} \pm t^*\text{SE}_{\hat{\mu}}$. **Prediction intervals** for an individual future response y have a similar form with a larger standard error, $\hat{y} \pm t^*\text{SE}_{\hat{y}}$. Software often gives these intervals.

STATISTICS IN SUMMARY

Here are the most important skills you should have acquired from reading this chapter.

A. Preliminaries

1. Make a scatterplot to show the relationship between an explanatory and a response variable.

2. Use a calculator or software to find the correlation and the equation of the least-squares regression line.

3. Recognize which type of inference you need in a particular regression setting.

B. Inference Using Software Output

1. Explain in any specific regression setting the meaning of the slope β of the population regression line.

2. Understand software output for regression. Find in the output the slope and intercept of the least-squares line, their standard errors, and the regression standard error.

3. Use that information to carry out tests of H_0: $\beta = 0$ and calculate confidence intervals for β.

4. Explain the distinction between a confidence interval for the mean response and a prediction interval for an individual response.

5. If software gives output for prediction, use that output to give either confidence or prediction intervals.

C. Checking the Conditions for Regression Inference

1. Make a stemplot or histogram of the residuals and look for strong departures from Normality.

2. Make a residual plot and look for departures from a linear pattern or unequal spread/variability about the "residual = 0" line.

3. Ask whether the study design suggests that observations are independent.

Link It

In Chapters 4 and 5 we studied scatterplots, correlation, and the least-squares regression line as methods for exploring data. In Chapter 5 we mentioned that we must exercise caution in how we interpret any relationships we observe through such exploratory analyses. We also mentioned that such interpretations rest on the assumption that the relationship is valid in some broader sense. And we promised to explore this more carefully later in this book. In this chapter we did so by considering inference for regression. Inference for regression allows us to determine whether the relationship we observe in a scatterplot is valid for some larger population. It also allows us to attach margins of error to our estimates of the slope and intercept, as well as to predictions based on the least-squares regression line.

In Chapters 4 and 5 we discussed several cautions about correlation and the least-squares regression line. These same cautions apply to inference for regression. We discussed a systematic approach using the residuals and what we know of the study design to determine if the assumptions behind inference for regression are reasonable.

This chapter considers inference for relationships between two quantitative variables. The previous chapter considered inference for relationships between two categorical variables. In the next chapter, we consider inference for relationships between a quantitative and a categorical variable—in particular, for deciding whether the mean of a response is the same for more than two categories.

CHECK YOUR SKILLS

Florida reappraises real estate every year, so the county appraiser's Web site lists the current "fair market value" of each piece of property. Property usually sells for somewhat more than the appraised market value. Here are the appraised market values and actual selling prices (in thousands of dollars) of condominium units sold in a beachfront building in a 93-month period between 2003 and 2010:[7]

© Franz Marc Frei/Corbis

SELLING PRICE	APPRAISED VALUE	MONTH	SELLING PRICE	APPRAISED VALUE	MONTH	SELLING PRICE	APPRAISED VALUE	MONTH
825	626	0	1900	1016	26	875	806	64
590	492	1	1600	1040	28	1510	1241	70
1075	930	3	980	442	34	1375	813	70
890	790	9	940	771	37	560	496	73
845	648	13	850	715	47	1050	774	79
1100	942	14	1100	997	54	605	470	86
715	345	15	1164	953	59	675	545	88
1325	1032	19	1425	922	64	693	690	93
700	556	21	1865	1190	64			
1322	879	26	1450	610	64			

Here is part of the Minitab output for regressing selling price on appraised value, along with prediction for a unit with appraised value $800,000:

```
Predictor    Coef  SE Coef    T       P
Constant     86.0   156.8   0.55    0.588
appraisal  1.2699  0.1938   6.55    0.000

S = 235.410  R-Sq = 62.3%  R-Sq(adj) = 60.8%

Predicted Values for New Observations

New
Obs   Fit SE Fit     95% CI        95% PI
  1 1101.9   44.7 (1010.0, 1193.9) (609.4, 1594.5)
```

Exercises 25.16 to 25.24 are based on this information.

25.16 The equation of the least-squares regression line for predicting selling price from appraised value is

(a) price = 86.0 + 1.2699 × appraised value.
(b) price = 1.2699 + 86.0 × appraised value.
(c) price = 156.8 + 0.1938 × appraised value.

25.17 What is the correlation between selling price and appraised value?

(a) 0.789   (b) 0.623   (c) 0.388

25.18 The slope β of the population regression line describes

(a) the average selling price in a population of units when a unit's appraised value is 0.
(b) the average increase in selling price in a population of units when appraised value increases by $1000.

(c) the exact increase in the selling price of an individual unit when its appraised value increases by $1000.

25.19 Is there significant evidence that selling price increases as appraised value increases? To answer this question, test the hypotheses

(a) $H_0: \beta = 0$ versus $H_a: \beta > 0$.
(b) $H_0: \beta = 0$ versus $H_a: \beta \neq 0$.
(c) $H_0: \alpha = 0$ versus $H_a: \alpha > 0$.

25.20 Minitab shows that the P-value for this test is

(a) 0.588.   (b) 0.1938   (c) less than 0.001.

25.21 The regression standard error for these data is

(a) 0.1938.   (b) 156.8.   (c) 235.41.

25.22 Confidence intervals and tests for these data use the t distribution with degrees of freedom

(a) 28.   (b) 27.   (c) 26.

25.23 A 95% confidence interval for the population slope β is

(a) 1.2699 ± 0.3306.
(b) 1.2699 ± 322.3808.
(c) 1.2699 ± 0.3985.

25.24 Louisa owns a unit in this building appraised at $800,000. The Minitab output includes prediction for this appraised value. She can be 95% confident that her unit would sell for between

(a) $609,400 and $1,594,500.
(b) $1,010,000 and $1,193,900.
(c) $1,057,200 and $1,146,6900.

CHAPTER 25 EXERCISES

25.25 Genetically engineered cotton. A strain of genetically engineered cotton, known as Bt cotton, is resistant to certain insects, which re-

Softdreams/Dreamstime.com

sults in larger yields of cotton. Farmers in northern China have increased the number of acres planted in Bt cotton. Because Bt cotton in resistant to certain pests, farmers have also reduced their use of insecticide. Scientists in China were interested in the long-term effects of Bt cotton cultivation and decreased insecticide use on insect populations that are not affected by Bt cotton. One such insect is the mirid bug. Scientists measured the number of mirid bugs per 100 plants and the proportion of Bt cotton planted at 38 locations in northern China for the 12-year period from 1997 and 2008. The scientists reported a regression analysis as follows:[8]

number of mirid bugs per 100 plants

$$= 0.54 + 6.81 \times \text{Bt cotton planting proportion}$$
$$r^2 = 0.90, \quad P < 0.0001$$

(a) What does the slope $b = 6.81$ say about the relation between Bt cotton planting proportion and number of mirid bugs per 100 plants?

(b) What does $r^2 = 0.90$ add to the information given by the equation of the least-squares line?

(c) What null and alternative hypotheses do you think the P-value refers to? What does this P-value tell you?

(d) Does the large value of r^2 and the small P-value indicate that increasing the proportion of acres planted in Bt cotton causes an increase in mirid bugs?

Exercise 7.51 (page 187) gives data from a study of the "gate velocity" of molten metal that experienced foundry workers choose based on the thickness of the aluminum piston being cast. Gate velocity is measured in feet per second, and the piston wall thickness is in inches. A scatterplot (you need not make one) shows no outliers and a moderately strong positive linear relationship. Figure 25.13 displays part of the Minitab regression output. Exercises 25.26 to 25.28 analyze these data.

FIGURE 25.13

Minitab output for the regression of gate velocity on piston thickness in casting aluminum parts, for Exercises 25.26 to 25.28.

Session

Regression Analysis: Veloc versus Thick

The regression equation is
Veloc = 70.4 + 275 Thick

Predictor	Coef	SE Coef	T	P
Constant	70.44	52.90		
Thick	274.78	88.18		

S = 56.3641   R-Sq = 49.3%   R-Sq(adj) = 44.2%

Obs	Thick	Veloc	Fit	SE Fit	Residual	St Resid
1	0.248	123.8	138.6	32.8	-14.8	-0.32
2	0.359	223.9	169.1	24.8	54.8	1.08
3	0.366	180.9	171.0	24.3	9.9	0.19
4	0.400	104.8	180.3	22.2	-75.5	-1.46
5	0.524	228.6	214.4	16.8	14.2	0.26
6	0.552	223.8	222.1	16.4	1.7	0.03
7	0.628	326.2	243.0	17.0	83.2	1.55
8	0.697	302.4	262.0	19.7	40.4	0.77
9	0.697	145.2	262.0	19.7	-116.8	-2.21R
10	0.752	263.1	277.1	22.8	-14.0	-0.27
11	0.806	302.4	291.9	26.4	10.5	0.21
12	0.821	302.4	296.0	27.4	6.4	0.13

R denotes an observation with a large standardized residual.

Predicted Values for New Observations

New Obs	Fit	SE Fit	90% CI	90% PI
1	207.8	17.4	(176.2, 239.4)	(100.9, 314.8)

25.26 **Casting aluminum: Is there a relationship?** Figure 25.13 leaves out the t statistics and their P-values. Based on the information in the output, test the hypothesis that there is no straight-line relationship between thickness and gate velocity. State hypotheses, give a test statistic and its approximate P-value, and state your conclusion in the context of the problem.

25.27 **Casting aluminum: intervals.** The output in Figure 25.13 includes prediction for piston wall thickness $x^* = 0.5$ inch. Use the output to give 90% intervals for

(a) the slope of the population regression line of gate velocity on piston thickness.
(b) the average gate velocity for a type of piston with thickness 0.5 inch.

25.28 **Casting aluminum: residuals.** The output in Figure 25.13 includes a table of the x and y variables, the fitted values $\hat{y}$ for each x, the residuals, and some related quantities.

(a) Plot the residuals against thickness (the explanatory variable). Use vertical scale -200 to 200 so that the pattern is clearer. Add the "residual = 0" line. Does your plot show a systematically nonlinear relationship? Does it show systematic change in the spread/variability about the regression line?
ALUMRES

(b) Make a histogram of the residuals. Minitab identifies the residual for Observation 9 as a suspected outlier. Does your histogram agree?
(c) Redoing the regression without Observation 9 gives regression standard error $s = 42.4725$ and predicted mean velocity 216 feet per second (90% confidence interval 191.4 to 240.6) for piston walls 0.5 inch thick. Compare these values with those in Figure 25.13. Is Observation 9 influential for inference?

Table 4.1(page 102) gives 36 years' data on boats registered in Florida and manatees killed by boats. Figure 4.2 (page 103) shows a strong linear relationship. The correlation is $r = 0.953$. Figure 25.14 shows part of the Minitab regression output. Exercises 25.29 to 25.31 analyze the manatee data.

25.29 **Manatees: conditions for inference.** We know that there is a strong linear relationship. Let's check the other conditions for inference. Figure 25.14 includes a table of the two variables, the predicted values $\hat{y}$ for each x in the data, the residuals, and related quantities.
MANATRS

(a) Round the residuals to the nearest whole number and make a stemplot. The distribution is single peaked and symmetric and appears close to Normal.
(b) Make a residual plot, residuals against boats registered. Use a vertical scale from -20 to 20 to show

the pattern more clearly. Add the "residual = 0" line. There is no clearly nonlinear pattern. The spread/variability about the line may be a bit greater for larger values of the explanatory variable, but the effect is not large.

(c) It is reasonable to regard the number of manatees killed by boats in successive years as independent. The number of boats grew over time. Someone says that pollution also grew over time and may explain more manatee deaths. How would you respond to this idea?

25.30 **Manatees: Do more boats bring more kills?** The output in Figure 25.14 omits the t statistics and their P-values. Based on the information in the output, is there good evidence that the number of manatees killed increases as the number of boats registered increases? State hypotheses and give a test statistic and its approximate P-value. What do you conclude?

25.31 **Manatees: estimation.** The output in Figure 25.14 includes prediction of the number of manatees killed when there are 1,050,000 boats registered in Florida. Give 95% intervals for

(a) the increase in the number of manatees killed for each additional 1000 boats registered.
(b) the number of manatees that will be killed next year if there are 1,050,000 boats registered next year.

25.32 **Fidgeting keeps you slim: inference.** Our first example of regression (Example 5.1, page 122) presented data showing that people who increased their nonexercise activity (NEA) when they were deliberately overfed gained less fat than other people. Use software to add formal inference to the data analysis for these data.
FATGAIN

(a) Based on 16 subjects, the correlation between NEA increase and fat gain was $r = -0.7786$. Is this significant evidence that people with higher NEA increase gain less fat? (Report a t statistic from regression output and give the one-sided P-value.)
(b) The slope of the least-squares regression line was $b = -0.00344$, so that fat gain decreased by 0.00344 kilogram for each added calorie of NEA. Give a 90% confidence interval for the slope of the population regression line. This rate of change is the most important parameter to be estimated.
(c) Sam's NEA increases by 400 calories. His predicted fat gain is 2.13 kilograms. Give a 95% interval for predicting Sam's fat gain.

25.33 **Predicting tropical storms.** Exercise 5.55 (page 150) gives data on William Gray's predictions of the number of named tropical storms in Atlantic hurricane seasons from 1984 to 2012. Use these data for regression inference as follows.
STORMS2

FIGURE 25.14
Minitab output for the regression of number of manatees killed by boats on the number of boats (in thousands) registered in Florida, for Exercises 25.29 to 25.31.

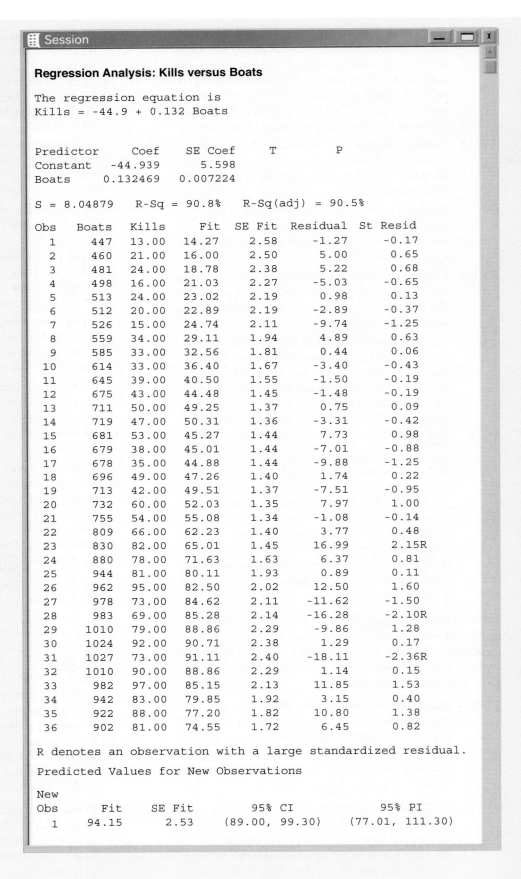

(a) Does Professor Gray do better than random guessing? That is, is there a significantly positive correlation between his forecasts and the actual number of storms? (Report a t statistic from regression output, and give the one-sided P-value.)

(b) Give a 95% confidence interval for the mean number of storms in years when Professor Gray forecasts 16 storms.

25.34 Coral growth. Sea surface temperatures across much of the tropics have been increasing since the mid-1970s. At the same time, the growth of coral has been decreasing. Scientists examined data on mean sea surface temperatures (SST) in degrees Celsius and mean coral growth in millimeters (mm) per year over a several-year period at locations in the Red Sea. Here are the data:[9] 📊 **CORAL**

SST	29.68	29.87	30.16	30.22	30.48	30.65	30.90
Growth	2.63	2.58	2.68	2.60	2.48	2.38	2.26

(a) Do the data indicate that coral growth decreases linearly as SST increases? Is this change statistically significant?

(b) Use the data to predict with 95% confidence the mean coral growth (mm per year) when SST is 30.0 degrees Celsius.

25.35 Predicting tropical storms: residuals. Make a stemplot of the residuals (round to the nearest tenth) from your regression in Exercise 25.33. Explain why your plot suggests that we should not use these data to get a prediction interval for the number of storms in a single year. 📊 **STORMS2**

25.36 Coral growth: residuals. Do the data in Exercise 25.34 on mean sea surface temperatures and coral growth in the Red Sea satisfy the conditions for regression inference? To examine this, here are the residuals: 📊 **CORALR**

SST	29.68	29.87	30.16	30.22	30.48	30.65	30.90
Residual	−0.067	−0.060	0.128	0.066	0.025	−0.024	−0.068

(a) **Linear relationship.** A plot of the residuals against the explanatory variable x magnifies the deviations from the least-squares line. Does the plot show any systematic deviation from a roughly linear pattern?

(b) **Normal variation about the line.** Make a histogram of the residuals. With only 7 observations, no clear shape emerges. Do strong skewness or outliers suggest lack of Normality?

(c) **Independent observations.** Why are the 7 observations independent?

(d) **Spread/variability about the line stays the same.** Does your plot in (a) show any systematic change in spread/variability as x changes?

25.37 Our brains don't like losses. Exercise 4.29 (page 115) describes an experiment that showed a linear relationship between how sensitive people are to monetary losses ("behavioral loss aversion") and activity in one part of their brains ("neural loss aversion"). 📊 **LOSSES**

(a) Make a scatterplot with neural loss aversion as x and behavioral loss aversion as y. One point is a high outlier in both the x and y directions. In Exercise 5.38 (page 147) you found that this outlier is not influential for the least-squares line.

(b) The research report says that $r = 0.85$ and that the test for regression slope has $P < 0.001$. Verify these results, using all the observations.

(c) The report recognizes the outlier and says, "However, this regression also remained highly significant ($P = 0.004$) when the extreme data point (top right corner) was removed from the analysis." Repeat your analysis omitting the outlier. Show that the outlier influences regression inference by comparing the t statistic for testing slope with and without the outlier. Then verify the report's claim about the P-value of this test.

25.38 Time at the table. Does how long young children remain at the lunch table help predict how much they eat? Here are data on 20 toddlers observed over several months at a nursery school.[10] "Time" is the average number of minutes a child spent at the table when lunch was served. "Calories" is the average number of calories the child consumed during lunch, calculated from careful observation of what the child ate each day. 📊 **TABTIME**

Time	21.4	30.8	37.7	33.5	32.8	39.5	22.8	34.1	33.9	43.8
Calories	472	498	465	456	423	437	508	431	479	454
Time	42.4	43.1	29.2	31.3	28.6	32.9	30.6	35.1	33.0	43.7
Calories	450	410	504	437	489	436	480	439	444	408

(a) Make a scatterplot. Find the correlation and the least-squares regression line. (Be sure to save the regression residuals.) Based on your work, describe the direction, form, and strength of the relationship.

(b) Check the conditions for regression inference. Parts (a) to (d) of Exercise 25.36 provide a handy outline. Use vertical limits −100 to 100 in your plot of the residuals against time to help you see the pattern. What do you conclude?

(c) Is there significant evidence that more time at the table is associated with more calories consumed? Give a 95% confidence interval to estimate how rapidly calories consumed changes as time at the table increases.

25.39 **DNA on the ocean floor.** We think of DNA as the stuff that stores the genetic code. It turns out that DNA occurs, mainly outside living cells, on the ocean floor. It is important in nourishing seafloor life. Scientists think that this DNA comes from organic matter that settles to the bottom from the top layers of the ocean. "Phytopigments," which come mainly from algae, are a measure of the amount of organic matter that has settled to the bottom. The data contains concentrations of DNA and phytopigments (both in grams per square meter) in 116 ocean locations around the world.[11] Look first at DNA alone. Describe the distribution of DNA concentration and give a confidence interval for the mean concentration. Be sure to explain why your confidence interval is trustworthy in the light of the shape of the distribution. The data show surprisingly high DNA concentration, and this by itself was an important finding. DNA

25.40 **Time at the table: prediction.** Rachel attends the nursery school of Exercise 25.38. Over several months, Rachel averages 40 minutes at the lunch table. Give a 95% interval to predict Rachel's average calorie consumption at lunch. TABTIME

*Exercises 25.41 to 25.45 ask practical questions involving regression inference without step-by-step instructions. Do complete regression analyses, using the **Plan, Solve,** and **Conclude** steps of the four-step process to organize your answers. Follow the model of Example 25.9 (page 579) and the subsequent discussion, and check the conditions as part of the **Solve** step.*

25.41 **Squirrels and their food supply.** The introduction to Exercises 7.24 to 7.26 (page 178) gives data on the abundance of the pinecones that red squirrels feed on and the mean number of offspring per female squirrel over 16 years. The strength of the relationship is remarkable because females produce young before the food is available. How significant is the evidence that more cones leads to more offspring? (Use a vertical scale from −2 to 2 in your residual plot to show the pattern more clearly.) SQUIRLS

25.42 **A big-toe problem.** Table 7.4 (page 186) and Exercises 7.47 and 7.49 describe the relationship between two deformities of the feet in young patients. Metatarsus adductus (MA) may help predict the severity of hallux abducto valgus (HAV). The paper that reports this study says, "Linear regression analysis, using the hallux abducto angle as the response variable, demonstrated a significant correlation between the metatarsus adductus and hallux abducto angles."[12] Do a suitable analysis to verify this finding. The study authors note that the scatterplot suggests that the variation in y may change as x changes, so they offer a more elaborate analysis as well. BIGTOE

25.43 **Beavers and beetles.** Exercise 5.53 (page 149) describes a study that found that the number of stumps from trees felled by beavers predicts the abundance of beetle larvae. Is there good evidence that more beetle larvae clusters are present when beavers have left more tree stumps? Estimate how many more clusters accompany each additional stump, with 95% confidence. BEAVERS

25.44 **Sulfur, the ocean, and the sun.** Sulfur in the atmosphere affects climate by influencing formation of clouds. The main natural source of sulfur is dimethylsulfide (DMS) produced by small organisms in the upper layers of the oceans. DMS production is in turn influenced by the amount of energy the upper ocean receives from sunlight. Exercise 4.30 (page 115) gives monthly data on solar radiation dose (SRD, in watts per square meter) and surface DMS concentration (in nanomolars) for a region in the Mediterranean. Do the data provide convincing evidence that DMS increases as SRD increases? We also want to estimate the rate of increase, with 90% confidence. SULFUR

25.45 **DNA on the ocean floor.** Another conclusion of the study introduced in Exercise 25.39 was that organic matter settling down from the top layers of the ocean is the main source of DNA on the seafloor. An important piece of evidence is the relationship between DNA and phytopigments. Do the data give good reason to think that phytopigment concentration helps explain DNA concentration? (Try vertical limits −1 to 1 to make the pattern of your residual plot clearer.) DNA

25.46 **(Optional) A lurking variable.** Return to the data on selling price versus appraised value for beachfront condominiums that are the basis for the Check Your Skills Exercises 25.16 to 25.24. The data are in order by date of the sale, and the data table includes the number of months from the start of the data period. Here are the residuals from the regression of selling price on appraised value (rounded): CONDRES

−55.90	−120.78	−192.01	−199.21	−63.88	−182.24
190.90	−71.54	−92.05	119.76	523.78	193.30
332.72	−125.09	−143.97	−252.09	−132.21	168.15
267.81	589.37	−234.53	−151.95	256.58	−155.85
−18.90	−77.84	−103.08	−269.22		

(a) Plot the residuals against the explanatory variable (appraised value). To make the pattern clearer, use vertical limits −600 to 600. Does the pattern you see agree with the conditions of linear relationship and constant standard deviation needed for regression inference?

(b) Make a stemplot of the residuals. Are there strong deviations from Normality that would prevent regression inference?

(c) Next, plot the residuals against month. Are the positive and negative residuals randomly scattered, as would be the case if the conditions for regression inference are satisfied? (*Comment:* Prices for beachfront property were rising rapidly during the first 36 months of this period. Because property is reassessed just once a year, selling prices might pull away from appraised values over time in this period, creating a pattern of many negative residuals followed by several positive residuals. As this example illustrates, it is often wise to plot residuals against important lurking variables as well as against the explanatory variable.)

25.47
standardized
residuals
(Optional) Standardized residuals. Software often calculates **standardized residuals** as well as the actual residuals from regression. Because the standardized residuals have the standard z-score scale, it is easier to judge whether any are extreme. Figure 25.13 (page 586) and the associated data include the standardized residuals for the regression of gate velocity on piston wall thickness.

(a) Find the mean and standard deviation of the standardized residuals. Why do you expect values close to those you obtain?

(b) Make a stemplot of the standardized residuals. Are there any striking deviations from Normality?

(c) The most extreme standardized residual is $z = -2.21$. Minitab flags this as "large." What is the probability that a standard Normal variable takes a value this extreme (that is, less than −2.21 or greater than 2.21)? Your result suggests that a residual this extreme would be a bit unusual when there are only 12 observations. That's why we examined Observation 9 in Exercise 25.28.

25.48 **(Optional) Tests for the intercept.** Figure 25.7 (page 575) gives Minitab output for the regression of blood alcohol content (BAC) on number of beers consumed. The t test for the hypothesis that the population regression line has *slope* $\beta = 0$ has $P < 0.001$. The data show a positive linear relationship between BAC and beers. We might expect the *intercept* α of the population regression line to be 0, because no beers ($x = 0$) should produce no alcohol in the blood ($y = 0$). To test

$$H_0: \alpha = 0$$
$$H_a: \alpha \neq 0$$

we use a t statistic formed by dividing the least-squares intercept a by its standard error SE_a. Locate this statistic in the output of Figure 25.7 and verify that it is in fact a divided by its standard error. What is the P-value? Do the data suggest that the intercept is not 0?

25.49 **(Optional) Confidence intervals for the intercept.** The output in Figure 25.7 (page 575) allows you to calculate confidence intervals for both the slope β and the intercept α of the population regression line of BAC on beers in the population of all students. Confidence intervals for the intercept α have the familiar form $a \pm t^*SE_a$ with degrees of freedom $n - 2$. What is the 95% confidence interval for the intercept? Does it contain 0, the value we might guess for α?

Exploring the Web

25.50 Predicting batting averages. As you did in Exercise 5.59 (page 152), go to www.mlb.com/ and find the batting averages for a diverse set of 30 players for both the 2012 and 2013 seasons. You can click on the "Stats" tab to find the results for the current season as well as historical data. You should select only players who played in at least 50 games both seasons. Find the least-squares regression line for predicting batting average in 2013 from that in 2012 based on your sample of 30 players. In 2012, the major league leader in batting was Buster Posey who had a batting average of .336. Find a 95% prediction interval for the 2013 batting average of someone who hit .336 in 2012. How does this prediction compare with Buster Posey's 2013 batting average?

25.51   Olympic medal counts. In Exercise 4.48 (page 120) you made a scatterplot of the Winter Olympics medal counts for 2010 and 2014. We investigate these medal counts further. Go the *Chance News* Web site at `www.causeweb.org/wiki/chance/index.php/Chance_News_61#Predicting_medal_counts` and read the article "Predicting Medal Counts." Next, search the Web (as you did in Chapter 4) and locate the Winter Olympics medal counts for 2010 and 2014 (I found Winter Olympics medal counts on Wikipedia). Find the equation of the least-squares regression line for predicting the 2014 medal counts from the 2010 counts. Compute 95% confidence intervals for the slope and intercept of your regression line. Are your results consistent with the comment in the *Chance News* article that states "we would have done well simply predicting that the Vancouver totals would match the Torino totals"?

One-Way Analysis of Variance: Comparing Several Means

Overview

In Chapter 21 the primary objective was inference for comparing two population means. This chapter represents the extension to more than two populations. Analysis of variance (ANOVA) is concerned with comparing group or treatment means in different samples by comparing variation *between* groups to variation *within* groups.

The early part of the chapter provides motivation for the method to be described. If there are I populations or treatments being compared, then there are $I(I-1)/2$ different pairwise comparisons of population means possible (1 vs. 2, 1 vs. 3, 2 vs. 3, etc.). When there are more than two populations or treatments being compared, application of the pairwise comparison methods of Chapter 21 breaks down for as few as three populations or treatments—we can't safely compare many parameters by comparing two at a time. This is the problem of multiple comparisons. The F test for ANOVA looks for evidence of any differences among the I population means. If necessary, we follow with an analysis to decide which of the means differ.

The ANOVA F test measures evidence for differences between population means by comparing two different kinds of variation in the samples: variation among the group sample means (measured by MSG, the mean square for groups), and variation among individuals within the same group (measured by MSE). If MSG is large compared with MSE, there is evidence of a difference between population means. These two sources of variation in the sample are central to the method. The ANOVA table provides a summary of these sources of variation, along with other components of the analysis—degrees of freedom, the F statistic and a P-value corresponding to the test.

If the hypothesis of equal means is rejected in favor of differences between means, the next step is to determine which group(s) differ. We use descriptive statistics and plots (typically side-by-side boxplots) to determine this. (If you have time in your class, you can investigate formal methods of comparison in Chapter 30, which is available online.)

Throughout the chapter, as with all inferential methods, we focus on conditions necessary for inference. Equal sample sizes are not necessary, although balanced designs offer some advantages (they provide greater power for the ANOVA F test, for example). Equality of population standard deviations is necessary, and rules of thumb for checking this based on the sample standard deviations are provided.

LEARNING OUTCOMES

- Recognize when testing the equality of several means is helpful in understanding data.
- Recognize that the statistical significance of differences among sample means depends on the sizes of the samples and on how much variation there is within the samples.
- Recognize when you can safely use ANOVA to compare means. Check the data production, the presence of outliers, and the sample standard deviations for the groups you want to compare.
- Explain what null hypothesis F tests in a specific setting.
- Locate the F statistic and its P-value on the output of analysis of variance software.
- Find the degrees of freedom for the F statistic from the number and sizes of the samples.
- If the test is significant, use graphs and descriptive statistics to see what differences among the means are most important.

Teaching Suggestions and Additional Examples/Activities for the Classroom

1. Use Examples to Compare Within-Group and Between-Group Variation

Start with an example asking students to compare, say, four treatments, but ask them to do it visually. A nice example is provided in the *IE* materials for Chapter 1. Here, provide two cases of data obtained from some hypothetical experiments for them to compare (times given in minutes)

Case 1: Treatment 1: 30 32 28 30 (mean = 30 minutes)
Treatment 2: 36 37 35 36 (mean = 36 minutes)
Treatment 3: 24 26 28 30 (mean = 27 minutes)

Case 2: Treatment 1: 24 22 36 38 (mean = 30 minutes)
Treatment 2: 30 34 38 46 (mean = 36 minutes)
Treatment 3: 20 24 30 34 (mean = 27 minutes)

If you plot the data for both cases on number lines, it is clear that the evidence for Treatment 3's superiority is stronger in Case 1 than in Case 2, even though in both cases the treatment averages are the same. Students can see that there are two different kinds of variation, and that if within-group variation is too large, we are less able to distinguish differences between groups. This is central to the method we will be discussing. Use the same example to discuss the problem of multiple comparisons. Why can't we simply run six different t tests for the six pairwise differences between population means? What would happen if we did?

2. Relate ANOVA Computations to Comparative Boxplots

You can discuss the formulas for computing the F statistic, but the focus should be on what is being compared. It is not productive to have students computing these statistics by hand. Rather, let software handle these tedious computations and instead focus on their meaning. The key is to relate statistical calculations in the ANOVA table to boxplots, so that students can see that large F statistics (and small P-values) correspond to greater relative variation between the groups and smaller relative variation within groups. You can, however, mention that under the null hypothesis F should be about 1 because the expected value of MSG is equal to the expected value of MSE under the null hypothesis. If the alternative hypothesis is true, F will be much greater than 1.

3. Use Examples of Interest to the Students

A class project or in-class activity can be done using data from the local grocery store (or downloaded). Visit the local store and record the number of grams of sugar per serving for different kinds of cereals. Record which shelf the brand is displayed on. Now test the hypothesis that mean sugar content is the same for all three (or four) shelves. It is likely that you will discover in your analysis that the shelf at eye-level for children has a greater mean sugar content per serving.

Another in-class example can be done using your students. If you have students at each class rank, treat your students as random samples from the ranks. Record the ages of all of your students, then compare mean ages for the different class ranks. In this example, within-group variation will be small, so it should be easy to detect the approximate one-year differences between groups. (Note that comparing ages assumes that you have no nontraditional students. If you do, talk about how the ages of the nontraditional students impact the analysis.)

Other Resources (LaunchPad)

StatClips
  ANOVA—Background and Introduction
  ANOVA—Inference

Snapshots Videos
  Introduction to ANOVA
  ANOVA Inference

EESEE Case Studies
  It's Tough to Crawl in March
  Stress Among Pets and Friends

Applet:
  One-Way ANOVA

Anthony Mercieca/Science Source

One-Way Analysis of Variance: Comparing Several Means

In this chapter we cover...

- Comparing several means
- The analysis of variance F test
- Using technology
- The idea of analysis of variance
- Conditions for ANOVA
- F distributions and degrees of freedom
- Some details of ANOVA

The two-sample t procedures of Chapter 21 compare the means of two populations or the mean responses to two treatments in an experiment. Of course, studies don't always compare just two groups. We need a method for comparing any number of means.

EXAMPLE 26.1   Comparing Tropical Flowers

FLOWER

STATE: Ethan Temeles and W. John Kress of Amherst College studied the relationship between varieties of the tropical flower *Heliconia* on the island of Dominica and the different species of hummingbirds that fertilize the flowers.[1] Over time, the researchers believe, the lengths of the flowers and the forms of the hummingbirds' beaks have evolved to match each other. If that is true, flower varieties fertilized by different hummingbird species should have distinct distributions of length.

Table 26.1 gives length measurements (in millimeters) for samples of three varieties of *Heliconia,* each fertilized by a different species of hummingbird. Do the three varieties display distinct distributions of length? In particular, are the mean lengths of their flowers different?

593

TABLE 26.1 FLOWER LENGTHS (MILLIMETERS) FOR THREE *HELICONIA* VARIETIES

H. BIHAI

47.12	46.75	46.81	47.12	46.67	47.43	46.44	46.64
48.07	48.34	48.15	50.26	50.12	46.34	46.94	48.36

H. CARIBAEA RED

41.90	42.01	41.93	43.09	41.47	41.69	39.78	40.57
39.63	42.18	40.66	37.87	39.16	37.40	38.20	38.07
38.10	37.97	38.79	38.23	38.87	37.78	38.01	

H. CARIBAEA YELLOW

36.78	37.02	36.52	36.11	36.03	35.45	38.13	37.10
35.17	36.82	36.66	35.68	36.03	34.57	34.63	

PLAN: Use graphs and numerical descriptions to describe and compare the three distributions of flower length. Finally, ask whether the differences among the mean lengths of the three varieties are *statistically significant*.

SOLVE (first steps): Figure 26.1 displays side-by-side stemplots with the stems lined up for easy comparison. The lengths have been rounded to the nearest tenth of a millimeter. Here are the summary measures we will use in further analysis:

Sample	Variety	Sample Size	Mean Length	Standard Deviation
1	*bihai*	16	47.60	1.213
2	red	23	39.71	1.799
3	yellow	15	36.18	0.975

```
     bihai                red                yellow
     34 |                 34 |               34 | 6 6
     35 |                 35 |               35 | 2 5 7
     36 |                 36 |               36 | 0 0 1 5 7 8 8
     37 |                 37 | 4 8 9         37 | 0 1
     38 |                 38 | 0 0 1 1 2 2 8 9   38 | 1
     39 |                 39 | 2 6 8         39 |
     40 |                 40 | 6 7           40 |
     41 |                 41 | 5 7 9 9       41 |
     42 |                 42 | 0 2           42 |
     43 |                 43 | 1             43 |
     44 |                 44 |               44 |
     45 |                 45 |               45 |
     46 | 3 4 6 7 8 8 9   46 |               46 |
     47 | 1 1 4           47 |               47 |
     48 | 1 2 3 4         48 |               48 |
     49 |                 49 |               49 |
     50 | 1 3             50 |               50 |
```

FIGURE 26.1

Side-by-side stemplots comparing the lengths in millimeters of samples of flowers from three varieties of *Heliconia*, from Table 26.1.

CONCLUDE (first steps): The three varieties differ so much in flower length that there is little overlap among them. In particular, the flowers of *bihai* are longer than either red or yellow. The mean lengths are 47.6 mm for *H. bihai*, 39.7 mm for *H. caribaea* red, and 36.2 mm for *H. caribaea* yellow. Are these observed differences in sample means statistically significant? We must develop a test for comparing more than two population means. ■

In the **SOLVE** step, we have chosen to draw side-by-side stemplots to compare the three distributions. The three distributions can also be compared using histograms or boxplots. Figure 26.2(a) provides comparative histograms and Figure 26.2(b) comparative boxplots. Because we want to compare the size of the responses for the three distributions as well as examining the general shape, it is best to align the response axis for the three graphs for ease of comparison. If the histograms in Figure 26.2(a) were side by side rather than above each other as in the figure, comparisons of the lengths for the three varieties would be more difficult. For comparison purposes, it is also best to use the percent scale for the histograms when the sample sizes differ. Figure 26.2(b) shows the comparative boxplots. Our impression of the distributions is similar for each of the three graphical displays. The flowers of *bihai* are longer than either red or yellow, and the spread for the lengths of the red flowers is somewhat larger than for the other two varieties. The choice of an appropriate graph is based on your software capabilities and to some extent personal preference. Although boxplots show less detail, they are useful for comparison as long as the sample sizes are not very small.

(a)

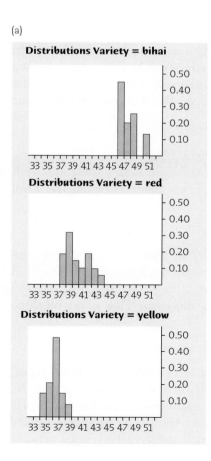

(b)

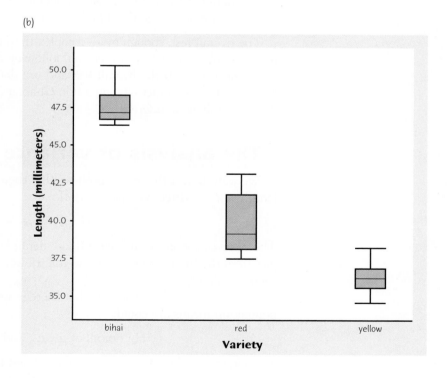

FIGURE 26.2

Histograms (a) and boxplots (b) comparing the lengths in millimeters of samples of flowers from three varieties of *Heliconia*, from Table 26.1.

Comparing several means

Call the mean lengths for the three populations of flowers μ_1 for *bihai*, μ_2 for red, and μ_3 for yellow. The subscript reminds us which group a parameter or statistic describes. To compare these three population means, we might use the two-sample *t* test several times:

- Test H_0: $\mu_1 = \mu_2$ to see if the mean length for *bihai* differs from the mean for red.
- Test H_0: $\mu_1 = \mu_3$ to see if *bihai* differs from yellow.
- Test H_0: $\mu_2 = \mu_3$ to see if red differs from yellow.

The weakness of doing three tests is that we get three *P*-values, one for each test alone. That doesn't tell us how likely it is that *three* sample means are spread apart as far as these are. It may be that $\bar{x}_1 = 47.60$ and $\bar{x}_3 = 36.18$ are significantly different if we look at just two groups but not significantly different if we know that they are the largest and the smallest means in three groups. As we look at more groups, we expect the gap between the largest and smallest sample mean to get larger. (Think of comparing the tallest and shortest person in larger and larger groups of people.) *We can't safely compare many parameters by doing tests or confidence intervals for two parameters at a time.*

multiple comparisons

The problem of how to do many comparisons at once with an overall measure of confidence in all our conclusions is common in statistics. This is the problem of **multiple comparisons.** Statistical methods for dealing with multiple comparisons usually have two steps:

1. An *overall test* to see if there is good evidence of *any* differences among the parameters that we want to compare
2. A detailed *follow-up analysis* to decide which of the parameters differ and to estimate how large the differences are

The overall test, though more complex than the tests we have met to this point, is reasonably straightforward. Formal follow-up analysis can be quite elaborate. We will concentrate on the overall test and use data analysis to describe in detail the nature of the differences. Companion Chapter 30, on the Web site, presents some details of follow-up inference.

The analysis of variance *F* test

We want to test the null hypothesis that there are *no differences* among the mean lengths for the three populations of flowers:

$$H_0: \mu_1 = \mu_2 = \mu_3$$

The basic *conditions for inference* (more detail later) are that we have random samples from the three populations and that flower lengths are Normally distributed in each population.

The alternative hypothesis is that there is *some difference*. That is, not all three population means are equal:

$$H_a: \text{not all of } \mu_1, \mu_2, \text{ and } \mu_3 \text{ are equal}$$

The alternative hypothesis is no longer one-sided or two-sided. It is "many-sided" because it allows any relationship other than "all three equal." For example, H_a includes the case in which $\mu_2 = \mu_3$ but μ_1 has a different value. The test of H_0

analysis of variance F test

against H_a is called the **analysis of variance *F* test.** Analysis of variance is usually abbreviated as ANOVA. The ANOVA *F* test is almost always carried out by software that reports the test statistic and its *P*-value.

EXAMPLE 26.2   Comparing Tropical Flowers: ANOVA

SOLVE (inference): Software tells us that for the flower length data in Table 26.1, the test statistic is $F = 259.12$ with *P*-value $P < 0.0001$. There is very strong evidence that the three varieties of flowers do not all have the same mean length.

The *F* test does not say *which* of the three means are significantly different. It appears from our preliminary data analysis that *bihai* flowers are distinctly longer than either red or yellow. Red and yellow are closer together, but the red flowers tend to be longer.

CONCLUDE: There is strong evidence ($P < 0.0001$) that the population means are not all equal. The most important difference among the means is that the *bihai* variety has longer flowers than the red and yellow varieties. ■

Example 26.2 illustrates our approach to comparing means. The ANOVA *F* test (done by software) assesses the evidence for *some* difference among the population means. Formal follow-up analysis would allow us to say which means differ and by how much, with (say) 95% confidence that *all* our conclusions are correct. We rely instead on examination of the data to show what differences are present and whether they are large enough to be interesting.

Apply Your Knowledge

26.1 Road Rage. "The phenomenon of road rage has been frequently discussed but infrequently examined." So begins a report based on interviews with 1382 randomly selected drivers.[2] The respondents' answers to interview questions produced scores on an "angry/threatening driving scale" with values between 0 and 19. What driver characteristics go with road rage? There were no significant differences among races or levels of education. What about the effect of the driver's age? Here are the mean responses for three age groups:

<30 yr	30–55 yr	>55 yr
2.22	1.33	0.66

The report says that $F = 34.96$, with $P < 0.01$.

(a) What are the null and alternative hypotheses for the ANOVA *F* test? Be sure to explain what means the test compares.

(b) Based on the sample means and the *F* test, what do you conclude?

26.2 Angry Women, Sad Men. What are the relationships among the portrayal of anger or sadness, sex and status conferred? Sixty-eight subjects were randomly assigned to view a videotaped job interview in which either a male or a female professional described feeling either anger or sadness. The targets being interviewed wore professional attire and were ostensibly being interviewed for a job while sitting at a table. The targets described an incident in which they and a colleague lost an account, and when asked by the interviewer how it made them feel, responded that the incident made them feel either angry or sad. The subjects were divided into four groups and evaluated one of the following four types of interviews:[3]

	Male Target	Female Target
Expressed anger	Group A	Group C
Expressed sadness	Group B	Group D

After watching the interview, subjects evaluated the target on a composite measure of status conferral that included items assessing how much status, power, and independence the target deserved in his or her future job. The measure of status ranged from $1 = $ none to $11 = $ a great deal.

(a) What are the null and alternative hypotheses for the ANOVA F test? Be sure to explain what means the test compares.

(b) Figure 26.3 is a bar graph displaying the means for the four groups. What is the approximate size of the mean difference in status conferred on angry men versus angry women? Which of these two groups has the higher mean status?

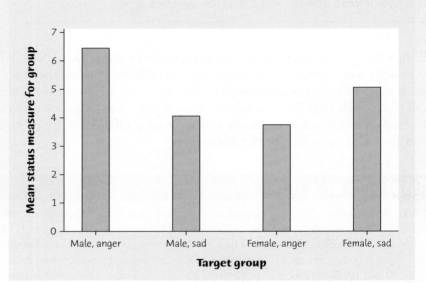

FIGURE 26.3
Bar graph comparing the mean status conferred for the four types of targets, for Exercise 26.2.

Using technology

Any technology used for statistics should perform analysis of variance. Figure 26.4 displays ANOVA output for the data of Table 26.1 from a graphing calculator, two statistical programs, and a spreadsheet program.

Minitab and Excel give the sizes of the three samples and their means. These agree with those in Example 26.1. Minitab also gives the standard deviations and Excel gives the variances. All four outputs report the F test statistic, $F = 259.12$, and its P-value. Minitab sensibly reports the P-value as 0 to three decimal places, while CrunchIt! reports that $P < 0.0001$. This is all we need to know about the P-value in practice. Excel and the graphing calculator offer the specific value 1.92×10^{-27}.

Texas Instruments Graphing Calculator

```
One-way ANOVA          One-way ANOVA
 F=259.1192995          ↑ MS=541.436183
 p=1.918818E-27          Error
 Factor                   df=51
  df=2                     SS=106.565761
  SS=1082.87237            MS=2.08952472
↓ MS=541.436183          Sxp=1.44551884
```

FIGURE 26.4
ANOVA for the flower length data: output from a graphing calculator, two statistical programs and a spreadsheet program.

Minitab

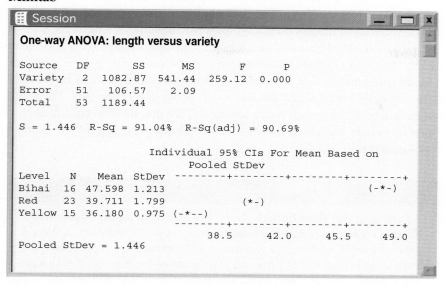

```
Session                                                      _ □ x

One-way ANOVA: length versus variety

Source    DF        SS       MS        F        P
Variety    2   1082.87   541.44   259.12   0.000
Error     51    106.57     2.09
Total     53   1189.44


S = 1.446   R-Sq = 91.04%   R-Sq(adj) = 90.69%

                      Individual 95% CIs For Mean Based on
                                 Pooled StDev
Level     N    Mean  StDev  --------+--------+--------+--------+
Bihai    16  47.598  1.213                               (-*-)
Red      23  39.711  1.799                (*-)
Yellow   15  36.180  0.975  (-*--)
                            --------+--------+--------+--------+
                               38.5     42.0     45.5     49.0

Pooled StDev = 1.446
```

CrunchIt!

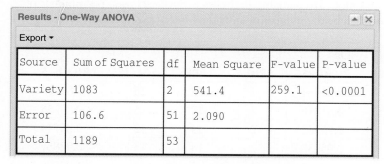

Results - One-Way ANOVA

Export ▾

Source	Sum of Squares	df	Mean Square	F-value	P-value
Variety	1083	2	541.4	259.1	<0.0001
Error	106.6	51	2.090		
Total	1189	53			

Excel

Microsoft Excel - ta25-01.dat

	A	B	C	D	E	F	G
1	Anova: Single Factor						
2							
3	SUMMARY						
4	*Groups*	*Count*	*Sum*	*Average*	*Variance*		
5	bihai	16	761.56	47.5975	1.471073		
6	red	23	913.36	39.7113	3.235548		
7	yellow	15	542.7	36.18	0.951257		
8							
9							
10	ANOVA						
11	*Source of variation*	*SS*	*df*	*MS*	*F*	*P-value*	*F crit*
12	Between Groups	1082.872	2	541.4362	259.1193	1.92E-27	3.178799
13	Within Groups	106.5658	51	2.089525			
14							
15	Total	1189.438	53				

Sheet4 / ta25-01 /

FIGURE 26.4
(Continued).

(This would be correct if the population distributions were exactly Normal. In practice, read such values simply as "P is very small.") There is very strong evidence that the three varieties of flowers do not all have the same mean length.

All four outputs report degrees of freedom (df), sums of squares (SS), and mean squares (MS). We don't need this information now. Minitab also gives confidence intervals for all three means that help us see which means differ and by how much. None of the intervals overlap, and *bihai* is much above the other two. These are 95% confidence intervals for each mean separately. We are *not* 95% confident that *all three* intervals cover the three means. This is another example of the peril of multiple comparisons.

Apply Your Knowledge

26.3   **Logging in the Rain Forest.** How does logging in a tropical rain forest affect the forest in later years? Researchers compared forest plots in Borneo that had never been logged (Group 1) with similar plots nearby that had been logged 1 year earlier (Group 2) and 8 years earlier (Group 3). Although the study was not an experiment, the authors explained why we can consider the plots to be randomly selected. The data appear in Table 26.2. The variable Trees is the count of trees in a plot; Species is the count of tree species in a plot. The variable Richness is Species/Trees, the number of species divided by the number of individual trees.[4] ▮▮▮ LOGGING

TABLE 26.2 DATA FROM A STUDY OF LOGGING IN BORNEO

GROUP	TREES	SPECIES	RICHNESS	GROUP	TREES	SPECIES	RICHNESS
1	27	22	0.81481	2	18	15	0.83333
1	22	18	0.81818	2	17	15	0.88235
1	29	22	0.75862	2	14	12	0.85714
1	21	20	0.95238	2	14	13	0.92857
1	19	15	0.78947	2	2	2	1.00000
1	33	21	0.63636	2	17	15	0.88235
1	16	13	0.81250	2	19	8	0.42105
1	20	13	0.65000	3	18	17	0.94444
1	24	19	0.79167	3	4	4	1.00000
1	27	13	0.48148	3	22	18	0.81818
1	28	19	0.67857	3	15	14	0.93333
1	19	15	0.78947	3	18	18	1.00000
2	12	11	0.91667	3	19	15	0.78947
2	12	11	0.91667	3	22	15	0.68182
2	15	14	0.93333	3	12	10	0.83333
2	9	7	0.77778	3	12	12	1.00000
2	20	18	0.90000				

(a)   Make an appropriate comparative graph of the variable Trees for the three groups. What effects of logging are visible?

(b)   Figure 26.5 shows Excel ANOVA output for Trees. What do the group means show about the effects of logging?

(c)   What are the values of the ANOVA F statistic and its P-value? What hypotheses does F test? What conclusions about the effects of logging on number of trees do the data lead to?

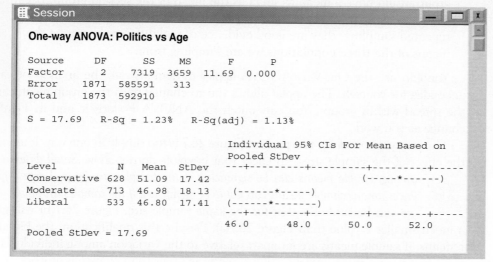

```
┌─────────────────────────────────────────────────────────────────────────────┐
│ ⬛ Microsoft Excel Book 1                                          _ □ X        │
├─────────────────────────────────────────────────────────────────────────────┤
│        A              B         C         D          E          F        G    │
│ 1  Anova: Single Factor                                                       │
│ 2                                                                             │
│ 3  SUMMARY                                                                    │
│ 4      Groups       Count      Sum     Average    Variance                    │
│ 5  Group 1            12       285       23.75     25.6591                     │
│ 6  Group 2            12       169     14.0833     24.8106                     │
│ 7  Group 3             9       142     15.7778     33.1944                     │
│ 8                                                                             │
│ 9                                                                             │
│ 10 ANOVA                                                                      │
│ 11 Source of variation  SS     df         MS          F      P-value   F crit │
│ 12 Between Groups     625.1566   2    312.57828   11.4257   0.000205  3.31583  │
│ 13 Within Groups      820.7222  30     27.3574                                 │
│ 14                                                                            │
│ 15 Total             1445.879   32                                            │
└─────────────────────────────────────────────────────────────────────────────┘
```

FIGURE 26.5

Excel output for analysis of variance on the number of trees in forest plots, for Exercise 26.3.

26.4 Political Views and Age. The University of Chicago's General Social Survey (GSS) is the nation's most important social science sample survey. The GSS asked a random sample of adults in 2012 their age and where they placed themselves on the political spectrum from extremely liberal to extremely conservative. The categories in the original survey included slightly liberal, liberal, and extremely liberal, but these have been combined into the single category liberal, and similarly with conservative.[5]

(a) Figure 26.6 gives the Minitab ANOVA output for these data. What do the mean ages say about the relationship between age and political views?

(b) What are the values of the ANOVA F statistic and its P-value? What hypotheses does F test? Briefly describe the conclusions you draw from these data.

```
┌──────────────────────────────────────────────────────────────────┐
│ ⬛ Session                                              — □ X        │
├──────────────────────────────────────────────────────────────────┤
│                                                                    │
│  One-way ANOVA: Politics vs Age                                    │
│                                                                    │
│  Source    DF      SS     MS      F      P                         │
│  Factor      2    7319   3659   11.69  0.000                       │
│  Error    1871  585591    313                                      │
│  Total    1873  592910                                             │
│                                                                    │
│  S = 17.69    R-Sq = 1.23%    R-Sq(adj) = 1.13%                     │
│                                                                    │
│                        Individual 95% CIs For Mean Based on        │
│                        Pooled StDev                                │
│  Level        N   Mean   StDev   ---+---------+---------+---------+-----  │
│  Conservative 628  51.09  17.42                    (-----*------)   │
│  Moderate     713  46.98  18.13    (------*-----)                   │
│  Liberal      533  46.80  17.41   (------*-------)                  │
│                                ---+---------+---------+---------+-----  │
│                                   46.0      48.0      50.0      52.0  │
│  Pooled StDev = 17.69                                              │
│                                                                    │
└──────────────────────────────────────────────────────────────────┘
```

FIGURE 26.6

Minitab output for the data on respondent's ages for three different political viewpoints, for Exercise 26.4.

The idea of analysis of variance

The details of ANOVA are a bit daunting (they appear in an optional section at the end of this chapter). The main idea of ANOVA is more accessible and much more important. Here it is: when we ask if a set of sample means gives evidence for

FIGURE 26.7

Boxplots for two sets of three samples each. The sample means are the same in (a) and (b). Analysis of variance will find a more significant difference among the means in (b) because there is less variation among the individuals within those samples.

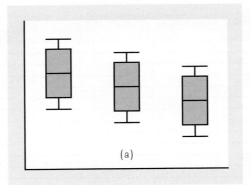

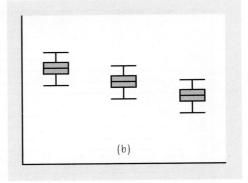

(a)   (b)

differences among the population means, what matters is not how far apart the sample means are but how far apart they are *relative to the variability of individual observations*.

Look at the two sets of boxplots in Figure 26.7. For simplicity, these distributions are all symmetric, so that the mean and median are the same. The center line in each boxplot is therefore the sample mean. Both sets of boxplots compare three samples with the same three means. Could differences this large easily arise just due to chance, or are they statistically significant?

■ The boxplots in Figure 26.7(a) have tall boxes, which show lots of variation among the individuals in each group. With this much variation among individuals, we would not be surprised if another set of samples gave quite different sample means. The observed differences among the sample means could easily happen just by chance.

■ The boxplots in Figure 26.7(b) have the same centers as those in Figure 26.7(a), but the boxes are much shorter. That is, there is much less variation among the individuals in each group. It is unlikely that any sample from the first group would have a mean as small as the mean of the second group. Because means as far apart as those observed would rarely arise just by chance in repeated sampling, they are good evidence of real differences among the means of the three populations we are sampling from.

You can use the *One-Way ANOVA* applet to demonstrate the analysis of variance idea for yourself. The applet allows you to change both the group means and the spread within groups. You can watch the ANOVA *F* statistic and its *P*-value change as you work.

This comparison of the two parts of Figure 26.7 is too simple in one way. It ignores the effect of the sample sizes, an effect that boxplots do not show. *Small differences* *among sample means can be significant if the samples are large. Large differences among sample means can fail to be significant if the samples are small.* All we can be sure of is that for the same sample size, Figure 26.7(b) will give a much smaller *P*-value than Figure 26.7(a). Despite this qualification, the big idea remains: if sample means are far apart relative to the variation among individuals in the same groups, that's evidence that something other than chance is at work.

The Analysis of Variance Idea

Analysis of variance compares the variation due to specific sources with the variation among individuals who should be similar. In particular, ANOVA tests whether several populations have the same mean by comparing how far apart the sample means are with how much variation there is within the samples.

It is one of the oddities of statistical language that methods for comparing means are named after the variance. The reason is that the test works by comparing two kinds of variation. Analysis of variance is a general method for studying sources of variation in responses. Comparing several means is the simplest form of ANOVA, called **one-way ANOVA.**

one-way ANOVA

The ANOVA *F* Statistic

The **analysis of variance *F* statistic** for testing the equality of several means has this form:

$$F = \frac{\text{variation among the sample means}}{\text{variation among individuals in the same sample}}$$

If you want more detail, read the optional section at the end of this chapter. The *F* statistic can take only values that are zero or positive. It is zero only when all the sample means are identical and gets larger as they move farther apart. Large values of *F* are evidence against the null hypothesis H_0 that all population means are the same. Although the alternative hypothesis H_a is many-sided, the ANOVA *F* test is one-sided because any violation of H_0 tends to produce a large value of *F*.

Apply Your Knowledge

26.5 ANOVA Compares Several Means. The *One-Way ANOVA* applet displays the observations in three groups, with the group means highlighted by black dots.

(a) What are the *F* and *P* value for these three samples? (The *P*-value is marked by a red dot that will move along the scale as you modify the samples.)

(b) Which group has the largest mean? If you grab its mean point with the mouse and begin to move it in either direction, the *F* value will change. How small can you make *F*? What did you do to this mean to make *F* small? Roughly how significant is your small *F*?

(c) Starting with tyour configuration at the end of (b), drag any one of the group means either up or down as far as they will go. What happens to *F*? What happens to the *P*-value? How large can you make *F*? What did you do to the three means to make *F* large?

26.6 ANOVA Uses Within-Group Variation. If the *One-Way ANOVA* applet is open from Exercise 26.5, click on New Samples. Otherwise, open the *One-Way ANOVA* applet which displays the observations in three groups, with the group means highlighted by black dots. You are going to use the Applet to investigate the effects described in Figure 26.7 (page 602).

(a) Use the mouse to slide the Standard Deviation at the top of the display to the right. You see that the group means do not change, but the spread of the observations in each group increases. What happens to *F* and *P* as the spread among the observations in each group increases? What are the values of *F* and *P* when the slider is all the way to the right? This is similar to Figure 26.7(a): variation within groups hides the differences among the group means.

(b) Leave the Standard Deviation slider at the extreme right of its scale, so that spread within groups stays fixed. Use the mouse to move the group means apart. What happens to *F* and *P* as you do this?

Conditions for ANOVA

Like all inference procedures, ANOVA is valid only in some circumstances. Here are the conditions under which we can use ANOVA to compare population means.

Conditions for ANOVA Inference

- We have *I* independent SRSs, one from each of *I* populations. We measure the same quantitative response variable for each sample.
- The *i*th population has a **Normal distribution** with unknown mean μ_i. One-way ANOVA tests the null hypothesis that all the population means are the same.
- All the populations have the **same standard deviation σ,** whose value is unknown.

The first two conditions are familiar from our study of the two-sample *t* procedures for comparing two means. As usual, the design of the data production is the most important condition for inference. Biased sampling or confounding can make any inference meaningless. *If we do not actually draw separate SRSs from each population or carry out a randomized comparative experiment, it may be unclear to what population the conclusions of inference apply.* ANOVA, like other inference procedures, is often used when random samples are not available. You must judge each use on its merits, a judgment that usually requires some knowledge of the subject of the study in addition to some knowledge of statistics.

Because no real population has exactly a Normal distribution, the usefulness of inference procedures that assume Normality depends on how sensitive they are to departures from Normality. Fortunately, procedures for comparing means are not very sensitive to lack of Normality. The ANOVA *F* test, like the *t* procedures, is **robust.** What matters is Normality of the sample means, so ANOVA becomes safer as the sample sizes get larger because of the central limit theorem effect. Remember to check for outliers that change the value of sample means and for extreme skewness. When there are no outliers and the distributions are roughly symmetric, you can safely use ANOVA for sample sizes as small as 4 or 5.

robustness

The third condition is annoying: ANOVA assumes that the variability of observations, measured by the standard deviation, is the same in all populations. The *t* test for comparing two means (Chapter 21) does not require equal standard deviations. Unfortunately, the ANOVA *F* for comparing more than two means is less broadly valid. It is not easy to check the condition that the populations have equal standard deviations. Statistical tests for equality of standard deviations are very sensitive to lack of Normality, so much so that they are of little practical value. You must either seek expert advice or rely on the robustness of ANOVA.

How serious are unequal standard deviations? ANOVA is not too sensitive to violations of the condition, especially when all samples have the same or similar sizes and no sample is very small. When designing a study, try to take samples of about the same size from all the groups you want to compare. The sample standard deviations estimate the population standard deviations, so check before doing ANOVA that the sample standard deviations are similar to each other. We expect some variation among them due to chance. Here is a rule of thumb that is safe in almost all situations.

Checking Standard Deviations in ANOVA

The results of the ANOVA F test are approximately correct when the largest sample standard deviation is no more than twice as large as the smallest sample standard deviation.

EXAMPLE 26.3   Comparing Tropical Flowers: Conditions for ANOVA

The study of *Heliconia* blossoms is based on three independent samples that the researchers consider to be random samples from all flowers of these varieties in Dominica. The stemplots in Figure 26.1 (page 594) show that the *bihai* and red varieties have slightly skewed distributions, but the sample means of samples of sizes 16 and 23 will have distributions that are close to Normal. The sample standard deviations for the three varieties are

$$s_1 = 1.213 \quad s_2 = 1.799 \quad s_3 = 0.975$$

These standard deviations satisfy our rule of thumb:

$$\frac{\text{largest } s}{\text{smallest } s} = \frac{1.799}{0.975} = 1.85 \quad \text{(less than 2)}$$

We can safely use ANOVA to compare the mean lengths for the three populations. ■

EXAMPLE 26.4   Thinking about Money Changes Behavior

STATE: Kathleen Vohs of the University of Minnesota and her coworkers carried out several randomized comparative experiments on the effects of thinking about money. Here's an outline of one of the experiments. Ask student subjects to unscramble 30 sets of five words to make a meaningful phrase from four of the five. The control group unscrambled phrases like "cold it desk outside is" into "it is cold outside." The "play money" group unscrambled similar sets of words, but a stack of Monopoly money was placed nearby. The "money prime" group unscrambled phrases that lead to thinking about money, turning "high a salary desk paying" into "a high-paying salary." Then each subject worked a hard puzzle, knowing that they could ask for help. Table 26.3 shows the time in seconds that each subject worked on the puzzle before asking for help.[6] Psychologists think that money tends to make people self-sufficient. If so, the two groups that were encouraged in different ways to think about money should take longer on the average to ask for help. Do the data support this idea?

MONEY

PLAN: Examine the data to compare the effect of the treatments and check that we can safely use ANOVA. If the data allow ANOVA, assess the significance of observed differences in mean times to ask for help.

SOLVE: Figure 26.8 compares the histograms of the data in the three groups. We expect some irregularity in small samples, but there are no outliers or strong skewness that would hinder use of ANOVA. The Minitab ANOVA output in Figure 26.9 shows that the group standard deviations easily satisfy our rule of thumb. The control group subjects asked for help much sooner (mean 186.1 seconds) than did subjects in the two money groups (means 305.2 seconds and 314.1 seconds). The three means are significantly different ($F = 3.73$, $P = 0.031$).

CONCLUDE: The experiment gives good evidence that reminding people of money in either of two ways does make them less willing to ask others for help. This is consistent with the idea that money makes people feel more self-sufficient. ■

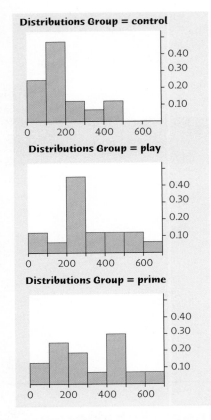

Distributions Group = control

Distributions Group = play

Distributions Group = prime

TABLE 26.3		TIME (SECONDS) UNTIL SUBJECTS ASK FOR HELP WITH A PUZZLE			
GROUP	**TIME**	**GROUP**	**TIME**	**GROUP**	**TIME**
Prime	609	Play	455	Control	118
Prime	444	Play	100	Control	272
Prime	242	Play	238	Control	413
Prime	199	Play	243	Control	291
Prime	174	Play	500	Control	140
Prime	55	Play	570	Control	104
Prime	251	Play	231	Control	55
Prime	466	Play	380	Control	189
Prime	443	Play	222	Control	126
Prime	531	Play	71	Control	400
Prime	135	Play	232	Control	92
Prime	241	Play	219	Control	64
Prime	476	Play	320	Control	88
Prime	482	Play	261	Control	142
Prime	362	Play	290	Control	141
Prime	69	Play	495	Control	373
Prime	160	Play	600	Control	156
		Play	67		

FIGURE 26.8
Histograms comparing the time until subjects asked for help with a puzzle, for Example 26.4.

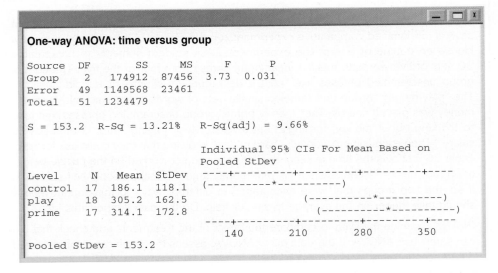

One-way ANOVA: time versus group

```
Source   DF       SS      MS     F      P
Group     2   174912   87456  3.73  0.031
Error    49  1149568   23461
Total    51  1234479

S = 153.2   R-Sq = 13.21%    R-Sq(adj) = 9.66%


                                Individual 95% CIs For Mean Based on
                                Pooled StDev
Level       N   Mean   StDev   ----+---------+---------+---------+----
control    17  186.1   118.1   (----------*----------)
play       18  305.2   162.5                  (----------*----------)
prime      17  314.1   172.8                    (----------*----------)
                                ----+---------+---------+---------+----
                                   140       210       280       350
Pooled StDev = 153.2
```

FIGURE 26.9
Minitab ANOVA output for comparing the three treatments in Example 26.4.

Apply Your Knowledge

26.7 Checking Standard Deviations. Verify that the sample standard deviations for these sets of data do allow use of ANOVA to compare the population means.

(a) The counts of trees in Exercise 26.3 (page 600) and Figure 26.5.

(b) The ages of Exercise 26.4 (page 601) and Figure 26.6.

26.8 Species Richness after Logging. Table 26.2 (page 600) gives data on the species richness in rain forest plots, defined as the number of tree species in a plot divided by the number of trees in the plot. ANOVA may not be trustworthy for the richness data. Do data analysis: make an appropriate graph to examine the distributions of the response variable in the three groups, and also compare the standard deviations. What characteristic of the data makes ANOVA risky? ▌▌ LOGGING

26.9 **Fertilizing Bromeliads.** Bromeliads are tropical flowering plants. Many are epiphytes that attach to trees and obtain moisture and nutrients from air and rain. Their leaf bases form cups that collect water and are home to the larvae of many insects. As a preliminary to a study of changes in the nutrient cycle, Jacqueline Ngai and Diane Srivastava examined the effects of adding nitrogen, phosphorus, or both to the cups. They randomly assigned 8 bromeliads growing in Costa Rica to each of four treatment groups, including an unfertilized control group. A monkey destroyed one of the plants in the control group, leaving 7 bromeliads in that group. Here are the numbers of new leaves on each plant over the 7 months following fertilization:[7] ▌▌ BROMLIAD

Nitrogen	Phosphorus	Both	Neither
15	14	14	11
14	14	16	13
15	14	15	16
16	11	14	15
17	13	14	15
18	12	13	11
17	15	17	12
13	15	14	

Analyze these data and discuss the results. Does nitrogen or phosphorus have a greater effect on the growth of bromeliads? Follow the four-step process as illustrated in Example 26.4.

F distributions and degrees of freedom

The ANOVA F statistic is

$$F = \frac{\text{variation among the sample means}}{\text{variation among individuals in the same sample}}$$

To find the P-value for this statistic, we must know the sampling distribution of F when the null hypothesis (all population means equal) is true. This sampling distribution is an **F distribution.**

F distribution

The F distributions are a family of right-skewed distributions that take only values greater than 0. The density curves in Figure 26.10 illustrate their shapes. A specific F distribution is determined by the *degrees of freedom* of the numerator and denominator of the F statistic. You may have noticed that all our software outputs include degrees of freedom, labeled either "df" or "DF." The optional section "Some Details of ANOVA" shows where the degrees of freedom come from. When describing an F distribution, always give the numerator degrees of freedom first. Our brief notation will be F(df1, df2) for the F distribution with df1 degrees of freedom in the numerator and df2 in the denominator. *Interchanging the degrees of freedom changes the distribution, so the order is important.*

Tables of F critical points are awkward because we need a separate table for every pair of degrees of freedom df1 and df2. Fortunately, software gives you P-values for the ANOVA F test without the need for a table.

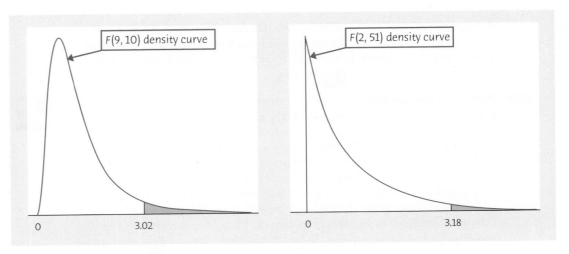

FIGURE 26.10
Density curves for two F distributions. Both are right-skewed and take only positive values. The upper 5% critical values are marked under the curves.

EXAMPLE 26.5   **Comparing Flowers: The *F* Distribution**

Look again at the software output in Figure 26.4 (page 598) for the flower length data. All four outputs give the degrees of freedom for the F test, labeled "df" or "DF." There are 2 degrees of freedom in the numerator and 51 in the denominator. P-values for the F test, therefore, come from the F distribution $F(2, 51)$ with 2 and 51 degrees of freedom. The right-hand curve in Figure 25.10 is the density curve of this distribution. The 5% critical value marked on that curve is 3.18, and the 1% critical value is 5.05. The observed value $F = 259.12$ of the ANOVA F statistic lies far to the right of these values, so the P-value is extremely small. ■

The degrees of freedom of the ANOVA F statistic depend on the number of means we are comparing and the number of observations in each sample. That is, the F test takes into account the number of observations. Here are the details.

Degrees of Freedom for the *F* Test

We want to compare the means of I populations. We have an SRS of size n_i from the ith population, so that the total number of observations in all samples combined is

$$N = n_1 + n_2 + \cdots + n_I$$

If the null hypothesis that all population means are equal is true, the ANOVA F statistic has the F distribution with $I - 1$ degrees of freedom in the numerator and $N - I$ degrees of freedom in the denominator.

EXAMPLE 26.6   **Degrees of Freedom for *F***

In Examples 26.1 and 26.2, we compared the mean lengths for three varieties of flowers, so $I = 3$. The three sample sizes are

$$n_1 = 16 \quad n_2 = 23 \quad n_3 = 15$$

The total number of observations is therefore

$$N = 16 + 23 + 15 = 54$$

The ANOVA F test has numerator degrees of freedom

$$I - 1 = 3 - 1 = 2$$

and denominator degrees of freedom

$$N - I = 54 - 3 = 51$$

These are the degrees of freedom given in the outputs in Figure 26.4 (page 598). ■

Apply Your Knowledge

26.10 Logging in the Rain Forest, Continued. Exercise 26.3 (page 600) compares the number of tree species in rain forest plots that had never been logged (Group 1) with similar plots nearby that had been logged 1 year earlier (Group 2) and 8 years earlier (Group 3).

(a) What are I, the n_i, and N for these data? Identify these quantities in words and give their numerical values.

(b) Find the degrees of freedom for the ANOVA F statistic. Check your work against the Excel output in Figure 26.5.

26.11 What Music Will You Play? People often match their behavior to their social environment. One study of this idea first established that the type of music most preferred by black college students is R&B and that whites' most preferred music is rock. Will students hosting a small group of other students choose music that matches the makeup of the people attending? Assign 90 black business students at random to three equal-sized groups. Do the same for 96 white students. Each student sees a picture of the people he or she will host. Group 1 sees 6 blacks, Group 2 sees 3 whites and 3 blacks, and Group 3 sees 6 whites. Ask how likely the host is to play the type of music preferred by the other race. Use ANOVA to compare the three groups to see whether the racial mix of the gathering affects the choice of music.[8]

© PhotoAlto/Alamy

(a) For the white subjects, $F = 16.48$. What are the degrees of freedom?

(b) For the black subjects, $F = 2.47$. What are the degrees of freedom?

Some details of ANOVA*

Now we will give the actual formula for the ANOVA F statistic. We have SRSs from each of I populations. Subscripts from 1 to I tell us which sample a statistic refers to:

Population	Sample Size	Sample Mean	Sample Std. Dev.
1	n_1	$\bar{x}_1$	s_1
2	n_2	$\bar{x}_2$	s_2
⋮	⋮	⋮	⋮
I	n_I	$\bar{x}_I$	s_I

You can find the F statistic from just the sample sizes n_i, the sample means $\bar{x}_i$, and the sample standard deviations s_i. You don't need to go back to the individual observations.

The ANOVA F statistic has the form

$$F = \frac{\text{variation among the sample means}}{\text{variation among individuals in the same sample}}$$

*This more advanced section is optional if you are using software to find the F statistic.

mean squares

The measures of variation in the numerator and denominator of F are called **mean squares.** A mean square is a more general form of a sample variance. An ordinary sample variance s^2 is an average (or mean) of the squared deviations of observations from their mean, so it qualifies as a "mean square."

Call the overall mean response $\bar{x}$. That is, $\bar{x}$ is the mean of all N observations together. You can find $\bar{x}$ from the I sample means by

$$\bar{x} = \frac{\text{sum of all observations}}{N} = \frac{n_1\bar{x}_1 + n_2\bar{x}_2 + \cdots + n_I\bar{x}_I}{N}$$

(This expression works because multiplying a group mean $\bar{x}_i$ by the number of observations n_i it represents gives the sum of the observations in that group.)

The numerator of F is a mean square that measures variation among the I sample means $\bar{x}_1, \bar{x}_2, \ldots, \bar{x}_I$. To measure this variation, look at the I deviations of the means of the samples from $\bar{x}$,

$$\bar{x}_1 - \bar{x}, \bar{x}_2 - \bar{x}, \ldots, \bar{x}_I - \bar{x}$$

MSG

The mean square in the numerator of F is an average of the squares of these deviations. We call it the **mean square for groups,** abbreviated as MSG:

$$\text{MSG} = \frac{n_1(\bar{x}_1 - \bar{x})^2 + n_2(\bar{x}_2 - \bar{x})^2 + \cdots + n_I(\bar{x}_I - \bar{x})^2}{I - 1}$$

Each squared deviation is weighted by n_i, the number of observations it represents.

MSE

The mean square in the denominator of F measures variation among individual observations in the same sample. For any one sample, the sample variance s_i^2 does this job. For all I samples together, we use an average of the individual sample variances. It is another weighted average, in which each s_i^2 is weighted by its degrees of freedom $n_i - 1$. The resulting mean square is called the **mean square for error,** MSE:

$$\text{MSE} = \frac{(n_1 - 1)s_1^2 + (n_2 - 1)s_2^2 + \cdots + (n_I - 1)s_I^2}{N - I}$$

"Error" doesn't mean a mistake has been made. It's a traditional term for chance variation. Here is a summary of the ANOVA test.

The ANOVA F Test

Draw an independent SRS from each of I Normal populations that have a common standard deviation but may have different means. The sample from the ith population has size n_i, sample mean $\bar{x}_i$, and sample standard deviation s_i.

To test the null hypothesis that all I populations have the same mean against the alternative hypothesis that not all the means are equal, calculate the **ANOVA F statistic**

$$F = \frac{\text{MSG}}{\text{MSE}}$$

The numerator of F is the **mean square for groups**

$$\text{MSG} = \frac{n_1(\bar{x}_1 - \bar{x})^2 + n_2(\bar{x}_2 - \bar{x})^2 + \cdots + n_I(\bar{x}_I - \bar{x})^2}{I - 1}$$

The denominator of F is the **mean square for error**

$$\text{MSE} = \frac{(n_1 - 1)s_1^2 + (n_2 - 1)s_2^2 + \cdots + (n_I - 1)s_I^2}{N - I}$$

When H_0 is true, F has the **F distribution** with $I - 1$ and $N - I$ degrees of freedom.

The denominators in the formulas for MSG and MSE are the two degrees of freedom $I - 1$ and $N - I$ of the F test. The numerators are called **sums of squares,** from their algebraic form. It is usual to present the results of ANOVA in an **ANOVA table.** Output from software usually includes an ANOVA table.

sums of squares

ANOVA table

EXAMPLE 26.7   ANOVA Calculations: Software

Look again at the four outputs in Figure 26.4 (page 598). The two software outputs give the ANOVA table. The calculator, with its small screen, gives the degrees of freedom, sums of squares, and mean squares separately. Each output uses slightly different language to identify the two sources of variation. The basic ANOVA table is

Source of Variation	df	SS	MS	F Statistic
Variation among samples	2	1082.87	MSG = 541.44	259.12
Variation within samples	51	106.57	MSE = 2.09	

You can check that each mean square MS is the corresponding sum of squares SS divided by its degrees of freedom df. The F statistic is MSG divided by MSE. ∎

Because MSE is an average of the individual sample variances, it is also called the *pooled sample variance,* written as s_p^2. When all I populations have the same population variance σ^2, as ANOVA assumes that they do, s_p^2 estimates the common variance σ^2. The square root of MSE is the **pooled standard deviation** s_p. It estimates the common standard deviation σ of observations in each group. The Minitab and calculator outputs in Figure 26.4 give the value $s_p = 1.446$.

pooled standard deviation

The pooled standard deviation s_p is a better estimator of the common σ than any individual sample standard deviation s_i because it combines (pools) the information in all I samples. We can get a confidence interval for any one of the means μ_i from the usual form

$$\text{estimate} \pm t^* SE_{\text{estimate}}$$

using s_p to estimate σ. The confidence interval for μ_i is

$$\bar{x}_i \pm t^* \frac{s_p}{\sqrt{n_i}}$$

Use the critical value t^* from the t distribution with $N - I$ degrees of freedom because s_p has $N - I$ degrees of freedom. These are the confidence intervals that appear in Minitab ANOVA output.

EXAMPLE 26.8   ANOVA Calculations: Without Software

We can do the ANOVA test comparing the mean lengths of *bihai*, red, and yellow flower varieties using only the sample sizes, sample means, and sample standard deviations. These appear in Example 26.1 (page 593), but it is easy to find them with a calculator. There are $I = 3$ groups with a total of $N = 54$ flowers.

The overall mean of the 54 lengths in Table 26.1 is

$$\bar{x} = \frac{n_1 \bar{x}_1 + n_2 \bar{x}_2 + n_3 \bar{x}_3}{N}$$

$$= \frac{(16)(47.598) + (23)(39.711) + (15)(36.180)}{54}$$

$$= \frac{2217.621}{54} = 41.067$$

The mean square for groups is

$$MSG = \frac{n_1(\bar{x}_1 - \bar{x})^2 + n_2(\bar{x}_2 - \bar{x})^2 + n_3(\bar{x}_3 - \bar{x})^2}{I - 1}$$

$$= \frac{1}{3 - 1}\left[(16)(47.598 - 41.067)^2 + (23)(39.711 - 41.067)^2\right.$$
$$\left. + (15)(36.180 - 41.067)^2\right]$$

$$= \frac{1082.996}{2} = 541.50$$

The mean square for error is

$$MSE = \frac{(n_1 - 1)s_1^2 + (n_2 - 1)s_2^2 + (n_3 - 1)s_3^2}{N - I}$$

$$= \frac{(15)(1.213^2) + (22)(1.799^2) + (14)(0.975^2)}{51}$$

$$= \frac{106.580}{51} = 2.09$$

Finally, the ANOVA test statistic is

$$F = \frac{MSG}{MSE} = \frac{541.50}{2.09} = 259.09$$

Our work differs slightly from the output in Figure 26.4 because of roundoff error. We don't recommend doing these calculations by hand, because tedium and roundoff errors cause frequent mistakes. ■

Apply Your Knowledge

The calculations of ANOVA use only the sample sizes n_i, the sample means $\bar{x}_i$, and the sample standard deviations s_i. You can therefore re-create the ANOVA calculations when a report gives these summaries but does not give the actual data. These optional exercises ask you to do the ANOVA calculations starting with the summary statistics. P-values require either a table or software for the F distributions.

26.12 Road Rage. Exercise 26.1 (page 597) describes a study of road rage. Here are the means and standard deviations for a measure of "angry/threatening driving" for random samples of drivers in three age groups:

Age Group	n	$\bar{x}$	s
Less than 30 years	244	2.22	3.11
30 to 55 years	734	1.33	2.21
Over 55 years	364	0.66	1.60

(a) The distributions of responses are somewhat right-skewed. ANOVA is nonetheless safe for these data. Why?

(b) Check that the standard deviations satisfy the guideline for ANOVA inference.

(c) Calculate the overall mean response $\bar{x}$, the mean squares MSG and MSE, and the ANOVA F statistic.

(d) Which F distribution would you use to find the P-value of the ANOVA F test? Software gives $P < 0.001$. Write a brief conclusion based on the sample means and the ANOVA F test.

26.13 **Angry Women, Sad Men.** Exercise 26.2 (page 597) describes a study in which subjects conferred status on men and women who were displaying either anger or sadness. Subjects were randomly assigned to the four treatments, with 17 subjects assigned to each treatment. The study report contains the following information about status measure conferred for each of the four groups:

Treatment	n	$\bar{x}$	s
Males expressing anger	17	6.47	2.25
Females expressing anger	17	3.75	1.77
Males expressing sadness	17	4.05	1.61
Females expressing sadness	17	5.02	1.80

(a) Do the standard deviations satisfy the rule of thumb for safe use of ANOVA?

(b) Calculate the overall mean response $\bar{x}$, the mean squares MSG and MSE, and the F statistic.

(c) Which F distribution would you use to find the P-value of the ANOVA F test? Write a brief conclusion based on the sample means and the ANOVA F test.

26.14 **Attitudes Toward Math.** Do high school students from different racial/ethnic groups have different attitudes toward mathematics? Measure the level of interest in mathematics on a 5-point scale for a national random sample of students. Here are summaries for students who were taking math at the time of the survey:[9]

Racial/Ethnic Group	n	$\bar{x}$	s
African American	809	2.57	1.40
White	1860	2.32	1.36
Asian/Pacific Islander	654	2.63	1.32
Hispanic	883	2.51	1.31
Native American	207	2.51	1.28

(a) The conditions for ANOVA are clearly satisfied. Explain why.

(b) Calculate the ANOVA table and the F statistic.

(c) Software gives $P < 0.001$. What explains the small P-value? Do you think the differences are large enough to be important?

CHAPTER 26 SUMMARY

Chapter Specifics

■ **One-way analysis of variance (ANOVA)** compares the means of several populations. The **ANOVA F test** tests the null hypothesis that all the populations have the same mean. If the F test shows significant differences, examine the data to see where the differences lie and whether they are large enough to be important.

■ The **conditions for ANOVA** state that we have an **independent SRS** from each population; that each population has a **Normal distribution;** and that all populations have the **same standard deviation.**

- In practice, ANOVA inference is relatively **robust** when the populations are non-Normal, especially when the samples are large. Before doing the F test, check the observations in each sample for outliers or strong skewness. Also verify that the largest sample standard deviation is no more than twice as large as the smallest standard deviation.

- When the null hypothesis is true, the **ANOVA F statistic** for comparing I means from a total of N observations in all samples combined has the **F distribution** with $I-1$ and $N-I$ degrees of freedom.

- ANOVA calculations are reported in an **ANOVA table** that gives sums of squares, mean squares, and degrees of freedom for variation among groups and for variation within groups. In practice, we use software to do the calculations.

STATISTICS IN SUMMARY

Here are the most important skills you should have acquired from reading this chapter.

A. Recognition

1. Recognize when testing the equality of several means is helpful in understanding data.
2. Recognize that the statistical significance of differences among sample means depends on the sizes of the samples and on how much variation there is within the samples.
3. Recognize when you can safely use ANOVA to compare means. Check the data production, the presence of outliers, and the sample standard deviations for the groups you want to compare.

B. Interpreting ANOVA

1. Explain what null hypothesis F tests in a specific setting.
2. Locate the F statistic and its P-value on the output of analysis of variance software.
3. Find the degrees of freedom for the F statistic from the number and sizes of the samples.
4. If the test is significant, use graphs and descriptive statistics to see what differences among the means are most important.

Link It

Analysis of variance is a general statistical method for studying sources of variation in a response. In this chapter, we have studied one-way ANOVA, which is a specific statistical technique designed to test the null hypothesis of the equality of the means of several populations. As such, it is an extension of the two-sample t test of Chapter 21, which tested the null hypothesis of the equality of *two* population means.

When we reject the null hypothesis of equality of two means with the two-sample t test, then we have evidence that the two means are different. Rejection of the null hypothesis with a one-way ANOVA is a more ambiguous conclusion—it is evidence of a difference in the means of the populations, but the result does not tell us *which* differences between the means are statistically significant. This is similar to the chi-square test in Chapter 24, where rejection of the null hypothesis indicates that there is a relationship between the two categorical variables, but says nothing about the nature of that relationship. Typically, after rejection of the null hypothesis in a one-way ANOVA, we perform a more detailed *follow-up analysis* to decide which of the means are different and to estimate how large these differences are. The companion Chapter 30, on the Web site, presents some details of this follow-up inference.

CHECK YOUR SKILLS

26.15 The purpose of analysis of variance is to compare

(a) the variances of several populations.
(b) the proportions of successes in several populations.
(c) the means of several populations.

26.16 The *F* distributions are

(a) a family of distributions with bell-shaped density curves centered at 0.
(b) a family of distributions that are right-skewed and take only values greater than 0.
(c) a family of distributions that are left-skewed and take values between 0 and 1.

An experiment to help determine if insects sleep gave caffeine to fruit flies to see if it affected their rest. The three treatments were a control, a low-caffeine dose of 1 mg/ml of blood and a higher dose 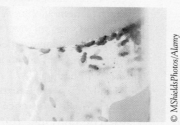 *of 5 mg/ml of blood. Nine fruit flies were assigned at random to the three treatments, three to each treatment, and the number of minutes of rest was measured over a 12-hour period.*

CONTROL	LOW DOSE	HIGH DOSE
450	466	265
413	420	330
418	435	389

Here is a partial Minitab output including the ANOVA table (several numbers have been omitted), along with the means and standard deviations of the rest times for the three groups:

```
Source      DF        SS        MS     F        P
Caffeine              22598                     0.027
Error                         1600
Total

Level       N        Mean     StDev
Control     3       427.00     20.07
Low         3       440.33     23.46
High        3       328.00     62.02
```

Exercises 26.17 to 26.22 are based on this study.

26.17 The degrees of freedom for the ANOVA *F* statistic comparing mean hours of rest are   **CAFFEIN**

(a) 2 and 6.   (b) 2 and 5.   (c) 3 and 7.

26.18 The null hypothesis for the ANOVA *F* test is

(a) that the population mean rest is the same for all three levels of caffeine.
(b) that the population mean rest is decreasing as the caffeine level gets larger.
(c) that the population mean rest is lowest for the high level of caffeine.

26.19 **(Optional)** The value of the ANOVA *F*-statistic for testing equality of the population means of the three caffeine levels is

(a) 4.73.   (b) 4.82.   (c) 7.06.

26.20 The conclusion of the ANOVA test is that

(a) there is strong evidence ($P = 0.027$) that the mean rest is not the same for all three treatments.
(b) there is strong evidence ($P = 0.027$) that the mean rest is lower in the high-caffeine treatment than in the other two.
(c) the data give no evidence ($P = 0.027$) to suggest that mean rest differs among the three treatments.

26.21 For this example, we notice

(a) ANOVA can be used on these data because ANOVA requires the sample sizes are equal.
(b) there is an extreme outlier in the data.
(c) the data show evidence of a violation of the assumption that the three populations have the same standard deviation.

26.22 To compare the treatments, we might use three 90% two-sample *t* confidence intervals to compare each pair of treatments: the control versus low dose, the control versus high dose, and the low dose versus high dose. The weakness of doing this is that

(a) we don't know how confident we can be that all three intervals cover the true differences in means.
(b) 90% confidence is okay for one comparison, but it isn't high enough for six comparisons done at once.
(c) we can't compare two treatments that use different doses of caffeine.

26.23 A company runs a three-day workshop on strategies for working effectively in teams. On each day, a different strategy is presented. Forty-eight employees of the company attend the workshop. At the outset, all 48 are divided into 12 teams of four. The teams remain the same for the entire workshop. Strategies are presented in the morning. In the afternoon, the teams are presented with a series of small tasks, and the number of these completed successfully using the strategy taught that morning is recorded for each team. The mean number of tasks completed successfully by all teams each day and the standard deviation follow:

DAY	*n*	$\bar{x}$	*s*
1	12	17.25	7.10
2	12	17.64	14.14
3	12	17.21	14.03

In this example, we notice

(a) the data show very strong evidence of a violation of the assumption that the three populations have the same standard deviation.

(b) ANOVA cannot be used on these data because the sample sizes are less than 20.

(c) the assumption that the data are independent for the three days is unreasonable because the same teams were observed each day.

CHAPTER 26 EXERCISES

Exercises 26.24 to 26.27 describe situations in which we want to compare the mean responses in several populations. For each setting, identify the populations and the response variable. Then give I, the n_i, and N. Finally, give the degrees of freedom of the ANOVA F statistic.

26.24 **Morning or evening?** Are you a morning person, an evening person, or neither? Does this personality trait affect how well you perform? A sample of 100 students took a psychological test that found 16 morning people, 30 evening people, and 54 who were neither. All the students then took a test of their ability to memorize at 8 A.M. and again at 9 P.M. The response variable is the score at 8 A.M. minus the score at 9 P.M.

26.25 **Does art sell products?** How does visual art affect the perception and evaluation of consumer products? Subjects were asked to evaluate an advertisement for bathroom fittings that contained an art image, a non-art image, or no image. The art image was Vermeer's painting *Girl with a Pearl Earring*, and the non-art image was a photograph of the actress Scarlett Johannsson in the same pose wearing the same garments as the girl in the painting and was taken from the motion picture *Girl with a Pearl Earring*. Thus the art and non-art image were a match on content. College students were divided at random into three groups of 39 each, with each group assigned to one of the three types of advertisements. Students evaluated the product in the advertisement on a scale of 1 to 7, with 1 being the most unfavorable rating and 7 being the most favorable. The paper reported a one way ANOVA on the product evaluation index had $F = 6.29$ with $P < 0.05$.[10]

26.26 **Test accommodations.** Many states require schoolchildren to take regular statewide tests to assess their progress. Children with learning disabilities who read poorly may not do well on mathematics tests because they can't read the problems. Most states allow "accommodations" for learning-disabled children. Randomly assign 100 learning-disabled children in equal numbers to three types of accommodation and a control group: math problems are read by a teacher, by a computer, by a computer that also shows a video, and standard test conditions. The researcher would like to compare the mean scores on the state mathematics assessment.

26.27 **Exercise and type 2 diabetes.** It is generally accepted that regular exercise provides health benefits to individuals with type 2 diabetes, although the exact exercise regimen (aerobic, resistance or both) is unclear. The subjects in this study were sedentary 30- to 75-year-old adults with type 2 diabetes and elevated hemoglobin A1c levels above 6.5%. The level of hemoglobin A1c correlates very well with a person's recent overall blood sugar levels. If the blood sugars have generally been running high during the previous few months, the level of hemoglobin A1c will be high. In a randomized controlled study, 41 subjects were assigned to a nonexercise control group, 73 to resistance training only, 72 to aerobic exercise only, and 76 to combined aerobic and resistance training. The weekly duration of exercise was similar for all three exercise groups, and subjects remained on the exercise regimens for 9 months. At the end of 9 months, the hemoglobin A1c levels of subjects was measured.[11]

26.28 **Don't handle the merchandise?** Although consumers often want to touch products before purchasing them, they generally prefer that others have not touched products they would like to buy. Can another person touching a product create a positive reaction? Subjects were given instructions to contact a sales associate at a university bookstore who would provide them with a shirt to try on. When meeting the sales associate, subjects were told there was only one shirt left and it was being tried on by another "customer." The other customer trying on the shirt was a confederate of the experimenter and was either an attractive, well-dressed professional female model or an average-looking female college student wearing jeans and a tee shirt. Subjects, who were either males or females, saw the confederate leaving the dressing room where the shirt was left for them to try on. There was also a control group of subjects who were handed the shirt directly off the rack by the sales associate. Thus there were five treatments: male subjects seeing a model, female subjects seeing a model, male subjects seeing a college student, female subjects seeing a college student, and the control group. Subjects evaluated the product on five dimensions, each dimension on a 7-point scale, with the five scores then averaged to give the subject's evaluation measure, with higher numbers indicating a more positive evaluation. Here are the sample sizes, means, and standard deviations for the five groups:[12]

TREATMENT GROUP	n	$\bar{x}$	s
Males seeing a model	22	5.34	0.87
Males seeing a student	23	3.32	1.21
Females seeing a model	24	4.10	1.32
Females seeing a student	23	3.50	1.43
Controls	27	4.17	1.50

(a) Verify that the sample standard deviations allow the use of ANOVA to compare the population means. What do the means suggest about the effect of the subject's gender and the attractiveness of the confederate on the evaluation of the product?

(b) The paper reports the ANOVA $F = 8.30$. What are the degrees of freedom for the ANOVA F statistic and the P-value? State your conclusions.

26.29 Plants defend themselves. When some plants are attacked by leaf-eating insects, they release chemical compounds that attract other insects that prey on the leaf-eaters. A study carried out on plants growing naturally in the Utah desert demonstrated both the release of the compounds and that they not only repel the leaf-eaters but also attract predators that act as the plants' bodyguards.[13] The investigators chose 8 plants attacked by each of three leaf-eaters and 8 more that were undamaged, 32 plants of the same species in all. They then measured emissions of several compounds during seven hours. Here are data (mean ± standard error of the mean for eight plants) for one compound. The emission rate is measured in nanograms (ng) per hour.

GROUP	EMISSION RATE (ng/hr)
Control	9.22 ± 5.93
Hornworm	31.03 ± 8.75
Leaf bug	18.97 ± 6.64
Flea beetle	27.12 ± 8.62

(a) Make a graph that compares the mean emission rates for the four groups. Does it appear that emissions increase when the plant is attacked?

(b) What hypotheses does ANOVA test in this setting?

(c) We do not have the full data. What would you look for in deciding whether you can safely use ANOVA?

(d) What is the relationship between the standard error of the mean (SEM) and the standard deviation for a sample? What are the four sample standard deviations? Do they satisfy our rule of thumb for safe use of ANOVA?

26.30 Can you hear these words? To test whether a hearing aid is right for a patient, audiologists play a tape on which words are pronounced at low volume. The patient tries to repeat the words. There are several different lists of words that are supposed to be equally difficult. Are the lists equally difficult when there is background noise? To find out, an experimenter had subjects with normal hearing listen to four lists with a noisy background. The response variable was the percent of the 50 words in a list that the subject repeated correctly. The data set contains 96 responses.[14] Here are two study designs that could produce these data:

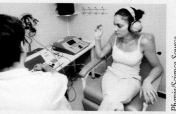
Phanie/Science Source

Design A. The experimenter assigns 96 subjects to 4 groups at random. Each group of 24 subjects listens to one of the lists. All individuals listen and respond separately.

Design B. The experimenter has 24 subjects. Each subject listens to all four lists in random order. All individuals listen and respond separately.

Does Design A allow use of one-way ANOVA to compare the lists? Does Design B allow use of one-way ANOVA to compare the lists? Briefly explain your answers.

26.31 More rain for California? The changing climate will probably bring more rain to California, but we don't know whether the additional rain will come during the winter wet season or extend into the long dry season in spring and summer. Kenwyn Suttle of the University of California at Berkeley and his coworkers randomly assigned plots of open grassland to three treatments: added water equal to 20% of annual rainfall either during January to March (winter) or during April to June (spring), and no added water (control). Here are some of the data, for plant biomass (in grams per square meter) produced by each plot in a single year.[15] 📊 MASS2003

WINTER	SPRING	CONTROL
264.1514	318.4182	129.0538
187.7312	281.6830	144.6578
291.1431	288.8433	172.7772
176.2879	382.6673	113.2813
141.7525	326.8877	142.1562
169.9737	293.8502	117.9808

FIGURE 26.11

Minitab ANOVA output for comparing the total plant biomass of grassland plots under different water conditions, for Exercise 26.31.

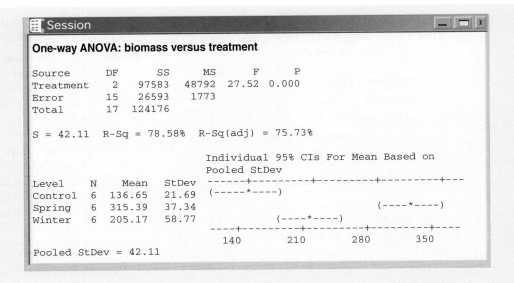

Figure 26.11 shows Minitab ANOVA output for these data.

(a) Make side-by-side stemplots of plant biomass for the three treatments, as well as a table of the sample means and standard deviations. What do the data appear to show about the effect of extra water in winter and in spring on biomass? Do these data satisfy the conditions for ANOVA?

(b) State H_0 and H_a for the ANOVA F test, and explain in words what ANOVA tests in this setting.

(c) Report your overall conclusions about the effect of added water on plant growth in California.

26.32 Can you hear these words? Figure 26.12 displays the Minitab output for one-way ANOVA applied to the hearing data described in Design A in Exercise 26.30. The response variable is "Percent," and "List" identifies the four lists of words. Based on this analysis,

is there good reason to think that the four lists are not all equally difficult? Write a brief summary of the study findings.

26.33 **College students need better sleep!** A random sample of 898 students between the ages of 20 and 24 at a large midwestern university completed a survey including questions about their sleep quality, moods, academic performance, physical health, and psychoactive drug use. Sleep quality was measured using the Pittsburgh Sleep Quality Index (PSQI), with students scoring less than or equal to 5 on the index classified as optimal sleepers, those scoring a 6 or 7 classified as borderline and those scoring over 7 classified as poor sleepers. The depression subscale of the Profile of Moods State (POMS) was used to assess how severely students experienced depression on a typical day, with high scores indicating greater levels of depression.

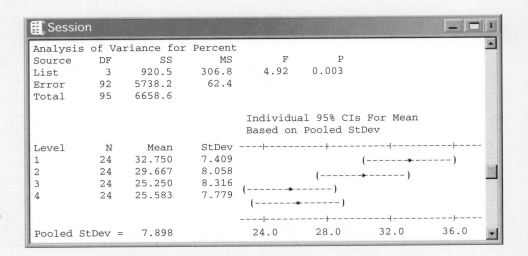

FIGURE 26.12

Minitab ANOVA output for comparing the percents heard correctly in four lists of words, for Exercise 26.32.

We want to know if there a significant difference in depression scores among the three classifications of sleep.[16] The full data set is too large to print here, but here are the first seven individuals: DEPRESED

Quality of sleep:	poor	poor	border	poor	poor	optimal	border
Depression score:	5	8	5	11	7	7	10

(a) Follow the four-step process in data analysis and ANOVA. Be sure to check the conditions for ANOVA and to include an appropriate graph that compares the depression scores for the three qualities of sleep.

(b) Explain why we can trust the ANOVA F test to give valid results for these data.

(c) This is an observational study. Explain why. In this study we can see why association does not prove causation. Explain how poor-quality sleep might lead to higher levels of depression. Then explain how higher depression scores might affect quality of sleep. You see that the cause-and-effect relationship might go in either direction.

26.34 **Do good smells bring good business?** Businesses know that customers often respond to background music. Do they also respond to odors? Nicolas Guéguen and his colleagues studied this question in a small pizza restaurant in France on Saturday evenings in May. On one evening, a relaxing lavender odor was spread through the restaurant; on another evening, a stimulating lemon odor; a third evening served as a control, with no odor. The three evenings were comparable in many ways (weather, customer count, and so on), so we are willing to regard the data as independent SRSs from spring Saturday evenings at this restaurant. Table 26.4 contains data on how long

TABLE 26.4 TIME (MINUTES) THAT CUSTOMERS REMAIN IN A RESTAURANT WHEN EXPOSED TO ODORS

				LAVENDER ODOR					
92	126	114	106	89	137	93	76	98	108
124	105	129	103	107	109	94	105	102	108
95	121	109	104	116	88	109	97	101	106
				LEMON ODOR					
78	104	74	75	112	88	105	97	101	89
88	73	94	63	83	108	91	88	83	106
108	60	96	94	56	90	113	97		
				NO ODOR					
103	68	79	106	72	121	92	84	72	92
85	69	73	87	109	115	91	84	76	96
107	98	92	107	93	118	87	101	75	86

(in minutes) customers stayed in the restaurant on each of the three evenings.[17] ODORS

(a) Make an appropriate graph comparing the customer times for each evening. Do any of the distributions show outliers, strong skewness, or other clear deviations from Normality?

(b) Do a complete analysis to see whether the groups differ in the average amount of time spent in the restaurant. Follow the four-step process in your work. Did you find anything surprising?

26.35 **Good weather and tipping.** Favorable weather has been shown to be associated with increased tipping. Will just the belief that future weather will be favorable lead to higher tips? The researchers gave 60 index cards to a waitress at an Italian restaurant in New Jersey. Before delivering the bill to each customer, the waitress randomly selected a card and wrote on the bill the same message that was printed on the index card. Twenty of the cards had the message "The weather is supposed to be really good tomorrow. I hope you enjoy the day!" Another 20 cards contained the message, "The weather is supposed to be not so good tomorrow. I hope you enjoy the day anyway!" The remaining 20 cards were blank, indicating that the waitress was not supposed to write any message. Choosing a card at random ensured that there was a random assignment of the diners to the three experimental conditions. Here are the percentage tips for the three messages:[18] TIPPING

Good weather report	20.8	18.7	19.9	20.6	22.0	23.4	22.8
	24.9	22.2	20.3	24.9	22.3	27.0	20.4
	22.2	24.0	21.2	22.1	22.0	22.7	
Bad weather report	18.0	19.0	19.2	18.8	18.4	19.0	18.5
	16.1	16.8	14.0	17.0	13.6	17.5	19.9
	20.2	18.8	18.0	23.2	18.2	19.4	
No weather report	19.9	16.0	15.0	20.1	19.3	19.2	18.0
	19.2	21.2	18.8	18.5	19.3	19.3	19.4
	10.8	19.1	19.7	19.8	21.3	20.6	

Do the data support the hypothesis that there are differences among the tipping percentages for the three experimental conditions? Does a prediction of good weather seem to increase the tip percentage? Follow the four-step process in data analysis and ANOVA. Be sure to check the conditions for ANOVA and to include an appropriate graph that compares the tipping percentages for the three conditions.

26.36 **Durable press fabrics are weaker.** "Durable press" cotton fabrics are treated to improve their recovery from wrinkles after washing. Unfortunately, the treatment also reduces the strength of the fabric. A study compared the breaking strength of untreated fabric with that of fabrics treated by three commercial durable press processes. Five specimens of the same fabric were assigned at random to each group. Here are the data, in pounds of pull needed to tear the fabric:[19] WEAKFAB

Untreated	60.1	56.7	61.5	55.1	59.4
Permafresh 55	29.9	30.7	30.0	29.5	27.6
Permafresh 48	24.8	24.6	27.3	28.1	30.3
Hylite LF	28.8	23.9	27.0	22.1	24.2

The untreated fabric is clearly much stronger than any of the treated fabrics. We want to know if there is a significant difference in breaking strength among the three durable press treatments. Analyze the data for the three processes and write a clear summary of your findings. Which process do you recommend if breaking strength is a main concern? Use the four-step process to guide your discussion. (Although the standard deviations do not quite satisfy our rule of thumb, that rule is conservative and many statisticians would use ANOVA for these data.)

26.37 Durable press fabrics wrinkle less. The data in Exercise 26.36 show that durable press treatment greatly reduces the breaking strength of cotton fabric. Of course, durable press treatment also reduces wrinkling. How much? "Wrinkle recovery angle" measures how well a fabric recovers from wrinkles. Higher is better. Here are data on the wrinkle recovery angle (in degrees) for the same fabric specimens discussed in the previous exercise: WRNKLFAB

Untreated	79	80	78	80	78
Permafresh 55	136	135	132	137	134
Permafresh 48	125	131	125	145	145
Hylite LF	143	141	146	141	145

The untreated fabric once again stands out, this time as inferior to the treated fabrics in wrinkle resistance. Examine the data for the three durable press processes and summarize your findings. How does the ranking of the three processes by wrinkle resistance compare with their ranking by breaking strength in Exercise 26.36? Explain why we can't trust the ANOVA F test.

26.38 Logging in the rain forest: species counts. Table 26.2 gives data on the number of trees per forest plot, the number of species per plot, and species richness. Exercise 26.3 analyzed the effect of logging on number of trees. Exercise 26.8 concludes that it would be risky to use ANOVA to analyze richness. Use software to analyze the effect of logging on the number of species. LOGGING

(a) Make a table of the group means and standard deviations. Do the standard deviations satisfy our rule of thumb for safe use of ANOVA? What do the means suggest about the effect of logging on the number of species?

(b) Carry out the ANOVA. Report the F statistic and its P-value and state your conclusion in context.

More rain for California? Exercise 26.31 describes a randomized experiment carried out by Kenwyn Suttle and his coworkers to examine the effects of additional water on California grassland. The experimental units are 18 plots of grassland, assigned at random among three treatments: added water in the winter wet season, added water in the spring dry season, and no added water (control group). Field experiments, unlike labora-

Suttle, K. B., M. A. Thomsen, and M. E. Power. "Species Interactions Reverse Grassland Responses to Changing Climate." *Science* 315.5812 (2007): 640–42.

tory experiments, are exposed to variations in the natural environment. The experiment therefore continued over 5 years, 2001 to 2005. Table 26.5 gives data on the total plant biomass (grams per square meter) that grew on each plot during each year.[20] The "Plot" column shows how the random assignment of 18 of the 36 available plots worked. Exercises 26.39 to 26.41 are based on this information.

26.39 Plot the means. Starting from the data in Table 26.5, you can calculate the mean plant biomass for each treatment in each year as follows: MASSMEAN

	YEAR				
	2001	**2002**	**2003**	**2004**	**2005**
Winter	132.58	203.33	205.17	223.58	332.84
Spring	257.69	388.85	315.39	299.54	289.66
Control	81.67	180.31	136.65	201.07	257.37

(a) Plot the means for each of the three treatments against year, connecting the yearly means for each treatment by lines to show the pattern over time. Use the same plot for all three treatments, with a different color for each treatment. From this plot, you can get an overall picture of the experiment's results.

(b) Across all 5 years, does more water in the wet season increase plant growth? What about more water in the dry season? Which seasonal addition of water has the larger effect?

(c) One-way ANOVAs comparing the mean plant biomass separately in each year find significant differences in 3 years and no significant difference in 2 years. Based on your plot, in which three years do you think the treatment means differ significantly?

TABLE 26.5 PLANT BIOMASS FOR THREE WATER CONDITIONS OVER FIVE YEARS

TREATMENT	PLOT	YEAR				
		2001	2002	2003	2004	2005
Winter	3	136.8358	228.0717	264.1514	254.6453	344.3933
Winter	8	151.4154	189.9505	187.7312	233.8155	203.3908
Winter	14	136.1536	209.0485	291.1431	253.4506	331.9724
Winter	20	121.6323	189.6755	176.2879	228.5882	388.1056
Winter	27	124.1459	188.0090	141.7525	158.6675	382.8617
Winter	32	125.2986	215.2174	169.9737	212.3232	346.3042
Spring	4	338.1301	422.7411	318.4182	517.6650	344.0489
Spring	11	291.8597	339.8243	281.6830	342.2825	261.8016
Spring	18	244.8727	398.7296	288.8433	270.5785	262.7238
Spring	22	234.6599	400.6878	382.6673	212.5324	316.9683
Spring	25	197.5830	326.9497	326.8877	213.9879	224.1109
Spring	35	239.0122	444.1556	293.8502	240.1927	328.2783
Control	6	73.4288	148.8907	129.0538	178.9988	237.6596
Control	7	110.6306	182.6762	144.6578	205.5165	281.1442
Control	17	95.3405	196.8303	172.7772	242.6795	313.7242
Control	24	83.0584	186.1953	113.2813	231.7639	258.3631
Control	28	30.5886	154.0401	142.1562	134.9847	235.8320
Control	33	96.9709	213.2537	117.9808	212.4862	217.5060

(d) In 2005, there were unusual late rains during the spring. How does the effect of this natural rainfall show up in your plot? (You see that it would not be wise to do an experiment like this in just one year.)

26.40 **The results for 2001.** Your work in Exercise 26.31 shows that there were significant differences in mean plant biomass among the three treatments in 2003. Do a complete analysis of the data for 2001 and report your conclusions. ▥▥ MASSALL

26.41 **Conditions for ANOVA.** Examine the data for the year 2004. The conditions for ANOVA inference are not met. In what way do these data fail to meet the conditions? (It is not very surprising that in 5 ANOVAs one will fail to satisfy our quite conservative conditions.) ▥▥ MASSALL

26.42 **Which test?** Example 26.4 (page 605) describes one of the experiments done by Kathleen Vohs and her coworkers to demonstrate that even being reminded of money makes people more self-sufficient and less involved with other people. Here are three more of these experiments. For each experiment, which statistical test from Chapters 17 to 24 would you use, and why?

(a) Randomly assign student subjects to money and control groups. The control group unscrambles neutral phrases, and the money group unscrambles money-oriented phrases, as described in Example 26.4. Then ask the subjects to volunteer to help the experimenter by coding data sheets, about 5 minutes per sheet. Subjects said how many sheets they would volunteer to code. "Participants in the money condition volunteered to help code fewer data sheets than did participants in the control condition."

(b) Randomly assign student subjects to high-money, low-money, and control groups. After playing Monopoly for a short time, the high-money group is left with $4000 in Monopoly money, the low-money group with $200, and the control group with no money. Each subject is asked to imagine a future with lots of money (high-money group), a little money (low-money group), or just their future plans (control group). Another student walks in and spills a box of 27 pencils. How many pencils does the subject pick up? "Participants in the high-money condition gathered fewer pencils" than subjects in the other two groups.

(c) Randomly assign student subjects to three groups. All do paperwork while a computer on the desk shows a screensaver of currency floating underwater (Group 1), a screensaver of fish swimming underwater (Group 2), or a blank screen (Group 3). Each subject must now develop an advertisement and can choose whether to work alone or with a partner. Count how many in each group make each choice. "Choosing to perform the task with a coworker was reduced among money condition participants."

Exploring the Web

26.43 **Confidence in the banking system.** The General Social Survey (GSS) is a socio-logical survey used to collect data on demographic characteristics and attitudes of residents of the United States. The survey is conducted face-to-face with an in-person interview by the National Opinion Research Center at the University of Chicago, of a randomly selected sample of adults (18+). SDA (Survey Documentation and Analysis) is a set of programs that allows you to analyze survey data and includes the GSS survey as part of its archive. Go to the website sda. berkeley.edu/ and click on *Archive*. Unless there is a more recent file, open the 1972–2012 cumulative data file (without the quick tables option).

(a) In the **Analysis** tab on the top of the page, click on *Comparison of means.* You are to do an ANOVA that examines how the mean age of the respondent varies with the person's confidence in the banking and financial systems. To do this, type in the dependent variable as "Age" and the row (treatment) variable as "Confinan." For the selection filter, type in "Year(2012)," or the most recent year available. For weight, change it to "noweight." Finally, in table options, the *only* boxes that should be checked are "Std dev," "N," and "ANOVA stats." Make sure checks are removed from the other boxes. Now click on *Run the table.*

(b) How many respondents are included in the analysis? What are the three means and standard deviations? In the ANOVA table, explain how the degrees of freedom were obtained. What are the F- and P-values? Write a brief report explaining the relationship between the average respondents age and their confidence in the banking system.

26.44 **Confidence in the banking system, continued.** This exercise is a continuation of the previous Web exercise. You are going to download the data set and reproduce the analysis, as well as provide some additional plots. First open the 1972–2012 cumulative data file following the instructions in the previous exercise.

(a) In the **Download** tab on the top of the page, click on *Customized Subset.* For the data file, if you highlight the CSV bubble, an Excel spread sheet will be downloaded. For the selection filter, again type in "Year(2012)" or the year used in the previous exercise. In the box for entering the names of individual variables, enter "Age" and "Confinan." Click *continue* on the bottom of the page. In the new window, click on *Create the Files* and in the next window click on *data files.* You can now either open or save the data file to your computer.

(b) Import the data into your statistical software package. You first need to "clean" the data a little because there are observations for which either the "Age" or "Confinan" variable are missing. For the "Confinan" variable, any value other than a 1, 2, or 3 is a missing value code. Delete these observations. For the "Age" variable, the missing value codes are 0, 98, and 99. Eliminate any observations with these values for "Age." You should now have the same number of observations as in the previous exercise.

(c) Draw comparative boxplots of the Age distribution for the three values of "Confinan." Describe the shapes of the three distributions. What information can you obtain from the boxplots that was not included in the output for the previous exercise?

(d) Reproduce the one-way ANOVA table using your software. Your results should agree with the previous exercise.

NOTES AND DATA SOURCES

Chapter 0 Notes

1. Parts of this essay are shared with David S. Moore, "Introduction: Learning from data," in Roxy Peck et al. (eds.), *Statistics: A Guide to the Unknown*, 4th ed., Thomson, 2006.

2. See, for example, Martin Enserink, "The vanishing promises of hormone replacement," *Science*, 297 (2002), pp. 325–326; and Brian Vastag, "Hormone replacement therapy falls out of favor with expert committee," *Journal of the American Medical Association*, 287 (2002), pp. 1923–1924. A National Institutes of Health panel's comprehensive report is *International Position Paper on Women's Health and Menopause,* NIH Publication 02-3284, 2002.

3. A. C. Nielsen, Jr., "Statistics in marketing," in *Making Statistics More Effective in Schools of Business*, Graduate School of Business, University of Chicago, 1986.

4. The data in Figure 0.2 were found online at `www.eia.gov/dnav/pet/hist/LeafHandler.ashx?n=PET&s=EMM_EPMR_PTE_NUS_DPG&f=W`.

5. FUTURE II Study Group, "Quadrivalent vaccine against human papillomavirus to prevent high-grade cervical lesions," *New England Journal of Medicine*, 356 (2007), pp. 1915–1927. We have simplified the conclusions so that students with as yet no statistics background can better follow the essay.

6. Rita F. Redburg, "Vitamin E and cardiovascular health," *Journal of the American Medical Association*, 294 (2005), pp. 107–109.

7. The data in Figure 0.3 were found online at `www1.eere.energy.gov/vehiclesandfuels/facts/2012_fotw741.html`.

8. Gerd Gigerenzer, "Dread risk, September 11, and fatal traffic accidents," *Psychological Science*, 15 (2004), pp. 286–287. The graph in Figure 0.3 was adapted from a graph in the article.

Chapter 1 Notes

1. Higher Education Research Institute 2011 Freshman Survey, at `www.heri.ucla.edu`.

2. *The Infinite Dial 2012: Navigating Digital Platforms*, at `www.arbitron.com`.

3. *Radio Today 2012: How America Listens to Radio*, at `www.arbitron.com`.

4. See Note 1.

5. Centers for Disease Control and Prevention, National Center for Health Statistics, *Births: Final data for 2010*, National Vital Statistics Reports, 61, No. 1, August 2012 at `www.cdc.gov`. These are the most recent data available

in May 2013, but the numbers change only slightly from year to year.

6. Marilyn M. Seastrom et al., "User's guide to computing high school graduation rates, Volume 1," from the National Center for Education Statistics, 2006, and the Department of Education Web site for the 2010 graduation rates at `ed.gov`.

7. Our eyes do respond to area, but not quite linearly. It appears that we perceive the ratio of two bars to be about the 0.7 power of the ratio of their actual areas. See W. S. Cleveland, *The Elements of Graphing Data*, Wadsworth, 1985, pp. 278–284.

8. From the 2012 Statistical Abstract at `census.gov/compendia/statab/2012/tables/12s0038.pdf`.

9. From the Gary Community School Corporation, courtesy of Celeste Foster, Purdue University.

10. The SAT Report on College & Career Readiness: 2012, from the College Board Web site at `www.collegeboard.org`.

11. From the Centers for Disease Control Web site at `cdc.gov/lyme/stats/index.html`. This is the chart of cases by age and sex, 2001–2010. The number of confirmed cases is from the chart of cases by symptom, 2001–2010.

12. The per capita total health expenditures in 2011 were obtained from the Country Data at the Global Health Observatory Data Repository of the World Health Organization at `http://apps.who.int/gho/data/`. All amounts are in international dollars at purchasing power parity. That is, the exchange rate between each currency, and the dollar is set not at the fluctuating market rate but at the rate that gives a dollar the same buying power in each country.

13. The U.S. Geological Survey maintains data for various water parameters at monitoring sites throughout the United States at `waterdata.usgs.gov/nwis`. The data can be graphed or downloaded. The data in Figure 1.12 are for USGS 254754080344300 SHARK RIVER SLOUGH NO.1.

14. College Entrance Examination Board, *Trends in College Pricing, 2012*, at `www.trends.collegeboard.org`. The averages are "enrollment weighted," so that they give average tuition over *students* rather than over *colleges*. The reported averages have been adjusted to constant 2012 dollars.

15. From the 2006 American Community Survey, at `factfinder.census.gov`.

16. *2012 DuPont Automotive Color Popularity Report*, at `www2.dupont.com`.

17. Keith N. Hampton et al., "Social networking sites and our lives," Pew Internet and American Life Project, 2011, at `www.pewinternet.org`.

18. Centers for Disease Control and Prevention, National Center for Health Statistics, *Deaths: Preliminary Data for 2011*, 61, No. 6, October 2012, at `www.cdc.gov/nchs`.

19. Sharon R. Ennis et al., "The Hispanic Population: 2010," 2010 U.S. Census Briefs at `www.census.gov`.

20. "2008 Student surveys: Complete results," *Macleans.ca*, February 19, 2008 at `oncampus.macleans.ca`.

21. Tom Lloyd et al., "Fruit consumption, fitness, and cardiovascular health in female adolescents: the Penn State Young Women's Health Study," *American Journal of Clinical Nutrition*, 67 (1998), pp. 624–630.

22. Data provided by Darlene Gordon from her PhD thesis, "Relationships among academic self-concept, academic achievement, and persistence with self-attribution, study habits, and perceived school environment," Purdue University, 1997.

23. Monthly stock returns from the Web site of Professor Kenneth French of Dartmouth, `mba.tuck.dartmouth.edu/pages/faculty/ken.french`. A fine point: the data are actually the "excess returns" on stocks, the actual returns less the small monthly returns on Treasury bills. The data is in the file Fama/French Benchmark Factors.

24. National Institutes of Health, Essential Fatty Acids Education site, `efaeducation.nih.gov`.

25. 2010 *Statistical Abstract of the United States*, Table 159, at `www.census.gov`.

26. As of the beginning of 2013, yearly data were available through 2009 at the United Nations Web site `unstats.un.org/unsd/mdg/SeriesDetail.aspx?srid=751&crid=`.

27. J. Ward Testa (editor), "Fur Seal Investigations, 2008–2009," *NOAA Technical Memorandum NMFS-AFSC-226*, (2011), p. 74. We would like to thank Rod Towell of NOAA Federal for supplying recent data not contained in the report.

28. David M. Fergusson and L. John Horwood, "Cannabis use and traffic accidents in a birth cohort of young adults," *Accident Analysis and Prevention*, 33 (2001), pp. 703–711.

29. From a plot in K. Krishna Kumar et al., "Unraveling the mystery of Indian monsoon failure during El Niño," *Science*, 314 (2006), pp. 115–119.

30. See Note 14.

31. U.S. Census Bureau, New Residential Construction page, at `http://www.census.gov/construction/nrc/`. Go to the *Search Database* link to download the data. These are monthly data that are not seasonally adjusted.

32. Ozone Hole Watch, at `ozonewatch.gsfc.nasa.gov/index.html`.

Chapter 2 Notes

1. From the 2003 American Community Survey, at the U.S. Census Bureau Web site, `www.census.gov`. The data are a subsample of the 13,194 individuals in the ACS North Carolina sample who had travel times greater than zero.

2. This study is available online at `www.dispatch.com/live/content/databases/index.html`.

3. This isn't a mathematical theorem. The mean can be less than the median in right-skewed distributions that take only a few values, many of which lie exactly at the median. The rule almost never fails for distributions taking many values, and most counterexamples don't appear clearly skewed in graphs even though they may be slightly skewed according to technical measures of skewness. See Paul T. von Hippel, "Mean, median, and skew: Correcting a textbook rule," *Journal of Statistics Education*, 13, No. 2 (2005), online journal.

4. National Association of College and University Business Officers, Commonfund study of endowments for 2012, at `www.nacubo.org`.

5. From the U.S. Census Bureau, `www.census.gov/const/uspricemon.pdf`.

6. U.S. Census Bureau, *Historical Income Tables: Households*, at `www.census.gov`.

7. The U.S. Department of Energy, `www.fueleconomy.gov/feg/download.shtml`.

8. We would like to thank Patricia Humphrey for supplying the test scores for students at Georgia Southern University.

9. From the Environmental Protection Agency, `www.epa.gov/radon/pubs/consguid.html`.

10. C. H. Cannon, D. R. Peart, and M. Leighton, "Tree species diversity in commercially logged Bornean rainforest," *Science*, 281 (1998), pp. 1366–1367. We thank Charles Cannon for providing the data.

11. Raymond Fisman and Edward Miguel, "Cultures of corruption: Evidence from diplomatic parking tickets," National Bureau of Economic Research Working Paper 12312, June 2006, at `www.nber.org`.

12. D. G. Jakovljevic and A. K. McConnell, "Influence of different breathing frequencies on the severity of inspiratory muscle fatigue induced by high-intensity front crawl swimming," *Journal of Strength and Conditioning Research*, 23, No. 4 (2009), pp. 1169–1174.

13. Current Population Survey 2012 Annual Social and Economic Supplement, at `www.census.gov`.

14. John J. Topoleski, *Retirement Savings and Household Wealth in 2010*, Congressional Research Service, April 2013.

15. T. Bjerkedal, "Acquisition of resistance in guinea pigs infected with different doses of virulent tubercle bacilli," *American Journal of Hygiene*, 72 (1960), pp. 130–148.

16. See Note 5 for Chapter 1.

17. Data for 1986 from David Brillinger, University of California, Berkeley. See David R. Brillinger, "Mapping aggregate birth data," in A. C. Singh and P. Whitridge (eds.), *Analysis of Data in Time*, Statistics Canada, 1990, pp. 77–83.

A boxplot similar to Figure 2.6 appears in David R. Brillinger, "Some examples of random process environmental data analysis," in P. K. Sen and C. R. Rao (eds.), *Handbook of Statistics*, Vol. 18, *Bioenvironmental and Public Health Statistics*, North Holland, 2000.

18. Paul E. O'Brien et al., "Laparascopic adjustable gastric banding in severely obese adolescents," *Journal of the American Medical Association*, 303 (2010), pp. 519–526. We thank the authors for providing the data.

19. The current roster is from `canadiens.nhl.com`, and their salaries were obtained from `usatoday.com/sports/nhl`.

20. Nicolas Guéguen and Christine Petr, "Odors and consumer behavior in a restaurant," *Journal of Hospitality Management*, 25 (2006), pp. 335–339. We thank Nicolas Guéguen for providing the data.

21. James A. Levine et al., "Inter-individual variation in posture allocation: Possible role in human obesity," *Science*, 307 (2005), pp. 584–586. We thank James Levine for providing the data.

22. Bruce Rind and David Strohmetz, "Effects of beliefs about future weather conditions on restaurant tipping," *Journal of Applied Social Psychology*, 31 (2001), pp. 2160–2164. We thank the authors for supplying the original data.

23. Information and data from the NHANES survey can be found at `cdc.gov/nchs/nhanes.htm`.

Chapter 3 Notes

1. See Note 9 for Chapter 1.

2. Cheryl D. Fryar et al., "Anthropometric reference data for children and adults: United States, 2007–2010," *Vital and Health Statistics*, Series 11, Number 252, (October, 2012), at `www.cdc.gov/nchs`. This report provides the means of various anthropometric measurements. Standard deviations were computed from the first and third quartiles assuming Normality. Instructions on measuring upper arm length are in the *National Health and Nutrition Examination Survey: Anthropometry Procedures Manual*, January 2007.

3. Monsoon rainfall from B. Parthasarathy, Indian Institute of Tropical Meterology, at `www.iges.org`. The data cover the years 1871 to 2000.

4. See Note 2.

5. All SAT facts are from the College Board Web site, `www.collegeboard.com`, and all ACT facts are from the ACT Web site, `www.act.org`.

6. See Note 2.

7. From the 2009–2010 Guide for the College Bound Student Athlete at `www.ncaastudent.org/NCAA_Guide`.

8. All MCAT facts are from the Medical College Admissions Test Web site, `www.aamc.org/students/mcat/`.

9. Detailed data appear in P. S. Levy et al., "Total Serum Cholesterol Values for Youths 12–17 Years," *Vital and Health Statistics*, Series 11, No. 155, National Center for Health Statistics, 1976.

10. See Note 21 for Chapter 2.

11. See Note 2.

12. See Note 22 for Chapter 1.

13. The data were provided by Nicolas Fisher.

14. See Note 8 for Chapter 2.

15. See Note 3.

16. See Note 23 for Chapter 2.

Chapter 4 Notes

1. Neal E. Cantin et al., "Ocean warming slows coral growth in the Central Red Sea," *Science*, 329 (2010), pp. 322–325.

2. Data for 2012 graduates from the College Board Web site, `www.collegeboard.org`.

3. Initial concerns were based on government data for 2005, presented in "An accident waiting to happen?" *Consumer Reports*, March 2007, pp. 16–19. Data for 2011 was found online at `www.transtats.bts.gov` (for delay percent by airline) and `web.mit.edu/airlinedata/www/default.html` (outsource).

4. The Florida Department of Highway Safety and Motor Vehicles (at `www.flhsmv.gov/dmv/vslfact.html`) gives the number of registered vessels. The Florida Wildlife Commission maintains a manatee death database at `research.myfwc.com/manatees`.

5. Based on T. N. Lam, "Estimating fuel consumption from engine size," *Journal of Transportation Engineering*, 111 (1985), pp. 339–357. The data for 10 to 50 km/h are measured; those for 60 and higher are calculated from a model given in the paper and are therefore smoothed.

6. A careful study of this phenomenon is W. S. Cleveland, P. Diaconis, and R. McGill, "Variables on scatterplots look more highly correlated when the scales are increased," *Science*, 216 (1982), pp. 1138–1141.

7. See Note 1.

8. Data for Figure 4.6(b) come from William Gray's Web site, at `hurricane.atmos.colostate.edu`. Data for Figure 4.6(c) were provided by Drina Iglesia, Purdue University, from a study reported in D. D. S. Iglesia, E. J. Cragoe, Jr., and J. W. Vanable, "Electric field strength and epithelization in the newt (*Notophthalmus viridescens*)," *Journal of Experimental Zoology*, 274 (1996), pp. 56–62. Data for Figure 4.6(d) are for the Wilshire 5000 stock index. As a fine point, plots (b), (c), and (d) are square with the same scales on both axes because both variables measure similar quantities in the same units.

9. This exercise is motivated by Scott Berry, "Statistical fallacies in sports," *Chance*, 19, No. 4 (2006), pp. 50–56, where scores from the 2006 Masters are analyzed. Masters scores for 2012 were found online at `www.masters.com/en_US/discover/past_winners.html`.

10. Andrew J. Oswald, et al., "Objective confirmation of subjective measures of human well-being: evidence from the U.S.A.," *Science*, 327 (2010), pp. 576–579.

11. From a graph in Naomi E. Allen et al., "Moderate alcohol intake and cancer incidence in women," *Journal of the National Cancer Institute,* 101 (2009), pp. 296–305.

12. From a graph in Magdalena Bermejo et al., "Ebola outbreak killed 5000 gorillas," *Science,* 314 (2006), p. 1564.

13. From a graph in Bernt-Erik Saether, Steiner Engen, and Erik Mattysen, "Demographic characteristics and population dynamical patterns of solitary birds," *Science,* 295 (2002), pp. 2070–2073.

14. From a graph in Sabrina M. Tom et al., "The neural basis of loss aversion in decision-making under risk," *Science,* 315 (2007), pp. 515–518.

15. From a graph in Sergio M. Vallina and Rafel Simó, "Strong relationship between DMS and the solar radiation dose over the global surface ocean," *Science,* 315 (2007), pp. 506–508.

16. From a graph in Camilla A. Hinde et al., "Parent-offspring conflict and coadaptation," *Science,* 327 (2010), pp. 1373–1376.

17. See Note 22 for Chapter 2.

18. From a graph in Martin Wild et al., "From dimming to brightening: Decadal changes in solar radiation at Earth's surface," *Science,* 308 (2005), pp. 847–850.

19. Brian J. Whipp and Susan A. Ward, "Will women soon outrun men?" *Nature,* 355 (1992), p. 25. An article in Scientific American in 2004 made a similar prediction that women would outrun men in the 100-meter dash in 2156. This article can be found online at www.scientificamerican.com/article.cfm?id=data-trends-suggest-women.

20. From a graph in Glenn J. Tattersall et al., "Heat exchange from the toucan bill reveals a controllable vascular thermal radiator," *Science,* 325 (2009), pp. 468–470.

21. From a graph in Naomi I. Eisenberger, Matthew D. Lieberman, and Kipling D. Williams, "Does rejection hurt? An fMRI study of social exclusion," *Science,* 302 (2003), pp. 290–292.

22. Justin S. Brashares et al., "Bushmeat hunting, wildlife declines, and fish supply in West Africa," *Science,* 306 (2004), pp. 1180–1183. The data used here are found in the online supplementary material. The published analysis omits data for 1999, an extreme low outlier, without explanation.

Chapter 5 Notes

1. From a graph in James A. Levine, Norman L. Eberhardt, and Michael D. Jensen, "Role of nonexercise activity thermogenesis in resistance to fat gain in humans," *Science,* 283 (1999), pp. 212–214.

2. See Note 1 for Chapter 4.

3. From a graph in Tania Singer et al., "Empathy for pain involves the affective but not sensory components of pain," *Science,* 303 (2004), pp. 1157–1162. Data for other brain regions showed a stronger correlation and no outliers.

4. Contributed by Marigene Arnold, Kalamazoo College.

5. Gannett News Service article appearing in the *Lafayette (Ind.) Journal and Courier,* April 23, 1994.

6. P. Goldblatt (ed.), *Longitudinal Study: Mortality and Social Organization,* Her Majesty's Stationery Office, 1990. At least, so claims Richard Conniff, *The Natural History of the Rich,* Norton, 2002, p. 45. The Goldblatt report is not available to us.

7. Laura L. Calderon et al., "Risk factors for obesity in Mexican-American girls: Dietary factors, anthropometric factors, physical activity, and hours of television viewing," *Journal of the American Dietetic Association,* 96 (1996), pp. 1177–1179.

8. *The Health Consequences of Smoking: 1983,* Public Health Service, Washington, D.C., 1983.

9. Data provided by Robert Dale, Purdue University.

10. G. L. Kooyman et al., "Diving behavior and energetics during foraging cycles in king penguins," *Ecological Monographs,* 62 (1992), pp. 143–163.

11. Chu, S., "Diamond ring pricing using simple linear regression," *Journal of Statistics Education,* 4 (1996), available online at www.amstat.org/publications/jse/v4n3/datasets.chu.html.

12. The last data pair is the height of one of the authors and his sister. The first 11 data pairs are from Karl Pearson and A. Lee, "On the laws of inheritance in man," *Biometrika,* 2 (1902), p. 357. These first 11 data also appear in D. J. Hand et al., *A Handbook of Small Data Sets,* Chapman & Hall, 1994. This book offers more than 500 data sets that can be used in statistical exercises.

13. From a presentation by Charles Knauf, Monroe County (N.Y.) Environmental Health Laboratory.

14. See Note 13 for Chapter 4.

15. Frank J. Anscombe, "Graphs in statistical analysis," *The American Statistician,* 27 (1973), pp. 17–21.

16. Debora L. Arsenau, "Comparison of diet management instruction for patients with non-insulin dependent diabetes mellitus: Learning activity package vs. group instruction," MS thesis, Purdue University, 1993.

17. Gary Smith, "Do statistics test scores regress toward the mean?" *Chance,* 10, No. 4 (1997), pp. 42–45.

18. From a graph in G. D. Martinsen, E. M. Driebe, and T. G. Whitham, "Indirect interactions mediated by changing plant chemistry: Beaver browsing benefits beetles," *Ecology,* 79 (1998), pp. 192–200.

19. P. Velleman, *ActivStats 2.0,* Addison Wesley Interactive, 1997.

20. From William Gray's Web site, hurricane.atmos.colostate.edu. Forecasts are those made each June.

21. Data for 1936–1999 are from a graph in Bruce J. Peterson et al., "Increasing river discharge to the Arctic Ocean," *Science,* 298 (2002), pp. 2171–2173. Data for 2000–2008 are from a graph in I. Ashik et al., "Arctic report card: Update for 2010," available online at http://www.arctic.noaa.gov/reportcard_previous.html. The graph is on page 41 of the report. Data for

2009–2010 are from a graph in K. R. Arrigo et al., "Arctic report card: Update for 2011," available online at `http://www.arctic.noaa.gov/reportcard_previous.html`. The graph is on page 157 of the report.
22. See Note 19 for Chapter 4.

Chapter 6 Notes

1. Lydia Said, "In U.S., half of women prefer a job outside the home," Gallup, September 7, 2012. Found online, `www.gallup.com`.
2. Rani A. Desai et al., "Video-gaming among high school students: Health correlates, gender differences, and problematic gaming," *Pediatrics,* 126 (2010), pp. 1416–1424.
3. From the October 2009 Current Population Survey, at `www.census.gov`.
4. Siem Oppe and Frank De Charro, "The effect of medical care by a helicopter trauma team on the probability of survival and the quality of life of hospitalized victims," *Accident Analysis and Prevention,* 33 (2001), pp. 129–138. The authors give the data in Example 6.4 as a "theoretical example" to illustrate the need for their more elaborate analysis of actual data using severity scores for each victim.
5. Found online at `www.math.kent.edu/darci/simpson/bballexamples.html`, a Web site maintained by Darci L. Kracht at Kent State University. Thanks to Patricia Humphrey at Georgia Southern University for bringing this example to our attention.
6. I. Westbrooke, "Simpson's paradox: An example in a New Zealand survey of jury composition," *Chance,* 11 (1998), pp. 40–42.
7. These data are from an April 20, 2010 report, "Teens and mobile phones," by Amanda Lenhart, Rich Ling, Scott Campbell, Kristen Purcell of the Pew Internet and American Life Project. Found online at `pewinternet.org/Reports/2010/Teens-and-Mobile-Phones.aspx`.
8. This General Social Survey exercise presents a table constructed using the search function at the GSS archive, `sda.berkeley.edu/archive.htm`. These data are from the 2012 GSS.
9. Gregory D. Myer et al., "Youth versus adult weightlifting injuries presenting to United States emergency rooms: Accidental versus nonaccidental injury mechanisms," *Journal of Strength and Conditioning Research,* 23 (2009), pp. 2054–2060.
10. Sanders Korenman and David Neumark, "Does marriage really make men more productive?" *Journal of Human Resources,* 26 (1991), pp. 282–307.
11. M. Radelet, "Racial characteristics and imposition of the death penalty," *American Sociological Review,* 46 (1981), pp. 918–927.
12. D. Gonzales et al., "Journal of the American Medical Association Varenicline, an $\alpha 4\beta 2$ Nicotinic Acetylcholine Receptor Partial Agonist, vs Sustained-Release Bupropion and Placebo for Smoking Cessation," *New England Journal of Medicine,* 340 (1999), pp. 685–691.
13. Michael Gurian, "Where have the men gone? No place good," *Washington Post,* December 4, 2005, at `www.washingtonpost.com`. The data are from the 2009 *Digest of Education Statistics* at the Web site of the National Center for Education Statistics, `nces.ed.gov`.
14. Nancy J. O. Birkmeyer, "Hospital complication rates with bariatric surgery in Michigan," *Journal of the American Medical Association,* 304 (2010), pp. 435–442.
15. The data for the University of Michigan Health and Retirement Study (HRS) can be downloaded from the Web site `ssl.isr.umich.edu/hrs/start.php`.
16. R. Shine, T. R. L. Madsen, M. J. Elphick, and P. S. Harlow, "The influence of nest temperatures and maternal brooding on hatchling phenotypes in water pythons," *Ecology,* 78 (1997), pp. 1713–1721.
17. Hannah Lund et al., "Sleep patterns and predictors of disturbed sleep in a large population of college students," *Journal of Adolescent Health,* 46 (2010), 124–132. We would like to thank the authors for supplying the data.

Chapter 7 Notes

1. Harry B. Meyers, "Investigations of the life history of the velvetleaf seed beetle, *Althaeus folkertsi Kingsolver,*" MS thesis, Purdue University, 1996.
2. Kristen Purcell et al., "Search engine use 2012," March 2012, Pew Internet and American Life Project at `www.pewinternet.org`.
3. See Note 8 for Chapter 3.
4. Data provided by Brigitte Baldi, University of California at Irvine.
5. From a graph in Stan Boutin et al., "Anticipatory reproduction and population growth in seed predators," *Science,* 314 (2006), pp. 1928–1930.
6. J. T. Dwyer et al., "Memory of food intake in the distant past," *American Journal of Epidemiology,* 130 (1989), pp. 1033–1046.
7. Data from a plot in Josef P. Rauschecker, Biao Tian, and Marc Hauser, "Processing of complex sounds in the macaque nonprimary auditory cortex," *Science,* 268 (1995), pp. 111–114. The paper states that there are $n = 41$ observations, but only $n = 37$ can be read accurately from the plot.
8. Mei-Hui Chen, "An exploratory comparison of American and Asian consumers' catalog patronage behavior," MS thesis, Purdue University, 1994.
9. "Dancing in step," *Economist,* March 22, 2001.
10. From a plot in Jon J. Ramsey et al., "Energy expenditure, body composition, and glucose metabolism in lean and obese rhesus monkeys treated with ephedrine and caffeine," *American Journal of Clinical Nutrition,* 68 (1998), pp. 42–51.

11. Centers for Disease Control and Prevention, *Cigarette Smoking among Adults–United States, 2006*, and related publications at www.cdc.gov.

12. Janice E. Williams et al., "Anger proneness predicts coronary heart disease risk," *Circulation*, 101 (2000), pp. 2034–2039.

13. From a graph in Peter A. Raymond and Jonathan J. Cole, "Increase in the export of alkalinity from North America's largest river," *Science*, 301 (2003), pp. 88–91.

14. From the Nenana Ice Classic Web site, www.nenanaakiceclassic.com. See Raphael Sagarin and Fiorenza Micheli, "Climate change in nontraditional data sets," *Science*, 294 (2001), p. 811, for a careful discussion.

15. Data for 2012 from the Organization for Economic Cooperation and Development website at oecd-library.org/statistics.

16. Louie H. Yang, "Periodical cicadas as resource pulses in North American forests," *Science*, 306 (2004), pp. 1565–1567. The data are simulated Normal values that match the means and standard deviations reported in this article.

17. Alan S. Banks et al., "Juvenile hallux abducto valgus association with metatarsus adductus," *Journal of the American Podiatric Medical Association*, 84 (1994), pp. 219–224.

18. Todd W. Anderson, "Predator responses, prey refuges, and density-dependent mortality of a marine fish," *Ecology*, 81 (2001), pp. 245–257.

19. From a graph in Craig Packer et al., "Ecological change, group territoriality, and population dynamics in Serengeti lions," *Science*, 307 (2005), pp. 390–393.

20. Peter H. Chen, Neftali Herrera, and Darren Christiansen, "Relationships between gate velocity and casting features among aluminum round castings," no date. Provided by Darren Christiansen.

21. Data compiled from a table of percents in "Americans view higher education as key to the American dream," press release by the National Center for Public Policy and Higher Education, at www.highereducation.org, May 3, 2000.

Chapter 8 Notes

1. From the *New York Times*/CBS News poll at www.nytimes.com. The methodological statement is similar for most polls listed.

2. Gary S. Foster and Craig M. Eckert, "Up from the grave: a sociohistorical reconstruction of an African American community from cemetary data in the rural Midwest," *Journal of Black Studies*, 33 (2003), pp. 468–489.

3. Pew Forum on Religion and Public Life, *Spirit and Power: A 10-Country Survey of Pentecostals*, October 2006, at www.pewforum.org.

4. The regulations that govern seat belt survey design can be found at www-nrd.nhtsa.dot.gov. Details on the Hawaii survey are in Karl Kim et al., *Results of the 2002 Highway Seat Belt Use Survey*, at www.state.hi.us/dot.

5. Donald L. McCabe, Linda Klebe Trevino, and Kenneth D. Butterfield, "Dishonesty in academic environments," *Journal of Higher Education*, 72 (2001), pp. 29–45.

6. For information about the response rates for the American Community Survey of households (there is a separate sample of group quarters), go to www.census.gov/acs/www/methodology/response_rates_data/.

7. Maeve Duncan and Joanna Brenner, "The demographics of social media users—2012," (2013). The Pew press release and the full report are at pewinternet.org/Reports/2013/Social-media-users.aspx.

8. For more detail on the limits of memory in surveys, see N. M. Bradburn, L. J. Rips, and S. K. Shevell, "Answering autobiographical questions: The impact of memory and inference on surveys," *Science*, 236 (1987), pp. 157–161.

9. The immigration questions are from the *New York Times/CBS News Poll* taken May 18 to 23, 2007, found at www.pollingreport.com. The responses on welfare are from a *New York Times/CBS News Poll* reported in the *New York Times*, July 5, 1992. Many other examples appear in T. W. Smith, "That which we call welfare by any other name would smell sweeter," *Public Opinion Quarterly*, 51 (1987), pp. 75–83. The example on the effect of question order is cited in Daniel Kahnemann et al., "Would you be happier if you were richer? A focusing illusion," *Science*, 312 (2006), pp. 1908–1910.

10. The research study "Health reform and the decline of physician private practice," (2010) found at www.physiciansfoundation.org.

11. You can go to www.pollingreport.com to see the results of many polling agencies compiled on a variety of issues.

12. Information from various articles in the special issue on cell phone surveys, *Public Opinion Quarterly*, 71, No. 5 (2007). The 2012 cell phone use numbers were obtained from the CDC Web site. www.cdc.gov.

13. See Mick P. Couper, "Web surveys: A review of issues and approaches," *Public Opinion Quarterly*, 64 (2000), pp. 464–494.

14. The Renfew Center Foundation, "New survey indicates there's more to makeup use than meets the eye," at wwww.renfrewcenter.com.

15. The American Association for Public Opinion Research, "AAPOR report on online panels," 2010, at www.aapor.org.

16. Rachel Sherman and John Hickner, "Academic physicians use placebos in clinical practice and believe in the mind-body connection," *Journal of General Internal Medicine*, 23 (2008), pp. 7–10.

17. The National Public Radio Fans on Facebook 2010 survey can be found at www.slideshare.net/nprresearch/npr-facebook-fans-survey-findings-overview.

18. From the Web site of the Gallup Organization, www.gallup.com. Individual poll reports remain on this site for only a limited time.

19. Information about area codes can be found on the North American Numbering Plan Administration Web site at, `www.nanpa.com`.

20. Robert C. Parker and Patrick A. Glass, "Preliminary results of double-sample forest inventory of pine and mixed stands with high- and low-density LiDAR," in Kristina F. Connoe (ed.), *Proceedings of the 12th Biennial Southern Silvicultural Research Conference*, U.S. Department of Agriculture, Forest Service, Southern Research Station, 2004. The researchers actually sampled every 10th plot. This is a systematic sample; see Exercise 8.43.

21. The questions are available from a Gallup poll, April 4–7, 2013, at `www.gallup.com/poll/1714/taxes.aspx`.

22. Bryan E. Porter and Thomas D. Berry, "A nationwide survey of self-reported red light running: Measuring prevalence, predictors, and perceived consequences," *Accident Analysis and Prevention*, 33 (2001), pp. 735–741.

23. Mario A. Parada et al., "The validity of self-reported seatbelt use: Hispanic and non-Hispanic drivers in El Paso," *Accident Analysis and Prevention*, 33 (2001), pp. 139–143.

24. Giuliana Coccia, "An overview of non-response in Italian telephone surveys," *Proceedings of the 99th Session of the International Statistical Institute*, 1993, Book 3, pp. 271–272.

25. Information about the Ontario College of Pharmacists obtained from their Web site at `www.ocpinfo.com`. The numbers reported are from the 2011–12 Annual Report.

26. The Health Care in Canada Survey can be found by going to survey reports and presentations at the Web site `www.hcic-sssc.ca`.

27. Clyde O. McDaniel, Jr., "Dating roles and reasons for dating," *Journal of Marriage and the Family*, 31 (1969), pp. 97–107.

28. The question is available from a Gallup poll, March 7–10, 2013, at `www.gallup.com/poll/161645/americans-concerns-global-warming-rise.aspx`.

29. The article can be found at `www2.macleans.ca/2010/07/16/sometimes-a-gaffe-is-more-than-a-gaffe/`.

Chapter 9 Notes

1. I. J. Goldberg et al., "Wine and your heart: A science advisory for healthcare professionals from the Nutrition Committee, Council on Epidemiology and Prevention, and Council on Cardiovascular Nursing of the American Heart Association," *Circulation*, 103 (2001), pp. 472–475.

2. J. E. Muscat et al., "Handheld cellular telephone use and risk of brain cancer," *Journal of the American Medical Association*, 284 (2000), pp. 3001–3007.

3. Hyunjin Song and Norbert Schwarz, "If it's hard to read, it's hard to do: Processing fluency affects effort prediction and motivation," *Psychological Science*, 19 (2008), pp. 986–988.

4. Hsin-Chieh Yeh et al., "Smoking, smoking cessation, and risk for type 2 diabetes mellitus: A cohort study," *Annals of Internal Medicine*, 152 (2010), pp. 10–17.

5. Charles A. Nelson III et al., "Cognitive recovery in socially deprived young children: The Bucharest Early Intervention Project," *Science*, 318 (2007), pp. 1937–1940.

6. The description of the factors and the response is based on a portion of the study by Alice Healy et al.,"Terrorism after 9/11: Reactions to simulated news reports," *American Journal of Psychology*, 122 (2009), pp. 153–165.

7. The description of the factors and the response is based on a study by M. K. Baloch and F. Bibi, "Effect of harvesting and storage conditions on the post harvest fruit quality and shelf life of mango, (*Mangifera indica* L.) fruit," *South African Journal of Botany*, 83 (2012), pp. 109–116.

8. See Note 18 for Chapter 2.

9. K. B. Suttle, Meredith A. Thomsen, and Mary E. Power, "Species interactions reverse grassland responses to changing climate," *Science*, 315 (2007), pp. 640–642. See Chapter 25 for an analysis of some data from this experiment.

10. Julie Mares et al., "Healthy diets and the subsequent prevalence of nuclear cataract in women," *Archives of Opthalmology*, 128 (2010), pp. 738–749.

11. Marielle H. Emmelot-Vonk et al., "Effect of testosterone supplementation on functional mobility, cognition, and other parameters in older men," *Journal of the American Medical Association*, 299 (2008), pp. 39–52.

12. David L. Strayer, Frank A. Drews, and William A. Johnston, "Cell phone-induced failures of visual attention during simulated driving," *Journal of Experimental Psychology: Applied*, 9 (2003), pp. 23–32.

13. See Note 12 for Chapter 2.

14. Alysha Fligner and Xiaoyan Deng, "The effect of font naturalness on perceived healthiness of food products," Senior thesis, The Ohio State University.

15. Sterling C. Hilton et al., "A randomized controlled experiment to assess technological innovations in the classroom on student outcomes: An overview of a clinical trial in education," manuscript, no date. A brief report is Sterling C. Hilton and Howard B. Christensen, "Evaluating the impact of multimedia lectures on student learning and attitudes," *Proceedings of the 6th International Conference on the Teaching of Statistics*, at `www.stat.aukland.ac.nz`.

16. S. Katherine Nelson et. al., "In defense of parenthood: Children are associated with more joy than misery," *Psychological Science*, 24 (2013), pp. 3–10.

17. Brad J. Bushman, "Violence and sex in television programs do not sell products in advertisements," *Psychological Science*, 16 (2005), pp. 702–707.

18. An Pan et al., "Red meat consumption and mortality," *Archives of Internal Medicine*, 172 (2012), pp. 555–563.

19. Margot Roosevelt, "Study finds traffic pollution can speed hardening of arteries," *LA Times*, February 14, 2010.

20. Jo Phelan et al., "The stigma of homelessness: The impact of the label 'homeless' on attitudes towards poor persons," *Social Psychology Quarterly*, 60 (1997), pp. 323–337.

21. Esther Duflo, Rema Hanna, and Stephan Ryan, "Monitoring works: Getting teachers to come to school," report dated November 21, 2007, at `econ-mit.edu/files/2066`.
22. John H. Kagel, Raymond C. Battalio, and C. G. Miles, "Marijuana and work performance: Results from an experiment," *Journal of Human Resources*, 15 (1980), pp. 373–395.
23. Shailja V. Nigdikar et al., "Consumption of red wine polyphenols reduces the susceptibility of low-density lipoproteins to oxidation in vivo," *American Journal of Clinical Nutrition*, 68 (1998), pp. 258–265. (There were in fact only 30 subjects, some of whom received more than one treatment with a four-week period intervening.)
24. The description of the factors and the response is based on a portion of the study by Brian Wnasik and Perre Chandon, "Can 'low-fat' nutrition labels lead to obesity," *Journal of Marketing Research*, 43 (2006), pp. 605–617.
25. Ian G. Williamson et al., "Antibiotics and topical nasal steroid for treatment of acute maxillary sinusitis," *Journal of the American Medical Association*, 298 (2007), pp. 2487–2496.
26. Based on Evan H. DeLucia et al., "Net primary production of a forest ecosystem with experimental CO_2 enhancement," *Science*, 284 (1999), pp. 1177–1179. The investigators used the block design.
27. E. M. Peters et al., "Vitamin C supplementation reduces the incidence of postrace symptoms of upper-respiratory tract infection in ultramarathon runners," *American Journal of Clinical Nutrition*, 57 (1993), pp. 170–174.
28. The study is described in Gina Kolata, "New study finds vitamins are not cancer preventers," *New York Times*, July 21, 1994. Look in the *Journal of the American Medical Association* of the same date for the details.
29. R. C. Shelton et al., "Effectiveness of St. John's wort in major depression," *Journal of the American Medical Association*, 285 (2001), pp. 1978–1986.

Data Ethics Notes

1. John C. Bailar III, "The real threats to the integrity of science," *Chronicle of Higher Education*, April 21, 1995, pp. B1–B2.
2. For more on the problems with open-access publishing see Gina Kolata, "Scientific articles accepted (personal checks, too)," *The New York Times*, April 7, 2013, and the March 28, 2013, issue of *Nature*.
3. See the details on the Web site of the Office for Human Research Protections of the Department of Health and Human Services, `www.hhs.gov/ohrp`.
4. The difficulties of interpreting guidelines for informed consent and for the work of institutional review boards in medical research are a main theme of Beverly Woodward, "Challenges to human subject protections in U.S. medical research," *Journal of the American Medical Association*, 282 (1999), pp. 1947–1952. The references in this paper point

to other discussions. Updated regulations and guidelines appear on the OHRP Web site (see Note 2).
5. Quotation from the *Report of the Tuskegee Syphilis Study Legacy Committee*, May 20, 1996. A detailed history is James H. Jones, *Bad Blood: The Tuskegee Syphilis Experiment*, New York: Free Press, 1993.
6. Dr. Hennekens's words are from an interview in the Annenberg/Corporation for Public Broadcasting video series *Against All Odds: Inside Statistics*. The lack of certainty that Dr. Hennekens refers to is now called "clinical equipoise" in discussions of ethics.
7. For more discussion of the ethics of the use of placebos in clinical trials see `http://www.ama-assn.org/ama/pub/physician-resources/medical-ethics/code-medical-ethics/opinion8083.page` on the American Medical Association Web site.
8. R. D. Middlemist, E. S. Knowles, and C. F. Matter, "Personal space invasions in the lavatory: Suggestive evidence for arousal," *Journal of Personality and Social Psychology*, 33 (1976), pp. 541–546.
9. For a review of domestic violence experiments, see C. D. Maxwell et al., *The Effects of Arrest on Intimate Partner Violence: New Evidence from the Spouse Assault Replication Program*, U.S. Department of Justice, NCH188199, 2001. Available online at `www.ojp.usdoj.gov/nij/pubs-sum/188199.htm`.
10. Joseph Millum and Ezekial J. Emanuel, "The ethics of international research with abandoned children," *Science*, 318 (2007), pp. 1874–1875. This paper has some useful comments on international research in general.

Chapter 10 Notes

1. Based on a news item "Bee off with you," *Economist*, November 2, 2002, p. 78.
2. Simplified from D. A. Marcus et al., "A double-blind provocative study of chocolate as a trigger of headache," *Cephalalgia*, 17 (1997), pp. 855–862.
3. Votes as of June 27, 2007, at `www.pbs.org/wgbh/nova/sciencenow`.
4. Simplified from Sanjay K. Dhar, Claudia González-Vallejo, and Dilip Soman, "Modeling the effects of advertised price claims: Tensile versus precise pricing," *Marketing Science*, 18 (1999), pp. 154–177.
5. K. J. Mukamal et al., "Prior alcohol consumption and mortality following acute myocardial infarction," *Journal of the American Medical Association*, 285 (2001), pp. 1965–1970.

Chapter 11 Notes

1. The Gallup Poll is based on telephone interviews conducted October 6–9, 2011. Each adult interviewed by Gallup had a known chance of being among those selected, but this chance depended on characteristics such as gender, age, race, Hispanic ethnicity, education, region, adults in the household, and phone status (cell phone only/landline only/both, cell phone mostly, and having an unlisted landline

number). Gallup used special weights to adjust for differences in the probability of being selected to obtain an estimate of the proportion of all adults who owned a gun at the time of the poll in the population of all U.S. adults. The actual estimate used by Gallup was close to 34% and uses the result from the sample to estimate what is true for the population.

2. Note that pennies have rims that make spinning more stable. The probability of a head in spinning a coin depends on the type of coin and also on the surface. See Exercise 21.3 for an account of 56% of heads in spinning a Belgian 1-euro coin. *Chance News 11.02* at `www.dartmouth.edu/~chance` reports about 45% heads in more than 20,000 spins of American pennies by Robin Lock's students at Saint Lawrence University.

3. The percentages were found at the GMAT Web site, `http://www.gmac.com/market-intelligence-and-research/research-library/gmat-test-taker-data/profile-documents/profile-of-gmat-candidates-2007-08-to-2011-12.aspx`.

4. Data for 2006 are from the Web site of Statistics Canada, `www.statcan.gc.ca`.

5. You can find a mathematical explanation of Benford's law in Ted Hill, "The first-digit phenomenon," *American Scientist*, 86 (1996), pp. 358–363; and Ted Hill, "The difficulty of faking data," *Chance*, 12, No. 3 (1999), pp. 27–31. Applications in fraud detection are discussed in the second paper by Hill and in Mark A. Nigrini, "I've got your number," *Journal of Accountancy*, May 1999, available online at `www.aicpa.org/pubs/jofa/joaiss.htm`.

6. Based on a November 2007 Gallup poll, found at `www.gallup.com/poll/1648/Personal-Health-Issues.aspx#2`.

7. Information from `gradedistribution.registrar.indiana.edu/reports.php`.

8. Thomas K. Cureton et al., *Endurance of Young Men*, Monographs of the Society for Research in Child Development, Vol. 10, No. 1, 1945.

9. Based on an April 2013 Gallup poll, found at `www.gallup.com/poll/15370/Party-Affiliation.aspx`.

10. Data from 2009 found online at `http://nhts.ornl.gov/tables09/fatcat/2009/household_HHVEHCNT.htm`.

11. See Note 16 for Chapter 1.

12. National population estimates for July 1, 2011, at the Census Bureau Web site `www.census.gov`. The table omits people who consider themselves to belong to more than one race.

13. Based on data from the *2012 Statistical Abstract of the United States*, Table 58, at `www.census.gov`.

Chapter 12 Notes

1. This is one of several tests discussed in Bernard M. Branson, "Rapid HIV testing: 2005 update," a presentation by the Centers for Disease Control and Prevention, at `www.cdc.gov`. The Malawi clinic result is reported by Bernard M. Branson, "Point-of-care rapid tests for HIV antibody," *Journal of Laboratory Medicine*, 27 (2003), pp. 288–295.

2. Robert P. Dellavalle et al., "Going, going, gone: Lost Internet references," *Science*, 302 (2003), pp. 787–788.

3. From the U.S. Department of Commerce Bureau of Economic Analysis at `www.bea.gov`. Motor vehicle sales information is included with the National Economic Accounts, and the numbers reported are based on sales from April 2012–March 2013.

4. Sales in 2006 from the Web site of the Entertainment Software Association, at `www.theesa.com`.

5. Information about Internet users comes from sample surveys carried out by the Pew Internet and American Life Project, at `www.pewinternet.org`.

6. Hal R. Arkes and Wolfgang Gaissmaier, "Psychological research and the prostate-cancer screening controversy," *Journal of Psychological Science*, 23(6) (2012), pp. 547–553.

7. W. Uter et al., "Inter-relation between variables determining constitutional UV sensitivity in Caucasian children," *Photodermatology, Photoimmunology and Photmeicine*, 20, (2004), pp. 9–13.

8. F. H. Schroder et al., "Screening and prostate-cancer mortality in a randomized European study," *New England Journal of Medicine*, 360(2009) pp. 1320–1328.

9. From the statistics page of the National Science Foundation Web site at, `www.nsf.gov/statistics`.

10. M. McCulloch et al., "Diagnostic accuracy of canine scent detection in early- and late-stage lung and breast cancers," *Integrative Cancer Therapies*, 5 (2006), pp. 30–39.

11. From the Internal Revenue Service Web site, at `www.irs.gov/taxstats`.

12. Projections from U.S. Department of Education, *Projections of Education Statistics to 2016*, December 2007, at `nces.ed.gov`.

13. Data provided by Patricia Heithaus and the Department of Biology at Kenyon College.

14. F. J. G. M. Klaassen and J. R. Magnus, "How to reduce the service dominance in tennis? Empirical results from four years at Wimbledon," in S. J. Haake and A. O. Coe (eds.), *Tennis Science and Technology*, Blackwell, 2000, pp. 277–284.

15. R. S. Gupta et al., "The Prevalence, Severity, and Distribution of Childhood Food Allergy in the United States," *Journal of Pediatrics*, 128 (2011), pp. e9–e17.

16. From the National Institutes of Health's National Digestive Diseases Information Clearinghouse, found at `wrongdiagnosis.com`.

17. The probabilities given are realistic, according to the fund-raising firm SCM Associates, at `scmassoc.com`.

18. B. Budowle et al., "Population Data on the Thirteen CODIS Core Short Tandem Repeat Loci in Afircan Americans, U.S. Caucasians, Hispanics, Bahamians, Jamaicans, and Trinidadians," *Journal of Forensic Sciences* (1999), pp. 1277–1286.

19. Marilyn vos Savant, "Ask Marilyn column," *Parade Magazine*, p. 16, September 9, 1990.

Chapter 13 Notes

1. Matthew A. Carlton and William D. Stansfield, "Making babies by the flip of a coin?" *American Statistician*, 59 (2005), pp. 180–182.

2. Richard F. MacLeod et al., "Damage to fresh tomatoes can be reduced" at `ucce.ucdavis.edu/files/repositoryfiles/ca3012p10-72079.pdf`

3. From Statistics Canada, Smoking 2011 at, `www.statcan.gc.ca`.

4. The survey question is reported in Susanah Fox and Maeve Duncan, "Tracking for Health" January, 2013 at `pewinternet.org`. In fact, the estimate in the survey was 60% saying "Yes."

5. Information obtained from the Centers for Disease Control website, at `www.cdc.gov`.

6. Information obtained from the Planned Parenthood website, at `www.plannedparenthood.org`.

7. Results for the full 2012 season, at `www.pgatour.com`.

8. From `gmauthority.com/blog/2012/10/gms-september-2012-sales-strongest-since-2008-by-the-numbers/`.

9. C. E. Finley et al., "Retention rates and weight loss in a commercial weight loss program," *International Journal of Obesity*, 31 (2006), pp. 292–298.

10. Associated Press news item dated December 9, 2007, found at `www.msnbc.msn.com`.

11. See `demonstrations.wolfram.com/MonteCarloEstimateForPi/` for an online demonstration of this idea.

Chapter 14 Notes

1. Data for U.S. searches in February 2008 from Hitwise, at `www.hitwise.com`.

2. U.S. Census Bureau, *Fertility of American Women: June 2004*, at `www.census.gov`.

3. From a Gallup Poll taken in 2003, `www.gallup.com`.

4. John Schwartz, "Leisure pursuits of today's young men," *New York Times,* March 29, 2004. The source cited is comScore Media Matrix.

5. From the Canadian Internet Use Survey at `www.statcan.gc.ca/daily-quotidien/100510/dg100510a-eng.htm`.

6. Probabilities from trials with 2897 people known to be free of HIV antibodies and 673 people known to be infected, reported in J. Richard George, "Alternative specimen sources: Methods for confirming positives," 1998 Conference on the Laboratory Science of HIV, found online at the Centers for Disease Control and Prevention, `www.cdc.gov`.

7. Higher Education Research Institute 2009 Freshman Survey, at `www.heri.ucla.edu`.

Chapter 15 Notes

1. Higher Education Research Institute 2010 Freshman Survey, at `www.heri.ucla.edu`.

2. Joseph H. Catania et al., "Prevalence of AIDS-related risk factors and condom use in the United States," *Science*, 258 (1992), pp. 1101–1106.

3. Jasmine Rogers, "More uninsured drivers on the road," *Marietta Times*, October 25, 2012.

4. See Note 3.

5. The data were obtained from the GSS Cumulative Datafile 1972–2012 at `sda.berkeley.edu/archive.htm`.

6. Strictly speaking, the formula $\sqrt{p(1-p)/n}$ for the standard deviation of $\hat{p}$ assumes that we draw an SRS of size n from an *infinite* population. If the population has finite size N, this standard deviation is multiplied by $\sqrt{1-(n-1)/(N-1)}$. This "finite population correction" approaches 1 as N increases. When the population is at least 20 times as large as the sample, the correction factor is between about 0.97 and 1. It is reasonable to use the simpler form $\sqrt{p(1-p)/n}$ in these settings.

7. From the article "Kellogg Survey: Most Americans Don't Eat Breakfast Daily," June 2011 at `detroitcbslocal.com`.

8. Joanne Brenner, "Pew Internet: Social Networking (full detail)," Pew Internet and American Life Project, 2012 at `pewinternet.org`.

9. "DrugFacts: High School and Youth Trends," December 2012 at `www.drugabuse.gov/publications/drugfacts/high-school-youth-trends`.

10. L. G. M. Bode, "Preventing surgical-site infections in nasal carriers of *Staphylococcus aureus*," *New England Journal of Medicine*, 362 (2010), pp. 9–17.

11. Maeve Duggan and Lee Rainie, "Cell phone activities 2012," Pew Internet and American Life Project, 2012 at `pewinternet.org`.

12. Susan Kleinman, "Online shopping customer experience study," 2012 at `comscore.com`.

13. Alex Leary, "Times/Bay News 9/Herald Florida Poll: Mitt Romney 51, Barack Obama 45," November 3, 2012. The article can be found online at `tampabay.com/news/politics/national/article1259531.ece`.

14. Keith Wagstaff, "Men are from Google+, Women are from Pinterest," 2012, at `techland.time.com`.

15. Anna Mancin, "Worldwide, Greeks most pessimistic about their lives," 2012, at `gallup.com`.

Chapter 16 Notes

1. We also need to be sure that the sample size is not too big a fraction of the population size. See Note 6 in Chapter 15.

2. Replacing p by $\hat{p}$ in the standard deviation can be justified on mathematical grounds. The argument begins by noting that $\frac{\hat{p}-p}{\sqrt{\hat{p}(1-\hat{p})/n}}$ has approximately a standard normal distribution for large n and then using a mathematical result that implies $\frac{\hat{p}-p}{\sqrt{\hat{p}(1-\hat{p})/n}}$ also has approximately a standard normal distribution for large n.

3. From a September 2012 Gallup Poll, at `http://www.gallup.com/poll/157958/americans-say-gov-not-favor-set-values.aspx`.

4. Found online at http://www.pollingreport.com/religion.htm.

5. Joseph H. Catania et al., "Prevalence of AIDS-related risk factors and condom use in the United States," *Science,* 258 (1992), pp. 1101–1106.

6. The quotation is from page 1104 of the article cited in Note 5.

7. G. A. Mauser and H. Taylor Buckner, "Canadian attitudes toward gun control: The real story," The Mackenzie Institute, 1997, at teapot.usask.ca/cdn-firearms/Mauser/gunstory.html.

8. Gregory D. Myer et al., "Youth versus adult weightlifting injuries presenting to United States emergency rooms: Accidental versus nonaccidental injury mechanisms," *Journal of Strength and Conditioning Research,* 23 (2009), pp. 2054–2060.

9. Gary Edwards and Josephine Mazzuca, "Three quarters of Canadians support doctor-assisted suicide," Gallup Poll press release, March 24, 1999, at www.gallup.com.

10. From the Gallup Web site, www.gallup.com. The poll was taken in July 2008.

11. This interval is proposed by Alan Agresti and Brent A. Coull, "Approximate is better than 'exact' for interval estimation of binomial proportions," *The American Statistician,* 52 (1998), pp. 119–126. Note in particular that the plus four interval is often more accurate than the Clopper-Pearson "exact interval" based on the binomial distribution of the sample count and implemented by, for example, Minitab.

    There are several even more accurate but considerably more complex intervals for p that might be used in professional practice. See Lawrence D. Brown, Tony Cai, and Anirban DasGupta, "Interval estimation for a binomial proportion," *Statistical Science,* 16 (2001), pp. 101–133. A detailed theoretical study that uncovers the reason the large-sample interval is inaccurate is Lawrence D. Brown, Tony Cai, and Anirban DasGupta, "Confidence intervals for a binomial proportion and asymptotic expansions," *Annals of Statistics,* 30 (2002), pp. 160–201.

12. BBC News, December 25, 2006, at news.bbc.co.uk.

13. From Alan Agresti and Brian Caffo, "Simple and effective confidence intervals for proportions and differences of proportions result from adding two successes and two failures," *The American Statistician,* 45 (2000), pp. 280–288. When can the plus four interval be safely used? The answer depends on just how much accuracy you insist on. Brown and coauthors (see Note 7) recommend $n \geq 40$. Agresti and Coull (see Note 7) demonstrate that performance is almost always satisfactory in their eyes when $n \geq 5$. Our rule of thumb $n \geq 10$ allows for confidence levels C other than 95% and fits our philosophy of not insisting on more exact results than practice requires. The big point is that plus four is very much more accurate than the standard interval for most values of p and all but very large n.

14. Gary Stoner et al., "Regression of rectal polyps in familial adenomatous polyposis patients with freezedried black raspberries," abstract and paper presented at the AACR meeting in 2008.

15. Lydia Saad, "In U.S., 11% of households report computer crimes, a new high," December, 2010, at www.gallup.com. The sampling scheme was more complex than an SRS, so the computation of the number in the sample reporting crimes and acting as if it were an SRS is oversimplified.

16. Michele L. Head, "Examining college students' ethical values," Consumer Science and Retailing honors project, Purdue University, 2003.

17. See Note 2 for Chapter 2.

18. Elizabeth Cohen, "Your top health searches, asked and answered," Pew Internet and American Life Project, 2010, at pewinternet.org. The cell phone sample used random digit dialing drawn through a systematic sampling from dedicated wireless 100-blocks and shared service 100-blocks with no directory-listed landline numbers, so acting as if we have an SRS is oversimplified.

19. See Note 22 for Chapter 8.

20. Bobby D. Rampey et al., *The Nation's Report Card: Trends in Academic Progress in Reading and Mathematics 2008* can be found on the Web site nces.ed.gov/nationsreportcard/ under "Long term trends."

21. Jon D. Miller, Eugenie C. Scott, and Shinji Okamoto, "Public acceptance of evolution," *Science,* 313 (2006), pp. 765–766. The information in the exercise appears in the supplementary online material.

22. A. Mantonakis et al., "Order in choice: Effects of serial position on preferences," *Psychological Science,* 20 (2009), pp. 1309–1312.

23. Laura Tutor, "The best drive through in America, 2002," on the Web site http://www.qsrmagazine.com/reports.

24. Francisco Lloret et al., "Fire and resprouting in Mediterranean ecosystems: Insights from an external biogeographical region, the Mexican shrubland," *American Journal of Botany,* 88 (1999), pp. 1655–1661.

25. The data were obtained from the GSS Cumulative Datafile 1972–2010—Quick Tables at sda.berkeley.edu/archive.htm.

Chapter 17 Notes

1. Daryl J. Bem, "Feeling the future: Experimental evidence for anomalous retroactive influences on cognition and affect," *Journal of Personality and Social Psychology,* 100 (2011) pp. 407–425. See also the rebuttal, Eric-Jan Wagenmakers, Ruud Wetzels, Denny Borsboom, and Han L. J. van der Mass, "Why psychologists must change the way they analyze their data: The case of psi: comment on Bem (2011)," *Journal of Personality and Social Psychology,* 100 (2011) pp. 426–432.

2. See Note 22 for Chapter 16.

3. Sam Dillon, "Incentives for advanced work let pupils and teachers cash in," *New York Times,* October 2, 2011.

4. In fact, *P*-values for two-sided tests are more accurate than those for one-sided tests. Our rule of thumb is a compromise to avoid the confusion of too many rules.

5. See Note 1 for Chapter 13.

6. Data found on the New Scientist Web site at `www.newscientist.com/article/dn1748-euro-coin-accused-of-unfair-flipping.html`.

7. Alexander Todorov et al., "Inferences of competence from faces predict election outcomes," *Science*, 308 (2005), pp. 1623–1626.

8. For a discussion of statistical significance in the legal setting, see D. H. Kaye, "Is proof of statistical significance relevant?" *Washington Law Review*, 61 (1986), pp. 1333–1365. Kaye argues: "Presenting the *P*-value without characterizing the evidence by a significance test is a step in the right direction. Interval estimation, in turn, is an improvement over *P*-values."

9. From a press release from the Harvard School of Public Health College Alcohol Study, April 12, 2001, at `www.hsph.harvard.edu/cas/`.

10. Warren E. Leary, "Cell phones: Questions but no answers," *New York Times*, October 26, 1999.

11. Poll published August 26, 2010, at `www.harrisinteractive.com/NewsRoom/HarrisPolls/tabid/447/mid/1508/articleId/555/ctl/ReadCustom%20Default/Default.aspx`. A note at the bottom of the page says: "Because the sample is based on those who agreed to participate in the Harris Interactive panel, no estimates of theoretical sampling error can be calculated."

12. Seung-Ok Kim, "Burials, pigs, and political prestige in Neolithic China," *Current Anthropology*, 35 (1994), pp. 119–141.

13. Gerardo Ramirez and Sian L. Bellock, "Writing about testing worries boosts exam performance in the classroom," *Science*, 331 (2011), pp. 211–213.

14. Kenneth A. Follett et al., "Pallidal versus subthalamic deep-brain stimulation for Parkinson's disease," *The New England Journal of Medicine*, 362, No. 22 (2010), pp. 2077–2091.

15. See Note 23 for Chapter 8.

16. Cara B. Ebbeling et al., "Effects of dietary composition on energy expenditure during weight-loss maintenance," *JAMA*, 307 (2012), pp. 2627–2634.

17. See Note 21 for Chapter 16.

18. Ted D. Adams et al., "Health benefits of gastric bypass surgery after 6 years," *JAMA*, 308 (2012), pp. 1122–1131.

Chapter 18 Notes

1. Shauna B. Wilson et al., "Dating across race: An examination of African American Internet personal advertisements," *Journal of Black Studies*, 37 (2007), pp. 964–982.

2. Based on data in Susannah Fox et al., "Health online 2012," January, 2013, at `pewinternet.org`.

3. From the 2011 Youth Risk Behavior Surveillance System at `apps.nccd.cdc.gov/youthonline/App/Default.aspx`. The data are from a complex multistage sample, so that acting as if we have SRSs is oversimplified.

4. The data was obtained from the GSS Cumulative Datafile 1972–2010 at `sda.berkeley.edu/archive.htm`.

5. Steven J. Frenda et al., "False memories of fabricated political events," *Journal of Experimental Social Psychology*, 49 (2013), pp. 280–286. We would like to thank the authors for providing the data and the photo in Figure 18.3.

6. This rule of thumb is quite conservative. It is in fact safe to arrange the data as a 2 × 2 table and apply the rule of thumb from Chapter 24 that all four *expected* counts must be 5 or greater. We give the conservative rule here because expected counts are messy to explain in the present context.

7. JoAnn K. Wells, Allan F. Williams, and Charles M. Farmer, "Seat belt use among African Americans, Hispanics, and whites," *Accident Analysis and Prevention*, 34 (2002), pp. 523–529.

8. Steiner Sulheim et al., "Helmet use and risk of head injuries in alpine skiers and snowboarders," *Journal of the American Medical Association*, 295 (2006), pp. 919–924.

9. Armando E. Giuliano MD et al., "Axillary dissection vs no axillary dissection in women with invasive breast cancer and sentinel node metastasis," *Journal of the American Medical Association*, 305 (2011), pp. 569–575. The sample sizes for the two groups and the proportions of patients in each group that are disease-free after 5 years have been chosen to match those in the paper.

10. The plus-four method is due to Alan Agresti and Brian Caffo. See Note 9 for Chapter 16.

11. Frances A. Campbell et al., "Adult outcomes as a function of an early childhood educational program: An Abecedarian Project follow up," *Developmental Psychology*, 48 (2012), pp. 1033–1043.

12. Modified from Richard A. Schieber et al., "Risk factors for injuries from in-line skating and the effectiveness of safety gear," *New England Journal of Medicine*, 335 (1996), Internet summary at `content.nejm.org`.

13. See Note 24 for Chapter 16.

14. See Note 4.

15. See Amanda Lenhart and Mary Madden, "Teens, Privacy and Online Social Networks," Pew Internet and American Life Project, 2007, at `www.pewinternet.org`.

16. W. P. T. James et al., "Effect of Sibutramine on cardiovascular outcomes in overweight and obese subjects," *New England Journal of Medicine*, 363 (2010), pp. 905–917.

17. François Gaudet et al., "Induction of tumors in mice by genomic hypomethylation," *Science*, 300 (2003), pp. 489–492.

18. Barbara Helmrich, "Window of Opportunity? Adolescence, Music and Algebra," *Journal of Adolescent Research*, 25 (2010), pp. 557–577.

19. Ninh T. Nguyen et al., "Improved bariatric surgery outcomes for medicare beneficiaries after implementation of the medicare national coverage determination," *Archives of Surgery*, 145 (2010), pp. 72–78.

20. Justin Dimick et al., "Bariatric surgery complications before vs after implementation of a national policy restricting coverage to centers of excellence," *Journal of the American Medical Association*, 309 (2013), pp. 792–799.

21. Arne L. Kalleberg and Kevin T. Leicht, "Gender and organizational performance: Determinants of small business survival and success," *The Academy of Management Journal*, 34 (1991), pp. 136–161.

22. See Maeve Duggan and Joanna Brenner, "The demographics of social media users—2012," Pew Internet and American Life Project, 2013, at `www.pewinternet.org`.

23. D. Gonzales et al., "Varenicline, an $\alpha 4\beta 2$ nicotinic acetylcholine receptor partial agonist, vs sustained-release bupropion and placebo for smoking cessation," *Journal of the American Medical Association* 296 (2006), pp. 47–55.

24. Data courtesy of Raymond Dumett, Purdue University.

25. Based on Alan G. Sanfey et al., "The neural basis of economic decision-making in the ultimatum game," *Science*, 300 (2003), pp. 1755–1758. The paper reports a chi-square test (equivalent to a two-sided z test). This analysis is incorrect for the paper's data, as there were in fact only 19 participants, each appearing twice in each row of the table given in the exercise. Exercise 18.31 therefore amends the data, assuming 76 participants, so that the elementary analysis is correct.

26. Ross L. Prentice et al., "Low-fat dietary pattern and risk of invasive breast cancer," *Journal of the American Medical Association*, 295 (2006), pp. 629–642.

27. Clive G. Jones et al., "Chain reactions linking acorns to gypsy moth outbreaks and Lyme disease risk," *Science*, 279 (1998), pp. 1023–1026.

28. See Note 11.

29. R. B. Turner et al., "A randomized trial of the efficacy of hand disinfection for prevention of rhinovirus infection," *Clinical Infectious Diseases*, 54 (2012), pp. 1422–1426.

Chapter 19 Notes

1. U.S. Census Bureau, *Income, Poverty, and Health Insurance in the United States: 2011*, Current Population Reports P60-243. Available online at `www.census.gov/prod/2012pubs/p60-243.pdf`.

2. Strictly speaking, the formula $\sigma/\sqrt{n}$ for the standard deviation of $\bar{x}$ assumes that we draw an SRS of size n from an *infinite* population. If the population has finite size N, this standard deviation is multiplied by $\sqrt{1-(n-1)/(N-1)}$. This "finite population correction" approaches 1 as N increases. When the population is at least 20 times as large as the sample, the correction factor is between about 0.97 and 1. It is reasonable to use the simpler form $\sigma/\sqrt{n}$ in these settings.

3. Earnings for all 73,366 households with earned incomes greater than zero were downloaded using the U.S. Census Bureau's Data Ferret software. The histograms in Figure 19.4 were produced from the downloaded data.

4. See Note 21 for Chapter 2.

5. Found online at `pages.stern.nyu.edu/adamodar/New_Home_Page/datafile/histret.html`. Sophisticates will note that for compounding over several years we want the geometric mean return, which was 9.38%.

Chapter 20 Notes

1. Note 2 for Chapter 19 explains the reason for this condition in the case of inference about a population mean.

2. See Note 12 for Chapter 2.

3. Margaret A. McDowell et al., "Anthropometric reference data for children and adults: U.S. population, 1999–2002," National Center for Health Statistics, Advance Data from Vital and Health Statistics, No. 361, 2005, at `www.cdc.gov/nchs`.

4. See Note 22 for Chapter 2.

5. From a graph in Benedetto De Martino et al., "Frames, biases, and rational decision-making in the human brain," *Science*, 313 (2006), pp. 684–687. We simplified the design a bit for easier comprehension: the starting amounts and gambles offered differed from trial to trial, though still matched in pairs; 32 very unbalanced "catch trials" were mixed with the 64 experimental trials to be sure subjects were paying attention, and all money amounts were in British pounds, not dollars.

6. R. A. Berner and G. P. Landis, "Gas bubbles in fossil amber as possible indicators of the major gas composition of ancient air," *Science*, 239 (1988), pp. 1406–1409. The 95% t confidence interval is 54.78 to 64.40. A bootstrap BCa interval is 55.03 to 62.63. So t is reasonably accurate despite the skew and the small sample.

7. See Note 2 for Chapter 2.

8. Alice P. Melis, Brian Hare, and Michael Tomasello, "Chimpanzees recruit the best collaborators," *Science*, 311 (2006), pp. 1297–1300. A Normal quantile plot does not show major lack of Normality, and a saddlepoint approximation that allows for skew gives $P = 0.0039$. So the t test is reasonably accurate despite the skew and small sample size.

9. See Note 7 for Chapter 7.

10. For a qualitative discussion explaining why skewness is the most serious violation of the Normal shape condition, see Dennis D. Boos and Jacqueline M. Hughes-Oliver, "How large does n have to be for the Z and t intervals?" *American Statistician*, 54 (2000), pp. 121–128. Our recommendations are based on extensive computer work. See, for example, Harry O. Posten, "The robustness of the one-sample t-test over the Pearson system," *Journal of Statistical Computation and Simulation*, 9 (1979), pp. 133–149; and E. S. Pearson and N. W. Please, "Relation between the shape of population distribution and the robustness of four simple test statistics," *Biometrika*, 62 (1975), pp. 223–241.

11. For more advanced users, a good way to ascertain if the t procedures are safe is to compare the 95% confidence

interval produced by t with the BCa interval from a bootstrap with at least 1000 resamples. For (b) the t interval is 29,428 to 32,254 and a BCa interval is 29,106 to 31,894. For (c), on the other hand, t gives 38.93 to 40.49, and BCa gives 38.97 to 40.44. These results confirm the judgment that t is safe for (c) but not for (b).

12. Table 1 of E. Thomassot et al., "Methane-related diamond crystallization in the earth's mantle: stable isotopes evidence from a single diamond-bearing xenolith," *Earth and Planetary Science Letters*, 257 (2007), pp. 362–371.

13. From the online supplement to Tor D. Wager et al., "Placebo-induced changes in fMRI in the anticipation and experience of pain," *Science*, 303 (2004), pp. 1162–1167.

14. TUDA results for 2009 from the National Center for Education Statistics, at `nationsreportcard.gov/tuda.asp`.

15. Ravi Mehta and Rui Zhu, "Blue or red? Exploring the effect of color on cognitive task performances," *Science*, 323 (2009), pp. 1226–1229.

16. Raul de la Fuente-Fernandez et al., "Expectation and dopamine release: Mechanism of the placebo effect in Parkinson's disease," *Science*, 293 (2001), pp. 1164–1166.

17. Robert R. Zarr and Dennis D. Leber, "Evaluation and Selection of Candidate Thermal Insulation Materials for NIST SRM 1450d, Fibrous-Glass Board," available online from the National Institute of Standards and Technology Web site, `www.nist.gov/manuscript-publication-search.cfm?pub_id5902936`.

18. J. D. Marshall et al., "Vehicle self-pollution intake fraction: Children's exposure to school bus emissions," *Environmental Science and Technology*, 39 (2005), pp. 2559–2563.

19. See Note 17 for Chapter 7.

20. Data provided by Drina Iglesia, Purdue University. The data are part of a larger study reported in D. D. S. Iglesia, E. J. Cragoe, Jr., and J. W. Vanable, "Electric field strength and epithelization in the newt (*Notophthalmus viridescens*)," *Journal of Experimental Zoology*, 274 (1996), pp. 56–62.

21. Matthias R. Mehl et al., "Are women really more talkative than men?," *Science*, 317 (2007), p. 82.

22. Data for Cohort 2 in Richard A. Morgan et al., "Cancer regression in patients after transfer of genetically engineered lymphocytes," *Science*, 314 (2006), pp. 126–129. The doubling time data are given in the paper, and the immune response data appear in the supplementary online material.

23. M. B. Laferty "OSU scientist gets a kick of out sports controversy," *The Columbus Dispatch*, November 21, 1993.

24. We thank Jason Hamilton, University of Illinois, for providing the data. The study is reported in Evan H. DeLucia et al., "Net primary production of a forest ecosystem with experimental CO_2 enhancement," *Science*, 284 (1999), pp. 1177–1179. No method for inference can be trusted with $n = 3$. In this study, each observation is very costly, so the small n is inevitable.

25. Michael W. Peugh, "Field investigation of ventilation and air quality in duck and turkey slaughter plants," MS thesis, Purdue University, 1996.

26. Harry B. Meyers, "Investigations of the life history of the velvetleaf seed beetle, *Althaeus folkertsi* Kingsolver," MS thesis, Purdue University, 1996. The 95% t interval is 1227.9 to 2507.6. A 95% bootstrap BCa interval is 1444 to 2718, confirming that t inference is inaccurate for these data.

27. J. Marcus Jobe and Hutch Jobe, "A statistical approach for additional infill development," *Energy Exploration and Exploitation*, 18 (2000), pp. 89–103. The comparison interval is the BCa interval based on 1000 bootstrap resamples.

28. See Note 2 for Chapter 2.

29. Ralf Bargou et al., "Tumor regression in cancer patients by very low doses of a T cell engaging antibody," *Science*, 321 (2008), pp. 974–977.

30. Data provided by Timothy Sturm.

31. Lianng Yuh, "A biopharmaceutical example for undergraduate students," manuscript, no date.

32. See Note 22 for Chapter 2.

Chapter 21 Notes

1. See Note 21 for Chapter 2.

2. Detailed information about the conservative t procedures can be found in Paul Leaverton and John J. Birch, "Small sample power curves for the two sample location problem," *Technometrics*, 11 (1969), pp. 299–307; Henry Scheffé, "Practical solutions of the Behrens-Fisher problem," *Journal of the American Statistical Association*, 65 (1970), pp. 1501–1508; and D. J. Best and J. C. W. Rayner, "Welch's approximate solution for the Behrens-Fisher problem," *Technometrics*, 29 (1987), pp. 205–210.

3. Kathleen G. McKinney, "Engagement in community service among college students: Is it affected by significant attachment relationships?" *Journal of Adolescence*, 25 (2002), pp. 139–154. To see the questions in the Inventory of Parent and Peer Attachments, go to `chipts.cch.ucla.edu/assessment/IB/List_Scales/inventory%20parent%20and%20peer%20attachment.htm`.

4. See Note 10 for Chapter 2.

5. P. A. Handcock, "The effect of age and sex on the perception of time in life," *The American Journal of Psychology*, 123 (2010), pp. 1–13.

6. See the extensive simulation studies in Harry O. Posten, "The robustness of the two-sample t-test over the Pearson system," *Journal of Statistical Computation and Simulation*, 6 (1978), pp. 295–311; and Harry O. Posten, H. Yeh, and Donald B. Owen, "Robustness of the two-sample t-test under violations of the homogeneity assumption," *Communications in Statistics*, 11 (1982), pp. 109–126.

7. Nicolas Guéguen and Christine Petr, "Odors and consumer behavior in a restaurant," *Journal of Hospitality Management*, 25 (2006), pp. 335–339. We thank Nicolas Gueguen for providing the data. Although the spending data are

quite discrete, a bootstrap BCa 95% confidence interval for the difference in means based on 1000 resamples is 2.394 to 4.826, close to the Option 1 95% interval 2.209 to 4.736. So the sample means are sufficiently Normal to allow use of t procedures.

8. Parmeshwar S. Gupta, "Reaction of plants to the density of soil," *Journal of Ecology*, 21 (1933), pp. 452–474.

9. Data provided by Samuel Phillips, Purdue University.

10. See Note 22 for Chapter 1.

11. The problem of comparing variability is difficult even with advanced methods. Common distribution-free procedures do not offer a satisfactory alternative to the F test because they are sensitive to unequal shapes when comparing two distributions. A recent survey of possible approaches is Dennis D. Boos and Cavell Brownie, "Comparing variances and other measures of dispersion," *Statistical Science*, 19 (2005), pp. 571–578.

12. See Note 21 for Chapter 20.

13. Michael A. Sayette et al., "Lost in the sauce, the effects of alcohol on mind wandering," *Psychological Science*, 20 (2009), pp. 747–752.

14. Eduardo Dias-Ferreira et al., "Chronic stress causes frontostriatal reorganization and affects decision-making," *Science*, 325 (2009), pp. 621–625. Many of the details appear in the supporting online material.

15. Angeline Lillard and Nicole Else-Quest, "Evaluating Montessori education," *Science*, 313 (2006), pp. 1893–1894. Many of the details appear in the supporting online material.

16. Mary K. Pawlik, "The effect of ginkgo biloba on the postlunch dip and chemosensory function," MS thesis, Purdue University, 2002.

17. Jennifer A. Whitson and Adam D. Galinsky, "Lacking control increases illusory pattern perception," *Science*, 322 (2008), pp. 115–117.

18. Atsushi Senju et al., "Mindblind eyes: An absence of spontaneous theory of mind in Asperger syndrome," *Science*, 325 (2009), pp. 883–885.

19. Wayne J. Camera and Donald Powers, "Coaching and the SAT I," *TIP* (online journal at www.siop.org/tip), July 1999.

20. See Note 22 for Chapter 2.

21. Sherri A. Buzinski, "The effect of position of methylation on the performance properties of durable press treated fabrics," CSR490 honors paper, Purdue University, 1985.

22. Fabrizio Grieco, Arie J. van Noordwijk, and Marcel E. Visser, "Evidence for the effect of learning on timing of reproduction in blue tits," *Science*, 296 (2002), pp. 136–138. The data in Exercise 18.48 are from a graph in this paper.

23. Kathleen D. Vohs, Nicole L. Mead, and Miranda R. Goode, "The psychological consequences of money," *Science*, 314 (2006), pp. 1154–1156. I thank Kathleen Vohs for supplying the data.

24. See Note 18 for Chapter 2.

25. Paul Kvam, "The effect of active learning methods on student retention in engineering statistics," *The American Statistician*, 54 (2000), pp. 136–140.

26. Data provided by Warren Page, New York City Technical College, from a study done by John Hudesman.

27. Ethan J. Temeles and W. John Kress, "Adaptation in a plant-hummingbird association," *Science*, 300 (2003), pp. 630–633. We thank Ethan J. Temeles for providing the data.

28. See Note 4 for Chapter 5.

Chapter 22 Notes

1. See www.cdc.gov/nchs/tutorials/NHANES/SurveyDesign/intro.htm.

2. See Note 22 for Chapter 8.

3. Poll published August 26, 2010, at www.harrisinteractive.com/NewsRoom/HarrisPolls/tabid/447/mid/1508/articleId/555/ctl/ReadCustom%20Default/Default.aspx. A note at the bottom of the page says: "Because the sample is based on those who agreed to participate in the Harris Interactive panel, no estimates of theoretical sampling error can be calculated."

4. Justin S. Brashares et al., "Bushmeat hunting, wildlife declines, and fish supply in West Africa," *Science*, 306 (2004), pp. 1180–1183. The data used here (and in Figure 1B of the article) are found in the online supplementary material.

5. Gabriel Gregoratos et al., ACC/AHA guidelines for implantation of cardiac pace-makers and antiarrhythmia devices: executive summary," Circulation, 97 (1998), pp.1325–1335.

6. From the commentary by Frank J. Sulloway, "Birth order and intelligence," *Science*, 316 (2007), pp. 1711–1712. The study report appears in the same issue, Petter Kristensen and Tor Bjerkedal, "Explaining the relation between birth order and intelligence," *Science*, 316 (2007), p. 1717.

7. C. Kopp et al., "Modulation of rhythmic brain activity by diazepam: GABAA receptor subtype and state specificity," *Proceedings of the National Academy of Sciences*, 101 (2004), pp. 3674–3679.

8. Bruce A. Cooper et al., "A randomized, controlled trial of early versus late initiation of dialysis," *New England Journal of Medicine*, 363, No. 7 (2010), pp. 609–619.

Chapter 23 Notes

1. See Note 20 for Chapter 20.

2. See Note 3 for Chapter 18.

3. Lee Rainie and Bill Tancer, "36% of online American adults consult Wikipedia," Pew Internet and American Life Project, 2007, at www.pewinternet.org.

4. Chi-Fu Jeffrey Yang, Peter Gray, and Harrison G. Pope, Jr., "Male body image in Taiwan versus the West," *American Journal of Psychiatry*, 16 (2005), pp. 263–269.

5. K. S. Oberhauser, "Fecundity, lifespan and egg mass in butterflies: Effects of male-derived nutrients and female size," *Functional Ecology*, 11 (1997), pp. 166–175.

6. Michael R. Dohm, Jack P. Hayes, and Theodore Garland, Jr., "Quantitative genetics of sprint running speed and swim-

ming endurance in laboratory house mice (*Mus domesticus*)," *Evolution,* 50 (1996), pp. 1688–1701.

7. Aaron S. Hervey et al., "Reaction time distribution analysis of neuropsychological performance in an ADHD sample," *Child Neuropsychology,* 12 (2006), pp. 125–140.

8. Maureen Hack, et al., "Outcomes in young adulthood for very-low-birth-weight infants," *New England Journal of Medicine,* 346 (2002), pp. 149–157. The exercises are simplified, in that the measures reported in this paper have been statistically adjusted for "sociodemographic status."

9. From the Web site of the Black Youth Project, blackyouthproject.uchicago.edu.

10. V. D. Bass, W. E. Hoffmann, and J. L. Dorner, "Normal canine lipid profiles and effects of experimentally induced pancreatitis and hepatic necrosis on lipids," *American Journal of Veterinary Research,* 37 (1976), pp. 1355–1357.

11. Jin Ha Lee and J. Stephen Downie, "Survey of music information needs, uses, and seeking behaviors: Preliminary findings," online Proceedings of the 5th International Conference on Music Information Retrieval, 2004, at ismir2004.ismir.net.

12. K. Carrie Armel and V. S. Ramachandran, "Projecting sensations to external objects: Evidence from skin conductance response," *Proceedings of the Royal Society of London, Series B,* 270 (2003), pp. 1499–1506.

13. Josh McDermott and Marc D. Hauser, "Nonhuman primates prefer slow tempos but dislike music overall," *Cognition,* 104 (2007), pp. 654–668. Failure to take account of repeated measures on the same subjects is one of the most common errors observed in statistical analysis.

14. James Otto, Michael F. Brown, and William Long III, "Training rats to search and alert on contraband odors," *Applied Animal Behaviour Science,* 77 (2002), pp. 217–232.

15. From the Merck Web site, www.merckvaccines.com/gardasilProductPage_frmst.html.

16. These data were originally collected by L. M. Linde of UCLA but were first published by M. R. Mickey, O. J. Dunn, and V. Clark, "Note on the use of stepwise regression in detecting outliers," *Computers and Biomedical Research,* 1 (1967), pp. 105–111. The data have been used by several authors. I found them in N. R. Draper and J. A. John, "Influential observations and outliers in regression," *Technometrics,* 23 (1981), pp. 21–26.

17. Jacqueline T. Ngai and Diane S. Srivastava, "Predators accelerate nutrient cycling in a bromeliad ecosystem," *Science,* 314 (2006), p. 963. We thank Jacqueline Ngai for providing the data.

18. Yvan R. Germain, "The dyeing of ramie with fiber reactive dyes using the cold pad-batch method," MS thesis, Purdue University, 1988.

19. Data provided by Marigene Arnold, Kalamazoo College.

20. Data provided by Corinne Lim, Purdue University, from a student project supervised by Professor Joseph Vanable.

21. Saiyad S. Ahmed, "Effects of microwave drying on checking and mechanical strength of low-moisture baked products," MS thesis, Purdue University, 1994.

22. Michael O. Finkelstein and Bruce Levin, "Statistical proof of discrimination in peremptory challenges," *Chance,* 17, No. 1 (2004), pp. 35–38.

23. G. S. Hotamisligil et al., "Uncoupling of obesity from insulin resistance through a targeted mutation in $aP2$, the adipocyte fatty acid binding protein," *Science,* 274 (1996), pp. 1377–1379.

24. Data simulated from a Normal distribution with $\mu = 98.2$ and $\sigma = 0.7$. These values are based on P. A. Mackowiak, S. S. Wasserman, and M. M. Levine, "A critical appraisal of 98.6 degrees F, the upper limit of the normal body temperature, and other legacies of Carl Reinhold August Wunderlich," *Journal of the American Medical Association,* 268 (1992), pp. 1578–1580.

25. J. F. Swain et al., "Comparison of the effects of oat bran and low-fiber wheat on serum lipoprotein levels and blood pressure," *New England Journal of Medicine,* 322 (1990), pp. 147–152.

Chapter 24 Notes

1. The National Longitudinal Study of Adolescent Health interviewed a stratified random sample of 27,000 adolescents, then reinterviewed many of the subjects six years later, when most were aged 19 to 25. These data are from the Wave III reinterviews in 2000 and 2001, found at the Web site of the Carolina Population Center, www.cpc.unc.edu.

2. Pennsylvania State University Division of Student Affairs, "Net behaviors November 2006," *Penn State Pulse,* at www.sa.psu.edu.

3. See Note 2 for Chapter 6.

4. All General Social Survey exercises in this chapter present tables constructed using the search function at the GSS archive, sda.berkeley.edu/archive.htm. Most concern data from the 2012 GSS.

5. There are many computer studies of the accuracy of chi-square critical values for X^2. Our guideline goes back to W. G. Cochran (1954). Later work has shown that it is often conservative in the sense that, if the expected cell counts are all similar and the degrees of freedom exceed 1, the chi-square approximation works well for an average expected count as small as 1 or 2. Our guideline protects against dissimilar expected counts. It has the added advantage that it is safe in the 2×2 case, where the chi-square approximation is least good. So our guideline is helpful for beginners—there is no single condition that is not conservative and applies to 2×2 and larger tables with similar and dissimilar expected cell counts. There are exact procedures that (with software) should be used for tables that do not satisfy our guideline. For a survey, see Alan Agresti, "A survey of exact inference for contingency tables," *Statistical Science,* 7 (1992), pp. 131–177.

6. Scott Keeter et al., "What's missing from national RDD surveys? The impact of the growing cell-only population," (2007) at www.pewresearch.org.

7. Leah Christian et al., "Assessing the cell phone challenge to survey research in 2010," at `www.pewresearch.org`.

8. Modified from Felicity Barringer, "Measuring sexuality through polls can be shaky," *New York Times*, April 25, 1993.

9. Based on a news item in *Science*, 305 (2004), p. 1560. The study, by Daniel Klem, appeared in the *Wilson Journal*.

10. David W. Eby et al., "The effect of changing from secondary to primary safety belt enforcement on police harassment," *Accident Analysis and Prevention*, 36 (2000), pp. 819–828.

11. See Note 2 from Chapter 16.

12. See Note 9 for Chapter 6.

13. See Note 12 for Chapter 6.

14. See Note 7.

15. See Note 17 for Chapter 6.

16. Virgilio P. Carnielli et al., "Intestinal absorption of long-chain polyunsaturated fatty acids in preterm infants fed breast milk or formula," *American Journal of Clinical Nutrition*, 67 (1998), pp. 97–103.

17. Adapted from M. A. Visintainer, J. R. Volpicelli, and M. E. P. Seligman, "Tumor rejection in rats after inescapable or escapable shock," *Science*, 216 (1982), pp. 437–439.

18. See Note 1.

19. Tom Reichert, "The prevalence of sexual imagery in ads targeted to young adults," *Journal of Consumer Affairs*, 37 (2003), pp. 403–412.

20. József Topáal et al., "Differential sensitivity to human communication in dogs, wolves and human infants," *Science*, 325 (2009), pp. 1269–1272. Many statistical software packages offer "exact tests" that are valid even when there are small expected counts.

21. U.S. Department of Commerce, Office of Travel and Tourism Industries, in-flight survey, 2007, at `tinet.ita.doc.gov`.

22. Andrew Steptoe et al., "Social isolation, loneliness, and all-cause mortality in older men and women," *Proceedings of the National Academy of Sciences*, 110 (2013), pp. 5797–5801. Two categories of depression were combined in the table.

23. See Note 2 for Chapter 10. I have simplified slightly: the table in the paper is exactly as in the exercise but contains data for 63 subjects plus data from one type of bar for 3 subjects who dropped out. Although the authors say that their chi-square refers to this table, they give a nonsignificant value that contradicts what the table shows.

24. See Note 15 for Chapter 6.

25. Two way tables from the Youth Risk Behavior Surveillance System can be constructed from the Web site `apps.nccd.cdc.gov/youthonline/App/Default.aspx`.

26. Data compiled from a table of percents in "Americans view higher education as key to the American dream," press release by the National Center for Public Policy and Higher Education, May 3, 2000, at `www.highereducation.org`.

27. See Note 14 for Chapter 6.

28. Data produced by Ries and Smith, found in William D. Johnson and Gary G. Koch, "A note on the weighted least squares analysis of the Ries-Smith contingency table data," *Technometrics*, 13 (1971), pp. 438–447.

Chapter 25 Notes

1. Samuel Karelitz et al., "Relation of crying activity in early infancy to speech and intellectual development at age three years," *Child Development*, 35 (1964), pp. 769–777.

2. From a graph in Naomi E. Allen et al., "Moderate alcohol intake and cancer incidence in women," *Journal of the National Cancer Institute*, 101 (2009), pp. 296–305. These data represent averages over large numbers of women and are an example of an ecological correlation (see page 137 in Chapter 5) and one must be careful not to interpret the data as applying to individuals.

3. From a graph in Stephen M. Fleming et al., "Relating introspective accuracy to individual differences in brain structure," *Science*, 329 (2010), pp. 1541–1543.

4. See Note 21 for Chapter 5.

5. Electronic Encyclopedia of Statistical Examples and Exercises (EESEE) at the text Web site, `www.whfreeman.com/bps`.

6. From a graph in Allison L. Perry et al., "Climate change and distribution shifts in marine fishes," *Science*, 308 (2005), pp. 1912–1915. The explanatory variable is the five-year running mean of winter (December to March) sea-bottom temperature.

7. Data for the building at 1800 Ben Franklin Drive, Sarasota, Florida, starting in February 2003. From the Web site of the Sarasota County Property Appraiser, `www.sarasotaproperty.net`.

8. Yanhui Lu et al., "Mirid bug outbreaks in multiple crops correlated with wide-scale adoption of Bt cotton in China," *Science*, 328 (2010), pp. 1151–1154.

9. See Note 1 for Chapter 4.

10. Based on Marion E. Dunshee, "A study of factors affecting the amount and kind of food eaten by nursery school children," *Child Development*, 2 (1931), pp. 163–183. This article gives the means, standard deviations, and correlation for 37 children, from which the data in the exercise are simulated.

11. From Table S2 in the online supplement to Antonio Dell'Anno and Roberto Danovaro, "Extracellular DNA plays a key role in deep-sea ecosystem functioning," *Science*, 309 (2005), p. 2179.

12. See Note 17 for Chapter 7.

Chapter 26 Notes

1. Ethan J. Temeles and W. John Kress, "Adaptation in a plant-hummingbird association," *Science*, 300 (2003), pp. 630–633. We thank Ethan J. Temeles for providing the data.

2. Elisabeth Wells-Parker et al., "An exploratory study of the relationship between road rage and crash experience in a representative sample of US drivers," *Accident Analysis and Prevention*, 34 (2002), pp. 271–278.

3. Victoria L. Brescoll and Eric L. Uhlmann, "Can an angry woman get ahead? Status conferral, gender and expression of emotion in the workplace," *Psychological Science*, 19 (2008), pp. 268–273. The description and data are based on study 1 in this article.

4. See Note 10 for Chapter 2.

5. The data from the General Social Survey for this exercise was constructed using the search function and download capabilities at the GSS archive, `sda.berkeley.edu/archive.htm`.

6. See Note 23 for Chapter 21.

7. See Note 17 for Chapter 23.

8. David B. Wooten, "One-of-a-kind in a full house: Some consequences of ethnic and gender distinctiveness," *Journal of Consumer Psychology*, 4 (1995), pp. 205–224.

9. John P. Thomas, "Influences on mathematics learning and attitudes among African American high school students," *Journal of Negro Education*, 69 (2000), pp. 165–183.

10. Henrik Hagvedt and Vanessa M. Patrick, "Art infusion: The influence of visual art on the perception and evaluation of consumer products," *Journal of Marketing Research*, XLV (2008), pp. 379–389.

11. Timothy Church et al., "Effects of aerobic and resistance training on hemoglobin A1c levels in patients with type 2 diabetes: A randomized controlled trial," *Journal of the American Medical Association*, 304 (2010), pp. 2253–2262.

12. Jennifer J. Argo et al., "Positive consumer contagion: Responses to attractive others in a retail context," *Journal of Marketing Research*, XLV (2008), pp. 690–701.

13. Data from the online supplement to André Kessler and Ian T. Baldwin, "Defensive function of herbivore-induced plant volatile emissions in nature," *Science*, 291 (2001), pp. 2141–2144.

14. The data and the full story can be found in the Data and Story Library at `lib.stat.cmu.edu`. The original study is by Faith Loven, "A study of interlist equivalency of the CID W-22 word list presented in quiet and in noise," MS thesis, University of Iowa, 1981.

15. See Note 9 for Chapter 9. We thank Kenwyn Suttle for providing these data for the year 2003.

16. See Note 17 for Chapter 6.

17. See Note 20 for Chapter 2.

18. See Note 22 for Chapter 2.

19. See Note 21 for Chapter 21.

20. See Note 9 for Chapter 9.

TABLES

Table entry for z is the area under the standard Normal curve to the left of z.

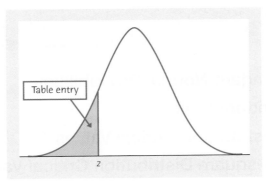

Table entry

TABLE A   STANDARD NORMAL CUMULATIVE PROPORTIONS

z	.00	.01	.02	.03	.04	.05	.06	.07	.08	.09
−3.4	.0003	.0003	.0003	.0003	.0003	.0003	.0003	.0003	.0003	.0002
−3.3	.0005	.0005	.0005	.0004	.0004	.0004	.0004	.0004	.0004	.0003
−3.2	.0007	.0007	.0006	.0006	.0006	.0006	.0006	.0005	.0005	.0005
−3.1	.0010	.0009	.0009	.0009	.0008	.0008	.0008	.0008	.0007	.0007
−3.0	.0013	.0013	.0013	.0012	.0012	.0011	.0011	.0011	.0010	.0010
−2.9	.0019	.0018	.0018	.0017	.0016	.0016	.0015	.0015	.0014	.0014
−2.8	.0026	.0025	.0024	.0023	.0023	.0022	.0021	.0021	.0020	.0019
−2.7	.0035	.0034	.0033	.0032	.0031	.0030	.0029	.0028	.0027	.0026
−2.6	.0047	.0045	.0044	.0043	.0041	.0040	.0039	.0038	.0037	.0036
−2.5	.0062	.0060	.0059	.0057	.0055	.0054	.0052	.0051	.0049	.0048
−2.4	.0082	.0080	.0078	.0075	.0073	.0071	.0069	.0068	.0066	.0064
−2.3	.0107	.0104	.0102	.0099	.0096	.0094	.0091	.0089	.0087	.0084
−2.2	.0139	.0136	.0132	.0129	.0125	.0122	.0119	.0116	.0113	.0110
−2.1	.0179	.0174	.0170	.0166	.0162	.0158	.0154	.0150	.0146	.0143
−2.0	.0228	.0222	.0217	.0212	.0207	.0202	.0197	.0192	.0188	.0183
−1.9	.0287	.0281	.0274	.0268	.0262	.0256	.0250	.0244	.0239	.0233
−1.8	.0359	.0351	.0344	.0336	.0329	.0322	.0314	.0307	.0301	.0294
−1.7	.0446	.0436	.0427	.0418	.0409	.0401	.0392	.0384	.0375	.0367
−1.6	.0548	.0537	.0526	.0516	.0505	.0495	.0485	.0475	.0465	.0455
−1.5	.0668	.0655	.0643	.0630	.0618	.0606	.0594	.0582	.0571	.0559
−1.4	.0808	.0793	.0778	.0764	.0749	.0735	.0721	.0708	.0694	.0681
−1.3	.0968	.0951	.0934	.0918	.0901	.0885	.0869	.0853	.0838	.0823
−1.2	.1151	.1131	.1112	.1093	.1075	.1056	.1038	.1020	.1003	.0985
−1.1	.1357	.1335	.1314	.1292	.1271	.1251	.1230	.1210	.1190	.1170
−1.0	.1587	.1562	.1539	.1515	.1492	.1469	.1446	.1423	.1401	.1379
−0.9	.1841	.1814	.1788	.1762	.1736	.1711	.1685	.1660	.1635	.1611
−0.8	.2119	.2090	.2061	.2033	.2005	.1977	.1949	.1922	.1894	.1867
−0.7	.2420	.2389	.2358	.2327	.2296	.2266	.2236	.2206	.2177	.2148
−0.6	.2743	.2709	.2676	.2643	.2611	.2578	.2546	.2514	.2483	.2451
−0.5	.3085	.3050	.3015	.2981	.2946	.2912	.2877	.2843	.2810	.2776
−0.4	.3446	.3409	.3372	.3336	.3300	.3264	.3228	.3192	.3156	.3121
−0.3	.3821	.3783	.3745	.3707	.3669	.3632	.3594	.3557	.3520	.3483
−0.2	.4207	.4168	.4129	.4090	.4052	.4013	.3974	.3936	.3897	.3859
−0.1	.4602	.4562	.4522	.4483	.4443	.4404	.4364	.4325	.4286	.4247
−0.0	.5000	.4960	.4920	.4880	.4840	.4801	.4761	.4721	.4681	.4641

Table entry for *z* is the area under the standard Normal curve to the left of *z*.

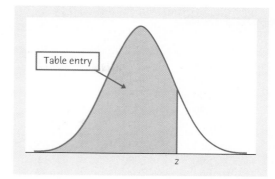

TABLE A STANDARD NORMAL CUMULATIVE PROPORTIONS (*CONTINUED*)

z	.00	.01	.02	.03	.04	.05	.06	.07	.08	.09
0.0	.5000	.5040	.5080	.5120	.5160	.5199	.5239	.5279	.5319	.5359
0.1	.5398	.5438	.5478	.5517	.5557	.5596	.5636	.5675	.5714	.5753
0.2	.5793	.5832	.5871	.5910	.5948	.5987	.6026	.6064	.6103	.6141
0.3	.6179	.6217	.6255	.6293	.6331	.6368	.6406	.6443	.6480	.6517
0.4	.6554	.6591	.6628	.6664	.6700	.6736	.6772	.6808	.6844	.6879
0.5	.6915	.6950	.6985	.7019	.7054	.7088	.7123	.7157	.7190	.7224
0.6	.7257	.7291	.7324	.7357	.7389	.7422	.7454	.7486	.7517	.7549
0.7	.7580	.7611	.7642	.7673	.7704	.7734	.7764	.7794	.7823	.7852
0.8	.7881	.7910	.7939	.7967	.7995	.8023	.8051	.8078	.8106	.8133
0.9	.8159	.8186	.8212	.8238	.8264	.8289	.8315	.8340	.8365	.8389
1.0	.8413	.8438	.8461	.8485	.8508	.8531	.8554	.8577	.8599	.8621
1.1	.8643	.8665	.8686	.8708	.8729	.8749	.8770	.8790	.8810	.8830
1.2	.8849	.8869	.8888	.8907	.8925	.8944	.8962	.8980	.8997	.9015
1.3	.9032	.9049	.9066	.9082	.9099	.9115	.9131	.9147	.9162	.9177
1.4	.9192	.9207	.9222	.9236	.9251	.9265	.9279	.9292	.9306	.9319
1.5	.9332	.9345	.9357	.9370	.9382	.9394	.9406	.9418	.9429	.9441
1.6	.9452	.9463	.9474	.9484	.9495	.9505	.9515	.9525	.9535	.9545
1.7	.9554	.9564	.9573	.9582	.9591	.9599	.9608	.9616	.9625	.9633
1.8	.9641	.9649	.9656	.9664	.9671	.9678	.9686	.9693	.9699	.9706
1.9	.9713	.9719	.9726	.9732	.9738	.9744	.9750	.9756	.9761	.9767
2.0	.9772	.9778	.9783	.9788	.9793	.9798	.9803	.9808	.9812	.9817
2.1	.9821	.9826	.9830	.9834	.9838	.9842	.9846	.9850	.9854	.9857
2.2	.9861	.9864	.9868	.9871	.9875	.9878	.9881	.9884	.9887	.9890
2.3	.9893	.9896	.9898	.9901	.9904	.9906	.9909	.9911	.9913	.9916
2.4	.9918	.9920	.9922	.9925	.9927	.9929	.9931	.9932	.9934	.9936
2.5	.9938	.9940	.9941	.9943	.9945	.9946	.9948	.9949	.9951	.9952
2.6	.9953	.9955	.9956	.9957	.9959	.9960	.9961	.9962	.9963	.9964
2.7	.9965	.9966	.9967	.9968	.9969	.9970	.9971	.9972	.9973	.9974
2.8	.9974	.9975	.9976	.9977	.9977	.9978	.9979	.9979	.9980	.9981
2.9	.9981	.9982	.9982	.9983	.9984	.9984	.9985	.9985	.9986	.9986
3.0	.9987	.9987	.9987	.9988	.9988	.9989	.9989	.9989	.9990	.9990
3.1	.9990	.9991	.9991	.9991	.9992	.9992	.9992	.9992	.9993	.9993
3.2	.9993	.9993	.9994	.9994	.9994	.9994	.9994	.9995	.9995	.9995
3.3	.9995	.9995	.9995	.9996	.9996	.9996	.9996	.9996	.9996	.9997
3.4	.9997	.9997	.9997	.9997	.9997	.9997	.9997	.9997	.9997	.9998

TABLE B RANDOM DIGITS

LINE							
101	19223	95034	05756	28713	96409	12531	42544	82853
102	73676	47150	99400	01927	27754	42648	82425	36290
103	45467	71709	77558	00095	32863	29485	82226	90056
104	52711	38889	93074	60227	40011	85848	48767	52573
105	95592	94007	69971	91481	60779	53791	17297	59335
106	68417	35013	15529	72765	85089	57067	50211	47487
107	82739	57890	20807	47511	81676	55300	94383	14893
108	60940	72024	17868	24943	61790	90656	87964	18883
109	36009	19365	15412	39638	85453	46816	83485	41979
110	38448	48789	18338	24697	39364	42006	76688	08708
111	81486	69487	60513	09297	00412	71238	27649	39950
112	59636	88804	04634	71197	19352	73089	84898	45785
113	62568	70206	40325	03699	71080	22553	11486	11776
114	45149	32992	75730	66280	03819	56202	02938	70915
115	61041	77684	94322	24709	73698	14526	31893	32592
116	14459	26056	31424	80371	65103	62253	50490	61181
117	38167	98532	62183	70632	23417	26185	41448	75532
118	73190	32533	04470	29669	84407	90785	65956	86382
119	95857	07118	87664	92099	58806	66979	98624	84826
120	35476	55972	39421	65850	04266	35435	43742	11937
121	71487	09984	29077	14863	61683	47052	62224	51025
122	13873	81598	95052	90908	73592	75186	87136	95761
123	54580	81507	27102	56027	55892	33063	41842	81868
124	71035	09001	43367	49497	72719	96758	27611	91596
125	96746	12149	37823	71868	18442	35119	62103	39244
126	96927	19931	36809	74192	77567	88741	48409	41903
127	43909	99477	25330	64359	40085	16925	85117	36071
128	15689	14227	06565	14374	13352	49367	81982	87209
129	36759	58984	68288	22913	18638	54303	00795	08727
130	69051	64817	87174	09517	84534	06489	87201	97245
131	05007	16632	81194	14873	04197	85576	45195	96565
132	68732	55259	84292	08796	43165	93739	31685	97150
133	45740	41807	65561	33302	07051	93623	18132	09547
134	27816	78416	18329	21337	35213	37741	04312	68508
135	66925	55658	39100	78458	11206	19876	87151	31260
136	08421	44753	77377	28744	75592	08563	79140	92454
137	53645	66812	61421	47836	12609	15373	98481	14592
138	66831	68908	40772	21558	47781	33586	79177	06928
139	55588	99404	70708	41098	43563	56934	48394	51719
140	12975	13258	13048	45144	72321	81940	00360	02428
141	96767	35964	23822	96012	94591	65194	50842	53372
142	72829	50232	97892	63408	77919	44575	24870	04178
143	88565	42628	17797	49376	61762	16953	88604	12724
144	62964	88145	83083	69453	46109	59505	69680	00900
145	19687	12633	57857	95806	09931	02150	43163	58636
146	37609	59057	66967	83401	60705	02384	90597	93600
147	54973	86278	88737	74351	47500	84552	19909	67181
148	00694	05977	19664	65441	20903	62371	22725	53340
149	71546	05233	53946	68743	72460	27601	45403	88692
150	07511	88915	41267	16853	84569	79367	32337	03316

Table entry for C is the critical value t* required for confidence level C. To approximate one- and two-sided P-values, compare the value of the t statistic with the critical values of t* that match the P-values given at the bottom of the table.

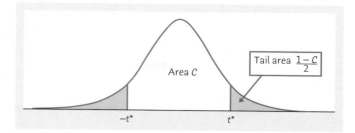

TABLE C   t DISTRIBUTION CRITICAL VALUES

DEGREES OF FREEDOM	CONFIDENCE LEVEL C											
	50%	60%	70%	80%	90%	95%	96%	98%	99%	99.5%	99.8%	99.9%
1	1.000	1.376	1.963	3.078	6.314	12.71	15.89	31.82	63.66	127.3	318.3	636.6
2	0.816	1.061	1.386	1.886	2.920	4.303	4.849	6.965	9.925	14.09	22.33	31.60
3	0.765	0.978	1.250	1.638	2.353	3.182	3.482	4.541	5.841	7.453	10.21	12.92
4	0.741	0.941	1.190	1.533	2.132	2.776	2.999	3.747	4.604	5.598	7.173	8.610
5	0.727	0.920	1.156	1.476	2.015	2.571	2.757	3.365	4.032	4.773	5.893	6.869
6	0.718	0.906	1.134	1.440	1.943	2.447	2.612	3.143	3.707	4.317	5.208	5.959
7	0.711	0.896	1.119	1.415	1.895	2.365	2.517	2.998	3.499	4.029	4.785	5.408
8	0.706	0.889	1.108	1.397	1.860	2.306	2.449	2.896	3.355	3.833	4.501	5.041
9	0.703	0.883	1.100	1.383	1.833	2.262	2.398	2.821	3.250	3.690	4.297	4.781
10	0.700	0.879	1.093	1.372	1.812	2.228	2.359	2.764	3.169	3.581	4.144	4.587
11	0.697	0.876	1.088	1.363	1.796	2.201	2.328	2.718	3.106	3.497	4.025	4.437
12	0.695	0.873	1.083	1.356	1.782	2.179	2.303	2.681	3.055	3.428	3.930	4.318
13	0.694	0.870	1.079	1.350	1.771	2.160	2.282	2.650	3.012	3.372	3.852	4.221
14	0.692	0.868	1.076	1.345	1.761	2.145	2.264	2.624	2.977	3.326	3.787	4.140
15	0.691	0.866	1.074	1.341	1.753	2.131	2.249	2.602	2.947	3.286	3.733	4.073
16	0.690	0.865	1.071	1.337	1.746	2.120	2.235	2.583	2.921	3.252	3.686	4.015
17	0.689	0.863	1.069	1.333	1.740	2.110	2.224	2.567	2.898	3.222	3.646	3.965
18	0.688	0.862	1.067	1.330	1.734	2.101	2.214	2.552	2.878	3.197	3.611	3.922
19	0.688	0.861	1.066	1.328	1.729	2.093	2.205	2.539	2.861	3.174	3.579	3.883
20	0.687	0.860	1.064	1.325	1.725	2.086	2.197	2.528	2.845	3.153	3.552	3.850
21	0.686	0.859	1.063	1.323	1.721	2.080	2.189	2.518	2.831	3.135	3.527	3.819
22	0.686	0.858	1.061	1.321	1.717	2.074	2.183	2.508	2.819	3.119	3.505	3.792
23	0.685	0.858	1.060	1.319	1.714	2.069	2.177	2.500	2.807	3.104	3.485	3.768
24	0.685	0.857	1.059	1.318	1.711	2.064	2.172	2.492	2.797	3.091	3.467	3.745
25	0.684	0.856	1.058	1.316	1.708	2.060	2.167	2.485	2.787	3.078	3.450	3.725
26	0.684	0.856	1.058	1.315	1.706	2.056	2.162	2.479	2.779	3.067	3.435	3.707
27	0.684	0.855	1.057	1.314	1.703	2.052	2.158	2.473	2.771	3.057	3.421	3.690
28	0.683	0.855	1.056	1.313	1.701	2.048	2.154	2.467	2.763	3.047	3.408	3.674
29	0.683	0.854	1.055	1.311	1.699	2.045	2.150	2.462	2.756	3.038	3.396	3.659
30	0.683	0.854	1.055	1.310	1.697	2.042	2.147	2.457	2.750	3.030	3.385	3.646
40	0.681	0.851	1.050	1.303	1.684	2.021	2.123	2.423	2.704	2.971	3.307	3.551
50	0.679	0.849	1.047	1.299	1.676	2.009	2.109	2.403	2.678	2.937	3.261	3.496
60	0.679	0.848	1.045	1.296	1.671	2.000	2.099	2.390	2.660	2.915	3.232	3.460
80	0.678	0.846	1.043	1.292	1.664	1.990	2.088	2.374	2.639	2.887	3.195	3.416
100	0.677	0.845	1.042	1.290	1.660	1.984	2.081	2.364	2.626	2.871	3.174	3.390
1000	0.675	0.842	1.037	1.282	1.646	1.962	2.056	2.330	2.581	2.813	3.098	3.300
z^*	0.674	0.841	1.036	1.282	1.645	1.960	2.054	2.326	2.576	2.807	3.091	3.291
One-sided P	.25	.20	.15	.10	.05	.025	.02	.01	.005	.0025	.001	.0005
Two-sided P	.50	.40	.30	.20	.10	.05	.04	.02	.01	.005	.002	.001

Table entry for *p* is the critical value χ^* with probability *p* lying to its right.

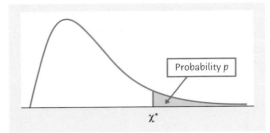

TABLE D CHI-SQUARE DISTRIBUTION CRITICAL VALUES

df	.25	.20	.15	.10	.05	.025	.02	.01	.005	.0025	.001	.0005
1	1.32	1.64	2.07	2.71	3.84	5.02	5.41	6.63	7.88	9.14	10.83	12.12
2	2.77	3.22	3.79	4.61	5.99	7.38	7.82	9.21	10.60	11.98	13.82	15.20
3	4.11	4.64	5.32	6.25	7.81	9.35	9.84	11.34	12.84	14.32	16.27	17.73
4	5.39	5.99	6.74	7.78	9.49	11.14	11.67	13.28	14.86	16.42	18.47	20.00
5	6.63	7.29	8.12	9.24	11.07	12.83	13.39	15.09	16.75	18.39	20.51	22.11
6	7.84	8.56	9.45	10.64	12.59	14.45	15.03	16.81	18.55	20.25	22.46	24.10
7	9.04	9.80	10.75	12.02	14.07	16.01	16.62	18.48	20.28	22.04	24.32	26.02
8	10.22	11.03	12.03	13.36	15.51	17.53	18.17	20.09	21.95	23.77	26.12	27.87
9	11.39	12.24	13.29	14.68	16.92	19.02	19.68	21.67	23.59	25.46	27.88	29.67
10	12.55	13.44	14.53	15.99	18.31	20.48	21.16	23.21	25.19	27.11	29.59	31.42
11	13.70	14.63	15.77	17.28	19.68	21.92	22.62	24.72	26.76	28.73	31.26	33.14
12	14.85	15.81	16.99	18.55	21.03	23.34	24.05	26.22	28.30	30.32	32.91	34.82
13	15.98	16.98	18.20	19.81	22.36	24.74	25.47	27.69	29.82	31.88	34.53	36.48
14	17.12	18.15	19.41	21.06	23.68	26.12	26.87	29.14	31.32	33.43	36.12	38.11
15	18.25	19.31	20.60	22.31	25.00	27.49	28.26	30.58	32.80	34.95	37.70	39.72
16	19.37	20.47	21.79	23.54	26.30	28.85	29.63	32.00	34.27	36.46	39.25	41.31
17	20.49	21.61	22.98	24.77	27.59	30.19	31.00	33.41	35.72	37.95	40.79	42.88
18	21.60	22.76	24.16	25.99	28.87	31.53	32.35	34.81	37.16	39.42	42.31	44.43
19	22.72	23.90	25.33	27.20	30.14	32.85	33.69	36.19	38.58	40.88	43.82	45.97
20	23.83	25.04	26.50	28.41	31.41	34.17	35.02	37.57	40.00	42.34	45.31	47.50
21	24.93	26.17	27.66	29.62	32.67	35.48	36.34	38.93	41.40	43.78	46.80	49.01
22	26.04	27.30	28.82	30.81	33.92	36.78	37.66	40.29	42.80	45.20	48.27	50.51
23	27.14	28.43	29.98	32.01	35.17	38.08	38.97	41.64	44.18	46.62	49.73	52.00
24	28.24	29.55	31.13	33.20	36.42	39.36	40.27	42.98	45.56	48.03	51.18	53.48
25	29.34	30.68	32.28	34.38	37.65	40.65	41.57	44.31	46.93	49.44	52.62	54.95
26	30.43	31.79	33.43	35.56	38.89	41.92	42.86	45.64	48.29	50.83	54.05	56.41
27	31.53	32.91	34.57	36.74	40.11	43.19	44.14	46.96	49.64	52.22	55.48	57.86
28	32.62	34.03	35.71	37.92	41.34	44.46	45.42	48.28	50.99	53.59	56.89	59.30
29	33.71	35.14	36.85	39.09	42.56	45.72	46.69	49.59	52.34	54.97	58.30	60.73
30	34.80	36.25	37.99	40.26	43.77	46.98	47.96	50.89	53.67	56.33	59.70	62.16
40	45.62	47.27	49.24	51.81	55.76	59.34	60.44	63.69	66.77	69.70	73.40	76.09
50	56.33	58.16	60.35	63.17	67.50	71.42	72.61	76.15	79.49	82.66	86.66	89.56
60	66.98	68.97	71.34	74.40	79.08	83.30	84.58	88.38	91.95	95.34	99.61	102.7
80	88.13	90.41	93.11	96.58	101.9	106.6	108.1	112.3	116.3	120.1	124.8	128.3
100	109.1	111.7	114.7	118.5	124.3	129.6	131.1	135.8	140.2	144.3	149.4	153.2

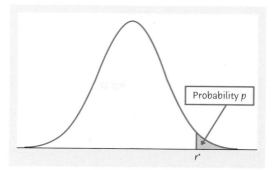

Table entry for p is the critical value r* of the correlation coefficient r with probability p lying to its right.

Probability p

r*

TABLE E   CRITICAL VALUES OF THE CORRELATION r

n	UPPER TAIL PROBABILITY p									
	.20	.10	.05	.025	.02	.01	.005	.0025	.001	.0005
3	0.8090	0.9511	0.9877	0.9969	0.9980	0.9995	0.9999	1.0000	1.0000	1.0000
4	0.6000	0.8000	0.9000	0.9500	0.9600	0.9800	0.9900	0.9950	0.9980	0.9990
5	0.4919	0.6870	0.8054	0.8783	0.8953	0.9343	0.9587	0.9740	0.9859	0.9911
6	0.4257	0.6084	0.7293	0.8114	0.8319	0.8822	0.9172	0.9417	0.9633	0.9741
7	0.3803	0.5509	0.6694	0.7545	0.7766	0.8329	0.8745	0.9056	0.9350	0.9509
8	0.3468	0.5067	0.6215	0.7067	0.7295	0.7887	0.8343	0.8697	0.9049	0.9249
9	0.3208	0.4716	0.5822	0.6664	0.6892	0.7498	0.7977	0.8359	0.8751	0.8983
10	0.2998	0.4428	0.5494	0.6319	0.6546	0.7155	0.7646	0.8046	0.8467	0.8721
11	0.2825	0.4187	0.5214	0.6021	0.6244	0.6851	0.7348	0.7759	0.8199	0.8470
12	0.2678	0.3981	0.4973	0.5760	0.5980	0.6581	0.7079	0.7496	0.7950	0.8233
13	0.2552	0.3802	0.4762	0.5529	0.5745	0.6339	0.6835	0.7255	0.7717	0.8010
14	0.2443	0.3646	0.4575	0.5324	0.5536	0.6120	0.6614	0.7034	0.7501	0.7800
15	0.2346	0.3507	0.4409	0.5140	0.5347	0.5923	0.6411	0.6831	0.7301	0.7604
16	0.2260	0.3383	0.4259	0.4973	0.5177	0.5742	0.6226	0.6643	0.7114	0.7419
17	0.2183	0.3271	0.4124	0.4821	0.5021	0.5577	0.6055	0.6470	0.6940	0.7247
18	0.2113	0.3170	0.4000	0.4683	0.4878	0.5425	0.5897	0.6308	0.6777	0.7084
19	0.2049	0.3077	0.3887	0.4555	0.4747	0.5285	0.5751	0.6158	0.6624	0.6932
20	0.1991	0.2992	0.3783	0.4438	0.4626	0.5155	0.5614	0.6018	0.6481	0.6788
21	0.1938	0.2914	0.3687	0.4329	0.4513	0.5034	0.5487	0.5886	0.6346	0.6652
22	0.1888	0.2841	0.3598	0.4227	0.4409	0.4921	0.5368	0.5763	0.6219	0.6524
23	0.1843	0.2774	0.3515	0.4132	0.4311	0.4815	0.5256	0.5647	0.6099	0.6402
24	0.1800	0.2711	0.3438	0.4044	0.4219	0.4716	0.5151	0.5537	0.5986	0.6287
25	0.1760	0.2653	0.3365	0.3961	0.4133	0.4622	0.5052	0.5434	0.5879	0.6178
26	0.1723	0.2598	0.3297	0.3882	0.4052	0.4534	0.4958	0.5336	0.5776	0.6074
27	0.1688	0.2546	0.3233	0.3809	0.3976	0.4451	0.4869	0.5243	0.5679	0.5974
28	0.1655	0.2497	0.3172	0.3739	0.3904	0.4372	0.4785	0.5154	0.5587	0.5880
29	0.1624	0.2451	0.3115	0.3673	0.3835	0.4297	0.4705	0.5070	0.5499	0.5790
30	0.1594	0.2407	0.3061	0.3610	0.3770	0.4226	0.4629	0.4990	0.5415	0.5703
40	0.1368	0.2070	0.2638	0.3120	0.3261	0.3665	0.4026	0.4353	0.4741	0.5007
50	0.1217	0.1843	0.2353	0.2787	0.2915	0.3281	0.3610	0.3909	0.4267	0.4514
60	0.1106	0.1678	0.2144	0.2542	0.2659	0.2997	0.3301	0.3578	0.3912	0.4143
80	0.0954	0.1448	0.1852	0.2199	0.2301	0.2597	0.2864	0.3109	0.3405	0.3611
100	0.0851	0.1292	0.1654	0.1966	0.2058	0.2324	0.2565	0.2786	0.3054	0.3242
1000	0.0266	0.0406	0.0520	0.0620	0.0650	0.0736	0.0814	0.0887	0.0976	0.1039

Chapter 0   Getting Started

0.1: (a) More than likely the individuals who chose to take vitamin E were more health conscious in general than those who didn't take it. They may also have been more affluent (i.e., had money available to purchase the vitamins). (b) In a randomized experiment, people of all types are randomly assigned to the treatments.

0.2: Most of this period had overall falling gas prices (about 1940 to 1975). There were instances of rapid price increases: 1931–32 (due to policies intended to "correct" the Depression), 1974–75 (the Arab oil embargo), roughly 1979–1980 (the overthrow of the Shah of Iran and Mideast tensions).

0.3: (a) The proportion of Americans who feel this way is very likely much lower. The "survey" was one of voluntary response (people were not randomly selected, but made their own decision to participate or not). Also, those who knew about the "poll" had just watched Mr. Schultz's long monologue. (b) As long as the poll was voluntary response, the sample size (868, 2500, or 100,000) doesn't matter.

0.4: (a) We can tell that this period was unusual because the dots for 2001 are at the top of the bars for October through December. (b) The terrorists didn't cause the fatal crashes. After the attacks of 9/11, people were reluctant to fly. With more car travel, there was more opportunity for car crashes.

Chapter 1   Picturing Distributions with Graphs

1.1: (a) Car makes and models. (b) Vehicle type (categorical), Transmission type (categorical), Number of cylinders (usually quantitative), City mpg (quantitative), Highway mpg (quantitative), Combined CO_2 (g/mile, quantitative).

1.2: Answers will vary. Possible categorical variables: Whether or not student plays on a sport team or club; Sex; Whether or not student smokes; Attitude about exercise. Possible quantitative variables: Weight (kilograms or pounds); Height (centimeters or inches); Resting heart rate (beats per minute); Body mass index (kg/m^2 or lb/ft^2).

1.3: (a) 100% − 71% = 29% of the radio audience listens to stations with other formats. (b)

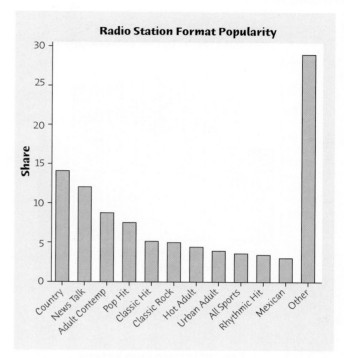

(c) No, the shares do not sum to 100%. If you include a wedge for "other," a pie chart would be reasonable.

1.4: (a) Individuals fall into more than one of the categories (pay for college expenses using more than one source). (b)

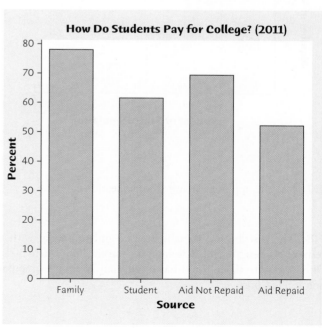

1.5: A pie chart would make it more difficult to distinguish between the weekend days and the weekdays. Some births are scheduled (induced labor, for example), and probably most are scheduled for weekdays.

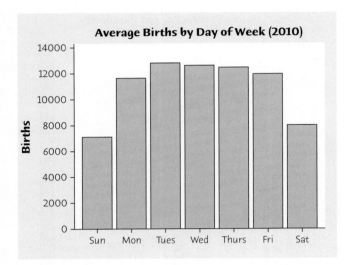

1.6:

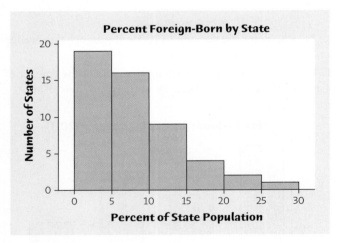

1.7: Use the applet to answer these questions.

1.8: The distribution is right-skewed and unimodal with a center between 5% and 10% and range from about 0% to 30%. Only one state (California) has more than 25% of its population foreign-born.

1.9: (a) There are two clear peaks in the distribution. Giving only one center would most likely be between these. (b) Young boys spend a lot of time outdoors playing and engaged in sports; their time outside in places where they would encounter ticks might well be less as they go through the college age and younger adulthood. With families and yard work, their time outside might increase. (c) This is wrong. Hiking in the woods at any age will make a person more likely to encounter the ticks that spread Lyme disease. (d) The histograms have the same shapes, but females have a slightly lower incidence rate at almost every age. Females possibly spend less time outdoors in areas where they would encounter ticks.

1.10: (a) The midpoint is 80.0%. The range is 59% to 88%.

```
5 | 9
6 | 2 3 7 8 8
7 | 1 1 2 4 4 4 5 6 6 6 6 7 7 7 8 8 8
8 | 0 0 0 1 1 2 2 3 3 3 3 3 3 3 4 4 6 6 6 6 6 6 6 7 7 8
```

(b) This plot is easier to read because there are fewer values per stem, but the shape is still left-skewed, and the skew seems even more pronounced than in the original.

```
5 | 9
6 | 2 3
6 | 7 8 8
7 | 1 1 2 4 4 4
7 | 5 6 6 6 6 7 7 7 8 8 8
8 | 0 0 0 1 1 2 2 3 3 3 3 3 3 4 4
8 | 6 6 6 6 6 6 7 7 8
```

1.11:

```
0 | 1 1 4 4
0 | 6 6 7 9 9 9 9
1 | 0 2 3 4 4
1 | 7
2 | 2
2 | 9
3 | 0 1 2 3
3 | 7 9
4 | 1 1 4
4 | 5 5 6
5 | 1
5 | 6 7
6 |
6 |
7 |
7 |
8 |
8 | 6
```

Distribution is somewhat right-skewed, with a single high outlier (United States) and two clusters of countries. Center is around 20 ($2000 spent per capita), ignoring the outlier. Range is from 1 ($100 spent per capita) to 86 ($8600 spent per capita).

1.12: (a)

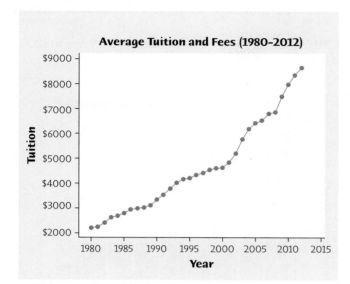

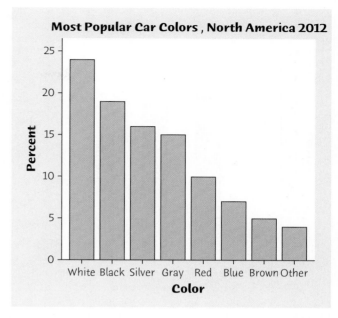

(b) Tuition has steadily climbed during the 30-year period, with sharpest absolute increases in the last five years. (c) Tuition has not decreased in this period, but there have been periods of small increases (1995–2000, for example). Very rapid increases were seen in 2000–2005 and 2008–2012. (d) It would be better to use percent increases rather than dollar increases.

1.26: (a)

1.13: (a)

1.14: (c)

1.15: (b)

1.16: (b) Zip codes are equivalent to town (or zone) names or identifications.

1.17: (b)

1.18: (b)

1.19: (c)

1.20: (b)

1.21: (b) Take the 26th ordered value.

1.22: (c)

1.23: (a) Students who have finished medical school. (b) 5; "Age" and "USMLE" are quantitative. The others are categorical. Age is in years, USMLE is points.

1.24: Brand name and model are identifiers. Categorical: type of freezer, Energy Star compliant. The rest are quantitative. Energy consumption is in kilowatts/yr; width, depth, and height are in inches; net capacities are in cubic feet.

1.25: "Other colors" should account for 4%.

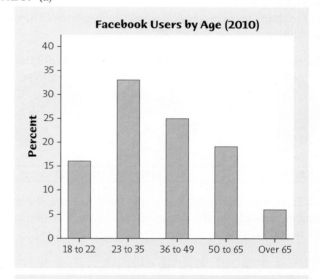

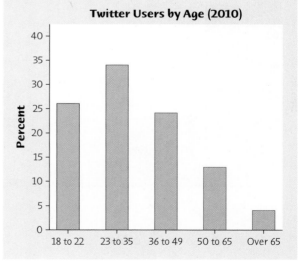

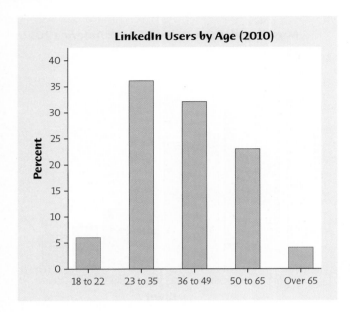

LinkedIn Users by Age (2010)

(b) Twitter has the largest proportion of users 18 to 22 years old and LinkedIn the smallest.

(c) Pie charts are appropriate because the percentages in each distribution should sum to 100%.

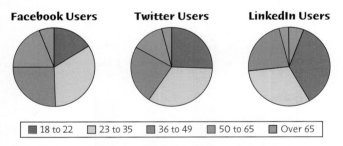

Facebook Users   **Twitter Users**   **LinkedIn Users**

■ 18 to 22   □ 23 to 35   ■ 36 to 49   ▨ 50 to 65   ▨ Over 65

1.27: (a) See below. (b) You would need to know the total number of deaths in this age group or the number of deaths due to "other" causes.

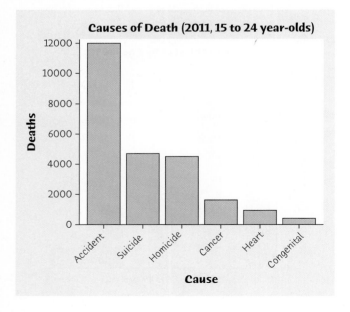

Causes of Death (2011, 15 to 24 year-olds)

1.28: Perhaps 60%–65% of the Hispanic population in the United States is Mexican and perhaps 10% is Puerto Rican.

1.29: (a) See below. (b) The percentages don't sum to 100%. Each university is its own "whole."

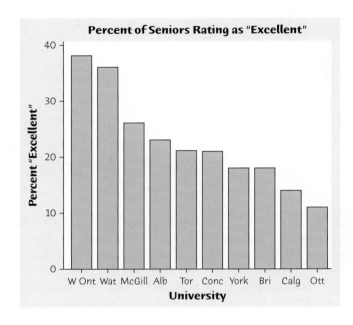

Percent of Seniors Rating as "Excellent"

1.30: This distribution is right-skewed, centered on 2 servings, with range from 0 to 8 servings. There are no outliers. About 12% (9 out of 74) consumed 6 or more servings, and about 35% (26 out of 74) ate fewer than 2 servings (which means 0 or 1 serving).

1.31: (a) Ignoring the outliers, the distribution is roughly symmetric, centered at about 110, with range 86 to 136. (b) 62/78 = 79.5%.

1.32: (a) Slightly left-skewed. (b) Somewhere between 0% and 2.5%. (c) Smallest: somewhere between −10% and −12.5%; largest: somewhere between 12.5% and 15%. (d) There are about 130 negative returns. This corresponds to about 39%.

1.33:

1. Are you male or female → Histogram (c).
2. Are you right-handed or left-handed → Histogram (b).
3. Heights → Histogram (d).
4. Time spent studying → Histogram (a).

1.34: (a) See following graph. (b) This is an extremely right-skewed distribution. Ratios greater than 1 correspond to an acid with more omega-3 than omega-6. Hence, this would be 7 of the 30 acids, or 23.3%. (c) Of the 7 healthier foods, 5 are types of fish. Furthermore, all the fish in the list have ratios higher than 1.

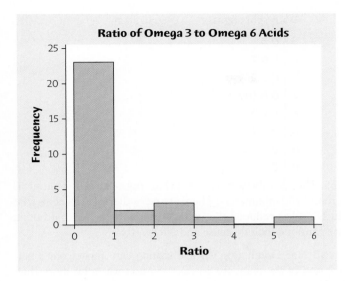

```
 0 | 1 1 1 2 3 3 3 5 5 7 9
 1 | 5 6 6 6 9 9
 2 | 7
 3 | 5 5 9
 4 | 0 0 4
 5 | 8 8
 6 | 0 3 7
 7 | 7 8
 8 | 2 7 9
 9 |
10 | 0 6
11 | 0
12 |
13 |
14 |
15 | 3
16 |
17 | 2
```

1.35: (a) States vary in population, so you would expect more nurses in California than in New Hampshire, for example. Nurses per 100,000 provides a better measure of how many nurses are available to serve a state's population. (b) See below. The District of Columbia, South Dakota, and Massachusetts are different from the others. Perhaps they could be considered outliers.

1.37: The shape of the distribution is symmetric; there are no visible outliers. The center is about 170 (the 12th observation). The data range from about 90 to about 250.

```
0 | 9
1 | 0 0 1 2
1 | 5 6 7 7 7 7 7 7 8 8 8 8 9
2 | 0 0 0 0
2 | 5
```

1.38: (a) It is natural for people to round to common multiples like 15 and 10.

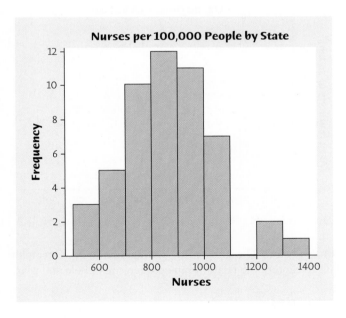

1.36: (a) Because the countries have varying populations, it is easier to compare them by emissions per person than by total emissions. (b) Use whole numbers as stems and tenths for the leaves. The United States and Canada are the extreme outliers in this data set. The distribution is right-skewed, with center about 3.5 and range of about 0.1 to about 17.

Women		Men
	0	0 3 3 3 3 4
9 6	0	6 6 6 7 9 9 9 9
2 2 2 2 2 2 2 1	1	2 2 2 2 2 2
8 8 8 8 8 8 8 8 8 7 5 5 5 5	1	5 5 5 8
4 4 4 0	2	0 0 3 4 4
7	2	
	3	0
6	3	

Both distributions are somewhat right-skewed. The women's distribution is shifted a bit to the right of the men, suggesting that they tend to study a bit more.

1.39: The decline in population is not described by the stem-plot made in Exercise 1.37.

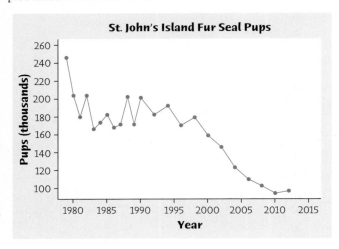

1.40: (a) Rates are appropriate because the group sizes are different.

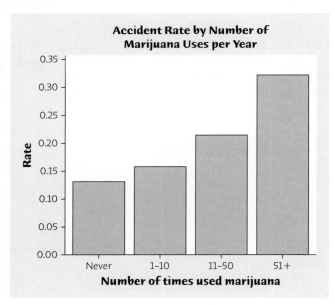

(b) Accident rates go up with the number of times drivers use marijuana. The data are "observational," and no cause-and-effect conclusion is reasonable.

1.41: Coins with earlier (lower) dates are older and rarer. There are more coins with larger dates (newer coins) than with smaller dates (older coins).

1.42: (a) Stems denote the hundreds place.

Without split stems:

```
6 | 0 3 5 5 7
7 | 0 1 2 4 4 8 8 9 9 9
8 | 1 1 3 6 6 7
9 | 0 6
```

With split stems:

```
6 | 0 3
6 | 5 5 7
7 | 0 1 2 4 4
7 | 8 8 9 9 9
8 | 1 1 3
8 | 6 6 7
9 | 0
9 | 6
```

(b) The distribution is somewhat right-skewed. Center is around 780 millimeters. There are no outliers. (c) It seems that El Niño has reduced the volume of monsoon rains, although we cannot conclude cause and effect.

1.43: (a) Graph (a). Vertical scaling can impact one's perception of the data. (b) In 2000, tuition was a bit more than $4000. Tuition rose to almost $9000 in 2012, so the increase is a bit less than $5000. Both plots describe the same data.

1.44: (a) Winter quarters are typically associated with lower housing starts. (b) and (c) Over the long run, housing starts have risen, except for the most recent years, which correspond with the 2007–2013 economic "crisis."

1.45: (a) See below. Since 2008 there has been no perceptible increase from year to year.

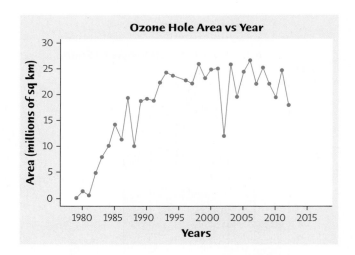

(b) See below. The midpoint is 19.5 million km². A stemplot fails to capture the relationship between size of hole and year.

```
0 | 0 1 1
0 | 5 8
1 | 0 0 1 2 4
1 | 8 9 9 9 9 9
2 | 0 2 2 2 2 3 3 4 4 4
2 | 5 5 5 5 6 6 7
```

1.46: Use the Applet to investigate this problem.

Chapter 2 Describing Distributions with Numbers

2.1: Mean *E. coli* level = 56.28 per ml. The mean is greater than most of the observations because of two high outliers.

2.2: Mean including the U.S. is $2604.29. Mean when the U.S. is excluded is $2427.71. The U.S. increases the mean by about $176.58.

2.3: Mean is 31.25 minutes. Median is 22.5 minutes. The mean is significantly larger than the median due to the right-skew in the distribution of times.

2.4: The mean is larger than the median because the distribution of home prices is right-skewed.

2.5: The histogram shows a right-skew. Hence, the mean is larger than the median. Here, the mean is 4.52 and the median is 3.55 tons per person.

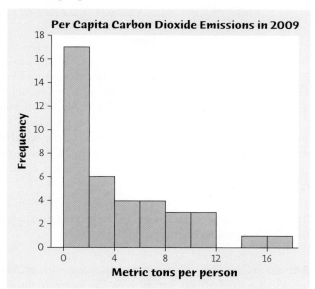

2.6: (a) and (b) A back-to-back stemplot is provided. The five-number summaries are tabulated below. (c) It seems that the offensive line players are slightly heavier; ignoring the outlier, the distribution of weights for the offensive linemen is symmetric, whereas the distribution for the defensive players is not symmetric and has fairly large gaps.

	MIN	Q_1	MED	Q_3	MAX
Offensive line	298	303.5	315	320.5	345 pounds
Defensive line	280	285.75	303.5	324.75	348 pounds

Offensive Linemen		Defensive Linemen
	28	0 5 8
8	29	
4 4 3	30	0 7
6 5 5	31	
5	32	4 5
	33	
5	34	8

2.7: (a) Minimum = 11, Q1 = 18, Median = 21, Q3 = 25, Maximum = 51. (b) The boxplot shows right-skew and several high outliers in the distribution of MPG values.

2.8: Q1 = 10 and Q3 = 30, so IQR = 30 − 10 = 20 minutes. Hence, Q1 − 1.5 × IQR = 10 − 1.5 × 20 = −20 minutes. No times can be negative, so no outliers are in the left tail. Q3 + 1.5 × IQR = 30 + 1.5 × 20 = 60 minutes. Hence, the "60" would not be considered an outlier, but it's close.

2.9: IQR = 25 − 18 = 7, so Q3 + 1.5 × IQR = 25 + 1.5 × 7 = 35.5. Seven values greater than 35.5 would be identified as potential outliers (40, 40, 40, 43, 45, 47, 51). Because Q1 − 1.5 × IQR = 18 − 1.5 × 7 = 7.5, there are no potential low outliers.

2.10: (a) $\bar{x}$ = (5.2 + 13.8 + 8.6 + 16.8)/4 = 44.4/4 = 11.1 picocuries per liter. (b) The standard deviation can be computed in steps:

x	5.2	13.8	8.6	16.8
$x - \bar{x}$	−5.9	2.7	−2.5	5.7
$(x - \bar{x})^2$	34.81	7.29	6.25	32.49

Hence, $s^2 = \dfrac{1}{n-1} \sum (x - \bar{x})^2$

$$= \dfrac{1}{4-1}(34.81 + 7.29 + 6.25 + 32.49) = 26.94667$$

So $s = \sqrt{s^2} = \sqrt{29.94667} = 5.19$ picocuries per liter.

2.11: Both data sets have the same mean and standard deviation (about 7.5 and 2.0, respectively). Stemplots reveal that Data A have a very left-skewed distribution, while Data B have a slightly right-skewed distribution.

2.12: (a) No. The distribution isn't symmetric. (b) Yes. The distribution is symmetric and mound-shaped with no severe outliers. (c) No. The distribution is strongly right-skewed.

2.13: Group 1: $\bar{x}$ = 23.75, s = 5.07. Group 2: $\bar{x}$ = 14.08, s = 4.98. Group 3: $\bar{x}$ = 15.78, s = 5.76.

2.14: Both groups (developing countries and developed countries) have right-skewed distributions for unpaid parking tickets. Comparing, developing countries' diplomats tend to have more unpaid tickets. National income alone, however, does not explain countries whose diplomats have more or fewer unpaid tickets.

2.15: (b)

2.16: (b)

2.17: (b)

2.18: (c)

2.19: (b)

2.20: (c)

2.21: (c)

2.22: (a)

2.23: (b)

2.24: (a)

2.25: The distribution is almost certainly right-skewed, so the mean is $61,691 and the median is $49,648.

2.26: In both cases (for single and married households), the distribution of account sizes is right-skewed. The median of $0 says that at least half of all singles have no retirement savings.

2.27: With 863 colleges, the median location is $(863 + 1)/2 = 432$, so the median is the 432nd (ordered) endowment. The 1st quartile, Q1, is found by taking the median of the first 431 endowments (when sorted). This would be the $(431 + 1)/2 = 216$th endowment. Similarly, Q3 is found as the 648th endowment (216 endowments above the median).

2.28: (a) Min = 23,000 pounds. Q1 = 30,350 pounds. Median = 31,950 pounds. Q3 = 32,700 pounds. Max = 33,700 pounds. (b) Notice that the Minimum is much farther from Q1 than the Maximum is from Q3. This suggests a long left tail, consistent with a left-skewed distribution.

2.29: The boxplots do not reveal the gap in the South between the rates for Georgia and the District of Columbia.

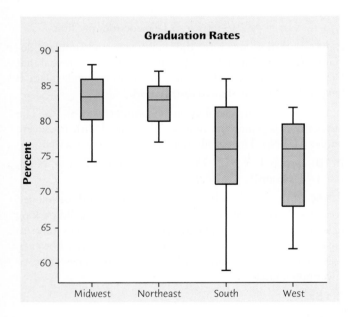

2.30: (a) Median = 2. Q1 = 1. Q3 = 4. (b) $\bar{x} = [(15)(0) + (11)(1) + (15)(2) + (11)(3) + (8)(4) + (5)(5) + (3)(6) + (3)(7) + (3)(8)]/74 = 194/74 = 2.62$ servings. This is larger than the median because the distribution is right-skewed.

2.31: (a) See top of next column. The distribution is strongly right-skewed, with center around 100 days and range from about 0 to 600 days. (b) Because of the extreme right skew, we should use the 5-number summary: 43, 82.5, 102.5, 151.5, 598 days. Notice that the median is closer to Q1 than to Q3.

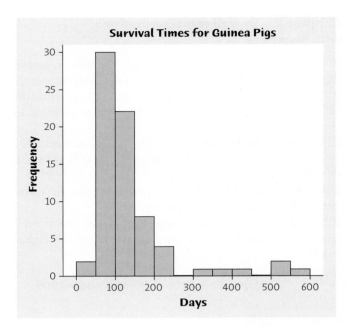

2.32: (a) If countries or years have very different numbers of babies born, it would be unreasonable to compare across years or across countries by counts. (b) 3,995,170 babies. (c) The distribution is left-skewed. (d) The median is the average of the 1,997,585th and 1,997,586th baby weights, and falls in the interval 3,000 to 3,499 grams. Q3 is in the interval 3,500 to 3,999 grams. Q1 is in the interval 2,500 to 2,999 grams.

2.33: (a) Symmetric distributions. (b) Removing the outliers reduces both means and both standard deviations.

2.34: (a) Mean (green arrow) moves along with moving point. Median (red arrow) points to middle point (rightmost nonmoving point). (b) Mean follows moving point. When moving point passes rightmost fixed point, median moves with it until moving point passes leftmost fixed point—then median stays there.

2.35: (a) The 6th observation must be placed at median for the original 5 observations. (b) No matter where you put the 7th observation, the median is one of the two repeated values above.

2.36: Both distributions are very similar: On weekdays more babies are born, and there is consistency from weekday to weekday, though Mondays appear to have slightly fewer births. On weekends, fewer births take place. Many more births take place in the United States.

2.37: The mean is 6.3%, far from the national percentage of 12.5%. You can't average averages. Some states, like California and Florida, are larger and should carry more weight in the national percentage.

2.38: More than half of all American households do not carry credit card debt.

2.39: (a) Pick any four numbers all the same: e.g., (4,4,4,4) or (6,6,6,6). (b) (0,0,10,10). (c) There is more than one possible answer for (a) but not for (b).

2.40: The TI-89 calculator used by the author reported $s = 1$ for the list 100,000,000,001 100,000,000,002 100,000,000,003. At some point, the calculator will fail, but for virtually any practical setting, a decent calculator will correctly compute.

2.41: Answers will vary. Start by ensuring that the median is 7, by "locking" 7 as the 3rd smallest value. Then, adjust the minimum or maximum accordingly to acquire a mean of 10 (so they sum to 50). One solution: 5 6 7 8 24.

2.42: Answers will vary. One solution: $(-100, 1, 2, 3, 4, 5, 6)$.

2.43: (a) Negative weight losses are weight *gains*. (b) See below. Gastric banding seems to produce higher weight losses, typically. (c) It's better to measure weight loss relative to initial weight. (d) If the subjects that dropped out had continued, the difference between these groups would be as great or greater because many of the "lifestyle" dropouts had negative weight losses (i.e., weight gains), which would pull that group down.

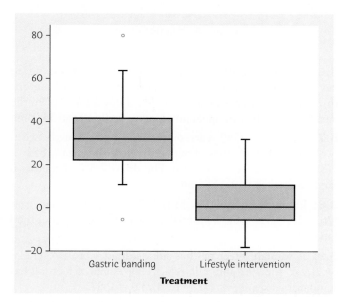

2.44: The distribution of Canadiens players' salaries is very right-skewed. Median salary is $1,750,000 (mean is $2,334,100, consistent with a strong right-skew). The middle half of players earn between $937,500 and $3,950,000, although a handful earn more than $5,000,000.

2.45: The distribution of average returns is left-skewed. Most years, average return is positive. Returns range from about −40% to 40%, with the median return about 16%.

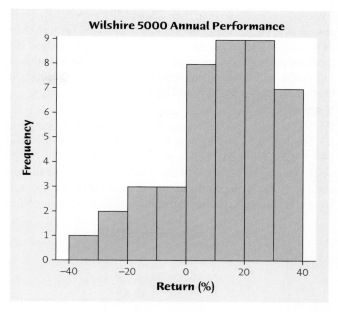

2.46: Comparing side-by-side boxplots, lavender seems to produce the highest customer expenditures.

2.47: Based on side-by-side boxplots, lean people spend relatively more time active, but there is little difference in the time these groups spend lying down.

2.48: Good weather forecasts generally yielded better tips, while there was little to no difference between a bad forecast and no forecast.

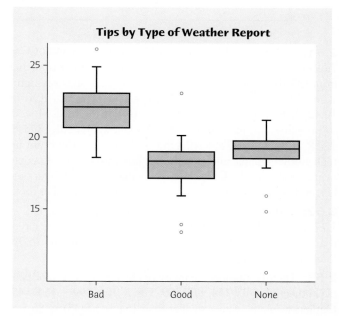

2.49: All age groups have right-skewed distributions with high outliers. The median increases slightly with increasing age (from 173 to 190 to 204). There is also an increase in variability with increasing age. (b) 25% or more of the individuals in each age group had total cholesterol levels above 200 ($Q_3 = 199$ for the people in their 20s and 218 and 229 for the other two). Unless their original cholesterol levels were *extremely* high, the 4 or

24 people on medication in their 20s and 30s, respectively, probably wouldn't affect these distributions a great deal because there were roughly 950 people in each group. For people in their 40s, those on medication were more than 10% of the group. If those 117 people had not been on medication, that distribution would likely show more variability and higher cholesterol readings.

2.50: (a) The five-number summary is 59, 74.5, 80, 83, 88. (b) The boxplot shows the same left-skew seen in the stemplot. (c) The *IQR* is $83 - 74.5 = 8.5$. Potential outliers are less than $74.5 - 1.5 \times 8.5 = 61.75$ or greater than $83 + 1.5 \times 8.5 = 95.75$. There are no high outliers, but the District of Columbia is the low outlier at 59%. Although the District of Columbia is not a state, it has a very high poverty rate; that could explain it being an outlier.

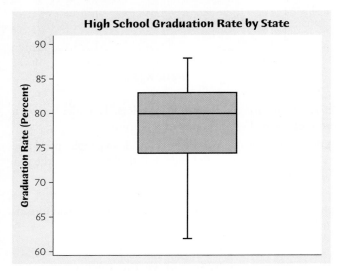

2.51: (a) Min = 0.0972. Q1 = 0.7475. Median = 3.5489. Q3 = 7.6977. Max = 17.2239. Notice that the maximum is farther from Q3 than the minimum is from Q1. This suggests right-skew. (b) IQR = $7.6977 - 0.7475 = 6.9502$. $1.5 \times$ IQR = 10.4253. Now Q1 $- 1.5 \times$ IQR = $0.7475 - 10.4253 < 0$, so no values are more than 1.5 IQRs below Q1. Also, Q3 $+ 1.5 \times$ IQR = $7.6977 + 10.4253 = 18.123$, so there are no high outliers. This rule is rather conservative—most people would easily call the United States's value (17.2239) a far outlier, and perhaps Canada would be considered an outlier too.

2.52: (a) Min = 1.3, Q1 = 3.8, Median = 6.3, Q3 = 12.7, Max = 26.9. (b) High end outliers will be values larger than $12.7 + 1.5(12.7 - 3.8) = 26.05$. California, at 26.9% foreign-born is an outlier.

2.53: The five-number summary of cholesterol levels for people in their 20s is 92, 154, 173, 199, 318. IQR = $199 - 154 = 45$. Outliers would be values smaller than $154 - 1.5(45) = 86.5$ or larger than $199 + 1.5(45) = 266.5$. Using this criterion, there are no low end outliers, but there are high end outliers.

Chapter 3   The Normal Distributions

3.1: Sketches will vary.

3.2: (a) It is on or above the horizontal axis everywhere, and the area beneath the curve is 1. (b) One-fifth of accidents

occur in the first mile, so the proportion is 0.20. (c) $(1.3 - 0.8) = 1/2$ mile. The area of this rectangle is $(1/2)(1/5)$, so the proportion is 0.10. (d) The part of the bike path more than a mile from either road is the 3-mile stretch from the 1-mile marker to the 4-mile marker. The area of this rectangle is $(3)(1/5)$, so the proportion is 3/5, or 0.6.

3.3: $\mu = 2.5$. The median is also 2.5 because the distribution is symmetric.

3.4: (a) Mean is C, median is B (the right-skew pulls the mean to the right). (b) Mean B, median B (this distribution is symmetric). (c) Mean A, median B (the left-skew pulls the mean to the left).

3.5: The vertical line at 35.8 indicates the mean; the horizontal line indicates the standard deviation.

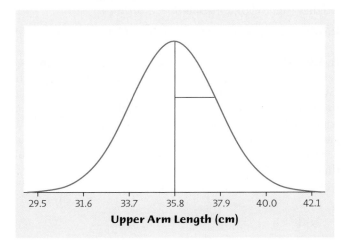

3.6: (a) 99.7% of all upper arm lengths are within 3 standard deviations of the mean, or between 29.5 and 42.1 centimeters. (b) This is the area 1 or more standard deviations above the mean. Hence, 16% of upper arm lengths are less than 33.7 cm.

3.7: (a) In 95% of all years, monsoon rain levels are between two standard deviations above and below the mean: $852 \pm 2(82) = 688$ to 1016 mm. (b) The driest 2.5% of monsoon rainfalls are less than 688 mm.

3.8: Idonna: $z = (670 - 514)/117 = 1.33$. Jonathan: $z = (26 - 21.1)/5.3 = 0.93$. Idonna's score is relatively higher than Jonathan's.

3.9: A woman 6 feet tall has standardized score $z = (72 - 64.2)/2.8 = 2.79$ (quite tall, relatively). A man 6 feet tall has standardized score $z = (72 - 69.4)/3.0 = 0.87$.

3.10: (a) 0.2236. (b) 0.7764. (c) 0.9265. (d) $0.9265 - 0.2236 = 0.7029$.

3.11: Let x be the monsoon rainfall in a given year. (a) $x \leq 697$ mm corresponds to $z \leq (697 - 852)/82 = -1.89$, for which Table A gives $0.0294 = 2.94\%$. (b) $682 < x < 1022$ corresponds to $(682 - 852)/82 < z < (1022 - 852)/82$, or $-2.07 < z < 2.07$. This proportion is $0.9808 - 0.0192 = 0.9616 = 96.16\%$.

3.12: (a) Let x be the MCAT score of a randomly selected student. Then $x > 30$ corresponds to $z > 0.75$, for which Table A gives 0.7734 as an area to the left. The answer is $1 - 0.7734 = 0.2266$. (b) $20 \leq x \leq 25$ corresponds to $-0.81 \leq z \leq -0.03$. Using Table A, the area is $0.4880 - 0.2089 = 0.2791$.

3.13: (a) Using Table A, looking for an area as close as possible to 0.6500, we find $z = 0.39$ (software gives $z = 0.3853$). (b) Now we want the value such that the proportion above is 0.20. This means that we want a proportion of 0.80 below. Using Table A, looking for an area as close to 0.8000 as possible, we find this value has $z = 0.84$ (software gives $z = 0.8416$).

3.14: Because the Normal distribution is symmetric, its median and mean are the same at 25.2. The first quartile has $z = -0.67$ (software gives $z = -0.6745$). The third quartile has $z = 0.67$. The first quartile is $25.2 - (0.67)(6.4) = 20.91$, and the third quartile is $25.2 + (0.67)(6.4) = 29.49$.

3.15: (c) Income and Sale prices of homes are economic variables that are usually right-skewed.

3.16: (a)

3.17: (b)

3.18: (b)

3.19: (b) $266 \pm 2(16) = 234$ to 298 days.

3.20: (c) 130 is two standard deviations above the mean, so 2.5% of adults have IQs of 130 or more.

3.21: (a) $z = \dfrac{132 - 100}{15} = 2.13$.

3.22: (b) $1 - 0.9265 = 0.0735$.

3.23: (b)

3.24: (c) About 98%. $z = 2.13$, and by Table A, the proportion below is 0.9834.

3.25: Sketches will vary but should be some variation on the one shown here: the peak at 0 should be "tall and skinny," while near 1, the curve should be "short and fat."

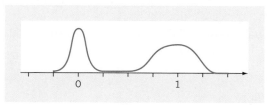

3.26: For mildly obese people: $373 \pm 2(67) = 239$ to 507 minutes. For lean people: $526 \pm 2(107) = 312$ to 740 minutes.

3.27: 70 is two standard deviations below the mean (that is, it has standard score $z = -2$), so about 2.5% (half of the outer 5%) of adults would have WAIS scores below 70.

3.28: (a) 0.0158. (b) $1 - 0.0158 = 0.9842$. (c) $1 - 0.9418 = 0.0582$. (d) $0.9418 - 0.0158 = 0.9260$.

3.29: (a) We want the proportion less than z to be 0.30, so $z = -0.52$. (Software gives $z = -0.5244$.) (b) If 35% are more than z, then 65% are less than or equal to z. $z = 0.39$. (Software gives $z = 0.3853$.)

3.30: (a) Let x be the length of a thorax for a randomly selected fruit fly. (a) $x < 0.6$ mm corresponds to $z < -2.56$. The area is 0.0052. (b) $x > 0.9$ mm corresponds to $z > 1.28$.

The area is $1 - 0.8997 = 0.1003$. (c) 0.6 mm $< x <$ 0.9 mm corresponds to $-2.56 < z < 1.28$. The area is $0.8997 - 0.0052 = 0.8945$.

3.31: $x < 5.0$ corresponds to $z < -0.80$, for which Table A gives 0.2119.

3.32: (a) $x > 135$ corresponds to $z > 2.48$. Table A gives $1 - 0.9934 = 0.0066 = 0.66\%$. (b) $x > 135$ corresponds to $z > 0.29$. Table A gives $1 - 0.6141 = 0.3859 = 38.59\%$.

3.33: $0.8725 < x < 0.8775$ corresponds to $\dfrac{0.8725 - 0.8750}{0.0012} < z < \dfrac{0.8775 - 0.8750}{0.0012}$, or $-2.08 < z < 2.08$, for which Table A gives $0.9812 - 0.0188 = 0.9624$. (Software gives 0.9628.)

3.34: Let x be the BMI for a randomly selected young man aged 20 to 29. (a) $z < -1.60$. So, 0.0548, or 5.48% are underweight. (b) $z > 0.62$. Hence, $1 - 0.7324 = 0.2676$, or 26.76% are obese.

For problems 3.35–3.38, let x denote the gas mileage of a randomly selected vehicle type from the population of 2013 model vehicles (excluding the high-mileage outliers, as mentioned).

3.35: Cars with better mileage than the Beetle correspond to $x > 25$, which corresponds to $z > 0.61$. This proportion is $1 - 0.7291 = 0.2709$, or 27.09%.

3.36: Looking for 0.1000 as a right-tail area in the table (0.9000 in the left tail) gives $z = 1.28$, so our vehicle would need mileage to be $21.9 + (1.28)(5.1) = 28.43$ mpg.

3.37: The first and third quartiles have $z = -0.67$ and $z = 0.67$, respectively. The first quartile is $21.9 - (0.67)(5.1) = 18.49$ mpg, and the third quartile is $21.9 + (0.67)(5.1) = 25.32$ mpg.

3.38: The first quintile is the mileage so that 20% of models have a lower mileage. This has $z = -0.84$ (find the number closest to 0.2000 in Table A as a left-tail area). Similarly, the second, third, and fourth quintiles have $z = -0.25$, $z = 0.25$ and $z = 0.84$, respectively. The first quintile is then $21.9 - (0.84)(5.1) = 17.62$ mpg. Similarly, the second, third, and fourth quintiles are, respectively, 20.63 mpg, 23.18 mpg, and 26.18 mpg.

3.39: (a) $z = \dfrac{(37.2 - 39.1)}{2.3} = -0.83$. From Table A, 20.33% is the proportion less than $z = -0.83$. Larry's percentile is 20.33. (Software gives 20.44%.)

3.40: A score of 1600 standardizes to $z = 2.76$, for which Table A gives a proportion of 0.9971 below. The proportion above 1600 (which are reported as 1600) is about 0.0029.

3.41: If x is the height of a randomly selected woman in this age group, we want the proportion corresponding to $x > 69.4$ inches. This corresponds to $z > 1.86$, which has proportion $1 - 0.9686 = 0.0314$, or 3.14%.

3.42: The distribution of weights of women is right-skewed. First, the mean weight is larger than the median weight. Another clue comes from the greater distance between the median and third quartile ($181.2 - 149.4 = 31.8$) than between the median and first quartile ($149.4 - 126.3 = 23.1$).

3.43: (a) Let x be a randomly selected man's SAT math score. $x > 750$ corresponds to $z > 1.83$. The proportion is $1 - 0.9664 = 0.0336$. (b) Let x be a randomly selected woman's SAT math score. $x > 750$ corresponds to $z > 2.22$. The proportion is $1 - 0.9868 = 0.0132$.

3.44: If the distribution is Normal, it must be symmetric about its mean—and in particular, the 10th and 90th percentiles must be equal distances below and above the mean—so the mean is 250 points. If 225 points below (above) the mean is the 10th (90th) percentile, this is 1.28 standard deviations below (above) the mean, so the distribution's standard deviation is $225/1.28 = 175.8$ points.

3.45. (a) About 0.6% of healthy young adults have osteoporosis ($z = -2.5$ gives 0.0062). (b) About 31% of this population of older women have osteoporosis ($z = -0.5$ gives 0.3085).

3.46. (a) There are two somewhat low IQs—72 qualifies as an outlier by the $1.5 \times IQR$ rule, while 74 is on the boundary. However, for a small sample, this stemplot looks reasonably Normal. (b) We compute $x = 105.84$ and $s = 14.27$ and find

$23/31 = 74.2\%$ of the scores in the range $x \pm 1s$, or 91.6 to 120.1

$29/31 = 93.5\%$ of the scores in the range $x \pm 2s$, or 77.3 to 134.4

For an exactly Normal distribution, we would expect these proportions to be 68% and 95%. Given the small sample, this is reasonably close agreement.

7	2 4
7	
8	
8	6 9
9	1 3
9	6 8
10	0 2 3 3 3 4
10	5 7 8
11	1 1 2 2 2 4 4 4
11	8 9
12	0
12	8
13	0 2

3.47: (a) $159{,}259/1{,}666{,}017 = 9.56\%$. (b) $\frac{159{,}259 + 54{,}167}{1{,}666{,}017} = 12.81\%$. (c) $x \geq 28$ corresponds to $z \geq \frac{28 - 21.1}{5.3} = 1.30$. From Table A, this is $1 - 0.9032 = 0.0968$, or 9.68%. (Software gives 9.65%.)

3.48: (a) The mean (5.43) is almost identical to the median (5.44), and the quartiles are similar distances from the median: $M - Q1 = 0.39$ while $Q3 - M = 0.35$. This suggests that the distribution is reasonably symmetric. (b) $x < 5.05$ corresponds to $z < -0.70$, and $x < 5.79$ corresponds to $z < 0.67$. Table A gives these proportions as 0.2420 and 0.7486. These are quite close to 0.25 and 0.75, which is what we would expect for the quartiles, so they are consistent with the idea that the distribution is close to Normal.

3.49: (a) A histogram (below) appears to be roughly symmetric with no outliers. (b) Mean = 544.42. Median = 540. Standard deviation = 61.24. Q1 = 500. Q3 = 580. The mean and median are close, and the distances of each quartile to the median are equal, consistent with a Normal distribution. (c) $z > -0.71$, or $1 - 0.2389 = 0.7611$, or 76.11%. (d) In fact, 1776 entering GSU students scored higher than 501, which represents $1776/2417 = 0.7348$, or 73.48%.

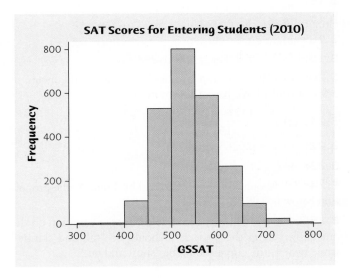

3.50: (a) Mean = 847.58 mm, and median = 860.8 mm. (b) The histogram shows a left skew.

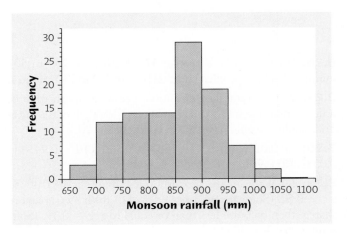

3.51: (a) $14/548 = 0.0255$ (2.55%) of the females in the data set weighed under 100 pounds. $z = \frac{100 - 161.58}{48.96} = -1.26$. From

Table A, 10.38% of the Normal (161.58, 48.96) distribution is less than 100. (b) $33/548 = 0.0602$ (6.02%) weighed more than 250 pounds. $z = \frac{250 - 161.58}{48.96} = 1.81$. From Table A, there should be 3.51% above 250. (c) No.

3.52: (a) The applet shows an area of 0.6827 between -1.000 and 1.000. (b) Between -2.000 and 2.000, the applet reports 0.9545. Between -3.000 and 3.000, the applet reports 0.9973.

3.53: Because the quartiles of any distribution have 50% of observations between them, we seek to place the flags so that the reported area is 0.5. The closest the applet gets is an area of 0.4978, between -0.671 and 0.671. Thus the quartiles of any Normal distribution are about 0.67 standard deviations above and below the mean.

Note: *Table A places the quartiles at about 0.67; other statistical software gives* ± 0.6745.

3.54: Placing the flags so that the area between them is as close as possible to 0.80, we find that the A/B cutoff is about 1.28 standard deviations above the mean, and the B/C cutoff is about 1.28 standard deviations below the mean.

Chapter 4   Scatterplots and Correlation

4.1: (a) Explanatory: time spent studying; response: grade. (b) Explore the relationship. (c) Explanatory: time spent online using Facebook; response: GPA. (d) Explore the relationship.

4.2: Explanatory: sea surface temperature; response: coral growth. Both variables are quantitative.

4.3: For example: weight, sex, other food eaten by the students, type of beer (light, imported, . . .).

4.4: The researchers suspect that lean body mass is explanatory, so it should be on the horizontal axis.

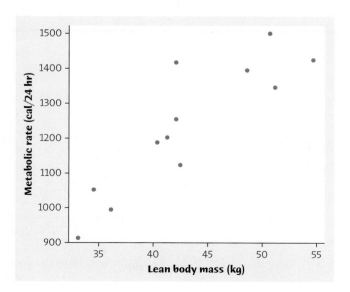

4.5: Outsource percent is the explanatory variable and should be on the horizontal axis. Delay percent is the response and should be on the vertical axis.

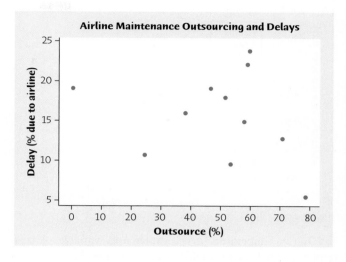

4.6: The scatterplot shows a positive direction, linear form, and moderately strong association.

4.7: There is an outlier (Frontier Airlines). Removing it, we would see almost no association between these variables. Without removing it, there is a very weak, negative association between the variables (which contradicts the suspicions described in Exercise 4.5).

4.8: (a) Below; speed is explanatory. (b) The relationship is curved—low in the middle, higher at the extremes. (c) Above-average (that is, bad) values of "fuel used" are found with both low and high values of "speed." (d) The relationship is very strong—there is little scatter around the curve, so the curve is very useful for prediction.

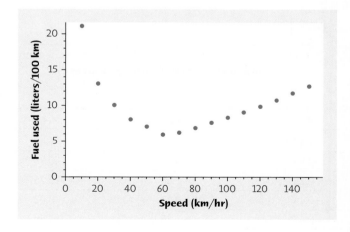

4.9: (a) Women are marked with filled circles, men with open circles. (b) For both men and women, the association is linear and positive. The women's points show a stronger association.

As a group, males typically have larger values for both variables (they tend to have more mass and tend to burn more calories per day).

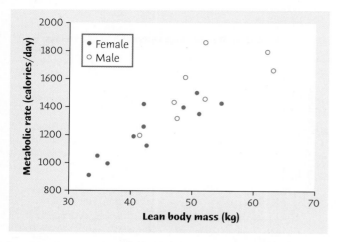

4.10: (a) Temperature is the explanatory variable. A scatterplot (below) shows a fairly strong, negative linear association between temperature and coral growth. (b) $\bar{x} = 30.28$ degrees, $s_x = 0.43$ degrees, $\bar{y} = 2.46$ mm, $s_y = 0.16$ mm. See the table below for the standardized scores. The correlation is $r = -5.24/6 = -0.873$. (c) Software gives a value of -0.8914. The more precision you carry at each step, the closer you'll get to that value.

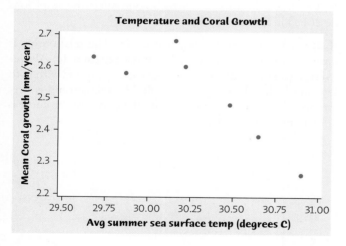

z_x	z_y	$z_x z_y$
-1.40	$+1.06$	-1.48
-0.95	$+0.75$	-0.71
-0.28	$+0.88$	-0.25
-0.14	-0.13	0.02
$+0.47$	-1.25	-0.59
$+0.86$	-0.50	-0.43
$+1.44$	-1.25	-1.80
		-5.24

4.11: r would not change; units do not affect correlation.
4.12: (a) $r = 0.8765$. (b) With Point A included, the correlation increases to 0.9273; with Point B, it drops to 0.7257. (c) Point A fits in with the positive linear association displayed by the other points, and even emphasizes (strengthens) that association. Point B deviates from the pattern, weakening the association.

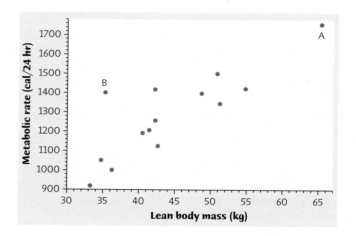

4.13: $\bar{x} = 50$ mph, $s_x = 15.8114$ mph, $\bar{y} = 26.8$ mpg, and $s_y = 2.6833$ mpg. Refer to the table of standardized scores below, then note that $r = 0/4 = 0$. The correlation is zero because these variables do not have a straight-line relationship; the association is neither positive nor negative.

z_x	z_y	$z_x z_y$
-1.2649	-1.0435	1.3199
-0.6325	0.4472	-0.2828
0	1.1926	0
0.6325	0.4472	0.2828
1.2649	-1.0435	-1.3199
		0

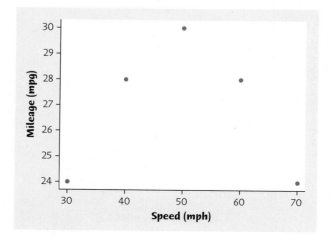

4.14: (a) We would expect that the price of a barrel of oil has an effect on the price of gasoline, rather than the reverse.

4.15: (a) The association should be positive (e.g., if oil prices rise, so do gas prices).

4.16: (b)

4.17: (a) 0.9. Without the outlier, there is a strong positive linear relationship.

4.18: (c)

4.19: (c) A correlation close to 0 might arise from a scatterplot with no visible pattern, but there could be a nonlinear pattern.

4.20: (c) Because we are not told how the x and y values vary together, we cannot tell whether the correlation will be -1 or $+1$.

4.21: (a)

4.22: (b) Correlation is unaffected by units.

4.23: (b)

4.24: (a) The lowest first-round score was 67, scored by one golfer. This golfer scored 73 in the second round. (b) Crenshaw scored 83 in the second round and 76 in the first round. (c) The correlation is very small but positive, so closest to 0.3. Knowing a golfer's first-round score would not be useful in predicting a second-round score.

4.25: (a) There is a slightly negative association between these variables. (b) There is general disagreement. (c) It does not appear that any of the values are obviously outside the general pattern. Perhaps one value (Rank = 8, BRFSS = 0.30) is an outlier, but this is hard to say.

4.26: (a) The scatterplot reveals a very strong positive linear relationship between wine intake and relative risk for cancer. We expect correlation to be close to $+1$. (b) $r = 0.9851$.

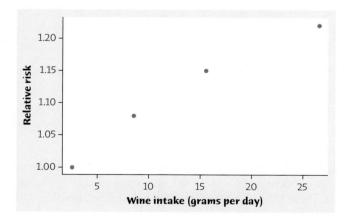

4.27: (a) The scatterplot suggests a strong positive linear association between distance and time with respect to the spread of Ebola. (b) $r = 0.9623$. (c) Correlation would not change, since it does not depend on units.

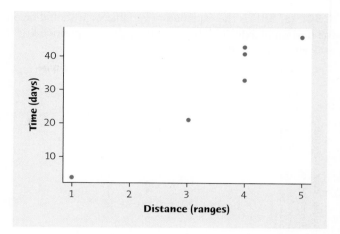

4.28: (a) The scatterplot shows a linear negative relationship. Correlation is an appropriate measure of strength: $r = -0.7485$. (b) The sparrowhawk is a long-lived territorial species.

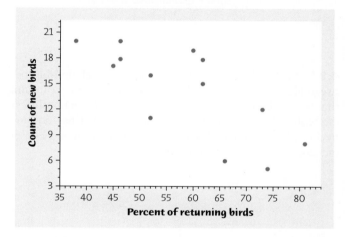

4.29: (a) See below. Neural activity is explanatory (and so should be on the horizontal axis). (b) The association is moderately strong, positive, and linear. The outlier is in the upper right corner. (c) For all points, $r = 0.8486$. Without the outlier, $r = 0.7015$. The correlation is greater with the outlier because it fits the pattern of the other points.

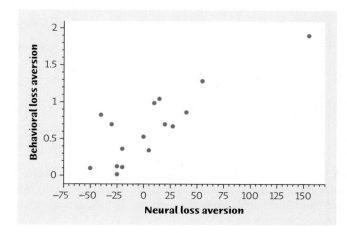

4.30: (a) SRD is the explanatory variable, so it should be on the horizontal axis. (b) The scatterplot shows a positive linear association. $r = 0.9685$.

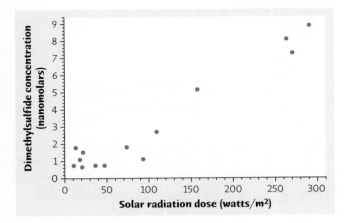

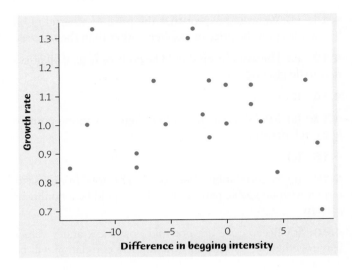

4.31: (a) See below. (b) There is a strong relationship between alcohol intake and relative risk of breast cancer (again, this is an observational study, so no causal relationship is established here). It seems that type of alcohol has nothing to do with the increase because the same pattern and rate of increase is seen for both groups.

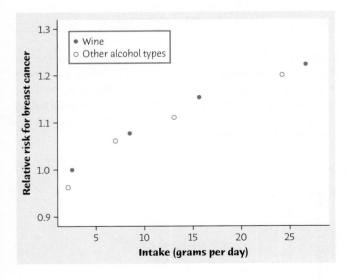

4.32: (a) See next column. (b) The scatterplot suggests that there is not a linear relationship. Here, $r = -0.1749$. (c) Neither theory is strongly supported, but the latter is more strongly supported.

4.33: (a) A plot follows that suggests that "Good" weather reports tend to yield higher tips. (b) The explanatory variable is categorical, not quantitative, so r cannot be used. Note that we can arrange the categories any way, and these different arrangements would suggest different associations.

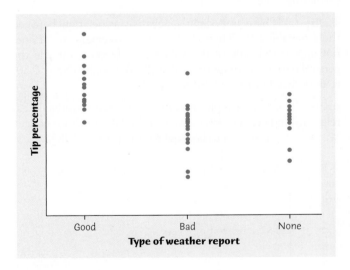

4.34: (a) Correlation would not change, as correlation does not depend on units. (b) Correlation would not change. (c) There would be a perfect positive linear relationship with $r = +1$.

4.35: (a) See next page. Changing the units has a dramatic impact on the plot. (b) Nevertheless, units do not impact correlation. For both data sets, $r = 0.8900$.

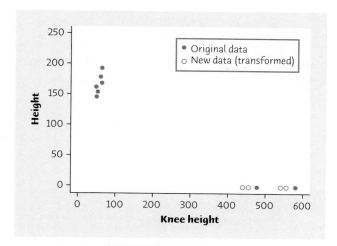

(b)

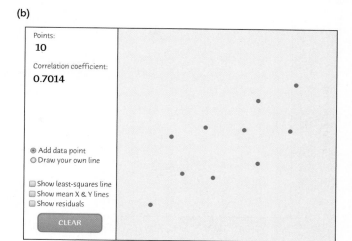

4.36: Explanations and sketches will vary but should note that correlation measures the strength of the association, not the slope of the line.

4.37: (a) Small-cap stocks have a lower correlation with municipal bonds, so the relationship is weaker. (b) She should look for a negative correlation.

4.38: The person who wrote the article interpreted a correlation close to 0 as if it were a correlation close to -1 (implying a negative association between teaching ability and research productivity). Professor McDaniel's findings mean there is little linear association between research and teaching ability. Also, remember that correlation is only meaningful if both variables are quantitative—and here there is no guarantee that this is the case.

4.39: (a) Because sex has a nominal scale, we cannot compute the correlation between sex and any other variable. Some writers and speakers use "correlation" as a synonym for "association," but this is not correct. (b) $r = 1.09$ is impossible, because r is restricted to be between -1 and 1. (c) Correlation has no units.

4.40: (a) The correlation will be closer to 1. (b) Answers will vary, but the correlation will decrease and can be made negative by dragging the point down far enough.

4.41: (a) Because two points determine a line, the correlation is always 1. (b) Sketches will vary; an example is shown at the top of the next column. Note that the scatterplot must be positively sloped. (c) One possibility is shown. (d) To have $r = 0.7$, the curve must be higher at the right than at the left. One possibility is shown.

(c)

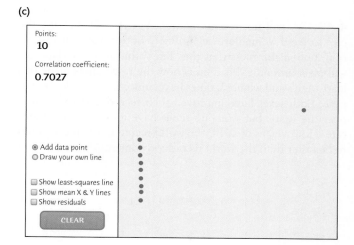

(d)

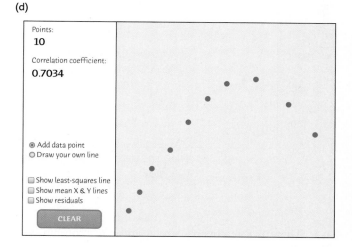

4.42: The plot suggests that sunlight has brightened overall, and the increase has been relatively steady. Correlation is a useful measure here, and $r = 0.9454$.

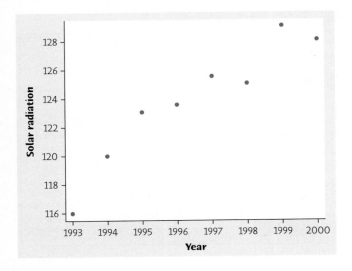

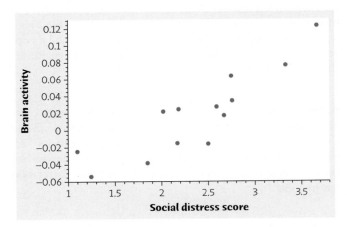

4.43: We will not use correlation, but we will examine the plot to see if women are beginning to outrun men. By inspection, one might guess that the "lines" that fit these data sets will meet around 1998. This is how the researchers made this leap. Men's and women's times have, indeed, grown closer over time. Both sexes have improved their record marathon times over the years, but women's times have improved at a faster rate. However, as of 2014, the world record for men continues to be faster than the world record for women.

4.45: A scatterplot shows a fairly strong, positive, linear association. There are no particular outliers; each variable has low and high values, but those points do not deviate from the pattern of the rest. $r = 0.8782$. Social exclusion does appear to trigger a pain response: higher social distress measurements are associated with increased activity in the pain-sensing area of the brain. However, no cause-and-effect conclusion is possible because this was not a designed experiment.

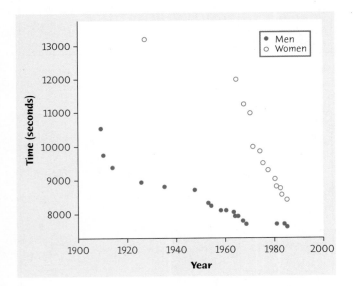

4.44: There is a reasonably strong linear relationship. $r = 0.9143$. When the outside temperature increases, a greater percentage of total heat loss is due to beak heat loss. That is, the beak plays a more important role in cooling down the toco toucan as the weather outside becomes hotter.

4.46: A scatterplot shows a moderately strong, positive, linear association. There are no clear outliers, although a few points fall slightly above (and one slightly below) the cluster. $r = 0.8042$. The positive association supports the idea that animal populations decline when the fish supply is low. The four years with the greatest fish supply were four of the five years in which biomass increased.

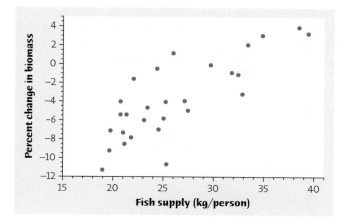

5.3: (a) $\bar{x} = 30.280$, $s_x = 0.4296$, $\bar{y} = 2.4557$, $s_y = 0.1579$, and $r = -0.8914$. Hence, $b = r\frac{s_y}{s_x} = (-0.8914)\frac{0.1579}{0.4296} = -0.3276$; $a = \bar{y} - b\bar{x} = 2.4557 - (-0.3275)(30.280) = 12.3754$. (b) Software agrees with these values to 3 decimal places because we rounded to the 4th decimal place.

5.4: (a) See below. (b) $\hat{y} = 201.2 + 24.026x$. (c) On average, metabolic rate increases by about 24 calories per day for each additional kilogram of body mass. (d) $\hat{y} = 1282.4$ calories per day.

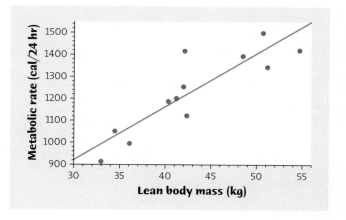

Chapter 5   Regression

5.1: (a) The slope is 1.033. On average, highway mileage increases by 1.033 mpg for each additional 1 mpg change in city mileage. (b) The intercept is 6.785 mpg. This is the highway mileage for a nonexistent car that gets 0 mpg in the city. Although this interpretation is valid, such a prediction would be invalid because it involves considerable extrapolation. (c) For a car that gets 16 mpg in the city, we predict highway mileage to be $6.785 + (1.033)(16) = 23.31$ mpg. For a car that gets 28 mpg in the city, we predict highway mileage to be $6.785 + (1.033)(28) = 35.71$ mpg. (d) The regression line passes through all the points of prediction. The plot was created by drawing a line through the two points (16, 23.31) and (28, 35.71), corresponding to the city mileages and predicted highway mileages for the two cars described in (c).

5.5: The farther r is from 0 (in either direction), the stronger the linear relationship between two variables.

5.6: (a) $\hat{y} = 1.0284 - 0.004498x$. The plot suggests a slightly curved pattern, not a strong linear pattern. A regression line is not useful for making predictions. (b) $r^2 = 0.0306$. The regression line does a poor job summarizing the relationship between difference in begging intensity and growth rate: Only 3% of the variation in growth rate is explained by the least-squares regression on difference in begging intensity.

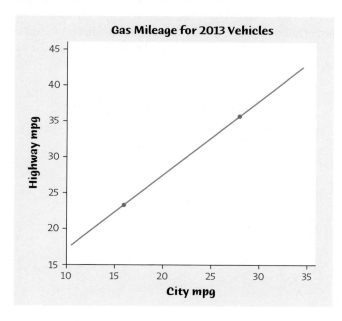

5.2: Weight = 80–5(days); intercept = 80 grams (the initial weight); slope = −5 grams/day.

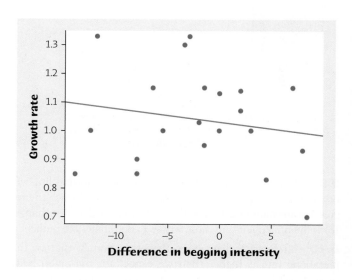

5.7: (a) The residuals are computed in the table below using $\hat{y} = 12.3754 - 0.3276x$. (b) They sum to zero, except for rounding error. (c) From software, the correlation between x and $y - \hat{y}$ is 0.000025, which is zero except for rounding.

x	y	$\hat{y}$	$y - \hat{y}$
29.68	2.63	2.652	−0.022
29.87	2.58	2.590	−0.010
30.16	2.60	2.495	0.105
30.22	2.48	2.475	0.005
30.48	2.26	2.390	−0.130
30.65	2.38	2.335	0.045
30.90	2.26	2.253	0.007
			0

5.8: (a) Below. (b) No; the pattern is curved, so linear regression is not appropriate for prediction. (c) $\hat{y} = 11.058 - 0.01466(10) = 10.91$, so the residual is $21.00 - 10.91 = 10.09$. The sum of the residuals is -0.01. (d) The first two and last four residuals are positive, and those in the middle are negative. Plot below.

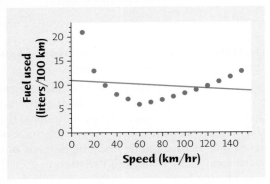

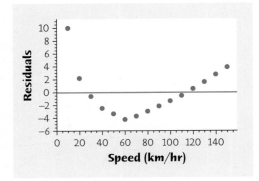

5.9: (a) Any point that falls exactly on the regression line will not increase the sum of squared vertical distances (which the regression line minimizes). Thus the regression line does not change. (b) Influential points are those whose x coordinates are outliers.

5.10: (a) Point A lies above the other points; that is, the metabolic rate is higher than we expect for the given body mass. Point B lies to the right of the other points; that is, it is an outlier in the x (mass) direction, and the metabolic rate is

lower than we would expect. (b) In the plot, the solid red line is the regression line for the original data. The dotted black line slightly above that includes Point A. The dashed red line includes Point B, the more influential point.

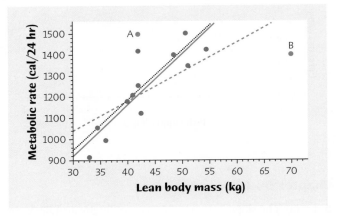

5.11: (a) In the plot, the outlier (Hawaiian Airlines) is the point at the lower right. Because this point is an outlier and falls outside the pattern suggested by the other data points, it is influential and will affect the regression line by "pulling" it. (b) With the outlier, $r = -0.243$. Without the outlier, $r = 0.038$. Notice that with the outlier, the correlation suggests a weaker linear relationship. (c) The two regression lines (one including the outlier, and the other without) are plotted. We see that the line based on the full data set (including the outlier) has been pulled down toward the outlier, indicating that the outlier is influential. Now, the regression line based on the complete (original) data set, including the outlier, is $\hat{y} = 18.667 - 0.0624x$. Using this, when $x = 78.6$, we predict 13.76% delays. The other regression line (fit without the outlier), is $\hat{y} = 16.223 + 0.0087x$, so our prediction would be 16.91% delays. The outlier impacts predictions because it impacts the regression line.

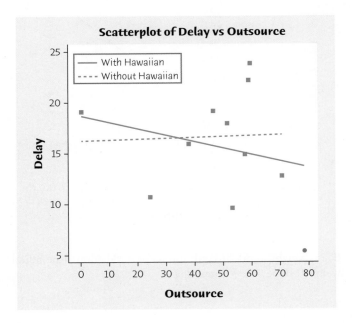

5.12: There is far more variability among individuals than among the averages. Correlation would be much smaller if it were calculated based on weights of individuals and their heights.

5.13: (a) The regression line is $\hat{y} = -44.939 + 0.1325x$ (or, $\widehat{Kills} = -44.939 + 0.1325\ Boats$). (b) If 900,000 boats are registered, then $x = 900$, and $\hat{y} = -44.939 + (0.1325)(900) = 74.3$ manatees killed. The prediction seems reasonable, as long as conditions remain the same, because "900" is within the space of observed values of x on which the regression line was based (this is not extrapolation). (c) If $x = 0$ (corresponding to no registered boats), then we would "predict" -44.939 manatees to be killed by boats. This is absurd, since it is clearly impossible for fewer than 0 manatees to be killed. This illustrates the folly of extrapolation … $x = 0$ is well outside the range of observed values of x on which the regression line was based.

5.14: A student's intelligence may be a lurking variable: stronger students (who are more likely to succeed when they get to college) are more likely to choose to take these math courses, whereas weaker students may avoid them.

5.15: Possible lurking variables include the IQ and socioeconomic status of the mother, as well as the mother's other habits (drinking, diet, etc.). These variables are associated with smoking in various ways and are also predictive of a child's IQ.

5.16: Socioeconomic status is a possible lurking variable: children from upper-class families can more easily afford higher education, and they would typically have had better preparation for college as well.

5.17: Answers will vary. One example would be that men who are married, widowed, or divorced may be more "invested" in their careers than men who are single. There is still a feeling of societal pressure for a man to "provide" for his family.

5.18: (b) The regression line seems to pass through the point (110, 7.5).

5.19: (b) Consider two points on the regression line, say (90,4) and (130,11). The slope of the line segment connecting these points is $\frac{11-4}{130-90} = 7/40$, which is close to 0.2.

5.20: (c)

5.21: (a)

5.22: (b) As the number of packs increases, average age at death decreases. Hence, correlation is negative, and so is the slope of the regression line.

5.23: (c)

5.24: (a)

5.25: (a) The slope of the line is positive.

5.26: (c)

5.27: (a)

5.28: (a) The slope is 0.0138 minutes per meter. (b) 5.45 minutes. (c)

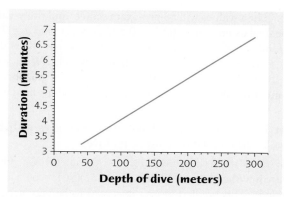

5.29: (a) Increasing the size of a diamond by an additional carat increases its price by 3721.02 Singapore dollars. (b) A diamond of size 0 carats would have a predicted price of 259.63 Singapore dollars. This is probably an extrapolation because the data set on which the line was constructed almost certainly had no rings with diamonds of size 0 carats.

5.30: (a) The regression equation is $\hat{y} = -0.126 + 0.0608x$. For $x = 2.0$, this formula gives $\hat{y} = -0.0044$. (b) $r^2 = 77.1\%$. (c) $r = \sqrt{r^2} = 0.878$; the sign is positive because it has the same sign as the slope coefficient.

5.31: (a) $\hat{y} = 0.919 + 2.0647x$. At 25 degrees Celsius, we predict beak heat loss of $\hat{y} = 0.919 + (2.0647)(25) = 52.54$ percent. (b) Because $r^2 = 0.836$, 83.6% of the total variation in beak heat loss is explained by the straight-line relationship with temperature. (c) $r = \sqrt{r^2} = \sqrt{0.836} = 0.914$. Correlation is positive here because the least-squares regression line has a positive slope.

5.32: Because we wish to regress husbands' heights on wives' heights, the women's heights will be the x-values, and the men's heights will be the y-values. (a) $b = (0.5)(\frac{3.1}{2.7}) = 0.574$, and $a = 69.9 - (0.574)(64.3) = 32.99$ inches. The regression equation is $\hat{y} = 32.99 + 0.574x$. (b) $\hat{y} = 32.99 + (0.574)(67) = 71.448$ inches. The plot, with this pair identified, is provided below. (c) We don't expect this prediction to be very accurate because the heights of men having wives 67 inches tall varies a lot.

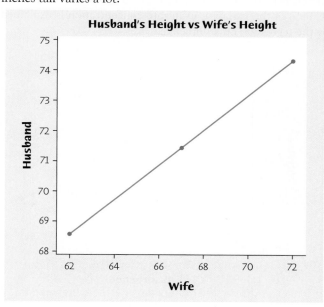

5.33: The x-values will be pre-exam scores, and the y-values will be final exam scores. (a) $b = 0.5 \times \frac{8}{40} = 0.1$, and $a = 75 - (0.1)(280) = 47$. The regression equation is $\hat{y} = 47 + 0.1x$. (b) $\hat{y} = 47 + (0.1)(300) = 77$. (c) Julie is right. With a correlation of $r = 0.5$, $r^2 = (0.5)^2 = 0.25$, so the regression line accounts for only 25% of the variability in student final exam scores.

5.34: $r = \sqrt{0.16} = 0.40$ (high attendance goes with high grades, so the correlation must be positive).

5.35: (a) $\hat{y} = 28.037 + 0.521x$. $r = 0.555$. (b) We predict Tonya to have height $\hat{y} = 28.037 + (0.521)(70) = 64.5$ inches (rounded). This prediction isn't expected to be very accurate because the correlation isn't very large. So $r^2 = (0.555)^2 = 0.308$. The regression line explains only 30.8% of the variation in sister heights.

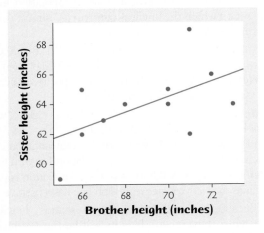

5.36: (a) A plot follows. The relationship between absorbence and nitrates is extremely linear. From software, $r = 0.99994 > 0.997$, so the calibration does not need to be repeated. From software, $\hat{y} = -14.522 + 8.825x$. For absorbence of 40: $-14.522 + (8.825)(40) = 338.478$ mg/liter. (c) We expect estimates of nitrate concentration from absorbence to be very accurate. $r^2 = (0.99994)^2 = 0.9999$, or 99.99% of the variation in nitrate concentration is explained by the regression on absorbence.

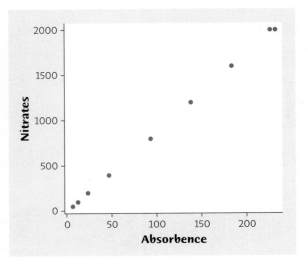

5.37: (a) $\hat{y} = 31.9 - 0.304x$. (b) The slope (-0.304) tells us that, on average, for each additional 1% increase in returning birds, the number of new birds joining the colony decreases by 0.304. (c) When $x = 60$, we predict $\hat{y} = 13.69$ new birds will join the colony.

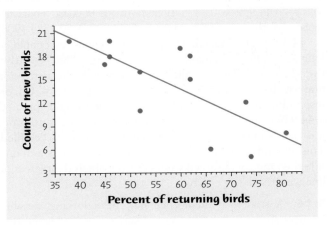

5.38: (a) The outlier in the upper-right corner is circled, because it is hard to see it with the two regression lines. (b) With the outlier omitted, $\hat{y} = 0.586 + 0.00891x$. (This is the solid line in the plot.) (c) The line does not change much because the outlier fits the pattern of the other points; r changes because the scatter (relative to the line) is greater with the outlier removed, and the outlier is located consistently with the linear pattern of the rest of the points. (d) The correlation changes from 0.8486 (with all points) to 0.7015 (without the outlier). With all points included, $\hat{y} = 0.585 + 0.0879x$ (nearly indistinguishable from the other regression line).

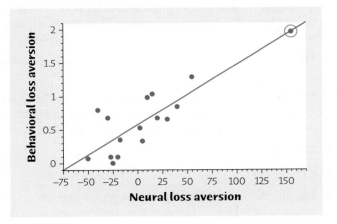

5.39: (a) To three decimal places, the correlations are all approximately 0.816, and the regression lines are all approximately $\hat{y} = 3.000 + 0.500x$. For all four sets, we predict $\hat{y} = 8$ when $x = 10$. (b) Plots below. (c) For Set A, the use of the regression line seems to be reasonable. For Set B, there is an obvious *non*linear relationship; we should fit a parabola or other curve. For Set C, the point (13, 12.74) deviates from the (highly linear) pattern of the other points. For Set D, the data point with $x = 19$ is a very influential point—the other points alone give no indication of slope for the line.

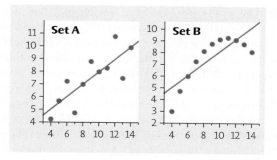

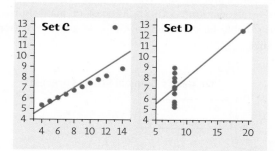

5.40: (a) The two unusual observations are indicated on the scatterplot. (b)

$$r_1 = 0.4819 \text{ (all observations)}$$
$$r_2 = 0.5684 \text{ (without Subject 15)}$$
$$r_3 = 0.3837 \text{ (without Subject 18)}$$

Both outliers change the correlation. Removing Subject 15 increases r because its presence makes the scatterplot less linear. Removing Subject 18 decreases r because its presence decreases the relative scatter about the linear pattern.

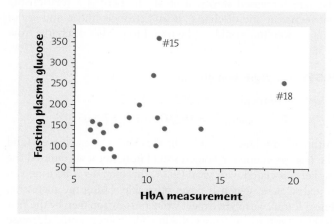

5.41: (a) $\hat{y} = 42.933 + 2.453x$. (b) $\hat{y} = 0.42933 + 0.002453x$. (c) Use the fact that 50 cm = 500 mm. When $x = 50$ cm, the first regression equation gives $\hat{y} = 165.583$ cm. Using the second equation, with $x = 500$ mm, $\hat{y} = 1.65583$ m. These are the same.

5.42: The equations are

$$\hat{y} = 66.4 + 10.4x \text{ (all observations)}$$
$$\hat{y} = 69.5 + 8.92x \text{ (without \#15)}$$
$$\hat{y} = 52.3 + 12.1x \text{ (without \#18)}$$

Although the equation changes in response to removing either subject, one could argue that neither one is particularly influential because the line moves very little over the range of x (HbA) values.

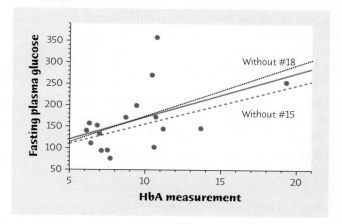

5.43: The correlation would be much lower because there is much greater variation in individuals than in the averages.

5.44: In this case, there may be a causative effect but in the direction opposite to the one suggested: People who are overweight are more likely to be on diets and so choose artificial sweeteners over sugar. (Also, heavier people are at a higher risk to develop Type 2 diabetes; if they do, they are likely to switch to artificial sweeteners.)

5.45: Responses will vary. For example, students who choose the online course might have more self-motivation, or have better computer skills (which might be helpful in doing well in the class).

5.46: For example, a student who in the past might have received a grade of B (and a lower SAT score) now receives an A (but has a lower SAT score than an A student in the past). Because of the grade inflation, we are not comparing students with equal abilities in the past and today.

5.47: Here is a (relatively) simple example to show how this can happen: suppose that most workers are currently 30 to 50 years old; of course, some are older or younger than that, but this age group dominates. Suppose further that each worker's current salary is his/her age (in thousands of dollars); for example, a 30-year-old worker is currently making \$30,000. Over the next 10 years, all workers age, and their salaries increase. Suppose every worker's salary increases by between \$4000 and \$8000. Then every worker will be making *more* money than he/she did 10 years before, but *less* money than a worker of that same age 10 years before. During that time, a few workers will retire, and others will enter the workforce, but that large cluster that had been between the ages of 30 and 50 (now between 40 and 60) will bring up the overall median salary despite the changes in older and younger workers.

5.48: We have slope $b = r\, s_y/s_x$ and intercept $a = \bar{y} - b\bar{x}$ and $\hat{y} = a + bx$. When $x = \bar{x}$, $\hat{y} = a + b\bar{x} = (\bar{y} - b\bar{x}) + b\bar{x} = \bar{y}$.

5.49: For a player who shot 80 in the first round: $\hat{y} = 51.42 + (0.314)(80) = 76.54$. For a player who shot 70 in the first round: $\hat{y} = 51.42 + (0.314)(70) = 73.4$. Notice that the player who shot 80 the first round (worse than average) is predicted to have a worse-than-average score the second round, but better than the first round. Similarly, the player who shot 70 the first round (better than average) is predicted to do better than average in the second round, but not as well (relatively) as in the first round. Both players are predicted to "regress" to the mean.

5.50: $\bar{y} = 46.6 + 0.41\bar{x}$. We predict that Octavio will score 4.1 points above the mean on the final exam: $\hat{y} = 46.6 + 0.41(\bar{x} + 10) = 46.6 + 0.41\bar{x} + 4.1 = \bar{y} + 4.1$.

5.51: See Exercise 4.41 for the three sample scatterplots. A regression line is appropriate only for the scatterplot of part (b).

5.52: (a) Drawing the "best line" by eye is a very inaccurate process; few people choose the best line. (b) Most people tend to overestimate the slope for a scatterplot with $r = 0.7$; that is, most students will find that the least-squares line is less steep than the one they draw.

5.53: The scatterplot shows a positive linear association. The regression line is $\hat{y} = -1.286 + 11.89x$. The straight-line relationship explains that $r^2 = 83.9\%$ of the variation in beetle larvae. The strong positive association supports the idea that beavers benefit beetles.

5.54: In the scatterplot, right-hand points are open circles, and left-hand points are filled circles. In general, the right-hand points lie below the left-hand points, meaning the right-hand times are shorter, so the subject is likely right-handed. There is no striking pattern for the left-hand points; the pattern for right-hand points is obscured because the points are squeezed at the bottom of the plot. Although neither plot looks particularly linear, we might nonetheless find the two regression lines: For the right hand, $\hat{y} = 99.36 + 0.0283x$ ($r = 0.305$, $r^2 = 9.3\%$), and for the left hand, $\hat{y} = 171.5 + 0.2619x$ ($r = 0.318$, $r^2 = 10.1\%$). Neither regression is particularly useful for prediction; distance accounts for only 9.3% (right) and 10.1% (left) of the variation in time.

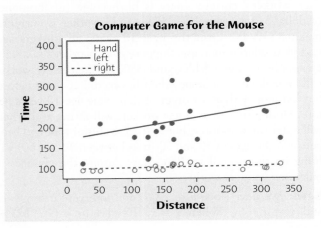

5.55: The scatterplot shows a reasonable, but not very strong linear relationship between Forecasted and Actual named storms. In the plot, the 2005 season is a noticeable outlier at the upper right. We might consider deleting this point and fitting the line again. Deleting the data point, we obtain $\hat{y} = 1.874 + 0.886x$. For reference, if the forecasts were perfect, the intercept of this line would be 0, and the slope would be 1. After deleting the 2005 season, $r = 0.647$, and $r^2 = 41.8\%$. Even after deleting the outlier, the regression line explains only 41.8% of variation in number of hurricanes. Predictions using the regression line are not very accurate. However, there is a positive association . . . so a forecast of many hurricanes may reasonably be expected to forebode a heavy season for hurricanes.

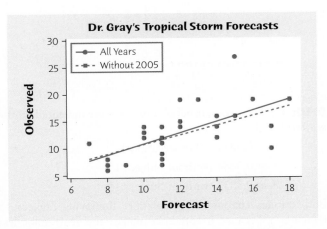

5.56: We see that during the recent 10 or 15 years, the volume of discharge has increased more rapidly, but before then, the rate increased slowly, if at all. If there is a relationship between year and discharge, it isn't strongly linear, and use of a regression line would not be useful to predict discharge from year.

5.57: The regression lines are:

For men: $\hat{y} = 66,072 - 29.535x$

For women: $\hat{y} = 182,976.15 - 87.73x$

Although the lines appear to fit the data reasonably well (and the regression line for women would fit better if we omitted the outlier associated with year 1926), this analysis is inviting you to extrapolate, which is never advisable. Using the regression lines plotted, we might expect women to outrun men by the year 2009. Omitting the outlier, the line for women would decrease more steeply, and the intersection would occur sooner, by 1995.

Chapter 6   Two-Way Tables

6.1: (a) This table describes 1808 people. $736 + 450 + 193 = 1379$ played video games. (b) The percent of boys earning A's and B's is $(736 + 205)/1808 = 0.5205 = 52.05\%$. We do this for all three grade levels. The complete marginal distribution for grades is

GRADE	PERCENT
A's and B's	52.05%
C's	32.85%
D's and F's	15.10%

Of all boys, 32.85% + 15.10% = 47.95% received a grade of C or lower.

6.2: (a) This table describes 19,558 thousand undergraduates. (b) There are 12,072 thousand undergraduates in the 18–24 age group. This represents 12,072/19,558 = 0.6172 = 61.72%. The marginal distribution for ages of undergraduates follows:

AGE GROUP	PERCENT
18 to 24	61.72%
25 to 34	21.95%
35 or older	16.33%

6.3: There are 1379 players. Of these, 736/1379 = 53.37% earned A's or B's. Similarly, there are 429 nonplayers. Of these, 205/429 = 47.79% earned A's or B's. Continuing in like manner, the conditional distribution of grades for players follows:

GRADES	PLAYERS	NONPLAYERS
A's and B's	53.37%	47.79%
C's	32.63%	33.56%
D's and F's	14.00%	18.65%

It doesn't look as if there's a big difference between these conditional distributions.

6.4: Shown in the table are the percents of women in each age group; for example, for the 18–24 age group, the proportion of women is 6432/(6432 + 5640) = 0.5328, or 53.28%. The data support our suspicion—the percent of women in the 25–34 age group (57.07%) is, indeed, larger than the percent of women in the 18–24 age group (53.28%). Notice that the percent of women in the 35 or older age group is particularly high, at 66.52%.

AGE GROUP	PERCENT FEMALE
18 to 24	53.28%
25 to 34	57.07%
35 or older	66.52%

6.5: Two examples are shown. In general, choose a to be any number from 10 to 50, and then all the other entries can be determined.

30	20
30	20

50	0
10	40

6.6: (a) Jamie made 155 field goals in 331 attempts. She made 155/331 = 0.468 = 46.8% of her field goal attempts.

Similarly, Lindsay made 47.6% of her field goal attempts. (b) The table below describes the percent of field goals made for each type of field goal, for each player. (c) While Jamie's overall field goal percentage is lower than Lindsay's, notice that Jamie's percentage is higher for both types of field goals. This is an example of Simpson's paradox.

	JAMIE	LINDSAY
Two-pointers	50.85%	50.59%
Three-pointers	37.11%	23.81%

6.7: (a) For Rotorua district, 79/8889 = 0.0089 or 0.9%, of Maori are in the jury pool, while 258/24,009 = 0.0107 or 1.07%, of the non-Maori are in the jury pool. For Nelson district, the corresponding percents are 0.08% for Maori and 0.17% for non-Maori. In each district, the percent of non-Maori in the jury pool exceeds the percent of Maori in the jury pool. (b) Combining the regions into one table:

	MAORI	NON-MAORI
In jury pool	80	314
Not in jury pool	10,138	56,353
Total	10,218	56,667

For the Maori, overall the percent in the jury pool is 80/10,218 = 0.0078, or 0.78%, while for the non-Maori, the overall percent in the jury pool is 314/56,667 = 0.0055, or 0.55%. Overall the Maori have a larger percent in the jury pool, but in each region they have a lower percent in the jury pool. (c) The reason for Simpson's paradox occurring with this example is that the Maori constitute a large proportion of Rotorua's population, while in Nelson they are small minority community.

6.8: (b)

6.9: (b)

6.10: (b) 150/612 = 0.245, or 24.5%

6.11: (a)

6.12: (c) 97/150 = 0.647, or 64.7%

6.13: (c)

6.14: (b) 97/468 = 0.207, or 20.7%

6.15: (b)

6.16: (b)

6.17: (b)

6.18: The two distributions are given below. For example, of the people that feel astrology is not at all scientific, the percent with JC degrees is 87/(87+198+111) = 0.2197, or 21.97%.

	NOT AT ALL SCIENTIFIC	VERY OR SORT OF SCIENTIFIC
Junior college	22.0%	33.6%
Bachelor's	50.0%	44.5%
Graduate	28.0%	21.9%

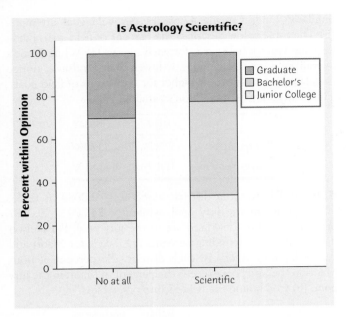

Loosely, adults who believe astrology is not at all scientific tend to have relatively more college education than adults who believe astrology is very or sort of scientific.

6.19: For each type of injury (accidental, not accidental), the distribution of ages is shown below.

AGES	ACCIDENTAL	NOT ACCIDENTAL
8–13	19.0%	4.0%
14–18	42.2%	35.8%
19–22	15.4%	20.8%
23–30	23.4%	39.4%

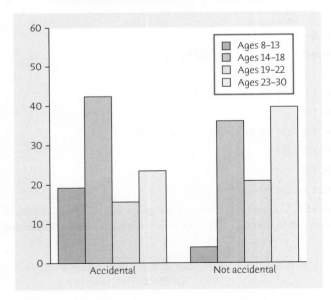

Among accidental weight-lifting injuries, the percentage of relatively younger lifters is larger, whereas among the injuries that are not accidental, the percentage of relatively older lifters is larger.

6.20: The table gives the two marginal distributions. The marginal distribution of marital status is found by taking, for example, $337/8235 = 0.041$, or 4.1% for the "Single" group. The marginal distribution of job grade is found by taking, for example, $955/8235 = 0.116$, or 11.6% for Grade 1. As rounded here, both sets of percents add to 100%.

Single	Married	Divorced	Widowed
4.1%	93.9%	1.5%	0.5%
Grade 1	Grade 2	Grade 3	Grade 4
11.6%	51.5%	30.2%	6.7%

6.21: The percent of single men in Grade 1 jobs is $58/337 = 0.172$, or 17.2%. The percent of Grade 1 jobs held by single men is $58/955 = 0.0607$, or 6.07%.

6.22: Divide the entries in the first column by the first column total; e.g., from Exercise 6.21, 17.2% is $0.172 = 58/337$. These should add to 100%, except for possible rounding error.

Job grade	1	2	3	4
% of single men	17.2%	65.9%	14.8%	2.1%

6.23: (a) We need to compute percents to account for the fact that the study included many more married men than single men, so we would expect their numbers to be higher in every job grade (even if marital status had no relationship with job level). (b) A table of percents is provided. Single and widowed men had higher percents of Grade 1 jobs; single men had the lowest (and widowed men the highest) percents of Grade 4 jobs.

	SINGLE	MARRIED	DIVORCED	WIDOWED
Grade 1	17.2%	11.3%	11.9%	19.0%
Grade 4	2.1%	6.9%	5.6%	9.5%

6.24: Answers will vary. One example would be that men who are married, widowed, or divorced may be more "invested" in their careers than men who are single. There is still a feeling of societal pressure for a man to "provide" for his family.

6.25: (a) The two-way table of race (White, Black) versus death penalty (Death penalty, No death penalty) follows.

	WHITE DEFENDANT	BLACK DEFENDANT
Death penalty	19	17
No death penalty	141	149

(b) For black victims: percentage of white defendants given the death penalty is $0/9 = 0$, or 0%; percentage of black defendants given the death penalty is $6/103 = 0.058$, or 5.8%. For white victims: percentage of white defendants given the death penalty is $19/151 = 0.126$, or 12.6%; percentage of black defendants given the death penalty is $11/63 = 0.175$, or 17.5%. For both victim races, black defendants are given the death penalty relatively more often than white defendants. Overall, referring to the table in (a), 19/160 = 0.119 or 11.9% of white defendants got the death penalty, while 17/166 = 0.102 or 10.2% of black defendants got the death penalty. This illustrates Simpson's paradox. (c) For white defendants, 151/160 = 0.9438 = 94.4% of victims were white. For black defendants, only 63/166 = 0.3795 or 37.95% of victims were white. The death penalty was predominantly assigned to cases involving white victims: 14.0% of all cases with a white victim, while only 5.5% of all cases with a black victim, had a death penalty assigned to the defendant. Because most white defendants' victims are white and cases with white victims carry additional risk of a death penalty, white defendants are being assigned the death penalty more often overall.

6.26: Examples will vary. Here is one very simple possibility. The key is to be sure that the three-way table has a lower percent of overweight people among the smokers than among the nonsmokers.

SMOKER	EARLY DEATH	
	YES	NO
Obese	1	0
Not obese	4	2

NONSMOKER	EARLY DEATH	
	YES	NO
Obese	3	6
Not obese	1	3

COMBINED	EARLY DEATH	
	YES	NO
Obese	4	6
Not obese	5	5

6.27: The percentages for each column are provided in the table. For example, for Chantix, the percentage of successes (no smoking in weeks 9–12) is 155/(155 + 197) = 0.4403, or 44.0%. Since we're comparing success rates, we'll leave off the row for "% smoking in weeks 9–12," since this is just 100% − % No smoking in weeks 9–12.

	CHANTIX	BUPROPION	PLACEBO
% No smoking in weeks 9–12	44.0%	29.5%	17.7%

A larger percentage of subjects using Chantix were not smoking during weeks 9–12, compared with results for either of the other treatments.

6.28: The table represents the responses of 516 men and 636 women. To find the conditional distributions, divide each entry in the table by its column total. These percents are given in that table; for example, 76/516 = 0.1473, or 14.73%. Men are more likely to view animal testing as justified if it might save human lives: over two-thirds of men agree or strongly agree with this statement, compared to slightly less than half of women. The percents who disagree or strongly disagree tell a similar story: 16% of men versus 30% of women.

RESPONSE	MALE	FEMALE
Strongly agree	14.7%	9.3%
Agree	52.3%	38.8%
Neither	16.9%	21.9%
Disagree	11.8%	19.3%
Strongly disagree	4.3%	10.7%

6.29: We compute, for example, the percentage of women earning associate's degrees: 519/823 = 0.631, or 63.1%. The table shows the percent of women at each degree level, which is all we need for comparison. Women constitute a substantial majority of associate's, bachelor's, and master's degrees, a scant majority of doctor's degrees, and slightly less than 50% of professional degrees.

DEGREE	% FEMALE
Associate's	63.1
Bachelor's	57.5
Master's	61.1
Professional	49.5
Doctor's	53.3

6.30: The table provides the percents of subjects with various complications for each treatment. For example, for subjects with gastric banding, 81/5380 = 0.0151 or 1.5% had non-life-threatening complications. Gastric bypass surgery carries the greatest risk for both non-life-threatening and serious complications. Gastric banding seems to be the safest procedure, with the lowest rates for both types of complications.

	NON-LIFE-THREATENING	SERIOUS	NONE
Gastric banding	1.5%	0.9%	97.6%
Sleeve gastrectomy	3.6%	2.2%	94.1%
Gastric bypass	6.7%	3.6%	89.7%

6.31: The table provides the percent of subjects with various health outlooks for each group. The outlooks of current smokers are generally bleaker than that of current nonsmokers. Much larger percentages of nonsmokers reported being in "excellent" or "very good" health, whereas much larger percentages of smokers reported being in "fair" or "poor" health.

	HEALTH OUTLOOK				
	EXCELLENT	VERY GOOD	GOOD	FAIR	POOR
Current smoker	6.2%	28.5%	35.9%	22.3%	7.2%
Current nonsmoker	12.4%	39.9%	33.5%	14.0%	0.3%

6.32: (a) The two-way table is provided. (b) In order of increasing temperature, the proportions hatching are 16/27 = 0.593, or 59.3%, 38/56 = 0.679, or 67.9%, and 75/104 = 0.721, or 72.1%. The percent hatching increases with temperature;

the cold temperature did not prevent hatching but made it less likely. The difference between the percents hatching at hot and neutral temperatures is fairly small and may not be big enough to be called significant.

	TEMPERATURE		
	COLD	NEUTRAL	HOT
Hatched	16	38	75
Did not hatch	11	18	29

6.33: Because the numbers of students who use (or do not use) medications is different, we find the conditional distributions of those who do and do not use medications. Those who use medications are less likely to have optimal sleep quality and more likely to have poor sleep quality than those who do not use medications. This is certainly a case where one would not want to ascribe causation—do those who use medications to stay awake have poor sleep quality because they use the medication, or do they use the medications to stay awake because they had poor sleep quality before using them?

	SLEEP QUALITY		
	OPTIMAL	BORDERLINE	POOR
Use medications	21.3%	30.5%	48.3%
Do not use medications	38.2%	26.7%	35.2%

Chapter 7 Exploring Data: Part I Review

7.1: (c)

7.2: Answers vary. Some suggestions for questions with a categorical response: "What is your class level (freshman, sophomore, etc.)?" or "Is this your first statistics class?" Some suggestions for questions with a quantitative response: "How many hours per week do you work at a paying job?" or "How many hours do you spend studying in a typical week?"

7.3: (c)

7.4: (d)

7.5: (b)

7.6: (a)

7.7: (b)

7.8: (b)

7.9: (c)

7.10: (b)

7.11: (a) centimeters; (b) centimeters; (c) centimeters; (d) grams^2.

7.12: (b)

7.13: (d) 5 years is 60 months, and there are 10 of 40 observations below 60.

7.14: (c)

7.15: (d)

7.16: His z-score is $z = \dfrac{32 - 25.2}{6.4} = 1.06$. From Table A, the area below $z = 1.06$ is 0.8554. He scored better than about 85.5% of all MCAT takers.

7.17: (a) 99.7% of all values are within 3σ of μ in a Normal distribution. This becomes $0.800 \pm 3(0.078) = 0.566$ to 1.034 mm. (b) 0.878 is 1 standard deviation above the mean; about 16% (15.87%) will have thorax lengths longer than 0.878 mm.

7.18: (a) $P(X > 20) = P(Z > 0.33) = 1 - 0.6293 = 0.3707$. (b) $Z = 1.64$ (or 1.65), so allow $19 + 1.65(3) = 23.95$, or about 24 minutes.

7.19: (a) Minimum $= 7.2$, $Q_1 = 8.5$, $M = 9.3$, $Q_3 = 10.9$, Maximum $= 12.8$. (b) $M = 27$. (c) 25% of values exceed $Q_3 = 30$. (d) Yes. Virtually all Torrey pine needles are longer than virtually all Aleppo pine needles. There is no overlap in the distributions, as seen by comparing, say, minimum for Torrey pine needles (21) to maximum for Aleppo pine needles (12.8).

7.20: (d)

7.21: (b)

7.22: (a)

7.23: (c)

7.24: (c)

7.25: (c)

7.26: (a)

7.27: (d)

7.28: (b)

7.29: (d)

7.30: (c)

7.31: (c)

7.32: (b)

7.33: (a)

7.34: The increased correlation suggests that the two types of stocks (American and European) now tend to rise together and fall together, which reduces the ability of one to hedge risk the other.

7.35: (a) No. (b) $r^2 = 0.64$, or 64%.

7.36: (a) $\hat{y} = 9.752 + 86.514x$. (b) If $x = 0.60$, $\hat{y} = 9.752 + (86.514)(0.60) = 61.66$ introspective ability. (c) For $x = 0.99$, $\hat{y} = 95.40$ introspective ability. Because $r^2 = (0.448)^2 = 0.201$, only 20.1% of the variation in introspective scores are explained by our regression model. Also, $x = 0.99$ is outside the range, so this might be extrapolation.

7.37: (a) 8.683 kg. (b) 10.517 kg. (c) Such a comparison would be unreasonable because the lean group is less massive and therefore would be expected to burn less energy on average. (d) See below. (e) It appears that the rate of increase in energy burned per kilogram of mass is about the same for both groups. Of course, the obese monkeys are more massive and therefore, on average, burn more energy, as computed in (a) and (b).

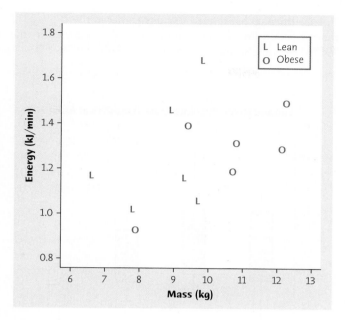

7.38: (a) See below. (b) There is a very strong negative linear relationship between Year and Percent of Smokers. There are no real outliers, but the rate of decline in smoking has varied over time. (c) $\hat{y} = 1017.009 - 0.497x$. (d) Over this time, on average, smoking declined by 0.497% per year (about 1% every two years). (e) $r^2 = (-0.98)^2 = 0.9604$, or 96.04%. (f) In the year 2010, we predict that $1017.009 - (0.497)(2010) = 18.04\%$ of adults smoke. The goal hasn't been achieved. (g) Such a prediction would be unreasonable because the year 2050 is far outside the range of years on which this model was based. In fact, the prediction is −1.8%, which is impossible.

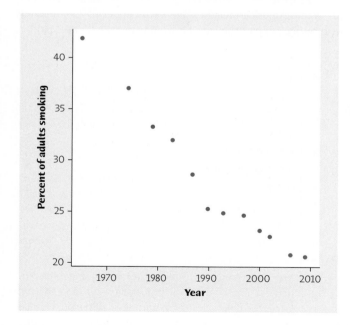

7.39: (a) $190/8474 = 0.0224$, or 2.24%. (b) $633/8474 = 0.0747$, or 7.47%. (c) $27/633 = 0.0427$, or 4.27%. (d) $4621/8284 = 0.5578$, or 55.78%. (e) The conditional distribution of

CHD for each level of anger is tabulated below. Angrier people are at greater risk of CHD.

LOW ANGER	MODERATE ANGER	HIGH ANGER
1.70%	2.33%	4.27%

7.40: (a) A graph shows that the distribution is slightly right-skewed; one observation is somewhat low, but not really an outlier. (b) Because of the slight right skew, we might expect the mean to be slightly larger, but the low observation will tend to counteract that. We find that $\bar{x} = 563.1$ and $M = 560$ km³ of water. (c) Because the distribution is not too skewed, one could choose the mean and standard deviation. Here, $s = 136.5$ km³ of water. Alternatively, use the five-number summary: Min = 290, $Q_1 = 445$, $M = 560$, $Q_3 = 670$, Max = 900 (all km³ of water).

```
2 | 9
3 |
3 | 6 9 9 9
4 | 1 2 2 2 2 3 4
4 | 5 6 7 8
5 | 0 0 1 1 4
5 | 5 5 6 6 8 8 9
6 | 0 0 0 1 3 4 4
6 | 7 7 8 8 8 9
7 | 0 1 1
7 | 7
8 | 0
8 | 8
9 | 0
```

7.41: The time plot shows a lot of fluctuation from year to year, but it also shows a recent increase: prior to 1972, the discharge rarely rose above 600 km³, but since then, it has exceeded that level more than half the time. A histogram or stemplot cannot show this change over time.

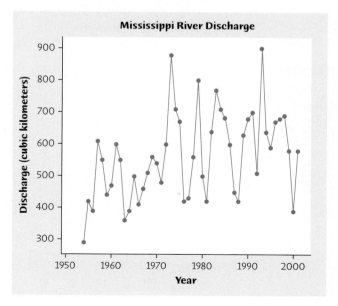

7.42: The distribution is roughly Normal. The two low numbers and one high number are not extreme enough to be called outliers. The mean, standard deviation and five-number summary (all in days) are $\bar{x} = 15.146$, $s = 5.989$, and Min $= 1$, $Q_1 = 11$, M $= 15.5$, $Q_3 = 19$, Max $= 31$. The median date is therefore May 4 or 5 (day 15 is May 4).

```
0 | 1 1
0 | 3
0 | 4 5 5
0 | 7 7
0 | 8 8 9 9 9
1 | 0 0 0 0 0 0 0 0 0 1 1 1 1 1 1 1 1 1
1 | 2 2 2 2 2 2 3 3 3
1 | 4 4 4 4 5 5 5 5
1 | 6 6 6 6 6 6 6 6 7 7 7 7 7 7
1 | 8 8 9 9 9 9 9 9 9 9
2 | 0 0 0 1 1 1 1
2 | 2 2 2 2 3 3 3 3 3 3
2 | 4 5 5
2 | 6 7
2 |
3 | 1
```

7.43: (a) The plot is provided. (b) The least-squares regression line is $\hat{y} = 168.96 - 0.0783x$. The slope is negative, suggesting that the ice breakup day is decreasing (by 0.0783 day per year). (c) The regression line is not very useful for prediction, as it accounts for only about 13.3% ($r^2 = 0.133$) of the variation in ice breakup time.

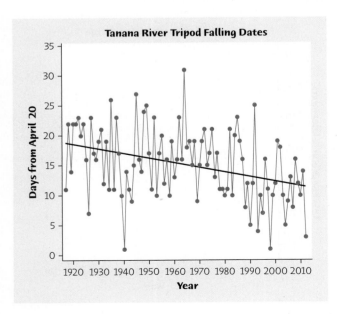

7.44: Breakup has tended to come earlier in the 1989–2012 time segment, but the minimum 1 day also occurred in the 1917–1940 time segment. 1990 was a high outlier.

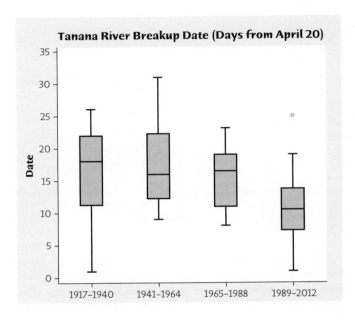

7.45: (a) and (b) The distribution is unimodal and left-skewed, with the exception of one high outlier. The high outlier is Luxembourg. (c) The mean and standard deviation are $\bar{x} = \$28,461$ and $s = \$11,455$. Some students may instead report the five-number summary, which is Min $= \$9,526$, $Q_1 = \$23,130$, M $= \$28,937$, $Q_3 = \$34,193$, Max $= \$65,171$. (d) While American teachers are in the upper 25% of teachers with respect to pay, they are far from being the highest paid.

```
0 | 9
1 | 0 1 1 4
1 | 5 7 7
2 | 3 4 4
2 | 5 6 6 6 7 7 8 9
3 | 0 0 0 0 0 2 2 4 4
3 | 6 6 7
4 | 3
4 | 5 6
5 |
5 |
6 |
6 | 5
```

7.46: Back-to-back stemplots show little difference overall. Both shapes are somewhat irregular, but neither is clearly higher or lower. Means and medians are also similar.

Cicada plants		Control plants
	1	
	1	3
4	1	4 4 5
7	1	7 7
9 9	1	8 9 9 9 9
1 1 1 1 0 0	2	0 1 1 1
3 3 3 3 3 3 2 2 2 2	2	2
5 5 4 4	2	4 4 4 4 4 4 5 5 5 5
7 7 7 7 6 6 6	2	6 6 6 6 6
9 9 9	2	8 9
1 1 0	3	
	3	
5	3	

The data give little reason to believe that cicadas make good fertilizer, at least on the basis of this response variable.

	$\bar{x}$	M
Cicada group	0.2426 mg	0.2380 mg
Control group	0.2221 mg	0.2410 mg

7.47: A stemplot is shown. The distribution seems to be fairly Normal apart from a high outlier of 50°. The five-number summary is preferred because of the outlier: Min = 13°, Q_1 = 20°, M = 25°, Q_3 = 30°, Max = 50°. (The mean and standard deviation are $\bar{x}$ = 25.4211° and s = 7.4748°.) Most patients have a deformity angle in the range of 15° to 35°.

1	3 4
1	6 6 7 8 8
2	0 0 0 1 1 1 1 2 3
2	5 5 5 5 6 6 6 6 8 8 8
3	0 0 0 1 2 2 2 4
3	8 8
4	
4	
5	0

7.48: The scatterplot suggests a positive linear association, albeit with lots of scatter, so correlation and regression are reasonable tools to summarize the relationship. r = 0.6821, $\hat{y}$ = 0.1205 + 0.008569x. The regression line explains r^2 = 46.5% of the variation in the proportion killed. The analysis provides weak support for the idea that the proportion of perch killed rises with the number of perch present.

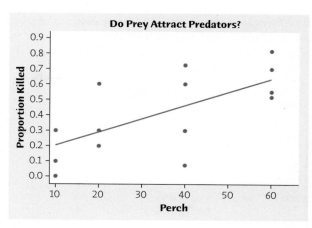

7.49: The scatterplot shows a moderate to weak positive linear association, with one clear outlier (the patient with HAV angle 50°). r = 0.3021, $\hat{y}$ = 19.723 + 0.3388x. MA angle can be used to give (very rough, imprecise) estimates of HAV angle, but the spread is so wide that the estimates would not be very reliable. The linear relationship explains only r^2 = 9.1% of the variation in HAV angle.

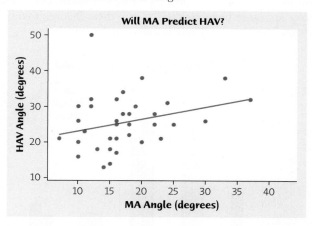

7.50: The scatterplot suggests a fairly strong negative linear association, so correlation and regression are reasonable tools to use here. r = −0.8035, $\hat{y}$ = 92.29 − 0.05762x; the equation explains r^2 = 64.6% of the variation in burned grassland. When wildebeest numbers are higher, the percent of grassland burned tends to be lower. Each additional 1000 wildebeest decrease burned area by about 0.058% on the average.

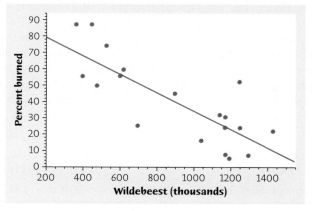

7.51: Software gives us the regression line ($\hat{y} =$ 70.44 + 274.78x), and a scatterplot shows a moderate positive linear relationship. The linear relationship explains about $r^2 = 49.3\%$ of the variation in gate velocity.

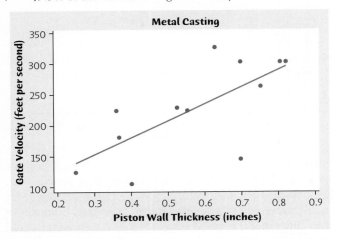

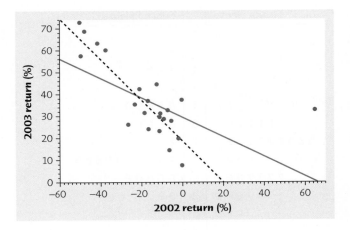

7.52: Some possible observations: all three groups were basically identical in the percent rating schools as "poor." Hispanics appear to be more likely to rate their children's schools as "excellent" but less likely to call them "good."

7.54: (a) $\hat{y}$ = 93.92 + 0.7783x. The third point (pure tone 241, call 485 spikes/second) is A (circled in plot). The first point (474 and 500 spikes/second) is B; it is marked with a square. (b) The correlation drops only slightly (from 0.639 to 0.610) when A is removed; it drops more drastically (to 0.479) without B. (c) When either point is removed, the slope decreases. Without A, the line is $\hat{y}$ = 98.42 + 0.6792x; without B, it is $\hat{y}$ = 101.1 + 0.6927x.

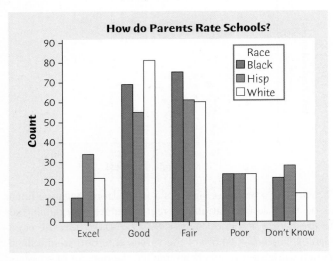

7.53: (a) The scatterplot (top of next column) of 2003 returns against 2002 returns shows (ignoring the outlier) a strong negative association. (b) The correlation for all 23 points is $r = -0.616$; with the outlier removed, the correlation is $r = -0.838$. (c) Regression formulas are given in the table below. The first line is solid in the plot, the second is the dashed line. The line for the 22 other funds is so far below Fidelity Gold that the squared deviation is very large. The line must pivot up toward Fidelity Gold in order to minimize the sum of squares for all 23 deviations. Fidelity Gold is very influential.

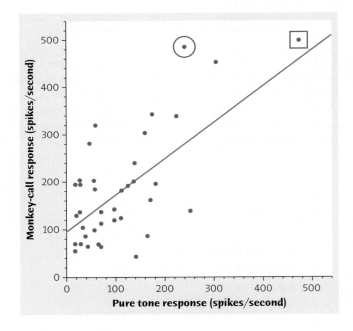

7.55: (a) Fish catch (on the horizontal axis) is the explanatory variable. The point for 1999 is at the bottom of the plot. (b) The correlations are given in the table. The outlier decreases r because it weakens the strength of the association. (c) The two regression lines are given in the table; the solid line in the plot uses all points, whereas the dashed line omits the outlier. The effect of the outlier on the line is small: there are several other years with similar changes in bushmeat biomass. Also, this year was not particularly extreme in the amount of fish caught.

	r	EQUATION
All 23 funds	−0.6156	$\hat{y}$ = 31.1167 − 0.4132x
Without Fidelity Gold	−0.8380	$\hat{y}$ = 21.4616 − 0.8403x

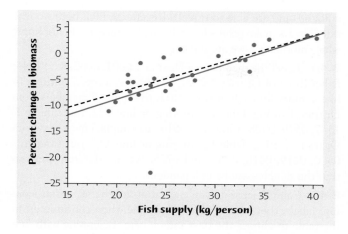

	r	EQUATION
All points	0.6724	$\hat{y} = -21.09 + 0.6345x$
Without 1999	0.8042	$\hat{y} = -19.05 + 0.5788x$

Chapter 8 Producing Data: Sampling

8.1: (a) (All) college students. (b) The 104 students at the researcher's college who returned the questionnaire.

8.2: Population: all the artifacts discovered at the dig. Sample: those artifacts chosen for inspection.

8.3: (a) All people who made credit card purchases. (b) The 137 people who returned the survey form.

8.4: It is a convenience sample; she is getting opinions from only students who are at the student center at a certain time of day. This might underrepresent some group: commuters, graduate students, or nontraditional students, for example.

8.5: Because all the students surveyed are enrolled in a special senior honors class, these students may be more likely to be interested in joining the club (and more willing to pay $35 to do so). The direction of bias is likely to overestimate the proportion of all psychology majors willing to pay to join this club. This is a convenience sample.

8.6: Number from 01 to 30 alphabetically (down the columns). With Table B, enter at line 122 and choose 13 = Crestview, 15 = Fairington, 05 = Brandon Place, and 29 = Village Square.

8.7: Number from 01 to 26 alphabetically (down the columns). With Table B, enter at line 134 and choose 16 = Ippolito, 18 = Jung, 13 = Gupta, 21 = Modur, and 04 = Bonds.

8.8: (a) Assign five-digit labels to each record, from 00001 to 55914. (b) With Table B, enter at line 120 and choose 35476, 39421, 04266, 35435, and 43742.

8.9: With the election close at hand, the polling organization wants to increase the accuracy of its results. Larger samples provide better information about the population.

8.10: The sample size for the general public is larger than the sample size for Pentecostals. Larger samples yield more information, which means more accuracy, which means a smaller margin of error.

8.11: Label the suburban townships from 01 to 30, alphabetically. With Table B, enter at line 105 and choose 29 = Wheeling, 07 = Elk Grove, 19 = Orland, 14 = New Trier, and 17 = Norwood Park. Next, label the Chicago townships from 1 to 8, alphabetically. With Table B, enter at line 115 and choose 6 = Rogers Park, 1 = Hyde Park, and 4 = Lake View.

8.12: Label the students in each class as shown in the table below. If Table B (starting at line 122) is used to choose the samples, the students selected are those listed in the table.

CLASS	LABELS	FIRST FIVE STUDENTS IN SAMPLE
Sophomores	001 to 989	138, 738, 159, 895, 052
Juniors	001 to 943	909, 087, 359, 275, 186
Seniors	001 to 895	871, 369, 576, 154, 580

8.13: (a) The population is all physicians practicing in the United States. The sample size is $n = 2379$. If the 2379 were randomly selected, we could draw conclusions, but there was too much nonresponse. (b) The nonresponse rate is $\frac{100{,}000 - 2379}{100{,}000} = 97.62\%$. We don't know the attitudes of the nonrespondents about health care reform. (c) They only received 2379 responses.

8.14: Anybody who answers "yes" to Question A would surely answer "yes" to Question B, but the converse is not true. Question A is slanted toward a more negative response on gays in the military.

8.15: (a) The sample was not randomly selected. (b) The true percentage is most likely lower.

8.16: Answers will vary. One possible answer follows. (a) "Do you approve of texting while driving?" (b) The cell-only group would be more supportive of texting while driving, so the sample percentage that favors making texting while driving illegal would decrease. (c) We're likely to overestimate the percentage of all adults that favor making texting while driving illegal. If we did not take into account the number of landline phones into a residence, a residence with multiple landlines would have a higher chance of being selected for the sample.

8.17: (a)

8.18: (b)

8.19: (b)

8.20: (b)

8.21: (b)

8.22: (a) Each member of the population needs a 3-digit label, and we need 440 of them (not 441, as in (b)).

8.23: (c) Notice that in (b) "07" appears in the sample twice.

8.24: (b)

8.25: (b)

8.26: The population is all adults, aged 18 and older, living in the United States. The sample consists of the 1,004 adults randomly selected.

8.27: The population is the 1000 envelopes stuffed during a given hour. The sample is the 40 envelopes selected.

8.28: Numbering from 01 to 35 alphabetically (down the columns), we enter Table B at line 131 and choose 05 = Burke, 32 = Vore, 19 = Kessis, 04 = Bower, 25 = Prince, 29 = Shoepf, 20 = Lu, and 16 = Heaton.

8.29: Using Table B, number the area codes 001 to 287. Then, enter at line 135, and pay attention to the instructions that if we use the table, we'll pick only 5 numbers. The selected area codes are 255, 100, 120, 126, 008.

8.30: (a) Assign labels 0001 through 1410. (b) Beginning at line 105, we choose plots 0769, 1315, 0094, 0720, and 0906.

8.31: Version A drew 55% saying the tax was fair. Perhaps the other version with the "too high" option prompted respondents with an option that more clearly reflected their views.

8.32: (a) False. Such regularity holds only in the long run. (b) True. All pairs of digits (there are 100, from 00 to 99) are equally likely. (c) False. Four random digits have chance 1/10,000 to be 0000, so this sequence will occasionally occur.

8.33: (a) The population is (something like) adult residents of the United States. (b) The nonresponse rate is 1169/2000 = 58.45%. (c) This question will likely have response bias; specifically, many people will give an inaccurate count of how many movies they have seen in the past year.

8.34: Online polls, call-in polls, and voluntary response polls in general tend to attract responses from those who have strong opinions on the subject and therefore are often not representative of the population as a whole.

8.35: The nonresponse rate was $\dfrac{45{,}956 - 5029}{45{,}956} = 89.1\%$.

8.36: (a) Assign labels 0001 through 5024, enter the table at line 104, and select: 1388, 0746, 0227, 4001, and 1858. (b) More than 171 respondents have run red lights. We would not expect very many people to claim they *have* run red lights when they have not, but some people will deny running red lights when they have.

8.37: People likely claim to wear their seat belts because they know they should; they may be embarrassed or ashamed to say that they do not always wear seat belts. Such bias is likely in most surveys about seat belt use (and similar topics).

8.38: (a) Each person has a 10% chance: 4 of 40 men, and 3 of 30 women. (b) This is not an SRS because not every group of 7 people can be chosen; the only possible samples are those with 4 men and 3 women.

8.39: The higher no-answer was probably the second period—more families are likely to be gone for vacations, or to be outside enjoying the warmer weather, and so on. Nonresponse of this type might underrepresent those who are more

affluent (and are able to travel). In general, high nonresponse rates always make results less reliable, because we do not know what information we are missing.

8.40: Label the members of District 1 0001, 0002, . . . , 1944. Label those of District 2 0001, 0002, . . . , 1566. Continue in like manner for each district. Now, to sample 5 members from District 1, using Table B, entering at line 122, our sample is 1387, 0529, 0908, 1369, and 0815. To sample 5 members from District 2, using Table B, entering at line 131, our sample is 0500, 0419, 0418, 0765, and 0705. We use different lines so that the samples will be independent.

8.41: Sample separately in each stratum; that is, assign separate labels, then choose the first sample, then continue on in the table to choose the next sample, etc. Beginning with line 102 in Table B, we choose:

FOREST TYPE	LABELS	PARCELS SELECTED
Climax 1	01 to 36	19, 27, 26, 17
Climax 2	01 to 72	09, 55, 32, 22, 69, 56, 52
Climax 3	01 to 31	13, 07, 02
Secondary	01 to 42	27, 40, 01, 18

8.42: (a) The sample size for the public is much larger, so the survey is more accurate for this group. (b) It's likely that people working in health-related fields have opinions that differ from those of the public. The researchers probably want to examine this.

8.43: (a) Because 200/5 = 40, we will choose one of the first 40 names at random. Beginning on line 120, the addresses selected are 35, 75, 115, 155, and 195. (Only the first number is chosen from the table.) (b) All addresses are equally likely; each has a 1/40 chance of being selected. This is not an SRS because the only possible samples have exactly one address from the first 40, one address from the second 40, and so on. An SRS could contain any 5 of the 200 addresses in the population.

8.44: (a) This design would omit households without telephones, those with only cell phones, and those with unlisted numbers. Such households would likely be made up of poor individuals (who cannot afford a phone), those who choose not to have landline phones, and those who do not wish to have their phone numbers published. (b) Those with unlisted landline numbers would be included in the sampling frame when a random-digit dialer (RDD) is used.

8.45: (a) In some families, the adult that answers the phone regularly may be systematically different from an adult that does not. (b) The population is adults 18 and older living in all 50 United States and the District of Columbia. There could be (and probably are) big differences between landline phone users and cellular phone users. The design is a stratified sample.

8.46: (a) The wording is clear, but will almost certainly be slanted toward a high positive response. (b) The question makes the case for a national health care system, and so will slant responses toward "yes." (c) This survey question is most likely to produce a response similar to: "Uhh . . . yes? I mean, no? I'm sorry, could you repeat the question?"

8.47: Answers will vary considerably.

8.48: In this case, the minister is terribly misguided. Critics of the proposal are worried that the sample will not be representative of the population—presumably because people who fill out the optional long-form questions will be systematically different from those who don't. Larger samples do not address such problems of bias.

Chapter 9 Producing Data: Experiments

9.1: This is an observational study: No treatment was assigned to the subjects; we merely observed cell phone usage (and presence/absence of cancer). The explanatory variable is cell phone usage, and the response variable is whether or not a subject has brain cancer.

9.2: This is an experiment: Each subject is (presumably randomly) assigned to a group, each with its own treatment (Arial or Brush font). The explanatory variable is the font, and the response variables are then perceived effort (in minutes) and willingness to make the exercise part of their daily routine.

9.3: This is an observational study, so it is not reasonable to conclude any cause-and-effect relationship. At best, we might advise smokers that they should be mindful of potential weight gain after smoking cessation and its accompanying ailments.

9.4: Subjects: the "healthy people aged 18 to 40." Factor: the pill given to the subject. Treatments: ginkgo or placebo. Response variable: the number (or fraction) of *e*'s identified by each subject.

9.5: Subjects: the students. Factors: type of attack, and prime used. Treatments: for the prime: *love-thy-neighbor* prime, or *eye-for-an-eye* prime; for the type of attack: on military target, or on cultural/educational target. Response variable: rating of U.S. reaction to attack.

		PRIME USED	
		LOVE-THY-NEIGHBOR	EYE-FOR-AN-EYE
Target	Military	1	2
	Cultural	3	4

9.6: (a) Factors: harvest time (treatments 80, 95, and 110 days after fruit setting) and storage temperature (treatments 20, 30, and 40 degrees centigrade. The treatment combinations are shown below. Response variable: time to ripening.

	STORAGE TEMPERATURE (°C)		
DAYS AFTER FRUIT SET	20	30	40
80	1	2	3
95	4	5	6
110	7	8	9

(b) If a tree were diseased, for example, its fruit would most likely not withstand any treatment and would indicate that the treatment was not good.

9.7: Making a comparison between the treatment group and the percent finding work *last year* is not helpful. Over a year, many things can change: the state of the economy, hiring costs (due to an increasing minimum wage or the cost of employee benefits), etc. (To draw conclusions, we would need to make the $500 bonus offer to some people and not to others during the same time period and compare the two groups.)

9.8: (a) The diagram is provided. The response variable is weight loss. (b) Label the students from 01 to 50, and pick 25 to receive Treatment 1 (the rest will receive Treatment 2). If using Table B, line 130, the sample is 05, 16, 48, 17, 40, 20, 19, 45, 32, 41, 04, 25, 29, 43, 37, 39, 31, 50, 18, 07, 13, 33, 02, 36, 23.

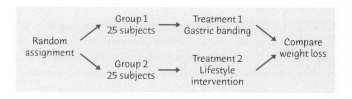

9.9: (a) Diagram below. (b) If using Table B, label 01 to 36 and take two digits at a time.

9.10: Assign 24/6 = 4 students to each treatment. The diagram is shown following. We assign labels 01 through 24, then use the first four 2-digit numbers in this range for Group 1, the next four for Group 2, etc. The table below shows the assignments. Note that with this many assignments, you will run through many lines of Table B. Once you've filled out members for 5 groups, the 6th group contains all the remaining, unassigned subjects.

Group 1: 20 Shi, 16 Kruger, 04 Baker, 18 Minor
Group 2: 07 Brower, 13 Greenberg, 02 Anthony, 05 Biery
Group 3: 19 Schwartz, 23 Truitt, 21 Stanley, 08 Carroll
Group 4: 10 Cote, 11 Delp, 15 Koster, 12 Disbro
Group 5: 14 Kessis, 09 Cohen, 24 Walsh, 22 Tory
Group 6: 01 Abramson, 03 Austen, 06 Blake, 17 Linder

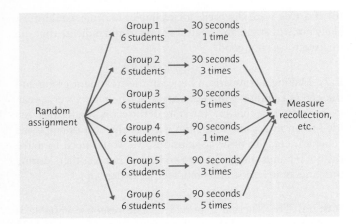

9.11: In a controlled scientific study, the effects of factors other than the nonphysical treatment (e.g., the placebo effect, differences in the prior health of the subjects) can be eliminated or accounted for, so that the differences in improvement observed between the subjects can be attributed to the differences in treatments.

9.12: If this year is considerably different in some way from last year, we cannot compare electricity consumption over the two years. For example, if this summer is warmer, the customers may run their air conditioners more. The possible differences between the two years would confound the effects of the treatments.

9.13: (a) The researchers simply observed the diets of subjects; they did not alter them. (That is, no treatments were assigned.) (b) Such language is reasonable because with observational studies, no "cause-and-effect" conclusion would be reasonable.

9.14: "Double-blind" means that the treatment (testosterone supplement or placebo) assigned to a subject was unknown to both the subject and those responsible for assessing the effectiveness of that treatment. "Randomized" means that patients were randomly assigned to receive either the testosterone supplement or a placebo. "Placebo-controlled" means that some of the subjects were given placebos. Even though placebos possess no medical properties, some subjects may show improvement or benefits just as a result of participating in the experiment; the placebos allow those doing the study to observe this effect.

9.15: In this case, "lack of blindness" means that the experimenter knows which subjects were taught to meditate. He or she may have some expectation about whether or not meditation will lower anxiety; this could unconsciously influence the diagnosis.

9.16: (a) Each swimmer swims one time using each breathing technique (B2 and B4). A coin is tossed to determine the order in which these techniques are used. (b) In a completely randomized design, the 10 male collegiate swimmers would be assigned randomly to the two treatments, 5 swimmers using technique B2 and the other 5 using technique B4. (c) If swimmers select their own technique, it would be an observational study.

9.17: (a) Explanatory variable is the font; the response is the degree of agreement with the statement. (b) *Completely*

randomized design: 50 students are randomly assigned to each of the two fonts and then rate the healthiness of the product. (c) *Matched pairs design:* each student sees both fonts in random order, then rates the product for each font.

9.18: For each block (pair of lecture sections), randomly assign one section to be taught using standard methods and the other to be taught with multimedia. Then (at the end of the term) compare final-exam scores and student attitudes. The diagram below is *part* of the whole block diagram; there would also be three other pieces like this (one for each of the other instructors). The randomization will vary with the starting line in Table B—or the randomization can be done by flipping a coin for each block.

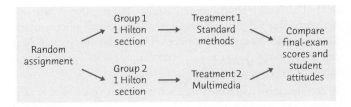

9.19: (a) Life satisfaction is observed, no treatments were imposed.

9.20: (c) Parents tend to have higher satisfaction in their lives than nonparents. Because this was a survey, we cannot make conclusions about cause-and-effect which is implied by both of the other options.

9.21: (b) All participating students had the same treatment.

9.22: (c)

9.23: (b)

9.24: (b) Both of the other two options would introduce lurking variables and confounding.

9.25: (a) Each of the 36 subjects needs a label, and the 18 subjects who will take the medicine need to be chosen at random.

9.26: (b) The communities are paired, then one is chosen to have the advertising campaign.

9.27: (a) The choice should be made randomly.

9.28: (b) This was a (matched-pairs) experiment, but to give useful information, the subjects should be chosen from those who might be expected to buy this car.

9.29: (a) This is an observational study; the subjects chose their own "treatments" (how much red meat to eat). The explanatory variable is red meat consumption, and the response variable is whether or not a subject dies. (There may have been other variables, but these were the only ones mentioned in the problem.) (b) Many answers are possible. For example, smoking is known to increase the risk of cancer. (c) Many answers are possible. For example, how many servings of fruits and vegetables were consumed along with the red meat?

9.30: Many answers are possible here. For example, it may take time to adjust to a new neighborhood and/or to be brought up to the academic performance of students already in the new neighborhood.

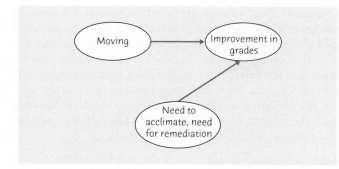

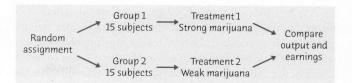

9.31: This is an experiment, because the treatment is selected (randomly, we assume) by the interviewer. The explanatory variable (treatment) is the level of identification, and the response variable is whether or not the interview is completed.

9.32: (a) Answers will vary. (b) Answers will vary.

9.33: (a) The researchers simply observe subjects who have chosen to live near highways and compare them with others who do not live near highways. (b) Answers will vary. For example, those who live near highways might have less money and be attracted to the (possibly) lower housing costs near a highway. Money (available for housing and health care) would be a variable that might be confounding. (c) Who would want their housing situation randomly assigned?

9.34: In the diagram below, equal numbers of subjects are assigned to each treatment.

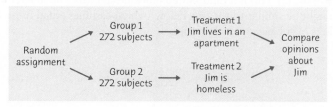

9.35: (a) Diagram below. (b) Assign labels 001 to 120. If using Table B, line 108 gives 090, 009, 067, 092, 041, 059, 040, 080, 029, 091.

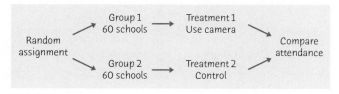

9.36: (a) A diagram is shown below. (b) Label the subjects from 01 through 30. From line 120, we choose subjects corresponding to the numbers 16, 04, 26, 21, 19, 07, 22, 10, 25, 13, 15, 05, 29, 09, 08 for the first group, and the rest for group 2. Hence, the marijuana group consists of Mattos, Bower, Williams, Sawant, Reichert, DeVore, Scannell, Giriunas, Stout, Kennedy, Mani, Burke, Zaccai, Fritz, and Fleming. All other subjects are assigned to the nonmarijuana group. (c) This could be a double-blind experiment, assuming that subjects can't distinguish between the types of marijuana smoked. Also, the persons measuring output and earnings of subjects don't know what kind of marijuana a subject smoked.

9.37: Use a completely randomized design; the diagram is provided. Labeling the men from 01 through 39, and starting on line 107 of Table B, we make the assignments shown in the table.

Group 1:   20, 11, 38, 31, 07, 24, 17, 09, 06
Group 2:   36, 15, 23, 34, 16, 19, 18, 33, 39
Group 3:   08, 30, 27, 12, 04, 35
Group 4:   02, 32, 25, 14, 29, 03, 22, 26, 10
Group 5:   Everyone else

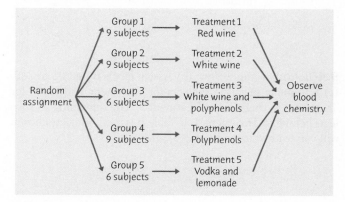

9.38: (a) There are two factors. The first factor is Type of granola, and it has two levels (regular and low-fat). The second factor is Serving size label, and it has three levels (2 servings, 1 serving, and no label). Hence, there are six treatment combinations (regular granola at 2 servings, regular granola at 1 serving, regular granola with no serving label, low-fat granola with 2 servings, low-fat granola with 1 serving, and low-fat granola with no serving label). At 20 subjects per treatment, there were 120 subjects in the experiment. (b)

| | GRANOLA TYPE | |
SERVING SIZE LABEL	REGULAR	LOW-FAT
2 servings	20 subjects	20 subjects
1 serving	20 subjects	20 subjects
No label	20 subjects	20 subjects

9.39: (a) The outline is given below. There are 40 subjects, so we assign 10 subjects to each of the four treatments. The four treatments are outlined:

	ANTIDEPRESSANT	NO DRUG
Stress management	1	2
None	3	4

(b) Assign labels 01 through 40 (in alphabetical order). The full randomization is easy with the Simple Random Sample applet: each successive sample leaves the population hopper, so that you need only click Sample three times to assign 30 subjects to three groups; the 10 subjects remaining in the hopper are the fourth group. Alternatively, line 125 of Table B gives the following subjects for Group 1: 21 Jiang, 37 Suarez, 18 Hersch, 23 Kim, 19 Hurwitz, 10 Devlin, 33 Richter, 31 Ramdas, 36 Smith, and 40 Xiang.

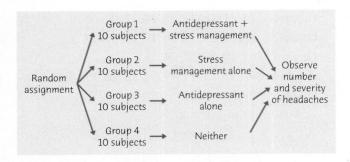

9.40: (a) Diagram below. (b) Assign labels from 001 to 240. (c) Randomly select 53 subjects for Treatment 1, then 64 for Treatment 2, then 60 for Treatment 3. The remaining 63 subjects belong to Treatment 4. If Table B is used, subjects chosen will vary with starting line.

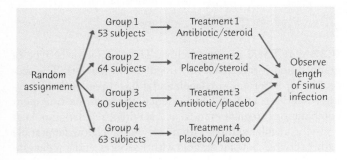

9.41: The factors are pill type and spray type. "Double-blind" means that the treatment assigned to a patient was unknown to both the patient and those responsible for assessing the effectiveness of that treatment. "Placebo-controlled" means that some of the subjects were given placebos. Even though placebos possess no medical properties, some subjects may show improvement or benefits just as a result of participating in the experiment; the placebos allow those doing the study to observe this effect.

9.42: "No significant difference" does *not* mean the groups are identical. Although there almost certainly were *some* differences in these variables between the four groups, those differences were no bigger than we might expect from true random allocation. For example, the proportions of smokers in the four groups were sufficiently similar that the effect of smoking on sinus infections would be nearly the same in each group.

9.43: (a) The subjects are randomly chosen Starbucks customers. Each subject tastes two cups of coffee, in identical unlabeled cups. One contains regular mocha Frappuccino, the other the new light version. The cups are presented in random order, half the subjects get regular then light, the other half light then regular. Each subject says which cup he or she prefers. (b) We must assign 10 customers to get regular coffee first. Label the subjects 01 to 20. Starting at line 141, the "regular first" group is: 12, 16, 02, 08, 17, 10, 05, 09, 19, 06.

9.44: The sketches requested in the problem are not shown here; random assignments will vary among students. (a) Label the circles 1 to 6, then randomly select three (using Table B, or simply by rolling a fair, six-sided die) to receive the extra CO_2. Observe the growth in all six regions, and compare the mean growth within the three treated circles with the mean growth in the other three (control) circles. (b) Select pairs of circles in each of three different areas of the forest. For each pair, randomly select one circle to receive the extra CO_2 (using Table B or by flipping a fair coin). For each pair, compute the difference in growth (treated minus control).

9.45: Each player will be put through the sequence (100 yards, four times) twice—once with oxygen and once without. For each player, randomly determine whether to use oxygen on the first or second trial. Allow ample time (perhaps a day or two) between trials for full recovery.

9.46: (a) This is a randomized block design. (b) The diagram might be similar to the one below (which assumes equal numbers of subjects in each group).

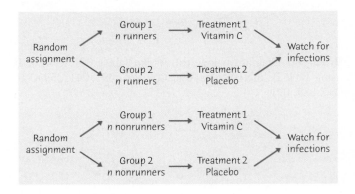

9.47: Diagram is shown below. The last stage ("Observe heart health") might be described in more detail.

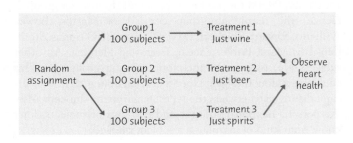

9.48: Divide the men and women into three groups of equal size. Diagram below.

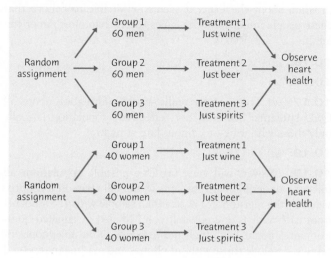

9.49: Any experiment randomized in this way assigns all the women to one treatment and all the men to the other. That is, sex is completely confounded with treatment. If women and men respond differently to the treatment, the experiment will be strongly biased. The direction of the bias is random, depending on the coin toss.

9.50: (a) The explanatory variable is the beta-carotene/vitamin(s) taken each day; the response variable is whether or not colon cancer develops. (b) Diagram is shown below; equal group sizes are convenient but not necessary. (c) Neither the subjects nor the researchers who examined them knew who was getting which treatment. (d) The observed differences were no more than what might reasonably occur by chance even if there is no effect due to the treatments. (e) Fruits and vegetables contain fiber; this could account for the benefits of those foods. Also, people who eat lots of fruits and vegetables may have healthier diets overall (e.g., less red meat).

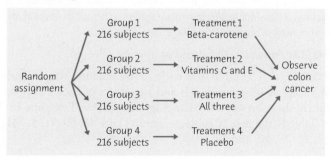

9.51: (a) "Randomized" means that patients were randomly assigned to receive either Saint-John's-wort or a placebo. "Double-blind" means that the treatment assigned to a patient was unknown to both the patient and those responsible for assessing the effectiveness of that treatment. "Placebo-controlled" means that some of the subjects were given placebos. Even though placebos possess no medical properties, some subjects may show improvement or benefits just as a result of

participating in the experiment; the placebos allow those doing the study to observe this effect. (b) Diagram below.

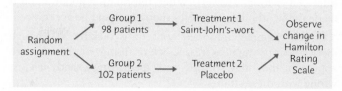

9.52: (a) We expect half of the sample to be made up of older students, so we expect 12.5 (half of 25) older students in the sample. (b) Results will vary, but probability computations reveal that more than 97.7% of samples will have 9 to 16 older employed subjects (and 99.6% of samples have 8 to 17 older employed subjects). In addition, if students average their 20 samples, nearly all students (more than 99%) should find that the average number of older employed subjects is between 11.3 and 13.7.

Note: X, *the number of older employed subjects in the sample, has a hypergeometric distribution with parameters* $N = 50$, $r = 25$, $n = 25$, *so that* $P(9 \leq X \leq 16) = 0.977$. *The theoretical average number of older employed subjects in the sample is 12.5.*

Data Ethics Solutions

As the text states, "Most of these exercises pose issues for discussion. There are no right or wrong answers, but there are more and less thoughtful answers." We have not tried to supply answers for exercises that are largely matters of opinion. For that reason, only seven solutions are provided here.

1. These five proposals are clearly in increasing order of risk. Most students will consider that (a) qualifies as minimal risk, and most will agree that (e) goes beyond minimal risk. Opinions will vary on where to "draw the line," of course.

2. (a) A nonscientist might raise different viewpoints and concerns from those considered by scientists. (b) Answers will vary.

3. It is good to state the purpose of the research plainly ("To study how people's political beliefs are related to risk of depression"). Stating the research *thesis* (that people with ultraliberal beliefs are at greater risk of depression) would cause bias.

6. They cannot be anonymous because the interviews are conducted in person in the subject's home. They are certainly kept confidential. More information can be found at the GSS Web site: www.norc.org/projects/General+Social+Survey.htm.

7. This offers anonymity because names are never revealed. (However, faces are seen, so there may be some chance of someone's identity becoming known.)

8. (a) Those being surveyed should be told the kind of questions they will be asked and the approximate amount of time required. (b) Giving the name and address of the organization may give the respondents a sense that they have an avenue to complain should they feel offended or mistreated by the pollster. (c) At the time that the questions are being asked, knowing who is paying for a poll may introduce bias, perhaps due to nonresponse (not wanting to give what might be considered a "wrong" answer). When information about a poll is made public, though, the poll's sponsor should be announced.

11. For example, informed consent is lacking. (Although most students might express shock or dismay at this research approach, they may struggle with identifying exactly *what* is wrong with these studies.)

Chapter 10   Producing Data – Part II Review

10.1: (c) Trees with active beehives, empty beehives, and no beehives.

10.2: (d)

10.3: (b)

10.4: (d)

10.5: (b)

10.6: (a) The researchers did not randomly assign how much the nurses drank.

10.7: (a) Although neither the subjects nor the technician fully knew what was happening, the exercise types were not randomly assigned.

10.8: (a) Label the students 0001, 0002, . . . , 3478. (b) Using Line 105, the selected students are those numbered 2940, 0769, 1481, 2975, and 1315. (c) There may be many, but certainly summer earnings is one of them.

10.9: Many answers are possible. One possible lurking variable is "student attitude about purpose of college" (students with a view that college is about partying, rather than studying, may be more likely to binge drink and more likely to have lower grades).

10.10: (a) Randomly select 32 of the 64 women, and assign these to the chocolate group. The other 32 women will be assigned to the carob group. Women, as well as the people evaluating them for impact of carob or chocolate, will not know to which group they were assigned (presuming they cannot tell the difference between carob and chocolate by taste). (b) Using Table B beginning at line 110, the first 5 women assigned to the chocolate group are those numbered 38, 44, 18, 33, and 46.

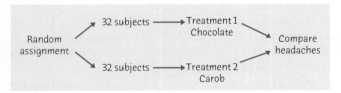

10.11: No doubt, Question A had 60% favoring a tax cut, whereas Question B had 22% favoring a tax cut.

10.12: (d) It's reasonable that just before lunch, the children would have been hungrier.

10.13: (b)

10.14: People who visit the *NOVA Science Now* Web site don't represent American adults broadly. Those people taking

the survey went out of their way to participate in this online poll, and they read pro and con arguments after watching a program about the issue. It seems reasonable to believe that these people understand the issues better than most American adults.

10.15: (a)

10.16: (c)

10.17: (c) The reporter really wants information about all OSU students; however, her "effective" population is really only those who walk on campus late at night.

10.18: (d)

10.19: Answers will vary. (a) One possible population: all full-time undergraduate students in the fall term on a list provided by the registrar. (b) A stratified sample with 125 students from each year is one possibility. (c) Mailed or emailed questionnaires might have high nonresponse rates. Telephone interviews exclude those without phones and may mean repeated calling for those who are not home. Face-to-face interviews might be more costly than your funding will allow. Some students might be sensitive about responding to questions about sexual harassment.

10.20: Placebos do work with real pain, so the placebo response tells nothing about physical basis of the pain. In fact, placebos work poorly in hypochondriacs.

10.21: Parents who fail to return the consent form may be less likely to place high priority and value on education, and therefore they may give their children less help with homework, and so forth. Including those children in the control group is likely to lower that group's score.

Note: *This is a generalization, to be sure; we are not saying that every such parent does not value education, only that the percent of this group that highly values education will almost certainly be lower than that same percent of the parents who return the form.*

10.22: (a) Increase. (b) Decrease. (c) Increase. (d) Decrease.

Note: *The first and third statements make an argument in favor of a national health insurance system, whereas the second and fourth suggest reasons to oppose it.*

10.23: Answers will depend on the site selected.

10.24: (a) The table shows the six treatments—three levels of Factor A (discount level) and two levels of Factor B (fraction of shoes on sale). (b) The diagram is shown. From line 111 of Table B, the first 10 subjects (group 1) are 48, 60, 51, 30, 41, 27, 12, 38, 50, and 59.

		Factor B Fraction of shoes on sale	
		50%	100%
Factor A	20%	1	2
Discount	40%	3	4
level	60%	5	6

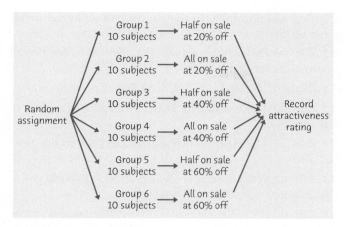

10.25: (a) The factors are storage method (three levels: fresh, room temperature for one month, refrigerated for one month) and preparation method (two levels: cooked immediately, or after one hour). There are therefore six treatments (summarized in the table). The response variables are the tasters' color and flavor ratings. (b) Randomly allocate n potatoes to each of the six groups, then compare ratings. (c) For each taster, randomly choose the order in which the fries are tasted.

	Cooked immediately	Wait one hour
Fresh	1	2
Stored	3	4
Refrigerated	5	6

10.26: (a) Randomly assign 15 students to the easy puzzles and 15 to the difficult puzzles. Compare the actual and estimated times. (b) Each student will work both types of puzzle, with the order randomized. The difference in the perception of time spent will be the response variable.

10.27: (a) This is an observational study; the subjects chose their own "treatments" (how much to drink). The explanatory variable is alcohol consumption, and the response variable is whether or not a subject dies. (b) Many answers are possible. For example, some nondrinkers might avoid drinking because of other health concerns. We do not know what kind of alcohol (beer? wine? whiskey?) the subjects were drinking.

Chapter 11   Introducing Probability

11.1: In the long run, of a large number of Texas Hold 'em games in which you hold a pair, the fraction in which you can make four of a kind will be about 2/245. It does *not* mean that exactly 2 out of 245 such hands would yield four of a kind.

11.2: (a) 0. (b) 1. (c) 0.99. (d) 0.45.

11.3: (a) There are 21 zeros among the first 200 digits of the table (rows 101–105) for a proportion of 0.105. (b) Answers will vary.

11.4: (a) In about 95% of all simulations, there will be between 2 and 8 heads, so the sample proportion will be between 0.2 and 0.8. (b) Shown is the theoretical histogram; a stemplot of 25 proportions will have roughly that shape.

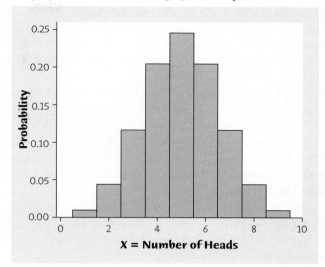

11.5: (a) S = {lives on campus, lives off campus}. (b) S = {All numbers between _____ and _____ years}. (Choices of upper and lower limits will vary.) (c) S = {all amounts greater than or equal to 0}, or S = {0, 0.01, 0.02, 0.03, . . .}. (d) S = {A, B, C, D, F}.

11.6: (a) See below. (b) Each of the 16 outcomes has probability 1/16.

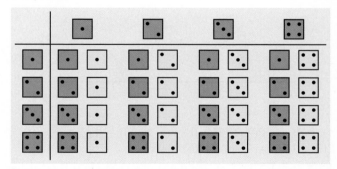

11.7: S = {3, 4, 5, 6, 7, 8, 9}. As all faces are equally likely and the dice are independent, each of the 16 possible pairings is equally likely, so (for example) the probability of a total of 5 is 3/16 because 3 pairings add to 4 (and then we add 1). The complete set of probabilities is shown in the table.

TOTAL	PROBABILITY
3	1/16
4	2/16
5	3/16
6	4/16
7	3/16
8	2/16
9	1/16

11.8: (a) 22%. This makes use of Rule 3 because (assuming there are no double majors) "undergraduate students in engineering" and "undergraduate students in science" have no students in common. (b) 46%. This makes use of Rule 4.

11.9: (a) Event B specifically rules out obese subjects, so there is no overlap with event A. (b) A or B is the event "The person chosen is overweight or obese." $P(A \text{ or } B) = P(A) + P(B) = 0.34 + 0.33 = 0.67$. (c) $P(C) = 1 - P(A \text{ or } B) = 1 - 0.67 = 0.33$.

11.10: (a) The given probabilities have sum 0.90, so $P(\text{other language}) = 0.10$. (b) $P(\text{not English}) = 1 - 0.08 = 0.92$. (c) $P(\text{neither English nor French}) = 0.02 + 0.10 = 0.12$.

11.11: Model 1: Not legitimate (probabilities have sum 6/7). Model 2: Legitimate. Model 3: Not legitimate (probabilities have sum 7/6). Model 4: Not legitimate (probabilities cannot be more than 1).

11.12: (a) $A = \{4, 5, 6, 7, 8, 9\}$, so $P(A) = 0.398$. (b) $B = \{2, 4, 6, 8\}$, so $P(B) = 0.391$. (c) A or B = $\{2, 4, 5, 6, 7, 8, 9\}$, so $P(A \text{ or } B) = 0.574$. This is different from $P(A) + P(B)$ because A and B are not disjoint.

11.13: (a) This is a legitimate probability model because the probabilities sum to 1. (b) $\{X < 4\}$ is the event that somebody lifts weights 3 or fewer days per week. $P(X < 4) = 0.91$. (c) $\{X \geq 1\}$. $P(X \geq 1) = 0.27$.

11.14: (a) 0.6. (b) 0.6. (c) 0.4.

11.15: (a) $\frac{1}{2}bh = \frac{1}{2}(2)(1) = 1$. (b) $P(X < 1) = 0.5$. (c) $P(X < 0.5) = 0.125$.

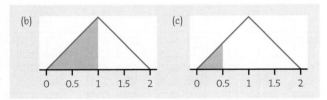

11.16: (a) $P(X \geq 35)$. (b) $P(X \geq 35) = P(Z \geq \frac{35 - 25}{6.4}) = P(Z \geq 1.56) = 1 - 0.9406 = 0.0594$.

11.17: (a) $X \geq 3$ means the student's grade is B or higher (B, B+, A− or A). $P(X \geq 3) = 0.41$. (b) "Poorer than B−" means any grade **lower** than B−. $P(X < 2.7) = P(X \leq 2.3) = 0.46$.

11.18: (a) The student runs the mile in 8 minutes or more. $P(Y \geq 8) = P(Z \geq \frac{8 - 7.11}{0.74}) = P(Z \geq 1.20) = 1 - 0.8849 = 0.1151$. (b) $\{Y < 6\}$. $P(Y < 6) = P(Z < \frac{6 - 7.11}{0.74}) = P(Z < -1.50) = 0.0668$.

11.19: (a) Answers will vary. (b) A personal probability might take into account specific information about one's own driving habits or about the kind of traffic one usually drives in. (c) Most people believe that they are better-than-average drivers (whether or not they have any evidence to support that belief).

11.20: (a) If Joe says $P(\text{Florida State wins}) = 0.1$, then he believes $P(\text{Duke wins}) = 0.3$ and $P(\text{North Carolina wins}) = 0.2$. (b) Joe's probabilities for Duke, Florida State, and North Carolina add up to 0.6, so that leaves probability 0.4 for all other teams.

11.21: (a) Probabilities express the *approximate* fraction of occurrences out of many trials.

11.22: (b) The set $\{0, 1, 2, 3, 4, 5\}$ lists all possible counts.

11.23: (b) This is a discrete model with a limited number of outcomes.

11.24: (b) The other probabilities add to 0.99, so this must be 0.01.

11.25: (c) $P(\text{Republican or Democrat}) = P(\text{Republican}) + P(\text{Democrat}) = 0.26 + 0.33 = 0.59$.

11.26: (b) $P(\text{not Republican}) = 1 - P(\text{Republican}) = 1 - 0.26 = 0.74$.

11.27: (b) There are 10 equally likely possibilities, so $P(\text{seven}) = 1/10$.

11.28: (c) "7 or greater" means 7, 8, or 9—3 of the 10 possibilities.

11.29: (b) 22% $(0.14 + 0.05 + 0.02 + 0.01 = 0.22$, or 22%) have 3 or more cars.

11.30: (c) $Y > 1$ standardizes to $Z > 2.56$, for which Table A gives 0.0052.

11.31: (a) There are 16 possible outcomes: {HHHH, HHHM, HHMH, HMHH, MHHH, HHMM, HMHM, HMMH, MHHM, MHMH, MMHH, HMMM, MHMM, MMHM, MMMH, MMMM}. (b) $S = \{0, 1, 2, 3, 4\}$.

11.32: (a) Legitimate. (b) Legitimate (even if the deck of cards is not!). (c) Not legitimate (the total is more than 1).

11.33: (a) $1 - 0.73 = 0.27$. (b) $P(\text{at least a high school education}) = 1 - P(\text{has not finished HS}) = 1 - 0.13 = 0.87$.

11.34: In computing the probabilities, we have dropped the trailing zeros from the land area figures. (a) $P(\text{area is forested}) = 4176/9094 = 0.4592$. (b) $P(\text{area is not forested}) = 1 - 0.4592 = 0.5408$.

11.35: (a) All probabilities are between 0 and 1, and they add to 1. (We must assume that no one takes more than one language.) (b) $0.43 = 1 - 0.57$. (c) $0.40 = 0.30 + 0.08 + 0.02$.

11.36: (a) The given probabilities add to 0.96, so other colors must account for the remaining 0.04. (b) $P(\text{silver or white}) = 0.23 + 0.18 = 0.41$, so $P(\text{neither silver nor white}) = 1 - 0.41 = 0.59$.

11.37: Of the seven cards, there are three 9's, two red 9's, and two 7's. (a) $P(\text{draw a 9}) = 3/7$. (b) $P(\text{draw a red 9}) = 2/7$. (c) $P(\text{don't draw a 7}) = 1 - P(\text{draw a 7}) = 1 - 2/7 = 5/7$.

11.38: The probabilities of 2, 3, 4, and 5 are unchanged (1/6), so $P(1 \text{ or } 6)$ must still be 1/3. If $P(6) = 0.2$, then $P(1) = 1/3 - 0.2 = 0.1333$ (or 2/15).

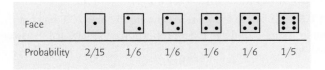

Face						
Probability	2/15	1/6	1/6	1/6	1/6	1/5

11.39: Each of the 90 guests has probability 1/90 of winning the prize. Because there are 42 women, the probability is 42/90 = 0.467.

11.40: (a) It is legitimate because every person must fall into exactly one category, the probabilities are all between 0 and 1, and they add up to 1. (b) 0.166 = 0.002 + 0.008 + 0.151 + 0.005. (c) 0.351 = 1 − 0.649.

11.41: (a) It is legitimate because every person must fall into exactly one category, the probabilities are all between 0 and 1, and they add up to 1. (b) 0.168. (c) 0.170—the sum of the numbers in the first column. (d) 0.548, the sum of the numbers in the third row.

11.42: (a) A corresponds to the outcomes in the first column and the third row. (b) $P(A) = 0.550$.

11.43: (a) 1 − 0.170 = 0.830. (b) 1 − 0.071 = 0.929.

11.44: (a) X is discrete because it has a finite sample space. (b) $\{X \geq 1\}$ (or $\{X > 0\}$). $P(X \geq 1) = 1 - P(X = 0) = 0.9$. (c) "no more than two nonword errors," or "fewer than three nonword errors." $P(X \leq 2) = P(X = 0) + P(X = 1) + P(X = 2) = 0.1 + 0.2 + 0.3 = 0.6$. $P(X < 2) = P(X = 0) + P(X = 1) = 0.1 + 0.2 = 0.3$.

11.45: (a) All 9 digits are equally likely, so each has probability 1/9:

Value of W	1	2	3	4	5	6	7	8	9
Probability	$\frac{1}{9}$	$\frac{1}{9}$	$\frac{1}{9}$	$\frac{1}{9}$	$\frac{1}{9}$	$\frac{1}{9}$	$\frac{1}{9}$	$\frac{1}{9}$	$\frac{1}{9}$

(b) $P(W \geq 6) = 4/9 = 0.444$, or twice as big as the Benford's law probability.

11.46: (a) There are 10 pairs. Just using initials: {(A, D), (A, M), (A, S), (A, R), (D, M), (D, S), (D, R), (M, S), (M, R), (S, R)}. (b) Each has probability 1/10 = 10%. (c) Mei-Ling is chosen in 4 of the 10 possible outcomes: 4/10 = 40%. (d) There are 3 pairs with neither Sam nor Roberto, so the probability is 3/10.

11.47: (a) {BBB, BBG, BGB, GBB, GGB, GBG, BGG, GGG}. Each has probability 1/8. (b) $P(X = 2) = 3/8 = 0.375$. (c) See table.

Value of X	0	1	2	3
Probability	1/8	3/8	3/8	1/8

11.48: The possible values of Y are 1, 2, 3, . . . , 12, each with probability 1/12. The first (regular) die is equally likely to show any number from 1 through 6. Half of the time, the second roll shows 0, and the other half it shows 6. Each possible outcome therefore has probability $(\frac{1}{6})(\frac{1}{2}) = 1/12$.

11.49: (a) This is a continuous random variable because the set of possible values is an interval. (b) The height should be 1/2 because the area under the curve must be 1. The density curve is illustrated. (c) 1/2.

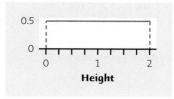

11.50: (a) 0.4. (b) 0.6.

11.51: (a) $P(0.49 \leq V \leq 0.53) = P(\frac{0.49 - 0.51}{0.009} \leq Z \leq \frac{0.53 - 0.51}{0.009}) = P(-2.22 \leq Z \leq 2.22) = 0.9868 - 0.0132 = 0.9736$. (b) $P(V \leq 0.49) = P(Z \leq \frac{0.49 - 0.51}{0.009}) = P(Z \leq 2.22) = 0.0132$.

11.52: $P(8.9 \leq x \leq 9.1) = P(\frac{8.9 - 9}{0.075} \leq Z \leq \frac{9.1 - 9}{0.075}) = P(-1.33 \leq Z \leq 1.33) = 0.9082 - 0.0918 = 0.8164$.

11.53: (a) Because there are 10,000 equally likely four-digit numbers (0000 through 9999), the probability of an exact match is 1/10,000. (b) There is a total of $24 = 4 \cdot 3 \cdot 2 \cdot 1$ arrangements of the four digits 5, 9, 7, and 4, so the probability of a match in any order is 24/10,000.

11.54: Note that in this experiment, factors other than the nickel's characteristics might affect the outcome.

11.55: (a)–(c) Results will vary, but after n tosses, the distribution of the proportion $\hat{p}$ is approximately Normal with mean 0.5 and standard deviation $1/(2\sqrt{n})$, whereas the distribution of the count of heads is approximately Normal with mean $0.5n$ and standard deviation $\sqrt{n}/2$, so using the 68–95–99.7 rule, we have the results shown in the following table. Note that the range for $\hat{p}$ gets narrower, whereas the range for the count gets wider.

n	99.7% RANGE FOR $\hat{p}$	99.7% RANGE FOR COUNT
40	0.5 ± 0.237	20 ± 9.5
120	0.5 ± 0.137	60 ± 16.4
240	0.5 ± 0.097	120 ± 23.2
480	0.5 ± 0.068	250 ± 329

11.56: (a) Virtually all answers will be between 62% and 88% (b) Answers will vary.

11.57: (a) With $n = 50$, the variability in $\hat{p}$ is larger. With $n = 100$, nearly all answers will be between 0.23 and 0.53. With $n = 400$, nearly all answers will be between 0.31 and 0.45.

Chapter 12   General Rules of Probability

12.1: It is unlikely that these events are independent. In particular, it is reasonable to expect that younger adults are more likely than older adults to be college students.

12.2: This would not be surprising: assuming that all the authors are independent (for example, none were written by siblings or married couples), $P(N = 0) = (1 - 0.096)^9 = 0.4032$.

12.3: If we assume that each site is independent of the others (and that they can be considered as a random sample from the collection of sites referenced in scientific journals), then P(all seven are still good) $= (0.87)^7 = 0.3773$.

12.4: A Venn diagram is provided. B is the event "the degree is a bachelor's degree," and W is the event "the degree was earned by a woman." (a) Because $P(W) = 0.59$, P(degree was earned by a man) $= P(\text{not } W) = 1 - 0.59 = 0.41$, or 41%. (b) $P(B \text{ and not } W) = 0.50 - 0.29 = 0.21$, or 21%.

(c) Because $P(B \text{ and } M) = 0.21$, but $P(B) \times P(M) = (0.50)(0.41) = 0.205$, $P(B \text{ and } M) \neq P(B) \times P(M)$. Hence, B and M are not independent. When using probabilities, "close" is not "equal."

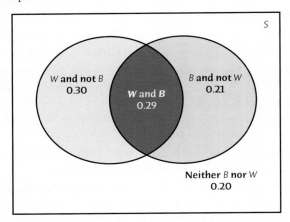

12.5: (a) A Venn diagram is provided. (b) The events are

$\{A \text{ and } B\}$	= {student is at least 25 and local}
$\{A \text{ and not } B\}$	= {student is at least 25 and not local}
$\{B \text{ and not } A\}$	= {student is less than 25 and local}
$\{\text{neither } A \text{ nor } B\}$	= {student is less than 25 and not local}

(c) $P(A \text{ and } B)$ is given. Subtracting this from the given probabilities for A and B gives $P(A \text{ and not } B)$ and $P(B \text{ and not } A)$. Those probabilities add to 0.90, so $P(\text{neither } B \text{ nor } W) = 0.10$.

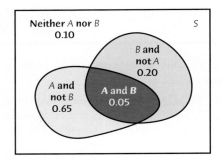

12.6: $P(W \mid B) = \frac{P(B \text{ and } W)}{P(B)} = 0.29/0.50 = 0.58$.

12.7: $P(B \mid \text{not } A) = \frac{P(B \text{ and not } A)}{P(\text{not } A)} = \frac{P(B) - P(B \text{ and } A)}{P(\text{not } A)} = \frac{0.2}{0.3} = 0.667$.

12.8: Let R be the event "game is a role-playing game," while S is the event "game is a strategy game." Then $P(\text{not } S) = 1 - 0.354 = 0.646$, and $P(R \mid \text{not } S) = \frac{P(R \text{ and not } S)}{P(\text{not } S)} = \frac{0.139}{0.646} = 0.2152$.

12.9: Let H be the event that an adult belongs to a club and T be the event that he/she goes at least twice a week. We have been given $P(H) = 0.10$ and $P(T \mid H) = 0.40$. Note also that $P(T \text{ and } H) = P(T)$ because one has to be a member of the club to attend. So $P(T) = P(H)P(T \mid H) = (0.10)(0.40) = 0.04$. About 4% of all adults go to health clubs at least twice a week.

12.10: Let A = {the teen is online}, B = {the teen has a profile}, and C = {the teen has commented on a friend's blog}. $P(A) = 0.93$, $P(B \mid A) = 0.55$, and $P(C \mid A \text{ and } B) = 0.76$. $P(A \text{ and } B \text{ and } C) = P(A)P(B \mid A)P(C \mid A \text{ and } B) = (0.93)(0.55)(0.76) = 0.3887$.

12.11: (a) and (b) These probabilities are provided below. (c) The product of these conditional probabilities gives the probability of a flush in spades by the general multiplication rule: we must draw a spade, and then another, and then a third, a fourth, and a fifth. The product of these probabilities is about 0.0004952. (d) Because there are four possible suits in which to have a flush, the probability of a flush is four times that found in (c), or about 0.001981.

$$P(\text{1st card } \spadesuit) = \tfrac{13}{52} = \tfrac{1}{4} = 0.25$$
$$P(\text{2nd card } \spadesuit \mid 1 \spadesuit \text{ picked}) = \tfrac{12}{51} = \tfrac{4}{17} = 0.2353$$
$$P(\text{3rd card } \spadesuit \mid 2 \spadesuit \text{s picked}) = \tfrac{11}{50} = 0.22$$
$$P(\text{4th card } \spadesuit \mid 3 \spadesuit \text{s picked}) = \tfrac{10}{49} = 0.2041$$
$$P(\text{5th card } \spadesuit \mid 4 \spadesuit \text{s picked}) = \tfrac{9}{48} = \tfrac{3}{16} = 0.1875$$

12.12: (a) $P(\text{woman}) = 254/863 = 0.2943$. (b) $P(\text{woman} \mid \text{full professor}) = 73/349 = 0.2092$. (c) Rank and sex are not independent; if they were, the probabilities in (a) and (b) would be equal.

12.13: (a) The tree diagram is shown below. (b) Two branches result in a positive test result. Multiply out the probabilities on those branches, and add to find $P(\text{Positive test}) = (0.063)(0.21) + (0.937)(0.06) = 0.06945$.

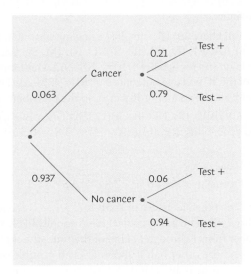

12.14: (a) The probability the child has blue eyes IF they have red hair is $P(\text{blue eyes} \mid \text{red hair}) = 0.473$; $P(\text{blue eyes AND red hair}) = (0.025)(0.473) = 0.0118$. (b) The probability of having freckles IF the child has both red hair and blue eyes is $P(\text{freckles} \mid \text{red hair and blue eyes}) = 0.857$; $P(\text{red hair AND blue eyes AND freckles}) = (0.025)(0.473)(0.857) = 0.0101$.

12.15: (a) $P(\text{no cancer} \mid \text{positive test}) = \frac{P(\text{no cancer and positive test})}{P(\text{positive test})} = \frac{0.05622}{0.06945} = 0.8095$.

(b) About 81% of positive test results come from people without prostate cancer; they will possibly suffer the "serious side effects" needlessly.

12.16: (a) $P(\text{red hair}) = 0.025$ is the probability a child has red hair, regardless of eye color or freckle status. $P(\text{blue eyes}) = (0.061)(0.030) + (0.461)(0.259) + (0.453)(0.562) + (0.025)(0.473) = 0.3876$ is the probability a child has blue eyes, regardless of hair color or freckle status. $P(\text{freckles}) = 0.061((0.030)(0.143) + (0.047)(0.182) + (0.923)(0.018)) + 0.461((0.259)(0.278) + (0.327)(0.232) + (0.414)(0.153)) + 0.453((0.562)(0.319) + (0.315)(0.302) + (0.123)(0.164)) + 0.025((0.473)(0.857) + (0.405)(0.767) + (0.122)(0.778)) = 0.2529$ is the probability a child has freckles, no matter what the hair or eye color. (b) $P(\text{freckles and red hair}) = 0.0203$ is the probability the child has red hair and freckles regardless of the eye color; it was the last row of the probability found in part (a). $P(\text{freckles} \mid \text{red hair}) = 0.0203/0.025 = 0.812$ is the probability of freckles IF the child has red hair, ignoring eye color. (c) Because $P(\text{freckles} \mid \text{red hair}) = 0.812$ is not equal to $P(\text{freckles}) = 0.2529$, freckles and hair color are not independent in the population of Germans of Caucasian descent.

12.17: (b) $(0.98)^3 = 0.9412$.

12.18: (b) $1 - (0.98)^3 = 0.0588$.

12.19: (a) $P(\text{at least one positive}) = 1 - P(\text{both negative}) = 1 - P(\text{first negative})P(\text{second negative}) = 1 - (0.1)(0.2) = 0.98$.

12.20: (b) $P(\text{female}) = 22{,}751/48{,}988 = 0.4644$.

12.21: (c) $1{,}778/7{,}996 = 0.2224$.

12.22: (b) $1{,}778/22{,}751 = 0.0782$.

12.23: (c) We want the fraction of engineering doctorates conferred to women. A (engineering doctorate) is what has been given.

12.24: (b) $P(W \text{ or } S) = P(W) + P(S) - P(W \text{ and } S) = 0.52 + 0.25 - 0.11 = 0.66$.

12.25: (c) $P(W \text{ and } D) = P(W)P(D \mid W) = (0.86)(0.028) = 0.024$.

12.26: (b) $P(D) = P(W \text{ and } D) + P(B \text{ and } D) + P(A \text{ and } D) = (0.86)(0.028) + (0.12)(0.044) + (0.02)(0.035) = 0.030$.

12.27: $(0.75)^8 = 0.1001$.

12.28: $P(\text{none are O-negative}) = (1 - 0.072)^{10} = 0.4737$, so $P(\text{at least one is O-negative}) = 1 - 0.4737 = 0.5263$.

12.29: (a) $\left(\frac{1}{20}\right)\left(\frac{9}{20}\right)\left(\frac{1}{20}\right) = 0.001125$. (b) The other (noncherry) symbol can show up on the middle wheel, with probability $\left(\frac{1}{20}\right)\left(\frac{11}{20}\right)\left(\frac{1}{20}\right) = 0.001375$, or on either of the outside wheels, with probability $= \left(\frac{19}{20}\right)\left(\frac{9}{20}\right)\left(\frac{1}{20}\right)$ (each). (c) Combining all three cases from part (b), we have $P(\text{exactly two cherries}) = 0.001375 + (2)(0.021375) = 0.044125$.

12.30: (a) $(0.80)^2 = 0.64$. (b) $P(\text{sighting on at least one day}) = 1 - P(\text{sighting on neither day}) = 1 - (0.20)^2 = 0.96$. (c) We want the number of trips, k, such that $1 - (0.20)^k \geq 0.99$. You can solve this algebraically for k, yielding $k \geq 2.86$,

which must be rounded up to 3. At least three trips are needed to ensure at least 0.99 probability of at least one sighting.

12.31: Let I be the event "infection occurs" and let F be "the repair fails." $P(I) = 0.03$, $P(F) = 0.14$, and $P(I \text{ and } F) = 0.01$. We want to find $P(\text{not } I \text{ and not } F)$. $P(I \text{ or } F) = P(I) + P(F) - P(I \text{ and } F) = 0.03 + 0.14 - 0.01 = 0.16$. Now observe that the desired probability is the complement of "I or F": $P(\text{not } I \text{ and not } F) = 1 - P(I \text{ or } F) = 0.84$.

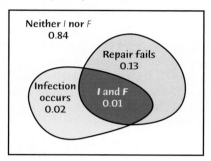

12.32: Let W be the event that a whale is seen. Let D be the event that a dolphin is seen. (a) $P(W) = 0.05 + 0.15 = 0.20$. (b) $P(W \text{ and not } D) = 0.05$. (c) Yes, because $P(W \text{ and } D) = 0.15 = P(W)P(D) = (0.20)(0.75) = 0.15$.

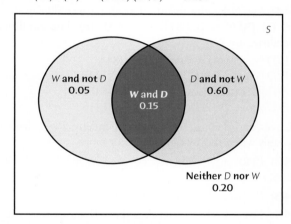

12.33: Let I be the event "infection occurs," and let F be "the repair fails." $P(I \mid \text{not } F) = \frac{P(I \text{ and not } F)}{P(\text{not } F)} = \frac{0.02}{0.86} = 0.0233$.

12.34: $P(D) = 0.4 = 0.1 + 0.1 + 0.2$.

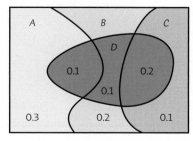

12.35: (a) $P(\text{positive} \mid \text{disease}) = 564/574 = 0.9826$. (b) $P(\text{no disease} \mid \text{negative}) = 708/718 = 0.9861$.

12.36: (a) $P(\text{income} \geq \$30{,}000) = 0.310 + 0.181 + 0.030 = 0.521$. (b) $P(\text{income} \geq \$75{,}000 \mid \text{income} \geq \$30{,}000) = \frac{P(\text{income} \geq \$75{,}000)}{P(\text{income} \geq \$30{,}000)} = \frac{0.211}{0.521} = 0.4050$.

12.37: (a) These events are not independent because P(pizza with mushrooms) $= 4/7$, but P(mushrooms | thick crust) $= 2/3$. Alternatively, note that P(thick crust with mushrooms) $= 2/7$, which is not equal to the product of P(mushrooms) $= 4/7$ and P(thick crust pizza) $= 3/7$. (b) With the eighth pizza, P(mushrooms) $= 4/8 = 1/2$ and P(mushrooms | thick crust) $= 2/4 = 1/2$, so these events are independent.

12.38: (a) P(two boys | at least one boy) $= \frac{P(\text{two boys})}{P(\text{at least one boy})} = \frac{0.25}{0.75} = \frac{1}{3}$ (b) P(two boys | older child is a boy) $= \frac{P(\text{two boys})}{P(\text{older child is boy})} = \frac{0.25}{0.50} = \frac{1}{2}$.

12.39: Let W be the event "the person is a woman" and M be "the person earned a master's degree." (a) $P(\text{not } W) = 1421/3560 = 0.3992$. (b) $P(\text{not } W \mid M) = 282/732 = 0.3852$. (c) The events "choose a man" and "choose a master's degree recipient" are not independent. If they were, the two probabilities in (a) and (b) would be equal.

12.40: Let W be the event "the person is a woman" and A be the event "the person earned an associate's degree." (a) $P(W) = 2139/3560 = 0.6008$. (b) $P(A \mid W) = 556/2139 = 0.2599$. (c) $P(W \text{ and } A) = P(W)P(A \mid W) = \left(\frac{2139}{3560}\right)\left(\frac{556}{2139}\right) = \frac{556}{3560}$. This agrees with the directly computed probability: $P(W \text{ and } A) = 556/3560 = 0.1562$.

12.41: Let D be the event "a seedling was damaged by a deer." (a) $P(D) = 209/871 = 0.2400$. (b) The conditional probabilities are

$P(D \mid \text{no cover}) = 60/211 = 0.2844$
$P(D \mid \text{cover} < 1/3) = 76/234 = 0.3248$
$P(D \mid 1/3 \text{ to } 2/3 \text{ cover}) = 44/221 = 0.1991$
$P(D \mid \text{cover} > 2/3) = 29/205 = 0.1415$

(c) Cover and damage are not independent; $P(D)$ decreases noticeably when thorny cover is 1/3 or more.

12.42: $P(\text{cover} > 2/3 \mid \text{not } D) = 176/662 = 0.2659$, or 26.59%.

12.43: $P(\text{cover} < 1/3 \mid D) = 136/209 = 0.6507 = 65.07\%$.

12.44: $P(\text{no offer}) = P(\text{not } A \text{ and not } B \text{ and not } C) = 0.10$.

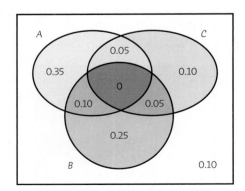

12.45: $P(A \text{ and not } B \text{ and not } C) = 0.35$.

12.46: $P(B \mid C) = 0.05/0.20 = 0.25$. $P(C \mid B) = 0.05/0.40 = 0.125$.

12.47: (a) P(doubles on first toss) $= 1/6$ because 6 of the 36 equally likely outcomes enumerated in Figure 11.2 involve rolling doubles. (b) We need no doubles on the first roll (which happens with probability 5/6), then doubles on the second toss. P(first doubles appears on toss 2) $= (5/6)(1/6) = 5/36$. (c) Similarly, P(first doubles appears on toss 3) $= (5/6)^2(1/6) = 25/216$. (d) P(first doubles appears on toss 4) $= (5/6)^3(1/6)$, etc. In general, P(first doubles appears on toss k) $= (5/6)^{k-1}(1/6)$. (e) P(go again within 3 turns) $= P$(roll doubles in 3 or fewer rolls) $= P$(roll doubles on 1st, 2nd or 3rd try) $= (1/6) + (5/6)(1/6) + (5/6)^2(1/6) = 0.4213$.

12.48: The probability of each outcome is the product of the individual branch probabilities leading to it. The total probability of the serving player winning a point is $0.4307 + 0.2080 = 0.6387$.

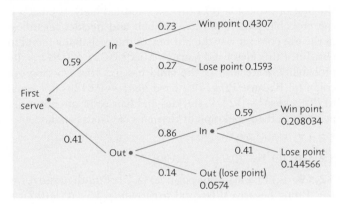

12.49: In the tree diagram, each "up-step" represents an allergic individual (and has probability 0.02), and each "down-step" is a nonallergic individual (and has probability 0.98). At the end of each of the 8 complete branches is the value of X. For example, any branch with 2 up-steps and 1 down-step has probability $0.02^2 \times 0.98^1 = 0.000392$ and yields $X = 2$. There are three branches each corresponding to $X = 2$ and $X = 1$ and only one branch each for $X = 3$ and $X = 0$. Because $X = 0$ and $X = 3$ appear on one branch each, $P(X = 0) = 0.98^3 = 0.941192$ and $P(X = 3) = 0.02^3 = 0.000008$. Meanwhile, $P(X = 1) = 3(0.02)^1(0.98)^2 = 0.057624$, and $P(X = 2) = 3(0.02)^2(0.98)^1 = 0.001176$.

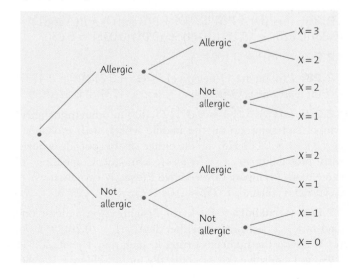

12.50: $P(\text{first serve in} \mid \text{server won point}) =$
$\frac{P(\text{first serve in and server won point})}{P(\text{server won point})} = \frac{0.4307}{0.6387} = 0.6743$, or 67.43%.

12.51: $P(X = 2 \mid X \geq 1) = \frac{P(X = 2 \text{ and } X \geq 1)}{P(X \geq 1)} = \frac{P(X = 2)}{P(X \geq 1)} =$
$\frac{0.001176}{1 - 0.941192} = 0.020$.

12.52: Let W, B, A, and L be (respectively) the events that this person is white, black, Asian, and lactose intolerant. We have been given

$$P(W) = 0.82 \qquad P(B) = 0.14 \qquad P(A) = 0.04$$
$$P(L \mid W) = 0.15 \qquad P(L \mid B) = 0.70 \qquad P(L \mid A) = 0.90$$

(a) $P(L) = (0.82)(0.15) + (0.14)(0.70) + (0.04)(0.90) = 0.257$, or 25.7%. (b) $P(A \mid L) = P(A \text{ and } L)/P(L) = (0.04)(0.90)/0.257 = 0.1401$, or 14%.

12.53: Let $R = \{\text{recent donor}\}$, $P = \{\text{pledged}\}$, and $C = \{\text{contributed}\}$. (a) The percent of calls resulting in a contribution can be found by considering all the branches of the tree that end in a contribution. $P(C) = (0.5)(0.4)(0.8) + (0.3)(0.3)(0.6) + (0.2)(0.1)(0.5) = 0.224$, or 22.4%. (b) $P(R \mid C) = \frac{P(C \text{ and } R)}{P(C)} = \frac{(0.5)(0.4)(0.8)}{0.224} = 0.7143$, or 71.4%.

12.54: In this problem, allele 29 is playing the role of A and 0.181 is the proportion with this allele ($a = 0.181$). Similarly, allele 31 is playing the role of B, and the proportion having this allele is $b = 0.071$. The proportion of the population with combination (29,31) is therefore $2(0.181)(0.071) = 0.025702$. The proportion with combination (29,29) is $(0.181)(0.181) = 0.032761$.

12.55: The proportion having combination (16,17) is $2(0.232)(0.212) = 0.098368$.

12.56: In Exercise 12.54, we found that the proportion of the population with allele (29,31) at loci D21S11 is 0.025702. In Exercise 12.55, we found that the proportion with allele (16,17) at loci D3S1358 is 0.098368. Assuming independence between loci, the proportion with allele (29,31) at D21S11 and (16,17) at D3S1358 is $(0.098368)(0.025702) = 0.002528$.

12.57: If the DNA profile found on the hair is possessed by 1 in 1.6 million individuals, then we would expect about 3 individuals in the database of 4.5 million convicted felons to demonstrate a match. This comes from (4.5 million)/(1.6 million) = 2.8125, which was rounded up to 3.

Chapter 13   Binomial Distributions

13.1: Binomial. (1) We have a fixed number of observations ($n = 15$). (2) It is reasonable to believe that each call is independent of the others. (3) "Success" means reaching a live person, "failure" is any other outcome. (4) Each randomly dialed number has chance $p = 0.2$ of reaching a live person.

13.2: Not binomial. We do not have a fixed number of observations.

13.3: Not binomial. The trials aren't independent. If one tile in a box is cracked, there are likely more tiles cracked.

13.4: Binomial. We have a fixed number of independent trials, each leading to success (smokes daily or occasionally) or failure, with the probability of success constant from trial to trial. We're counting the number of successes in our sample. The number in the sample who smoke daily or occasionally is binomial with $n = 1500$ and $p = 0.199$.

13.5: (a) C, the number caught, is binomial with $n = 10$ and $p = 0.7$. M, the number missed, is binomial with $n = 10$ and $p = 0.3$. (b) $P(M = 3) = \binom{10}{3}(0.3)^3(0.7)^7 = (120)(0.027)(0.08235) = 0.2668$. $P(M \geq 3) = 0.6172$.

13.6: Let N be the number of live persons contacted among the 15 calls observed. Then N has the binomial distribution with $n = 15$ and $p = 0.2$.

(a) $P(N = 3) = \binom{15}{3}(0.2)^3(0.8)^{12} = 0.2501$.

(b) $P(N \leq 3) = P(N = 0) + \cdots + P(N = 3) = 0.6482$.

(c) $P(N \geq 3) = P(N = 3) + \cdots + P(N = 15) = 0.6020$.

(d) $P(N < 3) = P(N = 0) + P(N = 1) + P(N = 2) = 0.3980$.

(e) $P(N > 3) = P(N = 4) + \cdots + P(N = 15) = 0.3518$.

13.7: (a) `5 choose 2 returns` 10. (b) `500 choose 2 returns` 124,750, and `500 choose 100 returns` 2.041694×10^{107}. (c) `(10 choose 1)*0.11*0.89^9 returns` 0.38539204407.

13.8: (a) $\mu = np = (15)(0.2) = 3$ calls. (b) $\sigma = \sqrt{np(1 - p)} = \sqrt{2.4} = 1.5492$ calls. (c) With $p = 0.08$, $\sigma = \sqrt{1.104} = 1.0507$ calls; with $p = 0.01$, $\sigma = \sqrt{0.1485} = 0.3854$ calls. As p approaches 0, the standard deviation decreases (that is, it also approaches 0).

13.9: (a) X is binomial with $n = 10$ and $p = 0.3$; Y is binomial with $n = 10$ and $p = 0.7$ (b) The mean of Y is $(10)(0.7) = 7$ errors caught, and for X the mean is $(10)(0.3) = 3$ errors missed. (c) The standard deviation of Y (or X) is $\sigma = \sqrt{10(0.7)(0.3)} = 1.4491$ errors.

13.10: Let X be the number of 1's and 2's; then X has a binomial distribution with $n = 90$ and $p = 0.477$ (in the absence of fraud). This should have a mean of 42.93 and standard deviation $\sigma = \sqrt{22.4524} = 4.7384$. Therefore, $P(X \leq 29) = P(Z \leq \frac{29 - 42.93}{4.7384}) = P(Z \leq -2.94) = 0.0016$. This probability is quite small, so we have reason to be suspicious.

13.11: (a) $\mu = (1520)(0.31) = 471.2$ and $\sigma = \sqrt{1520(0.31)(1 - 0.31)} = 18.0313$ students. (b) $np = (1520)(0.31) = 471.2 \geq 10$ and $n(1 - p) = (1520)(0.69) = 1048.8 \geq 10$, so n is large enough for the Normal approximation to be reasonable. The college wants 475 students, so $P(X \geq 476) = P(Z \geq \frac{476 - 471.2}{18.0313}) = P(Z \geq 0.27) = 0.3936$. (c) The exact binomial probability is 0.4045 (obtained from software), so the Normal approximation is 0.0109 too low.

13.12: (a) $\mu = (1000)(0.24) = 240$ and $\sigma = \sqrt{1000(0.24)(1 - 0.24)} = 13.5056$. (b) $np = 240$ and $n(1 - p) = 760$ are both more than 10. We compute $P(210 \leq X \leq 270) = P(\frac{210 - 240}{13.5056} \leq Z \leq \frac{270 - 240}{13.5056}) = P(-2.22 \leq Z \leq 2.22) = 0.9736$.

13.13: (b) He has 3 independent eggs, each with probability 1/4 of containing salmonella.

13.14: (b) $P(S \geq 1) = P(S > 0) = 1 - P(S = 0) = 1 - 0.4219 = 0.5781$.

13.15: (c) The selections are not independent; once we choose one student, it changes the probability that the next student is a business major.

13.16: (c) We must choose 2 of the 5 guesses to be correct; $\binom{5}{2} = 10$.

13.17: (b) This probability is $(0.25)(0.75)^3(0.25) = 0.0264$.

13.18: (a) Guessing correctly on 2 of the 5 means missing three times, so this probability is $\binom{5}{2}(0.25)^2(0.75)^3 = 0.264$.

13.19: (b) This is the event that a single digit is 8 or 9, so the probability is 0.20.

13.20: (a) Two lines of the table means that we have $2(40) = 80$ digits. This is the number of "successes" (8 or 9) in $n = 80$ independent trials with $p = 0.20$.

13.21: (a) $np = (80)(0.20) = 16$.

13.22: (a) No: there are more than two possible outcomes. (b) A binomial distribution is reasonable here; a "large city" will have a population much larger than 100 (the sample size), and each randomly selected juror has the same (unknown) probability p of opposing the death penalty. However, if the jurors are asked the question as a whole, for example, one person's answer could influence another, and independence would be lost. (c) In a Pick 3 game, Joe's chance of winning the lottery is the same every week, so assuming that a year consists of 52 weeks (observations), this would be binomial.

13.23: (a) A binomial distribution is *not* an appropriate choice for field goals made because given the different situations the kicker faces, his probability of success is likely to change from one attempt to another. (b) It would be reasonable to use a binomial distribution for free throws made because we have $n = 150$ attempts, presumably independent (or at least approximately so), with chance of success $p = 0.8$ each time.

13.24: (a) Binomial, $n = 10$ and $p = 0.30$. (b) $P(X = 1) = \binom{10}{1}(0.30)(0.70)^9 = 0.1211$. $P(X = 2) = \binom{10}{2}(0.30)^2(0.70)^8 = 0.2335$. (c) $P(X \geq 1) = 1 - P(X = 0) = 1 - (0.70)^{10} = 0.9718$.

13.25: (a) $n = 5$ and $p = 0.65$. (b) The possible values of X are the integers 0, 1, 2, 3, 4, 5. (c)

$P(X = 0) = \binom{5}{0}(0.65)^0(0.35)^5 = 0.00525$

$P(X = 1) = \binom{5}{1}(0.65)^1(0.35)^4 = 0.04877$

$P(X = 2) = \binom{5}{2}(0.65)^2(0.35)^3 = 0.18115$

$P(X = 3) = \binom{5}{3}(0.65)^3(0.35)^2 = 0.33642$

$P(X = 4) = \binom{5}{4}(0.65)^4(0.35)^1 = 0.31239$

$P(X = 5) = \binom{5}{5}(0.65)^5(0.35)^0 = 0.11603$

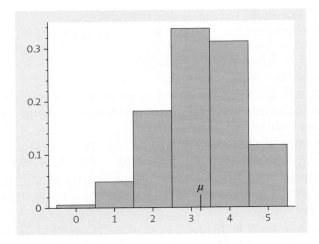

(d) $\mu = np = (5)(0.65) = 3.25$ years and $\sigma = \sqrt{5(0.65)(1 - 0.65)} = 1.0665$ years.

13.26: (a) $18/38 = 0.47368$. (b) X has the binomial ($n = 4$, $p = 0.47368$) distribution. (c) $P(\text{break even}) = P(X = 2) = \binom{4}{2}(0.47368)^2(1 - 0.47368)^2 = 0.37292$. (d) $P(\text{lose money}) = P(X < 2) = P(X = 0) + P(X = 1) = 0.07674 + 0.27625 = 0.35299$.

13.27: (a) All women are independent, and each has the same probability of getting pregnant. (b) Under ideal conditions, the number who get pregnant is binomial with $n = 20$ and $p = 0.01$; $P(N \geq 1) = 1 - P(N = 0) = 1 - 0.8179 = 0.1821$. In typical use, $p = 0.03$, and $P(N \geq 1) = 1 - 0.5438 = 0.4562$.

13.28: X, the number of wins betting on "red" 200 times, is binomial with $n = 200$ and $p = 0.47368$. $np = 94.736 > 10$ and $n(1 - p) = 105.264 > 10$. $\mu = np = 94.736$, $\sigma = \sqrt{200(0.47368)(1 - 0.47368)} = 7.06126$, so $P(X < 100) = P(X \leq 99) = P(Z \leq 0.60) = 0.7257$. The exact binomial probability is 0.7502. As the number of plays (n) increases, the probability of losing money will increase. For example, if $n = 400$, $P(X < 200) = P(Z \leq 0.95) = 0.8289$. The exact binomial probability is 0.8424.

13.29: (a) X, the number of women who get pregnant in typical use, is binomial with $n = 600$ and $p = 0.03$. $np = 18$ and $n(1 - p) = 582$ are both larger than 10. The mean is 18, and the standard deviation is 4.1785, so $P(X \geq 20) = P(Z \geq 0.48) = 0.3156$. The exact binomial probability is 0.3477. (b) Under ideal conditions, $p = 0.01$, so $np = 6$ is too small.

13.30: (a) We must assume that each drive is independent and that he has a 57% chance of hitting the fairway each time. Both of these assumptions are suspect. (b) The average number of fairways hit is $np = (14)(0.57) = 7.98$ fairways hit.

13.31: (a) If R is the number of red-blossomed plants out of a sample of 4, then $P(R = 3) = \binom{4}{3}(0.75)^3(0.25)^1 = 0.4219$. (b) With $n = 60$, the mean number of red-blossomed plants is $np = 45$. (c) If R is the number of red-blossomed plants out of a sample of 60, then $P(R \geq 45) = P(Z \geq 0) = 0.5000$ (software gives 0.5688 using the binomial distribution).

13.32: (a) X, the number of positive tests, has a binomial distribution with $n = 1000$ and $p = 0.004$. (b) $\mu = np = (1000)(0.004) = 4$. (c) To use the Normal approximation, we need np and $n(1 - p)$ to both be bigger than 10, and as we saw in (b), $np = 4$.

13.33: (a) Of 965,000 total vehicles in these top five name-plates, Impalas accounted for proportion $140,000/965,000 = 0.1451$. (b) If I is the number of Impala buyers in the 1000 surveyed buyers, then I has the binomial distribution with $n = 1000$, and $p = 0.1451$. Hence, $\mu = np = (1000)(0.1451) = 145.1$ and $\sigma = \sqrt{np(1 - p)} = \sqrt{1000(0.1451)(1 - 0.1451)} = 11.138$ Impala buyers. (c) $P(I > 150) = P(I \geq 151) = P(Z \geq 0.53) = 0.2981$.

13.34: Let D be the number of members that drop out in the first 4 weeks. Then D has the binomial distribution with $n = 300$ and $p = 0.25$, assuming customer results are independent. Let $S = 300 - D$ be the number still enrolled after 4 weeks. Then S has the binomial distribution with $n = 300$ and $p = 0.75$. (a) For D, $\mu = np = (300)(0.25) = 75$ and $\sigma = \sqrt{300(0.25)(1 - 0.25)} = 7.5$. (b) We use the Normal approximation to the distribution of S because $np = 225$ and $n(1 - p) = 75$ are both larger than 10. Now, $P(S \geq 210) = P(Z \geq -2) = 0.9772$.

13.35: (a) With $n = 100$, the mean and standard deviation are $\mu = 75$ and $\sigma = 4.3301$, so $P(70 \leq X \leq 80) = P(-1.15 \leq Z \leq 1.15) = 0.7498$ (software gives 0.7518). (b) With $n = 250$, we have $\mu = 187.5$ and $\sigma = 6.8465$, and a score between 70% and 80% means 175 to 200 correct answers, so $P(175 \leq X \leq 200) = P(-1.83 \leq Z \leq 1.83) = 0.9328$ (software gives 0.9428).

13.36: We have $\mu = 5000$ and $\sigma = 50$ heads, so using the Normal approximation, we compute $P(X \geq 5067$ or $X \leq 4933) = 2P(Z \geq 1.34) = 0.1802$. This is not unreasonable behavior for a fair coin.

13.37: (a) Answers will vary. (b) Each time we choose a sample of size 10, the probability that we have exactly 1 bad tomato is 0.3854; therefore, out of 20 samples, the number of times that we have exactly 1 bad tomato has a binomial distribution with parameters $n = 20$ and $p = 0.3854$.

13.38: The number N of new infections is binomial with $n = 20$ and $p = 0.80$ (for unvaccinated children) or 0.05 (for vaccinated children). (a) For vaccinated children, the mean is $(20)(0.05) = 1$, and $P(N \leq 2) = 0.9245$. (b) For unvaccinated children, the mean is $(20)(0.80) = 16$, and $P(N \geq 18) = 0.2061$.

13.39: The number N of infections among untreated BJU students is binomial with $n = 1400$ and $p = 0.80$, so the mean is 1120, and the standard deviation is 14.9666 students. 75% of that group is 1050, and the Normal approximation is safe: $P(N \geq 1050) = P(Z \geq \frac{1050 - 1120}{14.9666}) = P(Z \geq -4.68)$, which is very near 1.

13.40: Let V and U be (respectively) the number of new infections among the vaccinated and unvaccinated children.

(a) V is binomial with $n = 17$ and $p = 0.05$, with mean 0.85 infections. (b) U is binomial with $n = 3$ and $p = 0.80$, with mean 2.4 infections. (c) The overall mean is 3.25 infections.

13.41: Let V and U be (respectively) the number of new infections among the vaccinated and unvaccinated children. (a) $P(V = 1) = 0.3741$ and $P(U = 1) = 0.0960$. Because these events are independent, $P(V = 1$ and $U = 1) = P(V = 1)P(U = 1) = 0.0359$. (b) $P(2$ infections$) = P(V = 0$ and $U = 2) + P(V = 1$ and $U = 1) + P(V = 2$ and $U = 0) = P(V = 0)P(U = 2) + P(V = 1)P(U = 1) + P(V = 2)P(U = 0) = 0.1977$.

13.42: (a) and (b) The unit square and circle are shown; the intersection A is shaded. (c) The circle has area π, and A is a quarter of the circle, so the area of A is $\pi/4$. This is the probability that a randomly selected point (X, Y) falls in A, so T is binomial with $n = 2000$ and $p = \pi/4 = 0.7854$. (d) The mean of T is $np = (2000)(\pi/4) = 500\pi = 1570.7963$, and the standard deviation is 18.3602. (e) Because the mean of T is 500π, $T/500$ is an estimate of π.

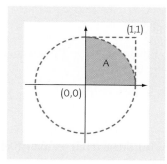

13.43: The number X of fairways Rory hits is binomial with $n = 24$ and $p = 0.57$. (a) $np = 13.68$ and $n(1 - p) = 10.32$, so the Normal approximation is (barely) safe. (b) The mean is $np = 13.68$, and the standard deviation is $\sqrt{24(0.57)(0.43)} = 2.4254$. Using the Normal approximation, $P(X \geq 17) = P(Z \geq 1.37) = 0.0853$. (c) With the continuity correction, $P(X \geq 17) = P(X \geq 16.5) = P(Z \geq 1.16) = 0.1230$. The answer using the continuity correction is closer to the exact answer (0.1215).

Chapter 14   Probability: Part III Review

14.1: (d)

14.2: (a) $S =$ {Male, Female}. (b) $S = \{6, 7, 8, \ldots, 19, 20\}$. (c) $S =$ {All values $2.5 \leq$ VOP ≤ 6.1 liters per minute}. (d) $S =$ {all heart rates such that heart rate > 0}.

14.3: (a) $1 - 0.66 - 0.21 - 0.07 - 0.04 = 0.02$.

14.4: (b) $1 - 0.66 = 0.34$.

14.5: $Y > 1$, or $Y \geq 2$. $P(Y \geq 2) = 1 - 0.26 = 0.74$.

14.6: $P(2 < Y \leq 4) = P(3 \leq Y \leq 4) = 0.16 + 0.15 = 0.31$.

14.7: (d) $1 - 0.33 = 0.67$.

14.8: All of the probabilities are between 0 and 1 (inclusive), and they sum to 1.

14.9: $P(X \leq 2)$ is the probability of women giving birth to 2 or fewer children during their childbearing years. $P(X \leq 2) = 0.193 + 0.174 + 0.344 = 0.711$.

14.10: (c) $P(X < 2) = P(X \leq 1) = 0.193 + 0.174 = 0.367$.

14.11: $P(X \geq 3) = 0.181 + 0.074 + 0.034 = 0.289$.

14.12: (b)

14.13: (a) The height of the density curve is $1/5 = 0.2$ because the area under the density function must be 1.

14.14: $P(1 \leq Y \leq 3) = 2/5 = 0.4$.

14.15: (b) This is a personal probability.

14.16: (b) 0.3707 because this is $P(Z > 1/3)$, or $P(Z > 0.33)$.

14.17: (a) $11{,}479/14{,}099 = 0.8142$. (b) $6457/(6457 + 1818) = 0.7803$.

14.18: (a) F and G.

14.19: (a) $1 - P(\text{failure}) = 1 - P(\text{both components fail}) = 1 - (0.20)(0.03) = 0.994$.

14.20: (b) $35\% - 15\% = 20\%$.

14.21: (c) $\binom{5}{3} = \binom{5}{2} = 10$.

14.22: (c) This is $P(\text{MMHHH}) = (0.6)^2(0.4)^3$, where M = miss and H = hit.

14.23: (a) This is a binomial distribution; $P(X = 3) = \binom{5}{3}(0.4)^3(0.6)^2 = 0.230$.

14.24: (c) This is a binomial random variable. $\mu = np = 1000*0.63 = 630$, $\sigma = \sqrt{np(1-p)} = 15.268$.

14.25: (a) This is a binomial distribution; $\binom{12}{8}(0.50)^8 (1 - 0.50)^4 = 0.1208$. (b) If $n = 500$ and $p = 0.50$, then $np = 250$ and $n(1 - p) = 250$, both of which are at least 10. The Normal approximation is reasonable. With mean $np = 250$ and standard deviation $\sqrt{np(1 - p)} = 11.18$, $P(X \geq 235) = P(Z \geq \frac{235 - 250}{11.18}) = P(Z \geq -1.34) = 0.9099$.

14.26: There are many possible answers; the key is that events A and B must be able to occur together.

14.27: (a) All probabilities are between 0 and 1, and their sum is 1. (b) Let R1 be Taster 1's rating and R2 be Taster 2's rating. $P(R1 = R2) = 0.03 + 0.08 + 0.25 + 0.20 + 0.06 = 0.62$. (c) $P(R1 > R2) = 0.19$. $P(R2 > R1) = 0.19$.

14.28: $P(A) = P(B) = \cdots = P(F) = 0.72/6 = 0.12$ and $P(1) = P(2) = \cdots = P(8) = (1 - 0.722)/8 = 0.035$.

14.29: (a) $n = 20$, $p = 0.25$. (b) $\mu = np = 5$. (c) $P(X = 5) = \binom{20}{5}(0.25)^5(0.75)^{15} = 0.2023$.

14.30: Approximately $N(1200, 15.492)$. $\mu = np = (1500)(0.80)$ and $\sigma = \sqrt{np(1 - p)} = \sqrt{1500(0.80)(0.20)} = 15.492$.

14.31: (a) See tree diagram following. (b) $P(\text{Positive test}) = (0.01)(0.9985) + (0.99)(0.0060) = 0.0159$.

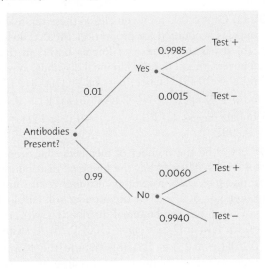

14.32: $P(\text{antibody} \mid \text{positive}) = P(\text{antibody and positive})/P(\text{positive}) = (0.9985)(0.01)/(0.0159) = 0.6280$.

14.33: $P(\text{home school} \mid \text{not regular public school}) = 0.006/(1 - 0.781) = 0.0274$.

Chapter 15 Sampling Distribution for a Proportion

15.1: Both are statistics; they were derived from the 11 patients.

15.2: 41% and 35% are parameters; 34% is a statistic.

15.3: Both are statistics.

15.4: (a) The population of interest is Americans aged 18 to 30 years old; p is the proportion of Americans 18 to 30 years old who pray at least once a week. (b) $\hat{p} = \frac{225}{377} = 0.5968$.

15.5: Answers will vary due to randomness.

15.6: (a) $p = 4/10 = 0.4$. (b) The first five digits of line 116 are 14459. We can't select student 4 twice, so the sample is students 1, 4, 5, and 9. These were F, F, M, and M. $\hat{p} = 2/4 = 0.5$. (c) The median of this sampling distribution is 0.5, and the mean is 0.45.

15.7: (a) "Unbiased" means that the average of many (actually, infinitely many) such sample proportions will be the actual population proportion. (b) Larger samples are more trustworthy because $\hat{p}$ has a smaller standard deviation; the results of such samples are closer to "truth."

15.8: (a) Because $np = (1500)(0.9) = 1350$ and $n(1 - p) = (1500)(1 - 0.9) = 150$, the sampling distribution of $\hat{p}$ will be approximately Normal with mean $\mu = 0.9$ and standard deviation $\sigma = \sqrt{\frac{(0.9)(0.1)}{1500}} = 0.00775$. $P(0.87 < \hat{p} < 0.93) = P(\frac{0.87 - 0.9}{0.00775} < Z < \frac{0.93 - 0.9}{0.00775}) = P(-3.87 < Z < 3.87) = 1 - 0 \approx 1$ (technology gives 0.9999). (b) For a sample of $n = 6000$, the mean of the sampling distribution is still $\mu = 0.9$, but the standard deviation will be $\sigma = \sqrt{\frac{(0.9)(0.1)}{6000}} = 0.00387$. $P(0.87 < \hat{p} < 0.93) = P(\frac{0.87 - 0.9}{0.00387} < Z < \frac{0.93 - 0.9}{0.00387}) = P(-7.75 < Z < 7.75) = 1 - 0 \approx 1$ (technology gives 1 as well).

15.9: (a) $\sigma = \sqrt{\frac{(0.24)(0.76)}{500}} = 0.0191$. (b) $n \geq (\frac{\sqrt{(0.24)(0.76)}}{0.01})^2 = \frac{(0.24)(0.76)}{(0.01)^2} = 1824$.

15.10: (b) 7.8% came from the sample of 60,000 households.

15.11: (c) 61.4% pertains to all registered voters.

15.12: (c) $\hat{p} = \frac{2140}{3000} = 0.7133$.

15.13: (a) This is the definition of "unbiased."

15.14: (b) The mean of the sampling distribution of $\hat{p}$ is p.

15.15: (b) $\sigma = \sqrt{\frac{(0.6)(0.4)}{117}} = 0.04529$.

15.16: (c) Because np and $n(1 - p)$ are both more than 10, $\hat{p}$ will be approximately Normal with $\mu = 0.08$ and $\sigma = \sqrt{\frac{(0.08)(0.92)}{500}} = 0.0121$.

15.17: (b) This is $P(Z > \frac{0.09 - 0.08}{0.0121}) = P(Z > 0.83) = 1 - 0.7967$.

15.18: (a) The population is surgical patients. p is the proportion of all surgical patients who will test positive for *Staphylococcus aureus*. (b) $\hat{p} = \frac{1251}{6771} = 0.185$.

15.19: (a) $\mu = 0.8$. $\sigma = \sqrt{\frac{(0.8)(0.2)}{1200}} = 0.0115$. (b) $P(0.78 < \hat{p} < 0.82) = P(\frac{0.78 - 0.8}{0.0115} < Z < \frac{0.82 - 0.8}{0.0115}) = P(-1.74 < Z < 1.74) = 0.9591 - 0.0409 = 0.9182$ (technology gives 0.9167).

15.20: Both are statistics; they came from the two classes of students.

15.21: (a) σ must be no larger than $0.02/3 = 0.00667$. (b) $n \geq (\frac{\sqrt{(0.8)(0.2)}}{0.00667})^2 = \frac{(0.8)(0.2)}{(0.00667)^2} = 3596.403$, or 3597.

15.22: In the long run, the gambler will lose money; he expects to win less than half the time (he'd need to expect to win at least half the time to stay even or come out ahead).

15.23: (a) The population is all voters in the district. (b) The sample is the 800 returned surveys. (c) $P(\hat{p} < 0.49$ or $\hat{p} > 0.59) = 2P(\hat{p} > 0.59) = 2P(Z > \frac{0.59 - 0.54}{0.0176}) = 2P(Z > 2.84) = 0.0045$.

15.24: (a) $\hat{p} = \frac{528}{800} = 0.66$. This is extremely unlikely to happen by chance variation. Exercise 15.23 found that the probability of differing from the 54% population proportion by more than 5% was 0.0045; this difference is 12%. More likely, the female head-of-household filled out the survey, speaking for all. (b) If women are more likely to support the bill, the bias would be in a positive direction.

15.25: The sample proportion will have $\mu = 0.67$ and $\sigma = \sqrt{\frac{(0.67)(0.33)}{1200}} = 0.01357$. Both $(1200)(0.67) = 804$ and $(1200)(0.33) = 396$ are more than 10, so the sampling distribution of $\hat{p}$ is approximately Normal. $P(\hat{p} > 0.70) = P(Z > \frac{0.70 - 0.67}{0.01357}) = P(Z > 2.21) = 0.0136$. There is only about a 1.3% chance that more than 70% of online shoppers in the sample will choose Economy Ground shipping, if the true proportion who would select that method is 67%.

15.26: (a) 51% is a statistic from the sample of 800 voters; 49.1% is a parameter, based on all Floridians who voted. (b) $\mu = 0.491$ and $\sigma = \sqrt{\frac{(0.491)(0.509)}{800}} = 0.01767$. (c) $\hat{p}$ is approximately Normal because $(800)(0.491) = 392.8$ and $(800)(0.509) = 407.2$ are both more than 10. $P(\hat{p} \geq 0.51) = P(Z \geq \frac{0.51 - 0.491}{0.01767}) = P(Z \geq 1.08) = 0.1401$ (technology gives 0.1412). There was a 14% chance of the sample survey result, if Romney's support was actually 49.1%.

15.27: (a) $\mu = 0.83$ and $\sigma = \sqrt{\frac{(0.83)(0.17)}{500}} = 0.01680$. (b) $P(0.80 < \hat{p} < 0.86) = P(\frac{0.80 - 0.83}{0.0168} < Z < \frac{0.86 - 0.83}{0.0168}) = P(-1.79 < Z < 1.79) = 0.9266$ (technology gives 0.9259).

15.28: $\mu = 0.25$ and $\sigma = \sqrt{\frac{(0.25)(0.75)}{500}} = 0.01936$. $P(\hat{p} < 0.30) = P(Z < \frac{0.30 - 0.25}{0.01936}) = P(Z < 2.58) = 0.9951$.

15.29: (a) The standard deviation of $\hat{p}$ must be no larger than $0.03/2 = 0.015$. (b) $n \geq (\frac{\sqrt{(0.83)(0.17)}}{0.015})^2 = \frac{(0.83)(0.17)}{(0.015)^2} = 627.111$; we must have a sample of at least $n = 628$.

15.30: (a) $\hat{p} = \frac{42}{300} = 0.14$. (b) We need $P(\hat{p} < 0.14)$ if the population proportion is 0.20. The distribution of the sample proportion will be approximately $N(0.2, \sqrt{\frac{(0.2)(0.8)}{300}} = 0.0231)$. $P(\hat{p} < 0.14) = P(Z < -2.60) = 0.0047$. (c) An alternative explanation is that there really are less than 20% white student cars on this campus.

Chapter 16   Confidence Intervals: The Basics

16.1: (a) Yes. There were $(0.52)(1017) = 529$ people who thought government should not favor any particular set of values and $(0.48)(1017) = 488$ who thought differently. (b) The 95% confidence interval is $0.52 \pm 1.96\sqrt{\frac{(0.52)(0.48)}{1017}}$, or 0.4893 to 0.5507.

16.2: $(0.84)(1000) = 840$ and $(1 - 0.84)(1000) = 160$, so the sample size condition is satisfied. The 95% confidence interval is $0.84 \pm 1.96\sqrt{\frac{(0.84)(0.16)}{1000}} = 0.8173$ to 0.8627.

16.3: (a) Answers will vary. (b) Once again, answers will vary. (c) By the time you have 500 SRSs, we'll expect that almost all students should have between 74.6% and 85.4% of the 80% intervals containing the true proportion. Similarly, for 1000 intervals, students should report between 76.2% and 83.8% of the intervals containing the true proportion.

16.4: (a) The 95% confidence interval will be $54\% \pm 4\%$, or 50% to 58%. (b) "95% confidence" means that in many repeated samples, intervals constructed in this manner contain the true proportion about 95% of the time.

16.5: $z^* = 1.44$. If using Table A, note that with 85% of the area in the middle of the distribution, there is 7.5% (0.0750) in either tail.

16.6: (a) The survey excludes residents of the northern territories, as well as those who have no phones or have only cell phone service. (b) Of the 1505 respondents, 1288 agreed with the statement and $(1505 - 1288) = 217$ disagreed or (presumably) were neutral; both of these counts are larger than 10, so the sample is large enough. $\hat{p} = \frac{1288}{1505} = 0.8558$ so $SE_{\hat{p}} = \sqrt{\frac{\hat{p}(1-\hat{p})}{n}} = 0.009055$; the 95% confidence interval is $0.8558 \pm (1.96)(0.009055) = 0.8381$ to 0.8736.

16.7: STATE: What proportion of weightlifting injuries in the 8 to 30 age group are accidental? PLAN: We will construct a 90% confidence interval for p, the proportion of all weightlifting injuries in the 18 to 30 age group that are accidental. SOLVE: We are told that the sample is random and will assume that the sample is close to an SRS. Because both the number of successes (1552) and failures $(4111 - 1552 = 2559)$ are much greater than 10, we may assume that the sampling distribution of $\hat{p}$ is approximately Normal. Here, $\hat{p} = \frac{1552}{4111} = 0.3775$ and $SE_{\hat{p}} = \sqrt{\frac{\hat{p}(1-\hat{p})}{4111}} = 0.0076$. Hence, a 90% confidence interval for p is given by $0.3775 \pm 1.645(0.0076) = 0.365$ to 0.390. CONCLUDE: With 90% confidence, the proportion of all weightlifting injuries in the 18 to 30 age group that are accidental is between 0.365 and 0.390 (36.5% to 39.0%).

16.8: $np = (0.002)(2673) = 5.346$. This is less than 10.

16.9: (a) The difference in the three intervals is the value of z^*. All have $\hat{p} = 0.39$ and $SE_{\hat{p}} = \sqrt{\frac{0.39(1 - 0.39)}{808}} = 0.0172$. The results are shown in the table below. (b) The margins of error are given in the table. Note that as the confidence level increases, the margin of error becomes larger.

CONFIDENCE LEVEL	z^*	INTERVAL	MARGIN OF ERROR
90%	1.645	0.3617 to 0.4183	0.0283
95%	1.960	0.3563 to 0.4237	0.0337
99%	2.576	0.3457 to 0.4343	0.0443

16.10: (a) With $n = 100$, the margin of error for a 95% confidence interval would be $1.96\sqrt{\frac{0.39(1 - 0.39)}{100}} = 0.0956$. (b) For $n = 400$, the margin of error is $1.96\sqrt{\frac{0.39(1 - 0.39)}{400}} = 0.0478$. Similarly, for $n = 1600$, the margin of error will be 0.0239. (c) When the sample size is quadrupled, the margin of error is cut in half.

16.11: (a) The 90% confidence interval is $0.84 \pm 1.645\sqrt{\frac{0.84(1 - 0.84)}{1000}} = 0.8209$ to 0.8591. (b) The margin of error is 0.0191. The 95% confidence interval computed in Exercise 16.2 was 0.8173 to 0.8627; this had a margin of error $0.8627 - 0.84 = 0.0227$. The 95% interval has a larger margin of error than a 90% interval. (c) With $n = 100$, we would have had 84 who believed in Santa and 16 who did not (both are at least 10). The confidence interval would be $0.84 \pm 1.96\sqrt{\frac{0.84(1 - 0.84)}{100}} = 0.7681$ to 0.9119. (d) Decreasing the sample size creates a wider interval.

16.12: (a) $\hat{p} = \frac{221}{270} = 0.8185$ and $SE_{\hat{p}} = \sqrt{\frac{\hat{p}(1-\hat{p})}{270}} = 0.02346$. Hence, the margin of error for 95% confidence is $(1.96)(0.02346) = 0.0460$ (4.6%). (b) For a ± 0.03 margin of error, we need $n = \left(\frac{1.96}{0.03}\right)^2 p^*(1 - p^*)$. Using the previous study as a pilot study, replace p^* with $\hat{p} = 0.8185$, yielding $n = 634.1$; we need at least $n = 635$.

16.13: $n = \left(\frac{z^*}{m}\right)^2 p^*(1 - p^*) = \left(\frac{1.645}{0.04}\right)^2 (0.75)(1 - 0.75) = 317.1$, so use $n = 318$.

16.14: (a) The sample size condition is met because $(0.42)(494) = 207$ and $(1 - 0.42)(494) = 287$; both the number of "successes" and "failures" are larger than 10. The 95% confidence interval is $0.42 \pm 1.96\sqrt{\frac{0.42(1 - 0.42)}{494}} = 0.3765$ to 0.4635. (b) Women may not want to admit that they are "very" or "somewhat" overweight.

16.15: (a) Missing people without landline telephones is not an error covered in the interval (undercoverage). (b) Nonresponse is not covered in the interval. (c) Random chance variation is the only error covered by the interval.

16.16: (a) Among the 14 observations, we have 11 successes and 3 failures. The number of successes and failures should both be at least 10 for the Normal approximation to be valid. (b) We add 4 observations: 2 successes and 2 failures. We now have 18 total observations with 13 successes and 5 failures. $\tilde{p} = \frac{11 + 2}{14 + 4} = \frac{13}{18} = 0.7222$. (c) Using the plus four method, $SE_{\tilde{p}} = \sqrt{\frac{\tilde{p}(1-\tilde{p})}{n + 4}} = \sqrt{\frac{0.7222(1 - 0.7222)}{18}} = 0.1056$. Hence, a 90% confidence interval for p is $0.7222 \pm 1.645(0.1056) = 0.5485$ to 0.8959. The confidence interval is quite wide (even with only 90% confidence) because the sample size is so small.

16.17: (a) The large-sample conditions are met because we had $113 > 10$ people who have experienced computer crime and $1025 - 113 = 912 > 10$ people who have not. $\hat{p} = \frac{113}{1025} = 0.1102$, and $SE_{\hat{p}} = \sqrt{\frac{\hat{p}(1-\hat{p})}{1025}} = 0.0098$. A 95% large-sample confidence interval for p is given by $0.1102 \pm 1.96(0.0098) = 0.0910$ to 0.1294, or 9.1% to 12.9%. (b) Using the plus four method, $\tilde{p} = \frac{113 + 2}{1025 + 4} = 0.1118$, and $SE_{\tilde{p}} = \sqrt{\frac{\tilde{p}(1-\tilde{p})}{1029}} = 0.0098$. A 95% confidence interval for p is given by $0.1118 \pm 1.96(0.0098) = 0.0926$ to 0.1310, or 9.3% to 13.1%. These intervals are virtually identical, but the plus four confidence interval is very slightly shifted to the right. This shift is very small because adding 2 successes and 2 failures to an already large sample size (1025) has little impact.

16.18: (a) The sample proportion is $\hat{p} = \frac{20}{20} = 1$, so $SE_{\hat{p}} = \sqrt{\frac{\hat{p}(1-\hat{p})}{n}} = 0$. The margin of error would therefore be 0 (regardless of the confidence level), so large-sample methods give the useless interval 1 to 1. (b) The plus four estimate is $\tilde{p} = \frac{20 + 2}{20 + 4} = 0.9167$, and $SE_{\tilde{p}} = \sqrt{\frac{\tilde{p}(1-\tilde{p})}{24}} = 0.0564$. A 95% confidence interval for p is then $0.9167 \pm 1.96(0.0564) = 0.8062$ to 1.0272. Because proportions can't exceed 1, we say that a 95% confidence interval for p is 0.8061 to 1.

16.19: (c) $z^* = 1.96$ is for 95% confidence and $z^* = 2.576$ is for 99%.

16.20: (a) $\hat{p} = \frac{70}{117} = 0.60$. The 95% confidence interval is $0.60 \pm 1.96\sqrt{\frac{0.60(1 - 0.60)}{117}}$.

16.21: (b) The value of z^* for 99% confidence is 2.576; for 95%, $z^* = 1.96$.

16.22: (b) Larger samples have smaller margins of error because n is in the denominator of the standard error.

16.23: (b) The 90% confidence interval is $0.80 \pm 1.645\sqrt{\frac{0.80(1 - 0.80)}{4500}}$.

16.24: (a) Less confidence means a smaller margin of error.

16.25: (b) Smaller samples have larger margins of error.

16.26: (c) $n = \left(\frac{z^*}{m}\right)^2 p^*(1 - p^*) = \left(\frac{2.576}{0.02}\right)^2(0.5)(0.5) = 4147.36$, so we would need at least 4148.

16.27: (c) There were only 8 who said it was very likely that they would seek a job out of state.

16.28: (a) This is what 95% confidence means.

16.29: (b) The margin of error for the 95% confidence interval is $1.96\sqrt{\frac{0.53(1 - 0.53)}{100}}$.

16.30: (a) The margin of error only covers random sampling variation.

16.31: (a) The conditions were met, and we have an SRS with 2012 who would ban texting and $2211 - 2012 = 199$ who would not. The sample proportion of those would ban texting while driving is $\hat{p} = \frac{2012}{2211} = 0.910$. The 95% confidence interval is $0.91 \pm 1.96\sqrt{\frac{0.91(1 - 0.91)}{2211}}$, or 0.8981 to 0.9219. (b) The individuals who have agreed to participate in the Harris Interactive Surveys do not represent an SRS from all adult Americans.

16.32: We have 19 successes and $172 - 19 = 153$ failures. Both are large enough so that conditions for large-sample confidence interval use are met. We estimate that $\hat{p} = \frac{19}{172} = 0.1105$, $SE_{\hat{p}} = 0.02391$, the margin of error is $1.96SE_{\hat{p}} = 0.04685$, and the 95% confidence interval is 0.0636 to 0.1573.

16.33: The sample proportion of those who would start smoking again is $\hat{p} = \frac{17}{166} = 0.1024$. The 95% confidence interval for the proportion of smokers who would start smoking again is $0.1024 \pm 1.96\sqrt{\frac{0.1024(1 - 0.1024)}{166}}$, or 0.0563 to 0.1485.

16.34: (a) The actual margin of error for 95% confidence is $1.96\sqrt{\frac{0.1024(1 - 0.1024)}{166}} = 4.6\%$. (b) If the sample proportion were 0.5, the margin of error would be 7.6%. (c) 10% is conservative; it is larger than either of these margins of error.

16.35: No. Once the sample has been taken and a confidence interval computed, nothing is random, so we cannot talk about probability.

16.36: No. We have no idea what future samples might show (conditions may very well change).

16.37: The mistake is in saying that 95% of other polls would have results close to the results of this poll. Other surveys should be close to the truth—not necessarily close to the results of this survey.

16.38: To construct a large-sample confidence interval, we require at least 10 successes (swimming areas with unsafe levels of fecal coliform) and at least 10 failures (swimming areas with safe levels of fecal coliform). Here, we have 13 successes and 7 failures.

16.39: (a) The survey excludes residents of Alaska and Hawaii and those who do not have cell phone service. (b) We have 422 successes and 2063 failures, so the sample is large enough to use large-sample inference procedure. We have $\hat{p} = \frac{422}{2485} = 0.1698$, and $SE_{\hat{p}} = 0.0075$. For 90% confidence, the margin of error is $1.645SE_{\hat{p}} = 0.0124$, and the confidence interval is 0.1574 to 0.1822, or 15.7% to 18.2%. (c) Answers will vary.

16.40: (a) The large-sample methods are safe because we have 880 trials, with 171 successes and $880 - 171 = 709$ failures. For the large-sample interval $\hat{p} = \frac{171}{880} = 0.1943$, $SE_{\hat{p}} = 0.01334$, the margin of error is $1.96SE_{\hat{p}} = 0.02614$, and the 95% confidence interval is 0.1682 to 0.2204. (b) It is likely that more than 171 respondents have run red lights.

16.41: (a) $n \geq \left(\frac{2.576}{0.015}\right)^2(0.2)(1 - 0.2) = 4718.77$; they will need a sample of at least $n = 4719$ people. (b) $E = 2.576\sqrt{\frac{(0.1)(0.9)}{4719}} = 0.0112$.

16.42: (a) PLAN: We construct a 99% confidence interval for the proportion of all 17-year-old students in 2008 who had at least one parent graduate from college. SOLVE: We are told to treat this sample as an SRS of 17-year-olds still in school. We are told that $\hat{p} = 0.46$, so we have $9600(0.46) = 4416$ students with a parent that graduated from college and 5184 who did not have a parent graduate from college. The number of successes and failures is very large (more than the required 10). We have $SE_{\hat{p}} = \sqrt{\frac{\hat{p}(1 - \hat{p})}{9600}} = 0.0051$, so the margin of error for 99% confidence is $2.576(0.0051) = 0.0131$, and a 99% confidence interval for the proportion is 0.4469 to 0.4731, or 44.69% to 47.31%. CONCLUDE: With 99% confidence, the proportion of 17-year-old students still in school with at least one parent that graduated from college is between about 0.447 and 0.473. (b) Answers will vary.

16.43: PLAN: We will give a 95% confidence interval for p, the proportion of American adults who think that humans developed from earlier species of animals. SOLVE: 594 answered "True," and $1484 - 594 = 890$ did not, so large-sample methods can be used. We have $\hat{p} = \frac{594}{1484} = 0.4003$, $SE_{\hat{p}} = 0.01272$, margin of error $1.96SE_{\hat{p}} = 0.02493$, and the 95% confidence interval is 0.3754 to 0.4252. CONCLUDE: We are 95% confident that the percent of American adults thinking that humans developed from earlier species of animals is between about 37.5% and 42.5%.

16.44: (a) PLAN: We will give a 95% confidence interval for p, the proportion of subjects who would select the first choice presented. SOLVE: We assume we have an SRS from the population. With 32 subjects, we have 22 successes (people that picked the first wine) and 10 failures (people that picked the second wine). With 10 failures, we can use either the large-sample or plus four. For the large-sample method, $\hat{p} = \frac{22}{32} = 0.6875$, $SE_{\hat{p}} = \sqrt{\frac{(0.6875)(1 - 0.6875)}{3632}} = 0.0819$. The large-sample method yields a confidence interval of 52.7% to 84.8%. If we use the plus-four method $\hat{p} = \frac{22 + 2}{32 + 4} = 0.6667$, $SE_{\hat{p}} = \sqrt{\frac{(0.6667)(1 - 0.6667)}{36}} = 0.0786$, and the 95% confidence interval is 0.5127 and 0.8207, or 51.27% and 82.07%. CONCLUDE: We are 95% confident that the proportion of subjects that would select the first wine is between about 52.7% and 84.8%. The exact interval depends on which method was used. (b) Answers will vary.

16.45: PLAN: We will give a 95% confidence interval for p, the proportion of Chick-fil-A orders correctly filled. SOLVE: We will assume that the 196 visits constitute a random sample of all possible visits. In our sample, we have 182 successes (correctly filled orders) and 14 failures (incorrectly filled orders). We have $\hat{p} = \frac{182}{196} = 0.9286$, $SE_{\hat{p}} = \sqrt{\frac{(0.9286)(1 - 0.9286)}{196}} = 0.0184$, margin of error $1.96SE_{\hat{p}} = 0.0360$, and the 95% confidence interval is 0.893 to 0.965. CONCLUDE: We are 95% confident the proportion of orders filled correctly by Chick-fil-A is between 0.893 and 0.965, or 89.3% to 96.5%.

16.46: PLAN: We obtain the sample size required to estimate the proportion of wine tasters that select the first choice to within ±0.05 with 95% confidence. SOLVE: In Exercise 16.44, $\hat{p} = 0.6875$. Hence, $n = \left(\frac{z^*}{m}\right)^2 p^*(1 - p^*) = \left(\frac{1.96}{0.05}\right)^2 (0.6875)(1 - 0.6875) = 330.14$; take $n = 331$. CONCLUDE: We require a sample of at least 331 wine tasters to estimate the proportion that would choose the first option to within 0.05 with 95% confidence.

16.47: Regardless of the level of confidence (the 95% confidence level has nothing to do with it), larger samples reduce margins of error (because the sample size, n, is in the denominator of the standard error).

16.48: The plus-four proportion is $\tilde{p} = \frac{13 + 2}{20 + 4} = 0.625$, and $SE_{\tilde{p}} = \sqrt{\frac{\tilde{p}(1 - \tilde{p})}{24}} = 0.0988$. The margin of error for 90% confidence is $1.645(0.0988) = 0.1626$, and the 90% confidence interval for the proportion of swimming areas with unsafe coliform levels is 0.4624 to 0.7879, or 46.2% to 78.8%.

16.49: PLAN: We will give a 90% confidence interval for the proportion of all *Krameria cytisoides* shrubs that will resprout after fire. SOLVE: We assume that the 12 shrubs in the sample can be treated as an SRS. Because the number of resprouting shrubs is just 5, the conditions for a large sample interval are not met. Using the plus-four method $\tilde{p} = \frac{5 + 2}{12 + 4} = 0.4375$, $SE_{\tilde{p}} = 0.1240$, the margin of error is $1.645SE_{\tilde{p}} = 0.2040$, and the 90% confidence interval is 0.2335 to 0.6415. CONCLUDE: We are 90% confident that the proportion of

Krameria cytisoides shrubs that will resprout after fire is between about 0.23 and 0.64.

16.50: (a) There were 277 who pray at least once a week and $411 - 277 = 134$ who do not; both of these counts are larger than 10, and we were told to assume an SRS. The sample proportion of those who pray at least once a week is $\hat{p} = \frac{277}{411} = 0.6740$. The margin of error is $2.576SE_{\hat{p}} = 0.0596$. The 99% large sample interval is 0.6740 ± 0.0596, or about 61.4% to 73.4%. (b) With the plus-four method, we have $\tilde{p} = \frac{277 + 2}{411 + 4} = 0.6723$ with a margin of error 0.0594, or about 61.3% to 73.2%. The plus-four interval is just slightly shifted toward 0.5.

Chapter 17 Tests of Significance: The Basics

17.1: (a) The sample is large. If the claim is true, $\sigma_{\hat{p}} = \sqrt{\frac{0.5(1 - 0.5)}{1800}} = 0.0118$. The sampling distribution will be approximately Normal (0.5, 0.0118). (b) $\hat{p} = \frac{963}{1800} = 0.535$. This result is almost three standard deviations above the mean of the sampling distribution and would be expected to rarely happen by chance; a result of $\hat{p} = \frac{918}{1800} = 0.51$ is less than one standard deviation above the mean and would be expected to occur by chance more than 16% of the time.

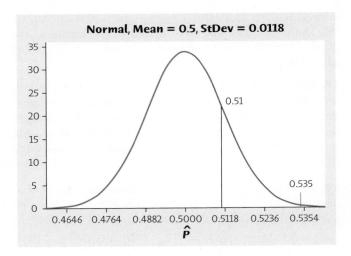

17.2: (a) If the claim is true the sample proportion will have an approximately Normal distribution with $\mu = 0.5$ and $\sigma = \sqrt{\frac{0.5(1 - 0.5)}{1038}} = 0.01552$. The conditions are met; we have a random sample (we do not know if it was a SRS), and we expect $(0.5)(1038) = 519$ each "successes" (people in favor of stricter laws on gun control) and "failures." The Normal distribution in the figure shows this sampling distribution. (b) The sample proportion $\hat{p} = 0.58$ is shown at the far right of the sampling distribution. A sample proportion of 0.51 is indicated as well by the vertical line. A sample proportion of 0.58 is far out in the right tail of this distribution (much farther than 3 standard deviations; it is unlikely to happen by chance). A sample proportion of 0.51 is less than one standard deviation above a claim of 0.5.

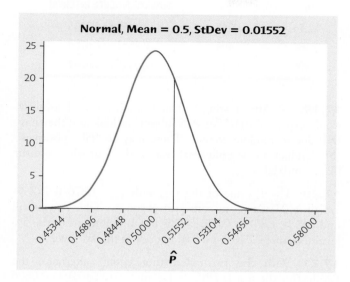

Normal, Mean = 0.5, StDev = 0.01552

17.3: H_0: $p = 0.5$ and H_a: $p > 0.5$. We chose the "greater than" (one-sided) alternative because these subjects were identified in advance as sensitive to negative emotions; as such, they should be expected to avoid the negative experience.

17.4: H_0: $p = 0.5$ and H_a: $p > 0.5$. The question is whether a majority (more than 50%) were in favor of stricter gun laws.

17.5: H_0: $p = 0.40$ and H_a: $p < 0.40$. The question of interest is whether this TA's students do worse than usual (where 40% of all students earn As and Bs).

17.6: H_0: $p = 0.723$ and H_a: $p \neq 0.723$. You want to know if your school district is *different*.

17.7: Hypotheses should be stated in terms of a parameter, not a sample value. $\hat{p}$ is the sample proportion.

17.8: (a) Approximately $N(0.5, 0.0068)$. (b) We want $P(\hat{p} \geq 0.517)$, which is equivalent to $P(Z \geq \frac{0.517 - 0.5}{0.0068}) = P(Z \geq 2.5) = 0.0062$.

17.9: (a) Approximately $N(0.5, 0.0884)$. (b) $P(\hat{p} \geq 0.6875) = P(Z \geq \frac{0.6875 - 0.5}{0.0884}) = P(Z \geq 2.12) = 0.0170$.

17.10: (a) The chance of obtaining a sample proportion $\hat{p} = \frac{15}{61} = 0.2459$ if the true proportion is $p = 0.15$ is 0.0183 (1.8%). This type of result would be expected to happen less than once in 50 trials, which is rare. (b) There were more than 10 each "successes" (5's on the AP exam) and failures. Whether we can view this particular class as a simple random sample (of all this teacher's students? of all AP Statistics students?) is questionable. (c) Answers will vary, but this was not a designed, randomized experiment.

17.11: (a) The P-value is 0.0015. This is statistically significant at both the 0.05 and 0.01 levels because $P < \alpha$. (b) If

there were only 918 correct selections, the P-value would be 0.1981. This is not significant at either the 0.05 or 0.01 levels because $P > \alpha$. (c) The first outcome is expected to happen less than twice in 1000 trials; this is rare. The second will happen almost 1/5 of the time; that is not rare.

17.12: (a) The P-value is 0.0000 (to four decimal places). Software says $P = 1.29 \times 10^{-7}$. This is statistically significant at both the 0.05 and 0.01 levels because $P < \alpha$. (b) With $\hat{p} = 0.51$ and the same $n = 1038$, we have $P = 0.2674$. This is not significant at either the 0.05 or 0.01 levels because $P > \alpha$. (c) The first sample proportion should happen less than once in a million trials (*very* rare!); the second should happen at least once in every four trials, which is not surprising at all.

17.13: STATE: We wonder if the proportion of times the Belgian euro coin spins heads is the same as the proportion of times is spins tails. PLAN: Let p be the proportion of times a spun Belgian euro coin lands heads. We test H_0: $p = 0.50$ vs. H_a: $p \neq 0.50$. SOLVE: The sample consists of 250 trials, we expect 125 "successes" (heads) and 125 "failures" (tails). Hence, the sample is large enough to use the Normal approximation to describe the sampling distribution of $\hat{p}$. We assume the sample represents an SRS of all possible coin spins. Here, $\hat{p} = \frac{140}{250} = 0.56$ $SE_{\hat{p}} = \sqrt{\frac{p_0(1 - p_0)}{n}} = \sqrt{\frac{0.50(1 - 0.50)}{250}} = 0.0316$. Hence, $z = \frac{\hat{p} - p_0}{SE} = \frac{0.56 - 0.50}{0.0316} = 1.90$, and $P = 0.0574$. CONCLUDE: We do not have sufficient evidence that the proportion of times a Belgian euro coin spins heads is different from 0.50.

17.14: STATE: We wonder if the proportion of times the "best face" wins is more than 0.50. PLAN: Let p be the proportion of times the "best face" wins. We test H_0: $p = 0.50$ vs. H_a: $p > 0.50$. SOLVE: The sample consists of 32 trials, we expect 16 "successes" (best face wins) and 16 "failures" (best face does not win). So, the sample is large enough to use the Normal approximation to describe the sampling distribution of $\hat{p}$. We assume the sample is an SRS. Here, $\hat{p} = \frac{22}{32} = 0.6875$ and $SE_{\hat{p}} = \sqrt{\frac{p_0(1 - p_0)}{n}} = \sqrt{\frac{0.50(1 - 0.50)}{32}} = 0.0884$. Hence, $z = \frac{\hat{p} - p_0}{SE} = \frac{0.6875 - 0.50}{0.0884} = 2.12$, and $P = 0.0170$. CONCLUDE: There is strong evidence that the proportion of times the "best face" wins is more than 0.50.

17.15: (a) The expected number of successes (heads) and the expected number of failures (tails) are both 5. These should be 10 or more. (b) A z test for a proportion can be used. (c) Under the null hypothesis, we expect only $200(0.01) = 2$ failures.

17.16: If H_0 is true, $\mu_{\hat{p}} = 0.5$ and $\sigma_{\hat{p}} = \sqrt{\frac{0.5(1 - 0.5)}{100}} = 0.05$. (a) With 58 of 100 scoring more than 514, $\hat{p} = 0.58$, and $z = \frac{0.58 - 0.5}{0.05} = 1.6$. The P-value for the test is 0.0548. This is not significant at the 5% level because $0.0548 > 0.05$. (b) With 59 of 100 scoring more than 514, $z = 1.8$ and $P = 0.0359$. This is significant at the 5% level because $0.0359 < 0.05$.

17.17: (a) For $n = 25$, $z = \frac{0.58 - 0.5}{0.1} = 0.8$, and $P = 0.2119$. For $n = 100$, $z = 1.6$ and $P = 0.0548$. For $n = 400$,

$z = 3.2$ and $P = 0.0007$. (b) The distribution curves are shown in the following figures.

(a)

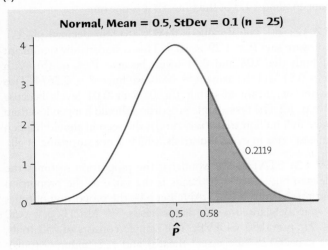

(b)

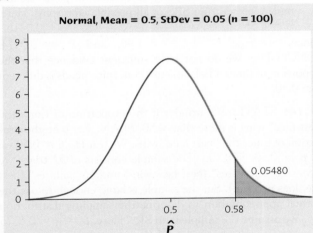

(c)

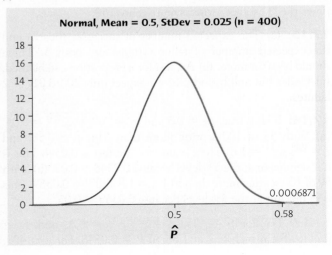

17.18: The confidence intervals are given below. Each interval was computed as $0.58 \pm 1.96 \sqrt{\frac{0.58(1 - 0.58)}{n}}$.

n	SE$_p$	95% CONFIDENCE INTERVAL
25	0.0987	0.3865 to 0.7735
100	0.0494	0.4832 to 0.6768
400	0.0247	0.5316 to 0.6284

17.19: (a) After testing 500 subjects at the 0.01 level, we would expect $(0.01)(500) = 5$ subjects to look as if they have ESP due to random chance. These may be "false positives." (b) Conduct a new (different) test on the four who were the likely candidates.

17.20: The P-value of the test with $z = 1.876$ is $P(Z > 1.876) = 0.0303$ (using software). If you use Table A, you will find $P(Z > 1.88) = 0.0301$. This test is significant at the 5% level (because $0.0303 < 0.05$), but not at the 1% level.

17.21: With the two-tailed (not equal) alternate, the P-value for the test is $2P(Z > 1.876) = 2(0.0303) = 0.0606$ (or 0.0602, using Table A). This test is not significant at either the 5% or 1% levels because $0.0606 > 0.05$.

17.22: (a) $z = \frac{0.43 - 0.5}{0.05} = -1.4$. (b) No. We would have had to have $|z| > 1.96$. (c) No. We would have had to have $|z| > 2.576$. (d) 1.4 lies between 1.282 and 1.645; this corresponds to $0.10 < P < 0.20$. This test does *not* give evidence against the null hypothesis.

17.23: (a) This is the definition of a P-value.

17.24: (b) To be significant at level α, we need $P < \alpha$.

17.25: (b) The null hypothesis always has "=" in it.

17.26: (c) The researcher thinks the noise will make the rats finish the maze faster, so more than half would finish within 18 seconds.

17.27: (c) The P-value for $z = 2.433$ is 0.0075 (assuming that the difference is in the correct direction; that is, assuming that the alternative hypothesis was H_a: $\mu > \mu_0$).

17.28: (a) $z = \frac{0.52 - 0.5}{0.0158}$.

17.29: (b) A P-value of 0.19 is not significant, so the test statistic will not be large.

17.30: (c) With a two-tailed alternate hypothesis, the test statistic can be either positive or negative and result in a rejection of the null hypothesis; $z = 2.807$ has area 0.0025 beyond it in the tail of the distribution.

17.31: (a) $z = 2.576$ has area 0.005 beyond it in the right-hand tail.

17.32: (a) The question is about "more than half."

17.33: (c) $z = \frac{\hat{p} - p_0}{\sqrt{\frac{p_0(1 - p_0)}{n}}} = \frac{0.53 - 0.50}{\sqrt{\frac{0.50(1 - 0.50)}{100}}} = 0.60.$

17.34: (a) All statistical methods are based on probability samples. We must have a random sample to apply them.

17.35: (b) Inference from a voluntary response sample is never reasonable. Online web surveys are voluntary response surveys.

17.36: (c) With only $n = 14$ women (presumably split between at least two treatments), any observed improvement may be due to the treatment, or may be due to another cause.

17.37: (a) The standard deviation of $\hat{p}$ decreases with increased sample size.

17.38: "$P = 0.03$" does *not* mean that there is a 3% chance that H_0 is true. It *does* mean that H_0 is not likely to be correct—but only in the sense that it provides a poor explanation of the data observed. It means that if H_0 is true, a sample as contrary to H_0 as our sample would occur by chance alone only 3% of the time.

17.39: If the presence of pig skulls were not an indication of wealth, then differences similar to or bigger than those observed in this study would occur at most 1% of the time by chance.

17.40: The person making the objection is confusing practical significance with statistical significance. In fact, a 5% increase isn't a lot in a pragmatic sense. However, $P = 0.03$ means that random chance does not easily explain the difference observed. That is, there does seem to be an increase in mean improvement for those that expressed their anxieties . . . but the significance test does not address whether the difference is large enough to matter. Statistical significance is not practical importance.

17.41: With a P-value of 0.50, the sample difference between the two study groups was as likely to occur as not. In other words, the observed difference was no less likely than "heads" on the toss of a fair coin.

17.42: In the sketch, the "significant at 1%" region includes only the dark shading ($z > 2.326$). The "significant at 5%" region of the sketch includes both the light and dark shading ($z > 1.645$). When a test is significant at the 1% level, it means that if the null hypothesis were true, outcomes similar to (or more extreme than) those seen are expected at most once in 100 repetitions of the study. When a test is significant at the 5% level, it means that if the null hypothesis were true, outcomes similar to (or more extreme than) those seen are expected at most five in 100 repetitions of the study. Thus, significance at the 1% level implies significance at the 5% level (or at any level higher than 1%). The converse is false; something that occurs "no more than 5 times in 100 repetitions" is not necessarily as rare as something that happens "no more than once in 100 repetitions."

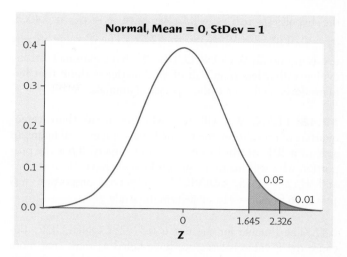

17.43: (a) The graduate student selected the alternative hypothesis after examining the data. The alternative hypothesis should be formulated before examining data and especially should not be motivated by data. (b) The correct P-value is $2P(Z > 1.9) = 2(0.0287) = 0.0574$.

17.44: Because a P-value is a probability, it can never be greater than 1. The correct P-value is $P(Z > 1.33) = 0.0918$.

17.45: STATE: We want to know if more than one-quarter (25%) of all adults in the U.S. in 2012 believed that a low-carbohydrate diet is more beneficial than a low-fat diet. PLAN: We will perform a test of the hypotheses H_0: $p = 0.25$ against H_a: $p > 0.25$. SOLVE: The sample proportion was $\hat{p} = 0.30$ from a random sample of $n = 1014$ adults nationwide. The standard deviation of $\hat{p}$ is $\sigma_{\hat{p}} = \sqrt{\frac{0.25(1 - 0.25)}{1014}} = 0.0136$. The test statistic is $z = \frac{0.30 - 0.25}{0.0136} = 3.68$. The P-value of the test is 0.0001. CONCLUDE: Because of the extremely small P-value, we have overwhelming evidence that more than 25% of American adults believed that a low-carbohydrate diet was more beneficial than a low-fat diet at the time of the survey in 2012.

17.46: PLAN: With p representing the proportion of songs loaded by Suzanne, we test H_0: $p = 0.50$ vs. H_a: $p \neq 0.50$. The test is two-sided because we wonder if the proportion loaded by Suzanne differs from that loaded by Ted. SOLVE: We assume that the 30 songs sampled are an SRS. With 30 songs sampled, we expect $30(0.50) = 15$ successes and $30(0.50) = 15$ failures, so conditions for use of the large sample z test are satisfied. We have $\hat{p} = \frac{22}{30} = 0.7333$, so $z = \frac{0.7333 - 0.50}{\sqrt{\frac{0.50(1 - 0.50)}{30}}} = 2.56$ and $P = 0.0106$. CONCLUDE: There is strong evidence that the proportion of songs downloaded by Suzanne differs from 0.50. It seems Suzanne downloaded more than Ted.

17.47: PLAN: We will test whether p, the proportion of American adults who think that humans developed from earlier species of animals is less than half of all Americans. The hypotheses are H_0: $p = 0.50$ and H_a: $p < 0.50$. SOLVE: We have an SRS with a very large sample size, so computing the hypothesis test is valid. We have $\hat{p} = \frac{594}{1484} = 0.4003$, $\sigma_{\hat{p}} = 0.0130$, $z = \frac{0.4003 - 0.5}{0.0130} = -7.67$. The P-value of the test is essentially 0. CONCLUDE: We have extremely strong evidence that less than half of all Americans think that humans developed from earlier species of animals.

17.48: PLAN: We will test whether more than three-quarters of those receiving gastric bypass surgery will maintain at least a 20% weight loss 6 years after surgery. If p is the proportion who will maintain such a loss, we have H_0: $p = 0.75$ and H_a: $p > 0.75$. SOLVE: The observed proportion was $\hat{p} = 0.76$ from $n = 418$ subjects in the study. $z = \frac{0.76 - 0.75}{\sqrt{\frac{0.75(1 - 0.75)}{418}}} = 0.47$. The P-value for the test is 0.3192. CONCLUDE: Because the P-value is large, we have failed to show that more than three-quarters of those undergoing gastric bypass surgery maintain a 20% weight loss 6 years after surgery.

17.49: PLAN: We will test whether less than half of all Americans supported Barack Obama at the time of the 2012 poll. We have H_0: $p = 0.50$ and H_a: $p < 0.50$. SOLVE: From the random sample of $n = 2551$ adults nationwide, $\hat{p} = 0.48$. $z = \frac{0.48 - 0.5}{\sqrt{\frac{0.5(1 - 0.5)}{2551}}} = -2.02$. The P-value of the test is 0.0217. CONCLUDE: At the time of this 2012 poll, we have good evidence that less than half of all Americans supported Barack Obama for president.

Chapter 18 Comparing Two Proportions

18.1: PLAN: Let p_M be the proportion of all males who have used the Internet to search for health information and p_F be the proportion of all females who have done so. We want a 95% confidence interval for the difference in these proportions. SOLVE: The samples were large with clearly more than 10 each "successes" and "failures." The sample proportions are $p_M = \frac{520}{1084} = 0.4797$ and $p_F = \frac{811}{1308} = 0.6200$. The standard error of the difference is $SE_{\hat{p}_F - \hat{p}_M} = \sqrt{\frac{0.62(1 - 0.62)}{1308} + \frac{0.4797(1 - 0.4797)}{1084}} = 0.0203$. The 95% confidence interval is $(0.62 - 0.4797) \pm 1.96(0.0203) = 0.1403 \pm 0.0398$, or 0.1005 to 0.1801. CONCLUDE: We are 95% confident that between 10% and 18% more women than men have looked for health information on the Internet.

18.2: PLAN: Let p_9 be the proportion of ninth graders who have used marijuana in the past 30 days and p_{12} be the proportion of twelfth graders who have done so. We want a 99% confidence interval for the difference in these proportions. SOLVE: The samples were large with clearly more than 10 each "successes" and "failures." The sample proportions are $p_9 = \frac{656}{3642} = 0.1801$ and $p_{12} = \frac{1010}{3606} = 0.2801$. The standard error of the difference is $SE_{\hat{p}_{12} - \hat{p}_9} = \sqrt{\frac{0.2801(1 - 0.2801)}{3606} + \frac{0.1801(1 - 0.1801)}{3642}} = 0.0098$. The 99% confidence interval is $(0.2801 - 0.1801) \pm 2.576(0.0098) = 0.10 \pm 0.0252$, or 0.0748 to 0.1252. CONCLUDE: Based on

these samples, we are 99% confident that between about 7.5% and 12.5% more twelfth graders use marijuana in a month than ninth graders.

18.3: PLAN: Let p_1 be the proportion of 18- to 29-year-olds who think claims about the environment are exaggerated and p_1 be the proportion of those 60 and older who think so. We want a 95% confidence interval for the differences in these proportions. SOLVE: The samples were large with clearly more than 10 each "successes" and "failures." The sample proportions are $p_1 = \frac{75}{251} = 0.2988$ and $p_2 = \frac{174}{376} = 0.4628$. The standard error of the difference is $SE_{\hat{p}_2 - \hat{p}_1} = \sqrt{\frac{0.4628(1 - 0.4628)}{376} + \frac{0.2988(1 - 0.2988)}{251}} = 0.0387$. The 95% confidence interval for the difference in proportions is $(0.4628 - 0.2988) \pm 1.96(0.0387) = 0.164 \pm 0.0759 = 0.0881$ to 0.2399. CONCLUDE: Based on these samples, we are 95% confident that between about 8.8% and 24% more people 60 and older believe that claims about the environment are exaggerated than people 18 to 29 years old.

18.4: (a) This is an observational study: no treatment was imposed; we simply observed drivers in the two cities. (b) PLAN: Let p_1 be the proportion of New York female Hispanic drivers who wear seat belts, and let p_2 be that proportion for Boston. The comparison of local laws suggests a one-sided alternative: H_0: $p_1 = p_2$ vs. H_a: $p_1 > p_2$. SOLVE: All counts $(183, 37, 68, 49)$ are greater than 5, so the significance test should be safe. We find $\hat{p}_1 = \frac{183}{220} = 0.8318$ and $\hat{p}_2 = \frac{68}{117} = 0.5812$. The pooled proportion is $\hat{p} = \frac{183 + 68}{220 + 117} = 0.7448$, and SE $= 0.04989$. The test statistic is therefore $z = \frac{0.8318 - 0.5812}{SE} = 5.02$, for which P is very small (<0.001). CONCLUDE: We conclude that female Hispanic drivers in Boston are less likely to wear seat belts than those in New York City.

18.5: STATE: Is helmet use less common among skiers and snowboarders with head injuries than skiers and snowboarders without head injuries? PLAN: Let p_1 and p_2 be (respectively) the proportions of injured skiers and injured snowboarders who wear helmets. We test H_0: $p_1 = p_2$ vs. H_a: $p_1 < p_2$. SOLVE: The smallest count is 96, so the significance testing procedure is safe. We find $\hat{p}_1 = \frac{96}{578} = 0.1661$ and $\hat{p}_2 = \frac{656}{2992} = 0.2193$. The pooled proportion is $\hat{p} = \frac{96 + 656}{578 + 2992} = 0.2106$. Then for the significance test, SE $= \sqrt{\hat{p}(1 - \hat{p})(\frac{1}{578} + \frac{1}{2992})} = 0.01853$. The test statistic is therefore $z = \frac{0.1661 - 0.2193}{SE} = -2.87$, and $P = 0.0021$. CONCLUDE: We have strong evidence (significant at $\alpha = 0.01$) that skiers and snowboarders with head injuries are less likely to use helmets than skiers and snowboarders without head injuries.

18.6: STATE: Is the proportion of disease-free patients after five years using ALND different from that for those using SLND alone? PLAN: Let p_1 and p_2 be (respectively) the proportions of patients using ALND and using SLND alone. We test H_0: $p_1 = p_2$ vs. H_a: $p_1 \neq p_2$. SOLVE: The smallest count is 70 (number of failures in the SLND alone group), so the large-sample significance testing procedure is safe. We find $\hat{p}_1 = \frac{345}{420} = 0.8214$ and $\hat{p}_2 = \frac{366}{436} = 0.8394$. The pooled proportion is $\hat{p} = \frac{345 + 366}{420 + 436} = 0.8306$. Then for the significance test, SE $= \sqrt{\hat{p}(1-\hat{p})(\frac{1}{420}+\frac{1}{436})} = 0.0256$. The test statistic is therefore $z = \frac{0.8214 - 0.8394}{SE} = -0.70$, and $P = 0.4840$. CONCLUDE: There is virtually no evidence of a difference in proportions disease-free five years later between patients on ALND and those using SLND alone. Random chance easily explains a difference in the observed sample proportions.

18.7: (a) One count is only 6, and the guidelines for using the large-sample confidence interval call for all counts to be at least 10. (b) For the plus-four method, the sample sizes are 55 and 110, and the success counts are 7 and 46. (c) We have $\tilde{p}_1 = \frac{6+1}{53+2} = 0.1273$ and $\tilde{p}_2 = \frac{45+1}{108+2} = 0.4182$. Hence, a plus-four 95% confidence interval for $p_1 - p_2$ is $(0.1273 - 0.4182) \pm 1.96 \sqrt{\frac{\tilde{p}_1(1-\tilde{p}_1)}{55} + \frac{\tilde{p}_2(1-\tilde{p}_2)}{110}} = -0.2909 \pm 0.1275 = -0.4184$ to -0.1634. Because the first population is for injured skaters with wrist guards, the proportion for skaters with wrist guards is seen to be lower than the proportion for skaters without wrist guards. With 95% confidence, among injured skaters, the difference between proportion with wrist guards and those without is between -41.8% and -16.3%. That is, it appears more injured skaters fail to wear wrist guards.

18.8: PLAN: We want a 95% confidence interval for the difference in *Xerospirea hartwegiana* shrubs that will resprout after being clipped and exposed to fire, or not exposed to fire. This was a designed experiment, with the shrubs randomly assigned to treatment (fire exposure) or control. SOLVE: The total number of shrubs exposed to each treatment was only 12; counts of resprouted shrubs were 12 and 8 ("successes") with 0 and 4 shrubs that did not resprout ("failures"). Because these counts are small, we need the plus-four confidence interval. Let $\tilde{p}_T = \frac{8+1}{12+2} = 0.6429$ be the estimate of resprouting for shrubs exposed to fire and $\tilde{p}_C = \frac{12+1}{12+2} = 0.9286$ be the estimate of resprouting for control shrubs. The plus-four 95% confidence interval for $p_C - p_T$ is $(0.9286 - 0.6429) \pm 1.96 \sqrt{\frac{\tilde{p}_C(1-\tilde{p}_C)}{14} + \frac{\tilde{p}_T(1-\tilde{p}_T)}{14}} = 0.2857 \pm 0.2849 = 0.0008$ to 0.5706. CONCLUDE: We are 95% confident that burning reduces the percent of shrubs that will resprout by between 0.08% and 57.1%. (Control shrubs are that much more likely to resprout.)

18.9: (a) The question is, "Is there a difference?"

18.10: (a) $\hat{p}_D = \frac{288}{407}$ and $\hat{p}_R = \frac{227}{293}$.

18.11: (b) $\hat{p} = \frac{288+227}{407+293}$.

18.12: (b) $z = -1.99$; P-value $= 0.0466$.

18.13: (a) The standard error is $\sqrt{\frac{0.708(1-0.708)}{407} + \frac{0.775(1-0.775)}{293}} = 0.0332$. The margin of error is $1.645(0.0332) = 0.0546$.

18.14: (a) $z = \frac{\hat{p}_1 - \hat{p}_2}{\sqrt{\hat{p}(1-\hat{p})(\frac{1}{n_1} + \frac{1}{n_2})}} = \frac{0.70 - 0.20}{\sqrt{0.45(1-0.45)(\frac{1}{10} + \frac{1}{10})}} = 2.25$, and the P-value is one-sided and less than 0.02.

18.15: (b) We have only three failures in the treatment group and only two successes in the control group.

18.16: (b) $\tilde{p}_1 = \frac{7+1}{10+2} = 0.667$, $\tilde{p}_2 = \frac{2+1}{10+2} = 0.25$, and the margin of error is $1.645 \sqrt{\frac{\tilde{p}_1(1-\tilde{p}_1)}{12} + \frac{\tilde{p}_2(1-\tilde{p}_2)}{12}} = 0.304$.

18.17: (a) The four counts are 117, 53, 152, and 165, so all counts are large enough. (b) Using the large-sample method, $\hat{p}_1 = \frac{117}{170} = 0.6882$, and $\hat{p}_2 = \frac{152}{317} = 0.4795$, and the 95% confidence interval is $\hat{p}_1 - \hat{p}_2 \pm 1.96 \sqrt{\frac{\hat{p}_1(1-\hat{p}_1)}{170} + \frac{\hat{p}_2(1-\hat{p}_2)}{317}} = 0.2087 \pm 0.0887 = 0.1200$. to 0.2974.

18.18: (a) For the subitramine group, $\hat{p}_1 = \frac{561}{4906} = 0.114$. For the control (placebo) group, $\hat{p}_2 = \frac{490}{4898} = 0.100$. (b) The counts are 561, 4345, 490, and 4408—easily large enough for use of the large-sample confidence interval procedure. (c) Using the large-sample method the 95% confidence interval is $\hat{p}_1 - \hat{p}_2 \pm 1.96 \sqrt{\frac{\hat{p}_1(1-\hat{p}_1)}{4906} + \frac{\hat{p}_2(1-\hat{p}_2)}{4898}} = 0.0143 \pm 0.0122 = 0.0021$ to 0.0265, or 0.2% to 2.7%. Using the plus-four method, $\tilde{p}_1 = \frac{561+1}{4906+2} = 0.1145$ and $\tilde{p}_2 = \frac{490+1}{4898+2} = 0.1002$. The 95% confidence interval is $\tilde{p}_1 - \tilde{p}_2 \pm 1.96 \sqrt{\frac{\tilde{p}_1(1-\tilde{p}_1)}{4908} + \frac{\tilde{p}_2(1-\tilde{p}_2)}{4900}} = 0.0145 \pm 0.0122 = 0.0023$ to 0.0267, or 0.2% to 2.7%. Note that the two methods produce the same confidence interval up to three decimal places of precision because the samples are so large.

18.19: (a) One of the counts is 0; for large-sample methods, we need all counts to be at least 10. (b) After we add two observations to each sample, the sample size for the treatment group is 35, 24 of which have tumors; the sample size for the control group is 20, 1 of which has a tumor. (c) $\tilde{p}_1 = \frac{23+1}{33+2} = 0.6857$ and $\tilde{p}_2 = \frac{0+1}{18+2} = 0.05$. The plus four 99% confidence interval is $\tilde{p}_1 - \tilde{p}_2 \pm 2.576 \sqrt{\frac{\tilde{p}_1(1-\tilde{p}_1)}{35} + \frac{\tilde{p}_2(1-\tilde{p}_2)}{20}} = 0.6357 \pm 0.2380 = 0.3977$ to 0.8737. We are 99% confident that lowering DNA methylation increases the incidence of tumors by between about 40% and 87%.

18.20: Let p_1 and p_2 be (respectively) the proportions of subjects in the treatment and control groups experiencing a primary outcome. We test $H_0: p_1 = p_2$ vs. $H_a: p_1 \neq p_2$. For the treatment group $\hat{p}_1 = \frac{561}{4906} = 0.1143$. For the control (placebo) group, $\hat{p}_2 = \frac{490}{4898} = 0.1000$. The pooled estimate is $\hat{p} = \frac{561+490}{4906+4898} = 0.1072$. Hence, $z = \frac{\hat{p}_1 - \hat{p}_2}{\sqrt{\hat{p}(1-\hat{p})(\frac{1}{4906} + \frac{1}{4898})}} = 2.29$ and $P = 0.0220$. We have strong evidence that the proportion of subjects on subitramine suffering a primary outcome differs from that on the placebo. (b) A comparison group is important because we want to learn about the difference in rate of primary outcome due to subitramine. A placebo should be used to blind patients to which group they are in and to account for any possible placebo effect.

18.21: (a) Let p_1 and p_2 be (respectively) the proportions of subjects in the music and no-music groups that receive a passing grade on the Maryland HSA. We test $H_0: p_1 = p_2$ vs. $H_a: p_1 \neq p_2$. For the music group $\hat{p}_1 = \frac{2818}{3239} = 0.870$. For the no-music group, $\hat{p}_2 = \frac{2091}{2787} = 0.750$. The pooled estimate is $\hat{p} = \frac{2818+2091}{3239+2787} = 0.815$. Hence, $z = \frac{\hat{p}_1 - \hat{p}_2}{\sqrt{\hat{p}(1-\hat{p})(\frac{1}{3239} + \frac{1}{2787})}} = 11.94$. An observed difference of $0.87 - 0.75 = 0.12$ in group proportions is much too large to be explained by chance alone, and $P < 0.0001$. We have overwhelming evidence that the proportion of music students passing the Maryland HSA is greater than that for the no-music group. (b) and (c) This is an observational study—people that choose to (or can afford to) take music lessons differ in many ways from those that do not. Hence, we cannot conclude that music causes an improvement in Maryland HSA achievement.

18.22: We estimate the overall proportion of ninth graders that passed the HSA test. As computed in Exercise 18.21, $\hat{p} = \frac{2,818 + 2,091}{3,239 + 2,787} = \frac{4,909}{6,026} = 0.815$. A 95% confidence interval for the proportion p is given by $\hat{p} \pm 1.96\sqrt{\frac{\hat{p}(1 - \hat{p})}{6,026}} = 0.815 \pm 0.010 = 0.805$ to 0.825, or 80.5% to 82.5%.

18.23: We have at least 10 successes and 10 failures for both samples. For the music group, $\hat{p}_1 = \frac{2,818}{3,239} = 0.870$. For the no music group, $\hat{p}_2 = \frac{2,091}{2,787} = 0.750$. $\hat{p}_1 - \hat{p}_2 \pm 1.96\sqrt{\frac{\hat{p}_1(1 - \hat{p}_1)}{3,239} + \frac{\hat{p}_2(1 - \hat{p}_2)}{2,787}} = 0.100$ to 0.140, or 10.0% to 14.0%.

18.24: (a) $\hat{p}_1 = \frac{270}{1847} = 0.1462$ is the proportion of patients with complications before the restriction, and $\hat{p}_2 = \frac{170}{1639} = 0.1037$ is the proportion with complications after the restriction. We can test $H_0: p_1 = p_2$ vs. $H_a: p_1 \neq p_2$ to assess the strength of the evidence that the proportions are different. We have $\hat{p} = \frac{270 + 170}{1847 + 1639} = 0.1262$ as the pooled proportion and find $z = \frac{\hat{p}_1 - \hat{p}_2}{\sqrt{\hat{p}(1 - \hat{p})\left(\frac{1}{3239} + \frac{1}{2787}\right)}} = 3.77$ with P-value 0.0002. This is extremely strong evidence that the proportions are different. (b) This study was observational; there were no assigned treatments because all "subjects" underwent bariatric surgery—they (or their records) were examined later for complications. (c) The possible other reasons for the decrease in complications are all lurking (confounding) variables; they weaken the case that the decline in complications is due to the restrictions. (d) Answers will vary, but the reason for a comparative control group is to eliminate lurking variables (as much as possible). We can (safely) assume that both groups experienced the same exposure to newer methods, increased surgeon experience, and so on. This allows us to focus on the factor of interest—the imposition of the restrictions and whether those improved results.

18.25: (a) To test $H_0: p_M = p_F$ vs. $H_a: p_M \neq p_F$, we find $\hat{p}_M = \frac{15}{106} = 0.1415$, $\hat{p}_F = \frac{7}{42} = 0.1667$, and $\hat{p} = 0.1486$. Then SE $= \sqrt{\hat{p}(1 - \hat{p})\left(\frac{1}{106} + \frac{1}{42}\right)} = 0.06485$, so $z = \frac{\hat{p}_M - \hat{p}_F}{\text{SE}} = -0.39$. This gives $P = 0.6966$, which provides virtually no evidence of a difference in failure rates. (b) We have $\hat{p}_M = \frac{450}{3180} = 0.1415$, $\hat{p}_F = \frac{210}{1260} = 0.1667$, and $\hat{p} = 0.1486$, but now SE $= \sqrt{\hat{p}(1 - \hat{p})\left(\frac{1}{3180} + \frac{1}{1260}\right)} = 0.01184$, so $z = \frac{\hat{p}_M - \hat{p}_F}{\text{SE}} = -2.13$ and $P = 0.0332$. (c) We are asked to construct two confidence intervals—one based on the smaller samples of part (a) and one based on the larger samples of part (b). First, for case (a), $\hat{p}_M = 0.1415$ and $\hat{p}_F = 0.1667$, so a 95% confidence interval for the difference is $\hat{p}_M - \hat{p}_F \pm 1.96\sqrt{\frac{\hat{p}_M(1 - \hat{p}_M)}{106} + \frac{\hat{p}_F(1 - \hat{p}_F)}{42}} = -0.156$ to 0.1056. We note that because there were only 7 business failures in those headed by women, that this interval is not really appropriate (even though the hypothesis test was). For case (b),

$\hat{p}_M = 0.1415$ and $\hat{p}_F = 0.1667$. The resulting confidence interval is $\hat{p}_M - \hat{p}_F \pm 1.96\sqrt{\frac{\hat{p}_M(1 - \hat{p}_M)}{3,180} + \frac{\hat{p}_F(1 - \hat{p}_F)}{1,260}} = -0.0491$ to -0.0013.

18.26: PLAN: Let p_M be the proportion of men who use social networking sites and p_F be the proportion of women who use them. We want to know if the proportions are different, so we will test $H_0: p_M = p_F$ vs. $H_a: p_M \neq p_F$. SOLVE: All sample counts are much larger than 5, so inference is appropriate. The sample proportions are $\hat{p}_M = \frac{525}{846} = 0.6206$ and $\hat{p}_F = \frac{679}{956} = 0.7103$. The pooled proportion is $\hat{p} = \frac{525 + 679}{1802} = 0.6681$. The test statistic is $z = \frac{0.6206 - 0.7103}{\sqrt{0.6681(1 - 0.6681)\left(\frac{1}{846} + \frac{1}{956}\right)}} = -4.04$. The P-value for the test will be < 0.0001. CONCLUDE: We have extremely strong evidence that the proportion of men who use social networking sites is different from the proportion of women who use social networking sites. It appears women are more likely to use those sites.

18.27: PLAN: Let p_W be the proportion of Whites who use social networking sites and p_H be the proportion of Hispanics who use them. We want to know if the proportions are different, so we will test $H_0: p_W = p_H$ vs. $H_a: p_W \neq p_H$. SOLVE: All sample counts are larger than 5 (the smallest is 43 for Hispanics who do not use social networking sites), so inference is appropriate. The sample proportions are $\hat{p}_W = \frac{866}{1332} = 0.6502$ and $\hat{p}_H = \frac{111}{154} = 0.7208$. The pooled proportion is $\hat{p} = \frac{866 + 111}{1332 + 154} = 0.6575$. The test statistic is $z = \frac{0.6502 - 0.7208}{\sqrt{0.6575(1 - 0.6575)\left(\frac{1}{1332} + \frac{1}{154}\right)}} = -1.75$. The P-value of the test is 0.0801. CONCLUDE: At the 0.05 level, we fail to reject H_0. This survey has failed to find a difference in the proportions of Whites and Hispanics who use social networking sites.

18.28: PLAN: We'd like a 90% confidence interval for the difference in the proportions of men and women who use social networking sites. SOLVE: All counts are much larger than 10; $\hat{p}_M = 0.6206$ and $\hat{p}_F = 0.7103$. The 90% confidence interval for the difference in proportions is $(0.7103 - 0.6206) \pm 1.645\sqrt{\frac{0.7103(1 - 0.7103)}{956} + \frac{0.6206(1 - 0.6206)}{846}} = 0.0897 \pm 0.0365$, or 0.0532 to 0.1262. CONCLUDE: Based on this survey, with 90% confidence, somewhere between 5.3% and 12.6% more women than men use social networking sites.

18.29: PLAN: We construct a 99% confidence interval for $p_1 - p_2$, where p_1 denotes the proportion of people on Chantix who abstained from smoking, and p_2 is the corresponding proportion for the placebo population. SOLVE: The sample counts are 155 and 61 (successes for treatment and control, respectively) and 197 and 283 (failures for the groups), so the large-sample procedures are safe. Using the large-sample method, $\hat{p}_1 = \frac{155}{352} = 0.4403$, and $\hat{p}_2 = \frac{61}{344} = 0.1773$, and the 99% confidence interval is $\hat{p}_1 - \hat{p}_2 \pm 2.576\sqrt{\frac{\hat{p}_1(1 - \hat{p}_1)}{352} + \frac{\hat{p}_2(1 - \hat{p}_2)}{344}} = 0.2630 \pm 0.0864 = 0.1766$ to 0.3494. CONCLUDE: We are 99% confident that the success rate for abstaining from smoking is between 17.7

and 34.9 percentage points higher for smokers using Chantix than for smokers on a placebo.

18.30: PLAN: Let p_1 be the proportion of deaths among African miners, and p_2 be the proportion among European miners. We test $H_0: p_1 = p_2$ vs. $H_a: p_1 > p_2$, where, presumably, these hypotheses were determined before looking at the data. SOLVE: The smallest success/failure count is $7 > 5$, so the conditions for a significance test are met. The sample proportions are $\hat{p}_1 = \frac{223}{33,809} = 0.006596$ and $\hat{p}_2 = \frac{7}{1541} = 0.004543$. The pooled proportion is $\hat{p} = \frac{223 + 7}{33,809 + 1541} = 0.006506$, and $SE = \sqrt{\hat{p}(1 - \hat{p})(\frac{1}{33,809} + \frac{1}{1541})} = 0.002094$. The test statistic is therefore $z = \frac{0.006596 - 0.004543}{SE} = 0.98$, for which $P = 0.1635$. CONCLUDE: We do not have enough evidence to conclude that the death rates are higher for European miners than for African miners.

18.31: PLAN: Let p_1 be the proportion of people that will reject an unfair offer from another person, and p_2 be the proportion for offers from a computer. We test $H_0: p_1 = p_2$ vs. $H_a: p_1 > p_2$. SOLVE: All counts are greater than 5, so the conditions for a significance test are met. The sample proportions are $\hat{p}_1 = \frac{18}{38} = 0.4737$ and $\hat{p}_2 = \frac{6}{38} = 0.1579$. The pooled proportion is $\hat{p} = \frac{18 + 6}{38 + 38} = 0.3158$, and $SE = \sqrt{\hat{p}(1 - \hat{p})(\frac{1}{38} + \frac{1}{38})} = 0.1066$. The test statistic is therefore $z = \frac{0.4737 - 0.1579}{SE} = 2.96$, for which $P = 0.0015$. CONCLUDE: There is very strong evidence that people are more likely to reject an unfair offer from another person than from a computer.

18.32: STATE: Is there evidence of a difference in the proportion of women with a family history of breast cancer between the treatment and control groups? PLAN: With p_1 the proportion for the treatment population, and p_2 the proportion for the control population, we test $H_0: p_1 = p_2$ vs. $H_a: p_1 \neq p_2$. SOLVE: All counts are large enough to use the significance testing procedure safely. $\hat{p}_1 = \frac{3396}{19,541} = 0.1738$, $\hat{p}_2 = \frac{4929}{29,294} = 0.1683$, and $\hat{p} = \frac{3,396 + 4,929}{19,541 + 29,294} = 0.1705$. Hence, $SE = \sqrt{\hat{p}(1 - \hat{p})(\frac{1}{19,541} + \frac{1}{29,294})} = 0.00347$. The test statistic is therefore $z = \frac{0.1738 - 0.1683}{SE} = 1.58$, for which $P = 0.1142$. CONCLUDE: We do not have enough evidence to reject the null hypothesis. Comparing treatment and control groups, there is little evidence of a difference in proportions of women with family history of breast cancer.

18.33: Let p_1 and p_2 be (respectively) the proportions of mice ready to breed in good acorn years and bad acorn years. We give a 90% confidence interval for $p_1 - p_2$. SOLVE: One count is only 7, and the guidelines for using the large-sample method call for all counts to be at least 10, so we use the plus-four method. We have $\tilde{p}_1 = \frac{54 + 1}{72 + 2} = 0.7432$, and $\tilde{p}_2 = \frac{10 + 1}{17 + 2} = 0.5789$, so the plus-four 90% confidence interval is $\tilde{p}_1 - \tilde{p}_2 \pm 1.645\sqrt{\frac{\tilde{p}_1(1 - \tilde{p}_1)}{74} + \frac{\tilde{p}_2(1 - \tilde{p}_2)}{19}} = 0.1643 \pm 0.2042 = -0.0399$ to 0.3685. CONCLUDE: We are 90% confident that the proportion of mice ready to breed in good acorn years is between 0.04 lower than and 0.37 higher than the proportion in bad acorn years.

18.34: PLAN: To answer the question about whether children who had early childhood education have a higher proportion of consistent employment, we will use a test of $H_0: p_E = p_C$ vs. $H_a: p_E > p_C$, where p_E is the proportion from the intervention group who received intensive early childhood education. If warranted by the test, we will estimate the difference with a 95% confidence interval for the difference in proportions. SOLVE: All counts are larger than 5 (the smallest is 13 for the intervention group who were not consistently employed), so a test is appropriate. If the interval is needed, we also note all counts are more than 10. The sample proportions are $\hat{p}_E = \frac{39}{52} = 0.75$ and $\hat{p}_C = \frac{26}{49} = 0.5306$, and $\hat{p} = \frac{39 + 26}{52 + 49} = 0.6436$. The test statistic is $z = 2.30$ with $P = 0.0107$. We have strong evidence of a difference in the proportions in the two groups who had consistent employment. How large is the difference? The 95% confidence interval is $(0.75 - 0.5306) \pm 1.96\sqrt{\frac{\hat{p}_1(1 - \hat{p}_1)}{52} + \frac{\hat{p}_2(1 - \hat{p}_2)}{49}} = 0.0367$ to 0.4021. CONCLUDE: We have strong evidence that children who receive intensive early childhood education are more likely to be consistently employed at age 30. Between about 3.7% and 40.2% more children with early education are likely to be consistently employed than those who do not receive the education, at 95% confidence.

18.35: (a) This is an experiment because the researchers assigned subjects to the groups being compared. (b) PLAN: Let p_1 and p_2 be (respectively) the proportions that have an RV infection for the HL+ group and a control group. We test $H_0: p_1 = p_2$ vs. $H_a: p_1 < p_2$. SOLVE: We have large enough counts to use large-sample significance testing procedure safely. Now $\hat{p}_1 = \frac{49}{49 + 67} = 0.4224$, $\hat{p}_2 = \frac{49}{49 + 47} = 0.5104$, and $\hat{p} = \frac{49 + 49}{116 + 96} = 0.4623$. Hence, $SE = \sqrt{\hat{p}(1 - \hat{p})(\frac{1}{116} + \frac{1}{96})} = 0.0688$. The test statistic is therefore $z = \frac{0.4224 - 0.5104}{SE} = -1.28$, for which $P = 0.1003$. CONCLUDE: We do not have enough evidence to reject the null hypothesis; there is little evidence to conclude that the proportion of HL+ users with a rhinovirus infection is less than that for non-HL+ users.

Chapter 19   Sampling Distribution for a Mean

19.1: Sketches will vary.

19.2: Five dice should settle down to 17.5 as the mean of the sum.

19.3: Although the probability of having to pay for a total loss for 1 or more of the 12 policies is very small, if this were to happen, it would be financially disastrous. On the other hand, for thousands of policies, the law of large numbers says that the average claim on many policies will be close to the mean, so the insurance company can be assured that the premiums they collect will (almost certainly) cover the claims.

19.4: (a) The population is the 12,000 students; the population distribution (Normal with mean 7.11 minutes and standard deviation 0.74 minute) describes the time it takes a randomly selected individual to run a mile. (b) The sampling distribution (Normal with mean 7.11 minutes and standard deviation 0.074 minute) describes the average mile time for 100 randomly selected students.

19.5: (a) The sampling distribution of $\bar{x}$ is $N(186$ mg/dl, 4.1 mg/dl). $P(183 < x < 189) = P(-0.73 < Z < 0.73) = 0.5346$. (b) With $n = 1000$, the sample mean has the $N(186$ mg/dl, 1.2965 mg/dl) distribution, so $P(183 < x < 189) = P(-2.31 < Z < 2.31) = 0.9792$.

19.6: (a) $10/\sqrt{4} = 5$ mg. (b) Solve $\sigma/\sqrt{n} = 2$, or $10/\sqrt{n} = 2$, so $\sqrt{n} = 5$, or $n = 25$. The average of several measurements is more likely than a single measurement to be close to the mean.

19.7: No: the histogram of the sample values will look like the population distribution, whatever it might happen to be. The central limit theorem says that the histogram of *sample means* (from many large samples) will look more and more Normal.

19.8: (a) $\mu_{\bar{x}} = 0.5$ and $\sigma_{\bar{x}} = 0.7/\sqrt{50} = 0.09899$. (b) Because this distribution is only approximately Normal, it would be quite reasonable to use the 68–95–99.7 rule to give a rough estimate: 0.6 is about one standard deviation above the mean, so the probability should be about 0.16. Alternatively, $P(\bar{x} > 0.6) = P(Z > \frac{0.6 - 0.5}{0.09899}) = P(Z > 1.01) = 0.1562$.

19.9: The central limit theorem says that, despite the skewness of the population distribution, the average loss among 10,000 policies will be approximately $N(\$75, \$3)$. $P(\bar{x} > \$85) = P(Z > \frac{85 - 75}{3}) = P(Z > 3.33) = 1 - 0.9996 = 0.0004$.

19.10: (b) The law of large numbers says that the mean from a large sample is close to the population mean.

19.11: (a) The mean of the sample means ($\bar{x}$'s) is the same as the population mean (μ).

19.12: (c) The standard deviation of the distribution of $\bar{x}$ is $\sigma/\sqrt{n}$.

19.13: (a) "Unbiased" means that the estimator is correct "on the average."

19.14: (c) The central limit theorem says that the mean from a large sample has (approximately) a Normal distribution.

19.15: (b) For $n = 6$ women, $\bar{x}$ has a $N(266, 16/\sqrt{6}) = N(266, 6.5320)$ distribution, so $P(\bar{x} > 270) = P(Z > 0.61) = 0.2709$.

19.16: In the long run, the gambler earns an average of 94.7 cents per bet. In other words, the gambler loses (and the house gains) an average of 5.3 cents for each $1 bet.

19.17: $\bar{x}$ has mean $\mu = 852$ mm, and standard deviation $\sigma/\sqrt{n} = 82/\sqrt{10} = 25.93$ mm.

19.18: (a) $P(20 < X < 30) = P(-0.78 < Z < 0.78) = 0.5646$. (b) If $n = 25$ students, the sampling distribution of $\bar{x}$ is $N(25, 6.4/\sqrt{25}) = N(25, 1.28)$. (c) $P(20 < \bar{x} < 30) = P(-3.91 < Z < 3.91) \approx 1$.

19.19: Let X be Shelia's measured glucose level. (a) $P(X > 140) = P(Z > 1.5) = 0.0668$. (b) If $\bar{x}$ is the mean of four

measurements (assumed to be independent), then $\bar{x}$ has a $N(122, 12/\sqrt{4}) = N(122$ mg/dl, 6 mg/dl) distribution, and $P(\bar{x} > 140) = P(Z > 3) = 0.0013$.

19.20: (a) Let $\bar{x}$ be the mean number of minutes per day that the 5 randomly selected mildly obese people spend walking. Then $\bar{x}$ has the $N(373, 67/\sqrt{5}) = N(373$ min., 29.96 min.) distribution. Now $P(\bar{x} > 420) = P(Z > \frac{420 - 373}{29.96}) = P(Z > 1.57) = 0.0582$. (b) Let $\bar{x}$ be the sample mean number of minutes per day for the 5 randomly selected lean people. $\bar{x}$ has the $N(526, 107/\sqrt{5}) = N(526$ min., 47.85 min.). $P(\bar{x} > 420) = P(Z > -2.22) = 0.9868$.

19.21: The mean of four measurements has a $N(122$ mg/dl, 6 mg/dl) distribution, and $P(Z > 1.645) = 0.05$ if Z is $N(0,1)$, so $L = 122 + 1.645(6) = 131.87$ mg/dl.

19.22: (a) For the emissions E of a single car, $P(E > 0.07) = P(Z > 2) = 0.0228$. (b) The average $\bar{x}$ is Normal with mean 0.05 g/mi and standard deviation $0.01/\sqrt{25} = 0.002$ g/mi Therefore, $P(\bar{x} > 0.07) = P(Z > 10) \approx 0$.

19.23: (a) The central limit theorem gives that $\bar{x}$ will have a Normal distribution with mean 8.8 beats per five seconds and standard deviation $1/\sqrt{12} = 0.288675$ beats per five seconds. (b) $P(\bar{x} < 8) = P(Z < -2.77) = 0.0028$. (c) If the total number of beats in one minute is less than 100, then the average over 12 five-second intervals needs to be less than $100/12 = 8.333$ beats per five seconds. $P(\bar{x} < 8.333) = P(Z < -1.62) = 0.0526$.

19.24: The mean NOX level for 25 cars has a $N(0.05$ g/mi, 0.002 g/mi) distribution, and $P(Z > 2.326) = 0.01$ if Z is $N(0,1)$, so $L = 0.05 + (2.326)(0.002) = 0.054652$ g/mi.

19.25: The central limit theorem says that over 40 years, $\bar{x}$ (the mean return) is approximately Normal with mean $\mu = 10.8\%$ and standard deviation $17.1\%/\sqrt{40} = 2.704\%$. Therefore, $P(\bar{x} > 10\%) = P(Z > -0.30) = 0.6179$, and $P(\bar{x} < 5\%) = P(Z < -2.14) = 0.0162$. Note: *We have to assume that returns in separate years are independent.*

19.26: If W is total weight, then the sample mean weight is $\bar{x} = W/22$. The event that the total weight exceeds 4500 pounds is equivalent to the event that $\bar{x}$ exceeds $4500/22 = 204.55$ lb. The central limit theorem says that $\bar{x}$ is approximately Normal with mean 190 lb and standard deviation $35/\sqrt{22} = 7.462$ lb. Therefore, $P(W > 4500) = P(\bar{x} > 204.55) = P(Z > 1.95) = 0.0256$.

19.27: We need to choose n so that $6.4/\sqrt{n} = 1$. That means $\sqrt{n} = 6.4$, so $n = 40.96$. Because n must be a whole number, take $n = 41$.

19.28: (a) 99.7% of all observations fall within 3 standard deviations, so we want $3\sigma/\sqrt{n} = 1$. The standard deviation of x must therefore be $1/3 = 0.33$ point. (b) We need to choose n so that $6.4/\sqrt{n} = 0.33$. This means $\sqrt{n} = 19.2$, so $n = 368.64$. Because n must be a whole number, take $n = 369$.

19.29: On the average, Joe loses 40 cents each time he plays (that is, he spends $1 and gets back 60 cents, on average).

19.30: (a) With $n = 14{,}000$, $\mu_{\bar{x}} = \$0.60$ and $\sigma_{\bar{x}} = \frac{\$18.96}{\sqrt{14{,}000}} = \$0.1602$. (b) $P(\$0.50 < \bar{x} < \$0.70) = P(-0.62 < Z < 0.62) = 0.4648$.

19.31: (a) With $n = 150{,}000$, $\mu_{\bar{x}} = \$0.40$ and $\sigma_{\bar{x}} = \frac{\$18.96}{\sqrt{150{,}000}} = \$0.0490$. (b) $P(\$0.30 < \bar{x} < \$0.50) = P(-2.04 < Z < 2.04) = 0.9586$.

19.32: (a) The estimate in Exercise 19.30 was 0.4648 (Table A), so the Normal approximation slightly underestimates the exact answer. (b) With $n = 3500$, the Normal approximation gives $P(\$0.50 < \bar{x} < \$0.70) = P(-0.31 < Z < 0.31)$, which is 0.2434 (Table A). This is just a bit smaller than the exact answer. (c) The probability that their average winnings fall between \$0.50 and \$0.70 is the same as the probability found in part (b) of the previous exercise, for which the Normal approximation gives 0.9586 (Table A), so the approximation differs from the exact value by only about 0.0043.

Chapter 20   Inference about a Population Mean

20.1: $s/\sqrt{n} = 63.9/\sqrt{1000} = 2.0207$ minutes.

20.2: $\bar{x} = 163$ beats per minute and $s/\sqrt{n} = 15/\sqrt{10} = 4.743$ beats per minute.

20.3: (a) $t^* = 2.132$. (b) $t^* = 2.479$.

20.4: $df = 30 - 1 = 29$. (a) $t^* = 2.045$. (b) $t^* = 0.683$.

20.5: (a) $df = 12 - 1 = 11$, so $t^* = 2.201$. (b) $df = 18 - 1 = 17$, so $t^* = 2.898$. (c) $df = 6 - 1 = 5$, so $t^* = 2.015$.

20.6: (a) The stemplot shows a slight skew to the right, but not so strong that it would invalidate the t procedures. (b) $t^* = 2.093$ (df = 19); $18.66 \pm 2.093\frac{10.2768}{\sqrt{20}} = 13.8504$ to 23.4696.

20.7: We are told to view the observations as an SRS. A stemplot shows some left-skewness; however, for such a small sample, the data are not unreasonably skewed. There are no outliers. $t^* = 1.860$ (df = 8); $59.5889 \pm 1.860\frac{6.2553}{\sqrt{9}} = 55.71\%$ to 63.47%. We are 90% confident that the mean percent of nitrogen in ancient air is between 55.71% and 63.47%.

20.8: (a) $df = 20 - 1 = 19$. (b) $t = 1.84$ is bracketed by $t^* = 1.729$ (with right-tail probability 0.05) and $t^* = 2.093$ (with right-tail probability 0.025). Because this is a one-sided significance test, $0.025 < P < 0.05$. (c) This test is significant at the 5% level because the $P < 0.05$. It is not significant at the 1% level because the $P > 0.01$. (d) From software, $P = 0.0407$.

20.9: (a) $df = 15 - 1 = 14$. (b) $t = 2.12$ is bracketed by $t^* = 1.761$ (with two-tail probability 0.10) and $t^* = 2.145$ (with two-tail probability 0.05). Because this is a two-sided significance test, $0.05 < P < 0.10$. (c) This test is significant at the 10% level because the $P < 0.10$. It is not significant at the 5% level because the $P > 0.05$. (d) From software, $P = 0.0524$.

20.10: $H_0\colon \mu = 78.1\%$ vs. $H_a\colon \mu \neq 78.1\%$. $t = \frac{59.5889 - 78.1}{6.2553/\sqrt{9}} = -8.88$. For df = 8, this is beyond anything shown in Table C, so $P < 0.001$ (software gives $P = 0.00002$). We have very

strong evidence $(P < 0.001)$ that Cretaceous air contained less nitrogen than modern air.

20.11: Let μ be the mean difference (monkey call minus pure tone) in firing rate. $H_0\colon \mu = 0$ vs. $H_0\colon \mu > 0$. We must assume that the monkeys can be regarded as an SRS. For each monkey, we compute the call minus pure tone differences; a stemplot of these differences shows no outliers or deviations from Normality. $t = \frac{70.378 - 0}{88.447/\sqrt{37}} = 4.84$ with df = 36. This has a very small P-value: $P < 0.0001$. We have very strong evidence that neural response to monkey calls is stronger than the response to pure tones.

20.12: Note that df = 36 is not contained in the table, so use df = 30 (round down to be conservative). $t^* = 2.750$; $70.378 \pm 2.750\frac{88.447}{\sqrt{37}} = 30.391$ to 110.365 spikes/second.

20.13: A stemplot suggests that the distribution of nitrogen contents is heavily skewed. Although t procedures are robust, they should not be used if the population being sampled from is this heavily skewed. In this case, t procedures are not reliable.

20.14: The stemplot of carbon-13 ratios suggests no strong skew, so use of t procedures is appropriate, assuming the sample is random. df = $24 - 1 = 23$, and $t^* = 1.714$. $-2.8825 \pm 1.714\frac{1.0360}{\sqrt{24}} = -3.2450$ to -2.5200.

20.15: (b) We virtually never know the value of σ.

20.16: (b) $t = \frac{8 - 10}{4/\sqrt{16}} = -2$.

20.17: (c) df = $25 - 1 = 24$.

20.18: (c) $P < 0.01$.

20.19: (a) 2.718. Here, df = 11.

20.20: (b)

20.21: (c) $85 \pm 3.250\frac{12}{\sqrt{10}}$.

20.22: (a) The t procedures are robust against mild skew, and they are used when σ is unknown.

20.23: (b) If you sample 64 unmarried male students and then sample 64 unmarried female students, no matching is present.

20.24: (c)

20.25: For the student group: $t = \frac{0.08 - 0}{0.37/\sqrt{12}} = 0.749$ (not 0.49, as stated). For the nonstudent group: $t = \frac{0.35 - 0}{0.37/\sqrt{12}} = 3.277$ (rather than 3.25, a difference that might be due to rounding error). From Table C, the first P-value is between 0.4 and 0.5 (software gives 0.47), and the second P-value is between 0.005 and 0.01 (software gives 0.007).

20.26: $t^* = 1.984$ (df = 100 from Table C) or $t^* = 1.9636$ (df = 653 with software). The 95% confidence interval for mean BMI is $\bar{x} \pm t^*\left(\frac{s}{\sqrt{654}}\right)$, computed in both cases as

$= 26.8 \pm 0.5756 = 26.2244$ to 27.3756 (using $t^* = 1.984$)

or

$= 26.8 \pm 0.5697 = 26.2303$ to 27.3697 (using $t^* = 1.9636$).

20.27: (a) The sample size is very large, so the only potential hazard is extreme skewness. Because scores range only from 0 to 500, there is a limit to how skewed the distribution could be.

(b) From Table C, we take $t^* = 2.581$ (df $= 1000$), or using software take $t^* = 2.5775$. For either value of t^*, the 99% confidence interval is $250 \pm 2.581 = 247.4$ to 252.6. (c) Because the 99% confidence interval for μ does not contain 243 and is entirely above 243, we would fail to reject H_0: $\mu = 243$ against the one-sided alternative hypothesis H_a: $\mu < 243$ at the 1% significance level.

20.28: (a) df $= 23 - 1 = 22$, so $t^* = 2.074$. A 95% confidence interval for the mean solution time is $11.58 \pm 2.074(\frac{4.37}{\sqrt{23}}) = 1158 \pm 1.89 = 9.69$ to 13.47 seconds. (b) We must assume that the 23 individuals in the neutral group can be regarded as an SRS from the population. Because the sample size is at least 15, we don't need to assume that the population is Normal. Indeed, t procedures can be used as long as the distribution of solution times for the neutral group is not heavily skewed, and as long as there are no outliers in the sample.

20.29: (a) A subject's responses to the two treatments would not be independent. (b) $t = \frac{-0.326 - 0}{0.181/\sqrt{6}} = -4.41$. With df $= 5$, $P = 0.0069$, significant evidence of a difference.

20.30: (a) A stemplot is provided. A "0" in the row with stem of "333" corresponds to a data value of 0.03330. Because the sample size is less than 15, we look to see if the data appear close to Normal. The stemplot is roughly symmetric with one peak and no outliers. (b) df $= 8$, and $t^* = 2.30600$ (using software to obtain t^* to 5 significant figures). $0.03339 \pm 2.30600(\frac{0.00027}{\sqrt{9}}) = 0.03318$ to 0.03360 Watts. (c) Because 0.0330 is not contained in the 95% confidence interval computed in (b), we would reject H_0: $\mu = 0.0330$ at the 5% significance level in favor of the two-sided alternative. There is strong evidence that the mean conductivity is different from 0.0330 watts.

```
330 | 0
331 |
332 | 0
333 | 0 0 0
334 | 0 0
335 |
336 |
337 | 0
338 |
339 | 0
```

20.31: (a) A stemplot suggests the presence of outliers. The sample is small, and the distribution is skewed, so use of t procedures is not appropriate. (b) We will compute two confidence intervals, as called for. In the first interval, using all 9 observations, we have df $= 8$ and $t^* = 1.860$. For the second interval, removing the two outliers (1.15 and 1.35), df $= 6$ and $t^* = 1.943$. The two 90% confidence intervals are

$$0.549 \pm 1.860(\tfrac{0.403}{\sqrt{9}}) = 0.299 \text{ to } 0.799 \text{ grams}$$

and

$$0.349 \pm 1.943(\tfrac{0.053}{\sqrt{7}}) = 0.310 \text{ to } 0.388 \text{ grams}.$$

(c) The confidence interval computed without the two outliers is much narrower. Using fewer data values reduces degrees of freedom (yielding a larger value of t^*). Also, smaller sample sizes yield larger margins of error. However both of these effects are offset by removing two values far from the others $-s$ reduces from 0.403 to 0.053 by removing them.

20.32: $t^* = 2.042$ (using df $= 30$ with Table C) or $t^* = 2.0262$ (using df $= 37$ with software). The confidence interval is nearly identical in both cases:

$$\bar{x} \pm t^*(\tfrac{s}{\sqrt{38}}) = 22.95° \text{ to } 27.89° \ (t^* = 2.042) \text{ or}$$
$$= 22.96° \text{ to } 27.88° \ (t^* = 2.0262).$$

We are 95% confident that the mean HAV angle among such patients is between 22.95° and 27.89°.

20.33: (a) The stemplot shows the high outlier mentioned in the text. (b) Let μ be the mean difference (control minus experimental) in healing rates. H_0: $\mu = 0$ vs. H_a: $\mu > 0$. $t = \frac{6.417 - 0}{10.7065/\sqrt{12}} = 2.08$. With df $= 11$, $P = 0.0311$ (using software). Omitting the outlier: $\bar{x} = 4.182$ and $s = 7.7565$, so $t = \frac{4.182 - 0}{7.7565/\sqrt{11}} = 1.79$. With df $= 10$, $P = 0.052$. Hence, with all 12 differences there is greater evidence that the mean healing time is greater for the control limb. When we omit the outlier, the evidence is weaker.

```
-1 | 3
-0 | 6
-0 |
 0 | 1 2
 0 | 5 7 8 9
 1 | 0 1 2
 1 |
 2 |
 2 |
 3 | 1
```

20.34: (a) Without the outlier, the mean is $\bar{x} = 24.76°$, and the standard deviation decreases to $s = 6.34°$. Using df $= 36$ with software, $t^* = 2.0281$, so a 95% confidence interval for the population mean becomes $24.76 \pm 2.0281(\tfrac{6.34}{\sqrt{37}}) = 22.65°$ to $26.87°$. (b) In Exercise 20.32, using all of the data, the 95% confidence interval was 22.96° to 27.88°. The confidence interval in Exercise 20.32 is wider because the presence of an outlier increases s.

20.35: (a) The sample has a significant outlier and indicates skew. We might consider applying t procedures to the sample after removing the most extreme observation (37,786). (b) If we remove the largest observation, the remaining sample is not heavily skewed and has no outliers. H_0: $\mu = 7000$ vs. H_a: $\mu \neq 7000$. With the outlier removed, $\bar{x} = 11,555.16$ and $s = 6,095.015$. $t = \frac{11,555.16 - 7000}{6095.015/\sqrt{19}} = 3.258$. With df $= 18$ with software, $P = 0.0044$ (this is a two-sided test). There is overwhelming evidence that the mean number of words per day of men at this university differs from 7000.

20.36: (a) A stemplot does not show any severe evidence of non-Normality, so t procedures should be safe. (b) $t^* = 1.812$ (df = 10). The 90% confidence interval is $1.1727 \pm 1.812(\frac{0.4606}{\sqrt{11}}) = 0.9211$ to 1.4243 days. There is no indication that the sample represents an SRS of all patients whose melanoma has not responded to existing treatments, so inferring to the population of similar patients may not be reasonable.

20.37: (a) A stemplot of differences shows an extreme right skew and one or two high outliers. The t procedures should not be used. (b) Some students might perform the test (H_0: $\mu = 0$ vs. H_a: $\mu > 0$) using t procedures, despite the presence of strong skew and outliers in the sample. If so, they should find $\bar{x} = 156.36$, $s = 234.2952$, and $t = 2.213$, yielding $P = 0.0256$.

20.38: The design described is matched pairs, and we are interested in the differences (helium filled − air filled) in punt distance. (a) A stemplot suggests a roughly symmetric, single-peak distribution for differences. Note that the randomization described means that we can treat the observations as a simple random sample from a population of all differences (for this kicker). Hence, t procedures seem to be appropriate here. (b) Let μ denote the mean difference (helium filled − air filled). H_0: $\mu = 0$ vs. H_a: $\mu > 0$. $t = \frac{0.462 - 0}{6.867/\sqrt{39}} = 0.420$. With df = 38, $P = 0.3384$, using software. There is virtually no evidence that the mean distance for helium-filled footballs is greater than that of air-filled footballs.

20.39: (a) H_0: $\mu = 0$ vs. H_a: $\mu > 0$, where μ is the mean difference (treated minus control). (b) $t = \frac{1.916 - 0}{1.050/\sqrt{3}} = 3.16$ with df = 2. Hence, $P = 0.044$. This is significant at the 5% significance level. (c) For very small samples, t procedures should only be used when we can assume that the population is Normal. We have no way to assess the Normality of the population based on these four observations. Hence, the validity of the analysis in (b) is dubious.

20.40: (a) Weather conditions that change day to day can affect spore counts. So the two measurements made on the same day form a matched pair. (b) Take the differences (kill room counts − processing counts). The 90% confidence interval for μ is $1824.5 \pm 2.353(\frac{834.1}{\sqrt{4}}) = 843.2$ to 2805.8 CFU/m³. The interval is so wide because the sample size is very small, but we are confident that the mean counts in the kill room are higher. (c) The data are counts, which are at best only approximately Normal, and we have only a small sample.

20.41: A stemplot reveals that these data contain two extreme high outliers (5973 and 8015). Hence, t procedures are not appropriate.

20.42: (a) H_0: $\mu = 0$ vs. H_a: $\mu > 0$, where μ is the mean loss in sweetness (sweetness before storage minus sweetness after storage). $t = \frac{1.02 - 0}{1.196/\sqrt{10}} = 2.697$ with df = 9. $P = 0.012$. There is strong evidence that storage reduces sweetness for this cola. (b) We have only 10 observations, and it isn't possible to assess Normality of the distribution of score differences. Indeed, a stemplot of these data reveals possible skew in this distribution.

```
−1 | 3
−0 | 4
 0 | 4 7
 1 | 1 2
 2 | 0 0 2 3
```

20.43: (a) From Table C, $t^* = 2.000$ (df = 60). Using software, with df = 63, $t^* = 1.998$. The 95% confidence interval for μ is $48.25 \pm 2.000(\frac{40.24}{\sqrt{64}}) = 38.19$ to 58.31 thousand barrels. (Using the software version of t^*, the confidence interval is almost identical: 38.20 to 58.30 thousand barrels.) (b) The stemplot confirms the skewness and outliers described in the exercise. The two intervals have similar widths, but the new interval is shifted higher by about 2000 barrels. Although t procedures are fairly robust, we should be cautious about trusting the result in (a) because of the strong skew and outliers. The computer-intensive method may produce a more reliable interval.

```
0 | 0 0 0 0 1 1 1 1 1 1 1 1 1 1
0 | 2 2 2 2 2 2 2 3 3 3 3 3 3 3 3 3 3 3 3
0 | 4 4 4 4 4 4 4 5 5 5 5 5 5 5
0 | 6 6 6 6 6 6 7
0 | 8 8 9 9
1 | 0 1
1 |
1 | 5
1 |
1 | 9
2 | 0
```

20.44: (a) Let μ represent the mean *E. coli* counts for all possible 100 milliliter samples taken from all Central Ohio swimming areas. H_0: $\mu = 130$ vs. H_a: $\mu > 130$; $t = \frac{56.28125 - 130}{77.28992/\sqrt{16}} = -3.815$ on df = 15, so $P = 0.9991$. There is no evidence that swimming areas in Central Ohio have mean *E. coli* counts greater than 130 bacteria per 100 ml. (b) A stemplot is provided. Note that stems are in units of 100, and data were rounded to the nearest 10. For example, "2 | 9" represents 290, which corresponds to the original sample value of 291, whereas "0 | 0" represents 0, which corresponds to the original sample value of 1. Due to extreme skew and the presence of outliers, t procedures should not be used here. Both tests provide similar P-values, and both tests reach the obvious conclusion, but this is not validation of the t-test.

```
0 | 0 1 1 1 2 2 2 3 3 4 5 5 5 9
1 | 9
2 | 9
```

20.45: Let μ be the mean percent of beetle-infected seeds. A stemplot shows a single-peak and roughly symmetric distribution. We assume that the 28 plants can be viewed as an SRS of the population, so t procedures are appropriate. Using df = 27, the 90% confidence interval for μ is 4.0786 ± 1.703

$\left(\frac{2.0135}{\sqrt{28}}\right) = 3.43\%$ to 4.73%. The beetle infects less than 5% of seeds, so it is unlikely to be effective in controlling velvetleaf.

20.46: H_0: $\mu = 0$ vs. H_a: $\mu > 0$, where μ represents the mean increase in T cell counts after 20 days on blinatumomab. A stemplot suggests that t procedures are reasonable, with no evidence of non-Normality and no outliers. df = 5 and $t = \frac{0.5283 - 0}{0.4574/\sqrt{6}} = 2.829$, $P = 0.0184$. We would reject H_0 at the 5% significance level. The data give convincing evidence that the mean count of T cells is higher after 20 days on blinatumomab.

20.47: A 95% confidence interval for the mean difference in T cell counts after 20 days on blinatumomab is $0.5283 \pm 2.517\left(\frac{0.4574}{\sqrt{6}}\right) = 0.5283 \pm 0.4801 = 0.0482$ to 1.0084 thousand cells.

20.48: (a) Fund and index performances are certainly not independent; for example, a good year for one is likely to be a good year for the other. (b) Let μ be the mean difference (Fund minus EAFE). H_0: $\mu = 0$ vs. H_a: $\mu \neq 0$. A stemplot shows no reason to doubt Normality. We must assume that the data we have can be viewed as an SRS. $t = \frac{-0.0754 - 0}{7.9979/\sqrt{24}} = -0.05$, for which $P > 0.5$ (software gives 0.9636). We have very little reason to doubt that $\mu = 0$; VIG Fund performance is not significantly different from its benchmark.

20.49: (a) For each subject, randomly select which knob (right or left) that subject should use first. (b) H_0: $\mu = 0$ vs. H_a: $\mu < 0$, where μ denotes the mean difference in time (right-thread time − left-thread time), so that $\mu < 0$ means "right-hand time is less than left-hand time on average." A stemplot of the differences gives no reason that t procedures are not appropriate. We assume our sample can be viewed as an SRS. $t = \frac{-13.32 - 0}{22.936/\sqrt{25}} = -2.90$. With df = 24 we find $P = 0.0039$. We have good evidence (significant at the 1% level) that the mean difference really is negative—that is, the mean time for right-hand-thread knobs is less than the mean time for left-hand-thread knobs.

20.50: H_0: $\mu = 0$ vs. H_a: $\mu \neq 0$, where μ denotes the mean difference in absorption (generic minus reference). We assume that the subjects can be considered an SRS. A stemplot of the differences looks reasonably Normal with no outliers, so the t procedures should be safe. $t = \frac{37 - 0}{1070.6/\sqrt{20}} = 0.15$. With df = 19 we see that $P > 0.5$ (software gives $P = 0.8788$). We cannot conclude that the two drugs differ in mean absorption level.

20.51: With df = 24, $t^* = 1.711$, so the confidence interval for μ is given by $-13.32 \pm 1.711\left(\frac{22.936}{\sqrt{25}}\right) = -13.32 \pm 7.85 = -21.2$ to -5.5 seconds. Now $\bar{x}_{RH}/\bar{x}_{LH} = 104.12/117.44 = 0.887$. Hence, right-handers working with right-handed knobs can accomplish the task in about 89% of the time needed by those working with left-handed knobs.

20.52: Let μ denote the average tip percentage for all patrons receiving a bad weather forecast. H_0: $\mu = 20\%$ vs. H_a: $\mu < 20\%$. We assume we may consider the sample to be an SRS taken from the population of all patrons receiving such a weather report. A stemplot of these data reveal no reason to suspect that t procedures are not appropriate. There are no outliers, and the data are roughly symmetric with one peak. $t = \frac{18.19 - 20}{2.105/\sqrt{20}} = -3.845$. With df = 19, $P = 0.0005$ (using software). There is overwhelming evidence that the mean tip percentage for patrons receiving a bad weather report is less than 20%.

20.53: (a) The margin of error for 90% confidence is $1.729\left(\frac{1.963}{\sqrt{20}}\right) = 0.759$, so the interval is $22.1 \pm 0.759 = 21.341$ to 22.859. (b) The test statistic is $t = \frac{22.1 - 22}{1.963/\sqrt{20}} = 0.227$, for which the two-sided P-value is greater than 50%. (c) The test statistic is $t = \frac{22.1 - 21}{1.963/\sqrt{20}} = 2.51$, for which the two-sided P-value is $0.02 < P < 0.04$. This is significant at the 10% level because $0.04 < 0.10$.

20.54: (a) No, because 33 falls in the 95% confidence interval, which is (27.5, 33.9). (b) Yes, because 34 does not fall in the 95% confidence interval.

20.55: (a) The test statistic is $z = \frac{0.667 - 0.5}{\sqrt{0.5(1 - 0.5)/33}} = 1.91$. The two-sided P-value is 0.0561. With this P-value we would fail to reject the null hypothesis that the proportion is 0.50 at the 0.05 level because $0.0561 > 0.05$. (b) The 95% confidence interval will be $0.667 \pm 1.96\sqrt{\frac{0.667(1 - 0.667)}{33}} = 0.5062$ to 0.8278. With this confidence interval, we would reject the null hypothesis that the proportion is 0.50 at the 0.05 level because 0.50 is not included in the confidence interval.

Chapter 21   Comparing Two Means

21.1: This is a matched pairs design. Each couple is a matched pair.

21.2: This involves two independent samples.

21.3: This involves a single sample.

21.4: This involves two independent samples.

21.5: (a) If the loggers had known that a study would be done, they might have (consciously or subconsciously) cut down fewer trees than they typically would, in order to reduce the impact of logging. (b) H_0: $\mu_1 = \mu_2$ vs. H_a: $\mu_1 > \mu_2$, where μ_1 is the mean number of species in unlogged plots and μ_2 is the mean number of species in plots logged 8 years earlier. We assume that the data come from SRSs of the two populations. Stemplots suggest some deviation from Normality and a possible low outlier for the logged-plot counts, but there is not strong evidence of non-Normality in either sample. With $\bar{x}_1 = 17.50$, $\bar{x}_2 = 13.67$, $s_1 = 3.53$, $s_2 = 4.50$, $n_1 = 12$, and $n_2 = 9$: SE $= \sqrt{\frac{s_1^2}{n_1} + \frac{s_2^2}{n_2}} = 1.813$ and $t = \frac{x_1 - x_2}{\text{SE}} = 2.11$. Using df as the smaller of $9 - 1$ and $12 - 1$, we have df = 8 and $0.025 < P < 0.05$. Using software, df = 14.8 and $P = 0.0260$. There is strong evidence that the mean number of species in unlogged plots is greater than that for logged plots 8 years after logging.

21.6: H_0: $\mu_1 = \mu_2$ vs. H_a: $\mu_1 \neq \mu_2$, where μ_1 is the mean time spent lying down for the lean group, and μ_2 is the mean time for the obese group. We assume that the data come from SRSs of the two populations. Stemplots do not indicate non-Normal data. We proceed with the t test for two samples. With $\bar{x}_1 = 501.6461$, $\bar{x}_2 = 491.7426$, $s_1 = 52.0449$, $s_2 = 46.5932$, $n_1 = 10$, and $n_2 = 10$: SE $= \sqrt{\frac{s_1^2}{n_1} + \frac{s_2^2}{n_2}} = 22.0898$ and $t = \frac{\bar{x}_1 - \bar{x}_2}{\text{SE}} = 0.448$. Using df as the smaller of $10 - 1$ and $10 - 1$, we have df $= 9$ and $P > 0.50$. Using software, df $= 17.8$ and $P = 0.6596$. There is no evidence to support a conclusion that lean people spend a different amount of time lying down (on average) than obese people.

21.7: $\bar{x}_1 = 17.50$, $\bar{x}_2 = 13.67$, and SE $= 1.813$. Using df $= 8$, $t^* = 3.355$. A 99% confidence interval for the mean difference in number of species in unlogged and logged plots is $\bar{x}_1 - \bar{x}_2 \pm t^*\text{SE} = -2.253$ to 9.913 species.

21.8: In this study, men underestimated their average life expectancy by 19.50%, whereas women underestimated their average life expectancy by 12.71%. If these samples can be viewed as SRSs, then under H_0: $\mu_1 = \mu_2$, a difference in sample means as great as the one observed ($12.71\% - 19.50\%$) is 2.177 standard errors below expected ($t = -2.177$), and a more extreme difference would have occurred by chance alone about 5.28% of the time under repeated sampling ($P = 0.0528$). There is somewhat strong evidence that men and women differ in their views on their own longevity.

21.9: (a) Back-to-back stemplots of the time data are shown below. They appear to be reasonably Normal, and the discussion in the exercise justifies our treating the data as independent SRSs, so we can use the t procedures. H_0: $\mu_1 = \mu_2$ vs. H_a: $\mu_1 < \mu_2$, where μ_1 is the is the population mean time in the restaurant with no scent, and μ_2 is the mean time in the restaurant with a lavender odor. With $\bar{x}_1 = 91.27$, $\bar{x}_2 = 105.700$, $s_1 = 14.930$, $s_2 = 13.105$, $n_1 = 30$, and $n_2 = 30$: SE $= \sqrt{\frac{s_1^2}{n_1} + \frac{s_2^2}{n_2}} = 3.627$ and $t = \frac{\bar{x}_1 - \bar{x}_2}{\text{SE}} = -3.98$. Using software, df $= 57.041$, and $P = 0.0001$. Using the more conservative df $= 29$ (lesser of $30 - 1$ and $30 - 1$) and Table C, $P < 0.0005$. There is very strong evidence that customers spend more time on average in the restaurant when the lavender scent is present. (b) Back-to-back stemplots of the spending data are below. The distributions are skewed and have many gaps. This may be due to pricing of menu items. We'll proceed, but be a bit cautious in conclusions. H_0: $\mu_1 = \mu_2$ vs. H_a: $\mu_1 < \mu_2$, where μ_1 is the the population mean amount spent in the restaurant with no scent and μ_2 is the mean amount spent in the restaurant with lavender odor. With $\bar{x}_1 = \$17.5133$, $\bar{x}_2 = \$21.1233$, $s_1 = \$2.3588$, $s_2 = \$2.3450$, $n_1 = 30$, and $n_2 = 30$: SE $= \sqrt{\frac{s_1^2}{n_1} + \frac{s_2^2}{n_2}} = \$0.6073$ and $t = \frac{\bar{x}_1 - \bar{x}_2}{\text{SE}} = -5.95$. Using software, df $= 57.998$ and $P < 0.0001$. Using the more conservative df $= 29$ and Table C,

$P < 0.0005$. There is very strong evidence that customers spend more money on average when the lavender scent is present. In this case, the possibly non-Normal distributions do not seem to affect the decision.

No scent		Lavender
9 8	6	
3 2 2	7	
9 6 5	7	6
4 4	8	
7 7 6 5	8	8 9
3 2 2 2 1	9	2 3 4
8 6	9	5 7 8
3 1	10	1 2 3 4
9 7 7 6	10	5 5 6 6 7 8 8 9 9 9
	11	4
8 5	11	6
1	12	1 4
	12	6 9
	13	
	13	7

No scent		Lavender
9	12	
	13	
	14	
9 9 9 9 9 9 9 9 9 9 9 9 9 9	15	
	16	
	17	
5 5 5 5 5 5 5 5 5 5 5 5	18	5 5 5 5 5 5 5 5 5 5 5
	19	
5	20	7
9	21	5 5 9 9 9 9 9 9 9
	22	3 5 5 8
	23	
	24	9 9
5	25	5 9

21.10: (a) The "compressed" stemplot shows no particular cause for concern. The "intermediate" stemplot shows slight skewness and the outlier described in the problem statement. (b) H_0: $\mu_1 = \mu_2$ vs. H_a: $\mu_1 < \mu_2$, where μ_1 is the population mean for the compressed soil and μ_2 is the mean for the intermediate soil. Here, with $\bar{x}_1 = 2.9075$, $\bar{x}_2 = 3.3360$, $s_1 = 0.1390$, $s_2 = 0.3190$, $n_1 = 20$, and $n_2 = 20$: SE $= \sqrt{\frac{s_1^2}{n_1} + \frac{s_2^2}{n_2}} = 0.0778$ and $t = \frac{\bar{x}_1 - \bar{x}_2}{\text{SE}} = -5.50$. Using either version of df (software df $= 25.96$, more conservative df $= 19$), $P < 0.0001$. There is very significant evidence that the mean penetrability for compressed soil is lower than that for intermediate soil.

Compressed		Intermediate	
26	8	2	9 9
27		3	0 1 1 1 1 1 1
27	6 8 8 8	3	2 3 3 3
28	1 2 2	3	4 4 4 5
28	6 6	3	6
29	0 2 4	3	8
29	6 8	4	
30	0	4	2
30	8 8		
31			
31	6 8		

21.11: We have two small samples ($n_1 = n_2 = 4$), so the t procedures are not reliable unless both distributions are Normal.

21.12: If we use software df = 25.96, then $t^* = 1.706$. If we use the more conservative df = 19, then $t^* = 1.729$. A 90% confidence interval for the mean difference in permeability between compressed and intermediate soils is $\bar{x}_1 - \bar{x}_2 \pm t^* SE$. So, -0.561 to -0.296 (if df = 25.96) or -0.563 to -0.294 (if df = 19).

21.13: Here are the details of the calculations:

$SE_F = \frac{12.6961}{\sqrt{31}} = 2.2803$

$SE_M = \frac{12.2649}{\sqrt{47}} = 1.7890$

$SE = \sqrt{SE_F^2 + SE_M^2} = 2.8983$

$df = \frac{SE^4}{\frac{1}{30}\left(\frac{12.6961^2}{31}\right)^2 + \frac{1}{46}\left(\frac{12.2649^2}{47}\right)^2} = \frac{70.565}{1.1239} = 62.8$

$t = \frac{55.5161 - 57.9149}{SE} = -0.8276.$

21.14: Let μ_1 denote the mean for men and μ_2 denote the mean for women. According to the output, $\bar{x}_1 = -19.50$, $\bar{x}_2 = -12.71$, $s_1 = 5.612$ and $s_2 = 5.589$. Hence, with $n_1 = 6$ and $n_2 = 7$:

$t = \frac{\bar{x}_1 - \bar{x}_2}{\sqrt{\frac{s_1^2}{n_1} + \frac{s_2^2}{n_2}}} = \frac{-21.50 - (-12.71)}{\sqrt{\frac{5.612^2}{6} + \frac{5.589^2}{7}}} = -2.179.$

Also,

$df = \frac{\left(\frac{5.612^2}{6} + \frac{5.589^2}{7}\right)^2}{\frac{1}{6-1}\left(\frac{5.612^2}{6}\right)^2 + \frac{1}{7-1}\left(\frac{5.589^2}{7}\right)^2} = 10.68$ rounded to two places.

21.15: Reading from the software output shown in the statement of Exercise 21.13, we find that there was no significant difference in mean Self-Concept Scale scores for men and women ($t = -0.8276$, df = 62.8, $P = 0.4110$).

21.16: (c) There is one sample and only one score from each member of the sample.

21.17: (a) We have two independent populations: females and males.

21.18: (b) Two measurements (one for each test) are being taken on each mouse.

21.19: (b)

21.20: (c) Here, df is the lesser of $(21 - 1)$ and $(21 - 1)$.

21.21: (b)

21.22: (c) If random digit dialing produces an SRS, then our samples are SRSs from their respective populations. Also, the samples are large enough to overcome problems of potential non-Normality.

21.23: (a) We suspect that younger people use social networks more than older people, so this is a one-sided alternative.

21.24: (b)

21.25: (a) $H_0: \mu_M = \mu_F$ vs. $H_a: \mu_M < \mu_F$. (b)–(d) The small table below provides a summary of t statistics, degrees of freedom, and P-values for both studies. The two-sample t statistic is computed as $t = \frac{\bar{x}_M - \bar{x}_F}{\sqrt{\frac{s_M^2}{n_M} + \frac{s_F^2}{n_F}}}$, and we take the conservative approach for computing df as the smaller sample size minus 1.

STUDY	t	df	TABLE C VALUES	P-VALUE
1	-0.248	55	$|t| < 0.679$	$P > 0.25$
2	1.507	19	$1.328 < t < 1.729$	$0.05 < P < 0.10$

Note that for Study 1 we reference df = 50 in Table C. (e) The first study gives no support to the belief that women talk more than men; the second study gives weak support, significant only at a relatively high significance level (say $\alpha = 0.10$).

21.26: (a) Call group 1 the Alcohol group and group 2 the Placebo group. Then, because SEM $= s/\sqrt{n}$, we have $s =$ SEM$\sqrt{n}$. So, $s_1 = 0.05\sqrt{25} = 0.25$ and $s_2 = 0.03\sqrt{25} = 0.15$. (b) Using the conservative Option 2, df = 24 (the lesser of 25 and 25, minus 1). (c) Here with $n_1 = n_2 = 25$, SE $= \sqrt{\frac{s_1^2}{n_1} + \frac{s_2^2}{n_2}} = 0.0583$. With df = 24, we have $t^* = 1.711$. A 90% confidence interval for the mean difference in proportions is given by $(0.25 - 0.12) \pm 1.711(0.0583) = 0.03$ to 0.23.

21.27: (a) Call group 1 the Stress group and group 2 the No stress group. Then, because SEM $= s/\sqrt{n}$, we have $s =$ SEM$\sqrt{n}$. Hence, $s_1 = 3\sqrt{20} = 13.416$ and $s_2 = 2\sqrt{51} = 14.283$. (b) Using the conservative Option 2, df = 19 (the lesser of 20 and 51, minus 1). (c) H_0: $\mu_1 = \mu_2$ vs. H_a: $\mu_1 \neq \mu_2$. With $n_1 = 20$ and $n_2 = 51$, SE $= \sqrt{\frac{s_1^2}{n_1} + \frac{s_2^2}{n_2}} = 3.605$ and $t = \frac{\bar{x}_1 - \bar{x}_2}{SE} = \frac{26 - 32}{3.605} = -1.664$. With df $= 19$, using Table C, $0.10 < P < 0.20$. There is little evidence in support of a conclusion that mean weights of rats in stressful environments differ from those of rats without stress.

21.28: (a) Parents who choose a Montessori school probably have different attitudes about education than other parents. (b) Over 55% of Montessori parents (30 out of 54) participated in the study, compared with about 22% of the other parents (25 out of 112). (c) H_0: $\mu_1 = \mu_2$ vs. H_a: $\mu_1 \neq \mu_2$, where μ_1 is the mean math score for Montessori children and μ_2 is the mean score for non-Montessori children. With $\bar{x}_1 = 19$, $\bar{x}_2 = 17$, $s_1 = 3.11$, $s_2 = 4.19$, $n_1 = 30$, and $n_2 = 25$: SE $= \sqrt{\frac{s_1^2}{n_1} + \frac{s_2^2}{n_2}} = 1.0122$ and $t = \frac{\bar{x}_1 - \bar{x}_2}{SE} = -1.976$. Using df $= 24$ under Option 2, $0.05 < P < 0.10$. Using software, df $= 43.5$ and $P = 0.0545$. There is moderate evidence of a difference in mean math scores between these two groups, but not quite enough evidence to reach such a conclusion at the 5% significance level.

21.29: (a) A placebo is an inactive pill that allows researchers to account for any psychological benefit (or detriment) the subject might get from taking a pill. (b) Neither the subjects nor the researchers who worked with them knew who was getting ginkgo extract; this prevents expectations or prejudices from affecting the evaluation of the effectiveness of the treatment. (c) SE $= \sqrt{\frac{0.01462^2}{21} + \frac{0.01549^2}{18}} = 0.0048$; $t = \frac{0.06383 - 0.05342}{SE} = 2.147$. This is significant at the 5% level: $P = 0.0387$ (df $= 35.35$) or $0.04 < P < 0.05$ (df $= 17$). There is strong evidence that those who take gingko extract average more misses per line read.

21.30: (a) Using the conservative two-sample procedures, df $= 17$ (the lesser of 18 and 18, minus 1). (b) Using the data summary provided in the problem description, $t = \frac{5.16 - 3.47}{\sqrt{\frac{3.5^2}{18} + \frac{2.0^2}{18}}} = 1.779$. (c) H_0: $\mu_1 = \mu_2$ vs. H_a: $\mu_1 \neq \mu_2$, where μ_1 is the mean for the lack-of-control group, and μ_2 is the mean for the in-control group. Using Table C, with df $= 17$, $0.05 < P < 0.10$. There is at best weak evidence of a difference in means between the in-control and lack-of-control groups.

21.31: Let μ_1 be mean for people with Asperger's syndrome and let and μ_2 be the mean for people without Asperger's syndrome. H_0: $\mu_1 = \mu_2$ vs. H_a: $\mu_1 \neq \mu_2$. Here $\bar{x}_1 = -0.001$, $\bar{x}_2 = 0.42$, $n_1 = 19$, and $n_2 = 17$. Because SEM $= s/\sqrt{n}$, we have $s = $ SEM$\sqrt{n}$. So, $s_1 = 0.15\sqrt{19} = 0.6538$ and $0.17\sqrt{17} = 0.7009$. SE $= \sqrt{\frac{s_1^2}{n_1} + \frac{s_2^2}{n_2}} = 0.2267$ and $t = \frac{\bar{x}_1 - \bar{x}_2}{SE} = -1.857$. Using the conservative version for df (Option 2), df $= 16$ and $0.05 < P < 0.10$. Using software, df $= 32.89$ and $P = 0.0723$. There is strong evidence that

the mean score for Asperger's syndrome population is different from that of the non-Asperger's population.

21.32: (a) The appropriate test is the matched-pairs test because a student's score on Try 2 is certainly dependent on his/her score on Try 1. Using the differences, we have $\bar{x} = 29$ and $s = 59$. (b) H_0: $\mu = 0$ vs. H_a: $\mu > 0$; $t = \frac{29 - 0}{59/\sqrt{427}} = 10.16$ with df $= 426$. $P < 0.0005$. It appears that coached students improve their scores on average. (c) Table C gives $t^* = 2.626$ for df $= 100$, while software gives $t^* = 2.587$ for df $= 426$. The confidence interval is $\bar{x} \pm t^*/s\sqrt{n}$. Using the conservative value of t^*, this yields 21.50 to 36.50 points. Using software, the confidence interval is 21.61 to 36.39.

21.33: (a) H_0: $\mu_1 = \mu_2$ vs. H_a: $\mu_1 > \mu_2$, where μ_1 is the mean gain among all coached students and μ_2 is the mean gain among uncoached students. SE $= \sqrt{\frac{59^2}{427} + \frac{52^2}{2733}} = 3.0235$ and $t = \frac{29 - 21}{3.0235} = 2.646$. Using the conservative approach, df $= 426$ is rounded down to df $= 100$ in Table C, and we obtain $0.0025 < P < 0.005$. Using software, df $= 534.45$ and $P = 0.0042$. There is evidence that coached students had a greater average increase than uncoached students. (b) $8 \pm t^*(3.0235)$ where t^* equals 2.626 (using df $= 100$ with Table C) or 2.585 (df $= 534.45$ with software). This gives either 0.06 to 15.94 points, or 0.184 to 15.816 points, respectively. (c) Increasing one's score by 0 to 16 points is not likely to make a difference in being granted admission or scholarships from any colleges.

21.34: This was an observational study, not an experiment. The students (or their parents) chose whether or not to be coached; students who choose coaching might have other motivating factors that help them do better the second time.

21.35: (a) Neither sample histogram suggests strong skew or presence of far outliers. t procedures are reasonable here. (b) Let μ_1 be the mean tip percentage when the forecast is good and μ_2 be the mean tip percentage when the forecast is bad. $\bar{x}_1 = 22.22$, $\bar{x}_2 = 18.19$, $s_1 = 1.955$, $s_2 = 2.105$, $n_1 = 20$, and $n_2 = 20$. H_0: $\mu_1 = \mu_2$ vs. H_a: $\mu_1 \neq \mu_2$. SE $= \sqrt{\frac{s_1^2}{n_1} + \frac{s_2^2}{n_2}} = 0.642$ and $t = \frac{\bar{x}_1 - \bar{x}_2}{SE} = 6.274$. Using df $= 19$ (Option 2) and Table C, we have $P < 0.001$. Using software, df $= 37.8$, and $P < 0.00001$. There is overwhelming evidence that the mean tip percentage differs between the two types of forecasts presented to patrons.

21.36: (a) Based on stemplots, the t procedures should be safe. Both stemplots and the means suggest that customers stayed (very slightly) longer when there was no odor. (b) H_0: $\mu_1 = \mu_2$ vs. H_a: $\mu_1 \neq \mu_2$, where μ_1 is the mean time in restaurant with no odor and μ_2 is the mean time in restaurant with lemon odor. $\bar{x}_1 = 91.2667$, $\bar{x}_2 = 89.7857$, $s_1 = 14.9296$,

$s_2 = 15.4377$, $n_1 = 30$, and $n_2 = 28$. We find SE $= 3.9927$ and $t = 0.371$. This is not at all significant: $P > 0.5$ (df $= 27$, using Table C with Option 2 for conservative df) or $P = 0.7121$ (df $= 55.4$, using software). We cannot conclude that mean time in the restaurant is different when the lemon odor is present.

No odor		Lemon
	5	6
	6	0 3
9 8	6	
3 2 2	7	3 4
9 6 5	7	5 8
4 4	8	3 3
7 7 6 5	8	8 8 8 9
3 2 2 2 1	9	0 1 4 4
8 6	9	6 7 7
3 1	10	1 4
9 7 6 6	10	5 6 8 8
	11	2 3
8 5	11	
1	12	

21.37: df $= 19$, $t^* = 2.093$. The 95% confidence interval for the difference in mean tip percentages between these two populations is $(22.22 - 18.19) \pm 2.093(0.642) = 2.69$ to 5.37 percent. Using df $= 37.8$ with software, the corresponding 95% confidence interval is 2.73% to 5.33%.

21.38: (a) We provide some summary statistics describing the two samples. The second line for Permafresh, denoted with *, has the low outlier of interest omitted.

	n	$\bar{x}$	s
Permafresh	5	29.54	1.1675
Permafresh*	4	30.025	0.4992
Hylite	5	25.20	2.6693

(b)

Permafresh		Hylite
	22	1
	23	9
	24	2
	25	
	26	
6	27	0
	28	8
9 5	29	
7 0	30	

(c) With summary statistics listed in the table, we find the following test statistics and P-values:

			CONSERVATIVE	SOFTWARE	
	t	df	P	df	P
All points	3.33	4	$0.02 < P < 0.04$	5.476	0.0181
Outlier removed	3.96	3	$0.02 < P < 0.04$	4.346	0.0142

The mild outlier in the Permafresh sample had almost no effect on our conclusion. Despite the small sample sizes, there is good evidence that the mean breaking strengths of the two processes differ.

21.39: (a) The Hylite mean is greater than the Permafresh mean. (b) Shown are back-to-back stemplots for the two processes, which confirm that there are no extreme outliers. (c) SE $= 1.334$ and $t = -6.296$. $0.002 < P < 0.005$ (using df $= 4$) or P-value $= 0.0003$ (using software, with df $= 7.779$). There is very strong evidence of a difference between the population means. As we might expect, the stronger process (Permafresh) is less resistant to wrinkles.

	n	$\bar{x}$	s
Permafresh	5	134.8	1.9235
Hylite	5	143.2	2.2804

Permafresh		Hylite
2	13	
5 4	13	
7 6	13	
	13	
	14	1 1
	14	3
	14	5
	14	6

21.40: The 90% confidence interval is $\bar{x}_1 - \bar{x}_2 \pm t^* SE$, where $t^* = 2.132$ (df $= 4$) or $t^* = 1.977$ (df $= 5.476$). This gives either

$4.34 \pm 2.778 = 1.562$ to 7.118 pounds (with df $= 4$) or
$4.34 \pm 2.576 = 1.764$ to 6.916 pounds (with df $= 5.476$).

21.41: The 90% confidence interval is $\bar{x}_1 - \bar{x}_2 \pm t^* SE$, where $t^* = 2.132$ (df $= 4$) or $t^* = 1.867$ (df $= 7.779$). This gives either

$-8.4 \pm 2.844 = -11.244$ to -5.556 degrees (with df $= 4$) or
$-8.4 \pm 2.491 = -10.891$ to -5.909 degrees (with df $= 7.779$).

21.42: (a) A stemplot is provided. Each data value is rounded to the nearest thousand, and stems are in units of ten thousand. The stemplots suggest that there is some skew in both populations, but the sample sizes should be large enough to overcome this problem. (b) $H_0: \mu_1 = \mu_2$ vs. $H_a: \mu_1 > \mu_2$.

$\bar{x}_1 = 16,496.1$, $\bar{x}_2 = 12,866.7$, $s_1 = 7914.35$, $s_2 = 8342.47$, $n_1 = 27$, and $n_2 = 20$. We find SE $= 2408.26$ and $t = 1.51$. With df $= 39.8$ (software), $P = 0.070$. Using Table C with the more conservative df $= 19$, $0.05 < P < 0.10$. There is some evidence that on average, women say more words than men, but the evidence is not particularly strong.

Women		Men
9 8 8 7 6	0	4 4 5 6 7 8 9
4 3 3 2 1 0 0 0	1	0 0 1 1 1 3 3 3
9 9 8 7 6 5	1	6 8
3 0	2	2
7 6 5 5 5	2	8
	3	
	3	8
	0	4

21.43: This is a two-sample t statistic, comparing two independent groups (supplemented and control). Using the conservative df $= 5$, $t = -1.05$ would have a P-value between 0.30 and 0.40, which (as the report said) is not significant. The test statistic $t = -1.05$ would not be significant for any value of df.

21.44: H_0: $\mu_1 = \mu_2$ vs. H_a: $\mu_1 \neq \mu_2$, where μ_1 is the mean days behind caterpillar peak for the control group, and μ_2 is the mean days for the supplemented group. $\bar{x}_1 = 4.0$, $\bar{x}_2 = 11.3$, $s_1 = 3.10934$, $s_2 = 3.92556$, $n_1 = 6$, and $n_2 = 7$; SE $= \sqrt{\frac{s_1^2}{n_1} + \frac{s_2^2}{n_2}} = 1.95263$, and $t = \frac{4.0 - 11.3}{\text{SE}} = -3.74$. The two-sided P-value is either $0.01 < P < 0.02$ (using df $= 5$) or 0.0033 (using df $= 10.96$ with software), agreeing with the stated conclusion (a significant difference).

21.45: These are paired t statistics: for each bird, the number of days behind the caterpillar peak was observed, and the t values were computed based on the pairwise differences between the first and second years. For the control group, df $= 5$, and for the supplemented group, df $= 6$. The control t is not significant (so the birds in that group did *not* "advance their laying date in the second year"), whereas the supplemented group t is significant with one-sided $P = 0.0195$ (so those birds did change their laying date).

21.46: H_0: $\mu_1 = \mu_2$ vs. H_a: $\mu_1 > \mu_2$, where μ_1 is the mean time for the treatment group, and μ_2 is the mean time for the control group. We must assume that the data come from an SRS of the intended population; we cannot check this with the data. A back-to-back stemplot shows some irregularity in the treatment times and skewness in the control times. We hope that our equal and moderate sample sizes will overcome any deviation from Normality. With $\bar{x}_1 = 314.0588$, $\bar{x}_2 = 186.1176$, $s_1 = 172.7898$, $s_2 = 118.0926$, $n_1 = 17$, and $n_2 = 17$, we find SE $= \sqrt{\frac{s_1^2}{n_1} + \frac{s_2^2}{n_2}} = 50.7602$, and $t = \frac{314.0588 - 186.1176}{\text{SE}} = 2.521$, for which $0.01 < P < 0.02$ (df $= 16$) or $P = 0.0088$ (df $= 28.27$). There is strong evidence that the treatment group waited longer to ask for help on average.

21.47: H_0: $\mu_1 = \mu_2$ vs. H_a: $\mu_1 > \mu_2$, where μ_1 is the mean weight loss for adolescents in the gastric banding group and μ_2 is the mean time for the lifestyle intervention group. We must assume that the data come from an SRS of the intended population; we cannot check this with the data. Stemplots for each sample show no heavy skew and no outliers. With $\bar{x}_1 = 34.87$, $\bar{x}_2 = 3.01$, $s_1 = 18.12$, $s_2 = 13.22$, $n_1 = 24$, and $n_2 = 18$ (note that not all subjects completed the study), SE $= \sqrt{\frac{s_1^2}{n_1} + \frac{s_2^2}{n_2}} = 4.84$ and $t = \frac{34.87 - 3.01}{\text{SE}} = 6.59$, for which $P < 0.0005$ (df $= 17$) or $P < 0.00001$ (df $= 39.98$ using software). There is strong evidence that adolescents using gastric banding lose more weight on average than those that use lifestyle modification.

21.48: H_0: $\mu_1 = \mu_2$ vs. H_a: $\mu_1 > \mu_2$, where μ_1 is the mean score for the Active group, and μ_2 is the mean score for the Traditional group. We must assume that the data come from an SRS of the intended population. Stemplots for each sample show no heavy skew and no outliers. $\bar{x}_1 = 3.6$, $\bar{x}_2 = 4.74$, $s_1 = 2.41$, $s_2 = 2.85$, $n_1 = 15$, and $n_2 = 23$. Of course, because $\bar{x}_1 < \bar{x}_2$, we will not conclude that $\mu_1 > \mu_2$. SE $= \sqrt{\frac{s_1^2}{n_1} + \frac{s_2^2}{n_2}} = 0.86$ and $t = \frac{3.6 - 4.74}{\text{SE}} = -1.32$, for which $P > 0.50$, regardless of df. Using software, df $= 33.42$ and $P = 0.9026$. There is no support for a conclusion that active learning yields higher average score than traditional learning.

21.49: H_0: $\mu_1 = \mu_2$ vs. H_a: $\mu_1 > \mu_2$, and find a 90% confidence interval for $\mu_1 - \mu_2$, where μ_1 is the mean for the treatment population, and μ_2 is the mean for the control population. We must assume that we have two SRSs and that the distributions of score improvements are Normal. Back-to-back stemplots of the differences ("after" minus "before") for the two groups; the samples are too small to assess Normality, but there are no outliers. With $\bar{x}_1 = 11.4$, $\bar{x}_2 = 8.25$, $s_1 = 3.1693$, $s_2 = 3.6936$, $n_1 = 10$, and $n_2 = 8$, we find SE $= 1.646$ and $t = 1.914$. With df $= 7$, $0.025 < P < 0.05$. With df $= 13.92$ (software), $P = 0.0382$. The 90% confidence interval is $(11.4 - 8.25) \pm t^*\text{SE}$, where $t^* = 1.895$ (df $= 7$) or $t^* = 1.762$ (df $= 13.92$): either 0.03 to 6.27 points or 0.25 to 6.05 points. We have fairly strong evidence that the encouraging subliminal message led to a greater improvement in math scores, on average. We are 90% confident that this increase is between 0.03 and 6.27 points (or 0.25 and 6.05 points).

21.50: (a) $t^* = 1.761$ (using df $= 14$) or $t^* = 1.692$ (using software). A 90% confidence interval for $\mu_1 - \mu_2$ is $(3.6 - 4.74) \pm t^*(0.86)$, or -2.65 to 0.37 (using df $= 14$) or -2.60 to 0.32 (using df $= 33.42$). (b) Now we want a 90% confidence interval for the mean change in score for the active class. That is, we construct a 90% confidence interval for μ_1. df $= 14$ and $t^* = 1.761$. $3.6 \pm 1.761 \frac{2.41}{\sqrt{15}} = 2.50$ to 4.70.

21.51: H_0: $\mu_1 = \mu_2$ vs. H_a: $\mu_1 \neq \mu_2$, and find a 95% confidence interval for $\mu_1 - \mu_2$, where μ_1 is the mean for the Red population and μ_2 is the mean for the yellow population. We must assume that the data come from an SRS. We also assume that the data are close to Normal. Back-to-back stemplots show some skewness in the red lengths, but the t procedures should be reasonably safe. With $\bar{x}_1 = 39.7113$, $\bar{x}_2 = 36.1800$, $s_1 = 1.7988$, $s_2 = 0.9753$, $n_1 = 23$, and $n_2 = 15$, we find SE = 0.4518 and $t = 7.817$. With either df = 14 or df = 35.10, $P < 0.0001$. The 95% confidence interval is $(39.711 - 36.180) \pm t^*\text{SE}$, where $t^* = 2.145$ (df = 14) or $t^* = 2.030$ (df = 35.1): either 2.562 to 4.500 mm or 2.614 to 4.448 mm. We have very strong evidence that the two varieties differ in mean length. We are 95% confident that the mean red length minus yellow length is between 2.562 and 4.500 mm (or 2.614 and 4.448 mm).

21.52: Answers will vary. A back-to-back stemplot of responses for men and women reveals that the distribution of claimed drinks per day for women is slightly skewed but has no outliers. For men, the distribution is only slightly skewed but contains four outliers. However, these outliers are not too extreme. In all problems, it seems use of t procedures is reasonable.

(a) We construct a 95% confidence interval for μ_w, the mean number of claimed drinks for women. $t^* = 1.990$ (df = 80 in Table A) or $t^* = 1.9855$ (df = 94, software), and SE = $2.1472/\sqrt{95} = 0.2203$. A 95% confidence interval for μ_w is $4.2737 \pm 1.990(0.223) = 3.84$ to 4.71 drinks. With 95% confidence, the mean number of claimed drinks for women is between 3.84 and 4.71 drinks. (b) We construct a 95% confidence interval for μ_m, the mean number of claimed drinks for men. $t^* = 1.990$ (df = 80 in Table A or software) and SE = $3.3471/\sqrt{81} = 0.3719$. A 95% confidence interval for μ_m is $6.5185 \pm 1.990(0.3719) = 5.78$ to 7.26 drinks. With 95% confidence, the mean number of claimed drinks for men is between 5.78 and 7.26 drinks. (c) H_0: $\mu_m = \mu_w$ vs. H_a: $\mu_m \neq \mu_w$. SE = $\sqrt{\frac{2.1472^2}{95} + \frac{3.3471^2}{81}} = 0.4322$ and $t = \frac{4.2737 - 6.5185}{\text{SE}} = -5.193$. Regardless of the choice of df (80 or 132.15), this is highly significant ($P < 0.001$). We have very strong evidence that the claimed number of drinks is different for men and women. To construct a 95% confidence interval for $\mu_m - \mu_w$, we use $t^* = 1.990$ (df = 80) or $t^* = 1.9781$ (df = 132.15). $(\bar{x}_1 - \bar{x}_2) \pm t^*\sqrt{\frac{s_1^2}{n_1} + \frac{s_2^2}{n_2}} = 2.2448 \pm 0.8601$ or 2.2448 ± 0.8549. After rounding either interval, we report that with 95% confidence, on average, sophomore men who drink claim 1.4 to 3.1 more drinks per day than sophomore women who drink.

Chapter 22 Inference in Practice

22.1: The most important reason is (c); this is a convenience sample consisting of the first 20 students on a list. This is not an SRS. Anything we learn from this sample will not extend to the larger population.

22.2: (a) $\bar{x} \pm t^*s/\sqrt{n} = 1.92 \pm 1.984\frac{1.83}{\sqrt{880}} = 1.80$ to 2.04 motorists. (b) The large sample size means that, because of the central limit theorem, the sampling distribution of $\bar{x}$ is roughly Normal, even if the distribution of responses is not. (c) Only people with listed telephone numbers were represented in the sample, and the low response rate (10.9% = 5,029/45,956)

means that even that group may not be well represented by this sample.

22.3: Any number of things could go wrong with this convenience sample. The day after Thanksgiving is widely regarded (rightly or wrongly) as a day on which retailers offer great deals—and the kinds of shoppers found that day probably don't represent shoppers generally. Also, the sample isn't random.

22.4: (a) "Power = 0.24" means that if really (unknown to the researcher) $\mu = 10.15$, then if we repeatedly sample $n = 6$ measurements, each time conducting the significance test described, we will correctly reject H_0 24% of the time. (b) This means that if $\mu = 10.15$, fully 76% of the time (100% − 24% = 76%) under repeated sampling, we will not reject H_0, even though we should.

22.5: (a) Increase power by taking more measurements. (b) If you increase α, you make it easier to reject H_0, and increase power. (c) A value of $\mu = 10.2$ is even farther from the stated value of $\mu = 10.1$ under H_0, so power increases.

22.6: The powers (obtained using the applet) are summarized in the table:

	(a)		(b)		(c)
n	POWER	μ	POWER	α	POWER
6	0.232	10.15	0.232	0.05	0.232
12	0.410	10.20	0.688	0.10	0.339
24	0.688	10.25	0.957	0.25	0.538

(a) As sample size increases (keeping everything else constant), power increases. (b) Keeping everything else constant, power is greater when the alternative considered is further away from 10.1. (c) Power increases when α increases, keeping everything else constant.

22.7: The table below summarizes power as σ changes. As σ decreases, power increases. More precise measurements increase the researcher's ability to recognize a false null hypothesis.

σ	0.10	0.05	0.025
Power	0.232	0.688	0.998

22.8: (a) H_0: The patient is healthy (or "the patient should not see a doctor"); H_a: The patient is ill (or "the patient should see a doctor"). A Type I error is a false positive—sending a healthy patient to the doctor. A Type II error means a false negative—clearing a patient who should be referred to a doctor. (b) One might wish to lower the probability of a false negative so that most ill patients are treated, especially for serious diseases that require fast treatment.

22.9: (a) All statistical methods are based on probability samples. We must have a random sample to be able to apply them.

22.10: (c) Especially with respect to heart rates, male athletes can't be considered to be representative of all male students.

22.11: (b) Inference from a voluntary response sample is never reasonable. Online web surveys are voluntary response surveys.

22.12: (c) Well-designed surveys incur error due to random chance. This random variation is the only source of error accounted for in the margin of error. All forms of bias are not accounted for and are errors in addition to those due to chance.

22.13: (a) There is no control group. Any observed improvement may be due to the treatment or may be due to another cause.

22.14: (a) The significance level (α) is the probability of rejecting H_0 when H_0 is true.

22.15: (b) The power of the test is the probability of rejecting H_0 when H_0 is false. In this case, if $\mu = 3$, then H_0 is false, and power is the probability of rejecting H_0.

22.16: (c) Power describes the test's ability to reject a false H_0.

22.17: We need to know that the samples taken from both populations (hunter-gatherers, agricultural) are random. Are the samples large? Recall that if the samples are very large, then even a small, practically insignificant difference in prevalence of color blindness in the two samples will be deemed statistically significant.

22.18: (a) The sample described is a random sample, but women shopping at a large suburban shopping mall don't represent the population of all women. (b) Because the sample is random, the sample is likely to represent the population of all women that shop at large suburban malls.

22.19: The effect is greater if the sample is small. With a larger sample, the impact of any one value is small.

22.20: (a) The distribution has a low outlier, which makes confidence interval methods unreliable. (b) The timeplot shows a decreasing trend over time, so we should not treat these 29 observations as a sample coming from a single population.

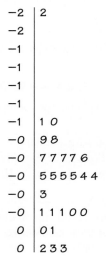

```
-2 | 2
-2 |
-1 |
-1 |
-1 |
-1 |
-1 | 1 0
-0 | 9 8
-0 | 7 7 7 7 6
-0 | 5 5 5 5 4 4
-0 | 3
-0 | 1 1 1 0 0
 0 | 0 1
 0 | 2 3 3
```

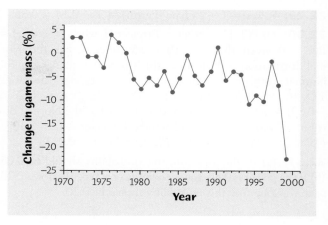

22.21: Opinion—even expert opinion—unsupported by data is the weakest type of evidence, so the third description is level C. The second description refers to experiments (clinical trials) and large samples; that is the strongest evidence (level A). The first description is level B: stronger than opinion, but not as strong as experiments with large numbers of subjects.

22.22: A low-power test may do a good job of *not* incorrectly rejecting the null hypothesis (that is, avoiding a Type I error), but it will often fail to reject H_0 even when it is false, simply because it is difficult to distinguish between H_0 and "nearby" alternatives.

22.23: (a) This test has a 20% chance of rejecting H_0 when the alternative is true. (b) If the test has 20% power, then when the alternative is true, it will fail to reject H_0 80% of the time. (c) The sample sizes are very small, which typically leads to low-power tests.

22.24: (a) The researchers conducted a two-sided test of hypotheses at the $\alpha = 0.05$ level of significance. (b) If there is, in fact a clinically significant difference in risk of death between the early and late intervention groups, then this test would detect that difference (reject the null hypothesis of no difference) 80% of the time, if the experiment was conducted repeatedly.

22.25: From the applet, against the alternative $\mu = 8$, power = 0.609.

22.26: (a) We reject H_0 at the 5% level when $z \geq 1.645$. (b) We reject H_0 when $(3.162)\bar{x} \geq 1.645$, or $\bar{x} \geq 0.5202$. (c) When $\mu = 0.8$, the power is $P(\bar{x} \geq 0.5202) = P(\frac{\bar{x} - 0.8}{1/\sqrt{10}} \geq \frac{0.5202 - 0.8}{1/\sqrt{10}}) = P(Z \geq -0.88) = 0.8106$.

22.27: (a) We reject H_0 at the 5% level when $z \geq 1.96$ or $z \leq -1.96$. Here, $z = \frac{\bar{x} - 10.1}{0.1/\sqrt{6}} = 24.4949(\bar{x} - 10.1)$. (b) We reject H_0 if $24.4949(\bar{x} - 10.1) \leq -1.96$ or if $24.4949(\bar{x} - 10.1) \geq 1.96$. Equivalently (solving for $\bar{x}$), we reject H_0 if $\bar{x} \leq 10.02$ or $\bar{x} \geq 10.18$. (c) When $\mu = 10.15$, the power is $P(\bar{x} \leq 10.02) + P(\bar{x} \geq 10.18) = P(Z \leq \frac{10.02 - 10.15}{0.1/\sqrt{6}}) + P(Z \geq \frac{10.18 - 10.15}{0.1/\sqrt{6}}) = P(Z \leq -3.18) + P(Z \geq 0.73) = 0.2334$.

22.28: The probability of committing a Type I error is $\alpha = 0.01$. The probability of a Type II error is $1 - \text{Power} = 1 - 0.78 = 0.22$.

22.29: Power $= 1 - P(\text{Type II error}) = 1 - 0.14 = 0.86$.

22.30: (a) $P(\text{Type I error}) = P(\text{reject } H_0 \text{ when } H_0 \text{ is true}) = P(\bar{x} > 0 \text{ given that } \mu = 0) = 0.5$, since $\bar{x}$ has the Normal distribution with mean 0. (b) If $\mu = 0.2$, then $\bar{x}$ has the Normal distribution with mean 0.2 and standard deviation $1/\sqrt{16} = 0.255$. $P(\bar{x} \le 0) = P(Z \le \frac{0 - 0.2}{0.25}) = P(Z \le -0.80) = 0.2119$. (c) If $\mu = 0.5$, then $\bar{x}$ has the Normal distribution with mean 0.5 and standard deviation 0.25. Hence, $P(\bar{x} \le 0) = P(Z \le \frac{0 - 0.5}{0.25}) = P(Z \le -2.00) = 0.0228$.

22.31: (a) In the long run, this probability should be 0.05. (b) If the power is 0.80, the probability of a Type II error is 0.20, so in the long run, this probability should be 0.20.

Chapter 23 Inference about Variables: Part IV Review

23.1: (c) The margin of error is $2.056(9.3)/\sqrt{27} = 3.7$.

23.2: The sample has to be an SRS taken from the population. Also, the sample should be free of outliers. The SRS condition is important for the validity of the procedure.

23.3: (b) $t = 2.023$, df $= 13$.

23.4: -2.79 to 14.21 micrometers.

23.5: (d) $\hat{p} = 1926/7028 = 0.274$.

23.6: (a) $\hat{p} = 1020/6889 = 0.148$.

23.7: (d) The standard error is 0.0068.

23.8: (c) The margin of error is $2.576(0.0068) = 0.0176$.

23.9: (a) The standard error is 0.0124. (b) A 95% confidence interval is 0.336 to 0.384.

23.10: (b) 0.3707 because this is $P(Z \ge 1/3)$, or $P(Z \ge 0.33)$.

23.11: (c) mean $= 100$, standard deviation $= 15/\sqrt{60} = 1.94$ (rounded).

23.12: (a) 0.0049; this is $P(Z \ge 2.58)$.

23.13: The answer in Exercise 23.10 would change because this refers to selecting one individual from the population distribution, which is now non-Normal. The answer in Exercise 23.11 would not change—the mean of $\bar{x}$ is 100 (the population mean), and the standard deviation of $\bar{x}$ is $\sigma/\sqrt{n} = 1.94$, regardless of the population distribution. The answer in Exercise 23.12 would, essentially, not change because of the central limit theorem and a sample of $n = 60$.

23.14: (b) The margin of error is $1.972(2.5)/\sqrt{200} = 0.349$.

23.15: (b) The margin of error is $2.005(3.2)/\sqrt{55} = 0.865$.

23.16: (d) $t = \dfrac{2.35 - 1.20}{\sqrt{\frac{2.5^2}{200} + \frac{3.2^2}{55}}} = 2.47$.

23.17: (a) df is the lesser of $(55 - 1)$ and $(200 - 1)$.

23.18: (a) $t = 10.417$.

23.19: With such large samples, t procedures are reasonable.

23.20: (b) The margin of error is $1.654(26.09)/\sqrt{162} = 3.39$.

23.21: (d) The margin of error is 3.53. The point estimate is $11.4 - 6.7 = 4.7$.

23.22: With 90% confidence, we estimate, that on average, female mouse endurance exceeds that of male mouse endurance by between 1.17 minutes and 8.23 minutes. This interval does not include 0, so female mice have significantly higher endurance at the 0.05 level.

23.23: (c) Plus four confidence intervals are reliable for samples of 5 or more in each group.

23.24: (c) Use $p^* = 0.5$ and $z^* = 2.576$.

23.25: Whether $n = 15$ or $n = 150$, the mean of $\bar{x}$ is 445 ms. If $n = 15$, the standard deviation of $\bar{x}$ is $82/\sqrt{15} = 21.17$ ms. If $n = 150$, the standard deviation of $\bar{x}$ is $82/\sqrt{150} = 6.70$ ms.

23.26: If the population we're sampling from is heavily skewed, then a larger sample is required for the central limit theorem to apply. Hence, if $n = 15$, the sampling distribution of $\bar{x}$ may not be approximately Normal, but if $n = 150$, it will surely be approximately Normal.

23.27: $P(\bar{x} > 450) = P(Z > 0.75) = 0.2266$. (Using software, this is 0.2276.)

23.28: (b) $\hat{p} = 225/757 = 0.297$.

23.29: (b) $\sqrt{\frac{0.297(1 - 0.297)}{757}} = 0.017$.

23.30: (b) $0.297 \pm 1.645(0.017)$.

23.31: (a) Subjects (babies) were not assigned to groups being compared.

23.32: (c) It seems reasonable that the researchers suspect that VLBW babies are less likely to graduate from high school.

23.33: (b) $\hat{p} = \dfrac{179 + 193}{242 + 233} = 0.78$.

23.34: (b) $z = \dfrac{0.7397 - 0.8283}{\sqrt{\hat{p}(1 - \hat{p})\left(\frac{1}{242} + \frac{1}{233}\right)}} = -2.34$.

23.35: (b) $t = \dfrac{87.6 - 94.7}{\sqrt{\frac{15.1^2}{113} + \frac{14.9^2}{106}}} = -3.50$.

23.36: (d) $t = \dfrac{86.2 - 89.8}{\sqrt{\frac{13.4^2}{38} + \frac{14^2}{54}}} = -1.25$, and the test is two-sided.

23.37: (a) $0.379 \pm 1.96\sqrt{\frac{0.379(1 - 0.379)}{378}}$.

23.38: (c) Hypotheses are in terms of population parameters (p), not sample statistics ($\hat{p}$).

23.39: (b) $z = \dfrac{0.379 - 0.41}{\sqrt{0.41(1 - 0.41)/348}} = -1.18$, so $P = 0.1190$.

23.40: (a) Random chance easily explains the observed difference.

23.41: (a) The mean will be $\mu_{\hat{p}} = 0.30$ with standard deviation $\sigma_{\hat{p}} = \sqrt{\frac{0.30(1 - 0.30)}{1500}} = 0.0118$. (b) $P(\hat{p} > 0.35) = P(Z > \frac{0.35 - 0.3}{0.0118}) = P(Z > 4.24) < 0.0001$ (software gives 1.1×10^{-5}).

23.42: (a) Using $p = 0.30$, we need $\sqrt{\frac{0.3(1 - 0.3)}{n}} < 0.01$. Doing the algebra, we have $n > \frac{0.21}{(0.01)^2} = 2100$. (b) Without a previous estimate, use $p^* = 0.5$. For 95% confidence, we would need $n \ge \left(\frac{1.96}{0.02}\right)^2 (0.5)(1 - 0.5) = 2401$.

23.43: $0.58 \pm 1.645\sqrt{\frac{0.58(1-0.58)}{634}} = 0.55$ to 0.61. A plus-four interval will agree to three decimal places and would also be appropriate.

23.44: $(0.45 - 0.23) \pm 1.645\sqrt{\frac{0.45(1-0.45)}{314} + \frac{0.23(1-0.23)}{567}} = 0.165$ to 0.275. The plus-four method would also be appropriate and will agree to three decimal places.

23.45: $H_0: p_b = p_w$ vs. $H_a: p_b \neq p_w$. $\hat{p} = \frac{0.72(634) + 0.68(567)}{(634 + 567)} = 0.701$, $z = \frac{0.72 - 0.68}{\sqrt{0.701(1-0.701)\left(\frac{1}{634} + \frac{1}{567}\right)}} = 1.51$ and $P = 2P(Z > 1.51) = 0.1310$. There is little evidence of a difference between the proportions of black and white young people who think that rap videos contain too many references to sex.

23.46: (c) $193 \pm 2.060\frac{68}{\sqrt{26}} = 165.5$ to 220.5.

23.47: (a) $t = \frac{193 - 174}{\sqrt{\frac{68^2}{26} + \frac{44^2}{23}}} = 1.174$, and df $= 22$, the lesser of $23 - 1 = 22$ and $26 - 1 = 25$.

23.48: (b) $(193 - 174) \pm 2.074\sqrt{\frac{68^2}{26} + \frac{44^2}{23}} = -14.6$ to 52.6.

23.49: We must assume that each sample is an SRS taken from its respective populations (clinic dogs and pet dogs). We must also assume that the populations (cholesterol levels of pet dogs and cholesterol levels of clinic dogs) are Normal.

23.50: The problem description makes clear that the sample of clinic dogs is not an SRS. The sample of pets also is not likely to represent an SRS.

23.51: Because each person was asked about several behaviors, this should be a paired t test.

23.52: Large-sample or plus-four confidence interval for a population proportion.

23.53: If the sample can be viewed as an SRS, a t confidence interval for a population mean.

23.54: This is the entire population of Chicago Cubs players. Statistical inference is not appropriate.

23.55: Matched pairs t test or confidence interval.

23.56: (a) Two-sample test or confidence interval for difference in proportions. (b) Two-sample test or confidence interval for difference in means. (c) Two-sample test or confidence interval for difference in proportions.

23.57: The response rate for the survey was only about 20% ($427/2100 = 0.203$), which might make the conclusions unreliable.

23.58: (a) This is a matched-pairs situation: the responses of each subject under both conditions (control and treatment) are not independent. (b) We need to know the standard deviation of the differences, not the two individual sample standard deviations.

23.59: Each of a monkey's six trials are not independent. If a monkey prefers silence, it will almost certainly spend more time in the silent arm of the cage each time it is tested.

23.60: (a) $\hat{p} = 80/80 = 1$, and the margin of error for 95% confidence (or any level of confidence) is 0: $z^* \sqrt{\frac{(1)(1-1)}{n}} = 0$. (b) $\tilde{p} = 82/84 = 0.9762$. The plus-four 95% confidence interval is $\tilde{p} \pm z^* \sqrt{\frac{\tilde{p}(1-\tilde{p})}{n+4}} = 0.9436$ to 1.0088. Ignoring the upper limit, we are 95% confident that the actual success rate is 0.9436 or greater.

23.61: (a) Let p_1 be the proportion of subjects on Gardasil who get cancer, and let p_2 be the corresponding proportion for the control group. We assume that we have SRSs from each population. Because there were no cases of cervical cancer in the Gardasil group, we should use the plus-four procedure. $\tilde{p}_1 = \frac{0+1}{8487+2} = 0.000118$, and $\tilde{p}_2 = \frac{32+1}{8460+2} = 0.003900$. A 99% confidence interval for $p_2 - p_1$ is then given by $\tilde{p}_2 - \tilde{p}_1 \pm 2.576\sqrt{\frac{\tilde{p}_1(1-\tilde{p}_1)}{8489} + \frac{\tilde{p}_2(1-\tilde{p}_2)}{8462}} = 0.0020$ to 0.0056. (b) Let p_1 denote the proportion in the Gardasil group with genital warts, and let p_2 be the corresponding proportion for the control group. Because we have fewer than 10 "successes" in the Gardasil group, conditions for using the large-sample interval are not met. However, we can use the plus-four interval. $\tilde{p}_1 = 0.000253$, and $\tilde{p}_2 = 0.011644$. A 99% confidence interval for $p_2 - p_1$ is then 0.0082 to 0.0145. (c) Gardasil is seen to be effective in reducing the risk of both cervical cancer (by between 0.0020 and 0.0056, with 99% confidence) and genital warts (by between 0.0082 and 0.0145, with 99% confidence).

23.62: $H_0: \mu = 12$ vs. $H_a: \mu > 12$, where μ denotes the mean age at first word, measured in months. We regard the sample as an SRS; a stemplot (not shown) shows that the data are right-skewed with a high outlier (26 months). If we proceed with the t procedures despite this, we find $\bar{x} = 13$ and $s = 4.9311$ months. $t = \frac{13 - 12}{4.9311/\sqrt{20}} = 0.907$ with df $= 19$ and $P = 0.1879$. (Note: If you delete the outlier mentioned above, $\bar{x} = 12.3158$, $s = 3.9729$, and $t = 0.346$, yielding $P = 0.3665$.) We cannot conclude that the mean age at first word is greater than one year.

23.63: $H_0: \mu_1 = \mu_2$ vs. $H_a: \mu_1 < \mu_2$, where μ_1 is the mean number of new leaves on plants from the control population, and μ_2 is the mean for the nitrogen population. $\bar{x}_1 = 13.2857$, $\bar{x}_2 = 15.6250$, $s_1 = 2.0587$, $s_2 = 1.6850$, $n_1 = 7$, $n_2 = 8$, SE $= \sqrt{\frac{2.0587^2}{7} + \frac{1.6850^2}{8}} = 0.9800$, $t = \frac{13.3857 - 15.6250}{SE} = -2.387$. With Option 2, df $= 6$ and $P = 0.0271$. Or, using Option 1, df $= 11.66$ and $P = 0.0174$. We have strong evidence that nitrogen increases the mean number of new leaves formed.

23.64: Let μ be the mean age at first word, measured in months. For df $= 19$, $t^* = 1.729$, and the 90% confidence interval is $13 \pm 1.729\frac{4.9311}{\sqrt{20}} = 11.09$ to 14.91 months. We are 90% confident that the mean age at first word for normal children is between 11 and 15 months.

23.65: $H_0: \mu_1 = \mu_2$ vs. $H_a: \mu_1 \neq \mu_2$. We view the data as coming from two SRSs; the distributions show no strong departures from Normality. $\bar{x}_1 = 48.9513$, $s_1 = 0.2154$ (cotton), $\bar{x}_2 = 41.6488$, and $s_2 = 0.3922$ (ramie). SE $= 0.1582$ and $t = 46.16$. With either df $= 7$ or df $= 10.87$ (software), $P \approx 0$.

There is overwhelming evidence that ramie is darker than cotton when dyed this way.

23.66: (a) The design is shown below. (b) H_0: $\mu_B = \mu_C$ vs. H_a: $\mu_B \neq \mu_C$. $\bar{x}_B = 41.2825$, $s_B = 0.2550$, $\bar{x}_C = 42.4925$, and $s_C = 0.2939$; $n_B = n_C = 8$. SE $= 0.1376$, and $t = \frac{x_B - x_C}{\text{SE}} = -8.79$. With df $= 7$ (of df $= 13.73$ from software), $P < 0.001$. There is overwhelming evidence that method B gives darker color on average. However, the magnitude of this difference may be too small to be important in practice.

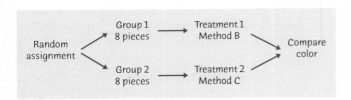

23.67: H_0: $\mu_1 = \mu_2$ vs. H_a: $\mu_1 \neq \mu_2$. We are told that the samples may be regarded as SRSs from their respective populations. Back-to-back stemplots show that t procedures are reasonably safe because both distributions are only slightly skewed, with no outliers and with fairly large sample sizes $\bar{x}_1 = 4.1769$, $s_1 = 2.0261$, and $n_1 = 65$ (parent allows drinking); $\bar{x}_2 = 4.5517$, $s_2 = 2.4251$, and $n_2 = 29$ (parent does not allow drinking). SE $= 0.5157$ and $t = \frac{x_1 - x_2}{\text{SE}} = -0.727$. This is close to zero, so we will certainly not reject the null hypothesis. Indeed, with df $= 46.19$ (software), $P = 0.4710$. There is no significant difference in the mean number of drinks between female students with a parent who allows drinking and those whose parents do not allow drinking.

23.68: Let p be the proportion of female students with at least one parent who allows drinking. We are told that the sample represents an SRS. Large-sample methods may be used. $\hat{p} = 0.6915$ and SE $= 0.04764$, so the margin of error is $1.96\text{SE} = 0.09337$, and the interval is 0.5981 to 0.7849. (Note: Using a plus-four estimate, the interval is 0.5916 to 0.7757.) With 95% confidence, the proportion of female students who have at least one parent who allows drinking is 0.598 to 0.785 (or 0.592 to 0.776).

23.69: (a) Stemplots are provided. The diabetic potentials appear to be larger. (b) H_0: $\mu_1 = \mu_2$ vs. H_a: $\mu_1 \neq \mu_2$, where μ_1 is the mean potential for diabetics, and μ_2 is the mean for the normal population. We assume we have two SRSs; the distributions appear to be safe for t procedures. $\bar{x}_1 = 13.0896$, $\bar{x}_2 = 9.5222$, $s_1 = 4.8391$, $s_2 = 2.5765$, $n_1 = 24$, $n_2 = 18$, SE $= 1.1595$, and $t = 3.077$. With Option 2, df $= 17$ and $0.005 < P < 0.01$. Or, using Option 1, df $= 36.6$, and $P = 0.0040$. We have strong evidence that the electric potential in diabetic mice is greater than the potential in normal mice. (c) If we remove the outlier, the diabetic mouse statistics change: $\bar{x}_1 = 13.6130$, $s_1 = 4.1959$, $n_1 = 23$. Now SE $= 1.065$ and $t = 3.841$. With df $= 16$, $0.001 < P < 0.002$. With df $= 37.15$,

$P = 0.0005$. With the outlier removed, the evidence that diabetic mice have higher mean electric potential is even stronger.

Diabetic		Normal
	0	
	0	
	0	4
7	0	6 7 7 7
9 8 8	0	8 8 8 8 9 9 9
1 0 0 0 0 0 0	1	0 0
3	1	2 3 3
5 4 4 4	1	4
7 6	1	
9 9 8 8	1	
	2	
2	2	

23.70: (a) We want to compare the proportions p_1 (microwave crackers that show checking) and p_2 (control crackers that show checking). We can do this either by testing hypotheses or with a confidence interval, but because the "microwave checked" count is only 3, significance tests are not appropriate. We will use the plus-four procedure and construct a confidence interval for $p_1 - p_2$. $\tilde{p}_1 = \frac{3 + 1}{65 + 2} = 0.0597$ and $\tilde{p}_2 = \frac{57 + 1}{65 + 2} = 0.8657$. SE $= \sqrt{\frac{\tilde{p}_1(1 - \tilde{p}_1)}{67} + \frac{\tilde{p}_2(1 - \tilde{p}_2)}{67}} = 0.05073$, and a 95% confidence interval is given by $(\tilde{p}_1 - \tilde{p}_2) \pm 1.96(0.0507) = -0.9054$ to -0.7066. We are 95% confident that microwaving reduces the percentage of checked crackers by between 70.7% and 90.5%. (b) Let μ_1 and μ_2 be the mean breaking pressures of microwaved and control crackers, respectively. H_0: $\mu_1 = \mu_2$ vs. H_a: $\mu_1 \neq \mu_2$, and construct a 95% confidence interval for $\mu_1 - \mu_2$. We assume the data can be considered SRS's from the two populations and that the population distributions are not far from Normal. SE $= 9.0546$ and $t = 6.914$, so the P-value is very small, regardless of whether we use df $= 19$ or df $= 33.27$. A 95% confidence interval for the difference in mean breaking pressures between these cracker types is 43.65 to 81.55 psi (using df $= 19$ and $t^* = 2.093$) or 44.18 to 81.02 psi (using df $= 33.27$ and $t^* = 2.0339$). There is very strong evidence that microwaving crackers changes their mean breaking strength. We are 95% confident that microwaving crackers increases their mean breaking strength by between 43.65 and 81.55 psi.

23.71: Let μ be the mean date on which the tripod falls through the ice. We assume that the data can be viewed as an SRS of fall-through times and that the distribution is roughly Normal. $n = 96$, $\bar{x} = 15.146$, and $s = 5.989$ days. df $= 95$. A 95% confidence interval is given by 13.932 to 16.359 days. We are 95% confident that the mean number of days for the tripod to or through the ice is 13.932 days to 16.359 days from April 20 or between May 3 and May 6.

23.72: Two of the counts are too small to perform a significance test safely.

23.73: (a) "SEM" stands for "standard error of the mean"; $SEM = s/\sqrt{n}$. (b) Two-sample t tests were done because there are two separate, independent groups of mice. (c) The observed differences between the two groups of mice were so large that it would be unlikely to occur by chance alone if the two groups were the same on average. $P < 0.005$ is stronger evidence than $P < 0.05$.

23.74: $\bar{x}_1 = 5.9$ (insulin), $\bar{x}_2 = 0.75$ (glucose) ng/ml, and the standard deviations are $s_1 = 0.9\sqrt{10} = 2.85$ and $s_2 = 0.2\sqrt{10} = 0.632$ ng/ml. $H_0: \mu_1 = \mu_2$ vs. $H_a: \mu_1 \neq \mu_2$. $SE = \sqrt{0.9^2 + 0.2^2} = 0.922$, $t = \frac{5.9 - 0.75}{SE} = 5.59$. With either df = 9 or df = 9.89, $P < 0.001$. The evidence is even stronger than the paper claimed.

23.75: (a) The stemplot confirms the description given in the text. (Arguably, there are two "mild outliers" visible in the stemplot, although the $1.5 \times IQR$ criterion only flags the highest as an outlier.) (b) Let μ be the mean body temperature. $H_0: \mu = 98.6°$ vs. $H_a: \mu \neq 98.6°$; the alternative is two-sided because we had no idea (before looking at the data) that μ might be higher or lower than 98.6°. Assume we have a Normal distribution and an SRS. The average body temperature in our sample is $\bar{x} = 98.203°$, $s = 0.803°$ so the test statistic is $t = \frac{98.203 - 98.6}{0.803/\sqrt{20}} = -2.21$. $P = 0.0396$ with df = 19. We have fairly strong evidence—significant at $\alpha = 0.05$, but not at $\alpha = 0.01$—that mean body temperature is not equal to 98.6°. (Specifically, the data suggests that mean body temperature is lower.)

23.76: Assume we have a Normal distribution and an SRS. With $\bar{x} = 98.203°$, $s = 0.803°$, and df = 19, our 90% confidence interval for μ is $98.203 \pm 1.729\left(\frac{0.803}{\sqrt{20}}\right) = 98.203 \pm 0.310$, or 97.89° to 98.51°. We are 90% confident that the mean body temperature for healthy adults is between 97.89° and 98.51°.

23.77: For the two-sided test $H_0: \mu = \$95,000$ vs. $H_a: \mu \neq \$95,000$ with significance level $\alpha = 0.01$, we can reject H_0 because $\$95,000$ falls outside the 99% confidence interval.

23.78: A low-power test has a small probability of rejecting the null hypothesis, at least for some alternatives. That is, we run a fairly high risk of making a Type II error (failing to reject H_0 when it is false) for such alternatives.

23.79: A Type I error means that we conclude the mean IQ is less than 100 when it really is 100 (or more). A Type II error means that we conclude the mean IQ is 100 (or more) when it really is less than 100.

Chapter 24 Two Categorical Variables: The Chi-Square Test

24.1: (a) The proportion of University Park campus students that do not use Facebook is 68/978 = 0.0695, which rounds to 0.070 and is represented as 7.0% in the table. (b) The bar graph reveals that students on the main (University Park) campus are much more likely to use Facebook at least daily, whereas Commonwealth campus students are more likely not to use it at all.

	UNIVERSITY PARK	COMMONWEALTH
Do not use Facebook	7.0%	28.3%
Use several times a month or fewer	5.6%	8.7%
Use at least once a week	22.0%	17.9%
Use at least once a day	65.4%	45.0%

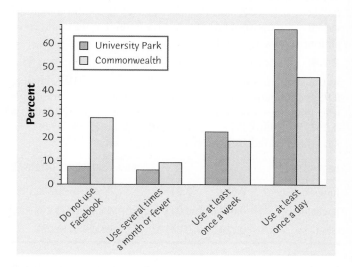

24.2: (a) For example, among those that played video games, the percentage that got A's and B's is 736/1379 = 53.4%. The table below summarizes. The accompanying bar graph reveals a larger percentage of A's and B's and a smaller percentage of D's and F's in the group of boys that played video games.

	A'S AND B'S	C'S	D'S AND F'S
Played games	53.4%	32.6%	14.0%
Never played games	47.8%	33.6%	18.6%

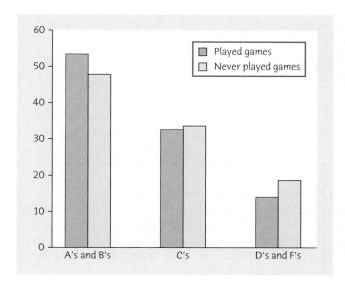

(b) We can't conclude that playing video games "causes" improved grades. This is an observational study—the boys who play video games may differ from those who do not in many ways.

24.3: (a) To test H_0: $p_1 = p_2$ vs. H_a: $p_1 \neq p_2$ for the proportions who do not use Facebook, we have $\hat{p}_1 = 0.0695$ and $\hat{p}_2 = 0.2834$. $\hat{p} = \frac{68 + 248}{978 + 875} = 0.1705$, SE = 0.01750, and $z = -12.22$, for which P is close to zero. (b) H_0: $p_1 = p_2$ vs. H_a: $p_1 \neq p_2$ for the proportions who use Facebook at least weekly, $\hat{p}_1 = 0.2198$ and $\hat{p}_2 = 0.1794$. $\hat{p} = 0.2008$, SE = 0.01864, $z = 2.17$, and $P = 0.0300$. (c) If we did four individual tests, we would not know how confident we could be in all four results when taken together.

24.4: (a) For the junior college sample, $\hat{p} = \frac{44}{44 + 31} = 0.587$, so the standard error is SE $= \sqrt{\frac{0.587(1 - 0.587)}{75}} = 0.057$. Hence, a 95% confidence interval for the proportion of all junior college graduates that think astrology is not at all scientific is $0.587 \pm 1.96(0.057) = 0.475$ to 0.699, or 47.5% to 69.9%. The three confidence intervals are displayed:

DEGREE HELD	$\hat{p}$	SE	95% CONFIDENCE INTERVAL
Junior college	0.587	0.057	47.5% to 69.9%
Bachelor's	0.663	0.035	59.4% to 73.2%
Graduate	0.724	0.045	63.6% to 81.2%

(b) As we construct more confidence intervals, each based on a different random sample, the chance that at least one of them fails to capture the parameter of interest increases.

24.5: (a) Expected counts are in the table provided. For example, $\frac{(131)(627)}{1537} = 53.44$. (b) Commonwealth students actually use Facebook several times a month or fewer more often than we would expect and use Facebook at least once a day less often than we would expect.

EXPECTED COUNTS	UNIVERSITY PARK	COMMONWEALTH
Several times a month or fewer	77.56	53.44
At least once a week	220.25	151.75
At least once a day	612.19	421.81

24.6: (a) The expected counts are shown in the table below. For example for boys who play video games, $\frac{(1379)(941)}{1808} = 717.72$.

EXPECTED COUNTS	A'S AND B'S	C'S	D'S AND F'S
Played games	717.72	453.06	208.22
Never played games	223.28	140.96	64.78

(b) By informal examination, there appear to be somewhat large differences between observed counts and expected counts, especially in a few of the cells. For example, for boys that get D's and F's and never play video games, the difference is $80 - 64.78 = 15.22$, which is fairly large relative to the expected cell count of 64.78.

24.7: (a) All expected counts are well above 5 (the smallest is 53.44). (b) H_0: there is no relationship between campus and Facebook use vs. H_a: there is a relationship between campus and Facebook use. Using software, $\chi^2 = 19.489$ and $P < 0.0005$. (c) The largest contributions come from the first row, reflecting the fact that monthly use is lower among University Park students and higher among Commonwealth students.

24.8: (a) Inspecting Figure 24.4, we see that all expected cell counts are more than 5, and all individual observed cell counts are at least 1, so conditions required for use of the chi-square test are satisfied. (b) $\chi^2 = 6.739$ and $P = 0.034$ by inspection of the output. Rejecting H_0 means we believe there is an association between playing video games and grades. (c) There is strong evidence of an association between grades and video game playing.

24.9: H_0: there is no relationship between education level and astrology opinion vs. H_a: there is some relationship between education level and astrology opinion. Examining the output provided in Figure 24.5, we see that all expected cell counts are greater than 5, and all observed cell counts are at least 1, so conditions for use of the chi-square test are satisfied. $\chi^2 = 3.618$ and $P = 0.160$. There is no evidence of an association between education level and opinion of astrology.

24.10: (a) The table is shown below. (b) Because there were three separate samples, this is a test of homogeneity. (c) H_0: The proportion of people who admit cocaine use is the same for all three interview methods H_a: The proportions are not the same (interview type makes a difference). Minitab output is provided below. All expected cell counts are greater than 5, so conditions for use of the chi-square test are satisfied. We see that $\chi^2 = 10.619$ on df $= 2$, and $P = 0.005$. We have strong evidence that the proportion of people who admit to cocaine use is related to the method in which the question was asked.

	No	Yes	All
Anon written	576	224	800
	602.7	197.3	800.0
Interview	600	200	800
	602.7	197.3	800.0
Phone	632	168	800
	602.7	197.3	800.0
All	1808	592	2400
	1808.0	592.0	2400.0

```
Cell Contents:          Count
                        Expected count

Pearson Chi-Square = 10.619, DF = 2, P-Value = 0.005
```

24.11: (a) There was one sample that was later categorized by two variables; this is a test of independence. (b) H_0: There is no relationship between age and how politically informed the person is H_a: There is a relationship between age and how politically informed the person is.

All expected counts are greater than 5, so the test is appropriate. $\chi^2 = 32.057$ on df $= 12$, and $P = 0.001$. Age and being informed about politics are related. In particular, we can see that 20- to 29-year-olds are less likely to be informed than older persons.

	Not at all	A little	Somewhat	Very	Extremely	Total
20–29	8	29	28	13	0	78
	5.73	17.57	31.40	17.57	5.73	
30–39	15	28	55	23	9	130
	9.55	29.28	52.33	29.28	9.55	
40–49	2	25	49	26	14	116
	8.52	26.13	46.70	26.13	8.52	
50 up	21	59	120	79	23	302
	22.19	68.02	121.57	68.02	22.19	
Total	46	141	252	141	46	626

Chi-Sq = 32.057, DF = 12, P-Value = 0.001

24.12: (a) df = $(3 - 1)(2 - 1) = 2$. (b) The largest critical value shown for df = 2 is 15.20; because the computed value (19.489) is greater than this, we conclude that $P < 0.0005$. (c) With $r = 4$ and $c = 2$, the appropriate degrees of freedom would be df = 3.

24.13: (a) df = $(2 - 1)(3 - 1) = 2$. (b) The computed value (6.739) is between the table values 5.99 and 7.38, so we conclude that $0.025 < P < 0.05$, which is consistent with output's reported $P = 0.034$. (c) Under the null hypothesis of no association, the mean value of χ^2 is df = 2. Our computed value is larger than this. The small P-value suggests that random chance does not easily explain the larger than expected value of χ^2.

24.14: $H_0: p_1 = p_2 = p_3 = \frac{1}{3}$ vs. $H_a:$ not all three are equally likely. The expected counts are each $53 \times \frac{1}{3} = 17.67$. $\chi^2 = \sum \frac{(\text{observed count} - 17.67)^2}{17.67} = \frac{(31 - 17.67)^2}{17.67} + \frac{(14 - 17.67)^2}{17.67} + \frac{(8 - 17.67)^2}{17.67} = 16.11$. df = 2. From Table D, $\chi^2 = 16.11$ falls beyond the 0.005 critical value, so $P < 0.005$. There is very strong evidence that the three tilts differ.

24.15: (a) If all days were equally likely, we would have $p_1 = p_2 = \cdots = p_7 = \frac{1}{7}$, and we would expect 100 births on each day.
(b) $\chi^2 = \frac{(84 - 100)^2}{100} + \frac{(110 - 100)^2}{100} + \cdots + \frac{(72 - 100)^2}{100} = 19.12$.
(c) df = $7 - 1 = 6$. From Table D, $\chi^2 = 19.12$ yields $0.0025 < P < 0.005$. Software gives $P = 0.004$. We have strong evidence that births are not spread evenly across the week.

24.16: The details of the computation are shown below. The expected counts are found by multiplying the expected frequencies by 803 (the total number of observations).

	EXPECTED FREQUENCY	OBSERVED COUNT	EXPECTED COUNT	$O - E$	$\frac{(O - E)^2}{E}$
16 to 29	0.328	401	263.384	137.616	71.9032
30 to 59	0.594	382	476.982	−94.982	18.9139
60 or older	0.078	20	62.634	−42.634	29.0203
		803			119.8374

The difference is significant: $\chi^2 = 119.84$, df = 2, and $P < 0.0005$ (using software, $P = 0.000$ to three decimal places).

24.17: (a) $15/33 = 0.455$, or 45.5% of subjects chose position 1. Similarly, the percentages for each position are 15.2% for position 2, 6.1% for position 3, and 33.3% for position 4. (b) If subjects are equally likely to select any position, then we would expect $33/4 = 8.25$ subjects in each position.

(c) $H_0: p_1 = p_2 = p_3 = p_4 = \frac{1}{4}$ vs. $H_a:$ the four order selection probabilities are not equally likely. We expect 8.25 subjects per cell under the null hypothesis, so all expected cell counts exceed 5. Also, we have at least one observation per cell. df = 3. Conditions for the chi-square test are satisfied. $\chi^2 = 12.45$. From Table D, $\chi^2 = 12.45$ yields $0.005 < P < 0.01$ ($P = 0.006$ from technology). There is strong evidence that positions are not selected with equal probability—some positions are more likely to be selected than others. (d) We see that the largest contributions to χ^2 are from the first and third positions. There is evidence of *both* primacy and recency effects.

24.18: $H_0: p_1 = p_2 = \cdots = p_{12} = \frac{1}{12}$ vs. $H_a:$ The 12 astrological sign birth probabilities are not equally likely. There are 1913 subjects in this sample. Under H_0, we expect $1913/12 = 159.42$ per sign, so a chi-square test is appropriate. $\chi^2 = \frac{(145 - 159.42)^2}{159.42} + \frac{(162 - 159.42)^2}{159.42} + \cdots + \frac{(157 - 159.42)^2}{159.42} = 12.31$. With df = 11, using Table D, $P > 0.25$ ($P = 0.341$ from technology). There is little evidence that some astrological signs are more likely in birth than others.

24.19: (a) $295 + 655 + 239 + 363 = 1552$.

24.20: (b) $655/(655 + 916) = 655/1571 = 0.4169$.

24.21: (a) For 23- to 30-year-olds, the percentage is 26.5%.

24.22: (b) There was only one sample.

24.23: (a) $(1571)(1552)/4111 = 593.09$.

24.24: (c) $(655 - 593.09)^2/593.09 = 6.463$.

24.25: (a) df = $(r - 1)(c - 1) = (4 - 1)(2 - 1) = 3$.

24.26: (b) This is the hypothesis of no association between "age" and "type of injury."

24.27: (b) This is the hypothesis of association between "age" and "type of injury."

24.28: (c) The largest entry in the table corresponding to df = 3 is 17.73.

24.29: (b) We assume that the sample is an SRS, or essentially an SRS from all weightlifting injuries.

24.30: (a) $\hat{p}_B = 0.2948$ and $\hat{p}_P = 0.1773$. The standard error is SE = 0.0325, so the large-sample 95% confidence interval for $p_B - p_P$ is $(\hat{p}_B - \hat{p}_P) \pm 1.96\text{SE} = 0.1175 \pm 0.0637 = 0.0538$ to 0.1812. (b) $H_0: p_C = p_B = p_P$ vs. $H_a:$ the three proportions are not the same. All expected cell counts are more than 5, so the guidelines for the chi-square test are satisfied. We have $\chi^2 = 56.992$, df = 2, and $P < 0.0005$. There is a significant difference in the proportions of those who did not smoke in weeks 9–12 among the groups. (c) This is a test of homogeneity; subjects were randomized into three separate treatments.

EXPECTED COUNTS	CHANTIX	BUPROPION	PLACEBO
No smoking in weeks 9–12	107.49	100.47	105.05
Smoked in weeks 9–12	244.51	228.53	238.95

24.31: (a) These were separate random samples, so this is a test of homogeneity. (b) H_0: the distribution of age groups is the same for landline and cell-only individuals vs. H_a: the distributions are different. All expected cell counts are more than 5, so the guidelines for the chi-square test are satisfied. We have $\chi^2 = 1032.892$, df = 3, and $P < 0.0005$. There is strong evidence of an association between age group and the type of telephone.

24.32: (a) 174 of the 871 students used OTC stimulants, $\hat{p} = 0.1998$. The sample counts are large, so we can use $\hat{p} \pm 1.96SE_{\hat{p}} = 0.1998 \pm 1.96\sqrt{\frac{0.1998(1 - 0.1998)}{871}}$, or 0.1732 to 0.2264. (b) The conditional distributions are given in the second row for each cell in the Minitab output below. It appears that those who use OTC medications are less likely to have optimal sleep and more likely to have poor sleep than those who do not use these medications. (c) H_0: there is no association between taking OTC medications to stay awake and sleep quality H_a: there is an association between OTC medications to stay awake and sleep quality. $\chi^2 = 18.504$, $P < 0.0005$. We conclude that there is an association between taking over-the-counter medications to stay awake and sleep quality.

	borderline	optimal	poor
No OTC	186	266	245
	77.82	87.79	74.47
	191.3	242.5	263.3
Uses OTC	53	37	84
	22.18	12.21	25.53
	47.7	60.5	65.7

Cell Contents: Count
                      % of Column
                      Expected count

Pearson Chi-Square = 18.504, DF = 2, P-Value = 0.000

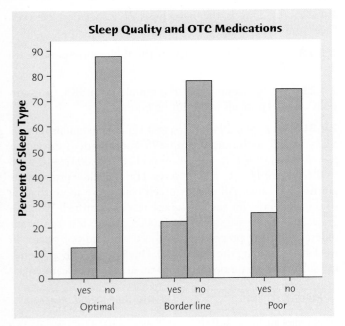

24.33: (a) The diagram is shown below. To perform the randomization, label the infants 01 to 77, and choose pairs of random digits. (b) See the table for the expected counts. $\chi^2 = 0.568$, df = 3, and $P = 0.904$. There is no reason to doubt that the randomization "worked."

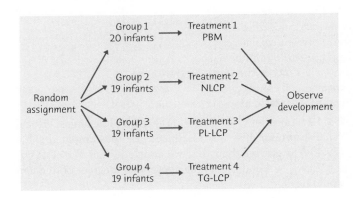

EXPECTED COUNTS	FEMALE	MALE
PBM	10.91	9.09
NLCP	10.36	8.64
PL-LCP	10.36	8.64
TG-LCP	10.36	8.64

24.34: (a) H_0: $p_G = p_{NG}$ vs. H_a: $p_G \neq p_{NG}$. $\hat{p}_G = \frac{36}{91} = 0.395604$ and $\hat{p}_{NG} = \frac{578}{2014} = 0.286991$, $\hat{p} = 0.291686$. SE $= 0.048713$. $z = \frac{\hat{p}_G - \hat{p}_{NG}}{SE} = 2.22965$ and $P = 0.0258$. (b) All expected cell counts exceed 5, and use of a chi-square test is appropriate. $\chi^2 = 4.971$ with df = 1. From software, $P = 0.0258$. (c) $z^2 = (2.22965)^2 = 4.971$. (d) We would use a one-sided z-test. The chi-square test is inherently two-sided—it tests for association generally, rather than for a particular direction of association.

EXPECTED COUNTS	FIGHT	NO FIGHT
Played games	26.54	64.46
Never played games	587.46	1426.54

24.35: (a) H_0: $p_1 = p_2$ vs. H_a: $p_1 < p_2$. (b) The z-test must be used because the chi-square procedure measures evidence in support of evidence of any association and is implicitly two-sided. $\hat{p}_1 = 0.3667$ and $\hat{p}_2 = 0.7333$. $\hat{p} = (11 + 22)/(30 + 30) = 0.55$, and SE $= 0.12845$, so $z = -2.85$ and $P = 0.0022$. We have strong evidence that rats that can stop the shock (and therefore presumably have better attitudes) develop tumors less often than rats that cannot (and therefore are presumably depressed).

24.36: (a) There was only one sample, so this is a test of independence. H_0: there is no relationship between sex and self-assessment of chances of being rich vs. H_a: there is some relationship between these factors. SOLVE: Examining the Minitab output in Figure 24.8, we see that conditions for use of the chi-square test are satisfied because all expected cell counts exceed 5. $\chi^2 = 43.946$ with df = 4, and $P < 0.0005$. There is evidence of a difference between how men and women assess their chances of being rich by age 30.

24.37: H_0: there is no relationship between sexual content of ads and magazine audience vs. H_a: there is some relationship between sexual content of ads and magazine audience. Examining the Minitab output in Figure 24.9, we see that conditions for use of the chi-square test are satisfied because all expected cell counts exceed 5. $\chi^2 = 80.874$ with df = 2, leading to $P < 0.0005$. Magazines aimed at women are much more likely to have sexual depictions of models than are the other two types of magazines.

24.38: (a) We compare the percentage of dogs in each "condition type" (I, II, III) making the specified number of errors. The table summarizes. We see that under Type I condition (social-communicative), dogs tend to make more errors, while under Type III condition (nonsocial), dogs tend to make fewer errors.

	0	1	2	3
Type I	0.0%	25.0%	25.0%	50.0%
Type II	41.7%	25.0%	8.3%	25.0%
Type III	66.7%	16.7%	16.7%	0.0%

(b) We have many cells with expected cell counts lower than 5. We also have two zeros in the table. (c) Software should warn users against using the chi-square test. Minitab does provide such a warning.

24.39: We need cell counts, not just percents. If we had been given the number of travelers in each group—leisure and business—we could have estimated the counts.

24.40: Some people are represented more than once; there are 6570 diseases for 5269 individuals in the low/average isolation group.

24.41: To do a chi-square test, each subject can be counted only once.

24.42: (a) $\hat{p} = \frac{404}{4310} = 0.0937$, and SE = 0.0044. A 99% confidence interval for the proportion of smokers is $0.0937 \pm 2.576(0.0044) = 0.0823$ to 0.1051, or 8.23% to 10.51%. (b) In a table or bar graph of conditional distributions, notice that smokers feel less optimistic about their own health than do nonsmokers: Smokers are more likely to view their health as Good, Fair, or Poor, whereas nonsmokers are more likely to view their health as Excellent or Very Good. (c) $\chi^2 = \frac{(25-47.7)^2}{47.7} + \frac{(484-461.3)^2}{461.3} + \frac{(115-156.7)^2}{156.7} + \cdots + \frac{(11-36.3)^2}{36.3} = 229.66$. With df = $(5-1)(2-1) = 4$, we have $P < 0.0005$. (d) Examining the terms of the chi-square statistic, or by comparing the differences between observed counts and expected counts (relative to expected), we see that the greatest contributions to χ^2 are due to differences at both extremes of the

conditional distributions: More than the expected number of smokers and fewer than expected nonsmokers were observed in the Poor and Fair categories, whereas the opposite trend occurs for the Very Good and Excellent categories.

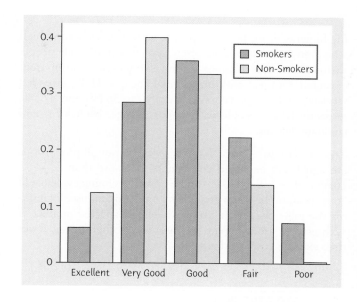

24.43: (a) H_0: there is no relationship between degree held and service attendance vs. H_a: there is some relationship between degree held and service attendance. Expected counts are shown in the table below. $\chi^2 = 14.19$ with df = 3, yielding P-value = 0.0027. There is strong evidence of an association between degree held and service attendance.

EXPECTED COUNTS	HIGH SCHOOL	JUNIOR COLLEGE	BACHELOR'S	GRADUATE
Attended services	437.3	55.7	129.2	61.8
Did not attend services	842.7	107.3	248.9	119.1

(b) Expected counts are shown in the table below. $\chi^2 = 0.73$ on df = 2. So, $P > 0.25$ (0.694 by software). In this table, we find no evidence of association between religious service attendance and degree held.

EXPECTED COUNTS	JUNIOR COLLEGE	BACHELOR'S	GRADUATE
Attended services	64.1	148.7	71.2
Did not attend services	98.9	229.3	109.8

(c) Expected counts are shown in the table on the next page. $\chi^2 = 13.40$ on df = 1. $P < 0.0005$ (0.0002 by software). There is overwhelming evidence of association between level of education (High School versus Beyond High School) and religious service attendance.

EXPECTED COUNTS	HIGH SCHOOL	BEYOND HS
Attended services	437.3	246.7
Did not attend services	842.7	475.3

(d) In general, we find that people with degrees beyond high school attend service more often than expected; while people with high school degrees attend services less often than expected. Of those with high school degrees, 31.3% attended services, and the percentages are 38.0%, 38.6% and 42.0%, respectively, for people with junior college, bachelor's, and graduate degrees.

24.44: H_0: $p_9 = p_{10} = p_{11} = p_{12}$ vs. H_a: not all proportions are equal. All expected cell counts are more than 5. $\chi^2 = 16.653$. df = 3, $P = 0.001$. There is overwhelming evidence of an association between grade level and condom use. Examining the table, we see that more than the expected number of 12th graders use condoms, whereas in all other grades, fewer than expected use condoms.

OBSERVED COUNTS EXPECTED COUNTS	CONDOM	NO CONDOM
9	281	463
	296.6	447.4
10	388	669
	421.4	635.6
11	597	938
	612.0	923.0
12	727	936
	663.0	1000.0

24.45: H_0: there is no relationship between race and opinion about schools vs. H_a: there is some relationship between race and opinion about schools. All expected cell counts exceed 5, so use of a chi-square test is appropriate. $\chi^2 = 22.426$, df = 8, and $P = 0.004$. We have strong evidence of a relationship between race and opinion of schools.

24.46: (a) This is an experiment in the sense that subjects were assigned to treatments—but it is not a randomized experiment, and it may well be the case that certain types of patients (e.g., high-risk patients) are more likely to be assigned to particular treatments. In this sense, it is easy to believe that patients within treatment groups differ in more ways than one. (b) H_0: there is no relationship between surgery type and complications, vs. H_a: there is some relationship between surgery type and complications. All expected cell counts exceed 5, so use of a chi-square test is appropriate. $\chi^2 = 318.668$, df = 4, and $P = 0.000$ to three decimal places. There is overwhelming evidence of an association between surgery type and complication.

24.47: H_0: there is no relationship between laundry habits and preference vs. H_a: there is some relationship between laundry habits and preference. To compare people with different laundry habits, we compare the percent in each class who prefer the new product.

	SOFT WATER, WARM WASH	SOFT WATER, HOT WASH	HARD WATER, WARM WASH	HARD WATER, HOT WASH
Prefer new product	54.3%	51.8%	61.8%	58.3%

The differences are not large, but the "Hard water, warm wash" group is most likely to prefer the new detergent. With all expected cell counts exceeding 5, a chi-square test is appropriate. $\chi^2 = 2.058$, df = 3, $P = 0.560$. The data provide no evidence to conclude that laundry habits and brand preference are related.

24.48: $\hat{p} = 373/1960 = 0.1903$. SE $= \sqrt{\frac{0.1903(1 - 0.1903)}{1960}} = 0.0089$. A 95% confidence interval for the proportion of Independents is $0.1903 \pm 1.96(0.0089) = 0.173$ to 0.208, or 17.3% to 20.8%.

24.49: We compare the percentages leaning toward each party within each education group. At each education level, we compute the percentage leaning each party. For example, among bachelor's degree holders, $166/(166 + 136) = 54.97\%$ lean Democrat, while the other 45.03% lean Republican.

At every education level, people leaning Democrat outweigh people leaning Republican. The difference is greatest at the "None" level of education, then decreases until the party support is nearly equal for bachelor's holders. Among graduate degree holders, Democrats strongly outnumber Republicans.

24.50: The conditional distributions are tabulated below and summarized in a bar graph. Expected cell counts are all greater than 5, so using a chi-square test is appropriate. $\chi^2 = 115.16$, df = 28. We see that $P = 0.000$ to three decimal places.

	None	High School	JrCollege	Bachelor	Graduate
Independent	101	186	26	36	24
	35.44	19.20	17.33	10.23	11.76
NearDemocrat	37	117	10	43	23528
	12.98	12.07	6.67	12.22	13.73
NearRepublican	20	76	1616	29	16
	7.02	7.84	10.67	8.24	7.84
Otherparty	5	22	5	14	8
	1.75	2.27	3.33	3.98	3.92
StrongDemocrat	47	60161	5335	60	53
	16.49	16.62	23.33	17.05	25.98
StrongRepublican	14	101	19	43	15
	4.91	10.42	12.67	12.22	7.35
WeakDemocrat	43	176	23	3863	34338
	15.09	18.16	15.33	17.90	18.63
WeakRepublican	18	64130	2216	64	22
	6.32	13.42	10.67	18.18	10.78

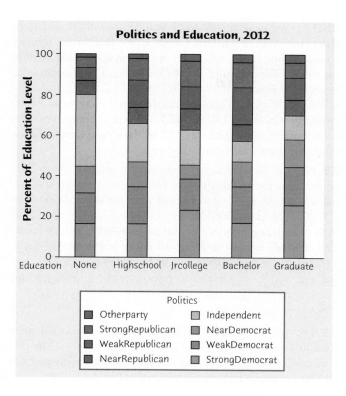

Chapter 25 Inference for Regression

25.1: (a) A scatterplot of the data is provided. From software, $r = 0.985$.

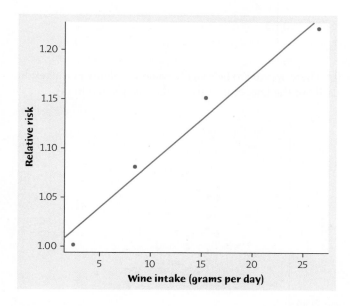

(b) We estimate that an increase in intake of 1 gram per day increases relative risk of breast cancer by 0.009. According to our estimate, wine intake of 0 grams per day is associated with a relative risk of breast cancer of 0.9931 (about 1). (c) $\hat{y} = 0.9931 + 0.0090x$. $s^2 = 0.00079/2 = 0.000395$. We estimate σ by $s = \sqrt{0.000395} = 0.01987$.

x	y	$\hat{y}$	RESIDUAL $y - \hat{y}$	$(y - \hat{y})^2$
2.5	1.00	1.0156	−0.0156	0.00024
8.5	1.08	1.0697	0.0103	0.00011
15.5	1.15	1.1328	0.0172	0.00030
26.5	1.22	1.2319	−0.0119	0.00014
			0	0.00079

25.2: (a) A scatterplot of the data is provided, with the least-squares regression line added. $r^2 = 0.201$. (b) $a = 0.49529$, $b = 0.002332$, and $s = 7.9$. (c) $\hat{y} = 0.49529 + 0.002332x$.

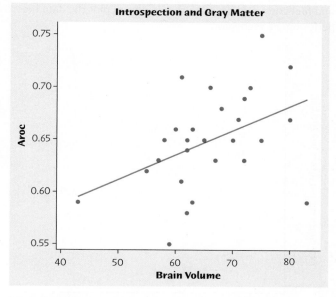

25.3: (a) A scatterplot of discharge by year is provided, along with the fitted regression line, which is requested in part (b). Discharge seems to be increasing over time, but there is also a lot of variation in this trend, and our impression is easily influenced by the most recent years' data. $r^2 = 0.215$. (b) $\hat{y} = -3362 + 2.6327x$, $s = 110.477$.

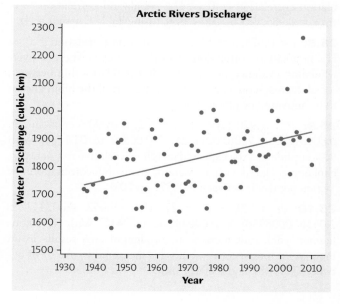

25.4: (a) $H_0: \beta = 0$ vs. $H_a: \beta > 0$; $t = \frac{b}{SE_b} = \frac{0.009012}{0.001112} = 8.104$. (b) df $= 4 - 2 = 2$; $0.005 < P < 0.01$ (0.0074 from technology).

25.5: $H_0: \beta = 0$ vs. $H_a: \beta > 0$. $t = 4.47$. Here, df $= 73$. Using Table C (df $= 60$), we obtain $P < 0.0005$. Using software, we obtain $P = 0.000$ (rounded to three decimal places). There is strong evidence of an increase in Arctic discharge over time.

25.6: $H_0: \beta = 0$ vs. $H_a: \beta \neq 0$. $t = \frac{b}{SE_b} = \frac{-0.01466}{0.02334} = -0.628$. df $= 15 - 2 = 13$. Using Table C we obtain $P > 0.50$ (software: $P = 0.5408$). There is little evidence of a straight-line relationship between fuel consumption and speed. However, there is a very strong (nonlinear) relationship between speed and fuel consumption. One should always plot the data before performing a regression.

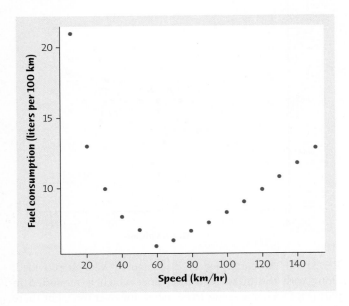

25.7: (a) $H_0: \beta = 0$ vs. $H_a: \beta > 0$. $t = 8.104$ with df $= 2$. $0.005 < P < 0.01$. This test is equivalent to testing H_0: population correlation $= 0$ vs. H_a: population correlation > 0. (b) $r = 0.985$. $0.005 < P < 0.01$ (using Table E with $n = 4$).

25.8: $r = 0.878$, $n = 13$. H_0: population correlation $= 0$ vs. H_a: population correlation > 0; $P < 0.0005$. There is overwhelming evidence of a positive linear relationship between social distress score and activity in the part of the brain known to be activated by physical pain.

25.9: $t^* = 2.920$ (df $= 4 - 2 = 2$, with 90% confidence). $0.009012 \pm 2.920(0.001112) = 0.00577$ to 0.01226. With 90% confidence, the expected increase in relative risk of breast cancer associated with an increase in alcohol consumption by 1 gram per day is between 0.00577 and 0.01226.

25.10: df $= 29 - 2 = 27$. $t^* = 2.052$. $0.002332 \pm 2.052(0.0008958) = 0.000492$ to 0.00417. With 95% confidence, each unit increase in Brodmann area results in an increase of introspective ability (as measured by Aroc) of between 0.000492 and 0.00417 units.

25.11: We have $b = 2.6327$ and $SE_b = 0.5893$. With 75 observations, df $= 73$. Using Table C, we look under the row corresponding to df $= 60$ (the nearest smaller value of df in the table). We obtain $t^* = 1.671$. Hence, a 90% confidence interval for β is given by $2.86327 \pm 1.671(0.5893) = 1.6480$ to 3.6174. With 90% confidence, the yearly increase in Arctic discharge is between 1.6480 and 3.6174 cubic kilometers. This confidence interval excludes "0," so there is evidence Arctic discharge is increasing over time.

25.12: (a) We should use a prediction interval. With 95% confidence, the relative risk of breast cancer for an individual woman drinking 10 mg of red wine per day is 0.98643 to 1.18000. (b) $\hat{\mu} = 1.08321$, $SE_{\hat{\mu}} = 0.01057$. $t^* = 2.920$. A 90% confidence interval for the mean relative risk of breast cancer in all women drinking 10 mg of red wine per day is $1.08321 \pm 2.920(0.01057) = 1.052$ to 1.114.

25.13: (a) $\hat{\mu} = 0.49529 + 0.002332(65) = 0.64687$. (b) $SE_{\hat{\mu}} = 0.00769$, df $= 29 - 2 = 27$, $t^* = 2.052$. $0.64687 \pm 2.052(0.00769) = 0.6311$ to 0.6626.

25.14: (a) A stemplot of the residuals is provided after rounding each residual to the nearest whole number. The distribution of residuals appears close to Normal, with perhaps one outlier in the right tail (a residual of "51").

```
-3 | 4
-2 | 4 3 3
-1 | 9 7 5 5 3 2
-0 | 9 9 9 9 7 6 6 6 3 2 2
 0 | 0 0 3 3 9
 1 | 0 1 1 1 4 8 9 9
 2 | 1 4
 3 | 3
 4 |
 5 | 1
```

(b) There appears to be roughly equal variability in the residuals about the line. Notice the outlier mentioned in (a).

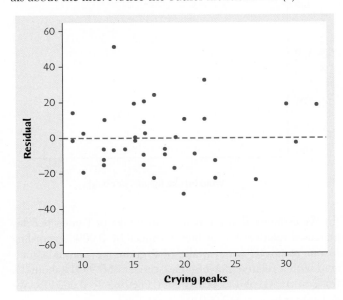

(c) The residuals sum to zero, so only $n - 1$ observations are independent.

25.15: (a) The residual plot provided does not suggest any deviation from a straight-line relationship between volume and Aroc score, although there are two residuals of larger magnitude present, both for Aroc scores slightly lower than 0.60.

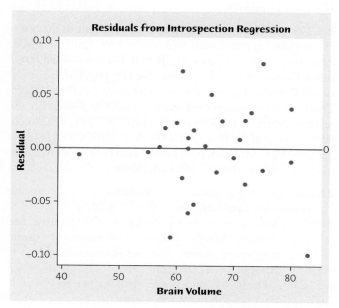

(b) A stemplot of residuals, provided below, does not suggest that the distribution of residuals departs strongly from Normality. There are two possible outliers, (the -0.099 and -0.083 could be viewed as separated from the -0.06), which agrees with the output provided by Minitab referenced in the problem statement.

```
-0 | 9 8
-0 | 6
-0 | 5
-0 | 3 2 2 2
-0 | 1 0 0 0
 0 | 0 0 0 0 0 0 1 1 1
 0 | 2 2 2 3 3
 0 | 5
 0 | 7
 0 | 8
```

(c) It is reasonable to assume that observations are independent because we have 29 different subjects, measured separately (d) It may be the case that variability is larger for smaller values of Aroc, but these happen to be the two outliers. It is difficult to make a definitive argument either way.

25.16: (a) From the output provided, $a = 86.0$ and $b = 1.2699$.

25.17: (a) $r = +\sqrt{r^2} = +\sqrt{0.623} = 0.789$.

25.18: (b) Individual price increases vary, so answer (c) is inappropriate. The regression line provides all predicted mean prices, given appraised value.

25.19: (a) This is a one-sided alternative because we wonder if larger appraisal values are associated with larger selling prices.

25.20: (c) Less than 0.001.

25.21: (c)

25.22: (c) df $= 28 - 2 = 26$.

25.23: (c) $t^* = 2.056$, so the margin of error is $2.056(0.1938) = 0.3985$.

25.24: (a) The prediction interval is appropriate.

25.25: (a) Scientists estimate that each additional 1% increase in the percentage of Bt cotton plants results in an average increase of 6.81 mirid bugs per 100 plants. (b) The regression model explains 90% of the variability in mirid bug density. (c) $H_0: \beta = 0$ vs. $H_a: \beta > 0$ (H_0: population correlation $= 0$ vs. H_a: population correlation > 0). $P < 0.0001$; there is strong evidence of a positive linear relationship between the proportion of Bt cotton plants and the density of mirid bugs. (d) We cannot conclude a causal relationship.

25.26: $H_0: \beta = 0$ vs. $H_a: \beta \neq 0$, $t = \frac{b}{SE_b} = \frac{274.78}{88.18} = 3.116$; df $= 12 - 2 = 10$. $0.01 < P < 0.02$ ($P = 0.0109$ from technology). There is strong evidence of a linear relationship between thickness and gate velocity.

25.27: df $= 10$, $t^* = 1.812$. (a) $b \pm t^* SE_b = 274.78 \pm 1.812(88.18) = 274.78 \pm 159.78 = 115.0$ to 434.6 fps/inch. (b) This is the "90% CI" in the output: 176.2 to 239.3.

25.28: (a) The scatterplot shows no obvious nonlinearity or change in variability. (b) A histogram is unimodal and slightly left-skewed, but not strikingly non-Normal. (c) There are some changes, but they might not be considered substantial: the regression standard error is about 25% smaller, the prediction for $x = 0.5$ inch is about 8 fps larger, and the confidence interval is narrower, due to the reduced standard error.

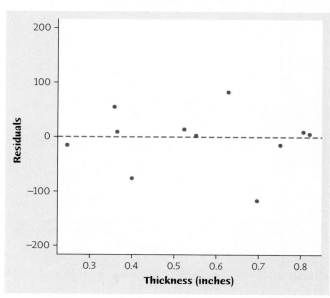

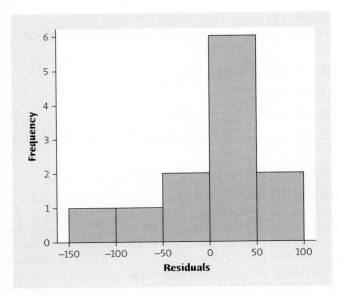

25.29: (a) There is little evidence of non-Normality in the residuals, and there don't appear to be any strong outliers. (b) The scatterplot confirms the comments made in the text. (c) Presumably, close inspection of a manatee's corpse will reveal nonsubtle clues when cause of death is from collision with a boat rotor. It seems reasonable that the number of kills listed in the table are mostly not caused by pollution.

```
-1 | 8 6
-1 | 2 0 0 0
-0 | 8 7 5
-0 | 3 3 3 2 1 1 1
 0 | 0 1 1 1 1 1 2 3 4
 0 | 5 5 5 6 6 8 8
 1 | 1 2 3
 1 | 7
```

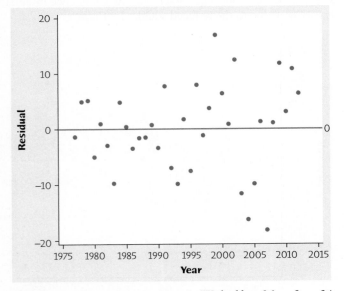

25.30: H_0: $\beta = 0$ vs. H_a: $\beta > 0$. With df $= 36 - 2 = 34$, $t = \frac{b}{SE_b} = \frac{0.132469}{0.007224} = 18.337$, $P < 0.0005$. There is overwhelming evidence that manatee kills increase with boats registered.

25.31: (a) This is a confidence interval for β. With df $= 34$, $t^* = 2.042$. $b \pm t^* SE_b = 0.132469 \pm 2.042(0.007224) = 0.11772$ to 0.14722 additional killed manatees per 1000 additional boats. (b) With 1,050,000 boats, we predict $\hat{y} = 94.15345$ killed manatees, which agrees with the output in Figure 25.14 under "Fit." A 95% prediction interval for the number of killed manatees if 1,050,000 boats are registered is 77.01 to 111.30 kills.

25.32: (a) The test for H_0: $\beta = 0$ is equivalent to the test of H_0: population correlation $= 0$. $t = -4.64$, and $P < 0.0005$ for testing H_0: $\beta = 0$ versus H_a: $\beta \neq 0$. For a one-sided test, P is half the size of Minitab's value. For H_0: population correlation $= 0$ against H_a: population correlation < 0, $P < 0.00025$. There is overwhelming evidence of a negative population correlation. (b) $SE_b = 0.0007414$. df $= 14$, $t^* = 1.761$. A 90% confidence interval for β is $-0.0034415 \pm 1.761(0.0007414) = -0.00475$ to -0.00213. (c) This question calls for a prediction interval. The interval is 0.488 to 3.769 kg.

25.33: (a) H_0: population correlation $= 0$ against H_a: population correlation is > 0. $t = 4.29$; df $= 27$; $P < 0.0005$. There is very strong evidence of a positive correlation between Gray's forecasted number of storms and the number of storms that actually occur. (b) $\hat{\mu} = 0.516 + 1.035(16) = 17.076$, and $SE_{\hat{\mu}} = 1.223$. df $= 27$, $t^* = 2.052$. The 95% confidence interval is given by $17.076 \pm 2.052(1.223) = 14.556$ to 19.586 storms.

25.34: (a) A scatterplot reveals a fairly strong negative linear relationship between SST and coral growth. H_0: $\beta = 0$ vs. H_a: $\beta < 0$; $t = -3.83$; df $= 7 - 2 = 5$; $P = 0.0122$. There is strong evidence of a negative linear relationship between SST and coral growth. (b) $\hat{\mu} = 11.6921 - 0.30305(30) = 2.6006$ mm/year. $SE_{\hat{\mu}} = 0.0385$. df $= 5$, $t^* = 2.571$. A 95% confidence interval for the mean coral growth per year when water temperature is 30 degrees is $2.6006 \pm 2.571(0.0385) = 2.5016$ to 2.6996 mm per year.

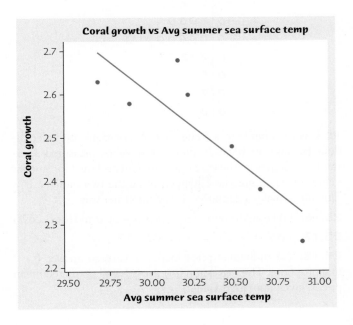

25.35: The stemplot is provided, where residuals are rounded to the nearest tenth. The plot suggests that the residuals do not follow a Normal distribution. Specifically, there are outliers on both ends of the distribution. This makes regression inference and interval procedures unreliable.

```
−0 | 8
−0 |
−0 | 4 4
−0 | 3 3 2 2 2 2
−0 | 1 1 0 0 0
 0 | 0 0 0 1 1 1 1
 0 | 2 2 2 3 3
 0 | 5
 0 | 6
 0 |
 1 | 1
```

25.36: (a) There is a potential outlier, but with only 7 observations, there is no evidence of a systematic departure from nonlinearity in the relationship between SST and coral growth. (b) There is some evidence (albeit difficult to detect with only 7 observations) that residuals are non-Normal. (c) It is not clear that observations are independent. (d) There may be a trend in the residual plot—negative residuals are associated with low and high temperatures, whereas large positive residuals are associated with moderate temperatures.

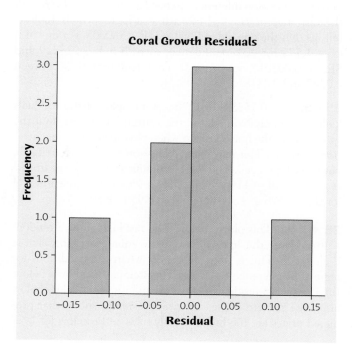

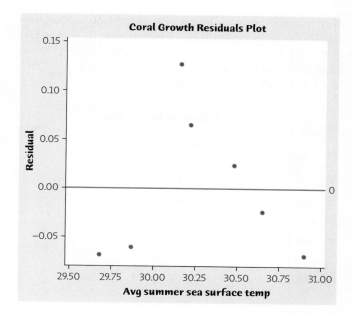

25.37: (a) Shown is the scatterplot with two (nearly identical) regression lines, one using all points and one with the outlier omitted. (b) The correlation for all points is $r = 0.8486$. For testing the slope, $t = 6.00$, for which $P < 0.0005$. (c) Without the outlier, $r = 0.7014$, the test statistic for the slope is $t = 3.55$, and $P = 0.004$. In both cases there is strong evidence of a linear relationship between neural loss aversion and behavioral loss aversion. However, omitting the outlier weakens this evidence somewhat.

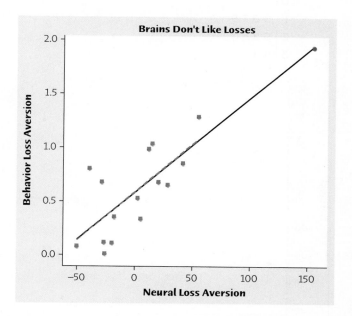

25.38: (a) $\hat{y} = 560.65 - 3.0771x$, $r = -0.6492$. Generally, the longer a child remains at the table, the fewer calories he or she will consume. This relationship is moderately strong and linear. (b) All the conditions for inference appear to be upheld, the stemplot of residuals is roughly Normal, and no overt patterns are seen in the plot of residuals against time. (c) The slope is significantly different from 0: $t = -3.62$, $P = 0.002$. $SE_b = 0.8498$. df $= 18$, $t^* = 2.101$. The 95% confidence interval for β is $-3.0771 \pm 2.101(0.8498) = -4.8625$ to -1.2917 calories per minute.

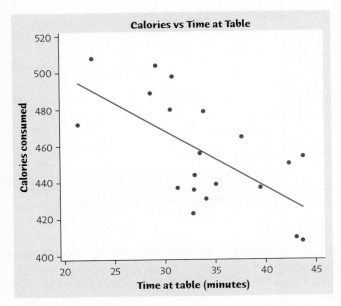

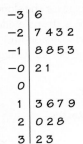

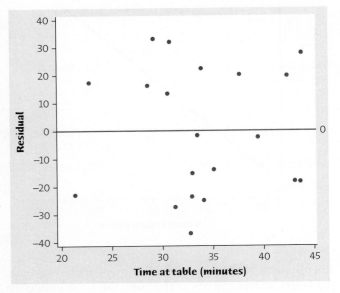

25.39: The distribution is skewed right but the sample is large, so t procedures should be safe. $\bar{x} = 0.2781$ g/m², $s = 0.1803$ g/m². $t^* = 1.984$ for df $= 100$ (rounded down from 115). The 95% confidence interval for μ is 0.2449 to 0.3113 g/m².

25.40: $\hat{y} = 560.65 - 3.0771(40) = 437.57$ calories. $SE_{\hat{y}} = s\sqrt{1 + \frac{1}{n} + \frac{(x^* - \bar{x})^2}{\Sigma(x - \bar{x})^2}} = 23.4\sqrt{1 + \frac{1}{20} + \frac{(40 - 34.01)^2}{758.07}} = 24.51$ calories. A 95% prediction interval is given by $437.57 \pm 2.101(24.51) = 386$ to 489 calories.

25.41: $\hat{y} = 1.4146 + 0.4399x$. The slope is significantly different from zero ($t = 4.33$, $P = 0.001$). To assess the evidence that more cones leads to more offspring, we should use the one-sided alternative, $H_a: \beta > 0$, for which P is half as large (so $P < 0.001$). The conditions for inference seem to be satisfied. One might also choose to find a confidence interval for β: df $= 14$, $t^* = 2.145$. A 95% confidence interval for β is $0.4399 \pm 2.145(0.1016) = 0.2220$ to 0.6578 offspring per unit of cone index.

25.42: $\hat{y} = 19.723 + 0.3388x$. For testing $H_0: \beta = 0$ vs. $H_a: \beta \neq 0$, $t = 1.90$, and $P = 0.065$. We have some evidence of a linear relationship between MA angle and HAV, but the evidence is not strong. Note that perhaps the researchers were really interested in testing $H_0: \beta = 0$ vs. $H_a: \beta > 0$ (as perhaps they felt that severe MA deformity is associated with larger MA angle). If so, then $P = 0.033$, half that for the two-sided test. We have strong evidence for such an assertion.

25.43: $\hat{y} = -1.286 + 11.894x$. An examination of the residuals does not suggest any severe violations of the conditions for regression inference. To test $H_0: \beta = 0$ vs. $H_a: \beta > 0$, $t = 10.47(\text{df} = 21)$, $P < 0.0005$. For df $= 21$, $t^* = 2.080$ for 95% confidence, so with b and $SE_b = 1.136$. We are 95% confident that β is between 9.531 and 14.257.

25.44: $\hat{y} = 0.1385 + 0.0282x$. An examination of the residuals does not suggest any severe violations of the conditions for regression inference. To test $H_0: \beta = 0$ vs. $H_a: \beta > 0$, the test statistic is $t = 14.03$ (df $= 13$), for which Table C tells us that the one-sided P-value is $P < 0.0005$. For df $= 13$, $t^* = 1.771$ for 90% confidence, so with $b = 0.028219$ and $SE_b = 0.002011$, we are 90% confident that β is between 0.0247 and 0.0318.

25.45: $\hat{y} = 0.1523 + 8.1676x$. A stemplot of the residuals looks reasonably Normal, but the scatterplot suggests that the spread about the line is greater when phytopigment concentration is greater. This may make regression inference unreliable, but we will proceed. The slope is significantly different from 0 ($t = 13.25$, df $= 114$, $P < 0.001$). A 95% confidence interval for β is $8.1676 \pm 1.984(0.6163) = 6.95$ to 9.39.

25.46: (a) This plot is below. The residuals show a random scatter about the line, with roughly equal variability across their range. This is what we expect when the conditions for regression inference hold. (b) The stemplot is shown below. (c) The plot of residuals against the month of the sale is below also. The pattern of steadily rising residuals shows that predicted prices are too high for early sales and too low for later sales. This is what we expect if selling prices are rising and appraised values aren't updated quickly enough to keep up.

(a)

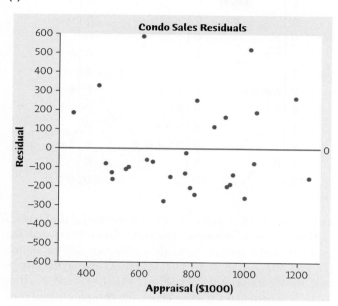

(b)

```
-2 | 6 5 3
-1 | 9 9 8 5 5 4 3 2 2 0
-0 | 9 7 7 6 5 1
 0 |
 1 | 1 6 9 9
 2 | 5 6
 3 | 3
 4 |
 5 | 2 8
```

(c)

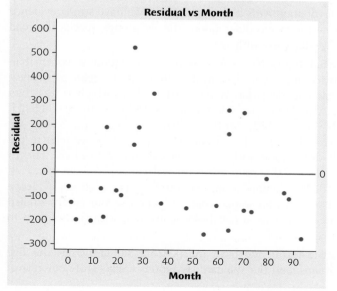

25.47: (a) $\bar{x} = -0.00333$, $s = 1.0233$. For a standardized set of values, we expect the mean and standard deviation to be (up to rounding error) 0 and 1, respectively. (b) The stemplot does not look particularly symmetric, but it is not strikingly non-Normal for such a small sample. (c) The probability that a standard Normal variable is as extreme as this is about 0.0272.

```
-2 | 2
-1 |
-1 | 4
-0 |
-0 | 3 2
 0 | 0 1 1 2 2
 0 | 7
 1 | 0
 1 | 5
```

25.48: $t = -1.00 = \frac{a}{SE_a} = \frac{-0.01270}{0.01264}$. $P = 0.332$, so we do not have enough evidence to conclude that the intercept α differs from 0.

25.49: df = 14, $t^* = 2.145$. $-0.01270 \pm 2.145(0.01264) = -0.0398$ to 0.0144. This interval does contain 0.

Chapter 26   One-Way Analysis of Variance: Comparing Several Means

26.1: (a) The null hypothesis is "all age groups have the same (population) mean road-rage measurement," and the alternative is "at least one group has a different mean." (b) The F test is quite significant, giving strong evidence that the means are different. The sample means suggest that the degree of road rage decreases with age. (We assume that higher numbers indicate *more* road rage.)

26.2: (a) H_0: $\mu_A = \mu_B = \mu_C = \mu_D$ vs. H_a: not all means are equal. (b) Referring to Figure 26.3, comparing Groups A and C, we see that the mean status for men expressing anger is about 6.3, whereas the mean status for women expressing anger is about 4. With both groups expressing anger, men receive higher mean status scores than women, and the mean difference is about 2.3. Notice that comparing Groups B and D, we see that women expressing sadness receive higher status scores than men expressing sadness, but the difference is relatively small.

26.3: (a) The stemplots appear to suggest that logging reduces the number of trees per plot and that recovery is slow (the 1-year-after and 8-years-after stemplots are similar).

(b) The means lead one to the same conclusion as in (a): the first mean is much larger than the other two. (c) H_0: $\mu_1 = \mu_2 = \mu_3$ vs. H_a: not all means are the same, $F = 11.43$, df = 2 and 30, $P = 0.0002$, so we conclude that these differences are significant: the mean number of trees per plot is significantly lower in logged areas.

Never logged		1 year ago		8 years ago	
0		0	2	0	4
0		0	9	0	
1		1	2244	1	22
1	699	1	57789	1	5889
2	0124	2	0	2	22
2	7789	2		2	
3	3	3		3	

26.4: (a) Examining the means provided in Figure 26.6, we see that liberals are notably younger than conservatives or moderates, and moderates are perhaps younger than conservatives (maybe this difference can be explained by chance). The mean ages of liberals, moderates, and conservatives are, respectively, 46.80, 46.98, and 51.09 years. (b) H_0: $\mu_L = \mu_M = \mu_C$ vs. H_a: not all means are the same. $F = 11.69$, df = 2 and 1871, $P = 0.000$ to three places. There is overwhelming evidence that the three means differ. We can see that, indeed, at least liberals differ in mean age from conservatives and moderates.

26.5: (a) Answers will vary due to randomness. (b) To make F small, move the mean so it is between the other two. How small this will be will vary. (c) As any mean becomes very different from the others, F becomes large, and P will become very small (<0.0001).

26.6: (a) As the standard deviation increases, F will become small and P will become large (close to 1). (b) As the group means move apart, F will become large again, and P will become small (close to 0).

26.7: (a) $s_1^2 = 25.6591$, $s_2^2 = 24.8106$, and $s_3^2 = 33.1944$. $s_1 = 5.065$, $s_2 = 4.981$, and $s_3 = 5.761$. The largest standard deviation (5.761) is not more than twice the size of the smallest standard deviation (4.981). Conditions are satisfied. (b) The three standard deviations are $s_L = 17.41$, $s_M = 18.13$ and $s_C = 17.42$. The ratio of largest to smallest standard deviation is $18.13/17.41 = 1.04$, which is less than 2. Conditions are satisfied.

26.8: The standard deviations (0.1201, 0.1472, 0.1134) do not violate our rule of thumb. However, the distributions appear to be skewed and have outliers, especially the 1-year-ago group.

Never logged		1 year ago		8 years ago	
4		4	2	4	
4	8	4		4	
5		5		5	
5		5		5	
6	3	6		6	
6	57	6		6	8
7		7		7	
7	5889	7	7	7	8
8	111	8	3	8	13
8		8	588	8	
9		9	01123	9	34
9	5	9		9	
10		10	0	10	000

26.9: Side-by-side stemplots show some irregularity but no outliers or strong skewness. ANOVA output shows that the group standard deviations easily satisfy our rule of thumb ($2.059/1.302 = 1.58 < 2$). The differences among the groups were significant at $\alpha = 0.05$: $F = 3.44$, df = 3 and 27, $P = 0.031$. Nitrogen had a positive effect, the phosphorus and control groups were similar, and the plants that got both nutrients fell between the others.

26.10: (a) $I = 3$; $n_1 = n_2 = 12$, and $n_3 = 9$; $N = 12 + 12 + 9 = 33$. (b) Numerator ("Between Groups") df: $I - 1 = 2$, denominator ("Within Groups") df: $N - I = 30$.

26.11: (a) $I = 3$ and $N = 96$, so df = 2 and 93. (b) $I = 3$ and $N = 90$, so df = 2 and 87.

26.12: (a) The sample sizes are quite large, and the F test is robust against non-Normality with large samples. (b) Yes (barely): the ratio is $3.11/1.60 = 1.94$, which is slightly less than 2. (c) We have $I = 3$ and $N = 1342$ $\bar{x} = 1.31$, SSG = 356.14, SSE = 6859.65, $F = 34.76$. (d) We compare to an F distribution with df = 2 and 1339. We have strong evidence that the means differ among the age groups; specifically, road rage decreases with age.

26.13: (a) No sample standard deviation is larger than twice any other. Specifically, the ratio of largest to smallest standard deviation is $2.25/1.61 = 1.40$, which is less than 2. Conditions are safe for use of ANOVA. (b) $\bar{x} = 4.8225$, MSG = 25.502, MSE = 3.507, $F = \frac{MSG}{MSE} = 7.272$. (c) We have df = $4 - 1 = 3$ and $68 - 4 = 64$, so we refer to the F distribution with 3 and 64 degrees of freedom. The P-value is 0.000 rounded to three decimal places. In fact, $P = 0.0003$ (obtained using software). There is strong evidence that the mean status scores between the four groups studied are not equal—a conclusion consistent with the solution to Exercise 26.2.

26.14: (a) We have independent samples from the five groups, and the standard deviations easily satisfy our rule of thumb ($1.40/1.28 = 1.09 < 2$). (b) The details of the computations, with $I = 5$ and $N = 4413$, are

$$\bar{x} = 2.459$$
$$= \frac{809 \times 2.57 + 1860 \times 2.32 + 654 \times 2.63 + 883 \times 2.51 + 207 \times 2.51}{4413}$$

$$\text{SSG} = 67.86 = 809(2.57 - \bar{x})^2 + 1860(2.32 - \bar{x})^2$$
$$+ 654(2.63 - \bar{x})^2 + 883(2.51 - \bar{x})^2 + 207(2.51 - \bar{x})^2$$

$$\text{MSG} = 16.97 = \frac{67.86}{5 - 1}$$

$$\text{SSE} = 8010.98 = 808 \times 1.40^2 + 1859 \times 1.36^2 + 653$$
$$\times 1.32^2 + 882 \times 1.31^2 + 206 \times 1.28^2$$

$$\text{MSE} = 1.82 = \frac{8010.98}{4413 - 5}$$

$$F = 9.34 = \frac{16.97}{1.82}$$

(c) The ANOVA is very significant ($P < 0.001$), but this is not surprising because the sample sizes were very large. The differences might not have practical importance. (The largest difference is 0.31, which is relatively small on a 5-point scale.)

26.15: (c)

26.16: (b)

26.17: (a) $I - 1 = 3 - 1 = 2$, and $N - I = 9 - 3 = 6$.

26.18: (a) The null hypothesis for an ANOVA test is always that the population means are equal.

26.19: (c) MSG $= 22,598/(3 - 1) = 11,299$. $F =$ MSG/MSE $= 11,299/1600 = 7.06$.

26.20: (a) The test measures evidence against the null hypothesis stated in Exercise 26.18. We conclude that not all means are the same.

26.21: (c) The largest standard deviation is 62.02 and the smallest is 20.07. Hence, the largest standard deviation is more than three times the smallest.

26.22: (a) This is the problem of multiple comparisons.

26.23: (c) We do not have three independent samples from three populations.

26.24: The populations are morning people, evening people, and people who are neither. The response variable is the difference in memorization scores. H_0: $\mu_1 = \mu_2 = \mu_3$ (all three groups have equal means) vs. H_a: not all means are equal. There are $I = 3$ populations; the sample sizes are $n_1 = 16$, $n_2 = 30$, and $n_3 = 54$, so the total sample size is $N = 100$. The degrees of freedom are therefore $I - 1 = 2$ and $N - I = 97$.

26.25: The populations are college students who might view the advertisement with art image, college students who might view the advertisement with a nonart image, and college students who might view the advertisement with no image. The response variable is student evaluation of the advertisement on the 1–7 scale. H_0: $\mu_1 = \mu_2 = \mu_3$ (all three groups have equal mean advertisement evaluation) vs. H_a: not all means are equal. There are $I = 3$ populations; the samples sizes are $n_1 = n_2 = n_3 = 39$, so there are $N = 39 + 39 + 39 = 117$ individuals in the total sample. There are then $I - 1 = 3 - 1 = 2$ and $N - I = 117 - 3 = 114$ df.

26.26: There are $I = 4$ populations: learning-disabled children with each of the three accommodations plus a control group. The response variable is the scores on the state math exam. H_0: $\mu_1 = \mu_2 = \mu_3 = \mu_4$ (all four groups have equal means) vs. H_a: not all means are equal. The sample sizes are $n_1 = n_2 = n_3 = n_4 = 25$, with a total sample size of $N = 100$. The degrees of freedom are therefore $I - 1 = 3$ and $N - I = 96$.

26.27: The response variable is hemoglobin A1c level. We have $I = 4$ populations: a control (sedentary) population, an aerobic exercise population, a resistance training population, and a combined aerobic and resistance training population. H_0: $\mu_1 = \mu_2 = \mu_3 = \mu_4$ (all four groups have equal mean hemoglobin A1c levels) vs. H_a: not all means are equal. Sample sizes are $n_1 = 41$, $n_2 = 73$, $n_3 = 72$, and $n_4 = 76$. The total sample size is $N = 41 + 73 + 72 + 76 = 262$. We have $I - 1 = 4 - 1 = 3$ and $N - I = 262 - 4 = 258$ df.

26.28: (a) The ratio of largest to smallest standard deviation is $1.50/0.87 = 1.72$, which is less than 2. ANOVA is safe to use for comparing means. Comparing the means provided, males seeing a model are clearly more positive in their evaluation of the product. Among female subjects, there is little difference between the impact of model or student confederate. (b) There are $I = 5$ populations being compared. $N = 22 + 23 + 24 + 23 + 27 = 119$ subjects in total. There are $I - 1 = 5 - 1 = 4$ and $N - I = 119 - 5 = 114$ df. With $F = 8.30$, using software $P = 0.000007$, so there is overwhelming evidence of a difference in population means.

26.29: (a) The graph suggests that emissions rise when a plant is attacked because the mean control emission rate is half the smallest of the other rates. (b) The null hypothesis is "all groups have the same mean emission rate." The alternative is "at least one group has a different mean emission rate." (c) The most important piece of additional information would be whether the data are sufficiently close to Normally distributed. (From the description, it seems reasonably safe to assume that these are more or less random samples.) (d) The SEM equals $s/\sqrt{8}$, so we can find the standard deviations by multiplying by $\sqrt{8}$; they are 16.77, 24.75, 18.78, and 24.38. However, this factor of $\sqrt{8}$ would cancel out in the process of finding the ratio of the largest and smallest standard deviations, so we can simply find this ratio directly from the SEMs: $\frac{8.75}{5.93} = \frac{24.75}{16.77} = 1.48$, which satisfies our rule of thumb.

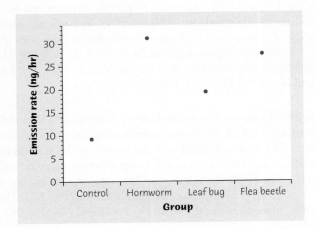

26.30: Only Design A would allow use of one-way ANOVA because it produces four independent sets of data. The data resulting from Design B would be dependent (a subject's responses to the first list would be related to that same subject's responses to the other lists), so ANOVA would not be appropriate for comparison.

26.31: (a) The means suggest that extra water in the spring has the greatest effect on biomass, with a lesser effect from added water in the winter. ANOVA is risky with these data; the standard deviation ratio is nearly 3 ($58.77/21.69 = 2.71$), and the winter and spring distributions may have skewness or outliers (although it is difficult to judge with such small samples). (b) H_0: $\mu_w = \mu_s = \mu_c$ vs. H_a: at least one mean is different. (c) ANOVA gives a statistically significant result ($F = 27.52$, df $= 2$ and 15, $P < 0.0005$), but as noted in (a), the conditions for ANOVA are not satisfied. Based on the stemplots and the means, however, we should still be safe in concluding that added water increases biomass.

26.32: The ANOVA test statistic is $F = 4.92$ (df $= 3$ and 92), which has $P = 0.003$, so there is strong evidence that the means are not all the same. In particular, list 1 seems to be the easiest, and lists 3 and 4 are the most difficult.

26.33: (a) STATE: Does sleep quality affect depression? PLAN: We'll have to assume these students are close to a random sample of college students, and that the observations (students) are independent of one another. SOLVE: We'll use side-by-side boxplots to examine the distributions. All three groups show outliers at the high end, but with such large samples (the smallest is 246), it is reasonable to believe the sample means have Normal distributions. The condition on standard deviations is satisfied because $4.719/2.560 = 1.84 < 2$. We have $F = 75.52$ with df 2 and 895, giving $P = 0.000$ (to three places). CONCLUDE: The mean depression scores for the three levels of sleep quality are not the same. From the output and graphs, it appears the mean depression score for poor sleepers is highest; the mean

depression score for optimal sleepers is lowest. (b) Assuming the students were randomly selected, the large sample size would lead us to believe these students are most likely representative of other college students. (c) Students were not randomly assigned to sleep conditions. Explanations about causation may vary, but this might well be a case of one condition (poor sleep) feeding the other (depression) in a "vicious cycle."

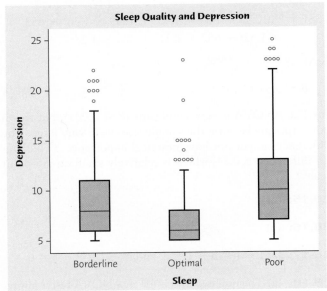

```
Source        DF            SS          MS         F        P
Sleep          2        2162.3      1081.1     72.52    0.000
Error        895       13343.7        14.9
Total        897       15506.0
S = 3.861   R-Sq = 13.94%   R-Sq(adj) = 13.75%
                         Individual 95% CIs For Mean
                         Based on Pooled StDev
Level    N   Mean  StDev   ----+-------+-------+-------+----
Border-
line   246  8.764  3.892         (---*---)
Optimal 309  7.013  2.560      (--*---)
Poor   343 10.656  4.719                 (---*---)
                           ----+-------+-------+-------+----
                           7.2     8.4     9.6    10.8
Pooled StDev = 3.861
```

26.34: (a) Stemplots are provided. There is some degree of left skew in the data corresponding to lemon odor, but it is not strong. There is not strong evidence of non-Normality in any of the population distributions based on these stemplots. There are no real outliers. (b) As discussed in part (a), there is little evidence of non-Normality in any of the three distributions. Also, the three standard deviations are reasonably close: the ratio of largest standard deviation to smallest standard deviation is $15.44/13.10 = 1.18$, which is less than 2. It is safe to apply ANOVA procedures. $F = 10.861$, df $= 2$ and 85, $P = 0.000$ to three decimal places. There is overwhelming evidence of a difference in the mean time customers spend in the restaurant, depending on the odor present. Lavender odor yields the longest mean time, while lemon odor reduces time spent on average compared with no odor at all.

```
Lavender                              Lemon
  7 | 6                                 5 | 6
  8 | 8 9                               6 | 0 3
  9 | 2 3 4 5 7 8                       7 | 3 4 5 8
 10 | 1 2 3 4 5 5 6 6 7 8 8 9 9 9       8 | 3 3 8 8 8 9
 11 | 4 6                               9 | 0 1 4 4 6 7 7
 12 | 1 4 6 9                          10 | 1 4 5 6 8 8
 13 | 7                                11 | 2 3
```

```
              No Odor
                6 | 8 9
                7 | 2 2 3 5 6 9
                8 | 4 4 5 6 7 7
                9 | 1 2 2 2 3 6 8
               10 | 1 3 6 7 7 9
               11 | 5 8
               12 | 1
```

26.35: First, we see that the ratio of largest standard deviation to smallest standard deviation is $2.388/1.959 = 1.219$, which is less than 2. There is some evidence of non-Normality, and perhaps one outlier in the "No Weather Report" group. We proceed, as the samples are reasonably large. $F = 20.679$, df $= 3 - 1 = 2$ and $60 - 3 = 57$. $P = 0.000$, to three decimal places. There is overwhelming evidence that the mean tip percentages are not the same for all three groups.

26.36: First, we note that ANOVA with all four groups shows large differences, as expected. The untreated mean is 58.56, roughly double the means for the other groups; the differences are highly significant ($F = 236.68$, $P < 0.0001$). It is reasonable to view the samples as SRSs from the three populations. The distributions show no severe deviations from Normality. The standard deviations do not satisfy our rule of thumb; the largest to smallest ratio is $2.669/1.167 = 2.29$. Because this is slightly more than 2, ANOVA is somewhat risky, but we proceed anyway. There is a highly significant ($F = 5.02$, df $= 2$ and 12, $P = 0.026$) difference among the mean breaking strengths for the three durable press treatments. Fabrics treated with Permafresh 55 have considerably higher strength than fabrics treated with Permafresh 48 or Hylite.

26.37: First, we note that the mean angle for untreated fabric is 79 degrees, showing much less wrinkle resistance than any of the treated fabrics. ANOVA on four groups gives $F = 153.76$ and $P < 0.0001$. A comparison of wrinkle recovery angle for the three durable press treatments is more interesting.

The ANOVA F-test cannot be trusted because the standard deviations violate our rule of thumb: $10.16/1.92 = 5.29$. This is much larger than 2. In particular, Permafresh 48 shows much more variability from piece to piece than either of the other treatments. Large variability in performance is a serious defect in a commercial product, so it appears that Permafresh 48 is unsuited for use on these grounds. The data are very helpful to a maker of durable press fabrics despite the fact that the formal test is not valid.

26.38: (a) Because $4.500/3.529 = 1.28 < 2$, ANOVA should be safe. The means suggest that logging reduces the number of species per plot and that recovery takes more than 8 years. (b) $F = 6.02$, df $= 2$ and 30, $P < 0.01$ (software gives 0.006). We conclude that these differences are significant; the mean number of species per plot really is lower in logged areas.

26.39: (a) See plot below. (b) There is a slight increase in growth when water is added in the wet season and a much greater increase when it is added during the dry season. (c) The means differ significantly during the first three years. (d) The year 2005 is the only one for which the winter biomass was higher than the spring biomass.

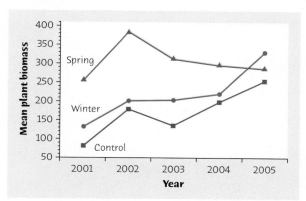

26.40: H_0: $\mu_w = \mu_s = \mu_c$ vs. H_a: at least one mean is different. It is reasonable to view the samples as SRSs from the three populations, but the standard deviation ratio is high ($49.59/11.22 = 4.42$), so ANOVA is risky. As with the 2003 data, the means suggest that extra water in the spring has the greatest effect on biomass, with a lesser effect from added water in the winter. ANOVA gives a statistically significant result ($F = 43.79$, df $= 2$ and 15, $P < 0.0001$). The combination of the (questionable) ANOVA results and the means supports the conclusion that added water in the spring increases biomass. The benefit of additional water in the winter is not so clear, especially when taking the plot of means in the solution to the previous exercise.

26.41: In addition to a high standard deviation ratio ($117.18/35.57 = 3.29$), the spring biomass distribution has a high outlier.

26.42: (a) This is a comparison of two means, so it requires a two-sample t test. (b) This is a comparison of three means, so it requires ANOVA. (c) This is a comparison of three proportions, so it requires a chi-square test of homogeneity.

INDEX